ENCYCLOPEDIA OF PHYSICS

EDITED BY

S. FLÜGGE

VOLUME LI

ASTROPHYSICS II: STELLAR STRUCTURE

WITH 197 FIGURES

SPRINGER-VERLAG
BERLIN · GÖTTINGEN · HEIDELBERG
1958

HANDBUCH DER PHYSIK

HERAUSGEGEBEN VON

S. FLÜGGE

BAND LI

ASTROPHYSIK II: STERNAUFBAU

MIT 197 FIGUREN

SPRINGER-VERLAG
BERLIN · GÖTTINGEN · HEIDELBERG
1958

ISBN 978-3-642-45910-8 ISBN 978-3-642-45908-5 (eBook)
DOI 10.1007/978-3-642-45908-5

Softcover reprint of the hardcover 1st edition 1958

Inhaltsverzeichnis.

Stellar Interiors.

By

Marshal H. Wrubel.

With 25 Figures.

A. Introduction.

I. Definitions.

The customary notation of physics is usually carried over to astrophysics. There are, however, some quantities peculiar to astrophysics.

For example, the physical properties of stars are frequently expressed in solar units where:

$$M_\odot = (1.991 \pm 0.002) \times 10^{33}\ \text{grams};$$

$$R_\odot = (6.960 \pm 0.001) \times 10^{10}\ \text{cm};$$

$$L_\odot = (3.86 \pm 0.03) \times 10^{33}\ \text{ergs/sec}.$$

The material of which stars are made is described in terms of

X = the fractional abundance of hydrogen, by mass;

Y = the fractional abundance of helium, by mass.

For some purposes, it is sufficient to group together all elements heavier than helium as

Z = the fractional abundance of "heavy elements" or "metals", by mass.

Z is used both to represent the gross abundance of elements beyond helium and also the atomic number of a particular element. In general, however, there is little danger of confusion.

It is also convenient to assign a value of

μ = mean molecular weight

to the material (Sect. 13). The perfect gas law then becomes

$$P = \frac{k}{\mu H} \varrho T, \tag{15.10}$$

where

H = mass of unit atomic weight.

From time to time the concept of stellar populations will be mentioned. In this connection, the reader is referred to the article on the Hertzsprung-Russell diagram by H. C. Arp in this volume.

Numerical values have generally been taken from Allen[1]. All logarithms are to the base 10 unless otherwise noted.

---

[1] C. W. Allen: Astrophysical Quantities. London: Athlone Press 1955.

II. Outline of the problem.

1. The scope of this article. It is the aim of the theory of the stellar interior to explain the observed masses, luminosities and radii of stars.

Part of this problem is the study of the formation of stars; that is, the circumstances under which a dark cloud of dust and gas can form a luminous star. Some progress has been made along these lines in recent years[1] but it will not be treated in this article. We will be concerned only with gaseous masses that are already stars.

We will, however, discuss to some extent the changes a star undergoes during its "lifetime". This is the study of stellar evolution, which is itself the subject of another article in this volume[2]. The theory of the stellar interior and stellar evolution have recently become so intricately entwined that it is quite impossible to discuss one and ignore the other. Nevertheless, our emphasis will be on the techniques of model construction from which a theory of stellar evolution may be devised, leaving the synthesis and speculation to the other chapter.

We will mainly be concerned with equilibrium models of stars. The processes which cause a star to evolve are in most cases sufficiently slow that the stars may be assumed to pass through a series of equilibrium configurations. These models may be thought of as representing a star at an instant of time. Ultimately, time must be introduced as an independent variable, but it is possible to build evolutionary sequences in an approximate way by estimating the likely changes in conditions and constructing equilibrium models accordingly. We thereby replace the time-dependent partial differential equations by ordinary differential equations and simplify the problem considerably.

Equilibrium models are also used as basic data in studying pulsating stars. These interesting objects are discussed in another part of this volume[3] and will not be treated here. It is worth mentioning, however, that those who work in stellar evolution must soon come to grips with this problem and explain why pulsation occurs at certain stages.

2. Historical resume and the status today. Although the masses, radii and luminosities of stars have always been the basic data of the theory of the stellar interior, the emphasis has been somewhat different in different generations. At first it was of interest to see if it were possible to construct gaseous spheres in hydrostatic equilibrium without much concern for the origin of the energy and assuming a particular form of the equation of state. The milestone of this era is R. EMDEN's „Gaskugeln" [*1*], and the models studied were of a type called polytropes (see Sect. 38). The next step forward was marked by A. S. EDDINGTON's classic "The Internal Constitution of the Stars" [*2*], in which the role of the radiative transport of energy was extensively discussed. Here EDDINGTON succeeded in establishing a theoretical basis for the observed relation between mass and luminosity. (In spite of the progress made since this book was written it remains an informative and delightful volume which no student of this subject should neglect.)

S. CHANDRASEKHAR summarized, in a complete and rigorous way, the progress to 1939 when his book, "An Introduction to the Study of Stellar Structure" [*3*], was published. Here many extensions of previous work appeared, as well as the detailed theory of white dwarfs.

[1] L. G. HENYEY, ROBERT LE LEVIER and R. D. LEVEE: Publ. Astronom. Soc. Pacific **67**, 154 (1955).

[2] G. R. BURBIDGE and E. M. BURBIDGE, p. 134.

[3] P. LEDOUX and TH. WALRAVEN, p. 353.

Simultaneous with the publication of CHANDRASEKHAR'S book, however, the entire subject took a new turn, for in a classic paper[1], BETHE established the nuclear origin of stellar energy. Thus for the first time not only the mode of energy transport but also its source could be studied.

In addition to this fundamental physical advance, an important observational result turned attention from the mass-luminosity relation to the Hertzsprung-Russell diagram. BAADE[2] persuasively showed that this diagram contains information about the types of stellar population and, largely through the work of SCHWARZSCHILD and HOYLE, the relation between the Hertzsprung-Russell diagram and stellar evolution has been elucidated[3]. SCHWARZSCHILD'S book on stellar structure, to be published soon, promises to be the next milestone in the subject.

The work immediately ahead is likely to be strongly influenced by the adoption of new techniques of high speed computation[4]. Our current knowledge of the detailed processes of absorption and energy production cannot be fully utilized if one is limited to laborious calculations by hand. The capacity of large electronic computers makes it possible to include a variety of physical effects and to vary parameters at will. As these devices become more powerful and as astrophysicists learn more of the necessary techniques, the complexity of the problems we can treat will increase. It is not impossible to hope that, aided by these devices, we may ultimately follow in detail the history of a star from the onset of energy production until it can no longer radiate.

This technical advance must be accompanied by improved physical theories. Our knowledge of the mechanism of convective transport is still rudimentary and the problem of the interactions between convection, rotation and magnetic fields are only beginning to be studied. Furthermore, lest the impression be given that the radiative opacity and nuclear processes are very accurately known, it should be pointed out that the most recently published opacities are only claimed to be accurate to within 10%[5] and the cross section of the $N^{14}(p, \gamma)$ reaction in the carbon cycle is still uncertain[6].

Therefore it is wise to bear in mind that the model stars that will be discussed in this article are to be viewed as explorations rather than as definitive answers; and for this reason we will emphasize techniques rather than numerical results.

3. Observational data. Let us consider the basic observational data that our models will be required to explain: masses, radii and luminosities. The accurate determination of these quantities for all types of stars is a difficult observational task but it is not our intention to go into detail. Masses are determined by gravitational interaction and the most accurate masses are determined from visual binaries[7]. This technique is limited to nearby stars, predominantly (if not exclusively) of BAADE'S Population I, and containing only a limited variety of spectral types.

---

[1] H. BETHE: Phys. Rev. **55**, 434 (1939).

[2] W. BAADE: Astrophys. Journ. **100**, 137 (1944).

[3] For a discussion of the pertinent observations see the article by H. C. ARP in this volume, p. 75.

[4] See, for example, C. B. HASELGROVE and F. HOYLE: Monthly Notices Roy. Astronom. Soc. London **116**, 515 (1956).

[5] GEOFFREY KELLER and ROLAND E. MEYEROTT: Astrophys. Journ. **122**, 32 (1955).

[6] E. M. BURBIDGE, G. R. BURBIDGE, W. A. FOWLER and F. HOYLE: Rev. Mod. Phys. **29**, 547 (1957).

[7] Cf. VAN DE KAMP'S contribution on visual binaries, Vol. L, this Encyclopedia.

Spectroscopic binaries[1] yield the mass function

$$f(m) = \frac{m_2^3 \sin^3 i}{(m_1 + m_2)^2}, \tag{3.1}$$

(where i is the unknown inclination of the orbit), which must be used together with suitable assumptions to get information concerning other spectral types. In this connection the data from very close binary systems may not be valid for single stars because of the possible interchange and loss of mass during the evolution of such systems[2].

To obtain luminosities, accurate photometry, together with reliable distances, is required. Once again, our data for the nearby stars are the most accurate, and the parallaxes of visual binaries can be combined with apparent magnitudes and masses to yield the remarkable mass-luminosity relation; that is, a plot of M versus L yields a line with little scatter. Fig. 1 is taken from a recent analysis by VAN DE KAMP[3]. The classical paper on this subject is by KUIPER[4] and includes data from spectroscopic binaries as well.

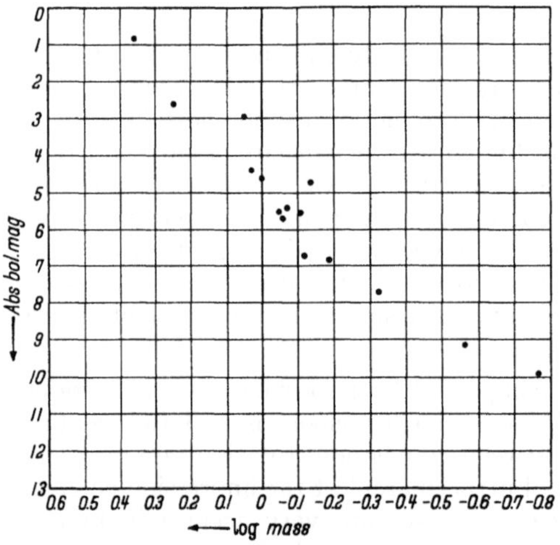

Fig. 1. The observed mass-luminosity relation for visual binaries, after VAN DE KAMP.

The mass-luminosity relation appears to be a phenomenon of the main sequence in the Hertzsprung-Russell diagram. In terms of stellar evolution, as we shall see, it is a relation among chemically homogeneous stars. A unique correlation between mass and luminosity does not exist for rapidly evolving stars which change in luminosity by several magnitudes while M remains virtually constant.

Stellar radii are deduced directly from eclipsing binaries. More often, however, the radius, R, is determined by combining the effective temperature, T_e, determined spectroscopically with the luminosity, L. The Stefan-Boltzmann relation is

$$L = 4\pi R^2 \sigma T_e^4, \tag{3.2}$$

where $\sigma = 5.6698 \times 10^{-5}$ erg cm^{-2} deg^{-4} sec^{-1}; and since L and T_e are assumed known, R may be determined.

The Hertzsprung-Russell (H-R) diagram may be thought of as a plot of $\log L$ vs. $\log T_e$ (T_e increasing to the left). The most recent papers tend to compare the properties of computed models with the observational H-R diagram in which absolute visual magnitude is plotted against Color Index, rather than to convert observations to L and R.

As indicated in the article by ARP in this volume, photoelectric photometry has yielded H-R diagrams of clusters to high accuracy. These are H-R diagrams

[1] Cf. STRUVE' and HUANG's contribution on spectroscopic binaries, Vol. L, this Encyclopedia.

[2] J. A. CRAWFORD: Astrophys. Journ. **121**, 71 (1955).

[3] P. VAN DE KAMP: Astronom. J. **59**, 447 (1954); see also K. AA. STRAND and R. G. HALL: Astrophys. Journ. **120**, 322 (1954).

[4] G. P. KUIPER: Astrophys. Journ. **88**, 472 (1938).

relative to the cluster and there is some uncertainty concerning the zero point. Nevertheless these observations are very important, particularly in connection with evolutionary sequences of models in which the loci of stars of different masses at a particular time are compared with the H-R diagrams of globular and galactic clusters (see Sects. 45 and 47).

It should be kept in mind that the H-R diagram is a two-dimensional projection of a three-dimensional function, the third dimension being mass. In spite of the rapid advance of the study of the H-R diagram of Population II, our knowledge of the masses of globular cluster red giants is still uncertain. (The currently accepted value is 1.2 solar masses.)

Each region of the H-R diagram presents its own particular problems. At one extreme we have the rapidly evolving, massive, bright, blue stars of "young" galactic clusters; at the other are the white dwarfs—feeble and spent. In the globular clusters we observe the effects of age on stars of roughly equal mass; and in the red dwarfs we have the conservative stars that hardly change over billions of years. Add to these the pulsating stars, magnetic stars and novae and one must admit that the term "star" comprises a wide variety of objects. We have made progress toward understanding a few of them.

B. The physical problem.

I. The differential equations of a star in equilibrium.

4. General remarks. The physical picture of a spherically symmetric star in equilibrium is mathematically expressed by four simultaneous, non-linear, ordinary differential equations of the first order. The physical variables involved are:

T = temperature,

P = total pressure,

ϱ = density,

$M(r)$ = mass interior to a sphere of radius r,

$L(r)$ = energy crossing a sphere of radius r per second,

$\varkappa$ = mass absorption coefficient (cm^2 per gram),

ε = energy produced per gram per second.

The four equations represent the radial gradients of P, $M(r)$, T and $L(r)$. Since there are four equations but more than four unknowns, we must have additional information before the system can be solved. The required relations between the unknowns are determined from the physical properties of the material, such as the "equation of state". These we shall call the "constitutive equations", and they usually involve the chemical composition. The constitutive equations are discussed at length in Sects. 13 to 29. We will be concerned with the differential equations in the sections which follow immediately.

Summarizing the results before deriving them, we have:

the pressure equation:

$$\frac{dP}{dr} = -\frac{GM(r)}{r^2}\varrho; \tag{4.1}$$

the mass equation:

$$\frac{dM(r)}{dr} = 4\pi r^2 \varrho; \tag{4.2}$$

the temperature gradient for radiative transport of energy:

$$\frac{dT}{dr} = -\frac{3\varkappa\varrho}{4ac}\frac{1}{T^3}\frac{L(r)}{4\pi r^2}; \tag{4.3}$$

or, for convective transport:

$$\frac{1}{T}\frac{dT}{dr} = \frac{\gamma-1}{\gamma}\frac{1}{P}\frac{dP}{dr}; \tag{4.4}$$

and the luminosity equation:

$$\frac{dL(r)}{dr} = 4\pi r^2 \varrho\,\varepsilon. \tag{4.5}$$

In the above equations, a is the radiation constant $= 7.568 \times 10^{-15}$ ergs cm^{-3} deg^{-4}; c is the velocity of light, and γ is the ratio of specific heats ($\frac{5}{3}$ for a perfect monatomic gas).

These equations must be solved subject to at most four boundary conditions. For further discussion of boundary conditions see Sect. 12.

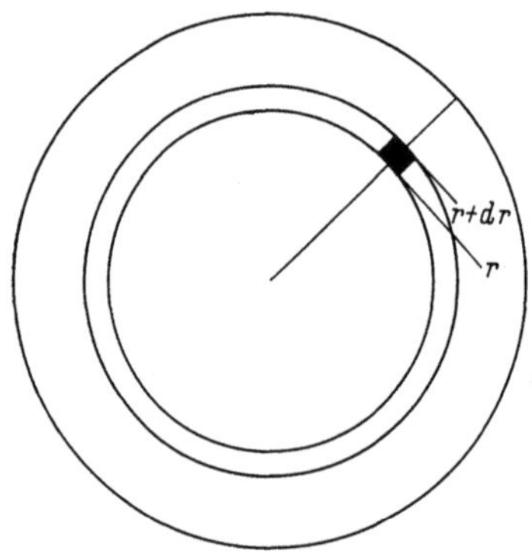

Fig. 2.

5. The pressure gradient. Two competing forces balance to maintain a nonrotating star in hydrostatic equilibrium. Gravity would collapse it if gas and radiation pressure did not suffice to keep it distended.

Consider a cylinder of material of unit cross-section, lying with its axis along a radius from r to $r + dr$ (Fig. 2). Its volume is dr and its mass is therefore $\varrho\, dr$. From potential theory the gravitational force on the cylinder will be due entirely to the mass interior to the sphere of radius r, denoted by $M(r)$. We may calculate it by assuming that the mass, $M(r)$, is located at a point at the center of the star. Thus the gravitational force is $\frac{-GM(r)\,\varrho\, dr}{r^2}$ and it is balanced by a pressure gradient across the cylinder, dP. Equating them, we obtain:

$$\frac{dP}{dr} = -\frac{GM(r)}{r^2}\varrho. \tag{5.1}$$

6. The mass gradient. The mass equation is easily derived by considering the mass of a spherical shell of radius r and thickness dr:

$$dM(r) = 4\pi r^2 \varrho\, dr. \tag{6.1}$$

The mass and pressure gradients may be combined by rewriting (4.1) as

$$\frac{r^2}{\varrho}\frac{dP}{dr} = -G\,M(r). \tag{6.2}$$

Differentiating, and substituting for the mass gradient, we find

$$\frac{d}{dr}\left(\frac{r^2}{\varrho}\frac{dP}{dr}\right) = -G\,4\pi r^2 \varrho. \tag{6.3}$$

In the form

$$\frac{1}{r^2}\frac{d}{dr}\left(\frac{r^2}{\varrho}\frac{dP}{dr}\right) = -4\pi G\,\varrho, \tag{6.4}$$

this equation will reappear in the study of convective zones, isothermal cores and white dwarfs.

In considering sequences of models evolving with time, mass is a more appropriate independent variable than radius. The mass of a given shell retains its identity, provided there is no mixing, although the shell may move outward or inward as the structure changes. In that case the appropriate form for the gradients in Eqs. (4.1) to (4.5) can be found by multiplying each by $(dM(r)/dr)^{-1}$.

7. General remarks concerning temperature gradients. In principle, there are three modes of energy transport available: radiation, convection, or conduction. In stellar interiors, however, conduction is unimportant except as the gas becomes degenerate. In that case it is quite effective[1].

Radiative energy transport will occur whenever there is a temperature gradient. Sometimes the gradient will be sufficient to drive all the energy produced in the interior outward against the resistance provided by the opacity of the material. The more opaque (i.e., absorptive) the material, the steeper the temperature gradient must be to drive a given amount of energy across a sphere of radius r per second.

In other cases, however, the temperature gradient necessary to drive all the energy may be too steep to be maintained (Sect. 9) and convection begins. Then the total energy is transported by a combination of radiation and convection.

Under certain circumstances, outlined in Sect. 9, convection is so efficient that radiative transport may be ignored.

8. The radiative gradient. Before deriving the appropriate equation it is necessary to discuss some of the macroscopic effects of the interaction of radiation and matter.

A field of radiation may be described in terms of the specific monochromatic intensity, I_ν, which is defined by the following construction. Imagine an infinitesimal plane surface, $d\sigma$, in the radiation field. Through a point on the surface draw a line in the direction (ϑ, φ); ϑ is the polar angle, measured with respect to the normal and φ is the azimuth measured on the surface. With this line as the axis, draw an infinitesimal cone with vertex angle $d\omega$ and vertex at the surface. Similar cones may be constructed through every point on $d\sigma$. The envelope of these cones forms a truncated cone called a *pencil* (Fig. 3). The energy, dE_ν, crossing the surface $d\sigma$ in the pencil $d\omega$, in time dt and in frequencies between ν and $\nu + d\nu$ is

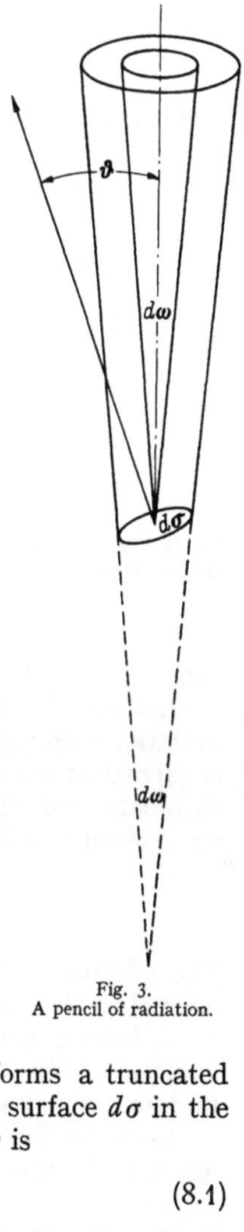

Fig. 3. A pencil of radiation.

$$dE_\nu = I_\nu \cos\vartheta \, d\sigma \, d\omega \, dt \, d\nu. \tag{8.1}$$

The factor, $\cos\vartheta$, arises from the projection of $d\sigma$ normal to the direction (ϑ, φ). The above equation defines the specific intensity, I_ν.

The radiation field in the stellar interior consists of energy flowing in all directions. There is, however, a slight preponderance of outward flow. This net flow of energy per unit area per second in a small frequency range $d\nu$ is called the monochromatic flux and, in terms of the specific intensity,

$$F_\nu = \int I_\nu \cos\vartheta \, d\omega. \tag{8.2}$$

[1] L. Mestel: Proc. Cambridge Phil. Soc. **46**, 331 (1950).

Both the monochromatic intensity and the monochromatic flux may be integrated over the spectrum to give the *integrated* intensity and *integrated* flux,

$$I = \int I_\nu \, d\nu \quad \text{and} \quad F = \int F_\nu \, d\nu. \tag{8.3}$$

F describes the outward flow per unit area. It is therefore related to $L(r)$, defined above. That is, the flux crossing any sphere of radius r concentric with the surface, is

$$F(r) = \frac{L(r)}{4\pi r^2}. \tag{8.4}$$

Due to the interaction of radiation and matter, the flow of radiation is hindered. This is expressed by means of the mass coefficient of absorption or opacity. Consider radiation traversing a cylinder of material of length ds and density ϱ,

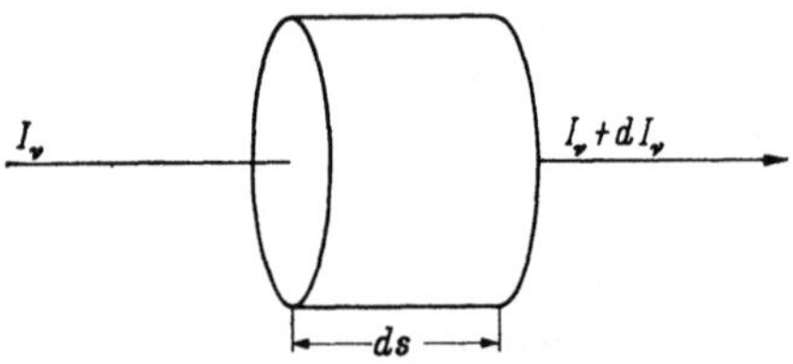

Fig. 4.

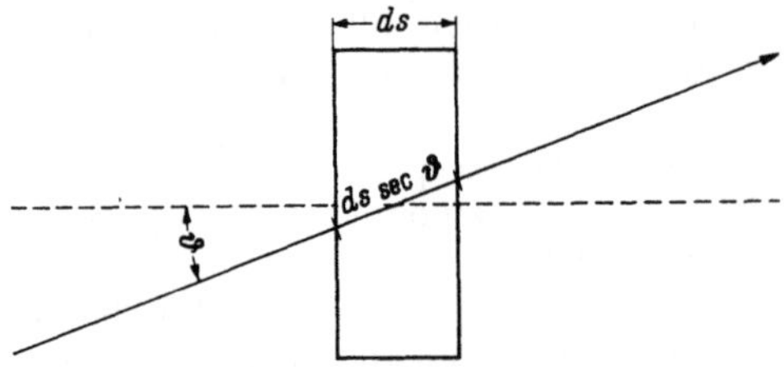

Fig. 5.

as in Fig. 4. If the radiation is described by the specific intensity I_ν on entering the cylinder, the intensity on leaving will be $I_\nu + dI_\nu$ where

$$dI_\nu = -\varkappa_\nu \varrho I_\nu \, ds. \tag{8.5}$$

(The units of the absorption coefficient, $\varkappa_\nu$, are clearly cm² per gram.)

Associated with each photon of energy E, is the momentum E/c. Thus the radiation field may be considered to have a pressure in the sense that momentum is carried across a surface placed in the radiation field. The pressure of monochromatic radiation is defined as the rate of transfer of momentum normal to an arbitrary unit surface. In terms of the specific intensity,

$$P_{r,\nu} = \frac{\int I_\nu \cos^2\vartheta \, d\omega}{c}. \tag{8.6}$$

The additional factor $\cos\vartheta$ comes from taking the component of momentum normal to the surface.

A force is actually exerted only when the momentum of the radiation field is altered. This occurs when the radiation is absorbed by material and the momentum of the absorbed radiation is communicated to the matter.

Consider Fig. 5. According to our previous discussion, $dI_\nu = \varkappa_\nu \varrho I_\nu \times$ (path length). In this case the path length is $dr \sec\vartheta$. Therefore

$$dI_\nu = -\varkappa_\nu \varrho I_\nu \, dr \sec\vartheta, \tag{8.7}$$

and

$$\frac{dP_{r,\nu}}{dr} = -\frac{\varkappa_\nu \varrho \int I_\nu \sec\vartheta \cos^2\vartheta \, d\omega}{c}, \tag{8.8}$$

$$= -\frac{\varkappa_\nu \varrho}{c} \int I_\nu \cos\vartheta \, d\omega; \tag{8.9}$$

or

$$\frac{dP_{r,\nu}}{dr} = -\frac{\varkappa_\nu \varrho}{c} F_\nu. \tag{8.10}$$

From this differential equation we will derive the radiative temperature gradient.

First it should be noted that, to quite a good approximation, the radiation pressure in the stellar interior can be taken as equal to the radiation pressure in an enclosure with black walls in thermodynamic equilibrium at temperature T. Under these conditions the radiation field is isotropic and the intensity is the Planck intensity:

$$I_\nu = B_\nu(T) = \frac{2h\nu^3}{c^2} \frac{1}{e^{h\nu/kT} - 1}. \tag{8.11}$$

From this it follows that

$$P_{r,\nu} = \frac{1}{c} \int I_\nu \cos^2\vartheta \, d\omega = \frac{4\pi}{3c} B_\nu(T). \tag{8.12}$$

It is true that the actual field is not precisely isotropic since there is a slight excess outwards. But this excess accounts for only a small percentage of the available energy and may be neglected in this case.

Combining (8.10) and (8.12) we find

$$\frac{4\pi}{3} \frac{\partial B_\nu(T)}{\partial r} = -\varkappa_\nu \varrho F_\nu. \tag{8.13}$$

Temperature has been introduced by means of the Planck function but the equation is still not in satisfactory form. For one thing, the *monochromatic* flux, F_ν, appears and we have no information concerning it in the deep interior. We want an expression involving the integrated flux, F, since this is a measure of the total energy flow. To achieve this, we rewrite (8.13) as

$$\frac{1}{\varkappa_\nu} \frac{4\pi}{3} \frac{\partial B_\nu(T)}{\partial T} \frac{dT}{dr} d\nu = -\varrho F_\nu d\nu. \tag{8.14}$$

Now we define the Rosseland mean opacity as

$$\frac{1}{\bar\varkappa} = \frac{\int\limits_0^\infty \frac{1}{\varkappa_\nu} \frac{\partial B_\nu}{\partial T} d\nu}{\int\limits_0^\infty \frac{\partial B_\nu}{\partial T} d\nu}. \tag{8.15}$$

Further, since

$$B = \int\limits_0^\infty B_\nu(T)\, d\nu = \frac{ac}{4\pi} T^4, \tag{8.16}$$

and

$$\int\limits_0^\infty \frac{\partial B_\nu}{\partial T} d\nu = \frac{d}{dT} \int\limits_0^\infty B_\nu \, d\nu = \frac{ac}{\pi} T^3, \tag{8.17}$$

we obtain

$$\int\limits_0^\infty \frac{1}{\varkappa_\nu} \frac{\partial B_\nu}{\partial T} d\nu = \frac{ac}{\pi \bar\varkappa} T^3. \tag{8.18}$$

Integrating (8.14) over all frequencies, we obtain

$$\frac{dT}{dr} = -\frac{3}{4ac} \frac{\bar\varkappa \varrho}{T^3} F; \tag{8.19}$$

or, using (8.4),

$$\frac{dT}{dr} = -\frac{3}{4ac} \frac{\bar\varkappa \varrho}{T^3} \frac{L(r)}{4\pi r^2}. \tag{8.20}$$

Eq. (8.20) is the radiative temperature gradient. In subsequent sections we will omit the bar and write $\varkappa$ for the Rosseland mean opacity.

9. The stability of the radiative gradient and the convective transport of energy. We will now establish a criterion due to K. SCHWARZSCHILD[1] which states that the radiative temperature gradient is unstable against convection when it is steeper than the local adiabat.

Consider a small element which somehow becomes hotter than its surroundings. Pressure equilibrium is established quite rapidly and the material expands. It is then at a lower density than the material in its vicinity and it experiences a buoyant force which drives it outward.

During its outward motion it does not readily exchange energy with its surroundings. We may therefore regard the element as a sort of balloon of hot gas enclosed by an elastic adiabatic wall.

The element comes to rest only if it encounters a region with which it is not only in pressure equilibrium but in temperature equilibrium as well. In that case it is indistinguishable from its surroundings and it experiences no buoyant force. This situation, which represents stability, is illustrated in Fig. 6a. The mass rising adiabatically does indeed encounter a region at the same temperature because the adiabat intersects the (less steep) radiative gradient.

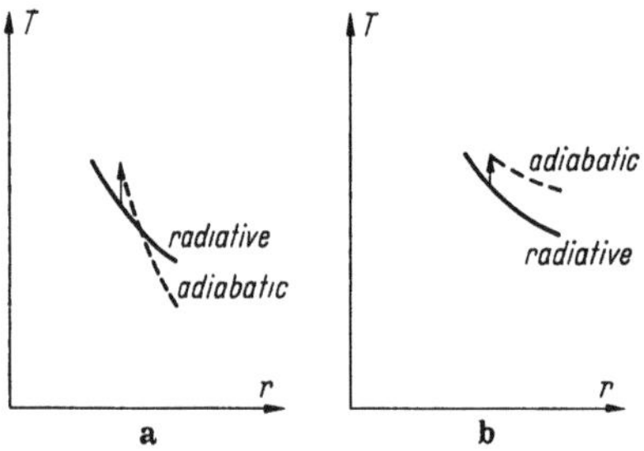

Fig. 6a and b. K. SCHWARZSCHILD's criterion: (a) stability; (b) instability.

On the other, hand the unstable situation is illustrated in Fig. 6b. When the radiative gradient is steeper than the adiabat, the mass element departs more and more from the local temperature as it rises, and it is continually driven outward.

The same argument holds for an element which sinks because it is cooler than its surroundings.

Eventually, even in the unstable case, the element loses its identity and mixes with its surroundings. On the average, before dissolution an element will travel a distance l, which is called the "mixing length", following PRANDTL[2]. Since, in the unstable case, a rising element will be hotter than its surroundings, it will contribute energy to the material with which it mixes; in this manner convection carries energy outward.

Thus, when the radiative gradient is unstable, we must consider energy transport by convection *as well as* by radiation. It should be emphasized that radiation will continue to carry energy, since any temperature gradient will drive radiant energy forward; but radiation need not carry all, or even a large part of the energy when convective transport is available. A temperature gradient will be set up such that the total flux will be carried partly by radiation and partly by convection. The radiative flux will be

$$F_r = -\frac{4ac}{3\varkappa\varrho} T^3 \frac{dT}{dr}, \tag{9.1}$$

where the magnitude of the actual gradient, $|dT/dr|$, will be less than the radiative gradient defined by (8.20).

---

[1] K. SCHWARZSCHILD: Göttinger Nachr. **1906**, 41.

[2] L. PRANDTL: Strömungslehre. Braunschweig: F. Vieweg & Sohn 1949; also, in translation: Fluid Mechanics. London: Blackie & Son 1952.

We must now evaluate the flux carried by convection[1].

First of all, the transport of energy must be proportional to the difference in temperature between a moving element and its surroundings. Let us first consider an element which rises adiabatically. Its temperature on moving a distance l (to terms of the first order) is

$$T_0 + l\left(\frac{dT}{dr}\right)_{\text{ad}},$$

where T_0 is the temperature at the point of origin. On the other hand, the material in its vicinity is at a temperature

$$T_0 + l\left(\frac{dT}{dr}\right),$$

where the gradient in this expression is the average gradient in the medium. Thus an element rising adiabatically differs from its surroundings by

$$\Delta T = l\left[\left(\frac{dT}{dr}\right)_{\text{ad}} - \left(\frac{dT}{dr}\right)\right]. \tag{9.2}$$

Consider a fixed surface of unit cross section, across which the rising element moves with velocity v. In one second, a volume equal to the velocity will cross this surface, carrying a mass ϱv. The rising material then mixes with its surroundings at constant pressure, transporting a flux of energy

$$F_c = c_P\,\varrho\, v\, l\left[\left(\frac{dT}{dr}\right)_{\text{ad}} - \left(\frac{dT}{dr}\right)\right], \tag{9.3}$$

where c_P is the specific heat at constant pressure[2].

In the central regions of the Sun, the flux is $\approx 10^{12.5}$ and the density is $\approx 10^{1.5}$. Furthermore, $c_P \approx 10^8$. By estimating the mixing length, BIERMANN[3] and COWLING[4] were able to show that the average gradient need not be very different from the adiabatic for convection to carry effectively all of the energy. It is therefore customary, when the radiative gradient becomes unstable in the *deep interior*, to replace it by the adiabatic gradient. In reality the gradient must be slightly steeper ("superadiabatic") but not sufficiently to affect the structure.

When convection occurs in the outer layers, however, its description is not so simple. In this case radiation may carry an appreciable fraction of the energy and the situation is further complicated by the ionization of hydrogen and helium which occurs within the convective zone.

A detailed procedure for treating this problem has been described by VITENSE[5] and UNSÖLD in the framework of stellar atmospheres. By this method, VITENSE has calculated a model for the Sun's atmosphere in which the total flux is carried by a combination of radiation and convection. The fraction carried by each is found in terms of four gradients, ∇, defined in terms of $d \ln T/d \ln P$ rather than dT/dr:

a) ∇_{rad}, a fictitious gradient which represents radiative equilibrium;

b) ∇, the average gradient of the medium;

---

[1] For a detailed discussion of the points which follow, see L. BIERMANN: Ergebn. exakt. Naturw. **21**, 1 (1945); T. G. COWLING: Monthly Notices Roy. Astronom. Soc. London **96**, 15 (1935); and for convection in the outer layers, as well as many additional references, A. UNSÖLD: Physik der Sternatmosphären, 2nd edit. Berlin-Göttingen-Heidelberg: Springer 1955.

[2] For a discussion of the appropriateness of c_P rather than c_V, see references above.

[3] L. BIERMANN: Z. Astrophys. **5**, 117 (1932). — Astronom. Nachr. **257**, 269 (1935); **258**, 257 (1936).

[4] T. G. COWLING: Monthly Notices Roy. Astronom. Soc. London **94**, 768 (1934).

[5] E. VITENSE: Z. Astrophys. **32**, 135 (1953).

c) ∇', the gradient for an average turbulent element; and

d) ∇_{ad}, the adiabatic gradient.

The turbulent elements are permitted to exchange energy with their surroundings so that $\nabla' \neq \nabla_{\text{ad}}$. In general

$$\nabla_{\text{rad}} > \nabla > \nabla' > \nabla_{\text{ad}}. \tag{9.4}$$

A complete discussion of this procedure belongs more appropriately to a treatise on stellar atmospheres[1]. Nevertheless, since convection in the outer layers may have considerable effect on the radius of a star, and this is a fundamental datum in the theory of stellar interiors, we will briefly discuss the adiabatic gradient of material undergoing ionization in the next section.

10. Adiabatic gradient. We will first consider the adiabatic gradient in material which is not undergoing an appreciable change of ionization. Partial ionization as well as radiation pressure will be introduced as we proceed.

From the first law of thermodynamics

$$dU = dQ - P\,dV, \tag{10.1}$$

and since in an adiabatic process $dQ = 0$, we have

$$dU = -P\,dV. \tag{10.2}$$

For a perfect gas

$$dU = C_V\,dT, \tag{10.3}$$

where C_V is the heat capacity at constant volume. Further,

$$PV = NkT = (C_P - C_V)\,T, \tag{10.4}$$

where N is the number of particles, k is the Boltzmann constant and C_P is the heat capacity at constant pressure; therefore

$$P\,dV = (C_P - C_V)\,dT - V\,dP. \tag{10.5}$$

Combining (10.2), (10.3) and (10.5) we obtain

$$C_V\,dT = -(C_P - C_V)\,dT + V\,dP; \tag{10.6}$$

or, rearranging terms and substituting for V,

$$C_P\,dT = \frac{(C_P - C_V)}{P}\,T\,dP. \tag{10.7}$$

Defining the "adiabatic exponent", γ, as the ratio of specific heats

$$\gamma = \frac{c_P}{c_V} = \frac{C_P}{C_V}, \tag{10.8}$$

and the "polytropic index"

$$n = \frac{1}{\gamma - 1}, \tag{10.9}$$

we obtain

$$\frac{1}{T}\,dT = \frac{\gamma - 1}{\gamma}\,\frac{1}{P}\,dP \tag{10.10}$$

$$= \frac{1}{n+1}\,\frac{1}{P}\,dP; \tag{10.11}$$

---

[1] Cf. the article by BARBIER in Vol. L, this Encyclopedia.

therefore,

$$\frac{d\log P}{d\log T} - 1 = n. \tag{10.12}$$

For a perfect monatomic gas, $\gamma = \frac{5}{3}$ and $n = \frac{3}{2}$ giving from (10.11)

$$\frac{d\log T}{dr} = 0.4\,\frac{d\log P}{dr}. \tag{10.13}$$

In a region where the temperature gradient is the adiabatic gradient, n is a constant equal to 1.5, and the gradients satisfy the relation

$$\frac{d\log P}{d\log T} - 1 = 1.5. \tag{10.14}$$

However, we may define an *effective polytropic index*, n_{eff}, which may vary through the star, by

$$n_{\mathrm{eff}} = \frac{d\log P}{d\log T} - 1. \tag{10.15}$$

SCHWARZSCHILD's criterion for *instability*, described in Sect. 9 as

$$\left|\left(\frac{dT}{dr}\right)_{\mathrm{rad}}\right| > \left|\left(\frac{dT}{dr}\right)_{\mathrm{ad}}\right|, \tag{10.16}$$

can easily be transformed to the equivalent and practical form

$$n_{\mathrm{eff}} < 1.5. \tag{10.17}$$

The relation

$$\frac{d\log P}{d\log T} = n + 1 \tag{10.18}$$

may be written in the integrated form

$$P T^{-(n+1)} = \mathrm{const.} \tag{10.19}$$

Using the perfect gas law to replace T by ϱ we obtain

$$P = K\varrho^{\frac{n+1}{n}} = K\varrho^{\gamma}. \tag{10.20}$$

An equation of this type, where n is a constant, is called a *polytropic equation of state.* It is encountered not only in perfect gases in adiabatic equilibrium but also, for example, in completely degenerate gases, as will be discussed in Sect. 15.

When radiation pressure is appreciable, the concept of an adiabatic exponent can be preserved but its definition is different. Let the gas pressure be a fraction, β, of the total pressure, the rest being radiation pressure; that is,

$$P = P_g + P_r; \tag{10.21}$$

$$P_g = \beta P \tag{10.22}$$

and

$$P_r = (1-\beta)\,P. \tag{10.23}$$

Then if we define a new adiabatic exponent, Γ_1, by requiring that

$$d\log P = \Gamma_1\, d\log\varrho, \tag{10.24}$$

it follows, after lengthy algebra[1], that

$$\Gamma_1 = \beta + \frac{(4-3\beta)^2(\gamma-1)}{\beta + 12(\gamma-1)(1-\beta)}, \tag{10.25}$$

where γ is the ratio of specific heats of the material alone.

[1] See [3], p. 57.

On the other hand, if we are interested in the logarithmic gradient of pressure versus *temperature*, we may define Γ_2 such that

$$d \log P = -\frac{\Gamma_2}{1-\Gamma_2} d \log T. \tag{10.26}$$

Then it follows that

$$\Gamma_2 = 1 + \frac{(4-3\beta)(\gamma-1)}{\beta^2 + 3(\gamma-1)(1-\beta)(4+\beta)}. \tag{10.27}$$

A third exponent, Γ_3, connects ϱ and T.

As expected, all three Γ's reduce to γ when $\beta = 1$ (vanishing radiation pressure), and $\frac{4}{3}$ when $\beta = 0$ (pure radiation).

We could also define an adiabatic exponent for a gas undergoing ionization. A completely unionized monatomic gas and a completely ionized gas would both be described by $\gamma = \frac{5}{3}$; but in a partially ionized gas, $\gamma < \frac{5}{3}$. [The gas would therefore also be more prone to convection by (10.9) and (10.17).]

An alternative procedure, due to UNSÖLD, is to calculate the *entropy* of the gas as a function of P and T and to determine $\left(\frac{d \ln T}{d \ln P}\right)_{\text{ad}}$ by numerical differentiation along lines of constant entropy[1].

11. Luminosity gradient. The energy produced in a shell of mass $4\pi r^2 \varrho\, dr$ is

$$d L(r) = 4\pi r^2 \varrho\, \varepsilon\, d r, \tag{11.1}$$

where ε is the energy produced per gram per second.

This is the most recent equation to be incorporated in detail into the system of differential equations because the nuclear processes responsible for ε were not known until 1939.

As we shall see later, in cases such as polytropes or under the special assumptions of the Cowling model, it is possible to determine a good deal about stellar structure without knowing ε. It is also possible, in approximate calculation, to avoid integrating the luminosity equation by assigning a temperature to the energy producing region sufficiently high to produce the necessary energy. (Such an approximation is used, for example, in hydrogen-burning shells.)

12. Boundary conditions and conditions at interfaces. Only particular solutions of the differential equations (4.1) to (4.5) are physically meaningful. These satisfy certain boundary conditions which may be stated as follows:

1. At $r = 0$ (the center of the star)

$$M(r) \to 0 \quad \text{and} \quad L(r) \to 0. \tag{12.1}$$

2. At the outer boundary

$$T \to T_0 \quad \text{and} \quad \varrho \to \varrho_0 \quad \text{simultaneously,} \tag{12.2}$$

where T_0 and ϱ_0 are photospheric values found from a consideration of the atmosphere of the star. The region producing the spectrum we observe directly is at a temperature roughly $T_0 = T_e$ and the density distribution in the atmosphere is such that, above this layer the atmosphere is virtually transparent, and below it, virtually opaque. The density ϱ_0 is therefore determined by the *atmospheric opacity*.

---

[1] UNSÖLD, footnote 1, p. 11. Also see R. C. TOLMAN: The Principles of Statistical Mechanics, p. 578. London: Oxford University Press 1938.

These refinements have often been ignored and replaced by the condition

$$P \to 0 \quad \text{as} \quad T \to 0. \tag{12.3}$$

Many models satisfying condition (12.3) do not appreciably violate condition (12.2) and are therefore acceptable. When models computed using (12.3) do not satisfy (12.2), however, they should be modified. These considerations will be important when we discuss red dwarfs, the globular cluster giants, and the Sun.

Because of singularities in the differential equations, the usual computational procedure is to integrate outward from the center and inward from the surface, beginning with series solutions that satisfy the boundary conditions. At some suitable point the solutions are matched subject to the appropriate fitting conditions. A simple example of a fitting procedure in a special case is described in Sect. 40.

Some models are constructed with abrupt discontinuities in chemical composition in the interior. At the interface between regions of different composition the temperature and pressure are required to be continuous and as a result the density is discontinuous. A technique for fitting solutions at interfaces of this type is given in Sect. 43.

II. The constitutive equations.

The physical quantities which appear in the differential equations of stellar structure are interrelated by certain macroscopic properties of the stellar material. We will call these relations *constitutive equations*, and they include the equations of state, the interaction with radiation, and the production of nuclear energy. These are, in turn, derived from the microscopic properties described by theoretical and experimental physics. Indeed, the numerical values we derive for such quantities as the central temperature of the Sun depend on nuclear cross sections which cannot be directly measured. In this respect, astrophysicists are at the mercy of nuclear physicists and the astrophysical results are only as reliable as the physical data.

13. Mean molecular weight. The interior of a star is a gaseous mixture of electrons, partially ionized atoms, bare nuclei, and radiation. The most abundant elements are hydrogen and helium, and one gram of stellar material contains X grams of the former and Y of the latter. The remaining material is a mixture of heavier elements contributing Z grams, where

$$X + Y + Z = 1. \tag{13.1}$$

In discussing the degree of ionization of stellar material we may consider the radiation field as closely equivalent to that in a black enclosure at temperature T and treat the material as though it were in thermodynamic equilibrium. The ratio between the number of ions of an element in two stages of ionization would then be given by the Saha equation:

$$\frac{N^{++} N_e}{N^+} = \frac{u^{++}}{u^+} \, \frac{2(2\pi m_e k T)^{\frac{3}{2}}}{h^3} \, e^{-\chi/kT}, \tag{13.2}$$

where

N^+, N^{++} = numbers of atoms in two successive stages of ionization;

u^+, u^{++} = partition functions for corresponding ions;

χ = ionization potential of lower stage;

N_e = number of electrons per unit volume;

m_e = mass of electron.

Eq. (13.2) should be multiplied by an additional factor to account for pressure ionization.

It is convenient to define a mean molecular weight, μ, such that the number of particles per gram of material is $(\mu H)^{-1}$ where H is the mass of unit atomic weight. (H^{-1} is AVOGADRO's number.) Alternatively, μ^{-1} is the number of particles per unit atomic weight.

Each hydrogen atom, when ionized, contributes two particles (an electron and a proton) for every unit of atomic weight (approximately). An ionized helium atom contributes three particles per four units of atomic weight. To sufficient accuracy for some purposes, the heavier elements may be assumed to contribute one particle for every two units of atomic weight. Therefore, noting (13.1),

$$\mu = \frac{1}{2X + \frac{3}{4}Y + \frac{1}{2}Z} = \frac{4}{2 + 6X + Y}. \tag{13.3}$$

To be more precise, one should take into account the degree of ionization (a function of temperature), and the abundance, α_i, of each element and define the mean molecular weight as

$$\mu = \frac{1}{\sum\limits_i \alpha_i \bar{n}_i}, \tag{13.4}$$

where $\bar{n}_i$ is the number of free particles per unit atomic weight contributed by element i. However, unless Z is large the approximate expression is sufficiently accurate and the temperature dependence of μ is usually neglected.

It is also convenient to define the number of *electrons* per gram of material as $(\mu_e H)^{-1}$. An approximation similar to the one derived for μ above, yields

$$\mu_e = \frac{2}{1 + X}. \tag{13.5}$$

14. Polytropic equations of state. The simplest type of equation of state to apply in the theory of the stellar interior is one which relates density and pressure without reference to temperature. If T does not explicitly occur in the equation of state, the pressure and mass gradient equations may be solved independently of the temperature gradient, permitting much to be determined concerning the stellar structure without a detailed knowledge of opacity or energy production.

An historically very important relation of this kind is the polytropic equation

$$P = K \varrho^{\frac{n+1}{n}}, \tag{14.1}$$

already encountered in Sect. 10.

We have seen how such a relation can arise in the case of a perfect gas undergoing an adiabatic process. There we defined the polytropic index, n, in terms of γ, the ratio of specific heats. This definition may be generalized to include any process in which the specific heat remains constant, i.e.,

$$dQ = c\,dT. \tag{14.2}$$

We may define a polytropic exponent,

$$\gamma' = \frac{c_P - c}{c_V - c}, \tag{14.3}$$

and corresponding polytropic index,

$$n = \frac{1}{\gamma' - 1}. \tag{14.4}$$

The corresponding equation of state for the process is again of the form (14.1). By these definitions an isothermal process is one of infinite heat capacity and, consequently, infinite polytropic index.

Equations of the form (14.1) occur not only in perfect gases but also as limiting cases in the behavior of completely degenerate gases. One significant difference, however, is worthy of mention at this point. In the polytropic relation for a perfect gas, the constant of proportionality is arbitrary and may be established, in any particular case, by specifying the density and pressure at a particular point. Since density can be varied at a fixed pressure by varying the temperature, the constant of proportionality in the polytropic equation is an implicit temperature factor.

The physical situation in a completely degenerate gas is entirely different since the relation between density and pressure is unaffected by temperature. Thus when relations of type (14.1) arise in discussing degenerate matter (Sect. 15), the constants of proportionality will be fixed physical constants and may be evaluated, once for all, numerically.

15. General equations of state. Over a large range of temperature and pressure the perfect gas law

$$P V = N k T \tag{15.1}$$

applies to stellar material. At high temperature and pressure, however, significant deviations occur due to the degeneracy of the electrons. The degeneracy of all other particles may be neglected (as shown below) and their partial pressures may be represented by the perfect gas law.

The theory of degeneracy is treated in detail elsewhere in these volumes so we will only summarize the results[1]. Degeneracy arises because electrons are subject to the Pauli exclusion principle, which implies that the number of particles with momenta between p and $p+dp$ in a volume V cannot exceed

$$N(p)\,dp = \frac{V\,8\pi\,p^2\,dp}{h^3}. \tag{15.2}$$

According to the Fermi-Dirac statistics the most probable distribution of electrons among available momentum states is

$$N(p)\,dp = \frac{V\,8\pi\,p^2\,dp}{h^3\,(\mathrm{e}^{-\psi+\vartheta E}+1)}, \tag{15.3}$$

where ψ is a parameter and

$$\vartheta = \frac{1}{kT}. \tag{15.4}$$

From this we may derive the following general results for the total number of particles and the pressure:

$$N = \frac{8\pi V}{h^3}\int\limits_0^\infty \frac{p^2\,dp}{\mathrm{e}^{-\psi+\vartheta E}+1}, \tag{15.5}$$

$$P = \frac{8\pi}{3h^3}\int\limits_0^\infty \frac{p^3}{\mathrm{e}^{-\psi+\vartheta E}+1}\,\frac{\partial E}{\partial p}\,dp, \tag{15.6}$$

in terms of the energy

$$E = m c^2\left[\left(1+\frac{p^2}{m^2 c^2}\right)^{\frac{1}{2}}-1\right]. \tag{15.7}$$

---

[1] See also [3], p. 357 and R. C. Tolman: The Principles of Statistical Mechanics, p. 388. London: Oxford University Press 1938.

The foregoing expressions are valid for all densities and temperatures, but in some regions of the $(\log \varrho, \log T)$ plane, less complicated expressions may be used. The behavior of the electron gas may be summarized in terms of Fig. 7, adapted from WARES[1].

In regions 1, 2, and 3, relativistic effects are negligible and we may write $E \approx p^2/2m$. In regions 5, 6, and 7, $E \approx pc$, the relativistic limit.

Degeneracy is negligible in regions 1, 7, and 8, since $\exp(-\psi + \vartheta E) \gg 1$. In regions 3, 4, and 5, degeneracy is complete. Here we may assume that all available momentum states up to a critical momentum, p_0, are occupied and there are no electrons with momenta greater than p_0.

We may summarize the equations of state of the electron gas as follows:

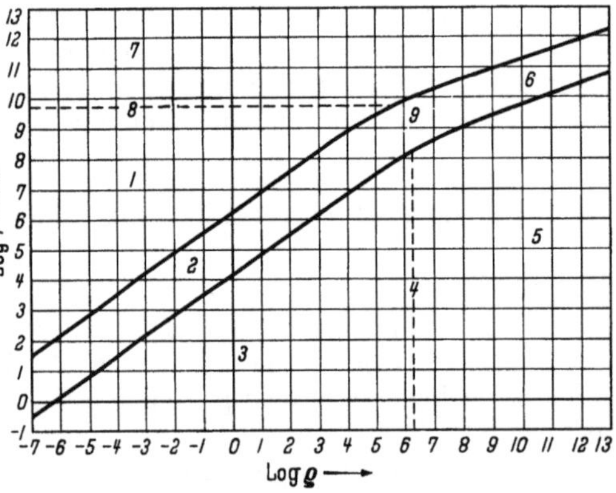

Fig. 7. Subdivision of the $(\log \varrho, \log T)$ plane, after WARES.

Regions 1, 7, 8; non-degenerate: The perfect gas law

$$P V = N k T \tag{15.8}$$

applies even for electrons. Since all particles are non-degenerate in this region and

$$\frac{N}{V} = \frac{\varrho}{\mu H}, \tag{15.9}$$

the total pressure becomes

$$P = \frac{k}{\mu H} \varrho T. \tag{15.10}$$

Region 2; transition from non-relativistic non-degeneracy to non-relativistic degeneracy: The equation of state for electrons is expressed in terms of the parameter, ψ:

$$\left. \begin{aligned} N_e &= \frac{2V}{h^3} \left(\frac{\pi m_e}{\vartheta} \right)^{\frac{3}{2}} F_{\frac{1}{2}}(\psi), \\ P_e &= \frac{N_e k T}{V} \, \frac{\frac{2}{3} F_{\frac{3}{2}}(\psi)}{F_{\frac{1}{2}}(\psi)}; \end{aligned} \right\} \tag{15.11}$$

where

$$F_w(\psi) = \int_0^\infty \frac{u^w}{e^{u-\psi} + 1} \, du. \tag{15.12}$$

Functions $F_{\frac{1}{2}}(\psi)$ and $\frac{2}{3} F_{\frac{3}{2}}(\psi)$. are tabulated by MCDOUGALL and STONER[2] from which columns 1, 2, and 3 of Table 1 are an excerpt.

This is a region of particular interest in recent astrophysical developments and merits special comment. As Fig. 7 indicates, an increase in temperature at a given density tends to remove degeneracy. Further, column 4 of Table 1 reveals that degenerate pressure is always greater than perfect gas pressure. These results are also seen near the border between region 1 and 2 where the approximation

$$P = \frac{N k T}{V} \left(1 + 0.1995 \, F_{\frac{1}{2}}(\psi) - \cdots \right) \tag{15.13}$$

is valid. Since $F_{\frac{1}{2}}(\psi) \propto N m^{-\frac{3}{2}}$, light particles show degeneracy effects at lower densities than heavy particles. Indeed, at stellar densities, as mentioned above, only the electrons are affected and even protons are not appreciably degenerate.

[1] G. WARES: Astrophys. Journ. **100**, 158 (1944).

[2] J. MCDOUGALL and E. C. STONER: Phil. Trans. Roy. Soc. Lond. **237**, 67 (1939).

Table 1. *Fermi-Dirac functions following McDougall and Stoner.*

ψ	$\frac{2}{3}F_{\frac{3}{2}}(\psi)$	$F_{\frac{1}{2}}(\psi)$	Ratio	ψ	$\frac{2}{3}F_{\frac{3}{2}}(\psi)$	$F_{\frac{1}{2}}(\psi)$	Ratio
−4.0	0.016179	0.016128	1.003	8.5	60.94678	16.80714	3.626
−3.5	0.026620	0.026480	1.005	9.0	69.71616	18.27756	3.814
−3.0	0.043741	0.043366	1.009	9.5	79.23141	19.79041	4.004
−2.5	0.071720	0.070724	1.014	10.0	89.51344	21.34447	4.194
−2.0	0.117200	0.114588	1.023	10.5	100.58256	22.93862	4.385
−1.5	0.190515	0.183802	1.036	11.0	112.45857	24.57184	4.577
−1.0	0.307232	0.290501	1.058	11.5	125.16076	26.24319	4.769
−0.5	0.489773	0.449793	1.089	12.0	138.70797	27.95178	4.962
0.0	0.768536	0.678094	1.133	12.5	153.11861	29.69679	5.156
0.5	1.181862	0.990209	1.194	13.0	168.41071	31.47746	5.350
1.0	1.774455	1.396375	1.271	13.5	184.60190	33.29308	5.545
1.5	2.594650	1.900833	1.365	14.0	201.70950	35.14297	5.740
2.0	3.691502	2.502458	1.475	14.5	219.75048	37.02649	5.935
2.5	5.112536	3.196598	1.599	15.0	238.74150	38.94304	6.130
3.0	6.902476	3.976985	1.736	15.5	258.69893	40.89206	6.326
3.5	9.102801	4.837066	1.882	16.0	279.63888	42.87300	6.522
4.0	11.751801	5.770726	2.036	16.5	301.57717	44.88535	6.719
4.5	14.88489	6.77257	2.198	17.0	324.52939	46.92862	6.915
5.0	18.53496	7.83797	2.365	17.5	348.51087	49.00235	7.112
5.5	22.73279	8.96299	2.536	18.0	373.53674	51.10608	7.309
6.0	27.50733	10.14428	2.712	18.5	399.62188	53.23939	7.506
6.5	32.88598	11.37898	2.890	19.0	426.78099	55.40187	7.703
7.0	38.89481	12.66464	3.071	19.5	455.02855	57.59313	7.901
7.5	45.55875	13.99910	3.254	20.0	484.37885	59.81279	8.098
8.0	52.90173	15.38048	3.440				

As the degenerate region is approached the pressure greatly exceeds the perfect gas pressure. In a mixture of equal numbers of degenerate electrons and non-degenerate protons, the pressure of the electrons predominates and other partial pressures may be neglected.

Regions 3, 4, 5; complete degeneracy: The equation of state is given in terms of a parameter, x, where

$$x = p_0/m\,c, \tag{15.14}$$

$$N_e = \frac{8\pi V}{3}\left(\frac{m_e c}{h}\right)^3 x^3, \tag{15.15}$$

$$P_e = A\,f(x), \tag{15.16}$$

$$A = \frac{\pi\, m_e^4\, c^5}{3h^3}, \tag{15.17}$$

$$f(x) = x(2x^2-3)(x^2+1)^{\frac{1}{2}} + 3\,\mathrm{Ar\,Sin}\,x. \tag{15.18}$$

These regions are of importance in the theory of white dwarfs, treated on p. 723 by SCHATZMAN.

In regions 3 and 5 the parameter x may be eliminated. In addition, neglecting all other pressures relative to the degenerate electron pressure and recalling the definition of μ_e, Eq. (13.5), the total pressure may be expressed in terms of the total mass density.

Region 3; complete non-relativistic degeneracy:

$$P = K_1\,\varrho^{\frac{5}{3}}, \tag{15.19}$$

$$K_1 = \frac{h^2}{5m_e}\left(\frac{3}{8\pi}\right)^{\frac{2}{3}}\frac{1}{(\mu_e H)^{\frac{5}{3}}}. \tag{15.20}$$

This corresponds to a polytropic relation with $n = \frac{3}{2}$.

Region 5; complete relativistic degeneracy:

$$P = K_2 \varrho^{\frac{4}{3}}, \tag{15.21}$$

$$K_2 = \frac{c\,h}{4}\left(\frac{3}{8\pi}\right)^{\frac{1}{3}} \frac{1}{(\mu_e H)^{\frac{4}{3}}}. \tag{15.22}$$

This corresponds to a polytropic relation with $n = 3$.

Region 6; transition region from relativistic non-degeneracy to relativistic degeneracy:

$$N_e = 8\pi V \left(\frac{kT}{h\,c}\right)^3 F_2(\psi), \tag{15.23}$$

$$P_e = \frac{1}{3}\frac{N_e}{V} kT \frac{F_3(\psi)}{F_2(\psi)}. \tag{15.24}$$

Region 9: In this region (15.5) and (15.6) must be solved by quadratures.

16. Radiative opacity. Opacity is the macroscopic description of the microscopic interaction of material and radiation. At the high temperatures of stellar interiors the material is highly ionized and most of the radiant energy consists of X-rays. Three sources of opacity are important in our problem: photoionization of bound electrons (bound-free); photoabsorption by free electrons in the neighborhood of ions (free-free); and scattering by free electrons. Photoabsorption which raises an electron from one bound state to another bound state (bound-bound or line absorption) is usually neglected.

The opacity at any frequency, $\varkappa_\nu$, is the sum of the contributions of each of these processes:

$$\varkappa_\nu = \varkappa_\nu\,(\text{bound-free}) + \varkappa_\nu\,(\text{free-free}) + \varkappa_\nu\,(\text{scattering}). \tag{16.1}$$

The Rosseland mean opacity, defined by (8.15), is calculated using the total opacity[1].

17. Bound-free contribution. Before discussing a more precise theory of this process we may easily obtain some heuristic results. In order for the contribution of bound-free processes to be important there must be an appreciable number of ions with bound electrons.

To a first approximation, the ionization is governed by the Saha equation

$$\frac{N^{(j-1)}}{N^{(j)}} = \frac{u_{j-1}}{u_j}\frac{G(T)}{N_e}\,e^{-\chi_j/kT}, \tag{17.1}$$

where

$$G(T) = \frac{2(2\pi m_e kT)^{\frac{3}{2}}}{h^3} = 4.830 \times 10^{15}\,T^{\frac{3}{2}}. \tag{17.2}$$

$N^{(k)}$ is the number of ions per unit volume with k bound electrons and u_k is the partition function.

At high temperatures, atoms with low ionization energies are completely stripped of electrons. The light elements H and He, which are abundant in stars, have no bound electrons at high temperatures and therefore do not contribute directly to bound-free opacity. They contribute indirectly, however, by supplying free electrons which affect the degree of ionization of the heavier elements.

---

[1] It should be noted that the Rosseland mean is a harmonic mean and therefore

$$\bar\varkappa \neq \bar\varkappa_1 + \bar\varkappa_2 + \bar\varkappa_3.$$

The outer shells of heavy atoms are also unoccupied. To estimate the general level of ionization we may define a dominant potential, ψ, such that

$$e^{\psi/kT} = G(T)/N_e. \tag{17.3}$$

Then, neglecting the ratio of partition functions,

$$\frac{N^{(j-1)}}{N^{(j)}} \approx e^{(\psi-\chi_j)/kT}. \tag{17.4}$$

Therefore, if the ionization energy of an ion having j bound electrons is approximately equal to the dominant potential, there are roughly as many ions with $j-1$ bound electrons as j bound electrons in the gas.

In particular, if any shell has an ionization potential appreciably less than ψ, it is stripped and does not contribute to bound-free absorption. Furthermore, if the K-shell of any element has $\chi \ll \psi$, that element does not contribute at all to the bound-free absorption. For this reason, the temperature dependence of the stellar opacity is sensitive to the abundance of elements whose K-shell ionization energy is close to the dominant potential. If these elements are abundant there is a decrease of opacity (per gram) when the temperature is sufficiently high to ionize the K-shell completely.

Approximating the ionization potential by the hydrogen-like expression

$$\chi = \frac{c\,R_\infty\,h\,Z^2}{n^2}, \tag{17.5}$$

where n is the principal quantum number, the condition

$$\chi = \psi \tag{17.6}$$

yields

$$Z^2 = 14.6\,n^2\,T\,[24.7 + \tfrac{3}{2}\log T - \log N_e], \tag{17.7}$$

where T is measured in millions of degrees.

Even if there are bound electrons in a shell of principal quantum number n, they cannot contribute to the opacity unless the incident photon has sufficient energy to remove an electron from this shell; that is, not unless

$$\nu \geq \nu_n, \tag{17.8}$$

where

$$h\,\nu_n \approx \chi_n. \tag{17.9}$$

[1]

Approaching ν_n from lower frequency, there is a sharp discontinuity in opacity upon crossing ν_n; this is called an *absorption edge*. Photons with energies larger than $h\nu_n$ can also cause ionizations from this shell but with smaller probability. The absorption coefficient varies approximately as ν^{-3} to the violet of an absorption edge.

For a given element, the frequency dependence of the bound-free absorption coefficient has a characteristic saw-toothed appearance due to the absorption edges of the K, L, M, and higher shells. The contribution of a particular edge to the total absorption depends upon temperature since it is proportional to the average number of bound electrons in that shell, $N_n^{(Z)}$, sometimes called the *occupation number*.

The detailed calculation of the bound-free absorption coefficient of a mixture is carried out in several steps. From the theory of KRAMERS [2], we find the classical

[1] The actual position of an absorption edge is affected by pressure.
[2] H. A. KRAMERS: Phil. Mag. **46**, 836 (1923).

absorption coefficient at frequency ν per electron in shell n per atom of atomic number Z. A correction, called the Gaunt factor, accounts for quantum mechanical effects.

To find the contribution of a particular shell, the previous result is multiplied by the occupation number of the shell. At a particular frequency, the contribution of a single element is the sum of contributions from all shells with absorption edges toward lower frequencies. Finally, the absorption coefficient of the mixture is the sum of the absorption coefficient of each element weighted by the abundance of that element. The result is usually expressed in cm² per gram.

Thus

$$\varkappa_\nu \text{ (bound-free)} = \sum_Z X_Z \varkappa_\nu^{(Z)} \text{ (bound-free)}, \tag{17.10}$$

where X_Z is the abundance of element Z and the Kramers opacity is

$$\varkappa_\nu^{(Z)} \text{ (bound-free)} = \frac{1}{AH} \frac{64\pi^4 Z^4}{3\sqrt{3}} \frac{m_e e^{10}}{c h^6} \frac{g}{n^5} \frac{1}{\nu^3} N_n^{(Z)}. \tag{17.11}$$

A is the mass number of element Z.

The Gaunt factor, g, depends on both n and ν; it is close to unity near an absorption edge. The detailed theory for hydrogen and helium has been developed by MENZEL and PEKERIS[1].

At low pressures and densities the occupation numbers may be found from[2]

$$N_n^{(Z)} = \frac{2n^2}{1 + \frac{G(T)}{N_e} e^{-\chi_n/kT}}. \tag{17.12}$$

Detailed calculations of occupation numbers under conditions appropriate to stellar interiors have recently been made by KELLER and MEYEROTT[3]. They considered an ion sphere in a medium of free electrons, protons and α-particles. The free particles perturb the outermost shells of the heavy ions beyond recognition and also effectively screen the ions from each other. Near the ions the medium is polarized. The physical quantity of importance is the potential in the vicinity of the ions. This is found by numerical integration of a generalized Thomas-Fermi equation. A comparison of the occupation numbers given by KELLER and MEYEROTT with those predicted from (17.12) are given in Table 2. We shall return to the Keller and Meyerott computations in Sect. 20.

Table 2. *Occupation numbers calculated by Keller and Meyerott compared with Eq. (17.12) for the K shell of oxygen, $T = 5.81 \times 10^6$ °K, $X = 0.485$.*

α_e	ϱ	KELLER-MEYEROTT	Eq. (17.12)	α_e	ϱ	KELLER-MEYEROTT	Eq. (17.12)
2	19.66	0.66	0.85	8	0.05118	0.0037	0.0038
4	2.828	0.170	0.192	10	0.006869	0.0005	0.0005
5	1.024	0.066	0.074	12	0.0009293	0.0001	0.00001
6	0.3771	0.028	0.028				

18. Free-free contribution. In the neighborhood of an ion of charge Ze, an electron may absorb or emit radiation by jumping from one hyperbolic orbit to

[1] D. H. MENZEL and C. L. PEKERIS: Monthly Notices Roy. Astronom. Soc. London **96**, 77 (1935).

[2] [3], p. 258.

[3] G. KELLER and R. E. MEYEROTT: Argonne National Laboratory Reports ANL-4771 and ANL-4856, Chicago 1952.

another. (The presence of the ion is necessary to conserve momentum.) The effect is again proportional to ν^{-3}, and in addition varies inversely as the velocity of the electron. For electrons obeying a Maxwellian velocity distribution, the theory gives

$$\varkappa_\nu^{(Z)} \text{ (free-free)} = \frac{1}{AH} \frac{16\pi^2 Z^2}{3\sqrt{3}} \frac{e^6}{hc(2\pi m_e)^{\frac{3}{2}}} \frac{N_e}{(kT)^{\frac{1}{2}}} \frac{g_{\rm ff}}{\nu^3}. \tag{18.1}$$

This expression also involves a Gaunt factor, $g_{\rm ff}$, which presents some difficulties that have not been fully resolved[1, 2].

Since free-free opacity does not require bound electrons, nuclei of H and He can play an active role. Indeed, the abundances of H and He are frequently so large that they provide almost all of the nuclei around which the free electrons swing, and heavier ions may be disregarded.

19. Electron scattering[3]. The classical theory of scattering by free electrons (Thomson scattering) is independent of the frequency of the incident photon. The coefficient per electron (cross section),

$$\sigma = \frac{8\pi e^4}{3 m_e^2 c^4} = 0.6653 \times 10^{-24} \text{ cm}^2, \tag{19.1}$$

is rather small but it provides a lower limit to the fogginess of ionized material. Expressed in terms of the mass absorption coefficient, this gives

$$\varkappa_\nu \text{ (scattering)} = \frac{8\pi e^4}{3 m_e^2 c^4} \frac{(1+X)}{2} \frac{1}{H}, \tag{19.2}$$

where (13.5) has been used to estimate the number of electrons.

For very high energy photons it is necessary to take into account the Compton effect. The scattering coefficient for photon energies much less than the rest mass energy of the electron is predicted by the Klein-Nishina formula to be

$$\sigma = \frac{8\pi e^4}{3 m_e^2 c^4} \left[1 - 2\frac{h\nu}{mc^2}\right], \tag{19.3}$$

so that even the minimum fogginess decreases toward the hard X-rays.

20. Rosseland mean opacities. For practical calculations (8.15) is evaluated in the form

$$\frac{1}{\varkappa} = \int_0^\infty \frac{W(u)}{\varkappa_u u^3} du, \tag{20.1}$$

where

$$u = h\nu/kT, \tag{20.2}$$

and the weighting function

$$W(u) = \frac{15}{4\pi^4} u^7 e^{2u} (e^u - 1)^{-3}. \tag{20.3}$$

The denominator of (20.1) is represented by a weighted sum over all shells of all important elements, plus an electron scattering term. It is convenient to write the contribution of a particular shell as

$$\varkappa_u(n, Z) \propto \frac{B(n, Z)}{u^3}, \tag{20.4}$$

[1] Keller and Meyerott: Astrophys. Journ. **122**, 32 (1955).
[2] R. M. Kulsrud: Astrophys. Journ. **119**, 386 (1954).
[3] Cf. R. D. Evans: Compton effect, Vol. XXXIV, this Encyclopedia.

where $B(n, Z)$ is a step function: it is zero for photons of insufficient energy to cause ionization ($u < u_n$); and it jumps discontinuously to a constant positive

Table 3. *Keller-Meyerott opacities for mixture given in legend to Fig. 8.*

$T \times 10^{-6}$	$\alpha_e = 6$		$\alpha_e = 8$		$\alpha_e = 12$	
	$\varkappa$	ϱ	$\varkappa$	ϱ	$\varkappa$	ϱ
0.116			320	0.000109	6.7	0.0000020
0.232			49	0.00030	1.49	0.0000057
0.581	70	0.0090	12.3	0.00122	0.74	0.000022
1.16	13.3	0.025	2.8	0.0034	0.49	0.000063
2.32	5.8	0.072	1.46	0.0097	0.44	0.000179
4.06	2.5	0.166	0.76	0.022	0.40	0.00042
5.81	1.08	0.28	0.52	0.038	0.39	0.00070
8.13	0.69	0.47	0.45	0.062	0.39	0.00116
11.6	0.52	0.80	0.41	0.109	0.38	0.0020
23.2	0.40	2.2	0.38	0.30	0.38	0.0057

value for $u \geq u_n$. Calculations of mean opacities have been carried out by MORSE[1], ZIRIN[2], and, most recently, KELLER and MEYEROTT[3]. The latter computations are the most complete to date. Several mixtures of elements were used and the results were given in terms of the temperature and a parameter α_e related to density. Some of these results are reproduced in Table 3.

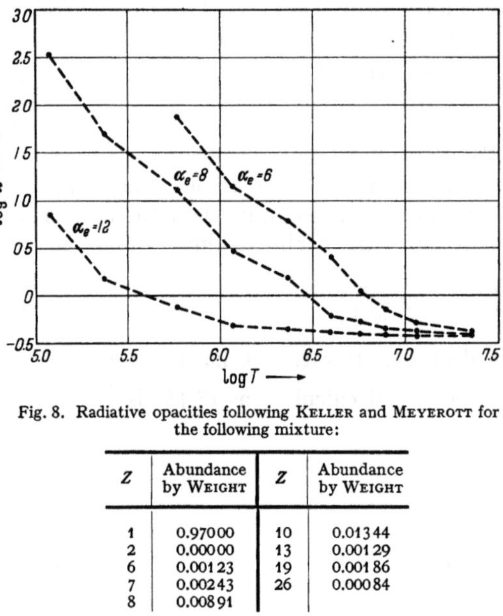

Fig. 8. Radiative opacities following KELLER and MEYEROTT for the following mixture:

Z	Abundance by WEIGHT	Z	Abundance by WEIGHT
1	0.97000	10	0.01344
2	0.00000	13	0.00129
6	0.00123	19	0.00186
7	0.00243	26	0.00084
8	0.00891		

Dotted lines connect tabulated points for $\alpha_e = 6, 8, 12$.

All of the above calculations are similar in general aspects, but they differ in details. At a constant density and composition they all exhibit the characteristic decrease of $\varkappa$ roughly as $T^{-3.5}$, with numerous fluctuations due to the absorption edges of various elements (Fig. 8). At high temperature the limit imposed by electron scattering is reached.

Numerous attempts have been made to represent absorption coefficient curves in terms of simple functions to facilitate calculations. The form

$$\varkappa = \varkappa_0 \varrho \, T^{-3.5}, \qquad (20.5)$$

where $\varkappa_0$ depends upon chemical composition, is known as Kramers opacity. This may be improved by dividing by the so-called "guillotine factor", τ, so that the opacity becomes

$$\varkappa = \varkappa_0 \frac{\varrho T^{-3.5}}{\tau}. \qquad (20.6)$$

[1] P. M. MORSE: Astrophys. Journ. **92**, 27 (1940).

[2] H. ZIRIN: Astrophys. Journ. **119**, 371 (1954). (L. ALLER has pointed out that these tables contain errors for which errata have not been published.)

[3] G. KELLER and R. E. MEYEROTT: Astrophys. Journ. **122**, 32 (1955).

The guillotine factor is found empirically from the detailed calculations, as a function of ϱ, T, and composition. It can therefore be regarded as a correction term to bring KRAMERS' expression into agreement with detailed results.

SCHWARZSCHILD has used a "modified Kramers opacity" of the form

$$\varkappa = \varkappa_0\, \varrho^{0.75}\, T^{-3.5}. \tag{20.7}$$

Others (for example, P. NAUR) have used a more general expression

$$\varkappa = \varkappa_0\, \varrho^{1-\alpha}\, T^{-3-s}, \tag{20.8}$$

where α and s are assigned different values in different regions of a star.

When computations are carried out on large-scale computers it is possible to use more sophisticated interpolation formulae. Nevertheless, for general surveys of various problems, these simple approximating functions are very useful.

21. Energy generation: general remarks. The two sources of stellar energy are gravitation and nuclear reactions. They each play a role at different stages of stellar evolution. A star has much more nuclear energy available than gravitational energy, but nuclear reactions can only occur if the temperature is high enough *and* if the reacting nuclei are sufficiently abundant. Thus in the early stages of star formation the gas cloud is too cool to produce nuclear reactions in hydrogen even though hydrogen is very abundant; and during later stages of evolution, although the central temperature is high enough, the local hydrogen supply may have been exhausted. Under these latter circumstances the star must fall back on its gravitational energy which, as we shall see, heats the star to the point where other nuclear reactions occur.

22. Gravitational contraction. A gas sphere contracting under its own gravitation decreases in potential energy. Only part of this energy is radiated; some is converted to internal energy and the star becomes hotter. The division of energy between released radiation and internal heating may be derived from the virial theorem (see [3], p. 49):

$$\frac{1}{2}\frac{d^2 I}{dt^2} = 2\mathcal{T} + \Omega, \tag{22.1}$$

where I is the moment of inertia of a configuration; $\mathcal{T}$, the kinetic energy of the constituent particles; and Ω, the potential energy of the system. If the second derivative of I may be neglected, we get the simpler result,

$$2\mathcal{T} + \Omega = 0, \tag{22.2}$$

on which our conclusions will be based.

This analysis will be restricted to the case of a perfect (though not necessarily monatomic) gas. The kinetic energy of an N particle system is

$$\left.\begin{aligned} \mathcal{T} &= \tfrac{3}{2} N\, kT, \\ &= \tfrac{3}{2} N\, HR\, T, \\ &= \tfrac{3}{2} N\, \mu\, H (c_P - c_V)\, T; \end{aligned}\right\} \tag{22.3}$$

where R is the gas constant per *mole* and c_P and c_V are heat capacities per *gram* at constant pressure and constant volume, respectively.

Further the internal energy is given by

$$U = N\, \mu\, H\, c_V\, T. \tag{22.4}$$

Therefore, using (10.8)

$$\mathcal{T} = \tfrac{3}{2}(\gamma - 1)\, U. \tag{22.5}$$

Substituting in the virial theorem, we obtain

$$3(\gamma - 1)\, U = -\Omega, \tag{22.6}$$

or

$$\Delta U = -\frac{1}{3(\gamma - 1)}\,\Delta\Omega. \tag{22.7}$$

In other words, a fraction, $\frac{1}{3(\gamma-1)}$, of the change of potential energy on contraction, $\Delta\Omega$, is transformed to internal energy. The remainder,

$$\frac{4 - 3\gamma}{3(\gamma - 1)}\,\Delta\Omega, \tag{22.8}$$

is radiated.

If instead of considering the contraction of a star as a whole, we consider the energy released by local changes of conditions, we can evaluate the energy production per gram per second due to contraction and expansion. From the first law of thermodynamics,

$$\varepsilon = \frac{\partial U}{\partial t} + P\,\frac{\partial V}{\partial t}; \tag{22.9}$$

where we are restricting our attention to one gram of material. The volume of one gram is the reciprocal of the density, and the internal energy, U, is $c_V T$; therefore:

$$\varepsilon = c_V\,\frac{\partial T}{\partial t} + P\,\frac{\partial\left(\frac{1}{\varrho}\right)}{\partial t}. \tag{22.10}$$

Writing the equation of a perfect monatomic gas as

$$\frac{P}{\varrho} = (c_P - c_V)\, T, \tag{22.11}$$

and noting (10.8), we have

$$\varepsilon = \frac{3}{2}\,\frac{\partial}{\partial t}\left(\frac{P}{\varrho}\right) + P\,\frac{\partial}{\partial t}\left(\frac{1}{\varrho}\right), \tag{22.12}$$

or

$$\left.\begin{aligned} \frac{2}{3}\,\frac{\varrho}{P}\,\varepsilon &= \frac{5}{3}\,\varrho\,\frac{\partial}{\partial t}\left(\frac{1}{\varrho}\right) + \frac{1}{P}\left(\frac{\partial P}{\partial t}\right), \\ &= \frac{\partial}{\partial t}\ln\left(\frac{P}{\varrho^{\frac{5}{3}}}\right), \\ &= \frac{\varrho^{\frac{5}{3}}}{P}\,\frac{\partial}{\partial t}\left(\frac{P}{\varrho^{\frac{5}{3}}}\right). \end{aligned}\right\} \tag{22.13}$$

Finally, we obtain

$$\varepsilon = \frac{3}{2}\,\varrho^{\frac{2}{3}}\,\frac{\partial}{\partial t}\left(\frac{P}{\varrho^{\frac{5}{3}}}\right). \tag{22.14}$$

In this form it is clearly seen that ε is the result of departures from the adiabatic relation between pressure and density.

23. The Helmholtz-Kelvin time scale. If gravitation were the only source of energy, stars would have relatively short lives. The luminosity would be

$$L = -\frac{3\gamma - 4}{3(\gamma - 1)}\,\frac{d\Omega}{dt}. \tag{23.1}$$

Assuming the star to be infinitely diffuse at some initial time and to radiate at a constant rate, $\bar{L}$, the total production of energy over a period of time, t, would be

$$\bar{L}\,t = -\frac{3\gamma - 4}{3(\gamma - 1)}\,\Omega, \tag{23.2}$$

where Ω would be the potential energy of the final configuration. Since, roughly

$$\Omega \approx -\frac{G M^2}{R}, \tag{23.3}$$

the total time the Sun could radiate at its present rate by contraction alone would be

$$t_{\text{sun}} = \frac{3\gamma - 4}{3(\gamma - 1)}\,\frac{G M_\odot^2}{R_\odot L_\odot}. \tag{23.4}$$

Assuming $\gamma = \frac{5}{3}$, we find a time scale of the order of 10^7 years; far too short to fit the geological evidence of several billion years.

24. Nuclear energy sources.—The proton-proton chain and the carbon-nitrogen cycle. The conversion of mass to energy as expressed by EINSTEIN's equation,

$$E = m c^2, \tag{24.1}$$

provides by far the largest fraction of the energy radiated by a star during its existence. The luminosity of the Sun is equivalent to the conversion of 4.3×10^{12} grams per second; assuming a constant luminosity for 5×10^9 years implies an original solar mass only 0.03% greater than at present. The nuclear time scale is therefore not inconsistent with the geological evidence.

The destruction of mass can occur only in certain types of reactions. For example, it is not possible to convert a proton entirely to energy in the stellar interior[1]. It is possible, however, to reap some gain by converting four hydrogen atoms to a helium atom. Each hydrogen atom has a mass of 1.00815 atomic mass units $= 1.6736 \times 10^{-24}$ grams and a helium atom has a mass of 4.00387 atomic mass units $= 6.647 \times 10^{-24}$ grams. From (24.1), one atomic mass unit is the equivalent of 1.492×10^{-3} ergs. Thus if the transmutation of hydrogen to helium can be accomplished, a difference of 0.0287 atomic mass units appears as energy (4.28×10^{-5} ergs). Not all of this energy is available to the star, however, since some is carried away by neutrinos which escape.

The fact that the mass of a nucleus differs from an integral number of atomic mass units[2] may be expressed in terms of the packing fraction

$$f(A) = \frac{M(A) - A}{A}, \tag{24.2}$$

where A is the mass number and $M(A)$ is the nuclear mass in atomic mass units. If it were possible to pack together A particles of unit atomic mass to form a nucleus of mass $M(A)$, the packing fraction would be proportional to the energy released per particle. Thus the energy available, per unit atomic mass, in converting from one element to another is proportional to the difference of their packing fractions. Naively it might be expected that the conversion of hydrogen to iron would release much more energy than the conversion of hydrogen to helium. The packing fraction of helium, however, is not very different from iron and the conversion of 56 protons to one iron nucleus yields only 20% more energy than the conversion of 56 protons to 14 α-particles (Fig. 9).

[1] Antiprotons, which could annihilate the protons, are not likely to be produced at such low energies.

[2] For details see H. WAPSTRA in Vol. XXXVIII, Part 1 of this Encyclopedia.

BETHE[1] systematically studied the possible reactions which could supply stellar energy. He concluded that the conversion of hydrogen to helium could be accomplished by two mechanisms: the proton-proton chain and the carbon cycle.

The proton-proton chain ($p-p$) involves six protons, but two of them are retrieved at the end. Reactions (24.3) each occur twice:

$$\left.\begin{aligned} &H^1 + H^1 \rightarrow D^2 + \beta^+ + \nu \quad \text{or} \quad H^1(p, \beta^+ \nu)\, D^2, \\ &D^2 + H^1 \rightarrow He^3 + \gamma \qquad \text{or} \quad D^2(p, \gamma)\, He^3. \end{aligned}\right\} \tag{24.3}$$

Then the two resulting He^3 nuclei combine to yield He^4 and two protons[2]:

$$\left.\begin{aligned} &He^3 + He^3 \rightarrow He^4 + 2H^1 \\ &\text{or} \quad He^3(He^3, 2p)\, He^4. \end{aligned}\right\} \tag{24.4}$$

Note added in proof: Recent results due to H. D. HOLMGREN and R. L. JOHNSTON have shown that the reaction $He^3(\alpha, \gamma)\, Be^7$ will compete with $He^3(He^3, 2p)\, He^4$ when the temperatures are high and helium is abundant. Statements made in the following sections concerning the proton-proton chain will have to be modified to take into account the following alternative reactions:

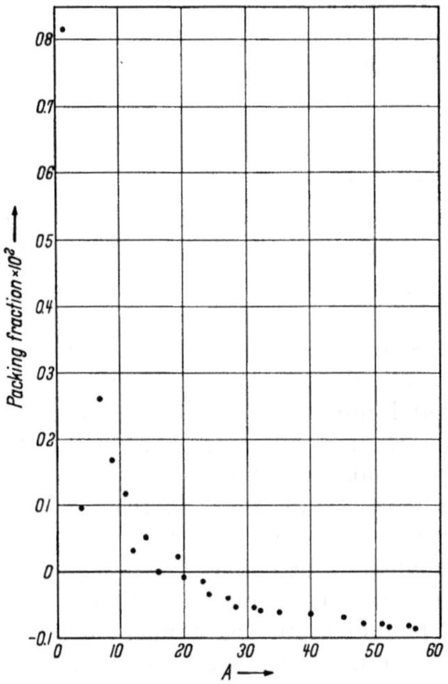

Fig. 9. Packing fractions for elements up to iron.

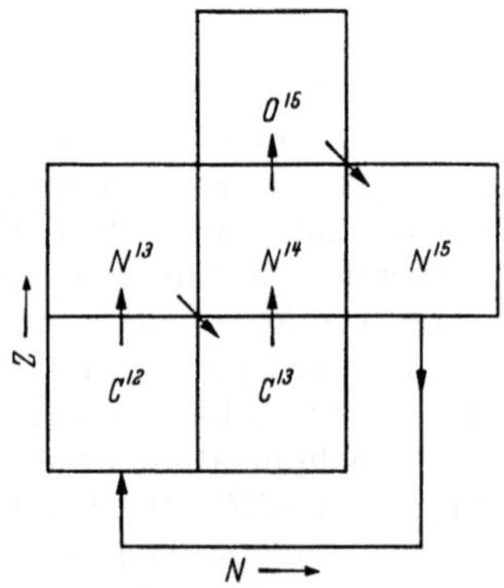

Fig. 10.

$$He^3(He^3, 2p)\, He^4$$

or

$$He^3(\alpha, \gamma)\, Be^7(e^- \nu)\, Li^7(p, \alpha)\, He^4$$

or

$$He^3(\alpha, \gamma)\, Be^7(p, \gamma)\, B^8(\beta^+ \nu)\, Be^{8*}(\alpha)\, He^4.$$

The carbon cycle involves C^{12} as a catalyst which reappears at the end of the cycle:

$$\left.\begin{aligned} &C^{12}(p, \gamma)\, N^{13} \\ &N^{13}(\beta^+ \nu)\, C^{13} \\ &C^{13}(p, \gamma)\, N^{14} \\ &N^{14}(p, \gamma)\, O^{15} \\ &O^{15}(\beta^+ \nu)\, N^{15} \\ &N^{15}(p, \alpha)\, C^{12}. \end{aligned}\right\} \tag{24.5}$$

[1] H. BETHE: Phys. Rev. 55, 434 (1939).

[2] This reaction was first proposed by C. C. LAURITSEN and by E. SCHATZMAN independently.

Schematically the carbon cycle is shown in Fig. 10. The two β-decay processes occur in a very short time compared with the (p, γ) and (p, α) reactions, except at temperatures greater than 10^8 degrees.

As we shall subsequently see, the proton-proton chain is predominantly responsible for energy production at low temperature and the carbon cycle at high temperature.

25. Thermonuclear reactions. We may regard reactions which build successively heavier nuclei by the absorption of protons as taking place in three steps:

1. collision with a proton;
2. penetration of the proton into the nucleus;
3. γ emission.

The number of collisions per second in the stellar interior is determined by the thermal motion of the nuclei; hence the term "thermonuclear reactions". Not all collision result in penetration; most protons are repulsed by the Coulomb barrier and get no closer than the point at which the Coulomb potential energy is equal to the particle's initial kinetic energy. Finally, not all penetrations result in transmutation to a heavier element; in some cases there is re-emission of the penetrating particle or shattering of the compound nucleus. We shall now consider each step in turn.

From kinetic theory[1] we find the number of collisions per unit volume per second between two types of nuclei with relative velocities between v and $v+dv$ to be

$$(\pi R_{12}^2 v)\, N_1 N_2\, 4\pi \left(\frac{m_1 m_2}{m_1+m_2}\right)^{\frac{3}{2}} \frac{1}{(2\pi k T)^{\frac{3}{2}}} \left[\exp\left(-\frac{m_1 m_2}{m_1+m_2}\,\frac{v^2}{2kT}\right)\right] v^2\, dv, \tag{25.1}$$

where R_{12} is the mean radius of the colliding nuclei, N_1 and N_2 are the number of particles of each type per unit volume and $m_1 m_2/(m_1+m_2)$ is the reduced mass. In the form given above, the first term in brackets may be regarded as the volume swept by a particle of radius R_{12} and velocity v in unit time and

$$4\pi N_2 \left(\frac{m_1 m_2}{m_1+m_2}\right)^{\frac{3}{2}} \frac{1}{(2\pi k T)^{\frac{3}{2}}} \left[\exp\left(-\frac{m_1 m_2}{m_1+m_2}\,\frac{v^2}{2kT}\right)\right] v^2\, dv \tag{25.2}$$

is the number of particles of type 2 encountered per unit volume with relative velocities between v and $v+dv$.

Transforming from relative velocity to energy we find the number of collisions per unit volume per unit time with relative energy between E and $E+dE$ to be

$$8 R_{12}^2 N_1 N_2 \frac{\pi^{\frac{1}{2}}}{(2kT)^{\frac{3}{2}}} \left(\frac{m_1+m_2}{m_1 m_2}\right)^{\frac{1}{2}} e^{-E/kT}\, E\, dE. \tag{25.3}$$

Instead of using the geometrical cross section, we use the maximum wave mechanical cross section for zero orbital angular momentum,

$$\pi \lambda\!\!\!^{-2} = \frac{\pi \hbar^2}{2E}\,\frac{m_1+m_2}{m_1 m_2}, \tag{25.4}$$

where $\hbar = h/2\pi$.

Penetrations are possible in spite of the Coulomb barrier because this obstruction is of finite thickness. In quantum mechanical terms, the wave function of the incident particle is attenuated in passing through the potential wall but not extinguished. The small amplitude inside the nucleus indicates a finite probability that the particle can penetrate.

[1] J. Jeans: An introduction to the kinetic theory of gases. Cambridge: Cambridge University Press 1940.

The probability of penetration per collision is given by the Gamow factor $\exp(-2G)$ where

$$2G = -\frac{4e(2Z_1Z_2R_{12})^{\frac{1}{2}}}{\hbar}\left(\frac{m_1m_2}{m_1+m_2}\right)^{\frac{1}{2}} + \frac{2^{\frac{3}{2}}\pi^2 Z_1Z_2e^2}{hE^{\frac{1}{2}}}\left(\frac{m_1m_2}{m_1+m_2}\right)^{\frac{1}{2}}. \tag{25.5}$$

That is, penetrations are least likely for small energies and large nuclear charges and most likely at high temperatures (large E) and when one of the constituents is a proton (small Z).

Multiplying the number of collisions by the probability of penetration we find the theoretical number of penetrations with relative kinetic energy between E and $E+dE$ per unit volume per second to be

$$N_1N_2\frac{h^2}{(2\pi kT)^{\frac{3}{2}}}\left(\frac{m_1+m_2}{m_1m_2}\right)^{\frac{3}{2}}\exp\left[-\frac{E}{kT}-2G\right]dE. \tag{25.6}$$

After penetration a compound nucleus is formed which may have several modes of decay. For a successful transmutation to a heavier nucleus to occur, no particle may be emitted; instead the excess energy is released by γ radiation. The likelihood of γ emission is usually expressed in terms of $\Gamma_\gamma = \hbar/\tau_\gamma$, where τ_γ is the mean life of the compound nucleus to γ emission. The probability of γ emission is

$$\frac{2\Gamma_\gamma R_{12}^2}{\hbar^2}\,\frac{m_1m_2}{m_1+m_2}, \tag{25.7}$$

where R is the radius of the compound nucleus and empirically

$$R_{12} = 1.4\times10^{-13}\left(A_1^{\frac{1}{3}}+A_2^{\frac{1}{3}}\right), \tag{25.8}$$

where A_1 and A_2 are the atomic numbers of the nuclei involved. The quantity Γ_γ has the dimension of energy and is referred to as the *gamma width* from its role in the dispersion theory of resonance cross sections.

If every gram of material contains x_1 and x_2 grams of elements 1 and 2, we may convert from N_1N_2 to x_1x_2 as follows

$$N_1 = \frac{\varrho x_1}{A_1H}; \quad N_2 = \frac{\varrho x_2}{A_2H}, \tag{25.9}$$

where A_1 and A_2 are the respective mass numbers. The number of reactions per *gram* per second with particles in the energy range E to $E+dE$ becomes

$$\varrho\frac{x_1x_2}{A_1A_2}\,\frac{\Gamma_\gamma R_{12}^2}{H^{\frac{5}{2}}}\,\frac{2^{\frac{3}{2}}\pi^{\frac{1}{2}}}{(kT)^{\frac{3}{2}}}\left(\frac{A_1+A_2}{A_1A_2}\right)^{\frac{1}{2}}\exp\left[-\frac{E}{kT}-2G\right]dE. \tag{25.10}$$

To find the total number of transmutations per gram per second, Eq. (25.10) must be integrated over all energies. The energy dependence appears in two exponential terms:

$$\exp\left(-y-2Q^3y^{-\frac{1}{2}}\right); \tag{25.11}$$

where we have made the transformation, ([3], p. 465),

$$\left.\begin{aligned} y &= E/kT,\\ Q^3 &= 5.34\times10^4\left(\frac{A_1A_2}{A_1+A_2}\right)^{\frac{1}{2}}\frac{Z_1Z_2}{T^{\frac{1}{2}}}. \end{aligned}\right\} \tag{25.12}$$

The term e^{-y} arises from the thermal collisions; it is a decreasing function of y. The other arises from the penetration factor, and increases with y. The product of the two functions (Fig. 11) is therefore small at both extremes and attains a maximum at $y = Q^2$ or at an energy

$$E_0 = 0.1222 \left(\frac{A_1 A_2}{A_1 + A_2}\right)^{\frac{1}{3}} Z_1^{\frac{2}{3}} Z_2^{\frac{2}{3}} T^{\frac{2}{3}} \text{ ev.} \tag{25.13}$$

Substituting values appropriate for carbon cycle reactions at 20×10^6 degrees, it is found that the maximum number of penetrations occur at energies in the range 25 to 30 kev. The maximum occurs at lower energies for the proton-proton chain. Since the maximum is sharp, we may associate with every stellar temperature an energy, E_0, which characterizes the energy of the nuclear reactions.

The evaluation of the integral,

$$\mathscr{I} = \int_0^\infty \exp\left(-y - 2Q^3 y^{-\frac{1}{2}}\right) dy, \tag{25.14}$$

may be accomplished approximately by expanding the integrand in the neighborhood of the maximum ([3], p. 465).

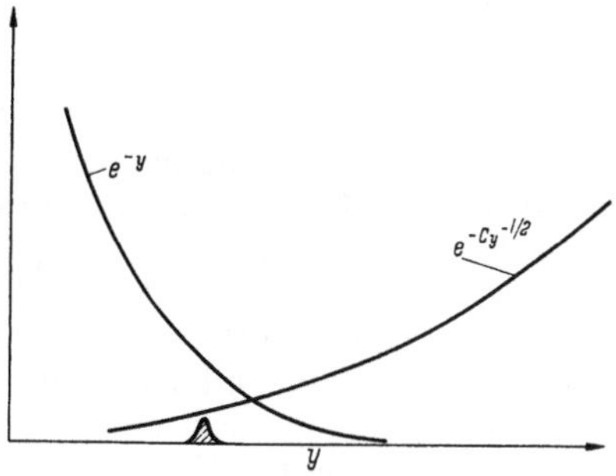

Fig. 11. Schematic diagram of the Maxwell and Gamow factors. The shaded area is the integrand in Eq. (25.14).

Thus, since Q is large,

$$\mathscr{I} \approx e^{-3Q^2} \int_{-3Q^2}^{\infty} \exp\left[-3(y - Q^2)^2/4Q^2\right] d(y - Q^2) = 2\left(\frac{\pi}{3}\right)^{\frac{1}{2}} Q\, e^{-3Q^2}. \tag{25.15}$$

From this the number of penetrations per gram per second is found to be proportional to

$$\varrho \frac{x_1 x_2}{A_1 A_2} \left(\frac{A_1 + A_2}{A_1 A_2}\right)^{\frac{1}{3}} \frac{Z_1^{\frac{1}{3}} Z_2^{\frac{1}{3}}}{T^{\frac{2}{3}}} \exp\left[-4.2483 \times 10^3 \left(Z_1^2 Z_2^2 A/T\right)^{\frac{1}{3}}\right]. \tag{25.16}$$

Experiments can be performed in the laboratory with particles of well-defined energy; therefore the term in (25.10) due to the thermal velocity distribution does not occur. For very low energies it can be shown[1] that the reaction cross section,

$$\sigma = \frac{\text{number of reactions per unit time per nucleus}}{\text{number of incident particles per unit area per unit time}} \tag{25.17}$$

becomes

$$\left.\begin{aligned} \sigma &= \frac{S_0}{E} \exp\left(-2\pi \frac{Z_1 Z_2 e^2}{\hbar v}\right), \\ &= \frac{S_0}{E} \exp\left(-31.281\, Z_1 Z_2 A^{\frac{1}{2}} E^{-\frac{1}{2}}\right); \end{aligned}\right\} \tag{25.18}$$

where E is measured in kev in the center-of-mass system of the two particles and A is the reduced mass in atomic units; i.e.,

$$A = \frac{A_1 A_2}{A_1 + A_2}; \tag{25.19}$$

S_0 is called the cross section factor and can be taken as approximately constant in non-resonant reactions.

[1] J. M. Blatt and V. F. Weisskopf: Theoretical Nuclear Physics. New York: John Wiley 1952.

Using the experimental value of S_0, the energy produced by a reaction per gram per second becomes[1]

$$\varepsilon = 7.567 \times 10^{26}\, \varrho\, x_1\, x_2\, S_0\, Q\, \frac{d Z_1^{\frac{1}{3}} Z_2^{\frac{1}{3}}}{A_1 A_2 A^{\frac{1}{3}}}\, \frac{f}{T^{\frac{2}{3}}}\, \mathrm{e}^{-\tau}\, \frac{\mathrm{erg}}{\mathrm{g\,sec}}; \tag{25.20}$$

where T is measured in million of degrees; Q is the energy liberated per reaction in Mev; f is the electron shielding factor which depends on both free and bound electrons; S_0 is in kev barns;

$$\tau = 42.483\, (Z_1^2 Z_2^2 A/T)^{\frac{1}{3}};$$

$d = 1$ for carbon cycle reactions and $\frac{1}{2}$ for the proton-proton reaction.

x_1 and x_2 are the abundances per gram as in (25.9). For further remarks concerning f and d see BOSMAN-CRESPIN, *et al.*

Apart from the factor f, the problem of stellar energy production becomes equivalent to the determination of S_0. In actuality, S_0 is not constant but has an energy (or, equivalently, temperature) dependence. Its determination experimentally is further complicated by the fact that existing laboratory techniques cannot be used to measure cross sections at the very low energies of astrophysical interest. As a result, the measurements are made at 100 kev or higher and extrapolated, using theoretical considerations, to energies in the vicinity of 25 kev.

Nuclear cross sections of light elements at low energies are marked by peaks or "resonances". An unobserved resonance in the 10 to 50 kev range may increase the cross section at astrophysical energies considerably and yet have little or no effect on the laboratory cross section at 100 kev. Extrapolations must therefore be used with care, particularly when the study of similar nuclei indicates the likelihood of a resonance at low energy.

26. Cross sections for hydrogen-burning reactions. Cross sections at stellar energies may be determined from (25.18):

$$\sigma = \frac{S_0}{E_0} \exp\left(-\, 31.281\, Z_1 Z_2 A^{\frac{1}{2}} E_0^{-\frac{1}{2}}\right) \text{ barns}, \tag{26.1}$$

where, (25.13):

$$E_0 = 1.220\, (Z_1^2 Z_2^2 A\, T^2)^{\frac{1}{3}} \text{ kev}. \tag{26.2}$$

Note that A is the reduced atomic mass number (25.19), and T is measured in millions of degrees.

The energy levels of many of the nuclei involved in the carbon cycle and the proton-proton chain may be found in Vol. XL of this Encyclopedia[2]. Three thorough summaries of cross sections have been made in recent years by FOWLER[3], BOSMAN-CRESPIN et al.[4], and BURBIDGE, BURBIDGE, FOWLER and HOYLE[5], and further discussion of these data will be found there.

$C^{12}(p, \gamma)\, N^{13}$: Fowler used the Thomas[6] modification of the Breit-Wigner dispersion formula to extrapolate measurements of the resonance curve of the

---

[1] D. BOSMAN-CRESPIN, W. A. FOWLER and J. HUMBLET: Bull. Soc. Roy. Sci. Liège **1954**, Nos. 9—10, 327.

[2] W. E. BURCHAM in Vol. XL of this Encyclopedia; see particularly pp. 62—96 and pp. 152—198.

[3] W. A. FOWLER: Mem. Soc. Roy. Sci. Liège **14**, 88 (1954).

[4] D. BOSMAN-CRESPIN, W. A. FOWLER and J. HUMBLET: Bull. Soc. Roy. Sci. Liège **1954**, Nos. 9—10.

[5] E. M. BURBIDGE, G. R. BURBIDGE, W. A. FOWLER and F. HOYLE: Rev. Mod. Phys. **29**, 547 (1957).

[6] R. G. THOMAS: Phys. Rev. **88**, 1109 (1952).

2.369 Mev level of N^{13} to stellar energies. The earlier measurements have been substantiated by very accurate measurements of low energy cross sections by LAMB and HESTER[1]. The best current value is

$$S_0 = (1.2 \pm 0.2) \text{ kev barns} \tag{26.3}$$

in the center of mass system.

BOSMAN-CRESPIN *et al.* refine this to include the temperature dependence of S_0:

$$S_0 = 1.09\,(1 + 0.017\,T^{\frac{2}{3}}) \text{ kev barns} \pm 17\%. \tag{26.4}$$

The energy release is 1.945 Mev.

$N^{13}(\beta^+\nu)\,C^{13}$: The measured mean lifetime of this process is 14.5 min. The energy released is 1.502 Mev plus a neutrino loss of 0.720 Mev.

$C^{13}(p, \gamma)\,N^{14}$: Fowler attributes the cross section at stellar energies predominately to the 8.06 Mev level of N^{14}. Nevertheless, a two-level formula employing the results of TEICHMANN[2] was used to obtain the value

$$S_0 = (6.1 \pm 2) \text{ kev barns.} \tag{26.5}$$

Recent preliminary measurements by LAMB and HESTER indicate a result larger by a factor of two.

The energy released by this process is 7.542 Mev.

$N^{14}(p, \gamma)\,O^{15}$: The rate of this reaction is the most uncertain in the carbon cycle. In his summary in 1953, FOWLER used the bound level of O^{15} at 6.84 Mev to extrapolate to stellar energies obtaining

$$S_0 = (33 \pm 11) \text{ kev barns.} \tag{26.6}$$

The cross section for $N^{14}(p, \gamma)$ at stellar energies is probably the smallest of any of the carbon cycle reactions. The rate of the entire cycle, therefore, depends upon the abundance of N^{14}. The mirror nucleus, N^{15}, indicates the possibility that an undetected state may exist in O^{15} in the stellar energy region; this might raise the value of S_0 by a very large factor. If this were true, $C^{12}(p, \gamma)$ would be the slowest reaction. Nevertheless, there is no experimental evidence to confirm this.

BOSMAN-CRESPIN *et al.* improved the value of S_0 given above by including temperature dependent terms:

$$S_0 = 36.2\,(1 - 0.016\,T^{\frac{2}{3}}) \text{ kev barns} \pm 33\%. \tag{26.7}$$

More recently, however, FOWLER has pointed out that experimental results of R. E. PIXLEY infer a much smaller non-resonant cross section factor at stellar energies. FOWLER combines PIXLEY's results with more recent measurements of LAMB and HESTER and suggests the value

$$S_0 = (3.0 \pm 0.6) \text{ kev barns.} \tag{26.8}$$

Indeed, on the basis of stellar abundance arguments he tends to regard this as the preferred value.

The energy release in this process is 7.347 Mev.

$O^{15}(\beta^+\nu)\,N^{15}$: The mean lifetime of this reaction is 3.0 min. The energy released is 1.729 Mev plus a neutrino loss of 0.976 Mev.

[1] W. A. S. LAMB and R. E. HESTER: Bull. Amer. Phys. Soc. **2**, 181 (1957).
[2] T. TEICHMANN: Phys. Rev. **77**, 506 (1950).

$N^{15}(p, \alpha)\, C^{12}$: FOWLER is quite confident of the value

$$S_0 = (1.1 \pm 0.3) \times 10^5 \text{ kev barns} \tag{26.9}$$

based on extrapolation of a simple dispersion curve through the resonance at 0.338 Mev.

The energy released is 4.961 Mev.

$H^1(p, \beta^+\nu)\, D^2$: A detailed discussion of this reaction was given by SALPETER[1]. The direct experimental determination of this cross section is impossible by present techniques because the predicted value is about 10^{-47} cm^2 $= 10^{-23}$ barns at 1 Mev. We must therefore rely upon theory alone. The results are uncertain to the extent that the Gamow-Teller β-decay constant is uncertain. The value adopted by SALPETER was

$$g = (7.5 \pm 1.5) \times 10^{-4} \text{ sec}^{-1}. \tag{26.10}$$

From this we may derive

$$S_0 = (4.4 \pm 0.8) \times 10^{-22} \text{ kev barns.} \tag{26.11}$$

An improved result by BOSMAN-CRESPIN *et al.* is

$$S_0 = 39.96\,(1 + 0.0123\, T^{\frac{1}{3}} + 0.00835\, T^{\frac{2}{3}}) \times 10^{-23} \text{ kev barns} \pm 20\%. \tag{26.12}$$

More recent experimental data indicate[2]

$$g = (5.35 \pm 0.15) \times 10^{-4} \text{ sec}^{-1}, \tag{26.13}$$

which would reduce the values calculated by BOSMAN-CRESPIN *et al.* by 29%.

The energy released by two such processes is 2×1.186 Mev and the neutrino loss is 0.257 Mev per process or 0.514 Mev per chain. This is the slowest reaction in the proton-proton chain.

$D^2(p, \gamma)\, He^3$: The cross section determined for $D^2(n, \gamma)$ has been used by SALPETER to give the result

$$S_0 = (8 \pm 2) \times 10^{-5} \text{ kev barns.} \tag{26.14}$$

The energy released by two processes is 2×5.494 Mev.

$He^3(He^3, 2p)\, He^4$: This once seemed to be the most likely termination of the proton-proton chain. FOWLER gives the value

$$S_0 = 2 \times 10^3 \text{ kev barns.} \tag{26.15}$$

The energy release is 12.847 Mev.

27. Representations of ε. We note the following total energy release in the two processes (see note in Sect. 24):

$$\left.\begin{aligned} Q &= 25.026 \text{ Mev for the carbon cycle;} \\ Q &= 26.207 \text{ Mev for the proton-proton chain.} \end{aligned}\right\} \tag{27.1}$$

The difference is due to different neutrino loss.

Using the cross section of the slowest reaction in each case [$N^{14}(p, \gamma)$ and $H^1(p, \beta^+\nu)$, respectively], and using the symbol ε_N in case of the carbon cycle, and ε_{pp} for the proton-proton chain, we find from(27.1), (25.20), and (26.7),

---

[1] E. E. SALPETER: Phys. Rev. **88**, 547 (1952).
[2] Footnote 5, p. 32.

according to the most recent cross section data,

$$\varepsilon_{\mathrm{N}} = 0.86\,\varrho\, x_{\mathrm{H}}\, x_{\mathrm{N}^{14}} f_{\mathrm{N}} \left(T^{-\frac{2}{3}} - 0.016\right) \times \exp\left(-152.28\, T^{-\frac{1}{3}}\right) \times 10^{28} \pm 20\,\% ; \tag{27.2}$$

$$\left.\begin{aligned} \varepsilon_{pp} = {} & 1.75\,\varrho\, x_{\mathrm{H}}^2 f_{pp}\, T^{-\frac{2}{3}} \times \left(1 + 0.012\, T^{\frac{1}{3}} + 0.008\, T^{\frac{2}{3}} + 0.00065\, T\right) \\ & \times \exp\left(-33.804\, T^{-\frac{1}{3}}\right) \times 10^{6} \pm 10\,\% \end{aligned}\right\} \tag{27.3}$$

in ergs per gram per sec, where T is in millions of degrees. The abundance of N^{14} is virtually the same as that of all the carbon and nitrogen isotopes involved:

$$x_{\mathrm{N}^{14}} \approx x_{\mathrm{CN}}. \tag{27.4}$$

In the event that the reaction $N^{14}(p,\gamma)\,O^{15}$ turns out to have a much larger cross section than currently believed, the rate of the carbon-nitrogen cycle would be determined by the $C^{12}(p,\gamma)\,N^{13}$ reaction and

$$\left.\begin{aligned} \varepsilon_{\mathrm{C}} = {} & (3.18 \pm 0.54)\,\varrho\, x_{\mathrm{H}}\, x_{\mathrm{C}^{12}} f_{\mathrm{C}} \times \left(T^{-\frac{2}{3}} + 0.017\right) \exp\left(-136.9\, T^{-\frac{1}{3}}\right) \\ & \times 10^{27} \text{ ergs per gram per sec}. \end{aligned}\right\} \tag{27.5}$$

Bosman-Crespin *et al.* have calculated the functions

$$E_{pp} = \varepsilon_{pp}/(\varrho\, x_{\mathrm{H}}^2 f_{\mathrm{H}}), \tag{27.6}$$

and

$$E_{\mathrm{N}} = \varepsilon_{\mathrm{N}}/(\varrho\, x_{\mathrm{H}}^2 f_{\mathrm{N}}), \tag{27.7}$$

using, in the latter case,

$$x_{\mathrm{CN}} = 0.005\, x_{\mathrm{H}}. \tag{27.8}$$

These results are given in Table 4 and Fig. 12, corrected for recent revisions of the cross sections.

Table 4. *The energy production functions E_{N} and E_{pp} together with the corresponding exponents. T is in million degrees.*

T	$\log E_{\mathrm{N}}$	n_{N}	$\log E_{pp}$	n_{pp}	T	$\log E_{\mathrm{N}}$	n_{N}	$\log E_{pp}$	n_{pp}
1	−40.5087	44.64	−8.4292	9.406	26	+2.3003	16.25	+0.3943	3.181
2	−27.0699	37.00	−5.5976	7.715	27	+2.5667	16.04	+0.4464	3.136
3	−20.5549	32.87	−4.2391	6.803	28	+2.8200	15.84	+0.4960	3.094
4	−16.4483	30.12	−3.3891	6.197	29	+3.0613	15.64	+0.5431	3.053
5	−13.5294	28.10	−2.7885	5.753	30	+3.2916	15.46	+0.5881	3.014
6	−11.3044	26.52	−2.3330	5.406	31	+3.5117	15.28	+0.6310	2.977
7	−9.5288	25.24	−1.9711	5.126	32	+3.7224	15.11	+0.6720	2.942
8	−8.0649	24.17	−1.6738	4.892	33	+3.9243	14.95	+0.7114	2.908
9	−6.8283	23.26	−1.4236	4.692	34	+4.1181	14.79	+0.7491	2.875
10	−5.7639	22.47	−1.2088	4.520	35	+4.3043	14.64	+0.7852	2.844
11	−4.8339	21.77	−1.0218	4.368	36	+4.4834	14.49	+0.8200	2.814
12	−4.0112	21.15	−0.8567	4.233	37	+4.6559	14.35	+0.8535	2.784
13	−3.2760	20.59	−0.7096	4.112	38	+4.8222	14.22	+0.8858	2.756
14	−2.6133	20.08	−0.5772	4.002	39	+4.9825	14.09	+0.9169	2.729
15	−2.0115	19.62	−0.4573	3.902	40	+5.1374	13.96	+0.9469	2.703
16	−1.4615	19.20	−0.3479	3.811	41	+5.2871	13.84	+0.9759	2.678
17	−0.9561	18.81	−0.2476	3.727	42	+5.4319	13.72	+1.0039	2.654
18	−0.4892	18.44	−0.1551	3.649	43	+5.5721	13.60	+1.0310	2.630
19	−0.0561	18.11	−0.0694	3.576	44	+5.7079	13.49	+1.0573	2.607
20	+0.3473	17.79	+0.0103	3.509	45	+5.8396	13.38	+1.0827	2.585
21	+0.7242	17.50	+0.0846	3.446	46	+5.9673	13.28	+1.1074	2.563
22	+1.0777	17.22	+0.1542	3.387	47	+6.0913	13.17	+1.1313	2.542
23	+1.4101	16.96	+0.2196	3.331	48	+6.2118	13.07	+1.1546	2.522
24	+1.7235	16.71	+0.2812	3.278	49	+6.3288	12.98	+1.1772	2.503
25	+2.0197	16.47	+0.3393	3.228	50	+6.4427		+1.1991	

Thus at low temperatures, energy production is largely by the proton-proton chain, but a switchover to the carbon cycle occurs at $T \approx 18$ million degrees.

The electron shielding factors may be estimated from the work of SCHATZMAN[1] and KELLER[2] to be

$$f_{\mathrm{H}} = 1,$$
$$f_{\mathrm{N}} = \exp\left(\frac{7.25}{T}\right). \tag{27.9}$$

Expressions (27.3) and (27.6) are not complicated by the standards of electronic computers. Nevertheless, for rough computations it is customary to employ approximate formulae of the form

$$\varepsilon = \varepsilon_0 \varrho T^{\nu}, \tag{27.10}$$

where

$$\nu = \frac{d \log \varepsilon}{d \log T}. \tag{27.11}$$

Because of the uncertainty of the shielding factors, BOSMAN-CRESPIN *et al.*, define, alternatively,

$$n = \frac{d \log E}{d \log T} = \nu - \frac{d \log f}{d \log T}. \tag{27.12}$$

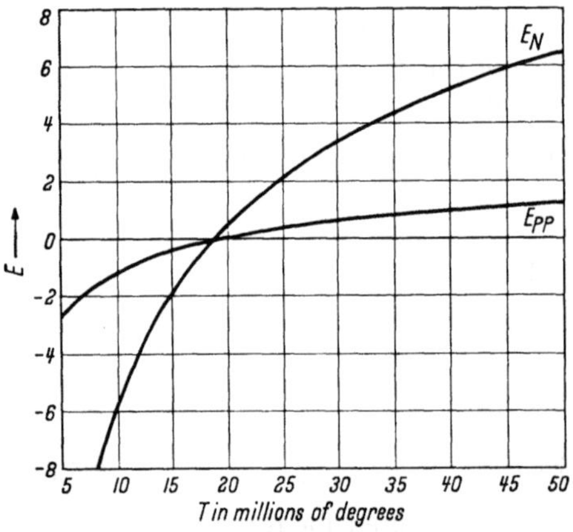

Fig. 12. E_{pp} and E_{CN} following BOSMAN-CRESPIN et al., with corrections.

Values of n may be easily calculated from Table 4. Both n_{pp} and n_{N} are monotonic decreasing functions of T. At 14×10^6 °K, $n_{pp} \approx 4$ and at 20×10^6 °K, $n_{\mathrm{N}} \approx 20$. In very approximate calculations some workers use

$$\left.\begin{aligned} \varepsilon &\propto T^4 \quad \text{for} \quad T < T'; \\ \varepsilon &\propto T^{18} \quad \text{for} \quad T > T'. \end{aligned}\right\} \tag{27.13}$$

This assumes energy production by one process *or* the other with constant exponents. In actuality, both processes are simultaneously operative and the total energy production in a given shell depends upon

$$\varepsilon = \varepsilon_{pp} + \varepsilon_{\mathrm{N}}. \tag{27.14}$$

It is entirely possible for the carbon cycle to predominate in the center of a star and yet to have the bulk of the luminosity produced by the proton-proton chain in the cooler regions, because the latter process operates in a much larger volume.

28. The time dependence of chemical composition. The gradual conversion of hydrogen to helium results in a slow increase of the molecular weight of the stellar material. If there is little or no mixing, the central regions will have a greater molecular weight than the envelope because the conversion is more rapid there. The consequences of chemical inhomogeneities will be studied in later sections.

The local rate of change of the hydrogen content in unmixed regions is

$$\frac{\partial X}{\partial t} = -\frac{4H}{Q} \varepsilon(X) = -\frac{\partial Y}{\partial t}, \tag{28.1}$$

[1] E. SCHATZMAN: Astrophys. Journ. **119**, 464 (1954).
[2] G. KELLER: Astrophys. Journ. **118**, 142 (1953).

where, as before, Q is the energy released per reaction, and the dependence of ε on X has been emphasized by writing $\varepsilon(X)$.

In convective regions it is assumed that mixing maintains chemical uniformity. In the case of a convective central core only the luminosity and mass of the entire core are involved:

$$\frac{\partial X_{\text{core}}}{\partial t} = -\frac{4H}{Q}\frac{L_{\text{core}}}{M_{\text{core}}}. \tag{28.2}$$

In general, however, the mass of the convective core is a time-dependent function and this introduces an added complication. For details concerning a numerical procedure see HASELGROVE and HOYLE[1].

29. Other nuclear reactions of astrophysical importance. The amount of hydrogen available as fuel for the nuclear reactions just described is limited. When it is exhausted other processes must take over. If nuclear energy sources are not sufficient, gravitational contraction will occur and the central regions will become hotter as a consequence. As the temperature rises, so does the average energy of a particle and the Coulomb barrier is more easily penetrated. Reactions between heavier nuclei are then possible. We shall discuss the subsequent events in terms of a theory of the origin of the elements in stars recently developed by HOYLE, FOWLER, CAMERON, GREENSTEIN, G. R. BURBIDGE and E. M. BURBIDGE[2].

If one postulates that originally all material was hydrogen then the initial nuclear reaction that produces appreciable stellar energy will be the proton-proton chain. When the hydrogen is exhausted the central temperature is raised by gravitational contraction until helium reactions are possible at $T \approx 150 \times 10^6$ °K. C^{12} may be formed as follows:

$$He^4 + He^4 \rightarrow Be^8 \tag{29.1}$$

$$Be^8 + He^4 \rightarrow C^{12} + \gamma. \tag{29.2}$$

Although Be^8 is not stable, there is enough present at these high temperatures to capture α-particles and form C^{12}. This reaction was suggested by SALPETER[3] and ÖPIK[4], and its importance in astrophysics was assured when laboratory results verified the low energy resonance predicted by HOYLE[5,6].

α-particle reactions may continue until the helium is exhausted, probably at Ne^{20} or Mg^{24}:

$$C^{12}(\alpha,\gamma)\,O^{16}(\alpha,\gamma)\,Ne^{24}(\alpha,\gamma)\,Mg^{24}. \tag{29.3}$$

If the stars remain stable, reactions may occur among the carbon, oxygen and neon nuclei until iron is formed. Up to this point the binding energy per nucleon has been increasing and each of the reactions produces energy. Beyond this point the formation of heavier nuclei will not yield additional energy.

---

[1] C. B. HASELGROVE and F. HOYLE: Monthly Notices Roy. Astronom. Soc. London **116**, 515 (1956).

[2] Summarizing accounts are to be found in W. A. FOWLER and J. L. GREENSTEIN: Proc. Nat. Acad. Sci. U.S.A. **42**, 173 (1956); W. A. FOWLER: Sci. Monthly **84**, 84 (1957). — Detailed discussions are found in W. A. FOWLER, G. R. BURBIDGE, E. M. BURBIDGE: Astrophys. Journ. **122**, 271 (1955) and Astrophys. Journ. Suppl. **2**, 167 (1955); F. HOYLE: Astrophys. Journ., Suppl. **1**, 121 (1954); the reference to Rev. Mod. Phys., footnote 5, p. 32; and A. G. W. CAMERON: Atomic Energy of Canada, Chalk River, CRL-41 (1957).

[3] E. E. SALPETER: Astrophys. Journ. **115**, 326 (1952).

[4] E. ÖPIK: Proc. Roy. Irish Acad. A **54**, 49 (1951).

[5] F. HOYLE: Astrophys. Journ., Suppl. **1**, 121 (1954).

[6] W. A. FOWLER, C. W. COOK, C. C. LAURITSEN, T. LAURITSEN and F. MOZER: Bull. Amer. Phys. Soc., Ser. II **1**, 191 (1956).

The prominent peak at iron, in the "cosmic" abundance curve[1] may be due to this culmination of nuclear energy processes at a temperature of about $3.5 \times 10^9\,^\circ$K.

On the other hand, suppose that instability sets in after the production of C^{12}, O^{16} and Ne^{20} and instead of proceeding as above, the star explodes, dispersing its matter into the interstellar medium. Stars that subsequently form will be enriched by these elements. These "second generation" stars need not rely on the proton-proton chain for energy production in the early stages, but, if the temperature is high enough, can produce helium via the carbon cycle.

Additional reactions produce C^{13}, O^{17} and Ne^{21}:

$$\left.\begin{array}{l} N^{15}(p,\gamma)\,O^{16}(p,\gamma)\,F^{17}(\beta^+\nu)\,O^{17}(p,\alpha)\,N^{14};\\ Ne^{20}(p,\gamma)\,Na^{21}(\beta^+\nu)\,Ne^{21}(p,\gamma)\,Na^{22}(\beta^+\nu)\,Ne^{22}(p,\gamma)\,Na^{23}(p,\alpha)\,Ne^{20}. \end{array}\right\} \tag{29.4}$$

Note that the reaction $N^{15}(p,\gamma)\,O^{16}$ is an infrequent alternative to the $N^{15}(p,\alpha)\,C^{12}$ reaction which recreates C^{12}. Further the neon-sodium reactions also form a cycle in which Ne^{20} is the catalyst[2].

The interesting property of C^{13}, O^{17} and Ne^{21} is that they will serve as sources of neutrons when the hydrogen is exhausted and the temperature rises[3]. The reactions are:

$$\left.\begin{array}{l} C^{13}(\alpha,n)\,O^{16};\\ O^{17}(\alpha,n)\,Ne^{20};\\ Ne^{21}(\alpha,n)\,Mg^{24}. \end{array}\right\} \tag{29.5}$$

By this time the medium is largely helium so that many α-particles are available. In addition, since He^5 is not stable, the neutrons will not be absorbed by the helium but instead will form heavier and heavier elements by combining with the nuclei near iron. This is the so-called *s-process* or slow neutron process, and it suffices to produce many elements up to the natural α-emitters. Evidence for the continual formation of elements in stars is the presence of technecium in the atmospheres of S-type stars. This element has a lifetime of 200000 years and must therefore be continually replenished to be observed.

The natural α-emitters, as well as other elements that cannot be built by the s-process, may be built up by rapid or *r-processes*, perhaps in the explosions of supernovae. Here the particles are packed together before they can fall apart, somewhat like the formation of Cf^{254} in the Bikini bomb tests.

Finally, to account for deuterium, lithium, beryllium and boron, which would be consumed by nuclear processes at relatively low temperatures, this theory postulates reactions on the surfaces of stars in which high-energy protons, accelerated in magnetic spots, are captured.

The verification of these hypotheses will depend upon a combination of experimental evidence from the nuclear physics laboratory and astrophysically determined abundance ratios.

III. Perturbations.

30. Rotation and mixing. The combination of hydrostatic equilibrium, radiative equilibrium and the assumption of *rigid rotation* leads to an interesting

[1] H. E. SUESS and H. C. UREY: article in this volume, p. 296.
[2] J. B. MARION and W. A. FOWLER: Astrophys. Journ. **125**, 221 (1957).
[3] A. G. W. CAMERON: Phys. Rev. **93**, 932 (1954). — Astrophys. Journ. **121**, 144 (1955).

result known as VON ZEIPEL's theorem[1]. There are two important consequences which we will derive:

1. A rotating star is brighter at the poles than at the equator.
2. The energy generation law must be of the form

$$\varepsilon = \text{const}\left(1 - \frac{\omega^2}{2\pi G\varrho}\right). \tag{30.1}$$

The effect of rotation is to add to the gravitational potential, V, a potential due to rotation so that the total potential becomes

$$W = V + \tfrac{1}{2}\omega^2 (x^2 + y^2), \tag{30.2}$$

where ω is the angular velocity of the rotation and the z-axis has been taken as the axis of rotation. The equation of hydrostatic equilibrium becomes

$$\operatorname{grad} P = \varrho \operatorname{grad} W. \tag{30.3}$$

Surfaces of constant W are called "level surfaces". Since, by (30.3), $dW = 0$ implies $dP = 0$, level surfaces are also surfaces of constant P, and P is a function of W only. Then (30.3) implies ϱ is a function of W only; and, since P depends only on ϱ and T, T is also constant on level surfaces. Further, if $\varkappa = \varkappa(\varrho, T)$, the opacity is also constant on level surfaces.

Since V satisfies the equation

$$\operatorname{div}\operatorname{grad} V = -4\pi G\varrho, \tag{30.4}$$

it follows from (30.3) and (30.4) that

$$\operatorname{div}\left(\frac{1}{\varrho}\operatorname{grad} P\right) = -4\pi G\varrho + 2\omega^2. \tag{30.5}$$

The condition of radiative equilibrium yields an equation of very similar form. Regarding the flux as a vector, the general equations are ([3], p. 211)

$$F_\nu = -\frac{c}{\varkappa_\nu \varrho}\operatorname{grad} P_{r,\nu}, \tag{30.6}$$

where $P_{r,\nu}$ is the pressure due to monochromatic radiation, and

$$\operatorname{div} F_\nu = \varepsilon_\nu \varrho. \tag{30.7}$$

Combining these equations we obtain, after defining a suitable mean opacity and integrating over frequency:

$$\operatorname{div}\left(\frac{1}{\varkappa\varrho}\operatorname{grad} P_r\right) = -\frac{\varepsilon\varrho}{c}. \tag{30.8}$$

Since P and P_r are both constant on a level surface, we may regard P_r as a function of P. Thus the left hand side of (30.8) becomes

$$\sum_{xyz}\frac{\partial}{\partial x}\left(\frac{1}{\varkappa}\frac{dP_r}{dP}\frac{1}{\varrho}\frac{\partial P}{\partial x}\right) = \frac{d}{dP}\left(\frac{1}{\varkappa}\frac{dP_r}{dP}\right)\frac{1}{\varrho}\left(\frac{dP}{dn}\right)^2 + \frac{1}{\varkappa}\frac{dP_r}{dP}\sum_{xyz}\frac{\partial}{\partial x}\left(\frac{1}{\varrho}\frac{\partial P}{\partial x}\right), \tag{30.9}$$

where

$$\left(\frac{dP}{dn}\right)^2 = \left(\frac{\partial P}{\partial x}\right)^2 + \left(\frac{\partial P}{\partial y}\right)^2 + \left(\frac{\partial P}{\partial z}\right)^2. \tag{30.10}$$

---

[1] H. VON ZEIPEL: Monthly Notices Roy. Astronom. Soc. London **84**, 665 (1924).

The rightmost sum in (30.9) is $\operatorname{div}(\varrho^{-1}\operatorname{grad} P)$; therefore, using (30.5),

$$\frac{d}{dP}\left(\frac{1}{\varkappa}\frac{dP_r}{dP}\right)\frac{1}{\varrho}\left(\frac{dP}{dn}\right)^2 = -\frac{\varepsilon\varrho}{c} + \frac{1}{\varkappa}\frac{dP_r}{dP}(4\pi G\varrho - 2\omega^2). \tag{30.11}$$

The right-hand side of this equation is constant along a level surface. Therefore, either dn is constant along a level surface or the coefficient of $(dP/dn)^2$ is zero. The former case implies that the level surfaces are parallel, which is not true in the case of rotation. Therefore we adopt the latter condition:

$$\frac{d}{dP}\left(\frac{1}{\varkappa}\frac{dP_r}{dP}\right) = 0, \tag{30.12}$$

or

$$\frac{1}{\varkappa}\frac{dP_r}{dP} = C. \tag{30.13}$$

Since the right hand side of (30.11) is now zero we may use (30.13) to find

$$\frac{\varepsilon\varrho}{c} = C(4\pi G\varrho - 2\omega^2), \tag{30.14}$$

or

$$\varepsilon = \text{const}\left(1 - \frac{\omega^2}{2\pi G\varrho}\right). \tag{30.15}$$

Further, since the net flux across the level surface is

$$F_n = -\frac{c}{\varkappa\varrho}\frac{dP_r}{dn}, \tag{30.16}$$

$$= -\frac{cC}{\varrho}\frac{dP}{dn}, \quad \text{by (30.13)}; \tag{30.17}$$

$$= -cC\frac{dW}{dn}, \quad \text{by (30.3)}; \tag{30.18}$$

and the flux is thus proportional to the effective gravity. The effective temperature should therefore be greater at the poles of a rotating star than at the equator. Further, according to an argument given by MILNE[1], the average effective temperature of the entire star is lowered because of a general expansion superposed on the ellipticity[2].

The condition (30.15) is, of course, physically unreasonable. One concludes therefore that stars do not rotate as rigid bodies. EDDINGTON suggested that if (30.15) were not satisfied the poles would heat up relative to the equator, the pressure would no longer be constant on level surfaces and meridional circulation currents would ensue ([2], p. 285). If currents of this type had large radial velocities, stars would be well-mixed and inhomogeneities would not develop in the course of evolution. As shown by SWEET[3] and MESTEL[4], however, EDDINGTON's original estimates of these velocities were too high. SWEET found a velocity of 3×10^{-10} cm/sec for the sun, which implies negligible mixing, and MESTEL pointed out that the gradient in composition set up by the circulation would itself damp further motion.

[1] E. A. MILNE: Handbuch der Astrophysik, Vol. III/1, p. 236. 1930.

[2] See also P. A. SWEET and A. E. ROY: Monthly Notices Roy. Astronom. Soc. London **113**, 701 (1953).

[3] P. A. SWEET: Monthly Notices Roy. Astronom. Soc. London **110**, 548 (1950).

[4] L. MESTEL: Monthly Notices Roy. Astronom. Soc. London **113**, 716 (1953).

The problem is further greatly complicated by the presence of magnetic fields. Two relevant summaries of magnetohydrodynamic phenomena have recently been made by COWLING[1].

31. Apsidal motion. Two mass points moving under their mutual gravitational attraction describe ellipses around their common center of mass. If we use this approximation to describe an eclipsing binary system, the light curve has two minima; the primary minimum occurs when the brighter component is eclipsed and the secondary minimum when the fainter is eclipsed. If the mass point theory holds, the time of the secondary minimum relative to the primary minimum should remain unchanged.

In several cases, for example, Y Cygni, it is observed that there is a periodic oscillation of the time of occurrence of the secondary minimum relative to the primary. This may be interpreted geometrically as the rotation of the major axis of the ellipse which describes the motion of one star around the other. Since this axis is known as the *line of apsides*, its periodic rotation is called *apsidal motion*.

The physical explanation lies in the fact that the mass point representation fails when the density distribution of one star is appreciably distorted by the tidal effects due to the other. The problem is related to the motion of a satellite around an oblate planet. It differs, however, because the satellite does not affect the oblateness of the planet, whereas the distortion of the star is *caused* by the companion whose motion is affected. The magnitude of the effect depends upon the equilibrium density distribution of the star in the absence of the disturber.

That oblateness causes a rotation of the line of apsides is a result of classical celestial mechanics[2]. The definitive treatment of the stellar case was given by STERNE[3] whose results encompass earlier work by RUSSELL[4] and COWLING[5]. In this analysis the distorting potential due to star 2 which alters the structure of star 1 is expanded in spherical harmonics about the center of mass of star 1:

$$V_t = \frac{m_2 G}{R} \sum_{2}^{\infty} \left(\frac{r}{R}\right)^n P_n(\cos\vartheta), \qquad (31.1)$$

where

m_2 = mass of star 2;
R = distance between centers of mass;
r = distance of an arbitrary point from the center of mass of star 1;
ϑ = angle at the center of star 1 between the point at r and the center of star 2.

(The omitted terms with $n=0$ and 1 cause, respectively, no force, and the orbital acceleration of the entire star.)

STERNE has obtained results up to fourth-order harmonic distortions, including the rotation of each star. The essentials may perhaps most easily be seen if we examine the second-order results for the particular case of negligible distortion of star 2 and a rotation period for star 1 equal to the orbital period.

---

[1] T. G. COWLING: Magnetohydrodynamics. New York: Interscience 1957. — Solar Electrodynamics, in: The Sun, G. P. KUIPER, Ed. Chicago, Ill.: Chicago University Press 1953.
[2] See, for example, F. R. MOULTON: An Introduction to Celestial Mechanics, p. 333. New York: Macmillan 1914.
[3] T. E. STERNE: Monthly Notices Roy. Astronom. Soc. London **99**, 451, 662, 670 (1939).
[4] H. N. RUSSELL: Monthly Notices Roy. Astronom. Soc. London **88**, 641 (1928).
[5] T. G. COWLING: Monthly Notices Roy. Astronom. Soc. London **98**, 734 (1938).

The equilibrium density distribution of star 1 enters the result in terms of a function, $k_{2,1}$, which is defined in terms of another function η_2:

$$k_{2,1} = \left[\frac{3-\eta_2(r)}{4+2\eta_2(r)}\right]_{r=a_1} \tag{31.2}$$

where a_1 is the average radius of star 1. The function η_2 is found by solving the equation

$$r\frac{d\eta_2}{dr} + \eta_2^2 - \eta_2 - 6 + \frac{6\varrho}{\varrho_m}(\eta_2+1) = 0, \tag{31.3}$$

subject to the condition that $\eta_2 = 0$ at $r = 0$. ϱ_m is defined as the mean density interior to r and the function ϱ/ϱ_m is to be found from the equilibrium model.

It then follows that the ratio of the orbital period to the apsidal period will be

$$\frac{P}{P'} = k_{2,1}\frac{a_1^5}{A^5}\left[\frac{m_2}{m_1}\left(15f_2(e) + g_2(e)\right) + g_2(e)\right], \tag{31.4}$$

where A = semi-major axis of the relative orbit. The two functions of the eccentricity, e, are

$$\left.\begin{aligned} f_2(e) &= (1-e^2)^{-5}\left(1+\tfrac{3}{2}e^2+\tfrac{1}{8}e^4\right),\\ g_2(e) &= (1-e^2)^{-2}, \end{aligned}\right\} \tag{31.5}$$

or $e < 1$.

For a completely homogeneous star (polytrope with $n = 0$), the value of $k_{2,1}$, is a maximum and equal to 0.75. For EDDINGTON'S standard model (a polytrope with $n = 3$), $k_{2,1} = 0.0144$. Recently extensive tables have been published of apsidal motion constants for polytropes including seventh-order harmonics by BROOKER and OLLE[1].

One may assign an apsidal polytropic index to a star, defined as the index of a polytrope having the same value of $k_{2,1}$. Observationally these values lie between $n_{ap} = 3$ and $n_{ap} = 4$. This has been used[2] as a check on theoretical models since $k_{2,1}$ can be computed from the equilibrium density distribution and used to find n_{ap}. If the apsidal polytropic index of the theoretical model lies outside the observed range, the model is open to serious question.

32. Pulsation. For a discussion of pulsation see the article by P. LEDOUX and TH. WALRAVEN in this volume, p. 353.

It is pertinent to note here that studies of pulsation usually begin with the properties of equilibrium models. Unfortunately, models for stars in the region of the H-R diagram where pulsation occurs are still quite uncertain.

C. Particular solutions.

I. Preliminary results.

33. The Vogt-Russell theorem. This theorem states that the entire structure of a star in equilibrium is uniquely determined by the distribution of chemical composition and the total mass, provided the equation of state, the opacity and the energy production are functions only of the local physical variables and the composition.

---

[1] R. A. BROOKER and T. W. OLLE: Monthly Notices Roy. Astronom. Soc. London **115**, 101 (1955).

[2] G. KELLER: Ph. D. Dissertation, Columbia University 1948. — R. S. KUSHWAHA: Astrophys. Journ. **125**, 242 (1957).

Consider Eqs. (4.1) to (4.5). Since either (4.4) or (4.5), but never both, are used simultaneously, the differential equations of equilibrium form a fourth order system. If the composition is specified, point by point in the star, $\varkappa$, ϱ, and ε may all be expressed in terms of P and T. The system of four equations now involves only four unknowns and a particular solution involves four constants to be determined from four boundary conditions.

The physics of the problem, however, requires only three independent boundary conditions: Eqs. (12.1) and (12.2). The requirement that ϱ and T tend simultaneously to specified values is only one condition since the point at which this occurs is arbitrary. This ambiguity can be removed by assigning the total mass. The boundary is required to be at the point where $M(r) = M$ and this provides the fourth condition. Thus when the run of composition and the total mass are assigned, the solution is unique.

Apart from the comfort this theorem provides, it permits us to make some important statements. It implies, in somewhat different terms, that mass and composition specify L and R uniquely. In particular if stars of different masses were formed from the same chemically homogeneous material, there would be a unique relation between their luminosities and radii. Since the coordinates of the H-R diagram involve only L and R, the locus of points of equivalent composition and differing mass should lie along a smooth curve. Indeed, we may interpret the "main sequence" in just these terms, as a sequence of masses having equivalent homogeneous composition. From the mass-luminosity relation, the more massive stars are those with higher luminosities. Slightly different mixtures of elements would result in roughly parallel main sequences.

From the point of view of stellar evolution the main sequence represents a star in the prime of life. Although nuclear energy production is going on, very little of the total fuel has been consumed. Stars of constant mass depart from the main sequence only as their compositions change due to nuclear processes. These changes will be most rapid for the most luminous (hence most massive) main sequence stars. Observations of H-R diagrams for galactic clusters confirm this point of view. Indeed, theory and observation may be combined to find the ages of galactic clusters[1].

34. Integral theorems. Estimates of the orders of magnitude for physical quantities we may expect to find in stellar interiors may be obtained without much difficulty.

The pressure and mass gradient equations alone permit us to find a lower bound for the central pressure. We simply compare the actual central pressure with the central pressure in a star of the same radius and mass but of a uniform density equal to the mean density, $\bar{\varrho}$, of the actual star. To form a star of uniform density $\bar{\varrho}$ from a star in which the density decreases outward, mass must be moved outward. This has the effect of lowering the central pressure; therefore, the central pressure in the star of uniform density $\bar{\varrho}$, is a lower limit to the actual central pressure.

This may be easily demonstrated. Eliminating the density between Eqs. (4.1) and (4.2), we find

$$\frac{dP}{dr} = -\frac{GM(r)}{4\pi r^4}\frac{dM(r)}{dr}, \tag{34.1}$$

which gives, on integration,

$$P_c = \frac{G}{4\pi}\int_0^M \frac{M(r)\,dM(r)}{r^4}. \tag{34.2}$$

---

[1] H. L. Johnson and W. A. Hiltner: Astrophys. Journ. **123**, 267 (1956).

Defining the mean density interior to r as

$$\bar{\varrho}(r) = \frac{3M(r)}{4\pi r^3}, \tag{34.3}$$

and eliminating r in the integrand, we obtain

$$P_c = \frac{G}{4\pi}\left(\frac{4\pi}{3}\right)^{\frac{4}{3}} \int_0^M \bar{\varrho}^{\frac{4}{3}}(r)\, M^{-\frac{2}{3}}(r)\, dM(r). \tag{34.4}$$

Clearly we obtain a lower limit to this integral if we take $\bar{\varrho}(r) = \text{const} = \bar{\varrho}$.

Therefore

$$P_c \geqq \frac{G}{4\pi}\left(\frac{4\pi}{3}\right)^{\frac{1}{3}} \bar{\varrho}^{\frac{4}{3}} \frac{3}{2} M^{\frac{2}{3}}. \tag{34.5}$$

But, recalling that

$$\bar{\varrho} = \frac{3M}{4\pi R^3}, \tag{34.6}$$

we finally obtain

$$P_c \geq \frac{3G}{8\pi} \frac{M^2}{R^4}. \tag{34.7}$$

In solar units of M and R this becomes

$$P_c \geq 1.326 \times 10^9\, M^2/R^4 \text{ atmospheres}. \tag{34.8}$$

By the same token we may establish an upper limit by considering a star of the same mass but a uniform density equal to the central density, ϱ_c, of the actual star:

$$P_c \leq \frac{1}{2} G \left(\frac{4\pi}{3}\right)^{\frac{1}{3}} M^{\frac{2}{3}} \varrho_c^{\frac{4}{3}}. \tag{34.9}$$

By assuming the equation of state for a perfect gas, and neglecting radiation pressure, a lower limit to the mean temperature may be obtained in an analogous way:

$$\overline{T} = \frac{1}{M} \int_0^M T\, dM(r) \geq \frac{1}{5} \frac{\mu H}{k} \frac{GM}{R}. \tag{34.10}$$

Of considerable interest is CHANDRASEKHAR's equation establishing a limit to the importance of radiation pressure. Expressing the gas pressure as a fraction, β, of the total pressure, we obtain

$$\left.\begin{aligned} P_g &= \frac{k}{\mu H} \varrho T = \beta P, \\ P_r &= \frac{a}{3} T^4 = (1-\beta) P. \end{aligned}\right\} \tag{34.11}$$

Dividing one equation by the other and solving for T we find

$$T = \left(\frac{3k}{a\mu H} \frac{1-\beta}{\beta}\right)^{\frac{1}{3}} \varrho^{\frac{1}{3}}. \tag{34.12}$$

Substituting in the pressure equation, we obtain

$$P = \varrho \frac{kT}{\mu H \beta} = \left[\left(\frac{k}{\mu H}\right)^4 \frac{3}{a} \frac{1-\beta}{\beta^4}\right]^{\frac{1}{3}} \varrho^{\frac{4}{3}}. \tag{34.13}$$

We have previously established an upper limit to the central pressure in terms of the central density. Therefore,

$$\left[\left(\frac{k}{\mu_c H}\right)^4 \frac{3}{a} \frac{1-\beta_c}{\beta_c^4}\right]^{\frac{1}{3}} \varrho_c^{\frac{4}{3}} \leqq \frac{1}{2} G \left(\frac{4\pi}{3}\right)^{\frac{1}{3}} M^{\frac{2}{3}} \varrho_c^{\frac{4}{3}}, \tag{34.14}$$

where the subscript, c, refers to central values. Alternatively,

$$M \geqq \left(\frac{6}{\pi}\right)^{\frac{1}{2}} \left[\left(\frac{k}{\mu_c H}\right)^4 \frac{3}{a} \frac{1-\beta_c}{\beta_c^4}\right]^{\frac{1}{2}} \frac{1}{G^{\frac{3}{2}}}. \tag{34.15}$$

If we define β^* by

$$M = \left(\frac{6}{\pi}\right)^{\frac{1}{2}} \left[\left(\frac{k}{\mu_c H}\right)^4 \frac{3}{a} \frac{1-\beta^*}{\beta^{*4}}\right]^{\frac{1}{2}} \frac{1}{G^{\frac{3}{2}}}, \tag{34.16}$$

we may conclude

$$\frac{1-\beta_c}{\beta_c^4} \leqq \frac{1-\beta^*}{\beta^{*4}}, \tag{34.17}$$

or equivalently

$$1-\beta_c \leqq 1-\beta^*. \tag{34.18}$$

The mass and mean molecular weight alone define β and thus M and μ provide an upper limit to the importance of radiation pressure. Table 5 shows that radiation pressure is important only in massive stars[1].

Table 5. *Upper limit to the importance of radiation pressure (after Chandrasekhar).*

$1-\beta^*$	$\frac{M}{M_\odot}\mu_c^2$	$1-\beta^*$	$\frac{M}{M_\odot}\mu_c^2$	$1-\beta^*$	$\frac{M}{M_\odot}\mu_c^2$
0.01	0.56	0.35	7.67	0.70	50.92
0.05	1.36	0.40	9.62	0.75	75.89
0.10	2.14	0.45	12.15	0.80	122.5
0.15	2.94	0.50	15.49	0.85	224.4
0.20	3.83	0.55	20.06	0.90	519.6
0.25	4.87	0.60	26.52	1.00	∞
0.30	6.12	0.65	36.05		

We may also note in passing, that when β is a constant (that is, when the gas pressure is always the same fraction of the total pressure), Eq. (34.13) becomes

$$P = K \varrho^{\frac{4}{3}} \tag{34.19}$$

which we recognize, from Sect. 14, to be a polytropic equation of state with $n=3$. The stellar model constructed using (34.19) is of some historic importance because it was extensively studied by EDDINGTON and is known as the "*standard model*". It is but one of the class of polytropes that will be discussed in Sect. 38.

35. Homologous stars. With an assigned mass and composition, the equations of equilibrium may be solved to yield L, R and all the physical variables as functions of r. Let us suppose that such a solution is available; is it necessary to integrate the equations all over again to obtain a solution for, say, a different mass? In certain simple cases it is not; the results for one star may be transformed so as to apply to another by using simple factors, known as *homology relations*. One solution may therefore be used to derive the properties of a whole family of *homologous stars*. Even in the case of more sophisticated stellar models, the homology relations, though not exact, may be used to estimate the properties of stars with parameters slightly different from those used in detailed computations.

[1] See also EDDINGTON'S quartic: [2], p. 117.

We shall derive homology relations, following STRÖMGREN, for stars composed of a perfect gas, obeying an opacity law of the form

$$\varkappa = \varkappa_0\, \varrho^{1-\alpha}\, T^{-3-s}, \tag{35.1}$$

and an energy generation law of the form

$$\varepsilon = \varepsilon_0\, \varrho\, T^{\nu}. \tag{35.2}$$

Let us compare the structure of a star of mass $M^{(1)}$ with that of a star of mass $M^{(0)}$ whose structure is known. Define the constant C_M such that

$$M^{(1)} = C_M\, M^{(0)}. \tag{35.3}$$

If the composition of the two stars differs then we may also define C_μ, $C_{\varkappa_0}$ and C_{ε_0} by means of the relations

$$\left.\begin{aligned} \mu^{(1)} &= C_\mu\, \mu^{(0)};\\ \varkappa_0^{(1)} &= C_{\varkappa_0}\, \varkappa_0^{(0)};\\ \varepsilon_0^{(1)} &= C_{\varepsilon_0}\, \varepsilon_0^{(0)}, \end{aligned}\right\} \tag{35.4}$$

where $\varkappa_0^{(1)}$ is the coefficient of the opacity equation in the star of mass $M^{(1)}$, etc.

Let us assume that the solution for star (1) would have a radius

$$R^{(1)} = C_R\, R^{(0)} \tag{35.5}$$

then point $r^{(1)}$ in star (1) will be called homologous to a point $r^{(0)}$ in star (0) if, at these points,

$$\left.\begin{aligned} r^{(1)} &= C_R\, r^{(0)};\\ M(r^{(1)}) &= C_M\, M(r^{(0)});\\ P^{(1)} &= C_P\, P^{(0)};\\ T^{(1)} &= C_T\, T^{(0)};\\ L(r^{(1)}) &= C_L\, L(r^{(0)});\\ r^{(1)} &= C_R\, r^{(0)}. \end{aligned}\right\} \tag{35.6}$$

We shall now show that the four differential equations of equilibrium will permit us to find four relations between C_M, C_μ, $C_{\varkappa_0}$, C_{ε_0}, C_R, C_P, C_T, and C_L. Equivalently, we will be able to describe star (1) in terms of star (0). In the course of the derivation, a theoretical mass-luminosity relation is derived.

Since the structure of star (0) is known, the physical variables are solutions of the equations

$$\frac{dP^{(0)}}{dr^{(0)}} = -\frac{G M(r^{(0)})}{r^{(0)2}}\, \varrho^{(0)} = -\frac{G M(r^{(0)})}{r^{(0)2}}\, \mu^{(0)} \frac{H}{k} \frac{P^{(0)}}{T^{(0)}}; \tag{35.7}$$

$$\frac{dM(r^{(0)})}{dr^{(0)}} = 4\pi\, r^{(0)2}\, \mu^{(0)} \frac{H}{k} \frac{P^{(0)}}{T^{(0)}}; \tag{35.8}$$

$$\frac{dT^{(0)}}{dr^{(0)}} = -\frac{3}{4ac}\, \varkappa_0^{(0)}\, \mu^{(0)(2-\alpha)} \left(\frac{k}{H}\right)^{2-\alpha} P^{(0)2-\alpha}\, T^{(0)(-6-s-\alpha)}\, \frac{L(r^{(0)})}{4\pi r^{(0)2}}; \tag{35.9}$$

$$\frac{dL(r^{(0)})}{dr^{(0)}} = 4\pi\, r^{(0)2}\, \varepsilon_0^{(0)} \left(\frac{\mu^{(0)} H}{k}\right)^2 P^{(0)2}\, T^{(0)\nu-2}. \tag{35.10}$$

The physical variables for star (1) must obey a similar set of equations in which (1) is substituted for (0): i.e.,

$$\frac{dP^{(1)}}{dr^{(1)}} = \frac{G M(r^{(1)})}{r^{(1)2}}\, \mu^{(1)} \frac{H}{k} \frac{P^{(1)}}{T^{(1)}}, \tag{35.11}$$

etc. Using (35.3) to (35.6), however, (35.11) may be written as

$$\frac{C_P}{C_R}\frac{dP^{(0)}}{dr^{(0)}} = -\frac{G C_M M(r^{(0)})}{C_R^2 r^{(0)2}} C_\mu \mu^{(0)} \frac{H}{k}\frac{C_P}{C_T}\frac{P^{(0)}}{T^{(0)}}, \tag{35.12}$$

which, by virtue of (35.7) reduces to

$$\frac{C_P}{C_R} = \frac{C_M}{C_R^2} C_\mu \frac{C_P}{C_T}, \tag{35.13}$$

or

$$C_T = C_\mu \frac{C_M}{C_R}. \tag{35.14}$$

This is the first of our relations. Using (35.3) to (35.6) we may write

$$T^{(1)} = \mu^{(1)} \frac{M^{(1)}}{R^{(1)}} \left[\frac{T^{(0)}}{\mu^{(0)}} \frac{R^{(0)}}{M^{(0)}}\right]. \tag{35.15}$$

Since all quantities with zero superscripts are known, the bracket on the right may be evaluated. Thus the temperature at any point in star (1) may be found from the temperature at the homologous point in star (0). In particular, since the centers of the two stars are homologous points,

$$T_c^{(1)} \propto \mu^{(1)} \frac{M^{(1)}}{R^{(1)}}, \tag{35.16}$$

where the constant of proportionality is determined by the parameters of star (0).

We may manipulate the mass equation in an identical way to obtain

$$C_M = C_R^3 C_\mu C_P / C_T. \tag{35.17}$$

Using the previous relation (35.14) we obtain

$$C_P = C_M^2 / C_R^4 \tag{35.18}$$

or

$$P^{(1)} \propto M^{(1)} / R^{(1)4}. \tag{35.19}$$

The relations (35.14) and (35.19) may be combined with the radiative temperature gradient to yield a very important theoretical relation between M, L, R, μ and $\varkappa_0$. In this equation the primary dependence of L is on M so that, although this is a mass-luminosity-radius-composition law, it is, in essence, a mass-luminosity law [we will omit the superscripts (1)]:

$$L \propto \frac{\mu^{7+s}}{\varkappa_0} \frac{M^{5+s+\alpha}}{R^{s+3\alpha}}. \tag{35.20}$$

Finally, the energy production equation yields an additional relation

$$L \propto \frac{\mu^\nu \varepsilon_0 M^{2+\nu}}{R^{3+\nu}}. \tag{35.21}$$

Thus, if the mass and the composition (equivalently, μ, $\varkappa_0$, and ε_0) of star (1) are specified, Eqs. (35.20) and (35.21) simultaneously determine the radius and luminosity, while (35.14) and (35.19) give the pressure and temperature in terms of the model of star (0).

It might also be pointed out that the period-mean density relation for Cepheids is also a form of homology relation.

36. Alternate variables. Several different variables are used in place of the physical variables in the current literature. As an example and a guide, we will define a few of them below and show how the differential equations for the pressure and mass gradients (assuming a perfect gas) are transformed on using them. In the physical variables, we have

$$\frac{dP}{dr} = -\frac{GM(r)}{r^2}\frac{\mu H}{k}\frac{P}{T}, \tag{36.1}$$

and

$$\frac{dM(r)}{dr} = 4\pi r^2 \frac{\mu H}{k}\frac{P}{T}. \tag{36.2}$$

α) Schwarzschild variables[1]: The following equations define the dimensionless variables p, t, q and x.

$$\left.\begin{aligned} P &= \frac{p}{4\pi}\frac{GM^2}{R^4}, \\ T &= \frac{t\mu H}{k}\frac{GM}{R}, \\ M(r) &= qM, \\ r &= xR. \end{aligned}\right\} \tag{36.3}$$

Note the similarity of the pressure and temperature definitions to the homology relations (35.16) and (35.19). In terms of these variables, Eqs. (36.1) and (36.2) become:

$$\left.\begin{aligned} \frac{dp}{dx} &= -\frac{pq}{x^2 t}, \\ \text{and} \quad \frac{dq}{dx} &= \frac{p x^2}{t}. \end{aligned}\right\} \tag{36.4}$$

These may be further transformed as follows[2]:

$$\left.\begin{aligned} \lambda &= \log p, \\ \psi &= \log q, \\ \tau &= \log t, \\ y &= \log x; \end{aligned}\right\} \tag{36.5}$$

and the differential equations become

$$\left.\begin{aligned} \log\left(-\frac{d\lambda}{dy}\right) &= \psi - y - \tau, \\ \log\left(\frac{d\psi}{dy}\right) &= \lambda + 3y - \psi - \tau. \end{aligned}\right\} \tag{36.6}$$

β) Bondi variables[3]:

$$\left.\begin{aligned} S &= \frac{4\pi P r^4}{GM(r)^2}, \\ Q &= \frac{Pr}{GM(r)\varrho}, \\ N &= 1 - \frac{d\log\varrho}{d\log P}. \end{aligned}\right\} \tag{36.7}$$

---

[1] M. SCHWARZSCHILD: Astrophys. Journ. **104**, 203 (1946).
[2] R. HÄRM and M. SCHWARZSCHILD: Astrophys. Journ., Suppl. **1**, 319 (1955).
[3] C. M. BONDI and H. BONDI: Monthly Notices Roy. Astronom. Soc. London **109**, 62 (1949).

From the equations of equilibrium

$$\left.\begin{aligned} S &= -\frac{d\log M}{d\log P}, \\ Q &= -\frac{d\log r}{d\log P}. \end{aligned}\right\} \tag{36.8}$$

These relations may be used to eliminate r, $M(r)$ and P in $d\log S/d\log P$ and $d\log Q/d\log P$.

Using, in addition, the definition of N, we obtain

$$\left.\begin{aligned} \frac{P\,dS}{S\,dP} &= 1 - 4Q + 2S, \\ \frac{P\,dQ}{Q\,dP} &= N - Q + S, \end{aligned}\right\} \tag{36.9}$$

as the relations equivalent to (36.1) and (36.2).

γ) Rosseland variables[1]:

$$\left.\begin{aligned} T &= \tau\, T_c, \\ \varrho &= \sigma\, \varrho_c, \\ M(r) &= \psi\, M_0, \\ r &= x\, r_0. \end{aligned}\right\} \tag{36.10}$$

The equations of equilibrium become

$$\left.\begin{aligned} -\frac{d(\sigma\tau)}{dx} &= \pi\,\frac{\psi\sigma}{x^2}, \\ \frac{d\psi}{dx} &= x^2\sigma; \end{aligned}\right\} \tag{36.11}$$

where

$$\pi = \frac{G M_0}{r_0}\,\frac{\mu H}{k T_c}, \tag{36.12}$$

and

$$M_0 = 4\pi r_0^3 \varrho_c. \tag{36.13}$$

δ) Core variables: When the central regions are convective or isothermal (in the latter case with or without degeneracy) it is convenient to define special variables for the core. Usually the radius is replaced by the dimensionless variable, ξ. The dependent variable, called ϑ, ψ, or x, depending upon the circumstances, arises from the appropriate equation of state relating ϱ and P. Each of these cases will be discussed in later sections.

II. The properties of particular models.

37. A survey of selected stellar models. In succeeding sections we will discuss various model stars, beginning with the simplest physical assumptions and proceeding to more sophisticated recent calculations. This section is a descriptive summary of the context in which these models are to be considered. The reader may wish to return to this section after reading some of the details concerning individual models, in order to unify the outlook.

---

[1] S. ROSSELAND: Z. Astrophys. **4**, 255 (1932); A. REIZ: Astrophys. Journ. **120**, 342 (1954).

Historically the first, and still the simplest, theory of stellar structure was that of the polytropes. These models obey an equation of state of the form (14.1) throughout. Since temperature does not explicitly occur in this relation between density and pressure, Eqs. (4.1) and (4.2) may be solved independently of the temperature and luminosity gradients, to provide considerable information about the structure. The range of polytropic index from 0 to 5 represents a range of density distributions from uniformity to infinite central condensation; as has been mentioned previously, $n=3$ is Eddington's "standard model". Because this range is likely to embrace the characteristics of more physically realistic stars and since tables of the properties of polytropes are available, they are often used to estimate the behavior of other models under rotation, apsidal motion and pulsation.

The special case $n=\infty$ represents the non-degenerate isothermal model.

Somewhat closer to reality are the composite models in which some zones are radiative and others convective and different approximate equations of state apply in different regions.

The Cowling model is the simplest composite configuration: the central core is convective and the remainder of the star (the "envelope") is in radiative equilibrium. The convective core may be described in terms of the polytrope $n=1.5$ and only the envelope requires new integrations. An additional assumption that the nuclear energy production takes place entirely in the core, simplifies the solution considerably because then the temperature gradient in the radiative envelope does not depend on a simultaneous luminosity gradient. The chemically homogeneous Cowling model may approximate a carbon cycle star on the main sequence.

If an appreciable fraction of energy is produced by the proton-proton chain, however, the energy sources may extend well into the radiative region. Indeed, the Naur-Osterbrock criterion shows there need not be a convective core at all.

For stars of solar mass or less, on the main sequence, the structure is affected by an exterior convective zone as shown in Osterbrock's models for red dwarfs.

Chemically homogeneous stars can only represent the earliest stages of evolution unless there is efficient mixing. Nuclear reactions, as discussed in Sect. 28, would be expected to cause an increase in molecular weight of the core relative to the envelope. In addition, the depletion of nuclear fuel should tend to make the core isothermal since there is no flux to be driven forward by a temperature gradient. The Schönberg-Chandrasekhar limit provides an upper bound to the mass that can exist in the non-degenerate isothermal core of a star in equilibrium.

The study of the paths of stars in the H-R diagram as they leave the main sequence is the most active aspect of the subject today. In this connection we will discuss the evolution of massive stars as described by Tayler and Kushwaha and the general survey of globular cluster evolution given by Hoyle and Schwarzschild, in which partial degeneracy, gravitational heating and external convection zones all play a role.

It is believed that ultimately stars become white dwarfs by some path, as yet unknown. The material becomes very dense and the electrons enter the domain of complete degeneracy. The theory of these objects has been developed by Chandrasekhar. There are many parallels with the theory of polytropes but the mass-radius relation and the limiting mass are characteristic of white dwarfs.

38. Polytropes. Stellar models obeying an equation of state of the form

$$P = K\varrho^{\frac{1+n}{n}} \tag{38.1}$$

with K and n constant throughout, are called complete polytropes of index n. Their properties are useful for illustrating some of the general concepts of stellar structure. Indeed, forty years ago they represented the foundations of the entire subject. The special case, $n=\frac{3}{2}$, is still important in stars with convective cores; and the polytropes $n=\frac{3}{2}$ and 3 represent extremes in the structure of white dwarfs; but by and large we use polytropes only as a guide and the importance attached to the polytrope $n=3$ is now a thing of the past.

The fundamental equation in the study of polytropes is the Lane-Emden equation. In Sect. 6 we have already seen that the pressure and mass gradients can be combined into an equation of the form

$$\frac{1}{r^2}\frac{d}{dr}\left(\frac{r^2}{\varrho}\frac{dP}{dr}\right)=-4\pi G\varrho. \tag{38.2}$$

Replacing P by (38.1) and defining the dimensionless variables

$$\left.\begin{aligned}\xi &= rA\\ \text{and}\quad \vartheta &= (\varrho/\varrho_c)^{1/n}=(P/P_c)^{1/(1+n)},\end{aligned}\right\} \tag{38.3}$$

where ϱ_c is the central density and P_c is the central pressure, we find the Lane-Emden equation

$$\frac{1}{\xi^2}\frac{d}{d\xi}\left(\xi^2\frac{d\vartheta}{d\xi}\right)=-\vartheta^n; \tag{38.4}$$

providing

$$A^2=\frac{4\pi G}{(n+1)K}\varrho_c^{\frac{n-1}{n}}. \tag{38.5}$$

If the polytrope also obeys the perfect gas law, it follows from (38.1) and (38.3), that

$$T=T_c\vartheta, \tag{38.6}$$

$$K=\frac{k}{\mu H}T_c\varrho_c^{-1/n}, \tag{38.7}$$

and

$$A^2=\frac{4\pi G}{(n+1)}\left(\frac{\mu H}{k}\right)^2\frac{P_c}{T_c^2}, \tag{38.8}$$

where T_c is the central temperature.

With these definitions it is clear that we must seek solutions which have the property $\vartheta=1$ at $\xi=0$. An additional requirement follows from an inspection of the term $\xi^2\frac{d\vartheta}{d\xi}$ which is the dimensionless form of

$$r^2\frac{dP}{dr}=-GM(r)\varrho. \tag{38.9}$$

Since the mass must go to zero as r^3 it follows that $\xi^2\frac{d\vartheta}{d\xi}$ must go to zero as ξ^3, or $\frac{d\vartheta}{d\xi}$ goes to zero as ξ. In other words,

$$\vartheta'=\frac{d\vartheta}{d\xi}=0 \quad\text{at}\quad \xi=0, \tag{38.10}$$

is the second condition on the solution of (38.4).

Table 6a. *Lane-Emden function—Polytrope $n = 1.5$.*
For the two last columns see Eqs. (38.21).

ξ	ϑ	$-\xi^2\vartheta'$	$\xi^4\vartheta$	U	V
0.0	1.0000000	0.00000	0.00000	3.00000	0.00000
0.1	0.9983346	0.00033	0.00010	2.99700	0.00833
0.2	0.9933533	0.00265	0.00159	2.98801	0.03336
0.3	0.9851007	0.00888	0.00798	2.97304	0.07511
0.4	0.9736505	0.02083	0.02493	2.95213	0.13370
0.5	0.9591039	0.04014	0.05994	2.92531	0.20924
0.6	0.9415881	0.06823	0.12203	2.89264	0.30191
0.7	0.9212547	0.10626	0.22119	2.85419	0.41195
0.8	0.8982765	0.15512	0.36793	2.81002	0.53966
0.9	0.8728456	0.21537	0.57267	2.76021	0.68541
1.0	0.8451698	0.28726	0.84517	2.70488	0.84970
1.1	0.8154699	0.37069	1.19393	2.64410	1.03312
1.2	0.7839768	0.46528	1.62565	2.57801	0.23643
1.3	0.7509276	0.57033	2.14472	2.50671	0.46057
1.4	0.7165631	0.68486	2.75275	2.43033	0.70670
1.5	0.6811243	0.80766	3.44819	2.34901	1.97629
1.6	0.6448499	0.93731	4.22609	2.26288	2.27116
1.7	0.6079733	1.07224	5.07785	2.17211	2.59357
1.8	0.5707202	1.21074	5.99119	2.07684	2.94641
1.9	0.5333066	1.35103	6.95011	1.97725	3.33330
2.0	0.4959368	1.49133	7.93499	1.87351	3.7589
2.1	0.4588015	1.62985	8.92282	1.76583	4.2290
2.2	0.4220770	1.76487	9.88741	1.65441	4.7516
2.3	0.3859239	1.89476	10.79973	1.53950	5.3366
2.4	0.3504866	2.01805	11.62831	1.42138	5.9978
2.5	0.3158926	2.13338	12.33955	1.30035	6.7535
2.6	0.2822524	2.23963	12.89826	1.17679	7.6297
2.7	0.2496598	2.33584	13.26795	1.05116	8.6631
2.8	0.2181919	2.42131	13.41129	0.92402	9.9082
2.9	0.1879094	2.49553	13.29048	0.79607	11.4487
3.0	0.1588576	2.55827	12.86747	0.66823	13.420
3.1	0.1310664	2.60954	12.10426	0.54170	16.057
3.2	0.1045515	2.64961	10.96302	0.41808	19.799
3.30	0.0793146	2.67901	9.40609	0.29964	25.589
3.35	0.0671724	2.68996	8.45999	0.24332	29.885
3.40	0.0553442	2.69860	7.39585	0.18963	35.853
3.45	0.0438268	2.70507	6.20892	0.13928	44.726
3.50	0.0326157	2.70959	4.89440	0.09321	59.340
3.55	0.0217058	2.71238	3.44739	0.052747	88.000
3.56	0.0195595	2.71275	3.14165	0.045496	97.396
3.57	0.0174249	2.71307	2.83037	0.038574	109.034
3.58	0.0153020	2.71334	2.51350	0.032009	123.827
3.59	0.0131907	2.71356	2.19101	0.025831	143.257
3.60	0.0110910	2.71373	1.86286	0.020082	169.92
3.61	0.0090028	2.71386	1.52900	0.014808	208.76
3.62	0.0069261	2.71395	1.18938	0.010075	270.61
3.63	0.0048607	2.71401	0.84397	0.0059726	384.54
3.64	0.0028067	2.71404	0.49272	0.0026423	664.14
3.65	0.0007639	2.71405	0.13559	0.00037830	2433.4

The solution of the Lane-Emden equation satisfying these conditions is called the Lane-Emden function, ϑ_n. Three solutions are known in closed form:

$$\left.\begin{aligned}\vartheta_0 &= 1-\tfrac{1}{6}\xi^2;\\ \vartheta_1 &= \sin\xi/\xi;\\ \vartheta_5 &= \frac{1}{(1+\frac{1}{3}\xi^2)^{\frac{1}{2}}}.\end{aligned}\right\} \qquad (38.11)$$

Other solutions are given in tabular form in a publication of the British Association for the Advancement of Science[1]. Table 6 gives some important quantities for the case $n=1.5$.

Table 6b. *The energy production integral for $n=1.5$ and $\nu=4$ and 20, respectively.*

ξ	$\int_0^\xi \xi^2\vartheta^7 d\xi$	$\int_0^\xi \xi^2\vartheta^{23} d\xi$	ξ	$\int_0^\xi \xi^2\vartheta^7 d\xi$	$\int_0^\xi \xi^2\vartheta^{23} d\xi$
0.0	0	0	1.2	0.23072	0.057915
0.1	0.000332	0.000326	1.3	0.25524	0.058280
0.2	0.002594	0.002435	1.4	0.27614	0.058433
0.3	0.008454	0.007347	1.5	0.29328	0.058491
0.4	0.019094	0.014956	1.6	0.30684	0.058511
0.5	0.035072	0.024197	1.7	0.31717	0.058517
0.6	0.056277	0.033600	1.8	0.32476	0.058518
0.7	0.081987	0.041884	1.9	0.33012	0.058519
0.8	0.111011	0.048322			
0.9	0.141875	0.052783	2.0	0.33378	0.058519
			2.1	0.33616	0.058519
1.0	0.17303	0.055561	2.2	0.33766	0.058519
1.1	0.20304	0.057121			

The series valid for small ξ is

$$\vartheta_n = 1-\frac{1}{6}\xi^2+\frac{n}{120}\xi^4-\cdots. \qquad (38.12)$$

The first zero of ϑ_n is called ξ_1 and represents the boundary of the polytrope.

As indicated by (38.6), ϑ represents the temperature in units of the central temperature. We will proceed to show how other properties of polytropes may be found in terms of ϑ and its derivatives.

The mass interior to r may be expressed as

$$\left.\begin{aligned}M(r) &= 4\pi\int_0^r r^2\varrho\,dr;\\ M(\xi) &= \frac{4\pi}{A^3}\varrho_c\int_0^\xi \xi^2\vartheta^n\,d\xi.\end{aligned}\right\} \qquad (38.13)$$

From the Lane-Emden equation we obtain, denoting $d\vartheta/d\xi$ by ϑ':

$$\left.\begin{aligned}M(\xi) &= \frac{4\pi}{A^3}\varrho_c\int_0^\xi \frac{d}{d\xi}\left[-\xi^2\vartheta'\right]d\xi\\ &= 4\pi\left[\frac{(n+1)K}{4\pi G}\right]^{\frac{3}{2}}\varrho_c^{(3-n)/2n}\left[-\xi^2\vartheta'\right]_\xi.\end{aligned}\right\} \qquad (38.14)$$

[1] British Assoc. for the Advancement of Science, Mathematical Tables, Vol. II. Emden Functions. London: Office of the British Association 1932.

The mass of the entire configuration may be found by evaluating (38.14) at $\xi = \xi_1$.

Another interesting relation connects the mean density and the central density. Define the mean density interior to ξ as

$$\bar{\varrho}(\xi) = \frac{M(\xi)}{\frac{4}{3}\pi\frac{\xi^3}{A^3}}. \tag{38.15}$$

From (38.14), it then follows that

$$\bar{\varrho}(\xi) = -\frac{3}{\xi}\left(\frac{d\vartheta}{d\xi}\right)\varrho_c. \tag{38.16}$$

In particular, the central density is related to the mean density of the entire star according to

$$\varrho_c = -\left[\frac{\xi}{3}\,\frac{1}{d\vartheta/d\xi}\right]_{\xi=\xi_1}\bar{\varrho}. \tag{38.17}$$

The quantity in brackets may be evaluated from the known solutions. Table 7 illustrates the wide range of density distribution that made the polytropes for $0 \leqq n \leqq 5$ so interesting in the development of studies of stellar structure.

Table 7. *The range of density variation in polytropic configuration (after Chandrasekhar).*

n	$\varrho_c/\bar{\varrho}$	n	$\varrho_c/\bar{\varrho}$	n	$\varrho_c/\bar{\varrho}$
0	1.0000	2.0	11.40	4.0	622.4
0.5	1.8361	2.5	23.41	4.5	6189
1.0	3.290	3.0	54.18	4.9	934800
1.5	5.991	3.5	152.9	5.0	∞

Finally, to evaluate the luminosity we integrate (4.5) using the known solution, ϑ_n:

$$\left.\begin{aligned} L &= 4\pi\int_0^R r^2\,\varepsilon\,\varrho\,dr,\\ &= 4\pi\,\varepsilon_0\int_0^R r^2\varrho^2\,T^\nu\,dr,\\ &= \frac{4\pi\,\varepsilon_0}{A^3}\,\varrho_c^2\,T_c^\nu\int_0^{\xi_1}\xi^2\,\vartheta^{2n+\nu}\,d\xi,\\ &= 4\pi\,\varepsilon_0\left(\frac{n+1}{4\pi G}\right)^{\frac{3}{2}}\frac{k}{\mu H}\,T_c^{1+\nu}\,P_c^{\frac{1}{2}}\int_0^{\xi_1}\xi^2\,\vartheta^{2n+\nu}\,d\xi;\end{aligned}\right\} \tag{38.18}$$

where we have assumed an energy generation law of the form

$$\varepsilon = \varepsilon_0\,\varrho\,T^\nu. \tag{38.19}$$

The Emden solution, ϑ_n, is but one of a family of solutions of the Lane-Emden equation, each member being finite and of vanishing slope at $\xi = 0$. A homology relation gives one member of the family, Φ, in terms of another, ϑ:

$$\Phi(\xi) = C^{\frac{2}{n-1}}\,\vartheta(C\,\xi), \tag{38.20}$$

where C is an arbitrary constant. A member of this family is called an E-solution.

If one defines the variables

$$\left.\begin{aligned} U &= \frac{d\log M(r)}{d\log r} = -\frac{\xi\vartheta^n}{\vartheta'};\\ V &= -\frac{d\log P}{d\log r} = -\frac{(n+1)\,\xi\vartheta'}{\vartheta}, \end{aligned}\right\} \tag{38.21}$$

which are invariant to homology transformation, the Lane-Emden equation takes on the form

$$\frac{d\log V}{d\log U} = \frac{U + \frac{V}{n+1} - 1}{3 - U - \frac{n}{n+1}V}. \tag{38.22}$$

In the (U, V) plane, all members of the family reduce to a single E-curve. The center of the star corresponds to the point (3,0) through which the E-curve goes with slope $-\frac{5(n+1)}{3n}$, Fig. 13[1].

The (U, V) plane is of particular interest when fitting zones together in composite models.

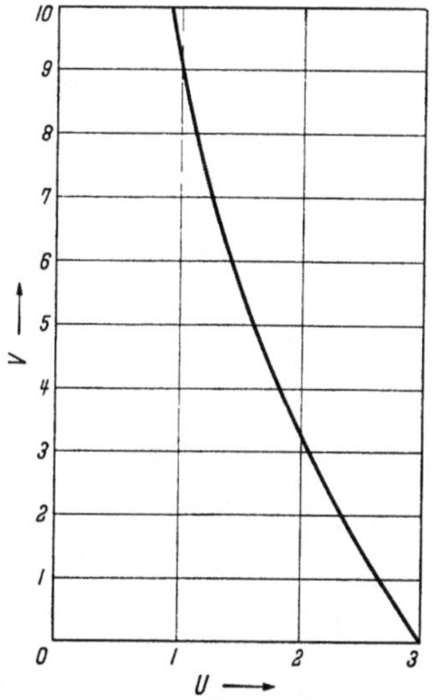

Fig. 13. The E curve in the (U, V) plane for $n = 1.5$.

39. The isothermal case. If the temperature is constant, the gas pressure is

$$P = K\varrho \tag{39.1}$$

where

$$K = \frac{k}{\mu H}T \tag{39.2}$$

and it is a constant. The isothermal case may therefore be considered to be a polytrope with infinite polytropic index. However it is more easily described in terms of a new variable ψ, defined by

$$\varrho = \varrho_c e^{-\psi}. \tag{39.3}$$

The appropriate definition of the dimensionless independent variable is

$$\left.\begin{aligned} r &= \xi/A\\ A^2 &= \frac{4\pi G\varrho_c}{K} \end{aligned}\right\} \tag{39.4}$$

and the analogue of the Lane-Emden equation becomes

$$\frac{1}{\xi^2}\frac{d}{d\xi}\left(\xi^2\frac{d\psi}{d\xi}\right) = e^{-\psi}. \tag{39.5}$$

Solutions describing the central regions of a star must obey the conditions

$$\psi = 0, \qquad \frac{d\psi}{d\xi} = 0 \quad \text{at} \quad \xi = 0. \tag{39.6}$$

A completely isothermal gas sphere has no finite boundary. Tables of the isothermal function have been computed by CHANDRASEKHAR and WARES[2].

[1] The E-curve divides the positive quadrant of the (U, V) plane into a region of M-solutions (between the E-curve and the origin) and a region of F-solutions. A thorough discussion will be found in CHANDRASEKHAR [3].

[2] S. CHANDRASEKHAR and G. W. WARES: Astrophys. Journ. **109**, 551 (1949).

The analogous homology invariant functions for the isothermal case are

$$U = \frac{\xi\, e^{-\psi}}{\psi'}; \qquad V = \xi\, \psi'. \tag{39.7}$$

The equation in the (U, V) plane becomes

$$\frac{d \log V}{d \log U} = -\frac{U-1}{U+V-3}, \tag{39.8}$$

and the corresponding E curve leaves the point (3, 0) with slope $-\frac{5}{3}$ and spirals around the point (1, 2), Fig. 14.

The above analysis is unaffected by radiation pressure, except that

$$P = K\varrho + \frac{a}{3} T^4. \tag{39.9}$$

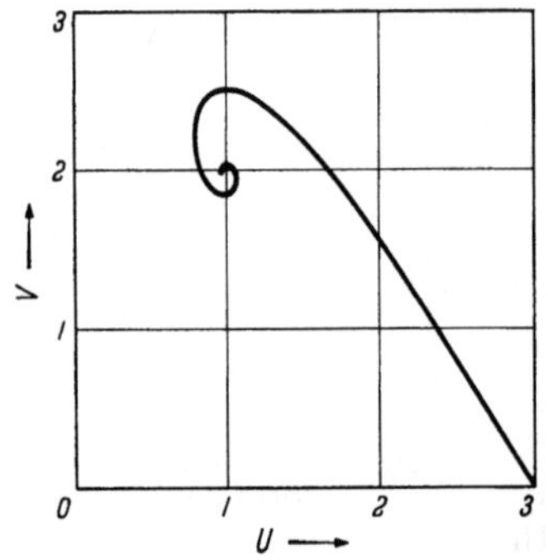

Fig. 14. The non-degenerate isothermal solution.

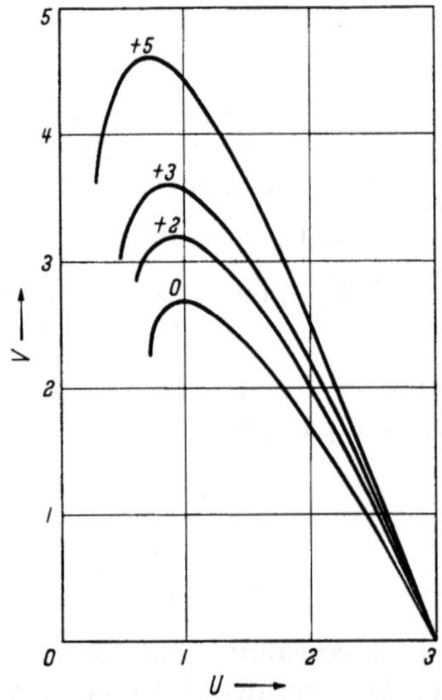

Fig. 15. Degenerate isothermal solutions. Each curve is marked by the value of ψ at the center.

When an isothermal region develops in the deep interior of a star, it is likely to be at high density and, consequently, the effects of degeneracy may become appreciable. In that case, the equation of state must be expressed in the parametric form (Sect. 15):

$$\varrho = \frac{2}{\sqrt{\pi}} G(T)\, \mu_e H\, F_{\frac{1}{2}}(\psi), \tag{39.10}$$

$$P = \frac{2}{\sqrt{\pi}} G(T)\, kT\, \frac{2}{3} F_{\frac{3}{2}}(\psi) + \frac{a}{3} T^4; \tag{39.11}$$

where $G(T)$ is defined by (17.2), F_w by (15.12) and Table 1 (Sect. 15), and μ_e by (13.5).

By making the transformation

$$r = \frac{h^{\frac{3}{2}}}{4\pi \mu_e H} (2m_e kT)^{-\frac{1}{4}} (2mG)^{-\frac{1}{2}} \xi, \tag{39.12}$$

the equation of hydrostatic equilibrium becomes[1]

$$\frac{1}{\xi^2} \frac{d}{d\xi}\left(\xi^2 \frac{d\psi}{d\xi}\right) = -F_{\frac{1}{2}}(\psi). \tag{39.13}$$

[1] G. W. WARES: Astrophys. Journ. **100**, 158 (1944).

This must be solved subject to the boundary conditions

$$\psi = \psi_0, \quad \frac{d\psi}{d\xi} = 0 \quad \text{as} \quad \xi \to 0. \tag{39.14}$$

ψ_0 is thus the value of ψ at the center of the configuration and from it the central density and central pressure may be determined by using (39.10).

The homology invariants become

$$\left.\begin{aligned} U &= F_{\frac{1}{2}}(\psi)\,\xi\left(-\frac{d\psi}{d\xi}\right)^{-1}; \\ V &= \frac{3}{2}\,F_{\frac{1}{2}}(\psi)\,F_{\frac{3}{2}}^{-1}(\psi)\,\xi\left(-\frac{d\psi}{d\xi}\right). \end{aligned}\right\} \tag{39.15}$$

Tables of the isothermal degenerate function are given by HÄRM and SCHWARZSCHILD[1] following WARES[2] for different values of ψ_0.

The behavior of some of these solutions in the (U, V) plane is shown in Fig. 15.

40. The Cowling model. From homology relations, the central temperature is proportional to the mass, (35.16). Further, at temperatures above 20×10^6 °K, the energy production by the carbon cycle predominates. We therefore conclude that for massive stars, energy is supplied by the carbon cycle.

These reactions are very sensitive to temperature so that the energy production is strongly concentrated to the center of the star. The total luminosity is effectively produced inside a relatively small sphere of radius r_0. The same energy then crosses all concentric spheres of larger radii including, ultimately, the surface. Although $L(r)$ = constant = L for all spheres with $r \geq r_0$, the flux, which is $L(r)/4\pi r^2$, must increase inward.

The radiative temperature gradient is proportional to the flux and therefore it becomes steeper going inward. Eventually it may become steeper than the local adiabat, convection will set in, and a convective core will form. If this occurs at a radius $r > r_0$, then all the energy will be produced in the convective core; while in the radiative envelope, the luminosity will be constant. These are the physical assumptions of the Cowling model.

As a result the problem is considerably simplified. In the radiative zone, $L(r)$ is constant and Eq. (4.5) need not be considered. In the convective core, the known solution for the polytrope $n = 1.5$ can be used. The calculation of the luminosity by integrating (38.18) from the center to the boundary of the core can be carried out separately.

For the purpose of simplicity we will assume a chemically homogeneous star and an opacity law of the form

$$\varkappa = \varkappa_0 \varrho\, T^{-3.5} \tag{40.1}$$

in the envelope. The envelope equations, in terms of the Schwarzschild variables, become

$$\left.\begin{aligned} \frac{dp}{dx} &= -\frac{pq}{t x^2}, \\ \frac{dq}{dx} &= \frac{p x^2}{t}, \\ \frac{dt}{dx} &= -\frac{C p^2}{x^2 t^{8.5}}. \end{aligned}\right\} \tag{40.2}$$

The constant

$$C = \frac{3\varkappa_0}{4ac(4\pi)^3}\left(\frac{k}{GH}\right)^{7.5} \frac{LR^{0.5}}{M^{5.5}} \frac{1}{\mu^{7.5}}, \tag{40.3}$$

[1] R. HÄRM and M. SCHWARZSCHILD: Astrophys. Journ., Suppl. 1, 319 (1955).

[2] Siehe Fußnote 1, S. 56.

is usually referred to as the mass-luminosity relation, [compare with (35.20)]. It should be noted, however, that it also involves composition (through $\varkappa_0$ and μ) and radius. If mass and composition are to result in a unique luminosity and radius, as implied by the Vogt-Russell theorem, one would expect an additional relation, similar to (35.21), between L, M, R and composition. That this turns out indeed to be the case will be seen below.

The system (40.2) is required to satisfy the boundary condition at the surface. We shall assume that the condition

$$p \to 0, \quad t \to 0, \quad \text{at} \quad x = 1, \tag{40.4}$$

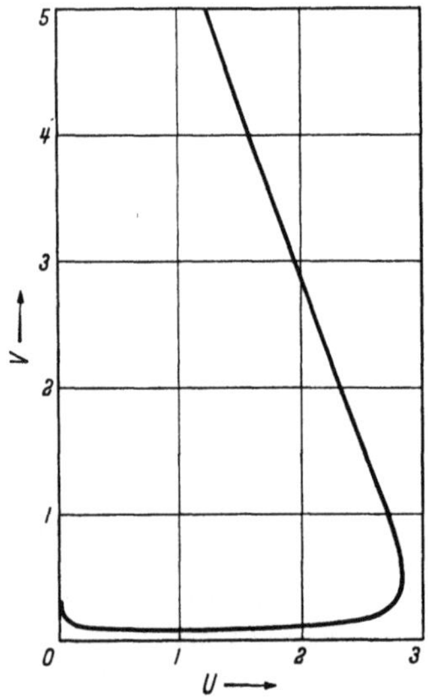

Fig. 16. Envelope solutions for $\log C = -6.0$ (after HÄRM and SCHWARZSCHILD).

is adequate. Further we may note that not only is

$$q = 1 \quad \text{at} \quad x = 1, \tag{40.5}$$

but $q \approx 1$ to good accuracy for some distance into the star. Eqs. (40.2) reduce to

$$\left.\begin{aligned} \frac{dp}{dx} &= -\frac{p}{t x^2}, \\ \frac{dt}{dx} &= -\frac{C p^2}{x^2 t^{8.5}}, \end{aligned}\right\} \tag{40.6}$$

from which x may be eliminated giving

$$\frac{dp}{dt} = \frac{1}{C} \frac{t^{7.5}}{p}. \tag{40.7}$$

We then may easily obtain the analytic solution valid near the surface:

$$\left.\begin{aligned} p^2 &= \frac{1}{C} \frac{2}{8.5} t^{8.5}, \\ t &= \frac{2}{8.5} \left(\frac{1}{x} - 1\right). \end{aligned}\right\} \tag{40.8}$$

These relations may be employed as long as $q \approx 1$; but when the mass of the exterior shells becomes appreciable we must continue inward by solving the complete system numerically. Solutions must be found separately for different values of the parameter, C.

The homology invariants U, V, and $(n+1) = \dfrac{d \log P}{d \log T}$ are known in terms of p, q, and t. In the (U, V) plane, these envelope solutions have a characteristic shape shown in Fig. 16 which is plotted from HÄRM and SCHWARZSCHILD's paper[1] for the case of Kramers opacity and $\log C = -6.0$. Further data for this case are given in Table 8.

The envelope solution is physically valid only as long as $n+1 > 2.5$. At the point where $n+1 = 2.5$ convection sets in and the envelope solution must join the E-curve for the core. At this point of fit, p, q, t and x must be continuous and hence, if the composition is continuous, the fit in the (U, V) plane must also be continuous.

For only one value of C will this condition be satisfied. C may therefore be regarded as an eigenvalue to be determined by fitting the envelope to the core.

[1] R. HÄRM and M. SCHWARZSCHILD: Astrophys. Journ., Suppl. 1, 319 (1955). See also J. G. GARDINER: Monthly Notices Roy. Astronom. Soc. London 111, 94 (1951).

Table 8. *Envelope solution for log C = − 6.0* (Härm and Schwarzschild).

$\log\left(\frac{1}{x}-1\right)$	$-\log q$	$\log t$	$\log p$	U	V	$n+1$
−1.00	0.0000	−1.62839	−4.2348	0.002	46.750	4.250
−0.95	0.0000	−1.57839	−4.0224	0.002	42.128	4.250
−0.90	0.0000	−1.52839	−3.8098	0.004	38.009	4.250
−0.85	0.0000	−1.47839	−3.5974	0.005	34.336	4.250
−0.80	0.0000	−1.42839	−3.3849	0.007	31.061	4.250
−0.75	0.0001	−1.37840	−3.1724	0.009	28.142	4.250
−0.70	0.0002	−1.32841	−2.9600	0.013	25.538	4.249
−0.65	0.0003	−1.27842	−2.7476	0.018	23.218	4.249
−0.60	0.0006	−1.22845	−2.5353	0.025	21.146	4.248
−0.55	0.0009	−1.17850	−2.3231	0.034	19.296	4.247
−0.50	0.0013	−1.12857	−2.1111	0.046	17.644	4.246
−0.45	0.0020	−1.07867	−1.8992	0.061	16.165	4.244
−0.40	0.0029	−1.02882	−1.6877	0.081	14.840	4.242
−0.35	0.0043	−0.97906	−1.4767	0.106	13.649	4.238
−0.30	0.0063	−0.92939	−1.2664	0.138	12.576	4.233
−0.25	0.0090	−0.87988	−1.0570	0.178	11.605	4.226
−0.20	0.0128	−0.83057	−0.8488	0.227	10.721	4.216
−0.15	0.0179	−0.78154	−0.6424	0.288	9.910	4.204
−0.10	0.0249	−0.73289	−0.4382	0.361	9.160	4.188
−0.05	0.0341	−0.68475	−0.2371	0.448	8.460	4.166
0.00	0.0462	−0.63725	−0.0398	0.550	7.800	4.139
+0.05	0.0619	−0.59058	+0.1525	0.668	7.169	4.104
0.10	0.0818	−0.54494	0.3388	0.801	6.562	4.060
0.15	0.1068	−0.50058	0.5178	0.950	5.974	4.006
0.20	0.1377	−0.45772	0.6881	1.112	5.401	3.940
0.25	0.1752	−0.41661	0.8485	1.286	4.843	3.860
0.30	0.2202	−0.37748	0.9977	1.466	4.303	3.765
0.35	0.2731	−0.34054	1.1348	1.649	3.783	3.654
0.40	0.3343	−0.30594	1.2590	1.831	3.290	3.524
0.45	0.4040	−0.27379	1.3701	2.006	2.829	3.377
0.50	0.4823	−0.24411	1.4680	2.170	2.405	3.212
0.55	0.5687	−0.21687	1.5532	2.319	2.023	3.030
0.60	0.6630	−0.19200	1.6262	2.450	1.684	2.836
0.65	0.7643	−0.16935	1.6882	2.562	1.389	2.629
0.70	0.8721	−0.14877	1.7401	2.654	1.137	2.416
0.75	0.9853	−0.13006	1.7833	2.726	0.924	2.200
0.80	1.1032	−0.11304	1.8190	2.777	0.748	1.986
0.85	1.2247	−0.09750	1.8483	2.808	0.603	1.778
0.90	1.3488	−0.08328	1.8722	2.817	0.485	1.581
0.95	1.4745	−0.07020	1.8916	2.804	0.391	1.398
1.00	1.6006	−0.05811	1.9075	2.767	0.315	1.232
1.05	1.7256	−0.04687	1.9205	2.703	0.256	1.084
1.10	1.8482	−0.03637	1.9312	2.607	0.210	0.956
1.15	1.9665	−0.02651	1.9401	2.477	0.174	0.847
1.20	2.0789	−0.01720	1.9476	2.312	0.146	0.759
1.25	2.1834	−0.00836	+1.9540	2.111	0.125	0.688
1.30	2.2782	+0.00009	1.9595	1.879	0.110	0.637
1.35	2.3619	0.00819	1.9645	1.627	0.100	0.600
1.40	2.4337	0.01600	1.9691	1.368	0.093	0.581
1.45	2.4936	0.02356	1.9735	1.117	0.089	0.575
1.50	2.5420	0.03091	1.9777	0.888	0.087	0.583
1.55	2.5801	0.03809	1.9820	0.688	0.088	0.602
1.60	2.6094	0.04512	1.9863	0.523	0.090	0.633
1.65	2.6316	0.05203	1.9908	0.390	0.095	0.675
1.70	2.6481	0.05884	1.9956	0.288	0.100	0.726

Table 8. (Continued.)

$\log\left(\frac{1}{x}-1\right)$	$-\log q$	$\log t$	$\log p$	U	V	$n+1$
1.75	2.6602	0.06559	2.0007	0.210	0.108	0.787
1.80	2.6690	0.07228	2.0062	0.152	0.116	0.856
1.85	2.6754	0.07895	2.0122	0.110	0.126	0.936
1.90	2.6800	0.08560	2.0187	0.079	0.138	1.023
1.95	2.6834	0.09226	2.0258	0.057	0.151	1.120
2.00	2.6857	0.09895	2.0336	0.040	0.166	1.225
2.05	2.6874	0.10567	2.0422	0.029	0.182	1.337
2.10	2.6886	0.11246	2.0517	0.021	0.201	1.458
2.15	2.6895	0.11933	2.0622	0.015	0.221	1.587
2.20	2.6902	0.12629	2.0737	0.011	0.243	1.721
2.25	2.6906	0.13338	2.0864	0.008	0.268	1.863
+2.30	2.6909	+0.14061	+2.1004	0.006	0.296	2.011

Schematically this is shown in Fig. 17, where it should be kept in mind that a third coordinate, $n+1$, not shown in the diagram, must also be continuous at the point of fit.

The inward integration gives p, q, t and x at the interface between core and envelope; the outward integration gives ξ and ϑ at the same point. From the dimensionless form of (38.6) and (38.3):

$$\left.\begin{aligned} t &= t_c\vartheta, \\ p &= p_c\vartheta^{1+n} \end{aligned}\right\} \tag{40.9}$$

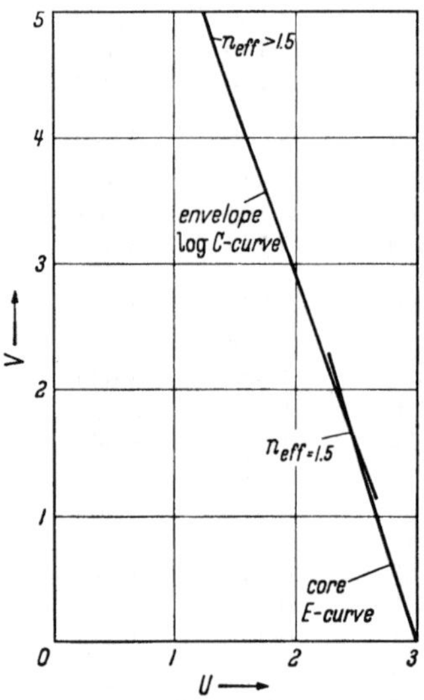

Fig. 17. Fitting a radiative envelope and a convective core in the (U, V) plane.

we may then evaluate p_c and t_c.

We may now turn our attention to the requirement that the total luminosity be produced in the core.

Assuming a single energy source for which

$$\varepsilon = \varepsilon_0 \varrho T^{\nu}, \tag{40.10}$$

the luminosity equation becomes

$$dL(r) = 4\pi r^2 \varepsilon_0 \varrho^2 T^{\nu} dr. \tag{40.11}$$

Defining

$$f = \frac{L(r)}{L}, \tag{40.12}$$

(40.11) becomes

$$df = D t_c^{1+\nu} p_c^{\frac{1}{2}} \xi^2 \vartheta^{2n+\nu} d\xi, \tag{40.13}$$

where [compare with (35.21)]:

$$D = \frac{\varepsilon_0}{4\pi}(n+1)^{\frac{3}{2}}\left(\frac{G\mu H}{k}\right)^{\nu}\frac{M^{2+\nu}}{R^{3+\nu}}. \tag{40.14}$$

This is the second mass-luminosity-radius-composition relation and it may be called the energy production constant.

If we require that the luminosity of the entire star is the luminosity of the core, (40.13) becomes:

$$1 = D t_c^{1+\nu} p_c^{\frac{1}{2}} \int \xi^2 \vartheta^{2n+\nu} d\xi, \tag{40.15}$$

where the integral is taken over the core. By using tables of this integral, such as Table 6b, and noting that t_c and p_c were already determined, (40.15) yields D.

If M and the composition are assigned, C and D may be regarded as two equations for L and R.

SCHWARZSCHILD, however, considered this problem somewhat differently[1]. Assuming the sun to be built on a Cowling model, he assigned M, L and R rather than M and composition. He then used C and D as two equations between the composition parameters X and Y (the hydrogen and helium content) which occur in μ, $\varkappa_0$, and ε_0. He regarded C as one relation in the (X, Y) plane and D as another. The point of intersection determined X and Y. This work was done when it was thought that the carbon cycle produced most of the solar energy. When it was shown, however, that the proton-proton reaction was the principal source, the Cowling model, as a solar model, had to be abandoned.

41. Criterion for a convective core. It was pointed out in the preceding section that energy production by the carbon cycle is likely to produce a convective core. On the other hand, if the proton-proton chain contributes greatly to the energy production, the situation is quite different. Since the proton-proton chain is not nearly so temperature-sensitive as the carbon cycle, we cannot argue that the energy producing region is very small. It is true that in the outer envelope the flux must increase inward; but once we enter the energy producing region, $L(r)$ is no longer constant but decreases, so that the temperature gradient need not be as steep as it was in the case of the carbon cycle.

Clearly the situation will depend on the opacity as well as the energy production since decreasing opacity will partially compensate for the increasing flux.

A useful criterion was established by NAUR and OSTERBROCK[2]. Their argument is as follows. At the boundary of the core the effective polytropic index is 1.5 and decreasing *inward*; therefore,

$$D = \frac{d \log (n+1)_{\mathrm{rad}}}{d \log r} \geqq 0. \tag{41.1}$$

Further,

$$n + 1 = \frac{d \log P}{d \log T}; \tag{41.2}$$

and from (4.1) and (4.3)

$$\frac{d \log P}{d \log r} \propto \frac{M(r)}{r T}, \tag{41.3}$$

and

$$\left(\frac{d \log T}{d \log r}\right)_{\mathrm{rad}} \propto \frac{P^{2-\alpha} T^{-9-s+\alpha} L(r)}{r}, \tag{41.4}$$

where an opacity law of the form

$$\varkappa = \varkappa_0 \varrho^{1-\alpha} T^{-3-s} \tag{41.5}$$

has been assumed. Noting that, at the boundary of the core

$$\left.\begin{aligned} \frac{d \log T}{d \log r} &= \frac{d \log P}{d \log r} \frac{1}{n+1}, \\ &= \frac{2}{5} \frac{d \log P}{d \log r}, \end{aligned}\right\} \tag{41.6}$$

and using the homology invariant quantities U, V, and

$$W = \frac{d \log L(r)}{d \log r}, \tag{41.7}$$

[1] M. SCHWARZSCHILD: Astrophys. Journ. **104**, 203 (1946).
[2] P. NAUR and D. E. OSTERBROCK: Astrophys. Journ. **117**, 306 (1953).

it is easily seen that

$$D = U - W - (1.2 + 0.4s + 0.6\alpha)\, V \gtrsim 0. \tag{41.8}$$

By using the solution of the Lane-Emden equation, $U - W$ may be plotted against V. This function is zero at the center of the star where V is also zero, and is an increasing function of V. Near the origin

$$\left(\frac{d(U-W)}{dV}\right)_{V=0} = \frac{6\nu}{25} + \frac{9}{25}, \tag{41.9}$$

where an energy generation formula of the form

$$\varepsilon = \varepsilon_0 \varrho T^{\nu} \tag{41.10}$$

has been used. A line passing through the origin with slope (41.9) would always lie above the computed curve and represents the limiting case, for which the core would have zero extent. Thus the criterion for the existence of a convective core becomes

$$6\nu + 9 > 15\alpha + 10s + 30, \tag{41.11}$$

or

$$6\nu - 10s - 15\alpha > 21. \tag{41.12}$$

As we would have expected, the greater the ν, that is, the more temperature sensitive the energy generation, the more likely convection would be. On the other hand, large values of α and s indicate a rapid decrease in opacity inward which would forestall convection. For the modified Kramers law, $\alpha = 0.25$ and $s = 0.5$; convection is therefore impossible if ν is less than about 4.9.

42. Red dwarfs. The stars which lie along the main sequence below the sun are red dwarfs. They are less massive than the sun and have lower effective temperatures. For example, the average of the two components of Castor C yields $\log L/L_{\odot} = -1.20$, $\log M/M_{\odot} = -0.22$, $\log R/R_{\odot} = -0.20$, and $T_e = 3600°$ K. For such a star it would be expected that the proton-proton chain is the predominant nuclear energy source; and by the arguments of the previous section, there should be no convective core.

Models computed on the assumption of radiative equilibrium throughout, however[1], indicated either too high a luminosity or, alternatively, a low hydrogen content. But the red dwarfs evolve slowly, because of their low luminosity, and their composition ought to be characteristic of the cloud from which they originated; therefore they should probably be rich in hydrogen.

This inconsistency was resolved by OSTERBROCK[2] who followed up earlier indications by BIERMANN[3], ÖPIK[4] and STRÖMGREN[5] that convection in the outer regions may affect the internal structure. This convection arises not because of high flux (which is the situation in the core) but because of high opacity in the atmosphere. Although the outermost layers of the stellar atmospheres are in radiative equilibrium, the opacity, due to the negative ions of hydrogen, increases inward and the radiative gradient necessary to drive the flux against this resistance becomes too steep to be stable. An extensive convective zone will develop, the structure of which is further complicated because this region contains the

[1] For example, L. H. ALLER: Astrophys. Journ. **111**, 173 (1950), or ALLER et al.: Astrophys. Journ. **115**, 328 (1952).
[2] D. E. OSTERBROCK: Astrophys. Journ. **118**, 529 (1953).
[3] L. BIERMANN: Astronom. Nachr. **264**, 361 (1938).
[4] E. ÖPIK: Publ. Obs. Tartu **30**, No. 4 (1938).
[5] B. STRÖMGREN: Astronom. J. **57**, 65 (1952).

zone of partial ionization of hydrogen and helium. (This problem was already mentioned in Sect. 9.) The convective temperature gradient is less steep than the radiative, and the ultimate effect is to lower the central temperature and, correspondingly, the luminosity.

OSTERBROCK used a rough procedure for taking the outer convective zone into account. It was assumed that the entire outer region was described by the adiabatic gradient, with $\gamma = \frac{5}{3}$. This circumvents the difficulty of integrating detailed atmospheres, but it yields somewhat erroneous radii.

In this adiabatic zone

$$P = K T^{\frac{5}{2}} \tag{42.1}$$

so that the usual boundary condition that P and T tend simultaneously to zero is fulfilled regardless of K. In principle, the value of K can be found by calculating a detailed atmosphere and comparing P and T at the inner boundary of the convective zone. Since there is no entirely suitable theory of convection, any evaluation of K in this manner will be uncertain. Therefore, OSTERBROCK chooses to regard K as a free parameter.

In terms of the Schwarzschild variables, the adiabatic relation becomes

$$p = E\, t^{2.5} \tag{42.2}$$

where the parameter

$$E = 4\pi K \left(\frac{\mu H}{k}\right)^{2.5} G^{1.5} M^{0.5} R^{1.5} \tag{42.3}$$

is used in the actual calculations, rather than K.

To represent the envelope, the equations

$$\frac{dp}{dx} = -\frac{pq}{t x^2}, \qquad \frac{dq}{dx} = \frac{p x^2}{t} \tag{42.4}$$

are integrated inward, using the boundary conditions

$$q = 1, \quad p = 0 \quad \text{at} \quad x = 1. \tag{42.5}$$

The radiative solutions are begun at the center and integrated outward. Using an opacity law:

$$\varkappa = \varkappa_0\, \varrho^{0.5}\, T^{-3.5}, \tag{42.6}$$

and an energy generation law:

$$\varepsilon = \varepsilon_0\, \varrho\, T^{4.5}, \tag{42.7}$$

and defining

$$f = \frac{L(r)}{L}, \tag{42.8}$$

the temperature and luminosity gradients become

$$\left.\begin{aligned} \frac{dt}{dx} &= -C\,\frac{p^{1.5} f}{t^8 x^2}, \\ &\text{and} \\ \frac{df}{dx} &= D p^2 t^{2.5} x^2; \end{aligned}\right\} \tag{42.9}$$

where

$$\left.\begin{aligned} C &= \frac{3\varkappa_0}{4ac}\left(\frac{1}{4\pi}\right)^{2.5}\left(\frac{k}{\mu H G}\right)^{7.5}\frac{LR^2}{M^6}, \\ &\text{and} \\ D &= \frac{\varepsilon_0}{4\pi}\left(\frac{\mu H G}{k}\right)^{4.5}\frac{M^{6.5}}{LR^{7.5}} \end{aligned}\right\} \tag{42.10}$$

represent the usual mass-luminosity-radius-composition relations. Eqs. (42.8) are combined with (42.3) and solved subject to the boundary conditions at the center:

$$f = q = 0 \quad \text{at} \quad x = 0. \tag{42.11}$$

By a suitable change of variables it is possible to reduce the solutions for the core to a one-parameter family, labeled by the value of $(n+1)$ at the center: $(n+1)_c$. On proceeding outward, eventually $n+1$ drops to 2.5 and at this point, the radiative core is fitted to a convective envelope by requiring continuity in the (U, V) plane.

Thus to every value of $(n+1)_c$ there corresponds a value of E, C, D and x_1, the boundary of the convective zone. Some of the results of OSTERBROCK'S integrations are given in Table 9, and illustrated in Fig. 18. Note in particular that as E decreases the boundary between the convective and the radiative zones moves outward.

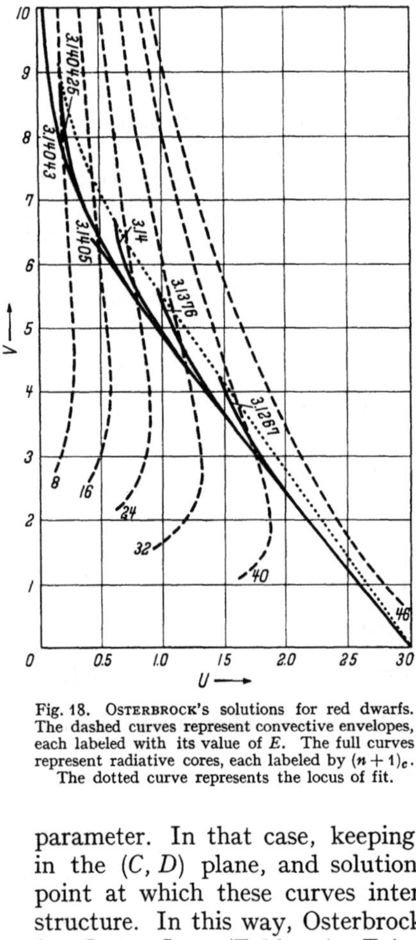

Fig. 18. OSTERBROCK'S solutions for red dwarfs. The dashed curves represent convective envelopes, each labeled with its value of E. The full curves represent radiative cores, each labeled by $(n+1)_c$. The dotted curve represents the locus of fit.

Table 9. *Mathematical properties of some of Osterbrock's red dwarf models.*

$(n+1)_c$	3.14	3.1403	3.14036	3.14041
E	21.43	16.82	14.69	11.05
x_1	0.655	0.676	0.686	0.705
q_1	0.873	0.913	0.930	0.953
C	-5.420	-5.385	-5.361	-5.309
D	$+0.066$	-0.180	-0.310	-0.570

If E were calculated from the atmosphere, the appropriate C and D could be found and combined with observed M, L and R to determine two composition parameters, such as X and Z. On the other hand, E may be treated as a free parameter. In that case, keeping X fixed and varying Z will give one curve in the (C, D) plane, and solutions for various values of E give another. The point at which these curves intersect, gives the appropriate composition and structure. In this way, Osterbrock constructed a table of possible compositions for Castor C av (Table 10); T_c is the central temperature, x_1 is the fractional radius of the radiative core and q_1 is the fractional mass in it; and T_1 is the temperature at the boundary of the core.

A fit can be obtained using quite reasonable compositions, down to spectral type $M0$. The most recent solar models have been constructed along similar lines.

Table 10. *Possible abundances of elements and physical properties of Castor* C av *after Osterbrock.*

X	Y	Z	E	x_1	q_1	T_1 million	T_c degrees	ϱ_c gm/cm³
0.90	0.091	0.009	18.7	0.669	0.900	2.2	7.8	81
0.80	0.184	0.016	19.4	0.666	0.894	2.4	8.3	79
0.70	0.271	0.029	19.9	0.663	0.888	2.6	8.9	76

43. Discontinuities in chemical composition. In the course of stellar evolution, discontinuities in chemical composition may arise. A suitable method for dealing with interfaces of this kind has been extensively used by Schwarzschild and his collaborators. The fitting is done in the (U, V) plane, as in the preceding sections, but the discontinuity in composition results in a discontinuity in the (U, V) curve.

Let us suppose that the discontinuity develops at some point within the radiative envelope, dividing it into an exterior region and an interior region, to be denoted by subscripts e and i, respectively. Such a situation was considered by LI HEN and SCHWARZSCHILD[1]. Let us consider, as they did, an opacity law

$$\varkappa = \varkappa_0 \varrho^{0.75} T^{-3.5}, \tag{43.1}$$

where

$$\varkappa_0 = \text{const}\, Z (1+X)^{0.75}; \tag{43.2}$$

and define

$$l = \frac{\mu}{\mu_e}, \qquad j = \frac{Z}{Z_e}\left[\frac{(1+X)\,\mu}{(1+X_e)\,\mu_e}\right]^{0.75}. \tag{43.3}$$

Then

$$l_i = \frac{\mu_i}{\mu_e}, \qquad j_i = \frac{Z_i}{Z_e}\left[\frac{(1+X_i)\,\mu_i}{(1+X_e)\,\mu_e}\right]^{0.75} \tag{43.4}$$

and $l_e = j_e = 1$. If the Schwarzschild variables are defined throughout the star in terms of μ_e, the equations of equilibrium become

$$\left.\begin{aligned} \frac{dp}{dx} &= -l\,\frac{pq}{t x^2},\\ \frac{dq}{dx} &= l\,\frac{p x^2}{t},\\ \frac{dt}{dx} &= -l j C\,\frac{p^{1.75}}{t^{8.25} x^2}; \end{aligned}\right\} \tag{43.5}$$

where

$$C = \frac{3\varkappa_{0,e}}{4ac}\left(\frac{k}{\mu_e H G}\right)^{7.5}\left(\frac{1}{4\pi}\right)^{2.75}\frac{L R^{1.25}}{M^{5.75}}, \tag{43.6}$$

and the homology invariant quantities are

$$\left.\begin{aligned} U &= l\,\frac{p x^3}{tq}\\ V &= l\,\frac{q}{tx}\\ n+1 &= \frac{1}{j}\,\frac{q t^{8.25}}{C p^{1.75}}. \end{aligned}\right\} \tag{43.7}$$

At the discontinuity, indicated by an asterisk,

$$\left.\begin{aligned} \left(\frac{U_i}{U_e}\right)^* &= \frac{l_i}{l_e} = l_i,\\ \left(\frac{V_i}{V_e}\right)^* &= \frac{l_i}{l_e} = l_i, \end{aligned}\right\} \tag{43.8}$$

and

$$\left(\frac{n_e+1}{n_i+1}\right)^* = \frac{j_i}{j_e} = j_i. \tag{43.9}$$

[1] LI HEN and M. SCHWARZSCHILD: Monthly Notices Roy. Astronom. Soc. London **109**, 631 (1949). Although this paper was important in the development of models for red giants, discontinuities of precisely this type probably do not actually occur.

It follows from (43.8) that

$$\left(\frac{U_i}{V_i}\right)^* = \left(\frac{U_e}{V_e}\right)^*; \tag{43.10}$$

which indicates that the jump in the (U, V) plane between the end of the exterior solution and the beginning of the interior solution occurs along a straight line that passes through the origin (Fig. 19). Furthermore, the values of $(n+1)$ at these end points must obey (43.9).

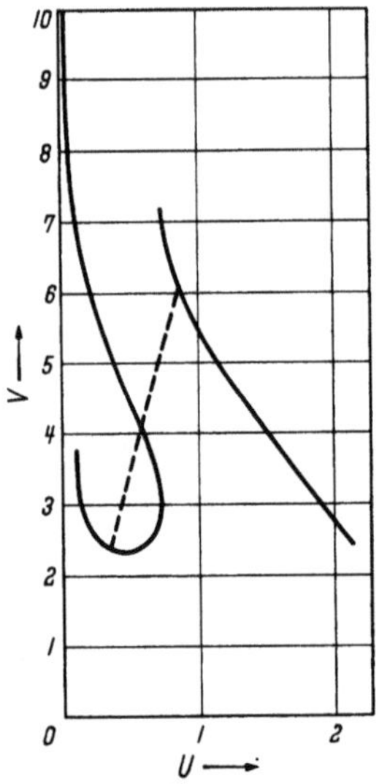

Fig. 19.

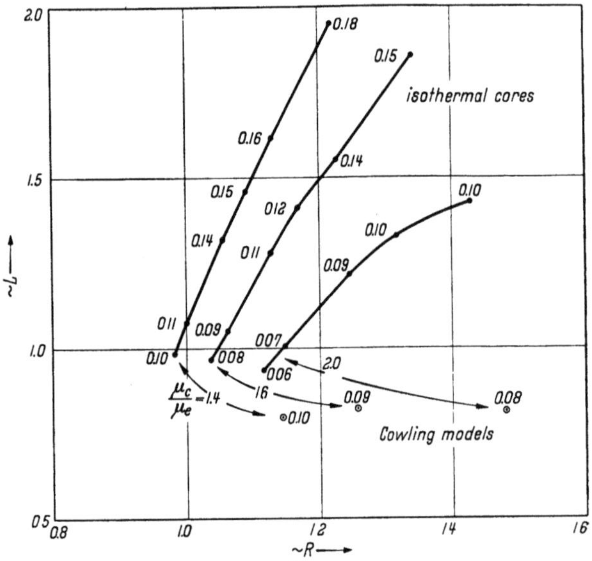

Fig. 20.

Fig. 19. The fitting conditions in the (U, V) plane for a discontinuity in the chemical composition (after LI HEN and SCHWARZSCHILD).

Fig. 20. The full curves represent non-degenerate isothermal cores with radiative envelopes, each curve for a different ratio of core-to-envelope composition. The open circles represent Cowling models for the same composition ratios. Each point is labeled by the fraction of mass contained in the isothermal or convective core. (After HARRISON.)

44. The Schönberg-Chandrasekhar limit. In 1942 SCHÖNBERG and CHANDRASEKHAR[1] and later HARRISON[2] constructed models motivated by the following scheme of stellar evolution. Assuming that mixing occurs only within the convective core, only the hydrogen in the core is available for energy production. Eventually this fuel is exhausted and the convection dies out. Since there is no further energy production in the core, no temperature gradient is necessary to drive energy outward and the core becomes isothermal. Energy production is then confined to a shell surrounding the isothermal core. As the hydrogen in *this* shell is exhausted, it too becomes isothermal, and the energy producing region moves outward. The mass of the isothermal region gradually increases by this process.

The Schönberg-Chandrasekhar and Harrison models consisted of two zones of differing chemical composition. The outer zone was always in radiative equilibrium and of uniform chemical composition. At the boundary between the zones there was a discontinuity in μ and the inner zone was also of uniform

[1] M. SCHÖNBERG and S. CHANDRASEKHAR: Astrophys. Journ. **96**, 161 (1942).
[2] M. H. HARRISON: Astrophys. Journ. **100**, 343 (1944); **105**, 322 (1947).

composition, but of higher mean molecular weight. In one series the inner zone was in convective equilibrium ($n = 1.5$) and in the second series the inner zone was isothermal ($n = \infty$). Degeneracy was neglected and Kramers opacity was assumed. The energy production was assumed confined to the convective core (if there was one) or to a very thin shell surrounding the isothermal core.

The convective core models indicated that the radius and mass of the convective region decreases with time as the ratio of molecular weight of the core to that of the envelope, μ_c/μ_e, increases. Even if the mass of the core had remained constant, the depletion of hydrogen would have taken place. If, in addition, the changing μ causes the mass included in the convection to decrease, the available hydrogen is even further reduced. The assumption is made that an isothermal core replaces the convective core at some μ_c/μ_e (the details are not discussed) and from then on the series of models with isothermal cores applies.

The result of the latter series that has attracted the most attention is the fact that for a given μ_c/μ_e there is a maximum mass of the isothermal region beyond which no equilibrium fit with a radiative envelope is possible. This maximum is a function of μ_c/μ_e having its largest value at $\mu_c/\mu_e = 1$ and decreasing (to $0.1\,M$) toward $\mu_c/\mu_e = 2$. This result is known as the Schönberg-Chandrasekhar limit.

In some theories of stellar evolution it is assumed that stars on the main sequence are chemically homogeneous ($\mu_c/\mu_e = 1$) and that the initial stages of evolution follow the outline given above. During this time the departure from the main sequence is slight. Once the Schönberg-Chandrasekhar limit is reached, however, new processes must set in and the departure from the main sequence is rapid.

Table 11 gives the main features of the Schönberg-Chandrasekhar and Harrison calculations and Fig. 20 illustrates their behavior in the (R, L) plane.

Table 11. *Summary of conditions when the maximum mass is in the isothermal core (after Harrison).*

$\mu_{core}/\mu_{envelope}$	1.0	1.4	1.6	1.8	2.0
Fractional mass of isothermal core	0.37	0.18	0.15	0.12	0.10
Fractional radius of isothermal core	0.12	0.09	0.08	0.07	0.06
Total luminosity in arbitrary units	4.4	1.9	1.85	1.5	1.4
Total radius in arbitrary units	1.2	1.2	1.3	1.3	1.4

These models leave some unanswered questions. In particular, during the initial stages, if the convective core shrinks, it will leave behind in the radiative zone a partially exhausted region which is likely to alter the structure considerably. We will return to this problem in discussing the calculations of Tayler and Kushwaha.

45. The evolution of early-type stars. Detailed calculations of the evolution of carbon cycle stars have recently been carried out by Tayler and Kushwaha[1], to find the evolutionary tracks in the H-R diagram. The early-type stars in galactic clusters (Population I) would be expected to follow such paths, with the most luminous stars moving most rapidly. The locus of points for stars of different masses at a given time should outline the H-R diagram, and differences

[1] R. J. Tayler: Astrophys. Journ. **120**, 332 (1954); Monthly Notices Roy. Astronom. Soc. London **116**, 25 (1956); and R. S. Kushwaha: Astrophys. Journ. **125**, 242 (1957). — These calculations differ primarily in the opacity law adopted. For these massive stars electron scattering must be included and radiation pressure is not altogether negligible.

in diagrams from cluster to cluster may be interpreted as being due to different ages.

In these calculations it is assumed that initially the stars are chemically homogeneous. As previously indicated, they will have convective cores due to the concentration of energy sources. Gradually the hydrogen in the core is consumed, but at the same time the core shrinks, leaving some of the hydrogen-deficient material behind in the radiative zone.

Fig. 21, from KUSHWAHA's integrations, illustrates the sequence of events. Initially, the star is homogeneous and is represented by a horizontal line in the (X, q) plane. The outer envelope continues at that level because the conversion of hydrogen to helium does not take place there. A sequence of horizontal lines of lower and lower hydrogen content represents the convective core, which is assumed to be well mixed. As the convective region encloses less and less mass it leaves behind an intermediate zone of varying composition connecting the convective region continuously with the inert envelope. This is the region that was at one time convective but has become radiative.

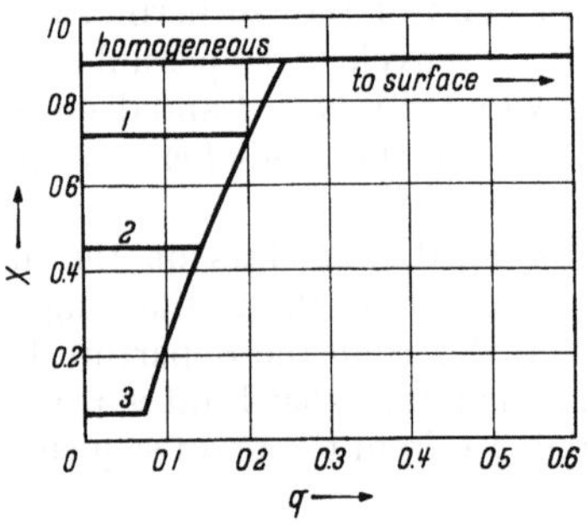

Fig. 21.

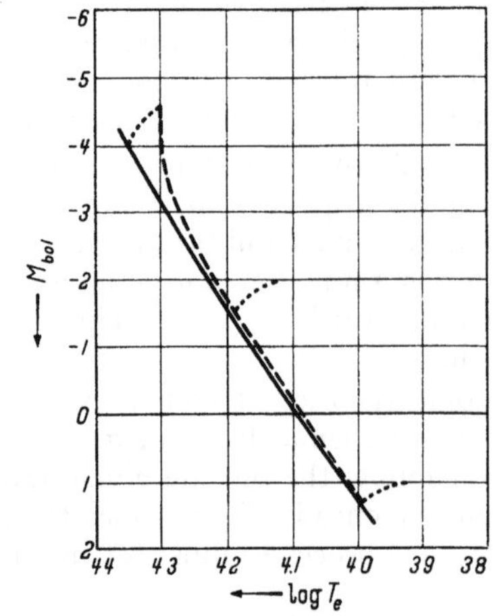

Fig. 22.

Fig. 21. The depletion of hydrogen in successive models (after KUSHWAHA).

Fig. 22. KUSHWAHA's evolutionary tracks. The full curve represents the age-zero main sequence. The dotted curves represent evolutionary tracks for 2.5, 5.0, and 10.0 solar masses. The dashed curve is the locus of points after 33.9×10^6 years.

Data concerning the sequence computed for $M = 5 M_\odot$ are given in Table 12. It should be noted that this is truly a sequence, because the composition of a

Table 12. *Characteristics of Kushwaha's models for the evolution of stars of five solar masses.*

	Homogeneous	1	2	3
$\log q_{core}$	-0.6978	-0.8101	-0.9529	-1.1868
X_{core}	0.90	0.620	0.370	0.077
$\log (L/L_\odot)$	$+2.4632$	$+2.5433$	$+2.6110$	$+2.6641$
$\log (R/R_\odot)$	$+0.3763$	$+0.4515$	$+0.5245$	$+0.6095$
M_{bol}	-1.47	-1.67	-1.84	-1.97
$\log T_e$	$+4.1880$	$+4.1704$	$+4.1509$	$+4.1216$
Spectral type	*B* 5	*B* 6	*B* 6	*B* 8
Bolometric correction	-1.57	-1.48	-1.37	-1.22
M_{vis}	$+0.10$	-0.19	-0.47	-0.75
$\log T_c$	7.3730	7.3941	7.4182	7.4694
$\log \varrho$	1.2894	1.2999	1.3370	1.4724
$t \times 10^{-9}$ years	0	0.0803	0.126	0.156

particular model depends upon the composition of the preceding model. X_c represents the hydrogen content of the core and q_f the fractional mass in the core.

Tracks in the H-R diagram are illustrated as dotted lines in Fig. 22. The solid line represents the locus of homogeneous models (age zero main sequence) and the dashed line is the locus after 33.9 million years.

46. Red giants. The upper right hand region of the H-R diagram, the domain of the cool stars of large radii, presented a difficult problem which was only recently resolved. The red giants in globular clusters are only perhaps twenty or thirty percent more massive than the sun; but their radii are factors ten or more larger. If these stars were homologous to the sun, relation (35.16)

$$T_c \propto \frac{\mu M}{R}, \tag{46.1}$$

would imply a very low central temperature; so low that known nuclear reactions could not account for their high luminosities. A homogeneous increase in μ throughout the star will not relieve this discrepancy, since the reasonable limits of pure hydrogen and pure helium give a range of only $\frac{1}{2}$ to $\frac{4}{3}$. But, as was pointed out by ÖPIK[1], HOYLE and LYTTLETON[2], and others, discontinuities in μ, such as might arise in the course of stellar evolution, could retain high central temperatures and produce large radii simultaneously.

As an example, we quote some results due to OKE and SCHWARZSCHILD[3]. They considered stars consisting of three regions, as follows:

1. envelope	radiative (zones 1–2)	hydrogen rich (zone 1)
2. intermediate zone		hydrogen poor (zones 2–3)
3. core	convective	

A sequence of models of this type was computed, with increasing mass contained in the hydrogen-poor regions (zones 2 and 3). Table 13 gives some of the characteristics calculated for $M = 2M_\odot$, with exterior composition: $X = 0.92$, $Y = 0.06$, $Z = 0.02$; and interior composition: $X = 0.01$, $Y = 0.97$, $Z = 0.02$. Instead of integrating the luminosity equation, all energy was assumed to be generated in the core, despite the low hydrogen content, and the central temperature was fixed at 30×10^6 °K. Fig. 23 illustrates the location of these models in the H-R diagram. Abrupt discontinuities tend to exaggerate the radii as compared with

Table 13. *Characteristics of the Oke-Schwarzschild models for* $M = 2M_\odot$.

	I	II	III	IV	V	VI
Fractional mass inside discontinuity	0.136	0.183	0.242	0.295	0.378	0.44
Fractional mass inside convective core	0.038	0.040	0.044	0.049	0.058	0.068
Fractional radius at discontinuity	0.068	0.058	0.045	0.030	0.010	0.002
Fractional radius of convective core	0.039	0.029	0.020	0.012	0.0025	0.0006
$R/R_\odot$	2.18	3.13	5.15	9.44	39.5	250
$\log L/L_\odot$	0.859	1.028	1.268	1.502	1.895	2.216
$\log T_e$	3.806	3.770	3.721	3.648	3.436	3.114

[1] E. ÖPIK: Publ. Obs. Tartu **30**, No. 4 (1938).

[2] F. HOYLE and R. A. LYTTLETON: Monthly Notices Roy. Astronom. Soc. London **102**, 218 (1942).

[3] J. B. OKE and M. SCHWARZSCHILD: Astrophys. Journ. **116**, 317 (1952).

stars having a continuous variation of μ[1]. Nevertheless, investigations such as this showed it was quite reasonable to construct red giants on the assumption of chemical inhomogeneity.

47. The evolution of stars in globular clusters. The advances in red giant theory previously described, made it possible to interpret, at least qualitatively, the H-R diagrams of globular clusters observed by ARP, BAUM and SANDAGE[2]. A preliminary reconnaissance was made by HOYLE and SCHWARZSCHILD[3].

Briefly, the concept is this: globular clusters represent groups of stars formed roughly at the same time as the galaxy. The most massive, hence most luminous

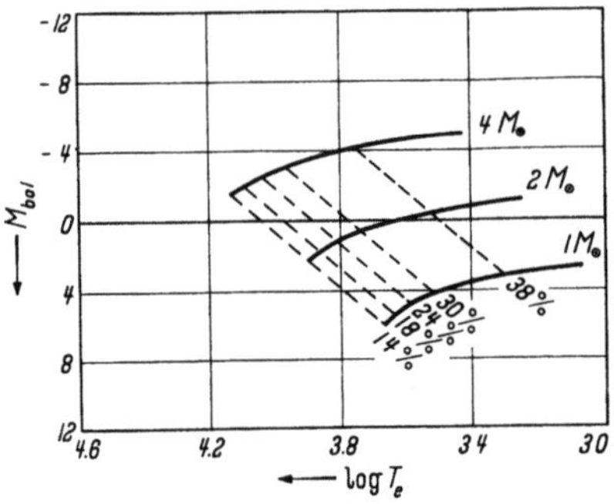

Fig. 23. The models of OKE and SCHWARZSCHILD. The dashed lines are labeled with the fraction of mass in the depleted core.

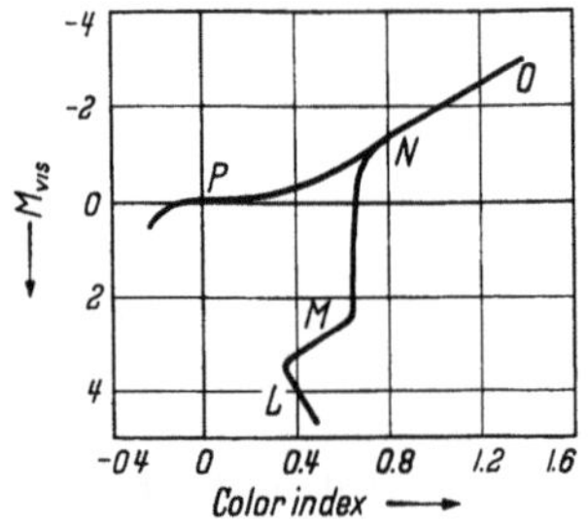

Fig. 24. Subdivision of the H-R diagram (after HOYLE and SCHWARZSCHILD).

stars in the original group, have long since evolved out of the diagram; the least massive stars have hardly evolved at all. The characteristic line which leaves the main sequence above $M_{vis} \approx 4$, culminates at the red giants and then retreats to form the blue horizontal branch, represents the locus of stars of the same age and roughly the same mass. Fig. 24 indicates a subdivision of the evolutionary track into regions in which different physical effects occur.

Table 14. *Characteristics of the Hoyle-Schwarzschild models from L to M in the* H-R *diagram.* ($M = 1.1\, M_\odot$.)

q_1	0.11	0.13	0.16	0.19
ψ_c	+1.28	+3.55	+5.54	+7.86
$\log (R/R_\odot)$	+0.07	+0.17	+0.33	+0.59
$\log (L/L_\odot)$	+0.45	+0.66	+0.85	+1.02
M_{vis}	+3.55	+3.03	+2.56	+2.18
Color index	+0.26	+0.26	+0.36	+0.65

HOYLE and SCHWARZSCHILD consider the evolution of a star whose initial composition is $X = 0.9$, $\mu = 0.533$. It is assumed that the Schönberg-Chandrasekhar description is adequate for the early stages and that subsequently the star develops an exhausted core of pure helium which is isothermal. Energy production (by the carbon cycle) then takes place in a very thin shell surrounding this core. At this stage the density of the core is sufficiently high for degeneracy to have an effect and the core is determined by solutions of Eq. (39.13), where ψ_0 describes the degeneracy at the center. The portion of the track from L to M is described by a series of such models with $M = 1.1\, M_\odot$ with the temperature of the shell, T_1, fixed at 20×10^6 °K. Some of the results are given in Table 14. As the fractional mass in the core, q_1, increases, the star moves along the diagram.

[1] R. HÄRM and M. SCHWARZSCHILD: Astrophys. Journ. **121**, 445 (1955).
[2] H. C. ARP, W. A. BAUM and A. R. SANDAGE: Astronom. J. **58**, 4 (1953).
[3] F. HOYLE and M. SCHWARZSCHILD: Astrophys. Journ., Suppl. **2**, 1 (1955).

When $q_1 > 0.19$, however, the points computed on this model deviate considerably from the observations. The difficulty was found to lie in the boundary conditions. These models, which were assumed to be radiative in the outer regions, satisfied (12.3):

$$P \to 0 \quad \text{as} \quad T \to 0; \tag{47.1}$$

but not the physically correct conditions

$$\varrho \to \varrho_{\text{photosphere}} \approx \frac{\mu H}{kT} \frac{GM}{\varkappa R^2} \quad \text{as} \quad T \to T_e, \tag{47.2}$$

where $\varkappa$ is the opacity coefficient for the *atmosphere*. (The quantity $R^2 kT/\mu HGM$ is an atmospheric scale height.)

Whereas the latter condition requires $\log \varrho = -7.5$ near the photosphere, the model computed for $q_1 = 0.22$ indicates $\log \varrho = -9.78$. In other words, the decline of density with temperature is too steep.

If the outer region is in radiative equilibrium, for the opacity law assumed, it follows that

$$\varrho \propto T^{3.25}. \tag{47.3}$$

If, on the other hand, the outermost regions are in convective equilibrium, the less steep relation

$$\varrho \propto T^{1.5} \tag{47.4}$$

would hold. HOYLE and SCHWARZSCHILD therefore assume that the stars adjust themselves to condition (47.2) by developing extended convection zones.

Beyond point M on the evolutionary track, therefore, they adopt an analysis similar to OSTERBROCK'S description of red dwarfs. The problem is considered in two parts: a model interior and a model atmosphere.

For the interior, the following structure was assumed: an exhausted degenerate core, an energy producing shell, a radiative intermediate zone, and a convective outer zone extending to the surface. In the convective region

$$P = K\, T^{2.5} \tag{47.5}$$

or, in Schwarzschild variables,

$$p = E\, t^{2.5}, \tag{47.6}$$

and the parameter

$$E = 4\pi\, K\, G^{1.5} \left(\frac{\mu H}{k}\right)^{2.5} M^{0.5} R^{1.5} \tag{47.7}$$

is used, instead of K, as the convection parameter. A set of models was computed using E and q_1 as independent parameters. (Beyond point N it was necessary to allow the temperature of the shell to rise to compensate for its decreasing density.) It was found that the luminosity of these models was largely determined by q_1, independent of R.

For the atmosphere, the following picture was adopted:

(1) an upper photosphere in radiative equilibrium;

(2) a lower photosphere in radiative equilibrium;

(3) a convective zone in two regions: (a) an upper region in which hydrogen is only partially ionized; (b) a lower region in which ionization is complete.

The lowest region, (3) (b), is to be fitted to the previous interior integrations.

Working downward through the atmosphere, the conditions at the bottom of the third zone may be found in terms of the effective temperature. The process is algebraically complicated but physically quite straightforward. The source of atmospheric opacity is assumed to be the negative hydrogen ion and all electrons are assumed to come from the ionization of hydrogen. Convection is ignored as a means of energy transport until the convective velocity approaches the velocity of sound. Further, it is arbitrarily assumed that convective transport, when it occurs, is 30% efficient.

At the base of the third zone, K is found by requiring the pressure and temperature to satisfy (47.5). This results in a relation between K (or, equivalently, E) M, R, and T_e:

$$f(E, M, R, T_e) = 0. \tag{47.8}$$

By examining the interior integrations, however, it is found that the range of E is rather limited if R is required to lie within reasonable bounds. By using $E = 15$ as an average (only its logarithm is actually involved) and assigning $M = 1.2 M_\odot$, relation (47.8) becomes

$$f(R, T_e) = 0. \tag{47.9}$$

Since R, T_e and L are related by

$$L = 4\pi R^2 \sigma T_e^4, \tag{47.10}$$

(47.9) may be considered to be

$$f(L, T_e) = 0. \tag{47.11}$$

This relation is the evolutionary track itself. The form of the track is therefore largely determined by the atmosphere, although the average value of E was found from the interior integrations.

Individual points on the track from M to O are identified with particular models by using L and the interior integrations to find q_1. Some of the results are given in Table 15.

Table 15. *Characteristics of the Hoyle-Schwarzschild models from M to P in the* H-R *diagram.* ($M = 1.2 M_\odot$.)

q_1	0.22	0.25	0.29	0.33	0.37	0.40	0.44	0.55
$\log R/R_\odot$	+0.55	+0.76	+0.96	+1.25	+1.48	+1.60	+1.79	+2.21
$\log L/L_\odot$	+1.028	+1.394	+1.728	+2.215	+2.606	+2.810	+3.130	+3.840
M_{vis}	+2.2	+1.3	+0.5	−0.7	−1.7	−2.1	−2.6	−3.9
Color index	+0.57	+0.62	+0.69	+0.81	+0.93	+1.00	+1.11	+1.30

Differences in the form of the track would be expected if the atmosphere were computed using different assumptions. HOYLE and SCHWARZSCHILD show that, if the electrons come from metals rather than hydrogen, the tracks are appreciably flatter. This is a tantalizing result, particularly in view of recent ideas concerning the greater abundance of metals in Population I objects.

The peak of the red giant sequence is described in the following terms: As the burning of the shell proceeds, more and more mass is added to the core, carrying with it gravitational energy that consequently heats the core[1]. The center of the core remains isothermal because of the efficient conductivity of highly degenerate matter. In the outer region of the core, however, there is a temperature gradient and the temperature falls to the hydrogen-burning shell. Thus the central core may be at 100×10^6 °K while the shell is at 20×10^6 °K.

At temperatures in excess of 100×10^6 °K, the helium within the core can produce energy by the Salpeter reaction. As pointed out by MESTEL[2], however, nuclear energy production in degenerate matter has some unusual consequences.

A star composed of a non-degenerate gas has a sort of "safety valve". If the energy production increases, the temperature rises and consequently the pressure rises. The star expands and the temperature decreases (Sect. 22). The nuclear reactions are quite sensitive to temperature so that this decrease in

[1] A. R. SANDAGE and M. SCHWARZSCHILD: Astrophys. Journ. **116**, 463 (1952).

[2] L. MESTEL: Monthly Notices Roy. Astronom. Soc. London **112**, 583 (1952).

temperature causes the energy production to decrease. The star thereby adjusts itself to conform to a stable situation.

The case of degenerate material is quite different because the pressure is determined almost entirely by the density and hardly at all by the temperature. An increase in energy production causes an increase in temperature which is not relieved by subsequent expansion. The rise in temperature, in turn, causes increased energy production. Relief can only come when the temperature is driven so high that the degeneracy is removed and the "safety valve" can take over.

Presumably that is what happens at the tip of a red giant sequence. Subsequent models have two energy sources; helium burning in the (now non-degenerate) core and hydrogen burning in a shell. Models of this type with completely radiative envelopes were also computed by HOYLE and SCHWARZSCHILD.

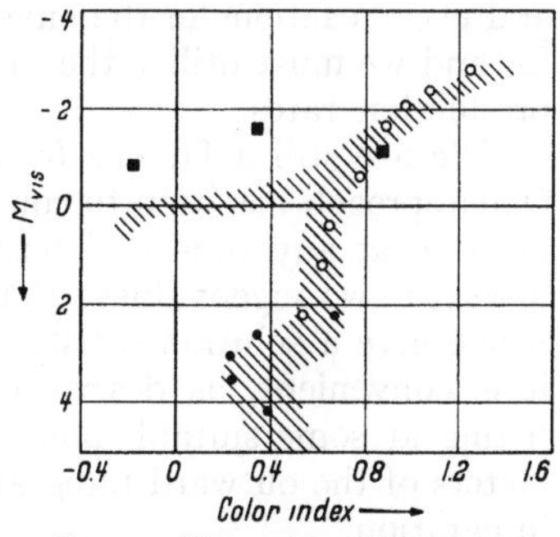

Fig. 25. The models of HOYLE and SCHWARZSCHILD in the H-R diagram.

Fig. 25 summarizes their evolutionary tracks. Much work remains to be done to verify and explore the details of this picture presented by HOYLE and SCHWARZSCHILD. It is now necessary to put in all the assumptions *ab initio* and allow a large scale computing machine to carry the models up to the red giant branch, and, if possible, beyond.

48. White dwarfs. Although we have little idea of the stages which immediately precede them, we believe the white dwarfs represent stars in the last phases of evolution. At some point the gas pressure can no longer support a distended star and it collapses to a very dense configuration. A separate article[1] in this volume is devoted to these objects; nevertheless it might be worthwhile to emphasize a number of important points here.

If we assume these stars obey the equation of state of completely degenerate matter (neglecting the outermost non-degenerate layers) we again find a relation between pressure and density independent of temperature, just as in the case of polytropes. Indeed, the extremes of vanishing density and infinite density are polytropic equations, as noted in Sect. 15. In the intermediate range, the equations of hydrostatic equilibrium reduce to (see [3], p. 417):

$$\frac{1}{\eta^2}\frac{d}{d\eta}\left(\eta^2\frac{d\Phi}{d\eta}\right) = -\left(\Phi^2 - \frac{1}{y_0^2}\right)^{\frac{3}{2}}, \tag{48.1}$$

to be solved subject to the conditions that

$$\Phi = 1, \quad \frac{d\Phi}{d\eta} = 0 \quad \text{at} \quad \eta = 0. \tag{48.2}$$

There is an obvious analogy with the Lane-Emden equation for polytropes; with a significant difference, however, in that this equation contains a parameter, y_0 (related to the central density of the configuration) and the equation must be solved separately for each value of y_0.

There are two significant results derived from this theory. First of all, there is a unique relation between mass and radius for an assigned composition; and secondly, there is an *upper limit* to the mass for which an equilibrium configuration is possible. The latter point is of considerable importance in the theory

[1] Cf. p. 723.

of stellar evolution since stars of larger mass must somehow lose their excess before they can become white dwarfs. Whether the mass loss is gradual or catastrophic is a subject for speculation.

49. The generalized problem. According to the Vogt-Russell theorem, the composition and mass determine the luminosity and radius of an equilibrium configuration. We have seen that some solutions are known under very special assumptions; but we would like to solve the more general problem: given an initial configuration, how does it evolve? We must therefore be prepared to solve for the luminosity and radius of any distribution of composition which may arise in the course of evolution. We may expect to encounter unusual distributions of energy sources, such as helium-burning cores, hydrogen-burning shells, and even reactions in the envelopes. Convection and mixing must be accounted for and we must utilize the most up-to-date data concerning opacity and energy production rates.

We are still a far cry from treating this problem in its most general form. One approach has been to consider a sequence of equilibrium models whose composition at any time is determined by past history. In this form the problem involves two eigenvalues; that is, one must search for the values of L and R which give a solution satisfying the physical boundary conditions. In practice, it is convenient, as described earlier, to integrate both inward and outward, fitting at some suitable point. It is convenient to use T_c and P_c as the two parameters of the outward integration and L and R as the parameters of the inward integration.

A procedure must be developed for improving the parameters so that they converge to a solution. To utilize high speed computers to their full advantage, this procedure should be fully automatic. Hoyle and Schwarzschild have been experimenting with different techniques. On the other hand, Henyey and his group prefer to treat the entire problem as one in partial differential equations, using a "relaxation" procedure to find solutions.

These are largely questions of numerical techniques. In addition we must develop improved physical theories of convection, pulsation and magneto-hydrodynamic phenomena and incorporate them in future attempts to follow stellar evolution in terms of stellar structure.

Acknowledgement.

The author is indebted to Dr. W. A. Fowler for his suggestions in the sections concerning nuclear energy production, and Dr. M. Schwarzschild for general comments.

It is also a pleasure to thank Mr. Vern L. Peterson for his assistance in assembling some of the tables and figures; Mr. Richard L. Sears for his comments; and particularly Miss Verona Sue Hughes for her careful preparation of the manuscript.

General references.

[1] Emden, R.: Gaskugeln. Leipzig: B. G. Teubner 1907.
[2] Eddington, A. S.: The internal constitution of the stars. Cambridge: Cambridge University Press 1926.
[3] Chandrasekhar, S.: An introduction to the study of stellar structure. Chicago, Ill.: Chicago University Press 1939.
[4] Schwarzschild, M.: Structure and evolution of the stars. Princeton, N. J.: Princeton University Press 1958.
Other recent surveys are:
[5] Chandrasekhar, S.: The structure, the composition, and the source of energy of the stars, in: J. A. Hynek, Astrophysics, p. 598. New York: McGraw-Hill 1951.
[6] Aller, L.: Astrophysics: Nuclear transformations, stellar interiors and nebulae. New York: Ronald 1954.
[7] Strömgren, B.: The Sun as a star, in: G. P. Kuiper, The Sun (The solar system, Vol. I), p. 36. Chicago, Ill.: Chicago University Press 1953.

The Hertzsprung-Russell Diagram.

By

H. C. Arp.

With 44 Figures.

Introduction.

1. Two of the fundamental quantities which describe a star are luminosity and temperature. A Hertzsprung-Russell diagram (H-R diagram) is a graphical representation of a group of stars which uses as coordinates their absolute magnitudes and spectral types. Spectral type is principally a temperature classification. Therefore the H-R diagram uses the observable characteristics of relative apparent magnitude and spectral type to show how the temperatures and luminosities of stars are related.

The great usefulness of the H-R diagram springs from the fact that stars exhibit relationships between their temperatures and luminosities. These sequences in the H-R diagram are used empirically to classify and identify various kinds of stars. Some of the major sequences, and the generally accepted terminology for the various regions in the H-R diagram are shown in Fig. 1.

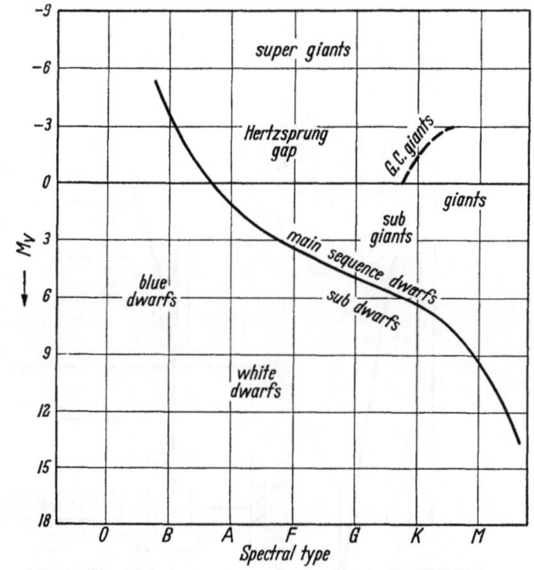

Fig. 1. Terminology of sequences and regions in the H-R diagram.

The reason for the existence of discrete sequences in the H-R diagram can be understood in terms of the Vogt-Russell theorem[1]. That theorem demonstrates that when a star of given mass and chemical composition is in equilibrium, its temperature and luminosity are determined. Symbolically, the following relations exist between the luminosity, mass, temperature and chemical composition[2] (L, M, T, μ).

$$\left.\begin{array}{l} L = f_1(M,\mu); \qquad T = f_2(M,\mu) \\ \text{yields} \quad L = f_3(T,\mu), \end{array}\right\} \text{Vogt-Russell Theorem}$$

$$\text{and, if } \mu = \text{const} \qquad L = f_4(T).$$

[1] S. Chandrasekhar: An Introduction to the Study of Stellar Structure. Chicago, Ill.: Chicago University Press 1939.

[2] μ is some parameter specifying the chemical composition, not necessarily the molecular weight.

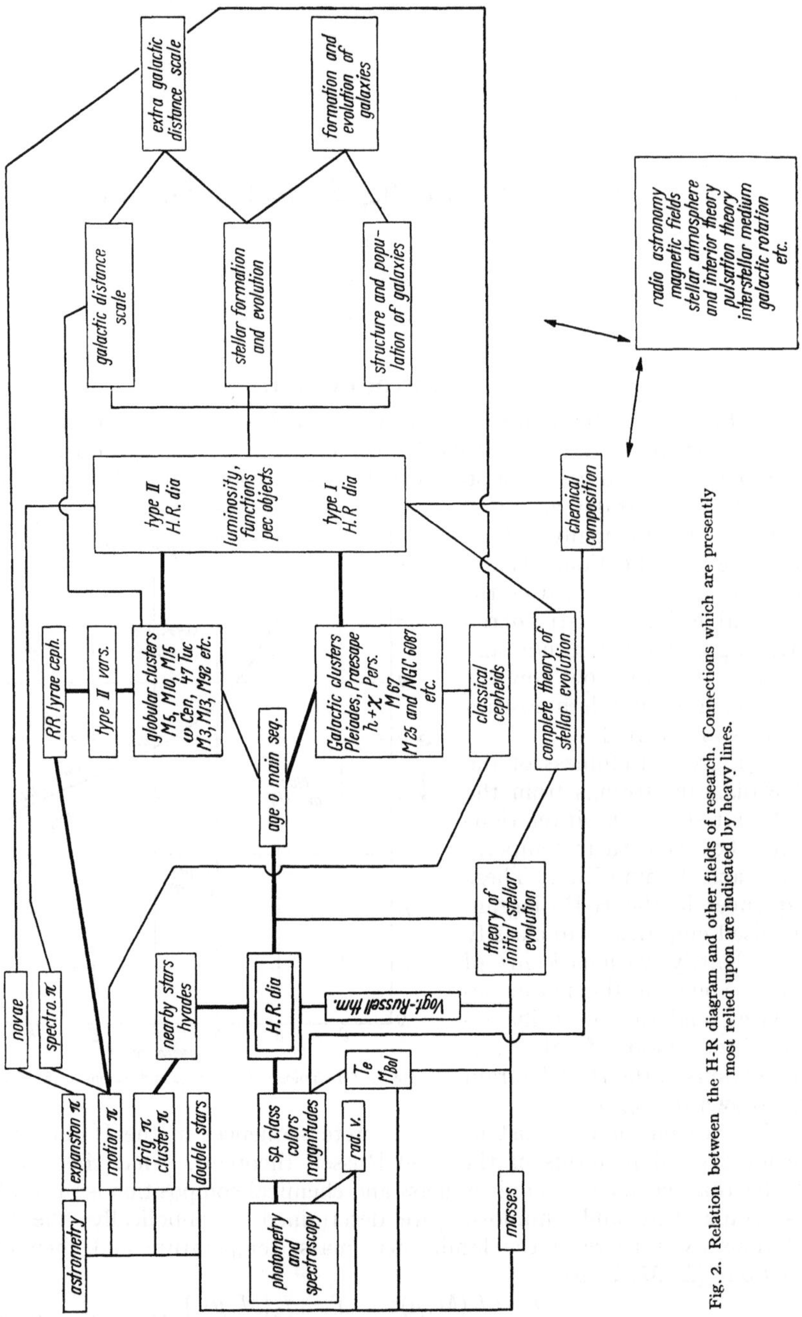

Fig. 2. Relation between the H-R diagram and other fields of research. Connections which are presently most relied upon are indicated by heavy lines.

If stars of only the same chemical composition are considered, then, to each mass there corresponds an L and T. Over the range of mass possible within this group of stars a functional relation between L and T results.

When different groups of stars in the same mass range delineate different sequences in the H-R diagram, it is concluded that the two groups must have different chemical composition, μ. Therefore classifications and identifications which are based on position in the H-R diagram refer to physically fundamental differences between stars so classified. Herein lies the power of the H-R diagram. Moreover, since differences in μ can only be caused by an initial chemical composition difference, or a subsequent change during the lifetime of a star (caused by evolution), the sequences in the H-R diagram may be used to study the conditions of stellar formation and stellar evolution.

The following article on Stellar Evolution, by G.R. and E.M. BURBIDGE, presents the interpretation of the H-R diagram in terms of specific theories of stellar evolution. In the present contribution, consequently, the observational data on the H-R diagrams for all kinds of stars will be presented but only as much of the generalized evolution picture will be utilized as is necessary for empirical discussion of the diagrams.

Of course many other fields of research in astronomy are intimately connected with the subject of the H-R diagram. Some of these interrelationships between the H-R diagram and knowledge gained from other basic fields of research are shown in Fig. 2. At various times the connections between different areas have been different and as knowledge in all divisions of astronomy advances, the connections which are presently judged to be the most trustworthy and accurate will undoubtedly change. Fig. 2 will be used in the present chapter, however, as a schematic in an attempt to systematize the large amount of diverse data which relates to the H-R diagram and furnish, it is hoped correctly, some central thread of continuity throughout the article.

A. Historical résumé.

2. The earliest history of the H-R diagram is directly connected with attempts to gain knowledge about evolution. Spectral classification as early as 1888 was influenced by LOCKYER'S conception of stellar evolution. It has always turned out, however, that, first, knowledge of the luminosities of the stars involved, and hence the distance scale of the stars in space, had to be obtained before significant progress on the evolution problem could be made. For example, Miss MAURY suggested in 1877 that the stars which showed sharp spectral lines, the *c* stars, represented a different line of evolution from ordinary stars. That suggestion spurred HERTZSPRUNG to initiate his fundamental studies of absolute magnitude. In 1905 and 1907 HERTZSPRUNG showed that, generally, bluish stars were intrinsically bright and that there were two main groups of red stars, one bright and one faint. At this time HERTZSPRUNG also demonstrated, from their proper motions, that the *c* stars were considerably more luminous than stars with broader spectral lines. He suggested the possibility of using certain spectral lines as luminosity criteria. This marked the beginning of spectroscopic parallaxes. By using these methods PANNEKOEK and others carried the investigation of absolute magnitudes further in the years immediately following.

That work was summarized in 1914, in a now famous lecture, by RUSSELL who discussed the relation between spectral type and other characteristics of stars. During the course of this review of the information then available, he presented the first H-R diagram. The diagram showed the relation between spectral type and absolute magnitude for nearby stars whose luminosities could be computed from their known trigonometric parallaxes. A similar diagram was presented using stars in binary systems whose absolute magnitudes could be computed from dynamical parallaxes. This latter diagram is shown in Fig. 3. It is striking how the essential form of this earliest diagram has been increasingly revealed as the characteristic one for nearby stars by all modern work, and how important this characteristic form has become in the present day analyses of stellar population and stellar evolution.

RUSSELL noted further, at that time, that the mass difference between red giants and dwarfs was not large and posed the question as to what caused their large difference in luminosity at the same temperature. He also mentioned the tentative appearance of a general correlation between the mass and luminosity of a star.

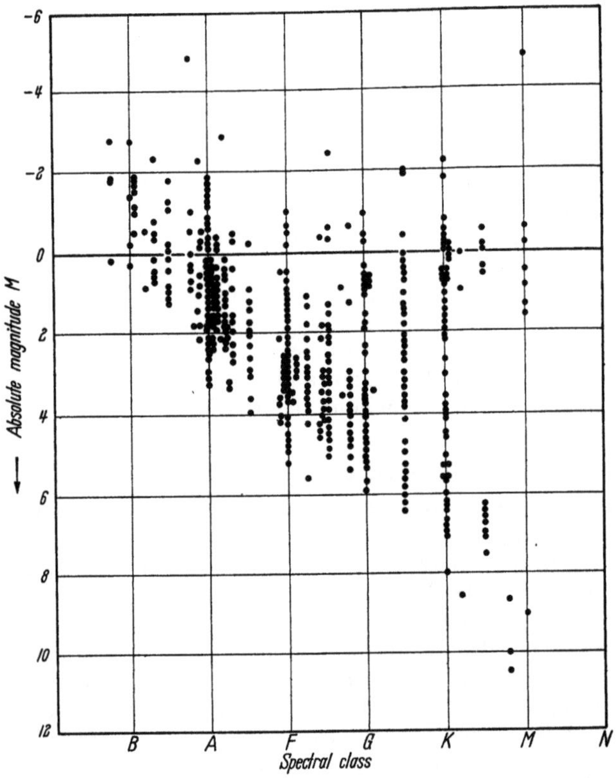

Fig. 3. Original H-R diagram presented by RUSSELL for stars with dynamical parallaxes.

In the years immediately following, work went forward principally on the spectroscopic side. ADAMS and KOHLSCHÜTTER investigated the use of spectroscopic criteria to determine absolute magnitudes, what was then, and still is today the more difficult of the two parameters in the H-R diagram to determine accurately. The luminosities were calibrated by means of proper motions and space motions. Mt. Wilson investigations subsequently carried the spectroscopic determination of absolute magnitudes further under ADAMS, JOY and STROMBERG. These researchers always confirmed and refined the original diagram given by RUSSELL. YOUNG and HARPER at Victoria found additional absolute magnitude criteria. The most recent knowledge of the luminosities of stars via their spectral characteristics comes through the work of R. E. WILSON, OORT, BLAAUW, GREENSTEIN and others. Parallaxes from the solution of space motions and galactic rotation have been used, and, most recently, photometric parallaxes. The culmination of this work has been the establishment of the MKK spectral classification system [6] with its luminosity classes defined from $M_v = -7$ mag. (I) to main sequence stars $M_v = -5$ to $+15$ (V).

Running parallel with the development of knowledge of the luminosities of the stars was the development of the ideas of the other major parameter of stellar classification, namely the temperature scale. Originally it was thought that the difference between spectra of stars was due to the differing abundance of the elements composing them. The early work of LOCKYER at the end of the 19th century, however, laid the ground work for the final proof by SAHA in 1922 that the spectral sequence was primarily one of descending temperature. Previous to this, of course, PICKERING and his co-workers at Harvard had empirically established the spectral classification scale of *O* through *M* which has persevered, essentially unchanged to the present day. An ever present clue to the validity of interpreting the spectral sequence as a temperature sequence was present in the correlation of the colors of the stars with this sequence.

Almost all the early workers conciously made color estimates by visual comparison, but the first quantitative measures of color seem to have sprung from the difference between photographic and photovisual magnitudes (the color index defined) principally in the magnitude scales set up in the North Polar sequence by Harvard, Mt. Wilson, Potsdam and Greenwich. In the subject of color indices HERTZSPRUNG also contributed work of fundamental importance, since shortly after RUSSELL's presentation of the first H-R diagram, his work on determining the effective wavelength of stars in the Pleiades led to what was essentially the first color-magnitude diagram. In 1922 he published a color-magnitude diagram for stars near the Sun which showed again the, now typical, main sequence and giant branches.

As the idea of the spectral class and color index both representing a temperature scale began to take form, it became obvious that the color index of very faint stars could be measured, much fainter stars than could be observed spectroscopically. It was natural then to use the color index of stars as spectral class equivalents—or simply color classes—to get some idea of what were the actual spectral classes of very faint stars. Gradually, the color indices replaced the spectral classes as the most commonly used parameters indicative of temperature and today the color indices are predominantly used in the H-R diagram. The distinction that the term *H-R diagram* is applied only to diagrams containing spectral class as a parameter has been more or less kept. At the same time the term *color-magnitude diagram* has been fully accepted as conveying the same significance as the term H-R diagram and the terms are used today, in the broad sense, interchangeably.

The era of investigations of *clusters of stars* was ushered in by TRUMPLER after 1920. In the investigation of clusters the H-R diagram began to come into its present powerful form as a tool of analysis. Since, in all but the nearest clusters, the stars were all at effectively the same photometric distance, the narrow sequences which exist in the H-R diagrams of stars with a common origin became apparent. By using spectral types and apparent magnitudes to derive the H-R diagrams of galactic clusters. TRUMPLER established the classification of 1*b*, 1*a* and 2*a* through 2*f* on the basis of the form and placement of the sequences they contained. This classification of galactic clusters formed a survey and framework which was prerequisite for much of the following work.

In the period around 1930 SHAPLEY [2] began to apply the technique of *color-magnitude measurement* to stars in globular and galactic clusters. The consequent start into the investigation of clusters of stars photometrically marked another important step in the development of the H-R diagram in general. With the possibilities of measuring H-R diagrams now extended enormously by measuring large groups of stars in the farthest reaches of the galaxy, the major problems became twofold: First there was the problem of computing the amount of reddening and absorption to the objects being observed in order to obtain their intrinsic color indices and absorption-free apparent magnitudes. Secondly there was the problem of computing the distance modulus, some zero point in the cluster which would convert all the apparent magnitudes to absolute magnitudes.

Around 1932 another big break-through came when Stebbins and Whitford began energetically to apply *photoelectric techniques* to the measurement of stars [*4*]. Two things resulted: One, the second-order difference quantity of color index now became measurable with extreme accuracy, about a factor of 10 more discriminating than spectral classification. Two, trustworthy magnitude scales could be obtained with great accuracy over the entire magnitude range observable, and not just in special areas such as the North Pole, but right in the object being investigated so that uncertain photographic magnitude transfers were made obsolete. The advent of photographic plates which were more sensitive in photovisual wavelengths and the invention of iris-diaphragm photometers[1] which yielded measures on photographic star images which were nearly linearly correlated with magnitude over a range of 5 magnitudes or more, reduced the calibration problems of color-magnitude diagrams to the setting up of a few accurate photoelectrically measured standards in the desired region.

Although spectral classes had during the time been used less and less as abscissae in the H-R diagram they, because they were essentially unaffected by reddening, became increasingly important, through the spectral class-normal color index relation, in judging reddening corrections to distant systems. Also, when they could be meaningfully applied, the spectroscopic luminosity criteria were important in giving rough checks on the moduli of these systems. For example the spectroscopic parallaxes to important clusters like the Pleiades and h and χ Persei were for many years the most accurate parallaxes available and not very different from the newest photometric parallaxes.

Finally the development of multicolor photometric systems, notably the U, B, V system of Johnson, furnished a method independent of spectroscopic observation for determining reddening and absorption corrections to various systems.

Previous to these latest photometric techniques which are now being used in the investigation of the H-R diagram, however, in 1944, a conceptual discovery of fundamental importance was made. When Baade resolved the brightest stars in some of the elliptical nebulae belonging to the local group of galaxies he was able to establish the fact that the brightest stars in these systems were unlike the stars in an ordinary H-R diagram. Rather, he pointed out they resembled the kind of stars which occurred in globular clusters. On this basis he established the division of the kinds of stars into *two distinct population types, I and II*. This distinction has proved enormously fruitful in empirical analyses of stars and stellar systems. Although the basic reasons for the difference between the populations is still imperfectly understood, the distinction has become better investigated with additional observation and is one of the subjects to which particular attention will be paid in the succeeding sections.

B. Spectroscopy and photometry.

For the purposes of the H-R diagram the radiative characteristics of stars need only be measured in such a way as to yield definite and empirically repeat-

[1] Eichner et al.: Astronom. J. **53**, 25 (1947).

able parameters which are related in some way to the star's luminosity and temperature. Consequently one whole category of effort has gone into setting up standard systems and measuring procedures so that observations by different researchers may be accurately compared.

Achieving a standard photometric and spectroscopic system has a further advantage. This comes about because of the need to make the H-R diagram interpretable in terms of astrophysical theory. For those purposes the quantities of bolometric magnitude and effective temperature must be derived and since they are usually derived semi-empirically, again, the existence of standard systems facilitates the conversion from the observations to these astrophysical parameters.

3. Spectral classification. As mentioned previously, the spectral classification scheme of *O* through *M*, *N*, *R* and *S* which is based on the relative strengths of helium, hydrogen, metal and molecular absorption lines, has for more than 30 years furnished an accurate stellar temperature scale. There are several modern systems of spectral classification which are closely similar to the original Harvard classification and to each other. The MKK system [*6*], however, is most used today because of the extensive work of W. W. Morgan and collaborating spectroscopists who have accurately defined the classification criteria and carefully separated the luminosity criteria in "An Atlas of Stellar Spectra". Of perhaps even more importance, of course, is the fact that almost universal use of this system by various researchers enables their results to be compared on the same system.

4. Color indices. Broad-band photometry has historically compared the integrated brightness of the spectral region between about 3600 and 5000 Å (photographic) to the region between 5000 and 6300 Å (photovisual). The color indices so obtained are directly related to the color temperature of a star. As in the case of spectral classification, the most important task has been to set up a standard system whereby all observers measure the same quantities and therefore may intercompare results. The diversity between various photometric systems used has always been greater than among systems of spectral classification. For example in 1952 Johnson [*7*] demonstrated that, unless light shortward of 3800 Å were excluded from the photographic magnitudes, non-linear transformations between various systems would result. In addition, because stars of higher luminosity class have different amounts of ultraviolet radiation than do main sequence stars, multi-valued, non-linear transformations would also result. As a consequence Johnson established the B, V system [*8*] in which the light shortward of 3800 Å is excluded from the B magnitudes by means of 2 mm of Schott GG13 filter. The V magnitudes are essentially equivalent to the old International photovisual (Ipv) as far as color dependence goes. The B, V system is presently used by the majority of observers and recent advances in knowledge of the H-R diagram are in great part due to the measurement on, or transformation of the results onto this common system.

Fig. 4 and Table 1 show the best present relation between B, V color indices and MKK spectral classes.

5. Apparent magnitudes. Magnitude scales are inextricably enmeshed with the subject of color indices. When once a standard color system is established, however, the magnitude scale problem reduces, particularly if V magnitudes are used as the basic scale, simply to one of determining the zero point of a particular system. By relying upon the accuracy and linearity of the photoelectric process, the color scale and zero point of apparent magnitude may be transferred to any desired region. This is done, in practice, principally through the use of Johnson-Morgan standard stars. This system of standards has the advantage of having more numerous stars, more easily accessible and covering a greater range of color index then those in the North Polar Sequence. The stars are bright, minimizing measuring errors and enabling system transformations to be performed with maximum accuracy. They also all have accurately determined spectral and luminosity classifications on the MKK system. Other photoelectric sequences exist principally in the Selected Areas[1] and, in the southern hemisphere, the E regions[2]. These sequences are at present expressible to within a few hundredths of a magnitude in the B, V system and work continues to connect them even more accurately.

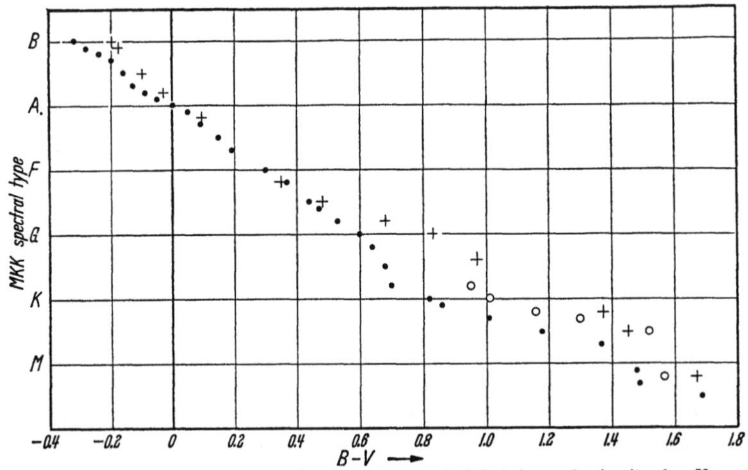

Fig. 4. Relation between B-V color index and MKK spectral type. Filled circles are luminosity class V, open circles III, and crosses I. Data from Table 1.

6. Effective temperature of stars. The effective temperature of a star is defined as that temperature which satisfies STEFAN's law, given the true luminosity and radius of the star. In order to proceed, for the use of astrophysical calculations, to this quantity of effective stellar temperature, the spectra, which are governed by excitation temperature, and the color indices, which are indicative of color temperature, must be related to the mean black body temperature of the photosphere[3]. Of course the stars do not radiate precisely as black bodies STEBBINS and WHITFORD [*11*] showed, however, that the observed colors of the stars agree rather closely with the colors of a black body at a suitable temperature.

Therefore the color temperatures can be used to interpolate between stars of different temperature whose effective temperatures are considered to be known with the most accuracy. KUIPER [*9*] originally set up his temperature scale in this way, using the radiometric measures of PETTIT and NICHOLSON, the effective temperature of the Sun and a few well-determined binary systems as the fundamental determinations. The more precise six-color photoelectric measures of STEBBINS and WHITFORD have enabled a small revision of this basic scale. KUIPER originally based his effective temperature scale for the

[1] STEBBINS, WHITFORD and JOHNSON: Astrophys. Journ. **112**, 469 (1950) and Mt. Wilson Program (to be published).

[2] Cape Mimeograms Nos. 1, 2 and 3. (Royal Observatory, South Africa.)

[3] For a discussion of these various temperatures see [*3*] p. 65.

Table 1. *Mean radiative parameters for MKK spectral types.*

Luminosity class V.

Sp. Type	B-V	U-B	M_v	T_e	B.C.
O 5				44600	−4.31
O 9			−4.6		
O 9.5			−4.4	36800	−3.68
B 0	(−0.32)	(−1.13)	−4.0		
B 1	−0.28	−1.00	−3.3	27300	−3.00
B 1.5				26360	−2.83
B 2	−0.24	−0.86	−2.7	22700	−2.63
B 3	−0.20	−0.71	−2.1		
B 5	−0.16	−0.56	−1.4	15600	−1.58
B 6					−1.42
B 6.5			−1.1	14000	
B 7	−0.13	−0.47			
B 8	−0.09	−0.29	−0.6	12800	−1.04
B 9	−0.05	−0.16	0.1	11800	
A 0	0.00	0.00	0.2	11000	−0.72
A 1	0.05	0.05		10300	
A 2			1.1	9700	−0.52
A 3	0.09	0.07	1.7	9100	
A 5	0.15	0.09	2.1	8700	−0.31
A 7	0.19	0.08	2.5	8100	
F 0	0.30	0.02	2.9	7600	
F 2	0.37	0.00	3.1	7000	−0.04
F 5	0.44	0.00	3.4	6600	−0.04
F 6	0.47	−0.02	3.7	6390	
F 8	0.53	0.02	4.0	6150	−0.05
G 0	0.60	0.06	4.3	6000	−0.06
G 2	0.64	0.16	4.6	5730	−0.07
G 5	0.68	0.21	5.0	5520	−0.10
G 8	0.70	0.24	5.5	5320	−0.10
K 0	0.82	0.48	5.9	5120	−0.11
K 1	0.86	0.54		4920	
K 2			6.3	4760	
K 3	1.01	0.89	6.8	4610	−0.15
K 4					−0.31
K 5	1.18	1.12	7.7	4400	−0.55
K 6				4000	
K 7	1.37	1.26			−1.00
M 0			9.1	3600	
M 0.5					−1.43
M 1	1.48	1.21	9.6	3400	−1.70
M 2			10.0	3200	−2.03
M 3	1.49	1.10	10.5		(−2.35)
M 4			11.2		(−2.7)
M 5	1.69	1.24	12.2		(−3.1:)
M 6			13.3		

Luminosity class III.

Sp.Type	B-V	U-B	T_e
F 5			6470
F 6	0.48	0.10	6020
F 8			5620
G 0			5300
G 2			4990
G 5			4650
G 8	0.95	0.72	4440
K 0	1.01	0.86	4200
K 1			4000
K 2	1.16	1.20	3810
K 3	1.30	1.44	3660
K 5	1.52	1.84	3550
M 0			3340
M 1			3200
M 2	1.57	1.86	3090
M 3			2980
M 4			2850
M 5			2710
M 6			2600:

* Luminosity class I.

Sp.Type	B-V	U-B	T_e
	(Ia, Iab, Ib)		(Ib)
B 0	−.20		
B 1	−.18	(−1.16)	
B 5	−.10	− .92	
B 8	−.03	− .64	
A 2	0.09	− .12	
F 2	0.35	0.32	
F 5	0.48	0.39	6200
F 6			5800
F 8	0.68	0.50	5300
G 0	0.83	0.63	5000
G 2			4600
G 4	0.97	0.81	
G 5			4290
G 8			4000
K 0			3820
K 1			3700
K 2	1.37	(1.60)	3590
K 3			3430
K 5	1.45	(1.74)	3320
M 0			3210
M 1			3100
M 2	1.67		3050

Explanation of Table 1 (luminosity class V).

Column 1 is the MKK spectral type. Columns 2 and 3, the B-V and U-B colors are from Johnson and Morgan [*8*]. Column 4, the absolute photovisual magnitude, is for the luminosity class V main sequence from Morgan [*3*] corrected by 0.1 mag. from HR mags., not the age zero main sequence. The absolute magnitudes of the *O* stars are from Johnson and Hiltner's calibration of h and χ Persei [*22*]. The effective temperature scale in Column 5 is from Underhill and McDonald [10] from the hottest stars through $B2$, and from Morgan thereafter. The bolometric corrections in Column 6 are to the photovisual magnitudes. From the hot stars through $B2$ the corrections are from Underhill and McDonald. From $B5$ through $A7$ they are theoretical corrections on black body assumption, from Kuiper [9]. Later than $F7$ the bolometric corrections are empirical corrections from the same source. From $K3$ through $M1$, Kuiper's original spectral classes have been reduced to the MKK system.

Explanation of luminosity class I.

The spectral type-color index relation, and the two-color index relation is a preliminary one obtained from work of H. L. Johnson and unpublished results of the author obtained in the southern hemisphere. The T_e scale is probably somewhat uncertain because of uncertain reddening corrections for these Ib stars.

cool stars on radiometric measures and binary stars and these are still the best today. His temperature scale for hot stars was based on Pannekoek's work with the maxima of certain spectral series. Recent work by A. B. Underhill [10], however, using a model atmosphere to compute the effective temperature and predict the spectral class, seems to offer a better determination of the T_e scale in very hot stars.

Correlation of the known spectral types then, with the stars which define this temperature scale gives a completely empirical relation between effective temperature and spectral type. Table 1 lists the best present determinations of effective temperature as a function of MKK spectral class. Since this table also contains the best present values for the normal color indices of the MKK spectral type it also exhibits the relation between intrinsic B-V color indices and the effective temperature.

7. Bolometric magnitudes. If the brightness of a star is known over some wavelength region of broad-band photometry, its total luminosity must be derived on the basis of some kind of calculation or assumption about how the energy curve of the star behaves in wavelength regions not observed. In some cases, however, namely for stars later than A, Pettit and Nicholson have made radiometric measures which include the majority of the light emitted by the star. The bolometric or total magnitudes for these stars may then be derived and compared to their, say, photovisual magnitudes. In this way an empirical set of corrections may be built up for the conversion of magnitudes in the photovisual to bolometric magnitudes.

The bolometric corrections for non-main sequence stars, such as giants and supergiants, are not the same as for main sequence stars of the same spectral class. This is principally due to the fact that they have different spectral class-T_e relationships. Kuiper gives different bolometric corrections for different luminosities at later spectral classes. Since his luminosity divisions, do not correspond to the MKK divisions, however, it is probably more convenient, within the accuracy of the determinations, to proceed from the T_e to the bolometric corrections for non-main sequence stars.

Originally, Kuiper's bolometric corrections for hot stars were made on the assumption of black body radiation with the T_e's pertaining to the spectral types as he had derived them at that time. Again, however, the model atmosphere calculations of Underhill and McDonald enable a more accurate estimate of the bolometric corrections for the very hot stars. Table 1 shows the region where Miss Underhill's bolometric corrections have been adopted.

8. Spectroscopic luminosity criteria. The spectroscopic luminosity criteria set up in the MKK system have been calibrated in terms of absolute magnitude. This has been done for main sequence stars later than spectral class F by using

Table 2. *Absolute photovisual magnitudes (M_v) of MKK luminosity classes.*

The blocked-in area represents calibrations obtained in the h and χ Persei association [22]. The remainder of the Table represents the calibrations obtained by standard methods as given by MORGAN [3].

Sp. Type	IV	III	II	Ib	Ia	Iab
*O*9	−5.4	−6.4				
*O*9.5	−5.0	−5.7	−6.2	−6.6		
*B*0	−4.6	−5.0	−5.4	−6.2		−6.9
*B*0.5	−4.2	−4.7	−5.2	−5.8		−6.9
*B*1	−3.8	−4.4	−4.9	−5.8	−6.5	−6.9
*B*1.5	−3.4	−4.0	−4.8	−5.8		−6.9
*B*2	−2.9	−3.6	−4.7	−5.8		−6.9
*B*3	−2.5	−3.3	−4.6	−5.8		−6.9
*B*5	−2.2	−3.2	−4.5	−5.7		−6.9
*B*6						−6.9
*B*6.5	−2.0	−3.1	−4.3	−5.6		
*B*7						−6.9
*B*8	−1.7	−3.0	−4.3	−5.5	−6.5	−6.9
*B*9	−1.0	−2.0	−3.8	−5.5	−6.5	−6.9
*A*0	−0.4:	−1.1	−3.0:	−4.8	−6.5	−6.9
*A*1					−6.5	−6.9
*A*2	0.2:	−0.7	−2.7:	−4.7	−6.5	−6.9
*A*3	1.0:	−0.3	−2.5:	−4.6	−6.5	−6.9
*A*5	1.4:	0.0	−2	−4.5	−6.5	−6.9
*A*7	1.7:	0.3	−2	−4.5		
*F*0	2.0:	0.6	−2	−4.5		
*F*2	2.5	0.8	−2	−4.5		
*F*5	2.7	1.0	−2	−4.5		
*F*6	2.9	1.0	−2	−4.5		
*F*8	3.1	1.0	−2	−4.5		
*G*0	3.2	0.7	−2	−4.5		
*G*2	3.3	0.4	−2	−4.5		
*G*5	3.4	0.2	−2.0	−4.5		
*G*8	3.4	0.4	−2.1	−4.5		
*K*0	3.4	0.2	−2.1	−4.5		
*K*2		0.0	−2.2	−4.5		
*K*3		0.1	−2.3	−4.5		
*K*5		0.3	−2.4	−4.5		
*M*0		−0.4	−2.4	−4.5	−5.3	
*M*1		−0.4	−2.4	−4.5	−5.3	
*M*2		−0.4	−2.4	−4.5	−5.3	
*M*3		−0.5	−2.4:		−5.3	
*M*4		−0.5			−5.3	

trigonometric parallaxes. For earlier spectral classes and stars more luminous than about class IV the calibrations have been made by recourse to proper motions and distances derived from solutions of galactic rotation. Very recently calibration has been effected by photometric cluster parallaxes.

The luminosity classes are separated by from 1 to 2 magnitudes. Therefore the average accuracy with which a luminosity classification yields an absolute magnitude is in the neighborhood of ± 0.5 mag. In order to gain more accuracy than this, the luminosity divisions would have to be capable of being expressed in finer gradations. With even the present divisions, however, luminosity classifications are limited to moderately bright stars where the necessary spectroscopic dispersion may be obtained.

Table 1 shows the best present relation between absolute photovisual magnitude and MKK spectral class. As explained in a later section, the age zero main sequence derived recently is, at any given spectral class, fainter than the average of the partially evolved stars which were used to calibrate the original luminosity class V main sequence (see Fig. 8). Although the precision of the criteria do not perhaps, as yet, make the distinction significant, the age zero main sequence must, then, strictly speaking, refer to a luminosity class slightly fainter than V. This difference is illustrated by comparing Table 1 with Table 5.

Table 2 lists the absolute photovisual magnitudes of the luminosity classes brighter than V. The blocked-in area of the Table represents calibrations recently obtained in h and χ Persei [*22*]. The remainder of the Table is from MORGAN'S 1952 calibrations [*3*].

9. Determinations and uses of spectra and colors. In the practical case, the accuracy of spectral classification is dependent upon the dispersion and widening at which the spectroscopy may be done which in turn rests upon the brightness of the star and the aperture of the telescope. An order-of-magnitude estimate of the relative faintness at which colors and spectra may be observed can be estimated in the following way: The spectrum classification rests on observation of absorption lines which are at most of the order of 1 Å in width. Photometry on the other hand utilizes effective band widths of the order of 700 Å or more. For equal exposure times this is a factor of roughly seven magnitudes. Of course, other considerations such as the consequent reduction of limiting sky background in spectra, favors the spectra. It is interesting to note, however, that in practice, say at the Newtonian focus of the 100-inch telescope, spectra for classification purposes are not usually taken much fainter than 13th magnitude, whereas color indices can be measured at fainter than 19th magnitude in a shorter time. The greater quantum efficiency of the photoelectric process offers promise, however, that in the future the measurement of spectra will be pushed to the noise level of the sky in shorter exposure times.

The relative advantages of spectra and color indices go much deeper than this of course. Over the spectral range from $B0$ to $M5$, there is a range of 2.01 mag. in color index. Since color indices over this range are commonly measured with an average accuracy of ± 0.01 mag. or better, this corresponds to more than 200 meaningful gradations over the range. In comparison, spectral classification gives around 53 separate classes in this range if we assume the spectral classes can be accurate to 0.1 of a class.

Spectra have unique advantages, however, since the types are semi-empirical depending on the appearance of the spectrum alone and in the two-dimensional MKK system requiring no additional information in order to place a star in the H-R diagram. In addition the spectra have the important advantage of being essentially independent of interstellar reddening and give more direct information on chemical abundances and the nature of peculiarities in stars.

10. Interstellar reddening and absorption. The existence of atomic, molecular and particle scattering of light in space has long been a troublesome problem in astronomy[1]. It particularly effects the H-R diagram because the measured apparent magnitudes must first be corrected for obscuration before luminosities can be derived even if an accurate distance to the star is available. Also observed

---

[1] See for example, J. L. GREENSTEIN, Interstellar Matter, [*3*] p. 526.

color indices must be corrected for the associated reddening affects before meaningful color-magnitude diagrams can be constructed.

Appreciable obscuration is encountered less than 400 parsecs from the Sun, particularly in the plane of the Milky Way. Until recently there were only two ways of approaching this problem. The first was by knowing the distance to stars of known luminosity to compute the amount by which the apparent magnitude was depressed. This led to essentially statistical values for absorption in space, for example 0.8 mag. obscuration per kiloparsecs in the plane of the galaxy. Patchiness is a characteristic of the interstellar medium, however, so that this method was unsatisfactory in detailed investigations. A more powerful tool is the estimate of the reddening of objects by comparing their measured color indices to the normal color indices, the normal color index for their spectral class being obtained by means of the normal color index-spectral type relation. The reddening so derived can then be converted also into absorption over a given wavelength region by the reddening-to-absorption ratio calculated by theory and observation. This technique has also been applied to objects whose normal spectral type-color index relation is uncertain, or whose spectra are unavailable by measuring stars in the neighborhood of the known object, that is objects in approximately the same region of space.

11. Multi-color photometry. The accuracy of the photoelectric method has recently enabled an independent method of obtaining reddening and absorption corrections. By measuring a third wavelength band in the ultraviolet where the interstellar absorption of light is relatively the greatest, the different reddening in the ultraviolet-blue color index relative to the blue-yellow color index may be empirically determined and calibrated. W. BECKER was the first to use this method for computing reddening and in 1953 JOHNSON [*8*] incorporated an ultraviolet color index U-B, into his standard system. Fig. 5 shows the relation between the U-B and B-V color indices for unreddened stars of various luminosity classes.

The general form of the curve is easily understood in terms of the nature of the stellar energy envelope. There is a general linear relation between the U-B and B-V color for stars of medium color index which is just the ratio of the slopes of their energy curves at about 3800 and 5000 Å. For very red and very blue stars the energy maximum moves outside the interval of observation and the U-B, B-V curve approaches a constant ratio at either of these extremes. There is a depression in U-B evident in Fig. 5 however, which reaches a maximum at about spectral class *A* 0. This is obviously due to the large amount of energy chopped out of the spectrum in the U region by the strong hydrogen lines in their confluence at the Balmer limit. For the stars of high luminosity the hydrogen lines are sharper and absorb, in the total, less in this region. Therefore these stars in the two-color index diagram would be above the curve for main sequence objects earlier than about *A* 0. Later than *A* 0 however, the depression shortward of the Balmer jump[1,2] in the supergiants brings the U-B, B-V curve below that of the main sequence until about *G* 0 where the hydrogen lines become unimportant and the relations for all luminosity classes come close together.

The ratio between color excess in U-B to color excess in B-V was determined to be

$$\frac{E_{\text{U-B}}}{E_{\text{B-V}}} = 0.72 + 0.05\, E_{\text{B-V}}$$

[1] See diagrams L. H. ALLER: The Atmospheres of the Sun and Stars, p. 188. New York: Ronald Press Company 1953.

[2] See also interpretation of U-B, B-V measures as a function of temperature and electron pressure, Astrophys. Journ. **125**, 139 (1957).

by BLANCO[1] and by HILTNER and JOHNSON[2]. They determined this ratio by measuring the slopes of the reddening lines for reddened *B* stars, that is, the lines connecting reddened and unreddened stars of the same spectral type. The ratio between reddening in B-V and total absorption in V is:

$$A_V = 3.0\, E_{B\text{-}V}.$$

The accuracy of this ratio is estimated to be about ± 0.2. By analyzing the six-color data of STEBBINS and WHITFORD[3], BLANCO obtained a factor of 3.0. STEBBINS quotes a value ot 3.0 and OORT derived 3.3. JOHNSON and HILTNER, taking advantage of the fact that the h and χ Persei cluster is more heavily reddened on one side than the other, derived a value of 3.0.

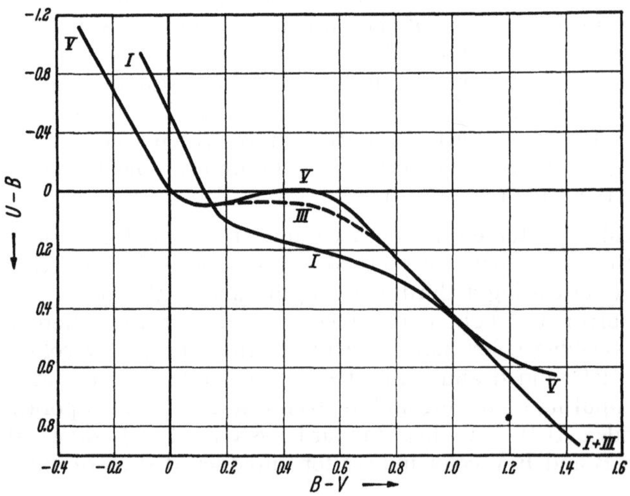

Fig. 5. Relation between unreddened U-B, B-V color indices for different luminosity classes. Data from Table 1.

This ratio, of course, might be different if the reddening law is different in various regions of space. Only in the region of the Trapezium in Orion and the rift in Cygnus is the reddening law known to be unusual. In those regions, however, although the results are still rather uncertain, the reddening to-absorption ratio seems to be also about 3.0.

Of course, successful use of the U-B, B-V curve to determine reddening corrections demands some assurance that the stars under consideration are known to intrinsically define one of the relations pictured in Fig. 5. In other words, if the reddening paths are assumed to have the slope of 0.72, it must also be known at what unreddened line they should be terminated. It is also true, of course, that where the two-color index curve has a secondary maximum at around B-V $= +0.5$ mag., that the backward extension of the reddening line for a single star may intersect the normal curve in more than on place, thereby leaving the reddening uncertain as judged solely from this criterion.

Fig. 5 illustrates that the normal relation between U-B and B-V is different for different luminosity classes. In addition white dwarfs are observed to have large ultraviolet excesses. Fig. 36 of Sect. D shows that the horizontal branches in globular clusters have ultraviolet deficiencies relative to the main

[1] V. M. BLANCO: Astrophys. Journ. **123**, 64 (1955).
[2] W. A. HILTNER and H. L. JOHNSON: Astrophys. Journ. **124**, 367 (1956).
[3] A. E. WHITFORD: Astrophys. Journ. **107**, 102 (1948).

sequence and that subgiant and fainter stars in globular clusters have ultraviolet excesses.

When dealing with stars whose physical properties are imperfectly understood, such as in globular cluster stars, we cannot rely too heavily on the empirical calibration by the kinds of stars used to define Fig. 5, to determine their true, unreddened U-B, B-V curve. But if by a combination of arguments, principally the reddening in the region of the stars we do known about, we can assign a fairly probable unreddened U-B, B-V curve to a group of stars about which we know little, the argument may be turned around. In this case some information may be gained about the energy envelope of the stars by examining the differences between the normal two-color index curves for the unknown group of stars compared to the known.

In general there seem to be two possible causes for different stars defining different normal sequences in the U-B, B-V plane. One, the relative energy distribution in the continuum in the U, B and V photometry bands are different. An example of this is the effect of the Balmer depression in supergiants. This, of course, requires deviation from black body radiation curves for one or both groups of stars. This cause seems to be the dominant effect for very blue, hot stars where the depression of the continuum by absorption lines is at a minimum. For cooler stars the presence of numerous absorption lines in different wavelength regions undoubtedly affects the ratio of U-B to B-V. If, then, we are dealing with stars of different chemical abundances in their atmospheres we could expect this blanketing effect to deviate the observed U-B, B-V relation, depending on the distribution and strength of the changed absorption lines in the different photometric bands.

12. Chemical abundance effects in the spectra. There are more direct means of observing chemical abundance differences in stars than effects in broad-band photometry. The more natural procedure is to measure the relative strengths of lines due to one element relative to lines caused by other elements, directly in the spectrum. This procedure is accomplished with maximum certainty by comparing stars of the same temperature and luminosity. If it is then observed that lines of one element are stronger in one star than in another, the most natural interpretation is that the relative abundance of that element is greater. Care must be exercised, however, to see to it that the results are not "tricks" of excitation. If all observable lines of a certain element, including lines of different ionization states, are strengthened the explanation in terms of abundance is strongly demanded. This is the case, for example, in the *S* stars where Merrill showed that ZrO, Zr I and Zr II are all strengthened relative to other lines.

Stars which do not fit the MKK spectral classification as defined, are listed as peculiar stars. The members of that group which are thought to be the results of chemical abundance differences are: (1) Wolf-Rayet, carbon and nitrogen stars, (2) carbon stars (very red), (3) *S* stars (very red), (4) some high-velocity stars (weak CN), (5) globular cluster giants (very weak metal lines).

C. The H-R diagram for galactic clusters.

13. Introduction. When once an H-R diagram has been constructed for a group of stars, by plotting apparent magnitudes against either colors or spectra, there is one basic step which must be undertaken. The absolute magnitudes of these stars must be determined as accurately as possible. In the case of a cluster or a group of stars which occupy a small volume of space relative to their distance from the observer, that is, are essentially all at the same distance, the problem resolves itself into one of determining the zero point of absolute magnitude, the *distance modulus to the system* which will convert all apparent magnitudes into absolute magnitudes. Since the stellar aggregates which exist in space near enough to the Sun to have geometrically determined distances contain only a relatively small number of intrinsically bright or unusual stars, information about the complete H-R diagram for all kinds of stars which occur in space must be acquired by a systematic process of building up distance scales and calibrating sequences and recognizable special stars relative to these nearby systems. As discussed in the Historical Resumé, Sect. 2, progress in achieving the complete H-R diagram has come through the continual extension of knowledge gained from nearby systems of stars to increasingly distant, and frequently richer, systems and subsystems, clusters, groups and associations. Today the H-R diagram is being extended into the nearby galaxies such as M 31 and the Magellanic Clouds.

There are indeed many interconnecting lines of evidence which lead to the present overall picture of the H-R diagram. The main line which will be discussed under the heading of galactic clusters is the one shown in the flow diagram of Fig. 2. That line proceeds from photometry and spectroscopy, to the H-R diagram for the nearby stars and the Hyades, to the age zero main sequence and thence into the galactic clusters in general and the type I H-R diagram. Other connecting lines of research, some of which are indicated by thin lines in Fig. 2 are important and will, no doubt, ultimately give as reliable calibrations as the line emphasized here. In fact accurate connections which give, independently, the same picture of the general H-R diagram are a necessity before we can feel fully confident that the final answer has been definitely obtained. These auxiliary checks on the main picture will be discussed separately at the ends of the sections on galactic clusters and at the end of the section on globular clusters.

I. Galactic clusters and the standard main sequence.

It has long been known that the most luminous stars can not long maintain their observed rate of energy dissipation and hence must eventually become less luminous as they age. In defining a standard sequence among such bright stars, therefore, care must be exercised and age effects considered. The presentation of a theory of initial stellar evolution by SCHÖNBERG and CHANDRASEKHAR in 1942 [*28*] which agreed with the available observational data was the basic contribution which has enabled an age zero main sequence to be built up from the fundamental main sequence defined by the H-R diagram of stars in the solar neighborhood.

14. The nearby stars. Table 3 presents the most recent list of stars which have trigonometric parallaxes greater than 0.″050. This list of 171 stars was compiled by R. H. STOY[1] who included only those stars for which two independent and accordant measures gave a distance less than 20 parsecs and which, further, had colors and magnitudes measured by either JOHNSON or at the Cape Observatory on the B, V system.

Fig. 6 exhibits the color-absolute magnitude diagram for these nearby stars. The error in magnitude due to an error in parallax is

$$\Delta M = 2.17 \frac{\Delta \pi}{\pi}.$$

[1] Compilation by R. H. STOY kindly made available in advance of publication.

Table 3. *Stars with trigonometric parallaxes greater than* 0″.050.

Yale No.	Sp. Type	V	B-V	π	Yale No.	Sp. Type	V	B-V	π
16	*F*2 IV	2.25	0.35	0″.072	1831	*G*5 IV	5.06	0.77	0″.050
49	*M*1 V	8.07	1.56	.278	1841	*G*0 V	5.36	0.57	.057
49a	*M*6 V	11.04	1.80	.278	1856	*G*0	5.17	0.59	.065
54	*G*0 V	4.23	0.56	.134	1864	*F*5 V	5.03	0.44	.067
69	*G*2 IV	2.81	0.61	.153	1902	*dF*8	5.60	0.55	.059
104	*G*5 V	5.58	0.70	.070	1946	*F*5	4.74	0.41	.051
110	*K*0 V	5.84	0.86	.100	1977	*G*8 V	5.97	0.76	.075
120	*G*0	5.90	0.53	.052	2042	*dG*8	6.37	0.78	.080
134	*K*0	2.06	1.03	.057	2067	*G*5	5.04	0.73	.058
155	*G*0 V	3.45	0.58	.182	2089	*K*1 V	6.55	0.93	.091
160	*G*	12.37	0.56	.236	2143	*A*7 V	3.12	0.18	.066
204	*dM*	9.57	1.50	.106	2153	*F*5 V	3.95	0.43	.070
219	*G*5 VI	5.12	0.69	.136	2174	*F*7 IV−V	4.78	0.48	.052
257	*dG*0	4.97	0.56	.064	2255	*F*6 V	4.54	0.47	.067
331	*F*8 V	4.08	0.54	.062	2268	*F*2 IV	3.57	0.35	.059
350	*G*2 V	4.94	0.63	.086	2280	*G*8 IV−V	5.47	0.75	.107
352	*K*0 V	5.07	0.88	.148	2390	*dM*0	6.59	1.38	.222
356	*K*1 V	5.20	0.84	.133	2420	*M*4.5: V	9.43	1.54	.211
365	*G*8 Vp	3.49	0.72	.275	2459	*F*8 V	4.82	0.51	.080
371	*dK*2	5.63	0.81	.111	2576	*M*2 V	7.47	1.51	.398
394	*A*5 V	2.62	0.14	.063	2631	*M*1 V	9.32	1.42	.113
405	*G*5 IV	3.70	0.85	.052	2645	*K*7 V	7.22	1.27	.085
445	*K*0 V	6.12	0.81	.083	2678	*K*0 V	5.97	0.80	.099
464	*G*0 V	4.87	0.61	.090	2701	*K*5 V	7.77	1.07	.075
479	*G*5 V	6.34	0.72	.091	2725	*G*5 V	4.91	0.64	.092
520	*K*3 V	5.82	0.99	.147	2738	*A*3 V	2.12	0.09	.076
520a	*dM*6	11.66	1.62	.147	2739	*F*8 V	3.63	0.54	.098
537	*F*5	4.82	0.42	.059	2745	*G*8 VI	6.49	0.75	.116
549	*F*7 V	4.12	0.48	.077	2762	*K*5 V	6.97	1.16	.096
557	*F*5	5.40	0.55	.071	2767	*dG*7	5.54	0.76	.078
599	*dK*0	6.01	0.88	.127	2824	*A*3 V	3.27	0.08	.052
647	*G*0 V	4.04	0.60	.084	2889	*K*0 V	7.12	0.82	.070
664	*F*8 IV	3.85	0.51	.070	2895	*G*0 V	4.29	0.59	.108
691	*G*5 V	4.82	0.68	.105	2924	*F*0 V	2.76	0.35	.101
703	*G*5 V	4.24	0.70	.156	2951	*M*0.5 V	8.49	1.41	.090
740	*dF*4	4.70	0.38	.052	3015	*G*0 V	4.30	0.56	.120
742	*K*2 V	3.74	0.88	.303	3039	*G*6 V	4.74	0.70	.115
788	*K*0 IV	3.52	0.92	.109	3144	*F*7 V	4.51	0.48	.056
827	*dF*3	4.21	0.42	.053	3161	*dG*5	6.44	0.68	.056
945	*K*1 V	4.42	0.81	.200	3175	*G*0 IV	2.70	0.59	.102
950	*K*3 V	7.62	1.14	.069	3206	*K*0 III−IV	2.15	0.98	.059
1070	*dG*1	5.49	0.63	.066	3242	*K*2 III p.	−0.06	1.23	.090
1077	*F*6 V	3.16	0.46	.125	3243	*K*3 V	6.66	1.04	.098
1129	*dK*5	6.21	1.09	.104	3274	*F*7 V	4.06	0.50	.067
1164	*F*8 V	4.69	0.51	.078	3309	*G*0 *K*5	−0.25	0.69	.751
1199	*G*0 V	4.71	0.67	.066	3375	*K*5 *M*1	5.65	1.13	.173
1255	*dM*1	7.96	1.48	.163	3462	*G*2 V	5.67	0.63	.059
1316	*F*6 V	3.56	0.50	.122	3536	*F*5 IV−V	4.64	0.39	.053
1339	*A*5 III	3.86	0.18	.055	3567	*G*6 V	6.03	0.70	.069
1543	*K*1 IV	3.97	1.09	.052	3570	*G*0 V	4.43	0.60	.091
1577	*A*1 V	−1.47	0.01	.375	3596	*F*9 V	4.60	0.56	.056
1606	*dK*6	6.56	1.08	.096	3602	*K*5	8.06	1.13	.055
1760	*F*0 V	4.16	0.31	.059	3604	*F*6 IV−V	3.85	0.48	.069
1805	*F*5 IV−V	0.34	0.40	.288	3669	*K*0 V	7.54	0.80	.063
1826	*K*0 III	1.15	1.00	.093	3712	*M*0 V	8.60	1.41	.088

Table 3. (Continued.)

Yale No.	Sp. Type	V	B-V	π	Yale No.	Sp. Type	V	B-V	π
3799	*G*0 IV	2.82	0.64	0″.110	4754	*G*8 V	3.54	0.75	0″.170
3815	*K*7 V	8.11	1.37	.118	4794	*M*0 V	7.96	1.44	.158
3878	*K*5 V	7.74	1.16	.083	4804	*K*0 V	5.72	0.91	.116
3880	*M*3–5:V	10.07	1.44	.096	4864	*G*5 V	6.59	0.73	.051
3913	*K*5 V	6.32	1.16	.172	4895	*A*	11.53	−0.06	.072
3919	*G*8 V	5.48	0.80	.125	4911	*dG*7	6.36	0.73	.065
3924	*K*3 *K*4	5.88	1.05	.137	4966	*K*0 IV	3.43	0.92	.071
3935	*dF*2	4.38	0.36	.058	5012	*dM*0.5	8.50	1.49	.137
3946	*G*2 V	5.39	0.62	.069	5077	*K*5 V	5.19	1.19	.292
3955	*K*7 V	7.54	1.36	.125	5117	*M*0 V	6.68	1.42	.255
4013	*G*1 V	5.23	0.61	.064	5139	*A*7 IV–V	2.41	0.23	.063
4027	*G*5 V	5.10	0.70	.071	5152	*F*8 V	4.23	0.47	.111
4029	*M*3.5 V	9.15	1.50	.203	5198	*K*2 V	7.17	0.89	.069
4060	*G*5 IV	3.42	0.75	.108	5262	*G*2 V	5.60	0.59	.068
4098	*M*5 V	9.53	1.74	.545	5314	*K*5 V	4.73	1.06	.285
4153	*F*7 V	5.04	0.52	.058	5345	*F*5 V	3.76	0.44	.074
4166	*K*7 V	8.37	1.31	.073	5395	*G*1 V	5.40	0.59	.077
4171	*K*2 V	6.40	0.87	.100	5415	*dG*0	5.35	0.66	.052
4215	*K*0 III–IV	3.26	0.94	.054	5432	*K*5 V	7.68	1.10	.070
4245	*F*7 V	3.58	0.50	.120	5546	*dM*5	10.16	1.60	.206
4293	*A*0 V	0.03	0.00	.123	5562	*K*5 V	6.47	1.13	.123
4330	*dM*3.5	8.90	1.54	.280	5563	*dM*2	8.66	1.51	.150
4330a	*dM*4	9.69	1.59	.280	5565	*A*3 V	1.18	0.12	.144
4433	*G*5	6.14	0.70	.056	5568	*dG*0	5.53	0.68	.073
4440		13.18	0.05	.066	5584	*M*2 V	7.36	1.30	.273
4494	*M*3.5 V	9.12	1.50	.168	5616	*K*3 V	5.56	1.01	.152
4607	*K*0 V	4.68	0.79	.179	5721	*K*3 V	7.09	1.01	.102
4665	*A*7 IV–V	0.75	0.23	.198	5724	*F*7 V	4.13	0.51	.064
4686	*dK*5	6.14	1.04	.067	5725	*K*1 IV	3.22	1.03	.064
4705	*G*8 IV	3.70	0.86	.070	5763	*M*2 V	8.98	1.49	.160
					5807	*G*2 V	5.75	0.56	.086

Therefore if we take the probable error of a trigonometric parallax to be ±0″.007, the maximum probable error of an absolute magnitude is ±0.3 mag. For all stars having parallaxes greater than 0″.050, the probable error in the magnitude is correspondingly less. The accuracy of the color indices is, of course, better than ±0.01 mag.

Table 3 and Fig. 6 then represent the best of the fundamental data on which the present absolute calibration of the H-R diagram rests.

15. The Hyades. The distance to this nearby galactic cluster may be obtained by the method of moving cluster parallaxes. Van Bueren[1] obtained a modulus of 3.03 mag. in 1952. Heckmann and Lubeck in 1956 [*24*] gave individual parallaxes of 97 members of the cluster. These individual parallaxes range from $m-M=2.56$ to 3.69 mag. and yield a modulus of the centroid equal to 3.08 mag.

Table 4 lists the individual moduli and measures by Johnson and Knuckles [*20*] of the color and magnitude of each star. Fig. 7 presents the color-absolute magnitude diagram for these 96 Hyades stars.

The accuracy of the individual parallax measures can be judged from the very narrow main sequence which is defined in Fig. 7. (The small number of points lying up to $\frac{3}{4}$ mag. above the main sequence are undoubtedly unresolved double stars. As will be seen, they are a common feature of accurately measured galactic

[1] H. G. van Bueren: Bull. Astron. Inst. Netherlands **11**, No. 432 (1952).

cluster color-magnitude diagrams.) Compare this Fig. 7 to Fig. 15 which was constructed using a mean modulus for all stars, and note the large reduction in scatter about the main sequence due to the use of individual moduli.

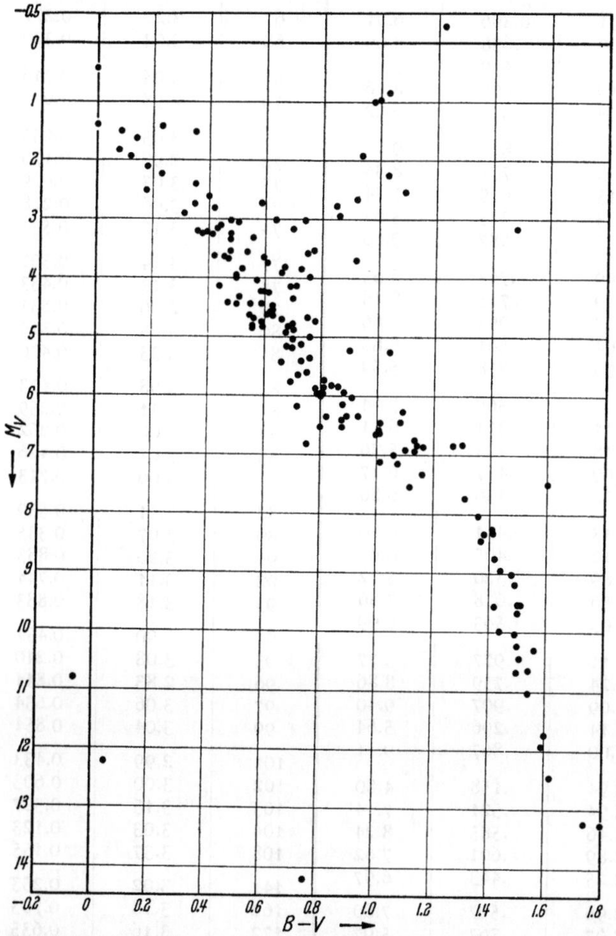

Fig. 6. Color-magnitude diagram for nearby stars. From Table 3 compiled by R. H. Stoy. The diagram plots all known stars which have a trigonometric parallax larger than 0".050 (two accordant determinations at different observatories) and color indices and magnitudes measured by either H. L. Johnson or the Royal Observatory at the Cape.

16. Theory of initial stellar evolution. Schönberg and Chandrasekhar [28] showed that in the absence of any significant mixing between the interior and exterior regions of a star, that the core becomes hydrogen exhausted in a time which is proportional to the luminosity of the star. When 12% of the mass of the star has become hydrogen exhausted, the star which has risen about 1.4 mag. and become slightly redder in the latter stages of its initial evolution, then begins to evolve very rapidly through non-equilibrium configurations to the right and up in the H-R diagram.

The result of this theory is first to explain the characteristic upcurving of the main sequences observed at the bright break-off point in the galactic cluster H-R diagrams. Secondly, the age of any particular cluster may be computed by substituting the luminosity of the break-off point of the cluster in the equation given by Sandage and Schwarzschild: $t = 1.1 \times 10^{10} M/L$ yrs. where the

Table 4. *The hyades stars.*

No.*	$m-M$	B-V	V	No.*	$m-M$	B-V	V
11+12	2.91	0.396	6.01	67	3.22	0.271	5.72
15	2.99	.658	8.09	68	3.03	0.320	5.90
16	3.69	.419	7.06	69	3.05	0.746	8.64
17	3.10	.696	8.46	70	2.98	1.011	3.52
20	3.05	.399	6.32	71	3.05	0.955	3.85
21	3.26	.816	9.15	72	2.99	0.179	3.41
22	3.03	.771	8.34	73	3.03	0.609	7.85
23	3.15	.679	7.54	74	3.06	0.228	5.03
24	3.43	.275	5.65	75	3.12	0 531	6.59
25	2.95	.987	9.60	77	3.14	0.502	7.05
26	2.99	.743	8.63	78	3.13	0.453	6.92
27	3.20	.715	8.46	79	2.96	0.831	8.96
28	2.98	.993	3.66	80	3.08	0.319	5.58
29	3.04	.561	6.88	81	3.33	0.470	7.10
30	2.96	.278	5.59	82	2.98	0.173	4.78
31	3.08	.566	7.44	83	3.11	0.259	5.48
32	3.11	.374	6.11	84	3.00	0.262	5.40
33	3.18	.223	5.26	85	3.03	0.426	6.51
34	3.09	.457	6.17	87	3.06	0.743	8.53
35	3.26	.436	6.80	88	3.11	0.539	7.78
36	3.13	.441	6.80	89	3.07	0.335	6.02
37	3.12	.405	6.61	91	3.20	0.883	8.94
38	2.89	.320	5.72	92	3.11	0.741	8.66
39	2.56	.678	7.86	93	3.18	0.883	9.40
40	3.03	.563	6.99	94	2.90	0.431	6.62
41	3.11	.977	3.77	95	3.08	0.240	6.66
42	3.24	.759	8.86	96	2.83	0.841	8.51
43	3.09	.907	9.40	97	3.06	0.634	7.94
45	3.14	.296	5.64	99	3.04	0 851	9.38
46	3.00	.867	9.11	101	2.99	0.433	6.65
47	2.94	.158	4.80	102	3.00	0.603	7.54
48	2.94	.521	7.14	103	3.15	0.307	5.79
49	3.46	.585	8.24	104	3.03	0.123	4.27
50	2.89	.601	7.62	108	3.37	0.155	4.68
51	3.25	.443	6.97	141	3.02	0.253	4.48
52	2.96	.597	7.80	162	3.18	0.705	7.84
53	3.07	.367	5.97	172	3.16	0.635	10.20
54	3.27	.137	4.22	173	2.83	1.237	10.49
55	3.04	.247	5.28	174	2.98	1.058	9.99
56	3.02	.048	4.30	175	3.37	1.031	10.30
57	3.09	.491	6.46	176	3.05	0.949	9.01
58	2.96	.680	7.53	177	2.93	1.123	10.57
59	3.01	.543	7.49	178	2.89	0.837	9.00
60	3.07	.268	4.29	179	2.92	0.934	9.52
62	3.20	.537	7.38	180	3.02	0.853	9.10
63	3.14	.632	8.06	181	3.14	1.167	10.33
64	3.17	.657	8.12	182	3.22	0.844	8.93
65	2.96	.535	7.42	183	3.19	0.910	9.69

* The numbers are those of Johnson and Knuckles (1955).

mass and luminosity (M and L) of the stars at the 12% limit are in solar units [*29*] (see also the discussion by Strömgren[1]). Thirdly, with the age of the cluster

[1] B. Strömgren: Astronom. J. **57**, 65 (1952).

so determined, the amount by which the remaining members of the observed main sequence have moved from their initial position may be computed. The result is a correction of from 1.4 to 1.0 mag. [30] for the brightest stars to the main sequence, decreasing for stars of lower and lower luminosity until the initial main sequence is reached.

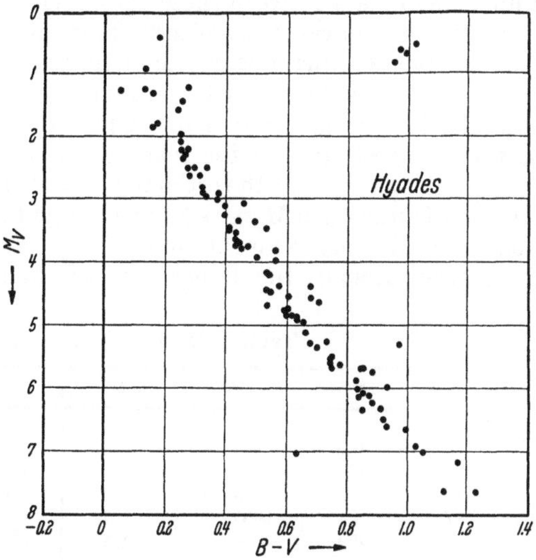

Fig. 7. Color-absolute magnitude diagram for Hyades. Colors and magnitudes from JOHNSON and KNUCKLES. Individual distance moduli from HECKMANN and LUBECK.

17. The age zero main sequence. The age zero main sequence is defined as the locus, in the color-absolute magnitude diagram, of stars which have completed their gravitational contraction but have not yet begun their evolutionary motion caused by changing chemical composition. By using the method outlined in the preceding section, JOHNSON [20], [22] has computed the age of the Hyades cluster to be 1×10^9 years. He has also computed the corrections to be applied to the Hyades main sequence in order to obtain the initial, or age zero, main sequence. By using the more populous and better defined Praesepe cluster as an interpolation sequence, he thereby establishes the age zero main sequence from about B-V = 0.2 to B-V = 1.5 mag. This age zero main sequence is estimated to be accurate within ± 0.2 mag. It is tabulated in Table 5 and shown schematically in the color-absolute magnitude diagram in Fig. 8.

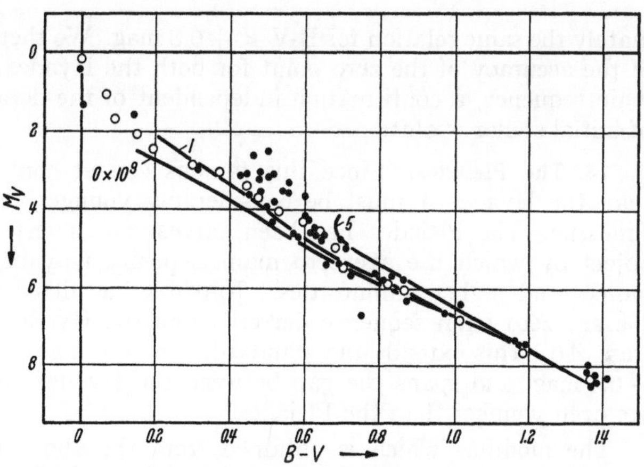

Fig. 8. Comparison of the computed main sequence for 0, 1 and 5×10^9 years with nearby stars. From JOHNSON and KNUCKLES. Filled circles represent individual nearby stars. Open circles represent KEENAN and MORGAN's calibration of luminosity class V.

JOHNSON has used the theory of initial stellar evolution as subsequently developed by HARRISON from the generalized Cowling model. He notes that by using the more recent theory of LEDOUX the age zero main sequence which is obtained lies only 0.1 mag. fainter than the one in Table 5, well within the estimated error of the determination.

There are two important ways in which we can check the accuracy of this derived age zero main sequence to this point and connect it firmly with the basic data from the H-R diagram of the nearby stars. First, comparison of the computed main sequence for age 0 and age 5×10^9 years to the color-absolute magnitude diagram of the nearby stars reveals that the solar neighborhood contains a mixture of stars between these ages. (This result first became evident with the derivation of the color magnitude diagram for M 67.) It is now seen, however, that the age zero main sequence coincides with the lower limit in the H-R diagram, of the envelope which contains the nearby main sequence. Fig. 8 shows how closely the age zero main sequence therefore corresponds to what must be the youngest stars in the solar neighborhood. The second test comes about through the fact within the estimated age of the universe, about 5 or 6×10^9 years, the stars fainter than $M_v\approx+7$ have not had time in which to evolve appreciably from the initial main sequence in the H-R diagram. Fig. 8 shows that the Hyades, the nearby stars and the derived age zero main sequence all converge to approximately the same relation for B-V $<+0.8$ mag. We therefore have a confirmation of the accuracy of the zero point for both the Hyades and the derived age zero main sequence, a confirmation independent of the details of any specific theory of initial stellar evolution.

Table 5. *Standard main sequence for age zero.*

MKK	B-V	M_v	MKK	B-V	M_v
B 8	−0.09	0.5	*F* 5	+0.44	4.0
B 9	−0.05	1.0	*F* 8	0.53	4.5
A 0	0.00	1.6	*G* 0	0.60	4.9
A 1	+0.05	1.8	*G* 2	0.64	5.1
A 2	+0.07	2.0	*G* 5	+0.68	5.3
A 3	+0.09	2.2	*G* 8	+0.70	5.5
A 5	0.15	2.5	*K* 0	+0.82	6.1
A 7	0.19	2.7	*K* 2	+0.86	6.3
F 0	0.30	3.2	*K* 3	+1.01	6.8
F 2	+0.37	3.6	*K* 5	+1.18	7.5
			K 7	+1.37	8.4

18. The Pleiades. Since this famous cluster contains brighter, bluer stars than the Hyades it must be considerably younger than either the Hyades or Praesepe. The Pleiades has been intensively investigated and form an ideal object by which the age zero main sequence may be extended to bluer color indices and higher luminosities. JOHNSON has fitted the Pleiades sequence to the age zero main sequence derived from the Hyades in the region of spectral class *A* 0. This extends the standard, age zero main sequence to almost B-V $=-0.1$ mag. and spans the gap between the Hyades and galactic clusters which are even younger than the Pleiades.

The modulus which is obtained from the above fitting procedure for the Pleiades is $m-M=(5.3\pm0.2)$ mag.[1] [*22*]. It is instructive to note that the original fitting of the Pleiades to the then standard main sequence yielded $m-M=5.8$ mag., and that evolution corrections have therefore changed the photometricaly derived modulus by this amount. This simply illustrates the change in the standard main sequence from JOHNSON (1953). It is interesting to note, that although his determination is of low weight, GRATTON[2] derives a modulus

[1] In most recent investigation, JOHNSON and MITCHELL [Astrophys. Journ. **125**, 414 (1957)] derive $m-M=(5.40\pm0.1)$ mag. by fitting at fainter magnitudes (corrected for ~0.1 mag absorption).

[2] L. GRATTON: Z. Astrophys. **14**, 48 (1938).

of $m-M=(5.0\pm0.2)$ mag. for the Pleiades by averaging what are predominantly geometrical distance criteria.

19. NGC 2362 and h and χ Persei. Given the age zero main sequence as computed from a combination of the nearby stars, Hyades, Praesepe and Pleiades, two more, even younger galactic clusters may than be fitted in the region between $B8$ and $A0$. The color-

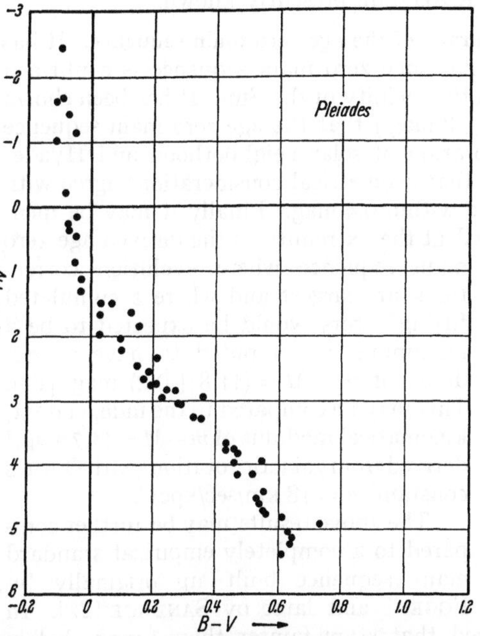

Fig. 9.

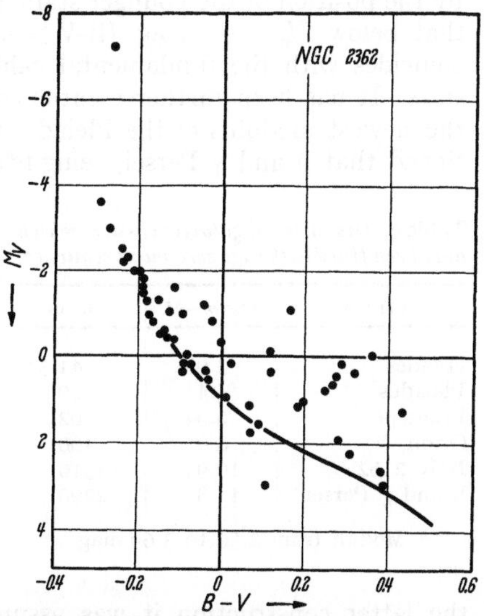

Fig. 10.

Fig. 9. Color-absolute magnitude diagram for the Pleiades. From JOHNSON and MORGAN.

Fig. 10. Color-absolute magnitude diagram for NGC 2362. From JOHNSON and HILTNER. Corrected for interstellar absorption and reddening. The solid line represents the age zero main sequence.

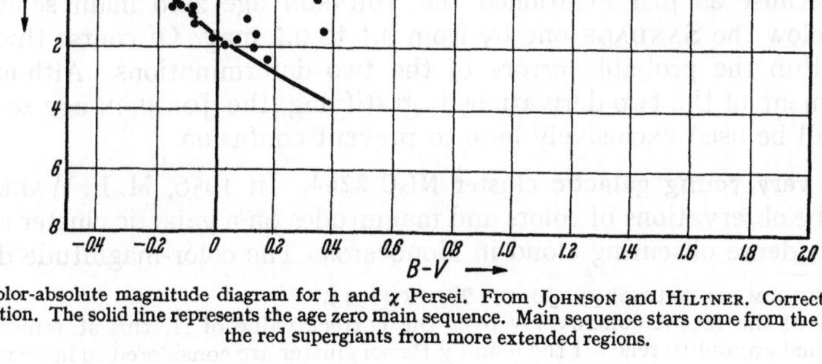

Fig. 11. Color-absolute magnitude diagram for h and χ Persei. From JOHNSON and HILTNER. Corrected for reddening and absorption. The solid line represents the age zero main sequence. Main sequence stars come from the nuclei of clusters, the red supergiants from more extended regions.

absolute magnitude diagram for NGC 2362 and h and χ Persei are shown in Figs 10 and 11. The age of these clusters is around 3×10^6 years. The fit yields a modulus of 10.9 mag. for NGC 2362 and 11.8 mag. for h and χ Persei. In each diagram the portion of the age zero main sequence which has been used to fit the clusters is shown by a full line. These last two clusters then represent the extension of the standard main sequence to the brightest, bluest, and earliest spectral type region of the H-R diagram presently known.

20. Independent checks on the accuracy of the age zero main sequence. It has already been mentioned how the derived age zero main sequence is confirmed by the position of the younger stars in the vicinity of the Sun. It has been shown that below $M_v = +7$ mag. (B-V $> +0.8$ mag.) that the age zero main sequence coincides with the fundamental calibration of solar neighborhood and Hyades stars. It has been further mentioned that geometrical considerations agree with the newest modulus of the Pleiades to within 0.4 mag. Finally it may be mentioned that h and χ Persei, being fitted at the extremity of the derived age zero main sequence where evolution corrections are largest and where accumulated fitting errors would be expected to be a maximum, is computed to have a modulus of $m-M = (11.8 \pm 0.2)$ mag. [22]. This may be compared to the independent, kinematical modulus of $m-M = 11.7$ mag.[1] derived from galactic rotation (with OORT's constant $A = 18$ km/sec/kpc)[2].

Table 6. *Distances of galactic clusters which have been fitted to the age zero main sequence.*

Clusters	true $m-M$	parsecs
Hyades	3.08	41.3*
Pleiades	5.4	120
Praesepe	6.04	162
Orion	8.0	400
NGC 2362	10.9	1510
h and χ Persei	11.8	2290

* Moduli from 2.56 to 3.69 mag.

The above results may be further compared to a completely empirical standard main sequence built up originally by EGGEN[3] and later by SANDAGE [27]. In the latter construction it was assumed that stars fainter than 3 mag. below the SCHÖNBERG-CHANDRASEKHAR limit had undergone no appreciable evolutionary motion in the H-R plane. Fitting the various clusters together only in these regions, SANDAGE derived a standard main sequence much like JOHNSON's. An independent check is derived from the cluster parallax of the 13 stars which compose the nucleus of the Ursa Major stream. The theory which JOHNSON used, however, predicts a few tenths of a magnitude motion, even this far below the break-off point. Consequently this "empirical" age zero main sequence lies somewhat above that of JOHNSON's in most parts of the H-R diagram. The derived true modulus of the Pleiades is $m-M = 5.45$ mag. from SANDAGE's empirical sequence. This compares favorably to JOHNSON and MITCHELL's most recent value of 5.4 mag. The difference is fortuitously small, however, because as just mentioned the JOHNSON age zero main sequence is generally below the SANDAGE one by from 0.1 to 0.2 mag. Of course this difference, is within the probable errors of the two determinations. Although the close agreement of the two derivations is gratifying, the JOHNSON age zero main sequence will be used exclusively here to prevent confusion.

21. The very young galactic cluster NGC 2264. In 1956, M. F. WALKER [25] presented the observations of colors and magnitudes on a galactic cluster situated in front of a dense obscuring cloud in Monoceros. The color-magnitude diagram

[1] W. P. BIDELMAN: Astrophys. Journ. **98**, 78 (1943).

[2] In view of the recent controversy over the correct value of A, this statement might be better turned around to read: If the h and χ Persei cluster are considered to have negligible peculiar motion, the present photometric distance modulus yields a value of $A = (17 \pm 2)$ km/sec/kps, close to the old Oort value of (20 ± 2) km/sec/kpc.

[3] O. J. EGGEN: Astronom. J. **60**, 401 (1955).

for that cluster, NGC 2264, is shown in Fig. 12. That diagram assumes great importance because WALKER has interpreted the stars which lie to the red of the age zero main sequence as stars which are still in the process of gravitational contraction on to the main sequence. If this interpretation is correct, it is the first observational confirmation of the predictions of SALPETER (1953) and HENYEY, LE LEVIER and LEVÉE (1955) that stars after their initial condensation, contract onto the main sequence from lower temperature regions of the H-R diagram. WALKER computes the age of the cluster to be between 2 and 5×10^6 years. The 1955 computations give a time scale of 3×10^6 years for the contraction of an A 0 star onto the main sequence.

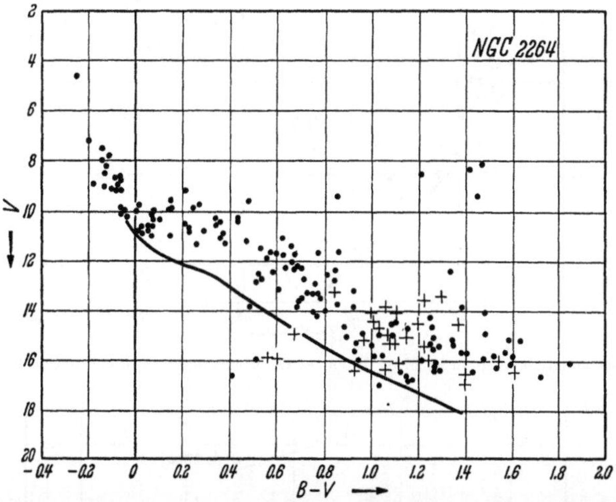

Fig. 12. Color-magnitude diagram of NGC 2264. Adapted from a diagram by WALKER. Full line represents age zero main sequence. Crosses represent stars which have either light variations or bright H α.

Spectroscopic luminosity classifications confirm that the stars later than A 0 in Fig. 12 lie above the main sequence. The major uncertainties in the interpretation of Fig. 12 are (1) possible inclusion of non-cluster members, (2) variable absorption within the cluster, (3) the fact that the stars which are interpreted as being still in the process of contraction lie above but parallel to the age zero main sequence rather than showing an increasing divergence toward fainter luminosities as predicted by the theory. WALKER's discussion indicates that considerations (1) and (2) have been adequately taken into account. However, both additional theoretical work and observational results on other similar clusters will undoubtedly improve our understanding of the contraction stages in the very earliest clusters. In this connection, JOHNSON and HILTNER [22] mention the fainter stars in NGC 2362 (see Fig. 10) as possibly being still in the gravitational contraction stage and WALKER indicated that NGC 6530[1] is another cluster similar to NGC 2264.

22. M 25 and NGC 6087. These two galactic clusters recently sprang into a position of extreme importance when it was discovered by J.B. IRWIN [26] that each of the clusters contained a classical cepheid. The membership of the cepheids in each cluster has been confirmed by radial velocity measures by M.W. FEAST and G. WALLERSTEIN. This discovery is of fundamental importance because it opens up for the first time an independent and basically accurate method for

[1] M.F. WALKER: Astronom. J. 62, 37 (1957).

measuring the zero point of the very important distance scale indicators, the classical cepheids. The values so far obtained for the absolute magnitude of the cepheids involved are discussed under the section on variable stars, Chap. D. In this section, only the color-magnitude diagrams of the clusters are presented below in Figs. 13 and 14.

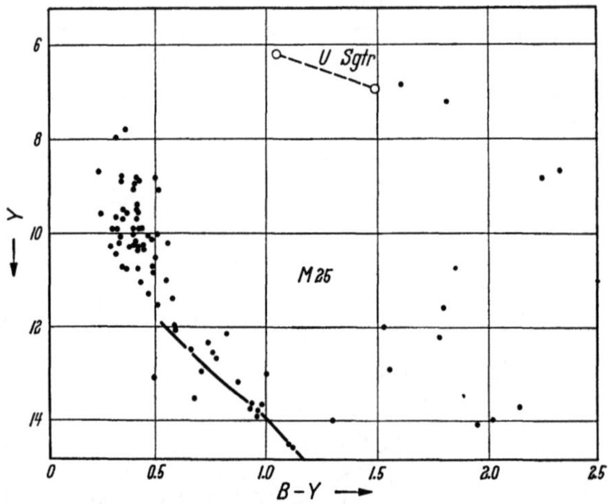

Fig. 13. Color-magnitude diagram for M25. From IRWIN. Colors and magnitudes not on B, V system.

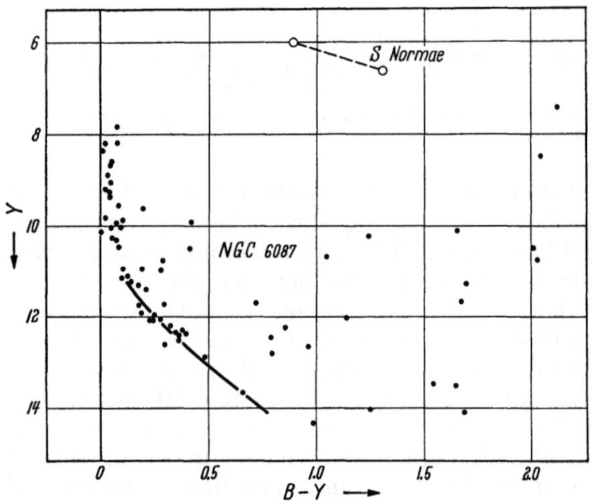

Fig. 14. Color-magnitude diagram for NGC 6087, from IRWIN. Colors and magnitudes not on B, V system.

The fitting procedure used by IRWIN was to match the cluster main sequences to the standard main sequence given by EGGEN. (The reddening was estimated from the two-color index curves.) This gave a true modulus, corrected for absorption, of $m-M=9.0$ mag. for M25 and $m-M=9.55$ mag. for NGC 6087. As has been previously mentioned, however, the Eggen main sequence is about 0.25 mag. brighter than the Johnson age zero main sequence at these colors. Therefore fitting to the main sequence which we have considered most fundamental in the preceding sections yields a modulus of $m-M=8.75$ mag. for M25 and $m-M=9.30$ mag. for NGC 6087.

II. Combining the galactic clusters in the H-R diagram.

In the previous section we have seen how the complete picture of the locus of stars which are so young that they have not evolved appreciably far in the H-R diagram, has been built up. Using this sequence, we may fit any cluster which contains at least a segment of the main sequence, into the general H-R diagram with an estimated accuracy of ± 0.2 mag. Except for the clusters listed in Table 6 which have been fitted to this age zero main sequence, however, most clusters were investigated earlier, before the evolutionary corrections had been applied. Consequently the published moduli for such clusters are usually around 0.5 mag. too large if the fitting was in the region of the Pleiades main sequence, or less if the fit was made in redder, less luminous regions where evolutionary effects are decreasing rapidly in size.

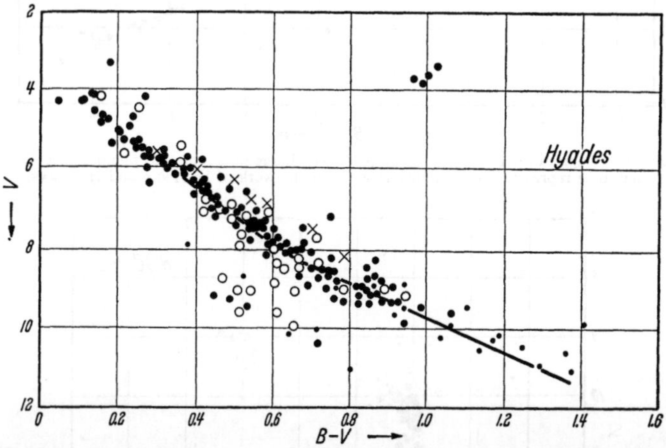

Fig. 15. Color-magnitude diagram for the Hyades. From JOHNSON and KNUCKLES. Filled circles: proper motions and radial velocities indicate membership. Open circles: proper motions indicate cluster membership, or proper motions and radial velocities indicate possible membership. Small dots: fainter stars, possible cluster members. Crosses: double stars.

It is of interest, nevertheless to inspect the color-apparent magnitude diagrams of a representative sample of galactic clusters and ascertain the form and definition of their sequences. Then, at the end we shall see that it is possible to fit them all together quite closely by a small overall revision of the original moduli. This enables a general discussion of the total H-R diagram as it is defined by galactic clusters.

23. Hyades, Praesepe, Coma Berenices, NGC 752. Fig. 15 shows the color-magnitude diagram for the Hyades. This plot differs from the plot of Fig. 7 in that many more stars are included, over a larger area of the cluster, and a mean modulus, not individual moduli, is used.

Fig. 16 shows the color-magnitude diagram for Praesepe. This is one of the most populous and best defined galactic clusters and as mentioned before, was used to interpolate the Hyades diagram is the fitting to the age zero main sequence. Because its main sequence is so well defined, JOHNSON was able to compute that the cosmic spread from an infinitely narrow main sequence could be no more than ± 0.03 mag., as an upper limit, in the region from B-V $= +0.6$ to 1.0 mag.

Fig. 17 shows the color-magnitude diagram for the Coma Berenicis cluster and Fig. 18 the color-magnitude diagram for NGC 752.

These four clusters are all of intermediate age in the order: Coma Berenicis, Hyades, Praesape and NGC 752. They demonstrate how the main sequence, in clusters of this age, breaks away from the standard main sequence at about $M_v = +3$ mag.

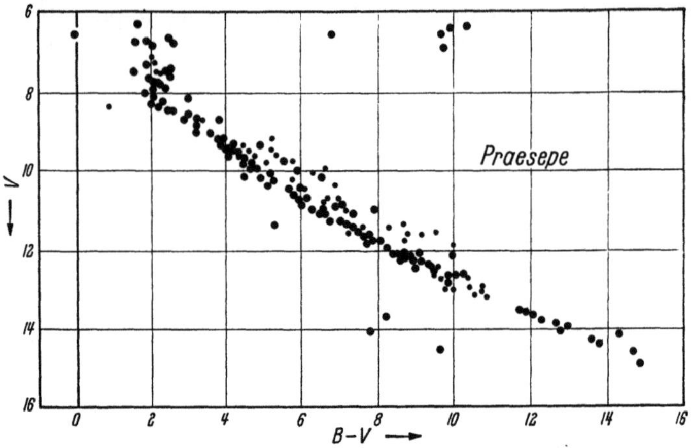

Fig. 16. Color-magnitude diagram for Praesepe. From Johnson. Large dots represent photoelectric observations, while the small dots represent values transformed from the work of Haffner and Heckmann.

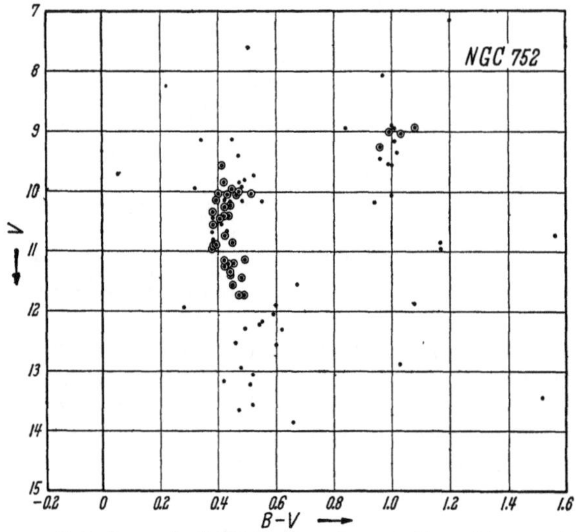

Fig. 17. Color-magnitude diagram for NGC 752. From Johnson. Large dots are certain members.

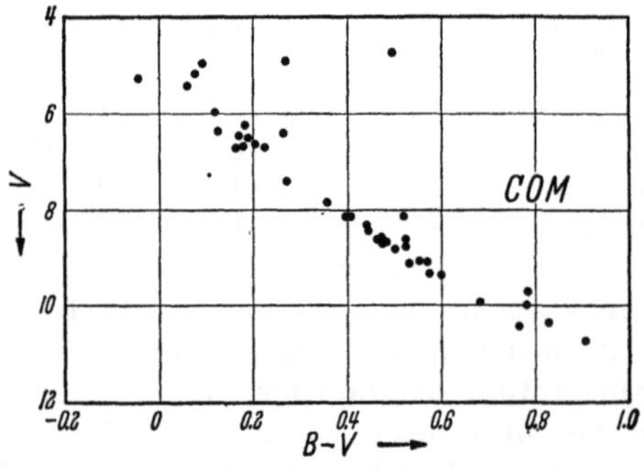

Fig. 18. Color-magnitude diagram for the star cluster in Coma Berenicis. From Johnson and Knuckles.

24. M 41 and M 11. The color-magnitude diagram for M 41 is shown in Fig. 19 and the colour-magnitude diagram for M 11, which is the nucleus of the Scutum

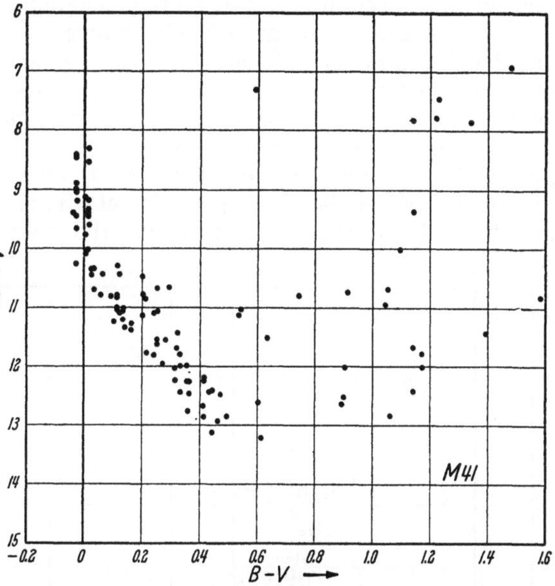

Fig. 19. Color-magnitude diagram for M 41 adapted from measures by Cox.

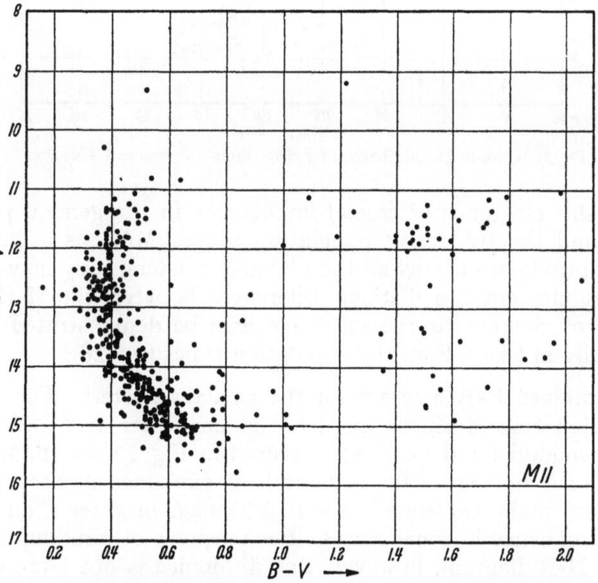

Fig. 20. Color-magnitude diagram for M 11. From JOHNSON and SANDAGE.

cloud, in Fig. 20. These two clusters are important because they fill the gap between the preceding group of clusters and the even younger Pleiades. Of even more importance are the red giants which they contribute in the region between $M_v = 0$ and -2 mag., a region which has heretofore been poorly represented.

25. M 67. This galactic cluster, because of its richness in faint stars, radial symmetry and the redness of its brightest stars, was at various times considered to be a globular cluster. The accurate color-magnitude diagram shown in Fig. 21 was obtained by JOHNSON and SANDAGE in 1955, however, and it shows convincingly that the M67 diagram is different from that of a globular cluster. Rather, the diagram demonstrates that this cluster is a natural continuation of the pattern of galactic clusters evolving away from the main sequence at lower luminosity as the age of the cluster increases. The age of this cluster is computed to be about 5×10^9 years. Since this is about the age of the galaxy itself, M67 then represents about the oldest possible galactic cluster. This completes the picture, then, of the galactic clusters over the range from the very young to the very old.

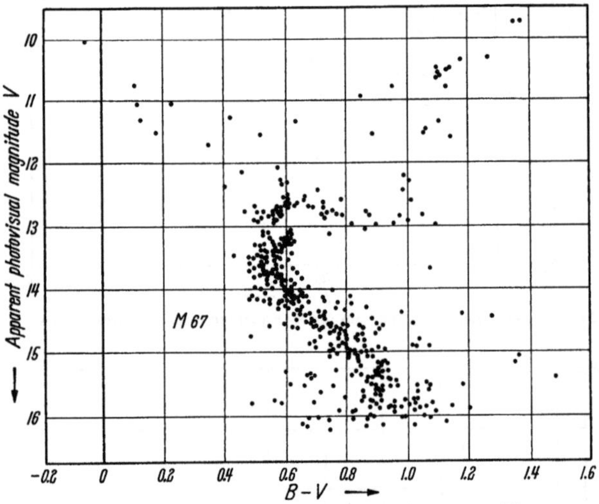

Fig. 21. Color-magnitude diagram for M67. From JOHNSON and SANDAGE.

Obviously this cluster is of crucial importance in the general picture of the H-R diagram and the subject of population types. M67, as a galactic cluster of the same approximate age as all the globular clusters, may now be compared to the globular clusters and distinct differences between the H-R diagrams of the two kinds of clusters, at the same age, can be demonstrated. This is discussed more fully in Chap. F on the population types (p. 128).

26. The combined H-R diagram for the galactic clusters. Fig. 22 shows the schematic combination of all the galactic clusters so far discussed. In order to obtain revised moduli for the clusters pictured in Fig. 22, SANDAGE fitted them to the "empirical" age zero main sequence. As explained in Sect. 17, this "empirical" age zero main sequence is about 0.25 mag. brighter than the Johnson age zero main sequence in some places. For purposes of building up the overall picture of the H-R diagram, however, this difference is not large enough to be significant. Therefore, taken together, these ten clusters show with striking clarity the total picture of the behavior of the galactic cluster color-magnitude diagrams in the general H-R diagram. From the youngest cluster, NGC 2362, down through the oldest, M67, the characteristic peeling away from the main sequence at successively fainter magnitudes is shown.

In summary we can say that Fig. 22 represents the culmination of the intensive investigation of galactic clusters which has taken place in the last decade: (1) It proves beautifully the correctness and usefulness of the theory of initial stellar evolution. (2) It gives a reliable method of estimating ages of galactic clusters. (3) It gives a complete picture of all possible configurations, as a function of age, of galactic cluster color-magnitude diagrams. (4) It furnishes a reliable, standard main sequence which enables the derivation of accurate distance moduli to galactic clusters. (5) It begins to define clearly the Hertzsprung gap in the H-R diagram. (6) It begins to give information on what regions of the H-R diagram red giants and supergiants occur in and from which (what age) clusters they come.

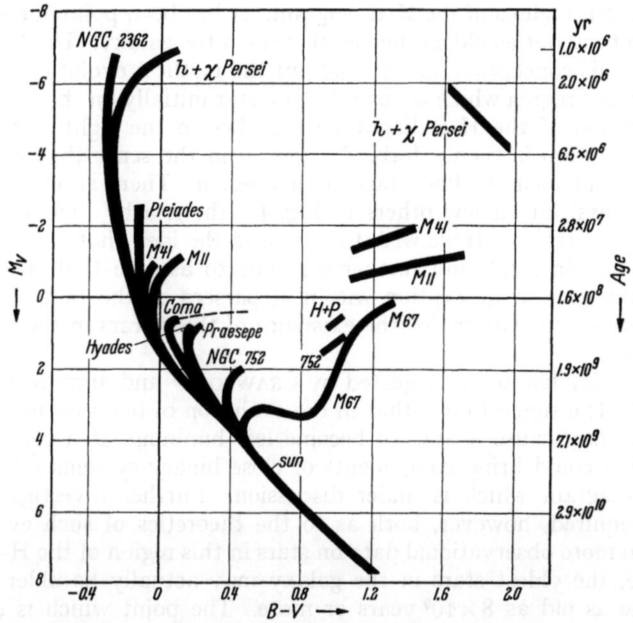

Fig. 22. Composite color-absolute magnitude diagram for 10 galactic clusters. Adapted from a diagram by SANDAGE.

27. Red giants and the funnel effect. Regardless of the luminosity of the break-off point on the main sequence which a particular cluster may have, Fig. 22 shows clearly the convergence in luminosity of the giant branches for redder color indices. The giants coming from fainter regions of the main sequence move to higher luminosities as they redden and the giants which spring from bright regions of the main sequence move to fainter luminosities as they redden. One consequence of this observed behavior of the giants is to render it difficult to estimate anything about the masses of stars in this red giant region. This is simply because stars passing through this region could have originally come from bright main sequence (massive) or faint main sequence (a little more than one solar mass). In fact SANDAGE estimates that there should be a mass spread among the luminosity class III giants with 70% having masses of about $1.5\ M_\odot$ while 30% should have masses like the *A* stars. The funnel effect in Fig. 22 explains the observed concentration of giants near $M_v = +0.5$ mag. on the basis that although the Hyades and M67 type giants move on different tracks they all intersect in the region of $M_v = +0.5$ and B-V $= 1.2$ mag. The kinematical properties of *K* III stars in the general field confirm this picture by showing that about one out of three *K* giants moves like an *A* star, and the rest like *F* stars ($M = 1.5\ \odot$).

28. The Hertzsprung gap. Fig. 22 pictures clearly the vacant region between the turn-off point of the main sequence and the red giants which occurs in most clusters. It is in this region where a star evolves rapidly away from the 12% break-off point, through non-equilibrium configurations (Sect. 16). It is also in this region that the classical cepheid variable stars occur and is, therefore, of great interest. Fig. 22 illustrates how the Hertzsprung gap narrows as the giant sequences become lower in absolute magnitude. Finally the gap disappears in M67. That is the M67 subgiant sequence cuts across below the Hertzsprung gap.

29. Forbidden regions of the H-R diagram. It has been pointed out that M 67 is believed to be about as old as the oldest stars in the galaxy. The M67 sequence in Fig. 22 on this conception, must represent the furthest evolutionary excursion into the red giant region which is possible for a star initially on the main sequence. The whole region of the H-R diagram which lies to the right of the main sequence-M67 diagram is then a forbidden region in the sense that we would not expect on present ideas to find stars in this region. There is one star in Fig. 6 (the nearby stars) and a few others in Fig. 44 (the nearby high velocity stars) which fall in this region. If we wish to persist in the idea that there are no stars older than those in M67, and further continue to assume that there is no appreciable reddening or absorption within 40 parsecs of the Sun, we must find some alternative explanation for the existence of these stars in the so called forbidden region.

One possibility has been suggested by CRAWFORD[1] and amplified by STRUVE and HUANG[2]. The suggestion is that in the evolution of binaries, mass exchange, or mass loss, may cause a star to become less luminous at a constant radius. Such evolution could bring components of close binary systems into the region of the H-R diagram which is under discussion. Further investigation of this problem is required, however, both as to the theoretics of such evolution and also to obtain more observational data on stars in this region of the H-R diagram.

Of course, the oldest stars in the galaxy may actually be older than M67, perhaps some as old as 8×10^9 years or more. The point which is being made here is that until the subject of close binary evolution is investigated in more detail, we are not necessarily forced to this conclusion by the observations of the nearby stars.

30. The problem of the subdwarfs. It can be seen from the color-magnitude diagrams of the Hyades and Praesepe clusters (Figs. 15 and 16) that there are a few stars in each of these clusters, particularly in the Hyades, which fall from one to two magnitudes below the main sequence in the region of $B\text{-}V = 0.4$ to 1.0 mag. Since these two clusters contribute by far the most stars in this region of the main sequence it could well be that other clusters would show a few such stars if a large enough sample of cluster stars could be observed in them. At the present time, however, the Hyades is the only galactic cluster which is suspected of containing subdwarfs.

The presence of subdwarfs in a galactic cluster causes severe difficulties with recent analyses which have tended to assign subdwarfs to low metal content populations such as the globular clusters. The only resolution of this paradox seems to be to postulate that the Hyades subdwarfs lie below the main sequence, not because of an initially low metal content like the true, unevolved subdwarfs, but that instead, they are far evolved, and this is only a transitory stage on their path to the dead star region of very faint white dwarfs. In this latter case,

[1] J. A. CRAWFORD: Astrophys. Journ. **121**, 71 (1955).
[2] SU-SHU HUANG and O. STRUVE: Astronom. J. **61**, 300 (1956).

their puzzling distribution towards the edges of the Hyades cluster [20] might reflect some kind of dynamical sorting due to their different mass.

As we see, the whole subject of Hyades subdwarfs, although undoubtedly very important, is a very uncertain one at this point. Their membership is being checked further by additional radial velocity measures and speculation about their nature is perhaps unwarranted until further date are available. Discussion of what are assumed to be a different kind of subdwarf will be made in connection with globular clusters and high velocity stars in Chaps. D and F, below.

31. The white and blue dwarfs and novae. There are three stars among the nearby stars shown in Fig. 6 which are tremendously underluminous for their color indices, falling in the range from $M_v = +10$ to $+14$ mag. Since these stars are so very faint, they are difficult to discover even if they are very close to the Sun. That selection effect has led consistently to estimates that, despite intensive searches, many more of these stars exist in a given volume of space than would be indicated by the ratios in Fig. 6.

Evidence that novae, in pre- and post-outburst states occur at $M_v = +5$ places the novating stars considerably higher in luminosity than the enormously underluminous stars just discussed. The accuracy with which the quiescent novae may be placed in the H-R diagram does not warrant their inclusion in any particular diagram. From the small amount of data available, however, they seem to be very blue stars in their non-outburst stages and it suffices to state that they fall in the blue dwarf region (Fig. 1). Their connection with the white dwarfs is uncertain therefore. As will be shown in the following section, the only stars known certainly to fall near the region of the quiescent novae in the H-R diagram, are the stars in the faint blue branch of globular clusters. This is not to say that the novae pertain particularly to the globular clusters, however, since the small amount of evidence available seems to indicate that the novae do not have the same space distribution as a globular cluster population.

D. The H-R diagram for globular clusters.

32. The designation of the two population types in 1944 gave tremendous impetus to the investigation of globular clusters. Prior to that date much work had been done on the variable stars in globular clusters but, relative to the galactic clusters, little was known about the H-R diagrams of a large sample of globular clusters[1]. The H-R diagram of globular clusters have been intensively investigated since then, however, and to date, results on the brighter regions of the color-magnitude diagrams for nearly a dozen globular clusters are available, results on fainter regions for about three globular clusters. Many more are being presently worked on.

Of course, the measurement of globular cluster color-magnitude diagrams represents a somewhat different problem from that which is involved in galactic clusters. There are not so many difficulties with spurious field stars in the diagrams. Enough faint cluster members may be measured over a small enough area, so that by their numbers, they overwhelm any field stars present. On the other hand this very high space density of stars in globular clusters makes their measurement difficult from a crowding standpoint and therefore requires instruments with large scales used in good seeing conditions. Because the globular clusters are generally quite distant from the Sun (5 to 10 kpc), and their brightest

[1] The first recognition of fundamental differences between color-magnitude diagrams of galactic and globular clusters seems to have been made by ten Bruggencate, „Sternhaufen", p. 129ff. Berlin 1927.

stars are $M_v = -3$ mag. or less, these stars invariably appear faint and require observation with telescopes of large aperture. This is particularly true in the range of the globular cluster main sequence which appears, in all clusters in which it has been observed so far, at fainter then 19th apparent magnitude. Also the photoelectric calibration of sequences so faint, and in such crowded regions, presents especially difficult problems and consequently work on the faint regions of globular cluster color-magnitude diagrams has progressed slowly.

It is hardly necessary to mention that the globular cluster stars are too faint to observe, with ease, spectroscopically. In addition, even when spectra are available, there is the problem of how to classify them, there being at present no definite system of classification for type II spectra[1,2]. It is here, as a result, that the H-R diagram properly called—with spectra as abscissae—cannot be constructed. And it is here consequently, that the more recent techniques of color-magnitude diagrams are exploited to their fullest.

I. Bright regions of the color-magnitude diagram.

33. When once the color indices and apparent magnitudes have been measured for a number of stars in a globular cluster, we can inspect the characteristic "inverted *y*" shape of the sequences in the color-magnitude diagram, and the distribution of stars along them. As shown in Fig. 23 the typical giant sequence approaches an asymptotic limit, about three magnitudes brighter than the horizontal or blue branch and at color indices about 1.5 mags. redder than the middle of the gap in the horizontal sequence. This giant branch, of course, as it becomes fainter becomes bluer and finally goes into the top of the vertical, subgiant branch at about $M_v = +1$.

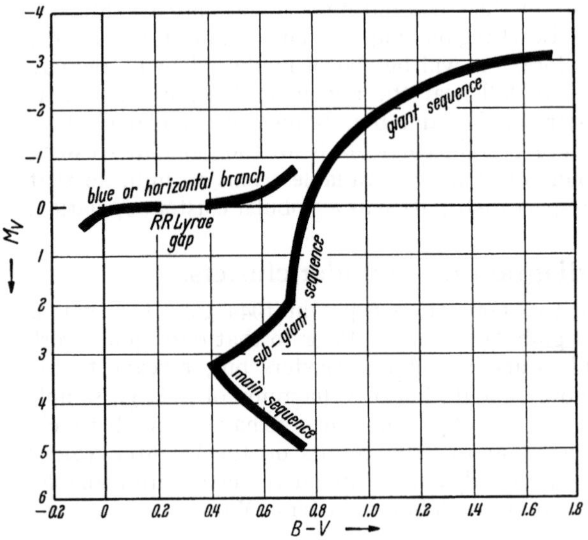

Fig. 23. Schematic color-magnitude diagram for M3. Terminology of the sequences is shown.

It would be completely impossible, however, to intercompare any of these globular cluster, color-magnitude diagrams without first taking two steps: One, we must evaluate any reddening and absorption which may be present. Secondly, unless we know the distance to the cluster by some other means, we must tentatively identify some point in the color-magnitude diagram as having the same absolute magnitude in all globular clusters. This is quite necessary since we could achieve no progress in obtaining systematically the characteristics of globular clusters if we did not assign some such unequivocal point of comparison.

---

[1] A. J. Deutsch, at present investigating spectral classification criteria of stars in globular clusters.

[2] D. M. Popper: Astrophys. Journ. **105**, 204 (1947). Spectra of bright stars in $M3$ and $M13$.

Unlike the galactic clusters, there is no main sequence in the globular clusters in the region of $M_v = +1$ or brighter, to which we may normalize each globular cluster color-magnitude diagram. In fact, the only point which we have any *a priori* reason for believing may remain invariant from cluster to cluster, is at the region of occurrence of the RR Lyrae cepheids.

34. The RR Lyrae cepheids as a zero point of color and magnitude. Historically, the convention of establishing the RR Lyrae cepheids as an invariant zero point among all globular clusters came about through SHAPLEY's early work on the period-luminosity relation for cepheids. Because of the firm belief in the existence of one, and only one, period-luminosity relation, the period was felt to be by far the most important parameter of a variable star. Therefore the fact that RR Lyrae cepheids all had the same period from cluster to cluster was taken to guarantee that they all had the same luminosity. As a working hypothesis it was assumed by SHAPLEY in his subsequent determination of the scale of the galaxy, that the RR Lyrae cepheids in all globular clusters were of the same absolute magnitude to within ± 0.1 mag.

The universal existence of one, unique period-luminosity relation was observationally refuted by BAADE in 1952. Some of the theoretical possibilities and implications which this discovery holds for the RR Lyrae cepheids as a zero point in globular clusters are discussed under variable stars in part E. In the following two paragraphs, however, only the observational evidence which relates to RR Lyrae cepheids as a zero point will be reviewed.

First, there are two independent lines of evidence which indicate that the RR Lyrae's all have closely the same intrinsic color index from cluster to cluster: (1) The nearest RR Lyrae stars in the general field have spectra which correspond with the color classes of the RR Lyrae cepheids in high galactic latitude (unreddened) globular clusters. (2) Of the six well-investigated globular clusters at high galactic latitudes, four turn out to have RR Lyrae cepheids at exactly the same color index (to within ± 0.05 mag.). The remaining two have reddenings of $E_{B\text{-}V} = 0.13$ mag. indicated. Now, using the same criterion for reddening, we find the two recently measured globular clusters (M 10 and M 22) at low galactic latitude, have reddenings of $E_{B\text{-}V} = 0.32$ and 0.40 mag. as would be expected. Observationally, therefore, the RR Lyrae cepheids in all globular clusters are shown to generally have the same mean intrinsic color indices to within about ± 0.1 mag. In fact, there are no observations which dispute the possibility that they may all have identical color indices to within a few hundredths of a magnitude, although this last has by no means been proved.

Secondly there is the observational evidence that the RR Lyrae cepheids are all of the same luminosity in different globular clusters: This evidence is indirect, but, when the zero point of absolute magnitude in each cluster is chosen as that of the RR Lyrae cepheids: (1) Characteristic features of all color-magnitude diagrams above $M_v \approx +1$ mag. occur at about the same absolute magnitude. (2) Cepheids of period greater than one day from different clusters, turn out to have the same absolute magnitude if they have the same period and same shape of light curve. (3) Long-period variables (around 100 days) turn out to have closely the same magnitudes.

In other words, stars which, as far as they have been tested, are identical in every respect, turn out to have identical luminosities when the globular cluster color-magnitude diagrams are fitted at the gap. In any particular cluster, of course, there may be no long period variables, no long-period cepheids, in fact in many cases no RR Lyrae cepheids, and in a few cases even the horizontal branch is missing. But for the case of the average globular cluster there are no observations in the bright region of the color-magnitude diagram at present which violate the hypothesis that all RR Lyrae cepheids are of the same luminosity to within ±0.1 mag. Contrary evidence from fainter regions is discussed in Sect. 44.

35. The RR Lyrae gap in the horizontal branch. It was shown originally by SCHWARZSCHILD [*32*] in M3, and later in other globular clusters, that the RR Lyrae cepheids at mean color index and luminosity, were bounded to within a few hundredths of a magnitude in color index by non-variable stars on either side of the gap. Fig. 24 shows the exactitude with which the gap in the horizontal branch defines the place of the RR Lyrae stars.

Inspection of the color-magnitude diagrams of the globular clusters shown in Figs. 25 through 32 reveals that wherever RR Lyrae cepheids have been measured they fall exactly in this gap. Inspection of all the diagrams reveals that the width of B-V = 0.2 mag. (B-V = 0.17 to 0.39 for M3) is satisfactory for the gap and that non-variable stars do not infringe into this area. Therefore, empirically, we may take the middle of the gap in the horizontal sequence as being identical with the mean of the RR Lyrae cepheids in a cluster. This result enables the zero point of a globular cluster to be determined more easily, by means of measures on non-variable stars.

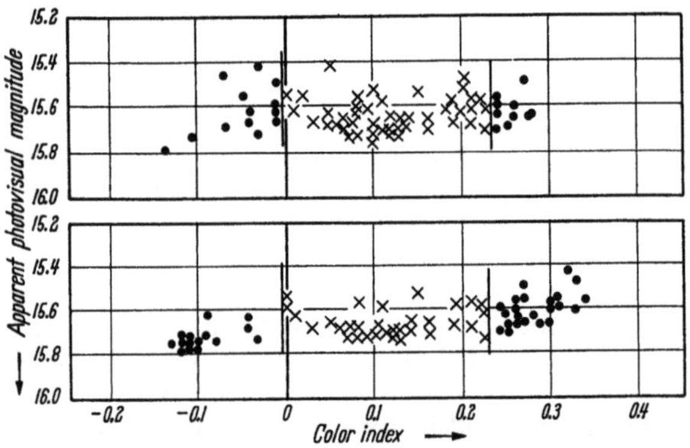

Fig. 24. Distribution of RR Lyrae cepheids, at mean light and color, in the horizontal branch of M3. Diagram from ROBERTS and SANDAGE. Filled circles represent non-variables, crosses represent RR Lyrae cepheids. Lower diagram represents a sample closer to the cluster center.

It is also apparent that a large number of RR Lyrae cepheids in a given cluster is usually associated with a strongly populated horizontal branch. Both M3 and M5 show this clearly. It appears necessary, however, to have a strong non-variable population on both sides of the variable gap, since a cluster like M2 has few RR Lyrae cepheids, even though it has a more populous blue side to its gap than M3. M3 is unusual in that the maximum of star density in the horizontal sequence, falls right in the RR Lyrae gap.

36. Fitting the color-magnitude diagrams together at the gap. The color-magnitude diagrams of the seven globular clusters which are exhibited in Figs. 25 through 32 were all measured in the same way, and on the same photometric system [*37*]. It is shown in Fig. 32 that, when the clusters are all fitted together at the gap, the remainder of the sequences all fall together, closely superposed in the color-absolute magnitude diagram. We can make the statement that the combination of all the globular clusters known today (including such recently measured clusters which are not shown here as NGC 4147[1], M22[2] and M53[3]) defines a mean color-absolute magnitude diagram which has a width of only about 0.25 mag. in color index.

[1] A. R. SANDAGE and M. F. WALKER: Astronom. J. **60**, 230 (1955).
[2] ARP and MELBOURNE: Unpublished.
[3] CUFFEY: Unpublished.

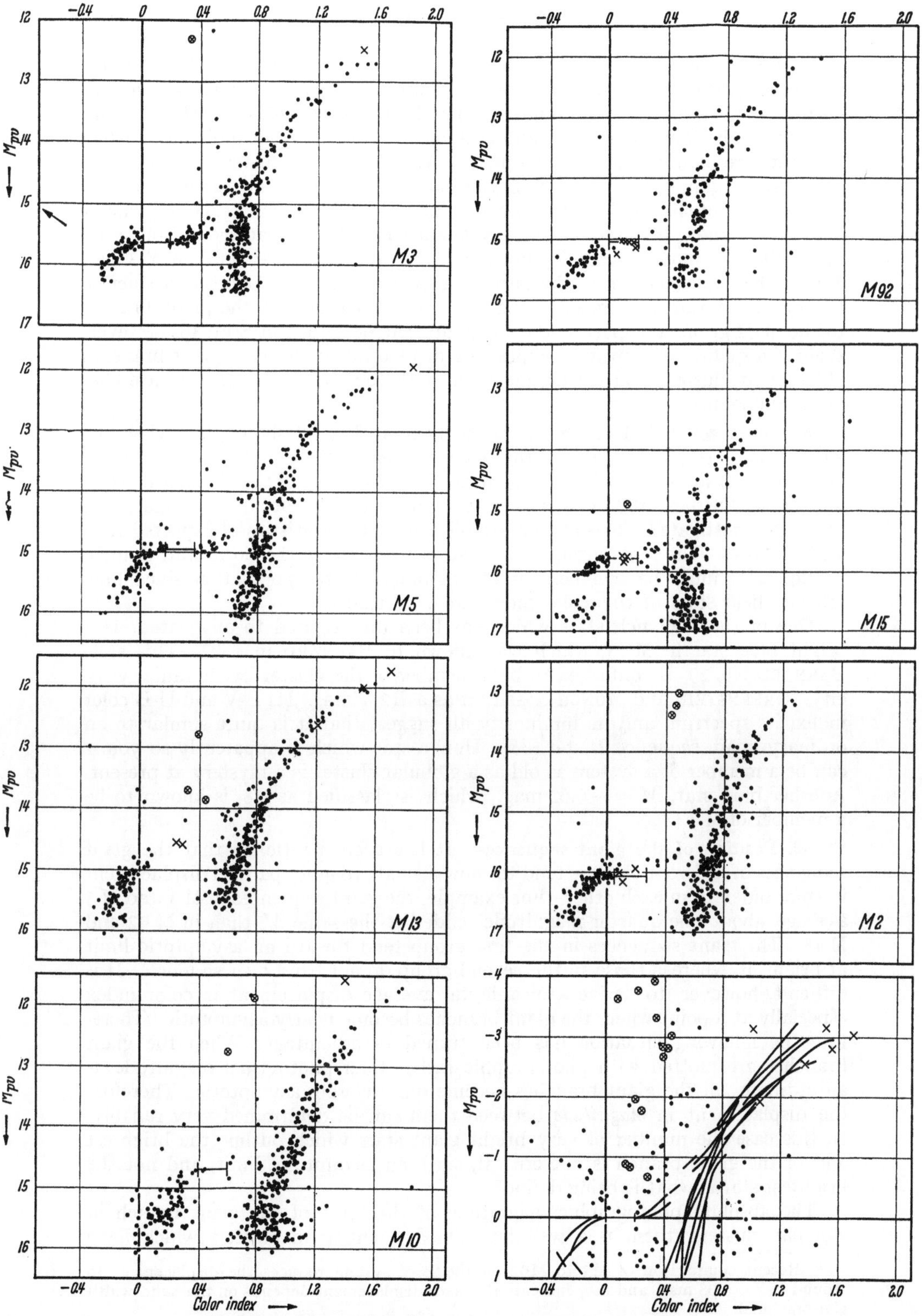

Fig. 25—32. Color-magnitude diagrams for seven globular clusters. From ARP.

The systematic accuracy of the photoelectric calibration of these color-magnitude diagrams, however, enables distinctions to be made within this narrow band spread of color indices. The RR Lyrae cepheids in both M 3 and M 92 are of the same color index and both of these high latitude globular clusters are therefore assumed to be unreddened. When the two clusters are fitted together at the gap, however, the M 92 giant and subgiant sequences show a parallel displacement of slightly less than 0.2 mag. in color index[1]. The remaining globular clusters show this same tendency toward parallel displacement from each other although none exceed by much the original difference between M 3 and M 92. The one exception to the parallel displacement is M 13 which cuts across the mean diagram of Fig. 33, going from the M 3 group at the giant branch, to the M 92 group and past at the subgiants. As described in Sect. 38, however, there is a quantitative way of expressing displacements between giant branches of different clusters and correlating them with the spectra which indicate chemical abundance differences.

37. Stars which fall off the major sequences. Fig. 32 shows that when the 2000 oddd stars which comprise the seven sequences are plotted together, there are about 100 stars which fall off the mean sequence range of the globular clusters in general. Of these 100 or so stars, not all are field stars accidently included in the field investigated. This can be established by the statistics of expected field star numbers. The loose group of blue stars lying above the horizontal branch, particularly, must be members of the globular cluster population since high latitude field stars of this color index are very rare.

One of the few such stars which has been investigated to any extent is a bright blue star in M 13, which was known to BARNARD in 1897. This star, BARNARD No. 29, is confirmed to be a member of the cluster by its radial velocity. GREENSTEIN and MÜNCH classify it as a *B* 2 V star. Its B-V and U-B color index, its spectrum and its luminosity all suggest that it is quite similar to an ordinary, main sequence *B* star [*38*]. How a star which is apparently so young can be a member of a system as old as a globular cluster is a mystery at present. Another blue star, $M_v = -0.62$ mag., which is classified as *O* 8, is known to be a member of M 3 [*35*].

38. Position of the giant sequences. It has been mentioned that the giant sequences in different clusters tend to show a small, roughly parallel displacement in color index from each other. For example, the giant sequences in M 3 and M 5 average about one-quarter magnitude redder at the same M_v than in M 92 and M 15. The giant sequences in the first group tend toward an asymptotic limit at bright M_v whereas those in the second group do not bend over so much. It is difficult, however, to derive a meaningful average displacement in color index especially at a point where the giant branches become nearly asymptotic. Therefore the following situation has been turned to advantage: When the giant branches are plotted with photographic rather than photovisual magnitude as an ordinate, all the giant branches become quite closely asymptotic. Therefore the displacement in *magnitude* between them can be determined very reliably. In this case the number of very bright giant stars which outline the brightest and of the giant branch is not critical, since an asymptotic limit, and not the brightest star is what is being defined.

The absolute photographic magnitude of the limit of the giant branch in globular clusters, then, is a well-determined quantity in clusters which have

[1] Recent comparison of M3 to M92 on the B, V system reduces the displacement to around B-V = 0.08 mag. and may indicate the color displacement depends on the band width photometered.

been carefully measured. The two groups of giant branch placements, typified on the one hand by M 3 and on the other by M 92, are clearly distinguished by this criterion in Table 7.

39. Correlations with mean period of RR Lyrae cepheids. In 1944 OOSTERHOFF pointed out that any globular cluster which contained an appreciable number of RR Lyrae cepheids, had, as a mean period for the *a* and *b* types, a value of either 0.51 to 0.55 days or 0.62 to 0.65 days. This remarkable bi-modal value has yet to receive a satisfactory explanation. Dividing the clusters on this basis, however, we see that there is an extremely good correlation of this mean period of the RR Lyrae *a* and *b* cepheids with almost every other characteristic of the cluster. These correlations are shown in Table 7.

Table 7. *Globular cluster characteristics which are correlated.*

Cluster	Giant branch M_{pg}	Mean period of RR Lyrae *a*, *b*	Metal lines in spectra	Longer period cepheids
M 3	−1.4	0.55 (124)	Weak	semi-classical
M 5	−1.4	0.54 (63)	Weak	semi-classical
M 13	−1.3	(3)	Weak	
M 4		0.51 (17)		
M 72		0.55 (21)		
Mean	−1.4	0.545	Weak	more nearly like classical cepheids
M 92	−1.7	0.63 (9)	Very Weak	
M 10	−1.7	(0)		RV Tauri — W Vir
M 2	−1.8	0.63 (11)		RV Tauri — W Vir
M 15	−2.0	0.65 (31)	Very Weak	
M 53	−1.8	0.62 (17)	Very Weak	
ω Cen		0.65 (77)	Weak	W Vir and RV Tauri
M 22	−1.3	0.63 (7)		
Mean	−1.7	0.643	Very Weak	W Vir and RV Tauri cepheids in this group

Table adapted from ARP [37] Giant branch M_{pg} of M 53 from CUFFEY (unpublished). Additional information on metal strength from integrated spectra by MORGAN.

M 22 because of a low giant M_{pg}, and ω Cen because of an integrated spectrum which places it in the M 3 group, are the only clusters which violate the correlation with the mean period of the RR Lyrae *a* and *b*'s in Table 7. M 22, ω Cen and M 13, however, all resemble each other strikingly in the nature of their variable stars (other than RR Lyrae) and the M 22 color-magnitude diagram resembles closely that of M 13. More observations on the quantities discussed in Table 7 are obviously needed, for a larger number of clusters, in order to understand better the relationships beginning to appear there.

40. Spectra and chemical abundance. In 1947 when POPPER observed the spectra of the brighter stars in M 3 and M 13, he confirmed that the CN absorption lines were weaker than in normal stars. This was expected on the basis that some of the high velocity stars showed this CN weakening and both the high velocity and globular cluster stars were at that time believed to be type II (see part F).

The spectral peculiarities of the M 3 and M 13 giants have been consistently interpreted as being due to a real deficiency of metals relative to hydrogen in these stars. When the giants in M 92 were observed spectroscopically, however, their spectra were so deficient in metallic lines, that they were at first classified as F stars even though their color classes corresponded to late K. It has since been estimated by GREENSTEIN[1] that the spectra are indeed K but that the metal lines on the spectra of M 92 are 60% weaker than those in the spectra of the M 3 giants. This makes the metal abundance in M 92 stars, altogether, very weak relative to normal, nearby stars.

Recently W. W. MORGAN[2] has classified the integrated spectra of globular clusters and separated out criteria of relative metal-to-hydrogen strength. His findings corroborate quite well the results derived by A. J. DEUTSCH[1] from the spectra of brighter individual cluster members. This might be expected since the brighter stars contribute heavily to the integrated light in a globular cluster [*37*]. MORGAN's method, however, is of great importance because it opens up the possibility of estimating the important parameter of chemical content for much more distant systems by using their integrated light.

The most important correlation in Table 7 seems to be that the clusters which have spectacularly weak metal lines have their giant branches placed at slightly higher luminosities in the H-R diagram. Although nothing can be said as to the size of the effect, this is in the same direction as the observed correlation between M 3 and the type I stars—namely that reduced metal content raises the luminosity of the giant branch[3].

II. Faint regions of the color-magnitude diagram.

Almost all interest in the fainter regions of the globular cluster color-magnitude diagram centers about the main sequence, which appears, as far as now known, somewhere between $M_v = +3.5$ and $M_v = +5$ mag. In an attempt to derive an accurate zero point for the RR Lyrae cepheids, BAADE suggested in 1953 that the main sequence in globular clusters should be fitted to the main sequence for nearby stars, in analogy to the procedures which were later used extensively in galactic clusters. In response to this suggestion, main sequences in three globular clusters have been measured. Almost from the start, however, it was clear that the globular cluster main sequence would probably have to be fitted to some subdwarf main sequence and not to the nearby star main sequence.

41. The observed main sequence. Fig. 33 shows the color-magnitude diagram for M 92. This cluster was the first one in which the main sequence was identified [*33*] and the photometric calibration extends only to the beginning of that main sequence. Fig. 34 exhibits the next globular cluster main sequence to be observed that of M 3 [*35*]. In M 3 a greater extent of the main sequence was obtained with better calibration. The measures plotted in Fig. 34 are the most recent redeterminations on the B, V system by JOHNSON and SANDAGE [*39*]. M 13 was the final globular cluster main sequence to be observed [*36*], and its diagram is shown in Fig. 35.

Because the color-magnitude diagram for M 3 has been investigated more intensively than that of any other cluster, the M 3 color-magnitude diagram has become a working prototype for globular clusters. Fig. 23 shows the schematicized diagram and the normal points on the mean M 3 sequences are listed in Table 8.

---

[1] Mt. Wilson and Palomar Observatories' Annual Report, Carnegie Institution of Washington Yearbook, 1954—1955 and 1955—1956.

[2] W. W. MORGAN: Publ. Astronom. Soc. Pacific **68**, 509 (1956).

[3] See F. HOYLE and M. SCHWARZSCHILD: Astrophys. Journ. Suppl. **1955**, No. 13.

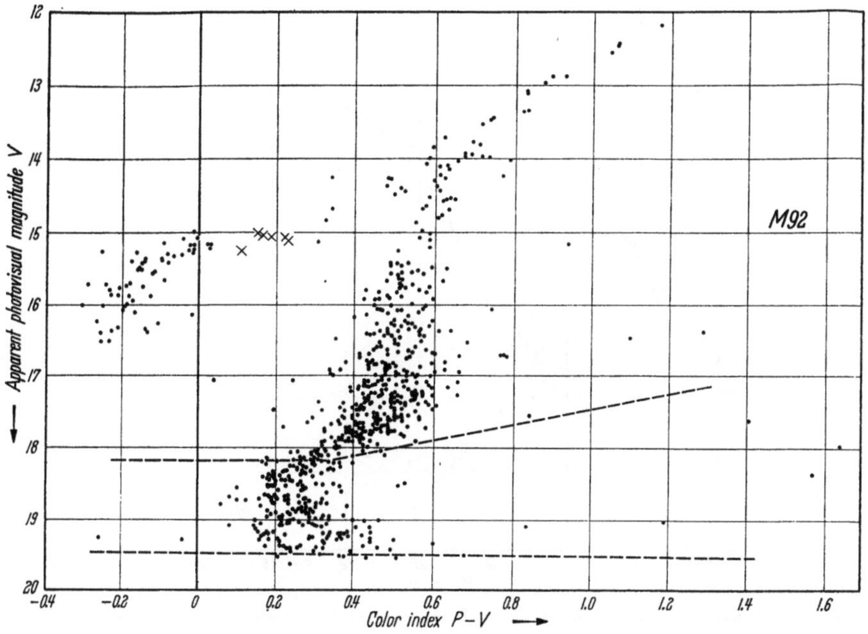

Fig. 33. Color-magnitude diagram for M92. From ARP, BAUM and SANDAGE. Area between dashed lines represents reduction in area sampled.

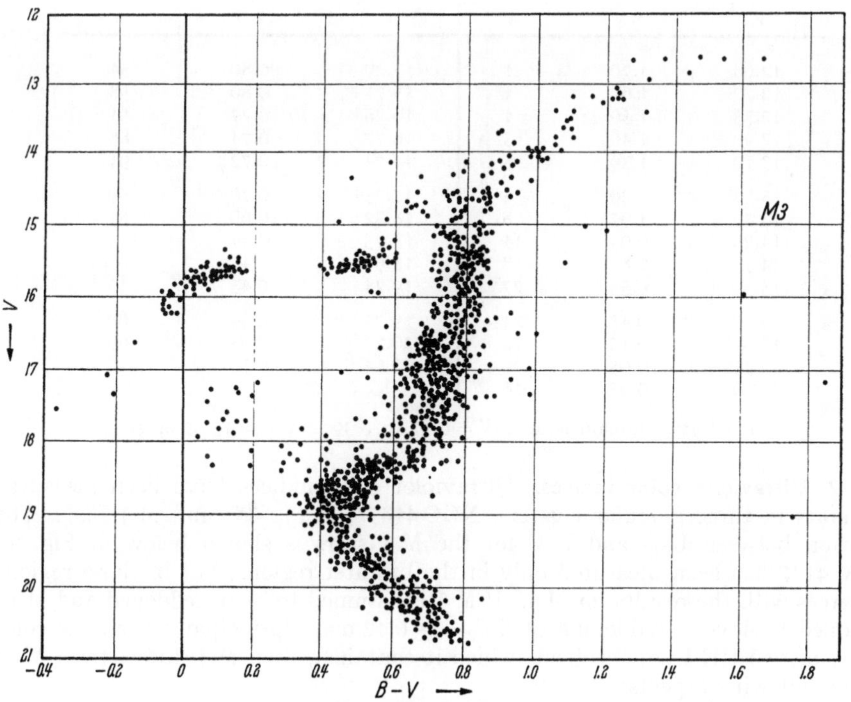

Fig. 34. Color-magnitude diagram for M3. From JOHNSON and SANDAGE.

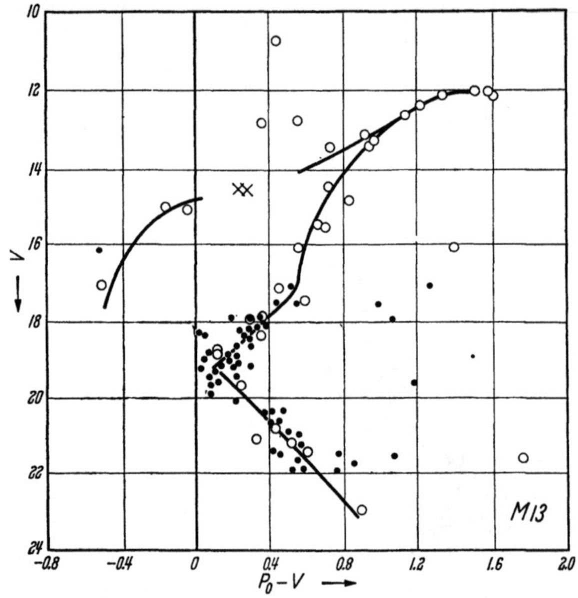

Fig. 35. Color-magnitude diagram for M13. From BAUM. Open circles represent photoelectric measures.

Table 8. *Mean color-magnitude diagram for M 3.*

Giants and horizontal branch stars			Subgiants and main sequence		
V	B-V	*n*	*V*	B-V	*n*
12.68	1.70	1	15.29	0.80	54
12.70	1.60	1	15.77	0.80	58
12.75	1.50	1	16.26	0.77	51
12.90	1.40	3	16.77	0.74	86
13.10	1.30	5	17.24	0.72	91
13.32	1.20	5	17.75	0.70	59
13.75	1.07	8	18.32	0.60	84
14.25	0.95	14	18.75	0.47	142
14.75	0.86	18	18.95	0.42	—
15.50	0.55	27	19.24	0.45	46
15.59	0.45	24	19.73	0.54	69
15.61	0.15	18	20.25	0.65	68
15.75	0.05	26	20.64	0.73	26
16.09	−0.05	16			

RR Lyrae domain is at B-V = 0.17 to 0.39 and $V = 15.67$ mag.

42. Ultraviolet color indices. Ultraviolet color indices have been measured for stars in three globular clusters, NGC 4147[1], M 13 [*38*] and M 3 [*39*]. The relation between U-B and B-V for the M 3 stars is shown below in Fig. 36. NGC 4147 has been measured only in the brighter regions, but in those regions it agrees with the results for M 3. If M 3 is assumed to be unreddened and M 13 assumed to have a reddening of $E_{\text{B-V}} = \pm 0.12$ mag. (principally from the color index of two RR Lyrae cepheids which it contains), then M 3 differs from M 13 in the following aspects:

[1] A. R. SANDAGE and M. F. WALKER: Astronom. J. **60**, 230 (1955).

1. The ultraviolet excess (relative to ordinary stars) for M 3 giants is about 0.25 mag. If there is any excess at all for the M 13 giants, it is less than 0.05 mag.

2. There is a small ultraviolet deficiency for M 3 horizontal branch star bluer than B-V = 0.1 mag. which becomes slightly larger toward bluer color indices. In M 13 there is a large ultraviolet deficiency for horizontal branch members which becomes very large for the bluest members.

M 3 is similar to M 13 in the following aspects:

1. In each cluster there is one very blue member which has a higher luminosity than the horizontal branch stars. In each cluster that star has the U-B and B-V relation of a normal luminosity class V star.

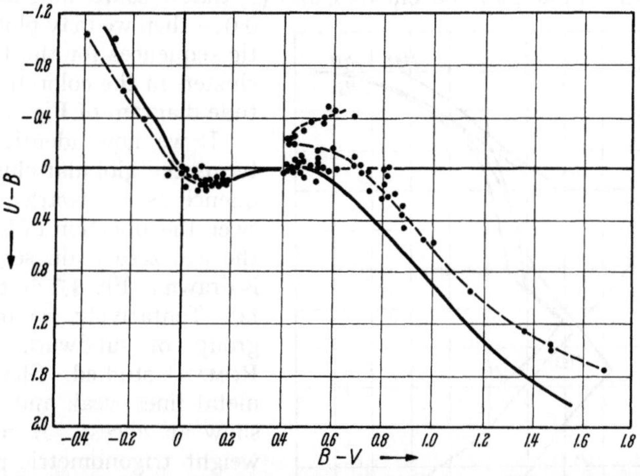

Fig. 36. The two-color diagram for M 3. Adapted from Johnson and Sandage. Dots represent mean points, photoelectric standards and individual stars (in the horizontal branch). Full line is relation for unreddened nearby stars. Dashed line is relation for M 3.

2. The subgiant stars in M 13 show an ultraviolet excess which, scale extrapolations indicate, becomes an increasingly large excess as fainter stars are reached. The measures in M 3 show clearly that the ultraviolet excess for its main sequence stars is larger than for its giants, and reaches 0.55 mag. excess.

43. Identification of globular cluster main sequence. Because of the far-reaching consequences for distance scales, caution has been exercised in actually identifying the main sequence in globular clusters. It must be admitted, nevertheless that, at present the strongest arguments point to the identification of the globular cluster main sequence as subdwarfs rather than as normal, main sequence dwarfs. There are three relevant points:

1. The globular cluster main sequence, as far as is presently known, shows a large ultraviolet excess. The main sequence dwarfs (by definition) do not. But the subdwarfs known in the vicinity of the Sun do show an ultraviolet excess[1].

2. In the vicinity of the Sun, the proportion of subdwarfs to dwarfs increases as stars of higher velocity are considered [*37*]. It is concluded, therefore, that the subdwarfs are more characteristic of the halo regions which the globular clusters inhabit.

3. The globular cluster stars are believed to have abnormally low metal abundance. This is true only of the subdwarfs and not of the dwarfs. (In fact this has been suggested as the explanation of the U-B excess [*39*].)

[1] N. G. Roman: Astronom. J. **59**, 307 (1954).

The identification of the globular cluster main sequence with the subdwarfs is thus seen to be very compelling. The formal fit has not yet been made, however, for several very good reasons. Foremost is the fact that the subdwarf sequence is not well enough defined to fit to. In fact there are indications that the subdwarfs do not define a unique sequence at all but that they fall in a curtain downward from the dwarfs as a function of metal abundance. Secondly the zero point of the RR Lyrae cepheids is left in a very uncertain position.

44. Zero point of absolute magnitude. If we continue with our two original assumptions:—(1) that the RR Lyrae stars are the same absolute magnitude and color index in all globular clusters, and (2) that absolute magnitude is $M_{gp} = 0.0$;—then we may plot the schematic sequences for the three globular clusters in the color-absolute magnitude diagram of Fig. 37.

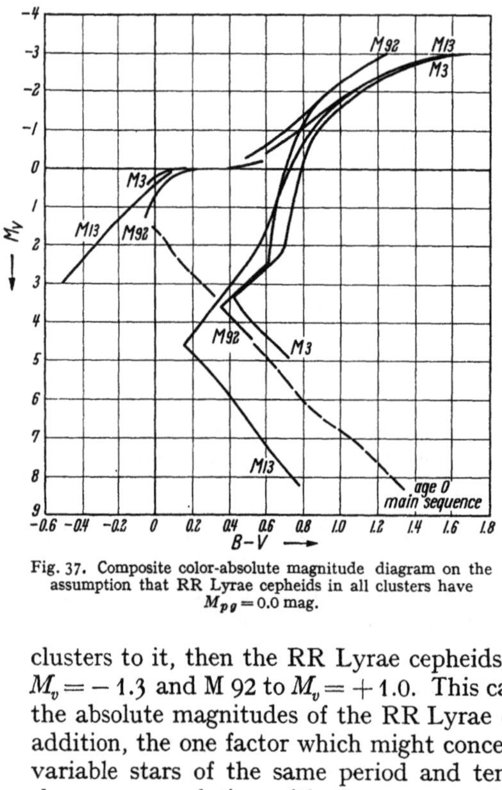

Fig. 37. Composite color-absolute magnitude diagram on the assumption that RR Lyrae cepheids in all clusters have $M_{pg} = 0.0$ mag.

If we now identify in general terms the globular cluster main sequence as subdwarfs, we must answer the question of how far below the age zero main sequence, which is drawn in Fig. 37, do the subdwarfs fall. Tentatively we might use the group of subdwarfs which N. G. ROMAN[1] studied. She found their metal lines weak and their U-B to show an excess. A mean of low-weight trigonometric parallaxes indicated that they fell somewhere between 0.5 and 1.5 mag. fainter than the normal dwarf sequence.

For the sake of argument, if we accept the appropriate main sequence to be 1.0 magnitude fainter than the one in Fig. 37 and fit the three clusters to it, then the RR Lyrae cepheids in M 3 go to $M_v = +1.5$, in M 13 to $M_v = -1.3$ and M 92 to $M_v = +1.0$. This causes a disturbingly large difference in the absolute magnitudes of the RR Lyrae cepheids in these various clusters. In addition, the one factor which might conceivably cause such differences between variable stars of the same period and temperature, the chemical composition, shows no correlation with these new variable star zero points or giant branch luminosities. That is, M 3 and M 92 which show the greatest spectroscopic difference in metal content fall rather close together in the main sequence region and elsewhere. It is M 13 which is apparently spectroscopically similar to M 3, which shows the greatest difference in sequence placement from M 3.

In summary, it might be fairly said that the observations at apparent magnitudes fainter than 19th are very difficult and consequently the observational picture is still not clear. For example M 13 is being reobserved to check the possibility of errors in the original measures and other clusters are being investigated in these regions. At present the indications are, however, that the RR Lyrae zero point will probably have to be changed to something fainter than $M_{pg} = 0.0$. We might guess that it could eventually be somewhere

[1] N. G. ROMAN: Astrophys. Journ. Suppl. **1955**, No. 18.

in the range of $M_{pg} = +0.5$ to $+1.0$ mag. But until some definite figure is derived the present convention will undoubtedly be kept. Of course, there is also the complicating possibility that RR Lyrae cepheids have different absolute magnitudes in different clusters. Further remarks on this will be made in part E below.

E. Variable stars in the H-R diagram.

45. It is usually assumed that the average temperature and luminosity of a variable star is the same as that which would obtain if there were no variation. Therefore, using the mean spectrum or color index and mean luminosity enables the variable stars to be placed in the H-R diagram in the same way as non-variables. In a few cases oscillations are performed about non-equilibrium, mean positions. If such variables have large amplitudes, as in U Gem stars or novae, the undisturbed or non-outburst measures are naturally the quantities desired in the H-R diagram.

In the usual case, though, the light curve of a variable star must be constructed over a representative portion of its cycle before its average luminosity and color index can be derived. In practice this procedure encounters several difficulties which are not inherent in the analysis of non-variable stars. There is, first, the observational difficulty that a number of observations must be made before an approximate value of the mean color, spectrum, or apparent magnitude may be obtained. The modern technique is to average *intensity* curves of light and color in order to escape asymmetry distortions.

Another circumstance which has militated against use of the H-R diagram in connection with variable stars, has been the emphasis on the period as the most significant parameter of a variable star. In other words, more attention has been paid to the period-luminosity relation than to the temperature-luminosity relation. Of course there is, among the type II cepheids, relatively little dependence of luminosity on mean color. There seems to be somewhat more dependence for the classical cepheids, but, unfortunately the intrinsic colors of the classical cepheids are, at this time still very uncertain. Basically the most dissatisfying situation of all is, of course, the fact that we are still quite unsure of the exact luminosities of either the type I or type II cepheids.

Because of the general uncertainty connected with the place of almost all the variables in the H-R diagram, an exhaustive treatment will not be undertaken here. Only the approximate location of some samples of various kinds of variables will be illustrated. But since the temperature-luminosity relationship is as fundamental for variable stars as for non-variable stars, it is clear that the place of the variables in the H-R diagram is a significant means of classifying and identifying them. It will be seen that in many cases the approximate placing of the variables in the H-R diagram is sufficient to define their group membership.

I. Mean spectral types and color indices.

The best known and most important variable stars have always been the cepheids. Perhaps next best known are the long-period variables. Other kinds of variables, of smaller amplitudes, are also known and will be mentioned below. As general information on the stars increases, however, it is clear that two things will happen: (1) The number of known variables will increase simply because more accurate techniques will enable discovery of variables of smaller and smaller amplitudes. (2) When smaller and smaller amplitudes are considered, the proportion of stars found to be variable will increase enormously. For example, in Sect. 50 evidence is put forward that indicates almost all stars brighter than

$M_v = -3$ and to the red of the main sequence are variables, at least by a few tenths of a magnitude. The variables of large amplitude, however, the so-called great sequence of cepheids, are intimately connected with a special region of the H-R diagram called the Hertzsprung gap.

46. Cepheids in globular clusters. Aside from the RR Lyrae cepheids, whose place in the H-R diagram was discussed in Sect. 34, globular clusters occasionally contain cepheids of period longer than one day. Table 9 gathers the best present data together for the majority of such cepheids known in globular clusters [*46*].

Table 9. *Cepheids in globular clusters.*

Star		Period	M_v	B-V	Sp. Type
M2	No. 1	$15^d.6$	−2.75	0.56	*F*8−*G*2
	No. 5	17.6−35.1	−2.90	0.58	*F*6−*G*0
	No. 6	19.3	−3.11	0.60	*F*6−*G*2
	No. 11	33.5−67.0	−3.91	0.50	*F*5−*G*3
M3	No. 154	15.3	−3.40	0.51	*F*5−*G*3
M5	No. 42	25.7	−3.77	0.47	*F*4−*G*3
	No. 84	26.6−53.2	−3.72	0.46	*F*5−*G*5
M10	No. 3	7.91	−2.06	0.41	
	No. 2	18.7−37.4	−2.99	0.47	*F*5−*G*2
M13	No. 1	1.46	−0.98	0.36	*A*2−*F*2
	No. 6	2.11	−0.82	0.47	*F*2−*F*5
	No. 2	5.11	−1.89	0.42	*F*0−*G*3
M15	No. 1	1.44	−1.01	0.34	*A*8−*F*0

Mean magnitudes and colors on the basis of the convention that the mean cepheids in each cluster are equal to $M_v = -0.1$ mag. B-V $= 0.31$ mag. In most cases B-V color indices were obtained from C. I. by the transformation[1]: B-V $= 0.84$ C. I. $+ 0.23$ spectral types from JOY [*42*].

As discussed later, the most uncertain quantity listed in Table 9 is the absolute magnitude. That absolute magnitude was derived on the usual convention that $M_{pg} = 0.0$ mag. for RR Lyrae cepheids. If a change is made in the accepted absolute magnitudes for the RR Lyrae stars, then, of course, a constant correction must be applied to column 3.

The W Virginis and RV Tauri sub-types, when taken together with the RR Lyrae cepheids, define the group called type II cepheids. With the above assumption about the zero point, these type II cepheids may be placed in the color-absolute magnitude diagram as shown in Fig. 38. Their place in the H-R diagram proper, again with open symbols, is shown in Fig. 39. The H-R diagram of Fig. 39 is plotted with M_{pg} as an ordinate instead of the usual M_v. (M_{pg} is the absolute magnitude on the old international photographic system.) This has been done because some variables, particularly classical cepheids, have only photographic luminosities and very uncertain color indices. Such variables therefore, may be placed more accurately in Fig. 39 than 38.

47. Classical cepheids. Relative to non-variable stars, the number of cepheids per unit volume of space is very small. Long period, high luminosity cepheids are particularly rare. In addition, these classical cepheids show a remarkable degree of concentration to the galactic plane and thus to the absorbing dust which lies in that plane. As a consequence even the nearest cepheids suffer

[1] H. C. ARP: Three-Color Photometry of Cepheids, A. J. in press.

appreciable reddening and absorption and, as a group, their intrinsic colors and magnitudes are very imperfectly known.

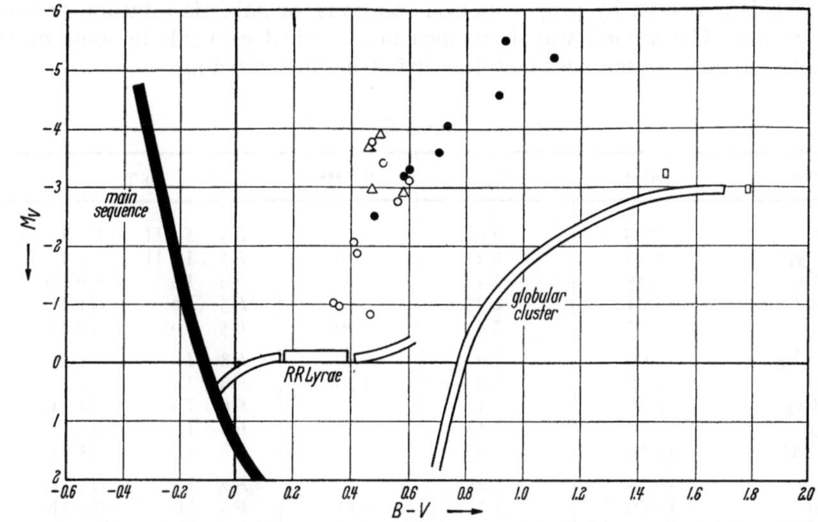

Fig. 38. Approximate place of variable stars in the color-absolute magnitude diagram. Open circles: globular cluster cepheids. Open triangles: globular cluster cepheids which show RV Tauri characteristics. Open rectangles: globular cluster long period variable (around 100 days). Filled circles: classical cepheids.

Spectral types are reliable data which are independent of reddening. The 18 classical cepheids listed in Table 10, however, represent about the extent of the classifications available on the MKK system [*41*].

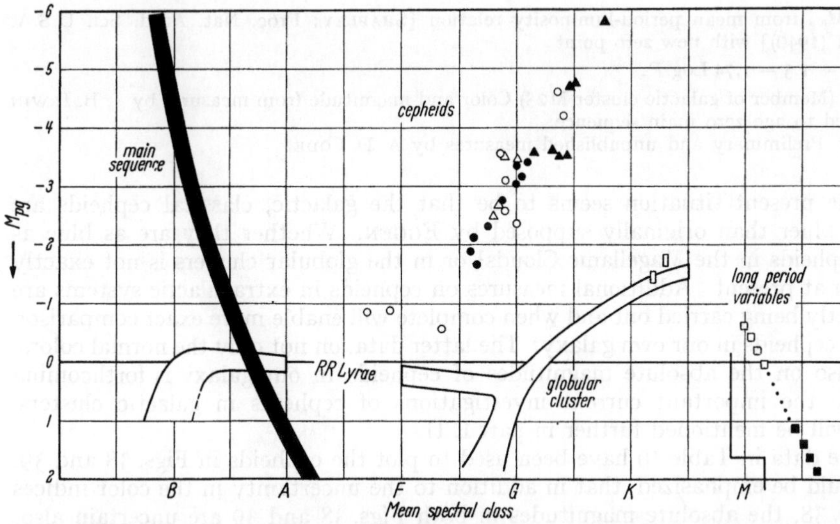

Fig. 39. Approximate place of variables in the H-R diagram. Symbols same as in Fig. 38 with the addition of: filled triangles: RV Tauri stars; open and filled squares: schematic representation of long period variables.

There are two ways in which the problem of intrinsic colors and luminosities may be attacked. One is to estimate the reddening in the same region of space as each individual cepheid. This has been done by A. D. Code[1] by using B stars.

[1] Unpublished material. See also D. W. N. Stibbs: Monthly Notices Roy. Astronom. Soc. London **115**, 323 (1955).

He gets the preliminary values listed in Table 10. The absorption-free apparent magnitudes may then be made to yield absolute magnitudes if the distance to the cepheids is known by proper motion solutions, or galactic rotation, or some other means. The second way is to measure classical cepheids in some extragalactic system to which we know the distance and reddening.

Table 10. *Classical cepheids.*

Star	Period	M_{pg}*	B-V***	Sp. Type
SU Cas	1ᵈ.95	−2.0		$F5$ I−II −$F7$ I−II
DT Cyg	2.50	−2.2	0.48	$F5.5$ I−II −$F7$ I−II
SZ Tau	3.15	−2.4		$F6$ Ib −$F9$ Ib
V386 Cyg.	5.24	−2.75		$F5$ Ib −$G1$ Ib
δ Cep	5.37	−2.8	0.60	$F5$ Ib −$G2$ Ib
MW Cyg	5.96	−2.9		$F8$ Ib −$G1$ Ib
η Agl	7.18	−3.0	0.70	$F6$ Ib −$G4$ Ib
VY Cyg	7.86	−3.1		$F6$ Ib −$G1$ Ib
S Sge	8.38	−3.15		$F6$ Ib −$G5$ Ib
BZ Cyg	10.14	−3.25		$F8$ Ib −$G5$ Ib
ζ Gem	10.15	−3.25		$F7$ Ib −$G3$ Ib
Z Lac	10.89	−3.3	0.73	$F6$ Ib −$G6$ Ib
TX Cyg	14.71	−3.5		$F5$ Ib −$G6$ Ib
SZ Cyg	15.11	−3.6		$F8$ Ib −$G8$ Ib
X Cyg	16.39	−3.6	0.91	$F7$ Ib −$G8$ Ib
CD Cyg	17.07	−3.6		$F8$ Ib −$K0$ Ib
T Mon	27.01	−4.0	1.10	$F7$ Ia−Ib−$K1$ Ia−Ib
SV Vul	45.13	−4.4	0.93	$F7$ Ia −($K0$)
U Sgr**	6.74	−2.70	0.58	

* M_{pg} from mean period-luminosity relation [SHAPLEY: Proc. Nat. Acad. Sci. U.S.A. **26**, 541 (1940)] with new zero point:

$M_{pg} = -1.5 - 1.74 \operatorname{Log} P$.

** (Member of galactic cluster M25) Color and magnitude from measures by J. B. IRWIN adjusted to age zero main sequence.

*** Preliminary and unpublished measures by A. D. CODE.

The present situation seems to be that the galactic, classical cepheids are much bluer than originally supposed by EGGEN. Whether they are as blue as the cepheids in the Magellanic Clouds[1] or in the globular clusters is not exactly known at present. Additional measures on cepheids in extragalactic systems are presently being carried out and when complete will enable more exact comparison to the cepheids in our own galaxy. The latter data, on not only the normal colors, but also on the absolute magnitudes of cepheids in our galaxy is forthcoming due to the important current investigations of cepheids in galactic clusters. This will be mentioned further in part E II.

The data in Table 10 have been used to plot the cepheids in Figs. 38 and 39, It should be emphasized, that in addition to the uncertainty in the color indices in Fig. 38, the absolute magnitudes in both Figs. 38 and 39 are uncertain also. The data at present can only show the approximate place of the classical cepheids in the H-R diagram.

48. RV Tauri Stars. The RV Tauri stars are defined by three properties: (1) They have spectra in the range of F through K. (2) They have periods from about 30 to 150 days. (3) They have light curves exhibiting a tendency for alter-

[1] S. C. B. GASCOIGNE and G. E. KRON: Publ. Astronom. Soc. Pacific **65**, 32 (1953).

nating deep and shallow minima. These variables have, at times, been suggested to be the transition between cepheids and long-period variables. This designation of the RV Tauri stars, however, does not seem tenable.

Properties (1) and (2) would indicate that they do indeed bridge the range between long period variables and cepheids. The luminosities of the RV Tauri variables indicates something quite different, however. Table 11 shows the spectral and luminosity classifications for eight RV Tauri stars including RV Tauri itself. These represent the only clues which we have to the luminosities of the RV Tauri stars. Although luminosity classes are never precisely accurate, and in particular they might be uncertain for stars as unusual as RV Tauri stars, the classifications by ROSINO [*43*] are strictly on the MKK system and should be meaningful enough to distinguish the 4 magnitude difference between the end of the cepheid sequence and the beginning of the long period variable sequence. As can be seen from Fig. 39, the RV Tauri stars do not fall near the long period variables, but form a natural extension of the cepheid sequence to higher luminosity[1].

Table 11. *RV Tauri variables.*

Star	Period	Sp. Type	
TW Cam	85.d6	*F*8 Ib	*G*8 Ib
SS Gem	89.3	*F*8 Ib	−*G*5 Ib
AC Her		*F*4*p* I	−*Rp*
U Mon	92.3	*F*8*e* Ib	−*K*0*p* Ib
R Sge	70.8	*G*0 Ib	−*G*8 Ib
R Sct	144	*G*0*e* Ia	−*K*0*p* Ib
RV Tau	78.6	*G*2*e* Ia	−*K*3*p*
V Vul	76.0	*G*1 Ia−Ib	−*G*8Ia−Ib
Semi-regular variables			
AG Agr	96.0	*G*0*e* Ib	−*K*0*ep*
UU Her	72−90	*F*2 Ib	−*cF*8
SV UMa	75.3	*G*1 Ib	−*K*3*p* Ia

Data from L. ROSINO, [*43*].

This conclusion is strengthened by the results for the cepheids in globular clusters. It was demonstrated there [*46*], that as the cepheids became longer in period, that there was a general transition from W Virginis characteristics to RV Tauri characteristics. That is, although the cepheids remained homogeneous in all their physical properties, including their luminosities, that the longer period light curves showed an increasing tendency to alternating deep and shallow minima. The classical cepheids appear to behave in an analogous way except the transition to RV Tauri behavior occurs at longer periods.

The evidence from the period-luminosity relation, Fig. 40, confirms this picture.

The first eight variables listed in Table 11 are numbers of JOY's type I group [*45*]. They have low velocities and are strongly flattened to the plane (although not so highly flattened as the classical cepheids). Therefore they appear to be geometrically allied to classical cepheids. Most of JOY's type I group are RV Tauri stars. In his group with large velocities and distributions away from the plane, which he calls type II, the predominant type is not RV Tauri, but rather semi-regular variables. Three of this latter group which were classified on the MKK system are listed in Table 11. The appended luminosity classes indicate they are generally similar to the group just discussed but beyond that, their exact connection requires further observational determination.

49. Long-period variables. The only data presently available, which allow us to place the long period variables in the H-R diagram at all, are due to R. E. WILSON and P. W. MERRILL [*40*]. The proper motions of the long period variables, checked by the mean of low weight trigonometric parallaxes, indicate that their

[1] JOY [*45*] also discusses the absolute magnitudes of the RV Tauri stars.

absolute photovisual magnitudes range from $M_v = -2.5$ for variables of periods around 150 days and spectra as early as $M\,1\,e$; to $M_v = +1$ for the longest periods, around 450 days, and spectra as late as $M\,8\,e$. Their position is indicated schematically in Fig. 39.

It is well known that, as we consider long period variables of progressively earlier spectral type (and hence shorter period and higher luminosity), the mean space velocity of these variables increases. This suggests that the higher space velocities may designate stars which pertain more to the extended spatial distribution of the globular cluster population. That this is indeed true can easily be seen by referring to Fig. 40. There, eight long period variables from three different globular clusters are plotted. It is seen that the long period variables

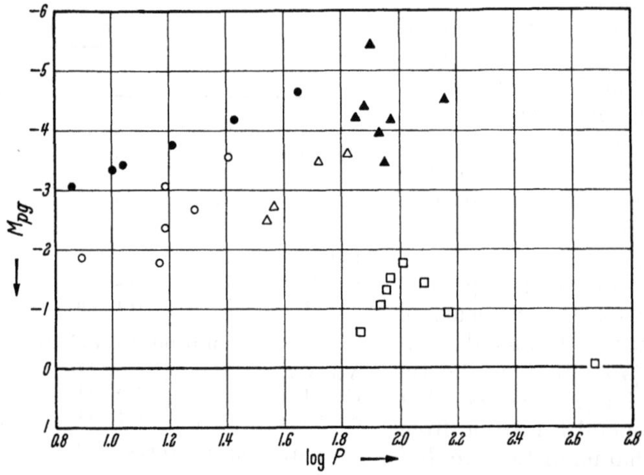

Fig. 40. Place of variables in period-luminosity diagram. Symbols as in Figs. 38 and 39. Long period variables at lower right are one each from M3 and M13 and the rest from ω Centauri.

in globular clusters are most commonly just about 100 days in period, the same as the long period variables in the general field which have the greatest space velocities. In addition the shape of the period-luminosity relation for these stars in globular clusters is the same as for the variables studied by WILSON and MERRILL.

The above identification of the 100-day long period variables in globular clusters with the extremum of the ordinary long period variables of high velocity seems quite likely to be correct. More of these variables are currently being investigated in globular clusters and the properties of the type II long period variables promises to be well known in the near future. The data available on the long period variables in the field, however, are very sparse. For instance, there are few mean apparent magnitudes available and almost no accurate or modern color indices for these stars. It is impossible to plot them for example in Fig. 38, and only the roughest, constant color estimate of B-V = +2 enables them to be approximately sketched in Fig. 39.

It will be important to remedy the lack of data on long period variable stars in the general field. For one thing, they apparently represent the first kind of star which runs the gamut continuously from the extreme of population I to population II. In this respect they could be very important as a population tracer in investigations of the structure of the galaxy. Secondly, if the photometric data on the stars in WILSON and MERRILL's catalogue can be improved, their distance determinations offer another, independent method of calibrating the zero point of absolute magnitude in globular clusters.

50. Supergiant variables. It has long been known that a number of the intrinsically brightest stars in the sky are at least slightly variable. Particularly the very red supergiants seem to be prone to some kind of oscillation. Examples in which some rough periodicity in the oscillation has been detected are RW Cyg and μ Cep (around 600 and 700 days respectively). Most of the red supergiant branch in h and χ Persei is irregularly variable. One of the best known of the latter group is S Persei which has a period of 835 days.

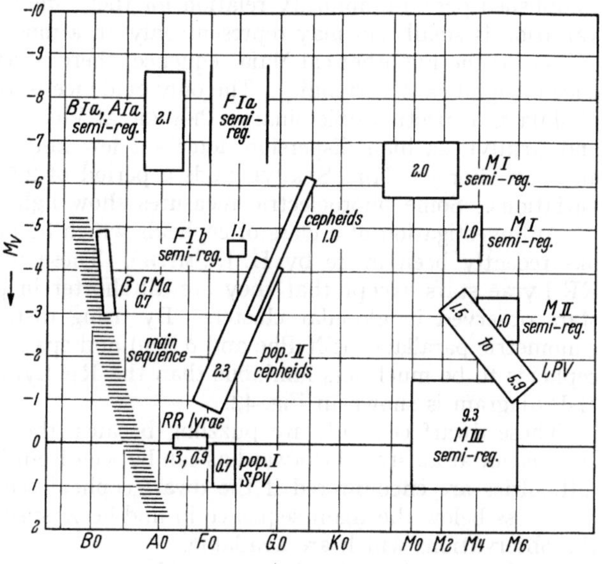

Fig. 41. Place of the supergiant variables in the H-R diagram. From H. A. Abt

Recent work by H. A. Abt[1] showed that when *A* and *F* type supergiants are examined, that they also tend to show some variations in radial velocity (and, where tested, usually in light also). In further surveying the upper part of the H-R diagram, Abt came to the important conclusion that probably all stars brighter than $M_v = +1$ and falling to the red of the main sequence (excluding globular cluster sequences) are at least somewhat variable. The designations which he has given to these variables and their location in the H-R diagram are shown in Fig. 41.

51. Dwarf cepheids and variables close to the main sequence. Fig. 42 shows the H-R diagram for variable stars as discussed by O. Struve[2].

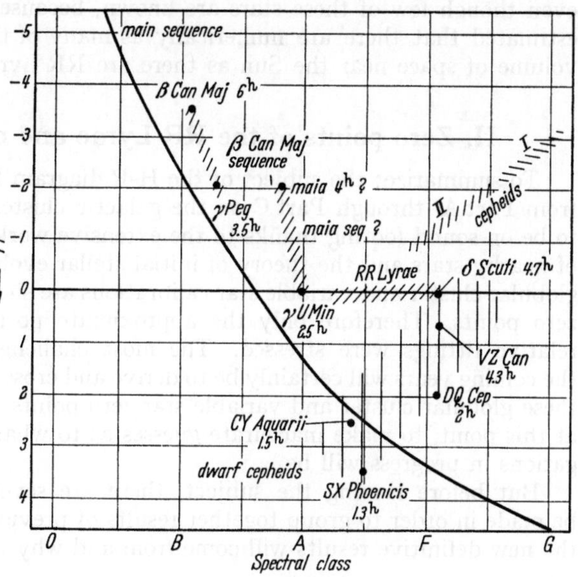

Fig. 42. Dwarf cepheids and other variables in the H-R diagram. Adapted from a diagram by O. Struve.

---

[1] H. A. Abt: The Variability of Supergiants. Astrophys. Journ. (in press).

[2] O. Struve: Sky and Telescope, September 1955, pp. 461–463.

The well known β Can Maj sequence, is the only definite group of variables that falls earlier than $A\,0$ in the H-R diagram. Even though there is a well established period-luminosity relation for these stars, the amplitude of the light variation is small and may represent only an atmospheric oscillation.

As for the hypothetical Maia sequence, there is severe doubt as to the existence of significant variability. The only evidence is W. S. ADAMS' announcement of Maia as a spectroscopic binary. PEARCE, at Victoria, says there is no variation, and STRUVE on high dispersion detects a few km/sec variation in the order of an hour. For γ U Min, STRUVE finds a period of 2.5 hours from radial velocity variations. Some photoelectric measures show light variations, others do not.

An investigation of a class of cepheids which he has called "dwarf cepheids" has recently been made by H. J. SMITH[1]. These variables resemble most the RR Lyrae stars except that they are all shorter in period than any RR Lyrae stars observed in globular clusters. By using distances determined from trigonometric parallaxes (SX Phe and δ Scu) and space motions, SMITH finds these cepheids to be much less luminous than the RR Lyrae stars. Their place in the H-R diagram is shown in Fig. 42.

These dwarf cepheids are puzzling because their relation to the RR Lyrae stars is not at all straightforward as can be seen from Fig. 42. Further theoretical difficulties are encountered if the dwarf cepheid sequence really does intersect and cross below the main sequence in middle A spectral classes as indicated by the observations which are available.

At the same time, this dwarf cepheid sequence appears to be a very important one. First, the average velocity of the dwarf cepheid is only 66 km/sec as against 163 km/sec for true RR Lyrae stars (period greater than 0.4 days). Therefore these fainter cepheids obviously tend more toward population type I characteristics and therefore may represent the type I analogue of RR Lyrae cepheids. Second, even though few of these stars are known, because of their faintness, it can be estimated that there are numerically as many of these dwarf cepheids per unit volume of space near the Sun as there are RR Lyrae cepheids.

II. Zero points of the RR Lyrae and classical cepheids.

To summarize; the subject of the H-R diagram has been elaborated in detail from Part A, through Part C on the galactic clusters. There the results appear to be on sound footing thanks to the extensive work on trigonometric parallaxes of nearby stars and the theory of initial stellar evolution. In Parts D and E the globular cluster and variable star calibrations are shown to depend on the cepheid zero points. Therefore only the approximate positions of the sequences, and relative fittings were stressed. The most challenging and important work in the coming years will certainly be to derive and cross check with extreme accuracy these globular cluster and variable star zero points. It would be no contribution, at this point, to make inaccurate guesses as to what the results of these investigations in progress will be.

But before leaving the subject, there are several comments which should be made in order to group together results of previous sections, indicating where the new definitive results will come from and why they will be important.

52. RR Lyrae stars. It has already been partially discussed how the conventional zero point of the RR Lyrae stars at $M_{pg} = 0.0$ mag. may eventually have to be revised towards fainter values. Comments have also been made on

[1] H. J. SMITH: Astronom J. **60**, 179 (1955).

the fact that, although there is some evidence that RR Lyrae stars have the same absolute magnitude in all clusters (features above $M_v = +1$ mag.), there is also some evidence that they may have different intrinsic magnitudes in some globular clusters (the main sequences in M3 and M13). These questions will be settled principally by faint observations in more globular clusters. Arguments like the following may ensue:

Consider a set of globular clusters all of the same age but of different initial chemical composition. Their far-evolved stars can cut through the Hertzsprung gap at different luminosities, depending on their particular chemical composition. But if the "period $\times \sqrt{\text{mean density}}$ = constant" law is valid, then a particular variable star with a certain color index and certain period has a unique luminosity. This comes about because the period-density relation imposes another condition on the mass, luminosity and temperature of the variable which must be satisfied. If we go through the same argument with the initial chemical composition held constant, and vary the age, the same result is obtained. From this we conclude that we would only expect it possible for RR Lyrae stars to have different luminosities in different clusters if *both* the initial chemical composition and age of the two clusters were different.

Whether this is true or not, and if true how much difference, in fact, an initial chemical composition and mass difference can make in the luminosities of RR Lyrae stars, remains to be seen. Even though such effects may turn out to be negligible between, say, globular clusters; considerations of this kind will undoubtedly be important in problems such as the dwarf cepheids and the differences between type I and type II cepheids.

53. Classical cepheids. The two strongest, original arguments for changing the classical cepheid zero point to $M_{pg} = -1.5$ mag. were: (1) The brightest type II stars observed by BAADE in the Andromeda Nebula gave a modulus of 23.9 mag. (now 24.6). (2) The RR Lyrae stars observed by THACKERAY in the Magellanic clouds give moduli between 18.6 and 18.7 mag. (now 19.2). The absolute magnitudes of the classical cepheids were then simply recomputed on the basis of these new distance moduli. It is apparent that if the zero point of the RR Lyrae stars is now moved, the moduli of both these systems will change, and the zero point of the classical cepheids along with them.

Of course, there have been independent checks on the new zero points. For example, A. BLAUUW and H. R. MORGAN[1] recomputed the absolute magnitudes of the nearest galactic cepheids by reanalyzing their space motions. However, they used the Eggen reddening values for the cepheids. If the increased values of absorption, believed appropriate today, are used, their value of $M_{pg} = -1.4$ mag. is made brighter. This is in the opposite direction from that indicated by the preliminary globular cluster results on the RR Lyrae variables. The masses, radii, temperatures and luminosities of the classical cepheids, with the new zero point now satisfy the theoretical requirements of the pulsation theory[2]. But the effect of evolution on the mass-luminosity relation has not been taken into account in these calculations, and some of the other quantities are quite uncertain as well.

The net result is that the zero point of the classical cepheids is quite uncertain at the present moment. Nevertheless the situation today is quite encouraging, principally due to the discovery by J. B. IRWIN of two classical cepheids which are members of galactic clusters[3]. That discovery spurred a search for other

[1] A. BLAUUW and H. R. MORGAN: Bull. Astron. Inst. Netherlands **1954**, No. 450, 95.
[2] M. SAVEDOFF: Bull. Astron. Inst. Netherlands. **1953**, No. 446, 48.
[3] See earlier section in galactic clusters.

cepheids which are members of galactic clusters. R. P. KRAFT and S. VAN DEN BERGH[1] have most recently computed a list of 10 additional classical cepheids, 4 likely and 6 possible members of other clusters.

The apparently normal cepheid in M25 gives an $M_{pg} = -1.15$ mag. for the cepheid zero point. The one in NGC 6087 is a *C* type, and gives a zero point 0.6 mag. fainter, as might be expected of a small-amplitude cepheid. When the accurate photometry of all of the cepheids which are now suspected of being cluster members is done, however, there should be just about enough information to give a reliable mean zero point for the classical cepheids. The most important feature of this approach is, of course, that it is observational and direct, tying the classical cepheids to the nearby stars in the same fundamental way in which the galactic clusters themselves have been calibrated.

F. Population I and II.

In 1944 BAADE [*47*] defined type II population as those kinds of stars which have the same H-R diagram as globular cluster stars. He defined population I as those kinds of stars which outline the ordinary H-R diagram, that is, the stars in the vicinity of the Sun, in galactic clusters and the spiral arms. The population concept has proved itself to be fundamentally valuable and the majority of astronomical papers have, since then, used these terms at least to some extent. It is now, perhaps, worthwhile to review this terminology in the light of the present, more complete knowledge of the H-R diagram.

54. Population type and chemical composition. Type I was defined as containing highly luminous *O* and *B* stars. Since such stars are very young this marked type I as partaking, at least to some extent, the characteristics of a young population. As knowledge of the globular clusters began to accumulate, it began to appear that the globular clusters, and therefore type II, were all very old. It was natural then for the population I to come to be regarded as young stars and population II as old stars.

Observations of their color magnitude diagrams, however, began to show the range of ages possible among galactic clusters. It was demonstrated that a related family of sequences existed in the H-R diagram, from the open clusters which contained bright *O* and *B* stars through to M67 in which the main sequence had "burned away" to $M_v = +3.5$ mag.

The derivation of the color-magnitude diagram for the galactic cluster M67 made it impossible to consider type I exclusively a young population any longer. Here was a galactic cluster as old as any globular cluster known, and of the same order of age as the galaxy itself.

There are several reasons why M67 must be, logically, considered as a type I population. First, since it is designated as a galactic cluster it must be type I by the original definition of population I which included "open" clusters. But there are other reasons besides this. For example, by its color-magnitude diagram it is not a globular cluster, therefore it must be something else besides population II. The M67 color-magnitude diagram does coincide with that of the majority of stars in the neighborhood of the Sun and the ordinary H-R diagram (main sequence, and luminosity class III, *G* and *K* giants). This again, by the original definition, marks it as type I. The really compelling identification, however, is the natural, continuous relation which M67 bears to the rest of the galactic clusters containing brighter and brighter main sequence objects, that is, the very conspicuous type I objects.

---

[1] R. P. KRAFT and S. VAN DEN BERG: Astrophys. Journ. (in press).

If we accept temporarily the view that M67 is best assigned to population I, then we may look at Fig. 43 and ask: What is the quantity which distinguishes type I from type II? Fig. 43 shows the distinct differences between M67 and two globular clusters. They are both of the same age but represent, as far as we now know, the extremes of population I and II, the limit of excursion in the H-R diagram of stars all of the same age.

Since they are both of the same age, present knowledge indicates that the masses of the stars involved cover the same range. Therefore, simple recourse to the Vogt-Russell theorem tells us that the only possible parameter which can explain their displacement in the H-R diagram, is a difference in their chemical composition.

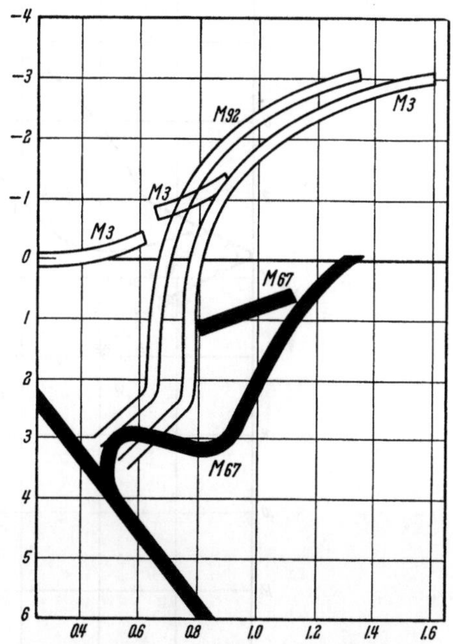

Fig. 43. Comparison of the galactic cluster M67 to the globular clusters M3 and M92 in the color-absolute magnitude diagram. From JOHNSON and SANDAGE.

How does this conclusion satisfy the observational facts? First, we know most of the nearby stars, stars known in galactic clusters (M67 included), all have normal spectra. In contrast, all the spectra available on globular cluster stars indicate that they are abnormal in the sense that they all show a weakening of metallic absorption lines. Secondly, the nearby stars define a U-B, B-V relation which we call normal. There is evidence which indicates that stars which have abnormally low metal abundances will deviate from this relation in the sense of having ultraviolet excesses[1]. Again, observationally, it is principally the type II population globular clusters which have these ultraviolet excesses, indicating that this is a population characterized by low metal content.

Taken all together, the foregoing evidence seems to indicate that the meaningful parameter which distinguishes population I from II is the chemical composition or, more specifically, the metal abundance relative to hydrogen. On this basis it is seen that age has no necessary connection with the population types. It is of interest to note, however, that we, at present, know no type II, or low metal content systems which are "young". It is uncertain, whether we could recognize such systems immediately. In fact this seems to be one of the most interesting problems for future investigation. Are there, anywhere, low metal content stars which are being, or have recently been formed?

As for the terminology employed, it may be felt that the words population I and population II contain too many misleading connotations of age, or incorrect geometrical correlations. Some recent papers have been written without using the population terminology at all but instead only referring to the parameter of metal abundance. So long as the populations are distinguished strictly in terms of chemical composition differences (which extends naturally from their

[1] This is due to the "blanketing" effect of metal lines in stars later than *A*. Originally suggested by STRÖMGREN and further substantiated in unpublished work by SANDAGE and others.

original definition), however, it does not seem important whether or not the present dual usage is kept or changed.

55. The high velocity stars. As an illustration of some of the difficulties which the population concept encountered with age inferences, Fig. 44 presents the color-magnitude diagram of the nearer high velocity stars. It is seen that these high velocity stars outline an M67 type color-magnitude diagram more closely than they do a globular cluster diagram. Originally [*47*] the high velocity stars

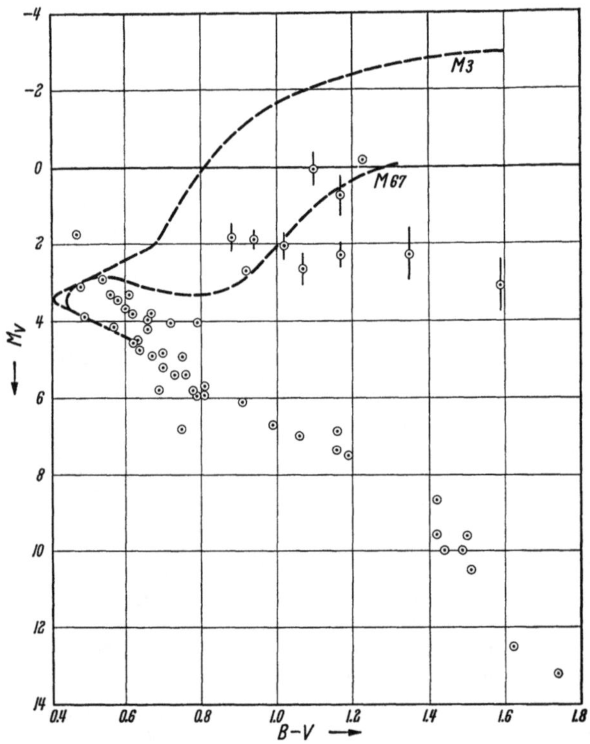

Fig. 44. Color-absolute magnitude diagram of high velocity stars. All high velocity stars with trigonometric parallaxes greater than 0".050. All high velocity stars brighter than $M_v = +4$ with trigonometric parallaxes greater than 0".030. Values mainly from catalogue by N. G. Roman. Probable errors indicated for giants.

were designated as population II, principally because they contained no *O* and *B* stars. But we now know that this can be simply an age effect and may not necessarily have anything to do with the chemical composition of a group of stars.

Now Popper, Schwarzschild, Chamberlain and Aller, the Burbidges and many others have shown that among the high velocity stars, there occur stars whose spectra show a weakening of CN absorption. Schwarzschild *et al.*[1] show that in the case of the high velocity star φ_2 Ori, the observations can be explained by a general reduction of a normal metal content by a factor of four. The rest of the sample of high velocity giants which he investigated, however, indicated lesser or no reduction in metal content. Against this we may weigh Aller and Chamberlain's estimate of a factor of ten reduction in the metal content of globular cluster stars [*48*].

[1] Martin and Barbara Schwarzschild, L. Searle and A. Meltzer: Astrophys. Journ. **125**, 123 (1957).

The general result seems to be that numerically, the majority of stars with space velocities greater than 63 km/sec define nearly a M67-type color-magnitude diagram and have metal contents which average not much lower than normal stars. In other words they are on the average, quite close to a high metal content, type I population. Among the high velocity stars, however, there are some stars (indications are that they have high z velocities bringing them from the halo regions of the globular clusters), which are indicated to be extreme type II stars. That is, they have very low metal content as indicated by their spectra and ultra-violet colors and fall on the type II sequences in the H-R diagram (for example, the subdwarfs). Therefore, there is in the group called high velocity stars, examples of all gradations between normal and very low metal content population types. These stars are relatively near the Sun and they therefore offer the best examples in which to study the range in population types.

Since the high velocity stars are high velocity because they enter the solar neighborhood from other parts of the galaxy their study will be fruitful in another respect. Namely, if it is possible to separate the high velocity stars which come from the central regions of our galaxy from those which come from the halo regions, it will be possible to tell more about the stellar content of the galaxy. At the present time, taking the fact that most of the high velocity stars have velocity vectors lying quite close to the plane of the galaxy, we might guess that most of the high velocity stars come from the nuclear regions of the galaxy. This, taken together with Fig. 44, would indicate that the stars toward the nucleus from our Sun, were predominantly a high metal content, nearly type I population, but one as old as M67. Recent photoelectric spectrum scans of the nucleus of the Andromeda Nebula by A.D. CODE[1], and the discovery of a group of nuclear globular clusters by W.W. MORGAN[2] which seem to have normal metal content, may offer some additional evidence that the nuclear regions of the galaxy are composed approximately of such stars.

Bibliography.

Historical and general.

[1] LUNDMARK, K.: Handbuch der Astrophysik, Vol. 5, part. 5, 1—210ff. Berlin: 1932. — An extremely detailed account of stellar astronomy from its earliest beginnings through to about 1930.

[2] SHAPLEY, H.: Star Clusters. New York-Toronto-London: McGraw-Hill Book Company 1930. — Shapley presents his own researches and sums up the information available at the time on clusters, variable stars and distance scales.

[3] HYNEK, J. A. (Editor): Astrophysics — A Topical Symposium. New York-Toronto-London: McGraw-Hill Book Company 1951. — An historical discussion by B. STRÖMGREN of developments of astrophysics during the last half centure (pp. 1—11). Classification of Stellar Spectra by P. C. KEENAN and W. W. MORGAN (pp. 12—28). Interpretation of Normal Stellar Spectra by L. H. ALLER (pp. 29—84).

[4] BEER, A. (Editor): Vistas in Astronomy, Vol. 1. London and New York: Pergamon Press 1955. — Discussion of the history of special fields in astronomy, particularly in Sects. 2 and 5.

[5] WHIPPLE, F. L. (Editor): New Horizons in Astronomy. Smithsonian Contributions to Astrophysics, Vol. 1, No. 1. 1956. — This Volume discusses the presently challenging problems in astronomy and advancements in instruments and techniques which would be desirable in order to effect their solution.

Spectroscopy and photometry.

[6] MORGAN, KEENAN and KELLMAN: An Atlas of Stellar Spectra. Chicago, Ill.: Chicago University Press 1943. — The definition of the MKK system.

[1] A. D. CODE: Carnegie Institution of Washington Yearbook, Mt. Wilson Annual Report 1955—1956.

[2] W. W. MORGAN: Publ. Astronom. Soc. Pacific **68**, 509 (1956).

[7] JOHNSON, H. L.: Astrophys. Journ. **116**, 272 (1952). — Exclusion of ultraviolet light from photographic (B) magnitudes.
[8] JOHNSON, H. L., and W. W. MORGAN: Astrophys. Journ. **117**, 313 (1953). — The presentation of the fundamental standards of the B, V system.
[9] KUIPER, G. P.: Astrophys. Journ. **88**, 429 (1938). — The fundamental work on the stellar temperature scale and bolometric corrections.
[10] UNDERHILL, A. B., and J. K. MCDONALD: Astrophys. Journ. **115**, 577 (1952). — Bolometric corrections and temperatures for early-type stars.
[11] STEBBINS, J., and A. E. WHITFORD: Astrophys. Journ. **102**, 318 (1945). — Six color photometry of stars.

Galactic clusters (Most recent work).

[12] JOHNSON, H. L., and W. W. MORGAN: Astrophys. Journ. **114**, 522 (1951). — Pleiades.
[13] JOHNSON, H. L.: Astrophys. Journ. **116**, 640 (1952). — Praesepe.
[14] JOHNSON, H. L., and W. W. MORGAN: Astrophys. Journ. **117**, 313 (1953). — Pleiades and others.
[15] JOHNSON, H. L.: Astrophys. Journ. **117**, 357 (1953). — NGC 752.
[16] JOHNSON, H. L.: Astrophys. Journ. **119**, 185 (1954). — M34.
[17] COX, A. N.: Astrophys. Journ. **119**, 195 (1954). — M41.
[18] SHARPLESS. S.: Astrophys. Journ. **119**, 200 (1954). — Orion.
[19] JOHNSON, H. L., and A. R. SANDAGE: Astrophys. Journ. **121**, 616 (1955). — M67.
[20] JOHNSON, H. L., and C. F. KNUCKLES: Astrophys. Journ. **122**, 209 (1955). — Hyades and Coma Berenices.
[21] JOHNSON, H. L., and W. W. MORGAN: Astrophys. Journ. **122**, 429 (1955). — h and χ Persei.
[22] JOHNSON, H. L., and W. A. HILTNER: Astrophys. Journ. **123**, 267 (1956). — h and χ Persei and NGC 2362.
[23] JOHNSON, H. L., A. R. SANDAGE and H. D. WAHLQUIST: Astrophys. Journ. **124**, 81 (1956). — M11.
[24] JOHNSON, H. L., and O. HECKMANN: Astrophys. Journ. **124**, 477 (1956). — Hyades.
[25] WALKER, M. F.: Astrophys. Journ. Suppl. **1956**, No. 23. — NGC 2264.
[26] IRWIN, J. B.: Proceedings of the National Science Foundation at Charlottesville, Virginia, April 1956. — NGC 6087 and M25.
[27] SANDAGE, A. R.: Astrophys. Journ. **125**, 435 (1957). — Red giants in the H-R diagram.

Initial stellar evolution.

[28] SCHÖNBERG, M., and S. CHANDRASEKHAR: Astrophys. Journ. **96**, 161 (1942). — The basic idea of initial evolution from the main sequence.
[29] SANDAGE, A. R., and M. SCHWARZSCHILD: Astrophys. Journ. **116**, 463 (1952). — The first extension of the theory past the 12% limit.
[30] SANDAGE, A. R.: Liège Symposium Volume, Les Processus nucléaires dans les astres, p. 254, 1954. — The application of the theory to globular clusters.

Globular clusters.

[31] HACKENBERG, O.: Z. Astrophys. **18**, 49 (1939). — One of the first modern color-magnitude diagrams of a globular cluster (M 92).
[32] SCHWARZSCHILD, M.: Circ. Harv. Coll. Obs. **1940**, No. 437. — First definition of RR Lyrae gap appears (M 3).
[33] ARP, H. C., W. A. BAUM and A. R. SANDAGE: Astronom. J. **58**, 4 (1953). — First color-magnitude diagram in which main sequence is identified (M 92).
[34] BAUM, W. A.: Astronom. J. **57**, 222 (1952). — Color displacement between M 3 and M 92-spectrum differences between M 3 and M 92.
[35] SANDAGE, A. R.: Astronom. J. **58**, 61 (1953). — Precision photometry down to $M_v = +6$ mag (M3).
[36] BAUM, W. A.: Astronom. J. **59**, 422 (1954). — MainSequence in M13.
[37] ARP, H. C.: Astronom. J. **60**, 317 (1955). — Color-magnitude diagrams for seven globular clusters above $M_v = +1$.
[38] ARP, H. C., and H. L. JOHNSON: Astrophys. Journ. **122**, 171 (1955). — Ultraviolet measures on M13.
[39] JOHNSON, H. L., and A. R. SANDAGE: Astrophys. Journ. **134**, 379 (1956). — Checking M3 and adding ultraviolet colors.

Variables in the H-R diagram.

[40] WILSON, R. E., and P. W. MERRILL: Astrophys. Journ. **95**, 248 (1942). — Date on motions, absolute magnitudes for long period variables.

[41] CODE, A. D.: Astrophys. Journ. **106**, 309 (1947). — Spectral classification of classical cepheids on MKK system.

[42] JOY, A. H.: Astrophys. Journ. **110**, 105 (1949). — Spectral types of variables in globular clusters (non-RR Lyrae variables).

[43] ROSINO, L.: Astrophys. Journ. **113**, 60 (1951). — Spectra of RV Tauri and semi-regular variables on the MKK system.

[44] EGGEN, O. J.: Astrophys. Journ. **113**, 367 (1951). — Extensive observations of light curves of classical cepheids.

[45] JOY, A. H.: Astrophys. Journ. **115**, 25 (1952). — Analysis of RV Tauri and Semi-regular variables in high and low velocity groups.

[46] ARP, H. C.: Astronom. J. **60**, 1 (1955). — Two-color observations of most of the cepheids available in globular clusters.

Zero points and population types.

[47] BAADE, W.: Astrophys. Journ. **100**, 137 (1944). — Original definition of population I and II.

[48] BAADE, W.: Michigan Symposium on Astrophysics. 1953. — Extensive discussion of population types and distance scales.

Stellar Evolution.

By

G. R. Burbidge and E. Margaret Burbidge.

With 32 Figures.

General introduction.

It can fairly be said that today the ideas of stellar evolution pervade practically all of the multitudinous branches of modern astrophysics. Therefore there are many ways in which a review article on such a subject could be attempted. Some might consider that the number of topics to be treated should be severely restricted. Others would wish to omit all topics which have less than an extremely sound base both in observation and in theoretical explanation. However, we have decided to be less selective in our discussion, if only so that we can indicate the large number of problems and approaches in this field which remain to be explored.

The plan of the article has been based on the idea that stellar evolution is the description of the birth, life and death of stars. Theoretical ideas concerning all of these stages will be described. However, because of the time scales involved observations of single stars must necessarily be restricted to a very short span of a star's evolution. To investigate by observation the whole evolutionary process it is necessary to turn to studies of clusters of stars which are coeval. A field of study which is only in its infancy at the present time is the evolution of galaxies taken as a whole

The theory of stellar evolution is intimately tied to the sources of nuclear energy in stars, so much so that the shortcomings of the early ideas were due in large part to their having been proposed long before the sources of energy were understood. Thus a brief historical sketch of early developments is given. The rapid advance of nuclear astrophysics in recent years has allowed us to outline the stages of chemical evolution in the star's structure.

In order adequately to discuss many topics it has been necessary to encroach to a limited extent on to the preserves of other authors of this Encyclopedia. In particular, some discussion of stellar structure and stellar models, of the H-R diagrams of clusters, and of the chemical compositions of stars has been required.

In A we discuss all of the evolutionary stages in the lives of individual stars. Here both theory and observation have been described together.

B is devoted to the facts of evolution which can be deduced from the observations of *groups* of stars. These range in age from the youngest galactic clusters and associations to the oldest globular clusters, and in numbers and assorted types of stars from the associations up to galaxies taken as a whole. Stellar populations are also briefly discussed.

In C we consider the processes of exchange of mass between stars and the interstellar medium which may affect or determine the course of a star's evolution.

D contains a discussion of the chemical evolution of stars.

In E, rotation, magnetism, and variability are discussed in as far as they affect stellar evolution.

A. Theory and observation of the evolution of individual stars.

I. Formation of stars.

1. Introduction. The theory of star formation is in a very incomplete state at the time that this review is being written. However, we shall outline the general ideas which have been put forward. Stars are fundamental constituents of galaxies and fit into a hierarchy of cosmical bodies, all of which may have condensed by essentially similar processes. Thus starting from the largest and most massive systems we have:

(i) Hyperclusters of galaxies, i.e., clusters of clusters of galaxies;
(ii) Clusters of galaxies;
(iii) Galaxies;
(iv) Star clusters;
(v) Stars.

From the observational standpoint only the existence of the first of these categories remains in some dispute. For example, ZWICKY[1] has claimed that the finite velocity of propagation of gravity $(=c)$ gives a natural limit to the size of systems and that this is not much greater than the dimensions of the greatest clusters so far discovered. However, if they do exist it is probable from estimates of their dimensions that they are the largest aggregates of matter which can exist in the universe. The next aggregate would be the universe as a whole.

Such a hierarchy of structure sizes might suggest a series of eddy sizes in a primordial turbulent gas. GAMOW [*1*] has proposed that the distribution of galaxies and clusters of galaxies reflects the initial conditions of turbulence in the primordial gas, and he has related these conditions to a particular model of the universe. However, the theory of von WEIZSÄCKER [*2*], which is based on general turbulence arguments for galaxies of stars, but does not discuss star formation in detail, and that of HOYLE [*3*], which relates the formation of galaxies and stars together, with no turbulent effects directly considered, are not tied to any particular cosmological models.

To begin condensation in a structureless initial gas mass it is generally argued that by some means an element of gas obtains a higher local density than its surroundings. Then, provided that it has not been considerably heated during this process, it will tend to contract under its own gravity, and if the Jeans criterion is satisfied in the most favorable conditions when its contraction is isothermal, it will ultimately form a protocluster of galaxies, a protogalaxy, protocluster of stars, or protostar. All of the theories which have been directed towards mechanisms of star formation have been based on this argument, and most of the discussion of this section is devoted to the consideration of the way in which an initial compression can be obtained, and then ways in which the fragments may eventually reach stellar dimensions. However, before we proceed to discuss these mechanisms, it is of interest to consider a general criticism of the basic idea which has been given by LAYZER [*4*]. The argument runs as follows.

Consider an element of gas in a protogalaxy. If it is in equilibrium its internal pressure is balanced in *part* by internal forces and in part by external forces due to the rest of the protogalaxy. The criterion for the relative importance of internal gravitation and the external forces is the relative magnitude of the internal kinetic energy and the internal potential energy, the internal gravitation dominating if these are of the same order. It is then argued that the ratio of

---

[1] F. ZWICKY: Proc. Third Berkeley Symposium on Mathematical Statistics and Probability, **3**, 113, 1956.

these two is given by $(R/r)^2 \alpha (\varrho_0/\varrho_c)^{\frac{1}{3}}$, where ϱ_c/ϱ_0 measures the extent of the compression, R/r measures the ratio of the dimensions of the whole system and the initially contracting element, and α is a measure of the fraction of the kinetic energy of the whole galaxy in the form of heat. LAYZER shows that for appropriate values of these quantities the ratio may be large. He thus concludes that the condensation will initially be at the mercy of the external forces and may be disrupted. It is of interest to consider the rate at which the tidal disrupting force will act, as compared with the rate at which the mass will contract under its own gravitation. The time taken for the high-density element to contract to a radius $r/2$ is

$$t_0 = (G\varrho_c)^{-\frac{1}{2}}. \tag{1.1}$$

On the other hand, the time taken for the whole galactic mass to move the high-density element through a distance $r/2$ is given by

$$t_1 = (G\varrho_0)^{-\frac{1}{2}} \left(\frac{r}{R}\right)^{\frac{1}{2}}. \tag{1.2}$$

Thus if

$$t_0 < t_1,$$

$$\frac{\varrho_c}{\varrho_0} > \frac{R}{r},$$

or if $r = 10^{-4} R$, $\varrho_c > 10^4 \varrho_0$. However, this is a more stringent condition than that for the condensation not to be destroyed by tidal disruption. For a spherical element, the gravitational acceleration at the far end of the diameter in the direction of the galactic center is $GM/(R+r)^2$, while that at the near end is $GM/(R-r)^2$. Thus the time taken for a gram of material at the far end to move a distance $\frac{r}{2}$ is $\left\{r\left(R+\frac{r}{2}\right)^2 \Big/ GM\right\}^{\frac{1}{2}} \approx t_1$, where M is the mass of the protogalaxy. Let us suppose that in this time a gram at the other end of the diameter moves a distance $x/2$.

To a first approximation the tidal distortion is then given by

$$\frac{1}{2}(x-r) = \frac{r}{2}\left\{\frac{(R+r/2)^2}{(R-r/2)^2} - 1\right\} = \frac{r^2 R}{(R-r/2)^2} \approx r\left(\frac{r}{R}\right). \tag{1.3}$$

Thus the approximate condition for the condensation not to be affected by tidal distortion is that $r/R \ll 1$, a condition which is always fulfilled for condensations within a galaxy.

Apart from this theoretical argument, we have incontrovertible evidence that stars are currently being born in our Galaxy; i.e., they are being formed in a locale in which the gravitational potential has remained essentially unchanged since the disk was formed, so that tidal disruption is a negligible effect, less important than the effect of galactic shear and rotation.

Arising from this general criticism, LAYZER has proposed an alternative theory, the hypothesis of gravitational clustering based on an idea of ISAAC NEWTON. This is essentially the inverse of the hierarchal system beginning with the largest aggregates and breaking them down into smaller and smaller groups. It is supposed that there are slight local irregularities in an initial distribution of matter. As the universe expands, these become more and more pronounced until finally self-gravitating systems are separated out. These then act as a second kind of density fluctuation and after further expansion they also form self-gravitating systems, and so on. Thus it is proposed that the initial building blocks are stars and planets which are formed at an early epoch in the expansion.

This ties the formation of stars to an evolutionary cosmological model. Further, it does not provide any explanation for the birth of the many generations of stars which are appreciably younger than the galaxies in which they lie. The details of star formation are not discussed.

2. Protostar formation in the absence of stars and dust. HOYLE [3] has considered an extragalactic cloud of atomic hydrogen with a density $\sim 10^{-27}$ g/cm³, and he has shown that in its early evolution a dichotomy of temperatures is to be expected, as follows. The heat energy is supposed to have been derived from the kinetic energy of the mass motions developed in the process of formation of this cloud. Excluding motions less than ~ 10 km/sec means that temperatures less than about 10^4 degrees are excluded. Thus the minimum energy required to heat the gas is $\sim 10^{12}$ ergs/g. Hydrogen becomes collisionally ionized at temperatures higher than 10000 degrees, so that even without radiation, hydrogen changes from being largely neutral to nearly completely ionized in the range 10000 to 25000 degrees, and the energy required to ionize is $> 10^{13}$ ergs/g. The time of formation of the cloud is $\sim (G\varrho)^{-\frac{1}{2}}$ seconds, which for the value of ϱ assumed is about 10^{17} seconds, and in this time the rate of loss of energy due to escaping radiation amounts to $\sim 3 \times 10^{13}$ ergs/g at 15000 degrees and $\sim 1.4 \times 10^{14}$ ergs/g at 25000 degrees. Hence the following stages can be distinguished, if we suppose that the hydrogen is heated up from very low temperatures:

(i) energy is absorbed in heating the neutral gas,
(ii) energy is absorbed in ionizing it, and then
(iii) energy is radiated by the ionized gas.

Now for stages (i) and (ii) about 10^{13} ergs/g is required. By the time that stage (iii) is reached, the energy supply must amount to $\sim 1.4 \times 10^{14}$ ergs/g. Since the radiation decreases with increasing temperature, the extra energy to heat the gas to $\sim 10^5$ degrees is only the difference in thermal energy between 25000 degrees and 10^5 degrees, and thus the plot of temperature against energy supply must be very flat in this region. When temperatures exceeding 3×10^5 degrees are reached, the energy required to heat the hydrogen begins to dominate again. Thus broadly speaking, the initial temperature of a condensing cloud will be in excess of about 300000 degrees. It is these two situations which must be considered in describing the condensation of galaxies and stars. The low temperature case (corresponding to initial mass motions of ~ 30 km/sec) is the one to be discussed first.

For a non-magnetic mass in mechanical equilibrium the virial theorem gives the relation between the thermal energy, U, gravitational energy, Ω, and the kinetic energy of mass motions, K,

$$2K + 3(\gamma - 1)\, U + \Omega = 0,$$

and the condition that the mass will contract is that the left-hand side is negative. Now in this model the mass motions have already been dissipated in heating the hydrogen, and if we put $\gamma = \frac{5}{3}$, the condition for contraction for a mass M of radius R is that approximately

$$\frac{GM^2}{R} > 3RMT$$

or

$$\varrho > 3 \times 10^{-21}\, T^3 \left(\frac{M_\odot}{M}\right)^2. \tag{2.1}$$

For the situation in which $\varrho = 10^{-27}$ g/cm³, and $T = 1.5 \times 10^4$ degrees (corresponding to the situation in which the hydrogen is about $\frac{2}{3}$ ionized), we find that the critical mass for a condensation is given by

$$M_c = 1.4 \times 10^{10} M_\odot . \tag{2.2}$$

No cloud less massive than this can condense to form a galaxy. If we now consider the condensation of a cloud with mass $> M_c$, we can consider two types of compression, isothermal and adiabatic. Under adiabatic conditions all of the gravitational energy is converted to thermal energy, so that if U_0 and Ω_0 are the initial values of U and Ω, $\delta U = \delta \Omega$ for each contraction step, and $U - U_0 = \Omega - \Omega_0$. For equilibrium $\Omega = 2U$, so that adiabatic compression leads to a rise in the thermal energy until hydrostatic equilibrium is reached, a situation which obtains for stellar masses.

In an isothermal contraction in which the dimension is decreased by a factor x, the gravitational energy released is of the order of $GM^2(1-x)/R$ while the thermal energy generated by direct compression is of the order of $3\,MRT \log_e x$, and to maintain isothermal conditions this energy must be removed from the gas. In order that the gravitational energy is dissipated in this way these two quantities must be nearly equal, and this demands that the condensation must be only just unstable against contraction, and it must not contract too far.

If $GM^2/R \gg 2U$, i.e., if very little of the gravitational energy is converted to heat, this gravitational energy must be dissipated into dynamical motions, and the dynamical pressures thus exerted must be almost sufficient (apart from the heat and viscous dissipation) to re-expand the cloud to its initial dimensions. Thus the condition that the gravitational potential energy greatly exceeds the thermal energy at equilibrium leads to the condition that there will be no permanent contraction of the cloud as a whole. However, within the cloud, density fluctuations will occur, which do satisfy the condition $GM^2/R \approx 2U$. These regions, therefore, can contract by factors of 2 or 3, forming sub-condensations within the cloud, and within these, further condensations can take place. A large cloud can in this way fragment very rapidly. Since at each stage the density will increase by a factor of x^3, and since the time scale of the contraction is of the order of $t_0 = (G\varrho_0)^{-\frac{1}{2}}$, the time required for a large number of stages is given by

$$t = t_0\left(1 + x^{-\frac{3}{2}} + x^{-3} + x^{-\frac{9}{2}} + \cdots\right) = t_0\left(1 - x^{\frac{3}{2}}\right)^{-1}. \tag{2.3}$$

If $x = 3$, the total time will increase over that for the initial step by only about 20%. The important conclusion derived from this result is that the condensation of galaxies, and the condensation of the first stars within them, are inextricably mixed together, as far as the initial time scale and the early dynamical effects are concerned. With the initial conditions set by HOYLE, this time is of the order of 10^9 years for galaxies with $M > 1.4 \times 10^{10} M_\odot$.

The fragmentation process must cease when the density of the object has reached such a point that the isothermal condition fails; the opacity becomes so high that the radiation is not able to escape freely. To determine the mass of the system at this point we can proceed as follows.

The opacity of the fragment arises partly from scattering and partly from absorption, and in the present case the scattering is unimportant. Absorption becomes important when the absorption in the radiation field becomes comparable with the energy radiated by the gas. Thus the transition to the adiabatic condition implies that the black-body distribution has been built up. The fragment then becomes similar to a star, radiating at a rate L, and for the transition we require that $d\Omega/dt > L$. The gravitational energy released per fragment in the

n-th step of the fragmentation is given by

$$\frac{G M_0^2}{R_0 x^{3n/2}} (x - 1),$$

the release occupying a time of the order of $x^{-3n/2} (G\varrho_0)^{-\frac{1}{2}}$, and the rate at which energy is radiated is given by $L x^{-3n/2} (G\varrho)^{-\frac{1}{2}}$. Therefore the transition comes when

$$L = \frac{G^{\frac{3}{2}} M_0^2 \varrho_0^{\frac{1}{2}}}{R_0}. \tag{2.4}$$

The outward flux at distance r from the center is given by

$$\frac{16\pi a c r^2 T^3}{3\varkappa\varrho} \frac{dT}{dr}. \tag{2.5}$$

If the fragments are composed of pure hydrogen $\varkappa$ becomes extremely small for temperatures below about 5000 degrees, since even in the presence of blackbody radiation the hydrogen would be predominantly neutral. Then the luminosity $L = 4\pi a c R^2 T^4$ at the surface. Thus for the plausible values of the initial mass of the cloud from which this fragmentation takes place, it is possible to calculate the final masses of the protostars if they have condensed from pure hydrogen. However, it is important to consider the effects when a small proportion of the mass is composed of non-hydrogenic material, since the opacity is considerably changed by such an admixture, because the low ionization potentials of the metals provide enough electrons at the temperature range 3000 to 5000 degrees for the formation of H^-. In this case HOYLE has argued as follows. The expression for the outward flux becomes, using a value of $\varkappa$ obtained from the tables of CHANDRASEKHAR and MÜNCH[1] and assuming $T = 4500°$,

$$L = \frac{160\pi a c R_0 T_0^3}{32 y \varrho_0^2 R} x^{-3n} \tag{2.6}$$

and equating this with the value when the adiabatic condition is reached, it is found that

$$x^{3n} = \frac{160\pi a c G^{\frac{1}{2}}}{27 y R^3} \frac{T_0}{\varrho_0^{\frac{3}{2}}} \left(\frac{G M_0}{R_0} \approx 3 R T_0\right) \tag{2.7}$$

where y is the number ratio of metals to hydrogen. Thus if $\varrho_0 = 10^{-27}$ g/cm³ and $T_0 = 10^4$ degrees, this gives

$$x^{3n/2} = 2.3 \times 10^9. \tag{2.8}$$

The mass of the fragments when this final stage is reached is given by $M_s = M_0/x^{3n/2}$, and if $M_0 = 3.6 \times 10^9 M_\odot$, $M_s \approx 1.6 M_\odot$. On the other hand, we have for a pure hydrogen star that

$$L = 4\pi a c T^4 R^2 = \frac{G^{\frac{3}{2}} M_0^2 \varrho_0^{\frac{1}{2}}}{R_0}. \tag{2.9}$$

Putting

$$R = R_0 x^{-3n/2}, \quad \text{and} \quad R_0^3 = \frac{3 M_0}{4\pi\varrho_0},$$

and substituting, we find that

$$x^{-3n} = \frac{G^{\frac{3}{2}} M_0 \varrho_0^{\frac{3}{2}}}{3 a c T^4}. \tag{2.10}$$

Thus if $M_0 = 3.6 \times 10^9 M_\odot$, $\varrho_0 = 10^{-27}$ g/cm³ and $T = 5 \times 10^3$ degrees,

$$x^{3n} = 1.10 \times 10^{20}, \quad \text{and} \quad x^{3n/2} = 1.05 \times 10^{10}.$$

[1] S. CHANDRASEKHAR and G. MÜNCH: Astrophys. J. **104**, 446 (1946).

Thus the masses of pure hydrogen protostars, M_s, are given by

$$M_s = \frac{3.6\times 10^9 M_\odot}{1.05\times 10^{10}} = 0.34 M_\odot . \tag{2.11}$$

These results are of considerable interest, but the numerical values that we obtained are not of especial significance. The following facts must be borne in mind. The critical masses which have been used by Hoyle have been derived for plausible values of the initial density and temperature, and have led to values of the masses of old stars which may be reasonable. However, in many cases the masses of elliptical systems are much greater than $3.6\times 10^9 M_\odot$. For example, the average masses of galaxies in the Virgo cluster are near $10^{11} M_\odot$, and these are mainly elliptical systems, and if their stars are completely analogous to the globular cluster stars, their average masses are near $1.5 M_\odot$. To reconcile these results with those derived above, it is necessary that there be a considerable degree of fragmentation taking place in an initial cloud of $\sim 10^{11} M_\odot$, which does not result in the production of a cluster of masses of $\sim 10^8 M_\odot$ but more likely a series of fragments that remain bound to each other, and then the stars which condense during the further fragmentation lose their identity with respect to a single fragment. Hoyle has treated the case of more massive systems ($\sim 5\times 10^{11} M_\odot$) in which he supposes that initially the high-temperature condition ($T \gtrsim 3\times 10^5$ degrees) held. In these cases he has shown that a large number of dwarf systems with masses $\sim 10^8 M_\odot$ will be formed. (In these, the later fragmentation stages are similar to those above.) He has supposed that the later dynamical evolution of these will lead to shrinkage, disintegration, and the escape of some systems. These might be identified with some clusters that have been discovered in our local group of galaxies, at least two of which have color-magnitude diagrams similar to that of a globular cluster [*5*], [*6*].

Furthermore, it is also probable that in these condensation processes, much gas will be left over, since the fragments will only condense in regions having higher than average densities. Thus this gas may condense under suitable dynamical conditions at a later epoch, and, for example, the presence of magnetic fields may also slow up this further condensation [*7*]. Neutral atomic hydrogen has been detected through 21-cm observations in both the Coma, Hercules, and Corona Borealis clusters of galaxies [*8*]; of these the Coma and Corona clusters are highly condensed and consist predominently of *S* 0 systems. This might well be primeval hydrogen contributing a considerable proportion (perhaps $\sim 20\%$ in the Coma cluster) of the total mass. In some galaxies of the local group a considerable amount of gas, comparable in mass to the stars, is present. Part of this may be a remnant of the original cloud. However, in general the fraction of the mass contained in gas within galaxies and perhaps within clusters as well is an index of the distance travelled along the evolutionary path of the galaxy or the cluster as a whole (cf. Chapter B V, p. 225).

This treatment has not taken into account the type of condensation which may occur if a fairly uniform contraction occurs with the development of dynamical motions and the turbulent dissipation of these motions. The effect of the contraction may well be to amplify the turbulence so that if sufficiently large density fluctuations were set up, the fragmentation and subsequent contraction would take place through them. This point demands quantitative investigation.

Another alternative is that the condensation becomes adiabatic at an earlier stage than has been suggested by Hoyle, due to the presence of molecules which absorb strongly in the infra-red. Then the condensation will be very slow and fragmentation may not occur appreciably. Other sources of opacity which have been neglected are the excitation of neutral hydrogen giving rise to the 21-cm radio radiation, and the extra ionization produced by cosmic ray particles (whose effect is hard to estimate at an early stage of galactic evolution).

These factors might lead to the condensation of stars, even those made of pure hydrogen, having masses far larger than those derived above. If all pure hydrogen stars are of very small masses considerable problems arise when the chemical evolution of the galaxy is considered (cf. Chap. D, p. 249).

In this treatment the angular momentum of a protogalaxy and the magnetic field which may exist in the initial cloud have both been neglected. Both of these must be important at all stages of the fragmentation process and also in determining the final mass beyond which fragmentation does not occur. The angular momentum problem is so severe (cf. Sects. 4, 5) that apparently it can only be resolved by supposing that the final protostars are members of bound multiple systems, which are later dissolved by tidal forces. Magnetic forces are important in that they will tend to halt contraction if the coupling between the magnetic field and the neutral gas by means of the plasma component is strong enough. The strength of the coupling is determined by the strength of the initial magnetic field and the degree of ionization of the gas. We shall return to this question in Sect. 6.

The final stage of the hierarchy described by HOYLE may be called a protostar. If we take as an example the final mass of $1.6 M_\odot$, the central density is $\varrho_0 x^{3n}$, which for the value $x^{3n} = 5.3 \times 10^{18}$ gives a central density of 5.3×10^{-9} g/cm³, while the central temperature is approximately 10^4 degrees. The radius of the protostar is about 5×10^{13} cm. This is very far from the conditions that exist when a star moves on to the main sequence. Some attempt at describing these intermediate conditions has been made by McVITTIE. However, before discussing this we shall consider the formation of stars when dust is already present, and then when other stars are also present.

3. Protostar formation through the concentration of dust. The existence of very young stars in *O*, *B* associations etc. shows that star formation must be currently going on in the spiral arms of our Galaxy. Attempts have been made to describe the effects which the dust and the stars have on the dynamical and temperature conditions in the interstellar gas, which may lead to the formation of new stars, or the rejuvenation of older stars.

Dust is a widespread constituent of the interstellar medium in the spiral arms in our own and other galaxies and it has been shown by LINDBLAD [*9*], TER HAAR [*10*], and KRAMERS and TER HAAR [*11*] that in a gaseous medium most of the atoms other than hydrogen and helium will tend to stick together, forming first molecules and then more complicated assemblies, i.e., dust grains will form. The only condition, therefore, for the formation of dust is that the galaxy has evolved sufficiently so that heavy elements, particularly the metals, are present. When dust is present the inelastic collisions between neutral hydrogen atoms and the grains are powerful cooling agents for the interstellar medium. The equilibrium temperature in the spiral arms where the dust is plentiful is $\leqq 100$ degrees. In regions where the temperature is below average and the density is greater, dust grains form and the temperature is reduced even further. Since the gas pressure tends to remain constant as the temperature decreases, the density will be further increased, and the rate of formation of dust grains will become even greater. These arguments, originally due to SPITZER [*12*], [*13*], suggest that the growth of dust is rapidly accelerated in limited regions, thus eventually forming what SPITZER has described as a "diffuse cloud", an idealized concept, since it is possible that the dust grains start to concentrate together before most of the atoms have condensed into grains.

When dust grains reach the size at which they might begin to absorb radiation effectively, different grains in the galactic plane will be forced towards each other by the radiation pressure of the general galactic field. This process as a basis for star formation was first proposed by SPITZER [12] and WHIPPLE [14]. An idealized situation obtains when we consider concentration of the dust in directions parallel to the galactic plane. The diffusion velocity V_d of a dust grain at distance R from the center of a diffuse cloud is given by WHIPPLE's equation

$$V_d = \frac{2 m_d g}{\pi a^2 n_H m_H V_H} = -\frac{dR}{dt}, \tag{3.1}$$

where m_d and a are the mass and radius of the grain, and where n_H, m_H and V_H are the number density, mass, and root mean square velocity of the neutral hydrogen atoms. The acceleration of a grain at distance R from the center of a diffuse cloud is given by

$$g = \frac{\pi^2 a^4 N_d (1-\gamma) Q^2 U R}{3 m_d},$$

where N_d is the number density of dust inside the cloud minus that outside the cloud, γ is the albedo, U is the energy density of radiation due to stars, and Q is the ratio of the effective scattering area of the particle to its geometrical cross section. If n_d is the number density of dust outside the cloud we can put

$$N_d = \frac{K}{R^2} - n_d.$$

Substituting in (3.1) for g and N_d and solving the resulting differential equation, we obtain

$$\left(\frac{R}{R_0}\right)^2 = 1 + \frac{N_d(0)}{n_d}\left(1 - e^{t/t_r}\right),$$

where $N_d(0) = N_d$ at $t=0$ and $R = R_0$. Here

$$t_r = \frac{3 n_H m_H V_H}{4U(1-\gamma) Q \pi a^2} = \frac{3}{2}\tau_r, \tag{3.2}$$

where τ_r is the time for the mass of the concentration to increase by a factor e. If t_c is the time at which R vanishes, which has been called the time of concentration, then

$$\frac{t_c}{t_r} = \log_e\left(1 + \frac{n_d}{N_d(0)}\right). \tag{3.3}$$

This equation shows the characteristics of any exponential build-up in that, for example, if $n_d = N_d(0)$, $t_c = 0.7 t_r$, whereas if $0.1 n_d = N_d(0)$, t_c is only increased to $2.4 t_r$. Under reasonable conditions ($T_H \approx 100$ degrees, $n_H = 1$, $U = 5.2 \times 10^{-13}$ erg/cm³, $(1-\gamma) Q = 1$, $\pi a^2 Q n_d = 3 \times 10^{-22}$), SPITZER and WHIPPLE have found that for $N_d(0)/n_d = 0.58$ [i.e., if the right-hand side in Eq. (3.3) becomes unity], $t_c = 3.9 \times 10^7$ years.

This is a time short compared with the age of the Galaxy, and modifications in the numbers due to new estimates of the interstellar density and temperature made since 1946 will not increase the time appreciably. It is a time just short enough for appreciable concentration to take place before a cloud is seriously distorted by galactic rotation. SPITZER has shown that effective limits on the size of the concentration, defined by R_0, can be made. If R_0 is too small V_d will be less than the thermal velocities of the grains, in which case only a slight concentration will be produced. The partial pressure of the grains will balance the radiative force.

Thus

$$\frac{R_0}{t_c} > \frac{3\,kT}{m_d},$$

and if $T = 100°$ K, and $a = 10^{-5}$ cm, $t_c = 3.9 \times 10^7$ years, then $R_0 \gg 10^{-3}$ psc. This value of 10^{-3} psc is an extreme lower limit, and SPITZER has concluded that the size of regions which can remain sufficiently cool for dust concentration for $\sim 10^7$ years must be considerably larger than this, so that concentrations with initial radii not much less than a parsec will arise.

On the other hand, if R_0 is too great the high inward velocities of the dust grains become so great that the gas is heated by collisions. Further, if the contraction velocity is greater than the gas kinetic velocity, the gas will be dragged inward and the concentration of the grains will be slowed down. Thus

$$\frac{R_0}{t_c} < \frac{3\,kT}{m_H},$$

or, for the values used by SPITZER, $R_0 < 64$ psc. This is of the same order as the radius of a spiral arm in gas and dust.

For sizes greater than this, the simple theory would break down and the effect might be that the initial concentration would begin to fragment. An upper limit on the density of the dust is about 2×10^{-26} g/cm³; this corresponds to the situation in which all of the elements that can be concentrated into grains are so ($\sim 2\%$ of the total mass in hydrogen gas with a mean density of 10^{-24} g/cm³), and in which it is assumed that the chemical composition of the gas and stars is the same. Thus the total mass of dust which can be concentrated in a single condensation lies in the range between 10^{-3} and $200\,M_\odot$.

This theory of dust concentration will only hold up to the point at which other factors, notably the opacity of the cloud and gravitational contraction, become important. Thus different methods must be used when the dust has been concentrated by a factor of about a hundred, so that the dust and gas densities become comparable. At this point both dust and gas begin to move together and continue to contract in this way until gravitational forces take over. In the case in which the gas and dust are mixed together, the usual Jeans condition for the gravitational instability can be used provided that the effective gravitational constant G_{eff} is used. If f_r/f_g is the ratio between the radiative and gravitational attraction of the two grains, and a fraction q of the matter is in the form of dust, then

$$G_{\text{eff}} = G\left(1 + q^2 \frac{f_r}{f_g}\right).$$

The usual condition for contraction in the absence of dust is that

$$\frac{G M_0}{R_0} \gtrsim 3\,RT.$$

Now it is found that the relation between the wavelength for instability in the presence and in the absence of dust is

$$L^2 = L_0^2 \frac{(1-q)^2}{(1+q^2 f_r/f_g)}.$$

Thus since

$$L_0^2 = 4\left(\frac{\pi}{3}\right)^2 R_0^2,$$

$$R^2 = R_0^2 \frac{(1-q)^2}{(1+q^2 f_r/f_g)}. \tag{3.4}$$

Therefore the condition for instability in the presence of dust becomes

$$\frac{GM(1+q^2 f_r/f_g)^{\frac{1}{2}}}{R_0(1-q)} \geq 3\,RT. \tag{3.5}$$

Since the critical mass $M \propto R^3$, and $M_0 \propto R_0^3$, this can be written

$$\frac{GM_0}{R_0}\,\frac{(1-q)^2}{(1+q^2 f_r/f_g)} \geq 3\,RT. \tag{3.6}$$

$T = 100°$ K and $\varrho_H = 1.6\times 10^{-24}$ g/cm³, the condition for contraction in the absence of dust is that $M_0 \geqq 10^5 M_\odot$. However, if dust is present we may follow SPITZER by putting $f_r/f_g = 1.4\times 10^4$ for icegrains with radii 10^{-5} cm. Then if $q = 0.1$, 0.5 and 0.9, $(1-q)\,(1+q^2 f_r/f_g)^{-\frac{1}{2}} = 7.6\times 10^{-2}$, 8.5×10^{-3}, and 9.4×10^{-4} respectively, and the critical masses become 580, 7.2, and $8.8\times 10^{-2} M_\odot$ respectively.

Thus as q increases a large cloud becomes unstable, and begins to contract as a whole, while local concentrations of dust will rapidly lead to condensation within the cloud, and a kind of fragmentation will proceed. The time for the protostar to contract to half its dimensions (i.e. an increase in density of order 10) is given by $(\varrho G_{\text{eff}})^{-\frac{1}{2}}$ which, for $q = 0.5$ and $\varrho = 3\times 10^{-24}$ g/cm³, is about 10^6 years. Contraction beyond a density of 10^{-23} g/cm³ through radiative gravitational processes is not possible, since at this point the gas pressure becomes comparable with the radiative pressure. Beyond this point the gravitational contraction process will take over, and the rate of contraction will be slowed up. However, it may be supposed that beyond this point the contraction will take place and some fragmentation may occur along the lines discussed by HOYLE, the breaking down into the fragments only ceasing when the isothermal contraction goes over to the adiabatic condition and the protostar becomes highly opaque.

This process of protostar formation leads naturally to a chemical composition which is more characteristic of the dust than of the interstellar medium as a whole. Since there are probably not extreme differences in composition between the youngest stars and the interstellar medium out of which they form, this suggests that if the protostars form out of a nucleus of dust they are later able to attract more gas which forms the bulk of the star. We shall later discuss the theory of accretion in this connection. The alternative to this explanation is to try and provide a theory which will give the necessary degree of compression to the gas and dust of the interstellar medium to form a protostar, without concentrating the dust alone.

Despite this difficulty it does appear that there is considerable concentration of dust relative to gas in some dark nebulae. Thus it is usually the case that $\varrho_d(\text{compressed})/\varrho_d(\text{original}) > \varrho_g(\text{compressed})/\varrho_g(\text{original})$.

This process of dust concentration will fail if the effect of the radiation pressure on the grains is very much less than that used in this treatment, and there is some evidence that the grains do have a very high albedo. The albedo depends both on the absorption and scattering and on whether the grain is metallic or dielectric. Thus instead of putting $(1-\gamma)\,Q = 1$ as was done originally, we might put $(1-\gamma)\,Q \approx 0.01$. Under these conditions t_c becomes about 3.9×10^9 years, so that the whole of the later discussion is invalidated, because any concentration will be dispersed by the effects of galactic shear long before any appreciable density increase is attained. However, at the present stage, our knowledge of grain physics is still very inadequate, and we have felt that a discussion of the older work in which all of the parameters favored rapid concentrations is not out of place.

4. Protostar formation through the direct influence of other stars. In a later section we shall describe the observations of globules which may be protostars in the process of formation. Bok originally proposed that these are dust concentrations, but more recently Oort [*15*] has suggested that they are mainly composed of gas with only a small admixture of dust. In the latter case the concentration may have been produced by a mechanism of protostar formation proposed by Biermann and Schlüter [*16*] and Oort and Spitzer [*17*].

The Biermann-Schlüter and Oort-Spitzer mechanism presupposes that a hot O-type star is initially present, associated with a complex of interstellar clouds. The theory is incomplete in that the presence of the O-type star is treated as an observational fact, and no theory concerning its origin is proposed.

It is supposed that the O-star is suddenly born in the middle of a cloud complex, an unreal picture, since the star will take $\sim 10^5$ years to contract on to the main sequence, but one which leads to a simple model. The sequence of events will then be as follows. The newly-born star will start to ionize the hydrogen surrounding it, and it is easily shown that the velocity of the ionization front is so rapid (at least one order of magnitude larger than the velocity of sound in the medium) that it can be said to ionize instantly up to approximately the radius of the Strömgren sphere for the star. (This rapid almost explosive ionization is a consequence of the initial assumption). Provided that the complex is large as compared with the Strömgren sphere, then the temperature inside the Strömgren sphere will be about 10000 degrees, which must be compared with the value of about 100 degrees outside it, and the inner pressure will be increased accordingly. Consequently the hot inner mass will expand into the cool shell.

Thus a compression region will proceed into the cool shell with a velocity of the order of that of sound in the hot gas (~ 10 km/sec). If the rate of radiation is sufficient to keep the temperature of the compressed region near to that in the cool shell, then the density will be increased by a factor ~ 100. The compression front will continue to expand outward after it has reached the outer edge of the complex. Its velocity of expansion is determined both by the rocket effect of the matter ionized from the inner surface of the shell and the pressure in the inner sphere, while the acceleration finally ceases through the braking action of the interstellar gas outside the cloud complex. Oort and Spitzer have calculated the minimum total masses which will escape complete evaporation through this process, and they amount to $\sim 10^3$ to $10^5 M_\odot$ for exciting stars ranging from $B0$ to $O5$. Since the cloud complexes are highly irregular it can be seen that in some cases the material will be completely ionized to its periphery, while in others masses of compressed neutral hydrogen will be ejected, to form separate clouds. This mechanism was developed primarily to explain the generation of motions of the interstellar clouds, and the velocities which are obtained are shown to be in reasonable agreement with the observed velocities of clouds.

For the condensation of protostars the critical masses from the Jeans criterion can be calculated. If we suppose that initially the mean densities lie in the range between 3.3×10^{-23} and 1.6×10^{-22} g/cm³ ($n_H = 20$ to 100), that the compression factor is 100, and that $T = 100$ degrees, then the critical masses for gravitational contraction are ~ 900 to $400 M_\odot$ while the radii (the condensations are supposed to be spherical) are ~ 1 psc. This simple argument may not be applicable, however, since the initial clouds are probably very far from being spherical (see the later discussion in this section). If these critical masses have any validity, it appears that the globules observed in the vicinity of a number of nebulae must be further advanced in protostar evolution than is the case when the clouds first begin to condense, since the majority of them are probably less massive, more dense, and certainly much smaller than they were at this stage. Thus several stages of fragmentation must again take place before stellar masses are achieved.

This process of protostar formation will give rise to an expanding system of stars, and it has been suggested that it is responsible for the birth of expanding *O* and *B* associations whose stellar velocity vectors have been shown in a number of cases to be traceable to a very small initial volume of space (cf. Sect. 36).

The formation of the luminous *O* star which starts this whole process has not been explained, and it is not impossible that an alternative explanation of the whole birth process may be forthcoming. Perhaps a more rapid, or even explosive, mechanism may be worth investigation. The very large amounts of kinetic energy which are found to be present in such complexes as the Orion nebula, amounting to some 10^{50} ergs [*18*], are reminiscent of the energies released in supernova outbursts. Thus ÖPIK [*19*] has suggested that the great arc in this region which amounts to some $10^5 M_\odot$ and is expanding with a velocity of ~ 10 km/sec is the result of a supernova outburst in which the supernova shell collided with the surrounding quiescent interstellar medium, and that a compression region has been driven outward. The momentum balance demands an initial mass for the supernova shell of at least $100 M_\odot$ if an ejection velocity as great as 10^4 km/sec is assumed. This is a very large supernova mass, and it has led SAVEDOFF [*18*] to reject the hypothesis and to explain the energy as being due to *O*-star formation. However, our knowledge of the range of supernova masses and velocities is very limited. If an explosion were responsible for this feature, then it is probable that the compression region was the place of origin of the high-velocity stars AE Aurigae, μ Columbae, and 53 Arietis which have apparently come from this vicinity (cf. Sect. 36).

As has just been described, the Biermann-Schlüter-Oort-Spitzer mechanism leads to a compression of cool gas by surrounding hot gas, i.e. a pressure is exerted on the cool gas at its boundary. The gravitational instability of a gas cloud under an external pressure at its boundary has recently been studied by EBERT [*19a*], BONNOR [*19b*], and MCCREA [*19c*]. It has been shown by these authors that the general form of the virial can be written

$$2U + \Omega - 3pV = 0,$$

where p is the pressure exerted over the boundary, and V is the volume enclosed by the boundary. If Ω is neglected this is simply BOYLE's law. If, on the other hand, we can neglect the external pressure, then we have the usual statement that for equilibrium

$$2U + \Omega = 0.$$

The system will contract if

$$2U + \Omega - 3pV < 0.$$

Since, by definition, the net pressure on the boundary is everywhere inward, it is clear that the term involving this pressure will make it more easy for the cloud to contract. This has been shown in some detail by EBERT who has given exact solutions for a sphere using EMDEN's equation, and by MCCREA using more approximate arguments. A convenient way of showing the result has been given by McCREA as follows. For a uniform sphere the critical mass above which the system will collapse is given by

$$M_c = 8.91\, T^2 \left(\frac{p}{k}\right)^{-\frac{1}{2}} M_\odot .$$

This must be compared with the critical mass in the absence of an external pressure, which is

$$M_c = 41.1\, T^{\frac{3}{2}}\, n_H^{-\frac{1}{2}}\, M_\odot .$$

The corresponding critical radii in the two cases are

$$R_c = 2.85\, T \left(\frac{p}{k}\right)^{-\frac{1}{2}} \text{psc}$$

and

$$R_c = 6.4\, T^{\frac{1}{2}}\, n_H^{-\frac{1}{2}}\, \text{psc}.$$

Some numerical examples have been described by these authors. The effect of an external pressure on the gravitational instability of a cylinder and a plane stratified layer has also been considered. The results show that in the latter case gravitational collapse will not occur in the way that it does for a spherical distribution of matter. This has led McCREA to conclude that if stars are formed out of material which originally had elongated form, or which lies in sheets, it must first break up through forces exerted by differential motions, or because of density fluctuations, so that the fragments have all of their dimensions comparable before gravitational collapse can occur.

5. Star formation through the concentration of pre-stellar nuclei. KRAT [*20*] has argued that the process of star formation consists of the concentration of many pre-stellar bodies into a larger body. Collisions between these pre-stellar nuclei which are envisaged as dark bodies with planetary structure (masses $\sim 10^{23}$ g) give rise to gas and dust which then condenses through dynamical instability. He has proposed that such processes are going on continually in star clusters which contain a large supply of invisible planetary material.

These pre-stellar nuclei have also been proposed by UREY [*21*] in considering the origin of the solar system. He has suggested that objects of considerable mass and size accumulate in a dust cloud at low temperature. No detailed theory of the way in which dust grains will be accreted into solid objects of large mass has yet been proposed. Once solid objects, perhaps of lunar size, have been formed, it is proposed that these will tend to be attracted together, and will then proceed rapidly to accrete gas and dust in the cloud (cf. Sect. 69). The collisions of these pre-stellar nuclei—the attraction may lead to an implosion—would probably lead to fragmentation and then recondensation into a gaseous star.

These ideas have further been discussed by HUANG [*22*]. He has supposed that all double and multiple stars are formed from pre-stellar nuclei, thus overcoming the angular momentum difficulties (which are discussed in Sect. 9). He has also supposed that no sharp distinction between the formation of multiple stellar systems, and single stars together with planetary systems, can be made, a point of view which has been put forward by KUIPER (Sect. 10) and by STRUVE [*23*].

HUANG has noted that a difficulty associated with the formation of pre-stellar nuclei is that if they are formed by random collisions between dust grains, the time scale for their formation is of the order of 10^9 years. He has therefore proposed that their growth will be accelerated if large dust concentrations can be formed by dynamical means, and has suggested that such concentrations will be formed in vortices in the boundary layers between colliding clouds. That the appropriate boundary layer conditions are formed in shearing collisions between dust clouds seems unlikely. These arguments are better translated into terms which apply more directly to the state of dust clouds. These will have large Reynolds numbers and we can suppose that the conditions are those of compressible turbulence. Consequently some compression in collisions between clouds, or the normal density fluctuations to be expected in a turbulent cloud, may lead to some acceleration in the dust concentration. The time scale for such processes remains unknown.

6. The influence of turbulence on protostar formation. Von WEIZSÄCKER [2] has considered protostar formation in the framework of a general cosmogony in which it is supposed that turbulence plays the dominant role. Thus it is supposed that the initial localized compression of the gas out of which the condensation begins is a density fluctuation characteristic of the turbulent gas. No detailed arguments for star formation in the general field have been put forward, but the importance of developing such a cosmogony using a theory of compressible hydromagnetic turbulence, since this will be more characteristic of astrophysical conditions, must be stressed. VON WEIZSÄCKER has considered in some detail the sequence of events in which the protostar out of which the solar system formed was developed. This and the later work based on these turbulence arguments will be described when we consider the evolution of the sun (Sect. 32; see also Sect. 10).

7. The mass-distribution function in star formation. None of the theories of star formation have been able to predict the distribution of masses among the protostars. However, by using the observed luminosity functions for different clusters and for the solar-neighborhood stars it has been possible to determine this function for stars *after they have arrived* on the main sequence. This problem is discussed in Sect. 54. It is found that the mass distribution function has the form

$$\xi(M) = k\,M^{-1.35}.$$

The $M^{-1.35}$ dependence is approximately followed for all of the clusters in which the function has been determined and also for the solar-neighborhood stars. It may well be universal. The value of k is different for different clusters, and is determined by the star density in the cluster. If it were possible to trace the evolution of the protostars back from the main sequence through their gravitational contraction stages and beyond in an unambiguous way, this observed mass-distribution might be expected to give valuable information concerning the process of star formation. However, a number of complicating factors are apparently at work.

As we shall show in Sect. 43, arguments may be advanced which suggest that considerable fragmentation takes place at a fairly early stage of the gravitational contraction of a protostar. Also, some astronomers believe that protostars form with their masses lying only within two narrow mass ranges, via the O-associations and the T-associations. In this case the mass-distribution function of stars on the main sequence has little to do with the original mechanism of star formation. However, on any other hypothesis the observed mass-distribution function on the main sequence must be intimately related to the process of star formation, and the constancy of the index -1.35 (perhaps better expressed as constant between the limits of -1.3 and -1.4) suggests that the same processes have occurred in clusters of many different ages.

8. Evidence for stars in the process of formation. Observational evidence for star formation has come from several directions. Objects which may be protostars at an early stage are the globules described by BOK and his associates, and the Herbig-Haro objects. Later stages of a star's contraction towards the main sequence may be represented by the T Tauri stars.

A number of small round dense nebulae or globules have been detected [*24*], [*25*], and it has been suggested by BOK that these are protostars being formed by the concentration of dust as we have described in the previous section. A number of these have been found in selected regions, particularly in the Rosette Nebula in Monoceros, part of which is shown in Fig. 1 (reproduction of a Palomar 48-inch Schmidt plate), the Southern Coalsack itself, and others. BOK has used the measures of the total photographic absorptions and dimensions of known obscuring clouds to estimate the total masses and densities of some of the globules. He has used radii of the dust grains of 10^{-5} cm. For three globules, including the Coalsack, he has made the following estimates:

	Globule I	Globule II	Coalsack
Diameter (parsec)	0.06	0.5	8
Density (g/cm³)	1.3×10^{-21}	5×10^{-23}	3×10^{-24}
Minimum mass ($M_\odot$)	0.002	0.05	13

It has further been argued that these globules are probably still growing, mainly because of the large effective gravity $G_{\text{eff}} \approx 50G$, so that, for example, the Coalsack could double its mass in a further 3×10^7 years. It is of considerable interest that some globules have been discovered in Messier 8, and these are very close in space to the young cluster NGC 6530 where the stars are mostly still contracting towards the main sequence and where T Tauri stars are found (cf. Sect. 42). It is not impossible, therefore, that this is a region where all stages of star formation are going on together.

It should be emphasized that the densities of the globules given by BOK are lower limits, and very much higher densities and thus larger total masses are entirely possible. Also, in some nebulosities there are large numbers of globules, very much smaller than those originally described by BOK, and there is probably a continuous sequence of sizes going down at least to the limit of resolution of the 200-inch telescope.

A recent study by BOK[1] by means of star counts in dust clouds in the Ophiuchus and Taurus regions has shown that very dense dust concentrations are present. For example, in Ophiuchus a dark cloud with a minimum total mass $\sim 30\,M_\odot$ has been found. The mean dust density in this cloud is about 300 times the average dust density in the interstellar medium. Also, 21-cm observations in some of the large complexes associated with these regions have shown that together with the dust concentrations there are often large volume concentrations of neutral atomic hydrogen[2], so that the total masses may be ~ 10 times the mass of the dust. However these concentrations have been achieved, it is clear that they are regions in which star formation on a large scale can be expected to occur.

---

[1] B. J. BOK: Astronom. J. **61**, 309 (1956).
[2] A. E. LILLEY: Astrophys. J. **121**, 559 (1955).

The Herbig-Haro objects are small knots of emission nebulosity, found in regions of the sky where heavy obscuration is present [26], [27]. For example, many of these objects are found in the region of the Orion Nebula. They often

Fig. 1. Part of the Rosette Nebula in Monoceros (NGC 2237) (48-inch Palomar Schmidt), showing small dark globules.

have a star-like nucleus. Their spectra show forbidden lines of [O I], [O II], and [S II], as well as hydrogen emission lines. The continuum radiation between the emission lines is about a factor 10 less strong than the radiation in the emission lines. That the Herbig-Haro objects are related to the T Tauri stars (cf. Sect. 35) is indicated by the fact that both occur in dense regions of interstellar matter, and also by the presence of the same forbidden emission lines in the nebulosity around T Tauri as appear in the Herbig-Haro objects. AMBARTZUMIAN [28] has suggested that the Herbig-Haro objects may represent an earlier evolutionary stage of the T Tauri stars.

A remarkable event has been described by HERBIG [29]. Direct photographs of a Herbig-Haro object in Orion taken in 1954 showed two star-like nuclei

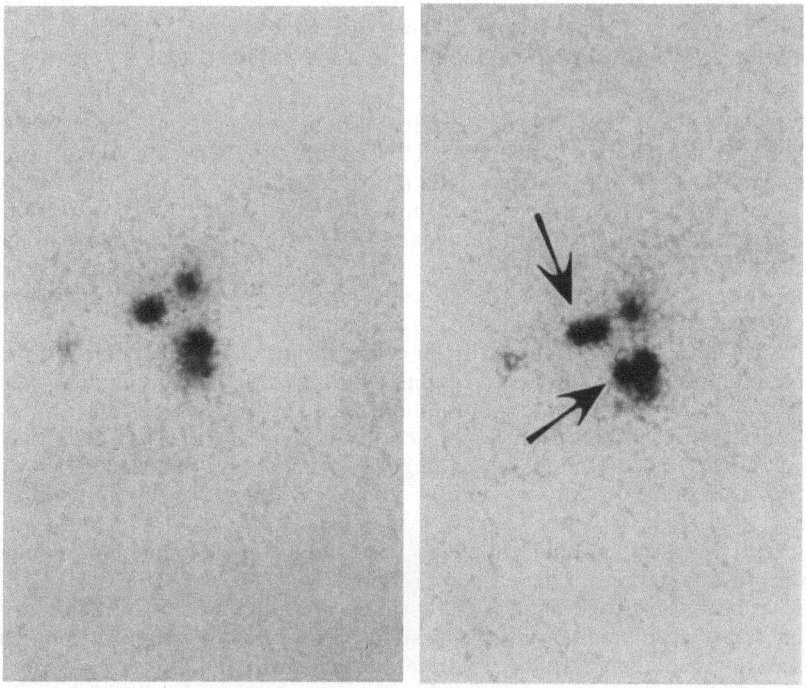

a b

Fig. 2a and b. Direct photographs of the Object Haro 12a = Herbig 2, in Orion. Te plates where obtained on 20 January 1947 and 20 December 1954 with the Crossley reflector (Lick Observatory) and blue-sensitive emulsions. North is at the top and east at the right. The scale can be obtained from the fact that the two brightest stars on the 1947 photograph are 8″ apart [29]. By courtesy: G. H. HERBIG.

which were not present on photographs taken in 1946 and 1947. This may perhaps be an observation of the birth of a star. This event is shown in Fig. 2. A difficulty associated with this interpretation is the rapidity with which the luminosity of the protostar appears to have grown and the condensation changed in 7 years, as compared with the rate of gravitational contraction.

FESSENKOV [30] has proposed that stars are currently being born in gas filaments. According to these ideas chains of stars like beads on a string are formed, the string being the filament of gaseous nebulosity. FESSENKOV and ROSHKOVSKY [30] have described observational evidence for a number of star chains in the Network Nebula in Cygnus, in NGC 6960, 6992 and 6995. The mean density along such filaments of nebulosity has been estimated by them to be about 10^{-19} g/cm^3. Now if we suppose that a protostar has already formed, the condition for another condensation to contract near to it is that the tidal force due to the nearby protostar does not disrupt it in the time necessary for it to contract, and this is a function of the distance apart of the two condensations and the density of the material. It has been argued that a separation of ~ 0.1 psc between the stars in the chain is consistent with the mean density of 10^{-19} g/cm³. The very existence of real *chains* of stars means that they must be very young since the galactic tidal forces will very rapidly disrupt such a configuration. However, considerable doubt has been cast on this interpretation of the observations. For example, the work of CHAMBERLAIN[1] gives mean densities in the Network Nebula of 10^{-21} g/cm³. Moreover, Mount Wilson photographs of some of the same regions do not show many of the star chains [*31*]. There is also the problem as to whether an apparently regular configuration of stars can be explained by accidental groupings. Thus the observations must be treated with reservation.

9. The removal of angular momentum. Whatever the initial state of the condensation, it is highly probable that it will contain a large amount of angular momentum, so that as it contracts the protostar will have a very high rotation. It is clear that as the contraction continues the effect of the angular momentum will be to halt contraction perpendicular to the axis of rotation and lead to the formation of a disk which is unstable in its equatorial region. This situation will develop rather rapidly and certainly will occur before the final fragmentation processes within the condensation have taken place.

The magnitude of this effect of the conservation of angular momentum in a contracting protostar is illustrated by the following example. The angular velocity of rotation of our galaxy amounts to about 0.04 km/sec parsec. Thus if a stellar mass initially contracted from a gas cloud with a radius of 2 parsecs into a star with a radius $5\,R_\odot$ without loss of angular momentum, the surface rotational velocity of the star would amount to about 10^7 km/sec, an entirely ridiculous result.

Several types of mechanism which will lead to loss of angular momentum from protostars and stars have been proposed.

It has been suggested that a fragmentation process in which a large proportion of the stars form bound multiple systems, in which the angular momentum becomes angular momentum of relative motion, might be envisaged. The high proportion of stars which are contained in binary systems supports this hypothesis. The best argument in favor of this idea is that it appears to be the only one which is capable of explaining the magnitude of the effect. It will be discussed further in Sect. 10. However, some other ideas have been proposed.

It has been suggested [*32*] that the effect of compression in interacting gas clouds may reduce the angular momentum rapidly. Thus two clouds with the same rotational sense will, in a collision, form a thin compressed layer which will try to rotate with an angular velocity of the same order as that of the clouds, while currents will flow in opposite directions on the two surfaces of the layer, and will probably vanish long before the collision is completed. The angular momentum in the layer can be reduced by a factor of the order of $(\varrho_c/\varrho)^{\frac{2}{3}}$ where ϱ_c and ϱ are the compressed layer density and the cloud density, respectively.

Another mechanism which has been proposed is that of non-magnetic braking due to the interaction of the surrounding interstellar medium on the star [*33*], [*34*]. This mechanism has been closely related to the formation of protostars in a turbulent medium in which a fast rotating protostar surrounded by a large rotating envelope is formed. It is then supposed that the interaction of a gaseous envelope on the central mass will be such that the central mass will lose angular momentum because of the drag of the envelope. This angular momentum will

---

[1] J. W. CHAMBERLAIN: Astrophys. J. **117**, 399 (1953).

be transferred to the envelope, and it will be transported away as the envelope is dissipated. Unfortunately it appears that this is a very weak mechanism under almost all astrophysical conditions, since it can transport away only a very small fraction (estimated by TER HAAR to be $\lesssim 10^{-4}$) of the total initial angular momentum.

A plausible mechanism which has been worked out in some detail is that of the magnetic braking of the rotation of a central mass by a surrounding ionized cloud [*35*], [*34*], [*36*]. A star containing a magnetic field will ionize the surrounding medium and will then, through the ionized cloud, be magnetically coupled to the interstellar gas; it will then be slowed down by this coupling. The rate of slowing will depend both on the initial magnetic field assumed for the star, and on the extent of the Strömgren sphere, i.e., on the luminosity of the star. LÜST and SCHLÜTER [*36*] have shown that, if a young star possesses a surface magnetic field of the order of 100 gauss, the effect of this braking is to reduce its rotation from the condition of limiting stability to negligible values in times $\lesssim 10^6$ years. However, the large amount of angular momentum which the protostar must lose even to reach this condition of limiting stability is not readily explicable in this way. At the very low temperatures which it must have if it originally condensed out of a gas and dust concentration, there will be very little ionization, and the coupling between the concentration and the surrounding medium will be very weak. Thus the condition for contraction in a magnetic field, which will be discussed in Sect. 11, and the conditions demanded for angular momentum to be lost by magnetic braking are, to a first approximation, mutually exclusive.

10. The formation of binary systems. We have previously pointed out in Sect. 9 that the problems associated with the angular momentum of contracting protostars are so severe that it appears that the solution lies in the idea that stars are formed in general in multiple systems which take up the net angular momentum of the initial clouds. The theories of HOYLE and SPITZER and others have to a large extent ignored the angular momentum problem, and the fact that the angular momentum may be an important factor in limiting the condensation also has not been discussed.

Until quite recently it was believed that the fission theory of binary star formation based on the work of DARWIN and JEANS might still be tenable [*37*]. However, a formidable array of arguments are now available which show that this theory cannot be correct. These are based both on the observational data [*38*], [*39*], on arguments relating to stellar structure and evolution, and on dynamical grounds directly refuting the arguments of JEANS [*40*].

Consequently the way appears to be open for the development of a theory based on the observations and theory described earlier in this article (Sects. 2 to 8) but taking into account the angular momentum problem encountered in the formation of single stars. Such an approach has been described by KUIPER [*39*], who has proposed that multiple stars are formed by independent condensations in random positions within protostars. He has pointed out that a general model might be expected to explain the two principal statistical properties of binary stars; i.e., the distribution of semi-major axes, a, and the distribution of mass ratios between the two components. The first is equivalent to the frequency distribution of the total angular momentum present in the original protostars, after allowance is made for components subtracted from multiple stars and from those wide binaries that are destroyed by stellar encounters. The second is just the mass division in the condensations. This, for main sequence stars, is apparently a random function, which is what might be expected. Whether or not this is consistent with the Salpeter mass function discussed in Sect. 54 remains to be tested.

The transfer of angular momentum in encounters between systems at an early stage of evolution has been considered by HUANG and STRUVE [41]. They have suggested that such an exchange might lead to a distribution of angular momentum over a coeval group of stars such that all will have values higher or lower, on the average, than the corresponding values of rotation for stars in the Galaxy as a whole.

KUIPER has considered that the interior region of the protostar, which he conceives as a type of globule similar to that described in Sect. 8, will be highly turbulent, since the Reynolds number is very high. The angular momentum of the system is then contained in the eddies, the major part in the largest outer eddies. Then, using resonable assumptions, it is possible to obtain a frequency distribution function for $\log_e a$, which is of the form

$$F(\log_e a) = \frac{2}{\sqrt{\pi}}\, x^3\, e^{-x^2},$$

where $x = A/A_p = \alpha (a/a_1)^{\frac{1}{2}}$, $a_1 = 1$ a.u., and α is a pure number, which is a function of β. The initial peripheral velocity is βv_c, v_c being the free circular velocity under attraction by the protostar. It is clearly very difficult to estimate β. KUIPER has shown that $0 < \beta \ll 1$; since the protostar does not expand and dissolve, $\beta \ll 1$. A is the total angular momentum and A_p is its most probable value. KUIPER has arbitrarily chosen $\beta = 0.1$. Comparison of this formula with the observational results then shows that the observed dispersion is higher than the computed one, i.e., there are more close and more wide binaries than are predicted theoretically. Variation in β alters the position of the maximum but not the dispersion. For $\beta = 0.1$, the maximum falls close to $\operatorname{Log} a = 1.4$, or $a = 25$ a.u. KUIPER has given a number of plausible arguments which explain the sense of the departure of theory from observation.

Some evolutionary effects are to be expected for all those binary systems in which the components are quite close (perhaps with $a \leqq 1$ a.u.). This will be discussed further in Sect. 71.

KUIPER's arguments are compatible with the observations of regularities among multiple stars. Thus the formation of three condensations will lead to instability and the ejection of one of them, unless two of the systems are very close together as compared with the third. Similarly, a quadruple system will reduce to a triple or a binary, unless the mutual distances are arranged in a hierarchy. Only in associations which have positive total energies, and are therefore dispersing, are such situations not obtained.

11. Protostar formation in the presence of a magnetic field. It has been shown that the Jeans criterion for gravitational instability is unaffected both by the presence of Coriolis force, or by the presence of a magnetic field, or by both acting together, [42], [43]. There is strong evidence for the existence of a galactic magnetic field with an average strength of about 10^{-5} gauss, and doubts have been expressed as to whether an initial instability can finally lead to a condensation of stellar dimensions in the presence of so much magnetic energy. The problem has been discussed by MESTEL and SPITZER [44]. In the presence of a magnetic field the virial condition is written

$$2U + \mathfrak{M} + \Omega = 0,$$

where

$$\mathfrak{M} = \frac{1}{8\pi} \int H^2 \, dv$$

is the magnetic energy. For a uniform sphere with magnetic field H,

$$\mathfrak{M} = \tfrac{1}{6} H^2 R^3 .$$

Thus the critical mass above which a cloud is able to condense (neglecting U) is given by

$$M \geq \frac{H^3}{(6G)^{\frac{3}{2}}} \left(\frac{3}{4\pi\varrho}\right)^2 . \tag{11.1}$$

Thus if $H = 7 \times 10^{-6}$ gauss and $\varrho = 1.6 \times 10^{-23}$ g/cm³,

$$M \geq 1.7 \times 10^5 M_\odot ,$$

or if $H = 10^{-6}$ gauss (somewhat less than its probable galactic strength), $M \geq 500 M_\odot$. This means that, as long as the magnetic field is frozen into the material, a mass smaller than this cannot fragment into stars. For a mass greater than this limit, contraction will lead to amplification of the magnetic field following the law

$$\frac{H}{H_0} = \frac{R_0^2}{R^2} ,$$

so that the magnetic energy increases at a rate proportional to R^{-1}; i.e., it has the same dependence as the gravitational energy, so that it dilutes gravity by a factor independent of the density.

This argument neglects the fact that the material can contract quite normally along the magnetic lines of force where the magnetic pressure is zero. This might suggest that in a spiral arm, in which we suppose that the magnetic field lines lie parallel to the axis of the "cylinder", the condensation along the arms will proceed more rapidly than transverse contraction, and only when the unstable element has formed a spheroid oblate to the direction of the field such that the effective gravity along the field is equal to the effective gravity (normal gravity minus magnetic impedance) across the field, normal condensation could take place.

However, the critical length given by the Jeans criterion for the contraction of a cylinder is

$$L \geq \left(\frac{\pi R T}{G \varrho}\right)^{\frac{1}{2}} ,$$

while the condition for the lateral instability is that

$$R > \left(\frac{2 R T}{\pi G \varrho}\right)^{\frac{1}{2}} ,$$

so that

$$L_c \approx R_c \frac{\pi}{\sqrt{2}} .$$

But to allow condensation without magnetic impedance we used condensation of a cylinder of volume $\pi R^2 L$, where $L \gg R$. Thus an initial length very much greater than the critical length is required, and, since this would be highly unstable against further break-up, this mechanism of protostar formation will not occur.

It is necessary, therefore, to consider in more detail whether the coupling between the magnetic field and the matter is so strong in the interstellar medium, that the conditions described above are completely valid. MESTEL and SPITZER have written down the equations of motion for the plasma and for the neutral gas and dust. These equations for the plasma are

$$\frac{\boldsymbol{j}\times\boldsymbol{H}}{c} + (n_i F_{iH} + n_e F_{eH}) - \nabla(p_i + p_e) + n_i m_i \nabla\varphi = n_i m_i \frac{d\boldsymbol{V}_i}{dt} \tag{11.2}$$

and for the natural gas

$$-(n_i F_{iH} + n_e F_{eH}) - \nabla p_H + n_H m_H \nabla\varphi = n_H m_H \frac{d\boldsymbol{V}_H}{dt}. \tag{11.3}$$

Here n_H, n_e, n_i are the number densities of neutral hydrogen, electrons, and ions, respectively, F_{eH} and F_{iH} are the mean forces on electrons and ions due to collisions between them and neutral atoms and molecules, p_H, p_i, and p_e are the partial pressures due to the neutral hydrogen, ions, and electrons, φ is the gravitational potential, and $\boldsymbol{V}_i$ and $\boldsymbol{V}_H$ are the mean drift velocities of the ions and neutral atoms. These equations show that the neutral gas is coupled to the magnetic field only through the friction represented by the term $(n_i F_{iH} + n_e F_{eH})$ between the neutral gas and the plasma.

Now in a normal interstellar cloud where $T \approx 100$ degrees, the density of the plasma is very small as compared with the neutral gas, so that the terms involving the partial pressure and gravitation of the plasma are negligible, and as a cloud contracts the tendency of the magnetic field to straighten out is counteracted by this frictional force. For initial conditions in which the cloud mass $= 10^3\, M_\odot$, $H_0 = 10^{-6}$ gauss, and $n_H(0) = 10$, and in which isotropic compression has occurred, it is found that the friction balances the magnetic force if

$$|\boldsymbol{V}_H - \boldsymbol{V}_i| \approx 1.5\times 10^3\, n_i^{-1}. \tag{11.4}$$

Since $V_H \approx 10^5$ cm/sec, the drift of the plasma relative to the neutral gas is small provided that $n_i \gg 1.5\times 10^{-2}$. Now if $n_i/n_H = 10^{-4}$, since $n_H = 2\times 10^4$ after compression, then $n_i = 2$, and the normal proportion of plasma is too high for any sensible drift to take place during the time (2.1×10^{13} secs) of free fall; i.e., $|\boldsymbol{V}_H - \boldsymbol{V}_i| \approx V_H$.

However, if we consider that the cloud is dense enough for the starlight to be absorbed by dust grains at its periphery, then the amount of ionization decays because of the capture of ions and electrons by the dust grains. The time constant for the decay has been estimated by MESTEL and SPITZER to be about 1.4×10^{12} secs, i.e., it is small compared with the time of free fall. Thus under the most favorable conditions the ion density can fall to a sufficiently low value so that the magnetic field can move the plasma through the infalling neutral gas, and so the field is not compressed by the contraction.

The condition that the magnetic force just balances the friction is rapidly reached and thereafter maintained, and as the cloud contracts the magnetic field tends to straighten itself out as it drags the plasma through the gas. If the situation is stabilized at a point in which starlight producing ionization balances the decay due to collisions with dust, then the magnetic field which remains in the condensation will be rather greater than its initial value. Very little of the magnetic energy released becomes converted to kinetic energy of mass motions.

Thus the condensation of a protostar from a cloud containing a magnetic field can take place if the conditions are such that the coupling between the magnetic field and gas as a whole is weak. However, increase in the degree of ionization, i.e., heating of the gas cloud, makes the condensation much more difficult. Though this is of no interest as far as HII regions in our Galaxy are concerned, since it is not believed that stars are formed in these regions, it is of some importance in considering the situation when the oldest stars condensed.

If we consider the initial conditions similar to those described earlier for condensation and fragmentation of a pure gas cloud (Sect. 2), we can put $\varrho_0 = 10^{-27}$ g/cm³ and $T = 25000$ degrees, so that in this case all of the hydrogen is ionized. The condition that the mass can contract is then that

$$\frac{G M^2}{R} > \frac{H^2 R^3}{6} + 3 R M T. \tag{11.5}$$

Thus, provided that the mass already fulfils the condition for gravitational contraction in the absence of a magnetic field, i.e. $M \gtrsim 10^{10} M_\odot$, it will contract in the presence of a magnetic field provided that

$$\left.\begin{aligned} &H_0 < \left(\frac{32G}{3}\right)^{\frac{1}{2}} (\pi R_0 \varrho_0) \\ \text{or} \quad & \\ &H_0 < 4.5 \times 10^{-7} \text{ gauss.} \end{aligned}\right\} \tag{11.6}$$

However, let us consider the situation if the mass of gas fragments. After n fragmentation processes, the mass of a fragment is $M_0/x^{3n/2}$, the radius of the fragment is $R_0/x^{3n/2}$, and the density is $\varrho_0 x^{3n}$, and the condition for contraction and break-up of the fragment is that

$$H_0 < \left(\frac{32G}{3}\right)^{\frac{1}{2}} (\pi R_0 \varrho_0)\, x^{-3n/2}. \tag{11.7}$$

To reach stellar masses it has been shown earlier that $x^{3n/2} \approx 10^9$, while to reach masses very much greater than stellar size, so that further fragmentation must be required, means that $x^{3n/2} \gtrsim 10^7$. Therefore the condition on the initial magnetic field in order that fragmentation down to masses of the order of $10^2 M_\odot$ is allowed is that

$$H_0 < 5 \times 10^{-14} \text{ gauss.} \tag{11.8}$$

Furthermore, it is clear that if the fragmentation did proceed, in the presence of a weak magnetic field, this would be amplified by the motions which take place when the Jeans condition is over-satisfied.

It appears, therefore, that if the gas is highly ionized in the initial configuration, it can contain only an extremely small seed field if it is to contract and fragment to masses approaching those of stellar size.

II. Gravitational contraction.

12. Gravitational contraction of protostars on to the main sequence. So far we have described the various proposals which have been suggested for the formation of protostars. Once these have formed, the stages of contraction and fragmentation will continue to take place, until the protostar becomes opaque and the adiabatic condition is fulfilled. Further contraction will then occur at a rate which is determined by the opacity of the stellar material, i.e., by the rate at which the gravitational energy can be released. It is this stage of the evolution, taking the star almost on to the main sequence, which we shall discuss here. However, before doing this it is necessary to describe briefly a paper by McVittie [*45*]. He has discussed in mathematical terms the stages of contraction from the initial state, in which the compression in the interstellar gas allows the Jeans criterion to be fulfilled, down to stellar dimensions.

McVittie has described the non-adiabatic contraction of a gas cloud to a complete polytrope, using a method of gas dynamics with gravitation. The assumption is made that the collapse follows a kind of homology transformation in which the change of temperature is not uniform. A large number of models are possible, depending on the kind of time dependence assumed for the inward gas velocity, and no important conclusions bearing on the most probable mode of contraction can be drawn. The conclusion that the initial temperature has to be very low for low-density contractions to proceed, and that there is therefore a dependence $T \propto \varrho^{\frac{1}{3}}$, is more easily derived from the simple statement of the Jeans criterion.

A wide range of initial densities and temperatures have been treated without considering the limitations imposed by the Jeans criterion. Any discussion of the contraction of a single mass to stellar dimensions is irrelevant if the initial conditions assumed are such that the critical mass is much greater than $100\,M_{\odot}$, since stars of masses greater than this probably do not exist. This imposes limits on the initial density considered, from 10^{-20} g/cm³ for a low-temperature limit of 30 degrees to 3×10^{-11} g/cm³ for $T\approx 10000$ degrees.

The phase of evolution after the protostar has become opaque to radiation is the so-called Kelvin-Helmholtz contraction phase. An estimate of the Kelvin-Helmholtz contraction time, t, can be made as follows [46][1]. Within the star

$$E = -(3\gamma - 4)\,U = \frac{(3\gamma-4)}{3(\gamma-1)}\,\Omega, \tag{12.1}$$

where E and U are the total and the internal energies. If the configuration contracts so that the potential energy Ω is changed by $\Delta\Omega$, a fraction $(3\gamma-4)/3(\gamma-1)$ is radiated while the internal energy gain is $\Delta U=[3(\gamma-1)]^{-1}\Delta\Omega$. Thus the luminosity is given by

$$L = -\frac{(3\gamma-4)}{3(\gamma-1)}\,\frac{\Delta\Omega}{\Delta t}. \tag{12.2}$$

Now the condensation has contracted from a condition with radius R_0 (given by the conditions of star formation) to a condition with radius R, in time t, and $R_0 \gg R$. If the configuration is a polytrope of index n (see article by M. WRUBEL in this volume, p. 12),

$$\Omega = \frac{3GM^2}{5-n}\left(\frac{1}{R}-\frac{1}{R_0}\right) \approx \frac{3GM^2}{(5-n)\,R}$$

so that

$$\int_0^t L\,dt = -\frac{(3\gamma-4)}{(\gamma-1)(5-n)}\,\frac{GM^2}{R} = \overline{L}\,t, \tag{12.3}$$

where $\overline{L}$ is the mean luminosity, or

$$t = \frac{(3\gamma-4)\,GM^2}{(\gamma-1)(5-n)\,R\overline{L}}. \tag{12.4}$$

Thus, for example, taking a model which is a crude approximation to the solar model, we put $n=3$, $\gamma=\frac{5}{3}$, and $\overline{L}$ equal to the observed luminosity of the Sun and find that $t=2.4\times 10^7$ years.

However, this ignores the fact that as the star contracts its luminosity is a function of its radius, the form of this function depending on the stellar model. It has been shown by THOMAS [47] that one of the conditions which must be fulfilled if a gas sphere is to contract homologously is satisfied if the opacity has the Kramers form and $\gamma=\frac{5}{3}$. Further conditions for the stability of such a contraction have been stated by him and by ROTH [48], who has made an exact integration. Such a model, neglecting radiation pressure, has been described by LEVÉE [49], who has shown that a homologously contracting gas sphere has

[1] See Sect. 23, p. 26 of M. H. WRUBEL's contribution to this volume.

an effective polytropic index ranging from 2.97 at the center to 3.25 at the surface, so that putting $n=3$ is a good approximation. However, the form of the mass-luminosity relation for a model with Kramers opacity is

$$L \propto M^{5.5} \mu^{7.5} R^{-0.5}. \tag{12.5}$$

It is easily shown that the rate of shrinkage onto the main sequence, dR/dt, is proportional to $L R^2$ for constant mass and chemical composition, and integration of the model of LEVÉE for the Sun gives

$$t = -2.52 \times 10^7 \int_{R_0/R}^{1} \frac{dR}{R^2} = 2.52 \times 10^7 \text{ years},$$

in good agreement with 2.4×10^7 years obtained above. However, if we put $L \propto R^{-\frac{1}{2}}$ from Eq. (12.5), we find that

$$t = -2.52 \times 10^7 \int_{R_0/R}^{1} \frac{dR}{R^{\frac{3}{2}}} = 5.04 \times 10^7 \text{ years}.$$

Thus the effect is to double the times obtained using the final luminosity L. When the mass, radius, and luminosity are measured in solar units and all models are supposed to be homologous, the total time for gravitational contraction is given by

$$t = 5.04 \times 10^7 \frac{M^2}{L} \text{ years} \tag{12.6}$$

for stars which approximate to the model with $n=3$. Similar calculations to these have been given by SANDAGE [*50*].

A number of tracks of gravitational contraction have recently been computed by HENYEY, LELEVIER and LEVÉE [*51*], using detailed opacity tables, for a series of masses and chemical compositions. The initial configurations have all been obtained by homology transformations from the model of LEVÉE. These authors have stated that their starting method necessarily leads to the introduction of certain transients which quickly disappear, and they also remark that a considerable latitude of choice (in starting models) leads after a few time-steps to very nearly the same results. Their starting models must always satisfy the conditions of dynamical and thermodynamical stability together with the third condition of THOMAS. It would not appear to be out of the question that protostar condensations, in moving from the isothermal to the adiabatic condition, begin to contract but reach configurations which violate the stability conditions. Their further development is not known but it may be that this instability leads to fragmentation.

Tracks given by HENYEY *et al.* are shown in Fig. 3. The luminosity slowly increases as the radius decreases, so that the tracks move from right to left. The approximate equation for the track is

$$\Delta M_{\text{bol}} = -2\,\Delta \operatorname{Log} T_e, \tag{12.7}$$

given by SANDAGE [*50*]. The tracks reach a maximum value of M_{bol} which approximately is the point at which thermonuclear energy release becomes incipiently significant. When this occurs the region very near to the center ceases to contract, and the rate of gravitational energy release is reduced, but the thermonuclear source is not able to make up this deficit completely. Thus the value of M_{bol} is reduced. At the point at which the thermonuclear source becomes important, convection may set in. An earlier investigation by HARRISON[1] suggested that even in the stage of purely gravitational contraction a convective core would appear, but this result is in error[2].

The time scales associated with each of the tracks in Fig. 3 are given in Table 1. The times t_1 and t_2 are, respectively, the time of maximum luminosity and the time at the end of the track (the end has been defined as the point at which changes on the original time scale become imperceptible).

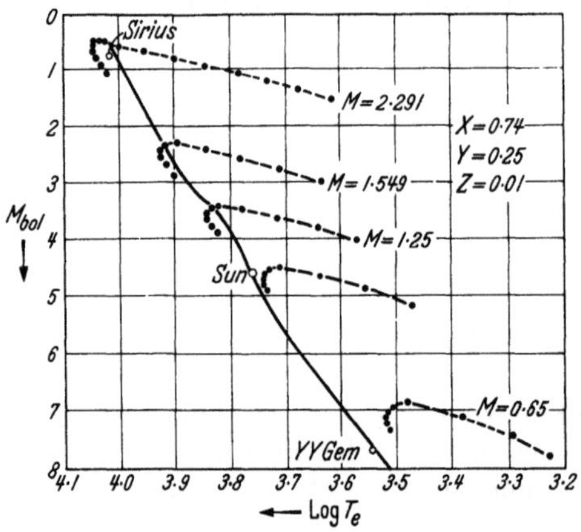

Fig. 3. Theoretical evolutionary tracks for gravitational contraction for different masses [51].

Tracks for a constant mass in which the amount of hydrogen has been varied have also been computed. They show that increasing the proportion of hydrogen reduces the maximum value of M_{bol} achieved in the gravitational contraction. Such tracks are shown in Fig. 4. Similar displacements are found when variations in Z, the proportion of the heavy elements, are put in.

Work foreshadowing that of HENYEY *et al.* on gravitational contraction and the color-magnitude diagrams which are to be expected for young clusters (cf. Sects. 41, 42, 43, 44β) was carried out by SALPETER [52].

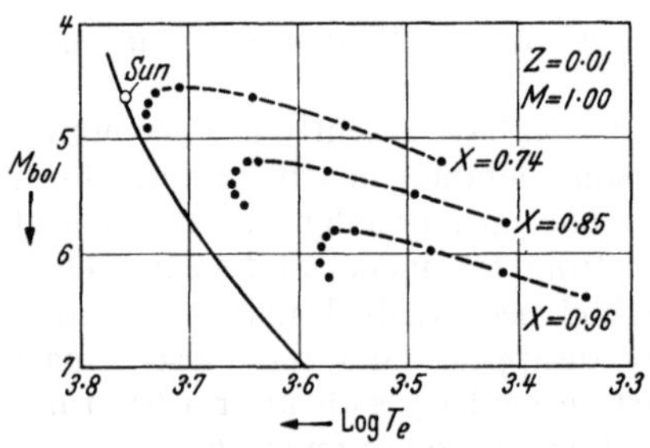

Fig. 4. Theoretical evolutionary tracks for gravitational contraction for different assumed proportions of hydrogen (X) [51].

Table 1. *Time scales for gravitational contraction of different masses.*
Times at maximum luminosity and at the end of the track are t_1 and t_2, respectively.

Mass ($M_\odot$)	t_1 (maximum luminosity) (years)	t_2 (end of track) (years)
0.65	7×10^7	1.5×10^8
1.00	1.6×10^7	3×10^7
1.25	8×10^6	1.4×10^7
1.549	4×10^6	8×10^6
2.291	1.8×10^6	3×10^6

III. Historical sketch of ideas concerning evolution on and off the main sequence.

13. The evolutionary scheme of RUSSELL and HERTZSPRUNG. So far we have described theories and observations which attempt to elucidate the problem of the formation of stars. In the previous section we have considered their final stage of evolution onto the main sequence. All of this work has been carried out over the last twenty years, and has to a large extent followed the development of ideas concerning the interstellar medium, though all of the condensation arguments are based on the fulfilment of the Jeans criterion for gravitational instability which was first propounded in 1901. The gravitational contraction theory must be attributed to the first theoretical physicists to attack the problem of the structure of a gaseous star, namely LANE, EMDEN, RITTER, and KELVIN.

[1] M. H. HARRISON: Astrophys. J. **102**, 216 (1945).
[2] Cf. E. J. ÖPIK: Contr. Baltic U. No. 35 (1947).

Information concerning these early developments has been given by CHANDRASEKHAR [46].

Already at the turn of the century the observers realised that the stars had a wide range of masses, densities, luminosities and surface temperatures. The classification of stellar spectra into "early" and "late" types were the first crude attempts at placing stars in a sequence of ages. However, with the exception of LOCKYER, the observers had ignored the theoretical developments which had been made up to that time.

LOCKYER [53] first advanced the idea that a star is hottest near the middle of its history and that the redder stars fall into two groups, one of rising and the other of falling temperature. It was essentially this idea which was developed by RUSSELL and HERTZSPRUNG in 1913 [54] in relating together the theoretical ideas and the current ideas of stellar classification. LANE's law states that if a star contracts, its internal temperature will rise so long as the material behaves as a perfect gas. Since at that time it was believed that the only available source of energy in a star was gravitational, it appeared that a star through radiating must continuously contract, and that the gravitational energy released must be sufficient both to replace the energy lost by radiation and also to raise the internal temperature. Of course, within the restrictions imposed this result is strictly correct, and it has been discussed in Sect. 12. However, RUSSELL and HERTZSPRUNG, after they had found the regularities in a plot of stars in a diagram of luminosity against temperature (the Hertzsprung-Russell diagram, hereinafter called the H-R diagram), attempted to explain them by using these theoretical ideas. RUSSELL proposed that the M-type giant stars represented an early stage of evolution. These stars then began to contract, continuously becoming hotter, and reaching a maximum temperature at the B-type stars, near the middle of the evolutionary sequence. It was argued that at this point the density of the star had increased to such an extent that the perfect gas laws no longer applied, so that the star would begin to cool like a solid or a liquid and travel down the main sequence.

The ascending and descending branches of the sequence were thus to be distinguished essentially by their densities, with low densities in the former and high ones in the latter. The theory gained wide acceptance in the decade following its advancement. Observational evidence in its favor was the determination of densities in eclipsing binary star systems by SHAPLEY and RUSSELL, and the realization that there was a real bifurcation of density between stars of solar type, for example, and giants.

However, by the early 1920's, and perhaps even earlier, a number of difficulties were arising so that this giant-dwarf theory could no longer be completely accepted. These arose for two major reasons: (i) the source of energy of the sun, and of stars in general, was recognized to be a problem; and (ii) the development of atomic physics and the early quantum theory led to the realization that the internal structure of stars under conditions of high temperature and density would demand a re-evaluation of the descending branch of the giant-dwarf theory.

14. Problem of energy sources. The energy problem and the time scale for the solar system constituted a particularly acute difficulty. EDDINGTON[1], basing his results on the figures then available, concluded that, on the contraction hypothesis, a reasonable upper limit to the age of the Sun was about 2×10^7 years.

[1] A. S. EDDINGTON: Internal Constitution of the Stars. Chap. 11. Cambridge: Cambridge University Press 1926.

On the other hand, he pointed out that the biological, geological, and physical arguments all lead to a very much greater age for the Earth. At that time, an age of 1.3×10^9 years had been set for the oldest sedimentary rocks from the uranium/lead and uranium/helium ratios. The situation for more massive stars was even more critical. Time scales $\lesssim 10^5$ years were given by the contraction hypothesis. For cepheid variables this implied that there would be changes in the period of pulsation (which is proportional to $\varrho^{-\frac{1}{2}}$ and hence to $R^{\frac{3}{2}}$) that should, for δ Cephei, for example, amount to a decrease of 40 seconds per year. The period of δ Cephei, which has been observed since 1785, does not change by any such amount, and this afforded a direct contradiction to the hypothesis that gravitational contraction was the only energy source.

Many speculations on other energy sources were made, though in 1919 RUSSELL[1] specified a series of conditions which in retrospect we see could be fulfilled only by the nuclear energy sources that we know today. Some of the speculations were concerned with the possibility that a stellar surface was heated by meteoritic bombardment, or even that radiation from the Sun was only emitted in some directions, where material was present to intercept it. Such ideas clearly failed abysmally, since they completely ignored the conditions for internal stability of stars, which rule them out immediately.

More interesting were the ideas concerning the possible "sub-atomic" sources. These fell into three categories: (i) radioactive sources; (ii) mass-annihilation of some kind; and (iii) the synthesis of heavy elements from lighter ones. Of these possibilities, radioactivity immediately ran into difficulties because it was realized that this was a spontaneous quantum effect which did not depend on the conditions in the stellar interior. Further, the energy release, if a terrestrial abundance of uranium and thorium was assumed, would not prolong the time scale sufficiently. Induced radioactivity was not considered at that time. JEANS was a strong advocate of some monatomic annihilation process, such as the annihilation of protons by electrons. He based his arguments on stability criteria which he had developed, and concluded that reactions between atoms, between atoms and electrons, or between atoms and radiation could all be ruled out because they would violate these criteria. He even argued that the synthesis of helium from four protons would lead to explosive instability, though he had not considered that this synthesis could take place through catalysts, or a number of successive reactions. Also he ruled out one of the conditions proposed by RUSSELL, which was that there was a threshold temperature below which the energy source would not operate. JEANS' source was similar to radioactivity in that it was independent of all external physical parameters. The time scale was simply the time taken to convert a solar mass to radiation, and this is of the order of 10^{13} years.

MILNE[2] also proposed that the mutual annihilation of protons and electrons in an exceedingly dense core with a temperature of at least 10^{10} degrees was the energy source. RUTHERFORD, at the Cavendish Laboratory, had already in 1919 produced the first nuclear transmutations. Those who strongly felt it was processes of this kind which were the source of energy in stars, and EDDINGTON and RUSSELL were the two most eminent authorities on this point, were inclined to use this experimental evidence. As EDDINGTON said in 1920, during his address to the British Association in Cardiff, "... what is possible in the Cavendish Laboratory may not be too difficult in the sun", though he was still assailed

[1] H. N. RUSSELL: Publ. Astronom. Soc. Pacific **31**, 205 (1919).

[2] E. A. MILNE: Monthly Notices Roy. Astronom. Soc. London **91**, 4 (1931).

by doubts and difficulties. For example, in 1924 he pointed out[1] that there was nothing in current knowledge of atomic physics that would support the idea of a critical temperature.

It is not necessary for us to discuss the developments in nuclear physics in the succeeding decades. In 1929 ATKINSON and HOUTERMANS[2] attempted to develop a theory of nuclear energy generation in stars, and ATKINSON[3] also tried to show that the elements could be synthesized in stars. This attempt was premature, because of the early stage of development of knowledge about nuclear reactions, but these authors argued that, while helium could not be synthesized directly from hydrogen, successive protons must be absorbed by nuclei which would eventually become α-unstable and eject helium nuclei. Thus they were using intermediate nuclei as catalysts. They were also using the barrier penetration theory of GAMOW to estimate the capture rates for protons. Further, they realized that, although temperatures in stellar interiors were such that the mean energies of particles were far smaller than those demanded to initiate nuclear reactions, a few particles in the high-energy tail of the Maxwellian distribution would be able to penetrate nuclei.

The problem was finally brilliantly solved in 1938 by BETHE [55] and VON WEIZSÄCKER [56], who showed that the carbon-nitrogen cycle was able to transmute hydrogen to helium in stellar interiors.

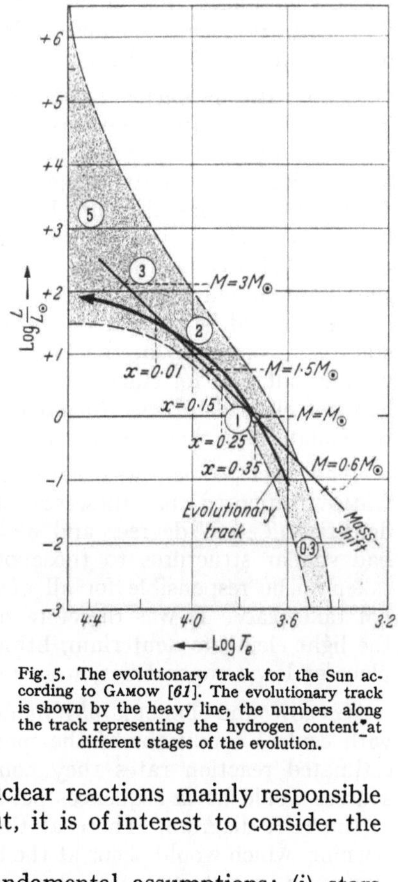

Fig. 5. The evolutionary track for the Sun according to GAMOW [61]. The evolutionary track is shown by the heavy line, the numbers along the track representing the hydrogen content at different stages of the evolution.

15. GAMOW's theory of stellar evolution.

The first attempt at a theory of stellar evolution using the assumption that stellar energy was nuclear in origin was made by GAMOW and his collaborators between 1937 and 1940 [57] to [61]. Since this attempt was made during the period in which the nuclear reactions mainly responsible for the energy supply were finally singled out, it is of interest to consider the theory in some detail.

The Gamow theory is based on three fundamental assumptions: (i) stars evolve gradually through a series of equilibrium configurations; (ii) these successive configurations are homologous; and (iii) nuclear reactions continue to take place until the entire amount of hydrogen in the star is exhausted. With these assumptions, GAMOW was able to compute the successive evolutionary configurations, starting from one on the main sequence. He showed that the effective temperature and luminosity increased so that the star would move upward along the main sequence until the whole of its hydrogen was exhausted. His evolutionary track for such a star is reproduced in Fig. 5.

[1] A. S. EDDINGTON: Internal Constitution of the Stars, p. 300. Cambridge: Cambridge University Press 1926.

[2] R. D'E. ATKINSON and F. G. HOUTERMANS: Z. Physik 54, 656 (1929).

[3] R. D'E. ATKINSON: Astrophys. J. 73, 250, 308 (1931).

Following the complete exhaustion of hydrogen, GAMOW supposed that the star would enter its final contractional stage. It would continue to move to the left in the H-R diagram, and after it reached a certain maximum luminosity, the interior would at first still be an ideal gas, but finally it would contract until a partially or totally degenerate configuration resulted and the star entered the white dwarf stage. Whether an instability would then develop depended on whether the mass of the configuration exceeded the Chandrasekhar limit. If not, then the star would remain a stable configuration, but if it did, then GAMOW envisaged that in the core neutrons would be produced; the star would then catastrophically collapse, releasing large amounts of gravitational energy, and an explosion would occur.

Two basic problems remained if GAMOW's theory were assumed correct. The first concerned the empirical mass-luminosity relation (Sect. 19). Since stars could evolve along the main sequence, a star of a given mass could have a variable luminosity and a unique mass-luminosity relation would not be expected. GAMOW evaded this difficulty by the following argument. The rate of hydrogen-burning is simply given by the luminosity of the star. Thus stars would spend most of their lives in the lower parts of the main sequence, where they were near their minimum luminosity. GAMOW concluded that the empirical mass-luminosity relation should be considered to be a statistical regularity arising because most stars were observed in the lower parts of their evolutionary tracks. He also pointed out that for stars in the lower half of the main sequence sufficient time had not elapsed for any to move upward appreciably, since the time scale for the Galaxy was then thought to be of order 10^9 years.

The other problem was that of the origin and energy sources of the red giants. GAMOW supposed that these stars had extremely low central temperatures and densities ($T \sim 10^6$ degrees and $\varrho \sim 5 \times 10^{-7}$ g/cm^3), since he assumed that they had similar structures to those of main sequence stars. Thus the CN-cycle, taken to be responsible for all of the main-sequence energy production, could not take place. It was therefore suggested that reactions between protons and the light elements deuterium, lithium, beryllium, and boron, which would take place at low temperatures, were responsible for the energy production.

GAMOW and TELLER [*59*] made calculations based on homologous models with central conditions of the magnitude described above, and with the then estimated reaction rates they concluded that stars existing on these energy sources would lie in sequences parallel to but to the right of the main sequence in the H-R diagram. Those at the extreme right would be due to deuterium-burning, which would occur at the lowest temperature, and so on. The argument was again that stars moved along these parallel tracks until they had exhausted whatever fuel they were burning. Then they contracted more rapidly until conditions suitable for the consumption of the next light element were reached. These arguments also led GAMOW to consider a possible origin for the pulsation phenomena in the red giant region. From the position in the H-R diagram where the various types of variable stars were at that time thought to lie, he concluded that this was the locus in each of the parallel tracks of the points where energy from nuclear sources began to play a lesser role than that from gravitational contraction. When this point was reached, and gravitational energy dominated, a tendency for the star to contract slightly would result in an increase in temperature and hence in the thermonuclear energy supply. This would make the star expand, and the central temperature would drop, and gravitational energy dominate again. It was supposed that stars would oscillate thus in an undamped pulsation that would continue until the isotope providing the energy source had been destroyed.

Thus the Gamow theory appeared to explain in a qualitative fashion the major features of the H-R diagram if all the specified conditions were fulfilled. Further developments of the theory of stellar evolution depended on the critical examination of these conditions, and we shall show in the later sections of this Chap. A that modern work suggests that none of the conditions (i), (ii), or (iii) above are fulfilled. The investigation of the applicability of these conditions, intensive study of the structure of the red giants, and knowledge of the rates of the reactions involving deuterium, lithium, beryllium, and boron (which were found to be of negligible importance) have all contributed to the development of the theory of stellar evolution off the main sequence which will be described in the following sections.

IV. Stars on the main sequence.

Observed masses and luminosities of solar-neighborhood stars.

Before turning to a description of the modern theory of the evolution of stars after they have contracted on to the main sequence, it is important to discuss briefly the structures of these stars on the main sequence, and the empirical mass-luminosity relation for the solar-neighborhood stars, if for no other reason than that stars spend the largest part of their lives on the main sequence. We shall also describe briefly the observational status of the giant, subgiant and subdwarf sequences in the mass-luminosity plane. A mathematical treatment of the appropriate stellar models has been given by M. WRUBEL in this volume.

The important parameters determining the structures of main-sequence stars are, whether the star is deriving its energy from the pp-chain or the CN-cycle, and whether the opacity of the stellar material is predominantly due to the free-free transitions (Kramers opacity) or to electron scattering. Since the central temperature is the parameter which determines whether the pp-chain or the CN-cycle will dominate, and since the pp-chain will operate at lower temperatures than the CN-cycle, it is found that the pp-chain dominates in the lower part of the main sequence, that there is a considerable region where both energy-producing mechanisms contribute in comparable amounts, and that the CN-cycle is dominant in the upper part of the main sequence. Estimates of the regions where the division can be made are given in Chap. D I. They are based on the pp-chain and CN-cycle rates given by FOWLER in the paper by BURBIDGE et al. [*62*].

16. Stars existing on the pp-chain. The Sun is shining predominantly through the pp-chain, the most recent calculations [*63*], [*64*], [*62*], [*65*] show that about 96% of its energy is coming from this chain and only about 4% from the CN-cycle. However, although the present studies of the solar models and the probable solar evolution will be described in Sect. 32, it may be mentioned here that the Sun has evolved away from the main sequence far enough so that it can no longer be considered to be a proto-type main-sequence model. For stars with masses less than the solar mass which are burning entirely on the pp-chain, the best models are the red-dwarf models recently investigated by OSTERBROCK [*66*] and LIMBER [*67*]. For these stars the weak temperature-dependence of the pp-chain means that there will be no convective core, and it was originally thought that the energy transport throughout these stars would be wholly radiative. However, models computed on this assumption (cf. [*68*] and [*69*]), led to computed luminosities which were higher than the observed luminosities. STRÖMGREN [*70*] pointed out that since these stars have lower temperatures and higher densities than stars higher up on the main sequence, the convective transport of energy may extend deep into the stellar interior. Since the temperature gradient is smaller in a convective zone than in the radiative zone, it follows that the central temperature will be lower than it would be if the star were in radiative equilibrium. Hence its luminosity would be expected to be smaller. This expectation was born out by the model calculations of OSTERBROCK. Thus for stars such as Castor C and α Cen B, which have masses of $0.6\, M_{\odot}$ and $0.9\, M_{\odot}$ and luminosities of $0.06\, L_{\odot}$ and $0.36\, L_{\odot}$, respectively, it is found that they have convective zones extending inward to a depth $\sim 30\%$ of their radii and central temperatures of 7.8 and 10.3×10^6 degrees, if a hydrogen content $X = 0.90$ is assumed. Furthermore, it appears that the values of the central temperature, the temperature at the bottom of the convective zone, and the depth of the convective zone are relatively insensitive to the assumed value of the hydrogen content. For the least massive of the stars—Krüger 60A—for which OSTERBROCK attempted to obtain a model ($M = 0.26\, M_{\odot}$, $L = 0.02\, L_{\odot}$), he failed to obtain a fit. However, recent studies by LIMBER [67] suggest that changes in the estimated bolometric corrections from those usually quoted in the literature are permissible and probable for stars on the main sequence in the range extending from $M4$ V to $M8$ V. He has carried out model calculations, extending OSTERBROCK's work, and has concluded that agreement between theory and observation may be reached by using completely convective models for these very low-mass stars.

17. Stars burning on the CN-cycle. Since the CN-cycle has a very strong temperature dependence, this means that stars burning totally on this cycle have a very concentrated energy-producing core. Thus in the radiative envelope, the flux of radiation will increase inward. Hence the radiative temperature gradient, which is proportional to the flux, will also increase inward, and it will eventually become steep enough to exceed the adiabatic and convection will set in. Provided that this takes place outside the energy-producing core we have a structure in which it can be supposed that all of the energy is produced in a convective core, while the outer envelope is radiative. This is the well-known Cowling model [71] which may be used to represent chemically homogeneous stars burning on the CN-cycle. Considerable mathematical simplification in treating this model is possible.

However, it is no longer permissible to use the Cowling model to represent stars over a wide span of the main sequence, as has often been done in the past, particularly since we have already seen that for a large range of low-mass stars the pp-chain is operating alone. KUSHWAHA [72] has recently constructed models, taking into account the effect of the electron-scattering opacity, for main-sequence stars of masses 2.5, 5, and $10\, M_{\odot}$ which are probably the best available at the present time. The mass-luminosity relation obtained by him will be compared with the observations in Sect. 19.

18. Stars in which both the *pp*-chain and the CN-cycle are important. For stars with masses in the range approximately $M_{\odot} < M < 3\, M_{\odot}$, both the energy-generating mechanisms are important. If both cycles are operating, the CN-cycle will be concentrated to the center, but the pp-chain will extend much further from the center. Thus the energy-producing region will be comparatively large. Only in the outer part of the star will the temperature gradient increase inward; once inside the energy-producing region the flux will decrease so that the temperature gradient will be smaller than it was in the case of the CN-cycle alone. An important factor is now the opacity. Criteria for the existence of convective cores in this case have been discussed by NAUR and OSTERBROCK [73].

19. Mass-luminosity relations for main-sequence stars. For chemically homogeneous stars in which the energy production is from the CN-cycle in a convective core, and the Kramers opacity is a good approximation, it is well known that the mass-luminosity-chemical composition relation has the form

$$L = \text{const} \frac{\tau}{Z(1+X)} R^{-0.5} M^{5.5} \mu^{7.5}.$$

Here τ is the so-called guillotine factor, and X, Z are the relative hydrogen and heavy element contents. μ is the mean molecular weight. Eliminating the radius, R, from this relation, it is found that [70]

$$L = \text{const} \left\{\frac{\tau}{Z(1+X)}\right\}^{46/45} (ZX)^{-1/45} \mu^{7.22} M^{5.13}.$$

Thus for stars of the same chemical composition, a mass-luminosity relation of the form

$$\operatorname{Log} L \approx 5 \operatorname{Log} M + \text{const} \tag{19.1}$$

should apply. The restrictions of this model mean that it will only be valid over a narrow mass range. For massive stars in the range 2.5—10 $M_\odot$ KUSHWAHA [72], by taking into account the electron-scattering opacity, has shown that a mass-luminosity relation of the form

$$\operatorname{Log} L \approx 3.5 \operatorname{Log} M + \text{const} \tag{19.2}$$

should be valid.

For low-mass stars a simple mass-luminosity relation does not appear to be easily derivable.

It is of interest now to turn to the observed mass-luminosity (M–L) relation. It should be borne in mind, however, that this is based on a small sample of stars whose ages cover a large range of time. If during this time the chemical evolution of the galaxy as a whole has led to an overall change in chemical composition, the oldest stars on the main sequence will have different values of X, Y, and Z from the youngest main-sequence stars, and the empirical relation will be one between mass, luminosity, and composition.

Stellar masses can be determined only for binary stars; detailed discussion of their derivation will be found in Vol. L of this Encyclopedia. The main-sequence stars with the lowest masses so far determined appear to be L 726–8 and Ross 614B. L 726–8 has a combined mass of about 0.08 $M_\odot$ [74]. LUYTEN has suggested that the components have roughly equal masses of 0.04 $M_\odot$. LIPPINCOTT [75] has shown that Ross 614B has a mass of 0.08 $M_\odot$. The most massive stars so far discovered are the members of the binary HD 47129 (PLASKETT's star), which both appear to have masses of about 90 $M_\odot$, and to be ejecting matter [76].

That an empirical relationship between mass and luminosity exists has long been known. References to early work are given by LUNDMARK [77] and KUIPER [78]. In 1937 PARENAGO [79] found the relation

$$\operatorname{Log} L = 3.3 \operatorname{Log} M$$

to hold over the range $-8 < M_{\text{bol}} < +10$, where L and M are the luminosity and mass in solar units. KUIPER [78] plotted the best data available in 1938, and showed that the majority of stars (mostly main-sequence) follow the approximate relation

$$\operatorname{Log} L = 4 \operatorname{Log} M$$

over the range $M_{bol} < +7$. The Sun falls on this line. For $M_{bol} > +7$ the relation is

$$\text{Log}\, L = \text{const} + 2\,\text{Log}\, M\,.$$

The bolometric corrections are uncertain in this region. RUSSELL and MOORE [80] found the relation

$$\text{Log}\, L = 3.82\,\text{Log}\, M - 0.24\,,$$

for all available binaries in the range $-7.8 < M_{bol} < +9.2$. According to this, the Sun is 0.6 magnitudes brighter, for its mass, than the average of the stars considered.

PARENAGO and MASSEVICH [81] found, as did KUIPER, two relations with different slopes. They are

$$\text{Log}\, L = 3.92\,\text{Log}\, M + 0.05$$

between spectral types O and $G4$, and

$$\text{Log}\, L = 2.29\,\text{Log}\, M - 0.39$$

between spectral types $G7$ and M. PETRIE [82], using spectroscopic binaries, found a quadratic relation between magnitude (with a small correction to take account of the dependence on radius) and Log M which can be approximated by the linear relation

$$\text{Log}\, L = 3.2\,\text{Log}\, M$$

for the range $-5 < M_{bol} < +3$. PLAUT [83] found that his determinations for a large number of main-sequence eclipsing binaries fit PETRIE's relation quite well, over the range $-8 < M_{bol} < +7$. KOPAL [84], also using main-sequence eclipsing binary components, found the relations

$$\text{Log}\, L = 2.8\,\text{Log}\, M$$

for the upper part of the main sequence and

$$\text{Log}\, L = 4.6\,\text{Log}\, M$$

for the lower part; the lines join at a mass near $2\,M_\odot$.

VAN DE KAMP [85] investigated a number of well-observed visual binaries, limited by observational selection to stars on or near the main sequence, and found the relations

$$\text{Log}\, L = 4\,\text{Log}\, M - 0.05$$

in the range $+1 < M_{bol} < +6$ and

$$\text{Log}\, L = 2\,\text{Log}\, M + \text{const}$$

in the range $M_{bol} > +6$. These equations are very similar to those obtained by KUIPER and PARENAGO and MASSEVICH. VAN DE KAMP found that the first equation has a moderate scatter while the second has very small dispersion. He also found indications that six stars lie on a third relation, parallel to the first but 1.5 magnitudes above it. These will be discussed in Sects. 20 and 22.

EGGEN [86] found a single equation

$$\text{Log}\, L = 3.1\,\text{Log}\, M$$

for main-sequence visual binary stars; this holds over the whole range $+1 < M_{bol} < +10$, without any break near $M_{bol} = +6$. That EGGEN's points are best fitted by a single equation, instead of two similar to VAN DE KAMP's, is really determined by one point, that for the faintest star, Krüger 60B. This star was used by both VAN DE KAMP and EGGEN; its mass is well determined, since the binary has an accurately determined parallax and orbit, and both workers adopted essentially the same value also for the visual magnitude. The difference in the plotted points (amounting to half a magnitude) lies in a discrepancy in the bolometric correction. Both used KUIPER'S scale, but for a star as faint as this a considerable extrapolation has to be made, and apparently different estimates were arrived at.

STRAND [87] has compiled data for the 37 visual binaries which he considers to be the best determined and finds two relations, given by

$$\operatorname{Log} L = 4.0 \operatorname{Log} M$$

in the range $-0.3 < M_{\text{bol}} < +7.5$, and

$$\operatorname{Log} L = 1.5 \operatorname{Log} M - 1.0$$

in the range $M_{\text{bol}} > +7.5$.

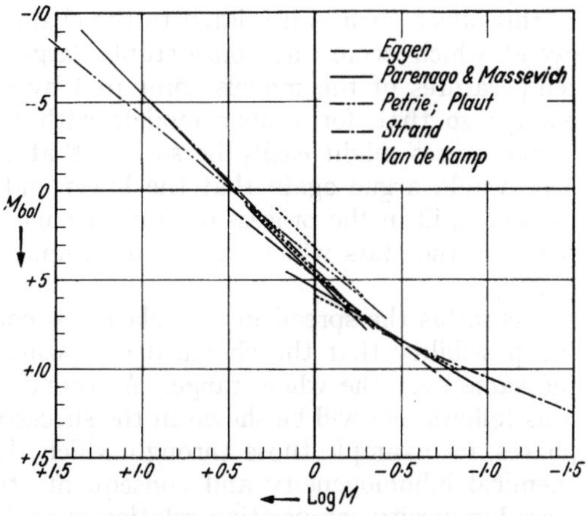

Fig. 6. Composite diagram showing the various mean mass-luminosity relations adopted by different authors (EGGEN [86], PARENAGO and MASSEVICH [81], PETRIE [82], PLAUT [83], STRAND [87], and VAN DE KAMP [85]).

A compilation of the different results obtained since 1950 is shown in Fig. 6. There is still considerable scatter in the positions of the various mean lines drawn through the different sets of selected points. In general, the visual binaries yield more accurate points than the spectroscopic and eclipsing binaries, so the lower part of the plot is better defined than the upper. Yet it is here that some disagreement still exists on whether or not there is a break and change of slope at a point somewhat fainter then the Sun. The majority agree that there is a break near $M_{\text{bol}} = +6$ to $+7$, in the sense that the slope becomes smaller for the fainter stars. In our opinion, the evidence for this break is satisfactory, yet it is clear that further checks should be made upon the size of the bolometric corrections for the fainter stars. A very recent study of these corrections by LIMBER [67] should enable the mass-luminosity relation in this mass range to be re-evaluated.

The fact that main-sequence stars over the whole available mass range cannot be represented by homologous models means that a single linear Log L–Log M relation would not be expected. For the brighter stars Eq. (19.2) gives a fairly good representation of the observations. The observed slope is never as steep as Eq. (19.1) would suggest. There is general agreement that the slope is less steep for the most massive stars than for the stars of intermediate mass; this effect is shown by KOPAL'S equations where the transition is at $2\,M_{\odot}$. The change in slope below the solar mass and the place at which it occurs have not yet been explained theoretically. Much more work needs to be done to correlate the observed and theoretical M–L relation. Models for a wide range of masses, using the tabulated opacities of KELLER and MEYEROTT[1], need to be calculated for different chemical compositions. Work along these lines with a large electronic computer has been begun by HOYLE, HASELGROVE, and BLACKLER [88].

[1] G. KELLER and R. MEYEROTT: Astrophys. J. 122, 32 (1955).

However, the following argument would suggest that a different M–L relation from that for bright stars must apply for the very faint stars. For, if a dependence of luminosity on mass of the form given either by Eqs. (19.1) or (19.2) held for such stars, then it is simply shown that the normal gravitational contraction time for stars with masses $< 0.1\,M_{\odot}$ would be in excess of 10^{10} years, a time that is currently believed to be of the order of the age of the Galaxy.

Of course, this argument might be used in reverse and it might be supposed that some of the least massive stars determining the lower end of the observed M–L relation are still in the process of gravitational contraction. The break and change of slope which occurs near $M = 0.5$ to $1\,M_{\odot}$ cannot, however, be attributed to this effect. Nor is it related to the change-over from the *pp*-chain to the CN-cycle[1] which occurs at considerably larger mass (cf. Chap. D I). The central temperatures of the models computed by OSTERBROCK for red dwarfs are low enough so that for similar models with masses $< 0.1\,M_{\odot}$, the corresponding temperatures might easily be so low that the *pp*-chain will no longer operate. This would argue again that the lower end of the main sequence is populated by stars still in the process of gravitational contraction. This must certainly be true for the stars which extend in luminosity down to $M_v = +20$ (cf. Table 6, p. 217).

As far as the spread in the relation is concerned, we have already mentioned the possibility that the chemical composition of the stars may not be exactly the same over the whole range. A second point which must be borne in mind is as follows. As will be shown in the succeeding sections, the initial evolutionary phases, for example those through which the Sun is now passing, lead to some chemical inhomogeneity and consequent structural changes. In this case, the mass-luminosity-composition relation is slightly different from that for a homogeneous model. This effect must lead to an intrinsic spread in the observed M–L relation.

20. Positions of subgiants in the mass-luminosity plane. In Chap. A V (p. 172), theoretical evolutionary paths will be discussed that carry stars off the main sequence into the subgiant and giant regions of the H-R diagram. However, from looking at the positions of such stars in the mass-luminosity plane, some preliminary deductions can be made about such evolutionary paths.

Subgiants are those stars that lie slightly above the main sequence in the H-R diagram, but below the giants. We shall divide them into two groups: those that are members of very close eclipsing and spectroscopic binaries, and those that are members of visual binaries (which are assumed to be identical with single stars). A large proportion of subgiant secondary components of close binaries have masses that are of the order of or less than $1 M_{\odot}$, while their radii and luminosities are greater than those of the Sun [*89*]. They therefore do not lie on the main-sequence M–L relation. There exists a well-established relation of the form [*89*]

$$\operatorname{Log} L = 0.33 \operatorname{Log} M + 2.06 \operatorname{Log} R + \text{const};$$

the luminosity also depends on the mass ratio of the components of the binary system [*90*]. The evolution of these stars is probably governed by mass loss from their surfaces and will be discussed in Chap. C II.

The visual binaries have large enough separations so that mass loss should not occur. Six stars lie on VAN DE KAMP's third M–L relation [*85*], which is parallel to the one satisfied by most of the brighter stars but 1.5 magnitudes

[1] H. BONDI: Monthly Notices Roy. Astronom. Soc. London **110**, 595 (1950).

above it. Of these, η Cas A probably lies slightly above the main sequence, while ζ Her A lies well above it, in the subgiant region. EGGEN [86] discussed the masses of six subgiant visual binaries; all probably lie on his M–L relation for the main sequence except for ζ Her A, which is above it. The best-determined subgiant, Procyon A, definitely lies on the established M–L relation [86], [87].

There is thus one well-determined case of a subgiant, ζ Her A, whose luminosity is not only too high for its color, but also too high for its mass. This star has probably started to evolve off the main sequence. The region on the main sequence where it should lie, according to its mass, is at spectral type $F9$ or $G0$. Its observed type is $G0$ IV. It may therefore be deduced from examination of the M–L plot alone that the evolution of a star of this mass ($1.1 M_{\odot}$ [1]) has moved it vertically upwards in the H-R diagram to the extent of 1.5 magnitudes. Evidence leading to the same conclusion, concerning evolution of stars of this mass off the main sequence, from a totally different point of view, will be discussed in Chap. B II and B III, and was applied to ζ Her A by SANDAGE [91].

The spectral type of Procyon A is $F5$ IV, and its mass is $1.8 M_{\odot}$. Since this star lies on the main-sequence M–L relation we may deduce that evolution has moved it horizontally off the main sequence in the H-R diagram. Similar conclusions can be drawn about the other five subgiants listed by EGGEN.

21. Positions of giants in the mass-luminosity plane. There are few reliable mass determinations for giants. For more than forty years it has been known that the solar-neighborhood giants lie near to the main-sequence M–L relation [54]. KUIPER [78] showed that both components of α Aur ($G5$ III, $G0$ III) lie on the main-sequence M–L relation; this was confirmed by EGGEN [86]. It may therefore be deduced that these yellow giants originated at a point on the main sequence where their value of M_{bol} was the same as the present value, i.e., at about spectral type $A3$ V, and that evolution has carried them horizontally in the H-R diagram. Indeed, EGGEN made a striking comparison between ζ UMa Aa, which has two nearly equal components and a combined spectral type $A3$ V, and α Aur: the masses of each are nearly the same. Different evidence leading to the same conclusion, concerning evolution of stars of this mass off the main sequence, will be discussed in Sect. 45.

22. Positions of subdwarfs in the mass-luminosity plane. The problem of the definition and life-history of subdwarfs is a vexing one at present. There may be a variety of species grouped together among stars thus designated, i.e., stars which lie below the main sequence in the H-R diagram. Some have been placed there on the basis of spectroscopic parallaxes and a discussion of these is given in Sects. 86 and 87. These are stars with a low abundance of the heavy elements. Since masses are not available for these stars, their positions in the M–L plane cannot be considered.

The most reliable masses of subdwarfs are those of the two components of 85 Peg.[2]. The components have almost equal masses, $0.82 M_{\odot}$ and $0.80 M_{\odot}$, while their bolometric luminosities differ by nearly two magnitudes. In the mass-luminosity plane [85] the fainter component lies near VAN DE KAMP's relation $\text{Log}\, L = 2\, \text{Log}\, M + \text{const}$, obeyed by stars fainter than $M_{\text{bol}} = +6$. The brighter component is one of the six stars which obey his third relation, parallel to and above the one defining the majority of stars brighter than $M_{\text{bol}} = +6$. Thus, while both stars are fainter than is normal for their temperatures, one is still

[1] A. A. WYLLER: Astronom. J. **60**, 39 (1955).
[2] A. A. WYLLER: Astronom. J. **61**, 76 (1956).

about 1.5 magnitudes brighter than its mass would predict according to the average mass-luminosity relation. The normal type for a main-sequence star of this mass is $K6$ V, with $M_{\mathrm{bol}} = +7$, while its observed type is $G2$ V. One highly speculative possibility for its evolutionary history is that the components may once have been much closer, so that mass-exchange and mass-loss became important, as discussed in Sect. 71; interaction between the stars and the ejected material might have increased the separation to its present value of about 20 a.u.

REIZ [*92*] has attempted to construct theoretical models for subdwarfs on the assumption that they contain only a negligible amount of heavy elements, i.e., $Z \approx 0$. He has considered homogeneous stars of masses ranging from about 0.7 to 1 $M_{\odot}$ which are burning on the *pp*-chain and are in radiative equilibrium throughout. In the absence of heavy elements the opacity is due to electron scattering and free-free transitions in hydrogen and helium alone. The models lie below the main sequence in the H-R diagram. Before this explanation is accepted it is necessary to establish that real sequences of subdwarfs exist, and also that they do have underabundances of the heavy elements (cf. Sects. 86 and 87). For example, the subdwarfs 85 Peg A and B do not appear to have such underabundances, although quantitative analyses have not been made, and evidently a different explanation is necessary here. Observationally it is not known at present, as will be discussed in Chap. B III, whether the main sequences of globular clusters (which do have low heavy-element content) lie below the solar-neighborhood main sequence. On the theoretical side it is necessary to consider a wide range of models and their further evolution after they leave their own main sequence.

V. Modern theories of evolution along and off the main sequence.

23. Mixing in main-sequence stars. An initially chemically homogeneous main-sequence star will begin to increase its molecular weight μ as it transmutes hydrogen to helium. If it can remain completely chemically homogeneous so that μ increases throughout the whole star uniformly, and if its mass remains constant, then from the mass-luminosity-composition relation and the mass-radius-composition relation it can be shown that the star will move upward and to the left of the main sequence in the H-R diagram, accelerating because equal luminosity increments will occur in decreasing intervals as the luminosity increases. In order that such a uniform increase can take place, it is necessary that mixing must take place at a rate which is fast compared with the rate of change of μ in the energy-generating core. There are two mechanisms by which such mixing might take place.

α) Convective mixing. In a zone in which convection is important it can be supposed that mixing is both rapid and efficient. In the absence of rotation this is deduced immediately from the properties of the turbulent mixing currents. Thus it is reasonable to suppose that in stars with convective cores, the cores are chemically homogeneous, while in stars in which there is a deep outer convective zone extending inward from the stellar atmosphere, the chemical composition of the atmosphere is characteristic of the whole of this outer zone. However, the only stars which may be convective throughout are the very faintest and least massive stars on the main sequence. It appears, therefore, that these are the only stars which could evolve upward and to the left in this manner. However, the luminosities of these stars are so small that the time taken to transmute a large proportion of their mass is very large; in the case of Krüger 60A this is about 10^{12} years. Consequently this mode of evolution is of no practical significance.

β) Rotational mixing. In the envelope of a non-rotating star in which radiative energy transport prevails, mixing is negligibly small since the diffusion effects are unimportant over time scales $\lesssim 10^{10}$ years [*93*]. However, in a rotating star currents will be set up in planes through the axis of rotation, the degree of mixing depending on the equatorial velocity. Investigations by SWEET and ÖPIK [*94*], [*95*] have shown that the estimate of the amount of mixing by EDDINGTON [*96*] was far too large. On the basis of the mixing time for a Cowling model calculated by these workers, STRÖMGREN [*70*] estimated that massive stars with abnormally high rotational velocities, i.e., the *Oe* and *Be* stars, would be well mixed while even some of the normal *O* and *B* stars might also be thoroughly mixed. However, more recently MESTEL [*97*] has pursued the subject further and has concluded that the conditions for mixing are even more stringent than those derived by SWEET and ÖPIK. He has concluded that no continuous mixing between core and envelope can take place in a uniformly rotating Cowling model since the non-spherical distribution of matter set up by the rotational currents themselves tend to choke back the motion. If the rotation increases inward sufficiently the star will build up a mixing zone of definite mass. If the inner region is constrained by a magnetic field to rotate uniformly the mass of the mixing zone is very sensitive to increases of angular velocity above a certain limit. In a star in which the angular velocity decreases outward, however, the mass of the mixing zone is likely to increase slowly with increasing angular momentum. Thus in starting from the Cowling model in calculations of stellar evolution, it is usually supposed that rotational mixing does not take place. However, MESTEL has also shown that the mass of the mixing zone is determined by a parameter which varies as the star's dimensions change. The efficiency of mixing increases as the star contracts and a star which is unable to mix in its initial state with a convective core may possess a mixing zone after contraction into a shell source state.

24. Evolution of stars down the main sequence. An idea basic to some theories of stellar evolution is that proposed by FESSENKOV [*98*] and developed by MASSEVICH, PARENAGO, SOROKIN, KRAT and others. This is that stars condense on to the main sequence in certain mass ranges and then evolve along it. The idea appears to have been developed in order to overcome a difficulty associated with the theoretical mass-luminosity-composition law for homogeneous stars, and the observed mass-luminosity relation. It has been pointed out that the very strong dependence of luminosity on μ means that, for example, a star which condenses with $\mu=0.53$ ($X=0.90$, $Y=0.09$, $Z=0.004$), as compared with a star with $\mu=0.62$ ($X=0.70$, $Y=0.28$, $Z=0.02$), will have a luminosity a factor of about 3 smaller. Thus there should be some scatter in the M–L relation if the stars remain thoroughly mixed, because as stars age, μ will steadily increase. Alternatively, there should also be scatter if there is a spread in chemical composition in gas clouds of the same age out of which stars condense. If the observed M–L relation is extremely tight, so that such scatter is precluded by the observations, then it has been argued that the solution is that stars eject mass, thus becoming less massive. It is then postulated that they eject mass at a rate which is proportional to their luminosity in order that they remain on the M–L relation. As will be discussed in Sects. 38 and 53, the formation of stars in two discrete mass ranges could then populate the main sequence.

A detailed discussion based on the general arguments has been given by Massevich [99]. She took standard models burning on the CN-cycle, with a convective core and radiative envelope (the basic models were those of Schwarzschild[1] and Harrison[2]). It was assumed that throughout the evolution the star remained thoroughly mixed. Three sets of evolutionary tracks along the main sequence were computed—each was derived from a slightly different initial model, but the three sets of tracks differed very little—and they were compared with suitable main sequence observations. The agreement between theory and observation was fairly good. For each track a limiting mass was obtained, essentially from the method of calculation, above which the calculations had no validity. The limits were 5, 8 and 15 $M_{\odot}$, respectively. Clearly, as the stars move along such an evolutionary track, the hydrogen/helium ratio decreases, but the heavy element composition remains constant. At the time that these calculations were carried out, it was thought that such a standard model would adequately represent the Sun (cf. Sect. 32 for the present status of solar models). Thus Massevich concluded from these arguments that the Sun had originally reached the main sequence as a star of between 5 and 8 $M_{\odot}$. The computed changes in mass and luminosity with time were plotted; the initial decrease in mass was large, but slowed down as the star decreased in luminosity, so that in the last 3×10^9 years it was practically unchanged. Its maximum lifetime was estimated to lie between 8.7 and 10×10^9 years. Over the range of applicability of the calculations, the relation between dM/dt and L was found to be approximately linear, as had initially been postulated by Fessenkov.

The rate of mass loss was considered further by Massevich [100], assuming that the stars between $O8$ and $G4$ are constrained to follow the empirical M–L relation. She found that a star of mass 14.2 $M_{\odot}$ with an initial hydrogen content $X=0.88$ would lose mass at an average rate of 0.0064 $M_{\odot}$ per 10^6 years until its mass is 2.53 $M_{\odot}$. The computed loss for a solar mass is then 0.07 $M_{\odot}$ per 10^9 years. If the rate of mass loss is written in the form

$$\frac{dM}{dt} = -\frac{k}{3} L,$$

this relation, together with a M–L relation of the form

$$L = M^4 \times \text{const},$$

leads to a theoretical luminosity function which is in agreement with the observed luminosity function for the solar neighborhood stars (see Sect. 53).

More recently Massevich [101] has considered various evolutionary tracks for completely mixed models with mass loss, for a range of stellar masses, for different assumed values of Z, and for a variety of assumed opacity laws. She has also considered ranges of models in which the mass has been kept constant.

At this point it is necessary to make some general criticisms of the postulates upon which evolution of stars down the main sequence is based, and upon which some models have been constructed.

The rate of mass loss can be calculated from the models and the observed mass-luminosity relation, once the form of the relation has been assumed. The basic justification for putting

$$\frac{dM}{dt} = -\frac{k}{3} L$$

[1] M. Schwarzschild: Astrophys. J. **104**, 203 (1946).
[2] M. H. Harrison: Astrophys. J. **108**, 310 (1948).

appears to be that it is possible in this way to populate the main sequence starting from only one or two discrete masses for condensing stars. If it is assumed that there is a whole spectrum of masses in the process of star formation, then this postulate is no longer necessary. There would appear to be no theoretical justification for assuming such a law, and no observational evidence is available that mass ejection is proportional to luminosity. The energy required to eject mass from the surface of the star must ultimately come from nuclear sources, and as is discussed in Chap. CII (p. 241) on mass loss, on this basis we would not expect in general that a linear relation between dM/dt and L would hold.

A more serious difficulty associated with the models just discussed is that they all depend on the assumption of complete mixing. As we described in Sect. 23, work carried out by SWEET, ÖPIK, and MESTEL has led conclusively to the result that mixing in radiative zones is never of importance in the time scales which are of interest.

Evolution of stars in which mass loss is assumed, and in which chemical inhomogeneities are developed, i.e., mixing does not occur, are described in the following Sect. 25.

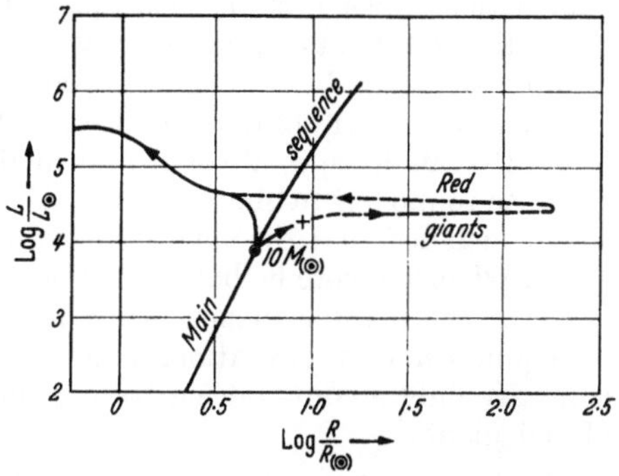

Fig. 7. Possible evolutionary track for a star with a growing energy-producing shell according to GAMOW [102]. Vertical cross indicates the place beyond which isothermal and radiative solutions cannot be fitted together according to SCHOENBERG and CHANDRASEKHAR.

25. Stellar models with inhomogeneities in chemical composition. The modern theory of the evolution of stars off the main sequence, worked out in recent years by SCHWARZSCHILD, HOYLE and their many collaborators, depends on the development in an originally homogeneous star of a core having a molecular weight, μ_c, different from that of the envelope, μ_e. Thus in this theory one of the basic postulates on which GAMOW's early theory is based (Sect. 15) has been discarded. These investigations take us right up to 1956. However, it is worth recalling that in the later sections of a paper written in 1943, GAMOW [*102*] anticipated in a qualitative discussion some of the developments that have come since. We show in Fig. 7 a diagram from that paper giving the schematic evolutionary track for a star as and after it moves off the main sequence.

During the last twenty years work on models with a chemical discontinuity between core and envelope has been carried out by a large number of workers in order primarily to elucidate the problem of the structure of the red giants [*103*] to [*115*], [*81*], [*116*] to [*122*]. From the evolutionary standpoint it is important to consider sequences of models which represent the transition from the main-sequence models to the red giants. Key investigations were therefore those of SCHOENBERG and CHANDRASEKHAR [*107*] and HARRISON [*109*]. They investigated the equilibrium configurations of stellar models having (i) convective cores in which the energy was generated, and radiative envelopes with different molecular weights, and (ii) isothermal cores and radiative envelopes with different molecular weights. They used Kramers opacity and neglected degeneracy. The major conclusions were as follows.

As the ratio of the molecular weights μ_c/μ_e increases with time, the radius and mass of the convective core both decrease. At some epoch the energy generation in the core ceases, the convection stops, and the core becomes isothermal, so that the series of models with isothermal cores must be used. The calculations then showed that for each assumed value of μ_c/μ_e there is a maximum mass of the isothermal core beyond which no fit can be made with the radiative envelope to make an equilibrium model. This critical mass is a function of μ_c/μ_e, and has its largest value for a homogeneous star, but diminishes to a lower limit, determined by these authors to be 0.1 of the mass of the star when $\mu_c/\mu_e=2$. This is known as the Schoenberg-Chandrasekhar limit. When a star has reached this limit its only nuclear energy production is confined to a thin shell between the isothermal core and the radiative envelope.

Following the work of SCHOENBERG and CHANDRASEKHAR, who did not take into account possible degeneracy in their models with isothermal cores, GAMOW and KELLER [*110*] considered isothermal cores with partial degeneracy and concluded that under certain conditions such models could give radii large enough to fit the red giants. The result was contested by HARRISON who computed a number of partially degenerate models which did not have large radii. More recently HAYASHI [*123*] and SCHWARZSCHILD, RABINOWICZ and HÄRM [*124*] found that under special conditions models with partially degenerate cores could give rise to models fitting well the radii and luminosities of red giants. However, they concluded that the conditions, which demand that the fraction of the mass in the exhausted core lie between very narrow limits (for smaller fractions the star lies near the main sequence, while for larger fractions it appears probable that no equilibrium configurations exist) are so stringent that partially degenerate cores are only a contributing factor, and not the main cause, for the large radii of red giants.

An important step forward was made by SANDAGE and SCHWARZSCHILD [*125*], who considered the quasi-equilibrium states through which an unmixed model passes after reaching the Schoenberg-Chandrasekhar limit. Their starting point is a model with an isothermal core which has reached the limiting mass, a shell source burning on the CN-cycle, and a radiative envelope. They then consider a series of models in which it is supposed that the hydrogen-poor core is contracting, so that an additional energy source is present in the core; this core is now in radiative equilibrium. It is found that as the cores contract the envelopes of these models greatly expand, while the shell source remains exceedingly thin, about 90% of the energy generation arising in a thickness of only 0.4% of the radius. As the core contracts only part of the gravitational energy is released as radiation flux, the remainder going to increase the internal energy of the core, so that the central temperature increases as the radius increases. This sequence of models suggests that stars originally on the main sequence will, after developing inhomogeneities in the cores, evolve in the following way. The core will grow until the Schoenberg-Chandrasekhar limit is reached. SANDAGE and SCHWARZSCHILD point out that the fraction, q, of the total mass at which this occurs is about 12% for a hydrogen content, X, of 0.6. During the evolution from the chemically homogeneous main-sequence model the luminosity will increase slightly, becoming about 1 magnitude brighter, the radius will increase by about 70% from the main-sequence model, while the effective temperature will remain nearly constant. Thus throughout this evolution the star will remain near the main sequence. However, the contraction of the core and the subsequent envelope expansion mean that the star will then move rapidly to the right of the main sequence in the H-R diagram.

The results of this theory are applied in Chaps. BII to BIV, when we consider the observations of color-magnitude diagrams and luminosity functions of star clusters. The limiting mass of 12% can be used in deducing the ages of clusters from the break-off point of the main sequence. Thus the age is given by

$$t = 0.007 c^2 \frac{X q M}{L} \tag{25.1}$$

where $Xq = 0.07$. This is approximately independent of the assumed hydrogen content of the envelope, X, since for varying X the jump in the molecular weight at the core-envelope boundary varies. It has been shown by HARRISON [109] that q varies in such a way that the product Xq is nearly constant.

This formula is only approximate, however, since it does not take into account the gradual brightening of the star as it goes to the Schoenberg-Chandrasekhar limit. In general the formula for the age can be written

$$t = 6.2 \times 10^{18} X M \int_0^q \frac{dq}{L} \tag{25.2}$$

and integrated from the starting point of the star on the main sequence. This relation has been used extensively to date a number of clusters, as discussed in Sect. 40.

The Sandage-Schwarzschild models suggest that the stars move extremely rapidly to the right after leaving the main sequence. They do not indicate that the envelope expansion ceases over the range considered, and this is in disaccord with the observations of star clusters. Neither do they indicate an increase in luminosity, i.e., a track both upward and to the right in the H-R diagram, which is demanded to explain the red giants in globular clusters. These authors proposed that a new physical process must occur which they had not taken into account, and in this connection they speculated both about the onset of helium-burning and about the possible effects of mixing. However, to elucidate the further evolution we must consider the work of HOYLE and SCHWARZSCHILD. Before doing this we shall consider briefly an independent approach to the early stages of evolution, using the mass-loss theory which has been developed in the U.S.S.R.

Evolutionary tracks for homogeneous and non-homogeneous models, both with constant mass and assuming mass loss, have been constructed by SOROKIN and MASSEVICH [126]. They consider three stages in the life of a single star. In its initial state it is a massive main-sequence star and it evolves down the main sequence by losing mass, remaining completely mixed throughout. Such evolution has been described in Sect. 24, where objections to it have also been discussed. As the star reaches the region of the Sun in the mass-luminosity diagram, the mass ejection becomes extremely small and the star enters a stage of "quiet evolution". In this stage, mixing apparently ceases and the star remains on the main sequence for a period estimated for $1\,M_\odot$ to be $6\text{–}8 \times 10^9$ years. The onset of this stage of quiet evolution is related in this theory to the break in the main sequence for the solar-neighborhood stars (Sect. 19). The third stage of evolution apparently begins when a limit, similar to the Schoenberg-Chandrasekhar limit, is approached. Several possibilities are investigated. In one it is supposed that as the energy sources in the core are exhausted, or as the convection recedes, catastrophic contraction takes place. This must presumably lead to violent ejection and re-organization of the star's structure. In a second model it is supposed that the contraction takes place non-catastrophically. The luminosity increases, mixing takes place, probably some mass is ejected, and the star begins its quiet evolutionary phase again. However, now the value of X/Y

will have decreased. A diagram showing these stages of evolution is reproduced in Fig. 8.

Further work along these lines was carried out by MASSEVICH [*127*], [*128*], [*101*], using non-homogeneous models with convective cores. She computed the times taken for stars of constant mass to reach limiting values of μ_c/μ_e, and values comparable with those of SANDAGE and SCHWARZSCHILD were obtained. Similar calculations were also carried out for stars which were also losing mass. In this case the time scales were found to be increased.

These authors do not appear to have discussed the interpretation of the color-magnitude diagrams for globular clusters (a discussion of galactic clusters is given in Sect. 44) on the basis of the concept of mass loss. It would seem that the turn-off point which is the characteristic dating point for these clusters on the basis of the arguments of SCHWARZSCHILD, SANDAGE, and others, would be reached, in this case, after all the stars originally above this point on the initial main sequence had lost mass and effectively ceased moving down the main sequence, and had then burned out their cores and moved to the right.

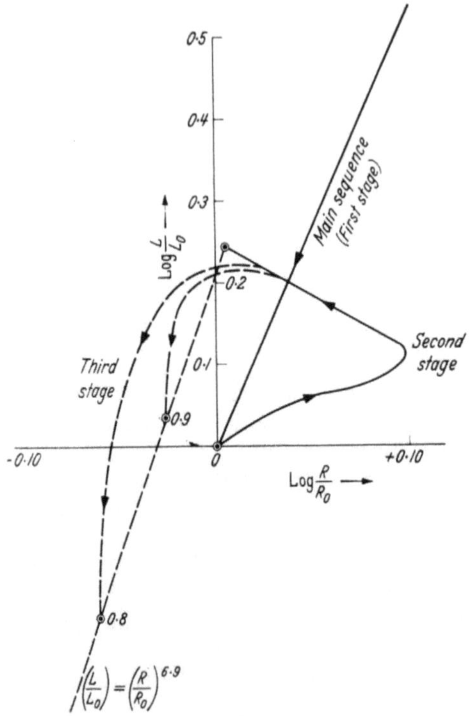

Fig. 8.

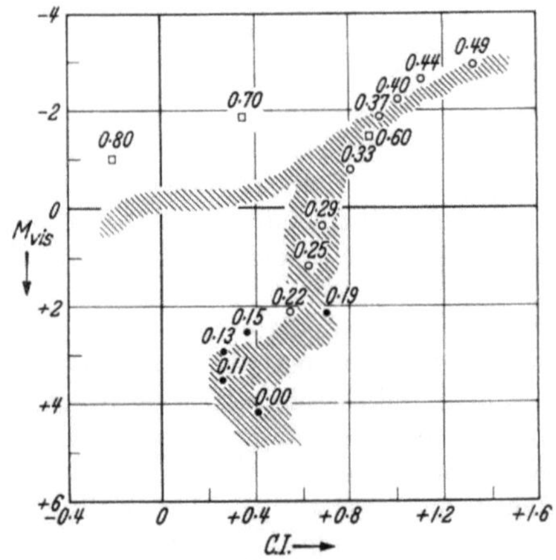

Fig. 9.

Fig. 8. Evolutionary track of a star of solar mass in the luminosity-radius plane according to SOROKIN and MASSEVICH [*126*]. First stage: evolution along the main sequence by ejection of mass. Second stage: "quiet evolution" (hydrogen-burning in a convective core). Third stage: non-statical change (contraction, accompanied by ejection of mass), after which "quiet evolution" returns.

Fig. 9. Hertzsprung-Russell diagram for the entire sequence of models following the evolution of population II stars according to HOYLE and SCHWARZSCHILD [*129*]. The numbers give the fraction of the mass in which hydrogen is burned out.

26. Evolutionary tracks of HOYLE and SCHWARZSCHILD. HOYLE and SCHWARZSCHILD [*129*] have attempted to construct evolving models which will explain the observed features of the color magnitude diagrams of globular clusters (see Figs. 9 and 11). Although they have termed their investigation a preliminary reconnaissance of the problem, it is at the same time the most ambitious attempt at understanding details of stellar evolution so far made. The major features are as follows. The tracks have been computed for stellar masses of 1.1 and

$1.2\,M_{\odot}$; these correspond approximately to the masses at the top of the truncated main sequence in the observed color-magnitude diagrams. Of course the evolutionary track of a single star is not identical with the line occupied by the stars in a given cluster, since this latter is the locus for stars of a range of masses at a given time, while the former is the locus for stars of a unique mass at different times. The difference between these two is small, and the reason is essentially that the time taken for stars to evolve after they have left the main sequence is short as compared with the time that the stars spend on the main sequence; consequently only a narrow mass range is demanded to populate the whole observed H-R diagram.

In the initial stage, HOYLE and SCHWARZSCHILD have constructed models ($M = 1.1\,M_{\odot}$) with isothermal partially degenerate cores, a thin energy-producing shell using the CN-cycle, and a radiative envelope; it is supposed that the core consists of essentially pure helium while the radiative envelope has a composition consisting predominantly of hydrogen. They have used the CN-cycle throughout, but the run of values of the shell temperature shows that initially the energy generation process must be due to the pp-chain, though the temperature soon increases so that the CN-cycle dominates. In this treatment they used Z (carbon and nitrogen) $= 0.005$, which they considered to be typical of the compositions of extreme population II stars. The sequence of models is shown in Fig. 9, where they are superposed on the mean diagram for M3 and M92. Models with this particular structure are represented by the closed circles; they are given up to a mass of the exhausted core $q = 0.19$. The next model obtained under those assumptions for $q = 0.22$ is not shown on this diagram, but it lay very far to the right of the other computed points, a result in agreement with the earlier work of SANDAGE and SCHWARZSCHILD [*125*]. HOYLE and SCHWARZSCHILD have attributed the small discrepancies between theory and observation in the range up to $q = 0.19$ to the approximations that have been made in this region. For example, since the star starts off on the pp-chain, the material is not sharply divided in composition at the core boundary; this does not occur until the CN-cycle dominates. Also, they have divided the outer envelope into two zones, an outer one in which the Kramers opacity alone is operating, and an inner one in which the electron scattering alone is at work. Since, ideally, contributions from both sources of opacity should be taken into account throughout the envelope, the result of this has been that they have slightly exaggerated the luminosities. Slight changes in luminosity can also be obtained if the assumed composition of the envelope (90% hydrogen and 10% helium) were changed.

The failure of these models beyond $q = 0.19$ was attributed to the simplification of the boundary condition at the surface, i.e., that the density and temperature send to zero at the surface. In fact, the condition which must be fulfilled at the surface is that

$$T \to T_{\text{eff}}, \qquad \varrho \to \frac{\mu}{kT}\,\frac{HGM}{R^2} \qquad \text{as} \quad r \to R,$$

i.e., T and ϱ approach the atmospheric values as the surface is reached. It has been argued that the envelope solutions which correspond to the idealized mathematical conditions at the surface differ negligibly in the interior of the star from the solutions that satisfy the boundary. However, in these models HOYLE and SCHWARZSCHILD have shown that this is true only for their models up to $q = 0.19$. Beyond this point the models fail badly in this respect. Stated in more physical terms, the models for $q > 0.19$ give a density which falls too soon as the surface is approached, i.e., when ϱ reaches the photospheric value, $T = T_1 > T_{\text{eff}}$. Now in a radiative zone the dependence of ϱ on T is of the form $\varrho \propto T^{3.25}$, whereas models taking into account the true surface condition require a dependence of ϱ on T which is smaller than this. More than this, in the models here studied there is a very thin convection zone immediately below the photosphere in which hydrogen is in a critical stage of ionization, and in this zone ϱ varies with T more rapidly than in a radiative zone, so that the discrepancy is even worse.

To overcome these difficulties it is proposed that the convection zone lying below the photosphere extends into the interior to a considerable depth. Below the hydrogen ionization level the temperature-dependence of the density is $\varrho \propto T^{\frac{3}{2}}$, a much slower decline than that in a radiative zone. Thus the star adjusts its photospheric density to the required value by adopting a structure with a sizeable outer convection zone, the parameter of adjustment being the depth of this zone.

Models have then been computed by elaborate techniques taking into account the effect of an outer convective zone on the structure of the star, and in this way it has been possible to fit the observed sequence in the globular clusters along the subgiant branch and up towards the giant branch. As the stars move along this sequence the calculations show that the temperature in the hydrogen-burning shell as it moves outward remains fairly constant at about 20×10^6 degrees. This situation holds until the luminosity of the stars is about $100 L_\odot$. At this stage the core begins to evolve in such a way that the density in the shell tends to decrease. This effect, together with the necessity of increasing the rate of energy generation, means that the temperature must rise to 40 to 50×10^6 degrees as the star reaches the giant branch.

This rise in temperature does not in itself affect the isothermal condition in the core. Departure from the isothermal condition occurs only when gravitational energy is released. The rate of this release parallels the luminosity. Since increased energy generation means that the helium core grows faster, towards the top of the giant branch (as distinct from much lower down as was supposed by SANDAGE and SCHWARZSCHILD), it is found that the rate of energy generation by the gravitational process is sufficient to maintain a temperature gradient between the center and the periphery of the exhausted core. Still, however, the release of gravitational energy is a negligible source of energy for the star as a whole.

The increase of temperature in the core has little effect on the star's structure as long as the inner parts of the core remain degenerate. However, for a large enough rise of the central temperature the helium-burning process

$$3\alpha \rightarrow C^{12}$$

will commence, as is discussed in Chap. D I (p. 249). For this to occur demands temperatures in excess of 10^8 degrees. When helium-burning begins in a degenerate zone we have the peculiar situation that there is no natural balancing process between energy generation and energy outflow as is the case for normal stellar matter. If energy production is inadequate to balance the outward flux, the shrinkage that occurs is accompanied by cooling and by a further falling behind of the energy generation. If energy generation exceeds the outward flux, the resulting expansion is accompanied by further heating which leads to more energy being generated. This situation is inherently unstable, and it can be stabilized only when the temperature has risen to such an extent that the material becomes non-degenerate.

In the Hoyle-Schwarzschild models it is supposed that it is at this point, where the increased energy generation leads to the creation of a non-degenerate core, that the stars cease to ascend the giant branch. Thus the tip of the giant branch is reached when helium-burning has begun. The time for a star to traverse any section of the evolutionary track can be obtained from the models; the total time to traverse the giant branch from $M_v = -0.7$ to -3 is $\sim 3 \times 10^7$ years. These times may be compared with the times estimated empirically from counts of numbers of stars in various parts of the H-R diagram of M 3, given in Sect. 56. The empirical times are slightly longer.

This, then, has traced the evolution of stars in globular clusters up to the top of the giant branch. The next problem is to consider how the stars evolve after helium-burning has commenced, and to explain why they move off and populate the horizontal branch and evolve into the RR Lyrae star region and beyond. HOYLE and SCHWARZSCHILD have given only a tentative discussion of the way in which the later stages of evolution may take place and these stages demand much more investigation. They have considered some models which now contain a non-degenerate core in which helium-burning is taking place; this core must contain an inner convective zone because of the strong dependence of helium-burning on temperature. Outside the helium-burning core there is a hydrogen-burning shell source, while the envelope is supposed to be radiative throughout. This latter postulate means that the true surface condition has again been neglected; the authors argue that if their models turn out to be to the right of the giant branch then their neglect of the surface condition is inappropriate, and only the calculated luminosity is of significance, while if they are to the left of the giant sequence all is well and good. In fact, as can be seen from Fig. 9, the model for which $q = 0.60$ lies on the giant sequence but considerably lower than the tip, indicating that the assumption of a non-degenerate structure by the core forces a star to retrace its steps down the giant branch. The models for $q = 0.70$ and $q = 0.80$ lie considerably off the horizontal branch, though they do show a trend towards rapid evolution to the left. However, when $q = 0.60$ the energy generation in the core by helium-burning is comparable with that in the shell by hydrogen-burning, but the energy generation through the conversion of 1 g of helium into carbon is only about 6×10^{17} ergs. Thus q cannot increase from 0.60 to 0.70 before all of the helium is consumed. So the later models using a helium-burning core and a hydrogen-burning shell are ruled out on energetic grounds. In fact, the helium in the center is rapidly exhausted and a helium-burning shell source is then present. The core will now consist of carbon. Further heating of the core could lead to generation of energy by reactions of the type discussed in Chap. D I. However, no models as complicated as these have yet been considered. Qualitative consideration of such evolution will be discussed in Chap. D.

The models of HOYLE and SCHWARZSCHILD are thus able to explain the evolution of stars as far as the giant branch in globular clusters, the chemical compositions of the models used being characteristic of these extreme population II systems (cf. Chaps. B III and D II). A plausible explanation is also given of the termination of the giant branch. The models are applicable only over a narrow mass range, near $1.1 - 1.2 M_\odot$ and any extrapolation of the evolutionary sequences very far outside this range will probably fail.

To conclude the discussion of the theoretical evolutionary tracks in globular clusters, it is necessary to mention the recent work of HASELGROVE and HOYLE [*130*] and the ambitious program of future work outlined by HOYLE. They have initiated a program of computation of evolutionary tracks using electronic computing machines, and have first of all used this method to rediscuss the problem of the age of stars in globular clusters with special application to M 3. They have computed a series of models using the EDSAC 1, starting with main-sequence models with $M = 1.25\ M_\odot$, $X = 0.9309$, $Y = 0.0666$ and Z (carbon and nitrogen) $=$ 0.0025, parameters which are reasonable for population II stars (this will be discussed in Chap. D II). No mixing between energy-producing core and envelope is assumed. The models develop a core with $q = 0.08$ which eventually is burned out and gives rise to an isothermal core and a shell source. (Their results depend on the old rate of the CN-cycle, which is now superseded, and the assumed abundance of C and N.) Evolution off the main sequence is then followed, the correct surface boundary condition being used, and the models move on to the giant branch. Fitting of the track with the observed main sequence in M3 gives them an age for this system of 6.5×10^9 years. HOYLE has recently outlined a program of work which will take the stars right to the tip of the giant branch and may allow him to consider the evolution onto the horizontal branch. The problem has been programmed for an IBM 704 machine. HASELGROVE and HOYLE [*130a*] have recently made further integrations of type II giant sequences for comparison with the observations of M 3. They carried the evolution through until the last two models were getting an appreciable energy yield from helium-burning. The models showed that until the helium core attained a temperature of about 80×10^6 degrees the star continued to increase in luminosity and radius, moving upward along the giant branch. However, when this temperature was reached, helium-burning became important (the energy yield overcoming the effect of compression), the helium zone became non-degenerate, and with the resulting change in structure the star moved rapidly back down the giant branch. The investigation was stopped at a point at which the luminosity due to helium-burning approached the total luminosity since there were indications that a further instability then ensued, and that mixing between the helium core and the hydrogen envelope might occur.

The value of the fractional mass of helium at a given luminosity ($\text{Log}\, L/L_\odot =$ 1.630) in the model can be compared with the fractional mass of helium obtained using the luminosity function for the stars in M 3. For M 3 SANDAGE (Sect. 56) [*251*] obtained 0.209, while in the models, HASELGROVE and HOYLE obtained 0.195.

An interesting problem, particularly when the chemical evolution of stars is considered, is that of the evolutionary path which will be followed by stars that have condensed out of pure hydrogen, so that the CN-cycle cannot operate at any stage until helium-burning has begun. In a recent paper HAYASHI [*131*] has claimed that pure hydrogen stars burning on the pp-chain in a shell can also explain the location of the giants in the globular-cluster H-R diagrams.

As a digression in their discussion of the evolution of globular cluster stars, HOYLE and SCHWARZSCHILD have considered the evolution of the giants of population I by carrying out similar calculations in which the chemical composition has been supposed different, i.e., they have increased the metal content (which was taken as very small for the globular cluster stars) to the normal value for population I stars. Although there are many uncertainties in such calculations, particularly since the characteristics of convective transport in the envelopes are largely unknown, they have obtained models which to some extent diverge from and tend to lie 2 to 3 magnitudes lower than, but roughly parallel to, the giant sequence in population II. This is in reasonable agreement with observation. Thus it appears that just this difference in composition is sufficient to explain the difference in luminosity between the giants of the two populations. Only in the photospheric regions does the metal content play a critical role. It greatly affects the extent of the envelope, thus changing the radius considerably for a given luminosity, but hardly affects the deep interior at all.

27. Evolution of massive stars off the main sequence. In considering the development of a chemical inhomogeneity between core and envelope and its role in the evolution of a star off the main sequence, the detailed way in which this inhomogeneity develops has so far been ignored. Essentially it has been supposed that it develops discontinuously. However, there are two possibilities when a star initially having a well-mixed convective core and a radiative envelope begins to evolve. The mass of the convective core may grow, or it may shrink. If it grows, then a discontinuity will be set up at the boundary, and the star will be sharply divided into two zones of different compositions. If it shrinks, convection will die out in the outer regions of the core, and since this material will have been processed to a lesser extent than material in the inner core, a continuous gradation of chemical composition will be set up. Thus at a given time the star will be divided into three zones: a central convective core with a uniform composition characteristic of its life so far, an intermediate radiative zone, inert from the point of view of nuclear energy generation but with a variable chemical composition reflecting the retreat of the convective core, and an outer radiative zone with a composition characteristic of the original material out of which the star condensed.

This intermediate zone was neglected in the calculations of SCHOENBERG and CHANDRASEKHAR [*107*], HARRISON [*109*], and SANDAGE and SCHWARZSCHILD [*125*], for the evolution of stars off the main sequence. However, it has been taken into account in the work of TAYLER [*132*] and of KUSHWAHA [*133*], and since it is apparent that for massive stars, where electron-scattering opacity is dominant, this intermediate zone is of importance, these models are the best available at present for investigating the way in which massive stars begin to evolve. For stars of smaller mass, Kramers opacity is more important and the extent of the intermediate zone will decrease. For example, in the original models of SCHOENBERG and CHANDRASEKHAR the initial mass of the convective core was 14.5% of the total mass, while the mass limit for the isothermal core was 10%.

Both TAYLER and KUSHWAHA have supposed that the CN-cycle is operating. In his first paper TAYLER assumed electron scattering opacity throughout while in the second he used only Kramers opacity. KUSHWAHA used a formula in which both opacity terms were added together, and this is probably the best approximation so far used. The major difference between these models and those in which the transition zone can be neglected (i.e., those with $M \approx 1 - 1.2\,M_\odot$) is as follows. In models such as those of SANDAGE and SCHWARZSCHILD, the star reaches the limiting core mass before gravitational contraction of the core and expansion of the envelope moves it rapidly to the right in the M_{bol}–Log T_e plane, having brightened by about a magnitude before abruptly leaving the vicinity of the main sequence. But massive stars with considerable intermediate zones will move almost immediately to the right off the main sequence and will only slowly increase in luminosity while doing this. Thus, for example, TAYLER has carried one of his models through until nearly 20% of the mass has been converted, but the isothermal core condition has still not been reached.

KUSHWAHA has considered masses of 2.5, 5, and $10\,M_\odot$. He has also attempted to take into account the fact that after a sufficient time has elapsed the decrease of hydrogen in the core as well as the general shrinkage of the core means that energy generation in the intermediate zone begins to be felt again. The theoretical evolutionary tracks for his models, which are quite similar to those of TAYLER, are shown in Fig. 10. The dotted line joins the points corresponding to the same age for the three stars. The starting models for the three stars were shown by him to lie quite close to the observed main sequence in this mass range. The dotted line should be compared with the observed H-R diagrams for such clusters as the Hyades and the Pleiades which are discussed in Sect. 45.

For the most massive star Kushwaha found that in the last inhomogeneous models the effective temperature reaches a minimum and then begins to increase again, thus forming an elbow in the evolutionary track which begins to move back towards the main sequence. Such an effect was also found for a less massive star by Henyey, Lelevier and Levée [*134*]. It is believed to be caused by the sharp reduction in the hydrogen content in the energy-generating core at this epoch, which necessitates a strong temperature increase and thus a shrinkage of the star.

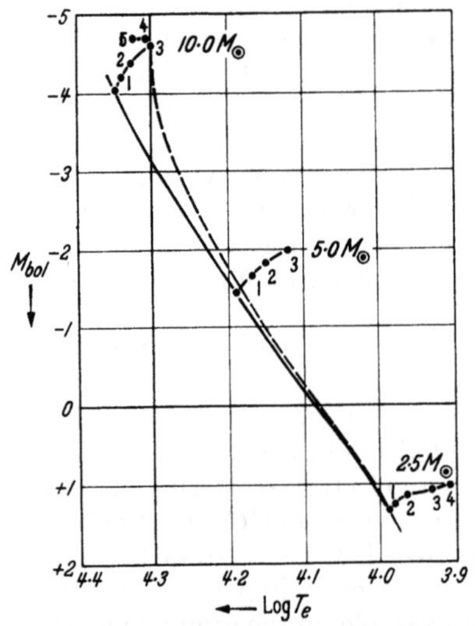

Fig. 10. The evolutionary tracks in the Hertzsprung-Russell diagram for three heavy stars with 10, 5, and 2.5 $M_{\odot}$, respectively, according to Kushwaha [*133*]. The line from the upper left to the lower right represents the main sequence. The solid curves show the evolutionary paths of each of the masses. The dotted line joins the points corresponding to the same age for the three stars. Black dots represent the different evolutionary models actually computed.

None of the tracks have been computed to the point where energy production in the core ceases and the isothermal condition is reached. Beyond this point it must be supposed that the release of gravitational energy and the development of a shell engery source involving both the intermediate zone and part of the unburned envelope, i.e., a thick shell source, will begin. The central temperature in the later stages of the most massive models of Kushwaha is already about 40×10^6 degrees (a rise of about 10×10^6 degrees over the initial models), so that after further contraction helium-burning will not be long delayed.

As will be discussed in Sect. 72, there is some observational evidence for mass loss by stars in the high-temperature giant and supergiant regions. Thus evolution with mass loss should be considered when tracks are calculated in the H-R diagram for massive stars. This has been done by Massevich [*187*] in considering the evolution in h and χ Persei (see Sect. 44). It is clear that, if we have a series of evolutionary tracks without mass loss for stars in the massive star region, covering a wide range of masses, it is possible to construct from them the evolutionary path for a star which is losing mass at a given rate, provided that it is supposed that the process of mass loss does not lead to an instability that becomes catastrophic. Since the tracks without mass loss after the stars leave the main sequence run roughly horizontally in the M_{bol}–Log T_e plane, according to Kushwaha and Tayler, the tracks with mass loss will tend to slope downwards from left to right in this plane, the slope of the curve being determined by the assumed rate of mass loss (which is probably variable).

It is generally believed that the next important stage of evolution of the massive stars is into the red supergiant region of the H-R diagram (Sects. 44 and 47). Construction of models of such stars would appear to be the next stage of the theoretical development.

VI. An empirical approach to evolution beyond the giant and supergiant stages.

After stars leave the red giant or supergiant regions of the H-R diagram, their evolution can be only schematically outlined, since computations have not yet carried stellar models as far as this. A number of factors may determine the course of this evolution:

(i) The exhaustion of nuclear fuels and the contraction of the stellar core which must occur to increase the temperature sufficiently to release the last amounts of nuclear energy (cf. Chap. D I).

(ii) The possible occurrence of mixing.

(iii) Steady ejection of mass, which may lead to more rapid evolution than would be the case if only nuclear transmutation were involved (cf. Chap. C II).

(iv) Catastrophic ejection of mass which may arise through a thermonuclear explosion and which might take the star immediately to a white dwarf configuration (cf. Chap. C II).

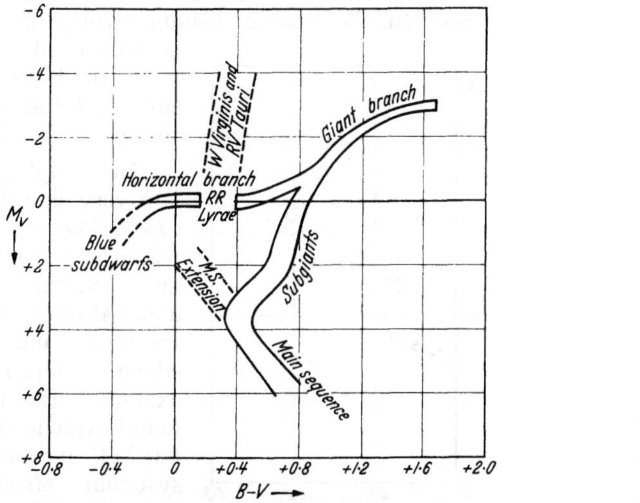

Fig. 11. Schematic color-magnitude diagram for a globular cluster. The ordinate is the absolute visual magnitude, and the abscissa is the B − V color index.

28. Horizontal branch stars. We must anticipate to some extent the discussion in Chap. B III and IV, where the observations in star clusters are used to formulate an empirical approach to stellar evolution. The existence of a well-populated sequence of stars forming the horizontal branch in the color-magnitude diagrams of globular clusters (cf. the article by H. C. ARP in this volume) suggests that this represents the evolutionary track, after leaving the giant branch, for stars with masses near $1.2 M_{\odot}$. Counts of numbers of stars in the horizontal branch, discussed in detail in Chap. B IV, suggest that the time spent here by a star is comparable with the time spent in the giant branch but that it is much shorter than the time spent on the main sequence.

A schematic H-R diagram of a typical globular cluster is shown in Fig. 11. In all the clusters for which accurate measures of colors and magnitudes are available, the giant branch bifurcates; one part is the subgiant branch leading up from the main sequence, while the other leads down, through a sparsely populated region, into the horizontal branch, which is at or near $M_v = 0$. There is a gap between the giant and horizontal branches in, for example, M 13 (see [*135*] and ARP's article in this volume).

In most globular clusters there is a short, sharply defined section of the horizontal branch which is filled exclusively with RR Lyrae variable stars. Quite wide variations exist in the relative numbers of stars on the blue and red sides of this "gap" where the variables occur. As SANDAGE [*136*] and ARP [*135*] showed, the number of variables in a cluster is strongly correlated with the relative populations of the blue and red sides of the variable star region, in the sense that clusters with the greatest number of variables (e.g. M 3) have almost equal stellar density on either side of this region, while M 13, with only a few variables, has no stars on the red side of the variable region. There is also a difference in the shape of the horizontal branch from cluster to cluster; for example, it is far from horizontal in M 13 and M 10, where the bluer stars are progressively fainter [*137*], [*135*]. There is variation also in the faint magnitude limit to which the blue end reaches. Assuming that the horizontal-branch stars represent a later evolutionary stage than the red giants, it is not clear whether a star, after leaving the tip of the red-giant branch, travels back down it, into the left-hand fork, along the horizontal branch from right to left, and down into the faint blue dwarf region, or whether it jumps from the red giant tip to the left end of the horizontal branch, and travels along this from left to right. That a star does travel along the horizontal branch is suggest by the continuity of star counts through the variable star gap in those clusters with an appreciable number of variables, but clusters like M 10 [*135*] do not necessarily suggest this. As discussed in Sect. 26, the onset of helium-burning or some other mechanism may lead to a structural adjustment which moves the star back towards the horizontal branch. Alternatively, it is possible that the star may first oscillate up and down the giant branch.

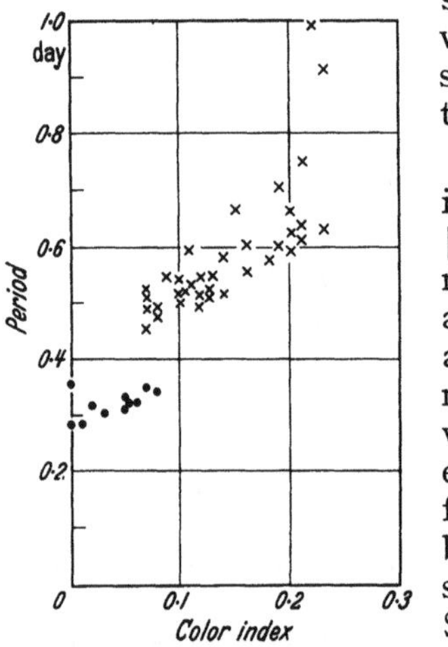

Fig. 12. Correlation between mean color index and period for variables in M 3, according to Roberts and Sandage [*139*].

The "gap" where the cluster-type variables occur is a very sharply defined region of the H-R diagram [*138*]. The extreme sharpness of the boundaries of this region was confirmed in M 3 and M 92 by Roberts and Sandage [*139*] and by Walker [*140*]. Roberts and Sandage found an excellent correlation between mean color index and period for the variables in M 3, which is reproduced in Fig. 12. Belserene [*141*] examined the periods of the variables in M 3 and found evidence for both abrupt and gradual changes, but both increases and decreases were found, with no systematic effect. The observations by Roberts and Sandage in most cases confirmed these apparent changes, but still on systematic effect was apparent. Hence, although a method is available here in principle for determining the direction of evolution, it has not led so far to a solution of the problem.

A third possibility must be borne in mind. Stars, after leaving the red giant branch, might undergo such structural changes as to move them very rapidly in the H-R diagram, but they might reach a stable configuration whose locus is defined by the horizontal branch. The position which they would take up on this branch might be defined by the amount of mass-loss experienced in the red giant stage, the amount of mixing, or the way in which mixing set in. Those stars finding themselves in a limited range of surface temperature might be unstable against pulsation (regular pulsation seems likely to be a phenomenon depending on the outer layers of a star and not on conditions within the core). After a period of structural stability, the stars might again become unstable and move rapidly off the horizontal branch, probably towards a white dwarf stage characterised by exhaustion of nuclear fuel and star death.

Although so far the existence of stars on a horizontal branch has been established only in globular clusters, there may be a region in old galactic clusters (population I) which is analogous to this. There is an indication of this in the color-magnitude diagram of M 67, where the evolving stars have masses near the evolving stars in globular clusters. This is discussed in Sects. 47 and 56, and is shown in Fig. 23 (p. 222). This sequence is much more sparsely populated than the corresponding region in globular clusters, and, if real, must therefore represent a more rapid evolution. It is not known yet whether this feature is a definite characteristic of old clusters like M 67, whether it is characteristic of both high- and low-mass stars in population I, or whether it is unique in M 67.

29. Hot subdwarfs. In Chap. D the chemical evolution of stars is considered, and observational evidence of the existence of stars which have used up all or most of their hydrogen is reviewed. It is shown there that many of the various types of stars in this category may be called hot subdwarfs from their positions in the H-R diagram. This is along the same lines as the evidence from globular clusters that the hot subdwarfs, at the end of the horizontal branch, may represent an advanced evolutionary stage, which is certainly subsequent to the red giant stage. However, this viewpoint involves considerable inference from, and extrapolation of, the direct evidence from observations, since no evolutionary tracks have been computed. Many of the hot subdwarf stars discussed in Chap. D II are solar-neighborhood stars, and are therefore not comparable to globular cluster stars, although the nuclei of planetary nebulae represent an older population and the stars in Sect. 89α may belong to the halo population. Further, the star *υ* Sgr, which must according to its chemical composition be near the end of its life, is an *A*-type supergiant, not a subdwarf.

The classical Wolf-Rayet stars may be classed as hot subdwarfs. They have very high temperatures and their luminosities[1] place them probably very slightly below the upper part of the main sequence. The spectra of classical WR stars indicate that considerable ejection of mass is occurring, although this interpretation of the observations is not universally accepted. As discussed in Sect. 89, these stars show evidence of having exhausted much of their hydrogen, while the products of hydrogen-burning and helium-burning have apparently mixed to the surface.

The classical WR stars show a high degree of concentration to the galactic plane (they belong to a flattened subsystem; see Chap. B V) and tend to be associated with *O* and *B* star aggregates. It may thus be inferred that they are recently formed stars, and their advanced stage of evolution implies that they must be massive stars which have a very short life time. However, a mass of only $10\,M_\odot$ for the WR component of V 444 Cyg has been determined[2]. The *O*-type star in V 444 Cyg is still more massive than the WR star, and yet for the latter to have evolved further, it would seem that it must originally have been more massive than the *O* star. Perhaps the discrepancy can be accounted for if the rate of mass loss from the WR star is high enough.

The planetary nebulae are rare in globular clusters. Their position in the H-R diagram would place them above the blue end of the horizontal branch, so their connection with the blue stars on the horizontal branch is not clear. Apparently they belong to an intermediate, not extreme, population II. Stars which eject material to form planetary nebulae must be unstable, and the fact that some nuclei of planetary nebulae have abnormal chemical compositions, even though the composition of the nebulae themselves seems normal (Sect. 89), suggests that they have reached a fairly late evolutionary stage.

A theoretical sequence in the H-R diagram for stars obtaining their energy from helium-burning was constructed by CRAWFORD [*142*]. The models were for a range of masses from 0.5 to $5\,M_\odot$, and had a small convective core. Central temperatures were close to 2×10^8 degrees. SALPETER'S[3] formula for the energy

---

[1] N. G. ROMAN: Astrophys. J. **114**, 492 (1951).
[2] O. C. WILSON: Astrophys. J. **91**, 379 (1940).
[3] E. E. SALPETER: Astrophys. J. **115**, 326 (1953).

generation from helium-burning was used and gave an approximate energy-generation law of the form

$$\varepsilon \sim \varrho\, T^{19.5}.$$

The opacity was taken to be due to electron scattering, so that the polytropic index in the envelope was 3.

The resulting sequence extends from a spectral type of about *O6* for a star of 0.5 $M_\odot$ towards hotter temperatures; it lies below the normal main sequence but converges toward it at the high-temperature end. It was suggested that the nuclei of planetary nebulae are contracting, having exhausted their hydrogen, until they reach temperatures at which helium-burning starts, when they take up positions on this curve. When the positions of stars are plotted in this diagram they are consistent with this idea, if the stars have masses $\sim 1\, M_\odot$. This diagram is reproduced in Fig. 13. The classical WR stars may also lie near the upper end of the curve, which has not been extended beyond 5 $M_\odot$.

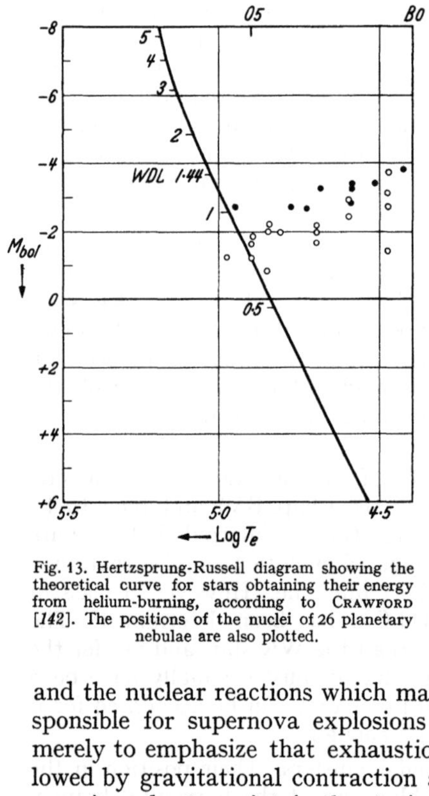

Fig. 13. Hertzsprung-Russell diagram showing the theoretical curve for stars obtaining their energy from helium-burning, according to CRAWFORD [*142*]. The positions of the nuclei of 26 planetary nebulae are also plotted.

30. Supernovae, novae, and explosive variables. When a star is using the last of its nuclear fuel, it must have a very high central temperature and pressure, and may easily become unstable. Supernova explosions seem likely to be the manifestations of such instability. Novae and nova-like variables of the SS Cygni and related types may represent lesser instabilities. Both supernovae and novae are discussed in detail in this volume by F. ZWICKY and C. PAYNE-GAPOSCHKIN. Further, estimates of the amount of mass ejected in catastrophic stellar out bursts are described in Chap. C II, and the nuclear reactions which may come at the end of a star's life and be responsible for supernova explosions are discussed in Chap. D I. Here we wish merely to emphasize that exhaustion of any kind of nuclear fuel is always followed by gravitational contraction and rise of the central temperature and pressure, in order to maintain the star's energy output, and this process must end in instability [*143*], [*62*].

Supernovae can be divided into two types, distinguished by their light curves and total light emitted [*344*] (cf. Sects. 73, 81, and 92). REAVES[1] has shownthat Type II supernovae tend to be found in spiral arms, and they have been reported so far only in spiral galaxies. This suggests an affiliation between Type II supernovae and population I. Type I supernovae have been found in elliptical galaxies and this suggests that they are associated with population II. Three supernovae in our Galaxy, all considered by BAADE to be of Type I, are between the arms, i.e., of disk population II (cf. Sect. 58).

Pre- and post-novae lie in the hot subdwarf region of the H-R diagram (see collected data and references in the book by PAYNE-GAPOSCHKIN [*144*], Chap. 11). It is not known observationally where pre-supernovae lie in the H-R diagram;

[1] G. REAVES: Publ. Astronom. Soc. Pacific **65**, 242 (1953).

theoretically it is to be expected that they may lie among stars whose chemical compositions show a high degree of exhaustion of nuclear fuels. The Crab Nebula, which is a post-supernova, consists of an amorphous mass of gas containing very high-energy electrons and positrons and a magnetic field[1,2], and an outer filamentary nebular structure. In the center is a star which is probably the remnant of the explosion. It has an absolute magnitude of $+4.8$[3] and a color index of $+0.14$[4]. The spectrum shows no lines at all.

The ex-nova DQ Herculis is a member of a very close binary system, and its companion may be an M-type dwarf [*145*].

First estimates for the radius and mass of the ex-nova were $0.10\,R_\odot$ and $0.006\,M_\odot$, respectively, but WALKER [*145a*] has shown that the light curve does not repeat exactly and the estimates, particularly of the mass, may need revision.

It is not known at present whether DQ Herculis is unique among novae in being a member of a close binary system. Many of the nova-like variables, in particular SS Cygni[5] and AE Aquarii[6], are also close binaries. The recurrent nova T Coronae Borealis is probably a spectroscopic binary ([*144*], p. 108).

A phenomenon which occurs in the hot subdwarfs MacRae $+43°$ 1 (which may be either a pre- or post-nova) is rapid light-variation[7] that may take place in minutes. This type of flickering variability has been found in a number of ex-novae, in stars thought from their spectra to be old novae[7], and in SS Cygni[8]. In DQ Herculis it has a periodicity of 1.18 minutes and a range of 0.07 magnitudes [*145*]. The cause of this flickering is not understood at present, but WALKER [*145a*] has suggested that in DQ Herculis it might be due to pulsation, since it is so regular. If so, then for a reasonable range of values of the surface temperature, the radius lies between 0.059 and $0.20\,R_\odot$, and the mass between 0.4 and $15\,M_\odot$.

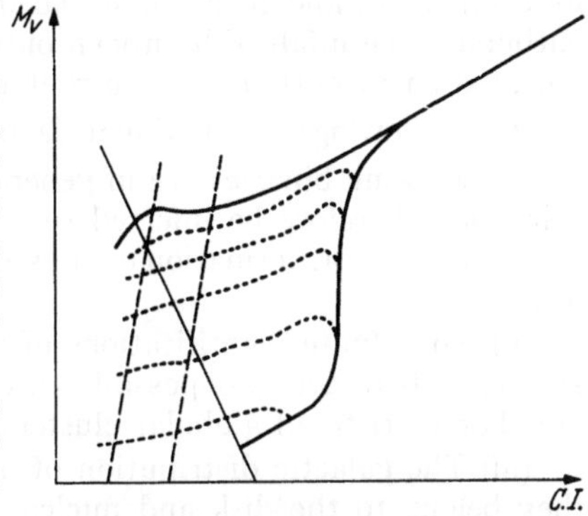

Fig. 14. A region of unstable stars in the Hertzsprung-Russell diagram as suggested by STRUVE and HUANG [*146*]. Different dotted curves show the evolutionary tracks on the H-R diagram of stars which are the components of close binary systems of different separations. The unstable zone is located between the two broken lines in the figure.

An evolutionary scheme for binaries of the type of DQ Herculis and SS Cygni has been proposed by STRUVE and HUANG [*146*]. It is supposed that in close binary systems considerable mass loss and mass exchange may occur. A discussion of the mass-loss part of this hypothesis is given in Chap. CII. This scheme is illustrated in Fig. 14, taken from the paper of STRUVE and HUANG. As has already been discussed, a single star of population II is believed to evolve so as to follow approximately the observed color-magnitude diagram of globular clusters (cf. Sects. 26 and 28). Close binaries, however, begin to lose mass as their radii increase to various limits determined by their separations. The stars may then follow tracks indicated by the dotted lines in Fig. 14; it was suggested

[1] J. H. OORT and T. WALRAVEN: Bull. astronom. Inst. Netherl. **12**, 285 (1956).
[2] G. R. BURBIDGE: Astrophys. J. **127**, 48 (1958).
[3] R. MINKOWSKI: Astrophys. J. **96**, 199 (1942).
[4] W. BAADE: Astrophys. J. **96**, 188 (1942).
[5] A. H. JOY: Publ. Astronom. Soc. Pacific **55**, 283 (1943).
[6] A. H. JOY: Astrophys. J. **120**, 377 (1954).
[7] M. F. WALKER: Publ. Astronom. Soc. Pacific. **66**, 71 (1954).
[8] G. GRANT: Astrophys. J. **122**, 566 (1955).

that such stars may not be too far from homologous models as regards their internal structure. The dashed lines represent an extrapolation of the region of instability in which the RR Lyrae variables lie. Thus, whereas single stars or well-separated binaries may evolve through the RR Lyrae stage, close binaries will manifest their instability in a different way. Those which lose considerable mass may become SS Cygni variables, while those which undergo the largest mass loss may become novae. It has also been proposed by SCHATZMAN [*147*] that the phenomena of pulsation, explosive variability, and nova outbursts might be related.

A somewhat different scheme has been proposed by CRAWFORD and KRAFT [*148*] for AE Aquarii. They suggest that matter is falling into the variable star from its companion, and it is the actual infall of matter which is responsible for the outburst. The infall of hydrogen onto the surface of a white dwarf might cause repeated small outbursts, recurrent novae, or full-scale nova outbursts.

The following observational facts may have a bearing on these ideas:

(i) Eclipsing binaries are in general not found in globular clusters and double stars may therefore be inferred to be absent. This may be connected with the conditions for star formation or subsequent stellar encounters in a dense aggregate of stars.

(ii) No definite identifications of novae in globular clusters have been made, although there are two possible cases ([*144*], p. 26), while RR Lyrae variables are characteristic of globular clusters.

(iii) The galactic distribution of novae, like planetary nebulae, suggests that they belong to the disk and nuclear regions of the Galaxy, rather than to the halo population [*149*]. In the nuclear region of the Galaxy, eclipsing binaries are frequent.

An evolutionary scheme for novae and related variables which is unconnected to the binary nature of these objects has been proposed by VORONTSOV-VELYAMINOV [*150*].

The role of angular momentum in a contracting star that is approaching the end of its life must be mentioned. Rotational instability as a factor in determining whether a star would become a nova or supernova was once considered by HOYLE [*143*]. Mass loss prior to an explosion may carry away angular momentum, but this has not been taken into account. Magnetic fields as factors in nova outbursts will be considered in Chap. E.

31. White dwarfs. The observations and theory of white dwarfs are described in Volumes L and LI of this Encyclopedia, by J. L. GREENSTEIN and E. SCHATZMAN respectively. White dwarfs will be discussed here only in so far as they represent a final stage in the evolution of stars.

White dwarfs are well known to be abnormally small and extremely dense. The theory of the structure of such stars (cf. [*46*], Chap. 11) shows that in their interiors the electron gas is degenerate. Whatever the chemical composition may be, gravitational separation of the elements may occur [*151*] so that observation of the spectra of white dwarfs may show only the elements of lowest atomic weight. Further, accretion of interstellar gas may mask the true composition of the star (cf. Sect. 69). The limiting mass for a stable configuration (CHANDRASEKHAR'S limit) is $5.75\ M_\odot/\mu_e^2$, where μ_e is the number of electrons per proton mass. Thus if a star exceeding this mass is to become a white dwarf it must eject mass.

It has been emphasized by MESTEL [*152*] that no unconsumed hydrogen can remain in the central regions of such a star, for the temperature would be high enough to liberate energy which would not be balanced by radiation; a white dwarf does not contain the self-balancing mechanism possessed by normal stars and could not remain in equilibrium. However, it is not known whether the star must be completely exhausted of nuclear fuel, i.e., be composed of the iron group of elements at the minimum of the packing fraction curve. It is conceivable that some energy sources might remain but be untapped, since temperatures $>10^9$ degrees are needed to liberate the last amounts of energy (cf. Chap. DI).

The evolution of stars after they have become white dwarfs, if they remain undisturbed, should follow a path in which they keep constant radii (they can contract no further), and radiate energy from their own internal heat supply. They should thus follow "cooling curves" appropriate to their radii (which are determined by their masses). Absolute magnitudes and colors of well-observed white dwarfs[1] [*153*] have been shown to lie very well along the computed curve for constant radius [*153*], [*154*]. The time for a star of mass M to cool from temperature T_1 to T_2, with corresponding luminosities L_1 and L_2, is given by [*152*]

$$t = 0.6\,\frac{RM}{A}\left\{\frac{T_2}{L_2} - \frac{T_1}{L_1}\right\}$$

where A is the mean atomic weight of its material. Cooling times given by this equation range from $10^8/A$ years to $10^{11}/A$ years. Thus, for example, if $A=4$ this gives the large age of 2.5×10^{10} years for some white dwarfs. If $A=56$, the age of a star such as VAN MAANEN 2 would be 2×10^9 years.

If a white dwarf does not remain undisturbed, for example, if it accretes material containing hydrogen, its evolution may not be quiet, but explosive instability may occur. We have already mentioned the suggestion that this might be the cause of nova outbursts in close binary systems; MESTEL [*152*] has proposed that such accretion by single stars might give rise to supernova outbursts.

From the motions of the white dwarfs they are deduced to belong to an intermediate galactic subsystem (cf. Sect. 59). White dwarfs can be counted only in a small sample of space, where a density of 0.15 per psc^3 has been found [*155*]. This suggests that the total number is large [*156*]. If a white dwarf represents a possible final stage in the life history of stars in any mass range, all those stars formed with masses greater than the Chandrasekhar limit must lose the excess mass before becoming white dwarfs. The large number of white dwarfs in the Galaxy thus indicates that a large proportion of the material of the Galaxy has been through at least one generation of stars.

The formation of white dwarfs in a galaxy eventually removes mass from the possibility of continuing exchange with the interstellar medium. Thus a conceivable end stage in the history of a galaxy would be the tying up of almost all of its mass in the form of white dwarfs. An approach to this stage might be indicated by a high ratio of mass to light (see discussion in Chap. BV).

VII. Evolution of the Sun.

32. Now that all of the different stages of stellar evolution have been discussed, it is of interest to give an outline of current ideas concerning the evolutionary track of the Sun from its condensation to its final demise.

The protostar stages of solar evolution are related to the origin of the solar system, a topic which has been investigated and speculated upon by a large

[1] D. L. HARRIS: Astrophys. J. **124**, 665 (1956).

number of investigators in the last several hundred years, but which lies outside the scope of this article. A review of such theories has been given by TER HAAR [*157*]. Early authors did not consider the origin of the Sun itself but were concerned with the ejection or capture of material to form the planets. However, the more recent developments begun by VON WEIZSÄCKER [*158*] and continued by TER HAAR [*157*], [*159*] and KUIPER [*160*] can be considered in the general framework of the formation of stars. Thus, by one of the processes described in Chap. AI, it is supposed that a cloud out of which the solar system forms becomes isolated and begins to contract. The nuclear region of the cloud becomes highly condensed, i.e., it is a well-advanced stage of the primitive Sun, but it is surrounded by the so-called solar nebula which has a mass $\sim 0.1\,M_\odot$. It is then supposed that a large part of this discoidal nebula is dissipated into space carrying the excess angular momentum with it, while the remainder is condensed into the planets.

The time scale for these processes is intimately connected with the amount of radioactive heating that occurred in the proto-planets. The chemical problems of the origin have been extensively discussed by UREY [*161*], [*21*], and by KUIPER [*160*]. UREY has raised a number of difficulties connected with the mechanisms that must be invoked to remove a large amount of material from the protoplanets, as envisaged by KUIPER, and has proposed a rather different sequence of events.

However, the scheme proposed by KUIPER is as follows:

(i) Gravitational contraction of the cloud until angular momentum arrests its progress.

(ii) Quiescent period of slow internal rearrangement of matter which by now has collapsed to a disk and is in rapid rotation; cooling of the cloud by radiation and the onset of gravitational instability in the densest parts.

(iii) The central star becomes luminous, i.e., it is getting near to the main sequence; the interior of the cloud is ionized; the angular velocity of the star is then reduced by magnetic braking in the way proposed by LÜST and SCHLÜTER [*36*].

(iv) The Sun reaches the main sequence and the nebula is dissipated.

There is considerable uncertainty as to the status of the protosun when the planetary condensations occur, but, for example, it has been supposed that the solar nebula lasts for a time $\sim 10^7$ years, a time during which the Sun will contract from a radius $\sim 5\times 10^{14}$ cm to 6×10^{12} cm; i.e., the Sun is still in its gravitational contraction phase with an effective temperature far lower than its main sequence value. It takes a total time of $\sim 5\times 10^7$ years to contract on to the main sequence. The gravitational contraction track for a solar mass has been calculated by HENYEY *et al.* [*51*] and is shown in Fig. 3 (p. 160).

The type of eddies which existed in the solar nebula and the way in which they condensed into the planets has been the subject of detailed discussion by a number of authors. The main point of interest for stellar evolution is that both VON WEIZSÄCKER and KUIPER have proposed that a cloud similar to the solar nebula may either give rise to a solar system such as our own, or it may give rise to a large number of small condensations, or most of the material may be captured by a single condensation in the nebula, thus giving rise to a binary system. This theory of binary star formation has been discussed in Sect. 10.

It is generally supposed that the Sun arrived on the main sequence about 5×10^9 years ago. This is compatible with the presently estimated age of the Earth of 4.55×10^9 years. We suppose that it was initially condensed with a homogeneous chemical composition. However, if it has not remained mixed throughout during the whole of its later life, it can be expected that by now it will contain a considerable chemical inhomogeneity.

A number of solar models have been constructed which attempt to represent the Sun as it is today by a homogeneous model in radiative equilibrium (e.g. [*162*] to [*165*]). Earlier models than these have been discussed by STRÖMGREN [*166*]. It has been shown earlier (Sect. 16) that effectively the major part of the energy generation in the Sun takes place through the *pp*-chain. These homogeneous models have not been successful. For example, the investigation of ABELL showed that with the assumptions which he had made it was impossible to obtain any sensible solution for the chemical composition, i.e., his model must contain 114% hydrogen. This was exactly the situation which HENYEY *et al.* [*51*] envisaged: "Since the evolution of a chemically inhomogeneous configuration is (in the direction of) higher luminosity and lower effective temperature, any attempt to represent a star, which, like the Sun, has suffered serious depletion of hydrogen in the central regions, by a homogeneous model will be subject to the following serious criticism. The effects of evolution will be taken up by making large fictitious modifications in the composition... Extravagant values for the hydrogen content may be encountered."

In addition to the need for taking into account the chemical inhomogeneity, the work of OSTERBROCK [*66*] has shown that for stars of masses as low as the solar mass an outer convective zone should be taken into account. The work of VITENSE [*167*] has shown that such a zone exists in the Sun. Models incorporating both of these corrections from the earlier models have been constructed by SCHWARZSCHILD, HOWARD, and HÄRM [*63*], the minor correction due to the CN-cycle being calculated by WEYMANN [*64*]. The method adopted by SCHWARZSCHILD *et al.* gives some insight into the solar evolutionary path up to the present time. They have first constructed a solar model which has a homogeneous composition but which contains an outer convective zone. This model represents the Sun immediately after it contracted on to the main sequence. From this model it is possible to compute the rate of transmutation at every point in the Sun. Having done this the authors are able, for a given time interval, to calculate the depletion of the hydrogen and hence the composition of the Sun today. They have assumed that the initial hydrogen content is that now present in the envelope, X_e, and have supposed that the hydrogen consumption in the initial model is a good approximation to the average over the time interval of 5×10^9 years.

With these new parameters they have then calculated an inhomogeneous solar model. In deriving final models they have had the same problem as that encountered by OSTERBROCK in considering red dwarfs. The mass-luminosity relation and the energy output equation are not sufficient to determine the three unknowns X_e, the envelope helium content Y_e, and a parameter E which represents the depth of the convection zone. They have therefore made three different assumptions about X_e and have calculated the other parameters accordingly. The results are shown in Table 2.

These are the best available published figures to date for the Sun as it is at present. Although it was assumed that the solar luminosity stayed roughly constant and equal to its initial value in order to make these calculations, it is possible to go back and make a detailed comparison between the two models—the initial and the present-day model—to determine the actual increase, since for both, a formula for the total luminosity independent of radius can be determined. SCHWARZSCHILD *et al.* have found, by doing this, that the Sun has increased in luminosity by about 20% or $\sim$0.5 magnitudes in the last 5×10^9 years.

Table 2. *Inhomogeneous solar models by* SCHWARZSCHILD, HOWARD and HÄRM [*63*].

	Model 1	Model 2	Model 3
X_e	0.60	0.70	0.80
Y_e	0.30	0.26	0.185
Z_e	0.10	0.04	0.015
Fraction of radius in radiative equilibrium	0.810	0.818	0.824
Fraction of mass in radiative equilibrium	0.995	0.996	0.997
Temperature at bottom of convective zone	1.46×10^6	1.27×10^6	1.12×10^6
Density at bottom of convective zone (g/cm³)	0.049	0.041	0.035
Central temperature	17.1×10^6	15.8×10^6	14.8×10^6
Central density (g/cm³)	122	127	132

Further unpublished work on solar evolution in which the change of radius of the model is taken into account has been carried by HOYLE [*356*] and SEARS [*357*]. HOYLE has given us some details of his evolutionary track, which is somewhat similar to that obtained by SEARS. HOYLE has used the Keller-Meyerott opacity tables but has found it possible to write them in algebraic form for machine computation, the departures from the exact (numerical) tables being less than 10%. It is assumed that energy generation is by the pp-chain, the CN-cycle contribution being assumed to be negligible. In order to obtain the correct present-day value of the radius (after 5×10^9 years of evolution), it was found necessary to increase the rate of energy generation by the pp-chain arbitrarily by a factor ~ 3. Since the computations were carried out, new experimental work on the reaction

$$\mathrm{He^3 + He^4 \to Be^7},$$

leading to completion of the pp-chain[1] through the reactions

$$\mathrm{Be^7 \to Li^7} + \beta^+ + \nu_+ \quad \text{and} \quad \mathrm{Li^7} + p \to \mathrm{He^4 + He^4},$$

as originally suggested by BETHE [*55*], has suggested that the energy-generation rate may be increased by a factor 2 over the values given in [*62*]. The overall factor in the Sun, when integrated over the energy-producing core, will be about 1.2. However, further work is needed to see whether this alternative route for the pp-chain has been effective over a large part of the Sun's history.

The correct surface condition was used in HOYLE's computations, essentially in the same way as in the work by HOYLE and SCHWARZSCHILD discussed in Sect. 26, except that the analytic approximation was replaced by a numerical integration. The evolutionary track for the Sun was followed (in about 20 steps) for about 11×10^9 years, at which point the luminosity had become more than three times the present value and the radius about 1.7 times the present value. This computation is to be preferred to the approximate treatment of the initial stages of evolution off the main sequence, using the Schoenberg-Chandrasekhar limit, given by SANDAGE [*168*].

SANDAGE [*168*] has considered the probable future evolution of the Sun, after it has passed through the stages just discussed, by using the observed evolutionary tracks of the stars in the galactic cluster M 67 (Sects. 46 and 56). The masses of these stars are near to $1.2 M_\odot$, and he has carried out homology transformations in order to obtain the evolutionary track for a star of solar mass. The predicted evolutionary track obtained in this way is shown in Fig. 15. If this comparison is valid, in approximately a further 6×10^9 years the Sun will have reached the Schoenberg-Chandrasekhar limit and will move very rapidly

[1] W. A. FOWLER: Astrophys. Journ. **127**, 551 (1958).

off the main sequence, having already brightened about a magnitude from its original luminosity. It will then move rapidly into the red giant branch and begin helium-burning and then move back on to the horizontal branch, probably passing through a pulsating phase. We do not know sufficient to determine whether it will undergo a supernova explosion. Eventually, some $11-12\times10^9$ years after its contraction on to the main sequence, it will enter the white dwarf phase of evolution and will very slowly cool. The increase of radius as it moves away from the main sequence and its increase in luminosity as it becomes a red giant will have considerable effects on the planetary system, and will destroy what little life remains on Earth after man's evolution has come to a catastrophic end.

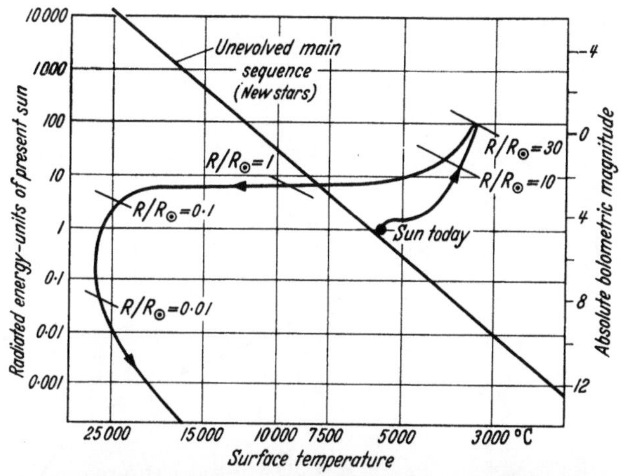

Fig. 15. Temperature-luminosity diagram showing the evolutionary track of the Sun, according to SANDAGE [168]. The radius of the Sun, in terms of its present value, is shown along the track.

It appears that the Sun is $1-2\times10^9$ years younger than the oldest stars so far dated in our Galaxy. A theory in which it is supposed that all of the elements were built in stars, starting from hydrogen, is described in Chap. D I. The present composition of the outer layers of the Sun, which must be original material out of which it first condensed, shows that this material must have passed through stars and undergone a number of nuclear processes before it condensed in the form of the Sun. It appears improbable that all of these processes could have taken place in the evolution of a single star. Thus the Sun is probably at least a third-generation star. In terms of the classification scheme described in Sect. 58 it is an old (disk) population I star.

B. Associations, clusters, and galaxies: Empirical approach to stellar evolution.

33. Introduction. In Chap. A, the evolution of individual stars has been discussed. Since such evolution is on too long a time scale for observation, it is mainly the statistical properties of numbers of stars which have led in the past to various evolutionary hypotheses. Historically, the statistical study that has been most closely connected with such hypotheses has been the plotting of absolute luminosities and spectral types (or colors, the abscissa being in either case the surface temperature on some scale) into the Hertzsprung-Russell diagram. This topic is treated in detail from the observational standpoint in the article by

H. C. ARP in this volume. We shall discuss here the connection between the regularities shown by the H-R diagram and evolutionary hypotheses.

Since the luminosity of stars varies as the mass raised to a power greater than one, while the energy sources of stars of a given composition will be directly proportional to their masses, a massive star will use up its initial energy supply faster than a small star and consequently must have a shorter total life. It follows that highly luminous, massive *O*- and *B*-type stars on the main sequence must

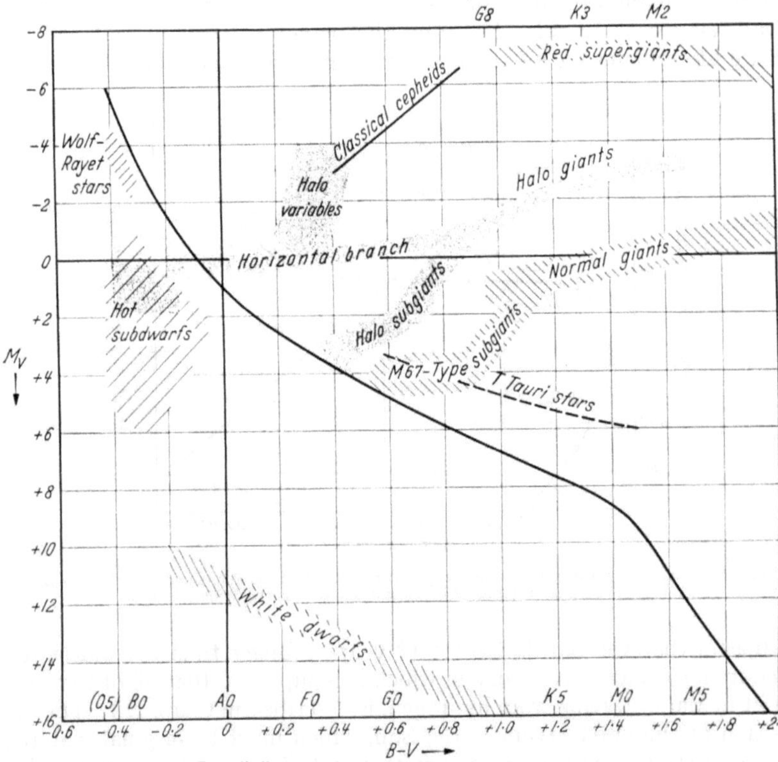

Fig. 16. Schematic Hertzsprung-Russell diagram, showing the locations of various sequences of stars. The ordinate is absolute visual magnitude, the abscissa is B − V color index. The spectral types at the bottom are for main-sequence stars; those at the top are for normal giants. The heavy black line represents the solar-neighborhood main sequence. Shaded areas represent solar-neighborhood stars; dotted areas represent globular-cluster stars. The dashed line gives the approximate location of the T Tauri stars.

be young, while dwarf *K*- and *M*-type stars may be older by many orders of magnitude. The presence of young massive stars is explained by supposing that star formation is continuously going on. If this is the case, then the H-R diagram of the solar neighborhood stars may represent an assembly of stars of different ages as well as different masses, and even the possibility of different compositions cannot be ruled out. This combination of possible variables complicates the interpretation of the data.

A star cluster, however, is likely to contain stars which had a common origin (chance formation of a cluster through gravitation being highly improbable), and thus are nearly coeval. It is also at least a reasonable working hypothesis to suppose that the material out of which a cluster formed was chemically homogeneous, so that no difference in composition would occur among the cluster stars. Thus two of the three variables present in a sample of field stars are eliminated, leaving only the stellar mass as a variable.

In the remainder of Chap. B, the data afforded by H-R diagrams or color-magnitude arrays of groups of stars in the Galaxy will be examined, in what we consider to be an age sequence, i.e., associations (10^5 to 10^7 years), very young ($\sim 10^6$ years), young (10^6 to 10^8 years), intermediate (10^8 to 10^9 years), and old galactic clusters ($\gtrsim 10^9$ years); globular clusters ($\gtrsim 5 \times 10^9$ years). In the case of associations, other properties that indicate their extreme youth are discussed, such as the nature of the T Tauri stars that give their name to the *T*-associations, and the expansion of associations.

The luminosity functions of star clusters are then described, i.e., counts of stars in different magnitude ranges. These give additional information besides that from the form of the H-R diagram alone. Finally, we turn to larger groups of stars and consider stellar evolution in the Galaxy as a whole and then in external galaxies, where necessarily stars in a range of ages instead of in coeval groups are studied together.

Fig. 16 shows a schematic H-R diagram, giving the locations of the various sequences of stars to be discussed.

I. Associations.

34. Definitions of *O*- and *T*-associations. That the *O* and early *B* stars are found in associations was shown by AMBARTZUMIAN [*169*] to [*171*]. He defined such associations as follows. They are groups containing $\lesssim 100$ stars of types *B* 0 or earlier, with diameters of 30 to 200 parsecs. They are often centered upon a cluster containing *O* stars, or a multiple star of the Trapezium type. In them the space density of the early-type stars, although much greater than the mean space density of such stars, is *less* by an order or half-order of magnitude than the mean space density of stars of all types in the neighborhood.

AMBARTZUMIAN showed that such associations are unstable groups, having a positive total energy, and must be in the process of disintegration through the general effect of the rest of the Galaxy. Hence, since the lifetimes of such groups cannot exceed a few million years, they must have formed recently. This work provided the first definite indications that star formation had occurred very recently in the Galaxy, and was presumably occurring at the present time. It thus resolved the paradox of the high-luminosity *O* and *B* stars, whose maximum ages before they must burn all their nuclear fuel at their present rate of energy output (a few times 10^7 years) were inconsistent with the known age of the Earth and therefore of the Sun, if all stars were supposed to have originated at the same time.

Another kind of association, the *T*-association, was also considered by AMBARTZUMIAN [*169*] to [*171*]. He put forward the theory that these also are groups of young stars, recently formed, and in the process of a disintegration that will feed their members into the general galactic field. The *T*-associations are groups of *T* Tauri stars, such as the group in the Taurus dark clouds discovered by JOY [*172*]. T Tauri stars are irregular variables (see Chap. E II) associated with dark or bright nebulosity, in whose spectra emission lines appear and may vary in strength. Their absorption-line spectral types are in the range *F* 8 to *M* 2. As more dense interstellar regions are surveyed for faint stars, so more *T*-associations are found. Sometimes *O*- and *T*-associations are found together, as in the Orion nebula [*173*] to [*176*], [*27*].

That there is a genetic relation between T Tauri stars and dense nebulae, rather than that the T Tauri stars are simply stars passing through nebulae and interacting with them, is demonstrated by the fact that their space density in, say, the central regions of the Taurus clouds, is about 5 to 15 times the space density of stars in the same absolute magnitude interval near the Sun [*177*], [*29*].

35. Properties of T Tauri stars. Further evidence that the T Tauri stars are young objects is as follows. Herbig [*26*] and Parenago [*176*] showed that they lie above the main sequence, and this has been beautifully confirmed by the accurate photometry of the very young clusters NGC 2264 and NGC 6530 [*178*] to [*180*]. The evolutionary interpretation of these and other young clusters is described in Sect. 43. Young stars which are still in the process of gravitational contraction and have not reached the main sequence lie above and to the right of it (Sect. 12). It is sufficient to point out here that their location is that to be expected from the prediction of Salpeter [*52*] and the calculated tracks of Henyey *et al.* [*51*].

An important observation by Joy [*172*] and Sanford[1] is that the spectra indicate that the upper layers of the T Tauri atmospheres where the emission lines originate are *rising*, relative to the underlying layers which give the absorption spectrum. At first sight this is surprising, since velocities indicating the *infall* of material into a contracting star might be expected. It is possible, however, that the observed velocities are those of shock fronts set up by disturbances arising after the infall of material. However, it must be admitted that at present there is no direct evidence for the infall of material. A T Tauri star may thus be nearly isolated from the interstellar medium and contracting in a self-consistent way, with the mass gain through accretion small or negligible.

The T Tauri stars might be expected to give information on such parameters of star formation as the mass distribution function and the amount of rotation. The absorption lines of the stars bright enough to be observed with moderate dispersion are wide and diffuse; if this is due to rotation, then the stars must be rotating faster than main-sequence stars of the same spectral types. The gravitational contraction tracks are from right to left and slightly upwards in the H-R diagram, so that stars now of types $F8$ to K will finally become main-sequence stars of types A to G. This would argue that the masses of the T Tauri stars are $\sim 2.5\, M_\odot$. Herbig [*181*] has pointed out that the rotations now observed will, if angular momentum is conserved, result in the correct order of magnitude for rotations of main-sequence stars. The rotations in NGC 2264 [*178*] set in at type $F8$ and steadily increase with advancing spectral type. This question of the angular momentum of contracting stars may be intimately connected with the explanation of the H-R diagrams of young clusters (Sect. 43).

36. Expansion of *O*-associations. Observational confirmation that the *O*-associations are unstable groups comes from studies of the motions of stars in associations, which show that they are indeed expanding [*182*] to [*185*]. If the measured expansions are assumed to have occurred at a constant velocity, then by tracing back the paths until minimum distances between the stars are reached, ages of 1.3×10^6 years, 4.2×10^6 years, 4.5×10^6 years, and $\sim 50\times 10^6$ years are derived for the ζ Per, Lacerta, II Cephei, and Cassiopeia-Taurus associations, respectively. These ages indicate how recently such star groups have been born. As an example, Fig. 17 shows the motions in the ζ Per association.

During the process of star formation, it is apparently possible for stars sometimes to acquire very high velocities which remove them rapidly from their place of origin. Blaauw and Morgan [*183*] showed that the stars AE Aur ($O9.5$ V) and μ Col ($B0$ V) have velocities that are almost equal in magnitude and op-

[1] R. F. Sanford: Publ. Astronom. Soc. Pacific **59**, 134 (1947).

posite in direction, and very high for early-type stars (128 and 127 km/sec, respectively). When the paths are traced back in space and time, they meet in a region very close to the Trapezium stars in the Orion nebula, 2.6×10^6 years ago. Other high-velocity early-type stars have been traced back to associations where they probably originated [*185*]; we note particularly 53 Ari, which apparently came from the Orion association 4.8×10^6 years ago.

The "expansion ages" of associations and individual stars are at first sight in good agreement with the maximum possible ages on the theory that these stars spend a time on the main sequence given by the dating procedure described in Sect. 40. A similar age is obtained under rather different assumptions by MASSEVICH [*187*]. However, MÜNCH has pointed out some discrepancies. Firstly [*188*], there are four high-luminosity $B1$ stars, very similar spectroscopically to ζ Per, but at such large distances from the galactic plane that the times elapsed since they could have been in the plane are about $3-4 \times 10^7$ years. These times are greater than the maximum ages given by Eq. (40.1), and an order of magnitude greater than the expansion age of ζ Per, which is so similar spectroscopically to these stars. This discrepancy can be avoided if it is supposed that the stars originated in high galactic latitudes [*187*]. Alternatively the variation in velocity with distance from the plane may be greatly different from that computed. Both of these possibilities seem unlikely.

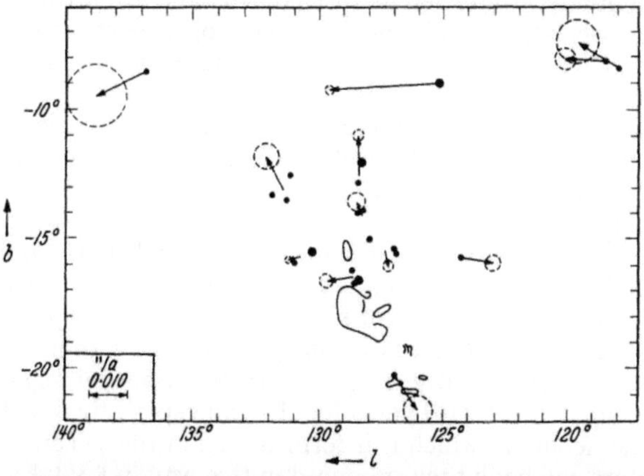

Fig. 17. Proper motions of the stars of the ζ Persei group and of surrounding O to $B\,5$ stars, relative to the mean proper motion of the ζ Persei group, according to BLAAUW [*182*]. The radii of the dotted circles represent the probable errors of the proper motions. BARNARD's dark regions are indicated.

The second discrepancy is in the age of 53 Ari [*189*], the high-velocity star which presumably originated in the Orion association 4.8×10^6 years ago. This is a β CMa variable star, as are also 12 Lac and 16 Lac which belong to the Lacerta association. The ages of these stars (cf. Chap. E II) may be an order of magnitude greater than that obtained from the expansion hypothesis. However, this may indicate only that the evolutionary arguments which have been proposed for the β CMa variables are in error. It should also be borne in mind that there may be a spread in the times of formation of stars in an association, as has been pointed out by MASSEVICH [*187*].

AMBARTZUMIAN has shown that the Trapezium-type multiple stars so often found in O-associations are unstable and must be disintegrating. SHARPLESS [*190*] pointed out that the observed systems might formerly have been much closer together. At that epoch a system consisting of one $O5$, one $O7$, and ten $B5$ stars, too close to be resolved, would be indistinguishable in spectrum and absolute magnitude from a single $O6$ star.

37. Expansion of *T*-associations. Knowledge of proper motions and radial velocities of T Tauri stars, which may eventually make possible a study of systematic motions in T-associations, are still lacking.

38. Rate of star formation in *O*- and *T*-associations. The postulate can be made that the *O*- and *T*-associations are the seats of all star formation in the Galaxy at the present time. Sometimes both types of association are found together, as in the Orion Nebula. There is some evidence suggesting that there is a real break in the main sequence of solar-neighborhood stars at or near spectral type *G*. The arguments come from two directions:

(i) the break in the mass-luminosity relation (Sect. 19);

(ii) different distributions of space velocity for stars above and below type *G* (Sect. 59).

As a consequence of this break it has been proposed by Parenago and Massevich [*81*] that the upper and lower parts of the main sequence have been populated by different mechanisms and from different starting points; i.e., the *O*-associations will populate the upper part and the *T*-associations the lower part. Whether there is a fundamental difference between modes of star formation in *O*- and *T*-associations, or whether they merely represent different parts of a continuous mass spectrum, it is instructive to look at estimates that have been made for the numbers of stars produced in associations.

Ambartzumian [*191*] has estimated that the total number of *O*-associations currently existing in the galaxy is about 10^4. Taking an average age for associations as 3×10^7 years, then the total number which have existed in a time 5×10^9 years is $\sim10^6$. If each association contains 10^2 to 10^3 stars, this gives 10^8 to 10^9 for the total number of stars which have been produced by *O*-associations.

Herbig [*181*] has estimated that the minimum present number of T Tauri stars in the Galaxy is 5×10^5. Taking what he regarded as a conservatively large estimate of 5×10^6 years for the time spent by a T Tauri star in association with the nebula in which it is formed, i.e., in the T Tauri stage, this means that 5×10^8 stars will have been produced in this way in 5×10^9 years. If the galactic distribution of the T Tauri stars were uniform, then the above rate would give about 30 stars within 10.5 parsecs of the Sun. Thus the estimated production rate by this process is galactically significant.

It has been suggested that the late-type dwarf stars with emission lines found in the solar neighborhood, which are not associated with nebulosity, may represent a later stage in the life of low-mass T Tauri stars which have still not reached the main sequence [*192*], [*181*]. For low-mass stars the contraction time scale is 10 to 100 times longer than the estimated time for the association to disperse. The dwarf *Me* stars tend to lie above the main sequence [*193*], and their velocity dispersion is only about half that of normal dwarf *M* stars (Sect. 59). This would be reasonable if *dMe* stars had formed in and escaped from nearby nebulosity and clouds, while the *dM* stars, being older, represent a less flat galactic subsystem. If this suggestion is correct, then the energy source for the emission lines must persist for a time after the star has left the nebulous region. However, the *dMe* stars in the solar neighborhood may be unconnected with T Tauri stars and may have another origin altogether.

Ambartzumian [*191*] has supposed that the *O*-associations feed stars into the flat subsystem of stars, while *T*-associations put stars into intermediate subsystems. There may not, however, be a clear-cut distinction. For example, some *O* and *B* stars, ejected from associations with high velocities, may eventually reach high galactic latitudes. Knowledge of these aspects of the problem is still rudimentary.

The overall estimates obtained by Ambartzumian and Herbig for the total numbers of stars formed in associations have been determined on the supposition that these rates have remained constant over the last 5×10^9 years. The results then show that the associations have contributed a large number, but nothing like the total number, of stars now present in the disk of our Galaxy. However, the number of stars which can be formed is ultimately determined by the amount of gas and dust available (Sect. 68), and there is some evidence suggesting that the present amount of gas is only $\sim 1\%$ of its initial value. Thus, whatever the detailed process of star formation is, it must be a decreasing function of the age of the Galaxy. We do not know what the most important modes of star formation were in the early life of our Galaxy, though we might suspect that processes other than associations dominated in this epoch. It seems clear, however, that it is unreasonable to extrapolate the formation of stars in associations over the whole life of the Galaxy. It is more reasonable to conclude that star formation in associations is the dominant mode of star formation at the present epoch, that it will continue to dominate, though it will decrease in magnitude as more of the gas is condensed into stars, but that it may not have had great significance in the early history of the Galaxy.

II. H-R diagrams of galactic clusters.

39. Early observational work and interpretations. A discussion of early observations of H-R diagrams of galactic clusters is given in the article by H. C. Arp in this volume. We refer here only to the description of galactic clusters by Trumpler [*194*], and the general deductions which he made from his observations, as follows. Clusters containing *O* and *B* main-sequence stars (i.e., massive stars) have very few or no red or yellow giant stars (the few being generally supergiants). Clusters whose main sequences do not extend upwards as far as types *O* and *B* most frequently contain an appreciable number of stars in the giant branch.

It was clear from these early observations that the H-R diagrams of individual clusters tended to have a smaller scatter in their main sequences than did the solar-neighborhood field stars. This feature, together with an apparent dispersion when the main sequences of several clusters were fitted together, led to an interpretation in terms of the well-known relation between luminosity, surface temperature, and hydrogen content (see Sect. 19). Kuiper [*195*] concluded that stars in each cluster had a common origin and therefore equal hydrogen content, but that the hydrogen content differed from one cluster to another. For the Hyades cluster, from four binaries whose masses were known, the low value of the hydrogen content $X = 0.13$ was found. Later, Parenago and Massevich [*196*] derived a value of the heavy-element content $Z = 0.12$ for the Hyades. The discussion to follow will show that both of these values are unacceptable today.

40. Modern observational work and interpretations; dating of clusters. The accumulation of accurate photometric observations and parallaxes for field stars and clusters in recent years has shown that below $M_v = +5$ main sequences of clusters and field stars, if plotted together in an H-R diagram, actually agree well and form a narrow sequence [*197*], [*193*]. It can therefore be postulated now that the hydrogen content in unevolved field stars and cluster stars in the solar neighborhood is approximately the same. The small differences in the heavy element content *Z*, which may be expected according to the arguments given in Chap. D I and II, will not materially affect *X*, since the presently accepted value of *Z* is only a small fraction of *X*. Spectroscopic observations show that there is little variation in the hydrogen content of stars, relative to the heavy elements, in different galactic clusters.

The dispersion in the upper end of the main sequences of clusters, shown by KUIPER'S composite H-R diagram, has, however, been substantiated by modern observations and can now be interpreted in terms of the theory of the early stages of stellar evolution (Sect. 25). Assuming that no mixing takes place between the core and the envelope, a star begins to move above and to the right of its original place on the main sequence as soon as it has converted an appreciable amount of the hydrogen in its core to helium. This takes place on a shorter time scale for the more luminous stars. Computed evolutionary tracks are shown in [*125*], [*132*]; see also Fig. 10. Composite H-R diagrams of clusters are shown in [*198*], [*199*]; see also Fig. 22 of ARP'S article, and Fig. 18 which gives a composite diagram in the M_{bol}, log T_e plane and is taken from [*200*]. The resemblance between the computed tracks and the upper ends of the cluster main sequences is to be noted.

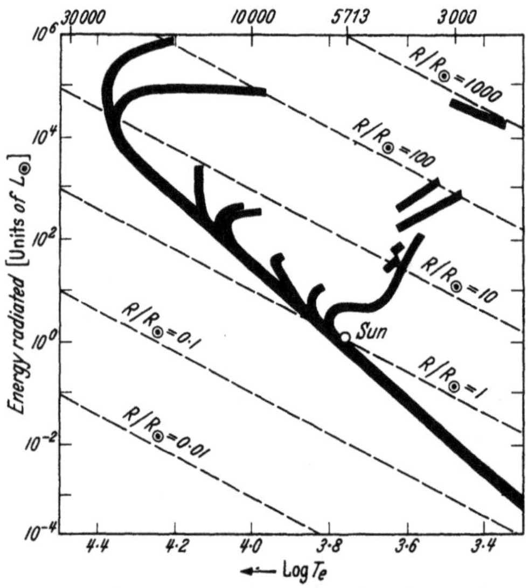

Fig. 18. A composite H-R diagram in the M_{bol}, Log T_e plane, compiled by SANDAGE [*200*] for the galactic clusters NGC 2362; h and χ Persei; Pleiades; M 41; M 11; Coma, Hyades, and Praesepe together; NGC 752; and M 67; in order of decreasing luminosity of the upper limit of the main sequence. The red giant branches belong to h and χ Persei, M 41, M 11, Hyades and Praesepe together, NGC 752, and M 67, again in order of decreasing luminosity. The ordinate gives luminosity in terms of the Sun. Lines of constant radii are dashed.

The abrupt end of all the observed main sequences is explained by the fact that, after a star reaches the Schoenberg-Chandrasekhar limit (Sect. 25), its structure changes rapidly. Since the time taken to reach this limit depends on the luminosity and mass, this provides a method of finding the age of stars that are just about to leave the main sequence [*70*] and hence a method of dating clusters [*125*]. A convenient formula for the time, τ, at which a star of mass M and luminosity L reaches the Schoenberg-Chandrasekhar limit has been obtained from the Sandage-Schwarzschild models [cf. Eqs. (25.1) and (25.2)] and is given by [*200*]

$$\tau = 1.10 \times 10^{10} \frac{L_\odot}{L} \frac{M}{M_\odot} \text{ years.} \tag{40.1}$$

For the more massive stars this method should be modified according to the evolutionary theory described in Sect. 27.

Tables 3 and 4 give the ages of a number of associations and galactic clusters as determined by various workers. MICZAIKA [*214*] used the formula given by STRÖMGREN [*70*], applied to the brightest stars in each cluster. LOHMANN [*201*] used TAYLER'S theory [*132*] to compute sequences of stars at different ages, starting from an "age zero" main sequence. VON HOERNER [*202*] used the position of the left-hand edge of the point of greatest curvature at the top of the main sequence. He used TAYLER'S theory and a mean, not an "age zero", main sequence. SANDAGE [*200*] used Eq. (40.1) and fitted the observed cluster sequences to an age zero main sequence. ARP'S article should be consulted for the derivation of an age zero main sequence; see also SANDAGE [*203*]. MASSEVICH [*204*],

[*205*], [*187*] used a different theory from the other workers, allowing mass loss to occur in unmixed stars (Sect. 25).

It will be noted that there is quite a range in the various age determinations for any one cluster. For the older clusters, the most accurate dating procedure is probably that of SANDAGE. It is certainly necessary to use, as he did, an age zero or unevolved main sequence for comparison with the individual clusters. Fitting the lower main sequence to this is probably less ambiguous than using the point of greatest curvature at the upper end of the main sequence, as was done by VON HOERNER. For the younger clusters, SANDAGE's assumption of homologous models probably leads to error, and the theory for more massive stars should be used, the clusters being compared with an unevolved main sequence. At the upper end of the main sequence it is more difficult to derive such an unevolved sequence, and modifications in the presently accepted one may yet be made. The validity of MASSEVICH's values depends on whether or not mass loss occurs in the way proposed.

Table 3. *Ages of associations.*

Name	Age (years)	
	VON HOERNER [*111*]	MASSEVICH [*187*]
γ Cyg	4×10^5	
II (P) Cyg	4×10^5	
I (h and χ) Per	4×10^5	$3-4\times10^5$
II Cep	7×10^5	
Orion	3×10^6	
I Gem	3×10^6	
Sco—Cen	$<4\times10^6$	
II (ζ) Per	5.5×10^6	
Lacerta	6.8×10^6	

Table 4. *Ages of clusters.*

Name	Age (years)				
	MICZAIKA [*214*]	LOHMANN [*201*]	VON HOERNER [*202*]	SANDAGE [*200*]	MASSEVICH [*187*], [*204*], [*205*]
h and χ Per	$>1\times10^6$		4.4×10^6	1×10^6	2×10^6
NGC 663			6.8×10^6		
NGC 2362		2×10^7	6.8×10^6	1×10^6	
IC 4665		6×10^7			
NGC 2264			1.5×10^6 ($<15\times10^6$)		
NGC 457			1.5×10^7		
Perseus moving cluster			2.0×10^7		
Pleiades	4×10^7	1.0×10^8	8.0×10^7	2×10^7	10^7
M 34			1.1×10^8		
NGC 7243			1.7×10^8		
M 41			1.7×10^8	6×10^7	
NGC 2516			1.7×10^8		
M 11		2.9×10^8	2.0×10^8	6×10^7	
M 39		3.0×10^8	2.3×10^8		
UMa stream	5×10^8		3.0×10^8		
Hyades	5×10^8	1.1×10^9	8.7×10^8	4×10^8	3×10^8
Praesepe	5×10^8	1.5×10^9	3.0×10^8	4×10^8	
Coma			5.9×10^8	3×10^8	
NGC 752		2.7×10^9	2.3×10^9	1×10^9	
M 67		3.8×10^9	4.6×10^9	5×10^9	

The present-day H-R diagram of those cluster stars which are not on the main sequence represents the locus of points through which stars of different masses will pass. Since the lifetime of an evolving star is short, relative to the time spent on the main sequence (Sect. 56), all evolving stars in the "snapshot" of a cluster at the present time will have come from only a small part of the main sequence, i.e., a small mass range. The sequences of stars not on the main sequence will thus very nearly represent the evolutionary track of a star having a mass slightly greater than that corresponding to the upper limit of the main sequence in that cluster. The number of stars in any part of the H-R diagram (off the main sequence) will be proportional to the time spent in that part of the evolutionary path.

DRAKE [*205a*] has recently attempted to investigate the amount of neutral atomic hydrogen in the vicinity of galactic clusters, using 21-cm techniques. His preliminary results suggest that there is a correlation between the amount of atomic hydrogen found and the age of the cluster, in the sense that the older the clusters are, the less gas there appears to be, associated with them. However, the question of the amount of molecular hydrogen which would be present as stars condense needs investigation.

We now consider in more detail the H-R diagrams of a few galactic clusters, and conclusions that can be drawn from them.

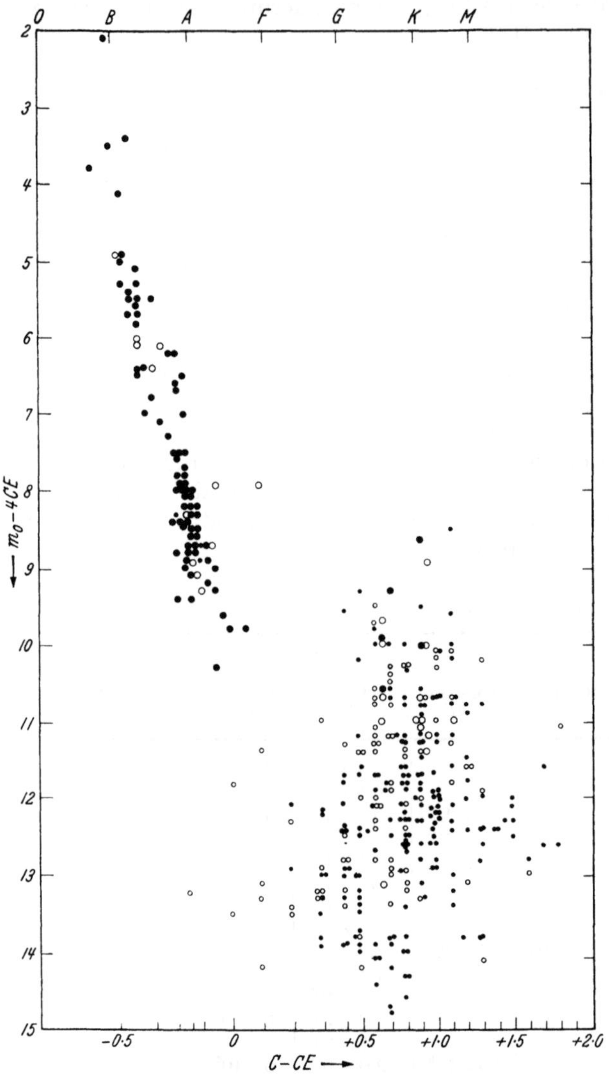

Fig. 19. The color-magnitude diagram of the Orion Nebula cluster constructed by PARENAGO [*206*]. The abscissa is the spectrum or the color index, corrected for the color excess (*C*−*C.E.*), and the ordinate is the photovisual magnitude corrected for absorption ($m_{pv} - 4\,C.E.$). The small symbols refer to the fainter stars for which only colors were determined. Closed and open circles refer to non-variable and variable stars, respectively.

41. The very young Orion Nebula cluster. 578 stars down to a limiting apparent magnitude of $15^{\mathrm{m}}.2$ have been found by PARENAGO to belong to a cluster in the region of the Orion association [*176*], [*206*]. Fig. 19 shows the color-magnitude diagram reproduced from PARENAGO's work; it is remarkable for the following features. The main sequence extends only between types *O* and *A* 5, and there is a definite break between this and the large scattered group of stars below. The latter lie above the normal position of the main sequence, and may be called subgiants. They include a great number of variables, presumably of T Tauri type.

The measures for the fainter stars were supplemented by measures from published prints of the field, which have low accuracy. However, it does not seem possible that any such systematic effect could have caused the break between the main sequence at *A* 5 and the group of subgiants, since this break occurs at about $M_{pv}=9.5$. A photoelectric investigation by JOHNSON [*207*] confirms the abrupt end of the main sequence at the same place as found by PARENAGO. Although JOHNSON measured few of the stars fainter than this, and gives no data from which it may be determined whether they are cluster members, he confirms their general location in the diagram. His measures also give a suggestion of a break between the position of these stars and the end of the main sequence.

42. The very young clusters NGC 2264 and NGC 6530. A photometric and spectrographic study of NGC 2264 by WALKER [*178*] has shown it to resemble the Orion Nebula cluster in that it has a normal main sequence extending only from $M_v=-5$ (type *O* 7) to $+1.5$ (type *A* 0), below which there is a group of stars lying above the normal main sequence. The H-R diagram is shown in Fig. 12 of ARP's article. This cluster differs from the Orion Nebula cluster in that there is no break between the main sequence proper and the group of subgiants. The latter are spread into a broad band parallel to the main sequence and about two magnitudes above it. Their spectra agree with the photometry in indicating luminosities that place them above the main sequence.

In addition, the majority of these stars fainter than $M_v=+3.5$ are T Tauri variables having hydrogen emission lines. WALKER concluded from the bright blue stars in this cluster that it is very young, so young in fact that the less massive stars may still be in the process of condensation, and may not yet have reached the main sequence. He gives 3×10^6 years as the time taken by an *A* 0 star to reach the main sequence by gravitational contraction from a starting configuration with surface temperature 4000°. The most luminous star in NGC 2264 is the *O* 7 star S Mon, and WALKER gives an upper limit to its age as about 5×10^6 years. The agreement here is good; however, when the contraction time of the *less* luminous stars is considered, a problem arises, which is discussed in the following section.

The color-magnitude diagram of NGC 6530 by WALKER [*179*] is very similar to NGC 2264, in that it has a normal main sequence extending only from *O* 5 to about *A* 0. Below this the stars lie above the main sequence, and in this region there are numerous T Tauri stars. The age of this cluster is concluded to be about 3×10^6 years, like NGC 2264.

43. Possible explanations of the H-R diagrams of the Orion Nebula cluster, NGC 2264, and NGC 6530. These three very young clusters are the only ones which at the time of writing have any detailed H-R diagrams constructed for them. Thus we can attempt to discuss the evolution of such clusters by using this first material. Two major features demand explanation. The first is the break in the main sequence in the Orion Nebula cluster, observed by both PARENAGO and JOHNSON, and shown in Fig. 19. In the other two clusters there does not appear to be a break. Second, in all three clusters an explanation is needed for the lower-luminosity stars which lie to the right of the main sequence. A number of different explanations can be considered.

(i) PARENAGO concluded that the H-R diagram of the Orion Nebula cluster supports his previous hypothesis, that there is a real break in the main sequence near solar type and that the upper part is populated by stars formed in *O*-associations and the lower part by stars formed in *T*-associations. The cluster is very young because of the *O*-type stars in it, and it was suggested that stars are probably being formed in this region at the present time. The theory of evolution down the main sequence through mass loss by completely mixed stars (Sect. 24) was applied, and an age of $\sim 10^7$ years was derived from this theory as the time taken for an O-type star to reach $A\,5$.

(ii) The second explanation is based on the assumption that all the stars in each of these clusters are coeval, and stars of all masses according to the mass-spectrum discussed in Sect. 54 began to contract towards the main sequence. The difficulty in explaining the diagram on the basis that the stars in the upper part have had time to reach the main sequence while those in the lower part are still gravitationally contracting has been discussed by Sandage [*50*], and is illustrated by his diagram reproduced in Fig. 20. Here the color-magnitude diagram of NGC 2264 has been plotted in the coordinates of M_{bol} and Log T_e. Computed contraction tracks for stars of various masses and their time scales are also shown. The ages derived by Walker from the upper end of the main sequence of NGC 2264 and from the contraction time of an $A\,0$ star are both 3×10^6 years, but if this is the age of the cluster, then below $A\,0$ one would expect a band of stars to be spread out more or less parallel to the contraction tracks and not, as is actually the case, parallel to the main sequence. In NGC 6530 these stars do populate a region which is funnel-shaped, widening towards the fainter stars, with its upper boundary nearly horizontal, like the contraction tracks. The fainter stars which lie near the main sequence, however, constitute just the same problem in this cluster as in NGC 2264. On the simple theory, there does not appear to have been time for the faintest stars to have contracted to their observed positions, which would take about 100 times as long as for the $A\,0$ stars to contract.

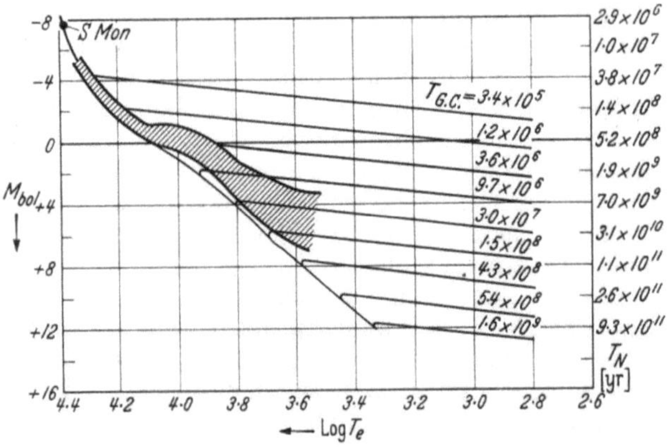

Fig. 20. Contraction paths in the M_{bol}, Log T_e plane computed by Sandage [*50*]. Contraction times $T_{g.c.}$ to reach the main sequence are written on the tracks. Times for the stars to remain on the main sequence at the given M_{bol} are on the right-hand ordinate. Walker's data for NGC 2264 are in the shaded area.

A way out of this difficulty may be found if it is supposed that fragmentation occurs. Let us consider the fragmentation of large contracting masses of gas (Sect. 2). According to Hoyle's theory, the last fragmentation of a cloud to masses of about 2 to 3 $M_\odot$ will give fragments of radius $\sim 10^{14}$ cm. This is considerably larger than the radius at the end of the tracks in Fig. 20, and in this theory the fragments would only just at this point become star-like, with the adiabatic condition superseding the isothermal. The fragments would still have to undergo the long contraction times indicated in Fig. 20. The locus of points from which stars would have reached their present positions in 3×10^6 years extends over to the right of Fig. 20 for present-day types $A\,0$ and earlier, but rapidly comes in close to their present positions for the less massive stars.

The number of stars later than type $A\,0$ is about 20 times the number of those earlier than $A\,0$. If there were a mechanism for the break-up of large contracting clouds that would result in stellar fragments with surface temperatures in the observed range, this could provide the answer. Stars with masses much in excess of $50 M_\odot$ are not observed; the complete break-up of only a few contracting gas masses of about $100 M_\odot$ or greater could provide the numbers of low-mass stars, but it is not clear that such fragments could rapidly become stellar, as would be necessary to explain the observations.

A possible mechanism of break-up, not considered by HOYLE, is rotation. It has recently been pointed out by PRENDERGAST [*208*] that the break in the main sequence observed in the Orion Nebula cluster at *A* 5 is fairly near to that at which there is a sharp drop in the rotational velocities of main-sequence solar-neighborhood stars, i.e., at *F* 5 (Sect. 94). This break, if measured in angular momentum units, is very sharp indeed. To explain the break it may be supposed that below a certain value of total angular momentum a protostar is able to contract without breaking up into separate condensations, while above this value fragmentation must take place. The appearance or non-appearance of a break in the observed number of stars would then be related to the ratios of the masses into which the protostars fragment and the original mass distribution. This hypothesis would suggest that all of the low-mass stars lying to the right of the main sequence were once members of binary or multiple systems and that most of them have been disrupted by tidal forces. A study of binary systems in these clusters is highly desirable. Also studies of the rotations of the stars in the Orion Nebula cluster would determine whether the break is coincident with a jump in rotational velocities for these cluster stars alone and not just close to, but not necessarily coincident with, the corresponding jump in the rotations of the general field stars. It would also be expected that the upper part of the main sequence of the Orion Nebula cluster would be populated by stars that have maximum values of rotation compatible with their masses.

(iii) A third possible explanation for the positions of the fainter stars in the H-R diagram might be forthcoming through the effect of dust grains. According to WHIPPLE and SPITZER, the presence of dust speeds up the contraction by a large factor over the time for pure gravitational contraction of gas alone (Sect. 3). The extent to which this could be important in low-mass condensations after they had become adiabatic with surface temperatures greater than a few hundred degrees is not clear. More work needs to be done in considering this possibility.

(iv) Finally, it is possible that the observed distribution of stars in the H-R diagram is due to a spread in starting-times for condensations of different masses. The presence of numerous small condensations would then have to be assumed to occur before and perhaps facilitate the start of a few larger ones. However, OORT and SPITZER and BIERMANN and SCHLÜTER have suggested the opposite; the presence of a massive *O* or *B* star might, through the jet action of ionization upon gas in its vicinity, cause local density increases and lead to the condensation of smaller stars in large numbers (Sect. 4). The sequence of events needed to explain these observations might be obtained if the ideas of KRAT and UREY are correct (Sect. 5), i.e., small solid condensations are present in large numbers and are gravitationally attracted, leading to the formation of stellar masses.

44. Young clusters. α) *h and χ Persei.* This double cluster contains very luminous *B*-type stars with absolute magnitudes as high as $M_v = -7$. BIDELMAN [*209*] showed, as had previously been hinted[1], that it is surrounded by a corona of *M*-type supergiants at the same distance as the cluster and therefore presumably physically connected with it, indicating a genetic relation between

[1] W. S. ADAMS, A. H. JOY and M. L. HUMASON: Astrophys. J. **64**, 225 (1926).

bright blue stars and M-type supergiants. Recently photometric and spectroscopic observations have been made on the brightest blue stars [*210*], [*211*], and on the red supergiants [*212*]. In the H-R diagram the high-temperature stars form a band extending from the main sequence, upwards and to the right to spectral type A 5 I a, where they come to an abrupt end. There is then a very wide Hertzsprung gap, and then the sequence of M-type supergiants.

MASSEVICH [*187*] discussed evolution in the double cluster and the association around it, considering only that part of the H-R diagram to the left of the Hertzsprung gap, and using her theory of evolution with mass loss but without mixing [*101*]. For the cluster, with its well-defined main sequence curving to the right at its upper end, she derives an age of 2×10^6 years. In the association a well-defined sequence is present at about the same position, indicating the same age, but in addition there are stars on the main sequence above this turn-off point, whose ages must be a factor of about 10 less than the age of the cluster. The measures are compared with a computed initial age zero main sequence, which agrees well with the initial main sequence of JOHNSON and HILTNER [*211*] but extends upwards further.

Also included in her H-R diagrams are TAYLER's [*132*] computed evolutionary tracks for a star of constant mass $10\,M_\odot$, and a track computed by MASSEVICH under the same assumptions for a constant mass of $20\,M_\odot$; these take stars off the main sequence into the B supergiant and giant region in times of 3×10^6 to 10^7 years. She concludes that both the cluster and the association were formed at about the same time (10^6 to 10^7 years ago), but that the process of star formation has continued in the outer association for a considerable time after it ceased in the nucleus. This is taken as possible support for the Oort-Spitzer-Biermann-Schlüter theory of star formation; the cluster formed first, and ionization from the hot stars in it has facilitated star formation in the region around it.

MASSEVICH postulates that the A-type supergiants, of which there are more in the association than in the cluster, are stars in the process of gravitational contraction and have not yet reached the main sequence. However, she does not discuss the M-type supergiants in the association, and the existence of these together with a wide Hertzsprung gap makes it more probable, in our opinion, that the A-type supergiants are evolving away from the main sequence, prior to a rapid structural readjustment that will take them into the M-supergiant region.

Finally, the question of the classical cepheids in the vicinity of h and χ Persei, to which BIDELMAN drew attention, should be mentioned. One of these, SZ Cas, is probably a member of the cluster. If so, it is not as luminous as the B and M supergiants. The evolutionary stage reached by the pulsating variable stars will be discussed in Chap. E II (p. 278).

β) Pleiades. The turn-off point in the Pleiades comes at a lower luminosity than that in h and χ Persei; the cluster is therefore somewhat older. There are no red giants in this cluster. MITCHELL and JOHNSON [*215*] have found what they believe to be a break in the main sequence near $M_v = +6.5$, and have suggested that this marks the place below which stars are still gravitationally contracting on to the main sequence. The break is very much smaller than that in the Orion Nebula cluster. It lies about 9 magnitudes below the brightest stars observed in the Pleiades. TAYLER's theory was used to derive an age of 1.5×10^8 years for this cluster, in good agreement with that obtained by LOHMANN [*201*]. If the break in the lower main sequence is real and is connected in some way with the gravitational contraction, then a similar age is derived from the time for contraction of stars at this luminosity.

45. Clusters of intermediate age. The observations on which the ages of the intermediate clusters in Table 4 are based are mainly by BECKER, COX, JOHNSON, SANDAGE, STOCK, WEAVER, and their associates; a full list of references is given by VON HOERNER [*202*]. A recent one may be added: the H-R diagram of NGC 1664 has been obtained by LARSSON-LEANDER [*213*] and an age of 1×10^9 years assigned.

Of the well-observed clusters in this group, M 41, M 11, Hyades, Praesepe, and NGC 1664 each have a Hertzsprung gap and then a sequence of red giants. The luminosities of the giants are approximately the same as the luminosity of the stars at the turn-off point on the main sequence, as may be seen in Fig. 18.

If white dwarfs represent the final stage of stellar evolution, then it is to be expected that in all clusters of intermediate and old age in which they are observationally accessible, white dwarfs will be found. There is evidence for their presence in the Hyades and probably in Praesepe and the Coma cluster [*216*], [*217*].

46. The old clusters NGC 752, NGC 7789, and M 67. The H-R diagrams of NGC 752 and M 67 are unusual for galactic clusters. The observations of NGC 752 [*218*] have so far shown no main-sequence stars above $M_v=+2.5$ or below $M_v=+4.5$. There is a Hertzsprung gap, and a few red giants. The upper limit to the main sequence is real, but there is a possibility that if photometry were carried to fainter stars, and if cluster membership could be checked, a lower extension of the main sequence might be found. The characteristic curve at the upper end, and the absence of main-sequence stars brighter than $+2.5$, show that this is a relatively old cluster. ROMAN [*219*] has found that the spectra show weak metallic lines, indicating a slightly lower than normal metal content. This problem is discussed in Chap. D II, where it is shown that a correlation of heavy-element content with age might be expected. ROMAN has computed models, and derived a minimum age of 1.5×10^9 years for the cluster, in reasonable agreement with the ages given in Table 4.

M 67 is a rich cluster in rather high galactic latitude. The main sequence contains practically no stars brighter than $M_v=+3.5$ [*220*]; it curves round into a subgiant sequence running approximately horizontally and then upwards into the red giant region. There is no Hertzsprung gap. The absence of a well-populated main sequence above $+3.5$, implying that all the brighter stars originally on the upper main sequence have passed through their evolution and have come to the end of their lives, shows that this is an old cluster. In fact, it is the oldest galactic cluster so far studied (see Table 4). The spectra, however, do not show weak-lined characteristics. This will be discussed further in Sect. 51 and Chap. D II. BAADE has reported, from inspection of his plates, that there may be a large number of white dwarfs in M 67 [*199*]. Since these are stars at the ends of their lives, there would be expected to be many in such a rich, old cluster as M 67. As discussed in Sect. 55, the number in any cluster can be predicted. Finally, there appears to be a thinly-populated sequence of stars which looks rather similar to the horizontal branch in globular clusters (Chap. B III) and may be the analogue in old galactic clusters of what is a characteristic and well-defined feature in globular clusters.

NGC 7789 is a rich cluster with a large number of yellow and red giants. An age of $1-2\times10^9$ years has been assigned to it [*220a*]. The giant branch slopes upwards in a similar way to that of M 67, but the break-off from the main sequence occurs higher up, and there is a well-defined Hertzsprung gap. The giant branch actually intersects that of M 41 in a composite H-R diagram; this effect will be discussed in Sect. 57.

Both M 67 and NGC 7789 contain a sparse distribution of stars on the main sequence above the break-off point. Such stars also appear in the globular cluster M 3 (cf. Sect. 50). Their evolutionary significance is not understood at present.

47. General conclusions from H-R diagrams of galactic clusters. From the H-R diagrams just described, together with the full discussion in the article by Arp, we may make the following empirical deductions concerning the course of stellar evolution.

(i) If clusters are very young, the lower-mass stars will not yet have reached to main sequence.

(ii) As clusters age, the brighter stars evolve off the main sequence, so that the upper extent of this can be used to indicate the age of clusters. Von Hoerner [*202*] noted that there should be a range of only 9.5 in bolometric magnitude on the main sequence in any one cluster, the upper end being limited by evolution off the main sequence, and the lower end by the time taken for gravitationally contracting stars to reach the main sequence. This conclusion appears to be confirmed by the recent observations of the Pleiades [*215*].

(iii) The Hertzsprung gap is wedge-shaped (see Fig. 18). It is widest in those clusters whose main sequences extend the furthest upwards, i.e., the youngest clusters, and disappears in the oldest (M 67). The existence of such a gap between the main sequence and the giant or supergiant region shows that stars above a certain mass near $1.5 M_\odot$ move very rapidly in their evolutionary paths after reaching the Schoenberg-Chandrasekhar limit. That evolution here is rapid, in the absence of degeneracy in the core, was shown by Sandage and Schwarzschild [*125*]. Reddish [*221*] suggested that whether or not there is a Hertzsprung gap may depend on whether or not the evolving stars are massive enough to have convective cores. This has also been discussed by Sandage [*200*].

(iv) After passing through the Hertzsprung gap, stars enter the giant or supergiant region. Evidence from most of the clusters does not indicate whether a star becomes stabilized at the left or right end of the giant sequences, but the diagram for M 67, where there is a continuous sequence of stars, strongly suggests that a star moves from left to right along the red-giant sequence.

(v) The luminosities of red giants are apparently related to the masses of the main-sequence stars from which they evolved. The most massive ones which were of types *O* and *B* become supergiants; less massive stars, as in M 41, become giants of luminosity class II; still less massive ones become normal (luminosity class III) giants with little dispersion in luminosity over quite a large mass range. In other words, after leaving the main-sequence region, stars of masses greater than about $2 M_\odot$ move approximately horizontally in the H-R diagram. This was already indicated by the positions of giants in the mass-luminosity plane (Sect. 21). The place of the five giants of types *G*5 III to *K*5 III in the very young cluster NGC 2264 in this scheme is not clear. Walker pointed out that they may be extreme cases of gravitationally contracting stars that have not reached the main sequence, or that possibly they may not be cluster members. Alternatively, they may be stars that have undergone considerable mass loss and have reached a giant configuration appropriate to lower-mass stars.

(vi) In every cluster the red giant branch comes to an abrupt end; these observations suggest either that a rapid change then takes place which removes them completely from the giant branch or that stars move back along the giant branch, presumably because of further structural changes. We saw in Sect. 28 that globular clusters provide some evidence for the subsequent history of population II stars, but of the galactic clusters so far observed, only M 67 contains enough stars to provide any clues to the later, apparently rapid, evolution of population I stars. In that cluster there is a sparse distribution of stars in a "horizontal branch", as may be seen in Fig. 21 of the article by ARP, and it is possible that this is the track of population I stars of masses near $1.2 M_{\odot}$, when they are nearing the ends of their lives.

48. Empirical deductions about evolution of field stars in the solar neighborhood. *α) The dissolution of galactic clusters.* Because of the effect of differential galactic rotation and shear, and also because the times of relaxation for all but the richest galactic clusters are small compared to 10^9 years[1,2] we should expect that there will be a steady loss of stars from clusters into the general field. The looser, poorer clusters dissipate faster than rich clusters. For example, MARKARIAN[3] has discussed the dissipation of clusters containing *O*-type stars.

OORT [*222*] has considered a problem posed by counts of the numbers of galactic clusters with main sequences extending up to various ranges of spectral type. On the basis of the dating procedure described in Sect. 40, time intervals, ΔT, between the ages corresponding to the various main-sequence limits were derived. Assuming a uniform rate of formation of clusters during the last 5×10^9 years, the number of clusters in each range should be proportional to ΔT, but the actually observed numbers of clusters depart from this by factors of more than 100 for the oldest ones, in the sense that there are too few old clusters. The data were taken from the list of TRUMPLER [*194*]. Yet the predicted times for complete dissolution of clusters are very much larger than the relaxation times, and are in general $\sim 3\times 10^9$ years.

At least three arguments which might explain this discrepancy can be put forward. Firstly, all the preceding arguments in this Chap. B II leading to the dating procedure may be erroneous. This appears highly unlikely. Secondly, clusters may, because of other factors, such as tidal effects due to interstellar matter and the acceleration of the rate of dissipation as the cluster is depleted, evaporate at a faster rate than that calculated. Thirdly, cluster formation may not have occurred at a uniform rate during the last 5×10^9 years. Finally, there may be an effect of observational selection in the available material.

The discrepancy is large, and is already apparent for clusters of ages 10^8 years. Further, if 3×10^9 years is the average time for total dissolution of a cluster, an estimate for the rate of star production in the general field by this means yields a smaller number than that estimated in Sect. 38 through the dissolution of associations, simply because of the very much longer lifetimes for clusters.

SPITZER [*222a*] has studied the disruption of galactic clusters by gravitational encounters with interstellar cloud complexes. He has shown that the effect is a full order of magnitude greater than the disruption resulting from passing field stars alone. Although conclusive results are not possible at present, this work strongly suggests that such an effect can account for the discrepancy pointed out by OORT. According to SPITZER's preliminary calculations, the disruption time for an extended cluster like the Hyades will be as little as 6×10^7 years, while that for the Pleiades probably does not exceed 10^9 years and may be considerably less. Only a relatively dense cluster, such as M 67, can apparently survive disruption for a period as long as 5×10^9 years.

[1] V. A. AMBARTZUMIAN: Ann. Leningrad State Univ. No. 22, 19 (1938).
[2] S. CHANDRASEKHAR: Principles of Stellar Dynamics. Chicago: Chicago University Press 1942.
[3] B. E. MARKARIAN: Soobs. Bjurakan Obs. **5** (1950).

However, at an epoch earlier than a few times 10^9 years ago, the possibility that the rate of formation of clusters may have been different from the rate in the more recent past should not be forgotten. When more gas and dust were present, the conditions for star formation and for the stability of clusters may have been different, and what we see now is the tail end of a decaying activity. This problem is connected with possible evolutionary stages of galaxies discussed in Chap. B V.

β) H-R diagram for solar-neighborhood field stars. Even though star formation may not in the past always have occurred in associations and clusters as we see them today, yet it does appear that at the present epoch at least, the general field is being populated by stars escaping from clusters. We should expect the stars in the solar neighborhood to be a mixture of stars of different masses and ages, and representable by a composite of the known populations of local galactic clusters of different ages. Furthermore, for those stars which really belong in the solar neighborhood, and have not wandered into the sun's vicinity from other parts of the Galaxy, we might expect the compositions to be approximately the same. The relatively few spectrophotometric analyses that have been carried out support this[1].

Table 5. *Subgiant stars lying below the* M 67 *sequence.*

Star	M_v	$P-V$ (Eggen)	Sp. Type	Parallax	Yale Cat. No.
β Hyi	+3.68	+0.52	G1 IV	0″.153 ± 0.007	69
ι Per	+3.67	+0.49	G0 V	0″.084 ± 0.005	647
δ Eri	+3.67	+0.85	K0 IV	0″.109 ± 0.004	788
λ Aur	+3.78	+0.505	G0 V	0″.066 ± 0.006	1199
31 Aql	+3.95	+0.665	G8 IV	0″.059 ± 0.005	4541

JOHNSON [198] plotted the H-R diagram of 45 stars with accurate trigonometric or moving cluster parallaxes and 40 brighter components of double stars for which he assumed that the fainter lay on the main sequence. Superimposed on the plot was a bounded area that included the H-R diagrams of a number of galactic clusters. This showed that the individually plotted stars mostly lay within bounded area.

Both EGGEN [223] and SANDAGE [224] have plotted the H-R diagrams of stars with well-determined trigonometric parallaxes, superimposed on cluster H-R diagrams. Eggen selected all stars of apparent visual magnitude brighter than 5,0, having $\pi > 0''.02$ with determinations at two or more observatories. Sandage selected all stars with $\pi > 0''.067$. In both cases the stars lie mostly within the area occupied by the galactic clusters, in particular the main sequence, the red giant region, and the region just above the main sequence, where lie stars of masses in the range of about $1.2-3\,M_\odot$ that have undergone some evolution but have not reached the Schoenberg-Chandrasekhar limit. It may be concluded that the evolution of stars in galactic clusters and in the general field is essentially the same.

An interesting feature, noted by both EGGEN and SANDAGE, is the group of stars lying to the right of the main sequence but *below* the position of the horizontal subgiant branch of M 67. The best-determined stars in this region, from [223], are given in Table 5. By the line of reasoning described in Sect. 40, it would seem as though these are stars that have evolved off points on the main sequence below the break-off point of M 67. Their ages would then be at least 8×10^9 years, which would make them the oldest stars so far dated in the Galaxy. Both authors have pointed out that more astrometric, photometric,

[1] L. H. ALLER: In this volume.

and spectroscopic work needs to be done on these stars, and other evolutionary histories should be considered before such a large age is accepted. In particular, the effects of mass loss as discussed in Chap. C II might be responsible for the positions of these stars in the H-R diagram.

III. Color-magnitude diagrams of globular clusters.

49. Early observations: Difference from solar-neighborhood stars. Early work on color-magnitude arrays by SHAPLEY and others (see for example [*225*]) revealed the fact that stars in globular clusters occupy a different area of the H-R diagram from that populated by galactic cluster and solar-neighborhood stars. A detailed study of earlier and modern work is given in the article by ARP in this volume. We are concerned here with the differences between H-R or color-magnitude diagrams of galactic and globular clusters, which have an important bearing on problems of stellar evolution.

Owing to the great distances of globular clusters, SHAPLEY'S studies did not extend to fainter absolute magnitudes than about +1, and it never became clear what was the relationship, if any, between the globular and galactic clusters. However, SHAPLEY'S studies showed that the absence of blue high-luminosity stars in the former, and the upward-inclined yellow and red giant branches extending to stars of similar color to those found in the galactic clusters and the solar neighborhood, but about three magnitudes brighter, were characteristic. It was these features which provided BAADE [*226*] with the clue leading to his resolution of stars in the disk of M 31 and its companions, and his demonstration of the existence of two stellar populations, which will be discussed in Sect. 58. The globular clusters belong to the extreme population II.

The observations of SHAPLEY and others revealed also the existence of a bifurcation of the characteristic color-magnitude diagram of globular clusters into a horizontal branch near $M_v = 0$, and a nearly vertical band of steadily increasing star density extending to fainter magnitudes and abruptly cut by the limiting magnitude to which the observations reached. Attempts to extrapolate the observations[1] to link them with the main sequence in the solar neighborhood were hampered by the fact that the measures extended only to $m_v = +17$.

50. Modern observations and interpretation: Dating of globular clusters. The tie-in between the two different kinds of color-magnitude diagrams was not provided until accurate photometric data extending to absolute magnitudes approximately as faint as that of the sun were obtained [*227*], [*228*]. It was then apparent that the red giant and subgiant branches in these diagrams lead into main sequences which are at approximately the same place as the population I main sequence, but which extend upwards only to $M_v = +3.5$ (see Figs. 25 to 35 in the article by ARP). The break-off point at this magnitude was taken to mean that all stars brighter than this had evolved off the main sequence. It was used by SANDAGE [*228*] to date the typical globular cluster M 3 by the method described in Sect. 40. An age of 5×10^9 years was found; the theory of HASELGROVE and HOYLE gives an age of 6.5×10^9 years (Sect. 26), which is probably more accurate.

The fact that the subgiant and giant branches of globular clusters form a continuous band of stars joining the main sequence, as in the old galactic cluster M 67, implies that the evolving stars, after leaving the main sequence, move continuously towards higher luminosities and lower surface temperatures, until

[1] H. L. JOHNSON and M. SCHWARZSCHILD: Astrophys. J. **113**, 630 (1951).

they reach a limiting magnitude and color beyond which their evolutionary paths must have an abrupt discontinuity. The theoretical tracks of Hoyle and Schwarzschild (Sect. 26) have succeeded in explaining these observed subgiant and giant branches. The time scale for stars in this stage is much shorter than that on the main sequence. Therefore, the color-magnitude diagram, which is actually the locus of points reached by stars in a small mass range, is very nearly represented by the actual computed tracks, as shown by Fig. 9, where the theory is compared with observations in M 3 and M 92.

The horizontal branch is a very characteristic feature of globular cluster diagrams, and presumably represents the evolutionary history of stars in this mass range after they reach the top of the giant branch. The theory of stellar evolution has not, at the time of writing, been carried further than the giant stage. The relation of the subsequent evolution to the horizontal branch of globular cluster diagrams is therefore not clear at present, nor is the evolutionary significance of the sharply defined region on the horizontal branch where the RR Lyrae variable stars occur. Some discussion of this and later stages was given in Chap. A VI.

A thinly-populated sequence of stars forms an extension of the main sequence in M 3 above the break-off point. It is not understood yet how stars can occur here, unless either they are younger stars, formed subsequently to the formation of the rest of the cluster, or they are stars which in some way have been well-mixed and have evolved as homogeneous stars (Sect. 23).

51. Effect of chemical composition. Fitting of main sequences of globular and galactic clusters. The first accurate color-magnitude diagrams of globular clusters preceded the work on M 67, and the comparison between them is interesting. The main sequences of M 67 and M 3 both break off at about the same point, implying that they have similar ages [*220*]. The fitting together of these diagrams may be revised somewhat in the near future, and Table 4 shows that Lohmann [*201*] derived a smaller age for M 67 than did von Hoerner [*202*] and Sandage [*200*]. However, the evolving stars probably have nearly the same masses in the two clusters. Of the different parameters mentioned in Sect. 33, namely age, mass, and composition, only the chemical composition is left as the possible cause of the difference.

Although no spectrophotometric determinations of chemical compositions of globular cluster stars have yet been made, work on spectral classification of individual red giants [*229*], [*230*], [*369*], and of the integrated light from clusters [*231*] has shown that the spectra differ from those of stars in the solar neighborhood. This difference can be explained by the globular cluster stars having a considerably lower abundance of the heavier elements (calcium and the iron group in particular), relative to hydrogen. The theoretical evolutionary tracks of Hoyle and Schwarzschild (Sect. 26), which agree so well with the observations of M 3 and M 92, were computed for a considerably lower abundance of the heavy elements than the accepted solar-system value. Hoyle and Schwarzschild showed that a higher abundance of these elements causes the giant branch of the track to lie at lower absolute magnitudes, and this is just what is observed in the comparison between M 67 and M 3.

In the discussion of the chemical evolution of stars (Chap. D II), in which the compositions of the halo population II stars are further described, reasons are given for expecting that a continuous gradation of the heavy-element content in stars might be found, in the sense that the youngest stars would have the largest amounts of these. The globular clusters are certainly old stellar systems, as would be demanded by this hypothesis, but M 67 is also old and yet its chemical composition is apparently similar to that of younger stars in the solar neighborhood.

It is not possible at present to determine the true age difference between M 67 and any of the globular clusters, because of uncertainty in locating the true position of the main sequences of globular clusters. This requires photometry at the limits of the best equipment available today. At the moment, fitting of globular clusters is usually done by assuming that the RR Lyrae stars on the horizontal branch always have $M_v = 0.0$. As discussed in Sect. 101, this may not be true. For the same reason, the exact nature of the small differences in color-magnitude diagrams of different globular clusters cannot be located at present. That there are small differences can be seen from Fig. 32 of the article by ARP, but it is not known at present whether the diagrams should be fitted together by means of the RR Lyrae variables or the main sequences, or possibly even the giant branch.

It is not entirely clear what the theoretical difference between the main sequences in the M_{bol}, Log T_e plane should be, as a function of the heavy-element content Z. Even if there is no appreciable difference in these coordinates for the range covered in the comparison, there will still be a shift in the color-magnitude diagrams, owing to different amounts of the blanketing effect in spectra with strong and weak lines. This effect is discussed in Chap. D II, in connection with the possibility that it may cause extreme population II stars to be spuriously called subdwarfs. There is no theoretical reason at present for expecting the globular cluster main sequences to lie as far below the solar-neighborhood main sequence as indicated by the preliminary photometry by BAUM *[232]* in M 13. A first check of this *[233]* indicates that the effect is not as large as found by BAUM.

Apart from the difficulty in determining the age difference between M 67 and the globular clusters by fitting the main sequences, or the color-magnitude diagrams in general, an entirely separate argument suggests that differences in age may exist for stars which turn off the main sequence at the same luminosity. This argument runs as follows.

As will be described in Sect. 75, a recent measurement of the $He^3(\alpha, \gamma) Be^7$ reaction and an analysis of the results by FOWLER *[357a]* shows that for stars slightly hotter than the Sun which contain abundant He^4 the pp-chain will be completed either by

$$He^3(\alpha, \gamma)\, Be^7(e^-, \nu)\, Li^7(p, \alpha)\, He^4$$

or

$$He^3(\alpha, \gamma)\, Be^7(p, \gamma)\, B^8(\beta^+, \nu)\, Be^8(\alpha)\, He^4.$$

In the case in which the chain is completed through B^8, the neutrino emitted in the decay to Be^8 has an average energy of 7.3 Mev. Together with the neutrino loss in $H^1(p, \beta^+\nu) H^2$ the total energy loss is 7.6 Mev; i.e. 7.6/28.7 or 28.4% of the energy is lost by neutrino emission. If the chain is completed through Li^7, the corresponding loss is only $1.1/26.7 = 4.1\%$. Thus to maintain a given luminosity a star burning on the pp-chain via B^8 will consume hydrogen at a rate 95.9/71.6 times as fast as one burning on the pp-chain via Li^7. Now the age determination for a cluster is obtained by calculating the time taken for stars of a given mass to consume a certain fraction of their mass. If the stars are burning on the pp-chain via B^8, therefore, we shall have to reduce the ages given by Eqs. (25.1), (25.2), and (40.1) by a factor 1.33. The relative rates of proton and electron capture by Be^7 as a function of stellar temperature and density are not yet available. However, there is some reason to believe that for stars slightly more massive than the Sun the pp-chain via B^8 may be important. This means that the chemical composition, or more precisely, the He^4/H ratio is an important and *unknown* factor in determining the ages of the clusters. (It must apparently remain unknown observationally since for stars in the mass range in which it is important the helium lines will not be visible in the spectrum.) However, we might suppose that in the globular clusters, by analogy with the metals, the initial He/H ratio is much smaller than it is in the disk population I stars and gas. In this case the pp-chain will first be completed through $He^3(He^3, 2p)\,He^4$ as is the case for stars less massive than the Sun. A core containing He^4 will gradually be produced. An important fact to determine now is whether the shell source where the energy production currently takes place will soon contain appreciable He^4, i.e. to what extent mixing with the inert core takes place. If little mixing occurs we can expect little change in the presently determined ages for the globular clusters. On the other hand, for galactic clusters we expect that the stars initially have a He/H ratio characteristic of disk population I. Thus for M 67 whose stars have a mass $\sim 1.2\,M_{\odot}$, the pp-chain via B^8 may be important. In this case the age of M 67, which has been estimated by SANDAGE [*200*] to be about 5×10^9 years, will be reduced to as low as 3.8×10^9 years, so that it might be significantly younger than M 3 or M 92.

It is clear that this effect must be taken into account when we are trying to determine the ages of clusters whose most luminous main-sequence stars have masses slightly in excess of that of the Sun. However, to conclude, it should be emphasized that the importance of the effect depends on, among other factors, the cross-section at low energies for the reaction $Be^7(p, \gamma)\,B^8$ which can be measured in the laboratory, and which is now being attempted, and the He/H ratio in these clusters which cannot be determined observationally.

IV. Luminosity functions of field stars and clusters.

52. Definition. Observations in solar-neighborhood field stars. Luminosity functions, i.e., counts of the number of stars in given magnitude and spectral ranges, for field stars in different regions of space and for stars in clusters, provide additional data on evolution, as well as information on the formation of stars.

The luminosity function $\varphi(M)$ is defined by

$$dN = \varphi(M)\,dM \tag{52.1}$$

where dN is the number of stars per cubic parsec with absolute magnitude lying between M and $M + dM$. In the remainder of this subdivision IV, we shall use the symbol $\mathfrak{M}$ to denote mass, to avoid confusion with M which represents magnitude. The function $\varphi(M)$ may be considered with respect to photographic or visual magnitudes, according to the data available, and may also be taken separately for different spectral ranges.

Two factors will combine to determine the observed luminosity function in any sample of stars:

(i) the distribution of masses with which the stars formed (original mass function) and,

(ii) the subsequent changes in brightness through evolution of the stars (a function of their age and mass).

The luminosity function for stars in the solar neighborhood has been studied by many workers, in particular [*234*] to [*238*]. There is good agreement in the general shape of the function, which increases towards fainter magnitudes, reaches a maximum near $M_v = +13$, and then decreases towards the limit to which LUYTEN's studies [*235*] have reached ($M_v = +20$). The recent work [*238*] indicates that the function is smoother than that originally given by VAN RHIJN [*234*].

Comparison of the luminosity function in different parts of the Galaxy is limited to regions fairly near the sun, because of the need for statistical completeness. Such a comparison has been made by MCCUSKEY [*238*]. The functions at distances of 100 to 600 parsecs show good general agreement. There is a slight tendency for the number of stars in the range $+4$ to $+7$ to increase with distance, and this is shown in Fig. 21; this confirms earlier work by OORT [*239*]. In theoretical discussions of the luminosity function, different workers have adopted slightly different mean values from the various observations. In Table 6 the adopted function covering the largest range in absolute magnitude, -6 to $+20$, by SANDAGE [*240*], is given in the second column [according to the usual practice, $\log\varphi(M)+10$ is given].

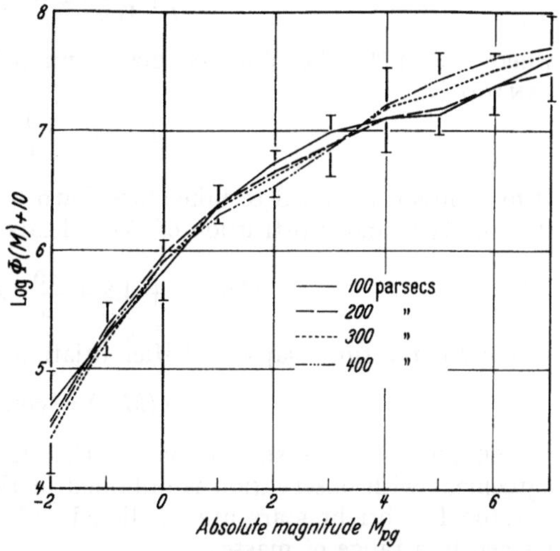

Fig. 21. MCCUSKEY's luminosity functions [*238*]. Variations in $\log\varphi(M)+10$. The average luminosity functions in the galactic plane for regions at 100, 200, 400, and 600 parsecs are shown. Vertical lines indicate the range in $\log\varphi(M)$.

Two different treatments of the observed luminosity function have been given. One is based on the theory of mass loss and mixing in main-sequence stars and will be discussed in Sect. 53. The other is based on the theory that stars as they evolve develop an inhomogeneity between core and envelope and evolve into the giant region, and is discussed in Sect. 54.

Table 6. *The adopted general luminosity function and the Salpeter "age zero" luminosity function.*

M_v	$\log\varphi+10$	$\log\psi+10$	M_v	$\log\varphi+10$	$\log\psi+10$
−6	1.29	4.71	+8	7.66	7.66
−5	2.43	5.59	+9	7.72	7.72
−4	3.18	6.08	+10	7.81	7.81
−3	3.82	6.41	+11	7.95	7.95
−2	4.42	6.68	+12	8.11	8.11
−1	5.04	6.92	+13	8.22	8.22
0	5.60	7.10	+14	8.21	8.21
+1	6.17	7.26	+15	8.12	8.12
+2	6.60	7.25	+16	7.98	7.98
+3	7.00	7.23	+17	7.76	7.76
+4	7.30	7.30	+18	7.40	7.40
+5	7.45	7.45	+19	6.58	6.58
+6	7.56	7.56	+20	5.28	5.28
+7	7.63	7.63			

53. Prediction of the form of the observed luminosity function according to the theory of mass loss by main-sequence stars; initial mass function. On the assumption that stars on the upper part of the main sequence were all originally formed at one particular mass, and have reached their present positions by mass loss through corpuscular radiation, as in the theory by FESSENKOV, MASSEVICH, and PARENAGO discussed in Sect. 24, it is possible to compute the expected luminosity function at the present day (if star formation is taken to be occurring at a constant rate).

PARENAGO [*241*] has taken the formula for the rate of loss of mass given in Sect. 24, which was derived from the theory of stellar structure together with the assumption of mixing and a uniform composition during the conversion of hydrogen to helium. This is given by

$$\frac{d}{dt}\left(\frac{\mathfrak{M}}{\mathfrak{M}_\odot}\right) = -\frac{k}{3}\frac{L}{L_\odot} \tag{53.1}$$

where $k = 10^{-10}$. The observed mass-luminosity relation (Sect. 19) taken by PARENAGO is

$$L = L_\odot \left(\frac{\mathfrak{M}}{\mathfrak{M}_\odot}\right)^4. \tag{53.2}$$

If mass loss is assumed to take stars down to the mass of the Sun, then integration of (53.1) and substitution of (53.2) leads to

$$M_{\text{bol}} = +4.8 + \frac{10}{3}\,\text{Log}\,(1 + kt) \tag{53.3}$$

where t is the time elapsed. Differentiation of (53.3) yields dM_{bol}/dt, and hence

$$\text{Log}\,(M_{\text{bol}}) = \text{const} + 0.3 M_{\text{bol}}. \tag{53.4}$$

MASSEVICH [*242*] has shown that, if the starting mass is taken as $8\mathfrak{M}_\odot$, good agreement with observation is obtained in the range of magnitude -3 to $+5$, but too few bright stars are predicted. She suggested that stars are actually formed in a range of masses.

The "initial" luminosity function is defined as the function describing the distribution of luminosities along the main sequence at the time of the star's arrival there; this can also be expressed in the form of the initial mass function. KAPLAN [*343*] has worked in the opposite direction to that considered by PARENAGO and MASSEVICH. If $\varphi(M)$ and $\psi(M)$ are the observed and initial luminosity functions, respectively, then on the assumption that both are time-independent, and using Eqs. (53.1) and (53.2), he derived

$$\varphi(M)\frac{dM}{dt} = \int\limits_{-\infty}^{M} \psi(M')\,dM'. \tag{53.5}$$

From this,

$$\psi(M) = K \cdot 10^{-0.3M}\,\varphi(M)\left[\frac{d\,\text{Log}\,\varphi(M)}{dM} - 0.3\right]. \tag{53.6}$$

The function ψ has a sharp maximum at $M = -5$, corresponding to spectral type $B1$ and mass about $10\mathfrak{M}_\odot$. KAPLAN concluded that about 80% of all new stars enter the main sequence with masses in the range 6 to $18\mathfrak{M}_\odot$.

The basic objection to this discussion of the luminosity function is the same as that given at the end of Sect. 24, where the theory of evolution of stars down the main sequence by mass loss was described. It is that the necessary assumption of complete mixing now appears to be unjustified.

54. SALPETER's luminosity function for unevolved main-sequence solar-neighborhood stars; initial mass function. On the entirely different theory of constant mass for stars on the main sequence, followed by evolution towards the Schoenberg-Chandrasekhar limit and then into the red giant region, SALPETER has computed an initial luminosity and mass function as follows [*244*].

Two factors were given in Sect. 52 as determining the observed luminosity function. According to the discussion in Chaps. BII and III, the upper limit to the age of known star clusters is about 6×10^9 years, and it is probable that this is also the upper limit to the age of most of the stars in the solar neighborhood. On the theory that stars do not change their mass during their life on the main sequence, the second factor therefore applies only to stars that have undergone appreciable evolution off the main sequence in 6×10^9 years, i.e., to stars with masses greater than about $1.2\times10^9\,\mathfrak{M}_{\odot}$.

In deriving initial luminosity or mass functions, it is important first to separate the observed stars into main-sequence stars and giants and supergiants (evolved stars). SALPETER has done this by choosing for each value of M_v a corresponding spectral type which would divide the stars so that stars earlier than this type would contain most of the main sequence stars. For each such dividing spectral type he has then obtained the fraction of stars which have types earlier than this. This fraction is then taken as the fraction of main-sequence stars at the given value of M_v. The observed (total) luminosity function adopted by SALPETER and the main-sequence function derived from it are shown in Fig. 22. SALPETER interpreted the change in slope near $M_v=+4$ as being due to the older stars brighter than this having evolved off the main sequence and most of them having come to the end of their life. By assuming that the rate of star formation has been uniform and constant since an epoch 6×10^9 years ago, and by using the Schoenberg-Chandrasekhar theory for the evolution of stars up to the point where they rapidly leave the main sequence, he derived an "initial" luminosity function, $\psi(M)$, which represents the relative frequency with which stars populate the main sequence at the time of their arrival there.

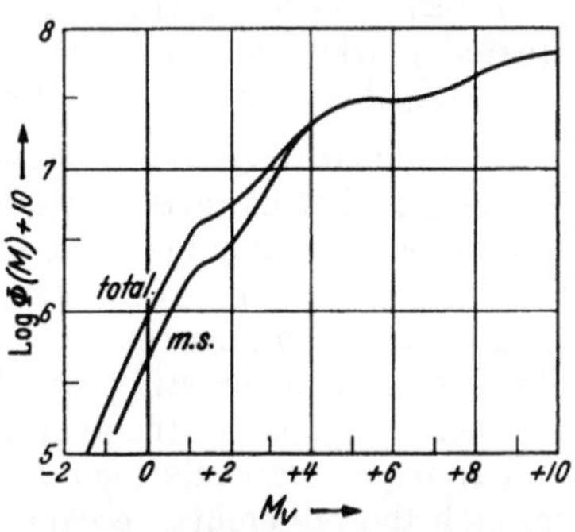

Fig. 22. The logarithm of the luminosity function, Log $\varphi(M)+10$, plotted against visual magnitude M_v, from SALPETER [244]. The curve marked *total* denotes the total observed function. The curve marked *m.s.* denotes the derived function for main-sequence stars only.

The derivation of $\psi(M)$ below is due to SANDAGE [240]. The life time onthe main sequence of a star is given by Eq. (40.1). With a constant rate of star formation, dN/dt, for a time T, we have:

$$\psi(M)=\frac{dN(M)}{dt}\,T. \tag{54.1}$$

For $\tau\leqq T$, only those stars formed less than τ years ago will still be present. Therefore

$$\varphi(M)=\frac{dN(M)}{dt}\,\tau \quad \text{for} \quad \tau\leqq T, \tag{54.2}$$

$$\varphi(M)=\psi(M) \quad \text{for} \quad \tau>T. \tag{54.3}$$

Hence for $\tau\leqq T$, from Eqs. (40.1), (54.1), and (54.2) we have that

$$\psi(M)=\varphi(M)\,\frac{\mathfrak{M}_L}{\mathfrak{M}}\,\frac{L}{L_L}, \tag{54.4}$$

where $\mathfrak{M}_L$, L_L are the mass and luminosity which give $\tau=T$ in Eq. (40.1). Table 6 gives in the third column $\text{Log}\,\psi(M)+10$, as calculated by SANDAGE from $\text{Log}\,\varphi(M)+10$ in the second column.

SALPETER defined the "initial" mass function, $\xi(\mathfrak{M})$, by

$$dN=\xi(\mathfrak{M})\,d\,(\text{Log}\,\mathfrak{M})\,\frac{dt}{T}, \tag{54.5}$$

where dN is the number of stars in the mass range $d\mathfrak{M}$ created in the time interval dt per cubic parsec. He derived $\xi(\mathfrak{M})$ from his initial luminosity function, and found it to have the form

$$\xi(\mathfrak{M}) \approx 0.03 \left(\frac{\mathfrak{M}}{\mathfrak{M}_\odot}\right)^{-1.35} \tag{54.6}$$

in the range 0.4 to 10 solar masses. From integration of this he found that the total mass of stars formed at any time during the last 6×10^9 years that were brighter than $M_v=+3.5$, i.e., stars that will have evolved off the main sequence and mostly come to the end of their life, is of the same order of magnitude as the total mass of stars fainter than $+3.5$ formed at any time, i.e., stars at present on the main sequence.

Integration of $\psi(M)$ in Table 6 gives 0.120 stars/psc³ for the number of stars created in the lifetime of the Galaxy in the solar neighborhood. If $T=6\times 10^9$ years, this gives an annual rate of 2×10^{-11} stars/parsec³ [*240*].

55. Comparison between SALPETER's initial luminosity function and observations in clusters. Let us take SALPETER's calculated initial luminosity function, $\psi(M)$, to represent the average mode of star formation in the solar neighborhood during the last 6×10^9 years. If the stars in a cluster or association formed simultaneously, then comparison between their observed luminosity functions, over that section of the main sequence still present in each cluster, should show whether star formation here is typical of the average. There are two provisos: that the assumptions underlying SALPETER's theory are justified, and that no modification in clusters through the preferential escape of stars in any mass range has occurred.

α) Galactic clusters. Comparisons have been made by WALKER [*178*], [*179*], SANDAGE [*199*], VAN DEN BERGH [*245*], and JASCHEK and JASCHEK [*246*]. In most clusters, down to about $M_v=+6$, there is good agreement, even in the very young clusters NGC 2264 and 6530. This is in striking contrast to the comparison of the observed general luminosity function, $\varphi(M)$, which has long been known to drop considerably below the cluster values at the bright end of the main sequence. This can be explained by the fact that there has been no depletion though evolution of the bright stars in clusters up to the end point of the main sequence in each case, while there has been such depletion in the general solar-neighborhood stars.

In M 67, the luminosity function has a maximum at $M_v=+4$ and declines toward zero at about $M_v=+9$. This has been interpreted [*247*] as being due to the escape of stars of low mass from the cluster. It has been shown that the present luminosity function for fainter stars can be predicted from an original (van Rhijn) function by the application of the theory of the escape of stars[1–3], if the cluster has an age of 180 relaxation times, or 4×10^9 years. This is of the same order of magnitude as the age ($4-6\times10^9$ years) predicted from the turn-off point of the main sequence.

In younger galactic clusters the luminosity function departs from the solar-neighborhood function for stars fainter than about $+6$. SANDAGE [*240*] has remarked that these clusters are not old enough for the escape of stars to explain the discrepancy, which is apparent even in the young cluster NGC 2264. VAN DEN BERGH [*245*] has suggested that it may be due to the inclusion of some high-velocity (population II) stars in the solar-neighborhood luminosity function, and that the stellar populations may differ in the relative numbers of faint stars

[1] V. A. AMBARTZUMIAN: Ann. Leningrad State Univ., No. 22, 19 (1938).

[2] L. SPITZER: Monthly Notices Roy. Astronom. Soc. London **100**, 396 (1940).

[3] S. CHANDRASEKHAR: Principles of Stellar Dynamics. Chicago, Ill.: Chicago University Press 1942.

they contain. It may also be partly due to the apparent loss of fainter stars in the statistical fluctuations of the background stars.

The agreement between $\psi(M)$ and the observed luminosity functions in galactic clusters for stars brighter than about $M_v = +6$ may be taken to imply that the conditions for star formation in this mass range in various clusters are generally similar to each other and to the conditions for formation of field stars (which may originally have been formed in clusters or associations). However, the discrepancy for the stars fainter than $M_v = +6$ may reflect a real difference in the conditions for formation of these lower-mass stars.

Finally, the function $\psi(M)$ has been used to predict the number of white dwarfs in certain clusters (on the assumption that all stars which were originally on the main sequence, but which have subsequently disappeared by exhaustion of their energy sources, have become white dwarfs) [*199*], [*240*]. The predicted numbers are given in Table 7, along with numbers of possible white dwarf members found in these clusters.

Table 7. *Predicted numbers of white dwarfs in certain galactic clusters, compared with numbers of possible white dwarf members observed.*

Cluster	Predicted	Observed	Reference
h Persei	0	—	—
Pleiades	2	—	—
Coma	9	15 possible	[*217*]
Hyades	23	6 + 36 possible	[*216*], [*217*]
Praesepe	20	12 possible	[*217*]
M 67	146	possibly large number	[*199*]

β) Globular clusters. Luminosity functions for M 3 [*248*], [*199*], M 92 [*249*], and NGC 4147 [*250*] have been obtained, and, apart from scale factors, are very similar. Direct comparison of the luminosity functions of globular clusters with $\psi(M)$ yields only a small region of overlap; the observed main sequences extend only from $+3.5$ to $+6$. Checks have been made [*248*], [*249*] and showed reasonable agreement. Sandage [*199*] has supposed that the fainter stars in M 3 are represented by the function given in Table 6. Integration of the function yields $M_v = -9.00$ for the total luminosity. Since the color index is 0.50, $M_{pg} = -8.5$, in good agreement with the observed value of -8.42. By using Kuiper's [*78*], [*237*] mass-luminosity relation for fainter stars, the masses are derived, giving the total mass (excluding the end-products of stars that have completed their life history) as $1.75 \times 10^5\, \mathfrak{M}_\odot$. By computing the number of "dead" stars from use of the original function $\psi(M)$ for stars brighter than $M_v = +3.5$, and by assuming that each "dead" star becomes a white dwarf of mass $1.44\, \mathfrak{M}_\odot$, the total mass of $2.45 \times 10^5\, \mathfrak{M}_\odot$ is derived.

There is no direct mass determination for M 3, but one is available for the cluster M 92, from the velocity dispersion of individual stars[1, 2]. If it is assumed that the mass of M 92 can be "scaled up" by using the ratio of the integrated luminosities of M 92 and M 3 and the mass-to-light ratio of M 92, a mass of $(2.4 \pm 1.2) \times 10^5\, \mathfrak{M}_\odot$ is derived. The good agreement of this value with that derived from the assumed luminosity function, and the agreement of the calculated and observed integrated magnitudes, have been taken to imply that $\varphi(M)$ and $\psi(M)$ in Table 6 give a good representation of the actual luminosity function in M 3. If this is the case, then the proportion of faint stars in the extreme halo population is the same as in the solar neighborhood.

56. Luminosity functions of M 67 and M 3 as a means of mapping evolutionary paths. The luminosity function and color-magnitude diagram of a cluster of stars

[1] O. C. Wilson and M. F. Coffeen: Astrophys. J. **119**, 197 (1954).
[2] M. Schwarzschild and S. Bernstein: Astrophys. J. **122**, 200 (1955).

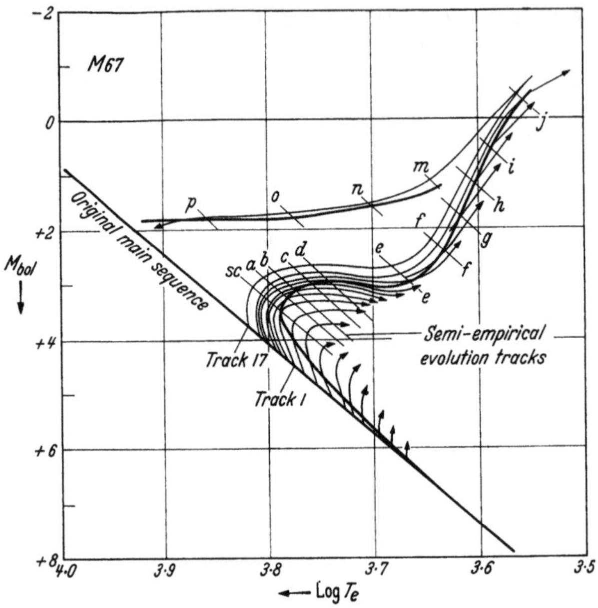

Fig. 23. Semi-empirical tracks of evolution for stars in M 67, from Sandage [*251*]. The various mapping points are shown as lines labelled *SC*, *a*, ..., *p*. The observed color-magnitude diagram for M 67 transformed to the $M_{\rm bol}$, Log T_e plane is shown as the heavy line cutting across the evolutionary tracks.

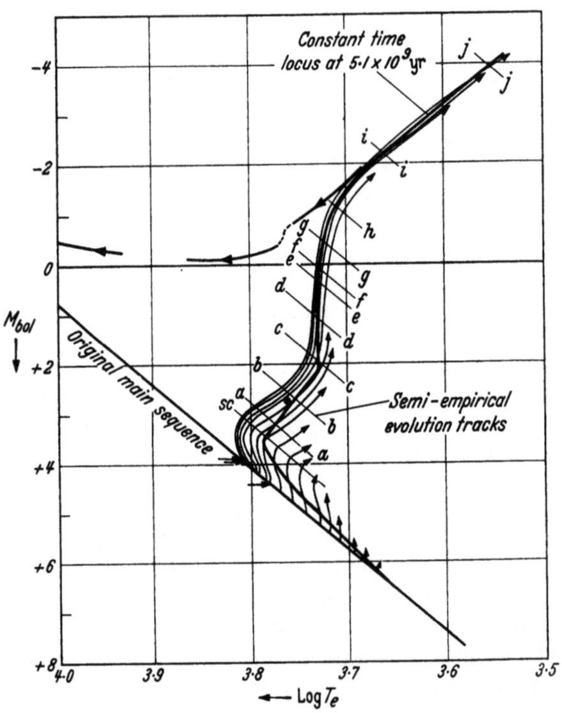

Fig. 24. Semi-empirical tracks of evolution for stars in M 3, from Sandage [*251*]. For details, see caption for Fig. 23.

can be used, if certain assumptions are made, to determine semi-empirically the evolutionary tracks beyond the Schoenberg-Chandrasekhar limit in the $M_{\rm bol}$, Log T_e plane for individual stars in the cluster, and to determine the time scale for traversing these tracks [*91*], [*251*]. This method will now be described.

In the preceding section it was seen that the luminosity function $\psi(M)$ (Table 6) is apparently a good representation of the function for the globular cluster M 3, but predicts more than the observed number of stars fainter than $M_v = +6$ in the galactic cluster M 67. As a starting assumption, therefore, van den Bergh's work [*247*] on the rate of loss of low-mass stars from M 67 is used to "correct" the observed luminosity function in order to derive the function unaffected by the escape of stars. The effects extend up to about $M_v = +4$.

The Schoenberg-Chandrasekhar (S-C) theory for the early evolution close to the main sequence is ten used as a starting point, giving the evolutionary tracks up to the limiting configuration when a star has consumed the hydrogen in a core containing a fraction q of the mass, where if the initial hydrogen content is X, $Xq = 0.07$. These tracks for M 67 and M 3, respectively, are shown in Figs. 23 and 24, running from the original main sequence up to the line marked SC, which indicates the locus of the limiting configuration. The heavy line in each figure gives the observed color-magnitude diagram converted to the $M_{\rm bol}$, Log T_e plane. The more massive a star, the faster it will reach the line SC. The line marked Track 1 in Fig. 23, and that marked by the lower arrow in Fig. 24, indicate the tracks of stars which have just reached the S-C limit in the age of each cluster respectively.

The extension of tracks beyond the S-C limit is done by means of the luminosity functions. At any evolutionary stage, homology is assumed for the evolving stars in the mass range under consideration; the locus of a homologous set of stars is parallel to the main sequence. Counts in the observed luminosity functions, $\varphi(M)$, for stars along the subgiant and giant branches have been compared with counts along the original main sequence (corrected in the case of M 67 for escape of stars), in order to proceed step by step up the evolutionary track. If there are N stars between two fairly close points on the giant branch, these will have come from a stretch of the main sequence containing N stars in the original, unevolved main sequence (corrected for escape of stars in M 67). The upper end of the stretch on the main sequence has been joined to the upper end of the corresponding stretch on the giant branch by a line drawn parallel to Track 1, but displaced above it and extrapolated slightly to meet the observed color-magnitude diagram. In this way, step by step, all the tracks shown in Figs. 23 and 24 were constructed by SANDAGE.

The times for stars to travel along the various marked segments of each track are given in Tables 8 and 9. The time for stars on Track 1 to reach the line SC is the age of the cluster, and

Table 8. *Times along the various evolutionary tracks for M 67 [251]. Times are in 10^9 years.*

Track No.	*MS–SC*	*SC–a*	*a–b*	*b–c*	*c–d*	*d–e*	*e–f*	*f–g*	*g–h*	*h–i*	*i–j*	*j–k*	*k–l*	*l–m*	*m–n*	*n–o*	*o–p*
1	5.1000																
2	4.6533	0.4467															
3	4.2429	0.4133	0.4438														
4	3.8695	0.3769	0.4067	0.4469													
5	3.7975	0.3684	0.3975	0.4387	0.0979												
6	3.7281	0.3616	0.3902	0.4307	0.0961	0.0933											
7	3.6612	0.3550	0.3830	0.4248	0.0948	0.0921	0.0891										
8	3.5941	0.3485	0.3761	0.4169	0.0931	0.0904	0.0880	0.0929									
9	3.5294	0.3421	0.3675	0.4075	0.0914	0.0887	0.0864	0.0917	0.0953								
10	3.4647	0.3359	0.3608	0.4000	0.0897	0.0871	0.0848	0.0900	0.0941	0.0929							
11	3.4000	0.3314	0.3558	0.3927	0.0881	0.0855	0.0832	0.0884	0.0924	0.0912	0.0913						
12	3.3377	0.3253	0.3493	0.3855	0.0865	0.0839	0.0817	0.0868	0.0907	0.0895	0.0896	0.0935					
13	3.2776	0.3178	0.3414	0.3784	0.0849	0.0824	0.0802	0.0851	0.0891	0.0879	0.0880	0.0918	0.0954				
14	3.2177	0.3120	0.3351	0.3714	0.0833	0.0809	0.0787	0.0836	0.0875	0.0863	0.0864	0.0901	0.0937	0.0933			
15	3.1598	0.3063	0.3290	0.3647	0.0818	0.0794	0.0773	0.0821	0.0859	0.0847	0.0848	0.0885	0.0920	0.0916	0.0921		
16	3.1022	0.3007	0.3230	0.3563	0.0796	0.0773	0.0756	0.0806	0.0843	0.0832	0.0827	0.0864	0.0903	0.0849	0.0904	0.1025	
17	3.0448	0.2952	0.3171	0.3498	0.0782	0.0759	0.0738	0.0789	0.0828	0.0817	0.0812	0.0849	0.0887	0.0833	0.0887	0.1006	0.0944

Table 9. *Times along the various evolutionary tracks for* M 3 [*251*]. *Times are in* 10^9 *years.*

Track No.	*MS–SC*	*SC–a*	*a–b*	*b–c*	*c–d*	*d–e*	*e–f*	*f–g*	*g–h*	*h–i*	*i–j*
1	5.1										
2	4.653	0.447									
3	4.243	0.408	0.449								
4	3.869	0.372	0.410	0.450							
5	3.696	0.355	0.391	0.430	0.228						
6	3.627	0.348	0.384	0.422	0.224	0.095					
7	3.594	0.345	0.380	0.418	0.222	0.094	0.046				
8	3.561	0.342	0.377	0.414	0.220	0.093	0.046	0.047			
9	3.529	0.339	0.373	0.410	0.218	0.092	0.045	0.047	0.046		
10	3.496	0.336	0.370	0.407	0.216	0.092	0.045	0.046	0.045	0.048	
11	3.465	0.333	0.367	0.403	0.214	0.091	0.044	0.046	0.045	0.048	0.046

the simple theory gives 5.1×10^9 years for both M 67 and M 3. The time to travel along the other tracks will be shorter, in proportion to the ratio of the luminosity on any track to the corresponding luminosity on Track 1. Subtraction of the time taken to reach the various marked lines in the figures yields the times taken to travel along any segment of the track.

The fraction of the star's mass which has been exhausted of hydrogen at each stage of the evolution of both clusters is given in Table 10, evaluated from the time spent by a star at each luminosity, the amount of energy emitted per gram of hydrogen converted to helium being known. The much smaller proportion exhausted in M 67 is a consequence of the lower luminosity of the giant branch in this cluster as compared with M 3, and is very interesting. Since we do not observe a well-populated sequence in M 67, subsequent to the giant branch, where the stars might be consuming the remaining 60% of their hydrogen, it has been suggested that a considerable amount may be ejected during the giant stage, by mass loss from the surface (cf. Chap. CII).

Table 10. *Fraction of mass burned at various points along the tracks in* M 67 *and* M 3.

M 67		M 3		
Point on Fig. 23	$X\,q$	Point on Fig. 24	$X\,q$	q for $X=0.9$
SC	0.070	*SC*	0.070	0.078
a	0.082	*a*	0.083	0.092
b	0.096	*b*	0.102	0.113
c	0.112	*c*	0.137	0.152
d	0.116	*d*	0.174	0.193
e	0.120	*e*	0.204	0.227
f	0.125	*f*	0.226	0.251
g	0.133	*g*	0.259	0.288
h	0.148	*h*	0.317	0.352
i	0.173	*i*	0.467	0.519
j	0.224	*j*	0.813	0.903
k	0.279			
l	0.313			
m	0.336			
n	0.351			
o	0.364			
p	0.375			

57. Observed and computed luminosity function for the *K* giants. In the H-R diagram for stars in the general field, the giants mostly lie in a fairly well-defined band, corresponding to luminosity class III on the Morgan, Keenan, and Kellman system, and do not scatter uniformly over the whole giant and supergiant domain. ROMAN [*252*] showed how sharp a maximum there is at class III in the stars of types *G*8 and later. On the basis of the theory accepted here, these stars have left the main sequence region after developing a chemical inhomogeneity between core and envelope, and have then moved into the giant region. The composite cluster diagram [*203*] reproduced in Fig. 22 of the preceding article by ARP gives an indication of the luminosity attained by giants coming from various regions of the main sequence. Thus, the supergiants in h and χ Persei have come from about $M_v = -6$, the luminosity class II giants in

M 41 have come from about $M_v = 0$, while the class III giants in the Hyades, Praesepe, NGC 752, and M 67 have come from a range of $M_v = +1$ to $+3.5$.

SANDAGE [*203*] has called this the "*funnelling effect*". Owing to the way in which the evolutionary tracks change between stars leaving the main sequence at $+1$ and at $+3.5$, stars from this whole range of main sequence are funnelled into a narrow band in the red giant region (see Fig. 25). The ratio of the numbers of stars in giant and main sequences of individual galactic clusters was used to obtain estimates for the ratio R of the time spent on the main sequence and in the giant region. This ratio was found to be approximately independent of absolute magnitude. Taking the luminosity functions $\varphi(M_v)$ or $\psi(M_v)$ for the main sequence, discussed in the previous paragraphs, and using the ratio R and the shapes of evolutionary tracks indicated by the clusters and interpolated for all starting points on the main sequence, it is then possible to evaluate empirically a predicted luminosity function in the giant region, which should agree with the observed function for field stars if the galactic cluster observations do give a true picture of evolution for all stars.

SANDAGE found good agreement between his computed function for $K0$ to $K2$ giants and the observed function in the solar neighborhood, both in the position of the maximum near $M_v = +1$ and in the width of the curve. Current ideas of stellar evolution can thus give a quantitative explanation for the K giants.

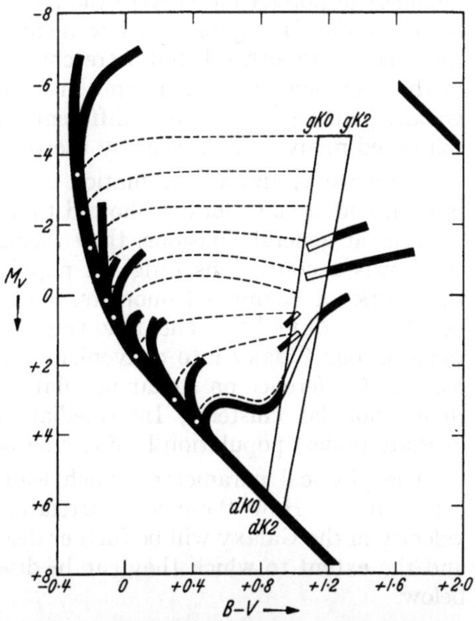

Fig. 25. Adopted evolutionary tracks for stars having different absolute magnitudes on the main sequence. Observed color-magnitude diagrams of different galactic clusters are shown by the filled areas; the adopted tracks are indicated by dashed lines starting at the various open circles on the main sequence. The region occupied by giant stars of various luminosities in the spectral range $K0-K2$ is shown, in order to indicate where the various evolutionary tracks intersect it.

V. Stellar evolution on the galactic scale.

58. Stellar populations. The concept of different stellar populations has repeatedly been used in differentiating between galactic and globular clusters, and indeed this concept is so familiar today that it permeates all discussion of stellar evolution. Before discussing stellar evolution on the galactic scale, however, we shall briefly recall BAADE's work and describe the situation as it appears at present.

In 1944 BAADE [*226*] succeeded in resolving the stars in the central region of the Andromeda Nebula and in its two elliptical companions, M 32 and NGC 205, by using red-sensitive plates and a fairly narrow-band red filter. He thus showed that these regions were populated by stars whose brightest members were red, high-luminosity giants. This population was different from that characterizing the solar neighborhood and spiral arms of galaxies, where the brightest members were high-luminosity O- and B-type stars. It was apparently similar to the population of stars in globular clusters, where also the brightest members were red giants.

Thus BAADE introduced the concept of two kinds of stars: population I was that found in spiral arms of galaxies, galactic clusters, and the solar neighborhood. Besides the *O* and *B* high-luminosity stars, the classical cepheids, *M* supergiants, and the carbon and *S* stars were among the typical members of population I, and the presence of dust was characteristic. Stars in the nuclei, halo regions, and between the arms of spiral galaxies, throughout elliptical galaxies, and in globular clusters were called population II. The most characteristic members of population II were the RR Lyrae variables. Dust was always absent in population II regions. A few members of population II might be found in the solar neighborhood, but were characterized by high space velocities relative to the Sun, since they had come into the Sun's vicinity from other parts of the Galaxy with galactic orbits different from the Sun's. These results were foreshadowed many years earlier by the work of OORT [*239*].

Later work, and a combination of studies of galactic structure, stellar distribution, and stellar motions showed that although populations I and II provided convenient general divisions, they themselves could be subdivided and actually there was a continuous range of populations. These conclusions are based on the work of many astronomers, notably OORT [*253*], PARENAGO [*254*], and VON WEIZSÄCKER [*2*]. The following is a schematic subdivision of the whole range in our Galaxy into convenient classifications, which was discussed at the Vatican Conference on stellar populations in 1957 [*255*]: Extreme population II (halo, globular clusters); Intermediate population II; Disk population; Intermediate (older) population I; Extreme population I.

The physical parameters which lead to this division are age and chemical composition. How these are correlated with the parameters of location and velocity in the Galaxy will be further discussed in the remainder of this Chap. B V, and the extent to which they can be disentangled will be discussed in Chap. D II below.

59. Spatial and velocity distributions of different kinds of stars as a clue to evolutionary linkage. The spatial distribution of stars in the Galaxy may be described by sets of ellipsoids or subsystems with varying degrees of flattening toward the equatorial plane of the Galaxy. If we consider the Galaxy in its initial stages of condensation into stars, and if this began soon after or at the same time as the Galaxy became differentiated from the rest of the Universe, then the first stars to condense would presumably have a roughly spherical distribution (like the system of globular clusters and halo stars). Stars, once formed, would remain "frozen" into the configuration possessed by the Galaxy at the time of their condensation, while the remaining matter would lose turbulent energy and begin to contract towards the equatorial plane defined by the rotation of the Galaxy. Thus stars forming at successive epochs would to a large extent delineate the shape possessed by the material out of which they formed, i.e., the shape of the Galaxy at the time of their formation; cf. [*2*].

Evidently the correlation of positions of stars in the H-R diagram with the ellipsoids or subsystems describing their spatial distribution is a useful clue in studying relative ages of and evolutionary links between various kinds of stars.

Stellar velocity distributions are strongly correlated with spatial distributions. Stars in highly flattened subsystems, like the young stars of high temperature and luminosity, have small velocities in the direction perpendicular to the galactic plane (the z-direction), and their total velocities are close to the circular velocity given by Keplerian motion at the corresponding distance from the galactic center. Stars in intermediate subsystems have larger z-velocities in order to maintain such a distribution against gravitational attraction by the main body of the Galaxy; their galactic orbits thus have an appreciable inclination to the galactic plane. Stars forming a spherical subsystem (globular clusters, halo stars) have the largest z-velocities.

Components of stellar peculiar motions in the galactic plane are correlated in magnitude with the z-component, but here an additional factor must be considered besides the peculiar motion of the star at the time of its formation. SPITZER and SCHWARZSCHILD [*256*] showed that the largest interstellar cloud complexes (of masses $10^6\ M_\odot$) could accelerate stars and cause an increase in velocity with age. Since the interstellar clouds lie close to the galactic plane, such an effect would be mainly apparent in the components of velocity in the plane, and would also be larger for stars that spend a larger proportion of their time in the plane. Whether or not this effect is important, the end result is the same: older stars, on the average, have larger peculiar motions, and stars of similar ages and past histories should have similar velocities.

The ordering of stars in different parts of the H-R diagram according to spatial distribution and velocities has been studied for many years by many workers. This work has been described in the book by TRUMPLER and WEAVER [*257*]. Reference may also be made to PAYNE-GAPOSCHKIN [*258*]. As an example, we reproduce in Table 11 the correlation between velocity and spectral type which has been obtained by PARENAGO [*254*]. Here σ_1, σ_2, and σ_3 are the components of the velocity ellipsoid, where σ_3 is in the direction which makes a fairly small angle with the direction of the galactic center. PARENAGO has also grouped the stars into subsystems according to spatial distribution as follows:

Table 11. *Components of the velocity ellipsoid for stars of different spectral types, after* PARENAGO [*254*].

Spectral type	σ_1 km/sec	σ_2 km/sec	σ_3 km/sec
$B0$ to $B5$	4.8	7.6	9.7
$B7$ to $A2$	5.5	10.5	16.2
$A3$ to $A8$	7.9	9.3	19.1
$A9$ to $dF1$	9.5	12.8	23.9
$dF2$ to $dF4$	11.7	17.0	26.8
$dF5$ to $dF7$	16.7	21.4	31.8
$dF8$ to $dG2$	22.6	27.5	46.0
$dG3$ to $dG7$	27.3	29.6	49.7
$dG8$ to $dK2$	22.9	30.5	52.0
$dK3$ to $dK6$	22.6	29.2	50.6
dM	22.5	25.6	45.7
Cepheid variables	5.4	8.6	12.5
gA to gF	10.3	14.2	26.8
Subgiants	23.7	27.1	42.5
$gG0$ to $gG8$	14.6	17.9	25.6
$gG9$ to $gK1$	15.7	20.5	30.5
$gK2$ to $gK5$	17.3	20.5	30.6
gM	16.3	22.5	31.2
Supergiants	7.7	10.2	12.7
High-velocity giants	100	60	50
Red variables	18.7	23.1	37.8

Very flat: interstellar gas, B stars, classical cepheids, supergiants, galactic clusters.

Rather flat: A, gA, gF, gG, gM, R and N stars.

Intermediate: dG to dM stars, red variables, long-period variables except for those with periods 150 to 200 days, subgiants, white dwarfs, planetary nebulae.

Spherical: long-period variables with periods 150 to 200 days, subdwarfs, high-velocity giants, globular clusters, RR Lyrae variables.

This grouping clearly shows that the youngest systems have a very flat distribution, while the oldest are spherical. Those of intermediate age lie between. The correlation between velocity dispersion and shape of subsystem is also clear, and the correlation of both with age is in agreement with our previous arguments.

However, the progression of velocities and shape of subsystem as the main sequence is descended is not yet fully explained. There is an approach towards equipartition of kinetic energies between stars of different masses, but complete equipartition is not attained [259] to [262]. Several factors may contribute to this. In the first place, it is conceivable that during star formation some kind of quasi-equipartition is set up in a condensing agglomerate, so that the stars of lowest mass might be formed with the largest peculiar motions. Secondly, stars fainter than $M_v = +4$ are on the average older than stars brighter than this, since the oldest brighter stars will have come to the end of their lives in times less than 6×10^9 years while the oldest fainter ones will still be in existence. Thus either the accelerating mechanism due to the interstellar clouds or the effects of decaying hydromagnetic turbulence and equatorial contraction of the Galaxy will result in larger velocities for the lower part of the main sequence; the latter would not necessarily give a continuous increase of velocities with decreasing mass, and indeed a levelling off is found. Thirdly, PARENAGO [254], [263], has interpreted the change in velocity dispersions as being due to a real break in the main sequence between types *G*4 and *G*7, stars in the two parts being thought to have different origins and evolutionary histories.

Table 12. *Velocity dispersions in* VYSSOTSKY'S *population groups* A *and* B.

Spectral type	Dispersion (km/sec)	
	population A	population B
A	11 ± 0.3	—
dG	17 ± 1	24 ± 2
gK	16 ± 2	27 ± 2
dM	18 ± 1	> 30

A link with the work on luminosity functions discussed in Sects. 52 and 55 can be seen. There it was shown that there is a slight deficiency of faint stars in some relatively young galactic clusters. This is in agreement with the velocity dispersions of stars at different points on the main sequence, and again the question is raised as to whether there is a real difference between conditions of star formation at an earlier epoch, when the Galaxy contained more interstellar matter, and recent times, when star formation is confined to the galactic plane and the amount of interstellar matter is a smaller fraction of the total mass of the Galaxy. In a discussion by VAN WIJK [264] of the place of origin of the high-velocity stars in the solar neighborhood, the possible effect of local density and amount of turbulence on the luminosity of the resulting stars is mentioned.

The correlation between the chemical composition of stars and their population characteristics has been mentioned in Sect. 51 and will be considered in more detail in Chap. D II. Extreme population II has a lower abundance of the heavier elements, relative to hydrogen, and there is in general a gradation of heavier-element content running through the various population types. This is reflected in the fairly good correlation between velocity and those spectral characteristics that depend on the heavier element content [265], [266]. VYSSOTSKY [267] has shown that stars of spectral types *dG*, *gK*, and *dM* can be divided into two groups, A and B, which are distinguished by spectroscopic criteria. The *dG* stars of population B show strengthening of CH and of the hydrogen lines and weakening of the metallic lines. The *gK* stars of population B show strengthening of CH and Ca I λ 4227. The *dM* stars of population A are the *dMe* stars, with hydrogen or Ca II emission. The two groups A and B have different velocity dispersions, as may be seen from Table 12 [267]. According to the velocities, population A is related to population I, flatter subsystems, and the spiral arm population; while population B is related to the less flat, intermediate subsystems. The very noticeable division at type *dM* confirms earlier work by DELHAYE [268] and suggests that the *dMe* stars are relatively young; the *dMe* stars with hydrogen emission as well as Ca II have a velocity dispersion of

15 km/sec, an even lower value than that in Table 12. The possible connection between these stars and T Tauri stars was discussed in Sect. 35.

Detailed work on the spatial and velocity distributions of particular classes of stars should continue to give information on the genetic relationships between these classes. For example, the *gK* stars are most closely correlated with the *F* main-sequence stars[1], and SANDAGE [*203*] pointed out that most of the *K*0 to *K*2 normal giants in the solar neighborhood should have evolved from the main sequence at type *F*, although a certain proportion should have come from type *A*.

The distribution of the *M* giants shows that they apparently belong to an intermediate subsystem and do not delineate spiral structure. They show a strong concentration towards the galactic center, which is more pronounced for the late-type *M* stars than for the earlier [*269*] to [*271*]. Thus the majority of the *M* giants are probably fairly old, and have therefore evolved from stars that are not very massive and lie on the main sequence perhaps in the late *A*—*F* range. The discovery of ordinary *M* giants in galactic clusters is awaited with great interest; their spatial distribution suggests that they might be found in clusters like M 67. So far they have been found in one cluster only, NGC 6940 [*271a*].

60. Spiral structure in the galaxy. *O*-associations, H II regions, individual *B* stars, and neutral atomic hydrogen all delineate spiral structure when their distributions are plotted spatially [*272*] to [*274*]; the 21-cm observations of spiral structure are discussed by J. H. OORT in Vol. LIII of this Encyclopedia. The 21-cm observations suggest that the neutral atomic hydrogen in our Galaxy comprises only $\sim 1\%$ of the total mass [*275*] and is very closely confined to the spiral arms. Since it seems unlikely that a comparable amount of interstellar matter is tied up in the form of molecular hydrogen and dust grains, the total mass of interstellar matter is therefore a small proportion of the mass of the Galaxy.

It has been made clear in the preceding sections that young stars may be expected to be confined to regions where interstellar gas and dust are located. This, with the exceptions noted in Sect. 36, is in general found to be the case. Therefore, we may conclude that star formation at the present time is confined to the spiral arms of our Galaxy [*276*], i.e., to a region containing only a small proportion of the total mass. Apart from the *O* and *B* stars, certain other groups of stars which belong to flat subsystems, and which are therefore relatively young, might be expected to delineate spiral structure. These are the *M* supergiants, the *S* and some carbon stars, and the classical cepheids.

The early *M*-type supergiants are in fact associated with H II regions and *O* and *B* stars, confirming the genetic relationship shown by the observations in h and χ Persei [*269*]. The *S* stars are also spiral arm objects, associated with *O* and *B* stars, and therefore probably represent a later evolutionary stage of massive stars [*277*], [*278*]. Carbon stars show clustering tendencies; there is apparently a separation between classes *R* and *N* [*269*], the *N* stars showing a tendency to spiral structure.

In the northern hemisphere, association between carbon stars and *O* and *B* stars has not been found, but such an association has been found in the southern hemisphere [*270*]. The distribution of the classical cepheids does not in general delineate spiral structure, except for some indications in the southern hemisphere [*279*]. Except for one probable member of h and χ Persei, the classical cepheids are not associated with *OB* stars [*191*], but are apparently correlated

[1] A N. VYSSOTSKY: Astronom. J. **56**, 62 (1951).

with stars of late *B* type. Some classical cepheids have been shown to be members of galactic clusters [*280*], [*281*].

61. Stellar evolution in external galaxies: General remarks. To determine whether stars in members of the local group and in other nearby galaxies have similar mass ranges, population characteristics, chemical compositions, and evolutionary tracks to those in our Galaxy, colors, magnitudes, and spectra of individual stars would ideally be required. However, such data are limited to stars of high luminosity, and less precise data can give much information. Integrated spectra studied by Morgan and Mayall [*282*], following the pioneer work of Humason and Mayall, are a very useful tool. Integrated colors have been studied by many workers of whom we will mention only Holmberg [*283*] and Stebbins and Whitford [*284*]. These colors show relatively little dispersion among elliptical nebulae, but there is a considerable range in the colors of spiral galaxies, indicating a range in the relative proportions of the stellar populations that are contributing to the light.

A search for the population indicators such as classical cepheids, *OB* stars, HII regions, and, in the nearer galaxies, RR Lyrae variables, can also be made. Another useful indicator of the stellar population of a galaxy, which may therefore throw light on conditions for star formation and stellar evolution, is the mass-to-light ratio [*285*] to [*287*]. This depends upon difficult observations with large telescopes in order to derive masses from the rotational velocities of galaxies, or else upon certain assumptions when the velocity dispersions in groups of nebulae are used.

Finally, 21-cm observations to yield the total mass and distribution of neutral atomic hydrogen may lead to tentative conclusions about the relative rates of present-day star formation in different galaxies. Some results of observations and their bearing on evolution in other galaxies are given in the succeeding paragraphs.

62. Stellar evolution in the Local Group: M 31. M 31 is thought to be similar to our Galaxy in form, content, and chemical composition. Both are *Sb* systems [*288*], [*289*]. In M 31, high-luminosity blue stars are confined to the spiral arms, where the interstellar dust and HII regions are also located, and thus the usual correlation between young stars and interstellar matter is observed. The amount of dust in the spiral arms decreases with increasing distance from the center [*276*]. An investigation of the relative ages of the young stars in the inner and outer arms would be interesting. A study of the novae in M 31 by Arp [*290*] showed that their distribution is intermediate between a highly flattened disk population and a nearly spherical globular cluster population.

The 21-cm observations made at Leiden suggest that the fraction of the mass which is neutral hydrogen is only a few percent [*291*], as in our Galaxy, although the results of Heeschen [*292*] lead to more neutral hydrogen than do the Leiden results, and we await clarification of this discrepancy. However, we conclude that probably the rate of star formation at present in M 31 is similar to that in our Galaxy and is also confined to the spiral arms.

Spectra of individual stars and integrated spectra do not indicate any difference in chemical composition between M 31 and our Galaxy. Morgan and Mayall [*282*] have constructed an H-R diagram that would reproduce the observed spectrum of the nucleus; it has a giant branch running from *G*8 to *M*, a Hertzsprung gap, and a main sequence containing *F*8 to *G*5 stars. Fainter main sequence stars could be present in the same proportions as in the solar neighborhood without contributing to the integrated spectra. The disk of M 31 has probably the same population as the nucleus. It is noteworthy that spectra indicating a deficiency of metals and carbon, oxygen, and nitrogen are not found: thus most of the light in M 31 comes from stars that are apparently "old population I" or disk population, and this is believed to be the case in our Galaxy also.

This conclusion is also reached from a study of the count-to-brightness ratio in M 31, i.e., the number of resolved stars brighter than a definite limit, divided by the total surface brightness [*293*]. This ratio is clearly quite sensitive to the stellar population, and gives information about a similar range of the luminosity function to that given by integrated spectra. The value for the main body (disk) of M 31 agrees well with that computed for the solar neighborhood by means of VAN RHIJN's luminosity function and differs from that computed for the globular cluster M 3. BAUM and SCHWARZSCHILD have therefore concluded that the main contribution to the light from the main body of M 31 is from old population I stars similar to those in our Galaxy. The color of M 31, +1.0 [*284*], is also consistent with this.

The most recent study of the mass-to-light ratio measured in solar units, f, in M 31 [*287*] yields the value of 16, and the data are consistent with this being constant over the nucleus and disk. The mass-to-light ratio gives information about the luminosity function far down the main sequence, where the stars contribute to the mass but not to the light. Using a value of $f=4$ in the solar neighborhood, SCHWARZSCHILD computed that 77% of the mass of M 31 is due to population II stars while 92% of the light is from old population I stars. For a mass-to-light ratio of 2.5 in the solar neighborhood, as given by GLIESE [*193*], the corresponding figures are 86% and 91%, respectively.

From these different methods of analysis, it seems probable that the evolutionary history of M 31 and its stars is very similar to that in our Galaxy.

63. Stellar evolution in the Local Group: M 33. Data for M 33, which is of type *Sc*, are not as extensive as for M 31. Red supergiants are found in the spiral arms in association with blue high-luminosity stars, in agreement with our picture of the evolution of these stars in our Galaxy. They are being studied at present by HUMASON and SANDAGE. Preliminary 21-cm observations [*294*] suggest that neutral atomic hydrogen contributes about one fifth of the mass of M 33. However, preliminary Leiden observations [*295*] do not confirm this and suggest a smaller fraction.

The color of M 33 is bluer than that of M 31; $C=0.63$ [*284*]. The integrated spectra also reveal an entirely different distribution of stellar population in the H-R diagram [*282*]. Although data for constructing an H-R diagram that would reproduce the spectrum are not yet available, the percentage of *A*- and *F*-type stars must be far higher in the main body of M 33 than in that of M 31.

The mass-to-light ratio, f, also indicates a difference in stellar population. Its value is 4 [*287*], similar to that in the solar neighborhood. Probably the mass and light are largely contributed by intermediate population I stars.

These facts all suggest that the present-day rate of star formation relative to the total mass is larger in M 33 than in M 31, so that a larger proportion of brighter main-sequence stars is present.

64. Stellar evolution in the Local Group: the Magellanic Clouds. Data on the Magellanic Clouds have been taken from a review article by BUSCOMBE, GASCOIGNE, and DE VAUCOULEURS [*296*], where full references are given. The Clouds were formerly classified as irregular, but primitive structure suggestive of barred spirals has been found, particularly in the Large Cloud. Many *OB* stars, HII regions, and bright red variables indicate that population I is well represented in both Clouds, while globular clusters containing RR Lyrae variables show that population II is also present, although it probably makes up a much smaller proportion than in M 31 and our Galaxy. Relatively few novae (intermediate population II) have been observed. Many galactic clusters are present in both clouds. Neutral hydrogen is present in both Clouds in large quantities [*297*], and the fraction of the total masses made up by this is probably of the order of a half in each case.

On the other hand, dust is observed in the Large Cloud but not in the Small Cloud; its absence in the latter case while neutral hydrogen and *O B* stars are present is surprising. Possibly the lower density of the Small Cloud has an effect on the formation of dust. Another possibility (not suggested in the review article quoted), which cannot be ruled out at present, is that the chemical composition of the Clouds, particularly the Small Cloud, is different from our Galaxy in having a smaller abundance of the elements heavier than helium.

The mass-to-light ratio, f, is 1.5 in the Small Cloud and 0.7 in the Large Cloud, and this low value suggests a higher proportion of extreme population I stars than is present in the solar neighborhood. The blue colors of both Clouds are also consistent with this; the color index on the (P, V) scale of EGGEN is $+0.18$ in the Large Cloud with no change between the central core and the outer regions, while in the Small Cloud it increases from -0.03 in the core to $+0.45$ in the outer parts. This latter result, which should be regarded as preliminary, is surprising since it indicates a higher proportion of bright blue stars in the central regions of the Small Cloud.

Luminosity functions for the brightest stars, as far as the data are available, seem to be similar to that for the bright stars in the solar neighborhood, but preliminary color-magnitude arrays (particularly in the Small Cloud) differ from anything found in our Galaxy in the large proportion of Ib supergiants of types *F–K*, which form a maximum in a band extending from color index -0.3 to $+1.5$.

The preponderance of population I and the high proportion of neutral hydrogen suggest that the present-day rate of star formation in the Clouds may be larger than in M 33, and very much larger than in M 31 and our Galaxy, although the absence of dust in the Small Cloud poses a problem. If there are evolutionary sequences in galaxies (as will be discussed in Sect. 68), then we might suggest that the Clouds are relatively young, and a chemical composition having a lower proportion of heavier elements would then not be surprising (see Chap. D II). The young and middle-aged stars in the Clouds might thus resemble the oldest stars in the Galaxy as regards chemical composition. Work is being carried out at present by ARP on color-magnitude diagrams in the Small Cloud, and should throw light on the evolutionary tracks of stars in clusters and in the general field. The latter may be, as in the solar neighborhood, a composite of evolutionary tracks for stars in a range of masses and ages. It will be interesting to know whether the apparent preponderance of *F–K* supergiants among the high-luminosity stars is real, and if so, from what part of the main sequence these stars might have come.

65. Stellar evolution in the Local Group: Elliptical galaxies. Resolution of the brightest stars in M 32, NGC 147, NGC 185, NGC 205, and the dwarf systems in Sculptor, Fornax, Draco, and Leo I and II shows that they are bright red giants as in the nucleus of M 31. The integrated colors, where available, are close to the mean value of $+0.86$ for elliptical galaxies as a group, with the exception of NGC 205, which has a color of about $+0.7$. Integrated spectra are not included in the lists by MORGAN and MAYALL [*282*], but other ellipticals (not in the local group) have spectra which indicate that most of the light comes from K giants. Again NGC 205 is an exception: probably F-type stars contribute largely to its light.

M 32, NGC 205, NGC 147, and NGC 185 were searched for novae in the same survey as that of M 31 [*290*], and none were found. The absence of dust in elliptical galaxies has been often commented upon since BAADE pointed out that this is correlated with systems being composed of population II stars. Once more, NGC 205 is an exception, since some dust clouds are present, and BAADE [*288*] has pointed out that near these there are some blue stars. This is a striking confirmation of the correlation between bright blue stars and the presence of dust.

Although dust is absent from most elliptical galaxies, interstellar gas may be present, as is shown by the frequent observation of the [OII] emission lines at λ 3727 in their spectra [*298*]. Recently, neutral hydrogen has been observed in M 32 [*299*]. The proportion of the total mass which is made up of this lies between 3×10^{-4} and 3×10^{-3}, which is considerably lower than the proportion in M 31 and our Galaxy. By arguing by analogy with globular clusters, where the detailed color-magnitude diagrams and theoretical and empirical evolutionary tracks enable the rate of star deaths and consequent ejection of gas to be estimated [*199*], it has been shown that the observed proportion of gas may be accounted for by ejection from stars; as will be shown in Chap. D II, the proportion of hydrogen to elements heavier than hydrogen may be much less than normal in this case. We may at least be certain of one fact, the present-day rate of star formation in elliptical galaxies must be very much less than that in the *Sb* systems M 31 and our Galaxy.

The mass-to-light ration in M 32 [*287*] is $f=200$, where the mass is derived upon the assumption that an asymmetry in velocities and form in M 31 is due to the gravitational pull of M 32. It is remarkable that this ratio is so much higher than that in the solar neighborhood and in the galaxies already discussed. Support for this high value is given by similarly high values found for other elliptical galaxies which are not members of the local group. Firstly, a direct determination in the nearby system NGC 3115 leads to a value of $f=100$. Secondly, study of the orbital motions of pairs of galaxies [*300*] gives an average value of $f=300$; while this is uncertain, more weight can be given to the factor of 6 by which f in average ellipticals exceeds f in average spiral galaxies. Thirdly, an uncertain and remarkably high value of $f=800$ has been determined in the Coma cluster of galaxies [*287*]; AMBARTZUMIAN [*300a*] has derived an even higher value. However, this depends on a number of uncertain factors: the small velocity sample available for calculating the kinetic energy may not be representative; there may be many more faint galaxies in the cluster than has been thought.

There have been suggested [*287*] three possible explanations for the high value of f in ellipticals, all of which imply different conditions for star formation than those in our Galaxy and in M 31. Firstly, the luminosity function might be similar in form to that in the solar neighborhood but shifted about 5 magnitudes fainter. The average stellar mass would then be about a quarter the average near the Sun. Secondly, the luminosity function might have a different form from that near the sun, with a much larger proportion of faint dwarfs. Thirdly, the excess mass might be contributed by white dwarfs, which might be much more numerous than near the Sun. This would imply a much higher rate of star deaths in earlier times, and consequently a larger number of massive stars. The original luminosity function might have been similar in shape to that in the solar neighborhood but shifted 7^m brighter. The average stellar mass would have been 6 times larger.

The high mass-to-light ratio in elliptical galaxies points to a difference in stellar content between them and globular clusters, where the ratio is only of order 1. The colors are different also; the average color index of globular clusters is +0.50 as compared with +0.86 for elliptical galaxies. The spectral energy distribution curve in M 32 is also different from the average in globular clusters [*301*], and the integrated spectra so far observed indicate a possible difference in chemical composition [*231*], [*282*] in that the elliptical nebulae do not indicate a lowered abundance of the metals. And yet both have been classified as "pure population II"; the brightest stars in both are similar, and the count-to-brightness ratio in one elliptical, NGC 205, is very similar to that in the globular cluster M 3 [*293*]. A preliminary color-magnitude diagram for the Draco system [*302*] looks very similar to that for globular clusters, and RR Lyrae variables have been found in that system. Possibly the stellar content of dwarf ellipticals like the Draco system is not the same as that in larger systems like M 32 [*240*].

A theoretical luminosity function for M 32 has been constructed so as to be consistent with the observed spectral energy distribution and to yield a mass-to-light ratio of the right order of magnitude [*303*]. Fainter than absolute magnitude +6 this theoretical function rises much more steeply than the function in the solar neighborhood, corresponding to the second possible explanation given above [*287*]; the proportion of faint dwarfs which contribute to the mass but not to the light is very much larger. Thus we see once again a hint that under different physical conditions, the mass distribution function in star formation may change and yield different luminosity functions (Sects. 52, 55, and 59).

66. Stellar evolution in the Local Group: Intergalactic star clusters. The National Geographic Society-Palomar Observatory Sky Survey plates have revealed the existence of objects which may be called intergalactic star clusters[1, 2]. These objects resemble poor globular clusters like NGC 4147, but they are at very great distances which place them right outside our Galaxy, in intergalactic space. It is not known at present whether they originated in the Galaxy and excaped from it, whether they have escaped from some other galaxy in the local group, or whether they had an independent origin. Radial velocity measures might enable the first possibility to be tested.

A study of the stellar population of these clusters, by means of color-magnitude diagrams, luminosity functions, and a search for variable stars will be particularly interesting in considering their possible origin. Some theoretical discussion concerning their possible origin has been given in Sect. 2.

A preliminary observational study of one cluster (the "11^h cluster") has been made from the Sky Survey prints, and a color-magnitude diagram derived [*5*]. A more detailed investigation of this cluster, and a preliminary study of another, the 10^h cluster, have been made from 200-inch Palomar plates [*6*]. By assuming that the horizontal branch is at $M_v = 0.0$, their distance moduli are found to be 20.5 and 20.6 magnitudes, respectively (125 kiloparsecs). The luminosity function of the 11^h cluster, when scaled up by a factor of 14, agrees well with that for the globular cluster M 3, down to the limit of observations at $M_v = +1$. From this the integrated apparent magnitude is found to be +14.3. Some strik-

[1] G. O. ABELL: Publ. Astronom. Soc. Pacific **67**, 258 (1955).
[2] A. G. WILSON: Publ. Astronom. Soc. Pacific **67**, 27 (1955).

ing differences are found, however, between this cluster and M 3. Its diameter is about twice as large, and there is little or no central concentration. The horizontal branch is well populated but very short; it extends only from the red giant branch to the place where the RR Lyrae variable star location usually starts in normal globular clusters. There are no RR Lyrae variables. In the 10^h cluster, however, the horizontal branch appears to be more normal and there is one RR Lyrae variable. In the 11^h cluster there are two red variables, with periods probably in the range 100 to 200 days, and these are among the brightest stars in the cluster[1].

Another of the clusters, from a preliminary uncalibrated study, has a typical population II giant branch, leading into the beginning of the main sequence, but no horizontal branch at all. More studies are clearly desirable. In spite of the general similarity to poor globular clusters in the Galaxy, there seems to be a difference in the evolutionary tracks of the stars which may be important.

67. Stellar evolution in more distant galaxies. Studies in more distant galaxies can be made by means of integrated spectra, colors (including spectral energy distribution curves), and, in some cases, the mass-to-light ratio. These may be correlated with the classification sequence according to galactic form (see the article by G. DE VAUCOULEURS in Vol. LIII of this Encyclopedia). As has already been mentioned, the elliptical and *S*0 galaxies show fairly little dispersion in color about the mean value of $+0.86$, while spiral galaxies range from quite blue colors for irregular, magellanic, *Sd*, and *Sc* systems to colors similar to those of ellipticals for the *Sa* and *Sb* systems, and within each subdivision there is a considerable dispersion.

The spectra show a similar correlation [*304*], [*282*]. Among those so far classified, the *A*-systems comprise irregular and normal and barred spirals of subdivision *c*; the *AF*-systems are all of *Sc* type; the *F*-systems contain *Sb* and *Sc*; the *FG* systems are mostly *Sb*; the *K*-systems contain subdivisions *a* and *b* of both normal and barred spirals, ellipticals, *S*0 systems, and one irregular system.

So far, no evidence of differences in chemical composition have been detected, but only large differences are likely to shown on low-dispersion spectra. There are, however, a few cases of lack of correlation between colors and spectra where the possibility of this being due to a remarkably low abundance of elements heavier then helium should not *a priori* be ruled out.

Most of the colors and spectra of galaxies outside the local group can be explained in the following way. The ellipticals and lenticulars contain mostly intermediate (not extreme) population II stars or old disk population I corresponding to subsystems of intermediate flattening in our Galaxy, together with some extreme population II. The spirals contain a mixture of young population I (flat subsystems) and old disk population I, together with some population II. The old population is contained in the nucleus and main body while the young population is located in the spiral arms. On going from subdivision *c* to *a*, i.e., from systems with small nuclei to those with large nuclei, the proportion of the old population increases.

The high mass-to-light ratio in ellipticals has already been mentioned, together with its implication that there is a different luminosity function for stars in these systems. At present the problems of stellar content, involving stellar formation and evolution, in different kinds of galaxies pose a large number of questions which still remain to be answered. How much extreme population I (similar to globular clusters in our Galaxy) is present in any class of galaxy?

[1] L. ROSINO: Mem. Soc. Astronom. Italiana **28**, 293 (1957).

The resolution of the brightest stars in the nucleus of M31, M32, and their globular clusters at the same absolute magnitude certainly suggests a fair proportion in these systems. In what respect do the globular clusters of spiral and elliptical systems differ from or resemble the globular clusters of our Galaxy or various parts of the galaxies themselves? Is there a difference in population between dwarf and giant ellipticals? Is there a range of chemical compositions and, if so, is it correlated with structural classification?

A final point concerns the colors of the most distant elliptical nebulae so far observed. The apparent excess reddening (Stebbins-Whitford effect) which was thought to be present implied that there was a real difference in stellar content that was correlated with distance, and which was therefore interpreted as an aging effect [*301*], [*305*]. The relatively short time scale spanned by the farthest measures did not, however, suggest that great changes in stellar content would be apparent, and it now appears probable that this excess reddening is not present [*306*].

68. Are there evolutionary sequences of galaxies? As was discussed in Sect. 2, it is generally supposed that a protogalaxy at the time of its formation as a separate unit is wholly composed of gas which contains some density fluctuations, and a certain amount of angular momentum and magnetic field. Stars begin to condense in this system, and since the rate of star formation must be closely connected with the amount of gas present, it is logical to assume that this rate is a function which decreases as the galaxy ages. If massive stars are formed at an early stage they will go through their life-history rapidly. At the end they must eject matter back into the galaxy in the form of gas (Chap. C II and D II), but a white dwarf remnant may be left which will play no further part in the interchange of matter between stars and the interstellar medium. Also, the low-mass stars which will continuously be formed as part of the natural mass-distribution will go through their life-histories so slowly that they will not have evolved appreciably in current estimates for the age of our sample of the universe. Thus, as time goes on, an increasing proportion of the mass of a galaxy will become tied up in the form of inert white dwarfs and low-mass red dwarfs.

It is natural to ask whether galaxies of different types form an evolutionary sequence or whether the structure of a galaxy is determined by some initial parameter, such as angular momentum or initial magnetic field. With regard to HUBBLE's original "tuning fork" classification diagram, he specifically remarked that the nomenclature "early" for the subdivision *a* and "late" for the subdivision *c* was not to be taken to have temporal implications, but that the sequence of classification was purely empirical. An early suggestion [*307*] was that a galaxy, shrinking under its gravitational attraction and possessing initial angular momentum, would tend to progress through configurations like those given in the sequence $E0 \rightarrow E7 \rightarrow Sa \rightarrow Sb \rightarrow Sc$. Individual galaxies might stop at any point in the sequence from want of angular momentum. However, the evidence on stellar populations today tends to argue against this.

That the sequence might be evolutionary in the opposite direction, $Irr \rightarrow Sc \rightarrow Sb \rightarrow Sa \rightarrow S0 \rightarrow E$, was suggested by SHAPLEY [*289*] because, in the light of present knowledge of stellar populations in different systems, there is a steadily decreasing proportion of young stars and interstellar matter, on the average, and an increasing proportion of old stars, in going this way along the sequence; see also [*308*]. Once a galaxy has come to be formed principally of small stars (elliptical system), one cannot see how it could ever again contain much gas and become "young" again, with the formation of massive stars. On the other hand, a young (spiral) system, with much gas and many massive stars, could become old, but whether it would, when old, look like an elliptical galaxy is another matter.

Theoretical arguments based upon the decay of initial turbulence, flattening towards the equatorial plane, and the transfer outwards of angular momentum and its eventual loss from the system also suggest that evolution would be in the sense suggested by SHAPLEY. They have been given by VON WEIZSÄCKER and his associates [*158*], [*2*] and references given therein, [*309*]. According to this theory, the time scale for a galaxy will depend on its mass, since the time scale for flattening connected with the decay of the original turbulence may be of the same order as one rotational period, while the time scale for loss of angular momentum (with the consequent growth of the nucleus of a galaxy) might be 10 or 20 times the rotational period. Thus galaxies born at the same instant of time might have different genetic "ages", the smaller ones becoming "old" more rapidly.

A critical discussion of this possible evolutionary sequence, with particular reference to the colors of galaxies, has stressed the large range in the colors of irregular and spiral galaxies of all subdivisions [*310*]. Galaxies of the whole range of structural types may have the same color and hence, by implication, the same stellar content. If the same stellar content in turn implies the same age, then there is not a simple one-to-one correlation between structural form and age. It should be remarked in passing that in certain peculiar galaxies with intense ultraviolet continuum radiation, the blue color index may be due to non-thermal radiation, unconnected with stellar radiation, and outside the scope of this chapter. Alternatively, it may be due in some cases to extremely hot stars, which are more luminous than any found in our galaxy, and which are presumably very massive. A recent investigation of one or two such galaxies, originally described by HARO [*310*], has been made by MÜNCH [*311*]. This has shown that the galaxies probably have small masses and their nuclear regions are apparently excited by these highly luminous, hot stars.

Haro indicated three possible interpretations of his comparison between colors and structural classification. First, as one extreme possibility, HUBBLE'S sequence might represent a simple and direct evolutionary progression, from irregular through spiral to elliptical systems. Second, at the opposite extreme, the initial conditions of formation might determine the structural form of a galaxy, and within that fixed structure it might undergo its specific evolutionary process, involving aging of the stars and reddening of the color. Third, a combination of the first two possibilities might represent the truth.

MORGAN [*315*] has devised a new system of classification of galaxies according to their structural features, in which the main criterion is the size of the nucleus. Although this criterion was one of those used by HUBBLE[1], it was sometimes outweighed by other criteria such as the form of the spiral arms. With this single criterion, a much better correlation is obtained between structural classification and integrated color, and, according to preliminary work, integrated spectral type also. Thus it appears that the size of the nuclear regions, relative to the whole galaxy, is a good indicator of the kind of stars composing the bulk of the galaxy. It is a very promising hypothesis that nuclear size is therefore an age indicator, by means of which galaxies can be ordered into an evolutionary sequence. However, other factors must also be borne in mind.

OORT [*312*] has given a qualitative discussion, arguing that evolution through the dichotomy between spiral and elliptical forms is determined largely by the initial angular momentum of the protogalaxy.

---

[1] E. HUBBLE: Realm of the Nebulae. New Haven: Yale University Press 1936.

The initial conditions of angular momentum, magnetic field, turbulent energy, and size of density fluctuations may govern the initial mass-function of the condensing stars and the speed of their formation. For example, perhaps spiral arms form in regions of high magnetic field; the persistence of bars in barred spirals despite rotation might suggest the presence of strong magnetic fields which inhibit differential rotation. If for some reason star formation has been partially inhibited in one galaxy as compared to another, e.g. by magnetic field conditions, then such a galaxy would retain more of its mass in the form of interstellar matter. Again, if the initial mass-functions vary in different galaxies, then this will clearly have an effect on the apparent age of a system as judged from its present-day stellar content.

Finally, the evolution of galaxies inside and outside dense clusters can be expected to be different because of the effect of collisions in the latter. These will sweep out gas and dust from the systems in collision, and the high rate of occurrence of $S0$ systems in dense clusters such as the Coma and the Corona clusters may be the result of such collisions [*313*]. Although the rate of collisions is considerably less than that originally suggested by Spitzer and Baade [*313*], because of the considerable increase in the distance scale, it still seems possible that galaxies which pass near to the center of such a cluster will undergo a number of collisions in making a single transit across the central region.

Evolution in galaxies is a speculative subject at present, and needs more study from many points of attack, both theoretical and observational. One important approach, currently being undertaken, is that of studying the stellar content of galaxies as deduced from their integrated spectra in relation to their structural features [*282*]. Further, the revision of the Hubble classification system by Sandage [*314*], the classification system devised by de Vaucouleurs (Vol. LIII of this Encyclopedia) and the classification according to the nuclear regions by Morgan [*315*], may all be expected to contribute to our understanding of galactic evolution.

C. Interchange of matter between stars and the interstellar medium.

I. Accretion of matter by stars.

69. Interchange of matter between gas and dust, and stars, may be an important factor in the evolutionary processes which take place in the stars. It is of importance to consider the situation in which a star is already present, and the gravitational attraction it exerts attracts and captures more mass. An early discussion of this problem was given by Eddington [*316*], and a more efficient process was proposed later [*317*], [*318*]. Here it was shown that capture can occur even if there is not a direct collision between an atom and a star, if the space density of the gas is high enough. Thus if we consider two parallel streams of gas passing symmetrically on either side of a star, they will be attracted and cross each other directly behind the star. In the crossing region collisions will occur in which atoms will be left with zero outward or transverse momentum, and the star's gravity then causes them to fall into its surface. For material of sufficiently small density and temperature, i.e., with the approximation that the heat generated would be rapidly radiated away, so that pressure effects can be ignored, the rate of gain of mass is given by

$$\frac{dM}{dt} = \alpha \frac{2\pi \varrho G^2 M^2}{v^3} \qquad (69.1)$$

where ϱ is the density of the interstellar gas, M is the mass of the star and v its velocity relative to the interstellar gas; α is a constant of the order of unity. On the other hand, in the limit in which the velocity is zero and pressure effects are taken into account the rate becomes [319]

$$\frac{dM}{dt} = \alpha' \frac{2\pi \varrho G^2 M^2}{c^3} \tag{69.2}$$

where c is the velocity of sound in the gas. In the case in which both velocity and pressure effects are important, it has been supposed by BONDI that the rate is approximately given by

$$\frac{dM}{dt} = \alpha \frac{2\pi \varrho G^2 M^2}{(v^2 + c^2)^{\frac{3}{2}}}. \tag{69.3}$$

Now it has been generally shown that for normal interstellar conditions, accretion is relatively unimportant. For example, let us suppose that $M = 10\,M_{\odot}$, $R = 6R_{\odot}$, $L = 10^{37}$ ergs/sec. Then from Eq. (69.1)

$$\frac{dM}{dt} \approx 10^{55} \frac{\varrho}{v^3}.$$

Thus if $\varrho = 10^{-23}$ g/cm³ and $v = 10$ km/sec,

$$\frac{dM}{dt} \approx 10^{14}\ \text{g/sec}.$$

On the other hand, the luminosity measures the rate of conversion of mass from hydrogen to helium, and thus the rate of evolution of the star given by

$$-\frac{dM}{dt} \approx 10^{18}\ \text{g/sec}.$$

Thus in these fairly normal conditions in a low-density interstellar cloud, the accretion of hydrogen would be negligible as far as the star's evolutionary rate is concerned. It will only become important when regions of high density are encountered by a star of very low velocity, in which case the rate of accretion will be primarily determined by Eq. (69.2). However, all of the observational arguments suggest, and the previously discussed theories concerning protostar formation demand, that stars are born in regions where the density is far greater than the mean. Thus it is important to decide: (i) whether accretion provides an alternative mechanism for producing stars which are apparently quite young, in regions of dense cloud complexes; in this case they will only be rejuvenated stars having cores which are far older; or (ii) whether accretion following initial condensation and contraction can increase the masses of protostars by a significant amount. This first problem has been studied by MCCREA [320].

(i) It is necessary to take into account the slowing down of the star in the dense medium into which it moves. The resistive force is given by [321]

$$F = -2\pi \varrho \frac{G^2 M^2}{v^2} \log_e \left(1 + \frac{s^2 v^4}{G^2 M^2}\right) = -2\pi \varrho \frac{G^2 M^2}{v^2} \beta \tag{69.4}$$

where s is the radius of the effective sphere of influence of the star. It has been shown by MCCREA that if x_1 and t_1 are the distance and the time elapsed to slow the star down to a velocity $\ll c$, then

$$t_1 \approx \frac{v_0^3}{(4\alpha + 3\beta)\, 2\pi \varrho G^2 M_0} \tag{69.5}$$

$$x_1 \approx \frac{v_0^4}{(5\alpha + 4\beta)\, 2\pi \varrho G^2 M_0} \tag{69.6}$$

where v_0 and M_0 are the initial velocity and mass, respectively. Eqs. (69.1) and (69.2) now show that the major part of the accretion will take place after the star has effectively been brought to rest. The time taken for the mass to grow to a value very much greater than the initial mass is given by

$$t_2 - t_1 = \frac{c^3}{\alpha' 2\pi G^2 \varrho M_0} . \tag{69.7}$$

Calculations based on these expressions are shown in Tables 13 and 14.

These results show that even in dense cloud complexes where the densities are as high as $10^3/\mathrm{cm}^3$, accretion will still not in general be an effective mechanism

Table 13. *Times and distances for reducing a star to rest relative to a cloud of density 2.5×10^{-21} g/cm³ ($T = 100°$).*

		$v_0 = 1$ km/sec	$v_0 = 2$ km/sec	$v_0 = 5$ km/sec
		Isothermal case ($\alpha = 2$)		
$M_0 = 2M_\odot$	t_1 (years)	3.1×10^6	2.0×10^7	2.6×10^8
	x_1 (parsec)	2.4	32	990
$M_0 = 5M_\odot$	t_1 (years)	1.5×10^6	9.3×10^6	1.1×10^8
	x_1 (parsec)	1.2	14	440
		Adiabatic case ($\alpha = 0$)		
$M_0 = 2M_\odot$	t_1 (years)	4.0×10^6	2.5×10^7	3.0×10^8
	x_1 (parsec)	3.1	38	1150
$M_0 = 5M_\odot$	t_1 (years)	2.0×10^6	1.2×10^7	1.4×10^8
	x_1 (parsec)	1.5	18	520

Table 14. *Times for large mass-increment in cloud of density 2.5×10^{-21} g/cm³ and mean molecular weight 1.4.*

		$T = 30°$ K	$T = 50°$ K	$T = 100°$ K
		Isothermal		
$M_0 = 2M_\odot$	$t_2 - t_1$ (years)	3.8×10^6	8.3×10^6	2.3×10^7
$M_0 = 5M_\odot$	$t_2 - t_1$	1.5×10^6	3.3×10^6	9.4×10^6
		Adiabatic		
$M_0 = 2M_\odot$	$t_2 - t_1$	3.7×10^7	8.0×10^7	2.3×10^8
$M_0 = 5M_\odot$	$t_2 - t_1$	1.5×10^7	3.2×10^7	9.0×10^7

for rejuvenating stars, since the initial velocities of stars entering such complexes will be high enough to carry them right through before they reach velocities so low that spherically symmetrical accretion can occur. This is based on the fact that regions of very dense gas rarely extend more than ~ 50 parsecs, while the peculiar velocities of stars relative to the gas in the Galaxy are predominently ~ 5 km/sec or greater. In the few cases in which $v < 1$ km/sec the mechanism will be effective. In the case of expanding associations the outward velocities are not explicable if an accretion theory in this form is assumed.

(ii) From Tables 13 and 14 it is possible to reach some conclusions regarding the importance of accretion following protostar formation by another mechanism. Thus, let us consider the formation of stars following the formation of a condensation either by the concentration of dust, or by the mechanism of ionization and compression by an *O*-type star. In these cases, the surrounding matter will fall into the initial condensation (spherically symmetrical accretion), and its amount of growth in a time t will be given by the entries in Table 14 for the time taken to increase the mass by a large amount in a time $t_2 - t_1$, where $t = t_2 - t_1$. For initial masses in the range 2 to 5 $M_\odot$ the times are of the order of 10^7 years. However, the rate of increase is proportional to $1/M_0$, so that for much larger masses, which are to be expected in an initial condensation, these times can become very short indeed. Furthermore, if we are trying to explain the birth of expanding associations in a cloud complex through the presence of an O-type star, spherically symmetrical accretion may still take place in the expanding compressed layer, since the material falling into the initial condensation can have had initially only a very small velocity component relative to this initial condensation, because they both form part of the outward-moving layer surrounding the O-type star.

In discussing the accretion theory we have neglected the effect of the radiation of a star on its rate of accretion, a point which has been considered by a number of authors [*322*] to [*324*]. Qualitatively this will have the following effect. Provided that the conditions of very low relative velocity and high density are satisfied, so that initially the rate of accretion is greater than the rate of mass depletion, the mass will grow and subsequent changes in the star's internal structure will mean that its luminosity will increase. Thus its output in the ultraviolet will increase and consequently the radius of the ionized sphere surrounding it will be increased. MESTEL [*322*] has shown that if the accretion rate is to be large, the radius of the ionized sphere must not be more than about 10^3 stellar radii. The absorbing power of a cloud is very much increased by the high densities near the star, so that a cloud whose mean density is quite high ($\gtrsim 10^{-21}$ g/cm³) is able to keep the ionized sphere sufficiently small for appreciable accretion to occur. When a star has accreted to a critical mass, the ionized sphere increases rapidly to the value given by STRÖMGREN and accretion effectively ceases. The effect of radiation is thus to restrict further the conditions under which accretion is a significant factor in a star's evolution. However, if accretion takes place on to a condensation which has been produced by gravitational collapse in a dense cloud, then the protostar will probably be so young that isothermal conditions will still prevail in the contraction. Thus the accreting mass will be very cool, and the effect of radiation can be neglected.

It is of interest to point out that accretion, while unimportant in the evolution of the main body of stars in the Galaxy, may, in the case of stars which are not mixing, have the effect of disguising their true chemical composition. For example, a white dwarf of mass $1 M_\odot$, $v = 10$ km/sec, moving in the disk of the Galaxy through gas of mean density 10^{-24} g/cm³, will accrete at a rate of 10^{11} g/sec. In a time of 5×10^9 years such a star will accrete about 1.5×10^{28} g or less than 10^{-5} of its initial mass. However, this is very much more than the normal fraction of the mass contained in the atmosphere, consequently the material accreted will gradually form a new atmosphere. Now the white dwarf may easily be composed almost entirely of helium, while the atmosphere which it has accreted will be composed predominantly of hydrogen. Consequently, white dwarfs may appear from their spectra to be rich in hydrogen, whereas in fact they are hydrogen-poor. Such effects may explain some of the white dwarf types described by GREENSTEIN in Vol. L of this Encyclopedia.

II. Mass loss from stars.

70. General arguments. In considering the life histories of stars it is easily proved that mass loss must be a very important factor in stellar evolution. For we have observational evidence from the extent of the main sequence that the masses of stars range up to $\sim 90\,M_\odot$. On the other hand, we have strong evidence that the final stage in the life of a star is reached when it becomes a white dwarf, with no energy sources, its only emission being due to its cooling. However, the masses of totally degenerate white dwarf configurations can never exceed the Chandrasekhar limit, which amounts to $1.44\,M_\odot$ for a pure helium white dwarf and $1.24\,M_\odot$ for the most extreme configuration, an iron white dwarf. Consequently all of this excess mass must be lost at some stage during the star's evolution.

Another general argument for mass loss, based on the observations of color-magnitude diagrams of stars in clusters, is as follows. In Sect. 56, a method was described for computing the fractional mass of hydrogen consumed by a star in the clusters M 67 and M 3 by the time that it has reached the tip of the giant branch. The low value of this fraction in M 67, about 40%, compared with about 80% in M 3, is a consequence of the lower luminosity of the giants in M 67. Before the stars in M 67 come to the end of their lives, they can be expected to have consumed almost all of their nuclear fuel (predominantly in the conversion of hydrogen to helium). Stars which have not used up all of their hydrogen should be luminous enough to have been measured in this cluster. However, subsequent to the giant branch no well-populated sequence of stars is observed at least down to $M_v = +9$. Where, then, are the stars that should be consuming the remaining 60% of their available energy supply? It seems that their non-appearance can only be explained by a process of mass loss. Thus the suggestion has been put forward that after stars leave the main sequence, but while they are on the giant branch, they are losing mass at a rapid rate. Mass loss taking place from the surface while the stars are unmixed will be extremely effective in removing hydrogen which has not been processed.

Turning from these general arguments, we shall describe in the following sections the conditions in single stars or in binary systems under which mass loss may occur. They may be divided into two groups. In the first of these we shall include steady, non-explosive ejection, while in the second we shall consider explosive ejection processes.

71. Loss of mass and exchange of mass in binary systems. This mode of instability and mass loss in close binary systems has been extensively discussed by STRUVE and his many collaborators, by KUIPER, KOPAL, WOOD, by KRAFT and CRAWFORD and others, [*146*], and other references given there, [*148*], [*325*] to [*327*]. The argument runs as follows.

In the restricted problem of three bodies we have that the equation of motion of a third infinitesimal body in the presence of two finite bodies rotating about their center of mass admits of the integral

$$v^2 = -2\Omega - C \tag{71.1}$$

where

$$\Omega = -\frac{1}{2}(x^2+y^2) - \frac{(1-\mu)}{r_1} - \frac{\mu}{r_2} \tag{71.2}$$

is the potential produced by the two bodies in the rotating frame of reference. Here units have been chosen such that the distance between the two masses is unity so that the masses are μ and $(1-\mu)$ respectively, and rotating axes have been chosen with the center of mass as the origin, the x-axis the line joining the two masses and the z-axis perpendicular to the plane of the motion. For a given value of C, if we put $v=0$, the equation will define a surface called the zero relative velocity surface whose equation is

$$x^2 + y^2 + \frac{2(1-\mu)}{r_1} + \frac{2\mu}{r_2} = C. \tag{71.3}$$

Once C is defined by the initial condition this surface defines the space to which the infinitesimal particle is restricted. The exact shape of the surface depends on the value of C. For large C the zero-velocity surfaces differ only slightly from separate spheres closed about each body. For decreasing values of C the ovals become more and more distorted and elongated until they touch each other at a single point on the x-axis, which is the first double point L_1. For smaller C, a single zero-velocity surface surrounds the two bodies, and for even smaller C, the dumb-bell increases in size and finally opens up first at L_2 and then at L_3, since further decrease of C makes the surface move away from the x-axis. Particles placed at the double points with zero relative velocity will remain at rest unless perturbed by external forces. Since particles will in general arrive at these points with a Maxwell-Boltzmann velocity distribution, the trajectories of some will enable them to escape.

Now if stars can be approximated to Roche models, i.e., if they are centrally condensed, these results are applicable to close binary systems, the zero-velocity surface being an equipotential surface, the potential being given by $\Omega = -C/2$. The difference between such a surface and one for a single star is that for a binary system, since it is non-spherical, material can escape more easily in certain directions than in others. Along the x-axis Ω has maxima at both L_2 and L_3, although along the y and z directions Ω has minima at these points. At L_1, Ω has a maximum in the x-direction, but minima in the y and z directions. Thus L_2 and L_3 control the mass flow out of the system, while L_1 provides a route for transferring mass from one star to the other. Mass flow via L_2 and L_3 does not mean that it will necessarily escape completely from the binary system, since at great distances the equipotential surfaces resemble those for a single star. Extra forces are needed for this to occur, and observational evidence is available in a number of cases which shows that the outer atmosphere surrounding the two stars is expanding, perhaps because of radiation pressure.

Proof that mass loss is currently occurring in some binary systems is obtained from the changes of period. This point has been discussed in detail by Wood [*327*]. For example, in the well-known case of β Lyrae, where the primary fills the zero-velocity surface, the period is increasing at the rate of about 9 sec per year [*325*] and this corresponds to an average rate of loss of mass given by

$$\frac{dM}{dt} = -2.82 \times 10^{22}\ \text{g/sec}.$$

Another argument, very closely related to evolutionary ideas, is as follows [*328*], [*148*]. Let us suppose that one star in the binary has evolved off the main sequence, and would in the normal course of events move on to the giant branch. It reaches a point at which it has expanded so that it fills the zero-velocity surface surrounding it. Further attempts to expand will then mean that the mass passing across the zero-velocity surface will be lost to the star, and will in general flow to the secondary where it may be trapped. This is a powerful mechanism of mass transfer between binary components. To estimate the rate at which mass loss takes place we can put the rate of increase of potential energy equal to k times the luminosity, so that the total energy output from the core is

$$\frac{dE}{dt} = L(1+k) \tag{71.4}$$

and

$$\frac{1}{k}\,\frac{GM}{R}\,\frac{dM}{dt} = L. \tag{71.5}$$

In general k is not known, but for the single-star models of SANDAGE and SCHWARZSCHILD, for a mass of $2M_\odot$, $k \approx 0.03$. On the other hand, the rate of mass loss may be obtained by writing

$$\frac{dM}{dt} = -4\pi R^2 \bar{\varrho} \frac{dR}{dt} \tag{71.6}$$

and numerically differentiating the solutions for given models to obtain dR/dt along the evolutionary tracks. By putting $k \approx 1$, STRUVE and HUANG have found that the rate of mass loss from the massive component of β Lyrae is comparable with the rate of mass loss determined from its change of period. Whether this is a reasonable comparison is not known, since we have no evolutionary tracks for such massive stars ($M \sim 20 M_\odot$), but it does appear as though this star is large enough to be about filling the zero-velocity surface.

CRAWFORD and KRAFT [*148*] have estimated the rate of mass transfer in the binary AE Aquarii from these arguments to be about 10^{18} g/sec. They have shown that the blue subluminous companion captures this material and that its luminosity may be essentially due to this in-falling material. Thus they have suggested that in such a system, this mechanism of mass transfer has been one of the most powerful factors in the evolution of the binary. Originally the now subluminous blue star was the more massive of the two stars. It evolved more rapidly and began to lose mass to its companion, and finally it contracted and would have become a white dwarf. However, the companion's evolution has been speeded up by its gain in mass, it is also now evolving off the main sequence, and is returning mass to the blue star.

CRAWFORD [*328*] has developed this mass transfer argument in connection with the general problem of the subgiant components of eclipsing binary systems. In a large number of systems which have main sequence primaries it has been shown by PARENAGO [*89*] that the secondary masses are on the average less than $1 M_\odot$ while the radii are considerably larger than $1 R_\odot$ (see Sect. 20). CRAWFORD has shown that in many cases the secondaries are so distended that they are filling the inner zero-velocity surface and that the mass transfer process is going on.

The idea that mass transfer can take place repeatedly in amounts sufficient to affect drastically both stars' evolutionary paths has been criticized by KOPAL [*329*]. He has pointed out that no cases have been detected where the stars are in the transient evolutionary stage in which the exchange of the roles of both components is in progress. Furthermore, he has argued that the total amount of mass transfer demanded to transform the primary into the secondary is about 80% of the total. This would demand a removal of mass right down to the core of the primary, and the total amount of nuclear energy released in the core to do this would need a large amount of rapid helium-burning which would be detectable. However, it is very difficult to see how this latter conclusion has been reached. Helium-burning produces only about 10% of the energy released in hydrogen-burning, per gram of material. Furthermore, helium-burning would not lead to a sudden increase in luminosity.

As an alternative, KOPAL has proposed that mass transfer takes place only to a very limited extent, so that the secondary is filled to its zero-velocity surface and perhaps gaseous streams are produced. This would not change the mass ratio appreciably, and the luminosity of the secondary would be only slightly increased. It is then supposed that the primary collapses away from its zero-velocity surface rapidly, while the secondary does this more slowly. It is then argued that the subgiants which do not fill their zero-velocity surface are at later evolutionary stages than those which do fill them. This asymmetrical cycle may then be repeated. However, the argument against this model is that, although the exhaustion of the hydrogen core gives rise to the expansion and the filling of the zero-velocity surface for the primary, there are no arguments connected with the internal structure of the primary or of the secondary which suggest that they should rapidly contract again. Furthermore, the amount of mass transfer demanded to fill the inner velocity surface of the secondary at the time that it can be ejected by the primary depends very critically on its mass and the age of the system. KOPAL has also argued that in many cases sufficient nuclear energy is not available to move material from the primary to the secondary. Since, except for the most extremely centrally condensed models, the available nuclear energy exceeds the gravitational potential energy by several orders of magnitude, this argument also needs further clarification.

Apart from these arguments which relate the evolution of one star to that of the other in a binary system, it should be emphasized that considerable loss of mass from the system as a whole will mean that both stars come more rapidly to the ends of their lives. Thus any process of violent ejection (localized disturbances like large-scale eruptive prominences), such as was originally proposed by WOOD [*327*], will be very effective in this respect.

In globular clusters, binary systems would be detected only if they were eclipsing binaries, and eclipsing variables are not in general found in globular clusters. This is explicable on the ground that tidal forces, the effect of near collisions in the central regions of such clusters, would destroy even the closest (most tightly bound) systems. Thus mass loss and transfer cannot be an important factor in the evolution of the stars in these clusters. In less condensed clusters, and in the field stars in the Galaxy, the effect will be present though small, since although a large proportion of the stars are physical binaries, only a small fraction of these are close enough for these processes to occur.

72. Steady mass loss from single stars. In many types of star there is some observational evidence that matter is being continuously ejected. General arguments have been given by BIERMANN [*330*]. For example, in the Sun there is a large body of evidence concerning the matter ejected at the time of large flares and other eruptions, and the high density of the interplanetary medium as compared with the density of normal interstellar matter is probably due to solar ejection. This ejection is attributed to electromagnetic processes.

Ejection of matter from the equatorial regions of fast-rotating stars was many years ago suggested by STRUVE and his associates to be responsible for the extended low-density atmospheres that give rise to emission lines and sharp shell absorption lines in the spectra of *Be* stars (see review article, [*331*]). The region giving rise to the emission lines usually has a ring-like structure lying in the star's equatorial plane. These extended tenuous atmospheres must dissipate, and indeed the outbursts and subsequent fading of emission lines and sharp absorption lines in stars like γ Cassiopeiae and Pleione show this very clearly. Thus a continuous loss of mass is occurring.

The rates of mass loss can be computed either by supposing that the shell is dispersed in an average time of perhaps 10 years, or else that it is continuously being lost at a low velocity ~ 5 km/sec. For stars such as Pleione[1] and 11 Camelopardalis[2] the extent of the outer shell is about $1 R_\odot$ and its mean density is $\sim 10^{-10}$ g/cm^3. If shells such as these are ejected in 10 years we find that $dM/dt \approx -3\times 10^{16}$ g/sec. On the other hand, if we suppose that ejection is steadily

[1] A. B. UNDERHILL: Astrophys. J. **110**, 166 (1949).

[2] G. R. BURBIDGE and E. M. BURBIDGE: Astrophys. J. **117**, 407 (1953).

taking place with an outward velocity in the shells of 5 km/sec, then $dM/dt \approx -10^{20}$ g/sec. Clearly these estimates are very uncertain. They may be compared with the rates of evolution of the mass through radiation, which are 2×10^{17} g/sec (Pleione) and 1.5×10^{18} g/sec (11 Cam). With such uncertainties it is not possible to decide whether mass loss is the most important mode of evolution for these main-sequence stars. Surveys of the *Be* stars made by MERRILL and others show that they comprise ~5 to 10% of the *B*-type stars in the Galaxy. It might be, therefore, that all stars of this type have a transient phase in which they eject matter, and that this speeds up their normal rate of evolution through radiation. Alternatively, it might be argued that only a small fraction of these stars, defined by their initial angular momentum, ever do eject matter while they are on the main sequence. This point is of importance in considering the chemical evolution of stars, since evolution by mass loss while the stars remain on the main sequence means that most of the ejected matter will remain in the form of hydrogen, so that the evolution will not enrich the interstellar medium in the heavier elements (see Chap. D II). More detailed work on the rate of ejection from *Be* stars might enable us to obtain a direct observational test of the hypothesis that has been made by some workers, which is based on rather different arguments, that the rate of mass loss is proportional to the luminosity of the star (Sect. 24).

Mass loss through rapid rotation may also be important in some giant stars. The *F* 5 II—III star π Pegasi has a large rotation and its spectrum shows evidence of a shell surrounding the star[1]. The equatorial velocity at which ejection can occur may be as low as 100 to 130 km/sec in this star. GREENSTEIN also pointed out that all stars in the range *A* 5 III to *F* 5 III lie in the Hertzsprung gap of the H-R diagram, and this might be connected with the ejection of matter through fast rotation.

There is some evidence suggesting that massive stars, after they have evolved into the supergiant or bright giant region of the H-R diagram, eject mass by processes other than rotation. This may be an extremely important effect. In this respect, the supergiants and bright giants may be divided into two classes, the higher-temperature and the lower-temperature groups, separated in the H-R diagram by the Hertzsprung gap. Among the former group is the well-known supergiant of type *B* 1, P Cygni, the outward velocity of the shell is about 100 km/sec, and if this shell is being continuously replenished it is losing mass at a rate of about 10^{21} g/sec [*332*]. On the other hand, P Cygni is evolving by radiation, i.e., transmuting hydrogen to helium, at a rate of 2×10^{20} g/sec. Thus its evolution is at present being mainly determined by the ejection of mass. Whether this situation will prevail through the remainder of the star's life remains an open question. If it did, then the star would lose most of its mass in the next two million years. Most of the higher-temperature group, above a certain luminosity, are probably losing mass, although at a lesser rate than in the case of P Cygni [*334*].

Many years ago it was pointed out [*333*] that the effective gravity at the surfaces of giant stars may be nearly zero, so that the outer layers might easily be lost. In the system of α Herculis, which consists of an *M*-type supergiant with a visual companion, a giant *G*-type spectroscopic binary, both immersed in a large envelope, mass is being ejected from the *M*-star into the envelope and from thence into the interstellar medium [*335*]. The rate of mass loss has been estimated to be about 10^{18} g/sec, while the rate of evolution by radiation is 10^{19} g/sec.

---

[1] J. L. GREENSTEIN: Astrophys. J. **117**, 269 (1953).

Thus in this case the normal evolution of the star is being slightly speeded up by mass loss. It has been suggested [*336*] that the ionization energy of the material in the convective zone of a late-type giant provides an energy source which is sufficient to account for the rate of ejection in a star like α Herculis. However, no indication has been given of the way in which this energy is converted into energy of mass motion. Such an explanation is not possible in the case of high-temperature supergiants like P Cygni, since even if there is a large reservoir of ionization energy it will not be made available in the atmosphere of a star with a black-body temperature $\gtrsim 15\,000$ degrees.

It does not appear that either radiation pressure or even the conditions which would prevail in a very high-temperature outer corona can in general be responsible for ejection at rates as large as those in the luminous supergiants. The effect of electromagnetic forces remains to be explored. Thus, although the energy necessary for escape must ultimately have come from nuclear transmutation, the physical causes for the ejection remain unknown.

It is perhaps of interest to point out that the velocity of escape $v_e \propto M$, so that the kinetic energy needed for escape is proportional to M. On the other hand, even for stars not on the main sequence the total energy output is proportional to M^x where in all cases $x > 1$. Thus the more massive the star, the smaller the fraction of its energy is required to eject matter at a given rate. Alternatively, this may be expressed by stating that any law of the rate of mass ejection can be expected to give much greater rates of ejection for massive than for low-mass stars, provided that the energy of ejection comes ultimately from nuclear sources.

The Wolf-Rayet stars which lie to the left of the main sequence have probably reached a very late stage of their chemical evolution. As was pointed out Sect. 29, their spectra indicate that considerable ejection of mass is occurring, although this interpretation of the observations is not universally accepted.

Other evidence for mass loss comes from the existence of planetary nebulae. It is generally supposed that these have been produced by ejection of matter from a central star, and since by interaction with the surrounding interstellar gas they will eventually be dispersed into the interstellar medium, the amount of mass loss from a single star is given by the mass of the planetary. Since planetaries are believed to have a galactic distribution of disk or intermediate population II character (see Sect. 58), it may be that the rate of dissipation is rather slow, since they spend a large part of their lives in an environment comparatively devoid of gas and dust. The rate of ejection of gas into the interstellar medium, i.e., the rate of dissipation of the planetary, has not been estimated. However, in one case the rate of ejection of the gas from the star to the planetary has been estimated. For the planetary NGC 4362 GURZADIAN [*337*] has obtained a value of $\sim 10^{20}$ g/sec. Stars which form the nuclei of planetaries have probably reached a late stage of chemical evolution, and it appears likely that the planetaries were ejected after the stars had left the main sequence.

The question of the effect of mass loss on the evolutionary path of a star in the H-R diagram was discussed in Sect. 25. Of particular interest, when steady slow mass ejection is taking place, is the situation which arises when an appreciable change in the chemical composition difference between core and envelope is produced by mass loss from the envelope, and not by transmutation in the interior.

73. Explosive mass loss from stars. A powerful mechanism of mass loss is that caused by instabilities in a star's interior which trigger explosions leading to the almost instantaneous ejection of large masses. Two types of explosion which are distinguished ultimately by their different rates and total amounts of energy released are the novae and the nova-like variables on the one hand, and the supernovae on the other hand. Early work was summarized in the report of the Paris conference [*338*], and more recently PAYNE-GAPOSCHKIN [*144*] has summarized all of the available material. In Table 15 we reproduce some of the integral properties of novae and supernovae given by PAYNE-GAPOSCHKIN, together with our modification of some of the quantities given by her (the latter being given in italics in the table). These latter changes will be discussed later. From the point of view of stellar evolution, the information of importance is the

Table 15. *Integral properties of supernovae, novae, and nova-like variables, in so far as it has been possible to estimate them. Mostly taken from* [*144*], *with some modifications.*

	Supernova Type I	Supernova Type II	Common nova	Recurrent nova	Symbiotic nova	U Gem star
L_{max} (M_{pg})	-16 to $-$*20*	-14 to $-$*18*	-7.8	-7.8	-7	$+5.5$
E_{vis} (ergs)	10^{49} to *10^{50}*	10^{47} to *10^{48}*	6×10^{44}	10^{44}	—	6×10^{38}
Mass ejected ($M_\odot$) .	10^{-1} to *1*	—	10^{-3}	5×10^{-6}	—	10^{-9}
v (km/sec)	10^4	5 to 7×10^3	1000	600	100	—
Number per year, our Galaxy . . .	0.005	0.025	50	—	—	—
Total in our Galaxy	2×10^7 ?	10^8 ?	2×10^6 ?	10^4 ?	10^5 ?	10^5 ?
Years between outbursts	∞	—	10^7 to 10^3 ?	20	3.5	0.2
Pre-nova: $M(M_\odot)$	*1* to *10*	*20* ?	0.2 to 2.0	—	—	—
$L(M_{pg})$	—	$-$*4* ?	3 to 4	0	0	$+9.5$
$L(M_{bol})$	—	—	0:	0:	-2:	$+7.5$:
$R(R_\odot)$	—	—	0.2	6	60	0.03
Post-nova: $M(M_\odot)$	—	—	0.2 to 2.0	—	—	—
$L(M_{pg})$	$+6$	—	3 to 4	0	0	9.5
$R(R_\odot)$	—	—	0.2	6	60	0.03
Spectrum	—	—	*O*−*B*	*F*:	*M*:	*B*−*A*

position of the pre- and post-nova or supernova in the H-R diagram, the population characteristics of each type of object, the total mass of the material ejected, and the frequency of occurrence of the explosions. The first and second points were discussed in Sect. 30, a few remarks concerning the second were made in Sect. 62, the third is discussed here, and the fourth is considered in Chap. D II.

The amount of mass ejected in nova explosions has been estimated by a number of authors from intensity measures in the spectra. Thus PAYNE-GAPOSCHKIN [*144*] has summarized the work of a number of authors [*339*] to [*343*], and concludes that the rate of mass loss from novae amounts to 10^{30} g for a common nova and $\sim 10^{28}$ g for a recurrent nova. The value for common novae is perhaps rather high. Although nova explosions are important events in the lives of stars from the point of view of their structural characteristics, it does not appear that they are important mechanisms of mass loss in as far as their final configurations are concerned. That is, nova explosions alone will not eject sufficient mass in most cases to enable the stars to achieve stable degenerate configurations, since even recurrent novae probably only explode $\sim 10^2$ times. Both the pre- and post-novae are subdwarfs which, although they have evolved far along their evolutionary tracks, have not yet reached the white dwarf stage. They may be stars which have already lost most of their mass by steady ejection processes and thus already have masses below the Chandrasekhar limit (see Sect. 30).

Supernovae have been divided into two categories, Types I and II [*344*]. There is no information on the amount of mass ejected in Type II supernovae and very little on that in Type I. The amount of radiant energy emitted is also in some doubt and in recent years revisions in the distance scale alone have been responsible for increases in the estimates of the absolute photographic magnitudes of supernovae at maximum. Thus the total amount of radiant energy emitted in typical supernovae is 10^{49} ergs and upward. This is comparable with the gravitational potential energy of the star. From these values it has been concluded that the total amount of mass ejected has a comparable total kinetic energy. However, attempts to determine a value have all of necessity been restricted to the Crab Nebula, the shell of the supernova of 1054 A.D., which is the only supernova remnant available for detailed investigation. The most recent determination is that of OSTERBROCK [*345*] who has obtained a mass of about 10^{32} g. This value is obtained from the density of free electrons and it is necessary to put in a value for the relative abundances of the elements to determine the mass. However, as will be discussed in Chap. D I, it is probable that a pre-supernova is already in an advanced state both of chemical evolution and of its evolution in the H-R diagram. Consequently, the mean atomic weight may be considerably greater than that assumed by OSTERBROCK and the mass will be correspondingly greater. It appears, therefore, that a mass of $\sim 10^{33}$ g is not unreasonable. A mass of stellar order is also to be expected if the explosion is so violent as literally to blow the star apart, whereas a mass of only a few percent would be expected for a less violent outburst. The possible causes of such a nuclear explosion will be discussed in Sects. 78 and 81.

The masses of pre-supernovae are very uncertain. PAYNE-GAPOSCHKIN [*144*] has given masses for Type I supernovae as large as $20 M_{\odot}$, and yet this type has been found in elliptical galaxies and has usually been classified as population II objects. However, high values of the mass may be demanded on other grounds (Sect. 81). It is not clear, therefore, whether the amount of mass ejected in a supernova outburst is sufficient to take a star from a mass above to one below the Chandrasekhar limit. If this is not the case, then steady mass ejection at an earlier stage of the star's evolution must always be the more important feature.

Mass loss in the explosive variables must also be mentioned. Well-known examples are SS Cygni, U Geminorum. Very little is known about the physics of these systems, but they are often binaries and it seems that the cause of their outburst may be related to their binary nature (see Sect. 30). The amount of mass loss per outburst from the U Geminorum stars has been estimated by GORDELATSE [*346*] and GREEP [*347*] to be about 10^{24} g. These stars are also at an advanced evolutionary stage.

D. Chemical evolution of stars.

I. Theory.

74. Introduction. Since nuclear transmutations are the dominant sources of stellar energy, all stars must evolve chemically as they age, since heavier nuclei are continuously being formed from lighter ones. Once stars have condensed on to the main sequence they immediately begin their chemical evolution. They spend the major parts of their lives in going through the first phase, i.e., in converting hydrogen to helium. However, in later stages further chemical evolution can occur, and this will continue, only being interrupted or finally stopped by processes of steady mass ejection or catastrophic explosion which may rapidly remove or use up prodigally the remaining nuclear fuel. Stars finally run out of fuel and/or become white dwarfs. In this last stage they will continue to radiate by cooling for very long periods.

Except at catastrophic phases a star possesses a self-governing mechanism in which the temperature is adjusted so that the outflow of energy through it is balanced by nuclear energy generation. The temperature required to give this adjustment depends on the nuclear fuel available. Hydrogen requires a lower temperature than helium, helium requires a lower temperature than carbon, and so on, the increasing temperature sequence ending at iron, since energy generation by fusion ends here. Thus throughout most of the star's life the temperature is adjusted to hydrogen-burning, but when hydrogen is exhausted in the central regions the temperature will eventually rise through release of gravitational energy so that helium burning will commence. In this way one set of reactions after another are brought into operation, the sequence always being accompanied by rising temperature. Since penetrations of Coulomb barriers occur more readily as the temperature rises, the sequence is one in which reactions take place between nuclei with increasing nuclear charges, and the nuclei will evolve towards configurations of greater stability so that heavier nuclei will be synthesized until iron is reached. Since for a star in equilibrium there is a very large temperature difference between the center and outer layers, it is clear that different layers will at the same time reach different stages of chemical evolution, provided that mixing does not take place.

Each stage of a star's evolution will correspond to a definite phase of element synthesis. These ideas have led recently to the development of a theory in which it is supposed that only hydrogen is primeval, and all of the elements have been synthesized in the stars [*143*], [*348*] to [*353*], [*62*], [*354*]. Previous theories of the origin of the elements have demanded special conditions either in pre-stellar matter at the beginning of the universe, or in special types of stars which are not presently known to exist. To produce all of the elements in stars it is found that eight different nuclear processes are demanded. In what follows we shall outline briefly what these are and relate them to the various stages of stellar evolution, while in Chap. D II the observational evidence for stars at different stages of chemical evolution will be discussed. The treatment given is mainly based on [*62*].

75. Hydrogen-burning[1]. The conversion of hydrogen to helium takes place either by the *pp*-chain, the CN-cycle, or the NeNa-cycle [*355*]. A considerable part of the hydrogen-burning takes place through the *pp*-chain or the CN-cycle while the stars are on the main sequence, and it is of interest to consider the rates of these processes. The latest available rates are given in [*62*].

The energy generation rate of the CN-cycle lies between ε_{C} and ε_{N}, where

$$\varepsilon_{\mathrm{C}} = (3.18 \pm 0.54)\, \alpha_{\mathrm{C}}\, x_{\mathrm{H}}\, x_{\mathrm{C}} \times 10^{27}\ \mathrm{ergs\ g^{-1}\ sec^{-1}} \tag{75.1}$$

where

$$\left.\begin{aligned} \alpha_{\mathrm{C}} &= \varrho\,(T_6^{-\frac{2}{3}} + 0.017)\exp(-136.9\,T_6^{-\frac{1}{3}})\exp(0.98\,\varrho^{\frac{1}{3}}/T_6) \\ &\approx 1.4 \times 10^{-25}\,\varrho \left(\frac{T_6}{15}\right)^{17.9} \exp(0.98\,\varrho^{\frac{1}{3}}/T_6) \quad \text{near} \quad T_6 = 15 \end{aligned}\right\} \tag{75.2}$$

and

$$\varepsilon_{\mathrm{N}} = (7.86 \pm 1.6)\, \alpha_{\mathrm{N}}\, x_{\mathrm{H}}\, x_{\mathrm{N}} \times 10^{27}\ \mathrm{ergs\ g^{-1}\ sec^{-1}} \tag{75.3}$$

where

$$\left.\begin{aligned} \alpha_{\mathrm{N}} &= \varrho\, T_6^{-\frac{2}{3}} \exp(-152.3\,T_6^{-\frac{1}{3}})\exp(1.11\,\varrho^{\frac{1}{3}}/T_6) \\ &\approx 2.5 \times 10^{-28}\,\varrho \left(\frac{T_6}{15}\right)^{19.9} \exp(1.11\,\varrho^{\frac{1}{3}}/T_6)\,. \end{aligned}\right\} \tag{75.4}$$

[1] See also p. 35 of M. H. WRUBEL's article, in this volume.

In these formulae, T_6 is the temperature in 10^6 degrees, ϱ is the density, and x_{H}, x_{C}, x_{N} are the concentrations by mass of hydrogen, carbon, and nitrogen. These are quite wide limits and they are due to the fact that it is still uncertain as to whether there is a resonance in the $N^{14}(p, \gamma)$ reaction. If there is, then the energy generation rate will be determined by the slowest reaction $\mathrm{C}^{12}(p, \gamma)$, and will be given by ε_{C}. If there is not, then it will be given by ε_{N}. Since the existence of the resonance has not been established, W. A. FOWLER is inclined to favor ε_{N}, which will be used in what follows. Other arguments favoring this have also been given in [62].

The rate of the pp-chain is given by

$$\varepsilon_{pp} = 1.75\, x_{\mathrm{H}}^2 \times 10^6 \alpha_{pp} \,\mathrm{ergs\, g^{-1}\, sec^{-1}}$$

where

$$\alpha_{pp} = \varrho f_{pp} T_6^{-\frac{2}{3}} (1 + 0.012\, T_6^{\frac{1}{3}} + 0.008\, T_6^{\frac{2}{3}} + 0.00065\, T_6) \exp(-33.804\, T_6^{-\frac{1}{3}}) \pm 10\%$$

$$\approx 2.0 \times 10^{-7} \varrho \left(\frac{T_6}{15}\right)^{3.95} \quad \text{near } T_6 = 15 \quad (f_{pp} \approx 1 \text{ for } T_6 \gtrsim 8).$$

These rates mean that in the Sun, the CN-cycle is only responsible for about 4% of the energy generation at the center, while the pp-chain is responsible for the remainder. Integrated over the whole core the CN-cycle is even less important because of its very strong temperature dependence. The two rates are equal when

$$\left(\frac{T_6}{15}\right)^{15.95} \exp(1.11\, \varrho^{\frac{1}{3}}/T_6) = 77.1 .$$

If we put $\varrho = 125\ \mathrm{g/cm^3}$, we find that $T_6 \approx 19.4$.

Thus for a main-sequence star the CN-cycle rate is equal to the pp-chain rate at the center for a central temperature $T_c \approx 19.4 \times 10^6$ degrees. A rough estimate of the mass at which this occurs is given if we use the condition that $T_c \propto M/R$, and suppose that the models are not very different from the Sun which has $T_c = 14.8 \times 10^6$ degrees [63]. We then find from the empirical mass-radius relation (and ignoring the small departure of the Sun from the main sequence), that $T_c = 19 \times 10^6$ degrees when $M = 2 M_{\odot}$. To determine at which point the CN-cycle dominates as the source of energy in the core, detailed models are demanded. We may guess that the situation occurs for $M \gtrsim 3 M_{\odot}$. More recent solar models computed by HOYLE [356] and SEARS [357] will probably lead to a somewhat higher central temperature for the Sun and a consequent revision in the mass at the point where the CN-cycle and the pp-chain energy generation rates become equal (cf. Sect. 32).

Very recently a change in the pp-chain energy generation rate from that given above has been announced by FOWLER [357a]. The rate given above has been calculated on the assumption that the last reaction in the chain is $\mathrm{He}^3(\mathrm{He}^3, 2p)\,\mathrm{He}^4$. However, HOLMGREN and JOHNSON[1] have successfully detected in the laboratory the reaction $\mathrm{He}^3(\alpha, \gamma)\,\mathrm{Be}^7$, and have measured its cross-section in the energy range 0.47 to 1.32 Mev. The cross-section is about 2500 times larger than was originally thought, and it means, when extrapolated to stellar energies, that in stars at high enough temperatures in which plenty of He^4 is available the pp-chain will go in one of the following ways. Either

$$\mathrm{H}^1(p, \beta^+ \nu)\,\mathrm{H}^2(p, \gamma)\,\mathrm{He}^3(\alpha, \gamma)\,\mathrm{Be}^7(e^-, \nu)\,\mathrm{Li}^7(p, \alpha)\,\mathrm{He}^4$$

or

$$\mathrm{H}^1(p, \beta^+ \nu)\,\mathrm{H}^2(p, \gamma)\,\mathrm{He}^3(\alpha, \gamma)\,\mathrm{Be}^7(p, \gamma)\,\mathrm{B}^8(\beta^+, \nu)\,\mathrm{Be}^{8*}(\alpha)\,\mathrm{He}^4.$$

[1] H. D. HOLMGREN and R. L. JOHNSON: Bull. Amer. Phys. Soc., Ser. II, 3, 26 (1958).

For stars less massive than the Sun, the older values will still be valid. For stars in which these alternative routes are important, the cooler ones will probably complete the chain through Li^7, and the hotter ones through B^8. At the highest temperatures at which the pp-chain is important it is probable that the increase in ε_{pp} over the value given above is of the order of 50%, while for the Sun, averaged over its life history, the increase is only about 20%. This will change slightly our estimates of the masses at which the energy generation rates of the pp-chain and the CN-cycle become comparable. However, these estimates are based on solar models which also now need further modification along the lines suggested by HOYLE and SEARS. In the case in which B^8 is produced, about 30% of the energy released is lost through neutrino emission. Thus in stars in which this route of the pp-chain is important, the times for evolution off the main sequence may also be changed. This is discussed in Sect. 51.

In stars which move off the main sequence after a chemical inhomogeneity between the core and envelope has been formed, hydrogen-burning will take place in a shell source surrounding an exhausted core. In the models of HOYLE and SCHWARZSCHILD (Sect. 26) such a shell source persists right to the tip of the giant branch, and in the last model which fits the observational sequence, hydrogen-burning and helium-burning are of approximately equal importance in energy generation. Provided that in the later stages of evolution some mixing occurs, hydrogen-burning may take place until the whole of the mass has evolved in this way.

Stars of masses much lower than the Sun are evolving at such a slow rate that little chemical evolution has taken place in that proportion of the mass of the galaxy which is tied up in faint stars. It is of interest, however, that none of these stars show evidence of being composed of pure hydrogen, as would be the case if they were some of the first stars to condense, and the protogalaxy were pure hydrogen. This already suggests that several generations of stars are required to synthesize all of the elements. Stars composed initially of pure hydrogen can evolve only by the pp-chain until a helium core has been produced and helium-burning has commenced so that C^{12} is available, at which point the CN-cycle can be started. By similar arguments to those given above we conclude that none of the stars on the upper part of the main sequence can be first generation stars. Provided that nuclei which are made by helium-burning, i.e., C^{12}, O^{16}, and Ne^{20}, are present already, hydrogen-burning in a gas containing these isotopes will synthesize the majority of the less abundant isotopes of carbon, oxygen, and neon, and also nitrogen. At temperatures near 15×10^6 degrees, an approximate expression for the equilibrium ratio of N^{14} to C^{12} as a function of temperature from the CN-cycle rates given in Eqs. (75.1) to (75.4) is

$$\frac{N^{14}}{C^{12}} = \left(\frac{200}{T_6}\right)^2. \tag{75.5}$$

The equilibrium ratio C^{12}/C^{13} is independent of temperature, the Coulomb barrier being the same for both isotopes, and is 4.6.

While a star is on the main sequence we cannot expect to see the results of the hydrogen-burning, unless it is totally convective (and hence of very small mass) so that the products of hydrogen-burning can mix to the surface. After a star leaves the main sequence the outward-burning shell source will leave behind it the results of hydrogen-burning, but large-scale mixing to the surface will not occur, according to the models of HOYLE and SCHWARZSCHILD, at least until a star has evolved on to the giant branch. This argument can be made only for stars in the mass range considered by HOYLE and SCHWARZSCHILD.

For more massive stars it is conceivable that mixing might occur soon after they leave the main sequence.

76. Helium-burning. The basic reactions in helium-burning are [*350*]:

$$He^4 + He^4 \rightleftharpoons Be^8,$$
$$Be^8 + He^4 \rightleftharpoons C^{12*} \rightarrow C^{12} + \gamma.$$

Be^8 is unstable to disintegration into two alpha-particles but only by about 95 kev, and SALPETER has shown that if $T \approx 10^8$ degrees and $\varrho \approx 10^5$ g/cm³, an equilibrium ratio $Be^8/He^4 \approx 10^{-9}$ is established. Thus it appears that in the central regions of red giants, temperature and density conditions such that helium-burning will take place are reached. It is also possible that very massive stars on the main sequence ($M \approx 100\, M_\odot$) might achieve conditions in the cores which would lead to incipient helium-burning. However, helium-burning is in general not to be expected until stars have left the main sequence and their hydrogen-exhausted cores have contracted. The reaction rate per alpha-particle is [*62*]:

$$p_{3\alpha}(He^4) = 2.37 \times 10^{-4} (\varrho\, x_\alpha)^2 \frac{\Gamma_\gamma}{T_8^3} \exp(-43.2/T_8)\ \mathrm{sec}^{-1}, \qquad (76.1)$$

where Γ_γ is the radiation width in ev and T_8 is in units of 10^8 degrees. The energy generation rate is given by

$$\varepsilon_{3\alpha} = 1.39 \times 10^{14} (\varrho^2 x_\alpha^3)\, f_{3\alpha} \frac{\Gamma_\gamma}{T_8^3} \exp(-43.2/T_8)\ \mathrm{ergs\ g^{-1}\ sec^{-1}}. \qquad (76.2)$$

With

$$\varrho = 10^5\ \mathrm{g/cm^3}, \quad x_\alpha = 1, \quad T = 10^8\ \text{degrees},$$
$$\varepsilon_{3\alpha} \approx 600\ \mathrm{ergs\ g^{-1}\ sec^{-1}}.$$

Further helium-burning reactions are

$$C^{12} + He^4 \rightarrow O^{16} + \gamma$$
$$O^{16} + He^4 \rightarrow Ne^{20} + \gamma$$
$$Ne^{20} + He^4 \rightarrow Mg^{24} + \gamma.$$

The relative rates of these reactions depend critically on the temperature, so that different astrophysical circumstances lead to different ratios of the abundances of these nuclei. However, apart from the slow rate of the last reaction all of the others are roughly comparable (within the experimental uncertainties) for temperatures in the range 1.0 to 1.3×10^8 degrees. The atomic abundance ratios are

$$C^{12}:O^{16}:Ne^{20}:Mg^{24} = 1:6:2:0.2$$

according to SUESS and UREY [*358*]. The corresponding ratios for young stars, according to ALLER (see his article in this volume) are 1:5:8:1. Both sets of ratios are reasonable values on the assumption that these isotopes were produced in a previous generation of stars by helium-burning. If these elements have later gone through a hydrogen-burning zone, much of the C^{12} may have been converted to N^{14}, and if the temperature was high enough, some of the O^{16} would have been destroyed. Furthermore, in the later evolutionary stages beyond the giants, processes which may deplete O^{16} and Ne^{20} and build more Mg^{24} may operate.

Helium-burning can occur at temperatures $\gtrsim 10^8$ degrees. Apart from one further nuclear process, the *s*-process, which will be discussed later (Sect. 80), all of the remaining nuclear processes to be described demand conditions of temperature and density more extreme than those required for helium-burning. The furthest evolved stellar models which have so far been constructed are those of HOYLE and SCHWARZSCHILD, and these have interior conditions where only hydrogen-burning and helium-burning can occur. Beyond this point it is widely believed that, as stars evolve into the horizontal branch regions and finally move to the white dwarf regions of the H-R diagram, they achieve the more extreme conditions of temperature and density for the advanced processes of chemical evolution to take place. Since no models which agree with observation are available, the discussion becomes more speculative. On the other hand, we can use as evidence that the processes do occur, the observed abundance anomalies in particular kinds of stars. These are described in Chap. DII.

77. The α-process. After a core of C^{12}, O^{16}, Ne^{20}, and a little Mg^{24} has been produced, all of the helium will be exhausted. It is probable that eventually this inner core will undergo gravitational contraction and heating since gravitation is a "built-in" mechanism of energy generation when a nuclear fuel becomes exhausted (provided that no mixing between core and envelope occurs). No important reactions occur among the nuclei until a temperature $\sim 10^9$ degrees is reached. At this point the gamma rays become sufficiently energetic so that the reaction

$$Ne^{20} + \gamma \rightarrow O^{16} + He^4 - 4.75 \text{ Mev}$$

will occur. The alpha particles released can be captured by the remaining Ne^{20}, producing more Mg^{24}. Thus the *net result* of these reactions is

$$Ne^{20} + Ne^{20} \rightarrow O^{16} + Mg^{24} + 4.56 \text{ Mev}.$$

Once an appreciable amount of Mg^{24} is formed, the reaction

$$Mg^{24} + He^4 \rightarrow Si^{28} + \gamma$$

will occur, also the further (α, γ) reactions that build S^{32}, A^{36}, and Ca^{40}.

These four-structure nuclei are built in decreasing abundance, partly because production of each nucleus does not begin until the previous one has been produced in considerable amounts, and partly because of the increasing repulsion of Coulomb forces. A slight variation of the α-process may also produce Ca^{44} and a little Ti^{48} [*62*]. It has been estimated that this whole α-process will occur when the temperature in the core of the star lies between 10^9 and 3×10^9 degrees. At temperatures near to 3×10^9 degrees interaction between the nuclei themselves begins to take place, and for temperatures above this, reactions begin to occur in such profusion that a statistical approach can be used.

78. The e-process. A further stage of chemical evolution is reached if the central regions of the star contract and heat up sufficiently so that temperatures exceeding about 3×10^9 degrees are reached, while the densities may be in the range 10^4 to 10^9 g/cm^3. Under these conditions, statistical equilibrium is achieved and the material moves to configurations with the maximum stability, i.e., the nuclei are transformed into the group of elements which forms the so-called iron peak in the abundance curve, a large proportion of the material being converted to Fe^{56}. Thus the star may evolve to a configuration with an iron core. Surrounding this, if no mixing takes place, will be a layer consisting of material made in the α-process, while outside this will lie a helium-burning zone, and finally a hydrogen-burning zone. To what extent this would represent a stable configuration is not known. However, the development of the large abundance of iron and the other elements in this region will take only a few seconds, while the α-process will take $\sim 10^2$ to 10^4 years. It may be that production of the elements by the *e*-process immediately precedes a supernova outburst.

The conditions under which the elements in the iron peak in the abundance curve have been synthesized can be deduced by calculating the relative abundances of the isotopes of the elements Ti, V, Cr, Mn, Fe, Co, and Ni as a function of temperature and density. The results may then be compared with the observational values, which are essentially those for the solar system (since only here have isotopic abundances been measured). It is found [*62*] that the best fit is obtained for $T = 3.78 \times 10^9$ degrees and $n_p/n_n \approx 300$. Here n_p and n_n are the number densities of free protons and neutrons, respectively, and their ratio is essentially a measure of the density in the core. Whether this means that stars, when they do evolve to this stage, always reach conditions which are approximated by these values of the parameters is unknown.

Further contraction of the stellar core beyond this stage means that the iron will rapidly be transformed into He^4 with an admixture of protons and neutrons, and arguments based on energy considerations have suggested that the explosions of supernovae are triggered through this process [*143*], [*62*]. This process will be considered further in Sect. 81. A sufficiently catastrophic explosion at this stage, or even at an earlier stage, may remove the outer layers of the star which are the least chemically evolved, so that the stellar stump which remains has little or no nuclear energy sources, and will become a white dwarf. Alternatively, if an explosion leaves behind it a stump which has expanded and cooled, the later stages of chemical evolution may never occur if the material becomes so cool that the nuclei are not energetic enough to react appreciably. Some white dwarfs may have chemical compositions characteristic of the α-process (see Sect. 90); a very speculative possibility is that white dwarfs like van Maanen 2, which has strong, very broadened lines of iron, may show evidence of the *e*-process. White dwarfs showing products of the α-process on their surfaces might have iron cores. Even if they were composed of material that had been mixed by an explosion, gravitational separation would lead to the heavier elements being concentrated to the center [*151*].

79. Synthesis of the remainder of the elements. In considering hydrogen-burning, pure helium-burning, the α-process, and the *e*-process, we have treated consecutively the stages through which stellar material may pass as the temperature is raised and contraction occurs. Although we have considered the case of hydrogen-burning in the presence of impurities, i.e., the CN-cycle and the NeNa-cycle, we have yet to consider the processes which may occur if, for example, we introduce hydrogen into a medium in which helium-burning is taking place, or if we suddenly heat up a region in which hydrogen has not been exhausted. Situations such as these can arise if mixing takes place between layers in a star which have chemically evolved to different extents; they may also occur when elements which have been synthesized in first-generation stars are condensed into second-generation stars, and so on.

Apart from considerations arising purely through the chemical evolution of stars, it is clear that, if all of the elements have been built in the stars, other processes of element synthesis are required, since the four processes so far discussed will synthesize only a selection of the most abundant isotopes up to $A \approx 60$, but none heavier than this. To build the remainder of the elements a *source of free neutrons* is required. This leads us to consider two processes by which these may be produced and captured in stars. If neutrons are available and are captured by nuclei already present, on a time scale sufficiently slow so that beta-decays can take place between neutron captures, then the majority of the nuclei between Ne^{22} and Ti^{50} will be built, if the initial nuclei produced in hydrogen- and helium-burning are already present. If, on the other hand, the nuclei which form the basic building material are those produced in the *e*-process, neutron capture on a long time scale will produce many of the nuclei up to and including Bi^{209}. This neutron capture chain will follow the neutron stability curve. It will end at Bi^{209} since captures beyond this point will lead to nuclei which α-decay in times very short compared with the average time for neutron captures in this region. It will pass through the especially stable and abundant heavy nuclei, those with closed shells of 50, 82, and 126 neutrons.

A second neutron-capture process is required to produce the remainder of the heavy nuclei which are predominantly those with excesses of neutrons, and the long-lived radioactive elements thorium and uranium. To synthesize these elements, a very large flux of neutrons which are captured very rapidly is required. In this case neutrons will be captured at a rate much faster than the natural beta-decay rates of nuclei on the main stability curve. Nuclei with very large neutron excesses will be synthesized, successive captures stopping only when neutrons can no longer be bound; further synthesis will then wait until beta-decay of the unstable nucleus has taken place. Synthesis by this process will proceed along a track roughly parallel to, but on the neutron-excess side of, the main stability line. After synthesis has stopped the neutron-rich nuclei will then decay to their stable forms, giving rise to the displaced peaks which are at the places of the 50, 82, and 126 closed shells of neutrons in the original (unstable) nuclei [*62*]. This rapid rate of captures also enables neutron-rich nuclei to be built in the trans-uranium region. It appears that the process is finally stopped near $A \sim 260$, at which point the spontaneous and neutron-induced fission rates have become exceedingly rapid.

These two different types of neutron-capture process have been designated the *s* (slow) and *r* (rapid) processes. Where they occur in a star's evolutionary track will be described in the next two sections.

80. The *s*-process. In the helium-burning layer a star will contain both He^4 and C^{12}. If some hydrogen is now introduced into this layer the reactions $C^{12}(p,\gamma)$ $N^{13}(\beta^+\nu_+)$ C^{13} will occur so that this can then be followed by the exothermic reaction

$$C^{13} + He^4 \rightarrow O^{16} + n + 2.20 \text{ Mev}$$

which will provide a source of free neutrons. Other exothermic (α, n) reactions which may occur are

$$O^{17} + He^4 \rightarrow Ne^{20} + n + 0.60 \text{ Mev}$$

$$Ne^{21} + He^4 \rightarrow Mg^{24} + n + 2.58 \text{ Mev}$$

$$Mg^{25} + He^4 \rightarrow Si^{28} + n + 2.57 \text{ Mev}$$

and the special case

$$Mg^{26} + He^4 \rightarrow Si^{29} + n + 0.04 \text{ Mev}.$$

The (α, n) reactions involving nuclei heavier than Mg^{26} are all endothermic. Of these reactions, $C^{13}(\alpha, n)$ was first speculated upon by GREENSTEIN [*359*] and proposed independently in some detail by CAMERON [*360*], [*351*], while $Ne^{21}(\alpha, n)$ was proposed by FOWLER *et al.* [*352*]. The other reactions are thought to be of little importance in the *s*-process since O^{17} is largely destroyed in hydrogen-burning, whereas Mg^{25} and Mg^{26} are not produced in hydrogen-burning and, in fact, are some of the nuclei which are built by the *s*-process.

Both $C^{13}(\alpha, n)$ and $Ne^{21}(\alpha, n)$ present problems when the element-building process is considered in detail. Both reactions probably occur in stars which have helium-burning cores and have reached the giant stage in their evolution. The number of neutrons which can be produced by the $C^{13}(\alpha, n)$ reaction is determined essentially by the abundance of C^{13}. If this is the equilibrium abundance produced in previous CN-cycle hydrogen-burning, then $C^{12}/C^{13}=4.6$. Now the "normal" abundance ratio $C^{12}/Fe^{56}=6.4$, so that only about 1.4 neutrons per iron nucleus will be made available, and this will be enough only to build nuclei slightly heavier than Fe^{56}. Furthermore, if neutron production occurs in a region where the relative abundances of the carbon and nitrogen have been determined by previous hydrogen-burning, the ratio N^{14}/C^{13} will be quite large, and hence the reaction $N^{14}(n, p)\, C^{14}$ will take a large proportion of the neutrons produced. CAMERON has suggested that these difficulties in the nuclear physics can be overcome if there is considerable mixing between core and envelope, so that C^{13} is made by (p, γ) reactions on C^{12} which has been produced in helium-burning, but little N^{14} is produced by $C^{13}(p, \gamma)\, N^{14}$.

However, as has been discussed previously, the later stages of stellar evolution do not, according to HOYLE and SCHWARZSCHILD, admit of large-scale mixing processes between core and envelope. On the other hand, it must be remembered that those models were only applicable in a small range of stellar masses, and that the later models were rather tentative. The observations (Sect. 91) appear to suggest that the isotopes produced in the *s*-process do appear over-abundant on the *surfaces* of some stars in the giant stages of their evolution; these stars may be much more massive than the stars for which the Hoyle-Schwarzschild models are valid. The fact that the anomalies are observed shows that mixing has taken place between the deep interior and the atmosphere either during the synthesis by the *s*-process or after it ceased. The incipient variability of many of these stars may also be taken as an indication of the occurrence of mixing. If this destroys the chemical inhomogeneity, it would appear to demand drastic changes in the star's structure and may well be the main factor in moving massive stars in their evolution off the giant branch. The situation clearly involves so many unknown factors that at the present time the $C^{13}(\alpha, n)$ reaction cannot be ruled out on these grounds. It is bound to come into operation and to synthesize elements at whatever point a star having a helium-burning or helium-exhausted core does begin core-envelope mixing.

The $Ne^{21}(\alpha, n)$ reaction was originally proposed as a neutron source which might overcome the mixing difficulties. Thus it is supposed that Ne^{20} which has been synthesized in a star of an earlier generation is converted to Ne^{21} in the hydrogen-burning shells of red giants at 30 to 50×10^6 degrees. As the shell burns outwards, the Ne^{21} will interact in the helium core to produce neutrons. Since the abundance ratio $Ne^{20}/Fe^{56}=14$, this means that, provided iron is present, having been synthesized in an *e*-process in a star of an earlier generation, about 14 neutrons per iron nucleus will become available, and this is sufficient for considerable heavy-element synthesis. This argument requires that the reaction $Ne^{20}(p,\gamma)\, Na^{21}(\beta^+\nu_+)\, Ne^{21}$ should produce Ne^{21} faster than it is destroyed by the $Ne^{21}(p,\gamma)\, Na^{22}$ reaction, and it appears that this condition will be fulfilled [*62*]. Furthermore, it is necessary that the Ne^{21} be produced before the hydrogen mixed with the Ne^{20} is exhausted by normal hydrogen-burning by the CN-cycle or the *pp*-chain. This will be possible provided that there is only a small concentration of carbon, nitrogen, and oxygen, about 8% of the normal abundance of these elements. Another condition is that the N^{14} which will rapidly destroy neutrons should be itself depleted by the $N^{14}(\alpha, \gamma)\, F^{18}$ reaction before the $Ne^{21}(\alpha, n)\,Mg^{24}$ reaction begins. This again appears to be possible [62]. Finally, it is necessary that the Ne^{21} should react with the helium before this helium has been burned. This demands that a high temperature near 2×10^8 degrees is achieved, and also that the concentration of helium has dropped very considerably, i.e., that a large proportion of it has been consumed in helium-burning in the core.

There is some indication from the form of the abundance curve for isotopes built by the *s*-process that both neutron-producing reactions may be important at different times in the giant stages of evolution. BURBIDGE *et al.* [62] have presented the following schematic picture. It is supposed that the stellar core spends some 10^7 to 10^8 years at a comparatively low temperature (20 to 40×10^6 degrees) while the star slowly progresses into the giant stage, some 10^5 years at 50 to 100×10^6 degrees, and then only a brief period $\sim10^3$ years at $\sim200\times10^6$ degrees as it rapidly exhausts helium and builds the middle-weight elements. The $C^{13}(\alpha, n)$ reaction has a life-time of about 10^5 years at 75×10^6 degrees, so that the C^{13} will be consumed during these stages. Thus a few neutrons per Fe^{56} become available on a time-scale $\sim10^5$ years. A time scale $\sim10^5$ years is needed to produce the early part of the abundance distribution built by the *s*-process [62]. With the consumption of C^{13} in its central regions, the core now heats to greater temperatures and helium-burning begins in earnest. Towards the end of this helium-burning, instability may set in and mixing occur. New C^{13} is produced, Ne^{21} is burned, and a great flux of neutrons is produced in a final period $\sim10^3$ years. A time scale of this order is required to account for the *s*-process abundance distribution with $A \gtrsim 100$ where the product of the neutron-capture cross section and the abundance becomes relatively constant. Details are given in [62].

81. The *r*-process. It has been proposed that the *r*-process occurs in the course of a supernova explosion [353], which presumably occurs at a very late stage of stellar evolution after the star has left the giant branch. We have no idea from observations where the pre-supernovae lie in the H-R diagram.

That supernovae are the seat of the *r*-process was proposed when it was found that the exponential decay of the light-curve of supernovae of Type I with a half-life of 55 days corresponded to the half-life by spontaneous fission of Cf^{254} (56.2 ± 0.7 days). This observational evidence will be considered further in Sect. 93. We shall outline here a qualitative theory of the supernova explosion as described by BURBIDGE *et al.* [62]. An earlier theory [361] of supernova outbursts demanded that stellar neutron cores had developed, a much more extreme condition than that to be described here.

A collapse theory involving the liberation of neutrinos—the so-called urca process—has also been proposed [362]. These theories will be described by F. ZWICKY in this Encyclopedia. Many other theories of supernovae have been proposed, but since they do not involve advanced chemical evolution through nuclear activity they will not be considered here.

We suppose that a configuration has been reached in which the central core has temperatures and densities sufficient for the *e*-process to occur. As we have already pointed out, for T near 4×10^9 degrees the peak of the abundance curve is very sharp and narrow, the abundances falling away extremely rapidly as A either increases or decreases from 56 by a few units. At higher temperatures the peak becomes rather wider, ranging from copper on the upper side to vanadium on the lower side. Outside these limits the abundances fall away even more rapidly so that they can be neglected.

The only exception to this is in the case of He^4. From the equation of statistical equilibrium it is found that the relation between the abundance of Fe^{56} and He^4

is given by [62]

$$\operatorname{Log} n\,(\mathrm{He}^4) - \frac{1}{14}\operatorname{Log} n\,(\mathrm{Fe}^{56}) = 32.08 + 1.39 \operatorname{Log} T_9 - \frac{34.62}{T_9} + \frac{1}{7}\operatorname{Log}\frac{n_p}{n_n}. \quad (81.1)$$

Here T_9 is the temperature in 10^9 degrees. The conditions at this stage are such that the densities of free protons and free neutrons are comparable and small as compared to $n(\mathrm{He}^4)$ and $n(\mathrm{Fe}^{56})$, so that the last term can be neglected. Now let us suppose that at a fixed density $\varrho = 10^8$ g/cm³, the mass is equally divided between Fe^{56} and He^4; calculation shows that the abundances of all other nuclei will be negligible under these conditions. Then we find that $T = 7.6 \times 10^9$ degrees. Hence we may suppose that as T rises above 7×10^9 degrees there is a considerable conversion of iron into helium. The result is not very dependent on the particular value of 10^8 g/cm³ since, as has been shown by HOYLE, the critical temperature at which the masses become comparable increases only slowly with the temperature. However, for $T \sim 7 \times 10^9$ degrees the density cannot differ very much from 10^8 g/cm³. Thus a value much less than 10^6 g/cm³ would be most implausible, while a value appreciably greater than 10^8 g/cm³ is forbidden by the conditions that there must be an appreciable contribution to the pressure.

Now let us suppose that the temperature increases to 8.2×10^9 degrees. In this case the right-hand side of the equation increases by 0.4, and in order that the equation be satisfied $[\operatorname{Log} n\,(\mathrm{He}^4) - \frac{1}{14}\operatorname{Log} n\,(\mathrm{Fe}^{56})]$ must also increase by 0.4. However, $\operatorname{Log} n\,(\mathrm{He}^4)$ can only increase by 0.3, since at this point the whole core will be helium. Thus an increase in T of 0.6×10^9 degrees implies that the factor $\frac{1}{14}\log n\,(\mathrm{Fe}^{56})$ is decreased by a factor of 0.1; i.e., that the amount of iron drops by a factor of about 25. Thus the increase in temperature from 7.6 to 8.2×10^9 degrees causes a change from 50% He^4 and 50% Fe^{56} to $\sim$98% He^4 and $\sim$2% Fe^{56}.

To convert 1 g of iron to helium rquires 1.65×10^{18} ergs. This may be compared with the total thermal energy of 1 g of material at 8×10^9 degrees, which amounts to 3×10^{17} ergs. Thus the energy supply for the conversion of iron to helium must come from the gravitational energy, practically all of it going to nuclear energy, so that little is available to increase the thermal energy. But in the large-scale contraction which is required, a large increase in thermal energy would be demanded to maintain mechanical equilibrium. Consequently, we conclude that in this contraction mechanical stability is not maintained, but free fall takes place. This happens in $\sim$1 second, and this implosion is the trigger which sets off the thermonuclear explosion. The loss of energy by neutrino-emission—the urca process—will also take place at this stage of evolution. However, it is easily shown [62] that the rate of energy loss by this process is such that the stellar core will lose energy equal to its thermal energy in about 10^8 secs, i.e., it is far less catastrophic than the process described here.

The relationship between the implosion of the central regions and the explosion of the outer region is as follows. Material in the outer regions normally possesses a thermal energy per unit mass which is far less than its gravitational potential energy per unit mass. Hence any abnormal process which leads to the thermal energy suddenly becoming comparable to the gravitational energy must lead to a sudden heating of the outer material. In the outer envelope the material is less chemically evolved, so that a sudden heating will lead to the release of nuclear energy.

Whether the gravitational energy released will be sufficient to trigger a thermonuclear explosion will depend on its magnitude. It has been shown that a sudden heating to 10^8 degrees would be sufficient to trigger the explosion. This corresponds to a thermal energy $\sim 10^{16}$ ergs/g. Thus for gravitational energies greater than this, explosion will occur. This means that for a stellar mass of $1.5 M_\odot$ the radius must be less than 10^{10} cm if the potential energy is to exceed 10^{16} ergs/g. At this advanced stage of evolution such a shrinkage from the main-sequence configuration is reasonable, and it means that the pre-supernova will lie in the subdwarf region to the left of the main sequence.

As the envelope implodes, its temperature will increase to about 2×10^8 degrees as a result of the conversion of gravitational to thermal energy, while the density may rise to $\sim 10^5$ g/cm³. Under these conditions reactions of the type $C^{12}(p, \gamma) N^{13}$, $O^{16}(p, \gamma) F^{17}$, $Ne^{20}(p, \gamma) Na^{21}$, $Mg^{24}(p, \gamma) Al^{25}$ will take place very rapidly, and a mean energy of about 2 Mev per proton is released. If we take a composition characteristic of a star that is very far advanced in its chemical evolution, consisting of approximately equal abundances by number of hydrogen, helium, and light nuclei, the energy released is sufficient to raise the temperature to $\sim 10^9$ degrees.

At this high temperature alpha-particle reactions become important and a neutron-producing reaction of importance is $Ne^{21}(\alpha, n) Mg^{24}$, which follows very rapidly on the 23-second beta decay of Na^{21}. The mean reaction time is the order of 10^{-5} sec if $T \approx 10^9$ degrees and $\varrho \approx 10^5$ g/cm³. The (α, n) reactions on C^{13} and O^{17} are not operative under these conditions, since N^{13} requires 14 minutes in the mean to decay to C^{13} and F^{17} requires 100 sec to decay to O^{17}. The neutrons produced in this way are rapidly thermalized ($kT \approx 100$ kev) and they are captured primarily by the abundant nuclei in the iron group which were initially present in the envelope. The neutrons become available in the mean lifetimes of Na^{21} (33 sec), Mg^{22} (7 sec), and Al^{26} (10 sec). Thus a reasonable mean time for the neutron production during the imploding high-density stage is about 10 sec. There will be one neutron produced for each Ne^{20} nucleus present, and on the assumption that there are approximately equal number densities of hydrogen, helium, and the light nuclei C^{12}, O^{16}, and Ne^{20}, each of the latter having equal abundance, the number density of Ne^{20} is 10^{22}/cm³.

An alternative mechanism of neutron production which has not been discussed in detail, but which may be of importance, is as follows. In the break-up of Fe^{56} into He^4, four neutrons per Fe^{56} nucleus are produced. In the explosion these neutrons may be captured either by some outer core iron which has not been broken up, or by iron peak nuclei in the envelope, to form heavy elements.

The neutron captures will take place during the expanding stage following the implosion. By then the density has fallen to $\sim 10^2$ g/cm³ so that the neutron density is $\sim 10^{24}$ neutrons/cm³. It is assumed that the radiation temperature remains near 10^9 degrees during the capture process, and it has been shown by Burbidge *et al.* [62] that this means that captures will occur until the neutron binding energy is reduced to ~ 2 Mev. Further captures must wait until beta emission occurs in the neutron-rich nuclei with lifetimes of ~ 3 sec. Each beta decay is followed by about 3 captures so that the total time for ~ 200 captures (Fe^{56} to Cf^{254}) is about 200 sec. Thus the total capture time is long as compared with the total neutron production time and in the case where the neutrons are produced by the Ne^{21} reaction, we are justified in assuming that the initial neutron density is that given by the original Ne^{20} density.

It has been shown that under these conditions most of the heavy isotopes not built by the *s*-process can be synthesized. Whether such a method of synthesis is able to account for the amounts of these heavy isotopes in the Galaxy depends on the rate of supernova explosions and other factors.

82. The *p*-process. The proton-rich isotopes of a large number of heavy elements cannot be built by either the *s*- or the *r*-process. These isotopes have a much lower abundance than the other heavy nuclei and lie on a separate curve on the schematic abundance diagram shown in Fig. 26. If we are attempting to show that all of the isotopes have been synthesized in stars we have to consider at what stage in a star's chemical evolution these isotopes can be built.

The reactions which must be involved in synthesizing these isotopes are (p,γ) and (γ, n) reactions on material which has already been synthesized by the *s*- and the *r*-processes. The astrophysical circumstances in which these reactions can take place must be such that material of density $\sim 10^2$ g/cm^3, containing a normal or excessive abundance of hydrogen, is heated to temperatures $\sim 2-3 \times 10^9$ degrees. It has been suggested [*62*] that these circumstances may be reached in the envelope of a supernova of Type II, it being assumed that the conditions are similar to those given above as applying to supernovae of Type I, except that some *r*-process material is already available, and there is no deficiency of hydrogen in the outer envelope. Alternatively, it has been suggested that they might be reached in the outermost parts of the envelope of a supernova of Type I in which the *r*-process has taken place in the inner envelope.

It is supposed that, following the initiating (p, γ) reactions, quasi-statistical equilibrium is set up between (p, γ), and (γ, n) reactions. Calculations based on this supposition show in a qualitative fashion that the nuclei will be driven from the main neutron stability line to synthesize the proton-rich nuclei. The whole process can take place in a time which is within the estimated time scale of a supernova explosion, i.e., 10 to 100 secs.

83. The *x*-process. We have discussed different stages of chemical evolution which may be responsible for synthesizing all of the isotopes in stars with the exception of D, Li, Be, and B. It is well known that these elements are extremely unstable at comparatively low temperatures and are easily converted to helium by proton bombardment. When we were discussing the gravitational contraction phase in a star's evolution it was pointed out that, just before reaching the main sequence, the star would burn deuterium and then lithium, beryllium, and boron in its interior, before the *pp*-chain or the CN-cycle commences. Thus very early in a star's chemical evolution these elements are destroyed, except in the outer parts of the star.

On the other hand, in the context of a theory of element synthesis in stars, there must be a stage at which synthesis of these elements can occur. Moreover, such synthesis must take place under conditions such that these elements are preserved after they have been synthesized; i.e., it cannot take place in stellar interiors. It has been proposed [*363*], [*62*] that such synthesis takes place in stellar atmospheres, or perhaps in the outer parts of supernova envelopes.

If these processes occur in stellar atmospheres, then they are probably related to flare phenomena or other activity in which particles may be accelerated and nuclear reactions take place. Either low-energy fusion reactions, or high-energy spallation reactions, or both, may be important. Evidence that such processes do occur is given by the observations of SEVERNY [*364*] and GOLDBERG and MÜLLER [*365*] that deuterium is synthesized in solar flares. Other more indirect evidence is the large apparent abundance of Li in T Tauri. Whether these elements are preserved when made on the surface depends on the depth of the mixing zone. Unless the nuclei are directly ejected from the star, mixing in the deep interior will destroy them. Synthesis of these elements may take place at many stages in a star's evolution.

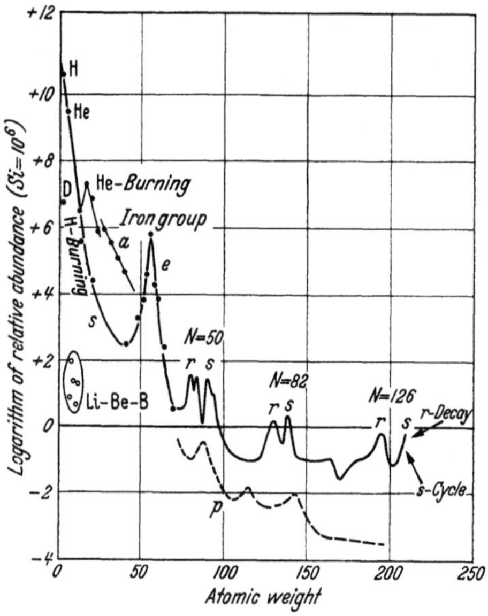

Fig. 26. Schematic curve of atomic abundances as a function of atomic weight based on the data of Suess and Urey [358], and drawn up by Burbidge *et al.* [62]. Suess and Urey have employed relative isotopic abundances to determine the slope and general trend of the curve. There is still considerable spread of the individual abundances about the curve illustrated, but the general features shown are now fairly well established. Note the over-abundances relative to their neighbors of the α-particle nuclei $A = 16$, 20, ..., 40, the peak at the iron group nuclei, and the twin peaks at $A = 80$ and 90, at 130 and 138, and at 194 and 208.

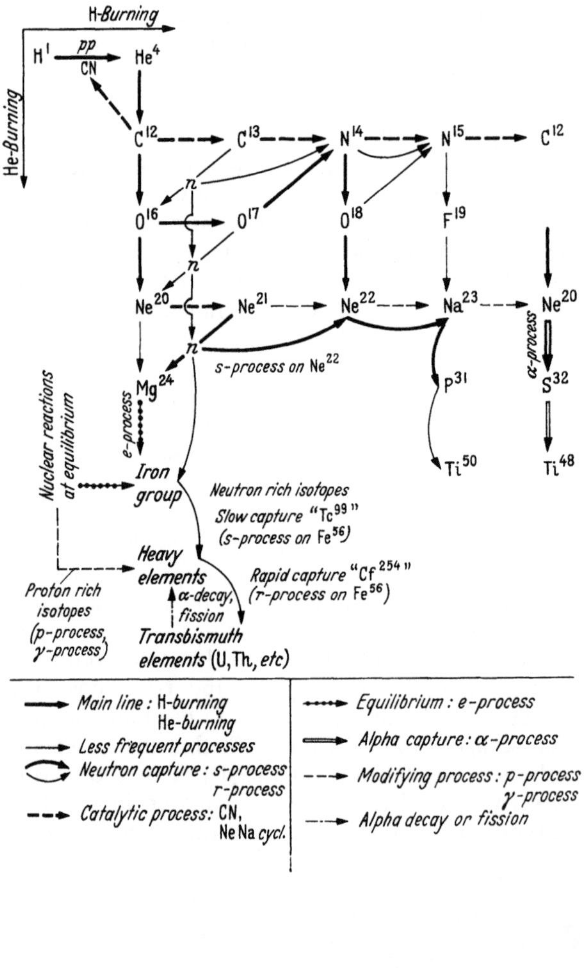

Fig. 27. A schematic diagram of the nuclear processes by which the synthesis of the elements in stars takes place, after Burbidge *et al.* [62]. Elements synthesized by interactions with protons (hydrogen-burning) are listed horizontally. Elements synthesized by interactions with alpha-particles (helium-burning) and by still more complicated processes are listed vertically. The details of the production of all the known stable isotopes of carbon, nitrogen, oxygen, fluorine, neon, and sodium are shown completely. Neutron-capture processes by which the highly-charged heavy elements are synthesized are indicated by curved arrows. The production of radioactive Tc^{99} is indicated as an example for which there is astrophysical evidence of neutron captures at a slow rate over long periods of time in red giant stars. Similarly Cf^{254}, produced in supernovae, is an example of neutron synthesis at a rapid rate. The iron group is produced by a variety of nuclear reactions at equilibrium in the last stable stage of a star's evolution.

To sum up the results of Sects. 75 to 83, Fig. 26 shows a schematic abundance curve. The various regions are indicated where the different processes demanded to synthesize all of the elements operate. In Fig. 27 the different kinds of nuclear process which occur are shown schematically.

84. Nova explosions. Nova explosions are apparently triggered in the outer parts of the star. Thus although the pre-novae are believed to be in an advanced stage of chemical evolution, it is not likely that they are effective in producing conditions under which further radical chemical evolution takes place. BIERMANN [*366*] has proposed that the explosion arises as a result of the release of ionization energy in a zone of instability. This will clearly leave the material unchanged. On the other hand, SCHATZMAN [*147*], [*367*], has proposed a theory in which the explosion is triggered by one of the pp-chain reactions. By considering the various ways this chain goes at different densities, it has been argued that conditions may arise in which the star becomes vibrationally unstable, a shock front develops, and the outburst occurs. This argument appears to demand further investigation, since it depends on the pp-chain at fairly high densities ($\sim 10^4$ g/cm^3). If the pre-novae are hydrogen-exhausted to a large degree, as might be expected, it is not obvious that instabilities can arise through reactions in the pp-chain.

It is of interest, however, that the remnant of the supernova in the central region of the Crab Nebula is apparently still unstable, and is giving rise to some form of outward-travelling disturbance.

II. Observations.

85. Introduction. As has been noted previously, the evolutionary models so far computed do not take stars any further than to configurations with central densities of $\sim 10^5$ g/cm^3 and central temperatures of $\sim 10^8$ degrees. Of the processes discussed in Part I of this Chapter, these conditions are sufficient only for hydrogen-burning, helium-burning, and neutron-capture on a slow time scale (s-process) to occur. Even the instabilities possibly leading to mixing between the core and envelope, which may be necessary if sufficient neutrons are to be made available for the s-process to build the heaviest elements from iron, have not yet been investigated theoretically.

However, it has been shown that chemical evolution of stars must occur, whether or not this process is found to be able quantitatively to account for the production of the observed abundances of the elements, starting from pure hydrogen, in the total lifetime of the Galaxy. From studies of stellar spectra, evidence can be obtained that the chemical composition of stars is related to their ages and the age of the gas out of which they are condensed. Moreover, there is also evidence from their spectra that certain types of stars have reached various stages of chemical evolution. Reference may be made to the article by L. H. ALLER in this volume for the general situation as regards the determination of abundances in stars. In the succeeding sections we shall discuss the evidence, firstly for a relation between chemical composition and age, and secondly for hydrogen-burning, helium-burning, the α-process, and the s-process having occurred in individual stars. We shall also discuss the evidence from light curves of supernovae for the occurrence of the r-process.

The problem of whether chemical evolution and death of stars within the Galaxy can account for its average present-day chemical composition depends firstly upon there having been an adequate rate of star formation and evolution, and interchange of matter between stars and the interstellar medium. Secondly, it depends upon the attainment, towards the end of a star's life, of structural conditions that have not yet been adequately studied theoretically, although they have some basis in observation (supernovae, novae, explosive variables, faint high-temperature stars).

We may note that, if hydrogen is the primeval element, and if there is a continual conversion of this into heavier elements in stars, then this one-way process has apparently not progressed very far in the sample of the universe accessible to our observation, since hydrogen still accounts for about 75% by mass of the total (see [*62*], Table II,1). In the building of heavier elements from lighter ones, energy is liberated until the group of elements in the iron peak in the abundance curve is reached (the minimum of the packing-fraction curve). To build heavier elements, energy has to be supplied and in fact this is where the neutron-capture processes take over. Thus a galaxy which had reached its lowest energy state would consist entirely of the iron group of elements or of degenerate matter (white dwarfs). However, by far the greatest amount of energy is liberated in the first step, the conversion of hydrogen to helium. The energy that would result from the conversion of the whole mass of a galaxy (say $10^{11} M_\odot$) from hydrogen to helium is 1.2×10^{63} ergs. This is sufficient for such a galaxy (say its integrated magnitude is -20) to radiate at its current rate for $\sim 10^{12}$ years.

If the current hydrogen/helium ratio is ~ 10 in our Galaxy, this might suggest that the Galaxy has had to convert more material, and hence release more energy, in the past than it is doing today [*368*].

86. Aging effect in composition: Extreme halo population. If elements are continually being synthesized in stars, so that the heavy-element content of the Galaxy is continuously increasing relative to hydrogen, then there should be a consequence that we may call the "aging effect" in the chemical compositions of stars formed at different epochs. Old main-sequence stars should, on the average, contain a smaller proportion of elements heavier than hydrogen than do the young main-sequence stars. The size of such an effect would depend upon whether production by the stars within a galaxy accounts for the whole abundance distribution, as suggested in the work presented in Part I of this Chapter, or whether a galaxy already contains some heavy elements at the time of its condensation.

The ideal way of testing for such an aging effect is to examine the chemical compositions of individual stars in globular clusters, which belong to the extreme halo population and are the oldest systems so far dated in our Galaxy. Owing to the faintness of such stars, quantitative analyses have not yet been carried out, but work is being planned by several people at the Mount Wilson and Palomar and the McDonald Observatories. Qualitative examination of spectra of giants in globular clusters [*229*], [*230*], [*369*] shows a general weakening of the metallic lines, which suggests lowered abundances of these elements. Such composition differences have been repeatedly referred to in Chap. B, in discussing the differences between color-magnitude diagrams of globular and galactic clusters.

A most valuable observational program is the study of integrated spectra of globular clusters [*370*], [*231*] and their accurate classification in terms, not only of spectral class, but also of the degree of weakening of spectrum lines of various elements [*231*]. This work indicates a range of compositions: those clusters in the direction of the galactic nucleus which may actually be very near it, appear to have more normal compositions, as distinct from the most extreme cases like M 92, which appear to have very large underabundance factors in the metallic elements.

Although quantitative analyses of globular cluster stars are not yet available, such an anlysis has been made for two stars in the solar neighborhood, HD 19445 and HD 140283, whose very high velocities indicate that they actually belong in the extreme halo population [*371*]. The iron abundance derived was about $^1/_{10}$ normal, and that of calcium, about $^1/_{30}$ normal. A spectrum of HD 19445 is shown in Fig. 28, where it may be compared with the normal standard *F* 7 V star, ξ Peg. It will be seen that not only the lines of iron and calcium, but also scandium and strontium, are weakened [*62*]. Thus the aging effect is apparent in at least three of the processes discussed in Part I of this Chapter, the α-, *e*-, and *s*-processes.

We may note that nothing is known about the helium content in unevolved stars in the old, spherical, extreme halo population, since such stars are of later type than *A* 0, while highly evolved, horizontal-branch stars of high enough temperature probably have a helium content that is representative of their own evolutionary stage and not of the material out of which they were formed.

A possible indirect way of deriving abundances in main-sequence stars in globular clusters by means of two-color plots, *U*–*B* against *B*–*V*, may prove useful. There is a well-defined relationship for normal solar-neighborhood stars[1]. The stars in the globular cluster M 3 lie considerably above this line [*372*]; they have an "ultra-violet excess". So also do many of the stars in the catalogue of high-velocity stars [*265*]; the spectra of these stars show a general weakening of all lines except hydrogen. This suggests an explanation for the ultra-violet excess in terms of differing blanketing effects [*373*], [*372*]. Lines due to the metals are more numerous in the *B* spectral region than in the *V*, and still more numerous in the *U*, so that a weakening of all lines would tend to displace a star in the *B*–*V*, *U*–*B* plane in such a way that it will lie above the normal *B*–*V*, *U*–*B* relation.

An attempt to measure the magnitude of this effect [*373*] is being repeated for the star HD 19445 [*374*]. If it does prove capable of explaining the whole ultraviolet excess, then it may be possible to calibrate the two-color plot in terms of abundances of the metals in stars and to use this to deduce abundances in clusters that are too distant for spectrophotometric analysis.

87. Problem of the so-called subdwarfs: Location of main sequence as a function of chemical composition. The blanketing effect discussed in Sect. 86 also has a bearing on the problem of whether the main sequence in extreme halo population II stars lies below the solar-neighborhood main sequence in the H-R diagram or not (cf. Sects. 22 and 51). Let us consider HD 19445. The trigonometric parallax leads to an absolute magnitude of $+4.57 \pm 0.64$. When the spectrophotometric analysis [*371*] and colors [*265*] showed that the star had a spectral type near *F* 7, instead of *A* 4[2], it was seen not to lie nearly so far below the main sequence as had previously been thought.

Now recent work [*375*], [*376*] indicates that the star is actually nearer to type *G* 0 or *G* 2, from its excitation and ionization temperature. Thus the star actually lies very close to the normal main sequence in the H-R diagram, and the changed blanketing effect may be the cause of its having a color index nearer to an *F*-type star.

Spectroscopic parallaxes of stars like this, with abnormal chemical compositions, are not of any value until a new calibration has been set up for them, adequately taking this effect into account. The term "subdwarf", as applied

---

[1] H. L. JOHNSON and W. W. MORGAN: Astrophys. J. **117**, 313 (1953).

[2] W. S. ADAMS, A. H. JOY, M. L. HUMASON and A. M. BRAYTON: Astrophys. J. **81**, 187 (1935).

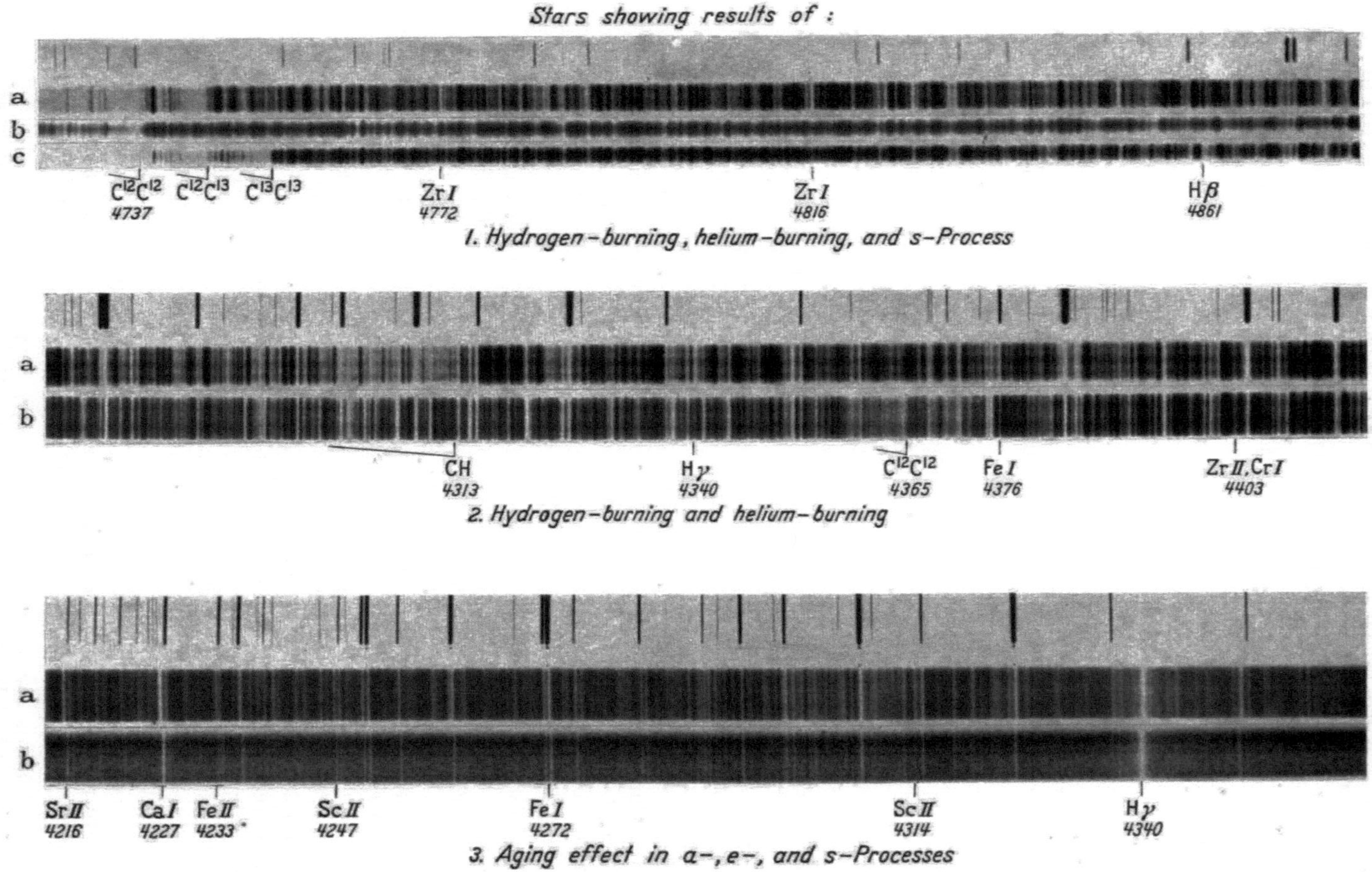

Fig. 28. Portions of the spectra of stars showing different aspects of element synthesis, reproduced from [62]. Upper: (a) Normal carbon star, X Cancri, which has $C^{12}/C^{13} \sim 3$ or 4. (b) Peculiar carbon star HD 137613, which shows no C^{13} bands, and in which hydrogen is apparently weak. (c) Normal carbon star, HD 52432, which has $C^{12}/C^{13} \sim 3$ or 4. Note that Zr I lines appear to be strongest in (a). Middle: (a) Normal carbon star, HD 156074, showing the CH band and Hγ (b) Peculiar carbon star, HD 182040, in which CH is not seen, although the weak band of C_2 at λ 4365 is visible. Hγ is also very weak, indicating that hydrogen has a low abundance. Lower: (a) Normal F-type star, ξ Pegasi. (b) Peculiar star HD 19445, which has a slightly lower temperature than ξ Pegasi, yet all lines but hydrogen are much weakened, showing that the abundances of α-, e-, and s-process elements are much lower than normal ("aging" effect). The

to stars of this sort, is misleading, and should be reserved for the peculiar stars which really do lie several magnitudes below the main sequence and which are probably highly-evolved stars.

The problem of the location of main sequences in the H-R diagram as a function of chemical composition is a very important one and needs more theoretical work. Apart from accurate trigonometric parallaxes of the few nearby members of the halo population, the measurement of accurate colors and magnitudes of sufficiently long sections of the main sequence in globular clusters may solve the problem, but observationally this is a difficult task [*232*], [*233*].

88. Aging effect in compositions: Intermediate cases. As there is a range of population types in the Galaxy intermediate between the extreme halo and extreme spiral arm, so a range of chemical compositions might be expected. Such a range has indeed been found [*377*] to [*384*], but although there is a general correlation between population type and composition, it is not a tight one. The problem is reminiscent of the attempt to correlate colors, spectral types, mass-to-light ratios, and structural forms of galaxies (Chap. B V).

Since all the analyses referred to above are for stars that are relatively close to the Sun, the population type has to be deduced by means of space motions. Two examples which show the scatter in the correlation are HD 161817, a star with a very high radial velocity (-363 km/sec) and abundances that are low only by factors of $\frac{1}{2}$ or $\frac{1}{4}$, and λ Bootis, which has abundances of Mg, Ca, Fe, and Sr that are low by factors of $\frac{1}{10}$ to $\frac{1}{30}$ but a space velocity of only 28 km per sec [*383*].

There is, however, reasonable correlation between space velocity and composition, and factors of about $\frac{1}{2}$ to $\frac{1}{4}$ of normal seem to be the average for the high-velocity *K* giants. Apparently elements involved in hydrogen-burning, helium-burning, α-, *e*-, and *s*-processes all share in the effect and differences between them tend to lie within the errors of determination. However, W Virginis, which is the prototype of this class of variable stars in the halo population II, has been found to have the ratio of strontium and scandium to iron lower than normal by a factor $\frac{1}{4}$ [*385*]. Strontium and scandium are produced in the *s*-process and iron in the *e*-process. In W Virginis, the departures from thermodynamic equilibrium in the atmosphere have prevented the determination of abundances relative to hydrogen.

The apparent range of composition among even the small sample of globular clusters so far surveyed is surprising, and suggests either a range of ages, or non-uniformity in the medium out of which they were formed. This is connected with another problem, the apparent similarity between the ages of the globular cluster M 3 and the old galactic cluster M 67 (see Sect. 51), as judged by the break-away point from the main sequence in the color-magnitude diagrams. Low-dispersion spectra of stars in M 67 [*386*], [*387*] do not show an abnormally low abundance of heavy elements. The calibration of the ages in the two clusters may not be correct yet, and uncertainty in the fitting of the main sequences makes this hard to check.

Possibly in the early history of the Galaxy, when plenty of interstellar gas was still present, star formation may have been much more rapid than today, during the decay of turbulence and flattening towards the disk. Then it is conceivable that there might have been a rapid rate of formation of elements, and a steep gradient of enrichment towards the equatorial plane, which might account for scatter in the compositions of globular clusters and other stars. Clusters would then have different compositions according to where they were formed in the Galaxy.

89. Evidence for element synthesis in individual stars: Hydrogen- and helium-burning. We now turn from the general aspects of the chemical evolution of stars to consideration of its occurrence in individual stars.

For a star to show on its surface evidence of element synthesis that has occurred in its interior, firstly the star must have reached a late stage of evolution, and a relatively high degree of exhaustion of nuclear fuel. Secondly, either the star must have undergone considerable mixing, or considerable mass loss to carry away its outer layers, or both. Anomalous abundances are observed in several general locations in the H-R diagram: giants, high-temperature subdwarfs, certain stars apparently near the main sequence, and white dwarfs. Anomalies in giants are found among cooler stars (*S* stars, carbon stars) but not exclusively so (e.g. R CrB stars, Ba II stars). The stars apparently near the main sequence comprise mostly stars where the anomalies are probably confined to the surface layers (peculiar *A* magnetic stars), and are not connected with chemical evolution in the sense of the using up of nuclear fuel. They are outside the scope of this chapter, and we shall not discuss the metallic-line stars where the anomalies may be only apparent.

Anomalies in the red giants may be connected with the onset of large-scale mixing and the occurrence of variable stars in this region. Once mixing has been achieved a star will no longer have the chemical inhomogeneity which led to its giant-type structure, so that it must adjust itself to a new model appropriate to its being homogeneous and having a higher helium content.

The effects of hydrogen-burning are the exhaustion of hydrogen and its replacement by helium, and the production of C^{13}, N^{14}, N^{15}, which are involved in the CN-cycle, in their equilibrium abundances. There may also be destruction of oxygen, if temperatures as high as 30×10^6 degrees are attained for as long as 10^5 years. The effects of helium-burning are the production of C^{12}, O^{16}, Ne^{20}. The following classes of stars show evidence of this sort.

(α) Complete hydrogen-exhaustion is shown by a group of stars in the hot subdwarf class; how far they lie below the main sequence in the *O* and *B* region is not certain, but they may not be very far. Examples are HD 124448 [*388*], HD 160641 [*389*], and HD 168476 [*390*]. These have no hydrogen lines but strong lines of helium, carbon, and, in some cases, nitrogen and neon. HD 168476 has no oxygen lines. Preliminary abundances in HD 160641 [*391*] are given in Table 16. These three stars probably have high space velocities.

In the category of complete hydrogen-exhaustion are also some white dwarfs, which have helium but no hydrogen lines.

(β) A fairly high degree of hydrogen exhaustion is shown by a group of stars which also lie in the hot subdwarf region, probably further below the main sequence than the HD 124448 group. Stars of this class seem to have fairly low velocities and may belong to population I. An example is HZ 44 [*392*]; the absolute magnitude is in the range $+3$ to $+5$, and the surface gravity seems high enough for the mass to be considerably larger than $1\, M_{\odot}$. Helium is very abundant and nitrogen is about 200 times as abundant as carbon, suggesting a low temperature for the last hydrogen-burning zone through which the surface material has passed. Abundances are given in Table 16.

The *A*-type stars υ Sgr [*393*] and HD 30353 [*394*] probably have very high luminosities, so that they lie in the supergiant region. Their masses are probably $\sim 10\, M_{\odot}$. They have very little hydrogen (see Fig. 29), and helium, nitrogen, and neon are probably greatly overabundant (although quantitative analyses have not been carried out). The great strength of sulphur and argon lines in

Table 16. *A comparison between the chemical compositions of young stars and evolved stars.* The second column gives the average "cosmic" abundances according to the compilation by SUESS and UREY [358]. The third column gives the average abundances in the young stars τ Scorpii and 10 Lacertae (for the elements C, N, O, Ne, and Si, the values are means of the determinations by TRAVING and ALLER, listed in [62]; the relative H/He abundance is that determined by TRAVING). The fourth column gives preliminary abundances in the hydrogen-exhausted star *HD* 160641 [391]. The fifth column gives preliminary relative abundances *by number* of the elements heavier than helium in *HZ* 44; the ratio H/He and the ratio of the sum of C, N, O, Ne, and Si to (H + He) are being studied by MÜNCH at the present time [392].

Element	Percentage abundance by mass			Relative abundance by number, referred to nitrogen as unity	Element	Percentage abundance by mass			Relative abundance by number, referred to nitrogen as unity
	SUESS and UREY	Young stars	HD 160641	HZ 44		SUESS and UREY	Young stars	HD 160641	HZ 44
H	75.5	58.6	0	—	O	0.65	0.48	0.75	1/16
He	23.3	39.8	95.4	—	Ne	0.32	0.73	3.00	1/5
C	0.080	0.10	0.31	1/200	Si	0.053	0.08	0.065	1/20
N	0.17	0.17	0.47	1					

v Sgr [395] may indicate the occurrence of the α-process also. It may be significant that both v Sgr and HD 30353 are spectroscopic binaries, so that mass loss and mass exchange may have played a part in stripping off their envelopes. If the high luminosities are correct, these stars can last only a short time in their present state (helium-burning yields only 10% of the energy per gram given by hydrogen-burning).

(γ) The classical Wolf-Rayet stars lie in the very high-temperature region, slightly below the normal main sequence. They belong to the extreme spiral-arm population I (see Sect. 29). The W-R stars of both the carbon and nitrogen sequences have been discussed as examples of a late evolutionary stage in massive population I stars [142], [396]. They are probably deficient in hydrogen and rich in helium [397]. In the *WC* stars, nitrogen is not seen at all, and they may be the result of helium-burning in stars which had exhausted almost all their hydrogen and had no hydrogen-burning zones. The *WN* stars have a ratio of nitrogen to carbon of about 20 [398]. Probably the results of helium-burning have passed out through a hydrogen-burning zone at about 40×10^6 degrees, according to the rates given in Part I of this chapter. The connection between W-R stars and the CN-cycle was first discussed in 1943 [399] and has been reviewed recently [400], [62].

The central stars of planetary nebulae, which belong to population II, have positions in the H-R diagram above the left-hand side of the horizontal branch. Their masses are about 1 $M_\odot$ [401]. These central stars may be of W-R type and are often apparently deficient in hydrogen, although the planetary nebulae themselves appear to have normal compositions [402]. The lack of a clearcut separation into *WC* and *WN* type suggests that in these stars the balance between the exhaustion of hydrogen, the attainment of conditions for the occurrence of helium-burning, and the onset of mixing, are not so critically defined as in the massive population I W-R stars.

(δ) The normal carbon stars are red giants, and from their spatial distribution they are population I objects. They have a C/O ratio several times greater than normal [403]. It is not certain to what extent this may be due to the destruction of oxygen in hydrogen-burning or to the formation of carbon in helium-burning, owing to the fact that the easily-formed CO molecule is spectroscopically

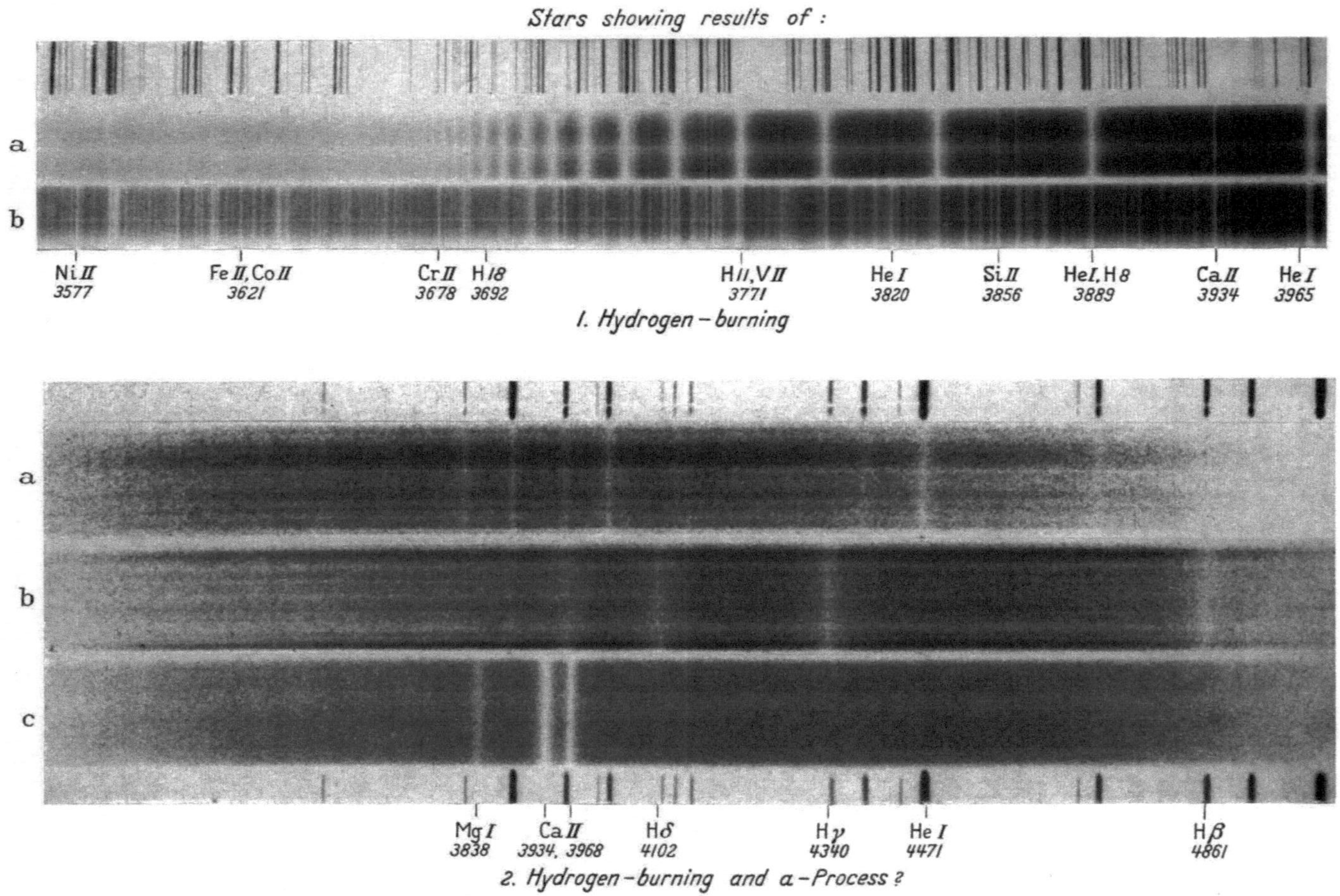

Fig. 29. Portions of the spectra of stars showing the results of hydrogen-burning and possibly the α-process, reproduced from [62]. Upper: (a) Normal *A*-type star, η Leonis, showing strong Balmer lines of hydrogen and a strong Balmer discontinuity at the series limit. (b) Peculiar star, υ Sagittarii, in which hydrogen has a much smaller abundance than normal. Lower: (a) White dwarf, L 1573−31, in which hydrogen is apparently absent. The comparison spectrum above the star is of a helium discharge tube; note the lines of helium in the star's spectrum. (b) White dwarf, L 770−3, which shows broad lines due to hydrogen only, for comparison with (a) and (c). (c) White dwarf, Ross 640, which shows only the two lines due to Ca II and a feature due to Mg I. All the spectrograms in this plate were obtained by J. L. Greenstein; the upper two are McDonald Observatory plates and the lower three are Palomar Observatory plates.

inaccessible. The occurrence of C^{13} in normal carbon stars in amounts that are near to the equilibrium ratio produced in the CN-cycle, i.e., $C^{12}/C^{13}=4.6$ [*404*], shows that the surface material has been through hydrogen-burning. It may be core material enriched in C^{12} by helium-burning, and then modified by a hydrogen-burning zone on its way to the surface. There should be a high abundance of nitrogen also. Certain *G* and *K* giants called the "4150 stars" [*265*] have strong CN bands in their spectra and may be related to the carbon stars, but these tend to be high-velocity stars and hence of population II type.

Some carbon stars of a peculiar type have weak or absent hydrogen lines and CH bands, and are presumably deficient in hydrogen. At the same time, these stars show no evidence for C^{13} so that the C^{12}/C^{13} ratio may be as large as the terrestrial one of 90. In these stars the C^{12} produced in the core may not have been mixed out to the surface until almost all the hydrogen had been exhausted and no hydrogen-burning zone was present (they may be like the *WC* stars in this respect). Examples of this type of star are R CrB [*405*] and HD 137613 [*406*], [*407*]. These stars should have a different abundance of nitrogen from normal carbon stars. Helium should be in high abundance, and a tentative identification of this element in R CrB has been suggested [*408*]. R CrB is of earlier spectral type than the normal carbon stars, and it may have a very high luminosity. Spectra of different kinds of carbon stars are shown in Fig. 28.

There is a different group of peculiar carbon stars whose high velocities place them in the extreme halo population II. These stars have excessively strong CH bands [*409*], indicating a high abundance of both carbon and hydrogen, and weakened metallic lines, in accordance with the aging effect. But some of them show no evidence for C^{13} [*404*], [*410*]. The absence of a product of hydrogen-burning cannot in this case be due to exhaustion of hydrogen; it may, therefore, be due to the structure and initial composition of the star and the temperatures of regions where different reactions occur. Perhaps the *pp*-chain has always dominated over the CN-cycle.

90. Evidence for element synthesis in individual stars: the α-process. Evidence for the α-process is very much less certain than that for hydrogen-burning and helium-burning. The possibility of its occurrence in v Sgr was mentioned in Sect. 89β. Two other possible examples are both white dwarfs. Ross 640 [*411*] has strong features attributed to the α-process elements magnesium and calcium (see Fig. 29). Although analyses of white dwarf atmospheres, with their very large surface gravities, have not been carried out, it seems hard to explain such a spectrum except by actual abundance peculiarities.

One peculiar and perhaps unique white dwarf, AC $+70°$ 8247, has no recognizable spectral features except very broad absorptions at $\lambda\lambda$ 3910, 4135, and 4470. It has been suggested [*412*] that these may be due to magnesium and silicon (both α-process elements). However, it has also been suggested that they may be due to helium under extreme pressure conditions [*413*], and the matter is not settled.

91. Evidence for element synthesis in individual stars: the s-process. The *s*-process can probably occur in red giants if suitable conditions for the liberation of neutrons are present [*360*], [*351*], [*352*], [*62*]. There are three groups of giant stars which have abundance anomalies that indicate the occurrence of the *s*-process in their interiors. These are the *S* stars, the BaII stars, and the carbon stars.

α) The *S* stars belong to the extreme spiral arm population I (see Sect. 60). Most of them probably have luminosities in the range of normal giants [*277*], as also is the case for two found in the Eta Carinae nebulosity [*278*]. However, one which lies near the young cluster NGC 6530 and is probably connected with it [*414*] would then have a magnitude between +1.5 and +2.4. A star in the same cluster which may be of *M–S* type would be of magnitude −0.1 to −1.0.

Elements with neutron magic-numbers appear more predominantly in the spectra of *S* stars than in *M* stars of the same temperature. It has been realized for some time that this must be a real atmospheric abundance effect, and not an effect due to the conditions of ionization, dissociation, or excitation. The *S* stars also probably have carbon abundances intermediate between the *M* and the *S* stars. Numerious references dealing with these spectroscopic features are listed in [*62*]. The discovery of the unstable element technetium in *S* stars [*415*], which is produced by neutron-capture processes, shows that nuclear activity and mixing to the surface must be occurring on a time scale not much longer than 2×10^5 years (the half-life of the longest-lived isotope of technetium). Portions of the spectra of the *S* star, R And, compared with a standard *M* star, are shown in Fig. 30.

The occurrence of an *S* star in NGC 6530, if it is confirmed that it is a cluster member, emphasizes that, although the *S* stars are apparently genetically old, having reached a fairly advanced stage of evolution, they are young in years. They must begin life as massive stars, and live through their span on the main sequence in a short time. Possibly they undergo considerable mass loss while they are *S* stars; their red giant structure and the frequent occurrence of light variability among them would seem likely to facilitate this.

Actual abundance determinations in *S* stars have been made only for a few elements so far. The elements which are probably overabundant are Sr, Y, Zr, Nb, Ru, Ba, La, Ce, Pr, Nd, and Sm, of which all but Ru and Sm have a magic neutron-number for their most abundant isotope.

β) The Ba II stars are *G* and *K* giants showing spectral peculiarities that link them closely with the *S* stars, and are easier to analyze spectrophotometrically, owing to their higher temperatures. A study of one such star, HD 46407 [*416*] has enabled a quantitative comparison to be made between those elements that are found to be overabundant and the predictions of the theory of the *s*-process discussed in Sect. 80. The agreement is good; all the elements with atomic weight greater than 75 whose principal isotopes are thought to be produced by the *s*-process and which are spectroscopically observable are found to be overabundant, with the possible exception of Yb, which needs further checking, both observationally and theoretically. These elements comprise Sr, Y, Zr, Nb, Mo, Ru, Ba, La, Ce, Pr, Nd, Sm, and W.

A few rare examples of the Ba II stars are found to have very high velocities, which place them as members of the extreme halo population II. HD 26 is an example. This star has weak metallic lines, in agreement with the aging effect.

As yet, technetium has not been identified in any Ba II stars.

γ) The normal carbon stars have apparent overabundances of those heavy elements which would be produced by the *s*-process, although these are not so striking as in the *S* stars [*417*], [*400*]. Technetium has been observed in 19 Piscium. It is hoped that eventually more abundance determinations in Ba II stars of both population groups, and in both normal and peculiar carbon stars, will be helpful in elucidating the evolutionary differences between *M*, *C*, *S*, and Ba II stars.

92. Supernovae and the *r*-process. Observational evidence for the occurrence of the *r*-process rests upon the shape of the light curves of supernovae of Type I

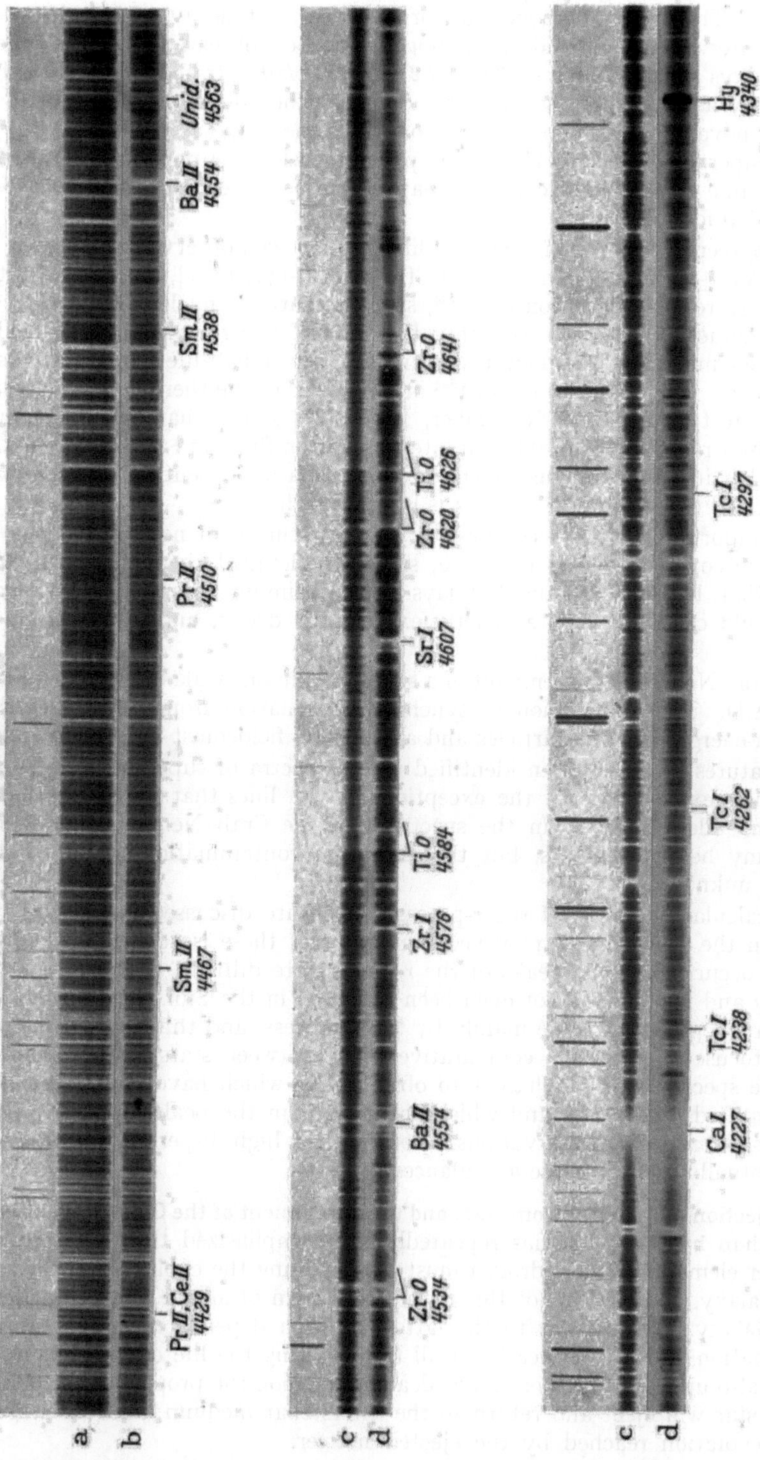

Fig. 30. Portions of the spectra of stars showing the results of the s-process, reproduced from [62]. Upper: (a) Normal G-type star, $\varkappa$ Geminorum. (b) Ba II star, HD 46407, showing the strengthening of the lines due to the s-process elements barium and some rare earths. Middle: (c) M-type star, 56 Leonis, showing TiO bands at $\lambda\lambda$ 4584 and 4626. (d) S-type star, R Andromedae, showing ZrO bands which replace the TiO bands. Lines due to Sr I, Zr I, and Ba II are all strengthened. Lower: (c) Another spectral region of the M-type star, 56 Leonis; note that Tc I lines are weak or absent. (d) R Andromedae; note the strong lines of Tc I. The spectrum of R Andromedae was obtained by P. W. Merrill, and the upper two spectra by E. M. and G. R. Burbidge (Mount Wilson plates).

(which appear to belong to population II), in which this process is suggested to take place. The strictly exponential form of the light decay, setting in 50 to 100 days after maximum and continuing in the best-observed case until more than 600 days after maximum [*353*], [*62*], [*419*], [*420*], strongly suggests radioactive decay as the energy source, provided that the bolometric light curves of such supernovae, so far unobtained, follow the same shape as the observed light curves. Supernovae of type II (which appear to belong to population I) do not have the same form of light curve, and appear to be somewhat different in their physical characteristics.

The agreement between the half-life of the well-observed supernova in IC 4182 ($55^d \pm 1^d$) and the half-life of Cf^{254} ($56\overset{d}{.}2 \pm 0\overset{d}{.}7$) [*421*], together with the large energy release by fission of Cf^{254}, suggest that this nucleus is responsible. Cf^{254} is unique in that it is the only beta-stable, neutron-rich nucleus which decays predominantly by spontaneous fission with a half-life between 10 days and 10^4 years. The observation of Cf^{254} in the debris of the thermonuclear hydrogen bomb test at Bikini in November, 1952 [*422*] shows that it will certainly be made by irradiation of U^{238} by an intense neutron flux, and we infer by extrapolation that it will be among the end products of a rapid neutron-capture chain beginning at Fe^{56}, if the neutron flux is large enough.

It is important that observations of the light curves of new supernovae, as they are discovered, should be made, since the detailed theory given in [*62*] suggests that, between 400 and 500 days after maximum, the radioactive energy input should cease to show a strictly exponential decay, and should begin to level off.

The Crab Nebula is the remnant of a supernova in our Galaxy which exploded in 1054 A.D. The observation of synchrotron radiation from this shows that very high-energy charged particles and a magnetic field must be present today.

No features have yet been identified in the spectra of supernovae of Type I during their explosion, with the exception of [OI] lines that appear 200^d after maximum. Identifications in the spectrum of the Crab Nebula itself do not include any heavy elements, but the extent of contamination by interstellar matter is unknown.

The calculated details of the *r*-process which are discussed in Sect. 81 rest only upon the solar-system relative abundances of these isotopes. Most of the elements occurring in the peaks of the *r*-process are difficult to detect spectroscopically and many have not even been observed in the Sun. Both isotopes of europium are probably made mainly by the *r*-process, and this may be the best element to use in making a comparative check between stars. Other elements which are spectroscopically harder to observe but which have almost certainly been identified in the Sun and which occur right in the peaks of the *r*-process are Os, Ir, Pt, and Au. Observations in cool stars on high-dispersion spectrograms may eventually yield relative abundances.

93. Ejection of material from stars and the enrichment of the Galaxy in elements heavier than hydrogen. It has repeatedly been emphasized that the synthesis of heavier elements from hydrogen must occur during the evolution of the stars in our Galaxy, whether or not this is the sole origin of all the heavier elements in the Galaxy. The answer to the latter question depends upon the rates of star formation and of star deaths at all times during the life of the Galaxy. It depends also upon the manner of star death, i.e., upon the proportion of its mass which a star will eject and return to the interstellar medium, and the stage of nuclear evolution reached by the ejected matter.

Some estimates of the present-day rate of ejection of matter by stars can be made (cf. [*62*] for a detailed discussion). Two kinds of ejection must be considered, steady and explosive.

α) Steady ejection. Ejection by red giants and supergiants, close binary systems, P Cygni and Wolf-Rayet stars, and planetary nebulae all contribute and the rates of ejection by these various kinds of stars were discussed in Chap. C II. DEUTSCH [*422a*] has estimated the total ejection by red giants and supergiants. Another way of estimating the amount of material which has been through a stellar form is by means of SALPETER'S mass function for solar-neighborhood stars (see Sect. 54). Integration shows that during the past 6×10^9 years, if star formation has been at a constant rate, 45% of the mass of the Galaxy has been condensed into stars which have by now gone through the whole of their evolutionary path. Apart from the fraction of this mass which is now in the form of white dwarfs, and that which came from the surface, unevolved, layers of stars, the rest represents material which has been processed by stars and enriched in heavier elements, and has been available for recondensation into stars.

Thus a proportion of the order of 10% of the mass of the Galaxy may have been processed by stars, taking the average rate of star formation and evolution given by SALPETER'S function. A small proportion of the material will have been ejected explosively; that part which has been ejected non-explosively may have been enriched in the products of hydrogen-burning, helium-burning, the *s*-process, and perhaps the *α*-process.

β) Explosive ejection. The rates of supernovae are thought to be 1 per 300 years for Type I and 1 per 50 years for Type II, per galaxy, as estimated by BAADE, MINKOWSKI, and ZWICKY. In 6×10^9 years this would lead to a total number of 2×10^7 and 1.2×10^8, respectively. The average mass ejection may be $\sim 10^{33}$ g; this depends on the original mass, which may be different for Types I and II, and on whether a white dwarf remnant is left afterwards. The ejected material (having a total mass of 2×10^{40} and 1.2×10^{41} g, respectively) may contain the results of the *e*-, *r*-, and *p*-processes, as well as other elements produced in the processes having longer time-scales.

The rate of novae in M 31 [*290*] is 26 ± 4 per year. If the rate is similar in our Galaxy, there will have been 1.6×10^{11} nova outbursts in 6×10^9 years. If 10^{28} to 10^{30} g is ejected in each outburst, then the total is of the same order of magnitude as that ejected by supernovae. It will probably contain results of hydrogen-burning, helium-burning, and perhaps the *s*-process and the *α*-process.

Thus it is not hard to account for the estimated total abundances of most of the heavier elements in the Galaxy, which are probably a factor 2 less, on the average, than in the Sun [*62*]. There is a problem, as we have seen, in accounting for the large amount of helium, if the abundance throughout the Galaxy is $\sim 10\%$ by number (see Sect. 85). However, since helium is not observable in the cool interstellar gas or in stars cooler than type *A* 0, the actual spatial abundance, even in the solar neighborhood, is very uncertain. Another problem is that a large proportion of our Galaxy is composed of stars which may be $4-6 \times 10^9$ years old. If the theory that all of the elements have been produced in stars is correct, then there are three possibilities. First, star formation and evolution may have been much more rapid at an early epoch in the history of our Galaxy. An evolutionary scheme for galaxies (see Sect. 68) would suggest that this has been the case. Second, our Galaxy may not originally have consisted of pure hydrogen, but may have condensed from gas which had already been processed by stars in other, earlier, galaxies. This would be expected in a steady-state cosmology. Third, the age of our Galaxy may be greater than 6×10^9 years (perhaps 10^{10} years).

E. Evolutionary aspects of stellar rotation, variability, and magnetism.

I. Rotation of single stars: Discussion of observations.

The problem of excess angular momentum in newly-forming stars was discussed in Sect. 9. The observed rotations of main-sequence stars of different masses and ages might be expected to contain clues to the conditions of stellar formation. Further, in so far as stars, once formed, are isolated systems which conserve angular momentum, the changing radii of stars as they evolve off the main sequence should lead to changes in the rotations. The distribution of rotations among giant stars should then be related to the distribution on the main sequence.

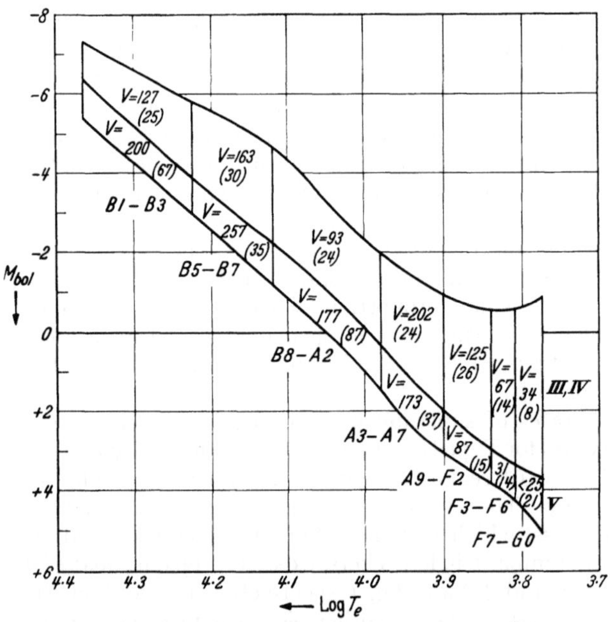

Fig. 31. The variation of stellar axial rotation across the M_{bol} versus Log T_e diagram, as determined from measured rotational velocities of 427 normal $B1-G0$ stars by SLETTEBAK [424]. The values of the mean true rotational velocities, v, are listed for each group of stars, and the numbers of stars discussed in each group are given in parentheses.

94. Rotations of main-sequence stars. The rotations were first observed in spectroscopic-eclipsing binaries. The pioneers in studying rotations of single stars from the broadening of spectral lines were STRUVE, SHAJN, ELVEY, and CARROLL. The correlation of size of rotation with spectral type [423] showed that there is a rather sharp drop at spectral type $F5$; large rotations are found in earlier but not in later types. The phenomenon of stellar rotation and its cosmogonic implications were discussed by STRUVE [331].

In recent years, standardized measures of rotations together with accurate spectral types have been accumulated, mainly through the work of SLETTEBAK, BÖHM, HUANG, HERBIG and SPALDING [424] to [427]. The distribution of the average values of the rotational velocities over the spectral range $B1$ to $G0$ and luminosity classes III, IV, and V is shown in Fig. 31 [424]. The mean of the observed quantity, $v \sin i$, where v is the equatorial velocity and i is the inclination of the axis, has been multiplied by $4/\pi$ to correct for the random spatial distribution of axes and thus to yield the true mean equatorial velocity[1].

The conclusions from these compilations are as follows:

(i) Maximum velocities occur at types $B5$ to $B7$; velocities decrease in both directions along the main sequence from this point.

(ii) Main-sequence stars later than $F5$ have small rotations.

(iii) There is no correlation between size of rotations and whether stars have spectra belonging to the strong- or weak-lined class.

[1] S. CHANDRASEKHAR and G. MÜNCH: Astrophys. J. **111**, 142 (1950).

(iv) There is no correlation between galactic latitude and size of rotation.

(v) Certain regions of the sky are found to contain higher than average rotations, i.e., the Pleiades cluster [*428*].

(vi) Rotations in the Scorpio-Centaurus cluster are correlated with space motions within the cluster [*429*].

(vii) In the general field there is no correlation between the magnitudes of radial velocity and rotational velocity, except for a possible sudden drop in rotational velocities for radial velocities >40 km/sec.

There is no satisfactory explanation for (i) and (ii), yet they suggest a correlation with mass and/or age of stars. The result (vii) should be considered in connection with the discussion of space motions, spectral types, and population types in the Galaxy, in Sect. 59. Both (iii) and (vii) imply that no well-defined correlation with population type is present, although the possible sudden drop in rotations for velocities >40 km/sec may be connected with the correlation between space motions and spectral type, as well as with the observed fact that very high-velocity stars have negligible rotations. The result (iv) implies that galactic rotation was not the prime factor governing the actual rotations acquired by stars. The results (v) and (vi) have been taken as lending support to the hypothesis that rotations are governed by the initial turbulent eddies in a condensing medium and interactions between them [*429*].

95. Rotations of giants. Both SLETTEBAK [*424*] and HERBIG and SPALDING [*427*] found that in the range $F0$ to $G0$, stars of luminosity classes III and IV have higher rotations than main-sequence stars. The considerable rotational velocities of giants of types A, F, and G have been noted and explained as follows [*430*]. Suppose that a star has radius R_0 when it is on the main sequence and a larger radius R at a later evolutionary stage. Suppose further that a star follows the Sandage-Schwarzschild evolutionary tracks during its evolution into a giant stage. It will then move approximately horizontally in the H-R diagram, and its moment of inertia at any stage can be calculated. As the star expands angular momentum is conserved and the value of $v \sin i$ as a function of R/R_0 can be calculated on either of two extreme hypotheses: that the star rotates as a rigid body, or that angular momentum is conserved in each thin spherical shell. The observed values of $v \sin i$ in giant stars, when plotted against R/R_0, were found to be consistent with the stars having evolved from an initial position fairly high on the main sequence, with the initial maximum $v \sin i = 225$ km/sec.

This test was repeated with more extensive observational material [*431*]. No stars of luminosity classes III and IV were found to be rotating faster than would be expected if they rotated as rigid bodies and had evolved along Sandage-Schwarzschild tracks. Further, initial main-sequence values of $v \sin i$ were computed for each star and the recovered distribution function was plotted and compared with the observed main-sequence distribution function. For stars in the ranges $A0$ to $A3$, $A4$ to $F0$, and $F1$ to $G0$ the agreement was good if the stars rotate as rigid bodies. However, the recovered function for classes $B0$ to $B9$ showed more small values of $v \sin i$ than are observed in the main-sequence function. Part of this effect may be due to observational limitations. The remainder, if significant, might be due to a change in the velocity distribution on the main sequence during the last 10^7 to 10^8 years (which seems unlikely) or to loss of mass by some early-type stars. It may be significant that the *Be* stars, discussed in Sect. 72 as examples of stars undergoing mass loss, have their greatest frequency in the middle of this spectral range, at type $B3$. Alternatively, the fastest-rotating early-type stars might not evolve into giants, and this might be the cause of the discrepancy.

Finally, study of the rotations of supergiants may be complicated by macroscopic eddies and currents in the stars' atmospheres[1,2]. An investigation is being made by ABT at Yerkes Observatory.

II. Stellar variability.

96. General. The phenomenon of intrinsic light and spectrum variation in stars indicates that the balancing conditions which hold for stars in statical equilibrium are not fully operating. For instance, in the case of pulsating variables there is departure, though slight, from hydrostatic equilibrium, and the energy output is not constant. Variability may be a phase through which stars pass, or it may be characteristic of a star right through its lifetime; such a star might evolve from one type of variability to another. In the light of all the foregoing discussion and the observational evidence for the locations of different types of variables in the H-R diagram, the former seems more probable. In that case, one might perhaps expect that variation would be characteristic of evolutionary phases when a star's structure is readjusting fairly rapidly, in a time short as compared to its total lifetime. This is not necessarily so in all cases, but it does seem likely for certain classes of variables.

A detailed discussion of stellar variability is given in this volume, by P. LEDOUX and T. WALRAVEN. As with so many other topics that bear on the problem of stellar evolution, we shall be concerned here only with those aspects that may be connected with evolution.

In considering in what stages of evolution and in what mass ranges the variables lie, their positions in the H-R diagram are of great importance. In particular the H-R diagrams of clusters in which there are variables are valuable, because in this case the evolving stars occur in a narrow mass range that can be determined, and counts of the number of stars in different regions may give an idea of the time spent by stars in different stages. The following sections will divide variable stars according to their approximate location in the H-R diagram. The treatment will not be exhaustive, but merely illustrative.

97. Variable stars in the gravitationally contracting phase. The T Tauri stars, which were discussed in Sect. 35, are irregular variables, probably stil gravitationally contracting, not yet having reached the main sequence. A number of proposals have been made to explain the irregular bursts of light accompanied by emission lines in the spectra. One is that this phenomenon is due to the release of energy through the infall of matter [*432*], [*433*], although it has been argued [*434*] that sufficient energy cannot be obtained in this way. Another proposal is that electromagnetic energy, gained through the condensation of an interstellar magnetic field, is released as matter falls into the star [*435*].

A third proposal is that the phenomenon is due to the release of nuclear energy by "pre-stellar" matter, i.e., matter in a hitherto unknown state out of which the stars condense [*436*]. All these suggestions would confine the seat of the disturbances to the outer layers of the star. AMBARTZUMIAN [*436*] pointed out that only energy release in or near the surface could result in such rapid variations; more deep-seated energy release would give slower light variations.

The consumption in the core of any deuterium, lithium, beryllium, or boron present in the newly-forming star would occur just before the star reached the main sequence but it would be unlikely to give rise to irregular, short-lived variations.

---

[1] O. STRUVE: Astrophys. J. **107**, 327 (1948).
[2] S. S. HUANG and O. STRUVE: Astrophys. J. **118**, 463 (1953).

The irregular variables without emission lines, which occur in obscured regions together with T Tauri stars, should be mentioned here. At present the evolutionary relationship of these to the T Tauri stars is not known, but their occurrence in the same regions indicates that they, too, are young stars.

It may be noted that flaring outbursts in T Tauri stars provide an additional way in which a gravitationally contracting star can lose energy, apart from its normal steady radiation. Thus such outbursts may help to shorten the time necessary for gravitational contraction, which is determined by the rate at which a star can lose energy by radiation. A more rapid mechanism of gravitational energy dissipation may help to resolve the difficulty associated with the positions in the H-R diagram of some of the low-mass stars in young clusters (see Sect. 43).

98. Variable stars on the main sequence. For almost the whole length of the main sequence, stars are constant in light, if we exclude such small-scale variations as the Sun's 22-year magnetic cycle. Two categories of stars do, however, show variations.

α) Be stars. The *Be* stars show non-periodic spectrum variations and often small-scale light variations. They were discussed in Sect. 72; they probably eject matter as a consequence of their fast rotation, and formation and dissipation of outer extended atmospheres probably accounts for the observed variations.

β) Dwarf M flare stars (UV Ceti stars). These stars appear to lie on the main sequence, and the basic cause of their sporadic variation is uncertain at present. They have been linked to the T Tauri stars and the same mechanism for both has been suggested [*436*], [*437*], [*192*].

It has been proposed [*438*] that the outbursts themselves may be flares which need cover only a small fraction of the star's surface to account for the large change in photographic light intensity, owing to the low intrinsic luminosities of these stars. If the flares are analogous to solar flares, then magnetic fields must be involved. It has been further suggested [*439*] that the phenomenon may be connected with the deep convective zones which have been shown to be present in the red dwarfs (see Sect. 16).

99. Variable stars just above the main sequence. *α) Magnetic stars of type A and F.* The stars of spectral classes *A* and *F* with magnetic fields [*440*] tend to lie slightly above the main sequence. All of the well-observed magnetic stars have been found to have variable fields, some regularly and some irregularly varying. Many are periodic spectrum variables, and many show small-period light variations. At this spectral class on the main sequence, stars have convective zones that set in right at the photosphere but extend only a short way into the star (see the discussion in [*363*]). A disturbance in the depth of the convective zone while a star in this region is just beginning to move slightly off the main sequence might be connected with the presence of strong magnetic fields. On the other hand, the presence of a magnetic field with consequent continual "hot spot" activity [*363*] over the star's surface might cause such a star to lie slightly above the main sequence.

β) β Canis Majoris or β Cephei stars. These stars are confined to a band in the H-R diagram extending from *B*1 II–III to *B*2 IV, shown in Fig. 32 reproduced from the papers by Struve [*441*]. The stars are pulsating (they show radial velocity variations as well as light variations). Two nearly commensurate periods of the order of a few hours, resulting in a long beat period, are present in some cases. Ledoux [*442*] has suggested that non-radial oscillations are occurring.

For the ten or so well-observed stars in this group, there is a tight correlation between color and period, indicated in Fig. 32. STRUVE [441] has suggested that this class of star may be one in which the time scale for evolutionary changes, in the form of changes in the period of variation, may be rapid enough to be observed. Main-sequence stars in the upper B region can have lifetimes of only a few times 10^7 years. Of the seven stars for which enough observations are available, four have increasing periods, to the extent of about 1 sec per century, and three are essentially constant. The increasing period, taken in conjunction with the color-period relation shown in the figure, suggest that the band in the H-R diagram occupied by these stars may actually represent an

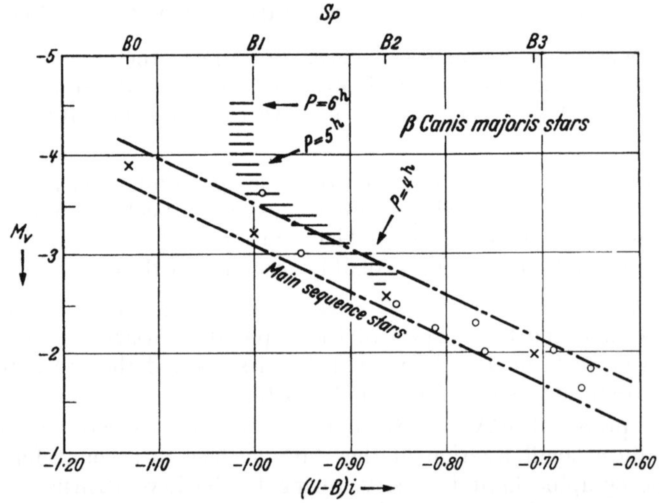

Fig. 32. Hertzsprung-Russell diagram showing the sequence of the β Canis Majoris or β Cephei stars, according to STRUVE [441]. Spectral type is shown at the top, and intrinsic $U-B$ color at the bottom, the open circles and crosses represent individual normal stars belonging to the main sequence. The positions of stars with periods of different length are indicated.

evolutionary sequence, going from just above the main sequence, upwards and diverging from the main sequence. For a star at the lower end of the sequence, with a period of two and a half hours, to reach the upper end, with a period of four hours, would take $\sim 10^6$ years if the period increases by 1 sec per century.

Struve pointed out that these stars should in this case have similar masses, while if the β CMa sequence represents a region of instability through which stars pass as they move off the main sequence, then a ratio of masses of 2:1 from the upper to the lower end might be expected. Assuming the relation $P\sqrt{\varrho}=$ const to hold, then if P_1, P_2, M_1, M_2 are the periods and masses at the upper and lower ends of the band, respectively, ratios of $P_1/P_2=1.8$ and 1.5, respectively, would be predicted for $M_1=M_2$ and $M_1/M_2=2$. The observed value is 1.7, and therefore the result is inconclusive.

Since this work was done, the theoretical evolutionary tracks for massive stars (Sect. 27) have been computed and tend to favor the second possibility, that stars move across and not along the β CMa region (compare Figs. 10 and 32).

100. Giant and supergiant variable stars. Most classes of variable stars fall into this category, and we consider the following subdivisions.

α) *Blue supergiants.* P Cygni represents a class of variable blue supergiants. Its present-day variations are small and irregular, but some 350 years ago the star brightened considerably and was called a nova. Its present-day absolute luminosity is very high ($M_v = -8.3$ has been assigned[1], although some workers have not set it quite as high). We have already pointed out (Sect. 72) that this star is evolving faster by mass loss than by radiation. It is probably passing through a short-lived evolutionary stage.

The bright blue variables discovered in M 31 and M 33 [*443*] also fall in this category. Their mean absolute photographic magnitude is -8.4 with small scatter, their color indices are blue, and their spectra are of *F*-type supergiants with *P* Cygni characteristics. Similar stars in the Large Magellanic Cloud, in addition to S Doradus, have also been studied [*444*].

Variations in radial velocity, probably due to pulsation, have been shown to be very common in *A*- and *F*-type I a supergiants [*445*]. All the stars in the sample investigated here proved to be variable.

Instability in such high-luminosity stars is not surprising, in view of the very large radii, the turbulence in the atmospheres, and the necessarily short lifetimes of such stars in this evolutionary stage.

β) Classical cepheids. These, perhaps the most studied of the periodic variables, form a band lying in the Hertzsprung gap, such that the coolest and brightest have the longest periods. Of all the variables, the cepheids probably come closest to pure radial pulsation.

The cepheids belong to a flat subsystem of the Galaxy, but do not delineate spiral structure in the way that *OB* stars and H II regions do so. They are apparently associated in space with the later *B*-type stars (Sect. 60), and probably represent a later stage of evolution of these. It is not known under what conditions *B* stars may evolve into cepheids. Nor is it known whether such stars evolve first into a normal red giant or supergiant configuration and then become cepheids, perhaps after losing some mass, or whether they become cepheids on their way to or instead of becoming red giants or supergiants.

Work is under way at present by several people on color-magnitude diagrams of clusters in which cepheids have recently been discovered. This may show where the cepheids originated on the main sequence. It is hoped that it will eventually show whether or not there is a relation between the period of cepheids and the luminosity of the brightest stars remaining on the main sequence of clusters. If a cepheid stage is subsequent to a red giant or supergiant stage, it may be governed by the amount of mass loss occurring then, and one might not expect a definite relation such as the one mentioned above.

Preliminary work on the clusters NGC 6664, NGC 6087, NGC 7788, and M 25, which contain one cepheid each, of periods 3.09, 9.75, 4.87, and 6.74 days, respectively, indicates that the ages of the latter three clusters, derived from the extent of the main sequence, according to SANDAGE's [*200*] dating method [see Eq. (40.1)], are about 1×10^8 years, and are not significantly different from each other. NGC 6664, containing the 3.09-day cepheid, is probably about 2×10^8 years old on the same dating system [*446*].

The present status of the pulsation theory of cepheids has been described by LEDOUX and WALRAVEN in this volume. Following earlier work of EDDINGTON, COWLING, LEDOUX and others, COX [*447*] has considered the pulsational stability of models which represent stars in the red giant stage of evolution. He has shown that the very high degree of central condensation in these stars makes them excessively stable to pulsations in which the whole star is involved. On the other hand, only a very small fraction of the normal energy supply would be necessary to maintain the pulsation. He has concluded that the region where the pulsations

[1] N. G. ROMAN: Astrophys. J. **114**, 492 (1951).

are generated and maintained must be sought in the outer 15% of the stellar radius, a region in which there is probably an extensive convective layer that must act as a moderator of the total energy output. This is essentially the same zone which EDDINGTON [*448*] deduced must be responsible for the well-known phase lag effect. The importance of the He II ionization zone in the outer layers of these stars has been pointed out by ZHEVAKIN [*449*] and COX and WHITNEY [*450*].

γ) W Virginis and RV Tauri stars. These variables, often called the population II cepheids, occur in globular clusters and hence belong to the halo population, although the velocities of some field RV Tauri stars suggest that they may occur in both spherical and intermediate subsystems. We group the two designations together, as suggested by PAYNE-GAPOSCHKIN [*258*]. Both show complications superimposed on a simple radial pulsation, in the form of discontinuous radial velocity curves and emission lines at some phases (see [*385*] and references given in Chap. 2 of [*258*]).

In the H-R diagram, these stars lie in the Hertzsprung gap outside the thickly populated principal sequences in globular clusters, and form a sparsely-populated sequence of their own. Their location may be seen in the composite diagram for seven globular clusters [*135*], reproduced in Figs. 32, 38, 39 of the article in this volume by H. C. ARP).

The most luminous members of this group are more luminous than the brightest red giants and considerably bluer in color. The sequence extends downwards in the direction of the RR Lyrae variable star location and indeed, if extrapolated, would pass through it.

ARP [*135*] suggested that these variables might possibly bear the same evolutionary relationship to the sparse scattering of stars above the horizontal branch and to the left of the giant branch as the more numerous RR Lyrae variables bear to the horizontal branch sequence of stars. That a certain combination of surface temperature and pressure defines the region of the pulsating variables is strongly indicated, but we do not know whether evolving stars pass across or along the variable region. Also, this may not be the whole story in the case of the W Virginis stars. A certain proportion of the scattered non-variable stars above the horizontal branch and to the left of the giant branch may be expected to be field stars, as ARP pointed out, and for stars brighter than $M_v = -2$ in this region the ratio of the density of variable to non-variable cluster stars appears to be greater than the corresponding ratio of RR Lyrae stars to non-variable horizontal branch stars. Their very existence in this region may be due to the same structural causes that give rise to the variability. More statistics are needed before conclusions can be drawn, and the cluster membership of all stars falling in this region needs to be checked. For example, spectra of four of the brighter non-variable stars have been obtained by WALLERSTEIN[1]. The velocities disagree with the cluster velocities and the spectral types are those of normal dwarfs, establishing that these stars are not cluster members.

δ) Long-period, semi-regular, and irregular variables. We consider these groups together under one designation, red giant variables (which includes both giants and supergiants). The stars occur in a wide range of population subsystems, the most luminous stars probably belonging to the flattest. Some are found in globular clusters; a division of the long-period variables can be made at periods of about 250 days, with those having shorter periods than this belonging to a more spherical population than those with longer periods. The significance of the division is

[1] G. WALLERSTEIN: Astrophys. Journ. **128**, July (1958).

not known; from the discussion in Chaps. B and D it may be one of mass or chemical composition.

In the H-R diagram these variables tend to occur at the right-hand side of the red giant and supergiant region. STEBBINS and HUFFER [*451*] showed that the majority of the *M*-type stars are in fact appreciably variable, the tendency being most marked for stars of high luminosity, or advanced spectral class, or both. WALKER [*140*] found that in the globular clusters M 3 and M 92 all the red giants brighter than about $M_v = -2.8$ are variable, with amplitudes that are small and increase with luminosity.

The carbon and *S*-type stars (Sects. 89, 91) lie in this region, and some of them are variable. It is possible, therefore, that variability may be connected with the onset of large-scale convection or mixing which will bring core material, containing the results of nuclear reactions, to the surface of the star [*221*]. Such mixing would at the same time destroy the chemical inhomogeneity that gives rise to the giant structure of these stars. That structural adjustment is ocurring in stars near the right-hand side of the red giant region is suggested by the frequency of variable stars here, and also by the very fact that giant sequences do not extend any further towards the right but come to a well-defined end. It is not known where these stars go if they leave the giant region, but the presumption is strong that they move to the left in the H-R diagram (Chap. A VI).

ε) R Coronae Borealis variables. The prototype of these irregular variables, R CrB itself, was mentioned in Sect. 89 *δ* as an example of an apparently carbon-rich, hydrogen-poor star which may have produced C^{12} in its core by helium-burning. Stars of this class would then be likely to have reached an advanced evolutionary stage. The cause of the light variation is not known; O'KEEFE[1] has suggested variable absorption by clouds of solid carbon particles ejected by the star. Alternatively, the cause might be intrinsic to the star, and connected with the evolutionary stage reached and the possible exhaustion of a nuclear fuel in part of the star. A related variable studied by HERBIG [*451a*], V 348 Sgr, appears to be rich in helium and carbon and to be surrounded by a small nebula containing hydrogen and oxygen (but peculiar excitation effects have not been ruled out as a possible cause of the unusual spectral features). The light variations in V 348 Sgr could be caused by sporadic increases in the surface temperature accompanied by the ejection of a shell of gas. Such stars may be the ancestors of planetary nebulae.

101. Variable stars in the horizontal branch and hot subdwarf regions. *α) RR Lyrae variables.* These characteristic members of the spherical population II were discussed in Sect. 28, in connection with the problem of evolution subsequent to the giant stage and the question of whether or not stars evolve along the horizontal branch and through the RR Lyrae region from one end to the other. The sharply-defined region of surface temperature in which the stars lie strengthens the suggestion that the phenomenon of pulsation has to do with the outer (probably convective) layers of the star. As with the W Virginis and RV Tauri stars, simple pulsation is complicated by motions shown by the discontinuous velocity curves.

The widespread use of these stars as distance indicators has led recently to a fresh assessment of the absolute magnitudes, usually taken as $M_v = 0.0$ for all these stars. The work of PARENAGO [*452*] and PAVLOVSKAYA [*453*] has suggested that $M_v = +0.5$. A possible connection between period and absolute magnitude has been considered by SANDAGE [*200*]. It has been suggested that the RR Lyrae

[1] J. A. O'KEEFE: Astrophys. Journ. **90**, 294 (1939).

stars in globular clusters may have a span of absolute magnitudes, and that this might be due to the stars having a range of ages, the oldest having the shortest periods and the lowest luminosities.

β) δ Scuti variables. Knowledge of a small group of *F*-type variables that have very short periods and small amplitudes, of which δ Scuti is the prototype, has been gradually growing through the work of EGGEN, WALRAVEN, HARLAN SMITH, and others. According to STRUVE [*454*] these stars lie in a band that crosses the main sequence approximately at type *F*, and is nearly at right angles to it. EGGEN [*455*] has suggested that such stars may be the population I analogues of RR Lyrae variables.

γ) Hot subdwarfs. The evolutionary significance of novae, recurrent novae, and explosive variables was considered in Sect. 30. Such catastrophic variables are likely to be stars that are near the ends of their lives. We draw attention again to the small flickering variations, with a time scale of minutes, that are observed in ex-novae, recurrent novae between outbursts, and explosive variables near minimum. Besides the suggestion by CRAWFORD and KRAFT [*148*] that material is being removed from the companion and falling into the high-temperature component of AE Aquarii, pulsation has been suggested in the case of DQ Herculis [*145*a].

III. Magnetic fields and stellar evolution.

102. We shall attempt here to discuss magnetic phenomena as they may affect stars in their lives, from birth to death.

A number of indirect arguments have led to the conclusion that a magnetic field with a mean strength $\sim 10^{-5}$ gauss exists throughout the disk of our Galaxy. Radioastronomical results also suggest that galactic magnetic fields exist in other galaxies and perhaps in intergalactic material as well. Since stars condense out of the interstellar gas and dust, it is to be expected, therefore, that all stars have magnetic fields.

The problem of the condensation of stars in the presence of a magnetic field has already been discussed (Sect. 11). No direct evidence for magnetic fields in stars in the gravitational contraction phase is available, but some of the phenomena in the T Tauri stars may be due to surface magnetic fields. The stars in which BABCOCK has detected large surface magnetic fields ($\sim 10^3$ to 10^4 gauss) comprise predominantly the peculiar *A* and *F* stars. Other types in which he has detected fields [*440*] are a few giant stars, RR Lyrae, and the classical cepheid FF Aquilae. The peculiar *A* and *F* stars lie slightly above the main sequence and it has previously been suggested that this position may be related to the appearance of large surface magnetic fields (Sect. 99α). On the other hand, the Sun also has reached a similar region in its evolutionary track, but only has a surface magnetic field ~ 1 gauss.

The success of the theory of stellar structure, in which magnetic forces have always been neglected, suggests that throughout a star's life on the main sequence, its internal magnetic field cannot be an important factor in its evolution. It was at one time proposed [*456*] that rotation and magnetism were intimately connected by a hitherto undiscovered law, so that the strength of the stellar magnetic field would be proportional to the total angular momentum of the star. Thus the magnetic field would vary along the main sequence in a prescribed way. However, there appears to be little theoretical, and no experimental [*457*], basis for such a relation.

However, it is known that at some stages of evolution, stars may have either convective cores, convective envelopes, or intermediate convective layers. In regions in which the energy is being transported by convection, the presence of a strong magnetic field may have the important effect of inhibiting the convection. This effect, while it has been studied in considerable detail by CHANDRASEKHAR and his collaborators [*458*], and has also been demonstrated in the laboratory [*459*] for conducting fluids of uniform density, has not been considered for even the simplest stellar model, because of the great mathematical complexity involved. However, qualitative arguments suggest that when a magnetic field is sufficiently strong, not complete inhibition, but a radical change both in the patterns of convection and the efficiency of energy transport will take place. Thus the depth of the convection zone may be altered. The structures of some of the red giants, which have surface magnetic fields $\sim 10^3$ gauss, should be investigated, taking this effect into account.

In the energy-generating cores of stars, another type of magnetic effect requires investigation. In degenerate cores, it is well known that the energy transport takes place through conduction, and the conduction electrons are also responsible for the electrical conductivity of the degenerate electron gas. If a strong magnetic field is present in the stellar core, it will be coupled to the heat-transporting electrons through the currents which they generate. This effect has been mentioned by BURBIDGE [*460*], but has not yet been investigated in detail. It may well be of considerable importance, since it offers a possible way of converting some nuclear energy, generated by the nuclear reactions, almost directly into magnetic energy in a star, via the electron component of the quanta, in which form all of the nuclear energy is converted and travels.

As has previously been mentioned, the prototype population II variable, RR Lyrae, has previously been shown by BABCOCK to have a surface magnetic field $\sim 10^3$ gauss. Now the extension of the virial theorem for a star to include the magnetic energy [*461*] shows that for sufficiently large internal magnetic fields, the periods of the natural modes of radial oscillation of the star may be altered. This effect has been investigated by BURBIDGE [*460*], who has shown that for central mean magnetic field strengths $\sim 10^6$ gauss, extending over the inner 10% of the radius, period changes $\sim 5\%$ or greater may be expected. This means that the same period-density relation cannot be expected to hold for highly magnetic and non-magnetic variables.

It has been shown that, if magnetic configurations in which the magnetic energy in a star becomes comparable to the potential energy ever develop, then instability will result [*462*], [*463*]. Using a solar-type model with a field distribution based on the work of COWLING and WRUBEL, the condition for dynamical instability has been numerically investigated [*464*]. It was found that a mean surface field, with an initial (primitive) field strength $\sim 10^8$ gauss in a star of solar mass would be sufficient for dynamical instability. The results are critically dependent on the model chosen, both as far as the density and temperature variation and the form of the field are concerned. It is clear, however, that none of the magnetic stars so far investigated have magnetic energies large enough for an outburst to result. The possibility cannot be excluded that the field can gradually be built up via nuclear processes in a degenerate core as the star evolves, and that it will eventually lead to instability.

The discovery of synchrotron radiation in supernova remnants provides indirect evidence that magnetic fields exist in these regions. In some of these, notably the Crab Nebula, the strength of the magnetic field has variously been estimated to lie between 10^{-3} and 10^{-2} gauss. Opinions are divided at the present time as to whether:

(i) The magnetic field of the star has been dispersed in the supernova explosion, reaching the values that it has today after the remnant has expanded for ~900 years, the total magnetic energy before and after the explosion remaining essentially constant.

(ii) The magnetic field in the remnant is the stellar field which was amplified in the explosion and the succeeding expansion to its present value.

(iii) The magnetic field in the remnant is an amplified interstellar field which has been built through the compression of the interstellar gas by the exploding gas shell of the supernova.

If (i) most nearly represents the real sequence of events, then it is probable that the magnetic field was closely connected with the triggering of the explosion itself. In any event, the total magnetic energy currently present amounts to 10^{48} to 10^{50} ergs, which is equal to or greater than any reasonable estimate of the initial gravitational energy of the pre-supernova. In our opinion the trigger of the explosion is not the magnetic field (cf. Sect. 81), and either (ii) or (iii) is more probable than (i). In this case the magnetic fields in supernova remnants have little to do with these catastrophic phases of stellar evolution.

Mustel [*465*] has proposed that the non-radial ejection of material in nova explosions is to be attributed to magnetic fields. In well-authenticated examples such as Nova Aquilae, the ejection in the form of two cones with a common axis demands, for a magnetic explanation, that large fields in the outer layers either are present in the pre-nova, or are generated by the explosion.

It is clear from the preceding arguments that stellar magnetic fields may be important in some phases of stellar evolution. However, in nearly all of the aspects mentioned, little detailed investigation has, as yet, been carried out.

References.

Chapter A.

[*1*] Gamow, G.: Kgl. danske Vidensk. Selsk. No. 10. (1953) — Proc. Nat. Acad. Sci. U.S.A. **40**, 480 (1954).
[*2*] Weizsäcker, C.F. von: Astrophys. J. **114**, 165 (1951).
[*3*] Hoyle, F.: Astrophys. J. **118**, 513 (1953).
[*4*] Layzer, D.: Astron. J. **59**, 170 (1954).
[*5*] Bergh, S. van den: Publ. Astronom. Soc. Pacific **68**, 449 (1956).
[*6*] Burbidge, E.M., and A.R. Sandage: Astrophys. J. **127**, 527 (1958).
[*7*] Burbidge, G.R.: In preparation.
[*8*] Heeschen, D.S.: Astrophys. J. **124**, 660 (1956). — Publ. Astronom. Soc. Pacific **69**, 350 (1957).
[*9*] Lindblad, B.: Monthly Notices Roy. Astronom. Soc. London **95**, 20 (1934). — Nature, Lond. **135**, 133 (1935).
[*10*] Haar, D. ter: Astrophys. J. **100**, 288 (1944).
[*11*] Kramers, H.A., and D. ter Haar: Bull. astronom. Soc. Netherl. **10**, 137 (1944).
[*12*] Spitzer, L.: Astrophys. J. **93**, 369 (1941); **94**, 232 (1941).
[*13*] Spitzer, L.: Centennial Symposia, Harvard College Observatory, 1946, p. 87.
[*14*] Whipple, F.L.: Astrophys. J. **104**, 1 (1946).
[*15*] Oort, J.H.: Bull. astronom. Soc. Netherl. **12**, 177 (1954).
[*16*] Biermann, L., u. A. Schlüter: Z. Naturforsch. **9**a, 463 (1954). — Gas Dynamics of Cosmic Clouds, Chap. 27. Amsterdam: North Holland Publishing Co. 1955.
[*17*] Oort, J.H., and L. Spitzer: Astrophys. J. **121**, 6 (1955). — Gas Dynamics of Cosmic Clouds, Chap. 28. Amsterdam: North Holland Publishing Co. 1955.
[*18*] Savedoff, M.P.: Astrophys. J. **124**, 533 (1956). — Menon, T.K.: Unpublished.
[*19*] Öpik, E.J.: Irish Astronom. J. **2**, 219 (1953).
[*19a*] R. Ebert: Z. Astrophys. **37**, 217 (1955).
[*19b*] W. B. Bonnor: Monthly Notices Roy. Astronom. Soc. London **116**, 351 (1956).
[*19c*] W. H. McCrea: Monthly Notices Roy. Astronom. Soc. London **117**, 562 (1957).

[20] KRAT, V.A.: Isv. Glavnoi Astronom. Obs. Pulkova **18**, 4, No. 145; **19**, 2, No. 149 (1952). Problems of Cosmology, Vol. 1, p. 34. Moscow: Academy of Sciences 1952.
[21] UREY, H.C.: Astrophys. J. **124**, 623 (1956).
[22] SU-SHU HUANG: Publ. Astronom. Soc. Pacific **69**, 427 (1957).
[23] STRUVE, O.: Stellar Evolution, Chap. 3. Princeton: University Press 1950.
[24] BOK, B. J., and E. REILLY: Astrophys. J. **105**, 255 (1947).
[25] BOK, B. J.: Centennial Symposia, Harvard College Observatory, 1946, p. 53.
[26] HERBIG, G.H.: Astrophys. J. **113**, 697 (1951). — J. ROY: Astronom. Soc. Canada **46**, 222 (1952).
[27] HARO, G.: Astrophys. J. **115**, 572 (1952); **117**, 73 (1953).
[28] AMBARTZUMIAN, V.A.: Soobs. Bjurakan Obs. No. 13 (1954).
[29] HERBIG, G.H.: Symposium on Non-Stable Stars, I.A.U. Monograph No. 3, p. 3. Cambridge: Cambridge University Press 1957.
[30] FESSENKOV, V.G., i D.L. ROSHKOVSKY: Astronom. J. USSR. **28**, 215 (1951); **29**, 382, 397 (1952); **30**, 3 (1953); **31**, 3 (1954). — Trans. Internat. Astronom. Union **8**, 702 (1954).
[31] STRUVE, O.: Sky and Telescope **13**, 181 (1954).
[32] OORT, J.H.: Gas Dynamics of Cosmic Clouds, p. 246. Amsterdam: North Holland Publishing Co. 1955.
[33] WEIZSÄCKER, C.F. VON: Z. Astrophys. **24**, 181 (1947).
[34] HAAR, D. TER: Astrophys. J. **110**, 321 (1949).
[35] ALFVÉN, H.: Ark. Mat. Astronom. Fys., Ser. A **28**, No. 6 (1942).
[36] LÜST, R., u. A. SCHLÜTER: Z. Astrophys. **38**, 190 (1955).
[37] STRUVE, O.: Stellar Evolution, Chap. 4. Princeton: University Press 1950.
[38] KUIPER, G.P.: Publ. Astronom. Soc. Pacific **47**, 121 (1935).
[39] KUIPER, G.P.: Publ. Astronom. Soc. Pacific **67**, 387 (1955).
[40] LYTTLETON, R.A.: The Stability of Rotating Liquid Masses, Chap. 10. Cambridge: Cambridge University Press 1953.
[41] SU-SHU HUANG et O. STRUVE: Ann. d'Astrophys. **17**, 85 (1954).
[42] CHANDRASEKHAR, S., and E. FERMI: Astrophys. J. **118**, 116 (1953).
[43] CHANDRASEKHAR, S.: Astrophys. J. **119**, 7 (1954).
[44] MESTEL, L., and L. SPITZER: Monthly Notices Roy. Astronom. Soc. London **116**, 503 (1956).
[45] MCVITTIE, G.C.: Astronom. J. **61**, 451 (1956).
[46] CHANDRASEKHAR, S.: An Introduction to the Theory of Stellar Structure, p. 453. Chicago: Chicago University Press 1939.
[47] THOMAS, L.H.: Monthly Notices Roy. Astronom. Soc. London **91**, 122, 619 (1931).
[48] ROTH, H.: Phys. Rev. **39**, 525 (1932).
[49] LEVÉE, R.D.: Astrophys. J. **117**, 200 (1953).
[50] SANDAGE, A.R.: Proc. Vatican Conference on Stellar Populations, 1957, p. 149.
[51] HENYEY, L.G., R. LELEVIER and R.D. LEVÉE: Publ. Astronom. Soc. Pacific **67**, 154 (1955).
[52] SALPETER, E.E.: Mém. Soc. Roy. Sci. Liège **14**, 116 (1954).
[53] LOCKYER, W.: Phil. Trans. Roy. Soc. Lond. **184**, 688 (1902). — Proc. Roy. Soc. Lond. **65**, 186 (1899).
[54] RUSSELL, H.N.: Observatory **36**, 324 (1913). — Pop. Astron. **22**, 275, 331 (1914).
[55] BETHE, H.A.: Phys. Rev. **55**, 434 (1939).
[56] WEIZSÄCKER, C.F. VON: Phys. Z. **39**, 633 (1938).
[57] GAMOW, G.: Phys. Rev. **53**, 595, 908 (1938).
[58] GAMOW, G., and E. TELLER: Phys. Rev. **53**, 608 (1938).
[59] GAMOW, G.: Phys. Rev. **55**, 718, 796 (1939).
[60] GAMOW, G., and E. TELLER: Phys. Rev. **55**, 791 (1939).
[61] GAMOW, G.: Nature, Lond. **144**, 575, 620 (1939).
[62] BURBIDGE, E.M., G.R. BURBIDGE, W.A. FOWLER and F. HOYLE: Rev. Mod. Phys. **29**, 547 (1957).
[63] SCHWARZSCHILD, M., R. HOWARD and R. HÄRM: Astrophys. J. **125**, 233 (1957).
[64] WEYMANN, R.: Astrophys. J. **126**, 203 (1957).
[65] OSTERBROCK, D.E.: Unpublished.
[66] OSTERBROCK, D.E.: Astrophys. J. **118**, 529 (1953).
[67] LIMBER, D.N.: Astrophys. J. **127**, 363, 387 (1958).
[68] ALLER, L.H.: Astrophys. J. **111**, 173 (1950).
[69] ALLER, L.H., J.W. CHAMBERLAIN, E.M. LEWIS, W.C. LILLER, J.K. MCDONALD, W.H. POTTER and N.E. WEBER: Astrophys. J. **115**, 328 (1952).
[70] STRÖMGREN, B.: Astronom. J. **57**, 65 (1952).
[71] COWLING, T.G.: Monthly Notices Roy. Astronom. Soc. London **96**, 42 (1936).

[72] Kushwaha, R. S.: Astrophys. J. **125**, 242 (1957).
[73] Naur, P., and D. E. Osterbrock: Astrophys. J. **117**, 306 (1953).
[74] Luyten, W. J.: Publ. Astronom. Soc. Pacific **66**, 337 (1954): **68**, 258 (1956).
[75] Lippincott, S. L.: Astronom. J. **60**, 379 (1955).
[76] Struve, O.: Sky and Telescope **17**, 18 (1957).
[77] Lundmark, K.: Handbuch der Astrophysik, **5**, Chap. 4, 1932.
[78] Kuiper, G. P.: Astrophys. J. **88**, 472 (1938).
[79] Parenago, P. P.: Astronom. J. USSR. **14**, 33 (1937).
[80] Russell, H. N., and C. E. Moore: Masses of the Stars. Chicago, Ill.: Chicago University Press 1940.
[81] Parenago, P. P., i. A. G. Massevich: Astronom. J. USSR. **27**, 137 (1950).
[82] Petrie, R. M.: Publ. Dom. Astrophys. Obs. Victoria **8**, 341 (1950).
[83] Plaut, L.: Publ. Astronom. Lab. Groningen No. 55 (1953).
[84] Kopal, Z.: Ann. d'Astrophys. **18**, 379 (1956).
[85] Kamp, P. van de: Astronom. J. **59**, 447 (1954).
[86] Eggen, O. J.: Astronom. J. **61**, 361 (1956).
[87] Strand, K. A.: J. Roy. Astronom. Soc. Canada **51**, 46 (1957).
[88] Hoyle, F., C. B. Haselgrove and J. Blackler: Private communication.
[89] Parenago, P. P.: Astronom. J. USSR. **27**, 41 (1950).
[90] Struve, O.: Mém. Soc. Roy. Sci. Liège **14**, 236 (1954).
[91] Sandage, A. R.: Mém. Soc. Roy. Sci. Liège **14**, 254 (1954).
[92] Reiz, A.: Astrophys. J. **120**, 342 (1954).
[93] Chapman, S.: Monthly Notices Roy. Astronom. Soc. London **82**, 292 (1922). — Eddington, A. S.: Internal Constitution of the Stars, p. 277. Cambridge: Cambridge University Press 1926.
[94] Sweet, P. A.: Monthly Notices Roy. Astronom. Soc. London **110**, 548 (1950).
[95] Öpik, E. J.: Monthly Notices Roy. Astronom. Soc. London **111**, 78 (1951).
[96] Eddington, A. S.: Monthly Notices Roy. Astronom. Soc. London **90**, 54 (1929).
[97] Mestel, L.: Monthly Notices Roy. Astronom. Soc. London **113**, 716 (1953).
[98] Fessenkov, V. G.: Astronom. J. USSR. **26**, 67 (1949). — Trans. Internat. Astronom. Union **8**, 702 (1952).
[99] Massevich, A. G.: Astronom. J. USSR. **26**, 207 (1949).
[100] Massevich, A. G.: Astronom. J. USSR. **28**, 36 (1951).
[101] Massevich, A. G.: Soobs. Astronom. Inst. Sternberg No. 99, 3 (1956).
[102] Gamow, G.: Phys. Rev. **65**, 20 (1944).
[103] Öpik, E. J.: Publ. Obs. Tartu **30**, Nos. 3 and 4 (1938); **31**, No. 1 (1943). — Armagh Obs. Contr. No. 2 (1949); No. 3 (1951).
[104] Gamow, G.: Astrophys. J. **87**, 206 (1938).
[105] Critchfield, C. L., and G. Gamow: Astrophys. J. **89**, 244 (1939).
[106] Chandrasekhar, S., and L. R. Henrich: Astrophys. J. **94**, 525 (1941).
[107] Schoenberg, M., and S. Chandrasekhar: Astrophys. J. **96**, 161 (1942).
[108] Hoyle, F., and R. A. Lyttleton: Monthly Notices Roy. Astronom. Soc. London **102**, 218 (1942); **109**, 614 (1949).
[109] Harrison, M. H.: Astrophys. J. **100**, 343 (1944); **103**, 192 (1946); **105**, 322 (1947).
[110] Gamow, G., and G. Keller: Rev. Mod. Phys. **17**, 125 (1945).
[111] Reiz, A.: Ann. d'Astrophys. **10**, 301 (1947).
[112] Ledoux, P.: Astrophys. J. **105**, 305 (1947). — Ann. d'Astrophys. **11**, 174 (1948).
[113] Massevich, A. G.: Astronom. J. USSR. **25**, 168 (1948). — Soobs. Astronom. Inst. Sternberg No. 30, 30 (1949).
[114] Li Hen and M. Schwarzschild: Monthly Notices Roy. Astronom. Soc. London **109**, 631 (1949).
[115] Bondi, C. M.: Monthly Notices Roy. Astronom. Soc. London **110**, 275 (1950).
[116] Bondi, C. M., and H. Bondi: Monthly Notices Roy. Astronom. Soc. London **110**, 287 (1950); **111**, 397 (1951).
[117] Gardiner, J. G.: Monthly Notices Roy. Astronom. Soc. London **111**, 102 (1951).
[118] Massevich, A. G., V. P. Matveyeva i. L. N. Toolenkova: Astronom. J. USSR. **28**, 432 (1951).
[119] Oke, J. B., and M. Schwarzschild: Astrophys. J. **116**, 3 (1952).
[120] Sooshkina, E. I.: Astronom. J. USSR. **30**, 180 (1953).
[121] Massevich, A. G.: Trudy Astronom. Inst. Sternberg **22**, 21 (1953).
[122] Resnikov, A. O.: Astronom. J. USSR. **31**, 60 (1954); **33**, 151 (1956).
[123] Hayashi, C.: Phys. Rev. **75**, 1619 (1949).
[124] Schwarzschild, M., I. Rabinowisc and R. Härm: Astrophys. J. **118**, 326 (1953).
[125] Sandage, A. R., and M. Schwarzschild: Astrophys. J. **116**, 463 (1952).
[126] Sorokin, V. S., i A. G. Massevich: Astronom. J. USSR. **28**, 21 (1951).

[127] MASSEVICH, A. G.: Astronom. J. USSR. **30**, 508 (1953).
[128] MASSEVICH, A. G.: Mém. Soc. Roy. Sci. Liège **14**, 170 (1954).
[129] HOYLE, F., and M. SCHWARZSCHILD: Astrophys. J. Suppl. **2**, 1 (1955).
[130] HASELGROVE, C.B., and F. HOYLE: Monthly Notices Roy. Astronom. Soc. London **116**, 515, 527 (1956).
[130a] HASELGROVE, C. B., and F. HOYLE: Monthly Notices Roy. Astronom. Soc. London (in the press, 1958).
[131] HAYASHI, C.: Progr. Theor. Phys. **17**, 737 (1957).
[132] TAYLER, R. J.: Astrophys. J. **120**, 332 (1954). — Monthly Notices Roy. Astronom. Soc. London **116**, 25 (1956).
[133] KUSHWAHA, R. S.: Astrophys. J. **125**, 242 (1957).
[134] HENYEY, L. G., R. LELEVIER and R.D. LEVÉE: Publ. Astronom. Soc. Pacific **67**, 341 (1955).
[135] ARP, H.C.: Astronom. J. **60**, 317 (1955).
[136] SANDAGE, A.R.: Astronom. J. **58**, 61 (1953).
[137] SAVEDOFF, M.P.: Astronom. J. **61**, 254 (1956).
[138] SCHWARZSCHILD, M.: Harvard Coll. Obs. Circ. No. 437 (1940).
[139] ROBERTS, M. S., and A.R. SANDAGE: Astronom. J. **60**, 185 (1955).
[140] WALKER, M.F.: Astronom. J. **60**, 197 (1955).
[141] BELSERENE, E.P.: Astronom. J. **57**, 237 (1952).
[142] CRAWFORD, J.A.: Publ. Astronom. Soc. Pacific **65**, 210 (1953).
[143] HOYLE, F.: Monthly Notices Roy. Astronom. Soc. London **106**, 343 (1946).
[144] PAYNE-GAPOSCHKIN, C.: The Galactic Novae. Amsterdam: North Holland Publishing Co. 1957.
[145] WALKER, M.F.: Astrophys. J. **123**, 68 (1956).
[145a] WALKER, M. F.: Astrophys. J. **127**, 319 (1958).
[146] STRUVE, O., and SU-SHU HUANG: Occ. Not. Roy. Astronom. Soc. London **3**, 161 (1957).
[147] SCHATZMAN, E.: Contr. Inst. d'Astrophys. No. 181 (1953). — Mém. Soc. Roy. Sci. Liège **14**, 163 (1954).
[148] CRAWFORD, J.A., and R.P. KRAFT: Astrophys. J. **123**, 44 (1956).
[149] KOPYLOW, I.M.: Symposium on Non-Stable Stars. I.A.U. Monograph No. 3, p. 71. Cambridge: Cambridge University Press 1957.
[150] VORONTSOV-VELYAMINOV, B.: Observatory **67**, 224 (1947).
[151] SCHATZMAN, E.: Ann. d'Astrophys. **8**, 143 (1945).
[152] MESTEL, L.: Monthly Notices Roy. Astronom. Soc. London **112**, 583, 598 (1952).
[153] LUYTEN, W. J.: Astrophys. J. **116**, 283 (1952).
[154] STRUVE, O.: Sky and Telescope **13**, 82 (1954).
[155] PARENAGO, P.P.: Astronom. J. USSR. **23**, 349 (1946).
[156] SCHATZMAN, E.: Trans. Internat. Astronom. Union **8**, 727 (1952).
[157] HAAR, D. TER: Kgl. danske Vidensk. Selsk. **25**, No. 3 (1948).
[158] WEIZSÄCKER, C.F. VON: Z. Astrophys. **22**, 319 (1944).
[159] HAAR, D. TER: Astrophys. J. **111**, 179 (1950).
[160] KUIPER, G.P.: Astrophysics, ed. J.A. HYNEK, Chap. 8. New York: McGraw-Hill 1951. — Proc. Nat. Acad. Sci. U.S.A. **39**, 1159 (1953); **40**, 1101 (1954). — J. Roy. Astronom. Soc. Canada **57**, 105, 158 (1955), and other references given there.
[160a] HOYLE, F.: Frontiers in Astronomy, p. 83. London: Heinemann 1955.
[161] UREY, H.C.: The Planets. New Haven: Yale University Press 1952. — Astrophys. J. Suppl. **1**, 147 (1954), and other references given there.
[162] EPSTEIN, I., and L. MOTZ: Astrophys. J. **117**, 311 (1953).
[163] NAUR, P.: Astrophys. J. **119**, 365 (1954).
[164] FRANK-KAMENETSKY, D.A.: Astronom. J. USSR. **32**, 139, 326 (1955).
[165] ABELL, G.O.: Astrophys. J. **121**, 430 (1955).
[166] STRÖMGREN, B.: The Sun, ed. G.P. KUIPER, Chap. 2. Chicago, Ill.: Chicago University Press 1953.
[167] VITENSE, E.: Z. Astrophys. **32**, 135 (1953).
[168] SANDAGE, A.R.: Amer. J. Phys. **25**, 525 (1957).

Chapter B.

[169] AMBARTZUMIAN, V.A.: Astrophysics and Stellar Evolution, ed. Armenian Acad. Sci., USSR. 1947.
[170] AMBARTZUMIAN, V.A.: Astronom. J. USSR. **26**, 3 (1949).
[171] AMBARTZUMIAN, V.A.: Trans. Internat. Astronom. Union **8**, 665 (1952).
[172] JOY, A.H.: Astrophys. J. **102**, 168 (1945); **110**, 424 (1949).

[173] Parenago, P.P.: Variable Stars **7**, 169 (1950).
[174] Haro, G.: Astronom. J. **55**, 72 (1950).
[175] Herbig, G.H.: Astrophys. J. **111**, 15 (1950).
[176] Parenago, P.P.: Astronom. J. USSR. **30**, 249 (1953).
[177] Haro, G., B. Iriarte and E. Chavira: Bol. Obs. Tonantzintla y Tacubaya No. 8 (1953).
[178] Walker, M.F.: Astrophys. J. Suppl. **2**, 365 (1956).
[179] Walker, M.F.: Astrophys. J. **125**, 636 (1957).
[180] Herbig, G.H.: Astrophys. J. **119**, 483 (1954).
[181] Herbig, G.H.: Proc. Vatican Conference on Stellar Populations, 1957, p. 127.
[182] Blaauw, A.: Bull. astronom. Soc. Netherl. No. 433 (1952).
[183] Blaauw, A., and W.W. Morgan: Astrophys. J. **119**, 625 (1954).
[184] Markarian, B.E.: Soobs. Bjurakan Obs. No. 11 (1953).
[185] Blaauw, A.: Astrophys. J. **123**, 408 (1956).
[187] Massevich, A.G.: Astronom. J. USSR. **34**, 176 (1957).
[188] Münch, G.: Publ. Astronom. Soc. Pacific **68**, 351 (1956).
[189] Münch, G., and E. Flather: Publ. Astronom. Soc. Pacific **69**, 142 (1957).
[190] Sharpless, S.: Astrophys. J. **119**, 334 (1954).
[191] Ambartzumian, V.A.: Isv. Acad. Sci. USSR., Phys. Ser. **14**, 15 (1950).
[192] Haro, G.: Symposium on Non-Stable Stars, I.A.U. Monograph No. 3, p. 26. Cambridge: Cambridge University Press 1957.
[193] Gliese, W.: Z. Astrophys. **39**, 1 (1956).
[194] Trumpler, R.J.: Publ. Astronom. Soc. Pacific **37**, 307 (1925). — Lick Obs. Bull. No. 420 (1930).
[195] Kuiper, G.P.: Harvard Bull. No. 903 (1936). — Astrophys. J. **86**, 176 (1937).
[196] Parenago, P.P., i A.G. Massevich: Astronom. J. USSR. **28**, 466 (1951).
[197] Johnson, H.L., and C.F. Knuckles: Astrophys. J. **122**, 209 (1955).
[198] Johnson, H.L.: Astrophys. J. **120**, 325 (1954).
[199] Sandage, A.R.: Astrophys. J. **125**, 422 (1957).
[200] Sandage, A.R.: Proc. Vatican Conference on Stellar Populations. 1957, p. 41.
[201] Lohmann, W.: Z. Astrophys. **42**, 114 (1957).
[202] Hoerner, S. von: Z. Astrophys. **42**, 273 (1957).
[203] Sandage, A.R.: Astrophys. J. **125**, 435 (1957).
[204] Massevich, A.G.: Astronom. J. USSR. **32**, 412 (1955).
[205] Massevich, A.G.: Astronom. J. USSR. **33**, 576 (1956).
[205a] Drake, F.: American Astronom. Soc. meeting, Indianapolis, Dec. 1957.
[206] Parenago, P.P.: Trudy Astronom. Inst. Sternberg **25** (1954).
[207] Johnson, H.L.: Astrophys. J. **126**, 134 (1957).
[208] Prendergast, K.V.: Private communication.
[209] Bidelman, W.P.: Astrophys. J. **98**, 61 (1943); **105**, 492 (1947).
[210] Johnson, H.L., and W.W. Morgan: Astrophys. J. **122**, 429 (1955).
[211] Johnson, H.L., and W.A. Hiltner: Astrophys. J. **123**, 267 (1956).
[212] Blanco, V.M.: Astrophys. J. **122**, 434 (1955).
[213] Larsson-Leander, G.: Stockholm Obs. Ann. **20**, No. 2 (1957).
[214] Miczaika, G.R.: Mém. Roy. Soc. Sci. Liège **14**, 275 (1954).
[215] Mitchell, R.I., and H.L. Johnson: Astrophys. J. **125**, 414 (1957).
[216] Humason, M.L., and F. Zwicky: Astrophys. J. **105**, 85 (1947).
[217] Luyten, W.J.: Astronom. J. **58**, 75 (1953); **59**, 224 (1954). — Harvard Announcement Cards, Nos. 1202 (1953); 1244 (1954).
[218] Johnson, H.L.: Astrophys. J. **117**, 357 (1953).
[219] Roman, N.G.: Astrophys. J. **121**, 454 (1955).
[220] Johnson, H.L., and A.R. Sandage: Astrophys. J. **121**, 616 (1955).
[220a] Burbidge, E.M., and A.R. Sandage: Astrophys. J. **128**, Sept. (1958).
[221] Reddish, V.C.: Observatory **74**, 68 (1954).
[222] Oort, J.H.: Proc. Vatican Conference on Stellar Populations, 1957, p. 63.
[222a] Spitzer, L.: Astrophys. J. **127**, 17 (1958).
[223] Eggen, O.J.: Astronom. J. **62**, 45 (1957).
[224] Sandage, A.R.: Proc. Vatican Conference on Stellar Populations, 1957, p. 287.
[225] Shapley, H.: Star Clusters. New York: McGraw-Hill 1930.
[226] Baade, W.: Astrophys. J. **100**, 137 (1944).
[227] Arp, H.C., W.A. Baum and A.R. Sandage: Astronom. J. **58**, 4 (1953).
[228] Sandage, A.R.: Astronom. J. **58**, 61 (1953).
[229] Popper, D.M.: Astrophys. J. **105**, 204 (1947).
[230] Baum, W.A.: Astronom. J. **57**, 222 (1952).

[231] MORGAN, W. W.: Publ. Astronom. Soc. Pacific **68**, 509 (1956).
[232] BAUM, W. A.: Astronom. J. **59**, 422 (1954).
[233] HILTNER, W. A., H. L. JOHNSON and A. R. SANDAGE: Unpublished.
[234] RHIJN, P. J. VAN: Groningen Publ. No. 47 (1936).
[235] LUYTEN, W. J.: Publ. Astronom. Obs. Univ. of Minnesota **2**, No. 7 (1939).
[236] BOK, B. J., and D. A. MACCRAE: Ann. New York Acad. Sci. **42**, 219 (1941).
[237] KUIPER, G. P.: Astrophys. J. **95**, 201 (1942).
[238] MCCUSKEY, S. W.: Astrophys. J. **123**, 458 (1956).
[239] OORT, J. H.: Groningen Publ. No. 40 (1926).
[240] SANDAGE, A. R.: Proc. Vatican Conference on Stellar Populations, 1957, p. 75.
[241] PARENAGO, P. P.: Astronom. J. USSR **28**, 93 (1951).
[242] MASSEVICH, A. G.: Astronom. J. USSR **33**, 216 (1956).
[243] KAPLAN, S. A.: Astronom. J. USSR. **30**, 391 (1953).
[244] SALPETER, E. E.: Astrophys. J. **121**, 161 (1955).
[245] BERGH, S. VAN DEN: Astrophys. J. **125**, 445 (1957).
[246] JASCHEK, C., and M. JASCHEK: Publ. Astronom. Soc. Pacific **69**, 337 (1957).
[247] BERGH, S. VAN DEN: Astronom. J. **62**, 100 (1957).
[248] SANDAGE, A. R.: Astronom. J. **59**, 162 (1954).
[249] TAYLER, R. J.: Astronom. J. **59**, 413 (1954).
[250] SANDAGE, A. R., and M. F. WALKER: Astronom. J. **60**, 230 (1955).
[251] SANDAGE, A. R.: Astrophys. J. **126**, 326 (1957).
[252] ROMAN, N. G.: Astrophys. J. **116**, 122 (1952).
[253] OORT, J. H.: Bull. astronom. Soc. Netherl. **9**, 185 (1941). — Astrophys. J. **116**, 233 (1952).
[254] PARENAGO, P. P.: Astronom. J. USSR. **27**, 150 (1950).
[255] Proc. Vatican Conference on Stellar Populations, 1957, pp. 426, 517, 533.
[256] SPITZER, L., and M. SCHWARZSCHILD: Astrophys. J. **114**, 385 (1951); **118**, 106 (1953). — OSTERBROCK, D. E.: Astrophys. J. **116**, 164 (1952).
[257] TRUMPLER, R. J., and H. F. WEAVER: Statistical Astronomy. Berkeley: University of California Press 1953.
[258] PAYNE-GAPOSCHKIN, C.: Variable Stars and Galactic Structure, Chap. 1. London: Athlone Press 1954.
[259] GONDOLATSCH, F.: Z. Astrophys. **24**, 330 (1947).
[260] MUMFORD, G. S.: Astronom. J. **61**, 224 (1956).
[261] DYER, E. R.: Astronom. J. **61**, 228 (1956).
[262] WEHLAU, A. W.: Astronom. J. **62**, 169 (1957).
[263] PARENAGO, P. P.: Astronom. News Letter No. 71 and Suppl. (1953).
[264] WIJK, U. VAN: Astronom. J. **61**, 279 (1956).
[265] ROMAN, N. G.: Astrophys. J. **112**, 554 (1950); **116**, 122 (1952). — Astronom. J. **59**, 307 (1954). — Astrophys. J. Suppl. **2**, 198 (1956).
[266] KEENAN, P. C., and G. KELLER: Astrophys. J. **117**, 241 (1953).
[267] VYSSOTSKY, A. N.: Publ. Astronom. Soc. Pacific. **69**, 109 (1957).
[268] DELHAYE, J.: C. R. Acad. Sci., Paris **237**, 294 (1953).
[269] NASSAU, J. J., and V. M. BLANCO: Astrophys. J. **120**, 464 (1954).
[270] SMITH, H. J., and E. v. P. SMITH: Astronom. J. **61**, 273 (1956).
[271] NASSAU, J. J.: Proc. Vatican Conference on Stellar Populations, 1957, p. 171.
[271a] VASELEVSKIS, S., and R. A. RACH: Astronom. J. **62**, 175 (1957).
[272] MORGAN, W. W., S. SHARPLESS and D. E. OSTERBROCK: Astronom. J. **57**, 3 (1952).
[273] MORGAN, W. W., A. E. WHITFORD and A. D. CODE: Astrophys. J. **118**, 318 (1953).
[274] WEAVER, H. F.: Astronom. J. **58**, 177 (1953).
[275] Bull astronom. Soc. Netherl. **13**, No. 475 (1957).
[276] BAADE, W.: Symposium Notes, University of Michigan Observatory, 1953.
[277] KEENAN, P. C.: Astrophys. J. **120**, 484 (1954).
[278] BLANCO, V. M., and L. MÜNCH: Bol. Obs. Tonantzintla y Tacubaya No. 12, 17 (1955).
[279] IRWIN, J. B.: Astronom J. (in the press).
[280] KRAFT, R. P.: Astrophys. J. **126**, 225 (1957).
[281] BERGH, S. VAN DEN: Astrophys. J. **126**, 323 (1957).
[282] MORGAN, W. W., and N. U. MAYALL: Publ. Astronom. Soc. Pacific **69**, 291 (1957).
[283] HOLMBERG, E.: Lund Medd. II, No. 128 (1950).
[284] STEBBINS, J., and A. E. WHITFORD: Astrophys. J. **115**, 284 (1952).
[285] OORT, J. H.: Astrophys. J. **91**, 273 (1940).
[286] HOLMBERG, E.: Lund. Medd. I, No. 180 (1952).
[287] SCHWARZSCHILD, M.: Astronom. J. **59**, 273 (1954).
[288] BAADE, W.: Publ. Univ. Michigan Obs. **10**, 7 (1950).

[289] SHAPLEY, H.: Publ. Univ. Michigan Obs. **10**, 79 (1950). — The Inner Metagalaxy, p. 183. New Haven: Yale University Press 1957. — Galaxies, p. 216. Philadelphia: Blakiston Company 1943.
[290] ARP, H. C.: Astronom. J. **61**, 15 (1956).
[291] HULST, H. C. VAN DE: J. RAIMOND and H. VAN WOERDEN: Bull. astronom. Soc. Netherl. **14**, 1 (1957).
[292] HEESCHEN, D. S.: Proc. Stockholm Conference on Galactic Structure, 1957.
[293] BAUM, W. A., and M. SCHWARZSCHILD: Astronom. J. **60**, 247 (1955).
[294] DIETER, N. H.: Publ. Astronom. Soc. Pacific **69**, 356 (1957). — Astronom. J. (in the press).
[295] OORT, J. H. et al.: Unpublished.
[296] BUSCOMBE, W., S. C. B. GASCOIGNE and G. DE VAUCOULEURS: Suppl. Austral. J. Sci. **17**, No. 3 (1954).
[297] KERR, F. J., J. V. HINDMAN and B. J. ROBINSON: Austral. J. Phys. **7**, 297 (1954).
[298] HUMASON, M. L., N. U. MAYALL and A. R. SANDAGE: Astronom. J. **61**, 97 (1956).
[299] HEESCHEN, D. S.: Astrophys. J. **126**, 471 (1957).
[300] PAGE, T.: Astrophys. J. **116**, 63 (1952).
[300a] AMBARTZUMIAN, V. A.: Symposium on Galactic Structure, I.A.U. Monograph No. 5, p. 4. Cambridge: Cambridge University Press 1958.
[301] STEBBINS, J., and A. E. WHITFORD: Astrophys. J. **108**, 413 (1948). — STEBBINS, J.: Monthly Notices Roy. Astronom. Soc. London **110**, 416 (1950).
[302] BAADE, W., and H. H. SWOPE: Astronom. J. (in the press).
[303] ROBERTS, M. S.: Astronom. J. **61**, 195 (1956).
[304] HUMASON, M. L.: Astrophys. J. **83**, 10 (1936).
[305] WHITFORD, A. E.: Astronom. J. **58**, 49 (1953). — Astrophys. J. **120**, 599 (1954).
[306] Ann. Rep. Washburn Observatory, Astronom. J. **61**, 352 (1956).
[307] JEANS, J. H.: Astronomy and Cosmogony, Chap. 13. Cambridge: Cambridge University Press 1929.
[308] PAYNE-GAPOSCHKIN, C.: The New Astronomy, Scientific American, p. 107. New York: Simon and Schuster 1955.
[309] LÜST, R.: Z. Naturforsch. **7**a, 87 (1952). — TREFFTZ, E.: Z. Naturforsch. **7**a, 99 (1952).
[310] HARO, G.: Bol. Obs. Tonantzintla y Tacubaya No. 14, 16 (1956).
[311] MÜNCH, G.: Astronom. J. (in the press).
[312] OORT, J. H.: Sci. Amer. **195**, 101 (1956).
[313] SPITZER. L., and W. BAADE: Astrophys. J. **113**, 413 (1951).
[314] SANDAGE, A. R.: Hubble Atlas of Galactic Forms, Carnegie Institution of Washington, 1958.
[315] MORGAN, W. W.: Publ. Astronom. Soc. Pacific **70**, August (1958).

Chapter C.

[316] EDDINGTON, A. S.: Internal Constitution of the Stars, p. 39. Cambridge: Cambridge University Press 1926.
[317] HOYLE, F., and R. A. LYTTLETON: Proc. Cambridge Phil. Soc. **35**, 405, 592 (1939); **36**, 325, 424 (1940). — Monthly Notices Roy. Astronom. Soc. London **101**, 227 (1941).
[318] BONDI, H., and F. HOYLE: Monthly Notices Roy. Astronom. Soc. London **104**, 273 (1944).
[319] BONDI, H.: Monthly Notices Roy. Astronom. Soc. London **112**, 195 (1952).
[320] MCCREA, W. H.: Monthly Notices Roy. Astronom. Soc. London **113**, 162 (1953).
[321] DODD, K. N., and W. H. MCCREA: Monthly Notices Roy. Astronom. Scc. London **112**, 205 (1952).
[322] MESTEL, L.: Monthly Notices Roy. Astronom. Soc. London **114**, 437 (1954).
[323] SCHATZMAN, E.: Gas Dynamics of Cosmic Clouds, p. 193. Amsterdam: North Holland Publishing Co. 1955.
[324] GURZADIAN, G. A.: Astronom. J. USSR. **26**, 104 (1949).
[325] KOPAL, Z.: Astrophys. J. **93**, 92 (1941).
[326] KUIPER, G. P.: Astrophys. J. **93**, 132 (1941).
[327] WOOD, F. B.: Astrophys. J. **112**, 196 (1950).
[328] CRAWFORD, J. A.: Astrophys. J. **121**, 71 (1955).
[329] KOPAL, Z.: Ann. d'Astrophys. **19**, 298 (1957).
[330] BIERMANN, L.: Gas Dynamics of Cosmic Clouds, p. 212. Amsterdam: North Holland Publishing Co. 1955.
[331] STRUVE, O.: Pop. Astron. **53**, 201, 259 (1945).
[332] PAGEL, B. E. J.: Monthly Notices Roy. Astronom Soc.. London (in the press).

[333] SHAJN, G. A.: Bull. Abastumani Astrophys. Obs. No. 7, 83 (1943).
[334] ABT, H. A.: Astrophys. J. (in preparation).
[335] DEUTSCH, A. J.: Astrophys. J. **123**, 210 (1956).
[336] HOYLE, F.: Astrophys. J. **124**, 482 (1956).
[337] GURZADIAN, G. A.: Dynamical Problems of Planetary Nebulae. Erivan, USSR. 1954.
[338] Les Novae et les Naines Blanches. Paris: Herman & Cie. 1939.
[339] AMBARTZUMIAN, V. A., u. N. KOSIREV: Z. Astrophys. **7**, 320 (1933).
[340] WHIPPLE, F., and C. PAYNE-GAPOSCHKIN: Harvard Coll. Obs. Circ. No. 413 (1936).
[341] GORDELATSE, S. G.: Bull. Abastumani Astrophys. Obs. **1**, 67 (1937).
[342] PAYNE-GAPOSCHKIN, C., and S. GAPOSCHKIN: Proc. Nat. Acad. Sci. U.S.A. **28**, 482 (1942).
[343] ALLER, L. H.: Astrophysics, Vol. II, p. 176. New York: Ronald Press 1955.
[344] MINKOWSKI, R.: Publ. Astronom. Soc. Pacific **53**, 224 (1942).
[345] OSTERBROCK, D. E.: Publ. Astronom. Soc. Pacific **69**, 227 (1957).
[346] GORDELATSE, S. G.: Bull. Abastumani Astrophys. Obs. **3**, 102 (1938).
[347] GREEP, P.: Diss. Utrecht.

Chapter D.

[348] HOYLE, F.: Astrophys. J. Suppl. **1**, 121 (1955).
[349] ALBADA, G. B. VAN: Astrophys. J. **105**, 393 (1947)
[350] SALPETER, E. E.: Astrophys. J. **115**, 326 (1952).
[351] CAMERON, A. G. W.: Astrophys. J. **121**, 144 (1955).
[352] FOWLER, W. A., G. R. BURBIDGE and E. M. BURBIDGE: Astrophys. J. **122**, 271 (1955).
[353] BURBIDGE, G. R., F. HOYLE, E. M. BURBIDGE, R. F. CHRISTY and W. A. FOWLER: Phys. Rev. **103**, 1145 (1956). — BAADE, W., G. R. BURBIDGE, F. HOYLE, E. M. BURBIDGE, R. F. CHRISTY and W. A. FOWLER: Publ. Astronom. Soc. Pacific **68**, 296 (1956).
[354] CAMERON, A. G. W.: Publ. Astronom. Soc. Pacific. **69**, 201 (1957).
[355] MARION, J. B., and W. A. FOWLER: Astrophys. J. **125**, 221 (1957).
[356] HOYLE, F.: Private communication.
[357] SEARS, R. L.: Astronom. J. (in the press).
[357a] FOWLER, W. A.: Astrophys. J. **127**, 551 (1958).
[358] SUESS, H. E., and H. C. UREY: Rev. Mod. Phys. **28**, 53 (1956).
[359] GREENSTEIN, J. L.: Modern Physics for Engineers, ed. L. Ridenour, p. 267. New York: McGraw Hill 1954.
[360] CAMERON, A. G. W.: Phys. Rev. **93**, 932 (1954).
[361] BAADE, W., and F. ZWICKY: Proc. Nat. Acad. Sci. U.S.A. **20**, 259 (1934).
[362] GAMOW, G., and M. SCHOENBERG: Phys. Rev. **59**, 539 (1941).
[363] FOWLER, W. A., G. R. BURBIDGE and E. M. BURBIDGE: Astrophys. J. Suppl. **2**, 167 (1955).
[364] SEVERNY, A. B.: Astronom. J. USSR. **34**, 328 (1957).
[365] GOLDBERG, L., O. C. MOHLER and E. MÜLLER: Astrophys. J. **127**, 302 (1958).
[366] BIERMANN, L.: Z. Astrophys. **18**, 344 (1939).
[367] SCHATZMAN, E.: Ann. d'Astrophys. **14**, 278, 294, 305 (1951).
[368] BURBIDGE, G. R.: Publ. Astronom. Soc. Pacific **70**, 83 (1958).
[369] DEUTSCH, A. J.: Principes Fondamentaux de Classification Stellaire, p. 32. Paris: Centre National de la Recherche Scientifique 1955.
[370] MAYALL, N. U.: Astrophys. J. **104**, 290 (1946).
[371] CHAMBERLAIN, J. W., and L. H. ALLER: Astrophys. J. **114**, 52 (1951).
[372] JOHNSON, H. L., and A. R. SANDAGE: Astrophys. J. **124**, 379 (1956).
[373] SCHWARZSCHILD, M., L. SEARLE and R. HOWARD: Astrophys. J. **122**, 353 (1955).
[374] BURBIDGE, E. M., A. R. SANDAGE and G. R. BURBIDGE: In preparation.
[375] BURBIDGE, E. M.: Unpublished.
[376] ALLER, L. H., and J. L. GREENSTEIN: In preparation.
[377] SCHWARZSCHILD, M. and B.: Astrophys. J. **112**, 248 (1950).
[378] MICZAIKA, G. R.: Z. Astrophys. **27**, 1 (1950).
[379] IWANOWSKA, W.: Bull. Astronom. Obs. Torun No. 9, 25 (1950).
[380] SCHWARZSCHILD, M., L. SPITZER and R. WILDT: Astrophys. J. **114**, 398 (1951).
[381] GRATTON, L.: Mém. Soc. Roy. Sci. Liège **14**, 419 (1954).
[382] WELLMAN, P.: Z. Astrophys. **36**, 194 (1955).
[383] BURBIDGE, E. M., and G. R. BURBIDGE: Astrophys. J. **124**, 116 (1956).
[384] SCHWARZSCHILD, M. and B., L. SEARLE and A. MELTZER: Astrophys. J. **125**, 123 (1957).
[384a] HACK, M.: Mem. Soc. Astronom. Italiana **28**, 339 (1957).
[385] ABT, H. A.: Astrophys. J. Suppl. **1**, 63 (1954).
[386] POPPER, D. M.: Astronom. J. **59**, 445 (1954).
[387] BURBIDGE, E. M., and G. R. BURBIDGE: In preparation.

[388] POPPER, D. M.: Publ. Astronom. Soc. Pacific **59**, 320 (1947).
[389] BIDELMAN, W. P.: Astrophys. J. **116**, 227 (1952).
[390] THACKERAY, A. D.: Monthly Notices Roy. Astronom. Soc. London **114**, 93 (1954).
[391] ALLER, L. H.: Mém. Soc. Roy. Sci. Liège **14**, 337 (1954).
[392] MÜNCH, G.: Astrophys. J. **127**, 642 (1958).
[393] GREENSTEIN, J. L., and W. S. ADAMS: Astrophys. J. **106**, 339 (1947).
[394] BIDELMAN, W. P.: Astrophys. J. **111**, 333 (1950).
[395] MORGAN, W. W.: Astrophys. J. **79**, 513 (1934).
[396] SALPETER, E. E.: Ann. Rev. Nucl. Sci. **2**, 41 (1953).
[397] SLJUSAREV, S. G.: Astronom. J. USSR. **32**, 346 (1955).
[398] ALLER, L. H.: Astrophys. J. **97**, 135 (1943).
[399] GAMOW, G.: Astrophys. J. **98**, 500 (1943).
[400] GREENSTEIN, J. L.: Mém. Soc. Roy. Sci. Liège **14**, 307 (1954).
[401] ALLER, L. H.: Gaseous Nebulae, p. 217. New York: John & Wiley Sons 1956.
[402] ALLER, L. H.: Astrophys. J. **125**, 84 (1957).
[403] BOUIGUE, R.: Ann. d'Astrophys. **17**, 104 (1954).
[404] McKELLAR, A.: Publ. Dom. Astrophys. Obs. **7**, 395 (1948).
[405] BERMAN. L.: Astrophys. J. **81**, 369 (1935).
[406] SANFORD, R. F.: Publ. Astronom. Soc. Pacific **52**, 203 (1940).
[407] BIDELMAN, W. P.: Astrophys. J. **117**, 25 (1953).
[408] HERBIG, G. H.: Astrophys. J. **110**, 143 (1949).
[409] KEENAN, P. C.: Astrophys. J. **96**, 101 (1942).
[410] BIDELMAN, W. P.: Vistas in Astronomy, ed. A. Beer, Vol. II, p. 1428. London: Pergamon Press 1957.
[411] GREENSTEIN, J. L.: Publ. Astronom. Soc. Pacific **68**, 501 (1956).
[412] BURBIDGE, G. R., and E. M. BURBIDGE: Publ. Astronom. Soc. Pacific **66**, 308 (1954).
[413] GREENSTEIN, J. L.: Third Berkeley Symposium on Statistics, University of California Press, 1956, p. 11.
[414] BLANCO, V. M., and G. GRANT: Private communication.
[415] MERRILL, P. W.: Science **115**, 484 (1952).
[416] BURBIDGE, E. M., and G. R. BURBIDGE: Astrophys. J. **126**, 357 (1957).
[417] BIDELMAN, W. P.: Mém. Soc. Roy. Sci. Liège **14**, 402 (1954).
[418] MERRILL, P. W.: Publ. Astronom. Soc. Pacific **68**, 70 (1956).
[419] BAADE, W.: Astrophys. J. **97**, 119 (1943); **102**, 309 (1945).
[420] MAYALL, N. U., and J. H. OORT: Publ. Astronom. Soc. Pacific **54**, 95 (1942).
[421] HUIZENGA, J. R., and J. DIAMOND: Phys. Rev. **107**, 1087 (1957).
[422] FIELDS, P. R., M. H. STUDIER, H. DIAMOND, J. F. MECH, M. G. INGHRAM, G. L. PYLE, C. M. STEVENS, S. FRIED, W. M. MANNING, A. GHIORSO, S. G. THOMPSON, G. H. HIGGINS and G. T. SEABORG: Phys. Rev. **102**, 180 (1956).
[422a] DEUTSCH, A. J.: Publ. Astronom. Soc. Pacific **68**, 308 (1956).

Chapter E.

[423] WESTGATE, C.: Astrophys. J. **79**, 357 (1934).
[424] SLETTEBAK, A.: Astrophys. J. **110**, 498 (1949); **119**, 146 (1954); **121**, 653 (1955). — SLETTEBAK, A., and R. F. HOWARD: Astrophys. J. **121**, 102 (1955).
[425] BÖHM, K. H.: Z. Astrophys. **30**, 117 (1952).
[426] SU-SHU HUANG: Astrophys. J. **118**, 285 (1953).
[427] HERBIG, G. H., and J. F. SPALDING: Astrophys. J. **121**, 118 (1955).
[428] STRUVE, O., Stellar Evolution, p. 123. Princeton: University Press 1950.
[429] SU-SHU HUANG and O. STRUVE: Ann. d'Astrophys. **17**, 85 (1954).
[430] OKE, J. B., and J. L. GREENSTEIN: Astrophys. J. **120**, 384 (1954).
[431] SANDAGE, A. R.: Astrophys. J. **122**, 263 (1955).
[432] STRUVE, O.: Stellar Evolution, p. 113. Princeton: University Press 1950.
[433] HOYLE, F.: Astrophys. J. **124**, 484 (1956).
[434] BÖHM, K. H.: Astrophys. J. **123**, 379 (1956).
[435] GREENSTEIN, J. L.: Publ. Astronom. Soc. Pacific **62**, 156 (1950).
[436] AMBARTZUMIAN, V. A., Soobs. Bjurakan Obs. No. 13 (1954). — Symposium on Non-Stable Stars, I.A.U. Monograph No. 3, p. 177. 1957. — KOURGANOFF, V.: Suppl. Nuovo Cim. **1**, No. 4 (1955).
[437] KHOLOPOV, P. N.: Symposium on Non-Stable Stars, I.A.U. Monograph No. 3, p. 11. 1957.
[438] GORDON, K. C., and G. E. KRON: Publ. Astronom. Soc. Pacific **61**, 210 (1949).
[439] BURBIDGE, G. R., and E. M. BURBIDGE: Observatory **75**, 212 (1955).

[440] BABCOCK, H. W.: Astrophys. J. Suppl. **3**, 141 (1958).
[441] STRUVE, O.: Publ. Astronom. Soc. Pacific **67**, 29, 135 (1955).
[442] LEDOUX, P.: Astrophys. J. **114**, 373 (1951).
[443] HUBBLE, E. P., and A. R. SANDAGE: Astrophys. J. **118**, 353 (1953).
[444] FEAST, M. W., A. D. THACKERAY and A. J. WESSELINK: Observatory **75**, 216 (1955).
[445] ABT, H. A.: Astrophys. J. **126**, 138 (1957).
[446] KRAFT, R. P.: Astronom. J. (in the press). — IRWIN, J. B.: Astronom. J. (in the press).
[447] COX, J. P.: Astrophys. J. **122**, 286 (1955).
[448] EDDINGTON, A. S.: Monthly Notices Roy. Astronom. Soc. London **101**, 182 (1941).
[449] ZHEVAKIN, S. A.: Astronom. J. USSR. **30**, 161 (1953); **31**, 141, 335 (1954).
[450] COX, J. P., and C. E. WHITNEY: Astrophys. J. (in the press).
[451] STEBBINS, J., and C. M. HUFFER: Publ. Washburn Obs. **15**, 143 (1930).
[451a] HERBIG, G. H.: Astrophys. J. **127**, 312 (1958).
[452] PARENAGO, P. P.: Variable Stars **10**, 193 (1954).
[453] PAVLOVSKAYA, E. D.: Variable Stars **9**, 349 (1954).
[454] STRUVE, O.: Symposium on Non-Stable Stars, I.A.U. Monograph No. 3, p. 93. 1957.
[455] EGGEN, O. J.: Astronom. J. **62**, 14, 45 (1957).
[456] WILSON, A. H.: Proc. Roy. Soc. Lond., Ser. A **104**, 451 (1923). — BLACKETT, P. M. S.: Nature, Lond. **159**, 658 (1947).
[457] BLACKETT, P. M. S.: Phil. Trans. Roy. Soc. Lond. **245**, 309 (1953).
[458] CHANDRASEKHAR, S.: Phil. Mag. **43**, 501, 1317 (1952). — Proc. Roy. Soc. Lond., Ser. A **217**, 306 (1953). — Phil. Mag. **44**, 233, 1129 (1953); **45**, 1177 (1954), and other references given there.
[459] FULTZ, D., and Y. NAKAGAWA: Proc. Roy. Soc. Lond., Ser. A **231**, 211 (1955). — NAKAGAWA, Y.: Proc. Roy. Soc. Lond., Ser. A **240**, 108 (1957).
[460] BURBIDGE, G. R.: Astrophys. J. **124**, 412 (1956).
[461] CHANDRASEKHAR, S., and D. N. LIMBER: Astrophys. J. **119**, 10 (1954).
[462] CHANDRASEKHAR, S., and E. FERMI: Astrophys. J. **118**, 116 (1953).
[463] PRENDERGAST, K. H.: Astrophys. J. (in the press).
[464] BURBIDGE, G. R.: Astrophys. J. **120**, 589 (1954).
[465] MUSTEL, E. R.: Symposium on Non-Stable Stars, I.A.U. Monograph No. 3, p. 57, 1957, and other references given there.

Die Häufigkeit der Elemente in den Planeten und Meteoriten.

Von

Hans E. Suess und Harold C. Urey.

Mit 1 Figur.

A. Einleitung.

1. Historisches. Seit mehr als 100 Jahren war die durchschnittliche Zusammensetzung der Erdkruste ein Gegenstand wissenschaftlichen Interesses. Systematische Untersuchungen, die sich auf eine große Anzahl einzelner Gesteinsanalysen gründeten, sind erstmalig 1889 von Clarke [*1*] und späterhin gemeinsam mit seinem Mitarbeiter Washington an der U.S. Geological Survey durchgeführt worden. Diese Untersuchungen gaben Aufschluß über geochemische Vorgänge, die zur An- bzw. Abreicherung gewisser Elemente in bestimmten Gesteinen und Mineralien führten. Diese Untersuchungen haben aber auch gezeigt, daß diese Schwankungen einer grundlegenden, durchschnittlichen Häufigkeitsverteilung der Elemente überlagert sind, die in keiner Weise mit ihren chemischen Eigenschaften in Zusammenhang gebracht werden kann.

Als ein Maß für die durchschnittliche chemische Zusammensetzung der Erde gelten die Analysenergebnisse der Meteoriten, besonders der Chondrite, da in ihnen eine quantitative Trennung der Metallphase von Silikat und Sulfid nicht stattgefunden hat, und es somit unwahrscheinlich erscheint, daß innerhalb dieser einzelnen Phasen erhebliche Konzentrationsverschiebungen stattgefunden haben. Den Höhepunkt dieser Untersuchungen bildete 1937 eine Arbeit von Goldschmidt [*2*], die die Grundlage aller weiteren Untersuchungen auf diesem Gebiet darstellt. Einen weiteren Beleg dafür, daß die auf diesem Wege gefundene mittlere chemische Zusammensetzung der Meteorite eine fundamentale Bedeutung besitzt, ergaben die spektralanalytischen Untersuchungen der Sonnen- und Sternatmosphären durch Unsöld [*3*]. Durch diese und zahlreiche weitere nachfolgende Arbeiten anderer Forscher[1] zeigte sich, daß die mittlere Zusammensetzung der Meteorite innerhalb der Fehlergrenzen der analytischen Methoden recht genau den nichtflüchtigen Bestandteilen der Atmosphären der zur Population I gehörenden Sterne unseres Milchstraßensystems entspricht.

2. Kernphysikalische Zusammenhänge. Diese Übereinstimmung der aus Meteoritenanalysen gewonnenen Daten mit den Ergebnissen spektralanalytischer Untersuchungen astronomischer Objekte weist darauf hin, daß die Häufigkeitsverteilung der Elemente mit kernphysikalischen Eigenschaften im Zusammenhang stehen müssen und daß die Materie, die uns umgibt, gleichsam die Asche darstellt, die nach dem Ablauf thermischer Kernprozesse in einem frühen Stadium der Entwicklung unseres galaktischen Systems zurückgeblieben ist. Bei einem Prozeß der Bildung schwererer Kerne wird die gebildete Menge der einzelnen Atomkernsorten in irgendeinem Zusammenhang stehen mit den kernphysi-

[1] Siehe den folgenden Artikel von L. H. Aller in diesem Bande.

kalischen Bestimmungsstücken und Eigenschaften dieser Kernsorten, wie etwa Massenzahl, Bindungsenergie, Neutronenüberschuß und Wirkungsquerschnitte für verschiedene Kernreaktionen. Gelingt es, einen solchen Zusammenhang aufzudecken, dann wird man ihn als einen Beweis dafür betrachten können, daß die für die Meteoriten ermittelten Häufigkeitsdaten in der Tat in guter Näherung der ursprünglichen Häufigkeitsverteilung der Elemente bei ihrer Entstehung entsprechen. Daß dies der Fall ist, ist 1947 von einem von uns [4] eingehend gezeigt worden.

Zur Deutung der Häufigkeitsverhältnisse der Elemente und der empirischen und halbempirischen Züge in der Häufigkeitsverteilung der Kernsorten wäre eine vollständige Theorie der Elemententstehung nötig. Derartige Theorien haben jedoch bisher kaum mehr gezeigt, als daß es möglich ist, die Existenz der schweren Kerne zu verstehen und ihr Häufigkeitsverhältnis in qualitativer Weise zu deuten. Der vorliegende Artikel beabsichtigt, das empirische Material so zusammenzufassen, daß es als Grundlage für weitere zukünftige Arbeiten dienen kann, die ein Verständnis der Elemententstehung zum Ziele haben.

B. Empirische Regeln für die relative Häufigkeit der Kernsorten.

3. Die Häufigkeitsregeln. Ein bekannter, nach dem Stande unseres Wissens beinahe selbstverständlicher Satz lautet: „Sämtliche stabilen Atomkernarten — und dazu noch einige wenige fast stabile mit ihren Folgeprodukten — kommen in der Natur vor.“ Ihre relative Häufigkeit zeigt jedoch Schwankungen um mehr als 15 Zehnerpotenzen. Die ersten Versuche zu einer Systematisierung dieser Schwankungen wurden 1917 von HARKINS [5] vorgenommen. Die Harkinssche Regel bringt eine der bedeutsamsten Eigenschaften der Häufigkeitsverteilung der Kernsorten zum Ausdruck. Eine Reihe von weiteren Regeln ist von MATTAUCH [6] formuliert worden. Die folgenden, von SUESS [4] aufgestellten Regeln beinhalten zum Teil die von HARKINS und von MATTAUCH gemachten Aussagen. Sie beabsichtigen eine quantitativere Betrachtungsweise der Zusammenhänge.

1. Kerne mit ungerader Massenzahl. Bei Kernen, deren Massenzahl A den Wert 50 übersteigt, bilden die Häufigkeitswerte eine glatte Funktion der Massenzahl. *Zusatz:* Existieren für eine Massenzahl zwei Isobare, dann gilt diese Regel für die Summe ihrer Häufigkeiten.

2. Kerne mit gerader Protonen- und gerader Neutronenzahl. Im Gebiete der Kerne, deren Massenzahl 90 übersteigt, bildet die Summe der Häufigkeitswerte der Isobaren eine glatte Funktion der Massenzahl. Im Gebiete der leichteren Kerne ($A < 90$) ändert sich die Häufigkeit der Kerne mit gleichem Neutronenüberschuß stetig mit der Massenzahl.

3. Relative Isobarenhäufigkeiten. Im Gebiete der schwereren Kerne ($A > 70$) ist das Isobar mit dem höheren Neutronenüberschuß, im Gebiete der leichteren Kerne ($A < 70$) das Isobar mit dem geringeren Neutronenüberschuß stets das häufigere.

4. Die Harkinssche Regel. Die Summe der Häufigkeiten der Kerne gerader Massenzahl ist stets größer als die Häufigkeit der Kerne der benachbarten ungeraden Massenzahlen. Der Unterschied nimmt mit zunehmender Massenzahl ab.

5. Ausnahmen. Diese Regeln besitzen vor allem Ausnahmen in Gebieten, in denen bei gewissen Neutronenzahlen (den sogenannten magischen Zahlen) Schalenabschlüsse im Kernbau auftreten.

Die Regeln von der stetigen Änderung der Häufigkeit mit der Massenzahl tritt augenfällig im Gebiet der Seltenen Erden zutage. Die Seltenen Erden besitzen so ähnliche chemische Eigenschaften, daß eine Verschiebung ihres Konzentrationsverhältnisses durch kosmochemische Vorgänge nicht anzunehmen ist. Die Regelmäßigkeiten, wie sie im Gebiet der Seltenen Erden unmittelbar zu erkennen sind, sind für das gesamte Gebiet des Periodischen Systems zu erwarten.

4. Abschätzung der Häufigkeiten. Bei der folgenden Diskussion der Häufigkeiten der einzelnen Elemente wird grundsätzlich angenommen, daß ihre isotope Zusammensetzung in irgendeiner Weise bedeutungsvoll sein muß und sich in ein geschlossenes Gesamtbild der Häufigkeiten aller Kernsorten einordnet. Es ist jedoch durchaus möglich, daß die für gewisse Massengebiete angenommene stetige Abhängigkeit der Häufigkeit von der Massenzahl nicht streng gilt und daß vielmehr die einzelnen Werte sich regellos um eine glatte Kurve gruppieren. Wir vermuten jedoch, daß dies im allgemeinen nicht der Fall ist, so daß durch das Häufigkeitsverhältnis isotoper Kernsorten die Neigung der Kurvenzüge streng festgelegt wird. Wir finden hierbei, daß die Linien für gerade und für ungerade Massenzahlen über weite Bereiche hin nahezu parallel verlaufen, was uns gleichfalls für die Richtigkeit unserer Annahme zu sprechen scheint.

C. Die empirischen Elementhäufigkeiten.

I. Allgemeines.

5. Daten und Normierung. Die folgende Diskussion schließt sich mit geringfügigen Änderungen der 1956 von SUESS und UREY [*14*] gegebenen an, in der die ersten Versuche von SUESS [*4*], zu einem geschlossenen Gesamtbild der Kernhäufigkeiten zu gelangen, einer eingehenden Revision unterzogen wurden. Während hierbei für eine Reihe von Elementen beträchtlich abweichende Werte ermittelt wurden, sind die wesentlichen Züge der Gesamtverteilung der Kernhäufigkeiten beibehalten worden.

Die atomaren Elementhäufigkeitswerte sind hier stets relativ zu einer Siliciumhäufigkeit gleich 10^6 angegeben. In der Tabelle GOLDSCHMIDTs wurde diese Häufigkeit gleich 100 gesetzt und BROWN [7] verwendet zur Definition seiner Häufigkeitseinheit einen Wert für Silicium gleich 10000. Die Wahl eines Bezugswertes von $Si = 10^6$ geschah zur Vermeidung großer negativer Exponenten. Die Elementhäufigkeiten sind in unseren Einheiten in Tabelle 5 angegeben. Die sich aus ihnen ergebenden Kernhäufigkeiten sind in Fig. 1a—c logarithmisch gegen die Massenzahl *A* aufgetragen und in Tabelle 6 angegeben (S. 317ff.).

6. Die Zusammensetzung der Sonne und der Planeten. Wir nehmen an, daß die Atmosphäre der Sonne alle Elemente in ihrem ursprünglichen Konzentrationsverhältnis enthält, sofern sie nicht durch Kernreaktionen im Sonneninneren Umwandlungen erfahren haben und deren Produkte durch Konvektion an die Sonnenoberfläche gelangt sind. Diese Möglichkeit besteht bei den leichtesten Kernen. Wir nehmen an, daß der Gehalt der Sonne an allen anderen Elementen der Zusammensetzung einer Urmaterie entspricht, aus der sich die Planeten und Meteoriten gebildet haben.

Der Stoffbestand des Planeten Jupiter und der anderen großen Planeten entspricht vermutlich der Zusammensetzung dieser Urmaterie, wenn man von der Möglichkeit eines geringeren Gehalts an Wasserstoff und Helium absieht. Während der Entstehung der terrestrischen Planeten sind jedoch flüchtige Bestandteile, wie Wasserstoff, die Edelgase, Kohlenstoff als CH_4, Stickstoff als NH_3, Sauerstoff als Wasserdampf, vermutlich auch Schwefel und die Halogene, sowie vielleicht auch andere Elemente zum überwiegenden Teil abgetrennt worden. Das Häufigkeitsverhältnis der Elemente in der Erdkruste weicht offensichtlich noch in anderer Hinsicht von dem der Urmaterie ab. Durch

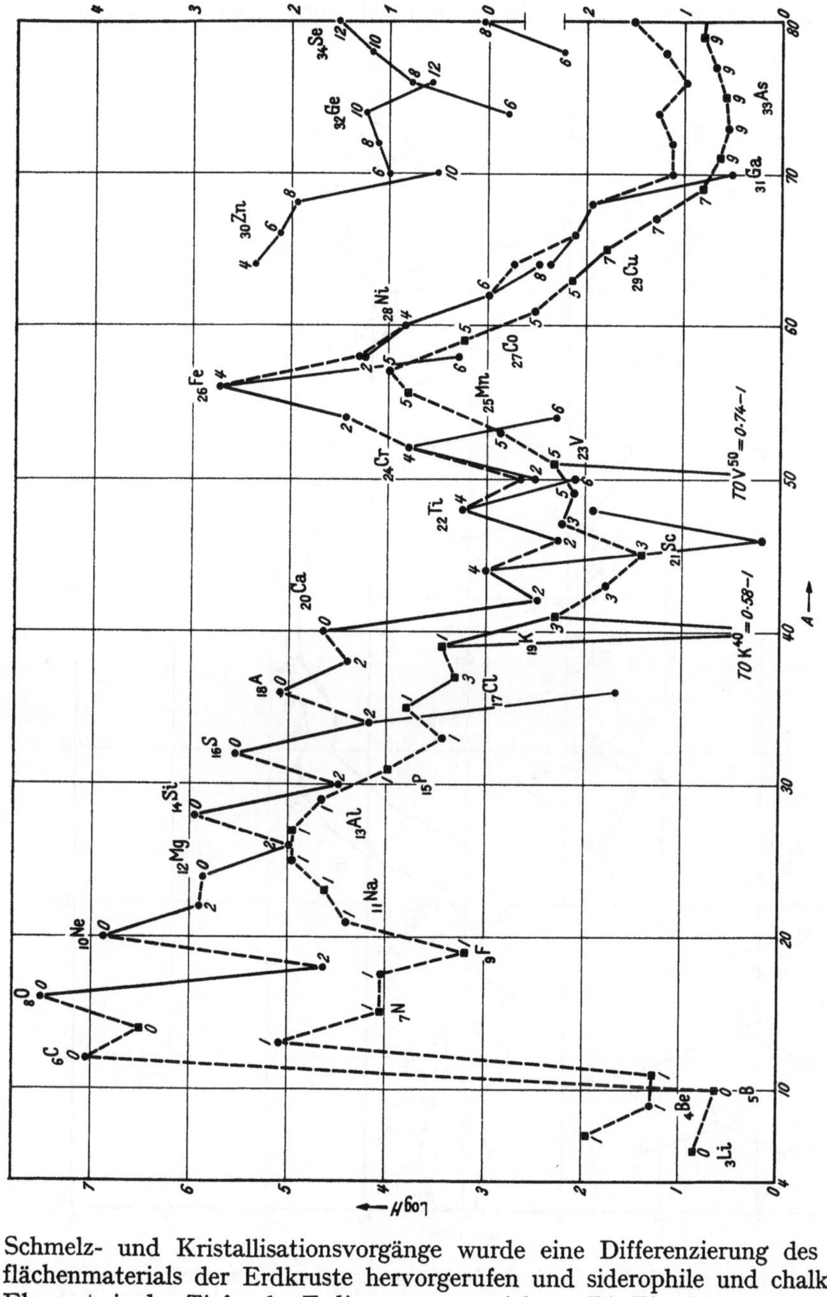

Fig. 1 a—c. Logarithmen der Kernhäufigkeiten nach Tabelle 6, aufgetragen gegen die Massenzahl *A*. Für jeden Punkt ist der Isotopenüberschuß angegeben. Die gestrichelten Linien verbinden die Werte für die Häufigkeitssummen von Isobaren. Die dünnen Linien verbinden Werte isotoper Kerne gerader Massenzahl. Der Maßstab dieser Werte ist für $A > 64$ um drei Einheiten versetzt und in der Figur rechts angegeben.

Schmelz- und Kristallisationsvorgänge wurde eine Differenzierung des Oberflächenmaterials der Erdkruste hervorgerufen und siderophile und chalkophile Elemente in den Tiefen des Erdinneren angereichert. Die Einwirkung des Wassers sorgte für weitere chemische Trennung. Die Abschätzung der mittleren chemischen Zusammensetzung des Oberflächenmaterials der Erde ist daher schwierig.

Trotzdem ist es möglich, einige Häufigkeitsdaten, die sich auf die Erdkruste beziehen, in unsere Betrachtungen einzubeziehen.

7. Der Stoffbestand der Meteorite. Eine bedeutend geringere chemische Differenzierung als das Material der Erde haben die Meteorite erfahren. In ihnen ist

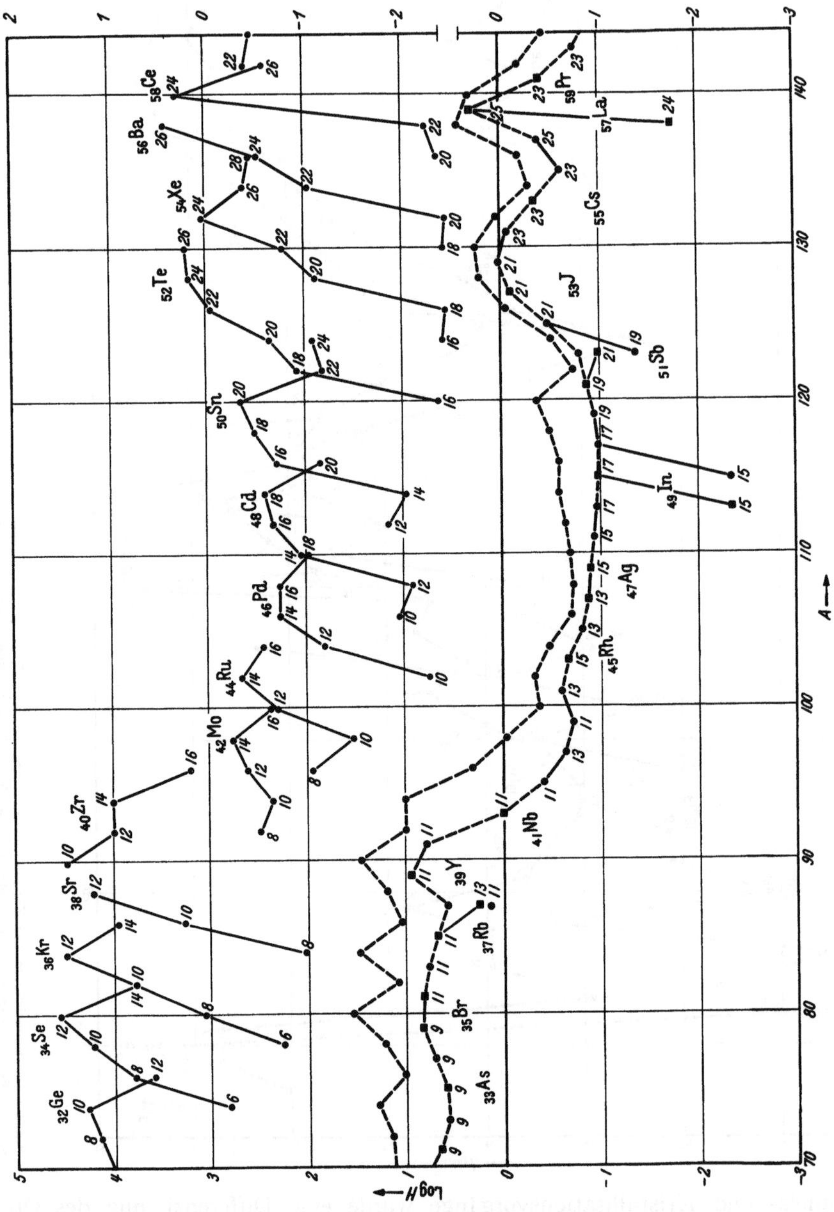

Fig. 1 b.

die Materie im wesentlichen in drei Phasen geschieden: dem Metall, Sulfid und Silikat. Je nach dem bevorzugten Auftreten eines Elementes in einer dieser Phasen wird es nach GOLDSCHMIDT [2] als siderophil, chalkophil oder lithophil bezeichnet. Diese Zuordnung kann jedoch nicht immer in eindeutiger Weise vorgenommen werden, da manche Elemente in wechselnder Menge sich zwischen zwei oder auch allen drei meteoritischen Hauptphasen verteilt vorfinden. Die wesentlichste Schwierigkeit bei der Angabe einer durchschnittlichen chemischen

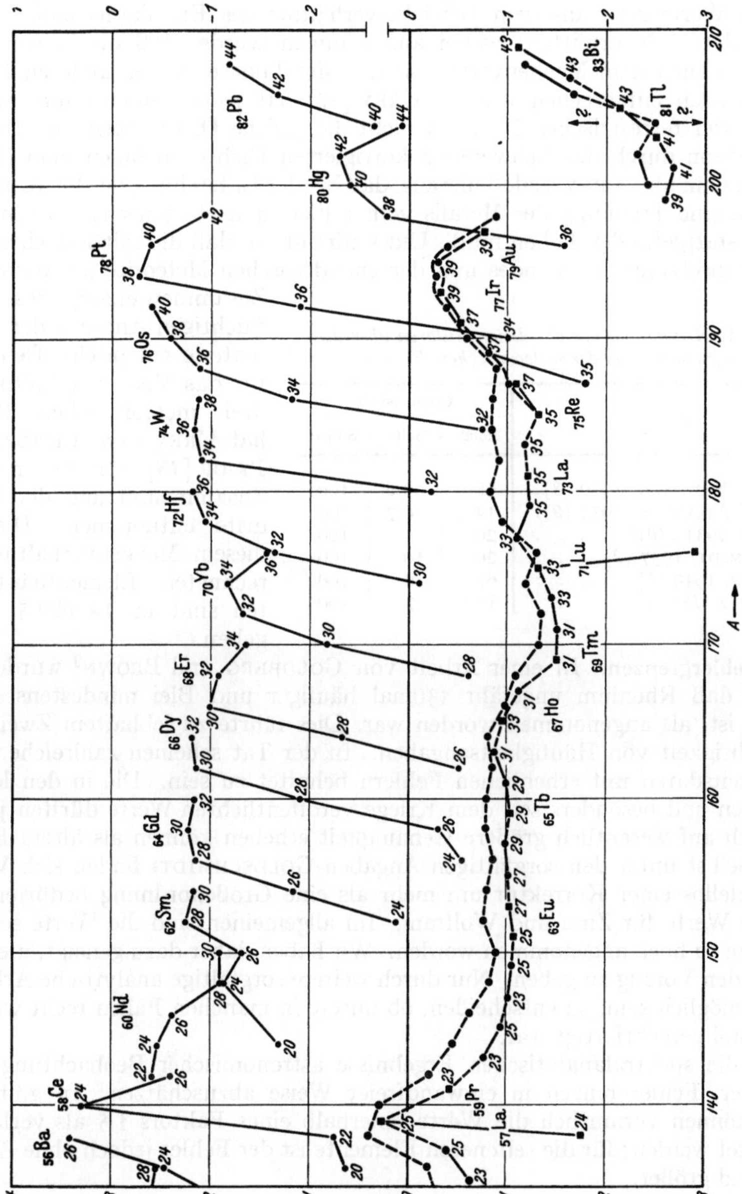

Fig. 1 c.

Zusammensetzung der Meteorite besteht in unserer Unkenntnis des Mengenverhältnisses dieser drei Hauptphasen. Eine Abschätzung dieses Mengenverhältnisses aus dem Bestand der auf der Erde auftreffenden Meteorite ist nicht ohne weiteres möglich, da Meteoreisen eine größere Wahrscheinlichkeit besitzen, die

Erdoberfläche zu erreichen als die chemisch und mechanisch wenig widerstandsfähigen Meteorsteine. In Tabelle 1 sind die Annahmen verschiedener Autoren über das Mengenverhältnis der drei meteoritischen Phasen angegeben.

Die von HARRISON BROWN [7] und die von I. und W. NODDACK [8] angegebenen Werte sind aus dem Gewichtsverhältnis des Eisenkerns zum Silikatmantel der Erde ermittelt, wobei angenommen wurde, daß dieses Gewichtsverhältnis auch dem Mengenverhältnis der Metallphase zu den anderen Phasen in Meteoriten entsprechen würde. UREY [11] hat jedoch darauf hingewiesen, daß die vier terrestrischen Planeten nicht die gleiche Dichte besitzen. Ihre für Kompression durch das Schwerefeld korrigierten Dichten nehmen vom Merkur zum Mars hin ab. UREY schloß hieraus, daß bei der Entstehung der terrestrischen Planeten eine Trennung des Metalls vom Silikat und ein teilweiser Verlust des Silikats stattgefunden haben muß. UREY nimmt an, daß die mittlere chemische Zusammensetzung des Mondes und der chondritischen Meteorite am besten der Zusammensetzung des nichtflüchtigen Anteiles der Solarmaterie entspricht. Den Wert für das Mengenverhältnis der drei meteoritischen Phasen hat UREY den Angaben von PRIOR [13] für die mittlere Zusammensetzung der Chondrite entnommen. Die mit diesem Mengenverhältnis berechneten Elementhäufigkeiten sind in Tabelle 5 angegeben.

Tabelle 1. *Annahmen verschiedener Autoren über das Mengenverhältnis der meteoritischen Phasen.*

Autor	Gewichtsteile		
	Metall	Sulfid	Silikat
I. und W. NODDACK, 1930 [8].	68	9,8	100
I. und W. NODDACK, 1934 [9].	14,6	6,7	100
FERSMAN, 1934 [10]	20	4	100
GOLDSCHMIDT, 1937 [2] . . .	20	10	100
H. BROWN, 1949 [7]	67	0	100
UREY, 1952 [11]	10,6	7	100

8. Fehlergrenzen. In einer Arbeit von GOLDBERG und BROWN[1] wurde 1950 gezeigt, daß Rhenium ungefähr 130mal häufiger und Blei mindestens 50mal seltener ist, als angenommen worden war. Dies führte zu lebhaftem Zweifel an der Richtigkeit von Häufigkeitsangaben. In der Tat scheinen zahlreiche ältere Häufigkeitsdaten mit erheblichen Fehlern behaftet zu sein. Die in den letzten 20 Jahren und besonders seit dem Kriege veröffentlichten Werte dürften jedoch Anspruch auf wesentlich größere Genauigkeit erheben können als ältere Ergebnisse. Selbst unter den sorgfältigen Angaben GOLDSCHMIDTs finden sich Werte, die zweifellos einer Korrektur um mehr als eine Größenordnung bedürfen, wie z.B. die Werte für Zinn und Wolfram. Im allgemeinen sind die Werte seltener Elemente zu hoch angenommen worden. Wir haben daher dazu geneigt, niederen Werten den Vorzug zu geben. Nur durch weitere sorgfältige analytische Arbeiten wird es möglich sein, zu entscheiden, ob unsere in manchen Fällen recht willkürliche Wahl gerechtfertigt war.

Für die spektralanalytischen Ergebnisse astronomischer Beobachtungen ist es schwer, Fehlergrenzen in einwandfreier Weise abzuschätzen. In günstigen Fällen können vermutlich die Werte innerhalb eines Faktors 1,5 als verläßlich betrachtet werden; für die selteneren Elemente ist der Fehler jedoch ohne Zweifel bedeutend größer.

II. Die Häufigkeit der leichteren Elemente bis Nickel.

a) Die Elemente von Wasserstoff bis Fluor.

9. Wasserstoff und Helium. Das Häufigkeitsverhältnis H/He bildet seit langem den Gegenstand astronomischer Untersuchungen. Über ihren gegen-

[1] E. GOLDBERG u. H. S. BROWN: Analyt. Chem. **22**, 308 (1950).

wärtigen Stand wird in dem Artikel von ALLER in diesem Bande ausführlich berichtet. Unsere Tabelle enthält den Wert 7,8 für das Verhältnis H/He.

Für das Verhältnis des Wasserstoffs zu Metallen nehmen wir das geometrische Mittel des Wertes von CLAAS[1] und UNSÖLD[2] und finden, normalisiert auf einen Mittelwert von Mg und Si für den Briggschen Logarithmus der Wasserstoffhäufigkeit 10,5 bezogen auf $\log[\mathrm{Si}] = 6$.

Für die Häufigkeit des Deuteriums relativ zum Wasserstoff in Meteoriten ist von BOATO[3] und von EDWARDS[4] ungefähr derselbe Betrag wie auf der Erde gefunden worden. Nach DE JAGER[5] soll auch auf der Sonne das Deuterium in ungefähr der gleichen Konzentration vorkommen, wenn auch diese Aussage lediglich auf der beobachteten Intensität der D_α-Linie beruht. Wir geben für H/D den Wert 7000 an, während das Durchschnittsverhältnis für irdischen Wasserstoff bei 6500 liegt.

Die Häufigkeit von He^3 in Solarmaterie ist unbekannt.

10. Lithium, Beryllium und Bor. Die Häufigkeit dieser drei Elemente ist etwa um einen Faktor 10^5 geringer als die der drei nächst schwereren Elemente Kohlenstoff, Stickstoff und Sauerstoff. Nach GREENSTEIN und RICHARDSON[6] ist in der Sonnenatmosphäre Li um einen weiteren Faktor 100 seltener als auf der Erde. Die geringe Häufigkeit dieser drei Elemente kann leicht verstanden werden als eine Folge thermischer Kernprozesse mit Protonen im Sterninneren. In unserer Tabelle sind für diese Elemente die von GOLDSCHMIDT [*2*] angegebenen Analysenwerte von Steinmeteoriten aufgenommen.

11. Kohlenstoff, Stickstoff, Sauerstoff und Neon. Unsere Kenntnis der Häufigkeit dieser Elemente beruht ausschließlich auf astronomischen Beobachtungen. Für unsere Tabelle übernehmen wir die von ALLER in diesem Bande angegebenen Daten, umgerechnet auf die von uns gewählte Normierung.

Die Häufigkeit von Neon ist wahrscheinlich geringer als die des Sauerstoffs, was verständlich wäre, da ein Abfall der Häufigkeitswerte mit dem Schalenabschluß im Kernbau bei 8 Protonen und 8 Neutronen plausibel erscheint.

12. Fluor. Die astronomischen Werte für Fluor sind sehr unsicher. Wir sind daher auf meteoritische und terrestrische Werte angewiesen. Eine Fluorbestimmung in Meteoriten durch I. und W. NODDACK [*9*] ergab 30 g/Tonne. Gesteinsanalysen von KORTINIK[7] deuten auf einen Durchschnittswert von etwa 700 g/Tonne für die Lithosphäre. Wir schätzen die Häufigkeit des Fluors zu etwa 200 g/Tonne Si und erhalten hieraus für die atomare Häufigkeit 1600 in unseren Einheiten.

b) Die Elemente von Natrium bis Nickel.

13. Allgemeines. Für die Elemente der Ordnungszahlen 11 bis 28 liegen sowohl für Gesteine als auch für Meteorite zahlreiche verläßliche Analysenergebnisse vor. Die älteren Daten sind 1937 von GOLDSCHMIDT [*2*] und 1947 von BROWN und PATTERSON [*15*] zusammengestellt und zur Angabe von Mittelwerten verwendet worden. Neuerdings wurde eine noch eingehendere und ausführlichere Auswertung älterer Analysen durch UREY und CRAIG [*16*] vorgenommen. Die von diesen Autoren erhaltenen Werte unterscheiden sich nicht wesentlich von den Angaben GOLDSCHMIDTs, doch zeigte sich, daß zwei wohl definierte

---

[1] W. J. CLAAS: Proc. Acad. Sci. Amsterd. **52**, 518 (1951).
[2] A. UNSÖLD: Z. Astrophys. **24**, 306 (1948).
[3] G. BOATO: Geochim. et Cosmochim. Acta **6**, 209 (1954). — Phys. Rev. **93**, 640 (1954).
[4] G. EDWARDS u. H. C. UREY: Geochim. et Cosmochim. Acta **7**, 154 (1955).
[5] C. DE JAGER: Mém. Soc. Roy. Sci., Liège, IV. ser. **13**, (III), 460 (1953).
[6] J. GREENSTEIN u. R. S. RICHARDSON: Astrophys. J. **113**, 536 (1951).
[7] S. KORTINIK: Geochim. et Cosmochim. Acta **1**, 89 (1951).

Gruppen von Chondriten existieren, die etwas verschiedene Mengen von Gesamteisen enthalten, und zwar 0.608 bzw. 0,849 Atome Eisen je Atom Silicium. Der Unterschied im Kobalt- und Nickelgehalt ist noch etwas größer. Da die terrestrischen Planeten bei ihrer Bildung Silikate verloren haben, nimmt UREY an, daß der niederere Eisenwert der Chondrite eher der Zusammensetzung der Solarmaterie entspricht als der höhere.

14. Natrium, Magnesium und Aluminium. Die Übereinstimmung der aus Chondritanalysen ermittelten Häufigkeitswerte dieser drei Elemente mit den astronomischen Daten ist befriedigend. Magnesium besitzt in Meteoriten sicherlich eine etwas geringere Häufigkeit als Silicium. Dies mag mit dem kernphysikalisch freilich nur schwach ausgeprägten Schalenabschluß bei N und $Z=14$ zusammenhängen. Die durch Meteoritenanalysen und astronomische Beobachtungen ermittelte Aluminiumhäufigkeit ist bedeutend geringer als der Verbreitung dieses Elements auf der Erdoberfläche entsprechen würde.

15. Phosphor, Schwefel, Chlor und Argon. Die astronomischen Daten für Schwefel und Chlor weisen eindeutig auf höhere Häufigkeitswerte, als die Analysen der Meteorite. Es ist möglich, daß diese Elemente bei der Bildung der Meteorite zum größeren Teil nicht zur Kondensation gelangt sind.

Für Phosphor ergaben die Meteoritenanalysen stark schwankende Werte. Dies gilt auch für die Bestimmungen, die 1956 von WIIK[1] ausgeführt wurden. Wir haben daher einen relativ hohen Phosphorwert gewählt, der den astronomischen Angaben nahekommt.

Das Schwefel-Silicium-Verhältnis bildet die Grundlage für die Ermittlung der Häufigkeiten chalkophiler Elemente, und ist daher äußerst bedeutsam. Sein Wert ist jedoch außerordentlich schwer abzuschätzen. Unser Schwefelwert liegt zwischen den astronomischen Angaben und den meteoritischen Werten. Er beträgt $3{,}75 \cdot 10^5$. Wir wählen diesen Wert, um mit dem bekannten Selen-Schwefel-Verhältnis einen passenden Wert für Selen zu erhalten.

Über die Chlorhäufigkeit liegen neuere Untersuchungen von BEHNE[2] und von SALPETER[3] vor. BEHNE wies in zwei Chondriten ungefähr 100 g/Tonne nach, und der Durchschnitt von SALPETERs Werten für Chondrite liegt bei 840 g/Tonne. SELIVANOV[4] gibt 100 g/Tonne für einen Chondriten an. Der ältere Wert der NODDACKS [*9*] beträgt 470 g/Tonne.

Die astronomischen Werte für Argon und Chlor in planetarischen Nebeln sind viel höher als die, die man durch einfache Interpolation der Isotopenhäufigkeiten und den meteoritischen Werten der Nachbarelemente erhalten würde. Wenn man die astronomischen Werte von A und Cl für die graphische Wiedergabe der Kernhäufigkeiten in diesen Massengebieten verwendet, dann ergibt sich ein auffallendes Bild, nämlich ein scharfer Sprung in den Häufigkeitslinien sowohl der geraden, als auch der ungeraden Massenzahlen bei $A=35$ bzw. 36, mit einem nachfolgenden stetigen Abfall von Cl^{37} über K^{39}, K^{41} zum Ca^{43} in der einen Linie und von A^{38} zu Ca^{42} in der anderen. Wir sehen keinen kernphysikalischen Grund für ein solches Verhalten der Kernhäufigkeiten und haben daher im Hinblick auf die Unsicherheit der spektralanalytischen Bestimmungen die Werte für A^{36} und A^{38} glatt interpoliert zwischen den Häufigkeitswerten der Schwefel- und Calciumisotope und den oben angegebenen Chlorwert verwendet.

---

[1] H. B. WIIK: Geochim. et Cosmochim. Acta **9**, 279 (1956).

[2] W. BEHNE: Geochim. et Cosmochim. Acta **3**, 186 (1953).

[3] E. SALPETER: Ric. Spectroscop. Lab. astrofis, specola Vaticana **2**, 1 (1952).

[4] L. S. SELIVANOV: C. R. Acad. Sci. USSR. **26**, 389 (1940).

16. Die Elemente von Calcium bis Nickel. Calcium ist in Meteoriten mit großer Genauigkeit bestimmt worden. Die neuesten Analysen von WIIK ergeben 49000 für die Häufigkeit dieses Elements.

Neuere Bestimmungen von Scandium in Meteoriten durch PINSON, AHRENS und FRANCK[1] bestätigen im wesentlichen den von GOLDSCHMIDT [*2*] angegebenen Wert von 6 g/Tonne. WIIK fand 8 g/Tonne in Chondriten. Die astronomischen Werte liegen um mehr als einen Faktor 2 höher. Aus diesem Grunde geben wir den höchsten Analysenwert für Meteorite in unserer Tabelle.

Titan ist mit vorzüglicher Genauigkeit in Meteoriten bestimmt worden. Nach WIIKs Angaben beträgt der Titangehalt einer Anzahl von Chondriten etwa 0,079% und ist nahezu konstant. Dies gibt 2440 für die atomare Häufigkeit des Titans.

Neuere Analysenergebnisse von WIIK für Chrom und Mangan in Chondriten stimmen gut mit den von UREY und CRAIG [*16*] angegebenen Daten überein. Ebenso sind die von UREY und CRAIG für die Chondritgruppe mit niedrigerem Eisengehalt ermittelten Werte für Eisen, Kobalt und Nickel hier übernommen. Auch diese Daten sind durch neue Analysen von WIIK bestätigt worden. Nur für Eisen wurde von WIIK ein geringfügig niedererer Wert ermittelt. Neue astronomische Beobachtungen führten zu einem noch niedrigeren Eisenwert.

III. Wichtige Häufigkeitsverhältnisse homologer Elemente.

17. Titan-Zirkon. Das Häufigkeitsverhältnis der schwereren zu den leichteren Elementen und somit der Gang der Häufigkeitslinien über den gesamten Bereich des Periodischen Systems hin ist durch einige wenige Verhältnisse der Häufigkeiten chemisch ähnlicher Elemente festgelegt. Am bedeutsamsten hierfür ist das Titan-Zirkon-Hafnium-Verhältnis, das in Meteoriten und in terrestrischen Gesteinen nicht wesentlich verschieden ist. GOLDSCHMIDT [*2*] gibt für das Gewichtsverhältnis Ti/Zr = 20 und Zr/Hf = 50 an. PINSON, AHRENS und FRANCK[1] fanden 33 g Zr/Tonne und WIIK[2] 790 g Ti/Tonne in Chondriten. Hieraus ergibt sich ein Gewichtsverhältnis Ti/Zr = 24, entsprechend einem Atomverhältnis von 45. Wir verwenden diesen Wert.

18. Zirkon-Hafnium. Das Zirkon-Hafnium-Verhältnis ist 1934 von HEVESY und WÜRSTLIN[3] an einer großen Zahl von Proben eingehend untersucht worden. Das Verhältnis ist praktisch konstant und nur selten finden sich in gewissen Mineralen Abweichungen im Sinne eines höheren Hafniumgehalts. Neuere, am U.S. Bureau of Mines ausgeführte Messungen[4] stimmen mit den älteren Befunden vorzüglich überein. Der Durchschnittswert von 68 Analysen beträgt 2,37% Hf relativ zur Summe Zr + Hf. Wir verwenden 55 für das Gewichtsverhältnis und 110 für das atomare Verhältnis von Zr/Hf.

19. Schwefel-Selen. Der Goldschmidtsche Wert für das Gewichtsverhältnis S/Se beträgt 6000 für die Litosphäre und 3300 für Meteorite. Neuere, 1938 von BYERS[5] ausgeführte Analysen ergaben 13 bzw. 10 g Se/Tonne in zwei Chondriten und ein Gewichtsverhältnis S/Se im Troilit des Eisens von Canyon Diablo

---

[1] W. H. PINSON, L. H. AHRENS u. M. L. FRANCK: Geochim. et Cosmochim. Acta **4**, 251 (1953).
[2] H. B. WIIK: Geochim. et Cosmochim. Acta **9**, 279 (1956).
[3] G. v. HEVESY u. K. WÜRSTLIN: Z. phys. Chem. **216**, 305 (1934).
[4] Nach einer freundlichen privaten Mitteilung von Dr. S. G. ENGLISH.
[5] H. BYERS: Industr. Engng. Chem., New Edit. **16**, 459 (1938).

von 4215. Wegen der Verschiedenheit in der Stabilität von H_2S und H_2Se sowie der Kohlenstoffverbindungen dieser beiden Elemente ist vermutlich Selen bei der Bildung der Meteorite weniger leicht abgetrennt worden als Schwefel. Wir wählen daher eine etwas höhere Selenhäufigkeit, als den Meteoritenanalysen entsprechen würde, nämlich 67,6.

Die Angaben über Tellurhäufigkeiten sind außerordentlich unsicher und wir sind auf eine Interpolation des Wertes für dieses Element angewiesen.

20. Chlor-Brom. Angaben über das Verhältnis Cl/Br scheinen verläßlicher zu sein, als die über die Absolutkonzentration von Br in Meteoriten. Im Meerwasser beträgt das Gewichtsverhältnis Cl/Br 292. Für Urgestein gibt SELIVANOV 1940[1] Werte von 100 bis nahezu 300 an. BEHNE[2] fand außerordentlich große Schwankungen in diesem Verhältnis für irdische Gesteine. Wir verwenden den Wert für Meerwasser aus dem sich mit unserem Wert für Chlor eine Bromhäufigkeit von 13,4 ergibt. Der Jodgehalt der Ozeane kann nicht in gleicher Weise für unsere Abschätzung herangezogen werden, da Jod in hohem Maße von lebenden Organismen angereichert wird und durch abgestorbenes Material in die Sedimente gelangt.

21. Kalium-Rubidium. Das Verhältnis K/Rb ist in den letzten Jahren eingehend untersucht worden. AHRENS, PINSON und KEARNS[3] fanden 1952 für das Gewichtsverhältnis K/Rb etwa 100. EDWARDS und UREY[4] geben im Durchschnitt 180 für einige Meteoritenproben an. HERZOG und PINSON[5] geben 1955 an, daß der wahrscheinlichste Wert bei 200 liegen dürfte. Die analytisch bestimmte Rubidiumhäufigkeit entspricht 3,8 g/Tonne Meteorit bzw. 6,5 Atomen Rb pro 10^6 Atome Si.

22. Siderophile Elemente. Die Elemente Ni, Co, Pd, Au und Ga sind 1951 mit großer Genauigkeit von GOLDBERG, UCHIYAMA und BROWN[6] in 45 Eisenmeteoriten bestimmt worden. Nimmt man an, daß diese Elemente ausschließlich in der Metallphase vorkommen, dann müssen die so gefundenen relativen Häufigkeitswerte ihrem kosmischen Vorkommen entsprechen. Das von diesen Autoren bestimmte Gewichtsverhältnis Ni/Pd beträgt $2{,}24 \cdot 10^4$. Dieser Wert ist der genaueste und verläßlichste, den wir kennen. Er ergibt mit unserem Nickelwert eine atomare Palladiumhäufigkeit von 0,675. Der Goldschmidtsche Wert von 2,5 beruht auf der Annahme eines weitaus höheren Anteiles der Metallphase in Meteoriten.

In gleicher Weise gibt das von GOLDBERG *et al.* angegebene Ni/Au-Verhältnis einen wertvollen Fixpunkt für den Verlauf der Häufigkeitslinien im Gebiet der schweren Kerne.

23. Die Edelgase. Wie erwähnt, liegen astronomische Daten für die Häufigkeit von Helium und Neon in den Atmosphären von Fixsternen vor. Für das Vorkommen von Argon können nur qualitative Angaben gemacht werden. Für die schwereren Edelgase liegen keinerlei Bestimmungen vor. Es können jedoch aus dem Konzentrationsverhältnis der Edelgase in der Erdatmosphäre wertvolle Schlüsse in bezug auf ihre kosmische Häufigkeit gezogen werden. Nach SUESS[7]

---

[1] L. S. SELIVANOV: C.R. Acad. Sci. USSR. **26**, 389 (1940).
[2] W. BEHNE: Geochim. et Cosmochim. Acta **3**, 186 (1953).
[3] L. H. AHRENS, W. H. PINSON u. M. M. KEARNS: Geochim. et Cosmochim. Acta **2**, 229 (1952).
[4] G. EDWARDS u. H. C. UREY: Geochim. et Cosmochim. Acta **7**, 154 (1955).
[5] L. HERZOG u. W. H. PINSON: Unveröffentlicht.
[6] E. GOLDBERG, A. UCHIYAMA u. H. BROWN: Geochim. et Cosmochim. Acta **2**, 1 (1951).
[7] H. E. SUESS: J. Geology **57**, 600 (1949).

und nach BROWN[1] kann man erwarten, daß das Verhältnis der irdischen zur kosmischen Häufigkeit der Edelgase eine glatte Funktion ihres Atomgewichts darstellt. Die Überlegungen zeigen, daß das Verhältnis Kr/Xe in der Sonne annähernd gleich dem in der Erdatmosphäre oder kleiner sein muß. Unser Wert für das Häufigkeitsverhältnis Kr/Xe entspricht dem Konzentrationsverhältnis in der Erdatmosphäre. Der Xenonwert bestimmt die unsichere Tellurhäufigkeit.

In der Erdatmosphäre sind He^3, He^4 sowie A^{40} radioaktiven Ursprungs. Diese Isotope können in unsere Betrachtungen nicht einbezogen werden.

IV. Die Häufigkeiten der mittelschweren und schweren Kerne unter Berücksichtigung der Häufigkeitsregeln.

a) Die Elemente von Kupfer bis Yttrium.

24. Allgemeines. Für die Elemente der Ordnungszahlen von 26 bis 40 gelingt es ohne Schwierigkeit, innerhalb der Fehlergrenzen der empirischen Daten, einen glatten Verlauf der Häufigkeit der Kerne ungerader Massenzahl als Funktion der Massenzahl zu erhalten. Mit den von uns gewählten Verhältnissen Cl/Br und S/Se besitzt die Häufigkeitslinie ein schwaches Maximum im Gebiet der Bromisotope. Das massenspektroskopisch ermittelte Häufigkeitsverhältnis der Isotopenpaare von Cu, Ga und Br bestimmt die Neigung der Linie an den betreffenden drei Punkten.

25. Kupfer. Ältere Kupferanalysen von Meteoriten zeigen überraschend große Schwankungen in ihren Ergebnissen, die mitunter einen Faktor 10 überschreiten. Auf Grund dieser älteren Analysen schätzt GOLDSCHMIDT [2] die atomare Kupferhäufigkeit auf 460. Neuere Messungen von WIIK[2] und von SANDELL[3] haben jedoch für Chondrite relativ konstante Werte ergeben, die einer Häufigkeit von 212 entsprechen. Wir übernehmen diesen Wert.

26. Zink. Dieses typisch chalkophile Element ist vorzugsweise im Sulfid der Meteorite erhalten. Seine Durchschnittshäufigkeit ist daher schwer abzuschätzen. Der von GOLDSCHMIDT angegebene Zinkwert ist kleiner als der von Kupfer und daher nicht mit unseren Häufigkeitsregeln verträglich. Der aus dem Sonnenspektrum ermittelte Wert ist um einen Faktor 3 höher. Wir verwenden einen um einen Faktor 1,3 höheren Wert als den von GOLDSCHMIDT angegebenen, um eine glatte Häufigkeitslinie für die Kerne ungerader Massenzahl zu erhalten und betrachten dies im Hinblick auf das Fehlen verläßlicher Meteoritenanalysen als gerechtfertigt.

27. Gallium. Für den Galliumgehalt in Eisenmeteoriten fanden GOLDBERG, UCHIYAMA und BROWN[4] ein sehr merkwürdiges Ergebnis. Danach fallen die Eisenmeteorite in bezug auf ihren Galliumgehalt in drei wohldefinierte Gruppen, die etwa 60, 20 und 2 g Ga pro Tonne enthalten. Eine befriedigende Erklärung dieses Befundes ist noch nicht gegeben worden, jedoch ist es wahrscheinlich, daß Gallium teilweise auch in der Silikatphase vorhanden ist. Eine neuere Analyse von SANDELL[5] stimmt gut mit den Angaben GOLDSCHMIDTs überein. Sie ergab 5,3 g Ga/Tonne Chondrit. Dies entspricht genau dem von uns verwendeten Wert von 11,4 für die atomare Häufigkeit.

---

[1] H. S. BROWN: In: The Atmospheres of the Earth and the Planets. Herausgeb. G. KUIPER. Chicago: University of Chicago Press 1952.

[2] H. B. WIIK: Private Mittlg.

[3] E. B. SANDELL: Geochim. et Cosmochim. Acta **8**, 221 (1955).

[4] E. GOLDBERG, A. UCHIYAMA u. H. BROWN: Geochim. et Cosmochim. Acta **2**, 1 (1951).

[5] H. ONISHI u. E. B. SANDELL: Geochim. et Cosmochim. Acta **9**, 78 (1956).

28. Germanium. Die sorgfältige Arbeit von GOLDSCHMIDT und PETERS[1] über den Gehalt von Meteoriten an Germanium ergab einen Durchschnittswert von 69 g Ge/Tonne, gemittelt über die drei meteoritischen Phasen. Das Germanium wurde vorzugsweise in der Metallphase gefunden. Wir nehmen zur Mittelwertbildung einen bedeutend geringeren Anteil von Metall an als GOLDSCHMIDT. Der Goldschmidtsche Wert entspricht einer atomaren Galliumhäufigkeit von 188, während ein glatter Verlauf der Häufigkeitslinie für ungerade Massenzahlen einen Wert von 50,5 erfordert. Wir betrachten diesen Wert als durchaus annehmbar.

29. Arsen. Neuere Analysen von SANDELL[2] ergaben 2,2 g/Tonne in Chondriten, entsprechend einer atomaren Häufigkeit von 4,0. Dieser Wert stellt einen Durchschnitt der Analysenergebnisse von 14 Chondriten dar und scheint sehr verläßlich zu sein. Die von I. und W. NODDACK [*9*] angegebenen Werte sind zweifellos viel zu hoch. Der Wert für Arsen ist bedeutsam, da sich aus ihm ein weiterer Hinweis dafür ergibt, daß bei der Bildung der Meteorite keine merkliche Abtrennung von Elementen stattgefunden hat, die weniger flüchtig sind als Quecksilber. Während nämlich Arsen und seine sämtlichen Verbindungen verhältnismäßig flüchtig sind, besitzen Kupfer und Gallium sowie die Verbindungen dieser Elemente eine sehr geringe Flüchtigkeit. Trotzdem liegen die Häufigkeitswerte der Isotopen ungerader Massenzahl dieser drei Elemente recht genau auf einer glatten Linie. Zink und seine Verbindungen sind gleichfalls relativ flüchtig, doch weicht unser interpolierter Wert nicht in unvereinbarer Weise von den empirischen Daten ab.

30. Selen, Brom, Krypton und Rubidium. Für die Häufigkeiten von Selen, Brom und Rubidium wählen wir die bereits besprochenen, aus den S/Se, Cl/Br bzw. K/Rb-Verhältnissen gewonnenen Werte. Der Wert für Krypton ist interpoliert.

31. Strontium. Mit Hilfe einer verbesserten Analysentechnik haben 1953 PINSON, AHRENS und FRANCK[3] aus einer Reihe von Meteoritenanalysen einen Durchschnittswert von 11 g Sr/Tonne ermittelt. SCHUMACHER[4] erhielt durch massenspektroskopische Messung nach dem Isotopenverdünnungsverfahren 12 g Sr/Tonne im Forest-City-Chondrit. Ein etwas höherer Wert für Strontium würde vielleicht besser in unser Häufigkeitsbild passen; die Analysenergebnisse sprechen jedoch nicht für einen solchen Wert, und wir übernehmen den empirischen Wert für unsere Tabelle.

32. Yttrium. Die Häufigkeiten von Yttrium und der Seltenen Erden sind wegen der Ungenauigkeit der chemischen Analysen nicht gut bekannt. Die Konzentration von Yttrium ist in sauren und basischen Gesteinen sowie in Meteoriten ungefähr dieselbe. Wir verwenden GOLDSCHMIDTs Wert von 5 g Y/Tonne und die entsprechende atomare Häufigkeit von 8,9[5].

33. Folgerungen. Mit den so gewählten Elementhäufigkeiten ergibt sich in diesem Gebiet eine stetige Abhängigkeit der Häufigkeiten der Kerne ungerader

---

[1] V. M. GOLDSCHMIDT u. C. PETERS: Nachr. Ges. Wiss. Göttingen, math.-phys. Kl. IV **1933**, 141.

[2] O. ONISHI u. E. B. SANDELL: Geochim. et Cosmochim. Acta **7**, 1 (1955). Siehe auch J. F. LOVERING, W. NICHIPORUK, A. CHODOS u. H. BROWN: Geochim. et Cosmochim. Acta **11**, 263 (1957).

[3] W. H. PINSON, L. H. AHRENS u. M. L. FRANCK: Geochim. et Cosmochim. Acta **4**, 251 (1951).

[4] E. SCHUMACHER: Helv. chim. Acta **239** (2), 531 (1956).

[5] Näheres bei K. RANKAMA u. TH. G. SAHAMA [*17*].

Massenzahl mit der Massenzahl. Wie erwähnt, besitzen diese Häufigkeiten ein flaches Maximum in der Gegend der Bromisotope ($A = 79$ und 81) und ein Minimum knapp vor dem Schalenabschluß im Kernbau bei 50 Neutronen. Dieser Verlauf mag vergleichbar sein mit jenem, der den Schalenabschluß bei 82 und 126 Neutronen vorhergeht und diese Schwankungen in viel ausgeprägterer Weise zeigt. Für die Kernsorten mit gerader Massenzahl ergibt sich jedoch keineswegs ein regelmäßiges Häufigkeitsbild, wenn man die für jede Massenzahl vorhandene Gesamthäufigkeit der Isobaren als Funktion der Massenzahl betrachtet. Faßt man jedoch Kerne mit gleichem Neutronenüberschuß zusammen, dann lassen sich auch für die Häufigkeiten der Kerne gerader Massenzahl gewisse Regelmäßigkeiten erkennen.

b) Die Elemente von Zirkon bis Zinn.

34. Allgemeines. Der Charakter der isotopen Zusammensetzung des Zirkons kann mit der des Nickels oder Neodyms verglichen werden. Im Gebiet dieser Elemente folgt auf einen Abschluß einer Neutronenschale eine rasche Abnahme der Kernhäufigkeiten mit zunehmender Massenzahl. Die Häufigkeiten der Elemente der Ordnungszahlen 40 bis 50 können ohne Schwierigkeit so gewählt werden, daß die Häufigkeitswerte der Isotope gerader Massenzahl, bzw. die Summe von Isobarenhäufigkeiten nach Regel 2 (S. 297) auf einer glatten Linie zu liegen kommen, wenn man ihre Logarithmen gegen die Massenzahl aufträgt. Gleichzeitig wird hierbei eine glatte Linie für die Häufigkeiten der Kerne ungerader Massenzahl erhalten.

35. Zirkon und Niob. Der Häufigkeitswert von Zirkon wird durch das gut bekannte Ti/Zr-Verhältnis bestimmt. Der Niobgehalt von Chondriten wurde von RANKAMA[1] zu 0,5 g/Tonne bestimmt. Dies entspricht einer atomaren Häufigkeit von 0,81. Der von uns angegebene Wert von 1,00 ist hiermit innerhalb der Fehlergrenzen identisch.

36. Molybdän, Ruthenium, Rhodium und Palladium. Als Fixpunkt zur Abschätzung des Verlaufes der Häufigkeiten in diesem und den folgenden Massengebiet verwenden wir die Palladiumhäufigkeit, die sich aus den bereits besprochenen, sorgfältigen Bestimmungen des Ni/Pd-Verhältnisses in Eisenmeteoriten ergibt. Diese Palladiumhäufigkeit paßt vorzüglich zu einer Molybdänbestimmung von KURODA und SANDELL[2], die für Chondrite 1,54 g/Tonne ergab. Dies entspricht einer atomaren Molybdänhäufigkeit von 2,42 und stellt etwa die Hälfte des von NODDACK [*8*] angegebenen Wertes dar.

Das Mengenverhältnis Ru:Rh:Pd wurde von GOLDSCHMIDT zu 10:5:9 geschätzt. Wir erhalten mit einem Verhältnis 10:1,44:5,7 glatte Häufigkeitslinien. Wir sind der Meinung, daß diese, wenn auch recht erhebliche Abweichung noch immer innerhalb der Fehlergrenzen der Goldschmidtschen Angaben liegt.

37. Silber. GOLDSCHMIDT gibt auf Grund seiner und der Noddackschen Analysen 3,2 für die atomare Silberhäufigkeit an. Ein glatter Häufigkeitsverlauf macht jedoch einen mehr als 10mal kleineren Silberwert nötig. Ein so niederer Wert scheint uns gerechtfertigt zu sein, da, wie wir von Dr. JOENSUU erfahren, bei den üblichen spektroskopischen Analyseverfahren Silberlinien sehr leicht gefunden werden, vermutlich wegen der Verbreitung von Silbermünzen in den Taschen und Händen der Analytiker. Der zu hohe Silberwert mag auch dadurch verursacht sein, daß die Menge der Troilitphase im allgemeinen überschätzt wird und Silber in hoher Konzentration im Troilit vorkommt.

---

[1] K. RANKAMA: Ann. Acad. Sci. fenn., Ser. A, III **13** (1948).

[2] P. K. KURODA u. E. B. SANDELL: Geochim. et Cosmochim. Acta **6**, 59 (1954).

38. Cadmium. Unsere atomare Häufigkeit des Cadmiums beträgt 0,89, d.h. die Hälfte des von I. und W. NODDACK [9] angegebenen Wertes für Chondrite. Genauere Cadmiumanalysen wären sehr erwünscht.

39. Indium. Für Indium geben ältere Daten 0,15 bis 0,2 g/Tonne in Meteoriten an. Neuerdings fand jedoch SHAW[1] den Indiumgehalt zweier Chondrite unterhalb seiner Nachweisbarkeitsgrenze von 0,02 g/Tonne. Unser interpolierter Wert von 0,1 entspricht einer Konzentration von 0,085 g/Tonne in chondritischen Meteoriten.

40. Zinn. GOLDSCHMIDT und PETERS[2] geben 1933 einen Gehalt von 100, 15 und 5 g/Tonne des Metalls, Troilits und Silikates der Meteorite an. Mit unserem Mengenverhältnis der meteoritischen Phasen würde dies einem Durchschnittsgehalt von ungefähr 15 g Sn/Tonne und einer atomaren Zinnhäufigkeit von 19 entsprechen. Die NODDACKS [9] geben 1934 eine atomare Zinnhäufigkeit von 50 an. Wir nehmen an, daß die älteren Analysenwerte für Zinn durchweg zu hoch sind. Zinn finden in Laboratoriumsgeräten verbreitete Verwendung, vor allem in den Apparaten zur Herstellung destillierten Wassers. Es ist bekannt, daß Zinn in Silikaten häufig zu hoch angegeben wird wegen der Verwendung gelöteter Siebe bei der Verarbeitung zerkleinerter Gesteinsproben[3]. Neuerdings hat SANDELL[4] für Zinn in Chondriten im Mittel etwa 1 g/Tonne gefunden. Die entsprechende atomare Häufigkeit von 1,33 paßt vorzüglich in den von uns angenommenen Verlauf der Häufigkeitslinien.

41. Der Häufigkeitsgang. Das Mengenverhältnis der Isotope ungerader Massenzahl des Rutheniums läßt unmittelbar erkennen, daß der glatte Abfall der Kernhäufigkeiten von den Isotopen des Zirkons zu höheren Massenzahlen hin durch mindestens eine Unregelmäßigkeit unterbrochen sein muß. Mit den hier angenommenen Elementhäufigkeiten nimmt diese Unregelmäßigkeit die Gestalt schwacher Maxima beider Häufigkeitslinien an. Es ist vielleicht denkbar, daß diese Maxima etwas mit dem Auffüllen der $g_{\frac{7}{2}}$-Neutronenschale bei 58 Neutronen zu tun haben[5], obgleich nach MAYER und JENSEN [18] in den Grundzuständen der Kerne gerader Massenzahl die $d_{\frac{5}{2}}$-Neutronenschale zuerst aufgefüllt werden sollte, die bei einer Neutronenanzahl 56 abgeschlossen ist. Es ist jedoch auch durchaus möglich, daß diese Unregelmäßigkeit nur die Häufigkeitslinie für die ungeraden Massenzahlen betrifft, wie früher [4] angenommen worden war.

c) Die Elemente von Antimon bis Barium.

42. Allgemeines. Aus der isotopen Zusammensetzung der Elemente gerader Ordnungszahl und ohne Kenntnis ihrer Häufigkeiten kann bereits ein ungefähres Bild des weiteren Verlaufes der Häufigkeitswerte gewonnen werden. Die isotope Zusammensetzung der Elemente Zinn und Tellur läßt auf einen Sprung zu niedrigeren Häufigkeitswerten in der Häufigkeitslinie für die geraden Massenzahlen bei $A = 120$ schließen. Die abnormale Isotopenhäufigkeit der Xenonisotope ungerader Massenzahl weist auf ein Maximum in den Häufigkeitslinien in der Gegend von $A = 130$. Von besonderem Interesse sind die beiden leichten, seltenen

---

[1] D. M. SHAW: Geochim. et Cosmochim. Acta **2**, 185 (1952).

[2] V. M. GOLDSCHMIDT u. C. PETERS: Nachr. Ges. Wiss. Göttingen, math.-phys. Kl. III **1933**, 36; Kl. IV **1933**, 37, 278.

[3] Hierauf hat uns Herr Dr. MICHAEL FLEISCHER, U.S. Geological Survey, freundlichst hingewiesen.

[4] H. ONISHI u. E. B. SANDELL: Geochim. et Cosmochim. Acta **12**, 262 (1957).

[5] Nach HUGHES, GARTH und EGGLER [Phys. Rev. **83**, 234 (1951)] besitzen Kerne, die 58 Neutronen enthalten, einen abnormal kleinen Einfangquerschnitt für schnelle Neutronen.

Isotope des Xenons und Bariums, deren relative Häufigkeit im Verein mit der der leichten Isotopen des Zinn, Tellur und Cer eine glatte, stetige Abhängigkeit der Häufigkeiten dieser Kernsorten von der Massenzahl nahelegt. Wir haben angenommen, daß eine derartige, stetige Abhängigkeit vorliegt und sie zur Abschätzung der Elementhäufigkeiten mitverwendet.

43. Antimon. Für Antimon in Meteoriten liegen nur spärliche Angaben von I. und W. NODDACK [*8*], [*9*] vor. Unser Wert von 0,246 entspricht genau dem Mittelwert der Noddackschen Angaben von 0,2 g Sb/Tonne, den man erhält, wenn man die Troilitphase vernachlässigt. ONISHI und SANDELL[1] findet 0,05 bis 0,2 g Sb/Tonne in Chondriten.

44. Tellur, Jod, Xenon und Cäsium. Für diese Elemente liegen kaum brauchbare Analysenergebnisse vor. GOLDSCHMIDT stellt fest [*2*], daß ein Se/Te-Verhältnis von 80, wie es sich aus den Angaben von I. und W. NODDACK [*9*] ergibt, „vielleicht die richtige Größenordnung" darstellt. Die hier angenommenen Werte entsprechen einem Se/Te-Verhältnis von 14,5. Für den mittleren Jodgehalt der Meteorite geben v. FELLENBERG 1,25 g/Tonne, und I. und W. NODDACK 0,035 g/Tonne an. Unser interpolierte Wert entspricht 0,66 g J/Tonne. Wie bereits erwähnt, ist für die Häufigkeit des Xenons durch den Kryptonwert eine obere Grenze gegeben. Kr/Xe ist entweder, wie in der Atmosphäre, 12,5, oder größer. Wir übernehmen den höchsten Wert für Xenon, d.h. ein Verhältnis Kr/Xe = 12,5, um nicht einen zu niedrigen Bariumwert annehmen zu müssen. Unser Wert für Cäsium läßt 0,40 g/Tonne in Chondriten erwarten, während von I. und W. NODDACK [*8*] 1930 ein Gehalt von 0,01 und 1934 ein solcher von 1,1 g/Tonne [*9*] angegeben wurde.

Neue Analysen von SMALES[2] (1957) geben ein Maximum von 0,10 g/Tonne an und zeigen, daß dieser Wert in den Chondriten beträchtlich schwankt. Dieses Resultat könnte ein Zeichen für einen gewissen Verlust an Caesium während der Bildung der Meteorite sein. Mehr Angaben über andere Elemente in dieser Gegend werden gebraucht.

45. Barium. Für die Bariummenge in Chondriten erhielten 1953 PINSON, AHRENS und FRANCK[3] im Durchschnitt 8 g/Tonne, entsprechend einer atomaren Bariumhäufigkeit von 8,8. Nimmt man jedoch an, daß für die leichten, seltenen Isotope der Elemente von Zinn bis Cer eine glatte Häufigkeitslinie erhalten werden soll, dann muß eine wesentlich niedrigere Bariumhäufigkeit von etwa 3,66 gewählt werden [*14*]. Anläßlich einer Untersuchung des Urangehaltes von Chondriten durch Neutronenaktivierung haben neuerdings TURKEVICH, HAMAGUCHI und REED[4] den Bariumgehalt dieser Chondrite bestimmt, da bei diesem Verfahren das sich bei der Uranspaltung bildende Ba^{140} gemessen wurde und bei der Auswertung der Ergebnisse für die Anwesenheit anderer Bariumaktivitäten korrigiert werden mußte. Hierbei ergaben sich in der Tat wesentlich niedrigere Bariumwerte, die mit unserem Häufigkeitswert von 3,66 vorzüglich übereinstimmen.

d) Die Seltenen Erden, Hafnium, Tantal und Wolfram.

46. Die Seltenen Erden. Wegen der Ähnlichkeit ihrer chemischen Eigenschaften sollten die Analysen von Meteoriten das ursprüngliche Mengenverhältnis

[1] H. ONISHI u. E. B. SANDELL: Geochim. et Cosmochim. Acta **8**, 213 (1955).

[2] A. A. SMALES: Private Mitteilung. Siehe auch B. M. GORDON, L. FRIEDMAN u. G. EDWARDS: Geochim. et Cosmochim. Acta **12**, 170 (1957).

[3] W. H. PINSON, L. H. AHRENS u. M. L. FRANCK: Geochim. et Cosmochim. Acta **4**, 251 (1953).

[4] H. HAMAGUCHI, G. W. REED u. A. TURKEVICH: Geochim. et Cosmochim. Acta **12**, 337 (1957); **13**, 248 (1958).

der Seltenen Erden mit großer Verläßlichkeit wiedergeben. Auch an der Oberfläche der Erde kann kaum eine weitgehende Trennung dieser Elemente voneinander stattgefunden haben. Nur Europium hat nach GOLDSCHMIDT und BAUER[1] die Tendenz, sich von den anderen Seltenen Erden zu trennen und dem Strontium und Blei in seinem geochemischen Verhalten zu folgen. Eine vollständige Analyse von Sedimentgesteinen für die Seltenen Erden ist 1935 von MINAMI[2] ausgeführt worden. Hierbei wurde keine abnormale Häufigkeit des Europiums gefunden und MINAMI schloß daraus, daß diese Sedimente die Seltenen Erden in einem Verhältnis enthalten, daß ihrem Durchschnittsvorkommen auf der Erdoberfläche entspricht.

Tabelle 2. *Häufigkeiten der Seltenen Erden relativ zur Häufigkeit des Lanthan.*

	La	Ce	Pr	Nd	Sm
Nach NODDACK (Meteorite)	1,00	1,12	0,46	1,57	0,57
Nach MINAMI (Sedimente)	1,00	2,46	0,30	1,25	0,22
Unser Wert	1,00	1,13	0,20	0,72	0,33
	Eu	**Gd**	**Tb**	**Dy**	**Ho**
Nach NODDACK (Meteorite)	0,13	0,81	0,25	0,95	0,27
Nach MINAMI (Sedimente)	0,05	0,31	0,043	0,21	0,052
Unser Wert	0,09	0,34	0,048	0,28	0,059
	Er	**Tm**	**Yb**	**Lu**	
Nach NODDACK (Meteorite)	0,76	0,14	0,72	0,23	
Nach MINAMI (Sedimente)	0,11	0,008	0,12	0,032	
Unser Wert	0,16	0,016	0,11	0,025	

Analysen von Meteoriten, die 1935 von I. NODDACK[3] ausgeführt wurden, führten zu wesentlich anderen Ergebnissen. Die Abweichungen erreichen für das Verhältnis der leichteren zu den schwereren Seltenen Erden mitunter einen Faktor 8. Es ist unwahrscheinlich, daß bei einer der beiden untersuchten Probenreihen eine Trennung der Seltenen Erden voneinander in einem solchen Ausmaß stattgefunden haben kann, und es ist wahrscheinlicher, daß eine der beiden Analysenreihen mit systematischen, experimentellen Fehlern behaftet ist. Beide Analysenreihen geben für die Kernhäufigkeiten als Funktion der Massenzahl relativ glatte Linien. Wir haben versuchsweise die Werte nach MINAMI unseren Daten zugrunde gelegt und sie nur in geringem Maße verändert um etwas glattere Linienzüge zu erhalten. Die Daten sind in Tabelle 2 wiedergegeben.

47. Hafnium. Als ein weiterer Fixpunkt im Verlauf der Häufigkeiten dient, wie erwähnt, das Hafnium, dessen Mengenverhältnis zu Zirkon sehr genau bekannt ist und hier zu 1/110 angenommen wurde.

48. Tantal. Für dieses Element gibt RANKAMA[4] als maximale Menge in Meteoriten 0,38 g/Tonne an. Die entsprechende atomare Häufigkeit beträgt 0,32. Unser interpolierte Wert beträgt 0,065. Wir haben dies mit Dr. RANKAMA besprochen, der diesen Wert für durchaus möglich findet.

[1] Siehe [2].
[2] E. MINAMI: Nachr. Ges. Wiss. Göttingen, math.-phys. Kl. IV, N.F. **1**, Nr. 14, 155 (1935).
[3] IDA NODDACK: Z. anorg. allg. Chem. **225**, 337 (1935).
[4] K. RANKAMA: Ann. Acad. Sci. fenn., Ser. A III **13** (1948).

49. Wolfram. Die älteren Abschätzungen der Elementhäufigkeiten [*4*] ergaben ein scharfes Maximum im Gebiet der Wolframisotope und einen Sprung zu bedeutend niedrigeren Häufigkeitswerten für die Isotope der folgenden Elemente. Für Wolfram in Urgestein liegen nunmehr neuere Messungen von SANDELL[1] vor, die um mehr als einen Faktor 10 niedrigere Werte ergaben, als die Bestimmungen von I. und W. NODDACK [*8*] und von HEVESY und HOBBIE[2] an ähnlichen Gesteinsproben. Wir schließen hieraus, daß auch die von NODDACK angegebenen Daten für Meteorite fehlerhaft waren, und daß daher kein verläßlicher analytischer Wert für den Wolframgehalt der Meteorite existiert. Wir haben für die Wolframhäufigkeit einen Wert von 0,49 interpoliert, der einem Gehalt von 0,59 g W/Tonne entspricht. Dies ist etwa $^1/_3$ des von SANDELL für Urgestein angegebenen Wertes.

e) Die Elemente von Rhenium bis Gold.

50. Rhenium. Sorgfältige Bestimmungen des Rheniums in fünf Eisenmeteoriten durch BROWN und GOLDBERG[3] ergaben 1949 einen Durchschnittsgehalt von 0,62 g Re/Tonne. Für die gesamte meteoritische Materie ergibt sich hieraus mit unserer Annahme eines mittleren Metallgehaltes von 10% ein Rheniumgehalt von 0,062 g/Tonne. Die thermodynamischen Eigenschaften des Rheniums sind jedoch nicht genau genug bekannt, um die Verteilung dieses Elements auf die meteoritischen Phasen mit Sicherheit vorhersagen zu können, doch lassen qualitative Angaben darauf schließen, daß ein Teil des Rheniums in der Silikatphase enthalten sein sollte. Unser Wert von 0,15 für die atomare Häufigkeit des Rhenium ist 2,3mal so groß als der, der sich aus dem obengenannten Wert für das Rhenium in der Metallphase ergibt.

51. Osmium, Iridium und Platin. Den Angaben GOLDSCHMIDTs [*2*] zufolge, sind diese Elemente in der Metall- und Troilitphase der Meteorite enthalten, und zwar in den in Tabelle 3 angeführten Mengen. Neuere Analysen liegen für diese Elemente nicht vor.

Tabelle 3.

	g/Tonne			Atomare Häufigkeit	
	Metall	Troilit	Durchschnitt	GOLDSCHMIDT	unsere Schätzung
Os	8	9	0,8	0,64	1,00
Ir	4	0,4	0,4	0,31	0,82
Pt	20	2	2,0	1,5	1,62

Die aus den Goldschmidtschen Angaben (Spalte 3 und 4, Tabelle 3) mit der Annahme von 10% Metall unter Vernachlässigung des Troilits berechneten Häufigkeiten stimmen mit den hier angenommenen Werten in befriedigender Weise überein, wie ein Vergleich der beiden letzten Spalten der Tabelle zeigt.

52. Gold. Dieses Element, das sicherlich nur im Metall in meßbaren Mengen vorkommt, ist in einer Reihe von Eisenmeteoriten von GOLDBERG, UCHIYAMA und BROWN[4] mittels Neutronenaktivierung bestimmt worden. Die ermittelten Mengen schwanken recht erheblich. Im Durchschnitt beträgt das gefundene Gewichtsverhältnis Ni/Au $5,8 \cdot 10^4$. Die atomare Goldhäufigkeit beträgt demnach mit unserem Nickelwert 0,140.

53. Der Häufigkeitsgang. Dieses Gebiet besitzt eine auffallende Ähnlichkeit mit jenem der Elemente von Zinn bis Barium. Die Häufigkeitslinien für beide

[1] E. B. SANDELL: Amer. J. Sci. **244**, 643 (1946). Vgl. auch S. LANDERGREN: Ark. Kemi, Mineral. Geol. A **19**, Nr. 25, 31 (1948).

[2] G. HEVESY u. R. HOBBIE: Z. anorg. allg. Chem. **212**, 134 (1933).

[3] H. S. BROWN u. E. GOLDBERG: Phys. Rev. **76**, 1260 (1949).

[4] E. GOLDBERG, A. UCHIYAMA u. H. BROWN: Geochim. et Cosmochim. Acta **2**, 1 (1951).

Kernsorten besitzen ausgeprägte Maxima im Gebiet der Massenzahlen $A = 193$ und 194. Auffallend ist ferner, daß der systematische Unterschied in den Häufigkeiten der Kerne gerader und ungerader Massenzahl in diesem Gebiet praktisch verschwindet.

f) Quecksilber, Thallium, Blei, Wismut, Thorium und Uran.

54. Allgemeines. Unsere Kenntnis des vom Standpunkt der Theorien der Elemententstehung wohl interessantesten Gebietes der Kernhäufigkeiten ist sehr unbefriedigend. Auch ist es keineswegs sicher, daß das Bild für die Häufigkeitsverteilung der Kernsorten in diesem Gebiete der schwersten Kerne einen ähnlichen Charakter besitzt, als im Gebiet der leichteren Massen. Abschätzungen durch Interpolation sind daher für diese Elemente höchst unzuverlässig.

55. Quecksilber. Dieses Element und seine Verbindungen sind zweifellos flüchtiger als die in normalen Chondriten vorkommenden Substanzen. Quecksilber mag daher zum größten Teile bei der Bildung der Meteorite nicht zur Kondensation gelangt sein. I. und W. NODDACK [*9*] berichten 1934, daß sie Quecksilber qualitativ im Troiliten des Eisens von Canyon Diablo nachweisen konnten. Die Häufigkeitsregeln würden erwarten lassen, daß die Quecksilberisotope ungerader Massenzahl etwa eine Häufigkeit besitzen dürften wie die der Thalliumisotope.

56. Thallium. I. und W. NODDACK geben 1934 für den Thalliumgehalt von Chondriten 0,15 g/Tonne an. Dies würde einer atomaren Häufigkeit von 0,108 entsprechen. SHAW[1] fand jedoch den Thalliumgehalt von zwei Chondriten und einem Achondriten unter seiner Nachweisbarkeitsgrenze von 0,01 g/Tonne. Als mittleren Wert für Urgestein gibt SHAW 1,3 g/Tonne an. Nach AHRENS[2] ist das Atomverhältnis Rb/Tl in pegmatitischen Mineralien bemerkenswert konstant und beträgt im Mittel 230. Für andere Gesteinsproben fand jedoch SHAW ein bedeutend größeres Rb/Tl-Verhältnis, und zwar etwa 780 für Granite. Dieses Verhältnis entspricht etwa einem Zehntel der von NODDACK angegebenen Thalliumhäufigkeit, d.h. einem Wert, der der von SHAW angegebenen oberen Grenze des Thalliumgehaltes der von ihm untersuchten Meteorite entspricht. Wir revidieren den von SUESS und UREY [*14*] übernommenen Noddackschen Wert und setzen für die atomare Thalliumhäufigkeit 0,0062.

57. Blei. Für die Bleihäufigkeit wurden 1934 von NODDACK relativ hohe Werte für alle drei meteoritischen Phasen angegeben und von GOLDSCHMIDT [*2*] übernommen. In neuerer Zeit sind von BROWN und Mitarbeitern eingehende Untersuchungen über Blei in Meteoriten und dessen isotope Zusammensetzung ausgeführt worden, hauptsächlich mit dem Ziele, durch Bestimmung von Urblei und radiogenem Blei das Alter der Erde und der Meteorite zu ermitteln. Es zeigt sich hierbei, daß von den älteren Angaben nur die von GOLDSCHMIDT angegebenen Troilitwerte einigermaßen zutreffen und daß für Metall und Silikat bedeutend niedrigere Bleiwerte angenommen werden müssen. Die gegenwärtig vorliegenden, das Blei in Meteoriten betreffenden Befunde sind in Tabelle 4 angegeben.

Spalte 1 der Tabelle gibt die Ergebnisse der Bleianalysen und Spalte 3, 4 und 5 die isotopische Zusammensetzung dieses Bleis nach PATTERSON, BROWN, TILTON und INGHRAM[3] an. In Spalte 6 ist die von TURKEVICH, HAMAGUCHI und REED

[1] D. M. SHAW: Geochim. et Cosmochim. Acta **2**, 118 (1952).
[2] L. H. AHRENS: J. Geology **56**, 578 (1948).
[3] C. PATTERSON, H. BROWN, G. TILTON u. M. INGHRAM: Phys. Rev. **92**, 1234 (1953). — Science, Lancaster, Pa. **121**, 69 (1955).

Tabelle 4. *Isotopische Zusammensetzung von Blei und Konzentration von Blei, Uran und Thorium in Meteoriten.*

	1	2	3	4	5	6	7
	Pb g/Tonne		$\frac{Pb^{206}}{Pb^{204}}$	$\frac{Pb^{207}}{Pb^{204}}$	$\frac{Pb^{208}}{Pb^{204}}$	U^{238} g/Tonne	Th^{232} g/Tonne
	gemessen	berechnet					
(1) Canyon Diablo (Troilit) .	18		9,41	10,27	29,16		
(2) Henbury (Troilit) . . .	5		9,50	10,30	29,26		
(3) Eisen (Durchschnitt) .			9,455	10,285	29,21		
(4) Forest City	0,4		19,27	15,95	39,05		
(5) Modoc	0,9		19,48	15,76	38,21		
(6) Chondrit (Durchschnitt)	0,65	0,08	19,375	15,855	38,63	0,0115	0,042
(7) Radiogenes Blei (6) — (3)			9,920	5,570	9,42		
(8) Nuevo Laredo	0,7	0,47	50,28	34,86	67,97	0,12	0,46
(9) Radiogenes Blei (8) — (3)			40,825	24,575	38,76		

durch Neutronenaktivierung ermittelte Urankonzentration angegeben. Nimmt man an, daß die isotope Zusammensetzung des Bleis aus dem Troilit der des Urbleis entspricht und ferner, daß die untersuchten Meteorite zu gleicher Zeit gebildet wurden und seither keine Verschiebung durch chemische Vorgänge im Mengenverhältnis von Blei zu Uran eingetreten ist, dann ergibt sich aus dem Isotopenverhältnis des Bleis dieser Proben und den bekannten Halbwertszeiten von U^{235} und U^{238} das Alter der Meteorite zu $4{,}55 \cdot 10^9$ Jahren. Altersbestimmungen nach der K^{40}- A^{40}-Methode durch WASSERBURG und HAYDEN[1] und nach der Rb^{87}- Sr^{87}-Methode durch SCHUMACHER[2] haben diesen Wert im wesentlichen bestätigt. Aus der analytisch bestimmten Uranmenge, der isotopen Zusammensetzung des Bleis der Proben und des Urbleis kann für das ermittelte Alter der Bleimenge in den Proben berechnet werden. Das Ergebnis der Rechnung ist in Spalte 2 angegeben. Die schlechte Übereinstimmung der gemessenen und berechneten Bleikonzentrationen zeigt, daß entweder eine unserer Grundannahmen nicht zutrifft, oder daß eine der Meßreihen mit erheblichen Fehlern behaftet ist. Die in Spalte 7 angegebene Thoriummenge ist aus der gemessenen Uranmenge und der isotopen Zusammensetzung des Bleis mit einem Alter von $4{,}55 \cdot 10^9$ Jahren berechnet. Die in unserer Tabelle angegebenen Häufigkeitswerte für Blei beruhen auf der Annahme, daß die Ergebnisse der direkten chemischen Bleibestimmungen zu hoch sind und daß die Voraussetzungen für die Berechnung des Bleiwertes zutreffen. Dieser Wert ist in guter Übereinstimmung mit neuen Messungen von POTRATZ (1957)[3] über den Thoriumgehalt in Chondriten. Auch MARSHALL (1957) hat gefunden, daß der Bleigehalt der Chondrite von Forest City nicht höher als 0,2 g/Tonne ist. Wir benützen einen etwas kleineren Wert von 0,14 g/Tonne oder eine atomare Häufigkeit von 0,1. Die Konzentration in Meteoriten kann sogar noch kleiner sein. Die Häufigkeit auf der Sonne ist nach den Schätzungen von ALLER um ungefähr zwei Größenordnungen höher, und wir wissen nicht, wie dieser Unterschied zu erklären ist. Auch liegt eine starke Abnahme der Häufigkeiten zwischen Gold und den schwereren Elementen vor. Weisen die Meteorite nicht die richtige Konzentration auf, so können die Kurven für Quecksilber, Thallium, Blei und Wismut höher liegen. Auch werden weitere Arbeiten möglicherweise zeigen, daß die Platingegend zu hoch ist. Es ist uns jedoch gegenwärtig kein Grund bekannt, warum Meteorite Blei und dessen Nachbarelemente verloren haben sollten.

[1] G. J. WASSERBURG u. R. J. HAYDEN: Phys. Rev. **97**, 86 (1955).
[2] E. SCHUMACHER: Helv. chim. Acta **39** (2), 531 (1956).
[3] H. A. POTRAZ: Private Mitteilung 1957.

58. Wismut. Über die Häufigkeit des Wismuts sind wir nur durch eine Angabe von I. und W. NODDACK [8], [9] unterrichtet. Wismut, als Endprodukt der vierten Zerfallsreihe muß sicher häufiger sein, als Pb^{207} und U^{235} zusammengenommen [1]. Der Noddacksche Wismutwert entspricht dieser Forderung und wir nehmen ihn in unsere Tabelle auf.

Tabelle 5. *Atomare Häufigkeiten der Elemente.* $H(\mathrm{Si}) = 10^6$.

Die Goldschmidtschen Werte sind auf Grund neuerer Analysen geringfügig revidiert; UREY's Angaben entsprechen den Analysenergebnissen für chondritische Meteorite. Die astronomischen Werte nach ALLER entsprechen den in dessen Beitrag zu diesem Bande angegebenen, umgerechnet auf unsere Normierung. Von den Werten nach SUESS und UREY sind die für die leichten Elemente diesen neuen astronomischen Angaben angeglichen und für Hg, Tl, und Pb niederere Werte eingesetzt. Nach neueren Meteoritenanalysen erscheint der hier angegebene Pb-Wert noch immer um einen Faktor 3 bis 4 zu hoch. Die Werte für Pb, Th und U gelten für die Zeit vor $4,5 \cdot 10^9$ Jahren, d.h. radiogenes Pb ist als U bzw. Th gerechnet.

Element	GOLDSCHMIDT [2]	BROWN [7]	UREY [11]	ALLER astronomisch	SUESS und UREY [14]	Element	GOLDSCHMIDT [2]	BROWN [7]	UREY [11]	ALLER astronomisch	SUESS und UREY [14]
1 H		$3,5 \cdot 10^{10}$		$3,2 \cdot 10^{10}$	$3,2 \cdot 10^{10}$	44 Ru	3,6	9,3	2,1	8,8	1,49
2 He		$3,5 \cdot 10^9$		$4,1 \cdot 10^9$	$4,1 \cdot 10^9$	45 Rh	1,3	3,5	0,71	11,5	0,214
3 Li	100		100	0,23	100	46 Pd	1,8	3,2	1,3	0,3	0,675
4 Be	20		16	1,0	20	47 Ag	3,2	2,7	0,35	0,01	0,26
5 B	24		20		24	48 Cd	2,6	2,6	1,9	1,8	0,89
6 C		$8,0 \cdot 10^6$		$11 \cdot 10^6$	$11 \cdot 10^6$	49 In	0,23	1,0	0,26	0,2	0,11
7 N		$1,6 \cdot 10^7$		$3 \cdot 10^6$	$3 \cdot 10^6$	50 Sn	29	62	1,33	1	1,33
8 O		$2,2 \cdot 10^7$		$3,1 \cdot 10^7$	$3,1 \cdot 10^7$	51 Sb	0,72	1,7	0,12		0,246
9 F	1500	9000	300		1600	52 Te	0,2		0,16		4,67
10 Ne		$9 \cdot 10^5$		$1,7 \cdot 10^7$	$8,6 \cdot 10^6$	53 I	1,4	1,8	1,5		0,80
11 Na	$4,42 \cdot 10^4$	$4,62 \cdot 10^4$	$4,38 \cdot 10^4$	$6,2 \cdot 10^4$	$4,38 \cdot 10^4$	54 Xe					4,0
12 Mg	$8,7 \cdot 10^5$	$8,87 \cdot 10^5$	$9,12 \cdot 10^5$	$4,8 \cdot 10^6$	$9,12 \cdot 10^5$	55 Cs	0,1	0,1	1,3		0,456
13 Al	$8,8 \cdot 10^4$	$8,82 \cdot 10^4$	$9,48 \cdot 10^4$	$5,0 \cdot 10^4$	$9,48 \cdot 10^4$	56 Ba	8,3	3,9	8,8	5,6	3,66
14 Si	10^6	10^6	10^6	$1,1 \cdot 10^6$	10^6	57 La	2,1	2,1	2,1		2,00
15 P	$5,8 \cdot 10^3$	$13 \cdot 10^3$	$5 \cdot 10^3$	$9 \cdot 10^3$	$1 \cdot 10^4$	58 Ce	5,2	2,3	2,3		2,26
16 S	$1,1 \cdot 10^5$	$3,5 \cdot 10^5$	$9,8 \cdot 10^4$	$43 \cdot 10^4$	$37,5 \cdot 10^4$	59 Pr	0,96	0,96	0,96		0,40
17 Cl	5000	17000	2100	$3 \cdot 10^5$	8850	60 Nd	3,3	3,3	3,3		1,44
18 A		$1,3 \cdot 10^4$		$1 \cdot 10^5$	$1,5 \cdot 10^5$	62 Sm	1,15	1,2	1,1		0,664
19 K	6900	6930	3160	2800	3160	63 Eu	0,28	0,28	0,28		0,187
20 Ca	57100	67000	49000	74000	49000	64 Gd	1,65	1,7	1,6		0,684
21 Sc	15	18	28	49	28	65 Tb	0,52	0,52	0,52		0,0956
22 Ti	4700	2600	2440	2900	2440	66 Dy	2,0	2,0	2,0		0,556
23 V	130	250	220	330	220	67 Ho	0,57	0,57	0,57		0,118
24 Cr	11300	9500	7800	4900	7800	68 Er	1,6	1,6	1,6		0,316
25 Mn	6600	7700	6850	6200	6850	69 Tm	0,29	0,29	0,29		0,0318
26 Fe	$8,9 \cdot 10^5$	$18,3 \cdot 10^5$	$6,0 \cdot 10^5$	$1,8 \cdot 10^5$	$6,0 \cdot 10^5$	70 Yb	1,5	1,5	1,5	(1)	0,220
27 Co	3500	9900	1800	1700	1800	71 Lu	0,48	0,48	0,48		0,050
28 Ni	$4,6 \cdot 10^4$	$1,34 \cdot 10^5$	$2,74 \cdot 10^4$	29000	$2,74 \cdot 10^4$	72 Hf	1,5	0,7	0,55		0,438
29 Cu	460	460	212	170	212	73 Ta	0,40	0,31	0,32		0,065
30 Zn	360	160	180	1000	486	74 W	14,5	17,0	13,0 ?		0,49
31 Ga	19	65	11,4	4,5	11,4	75 Re	0,12	0,41	0,05		0,135
32 Ge	190	250	65	60	50,5	76 Os	1,7	3,5	0,97		1,00
33 As	18	480	4,0		4,0	77 Ir	0,58	1,4	0,31		0,821
34 Se	15	25	24		67,6	78 Pt	2,9	8,7	1,5		1,625
35 Br	43	42	49 ?		13,4	79 Au	0,27	0,82	0,140		0,145
36 Kr					51,3	80 Hg	0,33		$<0,006$	(26)	0,017
37 Rb	6,8	7,1	6,5	5,2	6,5	81 Tl	0,17		0,11		0,108
38 Sr	40	41	18,9	19,5	18,9	82 Pb	9,1	$<2,0$	0,122	(15)	0,12
39 Y	9,7	10	8,9	30	8,9	83 Bi	0,11	0,21	0,21		0,078
40 Zr	140	150	54,5	4,5	54,5	90 Th	0,59		0,033		0,033
41 Nb	6,9	0,9	0,8	3	1,00	92 U	0,23	0,02	0,0178		0,0178
42 Mo	9,5	19	2,42	2,8	2,42						

[1] H. E. SUESS: Experientia, Basel **5**, 278 (1949).

Tabelle 6. *Atomare Häufigkeit der Kerne.*

Berechnet aus den in Tabelle 5 (letzte Spalte) angegebenen Elementhäufigkeiten und der isotopischen Zusammensetzung der Elemente.

Element	*A*	*N*	*I*	Log *H*	*H*
1 H				10,50	3,20 × 10^{10}
	1	*0*	*−1*	*10,50*	*3,20 ×* 10^{10}
	2	1	0	6,50	4,5 × 10^{6}
2 He					
	3	*1*	*−1*	...	...
	4	2	0	9,61	4,1 × 10^{9}
3 Li				2,00	100
	6	3	0	0,87	7,4
	7	*4*	*1*	*1,97*	*92,6*
4 Be	9	5	1	1,30	20
5 B				1,38	24
	10	5	0	0,65	4,5
	11	*6*	*1*	*1,29*	*19,5*
6 C				7,04	1,1 × 10^{7}
	12	6	0	7,02	1,09 × 10^{7}
	13	*7*	*1*	*5,07*	*1,22 ×* 10^{5}
7 N				6,48	3,0 × 10^{6}
	14	7	0	6,48	2,99 × 10^{6}
	15	*8*	*1*	*4,04*	*1,1 ×* 10^{4}
8 O				7,49	3,1 × 10^{7}
	16	8	0	7,49	3,08 × 10^{7}
	17	*9*	*1*	*4,06*	*1,16 ×* 10^{4}
	18	10	2	4,80	6,32 × 10^{4}
9 F	*19*	*10*	*1*	*3,20*	*1600*
10 Ne				6,93	8,6 × 10^{6}
	20	10	0	6,89	7,74 × 10^{6}
	21	*11*	*1*	*4,41*	*2,58 ×* 10^{4}
	22	12	2	5,92	8,36 × 10^{5}
11 Na	*23*	*12*	*1*	*4,64*	*4,38 ×* 10^{4}
12 Mg				5,96	9,12 × 10^{5}
	24	12	0	5,86	7,21 × 10^{5}
	25	*13*	*1*	*4,96*	*9,17 ×* 10^{4}
	26	14	2	5,00	1,00 × 10^{5}
13 Al	*27*	*14*	*1*	*4,98*	*9,48 ×* 10^{4}
14 Si				6,00	1,00 × 10^{6}
	28	14	0	5,96	9,22 × 10^{5}
	29	*15*	*1*	*4,67*	*4,70 ×* 10^{4}
	30	16	2	4,49	3,12 × 10^{4}
15 P	*31*	*16*	*1*	*4,00*	*1,00 ×* 10^{4}
16 S				5,57	3,75 × 10^{5}
	32	16	0	5,55	3,56 × 10^{5}
	33	*17*	*1*	*3,44*	*2,77 ×* 10^{3}
	34	18	2	4,19	1,57 × 10^{4}
	36	20	4	1,71	51
17 Cl				3,95	8850
	35	*18*	*1*	*3,82*	*6670*
	37	*20*	*3*	*3,34*	*2180*
18 A				5,18	1,50 × 10^{5}
	36	18	0	5,10	1,26 × 10^{5}
	38	20	2	4,38	2,4 × 10^{4}
	40	22	4		

Element	*A*	*N*	*I*	Log *H*	*H*
19 K				3,50	3160
	39	*20*	*1*	*3,47*	*2940*
	40	21	2	0,58 − 1	0,38
	41	*22*	*3*	*2,34*	*219*
20 Ca				4,69	4,90 × 10^{4}
	40	20	0	4,68	4,75 × 10^{4}
	42	22	2	2,50	314
	43	*23*	*3*	*1,80*	*64*
	44	24	4	3,02	1040
	46	26	6	0,20	1,6
	48	28	8	1,94	87,7
21 Sc	*45*	*24*	*3*	*1,43*	*28*
22 Ti				3,39	2440
	46	24	2	2,29	194
	47	*25*	*3*	*2,28*	*189*
	48	26	4	3,25	1790
	49	*27*	*5*	*2,13*	*134*
	50	28	6	2,11	130
23 V				2,34	220
	50	27	4	0,74 − 1	0,55
	51	*28*	*5*	*2,34*	*220*
24 Cr				3,89	7800
	50	26	2	2,54	344
	52	28	4	3,81	6510
	53	*29*	*5*	*2,87*	*744*
	54	30	6	2,31	204
25 Mn	*55*	*30*	*5*	*3,84*	*6850*
26 Fe				5,78	6,00 × 10^{5}
	54	28	2	4,55	3,54 × 10^{4}
	56	30	4	5,77	5,49 × 10^{5}
	57	*31*	*5*	*4,13*	*1,35 ×* 10^{4}
	58	32	6	3,30	1980
27 Co	*59*	*32*	*5*	*3,25*	*1800*
28 Ni				4,44	2,74 × 10^{4}
	58	30	2	4,27	1,86 × 10^{4}
	60	32	4	3,86	7170
	61	*33*	*5*	*2,53*	*342*
	62	34	6	3,00	1000
	64	36	8	2,50	318
29 Cu				2,33	212
	63	*34*	*5*	*2,16*	*146*
	65	*36*	*7*	*1,82*	*66*
30 Zn				2,69	486
	64	34	4	2,38	238
	66	36	6	2,13	134
	67	*37*	*7*	*1,30*	*20,0*
	68	38	8	1,96	90,9
	70	40	10	0,52	3,35
31 Ga				1,06	11,4
	69	*38*	*7*	*0,84*	*6,86*
	71	*40*	*9*	*0,66*	*4,54*

Tabelle 6. (Fortsetzung.)

Element	A	N	I	Log H	H
32 Ge				1,70	50,5
	70	38	6	1,02	10,4
	72	40	8	1,14	13,8
	73	*41*	*9*	*0,58*	*3,84*
	74	42	10	1,27	18,65
	76	44	12	0,59	3,87
33 As	*75*	*42*	*9*	*0,60*	*4,0*
34 Se				1,83	67,6
	74	40	6	0,81	0,649
	76	42	8	0,80	6,16
	77	*43*	*9*	*0,70*	*5,07*
	78	44	10	1,20	16,0
	80	46	12	1,53	33,8
	82	48	14	0,78	5,98
35 Br				1,13	13,4
	79	*44*	*9*	*0,83*	*6,78*
	81	*46*	*11*	*0,82*	*6,62*
36 Kr				1,71	51,3
	78	42	6	0,24 − 1	0,175
	80	44	8	0,06	1,14
	82	46	10	0,77	5,90
	83	*47*	*11*	*0,76*	*5,89*
	84	48	12	1,47	29,3
	86	50	14	0,95	8,94
37 Rb				0,81	6,5
	85	*48*	*11*	*0,67*	*4,73*
	87	*50*	*13*	*0,25*	*1,77*
38 Sr				1,28	18,9
	84	46	8	0,03	0,106
	86	48	10	0,26	1,86
	87	*49*	*11*	*0,12*	*1,33*
	88	50	12	1,19	15,6
39 Y	*89*	*50*	*11*	*0,95*	*8,9*
40 Zr				1,74	54,5
	90	50	10	1,45	28,0
	91	*51*	*11*	*0,79*	*6,12*
	92	52	12	0,97	9,32
	94	54	14	0,98	9,48
	96	56	16	0,18	1,53
41 Nb	*93*	*52*	*11*	*0,00*	*1,00*
42 Mo				0,38	2,42
	92	50	8	0,56 − 1	0,364
	94	52	10	0,35 − 1	0,226
	95	*53*	*11*	*0,58 − 1*	*0,382*
	96	54	12	0,60 − 1	0,401
	97	*55*	*13*	*0,37 − 1*	*0,232*
	98	56	14	0,76 − 1	0,581
	100	58	16	0,37 − 1	0,234
44 Ru				0,17	1,49
	96	52	8	0,93 − 2	0,0846
	98	54	10	0,52 − 2	0,0331
	99	*55*	*11*	*0,28 − 1*	*0,191*
	100	56	12	0,28 − 1	0,189
	101	*57*	*13*	*0,40 − 1*	*0,253*
	102	58	14	0,67 − 1	0,467
	104	60	16	0,43 − 1	0,272

Element	A	N	I	Log H	H
45 Rh	*103*	*58*	*15*	*0,33 − 1*	*0,214*
46 Pd				0,83 − 1	0,675
	102	56	10	0,73 − 3	0,0054
	104	58	12	0,80 − 2	0,0628
	105	*59*	*13*	*0,18 − 1*	*0,1536*
	106	60	14	0,26 − 1	0,1839
	108	62	16	0,26 − 1	0,180
	110	64	18	0,96 − 2	0,0911
47 Ag				0,41 − 1	0,26
	107	*60*	*13*	*0,13 − 1*	*0,134*
	109	*62*	*15*	*0,10 − 1*	*0,126*
48 Cd				0,95 − 1	0,89
	106	58	10	0,04 − 2	0,0109
	108	60	12	0,90 − 3	0,0079
	110	62	14	0,04 − 1	0,111
	111	*63*	*15*	*0,06 − 1*	*0,114*
	112	64	16	0,33 − 1	0,212
	113	*65*	*17*	*0,04 − 1*	*0,110*
	114	66	18	0,41 − 1	0,256
	116	68	20	0,83 − 2	0,068
49 In				0,04 − 1	0,11
	113	*64*	*15*	*0,66 − 3*	*0,0046*
	115	*66*	*17*	*0,02 − 1*	*0,105*
50 Sn				0,12	1,33
	112	62	12	0,13 − 2	0,0134
	114	64	14	0,96 − 3	0,0090
	115	*65*	*15*	*0,67 − 3*	*0,00465*
	116	66	16	0,28 − 1	0,189
	117	*67*	*17*	*0,01 − 1*	*0,102*
	118	68	18	0,50 − 1	0,316
	119	*69*	*19*	*0,06 − 1*	*0,115*
	120	70	20	0,64 − 1	0,433
	122	72	22	0,80 − 2	0,063
	124	74	24	0,90 − 2	0,079
51 Sb				0,39 − 1	0,246
	121	*70*	*19*	*0,15 − 1*	*0,141*
	123	*72*	*21*	*0,02 − 1*	*0,105*
52 Te				0,67	4,67
	120	68	16	0,62 − 3	0,00420
	122	70	18	0,06 − 1	0,115
	123	*71*	*19*	*0,62 − 2*	*0,0416*
	124	72	20	0,34 − 1	0,221
	125	*73*	*21*	*0,52 − 1*	*0,328*
	126	74	22	0,94 − 1	0,874
	128	76	24	0,17	1,48
	130	78	26	0,20	1,60
52 I	*127*	*74*	*21*	*0,90 − 1*	*0,80*
54 Xe				0,60	4,0
	124	70	16	0,58 − 3	0,00380
	126	72	18	0,55 − 3	0,00352
	128	74	20	0,88 − 2	0,0764
	129	*75*	*21*	*0,02*	*1,050*
	130	76	22	0,21 − 1	0,162
	131	*77*	*23*	*0,93 − 1*	*0,850*
	132	78	24	0,03	1,078
	134	80	26	0,62 − 1	0,420
	136	82	28	0,55 − 1	0,358

Tabelle 6. (Fortsetzung.)

Element	*A*	*N*	*I*	Log *H*	*H*
55 Cs	*133*	*78*	*23*	*0,66 − 1*	*0,456*
56 Ba				0,56	3,66
	130	74	18	0,57 − 3	0,003 70
	132	76	20	0,55 − 3	0,003 56
	134	78	22	0,95 − 2	0,0886
	135	*79*	*23*	*0,38 − 1*	*0,241*
	136	80	24	0,45 − 1	0,286
	137	*81*	*25*	*0,62 − 1*	*0,414*
	138	82	26	0,42	2,622
57 La				0,30	2,00
	138	81	24	0,25 − 3	0,0018
	139	*82*	*25*	*0,30*	*2,00*
58 Ce				0,35	2,26
	136	78	20	0,64 − 3	0,0044
	138	80	22	0,75 − 3	0,005 66
	140	82	24	0,30	2,00
	142	84	26	0,40 − 1	0,250
59 Pr	*141*	*82*	*23*	*0,60 − 1*	*0,40*
60 Nd				0,16	1,44
	142	82	22	0,59 − 1	0,39
	143	*83*	*23*	*0,24 − 1*	*0,175*
	144	84	24	0,54 − 1	0,344
	145	*85*	*25*	*0,08 − 1*	*0,119*
	146	86	26	0,39 − 1	0,248
	148	88	28	0,91 − 2	0,0824
	150	90	30	0,91 − 2	0,0806
62 Sm				0,82 − 1	0,664
	144	82	20	0,32 − 2	0,0108
	147	*85*	*23*	*0,00 − 1*	*0,100*
	148	86	24	0,87 − 2	0,0748
	149	*87*	*25*	*0,96 − 2*	*0,0920*
	150	88	26	0,69 − 2	0,0492
	152	90	28	0,25 − 1	0,176
	154	92	30	0,17 − 2	0,150
63 Eu				0,27 − 1	0,187
	151	*88*	*25*	*0,95 − 2*	*0,0892*
	153	*90*	*27*	*0,99 − 2*	*0,0976*
64 Gd				0,83 − 1	0,684
	152	88	24	0,14 − 3	0,001 37
	154	90	26	0,17 − 2	0,0147
	155	*91*	*27*	*0,00 − 1*	*0,101*
	156	92	28	0,15 − 1	0,141
	157	*93*	*29*	*0,03 − 1*	*0,107*
	158	94	30	0,23 − 1	0,169
	160	96	32	0,17 − 1	0,149
65 Tb	*159*	*94*	*29*	*0,98 − 2*	*0,0956*
66 Dy				0,74 − 1	0,556
	156	90	24	0,46 − 4	0,000 29
	158	92	26	0,70 − 4	0,000 502
	160	94	28	0,10 − 2	0,0127
	161	*95*	*29*	*0,02 − 1*	*0,105*
	162	96	30	0,15 − 1	0,142
	163	*97*	*31*	*0,14 − 1*	*0,139*
	164	98	32	0,19 − 1	0,157

Element	*A*	*N*	*I*	Log *H*	*H*
67 Ho	*165*	*98*	*33*	*0,07 − 1*	*0,118*
68 Er				0,50 − 1	0,316
	162	94	26	0,50 − 4	0,000 316
	164	96	28	0,67 − 3	0,004 74
	166	98	30	0,02 − 1	0,104
	167	*99*	*31*	*0,88 − 2*	*0,770*
	168	100	32	0,93 − 2	0,0850
	170	102	34	0,65 − 2	0,0228
69 Tm	*169*	*100*	*31*	*0,50 − 2*	*0,0318*
70 Yb				0,34 − 1	0,220
	168	98	28	0,48 − 4	0,000 30
	170	100	30	0,82 − 3	0,006 66
	171	*101*	*31*	*0,50 − 2*	*0,0316*
	172	102	32	0,68 − 2	0,0480
	173	*103*	*33*	*0,55 − 2*	*0,0356*
	174	104	34	0,84 − 2	0,0678
	176	106	36	0,44 − 2	0,0278
71 Lu				0,70 − 2	0,050
	175	*104*	*33*	*0,69 − 2*	*0,0488*
	176	105	34	0,11 − 3	0,0013
72 Hf				0,68 − 1	0,438
	174	102	30	0,90 − 4	0,000 78
	176	104	32	0,35 − 2	0,0226
	177	*105*	*33*	*0,91 − 2*	*0,0806*
	178	106	34	0,07 − 1	0,119
	179	*107*	*35*	*0,78 − 2*	*0,0604*
	180	108	36	0,19 − 1	0,155
73 Ta	*181*	*108*	*35*	*0,81 − 2*	*0,065*
74 W				0,69 − 1	0,49
	180	106	32	0,78 − 4	0,0006
	182	108	34	0,11 − 1	0,13
	183	*109*	*35*	*0,84 − 2*	*0,070*
	184	110	36	0,17 − 1	0,15
	186	112	38	0,14 − 1	0,14
75 Re				0,13 − 1	0,135
	185	*110*	*35*	*0,70 − 2*	*0,0500*
	187	*112*	*37*	*0,93 − 2*	*0,0850*
76 Os				0,00	1,00
	184	108	32	0,26 − 4	0,000 18
	186	110	34	0,20 − 2	0,0159
	187	*111*	*35*	*0,22 − 2*	*0,0164*
	188	112	36	0,12 − 1	0,133
	189	*113*	*37*	*0,21 − 1*	*0,161*
	190	114	38	0,42 − 1	0,264
	192	116	40	0,61 − 1	0,410
77 Ir				0,91 − 1	0,821
	191	*114*	*37*	*0,50 − 1*	*0,316*
	193	*116*	*39*	*0,70 − 1*	*0,505*
78 Pt				0,21	1,625
	190	112	34	0,00 − 4	0,0001
	192	114	36	0,10 − 2	0,0127
	194	116	38	0,73 − 1	0,533
	195	*117*	*39*	*0,74 − 1*	*0,548*
	196	118	40	0,62 − 1	0,413
	198	120	42	0,07 − 1	0,117

Tabelle 6. (Fortsetzung.)

Element	A	N	I	Log H	H
79 Au	*197*	*118*	*39*	*0,16 − 1*	*0,145*
80 Hg				0,23 − 2	0,017
	196	116	36	0,43 − 5	0,000027
	198	118	38	0,23 − 3	0,0017
	199	*119*	*39*	*0,46 − 3*	*0,0029*
	200	120	40	0,60 − 3	0,0040
	201	*121*	*41*	*0,35 − 3*	*0,0022*
	202	122	42	0,71 − 3	0,0051
	204	124	44	0,07 − 3	0,0012
81 Tl				0,08 − 3	0,0062
	203	*122*	*41*	*0,00 − 3*	*0,0010*
	205	*124*	*43*	*0,58 − 3*	*0,0038*

Element	A	N	I	Log H	H
82 Pb[1]				0,08 − 1	0,12
	204	122	40	0,36 − 3	0,0023
	206	124	42	0,34 − 2	0,022
	207	*125*	*43*	*0,38 − 2*	*0,024*
	208	126	44	0,83 − 2	0,068
83 Bi	*209*	*126*	*43*	*0,90 − 2*	*0,078*
90 Th[2]	232	142	52	0,52 − 2	0,033
92 U[2]				0,25 − 2	0,0178
	235	*143*	*51*	*0,61 − 3*	*0,0041*
	238	146	54	0,14 − 2	0,0137

59. Uran und Thorium. Im Chondriten von BEDDGELERT haben CHACKETT, GOLDEN, MERCER, PANETH und REASBECK[3] 0,106 g/Tonne Uran und 0,335 g/Tonne Thorium gefunden. UREY[4] wies 1955 darauf hin, daß derart hohe Uran- und Thoriummengen im Verein mit dem bekannten Kaliumgehalt eine so hohe Wärmeproduktion zur Folge hätten, daß es unmöglich wäre, den Wärmehaushalt der Erde, des Mondes, oder des Planeten Mars zu verstehen. Diese Schwierigkeit besteht nicht, wenn man annimmt, daß die 1950 von DAVIS[5] und 1953 von DE JAGER[6] bestimmten Mengen von 0,01 bis 0,03 g U/Tonne in Meteoriten dem durchschnittlichen Uranvorkommen entsprechen.

Die Ergebnisse von TURKEWICH, HAMAGUCHI und REED und die unpublizierten Werte von POTRATZ (1957) über Thorium (Tabelle 4) liegen im Bereich dieser Zahlen. Unsere Häufigkeitswerte entsprechen den in Tabelle 4 angegebenen Konzentrationen.

D. Zur Deutung der Häufigkeitsverteilung der Elemente.

60. Theorien der Elemententstehung. Die Häufigkeitsverteilung der Kernsorten, wie sie sich aus der vorhergehenden Diskussion ergibt, kann nun mit den Erwartungen der verschiedenen Theorien der Elemententstehung verglichen werden. Es ist mehrfach von verschiedenen Forschern darauf hingewiesen worden, daß ein Versuch, die Häufigkeitsverteilung als das Ergebnis eines eingefrorenen thermodynamischen Gleichgewichtes aufzufassen, zu keiner brauchbaren Arbeitshypothese führt[7]. Zur Deutung muß daher die Kinetik von Kernreaktionen herangezogen werden. Die naheliegendste und am eingehendsten untersuchte Reaktion, die zum Aufbau schwererer Kerne führen kann, ist Neutroneneinfang mit nachfolgendem Beta-Zerfall. ALPHER und HERMAN haben in drei vollständigen, zusammenfassenden Berichten [*19*] die Bedeutung dieser

---

[1] Urblei. Auf Grund der Meteoritenanalysen ist der hier angegebene Bleiwert um einen Faktor 3 bis 4 zu hoch.

[2] Th uud U vor $4,5 \cdot 10^9$ Jahren.

[3] K. F. CHACKETT, J. GOLDEN, E. R. MERCER, F. A. PANETH u. P. REASBECK: Geochim. et Cosmochim. Acta **1**, 3 (1950).

[4] H. C. UREY: Proc. Nat. Acad. Sci. U.S.A. **41**, 127 (1955).

[5] G. L. DAVIS: Amer. J. Sci. **248**, 107 (1950).

[6] C. DE JAGER: Mém. Soc. Roy. Sci., Liège, IV. ser. **13** (III), 460 (1953).

[7] H. C. UREY u. C. A. BRADLEY: Phys. Rev. **38**, 718 (1931). — H. J. D. JENSEN u. H. E. SUESS: Naturwiss. **34**, 131 (1947).

Reaktion für die Elemententstehung im Rahmen einer geschlossenen Kosmologie eingehend behandelt und mit den Vorstellungen anderer Autoren verglichen.

Allen diesen Versuchen lag das Bestreben zugrunde, der uns umgebenden Materie eine im wesentlichen einheitliche kernphysikalische Geschichte zuzuschreiben, und sie als das Produkt einer Reihe zeitlich aufeinanderfolgender Kernprozesse aufzufassen. Die anscheinend unüberwindlichen Schwierigkeiten, die sich einer solchen Auffassung entgegenstellen, haben nunmehr dazu geführt, einen physikalisch wesentlich anspruchsloseren Weg zu versuchen, nämlich den, unsere Materie als ein Gemisch von Produkten aufzufassen, die räumlich und zeitlich getrennt voneinander unter verschiedenen physikalisdhen Bedingungen gebildet worden sind. So werden, wie BURBIDGE, FOWLER und HOYLE[1] annehmen, bei Supernovae-Ausbrüchen schwerere Kernsorten gebildet und in den interstellaren Raum ausgestoßen. Das Mengenverhältnis, in dem diese Kernsorten gebildet werden, hängt ab von der Vorgeschichte des Sternes und von den Bedingungen, unter dem sich der Ausbruch vollzieht. Die sich aus interstellarer Materie neu kondensierenden Sterne enthalten ein Gemisch der bei verschiedenen Ausbrüchen entstandenen Produkte.

Wir hoffen, daß die hier gegebene Analyse der Häufigkeitsverteilung der Elemente dazu beitragen wird, zu entscheiden, ob eine quantitative Berechnung der möglichen astronomischen und kernphysikalischen Vorgänge eine solche oder eine ähnliche Deutung zuläßt. Im folgenden sei auf einige charakteristische Züge und Einzelheiten in der Häufigkeitsverteilung hingewiesen, deren Deutung von einer Theorie der Elemententstehung gefordert wird.

61. Der Verlauf der Häufigkeitslinien und die Summenregel für isobare Kernhäufigkeiten. Die Häufigkeiten der Elemente und deren Isotope sind in Tabelle 6 angegeben und in Fig. 1 aufgetragen. Der Charakter der Häufigkeitsverteilung unterscheidet sich in eindrucksvoller Weise im Gebiet der leichteren Kerne ($A < 90$) von dem im Gebiet der schwereren Kerne ($A > 90$). Im Gebiet der leichteren Kerne besitzt die Linie für die Summe der Isobarenhäufigkeiten der Kerne gerader Massenzahl einen zickzackförmigen unregelmäßigen Verlauf und die Häufigkeitswerte werden wesentlich vom Neutronenüberschuß mitbestimmt. Bei Massenzahlen, wo zwei stabile Isobare vorkommen, besitzt das Isobar mit dem geringeren Neutronenüberschuß die höhere Häufigkeit. Beim Übergang zu höheren Massenzahlen werden die Linien für die Summen der Isobarenhäufigkeit zunehmend glatter. Das Isobar mit dem höheren Neutronenüberschuß besitzt dann die größere Häufigkeit.

Eine stetige Abhängigkeit der Häufigkeiten von der Massenzahl muß nach jeder Theorie erwartet werden, bei der die Kernumwandlungen bei hohen Temperaturen ($kT \sim 1$ Mev) vor sich gehen, da in diesem Fall nicht nur die Grundzustände, sondern auch zahlreiche andere Niveaus bei der Reaktion mitspielen. Effekte, die von einer Änderung der Eigenschaften des Grundzustandes herrühren, werden durch das Mitwirken der höheren Niveaus weitgehend ausgeglichen. Ebenso werden Reaktionen, die mit einer Änderung der Massenzahl verbunden sind, zu einem glatten Verlauf der Häufigkeitslinien führen, wenn sie im Gebiet der beta-unstabilen Kerne an den Hängen des Energietales verlaufen.

Die möglichen Reaktionen, die zur Verteilung der Kernhäufigkeiten auf die verschiedenen Massenzahlen geführt haben können, lassen sich in zwei Typen gliedern:

1. Reaktionen, die primär zur Bildung von Kernen an der neutronenreichen Seite des Energietales führen, wie (n, γ)-Prozesse oder Kernspaltung.

[1] E. M. BURBIDGE, G. R. BURBIDGE, W. A. FOWLER u. F. HOYLE: Rev. Mod. Phys. **29**, 547 (1957).

2. Reaktionen, die auf der neutronenarmen Seite des Energietales verlaufen, wie (p,γ)- oder (γ,n)-Prozesse.

Zweifellos haben (n,γ)-Prozesse bei der Bildung der schwereren Kerne eine sehr wesentliche Rolle gespielt. Ein bei hoher Neutronenkonzentration innerhalb weniger Minuten verlaufender Neutronenaufbauprozeß durch (n,γ)-Reaktionen erklärt zwanglos den glatten Verlauf der Häufigkeitslinien und das Überwiegen der neutronenreichen Isobaren. Bei niederen Neutronenkonzentrationen, bei denen die Neutroneneinfänge langsamer verlaufen, als die entsprechenden β^--Umwandlungen, können auch gewisse abgeschirmte[1] Isobare entstehen, nämlich solche, die eine höhere Bindungsenergie besitzen, als ihre abschirmenden Partner. Nicht durch Neutronenaufbau gebildet werden können jene Kerne, die ein stabiles Isobar mit höherem Neutronenüberschuß besitzen, das energetisch günstiger liegt[2]. Diese Kerne müssen durch Reaktionen des Typus 2 entstanden sein. Auch hier wird man zwischen Reaktionen zu unterscheiden haben, die sich am Hange des Energietales vollziehen, und solchen, die in unmittelbarer Nachbarschaft stabiler Kernsorten verlaufen. Die erste Art dieser Reaktionen wird zu einem glatten, ausgeglichenen Verlauf der Häufigkeiten in Abhängigkeit von der Massenzahl führen, wie sie im Gebiet der leichten Isotope der Elemente Sn, Te, Xe, Ba und Ce in der Tat vorzuliegen scheint. Die zweite Art wird zu einem unregelmäßigen Mengenverhältnis der gebildeten Kerne führen, das von den individuellen Eigenschaften der an der Bildung beteiligten Kerne abhängt.

62. Kernhäufigkeiten und der Schalenbau der Atomkerne. Bereits 1933 hatte ELSASSER[3] darauf aufmerksam gemacht, daß die Häufigkeit von Kernsorten, die eine bestimmte Anzahl von Neutronen oder Protonen enthalten, besonders groß ist. Diese sog. magischen Neutronen- und Protonenzahlen sind die folgenden:

$$2,\ 8,\ 20,\ 28,\ 50,\ 82,\ 126.$$

Diese Zahlen gehören zu zwei verschiedenen arithmetischen Reihen:

$$2,\ 8,\ 20,\ 40,\ 70,\ 112\ldots$$
$$2,\ 6,\ 14,\ 28,\ 82,\ 126\ldots$$

Die erste Reihe tritt im Gebiet der niedrigen Massenzahlen besonders hervor, während die zweite Reihe bei Massenzahlen größer als 40 eine ausgeprägte Bedeutung besitzt. Wie man heute weiß, bedeuten diese magischen Zahlen Schalenabschlüsse im Bau der Atomkerne* [*18*].

Nach HUGHES und SHERMAN[4] besitzen Kerne mit abgeschlossener Neutronenschale einen besonders kleinen Wirkungsquerschnitt für den Einfang schneller Neutronen. Dies muß bei einer Bildung der Kerne durch Neutronenaufbau zu einer besonders großen Häufigkeit solcher Kerne führen, da ihre Umwandlung in Kerne der nächsthöheren Massenzahl langsamer verläuft, als bei anderen Kernsorten. Die breiten Maxima in den Häufigkeitslinien in der Gegend von $A=130$ und 194 können daher als durch Schalenabschlüsse bedingte Effekte gedeutet werden, wenn ein sehr rascher Neutronenaufbau im Gebiet der β^--unstabilen Kerne mit 50 bzw. 82 Neutronen stattgefunden hat. Ein langsamer Neutronenaufbau im Gebiet der stabilen Kerne kann die Häufigkeitsspitzen bei $A=138$ und 139 erklären.

Im Gebiet der leichteren Kerne existieren von Schalenabschlüssen herrührende Effekte bei $N=8$, 14 und 20, obgleich Unsicherheiten in den Häufig-

[1] In der englischen Literatur als "shielded isobars" bezeichnet werden jene Kernsorten, die wegen der Existenz eines stabilen Isobars mit höherem Neutronenüberschuß nicht durch eine Reihe aufeinanderfolgender β^--Zerfälle gebildet werden können.

[2] Diese Kernsorten werden in der englischen Literatur als "excluded nuclei" bezeichnet.

[3] W. ELSASSER: Nature, Lond. **131**, 764 (1933). — J. Phys. Radium **5**, 625 (1934).

[4] D. J. HUGHES u. D. SHERMAN: Phys. Rev. **78**, 632 (1950).

* Vgl. die Bände dieses Handbuches über Kernphysik, insbesondere Bd. XXXIX.

keitswerten und ihr rascher Wechsel mit der Massenzahl den Charakter dieser Effekte nicht deutlich erkennen lassen. Die Häufigkeit von Mg^{30} und A^{38}, Kerne, die 14 bzw. 20 Neutronen enthalten, ist deutlich erhöht.

Der Einfluß abgeschlossener Protonenschalen auf die Häufigkeitsverteilung kann nur von untergeordneter Bedeutung sein. Das Häufigkeitsbild enthält keine eindeutigen Merkmale eines solchen Einflusses.

63. Die Eisenspitze und das Gebiet der niedrigeren Massenzahlen. Während im Gebiet der schwereren Kerne die Bedeutung von Neutronen-Aufbaureaktionen klar zutage tritt, sind wir noch sehr im Unklaren darüber, welche Reaktionen das Mengenverhältnis der leichteren Kernsorten bestimmt haben. Eine Reihe von Anzeichen sprechen dafür, daß Reaktionen vom Typus 2, d.h. solche, die auf der neutronenärmeren Seite des Energietales verlaufen, eine entscheidende Rolle gespielt haben müssen. Ob die Eisenspitze mit der überragenden Häufigkeit von Fe^{56} einfach eine Folge der besonders großen Bindungsenergien der Kerne dieses Gebietes sein kann, oder ob ihr Bestehen mit dem Schalenabschluß bei 28 Neutronen und 28 Protonen in Zusammenhang steht, kann noch nicht entschieden werden. Es ist zu hoffen, daß die vorliegenden Ansätze[1] zur Lösung dieser Fragen führen werden.

64. Die Häufigkeit der Kerne ungerader Massenzahl. Der systematische Häufigkeitsunterschied der Kerne gerader und ungerader Massenzahl, der durch die Regel von HARKINS [*5*] zum Ausdruck gebracht wird, kann nicht in einfacher Weise als eine Folge des Unterschieds in den mittleren Bindungsenergien der beiden Kernsorten gedeutet werden. Er stellt ein weiteres, wohldefiniertes Problem dar, dessen Lösung von einer zukünftigen Theorie erwartet werden muß.

Literatur.

[*1*] CLARKE, F. W.: Bull. Phil. Soc. Washington **11**, 131 (1889).
[*2*] GOLDSCHMIDT, V. M.: Geochemische Verteilungsgesetze der Elemente IX. Skr. norske Vidensk.-Akad. Oslo, mat.-naturw. Kl. **1937**, Nr. 4. — Geochemistry. London: Oxford University Press 1954.
[*3*] UNSÖLD, A.: Z. Astrophys. **21**, 22 (1941); **24**, 306 (1948).
[*4*] SUESS, H. E.: Z. Naturforsch. **2**a, 311, 604 (1947). — Experientia, Basel **5**, 226 (1949). Vgl. auch H. E. SUESS u. H. J. D. JENSEN, LANDOLT-BÖRNSTEIN, Physikalische Tabellen, Bd. 3.
[*5*] HARKINS, W. D.: J. Amer. Chem. Soc. **39**, 856 (1917).
[*6*] MATTAUCH, J.: Phys. Z. **91**, 361 (1934).
[*7*] BROWN, HARRISON: Rev. Mod. Phys. **21**, 625 (1949).
[*8*] NODDACK, I. u. W.: Naturwiss. **35**, 59 (1930).
[*9*] NODDACK, I. u. W.: Svensk kem. Tidsskr. **46**, 173 (1934).
[*10*] FERSMAN, A. E.: Geochemie. Leningrad 1934.
[*11*] UREY, H. C.: The Planets. New Haven: Yale University Press 1952.
[*12*] UREY, H. C.: Phys. Rev. **88**, 248 (1952).
[*13*] PRIOR, G. T.: Miner. Mag. **23**, 33 (1933).
[*14*] SUESS, H. E., and H. C. UREY: Rev. Mod. Phys. **28**, 53 (1956).
[*15*] BROWN, HARRISON, and C. J. PATTERSON: J. Geology **55**, 405, 508 (1947).
[*16*] UREY, H. C., u. H. CRAIG: Geochim. et Cosmochim. Acta **2**, 269 (1952).
[*17*] RANKAMA, K., and TH. G. SAHAMA: Geochemistry. Chicago, Ill.: Chicago University Press 1950.
[*18*] MAYER, M. G., and H. J. D. JENSEN: Elementary Theory of Nuclear Shell Structure. New York: J. Wiley & Sons; London: Chapman & Hall 1955.
[*19*] ALPHER, R. A., and R. C. HERMAN: Rev. Mod. Phys. **22**, 153 (1950). — Phys. Rev. **84**, 60 (1951). — Annual Rev. Nucl. Sci. **2**, 1 (1953).

[1] Siehe z.B. F. HOYLE: Monthly Notices Roy. Astronom. Soc. London **109**, 343 (1946). — W. A. FOWLER u. J. L. GREENSTEIN: Proc. Nat. Acad. Sci. U.S.A. **42**, 173 (1956). — W. A. FOWLER, G. R. BURBIDGE u. E. M. BURBIDGE: Astrophys. Journ. **122**, 271 (1955); Suppl. **2**, 167 (1955). — A. G. W. CAMERON: Astrophys. Journ. **121**, 144 (1955). — Atomic Energy of Canada, Report CRP-652, 1956.

The Abundances of the Elements in the Sun and Stars.

By

LAWRENCE H. ALLER.

With 5 Figures.

I. Compositions of normal stars.

1. Introduction. The qualitative analysis of the Sun and stars was carried out in the latter half of the 19th century, particularly by KIRCHHOFF, LOCKYER and HUGGINS, with considerable success. They compared the solar and stellar spectra directly with laboratory spectra, identified the more abundant elements, and demonstrated the similarity in the qualitative compositions of the Earth and the stars. One of the more interesting incidents of this period was the discovery of helium in the Sun before it was found on the Earth. Helium's astrophysical significance lies not only in the fact that it is the second most abundant element cosmically, but also that it is the ash of the energy-generating proton-proton and carbon-cycle nuclear reactions.

Technological advances, such as the development of the Rowland diffraction grating, combined with an increasingly more complete knowledge of atomic and molecular spectra made continued progress possible. Of the approximately eighty-eight differing elements occurring in nature, the following have been identified in the Sun, stars, and nebulae: H, He, Li, B, Be, C, N, O, F, Ne, Na, Mg, Al, Si, P, S, Cl, A, K, Ca, Sc, Ti, V, Cr, Mn, Fe, Co, Ni, Cu, Zn, Ga, Ge, Kr, Sr, Y, Zr, Cb (Nb), In, Mo, Tc, Ru, Rh, Pd, Ag, Sn, Sb, Ba, Rb, La, Ce, Pr, Nd, Sm, Eu, Gd, Tb, Dy, Tm, Yb, Lu, Hf, W, Os, Ir, Pt, Pb, probably Cd, Au, Th, and possibly Er and Ta. In the Sun the elements Rb, Cs and In are observed only in sunspots, while boron and fluorine are represented only in compounds such as BH or MgF. Not found, although their strongest lines fall in the accessible part of the solar spectrum, are Re, Tl, Bi, and U, while the strongest lines of Hg, Te, Se, I, Br, Xe fall in the inaccessible parts of the spectrum of the Sun. These elements are presumably too rare to be found in other celestial objects. Holmium and arsenic have not been found, while the radio-active elements of relatively short life, Tc, Pm, Po, At, up through Cf, $Z = 100$, are not to be expected in the Sun and similar stars. Technetium, however, has been found in the *S*-type stars.

A cursory examination of the spectroscopic data suggests that hydrogen and helium must be very abundant; hydrogen persists in stellar spectra from the hottest to the coolest. It is clear that iron must be one of the most abundant metals since its lines appear in so many celestial sources, extended envelopes, nebulae, etc., as well as in normal stellar atmospheres.

The quantitative phase of the analysis was not possible until the necessary physical theory, the Saha ionization equation, atomic and molecular line strengths, and the theory of the broadening of spectral lines were known. Early attempts at quantitative analyses were made by Miss PAYNE[1], by MENZEL[2], and by UNSÖLD[3], but the first extensive study was that by H. N. RUSSELL[4] who investigated the composition of the solar atmosphere. MENZEL applied a similar analysis for the solar chromosphere. RUSSELL and MENZEL[5] recognized one of the most fundamental differences between terrestrial and stellar abundances, i.e., there are more H atoms in the universe than atoms of all other elements put together!

2. The observational data. We must now examine the character and limitations of the observational data available for abundance determinations. The accuracy of the derived abundances will depend ultimately on the accuracy of the measurements of the line intensities and profiles and on the correctness of the identifications of the lines that are utilized.

Identifications pose no problems for the stronger lines of abundant elements. The lines of the rarer elements, which are often very weak, are frequently blended with or masked by other lines. Hence a very detailed knowledge of the correct identifications of all lines is necessary. The spectra of some ions, e.g., Fe IV, and of certain of the rare earths, are poorly known. CHARLOTTE MOORE SITTERLY's[6] Multiplet Table of Astrophysical Interest includes most of the lines that are found in stellar spectra.

The Revised Rowland Table, a new edition of which is now in preparation, gives the wavelengths, rough intensities, and identifications of many solar lines. MERRILL has given a bibliography of identifications in the spectra of other stars together with a history of the occurrence of the lines of the various elements in the stars. The identifications of the lines in sunspot spectra are very incomplete; many of these lines are due to fragmentary molecules whose spectra are inadequately known.

Since intensity measurements of the highest accuracy are required, great care is necessary in the measurement of line profiles and equivalent widths. The highest dispersion possible for a given class of objects should be employed. The measured equivalent widths for a given star will often depend on the dispersion employed in the sense that the lower the dispersion the stronger will the observer usually measure the line. In prism instruments the changing dispersion with wavelength may introduce systematic errors, whereas scattered light in grating instruments often causes trouble.

Naturally, very high resolution may be employed in solar work as compared with stellar work. With the vacuum spectrograph at the McMath-Hulbert Observatory of the University of Michigan, it is possible to observe the λ 3933 K line of Ca II with a dispersion of 0.088 Å/mm and a resolving power of about 800000 in the seventh order. By contrast, with the Mt. Wilson 100-inch coude telescope, a dispersion of 2.8 Å/mm is obtained in the second order. Eventually, a dispersion of about 1 Å/mm will be available, but it will be possible to observe then only the very brightest stars. The solar spectrum can be studied in much greater detail than can that of any other star.

For most lines of most elements it is necessary to employ the equivalent widths or "total intensities"; they have the advantage that they are easy to measure, are not affected by the instrumental profile, and are expressed as single

[1] C. H. PAYNE: Stellar Atmospheres. Harvard Observatory Monographs No. 1. Cambridge, Mass. 1925.
[2] D. H. MENZEL: Harvard Circ. No. 258, 1924.
[3] A. UNSÖLD: Z. Physik **46**, 765 (1927).
[4] H. N. RUSSELL: Astrophys. Journ. **70**, 11 (1929).
[5] D. H. MENZEL: Publ. Lick Obs. **17**, 1 (1931).
[6] CHARLOTTE MOORE (Mrs. SITTERLY): Princeton Observatory Contributions, No. 20, 1945.

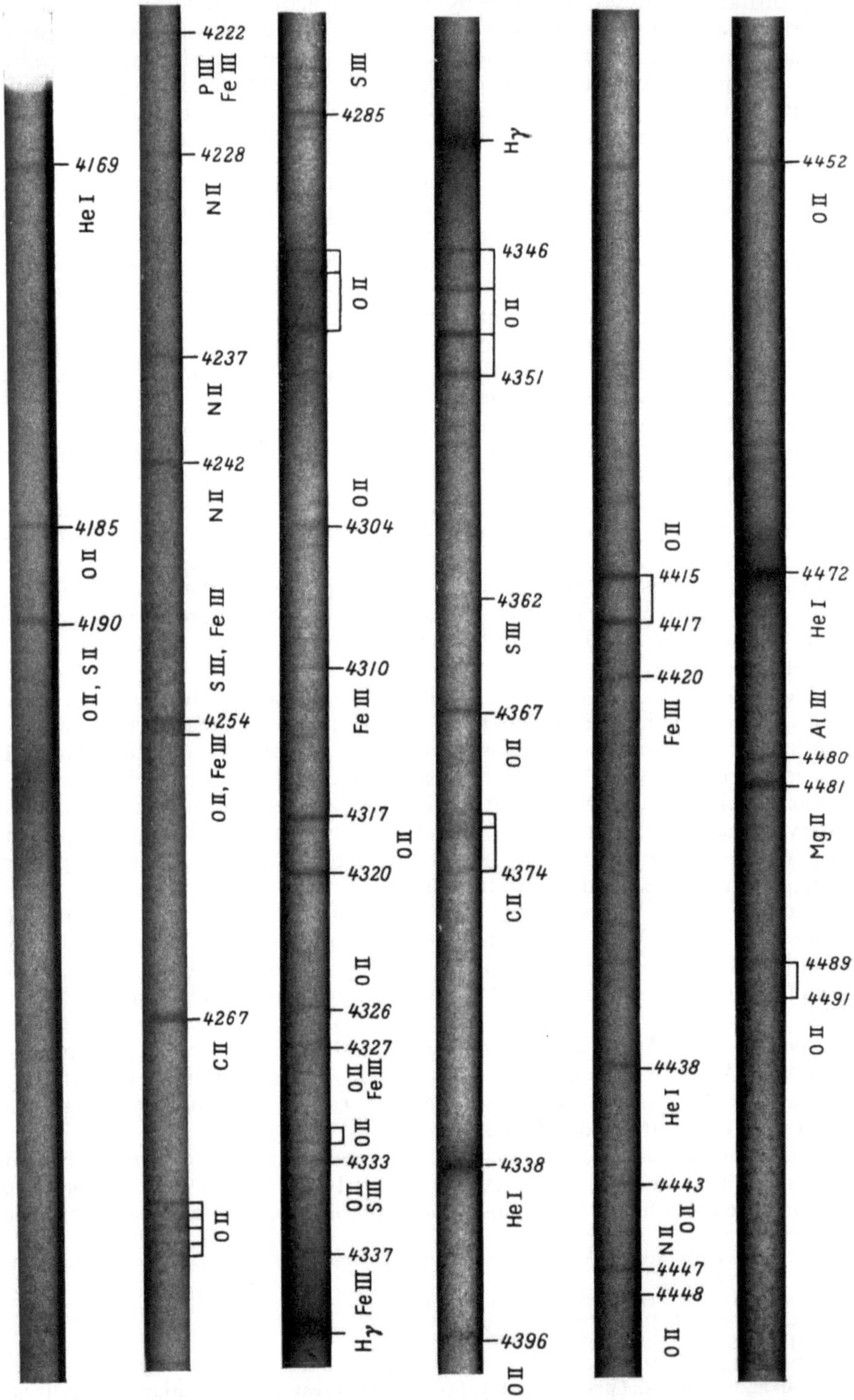
4169
He I
4185
O II
4190
O II, S II
4222
P III
Fe III
4228
N II
4237
N II
4242
N II
S III, Fe III
4254
O II, Fe III
4267
C II
O II
S III
4285
O II
O II
4304
4310
Fe III
4317
O II
4320
O II
4326
4327
O II
Fe III
O II
4333
O II
S III
4337
Hγ Fe III
Hγ
4346
O II
4351
4362
S III
4367
O II
4374
C II
4338
He I
4396
O II
O II
4415
4417
4420
Fe III
4438
He I
4443
O II
N II
4447
4448
O II
4452
O II
4472
He I
Al III
4480
4481
Mg II
4489
4491
O II

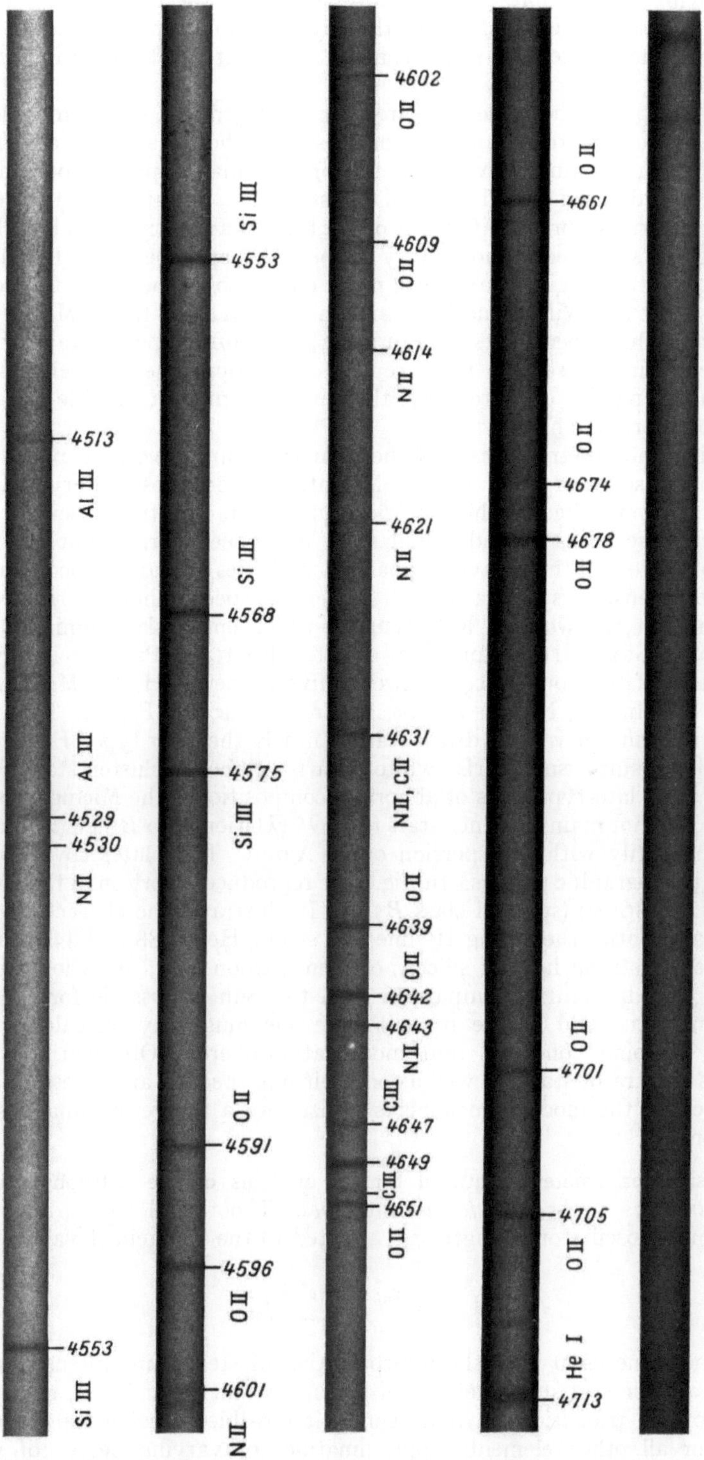

Fig. 1. The spectrum of the hot star HD 36959. Notice the numerous lines of the permanent gases, H, He, N, and O, and of other abundant elements such as C, Al, Si, and Fe. Photographed at the Mt. Wilson Observatory 1955.

numbers giving the amount of light subtracted from the continuum. The disadvantage is that the equivalent width may hide some information (which is carried in the line profile) and which is necessary to interpret the structure of the atmosphere.

Actual line profiles may be measured for hydrogen and strong metallic lines in the Sun and for hydrogen and helium lines in the hotter stars. If an adequate theory of line-broadening is available, the profiles may supply important clues concerning the abundance and atmospheric structure. Great care must be taken in the photometric calibration if results of the highest accuracy are to be achieved. Most high-dispersion spectrophotometry has been done photographically, although direct photoelectric tracings have been made on the solar spectrum, for example, by ROGERSON at Mt. Wilson and by PIERCE and MOHLER at the McMath-Hulbert Observatory. The experiments by HILTNER and CODE demonstrate that it will be possible to make accurate measurements of critical lines in stellar spectra. Here, as in other fields of astronomy, the development of a suitable image tube would be of enormous help.

Extensive photometric atlases of the solar spectrum have been published by the Utrecht Observatory and by the McMath-Hulbert Observatory. HILTNER and WILLIAMS published a photometric atlas of the spectra of several bright stars. Many researches on individual stars have been carried out by various investigators. Most of these have emphasized variables or stars of special interest. Among the extensive, systematic high dispersion spectrophotometric programs we mention first the work of K. O. WRIGHT which emphasizes luminous *G* and *K* type stars GREENSTEIN'S program at Mt. Wilson and Palomar involves the determination of the abundances of twenty-five elements, H, Na, Mg, Si, S, Ca, Sc, Ti, V, Cr, Mn, Fe, Ni, Co, Zn, Sr, Y, Zr, Ba, La, Ce, Pr, Sm, Eu and Gd, relative to the Sun for various dwarf stars of nearly the solar type. His programs also include *F* stars, subdwarfs, white dwarfs, globular cluster stars, of high luminosity; and late-type stars of abnormal composition. The Michigan program includes mostly hot main sequence stars from *O* 5 (λ Orionis) to *B* 8 IV (α Sculptoris) observed primarily with a dispersion of 2.8 Å/mm. The plates cover both the visual and photographic regions. In Fig. 1 we reproduce a portion of the spectrum of the star HD 36959 (spectral class *B* 1 V). It illustrates the characteristic lines of a hot star. Notice the strong Hγ line, the strong He I 4388 and 4471 lines and the moderately strong lines of silicon, oxygen, carbon, etc., in various stages of ionization. No differential comparison with the Sun is possible for these high temperature stars, and at the present time one must rely on calculated line strengths, damping constants and model atmospheres. Observations on the continuous spectra of stars, as well as on their line spectra, are necessary to provide a check on the model atmospheres utilized for a theoretical analysis of the observations.

3. Physical parameters required for an analysis of the astrophysical data. α) *Line strengths or f-values for atomic lines.* Theoretical work: The Ladenburg f-value or oscillator strength, $f_{n'n}$, related to the Einstein A-values, by the expression

$$A_{n n'} = \frac{g_{n'}}{g_n} \cdot \frac{8\pi^2 e^2 \nu^2}{m c^3} f_{n' n} \tag{3.1}$$

is a parameter necessary for the interpretation of stellar line intensities. Here g_n and $g_{n'}$ are the statistical weights of the upper and lower levels, respectively.

Only for the transitions in hydrogen is it possible to calculate the f-values exactly; for all other elements approximations of varying degree of validity

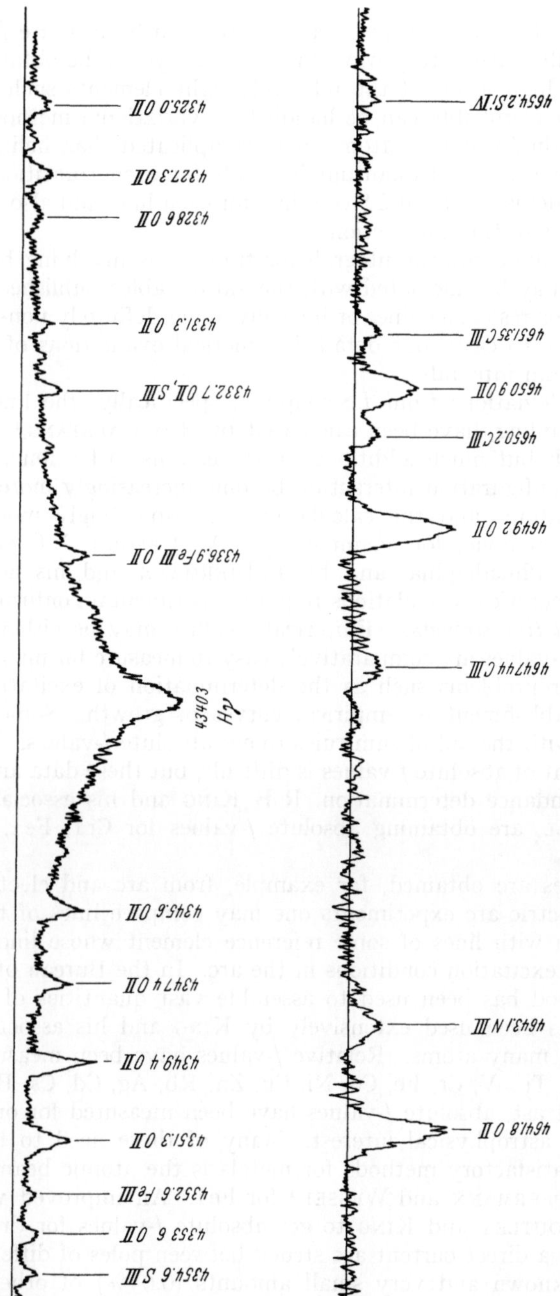

Fig. 2a and b. Tracings of portions of the spectrum of HD 36959 (a) Region of Hγ. (b) Region λ 4640 to 4650. Notice the broad shallow "wings" of the hydrogen line profile. For a given star the theoretical model atmosphere must have a temperature and density distribution such as to reproduce the observed profiles of the Balmer lines. For the weaker lines of other elements, the profiles cannot be measured and one can use only the equivalent widths.

must be employed. It should be possible to obtain fairly accurate *f*-values for helium, although fully satisfactory wave-functions are yet to be obtained for the higher levels. The high levels of the relatively light elements such as carbon, nitrogen, and oxygen probably can be handled by variational methods.

In order to obtain *f*-values in atoms more complicated than helium it is necessary to calculate the radial quantum integrals for each transition array involved, to find the deviation from *LS* coupling for each line, and also to estimate the amount of configuration interaction.

Approximate radial quantum integrals for transitions involving high, nearly hydrogenic, levels may be calculated with the aid of tables published by BATES and DAMGAARD[1]. For resonance lines or lines involving definitely non-hydrogenic levels it is necessary to carry out detailed numerical evaluations of the appropriate radial quantum integrals.

The effects of deviations from *LS* coupling, specifically, the line strengths in intermediate coupling, have been calculated by ROY GARSTANG[2] for lines of O II, Ne II and S II, but much additional work remains to be done.

The effects of configuration interaction become increasingly more important in more complex atoms and the calculations correspondingly more difficult. Much work has been done, for example, by C. W. UFFORD, L. C. GREEN, and their associates in Philadelphia, and by L. BIERMANN and his associates in Göttingen. The theoretical calculations require experimental confirmation.

β) Experimental line strengths. Empirical *f*-values may be either *relative* or *absolute*. Relative *f*-values are comparatively easy to measure for many elements. They are useful for problems such as the determination of excitation temperature and the establishment of empirical curves of growth. Sometimes they can be calibrated with the aid of sum rules to get absolute *f*-values. The experimental measurement of absolute *f*-values is difficult, but these data are necessary for the actual abundance determination. R. B. KING and his associates, DAVIS, ROUTLEY and BELL, are obtaining absolute *f*-values for Cr I, Fe I, Cu I, Co I, and Mn I[3].

Relative *f*-values are obtained, for example, from arc and electric furnace studies. In the electric arc experiments one may compare lines of the element under investigation with lines of some reference element whose line intensities are used to fix the excitation conditions in the arc. In the Bureau of Standards program this method has been used to assemble vast quantities of data. The electric furnace has been used extensively by KING and his associates to get relative *f*-values in many atoms. Relative *f*-values have been measured for Li, Na, Mg, K, Ca, Ti, Ti^+, V, Cr, Fe, Co, Ni, Cu, Zn, Rb, Ag, Cd, Cs, Ba, and Hg.

By way of contrast, absolute *f*-values have been measured for only about a dozen elements of astrophysical interest. Many of these need to be checked. One of the most satisfactory methods for metals is the atomic beam technique introduced by KOPFERMANN and WESSEL[4] for Fe I. An improved version was used by DAVIS, ROUTLEY and KING to get absolute *f*-values for Cr I and Ca I. C. W. ALLEN[5] used a direct current arc struck between poles of different copper alloys containing known and very small amounts (0.05%) of other elements. He tried to obtain absolute *f*-values but the method appears to be subject to systematic errors, at least for certain elements.

---

[1] D. R. BATES and A. DAMGAARD: Phil. Trans. Roy. Soc. Lond. A **242**, 101 (1949).

[2] R. H. GARSTANG: Monthly Notices Roy. Astronom. Soc. London **110**, 613 (1950).

[3] See for example, Proceedings of Conference on Stellar Atmospheres, Indiana University 1954, ed. M. WRUBEL.

[4] H. KOPFERMANN and G. WESSEL: Z. Physik **130**, 100 (1951).

[5] C. W. ALLEN: Monthly Notices Roy. Astronom. Soc. London **117**, 622 (1957).

A technique admirably suited to elements that form convenient liquid compounds is the whirling fluid arc which was developed at the laboratory of LOCHTE-HOLTGREVEN[1] in Kiel. Line strengths for C, N, and O have been obtained and since the apparatus can be operated at high pressures, important data on line broadening can be obtained.

A third technique involves the luminous shock tube developed by KANTROWITZ at Cornell and by E. B. TURNER in LAPORTE'S Laboratory at Michigan[2]. A shock wave is produced under controlled conditions and allowed to strike a rigid wall where it is reflected. After the shock wave is reflected, its energy is dissipated in the form of heat and the gas may be heated to incandescence. The pressures and temperatures may be computed from the hydrodynamics of the problem. Hence from a measurement of the intensities of the emitted lines, absolute f-values can be obtained.

In this way, LOWELL DOHERTY found f-values for Ne I in good agreement with those found by LADENBURG. The technique is well suited to permanent gases and simple gaseous compounds.

Molecular f-values, both relative and absolute, are of great importance for the analysis and interpretation of the spectra of the cooler stars. Vibrational transition probabilities have been published by JARMAIN, FRASER, and NICHOLLS[3] for bands of N_2, NO, O_2, O_2^+, OH, CO, and CO^+. Among recent and current investigations we may also mention FLOYD'S and KING'S[2] measurements of the relative transition probabilities of the violet CN bands and PHILLIPS'[2] program for relative f-values. He studies as many molecules of astrophysical interest as can be produced in the electric furnace, in particular, TiO, CN, C_2, ZrO, VO, YO, MgH, and CaH.

Mention must also be made of relative f-value determined empirically from solar and stellar curves of growth. MENZEL, BAKER, and GOLDBERG[4] and later K. O. WRIGHT[5] and BARBARA BELL[6] have obtained relative f-values from the solar curve of growth. GREENSTEIN[7] obtained empirical line strengths from his curves of growth for τ Ursae Majoris and v Sagittarii. Some astrophysicists, desperately in need of f-values, have even used emission lines from the solar chromosphere and bright line stars, but because of the huge range of density and temperature in these objects, such procedures are not to be recommended. Line strengths should be found from laboratory data—not from the stellar spectra.

γ) Damping constants and theories of line broadening. The interpretation of the profiles and equivalent widths of strong lines requires a knowledge not only of the f-value for the lines but also of their damping constants. The radiation damping constant can be calculated from the f-values and the radiation field. The collisional damping constant depends on the temperatures, density, and character of the perturbing ions or atoms. The strong metallic lines in the Sun and dwarf stars are broadened by the quadratic Stark effect and by the VAN DER WAAL'S interactions with neutral hydrogen atoms. The broadening coefficients must be found from the laboratory data.

For lines that obey the Lindholm theory of broadening, the collisional damping constant can be found if the wavelength displacement of the line is known as

---

[1] W. LOCHTE-HOLTGREVEN: Observatory **72**, 142 (1952).
[2] See footnote 3, p. 330.
[3] W. R. JARMAIN, P. A. FRASER and R. W. NICHOLLS: Astrophys. Journ. **122**, 55 (1955).
[4] D. H. MENZEL, J. G. BAKER and L. GOLDBERG: Astrophys. Journ. **87**, 81 (1938).
[5] K. O. WRIGHT: Publ. Dom. Astrophys. Obs. **8**, 1 (1948).
[6] BARBARA BELL: Harvard Obs. Spec. Rep. **35** (1951).
[7] J. L. GREENSTEIN: Astrophys. Journ. **95**, 161 (1942); **107**, 151 (1948); **109**, 121 (1949).

a function of density. Collisional damping constants differ not only from one atom to another but also from one type of level to another; e.g., Carter[1], Barbara Bell, and Pierce and Goldberg showed that Fe I lines arising from odd terms have higher damping constants than Fe I lines arising from metastable even terms.

Damping constants may be found experimentally with the aid of Lochte-Holtgreven's whirling fluid arc, with the shock tube, and with the electric furnace. Line broadening data on H, He and other light gases have been obtained at Kiel, but the results have so far been published only for H. In the luminous shock tube the pressure and density behind the shock may be calculated with the Rankine-Hugoniot relations with appropriate corrections for the ionization of the gas. The electric furnace or arc may be used with helium as the broadening gas in order to simulate the density broadening in the Sun and stars.

Hydrogen, which is subject to the first order Stark effect, therefore, shows different type of broadening than most other gases. Experimental studies have been made by Griem[2] at Kiel and by Turner and Doherty[3] at Michigan while the most satisfactory theory is that by A. C. Kolb[4] who finds that the broadening contributions of the electrons as well as those of the ions must be included and that non-adiabatic effects are important. Ionized helium shows the same type of broadening as hydrogen, whereas neutral helium shows both first and second-order Stark effects. The determination of the damping constants for helium presents a most difficult experimental problem that may be solved by shock tube techniques.

δ) Continuous absorption coefficient. A fundamental parameter in the interpretation of absorption line intensities is the ratio of the line absorption coefficient l_ν to the continuous absorption coefficient $\varkappa_\nu$. The hydrogen/metal ratio in the Sun is found from the intensities of metallic lines produced in an atmosphere whose continuous absorption is due primarily to the negative hydrogen ion[5]. The intensity at a point in the profile of a metal line thus depends on l_ν (metal)/$\varkappa_\nu$ (H^-).

The absorption coefficient for the negative hydrogen ion has to be calculated from wave functions that are difficult to determine at distances from the nucleus important for the calculation of transition probabilities and may be subject to errors of the order of 10 to 15% in the infrared. In the hotter stars where the absorption is contributed primarily by atomic hydrogen, fairly reliable absorption coefficients are available for the bound-free transitions; those for the free-free transitions are not quite as satisfactory.

In the hot stars of low surface gravity, electron scattering is the principal source of opacity. In cool stars of low surface gravity Rayleigh scattering by atomic or molecular hydrogen may play an important role. Continuous absorption by metals is probably not important except in stars of abnormal composition, while H_2^+ may be an important contributor to the opacity in A and F stars.

Other, as yet unidentified, sources of opacity occur in cooler stars than the Sun. P. C. Keenan, from a study of the CN bands, suggests that stars with effective temperatures less than 3500° K have another source of opacity which has a less pronounced maximum in the infra-red than the negative hydrogen ion.

---

[1] Carter: Phys. Rev. **76**, 962 (1949).
[2] H. Griem: Z. Physik **137**, 280 (1954).
[3] See footnote 3, p. 330.
[4] A. C. Kolb: Thesis, Univ. of Michigan 1956.
[5] S. Chandrasekhar and F. Breen: Astrophys. Journ. **104**, 430 (1946). — S. Geltman: Phys. Rev. **104**, 346 (1956).

SWINGS[1] suggests that in stars cooler than spectral class $M2$, the vibration-rotation bands of polyatomic molecules may be important.

In some stars of abnormal composition the continuous absorption may be produced by helium; hence an accurate theory of the continuous absorption by helium is needed.

ε) Dissociation potentials for molecules. The use of molecular bands to obtain abundances requires not only a knowledge of absolute f-values, but also accurate dissociation potentials. The latter are often difficult to determine and are not known for many astrophysically important molecules such as CO, CN, N_2, and NO. In the investigation of the cooler stars it is also necessary to have accurate dissociation potentials for TiO, ZrO, VO, and LaO, and eventually other molecules as well when the far infra-red and ultra-violet are more fully explored[2].

4. Methods of analysis of the spectroscopic data. *α)* RUSSELL's fundamental investigation of the composition of the Sun, which was based on a calibration of the Rowland intensity scale, established the general character of stellar atomic abundances. He obtained the relative abundances of 56 elements and 6 compounds. The abundances of H and He were not securely established, but his work and that of MENZEL on the chromosphere demonstrated the overwhelming preponderance of hydrogen by numbers of atoms. RUSSELL estimated that O was about as abundant by weight as all the metals put together and concluded that Na, Mg, Si, K, Ca, and Fe contributed about 95% of the total mass. He also noticed the abundance differences between elements of even and odd atomic number, the former averaging about ten times as abundant as the latter. Heavy metals from barium onward are but little less abundant than those that follow strontium. The metals from Na to Zn, inclusive, are far more common than the rest. C. DE JAGER[3] estimated the error in $\text{Log} N$ to be ± 0.55. RUSSELL's estimates were the only ones available for many elements until recently, while for some, ignorance of partition functions, etc., do not permit us even yet to make a substantial improvement over his values.

β) The curve of growth[4]. RUSSELL's procedure utilized eye estimates of intensity and allowed for saturation effects (curve of growth phenomenon) only in an approximate way. The introduction of the curve of growth, which employed quantitative data in the form of equivalent widths constituted an important step forward.

Early workers with the curve of growth simply plotted $\log W$, where W is the equivalent width of the line, against $\log Nf$, the number of atoms active in producing the line multiplied by the oscillator strength. When the number of atoms is small the equivalent width is directly proportional to Nf, since the amount of energy subtracted from the outgoing flux of radiation at this wavelength is small compared with the total amount of radiation. As Nf increases, "saturation effects" set in, the optical depth at the center of the line becomes very large and W grows slowly. The precise value of Nf at which this "flat portion" of the curve of growth occurs depends on the wavelength of the radiation and on the speed of the atoms. The higher the atomic speeds, the wider the wavelength interval over which absorption can take place and the greater the value of Nf

[1] P. SWINGS: Handbuch der Physik, Vol. L, p. 126. 1958.

[2] See for example, G. HERZBERG: Coll. Liege, 1956, p. 307.

[3] C. DE JAGER: Utrecht Obs. Reprint, No. 17, 1953.

[4] Discussion of the curve of growth may be found in the texts cited in the bibliography. See also D. BARBIER, Handbuch der Physik, Vol. L, p. 360. 1958. — A simple qualitative description may be found in L. GOLDBERG and L. H. ALLER: Atoms, Stars, and Nebulae, p. 103. Cambridge: Harvard Press 1943.

and W at which saturation will occur. Lines on the flat part of the curve of growth exhibit a bell-shaped profile. Finally, if Nf is made large enough, the "damping wings" or natural broadening of the line becomes important and W increases in proportion to $\sqrt{Nf}$[1]. The greater the value of the damping constant, the smaller the flat portion of the curve of growth. In the example depicted in Fig. 3, the required damping constants are large, so that the flat portion of the curve of growth is small.

The original formulation, by MINNAERT and SLOB[2], was improved by MENZEL[3] and by UNSÖLD for the Schuster-Schwarzschild model. In this model one supposes that the photosphere radiates a pure continuous spectrum and the lines are formed in an overlying stratum that absorbs monochromatically only. In the MENZEL notation one plots the quantity calculated from the equivalent width as ordinate against the optical depth at the center of the line as abscissae, viz.:

$$\log\left(\frac{W}{\lambda}\,\frac{c}{v}\right) \quad \text{vs.} \quad \log X = \log\left(N\,\frac{\pi e^2}{m c}\,f\,\frac{1}{\sqrt{\pi}}\,\frac{\lambda}{v}\right) \tag{4.1}$$

where c is the velocity of light and v is the most probable velocity of the atoms;

$$v = \sqrt{v^2_{\text{kinetic}} + \bar{v}^2_{\text{turbulence}}} \tag{4.2}$$

if turbulence is present in the stellar atmosphere. N is the number of atoms above the photosphere capable of absorbing the line. Now the number of atoms in the lower level j is

$$N_j = N_0\,\frac{g_j}{\boldsymbol{B}(T)}\,e^{-\chi_j/kT} \tag{4.3}$$

where g_j is its statistical weight and χ_j its excitation potential. $\boldsymbol{B}(T)$ is the partition function for the atoms. N_0 is the total number above the photosphere in the given stage of ionization. Hence the abscissa may be written in the form (where the symbol log stands for the logarithm to the base 10):

$$\operatorname{Log} X = \operatorname{Log} N_0 + \operatorname{Log}(g f \lambda) - \frac{5040}{T}\chi + \operatorname{Log}\frac{\text{const}}{\boldsymbol{B}(T)} - \operatorname{Log} v. \tag{4.4}$$

The theoretical curve of growth gives a uniquely defined relation between $\log W/\lambda$ and $\log X$. To obtain an empirical curve of growth we first compute

$$\log Y = \log\frac{X}{N_0} + \frac{5040}{T}\chi$$

for each line. Then we group the lines according to χ, the excitation potential of the lower level. The differential shifts required to fit the empirical $\log Wc/\lambda v$ vs. $\log Y$ plots to the theoretical curve give the excitation temperature and $\log N_0$. Turbulence, if present, requires a vertical shift of the empirical plot to fit it to the theoretical curve. We must know the absolute f-values and the damping constants in order to use the curve of growth to determine the abundance of the element in question in the given stage of ionization.

The electron density as well as the temperature must be found in order that we may use the Saha equation to evaluate the total concentrations of all the atoms of the given element.

If we wish to compare the numbers of atoms above the photosphere in a given stage of ionization in the Sun with the corresponding quantities for a star of

[1] See footnote 4, p. 333.

[2] M. MINNAERT and C. SLOB: Proc. Amst. Acad. **34**, No. 4 (1931).

[3] D. H. MENZEL: Astrophys. Journ. **84**, 462 (1936).

nearly the same spectral class we can easily do so with the aid of the curve of growth. If X_* and $X_\odot$ are the respective abscissae for a given line in the star and in the Sun, respectively, we have that the $gf\lambda$ values cancel and

$$\text{Log}\,\frac{X_*}{X_\odot} = \text{Log}\,\frac{N_*}{N_\odot} - 5040\,\chi\left(\frac{1}{T_*} - \frac{1}{T_\odot}\right) + \text{Log}\,\frac{\boldsymbol{B}_\odot(T)}{\boldsymbol{B}_*(T)}\,\frac{v_\odot}{v_*}. \tag{4.5}$$

The temperature $T_\odot$ and electron pressure $P_{\varepsilon\odot}$ in the solar atmosphere are presumed known. By comparing, for example, the ratio N_* (Fe II)/$N_\odot$ (Fe II) and the ratio

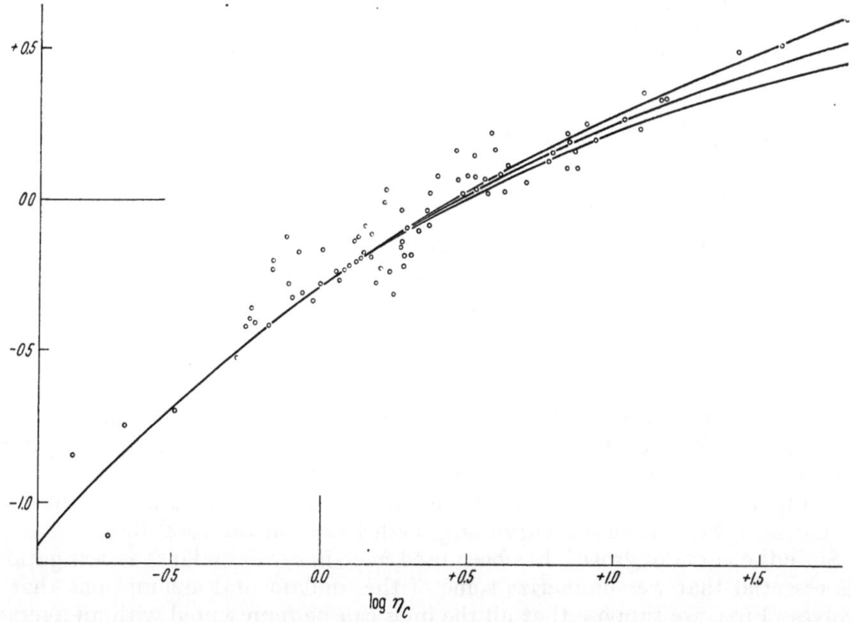

Fig. 3. Curve of growth for ε Canis majoris. Here Log $\frac{W}{\lambda}\frac{c}{v}$ is plotted against Log η_c [see Eq. (4.7)]; the theoretical curves are due to WRUBEL. See Astrophys. J. **109**, 71 (1949). The observational data were obtained at the McDonald Observatory and at the Mt. Wilson Observatory. Starting with the highest, the curves refer to log $a = -1.0$, -1.4 and -1.8 where $a = \Gamma/4\pi\,\Delta\nu_0$; Γ is the damping constant; $\Delta\nu_0 = \nu_0\,v/c$.

N_* (Fe I)/$N_\odot$ (Fe I) it is possible to solve for the quantity N_* (Fe II)/N_* (Fe I) since the corresponding ratio is known for the Sun. We find a relation between T and P_ε that the stellar atmosphere must follow from the Saha ionization equation[1]. Then we can fix T from the color of the star and evaluate P_ε. The total quantity of material "above the photosphere" will depend on the coefficient of continuous absorption $\varkappa_\lambda$. Since this absorption is due mostly to the negative hydrogen ion we can allow for its effect in computing the relative abundances.

Different forms of the curve of growth, variously dependent on the underlying physical assumptions, have been proposed. MENZEL and UNSÖLD used the Schuster-Schwarzschild model with a simplified theoretical formula in the one instance, and an empirical formula due to MINNAERT in the other. WRUBEL published curves for the Milne-Eddington model and an exact formula for pure scattering. Analogous curves for pure absorption have been discussed by the present writer; more complete and accurate curves have been computed by K. HUNGER and by WRUBEL. In the Milne-Eddington model, the ratio of line

[1] See for example, L. H. ALLER: Atmospheres of the Sun and Stars, p. 97. New York: Ronald Press. Co. 1953.

to continuous absorption is assumed independent of optical depth, i.e., the line and continuous spectra are formed strictly in the same layers. As abscissa for the curve of growth one employs

$$\mathrm{Log}\,\eta_c = \mathrm{Log}\,\frac{l_c}{\varkappa_{\mathrm{eff}}} \tag{4.6}$$

where l_c is the line absorption coefficient and $\varkappa_{\mathrm{eff}}$ is the sum of the continuous absorption coefficient and electron scattering coefficient, each computed per gram of stellar material.

Fig. 3 gives the curve of growth for ε Canis Majoris; the theoretical curve is one due to WRUBEL. Here the ordinates are Log $W c/\lambda v$, where v is given by Eq. (4.2), while the abscissae are

$$\mathrm{Log}\,\eta_c = \mathrm{Log}\,\frac{\pi\,\varepsilon^2}{m\,c}\,\frac{\lambda}{v\sqrt{\pi}} + \mathrm{Log}\,N_{r,s}\,\boldsymbol{f} - \vartheta(\chi - \chi_0) - \mathrm{Log}\,\varkappa_{\mathrm{eff}} + \Delta\,\mathrm{Log}\,\eta. \tag{4.7}$$

Here $\vartheta = \frac{5040}{T}$, χ is the excitation potential of the level in question, χ_0 that of the configuration (or term), $N_{r,s}$ is the number of atoms in the s-th configuration (or term) of the r-th stage of ionization, $\boldsymbol{f}$ is the Unsöld f which is related to the conventional Ladenburg f by the relation:

$$\boldsymbol{f}(J, J') = \frac{(2J'+1)}{g_s}\,f(J, J') \tag{4.8}$$

where g_s is the statistical weight of the whole configuration (or term) and $(2J'+1)$ that of the lower level from which the line arises. The last term, Δ Log η, take into account the variation of the theoretical curve with wavelength.

In the Sun one can use a curve of growth computed for the center of the disk. For the stars we must use a curve of growth based on the total flux.

Since the curve of growth has been used in so many abundance investigations it is essential that we summarize some of the fundamental assumptions that it involves. First, we suppose that all the lines can be represented with an average temperature and an average electron pressure, e.g., UNSÖLD obtained for the Sun $T = 5676°$ K and Log $P_\varepsilon = 1.51$, whereas K. O. WRIGHT got $T = 4700$ to $5275°$ K and Log $P_\varepsilon = 1.26$. Second, we get different results depending on what curve of growth is used, even though the same observational data are employed[1]. Third, usually all lines are assumed to follow the *same* curve of growth, although different lines are allowed to have different damping constants. In solar type stars and in the Sun K. O. WRIGHT and later BARBARA BELL attempted to establish individual empirical curves for different ions and the damping constants there to pertaining. Fourth, in the curves of growth computed for the Milne-Eddington model it is usually also assumed that the Planck function $B_\nu(T)$ is a linear function of the optical depth τ_ν at the wavelength in question. In Sects. 4γ and 5 we shall compare the curve-of-growth results with those obtained by more precise methods.

Molecular line and band intensities are useful for abundance studies because many elements are represented in the Sun or particular stars only by molecular bands. For example, molecules found in the Sun include BH, OH, CH, C_2, NH, CN, CO, O_2, SiH, MgH, MgO, MgF, AlO, CaH, ScO, TiO, ZrO, YO, and SrF.

[1] For example, in ε Canis Majoris [Astrophys. Journ. **123**, 117 (1956)] the following log N's (relative to oxygen) are found by the Wrubel, pure absorption, and Unsöld curves of growth, respectively: carbon −0.44, −0.74, −0.46; nitrogen −0.16, −0.26, and −0.20; Mg −0.46, −0.49 and −0.47; aluminum −2.02, −2.05, and −2.04; and silicon −0.96, −1.20, −1.02.

Attempts to estimate abundances from molecular data have been made by HUNAERTS[1]. From the bands of CN, CH, and OH he obtained concentrations for C and O that appeared to be too large. Some of the difficulties doubtless could be traced to uncertainties in the f-values and dissociation constants (for CN at least), but some of the trouble can be traced to the structure of the solar atmosphere. Thus, G. NEWKIRK'S[2] work on CO shows that a large dissociation potential plus a low boundary temperature is required. Alternatively, we may suppose that the plane parallel stratification approximation fails and alternate cool and hot areas exist. At any rate, the same temperature and gas pressure usually cannot be used for both atomic and molecular lines in a curve of growth analysis.

γ) *Use of model atmospheres. Pecker's method.* A stellar absorption line is not formed at a single level, but rather over a range of temperature, electron pressure, and gas pressures. Layers at different optical depths contribute by differing amounts to the production of the line, depending on the fraction of atoms capable of absorbing the line in question, the optical depth of the layer concerned and whether the line is formed by scattering or by absorption.

The first step is the construction of the model atmosphere, viz., the basic relationship between temperature and pressure or between temperature and the optical depth at some particular wavelength, τ_λ, viz., $T(P_\varepsilon)$ or $T(\tau_\lambda)$.

We suppose that the atmosphere is in hydrostatic equilibrium, viz.,

$$dP = -g\varrho\,dx. \tag{4.9}$$

The element of optical depth at the standard wavelength λ_0 is

$$d\tau_0 = -\varkappa_0\varrho\,dx \tag{4.10}$$

where $\varkappa_0$ is the mass absorption coefficient λ_0. The optical depth at any other wavelength is related to τ_0 by

$$d\tau_\lambda = \frac{\varkappa_\lambda}{\varkappa_0}\,d\tau_0. \tag{4.11}$$

If radiation pressure can be neglected, the equation of hydrostatic equilibrium becomes

$$\frac{dP}{d\tau_0} = -\frac{g}{\varkappa_0}. \tag{4.12}$$

Since $\varkappa_0$ depends on the temperature, electron pressure and the ratio of hydrogen to the metals, we must have some preliminary notion about the composition and we must know the dependence of T on τ_0.

The dependence of T on τ_0 may be obtained for the Sun from a study of the limb darkening as a function of wavelength and the energy distribution at the center of the solar disk. Various models of the solar atmosphere have been proposed; the dependence of temperature on optical depth seems well-established except in the outermost layers.

No direct empirical determination of the temperature distribution in stellar atmospheres is possible. In the hotter stars we may rely on the condition of radiative equilibrium, i.e., that the flux of energy is constant with depth.

The model atmosphere gives the temperature and pressure as a function of the optical depth τ_0 at some particular wavelength, λ_0. Hence the fraction of all the atoms of any particular element existing in a given energy state in a

[1] J. HUNAERTS: Trans. Internat. Astronom. Union 7, 462 (1950).
[2] G. NEWKIRK: Astrophys. Journ. 125, 571 (1957).

specified stage of ionization may be evaluated as a function of optical depth τ_0. Consider first the problem for the solar atmosphere, where W_λ^μ denotes the equivalent width of a line at wavelength λ, observed at a point on the Sun's disk such that the emergent ray makes an angle ϑ with the outward normal. Then

$$\frac{W_\lambda^\mu}{\lambda} = \int_{-\infty}^{+\infty} \frac{\overset{*}{M}}{\mu} \frac{\sqrt{\pi}}{c} Z(x)\, \Psi\left(\frac{Y}{\mu}; a\right) g_\lambda^\mu(x)\, dx. \tag{4.13}$$

Here

$$\mu = \cos\vartheta; \tag{4.14}$$

$\overset{*}{M}$ contains the abundance, the f-value, and several numerical constants.

$$M^* = \frac{N_i}{N_H} g_{r,s} f_{r,s} \lambda [A] \frac{10^{-8}}{Mod} \frac{\varepsilon^2 \sqrt{\pi}}{mc} \frac{1}{m_H \mu_0} 10^{\Theta_0 \cdot \Delta\chi}$$

where N_i/N_H is the abundance of the i-th element with respect to hydrogen, Mod is the modulus of natural logarithms, $g_{r,s}$ is the statistical weight of the s-th level in the r-th stage of ionization, m_H is the mass of the hydrogen atom, μ_0 is the molecular weight of the un-ionized material. Also $\vartheta_0 = 5040/T_a$ where T_a is conveniently chosen as the excitation temperature of the atmosphere, and $\Delta\chi$ depends on the state of ionization as follows: $\Delta\chi = (\chi_{r-1} - \chi_{r-1,s})$ for $r-1$, s; $\Delta\chi = -\chi_{r,s}$ for r, s, and $\Delta\chi = -(\chi_{r+1,s} + \chi_r)$ for $r+1$, s). Here χ_{r-1}, χ_r, and χ_{r+1} are the successive ionization potentials. The element of integration is

$$x = \operatorname{Log} \tau_0. \tag{4.15}$$

The integrand contains three depth dependent factors, Z, Ψ, and g_λ. Now $Z(x)$ includes the Boltzmann and ionization factors which depend on the particular atom and level involved.

$$Z(x) = \frac{1}{\varkappa_{\text{eff}}^\lambda} \frac{n_{r,s}}{\Sigma n_r} \frac{10^{\chi - \Theta_0 \Delta\chi}}{g_{r,s}}$$

where $\varkappa_{\text{eff}}^\lambda$ is the absorption coefficient at the wavelength in question, $n_{r,s}/\Sigma n_r$ is the ratio of the number of absorbing atoms to all atoms of the element. For each term or configuration of interest, we compute $Z(x)$.

The quantity $Y(x)$ is defined in terms of it by

$$Y(x) = \frac{\overset{*}{M}}{v} \int_{-\infty}^{x} Z(y)\, dy \tag{4.16}$$

where v is defined by Eq. (4.2). Finally, $\Psi\left(\frac{Y}{\mu}, a\right)$ is defined in terms of Y by the definite integral

$$\frac{2}{\sqrt{\pi}} \int_0^\infty H(u, a)\, e^{-H(u,a)\, Y/\mu}\, du = \Psi\left(\frac{Y}{\mu}, a\right) \tag{4.17}$$

where

$$H(u, a) = \frac{a}{\pi} \int_{-\infty}^{+\infty} \frac{\exp(-y^2)}{a^2 + (v-y)^2}\, dy$$

defines the shape of the absorption coefficient as a function of

$$u = \frac{\lambda - \lambda_0}{\Delta\lambda_0} \quad \text{and} \quad a = \frac{\Gamma}{4\pi \Delta\nu_0} \tag{4.18}$$

where Γ is the sum of the natural and collisional damping constants, and $\Delta\lambda_0$ (or $\Delta\nu_0$) is the Doppler width. Ψ can be calculated once and for all as a function of Y/μ and a and tabulated. Finally, the weighting function $g_\lambda^\mu(x)$ depends only on the model atmosphere and wavelength. It decreases monotonically with depth in the star. The entire integrand is often referred to as the "contribution function". The Ψ function is called the "saturation function" and expresses the well-known fact that strong lines are formed in higher levels of the atmosphere than are weak lines. In fact, the principal contribution to the equivalent width of very strong lines appear to come primarily from the chromosphere. For abundance determinations it is better to use weak to moderately strong lines. In this Pecker method, each atom in each stage of excitation and ionization has its own special curve of growth. A single, all-inclusive curve can no longer be used.

A formula somewhat similar to Eq. (4.13) holds for the stars except that here one deals with the integrated flux rather than with the radiation from a particular point on the disk. G. ELSTE has given a convenient formulation of the Pecker theory for this problem; we have followed his notation in this article.

For the hydrogen lines and most of the helium lines, the curve-of-growth theory cannot be applied. These lines show broad wings, largely due to the interatomic Stark effect. Hence, detailed profile calculations must be carried out for each line. The calculations are laborious and require that the separate contributions of ions and electrons to the broadening coefficient be taken into account.

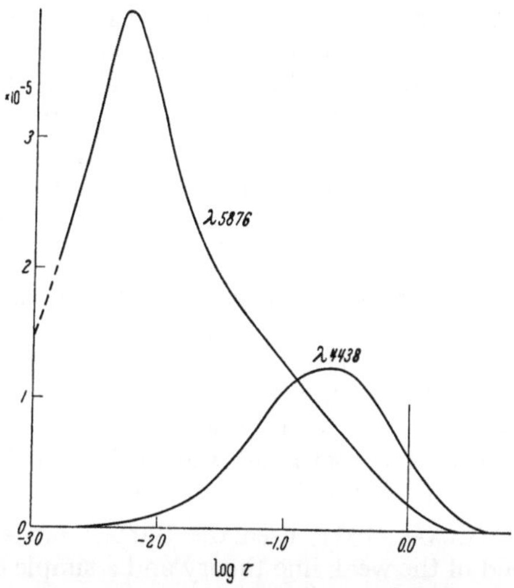

Fig. 4. The contribution function for helium lines in a B star. The ordinate is the integrand of the analogue of Eq. (4.13) for a stellar atmosphere. The abscissa is $x = \text{Log}\,\tau_0$ (τ_0 = optical depth at λ 5000) in the McDonald-Milligan model. For the strong "D_3" λ 5876 line, $\text{Log}\,a = -2.1$; for λ 4438, $\text{Log}\,a = -0.8$ [cf. Eq. (4.18)]. Notice that the main contribution to λ 5876 comes from layers with $\tau_0 < 0.05$ whereas the 4438 line is formed in the photospheric layers ($0.1 < \tau_0 < 1.0$).

5. Results obtained for the Sun. In his pioneer investigation of the quantitative composition of the Sun, RUSSELL used eye estimates of the line intensities. Later workers used equivalent widths and conventional curves of growth. Particular mention must be made of the work of UNSÖLD (1946). In 1940 STRÖMGREN introduced model atmosphere methods to determine the abundances of Mg, Ca, Na, and K with respect to hydrogen. Most subsequent work has employed the model atmosphere approach.

Many workers have calculated model solar atmospheres[1]. For example, G. MÜNCH obtained a purely theoretical model. K.H. BOHM, by a more elaborate treatment attempted to take into account the line spectrum as well. He derived a boundary temperature of about 3400° K but pointed out that the ionization temperature in these layers was appreciably higher. Actually, the centers of the strong lines used by K.H. BÖHM are formed in the chromosphere so that their

[1] See e.g., the discussion by M. MINNAERT, 1953, p. 88, "The Sun" ed. by G.P. KUIPER. Chicago: University of Chicago Press. See also the articles by L. GOLDBERG and by A. K. PIERCE: Handbuch der Physik, Vol. LII (in press).

central intensities may have no significance insofar as the structure of the photosphere is concerned. The Barbier model, adopted by CLAAS, deduced the temperature distribution $T(\tau_0)$, where τ_0 is the optical depth at λ 5000, from the observed limb darkening in the continuum. The boundary temperature of this model, 4900° K, seems to be too high. The de Jager model attempted to describe the center to limb variations of the Balmer lines as well as that of the continuous spectrum. The model atmosphere, obtained by A. K. PIERCE and the present writer and subsequently refined by JOHN WADDELL, is based on accurate limb darkening measurements in the continuum. The line spectrum is disregarded.

Recent data, particularly that obtained with the new vacuum spectrograph at the McMath-Hulbert Observatory, show that the cores of strong lines originate in the chromosphere whose complicated structure cannot easily be handled quantitatively. Model atmosphere methods are strictly appropriate only to the weak or moderately strong lines or to the wings of strong lines.

Most model solar atmospheres agree rather well with one another in the intermediate ranges of optical depth. They deviate from one another in the shallow layers ($\tau < 0.1$) and in the very deep layers ($\tau > 1.0$) where the limb darkening studies can give no information.

One of the big troubles in the model atmosphere methods as used originally by STRÖMGREN is that the damping constants are not known for the strong lines. Accordingly, MINNAERT proposed to use very weak lines for which the damping constants are unimportant and for which a simple theory of line formation may be employed. Many relatively rare elements are represented only by weak lines, but on the other hand, the f-values of the weak lines of more abundant elements are often not well enough known.

CLAAS (1951) used the Barbier model and studied sixteen elements with the aid of the weak line theory and a simple correction for saturation effects. WEIDEMANN also used a model atmosphere method for several elements. L. GOLDBERG, Miss E. MÜLLER, and the present writer have derived abundances for thirty elements in addition to those studied by CLAAS. Both the Claas model and the empirical model by PIERCE *et al.* have been used in the analysis which was carried out with the aid of the Pecker theory and several curves of growth based thereon and computed by G. ELSTE. Estimates had to be made for f-values for many transitions. UNSÖLD had used $\tilde{f}=1$ for transitions like $p^x s - p^x p$, $p^x p - p^x d$, $d^x s - d^x p$, $d^x p - d^x d$, etc. This approximation may be satisfactory for p electrons but for d electrons $\tilde{f}$ is probably in the neighborhood of 0.25. The question can be settled when the f-values are measured experimentally.

The results, which are discussed below under Sect. 7, are subject to numerous uncertainties. The equivalent widths of weak lines due to rare elements are difficult to measure accurately, often may be affected by serious or unknown blends, and hence may be sometimes in substantial error. The determinations deemed particularly uncertain are denoted by colons. Accurate experimental data on the f-values, and also the damping constants for the stronger lines are of paramount importance. Furthermore, we need refinements in the theory of the model solar atmosphere, e.g., correction for the failure of the hypothesis of plane-parallel stratified layers and for the influence of the chromosphere on lines of moderate to high intensity.

The finally adopted solar abundances are based on the Pierce-Aller empirical solar model atmosphere as extended by ELSTE both to deeper layers and to the chromospheric regions.

6. Results obtained for other stars. In Table 1 we enumerate some of the analyses that have been carried out for particular stars of population type I. Those of population type II have not been so completely studied.

Curve of growth analyses have been published by many investigators for early type stars, e.g., τ Scorpii by UNSÖLD, 55 Cygni by VOIGT, 10 Lacertae and a number of other objects by the present writer. The first attempt to apply model atmosphere methods to early type stars was made by the present writer in his investigation of γ Pegasi (1949). The model atmosphere employed was essentially that discussed by J.K. MCDONALD in her study of the Balmer lines. Later NEVEN and DE JAGER used an empirical model atmosphere intended to reproduce the profiles of the hydrogen lines. Since the temperature distribution they obtained is very different from that found by the present writer, the agreement between the abundances found in the two investigations may be largely fortuitous. More recently, TRAVING has discussed both τ Scorpii and 10 Lacertae by the method of model atmospheres while CAYREL has analyzed the supergiant ζ Persei. Traving used the same Unsöld curve of growth for all ions. That is, he assumed all lines to have the same intrinsic shape. He estimated the surface gravity by fitting the observed profiles of the hydrogen lines to the theoretical profiles computed by the HOLTSMARK theory.

G. ELSTE, JUN JUGAKU and the present writer analyzed τ Scorpii with the Pecker theory. The relative intensities of lines of the same element in two different stages of ionization, together with the profile of $H\gamma$ serve to fix the effective temperature and surface gravity of the model. The hydrogen line broadening theory developed by KOLB at the University of Michigan has been used. It takes into account the broadening influences of both the ions and the electrons. The required model differed from that of Traving in that the surface gravity had to be lower and the temperature was slightly different. Furthermore, different ions showed curves of growth with different shapes and the ionic concentrations were different from those found by TRAVING. Similar analyses for other stars are being carried out by the Pecker method. The big problem is to get a good model atmosphere that will reproduce both the line and continuous spectra.

With the aid of model atmospheres computed by Miss UNDERHILL[1] and by MILLIGAN, analyses have been carried out for several B stars. In particular a fairly detailed study is possible for the rich-line spectrum of γ Pegasi in which lines of fluorine, phosphorus, chlorine, and argon have been measured. See Table 2.

Most late B and A stars have been analyzed by curve of growth methods, although K. HUNGER has discussed Vega with the aid of a model atmosphere. Several investigators have calculated model atmospheres for A stars. One difficulty is that in late A and F stars the convective zone lies within the stellar atmosphere itself and the problem of the temperature distribution is a complicated one.

In stars of spectral classes A to G, the practice has often been to obtain abundances relative to the solar ones. For example, J.L. GREENSTEIN used empricial line strengths deduced from stellar or solar data in his analysis of τ Ursae Majoris, ϱ Puppis, α Persei, ϑ Ursae Majoris, and α Canis Minoris. He is using the same procedure in his analysis of stars of later spectral types, including unusual dwarfs and subdwarfs. He has employed ten stars of spectral class G with a luminosity range of 1000 and hopes to get the abundance ratios established to within 25%. For normal type I stars, the relative abundances of metals appears to remain pretty constant from one object to the next.

[1] A. B. UNDERHILL: Publ. Dom. Astrophys. Obs. Victoria **10**, 357 (1957).

Table 1. *Analyses of stellar atmospheres (O—G0)*. Type I Population.

Star	Spec.	Method of analysis	References
10 Lac	*O* 9 V	UCG, MAU	4, 14
τ Sco	*B* 0 V	UCG, EMA, MAU; MAP	1, 13, 11, 16
φ Ori	*B* 0 IV	UCG, WCG	11, 18
15 C Ma	*B* 1	WCG, MAP	11, 18
HD 36960	*B* 1	WCG, MAP	11, 18
ζ Persei	*B* 1 Ib	MA	20
ξ_1 C Ma	*B* 1 IV	UCG, WCG	11, 18
σ Sco	*B* 1 III	WCG	11
ε C Ma	*B* 2 II	CG	17
χ Ori	*B* 2 Ia	WCG	11
δ Ceti	*B* 2 IV	UCG, EMA	7, 13
γ Peg	*B* 2.5 IV	UCG, WCG, MAP, EMA	6, 7, 13
55 Cyg	*B* 3 Ia	UCG; MAU	12, 17
22 Ori	*B* 2 IV	WCG; MAP	11, 18
42 Ori	*B* 2 IV	WCG	11, 18
114 Tau	*B* 3	WCG; MAP	11, 18
i Her	*B* 3	UCG; EMA	7, 13
α Scl	*B* 8 IV	MA	18
β Ori	*B* 8 I	CG	19
α C Ma	*A* 0 V	MCG	3
α Lyr	*A* 0 V	MAU	15
γ Gem	*A* 1 IV	MCG	3, 10
α Cyg	*A* 2 Ia	CG	10
α C Mi	*F* 5 IV	MCG, ECG	5, 8
α Car	*F* 0 Ia	CG	2
α Per	*F* 5 Ib	CG	8, 9
ϱ Pup	*F* 6 II	CG	5
ϑ U Ma	*F* 6 IV	CG	5
γ Cyg	*F* 8 Ib	CG	5, 9

[1] A. UNSÖLD: Z. Astrophys. **21**, 22 (1941).
[2] J.L. GREENSTEIN: Astrophys. Journ. **95**, 161 (1942).
[3] L.H. ALLER: Astrophys. Journ. **96**, 321 (1942).
[4] L.H. ALLER: Astrophys. Journ. **104**, 347 (1946).
[5] J.L. GREENSTEIN: Astrophys. Journ. **107**, 151 (1948).
[6] G. MICZAIKA: Z. Naturforsch. **3**, 241 (1948).
[7] L.H. ALLER: Astrophys. Journ. **109**, 244 (1949).
[8] K.O. WRIGHT: Publ. D.A.O. Victoria **8**, 1 (1949).
[9] T.M. FOFANOVA: Isw. G.A.O. Pulkovo **18**, 68 (1950).
[10] W. BUSCOMBE: Astrophys. Journ. **114**, 73 (1951).
[11] L.H. ALLER: Chap. 2 of J.A. HYNEK, Astrophysics. New York-Toronto-London: McGraw-Hill 1952.
[12] H.H. VOIGT: Z. Astrophys. **31**, 48 (1952).
[13] L. NEVEN: Thesis Liège 1953. — NEVEN and DE JAGER: Bull. Astronom. Inst. Netherl. **12**, 103 (1954).
[14] G. TRAVING: Z. Astrophys. **36**, 1 (1955). — Z. Astrophys. **41**, 215 (1957).
[15] J. HUNGER: Z. Astrophys. **36**, 42 (1955).
[16] G. ELSTE, J. JUGAKU and L.H. ALLER: Publ. Astronom. Soc. Pacific **68**, 23 (1956). — Astrophys. Journ., Suppl. **25** (1957).
[17] L.H. ALLER: Astrophys. Journ. **123**, 117 (1956).
[18] In preparation.
[19] O. GINGERICH and C.H. PAYNE GAPOSCHKIN: Unpublished.
[20] R. CAYREL: Thesis Paris 1957.

The code for the method of analysis is:

CG = curve of growth; UCG = Unsold curve of growth;
WCG = Wrubel curve of growth; MCG = Menzel curve of growth;
ECG = empirical curve of growth; MA = model atmosphere;
MAU = model atmosphere and Unsold curve of growth;
MAP = model atmosphere and Pecker theory;
EMA = empirical model atmosphere.

Table 2. *Abundances of elements in the Sun, early type stars and gaseous nebulae.*

		Log N					Log N		
		Sun	Stars	Neb.			Sun	Stars	Neb.
1	H	12.00	12.00	12.00	44	Ru	1.36:		
2	He		11.15	11.24	45	Rh	0.70:		
3	Li	0.86			46	Pd	1.00:		
4	Be	2.35:			47	Ag	(− 0.50):		
5	B				48	Cd	1.76:		
6	C	8.56	8.24		49	In	0.80:		
7	N	7.98	8.28	8.44	50	Sn	(1.2):		
8	O	9.00	8.78	8.82	51	Sb	1.82		
9	F		5.5:	5.5	52	Te			
10	Ne		8.90	8.12	53	I			
11	Na	6.30			54	Xe			
12	Mg	7.28	8.02		55	Cs			
13	Al	6.21	6.44		56	Ba	2.26		
14	Si	7.60	7.50		57	La	[1.8]		
15	P	5.44	5.50		58	Ce	[2.4]		
16	S	7.17	7.65	7.82	59	Pr	[0.6]		
17	Cl		7.2:	6.55	60	Nd	[2.0]		
18	A		7.0:	6.90	62	Sm	[1.5]		
19	K	4.96			63	Eu	[1.4]		
20	Ca	6.38			64	Gd	[1.1]		
21	Sc	3.20			65	Tb			
22	Ti	4.96			66	Dy	[1.6]		
23	V	4.03			67	Ho			
24	Cr	6.00			68	Er	[0.1]		
25	Mn	5.30			69	Tm	[0.5]		
26	Fe	6.76			70	Yb	1.42		
27	Co	4.74			71	Lu	[1.0]		
28	Ni	5.80			72	Hf	[0.4]		
29	Cu	4.99			73	Ta	[0.0]		
30	Zn	4.53			74	W	[0.2]		
31	Ga	2.16:			75	Rh			
32	Ge	3.31:			76	Os	[0.5]		
33	As				77	Ir	[− 0.2]		
34	Se				78	Pt	[1.6]		
35	Br				79	Au			
36	Kr				80	Hg			
37	Rb	2.28			81	Tl			
38	Sr	2.80			82	Pb	2.68:		
39	Y	3.01:			83	Bi			
40	Zr	2.16							
41	Cb (Nb)	2.00:							
42	Mo	1.96:							

Column 1 gives atomic number. Column 2 the element. Column 3 solar abundance derived from GOLDBERG, MÜLLER and ALLER or from RUSSELL, indicated by square brackets. Determinations particularly uncertain, indicated by colons (:). Column 4: abundances found for early types stars. Column 5: abundances derived from planetary nebulae [Astrophys. Journ. **125**, 84 (1957)]. The abundances are normalized to log $N = 12.00$ for hydrogen.

7. Compilation of results. For the "standard" composition of the stars, the Sun serves as a natural reference, but many light elements are not observed there, although they are found in the hotter stars. We cannot, a priori, assume that the hot *B*-stars have the same composition as the Sun. Several lines of evidence indicate that they have been but recently formed out of the interstellar medium, i.e., out of material that has been cycled through nuclear reactions in previously existing stars. (See the contribution of G. and M. BURBIDGE on Stellar Evolution, in this volume.) Hence the ratio of the metals and of elements like carbon, nitrogen, and oxygen to hydrogen might be expected to be greater in these younger stars than in the Sun. Therefore, we shall in Table 2 list separately the data for the Sun, column 3; the hot stars, column 4; and the planetary nebulae, column 5, which are presumed to be remnants of the ancient type II population.

Hydrogen, helium, and carbon are represented in the spectra of the planetary nebulae by their recombination lines. Ionic concentrations of hydrogen and helium can be estimated with some assurance from the line intensities but no satisfactory theory exists for the recombination spectrum of carbon. The nitrogen

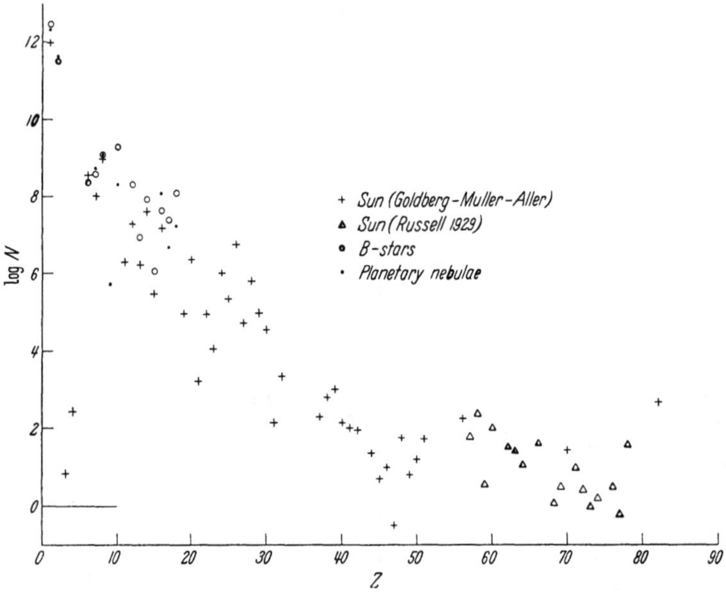

Fig. 5. The cosmic abundances of the elements. The logarithm of the number of atoms is plotted against the atomic number.

concentration is found from the forbidden [N II] lines which show a capricious behavior from one nebula to another. For all these elements it is troublesome to allow for the distribution of atoms among various stages of ionization, but the difficulties are particularly severe for nitrogen. Many other light elements are revealed only by their forbidden lines; the metals are poorly represented and exist in many different ionization stages.

The accuracy of the results (cf. Table 2) is very difficult to assess. It would be meaningless to quote probable errors from a comparison of results by various observers as different workers tend, for example, to use the same f-values, some of which are very poor. There is a further dependence on the model atmosphere and on the measured equivalent widths. Probably the relative abundance of elements such as C, N, O, Mg and Si are more accurately established than are the ratios of H to this group of elements. The hydrogen metal ratio in the Sun or the H/(C, N, O) ratio in hotter stars is established essentially from the ratio of the continuous to the line absorption.

The hydrogen/helium ratio is particularly difficult to establish. Miss ANNE UNDERHILL and later NEVEN and DE JAGER found a ratio of about 20:1 whereas UNSÖLD and his associates found a ratio of about 5:1. Much of the trouble lies in the circumstance that the broadening of most helium lines follows a very complicated theory, the necessary parameters for which have not been determined. As MISS UNDERHILL points out, the strong lines of HeI are formed predominantly in the stellar chromosphere where the theory of model atmospheres does not apply. Fig. 4, due to JUGAKU, shows the contribution function, i.e., the integrand of the analogue of Eq. (4.13) for a star, for a strong and a weak helium line formed in the Milligan-McDonald model atmosphere. The lines of ionized helium are broadened by the first order Stark effect and both the statistical broadening due to the ions and the phase shift broadening due to the electrons must be taken into account. KOLB has given an accurate theory for the line wings but the non-adiabatic terms are yet to be found for the hydrogen lines of interest. The damping constants for the HeI lines are not yet available.

At the moment the best estimate for the hydrogen/helium ratio in the early type stars agrees with that from the recombination lines of these elements in the Orion nebula where MATHIS finds an abundance ratio of about seven. A somewhat similar ratio is found for the planetary nebulae.

Some elements are represented in the Sun only by their compounds. An example in point is boron. The old estimate by RUSSELL is certainly too high and a new determination should be made. Incomplete term analyses preclude the accurate determination of the abundances of the rare earths in the Sun. Not only are f-values not available, but partition functions cannot be computed for these atoms and their ions. Hence no essential improvement over the old calculations by RUSSELL can be made. The best cosmic abundances of these elements must be obtained from the data of meteorites and the Earth's crust.

In Fig. 5 we plot the abundances of the elements as a function of Z, the atomic number.

II. Isotope abundances.

8. Abundances of isotopes may be found from atomic lines only for hydrogen or helium. The deuterium α line falls near the Hα line in a region affected by water vapor and the higher Balmer lines are not favorable. All indications point to a H^2/H^1 ratio that is much smaller than on the Earth. GREENSTEIN found a He^3/He^4 ratio in the Sun less than 1%. In the magnetic star, 21 Aquilae, the BURBIDGES found the abundance of He^3 to be comparable with that of He^4.

Isotope effects are most profitably investigated in molecular bands, such as those of C_2, CN, CH, TiO, ZrO, NH, OH, or MgH. From a study of the TiO bands, HERBIG found the titanium isotopes to have about the same ratio as on the Earth, while MCKELLAR noted that if the O^{17} and O^{18} isotopes were more abundant in the M stars than in the Earth's atmosphere the effects would be noted in the TiO bands. The isotope ratios of N, O, and Mg would appear to be about the same as on the Earth. There is no evidence for N^{15} in the stars. A study of ZrO would be particularly interesting since zirconium is an important element in the S stars and is apparently enriched there as a consequence of nuclear reactions.

Carbon shows the most exciting isotope effects. From a study of $G8-K5$ giant stars, J.D. SCHOPP[1] found no evidence that the C^{13}/C^{12} ratio differed from that found on the Earth, i.e., about (1/90). G. RIGHINI[2] found that the ratio in the Sun was at least as small as that found on the Earth or meteorites, and may possibly be even smaller. SHAJN and HASE studied the CN bands and MCKELLAR the C_2 bands in the carbon R and N stars. Among twenty-one R stars, three showed a terrestrial C^{13}/C^{12} ratio, whereas the others had a ratio of the order of $\frac{1}{3}$ or $\frac{1}{4}$. Curve of growth and saturation effects must be taken into account

[1] J. D. SCHOPP: Astrophys. Journ. **120**, 305 (1954).
[2] G. RIGHINI: Liège conference on molecules in celestial sources, 1956.

before a definitive result for this ratio can be realized, however. BIDELMAN suggested that among the carbon stars, "normal" C^{13}/C^{12} ratios may be connected with hydrogen deficiency or with a high velocity.

Isotope ratios and the relative abundances of light elements may serve as indicators of the degree of mixing between the interior and the surface. Thus, probably there is no deuterium in the Sun. The absence of C^{13} and the presence of Li^6 and Li^7 in the solar spectrum was shown by GREENSTEIN to be incompatible with deep mixing. Be^8 is abundant in the Sun. A solar Be/Li ratio of the order of 30 times the meteoritic value indicates that lithium is deficient while Be has a normal abundance as compared with other metals. GREENSTEIN concluded that mixing takes place down to strata where the temperature was about 3.2×10^6 °K but no deeper. Otherwise Be would be destroyed!

III. Composition differences between stars.

Composition differences are difficult to establish securely because stars that show abundance anomalies may also often have atmospheric conditions that make abundance determinations difficult to tie down. For example, if the coefficient of continuous absorption, $\varkappa_\lambda$, is not known, atmospheric models cannot be constructed.

9. Influence of population type I and II. SCHWARZSCHILD concluded (1950) that the only certain differences between high and low velocity stars is manifest in the CH band which indicates that the abundance of carbon relative to the metals is about 2.5 times larger in high velocity *F* stars than in low velocity dwarfs. Subsequently, SCHWARZSCHILD, SPITZER and WILDT suggested that the spectroscopic pecularities of the high velocity stars, i.e., the strengthening of the CH bands and the weakening of the CN bands and the metal lines, can be explained by a general reduction in the abundances of the heavy elements and a somewhat lesser reduction in the abundances of the medium-heavy elements relative to the low velocity stars. Presumably, the latter may have been formed from clouds with higher concentrations of grains.

From a comparison of high and low velocity *K* giants, L. GRATTON concluded that they were not homogeneous groups as far as chemical composition is concerned, but that the fluctuations were comparable with differences between the groups.

The subdwarfs represent an extreme type II population, similar perhaps in composition to the stars in the globular cluster M92. CHAMBERLAIN and the present writer found an enhancement of the hydrogen/metal ratio of the order of a factor of 10; the qualitative results seem to be borne out by a more recent study by GREENSTEIN and the writer who studied several such objects with better observational material. The BURBIDGES studied five stars which showed high velocity characteristics or a general weakening of the spectral lines similar to that observed in the subdwarfs. One of these stars, λ Boötis, had a composition very similar to that of the subdwarfs mentioned above, one had a normal composition and the others showed underabundance ratios intermediate between these two extremes.

The foregoing results apply to "run of the mill" stars; other stars of both types show pronounced effects of nuclear reactions and aging.

10. Wolf-Rayet stars and novae. The Wolf Rayet stars show broadened emission lines of high ionization and excitation. Hydrogen is generally weak. Some objects show C, O, and strong helium lines; others show strong N and

He lines, weak C and apparently weak O lines. In the nitrogen sequence of Wolf-Rayet stars, the He/N ratio seems to be about 20. In the carbon sequence the ratio He/C/O appears to be roughly 17/3/1.

FOWLER found that the N^{14}/C^{12} ratio produced in the carbon cycle is about 26 at 13×10^6 °K and 2 at 40×10^6 °K. Various workers since GAMOW have tried to correlate the compositions of Wolf-Rayet stars to the carbon cycle. For example, it has been suggested that the nitrogen sequence stars represent the residual of the operation of the CN cycle, whereas the *C* stars with weak N owe their compositions to the operation of the C^{12} building by the Salpeter process, which forms C, O, and Ne from the fusion of α-particles.

HD 45166 appears to be a Wolf-Rayet star with a "normal" composition or an *Of* star. The analysis is complicated by the variability of the spectrum. The compositions of ordinary *Of* stars have been determined by J.B. OKE[1].

Wolf-Rayet stars of population type II often occur as the central stars of planetary nebulae. In the carbon stars HD $+30°3639$ and the nucleus of NGC 40 relative abundances of He, C, and O were found to be roughly 10:6:1, whereas in the nucleus of NGC 6543 the C:N:O ratio seemed to be about 4:1:1. SWINGS and STRUVE emphasized the frequent occurrence of stars of intermediate composition among type II Wolf-Rayet stars.

The classical Wolf-Rayet stars (excepting HD 45166) appear to be rich in helium, deficient in hydrogen. On the other hand, the type II Wolf-Rayet stars may contain hydrogen in abundance as their surrounding envelopes (planetary nebulae) are composed mostly of this element.

NOVAE whose expanding envelopes necessarily have been analyzed only in an approximate fashion, appear to resemble ordinary stars and planetary nebulae nuclei rather than classical Wolf-Rayet stars.

11. Hydrogen deficient stars. It seems well established that certain stars are really deficient in hydrogen, probably as a consequence of the depletion of this element in the CN cycle or pp reaction. The data for a number of hydrogen deficient stars are listed in Table 3. To this table we should add σ Orionis *E*, in which there is no doubt that the helium to hydrogen ratio is very large. To this group we also may add certain of the Humason-Zwicky stars studied by GREENSTEIN and MÜNCH at Palomar. For example, HZ 44, which apparently has a luminosity similar to that of the Sun, shows weak hydrogen, strong helium, strong nitrogen and silicon lines and very weak carbon. Other stars may be enriched in carbon but no star heavily laden with oxygen and with virtually no C of N has yet been discovered.

The white dwarfs, which are discussed in Vol. L by J.L. GREENSTEIN, include a number of objects with abnormal helium/hydrogen ratios and other composition anomalies. The stars denoted as HZ 29 and L 1573-31 are pure helium objects, while Ross 640 shows lines of only Mg I and Ca II.

12. Cool stars of abnormal composition. The cool stars of abnormal composition fall into two categories; the "carbon group" and the "heavy metal" group.

The carbon group contains the stars of spectral classes *R* and *N*. Compounds of carbon dominate the spectra. Since the abundance of carbon is greater than that of oxygen, virtually all oxygen is tied up in CO. The C/O ratio is not fixed but varies from star to star in such a way as to produce difficulties in spectral

[1] J. B. OKE: Astrophys. Journ. **120**, 22 (1954).

Table 3. *Hydrogen deficient stars.*

Object	Characteristics of composition or spectrum	Spec.	Remarks	References
HD 160641	C/N/O/Ne = 6/7/10/30 He replaces H	*O* 9	1	W. P. BIDELMAN: Astrophys. Journ. **116**, 227 (1952)
HD 124448	HeI, CII strong	*B* 2	2	D. M. POPPER: Publ. Astronom. Soc. Pacific **59**, 320 (1947)
HD 168476	HeI, CII, NeI strong N, Mg, A, Si, S, A, Ca, V, Cr, Fe, Ni, Ti present O, H. missing	*B* 8		A. D. THACKERAY: Monthly Notices Roy. Astronom. Soc. London **114**, 93 (1954)
HD 135485	He/H = 10 × normal C, N, O, Si = normal, Mg ~ 1/10 normal	*B* 9		J. C. STEWART: Astrophys. Journ. **61**, 13 (1956)
υ Sag	Strong H, He, N, Ne, weak C	*c A pe*	3	J. L. GREENSTEIN: Astrophys. Journ. **91**, 438 (1940)
HD 30353	Very strong metallic lines	*A* 5	4	
Carbon H-deficient Stars	No enhancement of C^{13} with respect to C^{12}	*F—G*		W. P. BIDELMAN: Astrophys. Journ. **117**, 25 (1953)
R Cor Bor	No C^{13} bands	~*F* 8	5	Shajn, Obs. **64**, 255 (1942)

classification as were noted by KEENAN and MORGAN[6]. The C^{13}/C^{12} ratio is often greater than the normal terrestrial value. To this group can also be added the "lithium" stars, WZ Cassiopeiae (studied by McKELLAR) and T Arae (observed by FEAST).

The heavy metal group includes the stars of spectral class *S* and the so-called "barium" stars. The *S* stars are characterized by prominent lines of zirconium oxide. Some years ago WURM[7] suggested that since ZrO had a higher dissociation potential than TiO, the ZrO/TiO ratio would tend to be enhanced in stars of lower density, which would imply that the *S* stars are supergiants. The bands of YO and LaO are also strengthened in the *S* stars. Later, FUJITA[8] pointed out that an increase in the C/O ratio would decrease the pressure of free oxygen to the point where the ZrO/TiO ratio could be high even for normal zirconium and titanium abundances. On the other hand, the atomic lines of Zr as well as the bands of ZrO increase in intensity from the *M* to the *S* stars[9]. The neighboring elements, Y, Cb, Mo, and Ba, as well as the rare Earths are also strengthened. No juggling of physical conditions can account for this phenomenon. We deal here with an actual increase in the abundance of the heavy elements from the

[1] The temperature ~31500° K, $P_g \sim 1.3 \times 10^3$ dynes/cm². See L. H. ALLER: Liège Conference Memoirs, p. 354, 1953.

[2] No quantitative analysis available yet.

[3] Spectroscopic Binary, $L \sim 10^5 L_\odot$.

[4] A probable deficiency of H will explain the low opacity of the atmosphere and the great strength of the metal lines.

[5] HERBIG concludes that luminosity is not too great. The star is probably not as luminous as $M = -4.5$.

[6] P. C. KEENAN and W. W. MORGAN: Astrophys. Journ. **94**, 501 (1941).

[7] K. WURM: Astrophys. Journ. **91**, 103 (1940).

[8] FUJITA: Jap. J. Astronom. Geophys. **17**, 17 (1939); **18**, 45 (1940); **1941**, 177.

[9] P. W. MERRILL: Astrophys. Journ. **105**, 360 (1947). — P. C. KEENAN and L. H. ALLER: Astrophys. Journ. **113**, 72 (1951). — W. BUSCOMBE and P. W. MERRILL: Astrophys. Journ. **116**, 525 (1952). — W. P. BIDELMAN: Liege Memoires **14**, 402 (1954).

M to the *S* stars. Thus KEENAN suggested that the Ti/Si ratio was 27 in the *M* stars and 14 in the *S* stars whereas the Zr/Si ratio was 1.5 in the *M* stars and 8 in the *S* stars. Furthermore, continuous gradations of the C/O and Zr/Ti ratios appear to exist among these stars of unusual composition. For example, in W Cass, R Ori, and RC Mi, weak C_2 bands indicate that carbon is only slightly more abundant than O, but the stars show *S* characteristics in that the lines of Zr and its neighboring elements are strong.

Most striking, however, was MERRILL's observation of the lines of technetium in the *S* stars. The longest-lived isotope of this element has a half-life of only about 300000 years. GREENSTEIN estimated the technetium ratio in R Andromedae to be about 5×10^{-5}. Thus technetium is about as abundant in R Andromedae as zirconium is in the Sun and about a tenth as abundant as zirconium is in R Andromedae! No *S* stars are observed without the technetium lines and very possibly the lifetime of the *S* star is comparable with the lifetime of Tc itself. Technetium has also been found in the *N* stars. Presumably, the heavy metals of the zirconium row of the periodic table are manufactured in the stellar inteiior (possibly by neutron captures) and later mixed into the surface layers.

The barium and CH stars are related to the *S* stars in that they show heavy metals, although technetium is absent. All of the peculiar *G* and *K* stars show strong lines of Ba II, CH, C_2 and possibly CN[1]. This enhancement of Ba II represents an actual increase in the amount of barium present and is not an absolute magnitude effect. BIDELMAN[2] suggested that the Ba II and *S* stars are manifestations of the same abundance anomalies at different temperatures.

The most complete discussion of the problem from the standpoint of chemical compositions is the analysis of the Ba II star, HD 46407, by E.M. and G.R. BURBIDGE. They find that the elements from Na through Ge have normal abundances, but that most of the heavier elements beyond strontium have abundances of the order of ten times the normal one. They find abundance peaks at nuclei having a magic number of neutrons, e.g., Sr, Y, Zr, Ba, La, Ce, Pr, and Nd and have interpreted these results in terms of element building processes in stars.

An impressive quantity of evidence has now been accumulated to show that heavy elements are manufactured in stars by a variety of nuclear processes, jamming together of α particles, neutron capture, etc. Important contributions have been made by F. HOYLE, A. G. W. CAMERON, W. FOWLER, J. L. GREENSTEIN, and the BURBIDGES.

13. Magnetic stars and spectrum variables. Among the *A* stars, some objects are found which show remarkable spectral characteristics. These peculiar *A* stars are characterized by abnormally strong lines of Si II and Sr II and often a weak *K* line of Ca II. About 10% of these peculiar *A* stars show conspicuous variations in the strengths of certain absorption lines. The classic example is α^2 Canum Venaticorum in which lines of the ionized rare Earths vary strictly in phase with one another, whereas the line intensities of neutral chromium and of ionized chromium and strontium vary together in the mean but are 180° out of phase. Periods are known for thirteen stars, ranging from 0.52 days for HR 5313 to 20.3 days for 73 Draconis although one star, HR 7575, has a period of 226 days. In stars of longest period where the lines are sufficiently sharp to permit such measurements to be made, BABCOCK has found a variable magnetic field. The

[1] W. BIDELMAN and P.C. KEENAN: Astrophys. Journ. **114**, 473 (1951).
[2] W. BIDELMAN: Astrophys. Journ. **112**, 219 (1950).

changes in line intensities are often only approximately periodic. They show slow secular variations and also occasional abrupt changes.

Various explanations of the variable line intensities and simultaneous radial velocity changes have been sought. In one model, hydromagnetic oscillations of the star have been postulated. Since the magnetic field is frozen into the gas, the oscillations involve actual mass motions of the gas. The second model is an oblique rotator which, however, cannot have axial symmetry. The rare-earth and chromium-line enhancements are attributed to patchs on the surface in which these elements are concentrated.

The spectral peculiarities cannot be explained by differences in temperature and pressure, nor by abnormal ionization and excitation effects. Magnetic intensification of certain lines must play some role although it cannot cause the spectrum variations. The increase in the concentrations of the rare Earths in the atmospheres of these stars must represent an actual enhancement of these elements in the surface layers, possibly in small patches on the surface.

From an analysis of spectrograms secured by P. W. MERRILL, the present writer suggested an enhancement of the metal/hydrogen ratio in these stars by a factor of 10 or more. The BURBIDGES[1], using spectrograms of much higher dispersion, were able to make detailed estimates for a number of elements in α_2 Can Ven and in HD 133029. They found calcium to be underabundant by a factor of 20 in HD 133029 and by 50 in α_2 Can Ven. The abundances of Mn and silicon are enhanced by about 15 times, Cr is less enhanced and iron is affected only by about a factor of 2. Sr, Y, and Zr are enhanced about 25 times, while the greatest increase in abundance is found among the rare Earths where the average enrichment factor is about 600 times! Barium, on the other hand, appears to have a normal abundance! The abundance of lead is enhanced!

FOWLER and the BURBIDGES[2] suggested that these anomalous overabundances were not due to element synthesis in the interior but rather to the building of the heavy elements on the surface of the star itself. Charged particles are presumed to be accelerated by a changing magnetic field, by mechanisms similar to some that have been invoked to explain the generation of soft cosmic-ray radiation in the Sun. The synthesized atoms are not mixed with the interior but are concentrated in the shallow zone where convection currents occur. In the order of $5 \cdot 10^8$ years the observed enhancements of the heavy elements could be produced.

The peculiar A stars are not to be confused with the so-called metallic line stars. This type of object is characterized by intensity anomalies among the lines familiar in spectral classification work. Thus, a star classified as $A3$ on the basis of its Ca II K line would be listed as $F5$ from the great strength of the metallic lines.

GREENSTEIN[3], who studied τ Ursae Majoris and ζ Lyrae, noted that the differences between the spectra of metallic and normal F stars appears as an apparent abundance deficiency of elements whose second ionization potential lies between 2 and 16 eV. RUDKJÖBING[4] suggested that the differences are probably due to differences in atmospheric structure rather than to real abundance differences. Metallic line stars are often found in galactic clusters; perhaps they are associated in some way with the evolutionary development of stars.

---

[1] E.M. and G.R. BURBIDGE: Astrophys. Journ. **122**, 396 (1955).
[2] W.A. FOWLER and G.R. BURBIDGE: Astrophys. Journ., Suppl. **2**, No. 17, 167 (1955).
[3] J.L. GREENSTEIN: Astrophys. Journ. **109**, 121 (1949).
[4] M. RUDKJÖBING: Ann. d'Astrophys. **13**, 69, 164 (1950).

Bibliography.

I. Compositions of normal stars.

Sources of observational data

1. Identifications of lines.

The Sun. Revised Rowland Table, C. E. St. JOHN and others, Carnegie Instn. Publ. **1928**, No. 396, gives wavelengths, intensities (arbitrary scale) and identifications. BABCOCK, H. D., and C. E. MOORE prepared an extension to λ 6600 $-$ λ 13490. Carnegie Instn. Publ. No. 579. In collaboration with MINNAERT and Mrs. CHARLOTTE M. SITTERLY, H. D. BABCOCK is preparing a new edition which will give also the equivalent widths, W'_λ, of each solar line. MOHLER, O. C., Table of Solar Spectrum Wavelengths λ 11984 A to 25578 A, University of Michigan Press 1955, gives observed wavelengths and equivalent widths, wave numbers and identifications of solar and telluric lines.

The Stars. MERRILL, P. W., List of Chemical Elements in Astronomical Spectra, Carnegie Instn. Publ. **1956**, No. 610, gives tables of wavelengths of lines and identification in stellar spectra, occurrences of various elements, partial Grotrian Diagrams with principal transitions, and an extensive bibliography.

2. Measurements of line intensities.

The Sun. MINNAERT, M., G. F. W. MULDERS and J. HOUTGART, Photometric Atlas of Solar Spectrum, Utrecht Observatory, Holland, gives profiles of solar lines as measured on a true intensity scale at the center of the disk. MOHLER, O. C., A. K. PIERCE, R. R. MCMATH and LEO GOLDBERG: Photometric Atlas of Infrared Solar Spectrum from λ 8465 to λ 25242, University of Michigan Observatory 1951. PIERCE, A. K.: Proceedings Indiana University Conference on Stellar Atmospheres 1954 — bibliography of measurements of equivalent widths of lines in solar spectrums; listed element by element. See also ALLEN, C. W.: Memoirs of Commonwealth Observatory, Canberra, Australia, No. 5; 1, 2, 1934; No. 6, 1938; Astrophys. Journ. **88**, 125, 165 (1938) ($\lambda\lambda$ 800 to 11000); Monthly Notices Roy. Astronom. Soc. London **96**, 842 (1936); **109**, 343 (1949); PHILLIPS, J. G.: Astrophys. Journ. **96**, 61 (1942) (3530 to 3915).

The Stars. HILTNER, W. A., and R. WILLIAMS: Photometric Atlas of Stellar Spectra. University of Michigan 1946. Many measurements for individual stars are listed in the literature cited for Table 1.

Transition Probabilities, Damping Constants, and Dissociation Constants.
- ALLEN, C. W.: Astrophysical Quantities, p. 63. London: Athlone Press 1955.
- LANDOLT-BÖRNSTEIN: Zahlenwerte und Funktionen aus Physik, Chemie, Astronomie, Geophysik und Technik, Vol. 1, part I, p. 260. Berlin: Springer 1950.
- Proceedings: National Science Foundation Conference at Indiana University on Stellar Atmospheres, ed. by M. WRUBEL. 1954.

Texts and References for Methods of Analyses of Stellar Atmospheres.
- ALLER, L. H.: Atmospheres of the Sun and Stars. New York: Ronald Press Co. 1953.
- RUSSELL, H. N.: Astrophys. Journ. **70**, 11 (1929).
- UNDERHILL, A. B.: Publ. Copenhagen Obs. No. 151, 1950.
- UNSÖLD, A.: Physik der Sternatmosphären. 2nd edition. Berlin: Springer 1955.
- WOOLLEY, R., and STIBBS: Outer Layers of a Star. Oxford 1953.

Results of Analysis for the Sun.
- CLAAS, W. J.: Utrecht Recherches Astron. **12**, part I (1951) (model atmosphere).
- UNSÖLD, A.: Z. Astrophys. **24**, 306 (1948) (curve of growth).
- WEIDEMANN, V.: Z. Astrophys. **36**, 101 (1955).

The most detailed modern study with extensive bibliography is given in the chapter by L. GOLDBERG, E. MÜLLER and L. H. ALLER in: Handbook of the Solar System, vol. 4, ed. by G. P. KUIPER. Chicago, Ill.: Chicago University Press.

Results of Analyses for Stars and Gaseous Nebulae.
- For gaseous nebulae see: ALLER, L. H.: Gaseous Nebulae. London: Chapmann & Hall 1956; Astrophys. Journ. **125**, 84 (1957).
- In addition to previously cited references, results for normal stars are discussed in ALLER, L. H.: Chemical Composition of the Universe. New York: Interscience Press (in press).

II. Isotope abundances.

GREENSTEIN, J. L.: Astrophys. Journ. **113**, 531 (1951).
GREENSTEIN, J. L., and R. S. RICHARDSON: Astrophys. Journ. **113**, 536 (1951).
GREENSTEIN, J. L., R. S. RICHARDSON and M. SCHWARZSCHILD: Publ. Astronom. Soc. Pacific **62**, 15 (1950).
HERBIG, G. H.: Publ. Astronom. Soc. Pacific **60**, 378 (1948).
MCKELLAR, A.: Publ. Astronom. Soc. Pacific **61**, 199 (1949); **62**, 110 (1950).

III. Composition differences between stars.

See especially BURBIDGE, G. R. and E. M.: Stellar Evolution, in this volume.
ALLER, L. H.: Nuclear Transformations, Stellar Interiors and Nebulae, Chap. 4. New York: Ronald Press Co. 1954.
BURBIDGE, G. R. and E. M.: Astrophys. Journ. **124**, 116 (1956).
BURBIDGE, E. and G. R., W. A. FOWLER and F. HOYLE: Rev. Mod. Phys. **29**, 547 (1957).
GRATTON, L.: Liège Mémoires **14**, 419 (1954).
GREENSTEIN, L.: Liège Mémoires **14**, 307 (1954).
SCHWARZSCHILD, M. and B.: Astrophys. Journ. **112**, 248 (1950).
See especially the Liège Mémoires **14** (1954) in addition to papers cited in text.
See also MICZAIKA, G.: Mitt. Heidelberg **62** (1948).
See, for example, the review article by DEUTSCH, A. J.: Publ. Astronom. Soc. Pacific. **68**, 92 (1956).
Additional bibliography will be found in the references cited.

Variable Stars.

By

P. LEDOUX and TH. WALRAVEN.

With 51 Figures.

A. Introduction.

I. General remarks.

Historically, the name *variable stars* came into use to designate all those stars which vary in brightness but, more recently, thanks to spectroscopy, the class has been extended to include stars such as spectroscopic and magnetic variables in which the most conspicuous variation is not one of luminosity.

However, the title of this article excludes all variables in which the main variation is due to purely geometrical effects as in eclipsing binaries. Furthermore, due to their great importance on their own right, some of the most spectacular intrinsic variables, the novae and supernovae fall also outside the scope of this article and are treated as a separate class elsewhere in this volume.

If we could study the instantaneous luminosity of stars with all the precision we could wish for, it is probable that they would all show some kind of minor erratic variability. But in most cases, this would reflect only the unavoidable fluctuations of large gaseous assemblies of particles in turbulent motions under the action of various mechanical and electromagnetic fields of forces. To be significant, a variation must present a sizable amplitude and a certain amount of regularity, the two requirements being in a sense complementary: a very small variation can still be interesting if it is very regular.

These conditions, especially the first one which, in the past, were more or less forced upon astronomers by the methods of observation limit rather strongly the membership of the class. If, as we shall see, the total number of known variable stars might seem impressive, their proportion to other stars remains very small, of the order of 10^{-6} according to Mrs. PAYNE-GAPOSCHKIN [*1*]. Despite this and their relatively late discovery, no other group of stars has awakened more interest among professional and amateur astronomers alike nor provided more important information.

The association between the different classes of variable stars and specific characteristics of galactic structure (flat and spherical subsystems) and stellar populations (young and old) has made of them one of the most useful tools for the probing of the nature of stellar systems as well as valuable pointers of stellar evolution [*2*].

From the theoretical point of view, the interpretation of their variations has been and still is a perpetual challenge to the ingenuity and ability of astronomers and a stringent test for their theories. Let us recall that the epoch-making papers of EDDINGTON on the radiative equilibrium of gaseous stars originated in his attempt to tackle the cepheid problem [*3*]. There is little doubt that in the future too, the solution of the numerous problems which we shall encounter in the course of this article will either lead to or be associated with fundamental advances in the physics of stellar interiors and atmospheres. In some of the theories proposed in the past, for instance in ZÖLLNER's or NÖLKE's or JEAN's theories, the interpretation of variable stars was directly connected with a special phase of stellar evolution and it is likely that in this respect, their final explanation will reveal a deep meaning, which however remains difficult to foresee at present.

Anyway, some of their elements and especially the periods in the case of the most regular variables, could be very sensitive indicators of evolution. Also the comparison of the behaviour of variables belonging to young and old stellar populations will certainly help us to trace the significant physical differences between these populations.

II. Historical background and development.

Many interesting books [*4*] have been written on the discovery of variable stars, on the development of the methods of observations and on the attempts to explain their variations. Here we have to limit ourselves to a brief outline of this fascinating subject.

a) Discovery and observations.

1. Early history and visual observations. As far as the discovery of variable stars is concerned, perhaps the most striking fact is that, although quite a number of them are naked eye objects at maximum light, the first definite record of a variable star, except for a few references to novae, comes as late as 1596. In August of that year, FABRICIUS noticed a star in Cetus of about the second magnitude, the brightness of which went on declining regularly until sometime in October when it vanished from sight.

It is difficult to understand how such an object can have escaped the notice of astronomers since the time when HIPPARCHUS and PTOLEMY started classifying hundreds of stars down to the sixth magnitude. Especially since, according to PLINY[1], HIPPARCHUS' aim was not only to detect the appearance of new stars but also to verify whether some of them increased or decreased in intensity. On the other hand, if any such periodical variability had ever been recognized, how is it that no record of it has come down to us as its significance with respect to the ideas of perfection and immutability of the fixed stars prevailing in the ancient world, could hardly have been missed?

It is true that in a few instances, etymology (Algol, a name often considered as deriving from the Arabian "El Ghoul" meaning "changing spirit"[2]) or mythology (alleged disappearance of one of the seven pleiades towards the end of the second millennium B. C.[3]) suggest that stellar variability was noticed before. But apparently, it did not find its place among the body of rational knowledge of the time or perhaps a very simple explanation in terms of motion of approach or recession such as was favoured by the Greeks even for novae, deprived it of sufficient interest.

Anyway, even after FABRICIUS' discovery and the christening of the star as "the wonderful" (Mira Ceti), more than 60 years elapsed before its periodicity

---

[1] Cf. E. ZINNER: Astronom. Nachr. **238**, 65 (1930). In this article as well as in the following: Phys. Z. **31**, 440 (1930); Astronom. Nachr. **237**, 77 (1929), ZINNER has compared critically old and recent determinations of stellar magnitudes in view of establishing a possible progressive change in the brightness of some stars.

[2] Cf. however G. A. DAVIS: Sky and Telescope, Vol. XVI, p. 177, 1957, where another origin of the name is advocated.

[3] Cf. R. GRAVES: The Greek Myths, Vol. 1, pp. 154 and 220. Penguin Books 1955.

suggested by HOLWARDA in 1638, was definitely established around 1660. With its period of about 11 months, it became the first member of a loose group often referred to as the *long-period variables*. Apart from the eclipsing binary β Persei (Algol), no new variable star was recorded until 1784, when GOODRICKE discovered the light variations of δ Cephei which was to give its name to a class of extremely important and interesting variables: *the Cepheids*.

Before the end of the eighteenth century, examples of three more classes of intrinsic variables were discovered: *irregular* (α Herculis, HERSCHEL 1795) and *semi-regular variables* (R Scuti, PIGOTT 1795) and the peculiar star R Coronae Borealis (PIGOTT 1795).

However, for a long time the total number of known variable stars increased only very slowly and the first ARGELANDER'S catalogue in 1844 contained only

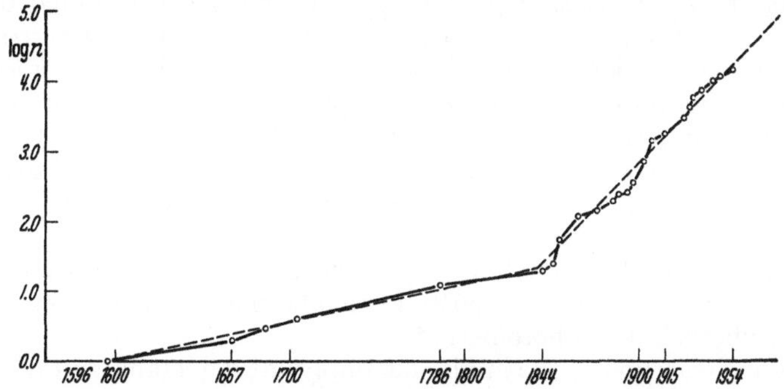

Fig. 1. Number of variable stars discovered as a function of time.

six more variables than PIGOTT'S catalogue of 1786. After HERSCHEL'S and ARGELANDER'S efforts to put the observations on a more systematic and quantitative basis, a new interest arose and the rate of discovery increased rapidly as illustrated in Fig. 1; in 1865, CHAMBERS could already list 113 variable stars.

While ARGELANDER'S system in the hands of HAGEN and PICKERING led to the establishment of sequences of comparison stars with well determined magnitudes and to a precise system of estimation in tenths of a magnitude, differential photometers of the wedge and polarizing types came into use and both contributed to a much greater precision of the light curves. But in this respect, the recent developments of photoelectric photometry have made all other methods obsolete.

However, it is still very important for statistical and reconnoitring purposes especially in the case of long-period variables, that a great many stars be followed visually as regularly as possible. Here, the help of amateur astronomers is invaluable and when their work is well coordinated, it can lead to such important publications as the recent "Studies of Long-Period Variables" by L. C. CAMPBELL [*5*].

2. Photography and photoelectric photometry. The advent of stellar photography towards the end of the nineteenth century and its application to the search for variable stars following the early work of PICKERING at Harvard allowed a continuous rise in the rate of discovery. In 1915, the number of stars definitely known to be variables amounted already to 1687 and the latest catalogue with its supplement [*6*] has brought this number to 13745.

In particular around 1900, photography enabled BAILEY [*7*] to discover numerous short period variable stars in globular clusters opening a new and important class of variable stars known as *cluster variables* or *RR Lyrae variables* after the name of the first member of this class discovered outside a globular cluster by Mrs. W. P. FLEMING in 1901.

But the photographic method was still to reap its most spectacular success in the work of Miss LEAVITT [8] who applying it to the Cepheids of the Magellanic Clouds was led in 1912, to the discovery of the period-luminosity relation which was later made the basis of an absolute scale of distances not only for our galaxy, but for the whole universe. The fact that a correction to the zero point of this relation, such as that introduced by BAADE [9] in 1952, implies drastic consequences not only for the theory of the Cepheids themselves, but also for the sizes and distances of extragalactic objects and for the relativistic time-scale of the Universe shows better than any argument, the fundamental importance of this relation.

However the advantages of photography did not remain limited to the field of discoveries, and soon the theoretical and experimental investigations of K. SCHWARZSCHILD[1] provided a sound basis for photographic photometry which enabled one to obtain many good photographic light curves. Their comparison[2] with the corresponding visual results reinforced the physical interest of observations in different colours and, for stars in general, it led PARKHURST[3] to a workable definition of a colour index which superseded all earlier attempts at stellar colorimetry.

It is only in recent years that a systematic and precise study of star-light in different colours has become possible thanks mainly to the sensitivity and precision of photoelectric photometry[4].

By lowering enormously the minimum amplitude detectable and the resolution time, this method should add to the ranks of variables many a star considered stable so far and reveal many unforeseen details in the light curves. Spectacular results have already been reached in both directions.

3. Spectroscopy. From 1870 onwards, the newly born science of spectroscopy was also to contribute fundamentally to our knowledge of variable stars.

In some cases such as the long-period variables, the existence of specific spectral characteristics easily recognized on prism-objective spectra increased the chances of discovery.

In addition, following the work of VOGEL[5] and BELOPOLSKI[6], it revealed in many variable stars a periodic or cyclic Doppler-shift of the lines providing precious informations on the motions attending the light variations. Conversely, displacements of spectral lines sometimes led to the discovery of new variables such as those belonging to the *β Cephei type*[7] (or *β Canis Majoris type*).

Finally, it brought all the resources of quantitative spectral analysis on which so much of Astrophysics is built, to bear on the problem of the physical nature of variable stars.

It has even disclosed a new class of variables *(spectrum variables)* in which the most conspicuous variations are changes in the strength of some of the spectral lines such as occur for instance in *α Canum Venaticorum*, the first member of

[1] K. SCHWARZSCHILD: Publ. der v. Kuffnerschen Sternw. **5** (1897).

[2] Cf. for instance an interesting comparison by A. KOHLSCHÜTTER: Astronom. Nachr. **183**, 265 (1910).

[3] J. A. PARKHURST: Astrophys. Journ. **36**, 169 (1912).

[4] A technique with took its first steps at the beginning of this century with STEBBIN's pioneering work: Science, Lancaster, Pa. **28**, 852 (1908).

[5] H. C. VOGEL: Astronom. Nachr. **123**, 289 (1889).

[6] A. BELOPOLSKI: Bull. Acad. Imp. Sci., St. Petersbourg, Ser. V **1**, 267 (1894); **7**, 367 (1897). — Astrophys. Journ. **6**, 393 (1897).

[7] E. B. FROST: Astrophys. Journ. **15**, 340 (1902).

the class discovered by LUDENDORFF[1] in 1906. In many of these stars[2], spectroscopy has recently revealed the existence of large varying magnetic fields. This new physical factor probably plays an important rôle even in more classical objects such as the RR Lyrae variables or even the classical Cepheids[3]. In other stars, such as the long-period variables, spectroscopy points to appreciable anomalies in the chemical abundances, one of the most striking being the presence in *S* type stars of relatively large amounts of the unstable element technetium[4] which might be significant for their variations.

4. Observational informations on absolute luminosity, temperature, radius, mass. Granted that the interstellar reddening should be taken into account as carefully as possible[5], spectroscopic and photometric observations provide us with rather direct evidence on the superficial temperatures of the variable stars and on the other hand, the period-luminosity relation or the spectral luminosity criteria enable us to estimate their absolute luminosities. From these data, their radii can be derived. But there is a fundamental parameter, the mass on which, until very recently, we had absolutely no observational information. Apparently while most other types of stars occur frequently in binaries, the rarity of variable stars in such close gravitational associations is very striking and has sometimes been used as an argument in favour of some kind of duplicity of the variable itself[6].

However, tables of double stars having one variable component have been compiled[7] and a few cases of real physical pairs are known. Recently, a very promising system has been discussed by G. THIESSEN[8] who succeeded in deriving the mass of the variable component, considered by the author as a typical Cepheid. It is to be hoped that this system and possibly others similar to it, will receive the greatest possible attention in the future.

b) Theory.

At first of course, no distinction was made between the different types of variables. Tentative explanations can be subdivided into two types, those that are mainly geometrical or mechanical and those that are mainly physical.

5. Precession of ellipsoidal bodies. Among the hypotheses of the first group, not counting the Ancients' suggestion of mere approach or recession which had to be given up as soon as the extreme remoteness of the stars was realized, the "purest" one is probably that due to MAUPERTUIS[9]. He attributed all light variations (even, in the case of novae) to the motion of the axis of symmetry of ellipsoids of different flattenings due to the action of planets on highly excentric orbits lying in planes making appreciable angles with the equator of the main body.

[1] H. LUDENDORFF: Astronom. Nachr. **173**, 1 (1906).
[2] H. W. BABCOCK: Astrophys. Journ. **105**, 105 (1947).
[3] H. W. BABCOCK: Publ. Astronom. Soc. Pacific **68**, 70 (1956) and communication at the Stockholm Symposium on "Electromagnetic Phenomena in Cosmical Physics", August 1956.
[4] P. MERRILL: Astrophys. Journ. **116**, 21 (1952).
[5] D. L. HARRIS III: Astrophys. Journ. **119**, 297 (1954); **123**, 549 (1956).
[6] C. D. PERRINE: Astrophys. Journ. **50**, 81 (1919). — F. HOYLE and R. A. LYTTLETON: Monthly Notices Roy. Astronom. Soc. London **103**, 21 (1943).
[7] L. PLAUT: Bull. astronom. Inst. Netherl. **7**, 181 (1934).
[8] G. THIESSEN: Z. Astrophys. **39**, 65 (1956).
[9] P. L. MOREAU DE MAUPERTUIS: Discours sur les différentes figures des astres où l'on essaie d'expliquer les principaux phénomènes du Ciel. Paris 1732.

6. Solid body rotation. Theories based on the uniform rotation of a body with non-uniform distribution of surface brightness constant in time such as that of RICCIOLI (1651) for novae and adapted to Mira Ceti by BOULLIAUD (1667) also fall into the first category. Starting with BRUNS[1], the rotation hypothesis has been the subject of many investigations.

On the theoretical side, it was extended by PICKERING[2] to cover the case where the surface departs from axial symmetry and later, by RUSSELL[3] for a convex surface in general. In an earlier investigation, GYLDEN[4] had shown that quite complicated light curves and cyclic changes in the period could be accounted for by the general theory of rigid body rotation when the axis of rotation does not coincide with one of the principal axes of inertia, as had already been suggested by CASSINI[5] in 1740.

Many of these results were checked or extended by experimental investigations[6]. Although practically all types of light curves could be reproduced, many of these possibilities applicable to solid asteroids or satellites had to be given up when the gaseous nature of the stars was recognized. Indeed this implied a spherical or a slightly flattened shape for the star with the rotation axis along the major principal axis of inertia and in that case, as pointed out by RUSSELL[7] and BOTTLINGER[8], the explanation of some of the light curves especially those of Cepheids with large asymmetry requires impossible physical assumptions as to the distribution of surface brightness and the limb darkening.

7. Differential rotation. Fairly early however, other theories introduced physical factors which slowly took more and more importance. Already ZÖLLNER's theory[9], grafted onto general views on the evolution of stars by cooling, had a more physical character with its progressive building up of superficial slag drifting towards the equator along trajectories deviated against the general rotation. This theory which was compatible with simultaneous variations in colour and luminosity, with some asymmetry in the light curves and with changes from cycle to cycle, in the periods and the form of the light curves, remained in favour for quite a long time especially for long-period variable stars. However the amplitude of the variations depended very much on the position of the axis of rotation with respect to the line of sight, a rather unfavourable feature considering the great similarity of the observed amplitudes for stars with approximately the same period.

8. Sunspot theory. A more drastic step towards a purely physical interpretation of the long-period variables was taken as early as 1852 by R. WOLF[10], who pointed

[1] H. BRUNS: Bemerkungen über den Lichtwechsel der Sterne von Algol typus. Sitzgsber. preuß. Akad. Wiss. Berlin **48** (1881).

[2] E. C. PICKERING: Variable Stars of short Period. Proc. Amer. Assoc. Adv. Sci. **16**, 257 (1881).

[3] H. N. RUSSELL: On the light-variations of Asteroids and Satellites. Astrophys. Journ. **24**, 1 (1906).

[4] H. GYLDEN: Versuch einer mathematischen Theorie zur Erklärung des Lichtwechsels der veränderlichen Sterne. Helsingfors Acta 1880.

[5] G. CASSINI: Eléments d'Astronomie, livre 1, p. 67. 1740.

[6] J. C. F. ZÖLLNER: Photometrische Untersuchungen mit besonderer Rücksicht auf die physische Beschaffenheit der Himmelskörper. Leipzig 1865. — P. GUTHNICK: Künstliche Lichtkurven. Astronom. Nachr. **209**, 1 (1919).

[7] H. N. RUSSELL: Proc. Amer. Assoc. Adv. Sci. **1913**.

[8] K. F. BOTTLINGER: Astronom. Nachr. **210**, 33 (1919).

[9] Cf. Ref. 6, p. 358.

[10] R. WOLF: Neue Untersuchungen über die Periode der Sonnenflecken und ihre Bedeutung. Bern 1852. For a short summary of common characteristics, cf. A. M. CLERKE: Problems in Astrophysics, p. 361. London 1903.

out the general similarity between their light curves and the frequency curve of sunspots which he had just established. Soon after, SECCHI[1] noticed that there was a close analogy between the spectra of sunspots and Mira-variables. This led to the idea that a periodic building up of spots on the star surface might be the cause of the variation rather than the rotation of a spotted surface as in ZÖLLNER's theory. Since the light variations attending the sunspot cycle are extremely small a more direct comparison was impossible. A detailed discussion by TURNER[2] showed that this comparison is satisfactory only if maximum light coincides with maximum spottiness and activity, a point which is difficult to establish definitely. The discovery that long-period variables are giants while the Sun is a typical dwarf and a more complete analysis of the complicated spectrum of these variables led to the neglect of this hypothesis. However a general theory of the origin of sunspots and solar activity capable of being extrapolated to widely different conditions might still be of significance for some types of stellar variability[3].

In any case, this attempt was important in focusing the attention on the possibility of purely intrinsic physical causes of the observed variations.

9. Veil theories. In fact, all subsequent theories were to keep this character. For instance, BRESTER[4] developed extensively the idea of a periodical evaporation and condensation into separate clouds of a layer of saturated vapour above the photosphere accompanied respectively by the rising and expansion or sinking and contraction of that layer. He suggested also that chemical reactions and a periodical displacement of dissociation equilibrium could play a rôle in the excitation of the bright lines. Many details appear now as unjustifiable and even the entire scheme is probably thermodynamically unsound but nevertheless BRESTER's investigations introduced many interesting ideas and they were the natural forerunner of MERRILL's veil theory[5]. Merrill assumes that the very outer layers of the star become periodically opaque due to some kind of condensation phenomena, the products of condensation (liquid droplets or solid dust) being destroyed later by the heat accumulating below this layer.

To this group, we may also add CAMPBELL's geyser-theory[6] with its periodical formation of a superficial crust broken at more or less regular intervals by the heated gases and vapours imprisoned under it. If it has to be rejected on physical grounds, the idea of a regular alternation between phases of quietness and eruptive activity is a valuable one. It was taken up again in 1923, by NÖLKE[7] in the more satisfactory form of an oscillation between states of radiative and convective equilibrium.

10. Double star theories. It seems that in the case of long-period variables the presence of small irregularities in their periods and light curves was always

[1] A. SECCHI: Etude spectrale de diverses régions du soleil, et rapprochements entre les spectres obtenus et ceux de certaines étoiles. C. R. Acad. Sci., Paris **68**, 959 (1869).

[2] H. H. TURNER: Monthly Notices Roy. Astronom. Soc. London **64**, 543 (1904); **67**, 322 (1907).

[3] For instance, S. ROSSELAND: Astrophysik, pp. 96—99. Berlin 1931.

[4] A. BRESTER: Essai d'une théorie du soleil et des étoiles variables. Delft 1889. — Essai d'une explication du mécanisme de la périodicité dans le soleil et les étoiles rouges variables. Verh. Kon. Akad. Wetensch., Amsterd. Sect. 9 **1908**.

[5] P. W. MERRILL: Publ. Astronom. Obs. Univ. Michigan **2**, 70 (1916). — Spectra of Long Period Variable Stars, pp. 98—100. Chicago, Ill.: Chicago University Press 1940. For a general review of the problem and further references, see O. STRUVE: Dust and Related Phenomena in Stars, 6th Internat. Astrophys. Symposium, Liège 1954 (Mém. Soc. Roy. Sci. Liège, Sér. IV, **15**).

[6] W. W. CAMPBELL: Stellar Motions, p. 303. New-Haven 1913.

[7] FR. NÖLKE: Astronom. Nachr. **218**, 143 (1923).

a favourable argument for a purely physical origin of their variability. However the much greater regularity of Cepheids and especially the early discovery[1] of a periodical variation in their radial velocity very analogous to that of spectroscopic binaries was, to contemporaries, compelling evidence of their double star nature.

Since any interpretation of their light variation in terms of eclipses was ruled out by the form of the light curve and the phase relationship between the light and velocity curves, a separate explanation of their variable brightness was needed. Such a theory based on very excentric orbits with short periastron distances had already been developed by KLINKERFUES[2] for variable stars in general. In agreement with current ideas at the time, KLINKERFUES considered the star as composed of a solid sphere surrounded by a fairly extensive gaseous layer which due to tidal distortion at periastron would become thinner in a direction at right angle to the apsidal line making the star look brighter in that direction.

Two later theories also brought the maximum luminosity in coincidence with periastron. One due to ROBERTS[3] suggested the presence of a dark companion brought to incandescence at periastron by the radiation of the bright component However, it is easy to verify that even if the companion would become as bright as the primary, it would not be sufficient to explain the observed amplitudes and furthermore, in that case, the spectral lines of the companion should also become visible.

In the other theory, EDDIE[4] suggested that it was mainly the primary which was heated up at periastron this time by tidal friction. Later PANNEKOEK[5] showed that the dissipation of mechanical energy involved, would bring about a yearly change in the period of revolution quite incompatible with the observations.

Anyway, as the number of known velocity curves increased, it became clear that maximum luminosity was not systematically associated with periastron but rather with the maximum radial velocity of approach. ALBRECHT[6] who was the first to verify the validity of this law also noted that generally there was a very close similarity even in details between light and velocity curves. As no constant relation between the orientation of the orbit and the line of sight could be expected, it was necessary in order to save the double star hypothesis, to introduce a new element with the maximum possible symmetry.

CURTISS[7] thought of a resisting medium pervading the system and enhancing the brightness of that side of the star which faces the direction of motion. This hypothesis was further developed by LOUD[8], but DUNCAN[9] raised serious objections concerning the dissipation of mechanical energy. Instead, he suggested that the atmosphere of the visible star was brushed back from its front by a very tenuous resisting medium associated with the other component. But it isn't at all clear whether, in the stated conditions, the physical process would take place as depicted by DUNCAN or would rather lead to the formation of a compressed layer in front of the primary.

---

[1] Cf. Ref. 6, p. 356.
[2] W. KLINKERFUES: Nachr. Ges. Wiss. Göttingen **1865**.
[3] A.W. ROBERTS: Astrophys. Journ. **2**, 283 (1895).
[4] L. A. EDDIE: Astrophys. Journ. **3**, 227 (1896).
[5] A. PANNEKOEK: Astronom. Nachr. **215**, 227 (1922).
[6] S. ALBRECHT: Astrophys. Journ. **25**, 330 (1907).
[7] R. H. CURTISS: Astrophys. Journ. **20**, 186 (1904).
[8] F. H. LOUD: Astrophys. Journ. **26**, 369 (1907).
[9] J. C. DUNCAN: Lick Obs. Bull. **151** (1908).

Probably the most detailed discussion of this type of hypothesis is found in LUIZET's monograph[1] where he showed that, with an appropriate choice of the orbital elements, it was possible to represent perfectly the light curves of a number of Cepheids on the simple hypothesis that their surface is divided into two hemispheres of unequal brightness by the meridian which, at each instant, contains the centre of gravity of the system. These elements did show a reasonable agreement with the orbital elements determined from the velocity curves in the few cases where these were known. However this was a purely formal success.

In all these double star theories of the light variations, the oversimplified physical background remained vague and unsatisfactory and soon even the interpretation of the velocity curves led to arguments against the binary nature of Cepheids. In some cases, there were secondary oscillations which were difficult to interpret in terms of gravitational perturbations by a third body[2]. The mass-function and $(a \cdot \sin i)$ were consistently very small[3] and periastron showed a preferential direction with respect to the line of sight.

After the discovery[4] by HERTZSPRUNG and RUSSELL of the giant nature of Cepheids, some of these dynamical difficulties summarized by H. SHAPLEY[5] became very acute especially the fact that the orbit of the companion had to dip deeply into the principal star at periastron.

On the other hand, as already pointed out by K. SCHWARZSCHILD[6] as early as 1900, the light variations are accompanied by a regular change in the spectral type which certainly implies a continuous physical change in the atmosphere. Furthermore, at any instant, the deviations with respect to the spectra of normal stars are small and little suggestive of the more or less catastrophic or at least unusual circumstances necessary to explain the light curves on any of the double star theories.

11. Free oscillations. The growing discontent with the binary hypothesis led to the rather timid suggestion at first, that perhaps a single star was involved and that some kind of free oscillation gave rise to the observed phenomena.

As a matter of fact, this idea had already been explicitely stated by RITTER in the 1880's in the course of a serie of remarkable memoirs [*10*] on dilatational pulsations. It was advocated also by MOULTON[7] in an interesting paper where he tried to derive from the equation of conservation of energy, the surface temperature changes and the light variation corresponding to a general deformation expressed in a series of Legendre polynomials. Unfortunately, his results are marred by a very rough evaluation of the kinetic energy of motion and by the approximation that the density remains uniform in the star, its mean variation only being taken into account. Furthermore, as we shall see later even if applied correctly, as in some recent attempts[8], this principle cannot be used in practice to determine the flux variation which is extremely small compared to other terms of the equation.

---

[1] M. LUIZET: Les céphéides. Ann. Univ. Lyon, Nouvelle Série, **1912**, Fasc. 33.

[2] Cf. for instance H. C. PLUMMER: Monthly Notices Roy. Astronom. Soc. London **73**, 665 (1913).

[3] H. LUDENDORFF: Astronom. Nachr. **184**, 373 (1910); **193**, 304 (1912).

[4] H. N. RUSSELL: Science, Lancaster, Pa. N. S. **37**, 662 (1913). — E. HERTZSPRUNG: Astronom. Nachr. **196**, 201 (1913).

[5] H. SHAPLEY: Astrophys. Journ. **40**, 448 (1914).

[6] K. SCHWARZSCHILD: Publ. der v. Kuffnerschen Sternw. **5** C, 100 (1900).

[7] F. R. MOULTON: Astrophys. Journ. **29**, 257 (1909).

[8] E. A. MILNE: Monthly Notices Roy. Astronom. Soc. London **109**, 517 (1949).

Another and better justified feature of MOULTON's paper, was the evaluation of the mean density of variable stars by means of the theoretical formula for the period of oscillation which, already, revealed the giant character of Cepheids. When this had been established independently, SHAPLEY[1] could reverse the procepure as an argument for the pulsation theory. But SHAPLEY as well as MOULTON used KELVIN's formula [*11*] for the periods of non-radial oscillations of liquid spheres although both of them were aware of RITTER's work on purely radial pulsations.

Thus it seems that the exact nature of the oscillations to be considered was still vague at the time and, in fact, there was a general feeling that non-radial oscillations, especially those corresponding to spherical harmonics of degree 2, could be more easily excited than any other type and, of course, this is certainly true if one thinks of external causes.

It is difficult to trace the precise reasons why a few years later, the emphasis was laid so strongly on purely radial pulsations. In his first paper on the subject, EDDINGTON[2] mentions mainly reasons of facility although he remarks also that observational evidence points towards a very high spatial symmetry of the phenomenon. In fact, the amplitudes appear to be confined to a limited range and the existence of a lower limit would seem difficult to reconcile with any axial symmetry since no correlation between the position of the axis and the line of sight could be expected. Of course, this applies as well to any rotational or double star theory as to non-radial oscillations of a simple type such as can be represented by low-order spherical harmonics. More complicated non-radial oscillations would escape this criticism but it is usually agreed that they would be rapidly damped out and furthermore, they are not very effective in producing large observable amplitudes. This argument would still be strongly reinforced by the existence for individual stars, of a relation between the amplitude and the period such as is suggested by EGGEN's recent work [*12*]. Of course, in this respect, any possible selection effect should be carefully discussed before definite conclusions are reached.

On the other hand, with the problems of internal stellar structure coming to the fore, it was natural at the time, especially in view of the existence of a period-luminosity relation to think more of an internal origin for the pulsations and this makes spherical symmetry much more likely. However, one should not forget that many of these stars may possess some rotation and perhaps magnetic fields tending to create departures from that symmetry.

Anyway, from EDDINGTON's papers [*13*] onwards, practically only radial oscillations have been considered and have been the subject of a considerable amount of theoretical work. However if, after some setbacks, the agreement between theoretical and observed periods is to-day reasonably satisfactory yet some of the early difficulties concerning the maintenance of the pulsation, the anharmonicity of the light curve and its phase-lag with respect to the velocity curve are still partly with us to-day.

12. Modified double star theories and forced oscillations. At first, these difficulties gave rise to many attempts to revive the double star theories[3] but

[1] See footnote 5, p. 361.

[2] A. S. EDDINGTON: Observatory **40**, 209 (1917).

[3] Cf. for instance: a) C. D. PERRINE: Astrophys. Journ. **50**, 81, 148 (1919). — Monthly Notices Roy. Astronom. Soc. London **81**, 442 (1921). — b) K. F. BOTTLINGER: Astronom. Nachr. **210**, 33 (1920). — c) J. G. HAGEN: Astronom. Nachr. **209**, 33, (1919); **211**, 247, 413 (1920); **225**, 175 (1925). — Monthly Notices Roy. Astronom. Soc. London **81**, 226 (1921). — d) J. HELLERICH: Astronom. Nachr. **215**, 291 (1922). — e) FR. NÖLKE: Astronom. Nachr. **217**, 65 (1922). — f) SHINZO SHINJO: Jap. J. Astronom. Geophys. **1**, 7 (1922).

although some positive arguments could be produced, none of these theories proved really fruitful and all were again characterized by more or less arbitrary assumptions without a sound physical basis. For instance, there is a curious analogy between the frequency distribution of these variables as a function of the period and that of true spectroscopic binaries[1]. But on the other hand, it is nearly always possible to find a case of a known double star where the proposed mechanism should work as well as for Cepheids but where no effect is observed.

However NÖLKE's investigation, referred to above, deserves some comments. While taking the giant character of the primary into account, he tried to avoid the difficulty of a companion moving deeply into it, at periastron by assuming with GUTHNICK[2] that the high observed excentricities are due to systematic internal motions in the primary which alter the observed radial velocity curve and that the real orbits are practically circular and very close to the surface of the main component. Let us note that, in a way, this is very close to introducing oscillations even if they are of the forced type. Going further, NÖLKE suggested that in the above conditions, the atmosphere of the primary distended by tidal effects would envelop completely the companion and with the help of a series of ad hoc hypotheses he thought that he could explain the main characteristics of Cepheid variation.

Most of these hypotheses could not be upheld now, but an interesting feature of this theory, and an important one according to NÖLKE, is that the existence of such close systems could be interpreted as a natural step in the evolution of rotating stars, a point of view which will again receive emphasis in JEANS' theory. Moreover, NÖLKE's fundamental idea was to find a new expression some years later in HOYLE and LYTTLETON's theory.

However, our increased knowledge of the physical properties of Cepheids has rather strengthened the difficulty concerning the relative size of the orbit and of the primary. For instance, adopting for δ Cephei, the radius $R=3.8\times 10^{12}$ cm and the mass $M=m_1=9\,M_\odot$ given by SAVEDOFF[3], the orbital elements (cf. [4b], p. 205) lead to a value of the semi-major axis of the relative orbit $a=a_1+a_2=2\cdot 10^{12}$ cm which is hardly more than half the radius R of the primary. This result corresponds to a mean value of $\sin^3 i$ equal $\frac{2}{3}$. But even for a smaller value of i, say $i=20°$ which is rather unlikely, a remains small of the order of 2.23×10^{12} cm. Decreasing the mass of the primary would make matters rather worse since for instance with $M=5\,M_\odot$, one gets $a=1.8\times 10^{12}$ cm. Thus if one wants to keep to the interpretation of the period as the time taken by the revolution of two centers of mass around each other, one will have to go one step further than NÖLKE and place them both rather deeply in a common envelope which will undergo forced oscillations.

This point of view has found expression in two theories. The first one due to JEANS who realized early the main short-comings of the radial pulsation theory[4], considers variable stars and especially Cepheids as illustrating the passage of a rotating star from a Jacobian ellipsoid to a double star through the stage of a pearshaped configuration. However, most of the usual objections to rotational theories and namely the enormous broadening of the spectral lines that they imply for objects of giant star dimensions apply here and, furthermore, Jacobian

[1] J. HELLERICH: Astronom. Nachr. **218**, 33 (1925) and O. STRUVE: Monthly Notices Roy. Astronom. Soc. London **86**, 63 (1925).

[2] P. GUTHNIK: Astronom. Nachr. **205**, 119 (1917); **208**, 173 (1919).

[3] M. SAVEDOFF: Bull. astronom. Inst. Netherl. **12**, 58 (1953).

[4] J. H. JEANS: Monthly Notices Roy. Astronom. Soc. London **86**, 90 (1926).

ellipsoids and fission occur only, if at all, in configurations of very small compressibility while the mean density of giants suggests that they are gaseous throughout.

In a later version [*14*], JEANS tried to avoid these objections by considering that only the central core of the star was going through this process and was surrounded by an extended atmosphere rotating much more slowly than the central core. He suggested that the motion of the core would be transmitted to this atmosphere through the action of gravitation and pressure in the form of a running wave travelling around the equator and developing a sharp front of high temperature as the amplitude increased to a finite value. He thought that this could account for the phase relation between light and velocity curves, but this has yet to be proved. Furthermore, apart from the undesirable axial symmetry that the phenomenon would exhibit, it would still be very difficult to justify the small compressibility required for fission, even in a small dense central core.

Later HOYLE and LYTTLETON[1] attempted to dress JEANS' theory in modern attire, considering a double star imbedded in an extended atmosphere acquired mainly by accretion, a theory which apart from its own difficulties has some in common with the pulsation theory.

13. Conclusions. In conclusion, we may say that despite its drawbacks, the pulsation hypothesis still appears as the most hopeful one, not only for the Cepheids but also for practically all regular or semi-regular variables and especially for those exhibiting a close relationship to Cepheids such as the RR Lyrae and long-period variables. In the latter case, one may add that interferometric measurements have revealed at least for one of them, Betelgeuse[2], a large cyclical variation of its angular diameter which it would be very difficult to interpret otherwise than by successive expansions and contractions of the star.

B. Observational data.

14. Classification. Any system of classification depends on the properties considered as providing significant criteria. In a complex and growing subject as that of variable stars, it is understandable that different authors be tempted to adopt different criteria according to their interests. Apart from intrinsic physical properties, the recent advances in our knowledge of galactic structure suggest that kinematical properties are also very important [*2*]. There are certainly correlations between the two types of properties, but nevertheless, they will probably lead to a two-dimensional classification. Furthermore, the discovery of new properties or a better interpretation may require revisions or refinements.

At the present time, there is a fair agreement in the literature, and the minor differences that are encountered often concern simply the names of the classes. But a comparison between a few standard books[3] will permit to establish the necessary correspondences. Here, we shall follow mainly the classification adopted in the General Catalogue [*6*].

Table 1 gives some characteristics of the most important classes.

---

[1] F. HOYLE and R. A. LYTTLETON: Monthly Notices Roy. Astronom. Soc. London **103**, 21 (1943).

[2] F. G. PEASE: Publ. Astronom. Soc. Pacific **34**, 346 (1922) also Mt. Wilson Reports, 1920—1928.

[3] Cf. for instance [*1*], [*4e*], [*4f*] and [*6*].

Table 1. *Main classes of periodic variables and some characteristics.*

Class	Number (1955)	Range of periods	Extreme periods				Spectral types	Absolute magnitude
Mira Ceti Stars	3099	150^d to 450^d	$90^{d}.6$	(T Cen)	1379^d	(BX Mon)	*Me, Re, Ne, Se*	2^M to -1^M
Long-Period variables	962	120^d to 410^d	70^d	(AU Tau)	700^d	(SW Gem)	*Me, Re, Ne, Se*	1^M to -2^M
Semi-regular variables	670	100^d to 200^d	42^d	(TX Tau)	810^d	(S Per)	*G, K, M, R, N, S*	
RV Tauri stars	71	60^d to 100^d	33^d	(SX Cen)	146^d	(R Sct)	*G, K*	-3^M
Classical Cepheids	536	2^d to 40^d	$1^{d}.13$	(BQ CrA)	$45^{d}.2$	(SV Vul)	*F* 6 to *K* 2	$-0^{M}.5$ to $-5^{M}.0$
RR Lyrae stars	2102	$0^{d}.3$ to $0^{d}.9$	$0^{d}.055$	(SX Phe)	$1^{d}.35$	(XX Vir)	*A* 2 to *F* 6	$0^{M}.0$ to $+0^{M}.5$
β Cephei	12	$0^{d}.2$	$0^{d}.15$	(γ Peg)	$0^{d}.25$	(β CMaj)	*B* 1 to *B* 2	$-3^{M}.5$ to $-4^{M}.5$

a) Cepheids and RR Lyrae stars.

15. Introduction. The study of the properties of the Cepheid variables has always been connected with the study of the structure of stellar systems. Originally Cepheid variables were found only among the stars in the Milky Way. Since 1895 BAILEY [*7*] discovered Cepheid variables in great numbers in several globular clusters. It soon appeared that these variables had all much shorter periods than the galactic Cepheids. They obviously represented a new class of variables and they were designated as cluster variables.

Soon afterwards, however, a variable of this type was discovered outside a globular cluster, and, in the course of time, while the photographic method of discovering variables was developed, many more followed. At last the number of galactic "cluster variables" exceeded that of the classical Cepheids. The name cluster variables thereby lost its significance and was replaced by RR Lyrae variables after the first discovered and brightest member of the class.

The name Cepheids was maintained for the galactic variables with periods longer than one day.

The relation between class and location in stellar systems was again emphasized by the discovery by Miss LEAVITT of many Cepheids in the Small Magellanic Cloud. The study of these variables was continued by SHAPLEY, with the result, that at present more than 2500 Cepheids are known in the Magellanic Clouds. Until very recently no RR Lyrae variables were known in these systems.

A relation between class and stellar dynamics was established when it was found that the galactic Cepheids move with low velocity in the galactic plane, whereas the RR Lyrae variables have high velocities and can move perpendicularly to the galactic plane. Such relations as just described have played an important rôle in the development of our present picture of the structure of galactic systems. According to BAADE [*9a*] each stellar system consists of two types of stars with different properties, designated as Population I and II.

In our Galaxy the objects belonging to Population I are confined within a narrow layer in the galactic plane, and preferably in the spiral arms; they participate in the rotation of the galaxy and show only small motions with respect to the Sun.

The stars of Population type II have a nearly spherical distribution and are concentrated towards the galactic nucleus and have high random velocities.

The RR Lyrae stars, whether in or outside a globular cluster, are Population II objects and the classical Cepheids both in our Galaxy and in the Magellanic Clouds belong to Population I.

Since from the period and the light curve the population type of a variable can be determined, they can serve as population indicators in unknown systems.

On the other hand, the study of the properties of the different classes of variables is facilitated by the possibility to select locations of a pure population type.

For example, our insight in the properties of the Cepheids has been much clarified by the separation into Population I and Population II Cepheids. Since it is known from their location in the galaxy, that the globular clusters are pure Population II systems, the Cepheids observed in them, with periods longer than one day, must be also Population II objects. These variables have periods which are concentrated between 1 and 2 days and between 13 and 19 days thus quite different from the distribution of periods of the galactic Cepheids [*15a, b*]. Moreover the light curves of these variables are perceptably different from those of most of the galactic Cepheids with corresponding periods.

However, the same type of light curve was also encountered among the galactic Cepheids. In general such variables are characterized by a high galactic latitude or high velocity, which confirms that they are Population II Cepheids.

This discovery led to the introduction of a new class of Cepheids, called W Virginis stars. Their properties are in seveıal respects different from those of the Population I Cepheids.

Of course, the question arises whether also among the RR Lyrae stars two population types are present. This has been suggested by IWANOWSKA[1], who discovered that RR Lyrae stars can be divided into two groups on the basis of differences in intensity of spectral lines. However, whereas the W Virginis stars can be distinguished from Population I Cepheids by their light curve, luminosity, colour and the occurrence of emission lines, such distinctions are only weakly present, or not yet observed between the groups of IWANOWSKA. Therefore, the existence of Population I RR Lyrae stars does not seem to be definitely established.

Recently a new group of variables has attracted attention. According to their periods, which are of the order of a tenth of a day, they should be assigned to the RR Lyrae class, but in some respects they are appreciably different from the RR Lyrae stars. What makes these variables interesting is that they are of low intrinsic luminosity but on the other hand several of them are rather bright stars, so that they constitute probably a numerous class of variables[2].

Their rapid variation and low range make their discovery on photographic plates difficult and this explains why only a limited number of these variables are known at the present. However, in recent years, the search with photoelectric photometres has led to the discovery of several such variables.

16. The form of the light curves and their relation with period. In his pioneer studies on RR Lyrae variables in globular clusters BAILEY discovered three types of light curves, which he designated by *a*, *b* and *c*.

The light curves of type *a* have a big amplitude, on the average 1.0 magnitude, and are very asymmetrical, showing a steep rising branch, a sharp maximum and a long flat minimum.

---

[1] W. IWANOWSKA: Torun Bull. **11** (1953).

[2] L. WOLTJER: Bull. astronom. Inst. Netherl. **470**, 62 (1956).

The light curves of type *b* are also asymmetrical but have a rather low amplitude, of about 0.6 magnitude.

The light curves of type *c* have a nearly sinusoidal shape and an amplitude of about 0.5 magnitude.

Later studies only confirmed the adequacy of BAILEY'S classification. No matter where they are, the RR Lyrae stars always show the same characteristic light curves.

In Fig. 2 some typical light curves are shown. They are reproduced from MARTIN'S exhaustive investigation of the variables in ω Centauri [*16*]. The weak hump preceding the maximum of the curve of type *c* is present in practically all well determined light curves of this type and the same is true for the details in the light curves of types *a* and *b*.

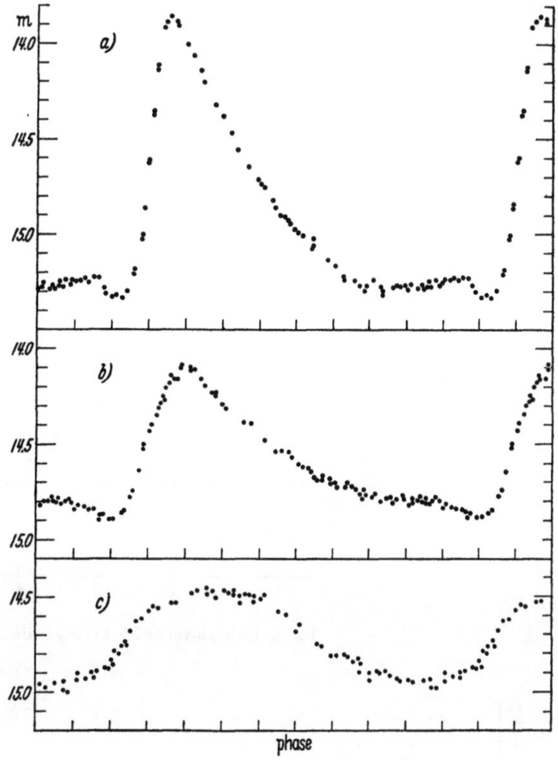

Fig. 2 a—c. Typical light curves of RR Lyrae variables. BAILEY'S types a, b, and c.

As an example for galactic RR Lyrae stars, two light curves of RR Lyrae are shown in Fig. 3. This star, like several other members of its class, has a light curve which varies periodically in shape. When the amplitude is large, the curve is typical for BAILEY'S class *a*, when the amplitude is low the curve resembles strongly type *b*. The fact that one and the same star may show both types proves that there exists only a gradual difference between type *a* and type *b*. This is confirmed by other relationships and it is only the difference between type *c* and the others which is clearly pronounced.

With respect to period the variables of type *c* are separated from the other types. Especially in globular clusters the division is sharp. In ω Centauri, for example, the type *c* variables, with one exception, have periods ranging from $0^d_.25$ to $0^d_.48$, all types *a* and *b* fall within the range $0^d_.47$ to $0^d_.90$. For different clusters the division may take place at slightly different periods.

For the galactic RR Lyrae type variables the ranges of periods of the different types overlap to some extent and in the galactic nucleus they overlap almost entirely.

The light curves of the very rapid RR Lyrae stars, with periods between $0^d_.05$ and $0^d_.2$ are not in conformity with the Bailey types. The curves are smoother and less asymmetrical than the curves of types *a* and *b*. In general, they show no humps; if these occur, as in AI Velorum, they are not at a fixed position in the light curve.

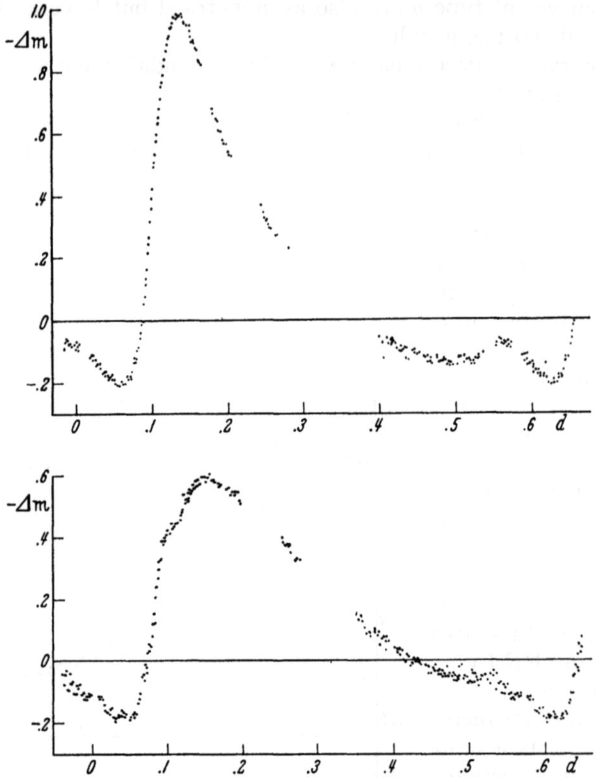

Fig. 3. Light curves of RR Lyrae at different times.

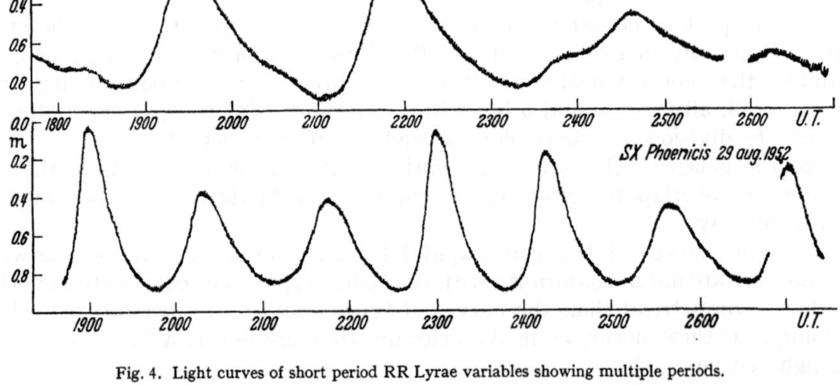

Fig. 4. Light curves of short period RR Lyrae variables showing multiple periods.

Several of these variables have a double periodicity, i.e. the light curve resembles the result of the interference of two different oscillations as shown in Fig. 4.

At periods of about one day only a limited number of variables is known and it is not clear whether they belong to the class of the RR Lyrae stars or to that of the Cepheids. The light curves do not fit the Bailey types and often have queer shapes, with sharp humps on the rising part or shoulders on the descending part, but very regular curves occur also. Nearly every variable with a period between one and three days has its own individual shape[1].

In the globular clusters, Cepheid variables occur with periods between one and three days which should be considered as Population II objects. On the

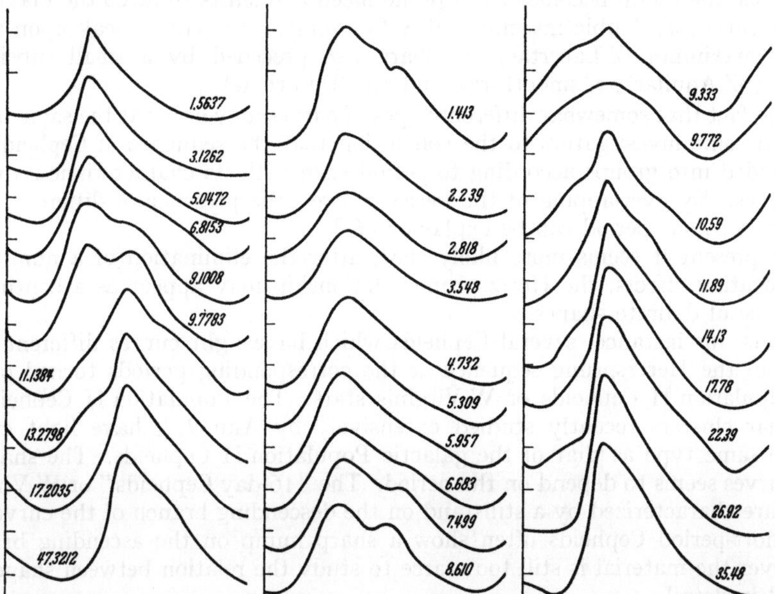

Fig. 5. Hertzsprung sequence showing variation of light curves of Cepheids with periods. The left-hand section refers to the Magellanic Cepheids and the other sections to galactic Cepheids (after Mrs. PAYNE-GAPOSCHKIN [*1*]).

other hand, in the outer parts of the Small Magellanic Cloud also a considerable number of variables have periods within this range[2]. However, it seems difficult to consider these as Population II objects since they satisfy the period-luminosity relation of Cepheids of type I. Thus in many respects the situation in the range of periods between one and three days is not clear.

In the case of the classical Cepheids, which have periods from 3 to 40 days it was first noticed by HERTZSPRUNG[3], that the light curves, if arranged according to period, show a systematic variation of shape.

This relationship was studied carefully by KUKARKIN and PARENAGO[4] for the galactic Cepheids and it was found that although the individual light curves at the same period are not always strictly identical, the average curves of a given period have a definite shape. The same shape was also found for other stellar systems such as M31, M33 and NGC 6822.

[1] Cf. Fig. 1, Vistas in Astronomy, Vol. II, p. 1146. Pergamon Press 1956.

[2] H. SHAPLEY and V. MCK. NAIL: Proc. Nat. Acad. Sci. U.S.A. **26**, 105 (1940).

[3] E. HERTZSPRUNG: Bull. astronom. Inst. Netherl. **3**, 115 (1926).

[4] B. V. KUKARKIN and P. P. PARENAGO: Astronom. USSR. **14**, 181 (1937).

From unpublished observations of 150 galactic Cepheids PAYNE-GAPOSCHKIN [1] derived a relationship which is practically identical to that of KUKARKIN and PARENAGO.

The relationship found by SHAPLEY and McK. NAIL[1] for the Cepheids in the Magellanic Clouds is quite similar.

In Fig. 5 the Hertzsprung relationship is shown. The left-hand section refers to the Magellanic Cepheids and the other sections to galactic Cepheids.

Mrs. PAYNE-GAPOSCHKIN has introduced the following symbols for the typical light curves of the sequence: *u*, smooth asymmetrical curve (δ Cepheid); *v*, smooth rise with hump on the descending branch (η Aquilae); while the period increases, the hump becomes more pronounced and shifts towards the maximum of the curve; *w*, double maximum (SX Velorum); *x*, a central peak upon a flat broad maximum (Z Lacertae); *y*, sharp rise preceded by a small subsidiary hump (SZ Aquilae); *z*, smooth rise and fall (U Carinae).

The fact that somewhat different types of curves may occur at the same period has led some investigators to the conclusion that the sequence of Cepheids can be divided into groups according to period each with its characteristic sequence of curves. By overlapping of the period ranges, the presence of different types of curves at one period can be explained [*4e*].

At present it seems more likely that, after the elimination of a number of complicating effects, the Hertzsprung relationship may appear as a continuous sequence of definite shapes.

Thus, for instance, several Cepheids which have light curves different from those in the Hertzsprung sequence at the corresponding periods turned out to be Population II Cepheids or W Virginis stars. The Population II Cepheids in globular clusters recently studied extensively by ARP [*17*], have light curves of the same type as that of the galactic Population II Cepheids. The shape of the curves seems to depend on the period. The "16-day Cepheids" or W Virginis stars are characterized by a stillstand on the descending branch of the curve and the short-period Cepheids often show a sharp hump on the ascending branch. However the material is still too scarce to study the relation between shape and period in detail.

From the foregoing it follows that the Hertzsprung sequence holds only for population type I Cepheids and that, in a study of its properties, the type II Cepheids must be carefully eliminated. A further complication arises from the fact that, at any period, some Cepheids may show variations in light which are less than normal. This was proved by EGGEN's photo-electric observations [*12*].

In analogy with BAILEY's types, EGGEN designated the Cepheids with low amplitude as type *C* and those with normal amplitudes as types *A* and *B*. Moreover, a given type of light curves is not strictly associated with one period but may be found within a small range of periods.

17. Correlation between amplitude, asymmetry and period. In Fig. 6 the amplitude-period diagram is shown for the RR Lyrae stars in the globular cluster ω Centauri.

The variables of type *c* have all more or less the same amplitudes scattered around 0.5 magnitude. At the period of nearly 0.5 day, the amplitude suddenly increases and toward longer periods it gradually decreases again while the light curves change from type *a* to type *b*. The greatest dispersion in amplitude occurs at periods just above the critical period. It is interesting that at the same periods also, the tendency of the variables to have unstable light curves is greatest.

[1] H. SHAPLEY and MCK. NAIL: Proc. Amer. Phil. Soc. **92**, 310 (1948).

Some cases are noted where the height of the maximum of the curve varied by 0.8 magnitude. It is not yet decided whether the variation is periodic and thus of the same nature as the variations shown by many galactic RR Lyrae stars.

The properties of the amplitude-period diagram just described are common to all globular clusters which are rich in variable stars, but quantitative differences exist. Notable differences are shown by the various clusters in the period at which the transition from type *c* to type *a* takes place.

The amplitude-period diagram for the galactic RR Lyrae stars is similar to that of the cluster variables, but less sharply defined. This might be interpreted as the result of the less homogeneous character of the observations but, to a certain extent, it might also be due to the mixing of stars of different origin or in different stages of evolution.

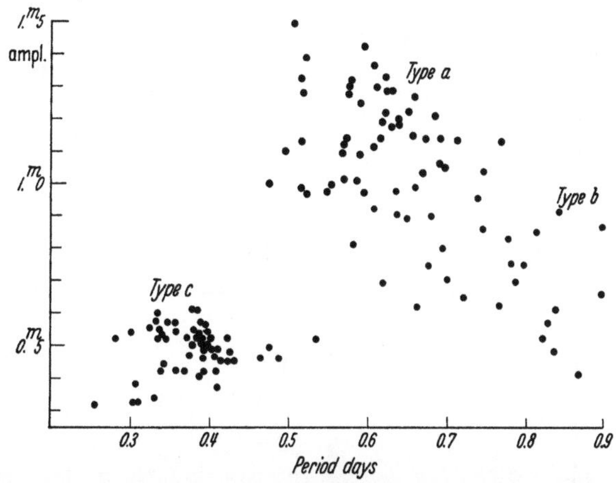

Fig. 6. The photographic amplitudes of the RR Lyrae variables in the globular cluster ω Centauri as a function of periods.

In the diagram for the RR Lyrae stars in the galactic nucleus the type *c* and type *a* variables no longer occupy separate regions.

The change in asymmetry is very spectacular in the transition from type *c* to type *a*. The type *c* variables show practically no asymmetry, the type *a* variables have the strongest asymmetry of all pulsating variables. Usually the asymmetry in time is measured by the quantity ε, which is the interval between minimum and maximum, divided by the period. For the type *c* variables, ε is 0.5; for type *a*, it reaches the value 0.1 and with increasing period ε decreases to about 0.2 for the type *b* variables, which have still appreciable asymmetry in time.

Also the asymmetry in magnitude is very pronounced for the type *a* variables, the maxima being much narrower and sharper than the minimum. This asymmetry can be described by the quantity ζ, defined as:

$$\zeta = m_{\frac{1}{2}P} - \tfrac{1}{2}(m_{\max} + m_{\min})$$

where $m_{\frac{1}{2}P}$ is the magnitude at which the interval between the rising and descending branches is half the period. For the type *c* variables ζ is zero, for the type *a* variables it suddenly jumps up to about $0^{\mathrm{m}}.4$ and decreases again to become nearly zero for the type *b* variables.

The quantities ε and ζ are very closely correlated with the amplitude for the curves of the types *a* and *b*, as is shown in Fig. 7 for the variables in ω Centauri [*16*].

When the amplitudes of the Cepheids are collected from the various sources in the literature and their relation with period is investigated only a broad band of scattered values is found and hardly any relation can be found. It is known now that this is due almost entirely to the inhomogeneity of the photographic observation by different observers with different instruments. In this respect the situation is worse than for the RR Lyrae stars in a cluster which at least are observed simultaneously.

It has been the experience that high quality observations though limited in number give better defined relationships than numerous observations of average quality.

Accurate photo-electric light curves of 32 classical Cepheids were obtained by EGGEN [*12*] with blue and yellow filters.

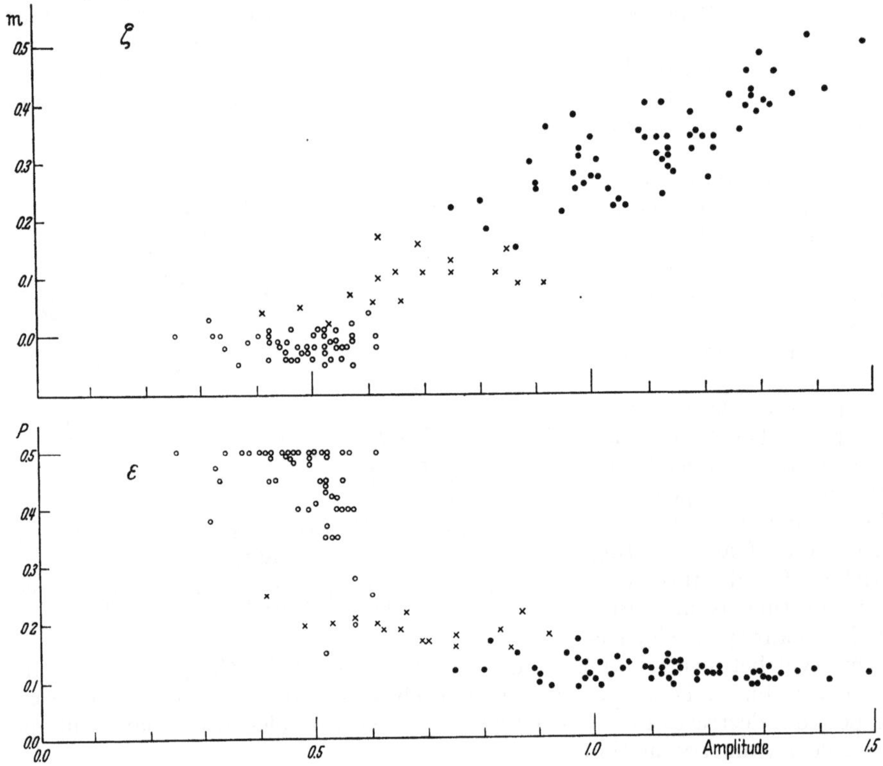

Fig. 7. Asymmetry in magnitude (ζ) and asymmetry in phase (ε) of RR Lyrae variables as a function of amplitudes. Dots: type *a*; crosses: type *b*; open circles: type *c*.

The amplitude A_{pg} of the curves obtained with the blue filter are plotted in Fig. 8 against the logarithm of the period in days. The open circles represent unpublished photo-electric observations of Southern Cepheids.

The figure shows that there does not exist a sharply defined relation between amplitude and period although the majority of the stars obeys a period-amplitude relation indicated by the dashed lines. For some variables the amplitude is considerably lower but no abnormally high amplitudes occur.

The variables with reduced amplitude were designated by EGGEN as type *C* and the other ones as types *A* or *B*. EGGEN considered all variables with periods of nearly 9 days observed by him as being of type *C* and interpreted the absence of types *A* or *B* at this period as a gap in the frequency distribution of these stars. However the frequency curve of the periods of all Cepheids together does not show a deficiency at this period.

Thus in EGGEN's hypothesis, the deficiency of types *A* and *B* at this period should be compensated by an excess of type *C* stars.

It would seem more plausible, to assume that the amplitude-period relation shows a depression at the period of 9 days as shown by the dashed lines in Fig. 8, which is connected with the rapid transition in shapes of the curves in the Hertzsprung sequence. The lowest amplitude occurs when the hump on the descending branch has reached or just surpassed the height of the original maximum of the light curve, and starts developing into the new maximum (see Fig. 5).

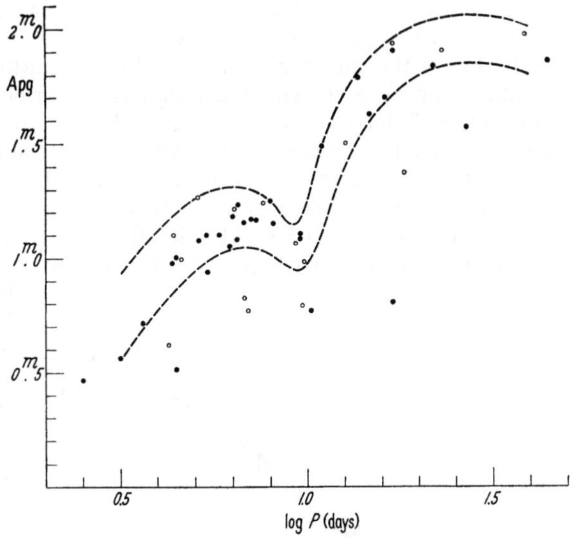

Fig. 8. Photo-electric amplitudes of Cepheids as a function of the period.

All light curves with periods of nearly 9 days have a maximum of which the average curvature is less than that of the minimum. This shows that also in those cases where the humps on the maximum are not plainly visible the character of the curve is still the same.

In general, the light curves of type *C* have shapes similar to those of the Hertzsprung sequence at the corresponding period.

A satisfactory interpretation of the remarkable progression of shapes has not yet been found. KUKARKIN and PARENAGO[1] and PAYNE-GAPOSCHKIN[2] have resolved standard light curves into harmonic components, but the results did not help very much to elucidate the problem.

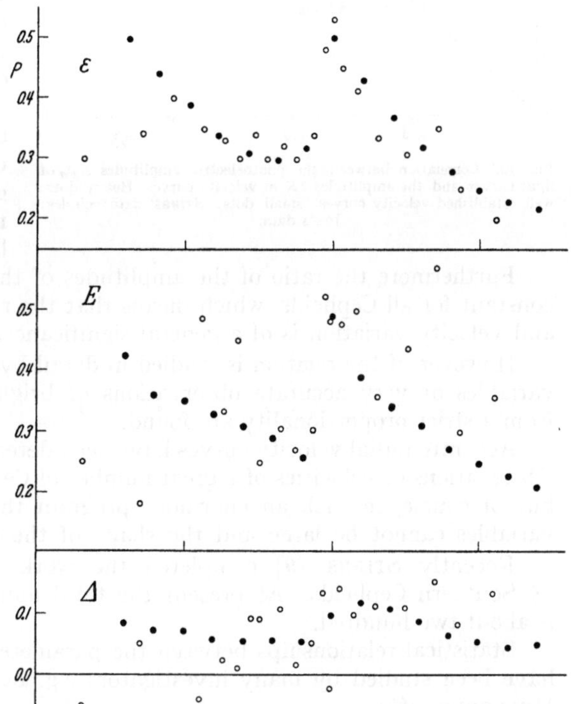

Fig. 9. Top: asymmetry in phase of light curves of Cepheids. Middle: asymmetry in phase of radial velocity curves. Bottom: phase lag of velocity curves with respect to light curves. Dots: standard curves of KUKARKIN and PARENAGO. Open circles: curves of C. PAYNE-GAPOSCHKIN.

---

[1] Cf. Ref. 4, p. 369.

[2] C. PAYNE-GAPOSCHKIN: Astronom. J. **52**, 218 (1947).

It is clear also that the use of the asymmetry as a parameter to describe the form has only a formal significance.

In Fig. 9 the quantity ε derived by KUKARKIN and PARENAGO[1] from their standard curves is shown as a function of period. The data given by PAYNE-GAPOSCHKIN[2] derived from entirely independent observations of 135 Cepheids, are shown for the purpose of comparison. At the critical period of 9 days where the light curves have two maxima of equal height a discontinuity in ε exists, but it is clear that this does not correspond to a sudden change in physical properties but is introduced by the definition of ε. On both sides of the jump the asymmetry increases with period. The Cepheids with the longest periods have the most asymmetrical curves. Let us note that this maximum asymmetry remains smaller than that of the RR Lyrae stars of type *a*.

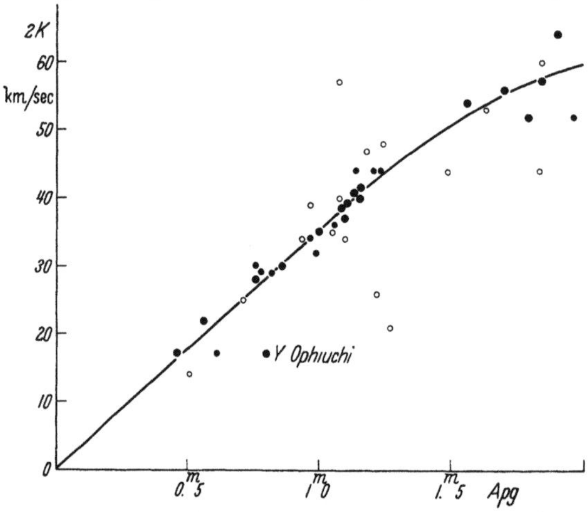

Fig. 10. Correlation between the photoelectric amplitudes A_{pg} of light curves and the amplitudes $2K$ of velocity curves. Heavy dots: well established velocity curves; small dots: STIBBS' data; circles: JOY's data.

18. Correlation between the radial velocity curves and the light curves. The light curves and the radial velocity curves of Cepheids are closely related. In first approximation the velocity is a linear function of the magnitude. The minimum radial velocity practically coincides in time with the maximum of light and also maximum velocity, with minimum light; in other words the velocity curve is the "mirror image" of the light curve.

Furthermore the ratio of the amplitudes of the two curves is approximately constant for all Cepheids, which means that the relation between light variation and velocity variation is of a general significance.

However, if the relation is studied in detail by using statistical data of many variables or very accurate observations of bright variables certain deviations from a strict proportionality are found.

Accurate radial velocity curves have been determined for only a few Cepheids. Observations of velocities of a great number of Cepheids were made by JOY [*18*], but of course, in such an enormous program the number of observations per variables cannot be large and the shape of the curves is not sharply defined.

Recently STIBBS [*19*] completed the work of JOY with observations of 55 Southern Cepheids. At present the total number of known velocity curves is about two hundred.

Statistical relationships between the parameters of light and velocity curves have been studied by many investigators e.g. by LUDENDORFF, ROBINSON and HELLERICH [*4b*].

Originally only weak correlations were found. This was due entirely to the inaccuracy of the observations for it appeared that, if only photo-electric light curves and accurate velocity curves are compared, extremely sharp relations

[1] Cf. Ref. 4, p. 369.

[2] C. PAYNE-GAPOSCHKIN: Harvard Ann. **113**, 153 (1954).

are found. This was demonstrated by EGGEN [12] and by PARENAGO [20] in the case of the relation between the total amplitude $2K$ (km/sec) of the radial velocity curve and the amplitude A_{pg} of the light curve.

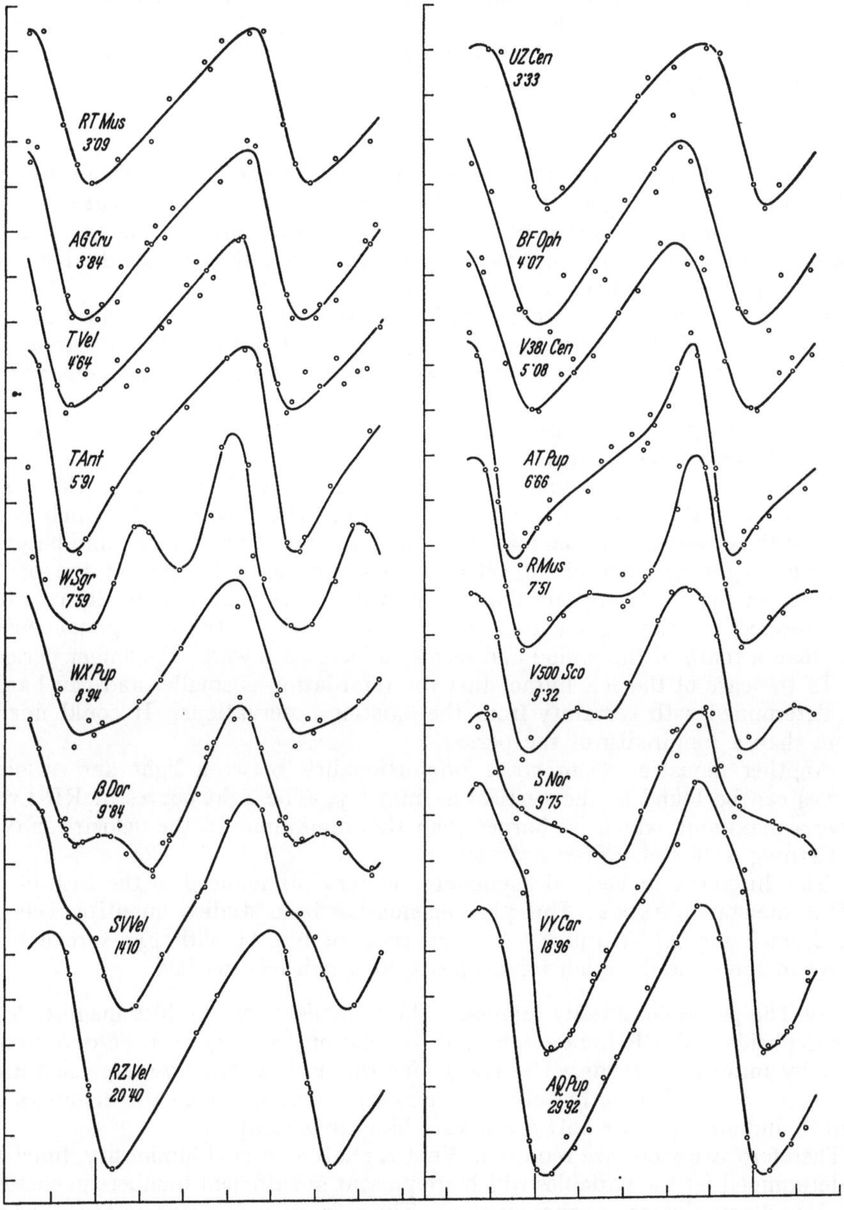

Fig. 11. Typical radial velocity curves of Cepheids with different periods. One sub-division of the vertical scale equals 10 km/sec.

It is seen in Fig. 10 that, especially for the more reliable observations the relation is remarkably close. The large deviations of some stars are probably real, as for Y Ophiuchi. However, in many cases, the radial velocities have been observed about twenty years earlier than the photo-electric light curves and variations in the amplitude may have occurred.

The smaller deviations may be caused by observational errors and improvement in the observational methods might reduce the scatter still more.

PARENAGO first drew attention to the fact that for big amplitudes the relation is no longer linear A_{pg} increasing more rapidly then $2K$. For the linear part ($A_{pg} < 1\overset{\text{m}}{.}2$), he derived the formula

$$2K = 35.3\, A_{pg}.$$

The shape of the radial velocity curve of a Cepheid resembles closely that of the light curve. For example, the asymmetry of the velocity curve varies in the same way with period as that of the light curve, as is shown in Fig. 9. Also the discontinuity at the period of 9 days is present. It is caused by the double minimum in the velocity curves at this period.

Fig. 11 reproduces a sequence of radial velocity curves as observed by STIBBS. A comparison of these curves with Fig. 5 shows that the velocity curves are subject to a gradual change of shape with period, which is of the same nature as that of the light curves.

In the foregoing, the close connection between the variations in light and velocity has been demonstrated.

Nevertheless there are some systematic differences. For example the variations do not take place in exact synchronism. The quantity Δ, which is the mean of the intervals between maximum light and minimum velocity and between minimum light and maximum velocity, expressed as a fraction of the period is shown in Fig. 9. It appears that the velocity curve is systematically retarded with respect to the light curve. The retardation is on the average somewhat less than a tenth of the period and seems to increase towards the longer periods.

In the case of the RR Lyrae stars the retardation is smaller and can hardly be determined with certainty from the existing observations. It could not be more than a hundredth of the period.

Another deviation from strict proportionality between light and velocity curves can be found in the vertical asymmetry. The light curves of RR Lyrae have a maximum which is sharper than the minimum. In the velocity curves, this strong asymmetry does not appear.

The difference in vertical asymmetry is very pronounced in the case of the RR Lyrae stars of type *a*. This phenomenon has been studied quantitatively for RR Lyrae[1] and AI Velorum[2]. A comparison of Fig. 11 with Fig. 5 reveals an effect in the same direction for Cepheids, although less marked.

19. The period-luminosity relation. The luminosity or absolute magnitude of the Cepheids and RR Lyrae stars and its relation to period can be determined only by indirect methods. The reason for this is that the absolute magnitude can be determined directly only for the nearby stars of which the distances are known and among these only a few variables are present.

Therefore two steps are required. First a relative period-luminosity function, is determined for the variables which are present in sufficient numbers in clusters, the Magellanic Clouds or other systems. The stars in such a system are all at the same distance from us and their apparent magnitudes are directly comparable. This gives period-luminosity relations of the correct form but with unknown position on the absolute magnitude scale.

---

[1] TH. WALRAVEN: Bull. astronom. Inst. Netherl. **1953**, No. 434.

[2] L. GRATTON: Bull. astronom. Inst. Netherl. **1952**, No. 444.

The next step is to correct the scale by comparison with the directly determined absolute magnitudes of variables or other objects in our neighbourhood. In this procedure it is assumed that the nearby objects have the same properties as those in the remote systems.

The first period-luminosity relation was established by means of Cepheids in the Magellanic Clouds. HERTZSPRUNG first tried to bring the relation on an absolute magnitude scale[1]. SHAPLEY devoted a considerable amount of work to the problem and derived a period-luminosity relation which for many years served as a tool for exploring the dimensions and distances of stellar systems [*21*]. SHAPLEY's relation, which is shown in Fig. 12 as a dotted line, includes the RR Lyrae stars.

In 1937, KUKARKIN[2] improved the form of the relation by showing that it consisted of two parts with different slopes, fitting together at about the 10 day period i.e. the critical period of the Hertzsprung sequence. KUKARKIN's relations can be written

$$\left.\begin{aligned} \Delta M_{pg} &= -1.03\,\Delta \operatorname{Log} P, \quad \operatorname{Log} P < 0.97, \\ \Delta M_{pg} &= -1.73\,\Delta \operatorname{Log} P, \quad \operatorname{Log} P > 0.97 \end{aligned}\right\} \tag{19.1}$$

if we limit ourselves to the relative correlations to avoid the difficult question of the zero point. Mrs. PAYNE-GAPOSCHKIN and GAPOSCHKIN[3] have also tabulated the bolometric magnitude-period relation which can be represented roughly by

$$\Delta M = -2.7\,\Delta \operatorname{Log} P. \tag{19.2}$$

Apart from a few dissenting voices[4], SHAPLEY's relation as modified by KUKARKIN was generally accepted[5] until 1952 when BAADE [*9a*], in his studies on extragalactic systems arrived at serious discrepancies between his results and SHAPLEY's relation, which led him to the conclusion that the zero-point of the magnitude scale of the classical Cepheids was wrong, these stars being brighter by $1^{\mathrm{m}}.5$ than hitherto supposed. The absolute magnitude of the RR Lyrae stars, however, needed no correction.

As an illustration of the intricacies of the problem, some of the arguments are mentioned. For example, assuming that the classical Cepheids in the Andromeda nebula M 31 obey SHAPLEY's period-luminosity relation the apparent magnitude of the RR Lyrae stars in this system should be $22^{\mathrm{m}}.4$, but although BAADE could easily observe such weak stars he found no RR Lyrae stars. From accurate photometry of stars in globular clusters it is known[6,7] that the RR Lyrae stars are placed in a sharply defined position in the colour-magnitude diagram. By comparison with the corresponding array of nearby stars of Population type II the absolute magnitude of the RR Lyrae stars is rather certainly fixed at $M_{pg} = 0.0$.

Furthermore the diagrams of the clusters show that the brightest stars in the clusters are $1^{\mathrm{m}}.5$ brighter than the RR Lyrae stars. In the Andromeda nebula, BAADE could observe such bright Population II stars but, in comparison with the Cepheids they appeared to be $1^{\mathrm{m}}.5$ weaker than expected.

---

[1] E. HERTZSPRUNG: Astronom. Nachr. **196**, 201 (1913).

[2] B. V. KUKARKIN: Astronom. J. USSR **14**, 125 (1937).

[3] C. PAYNE-GAPOSCHKIN and S. GAPOSCHKIN: Proc. Amer. Phil. Soc. **86**, 329 (1943) (H. R., Series II, No. 2).

[4] K. LUNDMARK: Vjschr. Astronom. Gesellsch. **68**, 369 (1933). — H. MINEUR: Ann d'Astrophys. **7**, 160 (1945).

[5] R. E. WILSON: Astrophys. Journ. **89**, 218 (1939).

[6] M. SCHWARZSCHILD: Circ. Harv. Coll. Obs. **1940**, No. 437.

[7] H. C. ARP, W. A. BAUM and A. R. SANDAGE: Astronom. J. **58**, 4 (1953).

The unescapable conclusion therefore is that the brightness of the Cepheids was underestimated by this amount in SHAPLEY's relation. The consequences of BAADE's discovery were enormous. The sizes and distances of the Andromeda nebula and other systems which were determined with the aid of the Cepheids and SHAPLEY's relation, had to be doubled by the correction.

Shortly later, BAADE's discovery was confirmed by THACKERAY[1] and WESSELINK who found some RR Lyrae stars in globular clusters in the Magellanic Cloud system. Their apparent brightness was about 2 magnitudes weaker than expected on the basis of the brightness of the Cepheids in the Clouds.

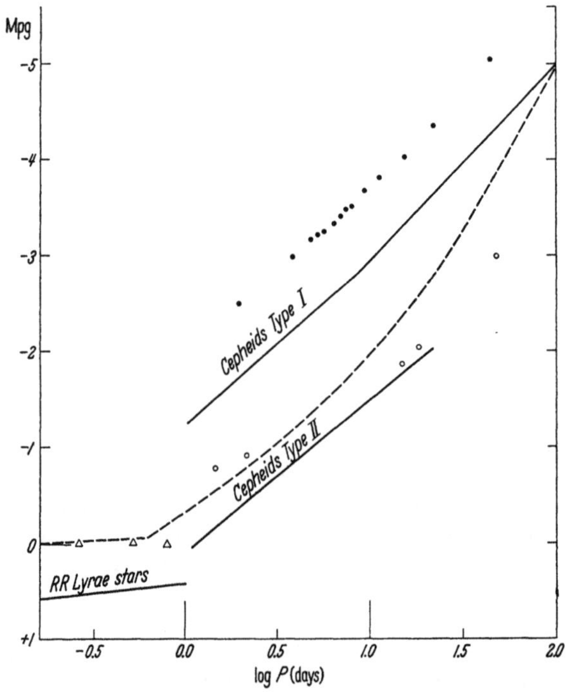

Fig. 12. Period-luminosity relation for RR Lyrae stars and Cepheids. Shapley 1930: dashed lines; PARENAGO 1955 [20]: full lines; C. PAYNE-GAPOSCHKIN 1955 [1], dots: Cepheids type I, open circles: Cepheids type II and triangles: RR Lyrae stars.

Moreover BLAAUW and MORGAN[2] from proper motions of 18 nearby Cepheids concluded that their luminosity is $1^{\mathrm{m}}.4$ greater than according to the old period-luminosity relation in perfect agreement with BAADE's results.

Unfortunately the discussions which brought forth the need for a correction, led at the same time to a better insight in the difficulties of the problem and subsequent rediscussions of the exact value of the correction gave inconsistent results.

For example, PARENAGO [20] in a careful discussion of all available material, based in trigonometrical parallaxes, proper motions, galactic rotation etc. derived a correction of $1^{\mathrm{m}}.0$ for the Cepheids and placed the absolute magnitude of the RR Lyrae stars at $M_{pg} = +0.5$. It seems that at present the zero point of the period-luminosity relation cannot be determined with an accuracy better than 0.5 magnitude. For this reason, the period-luminosity relation as determined by different authors is given in Fig. 12. The reasons for the uncertainty are the following:

a) the distances of the nearest variables are still considerable and cannot easily be determined from parallaxes or proper motion;

b) the effect of interstellar absorption on the apparent brightness is appreciable and difficult to estimate;

c) it is not known in how far the colour-luminosity arrays of the stars in our neighbourhood and in other location are comparable;

[1] A. D. THACKERAY: Nature, Lond. **171**, 693 (1953).

[2] A. BLAAUW and H. MORGAN: Bull. astronom. Inst. Netherl. **12**, 95 (1954).

d) in systems of different types of population different types of variables are prevailing;

e) it is not known whether the period-luminosity relations of variables in different systems are exactly identical.

With regard to the last point it might be remarked that the Hertzsprung relationship for the classical Cepheids in the Magellanic Clouds is not exactly identical with that of the galactic Cepheids (see Fig. 5). The critical period at which the light curves have double maxima of equal height and at which the discontinuity in asymmetry takes place is 9 days for the galactic Cepheids and nearly 10 days for the Cepheids in the Magellanic Clouds. The individual light curves of this type are encountered at periods ranging from 8.4 days to 9.7 days among the galactic variables and a similar dispersion in period exists for the Magellanic Cepheids. Mrs. PAYNE-GAPOSCHKIN [*1*] has discovered that in the

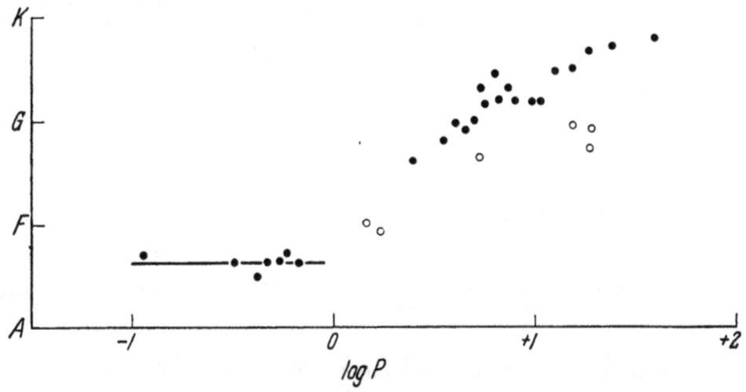

Fig. 13. Mean spectral type-period relation after PARENAGO [*20*]. Dots: RR Lyrae and type I Cepheids; open circles: type II Cepheids.

Magellanic Clouds this critical period is related to the apparent brightness of the stars (and consequently to the absolute brightness). From the shortest to the longest periods the brightness of the variables with this special shape increases systematically by more than half a magnitude. For adjacent curve types an analogous relation between period and brightness exists.

If this is correct, the hitherto unexplained large dispersion in the period-luminosity relation of the Cepheids in the Magellanic Clouds is real and the identification of this relation with that of the galactic Cepheids becomes uncertain.

20. The period-spectral type relation. The spectral type of the Cepheids and RR Lyrae stars is subject to variations similar to the variations in light and radial velocity.

The spectral type curve has roughly the same asymmetry as the light curve, as the earliest spectral type coincides in time with maximum light and the latest spectral type with minimum light.

Except for some of the brighter variables the spectral type has been determined only on low dispersion spectrograms. For many variables only a few determinations of the spectral type at unknown phases exist.

Nevertheless, PARENAGO could derive a satisfactory statistical relation between mean spectral type and period, using the spectral types of 248 variables given in the General Catalogue of Variable Stars (1948), which is shown in Fig. 13. It appears that the mean spectral type of the RR Lyrae stars is independent of period and is on the average *A*6. The mean spectral type of the classical

Cepheids changes with increasing period from $F6$ to $G8$ and shows a dip at the period of 9 days, where nearly every relation shows some peculiarity.

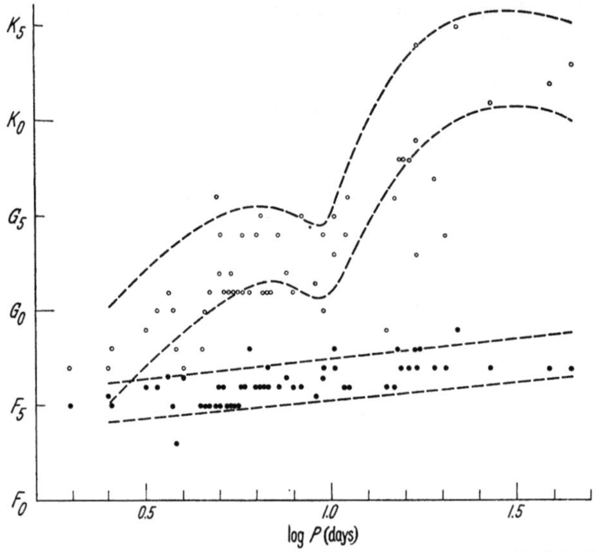

Fig. 14. Spectral types at maximum (dots) and at minimum (circles) of Cepheids as a function of the period.

The figure shows also the mean spectral types of Population II Cepheids derived from data for six W Virginis stars and fifteen variables in globular clusters, studied by JOY [*15a*]. The spectral types of the Population II Cepheids are about 0.7 of a spectral class earlier than that of the Population I Cepheids.

For a more limited number of variables, the spectral types have been determined at a sufficient number of different phases so that the earliest and latest types are known.

It was first remarked by STRUVE[1] and confirmed by CODE[2] that the earliest spectral type varies only very slowly with period, whereas the latest spectral type shifts considerably to later classes with increasing period.

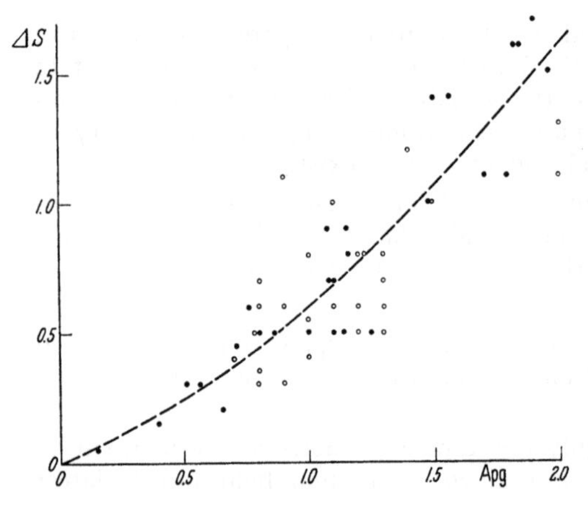

Fig. 15. Correlation between the range of spectral types ΔS and amplitudes A_{pg} of the light curves for Cepheids. Dots: photo-electric light curves; circles: photographic light curves.

This effect is clearly shown in Fig. 14 where the spectral types at maximum and at minimum light are plotted for individual classical Cepheids against the logarithm of period. The spectral types are those adopted by PARENAGO [*20*] who reduced the available data to the Yerkes classification system.

The earliest spectrum varies from $F5$ to $F7.5$ with an average dispersion of a tenth of a spectral class, which is not more than the accuracy of the classification.

The latest spectrum varies much more, in a way which reminds us of the relation between amplitude of the light curve and period (see Fig. 8). Certain variables show large deviations. The variables with exceptionally small variation in spectral type have also low amplitude light curves and must be classified as of EGGEN'S type *C*.

---

[1] O. STRUVE: Observatory **65**, 257 (1944). — Astronom. J. **102**, 232 (1945).
[2] A. D. CODE: Astronom. J. **106**, 309 (1947).

This shows that there is a correlation between the range of spectral type and the amplitude of the light curve which is demonstrated more explicitly in Fig. 15.

The spectral classification of the RR Lyrae stars meets with difficulties which render uncertain the spectral type-period relation for these stars.

It is interesting to note that, as was pointed out by STRUVE and BLAAUW[1] the spectral type at maximum of RR Lyrae remains constant at A_2 during successive cycles whereas the magnitude of the maximum varies strongly.

21. Secular variations of the periods. The periods of Cepheids and RR Lyrae stars are often given as numbers of seven or eight decimal places. Such amazingly precise values are of course not directly observed. The moment at which a characteristic feature of the light curve, for example the middle of the rising branch, arrives can be observed with a precision of at most a thousandth of a cycle. By combining observations of such moments separated by a long time interval which contains thousands of cycles a very accurate value of the period can be derived.

Strictly speaking by this method only the mean period is determined and the question of how precise and stable the real period is, remains open. Changes in the period can be discovered only when after a sufficient number of cycles, the difference between the observed phase and the phase predicted with the original period becomes noticeable. Thus, small periodic changes in the length of the period may escape attention, but secular changes in period must become observable in due time.

Suppose that for a variable star the period P_0 has been derived, then the arrival of future light curves can be predicted by adding any multiple value of the period, say P_0E. If however the period is subject to a secular change and has in reality the value $P = P_0 + \beta P_0 E$, then the light curves shall arrive at moments given by $P_0E + \frac{1}{2}\beta P_0 E^2$ and a growing difference $\frac{1}{2}\beta P_0 E^2$ between the observed and predicted moments arises.

The observations of the light curves of variable stars started more than half a century ago and, for several variables, deviations in phase were noted after some time, which could be described by a quadratic term in the formula, and which were often interpreted as caused by secular variations in the period. Many such quadratic formulae can be found scattered throughout the literature.

However in many cases the observations were not very continuous and often quadratic formulae have been derived for only a few isolated groups of observations. Often the quadratic formula was not confirmed by later observations. Periods which were lengthening became shorter again or otherwise. While the material increased, it became clearer and clearer that, in general, the variations in length of the periods are not of a secular nature but are rather quasi-periodical or irregular, and the question whether real secular changes in period take place cannot yet be answered.

Some observational results may serve to illustrate our present insight in the problem. For the sake of argument we shall speak of secular variations of period when the observed variations in phase can be described by a quadratic formula.

MARTIN [*16*], by comparing observations of the RR Lyrae stars of types *a* and *b* in the globular cluster ω Centauri made in 1931 to 1935 with the observations by BAILEY, of 1892 to 1898, found that, for 49 variables, the periods had lengthened and, for 19 variables, they had shortened. The mean value of β is 5×10^{-10}, i.e. on the average the periods have increased at a rate of 0.2 day per 10^6 years.

[1] O. STRUVE and A. BLAAUW: Astrophys. Journ. **108**, 60 (1948).

The spread in the values of β for the different variables is 8×10^{-10}, which makes the reality of the systematic effect questionable.

About one third of the type *c* variables showed large fluctuations in period, which are too large to be considered as secular changes, but the material was not sufficient to decide whether they were periodic.

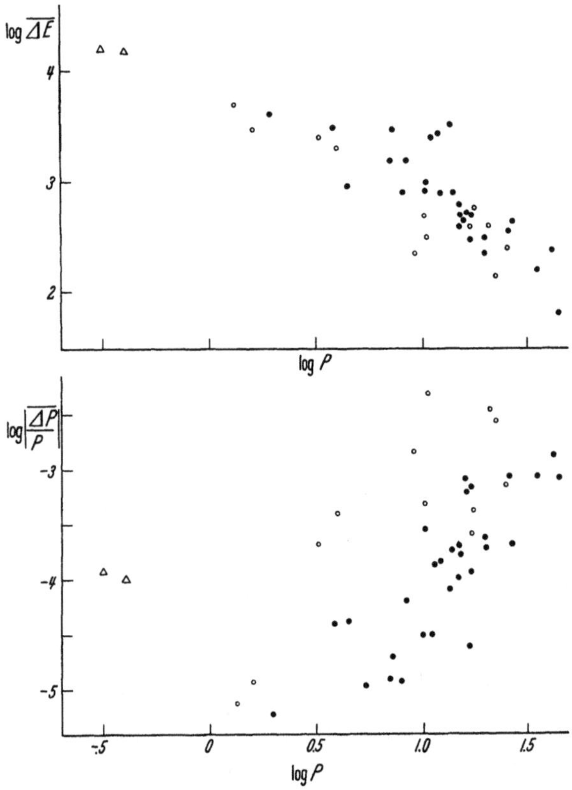

Fig. 16. Upper part of the figure shows the logarithm of the average number of cycles between successive changes in period as a function of period. Lower part gives the average relative change in period as a function of period. Dots: Cepheids type I; circles: Cepheids type II; triangles: RR Lyrae stars.

In other globular clusters the situation is similar. BELSERENE[1] found in M 3, 15 increasing periods and 9 decreasing. For M 15, OOSTERHOFF[2] gives the numbers 25 and 15.

Several galactic RR Lyrae stars have been observed more or less regularly at different observatories, especially at Budapest by BALASZ and DETRE[3]. DETRE [22] has discussed the changes in periods of 69 of these stars which have been observed for more than 25 years.

Of these variables, 19 showed no appreciable change in period. The deviations in phase of 24 variables could satisfactorily be represented by a quadratic formula, with β of the order of 10^{-9}, which might be due to secular variations. Of these, 20 variables have lengthening periods. In 11 cases no secular effects, but alternating changes are observed with periods ranging from 1.5 years to 46 years.

The changes of the remaining 15 variables can be described according to DETRE, as a superposition of a secular variation and periodic fluctuations; in 14 cases, the period is lengthening.

It should be remarked that often the interpretation of an inhomogeneous, interrupted series of observations is doubtful. For example PRAGER[4] noticed for RZ Cephei abrupt changes in period. But according to DETRE abrupt changes do not occur in any RR Lyrae type variable.

With regard to the explanation of the variations it may be important that, in some cases (RR Lyrae, RR Leo), DETRE has found a change in the slope of

[1] E. P. BELSERENE: Astronom. J. **57**, 237 (1952).
[2] P. TH. OOSTERHOFF: Leiden Ann. **17**, No. 4 (1941).
[3] Cf. Mitt. Budapest Obs. as from 1938 onwards.
[4] R. PRAGER: Harvard Bull. **1939**, No. 911.

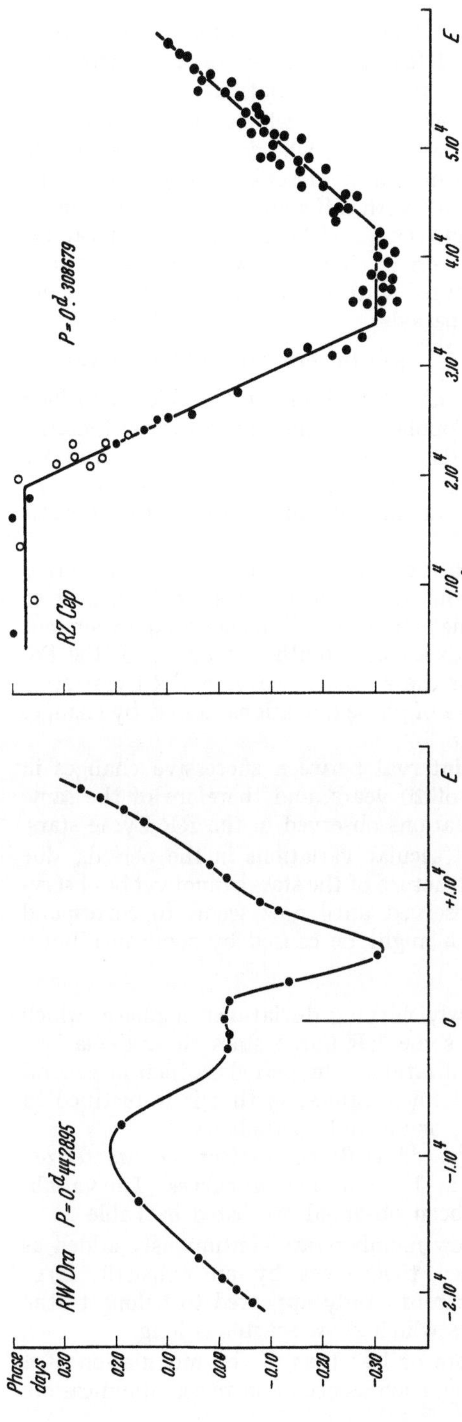

Fig. 17. Deviation of observed phase from computed phase for RW Draconis and RZ Cephei.

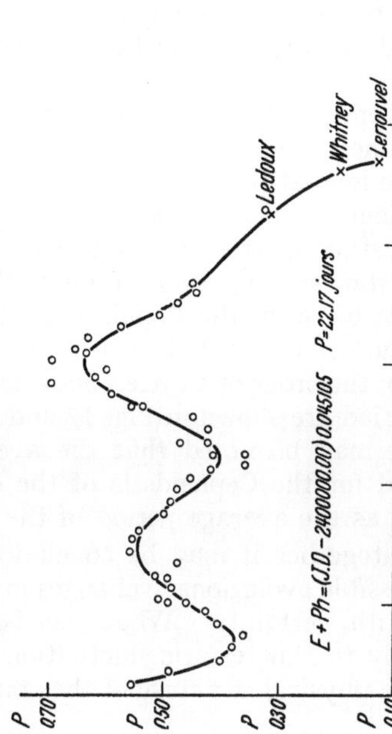

Fig. 18. Deviation of observed phase from computed phase for RU Camelopardalis.

the rising branch of the light curve taking place simultaneously with the variation in phase shift.

Also most of the classical Cepheids, which are sufficiently well observed show variations in period and, as for the RR Lyrae stars, more or less irregular fluctuations are the rule rather than regular secular variations.

A survey of the behaviour of about 40 Cepheids which show pronounced variations in period has been given by PARENAGO[1], who found that, as a rule the deviations in phase as a function of time can be described by a broken line, the successive portions representing intervals with different periods. The number of changes of period which occur varies between 1 and 13 for the different variables, and on the average it is 3. The numbers of changes towards longer periods and towards shorter periods are distributed entirely at random; there is no preference for lengthening or shortening periods.

The average relative change in period $\frac{\Delta P}{P}$ is of the order of 10^{-5} for the variables with shortest periods and increases to about 10^{-3} for the variables with long period (see Fig. 16). For the Cepheids of Population II the changes are systematically larger. Also the values for two RR Lyrae stars with strong variations (RZ Cephei, RR Geminorum) are shown. The average number of cycles ΔE, contained in the intervals varies between about 5000 and 100 from the shortest to the longest periods.

Hence it follows that the stronger relative changes in the value of the period are lasting shorter, in other words the total deviation in phase with respect to a constant mean period is roughly the same in all cases. The maximum excursions in phase are for the Population I Cepheids about a tenth of a cycle, for the Population II Cepheids half a cycle and for the strongest varying RR Lyrae stars it is of the order of a cycle. Some examples of phase deviations caused by changes in period are shown in Fig. 17 and Fig. 18.

It may be noted that the average interval between successive changes in period for the Cepheids is of the order of 20 years, and therefore of the same order as the average period of the fluctuations observed in the RR Lyrae stars.

Altogether it may be concluded that secular variations in the periods, due to possible evolutionary changes in the structure of the stars cannot yet be observed with certainty. What has been observed until now seems to correspond mainly to slow erratic fluctuations which might be caused by some instability in the physical structure of the stars.

22. Multiple periods. Besides the slowly varying deviations in phase, which were discussed in the previous section, some RR Lyrae stars show deviations which are subject to a regular periodic variation. The periods, which in general are of the order of a month have well defined values, so that it is justified to use in these cases the expression double, or multiple periodicity.

This phenomenon, which is also designated as Blazhko effect is characterized by a simultaneous variation in phase and in shape of the light curves. The variables in which a multiple periodicity has been observed are listed in Table 2.

The list is by no means complete; new members are continuously added as variables, which previously distinguished themselves by an unusually large scatter in the observations, after more careful study appeared to belong to the class, and the list of RR Lyrae variables which show scatter is long.

The phenomenon itself resembles more or less that of the modulation of a carrier wave by a wave of lower frequency, such as occurs in telecommunication.

[1] P. P. PARENAGO: Colloquium on Variable Stars, Budapest, 1956.

The phase of the light curve is modulated, in the case of RR Lyrae for example, with a total amplitude of 0.037 of the period. The amplitude of the light curve is subject to a modulation which, in the case of RR Lyrae, amounts to $0^{m}.4$. This is mainly a variation of the height of the maximum, the minimum varies much less. Also the shape of the curve varies during the cycle of modulation as shown in Fig. 19. The curve with large amplitude is also the most asymmetrical one and resembles BAILEY's type *a*, when the amplitude is small the curve is of BAILEY's type *b*.

The character of the changes is not always exactly the same for every variable. Whereas in RR Lyrae[1] the maximum retardation in phase is reached when the amplitude has its mean value and is increasing, in XZ Cygni[2] and RS Bootis[3] the maximum retardation in phase coincides with maximum amplitude.

In RV Ursae Majoris[4], the modulation in amplitude is large but for the phase it is small. SW Andromedae[5] shows practically no variations of amplitude but pronounced changes take place in the strength of the hump on the ascending branch, as is also shown by RR Lyrae.

In some cases, where the modulation is strong and is observed with sufficient continuity, it appears to be variable in strength in a period which is always nearly three times the period of the modulation. In Table 2, these cases are indicated by giving the value of the longer period in brackets.

Table 2. *Variables with multiple periods.*

	Period P_0 (days)	Period of modulation P_M (days)
Cepheids		
T Trianguli Australis .	2.568	6.979
TU Cassiopeiae . . .	2.139	5.230
RR *Lyrae stars*		
DL Herculis	0.592	49.2
RR Lyrae	0.567	41.0 (123.0)
RW Cancri	0.547	29.9 (91.1)
Y Leo Minoris	0.524	33.4
RZ Lyrae	0.511	115.7
XZ Draconis.	0.476	76
AR Herculis	0.470	31.5
RV Ursae Majoris . .	0.468	91
XZ Cygni	0.466	41.6 (148)
RV Capricorni	0.448	221.9
RW Draconis	0.443	41.6 (124)
SW Andromedae . . .	0.442	36.8
RR Geminorum . . .	0.397	~40
RS Bootis.	0.377	537
Very short period variables		
δ Scuti	0.194	0.838
VZ Cancri	0.178	0.716
AI Velorum	0.112	0.379
RV Arietis	0.093	0.316
DQ Cephei	0.079	0.375
SX Phoenicis	0.055	0.193

With respect to multiple periodicity, the very short-period variables constitute a separate group. As can be seen in Table 2 the modulation is very rapid, its period being only 3 to 4 times the fundamental period. In one night the complete cycle of modulation can be observed, as is shown in Fig. 4.

In AI Velorum the modulation is exceptionally strong and as a consequence it was possible to show that the variations in amplitude and phase have exactly the same character as those shown by a curve which is a superposition of two harmonic oscillations. However, the light curves such as observed cannot be described as a superposition of two waves, i.e., no wave-shapes can be formed which, if added, reproduce as well the high light curves as the flat light curves.

[1] TH. WALRAVEN: Bull. astronom. Inst. Netherl. **11**, 17 (1949).
[2] A. B. MULLER: Bull. astronom. Inst. Netherl. **12**, 11 (1953).
[3] P. TH. OOSTERHOFF: Bull. astronom. Inst. Netherl. **10**, 101 (1946).
[4] L. DETRE: Colloquium Budapest, 1956; also, Vistas in Astronomy, Vol. II, p. 1156.
[5] J. BALAZS and L. DETRE: Mitt. Budapest **1954**, No. 33.

This follows for example from the fact that, in a curve resulting from the combination of two arbitrary wave-shapes, the variations in height of the maxima and minima are symmetrical which is not observed in the light curves. To account for the difference, a systematic distortion in magnitude has to be applied.

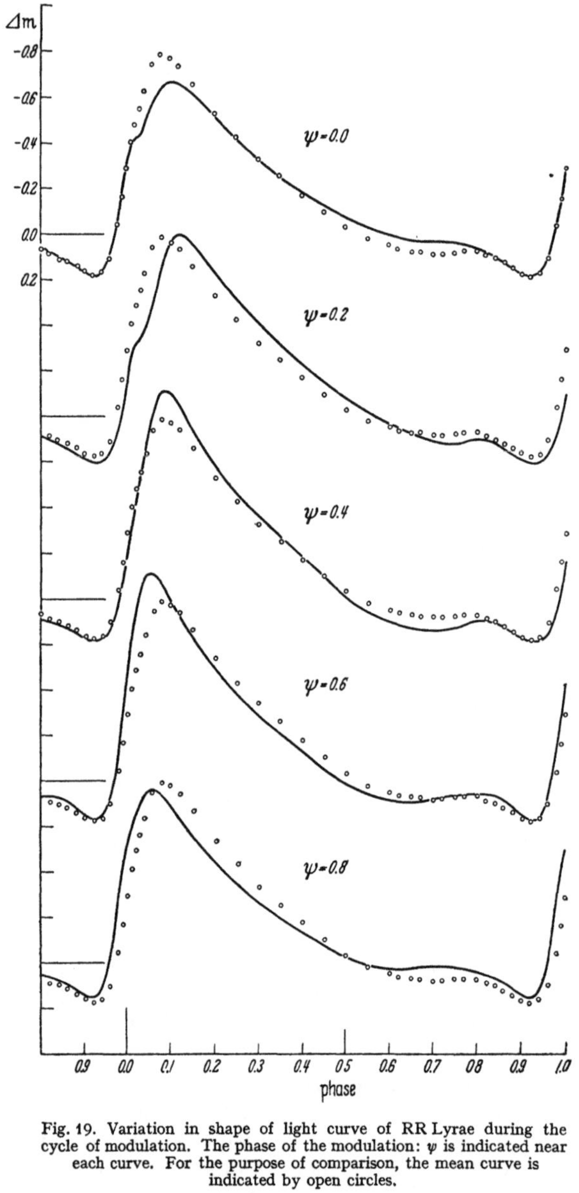

Fig. 19. Variation in shape of light curve of RR Lyrae during the cycle of modulation. The phase of the modulation: ψ is indicated near each curve. For the purpose of comparison, the mean curve is indicated by open circles.

Secondly the smoothness of the flat light curves can be explained only if the two components are sinusoidal. As sinusoidal components cannot produce the strong asymmetry in phase observed in the high light curves it must be assumed that the combined curve is subject also to a systematic distortion in phase.

The two distortions which thus present themselves in a natural way have been studied in a detailed analysis of the light curves of AI Velorum and SX Phoenicis [*23*]. The method can be applied also to RR Lyrae.

In Fig. 20a, the distortion in magnitude is shown for the three stars as a relation between the magnitude Δm in the light curve and the corresponding value u of the model curve:

$$u = a \sin \frac{2\pi t}{P_0} + b \sin \frac{2\pi t}{P_1}.$$

The distortion in phase derived from the comparison of the light curve with the model curve is shown in Fig. 20b and it appears that the phase-shift χ is a linear function of u. Fig. 20c illustrates how the combination of the different factors results in a curve which is in very close agreement with the observed light curves (cf. Fig. 4). All this makes it probable that in AI Velorum and similar stars, the modulation can be physically interpreted as an interference of two oscillations. The exact values of the two periods are, for AI Velorum, $P_0 = 0^{\mathrm{d}}.11157375$ and $P_1 = 0^{\mathrm{d}}.08630767$ and that of the period of modulation $P_M = 0^{\mathrm{d}}.379188$.

Still other periods can be found in AI Velorum. Small waves are seen in the light curves, which have an average amplitude of $0^{\rm m}.02$. Apparently they do not form a continuous wavetrain but it appears that the phases of the well observable waves always fit a definite periodicity. The length of this period is $P_2 = 0^{\rm d}.044402$. The occasional disappearance of the wave is caused by a modulation of the amplitude which takes place with the rhythm of the main period P_0. Also an interaction of P_2 with P_1 takes place. Altogether AI Velorum is a beautiful case of a variable with multiple periodicity.

The radial velocity curves of the variables with multiple periods are subject to changes which in nearly all respects resemble those of the light curves. This is in conformity with the generally observed close relationship between light and velocity variation. Only one difference is noticeable, the vertical asymmetry of the velocity curves is much less than that of the light curves. This effect was mentioned already in the case of

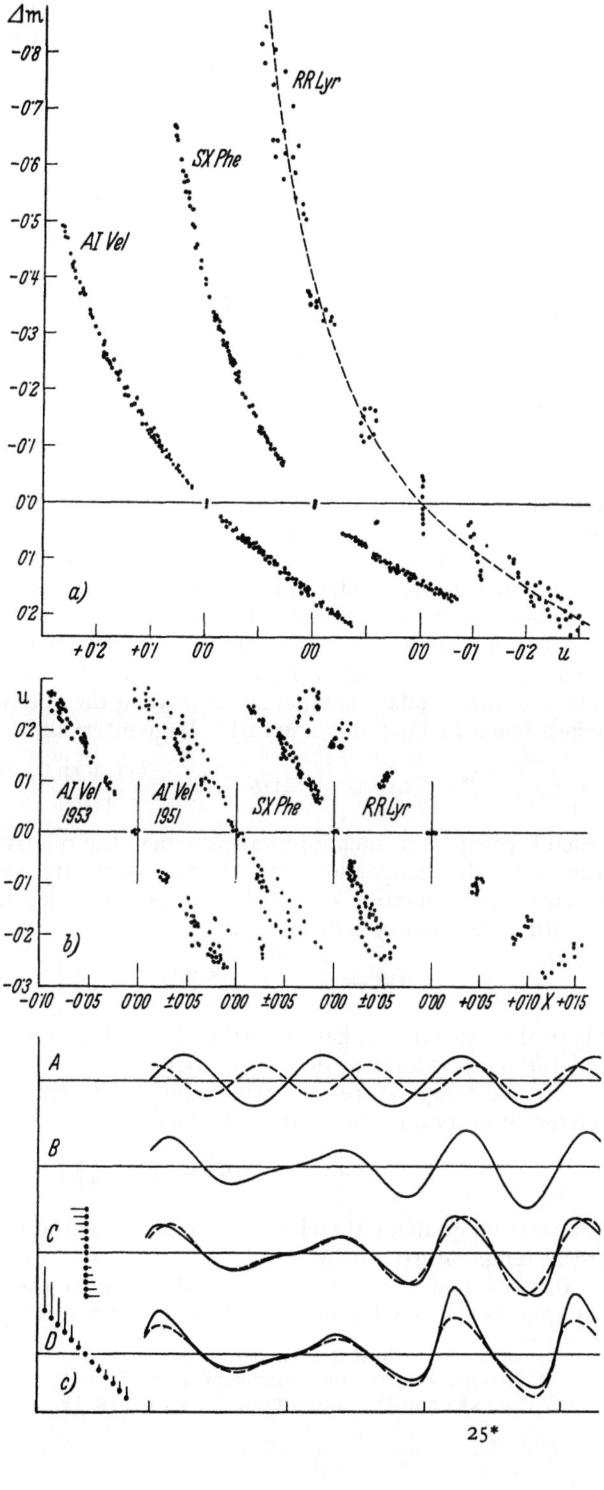

Fig. 20a–c. (a) The distortion in magnitude for three variables as derived from the comparison of the observed light curves with the sinusoidal model curves. (b) The distribution of phase derived in the same way. (c) Representation of observed light curve of AI Velorum as a combination of two oscillations; *A*: two harmonic curves; *B*: sum of curves in *A*; *C*: curve *B* after distortion in phase (resembles radial velocity curve); *D*: curve C after distortion in magnitude (resembles light curve).

the Cepheids but is even more clearly shown by the RR Lyrae variables with strong modulation. The vertical asymmetry of the light curves increases considerably with the amplitude and this is not shown by the velocity curves.

As was shown by GRATTON[1] the radial velocity curves of AI Velorum resemble the combination of two sine-curves subject to a distortion in phase only such as curve C of Fig. 20c.

The variations in spectral type and in colour are also subject to the effects of modulation.

Until recently no multiple periodicity had been observed for Cepheids. But in the case of TU Cassiopeiae which presents strong variations in amplitude and phase, OOSTERHOFF[2] has succeeded in interpreting these satisfactorily as the result of a double periodicity. The period of modulation is $P_M = 5\overset{\mathrm{d}}{.}23$ which is only about $2\frac{1}{2}$ times the period $P_0 = 2\overset{\mathrm{d}}{.}14$. A similar effect is present in T Trianguli Australis in which case $P_M = 6\overset{\mathrm{d}}{.}98$ and $P_0 = 2\overset{\mathrm{d}}{.}57$.

Since the same type of irregularity is shown by many Cepheids with P_0 of the order of 2 days one may expect that in their case too a double periodicity is present.

23. The continuous spectrum; spectrophotometry and multi-colour photometry. The determination of colour-indices is the first step towards the study of the continuous spectrum. For a more detailed investigation of the spectral intensity distribution two methods have come into practice: photographic spectrophotometry and multi-colour photoelectric photometry.

The spectrophotometric method consists in measuring the intensity of the continuous spectrum between the absorption lines at as many wavelengths as possible and comparing these intensities for different stars. If I_λ is the observed intensity in a star at wavelength λ and I_λ^* the intensity at the same wavelength observed in a standard comparison star, then the quantity $\log I_\lambda - \log I_\lambda^*$ appears to be a linear function of $1/\lambda$ in rather large intervals of wavelength. The quantity

$$\Delta\Phi = -2.3\,\frac{\Delta \operatorname{Log} I_\lambda/I_\lambda^*}{\Delta\,(1/\lambda)} \tag{23.1}$$

which is constant in such intervals is named the relative gradient of the star with respect to the comparison star. If the stars are radiating like black bodies, which is approximately true, the relative gradient can be determined theoretically from the laws of radiation, as

$$\Delta\Phi = \frac{C_2}{T}\left(1 - \mathrm{e}^{-C_2/\lambda T}\right)^{-1} - \frac{C_2}{T^*}\left(1 - \mathrm{e}^{-C_2/\lambda T^*}\right)^{-1} \tag{23.2}$$

where the constant C_2 has the value 14300 if λ is expressed in Ångström units and the temperatures in degrees centigrade.

For the temperatures we are dealing with, the factor between parentheses is close to unity and the relative gradient is

$$\Delta\Phi \approx \frac{C_2}{T} - \frac{C_2}{T^*}. \tag{23.3}$$

The relative gradient therefore is a measure of the difference between the reciprocal temperatures of the stars.

BECKER and STROHMEIER[3] have determined these gradients for a number of Cepheids and RR Lyrae stars. They found that they could represent the spec-

[1] L. GRATTON: Bull. astronom. Inst. Netherl. **1953**, No. 444.

[2] P. TH. OOSTERHOFF: Bull. astronom. Inst. Netherl. (in press).

[3] W. BECKER and W. STROHMEIER: Z. Astrophys. **19**, 249 (1940).

tral intensity distribution of Cepheids and stars of the same spectral type satisfactorily by means of two gradients, one from 6500 to 4800 Å and the other from 4800 to 3900 Å, and derived corresponding temperatures. The variation of the gradients in the course of the cycle is closely related with the variation of the colour-indices derived from monochromatic light curves. No large discrepancies were found between the spectral intensity distribution of the Cepheids and that of the non-variable stars.

BECKER[1] was able to show that the temperature derived from the gradients (colour-temperature) is different from the temperature which determines, according to the law of WIEN, the surface brightness of the stars (radiation-temperature). The difference, due to the fact that the stars do not radiate like black bodies is very large, as shown by Table 3.

Table 3. *Colour temperatures and radiation temperatures of Cepheids.*

Stars	Colour temperature			Radiation temperature		
	Max. °K	Min. °K	Ampl. °K	Max. °K	Min. °K	Ampl. °K
SU Cas	6050	5240	810	5700	5290	410
RT Aur	7500	5600	1900	6420	5450	970
α U Mi	6050	5760	290	5730	5550	180
T Vul	6240	5160	1080	5820	5240	580
δ Cep	6810	5320	1490	6070	5340	730
η Aql	6000	4440	1560	5680	4990	690
S Sge	6050	4450	1600	5720	4990	730
ζ Gem	6140	5050	1090	5770	5300	470

This important result reveals one of the most serious obstacles on the way to the interpretation of the variation in light and colour in terms of the variation in radius and temperature.

CANAVAGGIA[2], following the methods of CHALONGE and BARBIER determined two gradients in the continuous spectrum, respectively before (4800 to 4000 Å) and after the Balmer discontinuity (3600 to 3100 Å), for δ Cephei, η Aquilae and ζ Geminorum. She also determined the value of the Balmer discontinuity defined as $D = \log \frac{I_{3700^+}}{I_{3700^-}}$ where I_{3700^+} and I_{3700^-} represent the intensities of the continuous spectrum just above and just below the wavelength λ 3700. This quantity is a measure for the depression in the continuous spectrum caused by the continuous absorption band which extends from the limit of the Balmer series at λ 3647 towards shorter wavelengths.

The gradient in the visual region of the spectrum varies in synchronism with the light, indicating that the star is bluest at maximum light in qualitative agreement with BECKER'S and STROHMEIER'S results. The ultraviolet gradient, on the other hand, varies like the mirror image of the light curve, which, if interpreted as a variation of colour temperature, would make the temperature a minimum at maximum light. It is clear, in this case, that the variation of the gradient due to temperature changes is completely masked by the disturbing effect of the varying continuous Balmer absorption which is measured by the quantity D. In δ Cephei, D varies between 0.11 and 0.40 more or less in the same way as the brightness of the star. The maximum value (strong absorption) coincides with maximum light and the minimum of absorption with minimum light.

[1] W. BECKER: Z. Astrophys. **19**, 269 (1940).
[2] R. CANAVAGGIA: Ann. d'Astrophys. **12**, 21 (1949).

In a second paper CANAVAGGIA[1] discussed these results in terms of the variation of electron pressure and temperature.

Perhaps no photometric observations have been used so often as the six-colour photoelectric observations, by STEBBINS and collaborators, of δ Cephei [*24a*], η Aquilae [*24b*] and RR Lyrae [*24c*].

The effective wavelengths, situated at $\lambda\lambda$ 10300 (I), 7190 (R), 5700 (G), 4880 (B), 4220 (V) and 3530 (U), cover an enormous interval, from the infrared to the ultraviolet and, as a result, striking differences are shown by the various light curves.

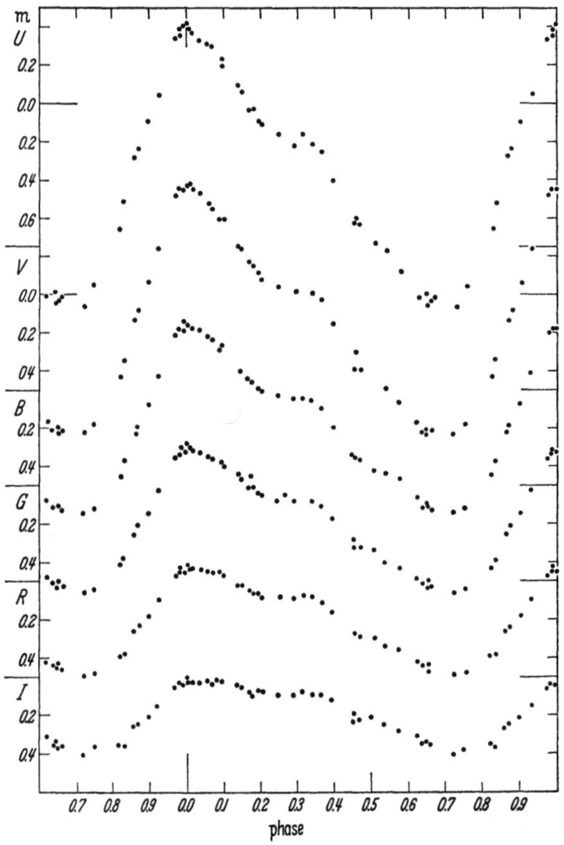

Fig. 21. Light curves in six colours of η Aquilae according to STEBBINS and WHITFORD.

With regard to the study of the continuous spectrum the advantage of the six-colour photometry is the reliability of the magnitude scales. An excellent comparison of the Cepheids with non-variable stars is possible because STEBBINS and WHITFORD have measured hundreds of such stars with the same photometer and have established moreover a careful calibration of the colours against the temperature. It is regrettable that hitherto six-colour observations of only three Cepheids are available.

The only disadvantage of the photoelectric observations as compared to the photographic spectrophotometry is that the influence of absorption lines is not eliminated. This makes a comparison of the observations with theoretically computed spectral intensity distribution difficult.

In Fig. 21 the six light curves of η Aquilae are shown. The variations in colour, shown by the large differences in amplitude, have given rise to several discussions of the variation in radius and temperature which will be reviewed in Sects. 112 and 113.

The amplitude of the ultraviolet light curve (U) is not as large as might be expected from the general increase of the amplitudes towards shorter wavelengths. This is caused by the continuous Balmer absorption, which is strongest at maximum light. In the case of RR Lyrae the amplitude of the curve U is even smaller than that of the curve V for the same reason. However, a special effect was

[1] R. CANAVAGGIA: Ann. d'Astrophys. **12**, 96 (1949).

discovered by HARDIE[1] in the ultraviolet light curve of RR Lyrae. This consists of a short lasting excess of light of about $0^m.1$, which occurs during the rise of brightness.

The six-colour observations are particularly useful for the study of interstellar reddening, as was pointed out by HARRIS[2] and CANAVAGGIA[3]. Whereas in two-colour observations the effect of space reddening cannot be distinguished from intrinsic reddening caused by a change in temperature, the two effects can be separated in the six-colour observations. It was found in this way that η Aquilae is more reddened, by $0^m.23$ on the $V-I$ scale, than δ Cephei. It follows then, that the weakening of the visible by interstellar absorption is larger for η Aquilae than for δ Cephei, by an amount of $0^m.3$. How large the absorption is for δ Cephei is not known.

Considering that usually δ Cephei and η Aquilae are used as standards, it becomes clear how uncertain our knowledge of the intrinsic colours and of the luminosities of the Cepheids is and more observations of the six-colour type are needed.

24. The line spectrum: comparison with non-variable supergiants, variations with phase, line asymmetry, double lines, emission lines, molecular bands, differential velocities, metallic lines, hydrogen lines, *H* and *K* lines. The investigation of the spectral lines of Cepheids was started in 1894 by BELOPOLSKI[4], who discovered the variability of the radial velocity of δ Cephei.

The variations of the intensity of lines in the spectrum of η Aquilae were discovered by DUNCAN[5].

Systematic studies of the variations in the spectra of Cepheids were carried out by ADAMS and JOY at the Mt. Wilson Observatory and by SHAPLEY and his collaborators at Harvard.

Each group developed its own system of classification with somewhat different criteria.

According to the Harvard classification the spectral type of the Cepheids varies appreciably with phase, the earliest type occurs at maximum light, the latest type at minimum light. It was found that the range of variation of spectral type is related to the amplitude of the light curve and that the mean type shifts towards a later class with increasing period. In short, most of the relationships described in Sect. 20 were established at Harvard.

Different results were found with the Mt. Wilson classification scheme. The spectral type as given by ADAMS and JOY[6] for the Cepheids is roughly the same for all periods and shows practically no variation with phase. At maximum light the average spectral type is $F9$ and at minimum it is $G0$.

However, instead of variations in type, variations in luminosity were indicated by the criteria. According to these criteria the variables should be more luminous at maximum light and by roughly the same amount as shown by the light curve.

ADAMS and JOY noticed that the hydrogen lines in the spectrum of the Cepheids are abnormally strong, especially at maximum light, as compared with the spectrum of non-variable stars.

If the spectra are classified by means of the hydrogen lines alone, then at maximum light type $F1$ is found and at minimum, $F7$.

---

[1] R. H HARDIE: Astrophys. Journ. **122**, 256 (1955).
[2] D. HARRIS: Astrophys. Journ. **119**, 297 (1954).
[3] R. CANAVAGGIA: C. R. Acad. Sci., Paris **238**, 2390 (1954).
[4] A. BELOPOLSKI: Bull. Acad. Imp. Sci., St. Petersbourg **1**, 267 (1894).
[5] J. C. DUNCAN: Lick Obs. Bull. **5**, 93 (1909).
[6] W. S. ADAMS and A.H. JOY: Comm. Mt. Wilson Obs. **1918**, No. 53.

In an excellent discussion of the spectra of Cepheid variables, STRUVE [25] gives a detailed description of the discrepancies and more or less solves the problem in the light of his own work.

STRUVE used the classification scheme developed at the Yerkes Observatory by MORGAN, KEENAN and KELLY [26] for the determination of the spectral types and their variations with phase in the case of seven Cepheids in Cygnus. Some of the results follow.

The spectral types vary appreciably with phases. At minimum light the spectra are almost indistinguishable from the spectra of non-variable standard stars of luminosity class Ib (supergiants). At maximum light the spectra do not match any of the spectra of standard stars; they are abnormal. This distinction between maximum and minimum appears also in the differences between radiative and colour temperature at those phases as given in Table 3. There also the minimum appears to be more normal.

The hydrogen lines at maximum are abnormally strong, as are also, but in a lesser degree, the lines of ionized titanium. The criteria indicate a luminosity class intermediate between Ia and Ib, that is they imitate higher luminosity at maximum light.

The changes in appearance of the CH molecular band with phase are striking and changes of the intensity of the metallic lines are certainly real. It is seen that STRUVE'S conclusions confirm the results of the Harvard classifications as well as most of ADAMS and JOY'S results. The essential point is that the spectrum of the Cepheids at maximum does not occur among the normal stars and, consequently, classification systems, calibrated for the normal stars, may give discordant results for the Cepheids.

The strengthening of the hydrogen lines near maximum light is convincingly shown in Fig. 22 which represents two microphotometer tracings of small dispersion spectra of δ Cephei[1]. At maximum the hydrogen lines dominate and the Balmer absorption beyond λ 3700 is strong while the metallic lines, excepting *H* and *K* of Ca II, are weak. At minimum the hydrogen lines are nearly submerged in the strong crowding metallic lines, and the *H* and *K* lines are very intense.

It should be remarked that due to the crowding of the lines, the spectrum between λ 3700 and the *H* and *K* lines is not used in normal classification work, which rests mainly on the region from the *H* and *K* lines up to λ 4800. For example, the ratio of the intensities of the lines Sr II, λ 4077 and H δ which shows an appreciable change in Fig. 22 is one of the criteria for luminosity in MORGAN and KEENAN'S system.

Also the spectra of the RR Lyrae stars deviate from the spectra of normal stars. In this case also, the hydrogen lines spoil the resemblance as they are too weak for the spectral types derived from the metallic lines. It is to be noted that the deviation is opposite to that for the Cepheids.

From a large number of spectra STRUVE and BLAAUW[2] estimated the spectral type and its variations with phase in the case of RR Lyrae. At maximum light the spectral type is the earliest and is of class *A* 2 according to the appearance of the metallic lines and the Ca II lines. The latest spectra occur shortly before minimum light and are of class *A* 8 when the amplitude of the light curve is low and of class *F* 1 when the amplitude is large. It is remarkable that in the light curves it is primarily the magnitude of maximum during the 41-day cycle (cf. Sect. 22) which changes and not that of the minimum, whereas the spectral type behaves just in the opposite way.

---

[1] TH. WALRAVEN: Hemel en Dampkring, p. 91. 1944.

[2] O. STRUVE and A. BLAAUW: Astronom. J. **108**, 60 (1948).

The spectral types derived from the appearance of the hydrogen lines show the same behaviour, but are systematically later by about half a spectral class.

In agreement with the results from the spectrum, STEBBINS [*24c*] found that the six-colour photometry of RR Lyrae suggests also that the continuous Balmer absorption is weaker than in ordinary stars of the same colour class.

The results of STRUVE and BLAAUW agree with earlier investigations by MÜNCH and TERRAZAS[1]. More recently the spectra of 15 RR Lyrae stars were investigated by IWANOWSKA[2] who distinguished two kinds of spectra. In group II, which according to IWANOWSKA represents Population II stars, the Balmer lines

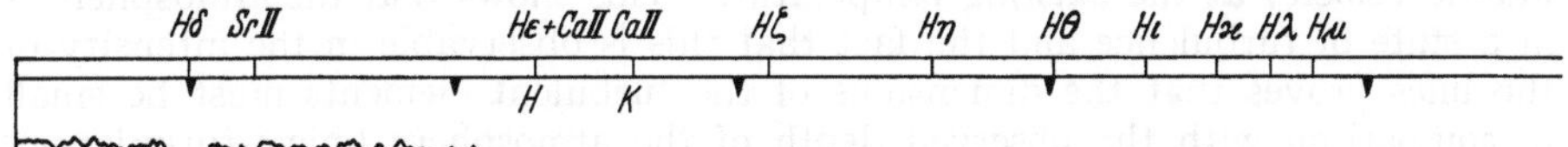

Fig. 22. Photometer tracings of the blue-ultraviolet part of the spectrum of δ Cephei. Top: maximum light; bottom: minimum light.

and continuum are very much weakened as compared with standard stars. The lines of Sc, V, Ti, Eu and some others are stronger than normal. These spectra present some analogy with the spectra of high-velocity stars.

The spectra of the other group have a more normal appearance. Most of the above-mentioned results are based on direct estimates or simple measurements in medium-dispersion spectrograms and the relation of the estimated quantities with the physical parameters of the star can be established only through an empirical calibration and comparison with stars for which the parameters are known already.

This problem can be solved more directly by measuring carefully the strength or the profiles of the absorption lines in large-scale spectrograms. The measured intensities, or equivalent widths, are converted by means of the curve of growth into numbers of absorbing atoms in various stages of ionization and excitation in the atmosphere of the star. By applying the theories of thermodynamical equilibrium, the temperature and the electron pressure are computed from the

[1] G. MÜNCH and L. R. TERRAZAS: Astrophys. Journ. **103**, 371 (1946).
[2] W. IWANOWSKA: Trans. Internat. Astronom. Union **8**, 814 (1952).

degree of ionization and excitation shown by these numbers of atoms. With the help of the model atmospheres the density of the gases and the gravity can be computed.

The theories involved in this work can be found in textbooks such as that of UNSÖLD [27] or ALLER [28]. The method has been applied to δ Cephei [29a], η Aquilae [29b], S Sagittae, T Vulpeculae [29c], RR Lyrae [29d] and W Virginis [29e]. In all cases, the curve of growth, i.e. the relation between the intensity of the lines and the numbers of atoms, indicates that the average velocity of the atoms is of the order of 5 km/sec, which is many times larger than the kinetic velocity at the existing temperature. This shows that the atmosphere is in a state of turbulence and the fact that this is observable in the intensity of the lines proves that the dimensions of the turbulent elements must be small in comparison with the observed depth of the atmosphere (micro-turbulence). The phenomenon is also shown in the spectra of many non-variable supergiants.

The SCHWARZSCHILDS and ADAMS, using large-dispersion spectrograms (2.9 Å per mm) of η Aquilae, concluded from the profiles of the lines that in addition turbulence on a larger scale (macro-turbulence) takes place, involving masses of gas, with an average speed of 12 km/sec. Since this turbulence has no effect on the intensities of the lines, the dimensions of the elements must be comparable to the observed depth of the atmosphere.

It was found for η Aquilae [29b] and for δ Cephei[1] that the micro-turbulence varies with phase being stronger by a few km/sec at minimum light.

In the case of one element these numbers give the relative population of different levels of excitation from which the temperature can be computed using BOLTZMANN'S formula. The temperatures derived in this way for different elements show a good agreement between themselves and their variation resembles the variation of the colour-temperature. On the other hand, the numbers of neutral and ionized atoms of the same element give the temperature and pressure and their variations by means of SAHA'S formula. This temperature variation also runs parallel to the variation of colour-temperature.

It appears that complicated details in behaviour of different lines can be accounted for completely by the differences in ionization and excitation potentials.

The remaining exceptions are all cases where it is obvious that a more refined theory is required. For instance, the hydrogen lines are formed at much higher levels than the metallic lines and must be discussed on the basis of the Stark effect. Other examples are provided by the Eu II lines which show an abnormally large variation in strength, and the molecular lines of CH and CN.

A very interesting discovery was made by STRUVE[2], and independently by SANFORD[3], in the spectrum of RR Lyrae. On some occasions emission lines in H_α and/or H_β appear at the phase about half way up the rising branch of the light curve but only during a short time, not more than 30 minutes. The emission is shifted to the violet side of the normal position of the line by about 20 km/sec. Immediately after the disappearance of the emission on the violet side of the H_α absorption line which is displaced 75 km/sec towards the red a second absorption appears, which rapidly gains strength, whilst the original line becomes weaker. The separation of the components is 150 km/sec. The original line gradually fades out and during the remainder of the cycle the new line gradually shifts towards the red until on the next following rising branch the phenomenon is

[1] J. HITOTUYANAGI and H. INABA: Sci. Rep. Tôhoku Univ. **36**, 321 (1952).
[2] O. STRUVE: Publ. Astronom. Soc. Pacific **59**, 192 (1947).
[3] R. F. SANFORD: Astrophys. Journ. **109**, 208 (1949).

repeated. The duplicity of the line lasts about 80 min; it ends about at maximum light. This observation means that for H_α the descending part of the radial velocity curve is missing and the transition from maximum to minimum velocity is discontinuous. The same is true for the Ca II lines but not for the other lines in the spectrum.

STRUVE noticed that the H_α emission line appears only at a certain phase of the 41-day modulation period, i.e. where the retardation in phase of the light curve is greatest. At the same time, the light curve shows a standstill in the middle of the rising branch, which is less pronounced or absent at other phases of the 41-day modulation period.

XZ Cygni, another RR Lyrae star extensively observed by STRUVE and VAN HOOF[1] shows doubling of the hydrogen line at median rising brightness, but no emission has been observed. According to MULLER[2] this variable shows no standstill or hump on the rising branch.

The phenomena observed in the spectrum of RR Lyrae present themselves more obtrusively in W Virginis, the prototype of the Population II Cepheids. The spectrum has been studied by SANFORD[3] and ABT [29e]. In Fig. 23 the light curve of this variable is shown, it is quite abnormal for Population I Cepheids with similar periods. The colour curve has a more normal appearance.

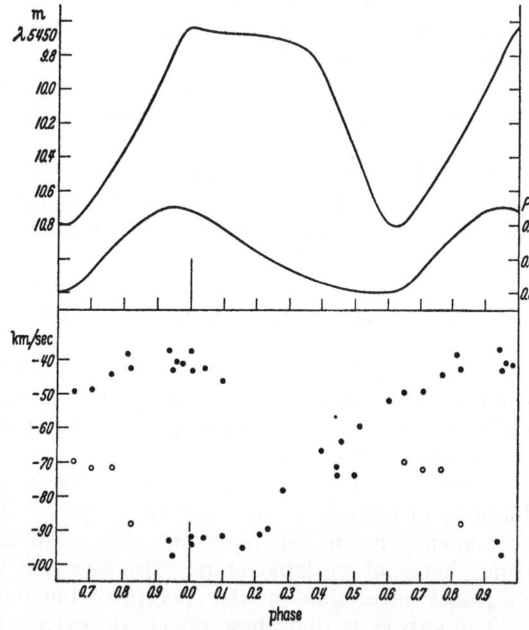

Fig. 23. Upper part. Light curve and colour curve of W Virginis. Lower part. Dots: radial velocity curve from absorption lines; open circles: radial velocity of emission lines.

The radial velocity curve is discontinuous, not only for the hydrogen lines but also for many other lines in the spectrum. The weakening line, displaced towards the red, and the new-born line which is violet-displaced are present simultaneously during about a fifth of a cycle at maximum brightness. As in the case of RR Lyrae the duplicity of the lines is preceded by emission, although in W Virginis the emission is much stronger and lasts longer. The emission lines already appear while the light is at minimum, attain maximum strength about halfway on the ascending branch and disappear shortly before maximum brightness.

According to ABT the complete history of the hydrogen line H_δ is as follows. Just before minimum light the absorption line is sharp and displaced redward with respect to the mean position by 20 km/sec. At minimum light an emission appears as a diffuse line, shifted about 10 km/sec to the violet. The emission line is divided by the sharp absorption line into two portions of which the one on the violet side is stronger. While the brightness of the star increases the

[1] A. VAN HOOF: Astrophys. Journ. **109**, 215 (1949).
[2] A. B. MULLER: Bull. astronom. Inst. Netherl. **12**, 11 (1953).
[3] R. F. SANFORD: Astrophys. Journ. **116**, 331 (1952).

emission becomes stronger and wider and shifts more towards the violet, while the narrow absorption line shifts redward. At maximum light, when the emission has become weaker already, weak absorption wings appear on both sides extending up to 9.5 Å from the centre of the line. It seems that now we have a superposition of three lines, a very wide absorption line and a less diffuse emission line both shifted towards the violet by about 25 km/sec, and the original sharp absorption line shifted by the same amount towards the red.

When the emission line disappears, the violetward widewinged absorption line and the redward sharp absorption line remain. The former becomes deeper and narrower, while the other fades out. During the remainder of the period the single line left over shifts towards the red while it becomes sharp.

The behaviour is not exactly the same for all cycles, likewise the shape of the light curve is variable but the observations do not allow to detect a certain regularity in these changes.

Another Cepheid of population type II with similar properties is RU Camelopardalis (period $22\overset{d}{.}1$). The shape of its light curve is similar to that of W Virginis, showing a strong hump on the descending branch and a sharp minimum. The period and the shape of the light curve are subject to appreciable changes[1] (cf. Fig. 17). Miss JEHOULET[2] described the appearance of wide hydrogen emission lines during the rise of brightness. The structure of these lines varies from cycle to cycle. Certain lines especially that of Sr II, λ 4077 are doubled at maximum light, the two components showing a relative displacement of 120 km/sec. The violet component is sharp. Many other lines are broadened at the same phase. Violet-displaced emission lines appear in the centres of the broad K line of Ca II at maximum light and later.

From the foregoing it follows that the formation of emission lines and the doubling of lines is a characteristic property of Population II variables. These phenomena give us an important clue as to the processes taking place in the atmospheres of variable stars. The question then arises why the doubling of lines does never show in the spectra of the Population I Cepheids.

The answer is that these effects do exist, although much less spectacularly, in the spectra of Population I variables. For example, the difference between the discontinuous velocity curves of some of the hydrogen lines and the Ca II lines which have large amplitudes and the continuous curves with lower amplitude of the weak metallic lines observed in RR Lyrae is simulated by a difference in amplitude of the velocity curves of the corresponding lines in the Cepheids.

In the last thirty years, all kinds of level effects sometimes contradictory have been reported[3]. The great number of such investigations arose from the desire to get some information on the type of wave motion in the atmosphere.

More recently the subject was taken up again by JACOBSEN[4] who reported well-pronounced level-effects in the velocity curves of δ Cephei and η Aquilae. In both cases, the velocity curves of the high level lines such as H_β, H_γ, H_δ, Ca I λ 4227 Sr II λ 4077, λ 4216 (Ca II is not observed) have a larger amplitude than the low level lines, mostly due to neutral metals. The differences are of the order of 5 km/sec. For both stars, JACOBSEN indicated a rate of increase of the

[1] F. LENOUVEL and D. JEHOULET: Ann. d'Astroph. **16**, 139 (1953).

[2] D. JEHOULET: Bull. Acad. Roy. Belg., Cl. Sci. **40**, 377 (1954).

[3] R. F. SANFORD: Astronom. J. **66**, 170 (1927); cf. Publ. Michigan **4** (1932) which contains papers by W. C. RUFUS, J. A. ALDRICH and D. W. LEE on the same subject. — V. HASE: Poulkovo Mitt. **11**, 6, 345 (1928). — H. A. BRUCK and H. E. GREEN: Monthly Notices Roy. Astronom. Soc. London **101**, 376 (1942).

[4] T. S. JACOBSEN: J. Roy. Astronom. Soc. Canada **43**, 142 (1949); **44**, 104 (1950).

amplitude which is 1 km/sec per 1800 km increase in height of level. The height of level in this case refers to the Sun, as given by MITCHELL[1] and not to the atmosphere of the variables.

The high level lines in this case are the same that show most clearly duplicity in the Population II variables.

On the other hand, several recent investigations[2] on large dispersion spectra failed to show any appreciable effects amongst lines originating in the main body of the atmosphere.

In a recent paper, JACOBSEN[3] has reviewed the problem again in the case of δ Cephei and η Aquilae and concluded that the level effect is practically unobservable except for the cores of the strong *H* and *K* lines of Ca II which show a greater asymmetry, about 50 to 80% greater amplitude than the other lines and lag slightly in phase.

Also a rudimentary counterpart of the emission lines can be observed in the Population I Cepheids. Hitherto no observations of hydrogen emission lines have been published but numerous observations have been made of weak emission lines in the centre of the Ca II *H* and *K*-lines. ADAMS and JOY[4] observed such emissions in the spectrum of ζ Geminorum. VAN HOOF[5] found emission in the spectrum of X Cygni twice during a cycle, first at the standstill on the rising branch of the light curve and secondly, just after maximum light. JACOBSEN[6] observed the emission in the spectrum of η Aquilae at the middle of the rising branch.

In JOY and WILSON's list[7] of stars with bright *H* and *K* lines, several Cepheids occur. In all cases, the emission line is displaced toward the violet and if it is strong it is divided in two parts of which that on the violet side is much stronger.

The phenomenon is accurately described by HERBIG[8] who found the emission in the spectrum of S Sagittae, again midway up the rising branch of the light curve.

The emission line is placed at the bottom of the very wide and deep absorption line. It is split up, as in the case of H_α in W Virginis, by a narrow absorption line displaced toward the red so that apparently two components of emission are left of which the violet one is much stronger.

In order to find out whether the occurrence of emission in the Ca II lines is a general characteristic of Cepheids, HERBIG[9] made a search for it and he reached the conclusion that the temporary presence of shortward displaced emission lines of Ca II is a general characteristic of Cepheids, with sufficiently large amplitudes of light variation, and periods between five and sixteen days.

Although many non-variable stars of spectral class *G* or later show bright Ca II lines, as can be judged from the long list compiled by JOY and WILSON, the systematic displacement towards the violet of these lines in the Cepheids and their short-lived appearance on the middle of the rising branch of the light curve suggests strongly that the cause of the phenomenon may be of the same nature as that present in the Population II variables.

---

1 S. A. MITCHELL: Astrophys. Journ. **38**, 407 (1913).
2 Cf. [*29b*] and also J. GRANDJEAN: Mém. Acad. Roy. Sci. Belg. **29**, 3 (1956).
3 T. S. JACOBSEN: Dom. Astrophys. Obs. **10**, 145 (1956).
4 W. S. ADAMS and A. H. JOY: Publ. Amer. Astr. Soc. **9**, 254 (1939).
5 A. VAN HOOF: Astrophys. Journ. **108**, 160 (1948).
6 T. S. JACOBSEN: Publ. Astronom. Soc. Pacific **62**, 269 (1950).
7 A. H. JOY and R. E. WILSON: Astrophys. Journ. **109**, 231 (1949).
8 G. H. HERBIG: Astrophys. Journ. **116**, 369 (1952).
9 G. H. HERBIG: Publ. Astronom. Soc. Pacific **64**, 302 (1952); cf. also R. P. KRAFT: Astronom. J. **62**, 21 (1957).

b) β Cephei stars.

25. Characteristics of velocity curve and light curve. The β Cephei stars form a small homogeneous group of bright variable stars with very interesting properties. Already in 1906, FROST[1] announced that β Cephei shows large variations in radial velocity, with a total amplitude of 34 km/sec taking place in the amazingly short period of $4^h 34^m 11^s$.

In 1913, GUTHNICK[2] discovered that the brightness of this star varies by 0.05 magnitude in the same period. Within a few years, three other bright stars with similar properties were discovered, β Canis Majoris (1910), σ Scorpii (1916) and 12 Lacertae (1915). In recent times, the name β Canis Majoris has often been used but since β Cephei was the first to be discovered, it is more correct historically to keep it as the prototype of the class.

On account of large and rapid variations in radial velocity, 22 stars were listed by HENROTEAU[3] in 1928 as being members or probable members of the group. At present, 11 stars are considered as really belonging to it.

Very little progress was made in the investigation of these stars until, in 1932, MEYER[4] discovered a double periodicity in the velocity variations of β Canis Majoris. Since then, our knowledge of the properties of the β Cephei stars has increased more rapidly due mainly to the work of STRUVE and his collaborators and, at present, a widespread interest is shown in this type of variable stars[5].

STRUVE [30] has given an excellent summary of the properties of the β Cephei stars in a paper entitled: "An Interesting Group of Pulsating Stars". This means that at present, they are considered as physical variables in the same sense as the Cepheids or RR Lyrae stars.

The numerical data given in STRUVE'S article are reproduced in Table 4, which clearly shows the homogeneity of the group.

The spectral types, the colour and the absolute magnitude M_v are confined within narrow limits and are even correlated with the periods although these vary only over a small range: 4 to 6 hours.

The amplitudes $2K$ of the radial velocity curves and the amplitudes Δm of the light curves vary appreciably, but it is possible that this means only that one star is pulsating more strongly than the other although MCNAMARA[6] has suggested a systematic relation between period and amplitude of the velocity curve.

Five stars of the list are subject to a modulation or periodic variation in the amplitude of the radial velocity curve which can be interpreted as resulting from the interference of two oscillations with nearly equal periods P_0 and P_0'. This was first discovered in the case of β Canis Majoris by MEYER[7] who resolved the two oscillations present in the velocity curve and found that both components were sinusoidal in shape. The two periods $P_0 = 6^h 0^m$ and $P_0' = 6^h 2^m$ produce a beat period of 49 days. The amplitudes of the components are respectively $2K_0 = 12$ km/sec and $2K_0' = 6$ km/sec. The observed amplitude of the velocity curve varies between the sum and the difference of these two values. The corresponding data for the four other variables are given in Table 4. In

[1] E. B. FROST: Astrophys. Journ. **15**, 340 (1902); **24**, 259 (1906).
[2] P. GUTHNICK: Astronom. Nachr. **196**, 357 (1913).
[3] F. C. HENROTEAU: Handbuch der Astrophysik, Vol. 6, p. 436. 1928.
[4] W. F. MEYER: Publ. Astronom. Soc. Pacific **46**, 202 (1934).
[5] One member of the class, 12 Lacertae has given rise recently, under the leadership of Dr. DE JAGER to one of the most effective collaboration program of observations ever realized in this field and it is to be hoped that this example will be followed.
[6] D. H. MCNAMARA: Astrophys. Journ. **122**, 95 (1955).
[7] W. F. MEYER: Publ. Astronom. Soc. Pacific **46**, 202 (1934).

Table 4. *β Cephei stars.*

Star	P_0	P_0'	$2K_0$ km/sec	$2K_0'$ km/sec	Δm_0	$\Delta m_0'$	Spectral Type	Color	M_v	Line profile	Rotational velocity
β CMa	6^h0^m	6^h2^m	12	6	0.03	—	*B* 1 II–III	−0.280	−4.7	variable	large
σ Sco	5^h44^m or 6^h7^m	5^h55^m	15	110	—	0.08	*B* 1 III		−4.3	variable	large
ξ^1 CMa	5^h2^m	—	36	—	0.01 or 0.045	—	*B* 1 IV	−0.280	−4.2	constant	small
BW Vul	—	4^h49^m	—	150	—	0.19–0.26	*B* 2 III	−0.270	−4.1	variable	average
12 DD Lac	4^h44^m	4^h38^m	15	36	0.042	0.074	*B* 2 III	−0.265	−4.1	variable	average
β Cep	4^h34^m	—	18–46	—	0.02–0.05	—	*B* 2 III	−0.275	−4.1	constant	small
15 Can Maj.	4^h26^m	—	7	—	0.01	—	*B* 1–*B* 2 III	—	—	constant	—
ν Eri	4^h16^m	4^h10^m	22	49	0.067	0.114	*B* 2 III	−0.255	−4.1	variable	average
16 EN Lac	4^h6^m	4^h4^m	9	30	0.035	0.055	*B* 2 IV	−0.260	−3.3	variable	small
δ Cet	3^h52^m	—	13	—	0.025	—	*B* 2 IV	−0.245	−3.3	constant	small
γ Peg	3^h38^m	—	7	—	0.015	—	*B* 2 IV	−0.240	−3.0	constant	small

the case of three stars a modulation is also observed in the light curve[1] which is correlated with the modulation of the velocity curve i.e. both curves have large or small amplitudes at the same time. In these cases, the light curve can be represented also as the sum of two components with periods P_0 and P_0' and amplitudes Δm_0 and $\Delta m_0'$. The relation between $2K$ and Δm for the stars which show no modulation neither in the light curve nor the velocity curve is somewhat different from the relation between $2K_0$ and Δm_0 and between $2K_0'$ and $\Delta m_0'$ for the components of the doubly periodic variables as shown in Fig. 24. The light curves of the β Cephei stars are always nearly sinusoidal in shape and they lag 90° in phase behind the radial velocity curves[2]. As was demonstrated by DE JAGER this holds also for each of the components of the doubly periodic variables. For each oscillation, maximum brightness occurs halfway on the descending branch of the velocity curve. Table 5 provides a comparison between the β Cephei stars and the RR Lyrae stars and Cepheids with double periodicity.

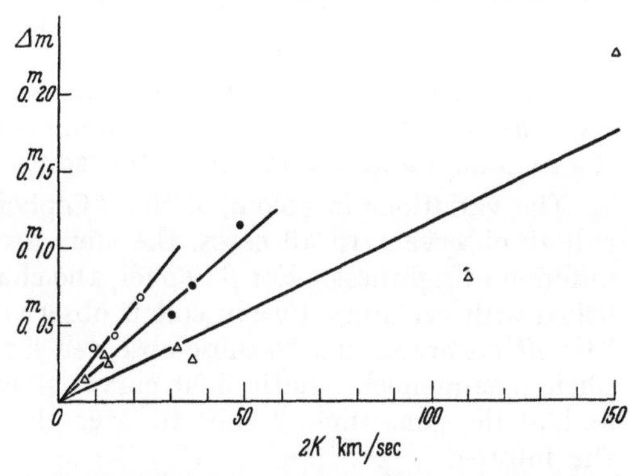

Fig. 24. Correlation between amplitude $2K$ of radial velocity curve and amplitude Δm of light curve of β Cephei stars. Triangles: variables with one period; open circles: component P_1; dots: components P_2 for doubly periodic variables.

The table shows that, in nearly every respect, the β Cephei stars are different from the Cepheids and RR Lyrae stars. The distinction in phase lag is especially noteworthy. The 180° phase lag in the case of the Cepheids and RR Lyrae stars means that, if the variations in radial velocity are interpreted as due to a radial pulsation, maximum and minimum brightness both occur when the radius has

[1] (a) C. DE JAGER: Bull. astronom. Inst. Netherl. **12**, 81, 88, 91 (1953). — (b) M. F. WALKER: Astrophys. Journ. **116**, 81 (1952). — (c) M. F. WALKER: Astrophys. Journ. **116**, 99 (1952).

[2] Cf. Ref. 1, (a), this page. — A. D. WILLIAMS and O. STRUVE: Publ. Astronom. Soc. Pacific **67**, 250 (1955). — A. D. WILLIAMS: Publ. Astronom. Soc. Pacific **66**, 200 (1954). — D. H. MCNAMARA: Astrophys. Journ. **122**, 95 (1955).

Table 5. *Comparison of doubly periodic variables.*

Property	Cepheids with periods of 2 to 3 days	RR Lyrae stars	β Cephei stars
Asymmetry of L-curve	$\propto \Delta m$ both in phase and in magnitude	strong, $\propto \Delta m$ both in phase and in magnitude	very weak
Asymmetry of V_R-curve	$\propto 2K$ in phase only	$\propto 2K$ in phase only	very weak
Phase lag of L-curve behind V_R-curve	180°	180°	90°
$2K/\Delta m$	$\approx$ 35 km/sec mag.	$\approx$ 64 km/sec mag.	$\approx$ 600 km/sec mag.
$\frac{2K_0'}{\Delta m_0'} \Big/ \frac{2K_0}{\Delta m_0}$	?	1	>1
$\frac{P_0'}{P_0}$	<1	<1	4 cases <1 1 case >1
$\frac{2K_0'}{2K_0}$ and $\frac{\Delta m_0'}{\Delta m_0}$	<1	<1	4 cases >1 1 case <1
$\frac{\text{Beat period}}{\text{Main period}}$	2.5 (TU Cas and TTrA)	60 to 1400 (RR Lyrae stars) 3.5 (AI Vel)	30 to 200

its mean value. On the other hand, the 90° phase lag in the case of the β Cephei stars means that maximum and minimum brightness coincide respectively with the minimum and maximum of the radius.

The variations in colour of the β Cephei stars are extremely small and difficult to observe[1]. In all cases, the stars are slightly bluer at maximum than at minimum brightness. For β Cephei, the character of these variations was established with certainty by six colour observations made by STEBBINS and KRON[2]. The effects are similar to those observed for Cepheids such as η Aquilae and δ Cephei. The monochromatic light curves show a systematic decrease in amplitude and, at the same time, a shift to later phases in passing from the ultraviolet to the infrared.

26. Line spectrum. Variable broadening, doubling of lines. The spectrum of the β Cephei stars is normal and of type $B1$ or $B2$. As it is often the case with B stars, the spectral lines may be broadened by rotation. The last column of Table 4 contains the rotational velocity as evaluated by STRUVE: small, average and large correspond to equatorial velocities of the order of 15, 30 and 60 km/sec respectively. Normal B stars may show much faster rotations ($\approx$ 200 km/sec). The variations in spectral type during the period are too small to be observed with certainty. In BW Vulpeculae which shows the largest variations in light and velocity, the hydrogen lines H_γ and H_δ are somewhat stronger near minimum light indicating a slightly lower temperature. The lines H_α and H_β show a weakening near minimum light which is probably caused by emission which fills up the absorption cores[3].

[1] Cf. Footnote 1, p. 399. — V. B. NIKONOV and NIKONOVA: Dokl. Akad. Nauk. USSR. **9**, 135 (1952). — D. H. MCNAMARA and A. D. WILLIAMS: Astrophys. Journ. **121**, 51 (1955).

[2] J. STEBBINS and G. E. KRON: Astrophys. Journ. **120**, 189 (1954).

[3] G. J. ODGERS: Publ. Dom. Astrophys. Obs. **10**, 215 (1956).

The most interesting aspect of the spectral variation is the variable broadening of the lines first noted by HENROTEAU in the spectrum of β Canis Majoris. It turned out that such variable line profiles are commonly found in the spectrum of β Cephei stars and are correlated with the presence of a double periodicity as can be seen in Table 4. The four stars with constant line profiles, ξ^1 Canis Majoris, β Cephei, δ Ceti and γ Pegasi, all have small rotational velocity and single periods.

It is very remarkable that the variations in the line profiles take place with the period P_0' only. Except for β Canis Majoris, this period corresponds also to the oscillation with the largest amplitudes $2K_0'$ and $\Delta m_0'$. It may be that the other oscillation also produces changes in line profiles which are too small to be detected.

Although the line broadening has not yet received a completely satisfactory explanation, it shows that the separation of the observed velocity curve and light curve into two components P_0 and P_0' has a real physical meaning.

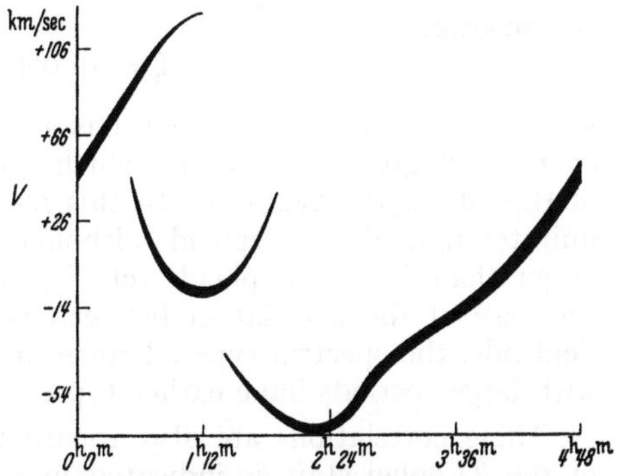

Fig. 25. Schematic velocity curve of BW Vulpeculae, after STRUVE. Thickness of the curves proportional to the intensity of line components.

In the case of BW Vulpeculae, large dispersion spectra[1] showed that the change in the line profiles usually observed as a broadening consists really of a splitting of the lines into two components. On small scale spectra or in the case of large rotational broadening, this is observed as a varying diffuseness of the line. Actually, the splitting occurs twice as shown in Fig. 25. First a weak component appears which moves towards the violet side of the spectrum increasing in intensity at the same time. The original line moves towards the red and weakens. When the latter line has faded out the phenomenon is repeated. The effect resembles the doubling of the lines in the W Virginis stars and RR Lyrae stars but, in these cases, it takes place only once per cycle.

The double discontinuity in the velocity curve of BW Vulpeculae is also observed in 12 Lacertae[2] and σ Scorpii[3]. The moment at which the doubling or broadening occurs is not the same for all spectral lines as was discovered by VAN HOOF[4]. It is first observed in the lines of Si III and O II; a few minutes later, the helium lines are affected and finally, the hydrogen lines. This suggests that the discontinuity in velocity is propagated outwards through the atmosphere and reaches successively the different levels where the lines of Si III and O II, the helium lines and finally, the hydrogen lines are formed. HUANG[5] found that the sum of the equivalent widths of the components in the double line stage is equal to the equivalent width of the lines when single. This would seem to

[1] O. STRUVE: Publ. Astronom. Soc. Pacific **66**, 329 (1954). — G. J. ODGERS: Publ. Dom Astrophys. Obs. **10**, 215 (1956) where reproductions and analysis of very high dispersion spectra obtained by DEUTSCH are included.

[2] O. STRUVE: Astrophys. Journ. **113**, 589 (1951). — O. STRUVE and V. ZEBERG: Astrophys. Journ. **122**, 134 (1955).

[3] SU-SHU HUANG and O. STRUVE: Astrophys. Journ. **122**, 103 (1955).

[4] O. STRUVE and A. VAN HOOF: Publ. Astronom. Soc. Pacific **65**, 158 (1953). — A. VAN HOOF, M. DE RIDDER and O. STRUVE: Astrophys. Journ. **120**, 179 (1954). — O. STRUVE: Publ. Astronom. Soc. Pacific **67**, 173 (1955).

[5] SU-SHU HUANG: Publ. Astronom. Soc. Pacific **67**, 22 (1955). — SU-SHU HUANG and O. STRUVE: Astrophys. Journ. **122**, 103 (1955).

indicate that the double lines are not formed in two layers at different depths but rather in different areas on the surface of the star. But in that case, the distribution of the velocities should depart from the spherical symmetry characteristic of purely radial pulsation.

27. The period-luminosity relation. As shown already by Table 4, both the luminosity and the colour of the β Cephei stars are correlated with the period. As the logarithm of the period increases from about -0.8 to -0.6, the luminosity increases from about $M_v = -3.0$ to $M_v = -4.7$. BLAAUW and SAVEDOFF[1] who investigated the question in some details propose the following relation

$$M_v = -10 - 9 \operatorname{Log} P_0 \tag{27.1}$$

D. H. MCNAMARA[2] and R. M. PETRIE[3] arrived also at the conclusion that a definite period-luminosity relation exists. A least-square solution led PETRIE to the relation

$$M_v = +0.4 - (18.1 \pm 2.3)\, P_0 \tag{27.2}$$

which, in the range considered, has a slope practically identical to (27.1). However (27.2) gives magnitudes which are systematically $0^m\!.4$ fainter than those of BLAAUW and SAVEDOFF. In this respect, the behaviour of the β Cephei stars imitates that of the Cepheids although the slope of the relation (27.1) is much larger than for the Cepheids [cf. Eq. (19.2)]. There is no such resemblance in the case of the correlation between period and spectral type. While for the Cepheids, the spectral type advances as the period increases, the β Cephei stars with larger periods have earlier types[4].

These correlations are also apparent on Fig. 51 (p. 572) where the position of the β Cephei stars is indicated in a spectral type-magnitude diagram. The variables with the shortest periods fall nearly on the main sequence, those with larger periods are more luminous and fall in MORGAN's luminosity classes II to III showing giant characteristics.

c) Long-period variable stars.

28. Distribution as a function of periods and spectral classes. Spatial distribution. The class of the long-period variables cannot be defined precisely. The borderlines between regular long-period variables, semi-regular and irregular variables or between variable and non-variable red giant stars are very ill-defined. As a consequence, in the literature and the catalogues of variables stars, a certain confusion exists in the names of the sub-classes and in the assignment of the individual stars to these classes.

In this and in the following sections we use the name long-period variables or red variables for general purposes. If more specification is required the nomenclature of the General Catalogue of Variable Stars is followed in which the variables are distributed in four groups.

M, Mira-type: variables with amplitude of light variation greater than 2.5 magnitudes, good regularity and typical spectral types.

LP, Long-period variables: same properties as the Mira-type stars but amplitudes smaller than 2.5 magnitudes.

[1] A. BLAAUW and M. SAVEDOFF: Bull. astronom. Inst. Netherl. **12**, 69 (1953).
[2] D. H. MCNAMARA: Publ. Astronom. Soc. Pacific **65**, 286 (1953).
[3] R. M. PETRIE: J. Roy. Astronom. Soc. Canada **48**, 185 (1954).
[4] Cf. D. H. MCNAMARA: Publ. Astronom. Soc. Pacific **65**, 155 (1953).

SR, Semi-regular variables: more or less periodic variations with appreciable irregularities.

I, Irregular variables: slow variations in brightness in which no periodicity can be recognized.

The known long-period variable stars are more numerous than any other type of variables. In the General Catalogue of Variable Stars more than 4000 variables are named and accepted in the classe *M* and *LP*.

If all known variables are distributed according to the period a broad single-moded distribution is found with a maximum at the period $P = 270$ days. There is no doubt that this distribution is affected by selection effects and does not represent the real distribution. If, for example, only the brightest variables are taken, an appreciably different distribution is obtained, which has a much flatter maximum shifted towards longer period.

Several investigations have born out that the distribution in the galactic system is different for various groups of long-period variables and depends on the mean period.

Together with the mean period also the mean spectral type and the luminosity vary. C. PAYNE-GAPOSCHKIN [*1*] found that the mean distance from the galactic plane is 530 parsecs for the variables with periods shorter than 250 days and 230 parsecs for the variables with longer periods.

A similar result was found by KUKARKIN[1] who selected the Mira-variables with well-defined light curves and established a correlation between the mean distance from the galactic plane and the curve-type, according to LUDENDORFF (cf. Sect. 29).

The mean distance increases from 110 parsecs for the curve-type γ to 500 parsecs for the types β_2 and β_3.

Since the curve-types are correlated with the periods, KUKARKIN'S result means that the variables with longer periods are more concentrated towards the galactic plane.

On the average the concentration of the long-period variables is intermediate between that of the classical Cepheids and that of the RR Lyrae stars for which the respective mean distances from the galactic plane are about 50 and 740 parsecs [*1*].

The variables with short periods are more concentrated towards the galactic centre than those with longer periods. Altogether it seems that of the long-period variables those with the shortest periods have a distribution which approaches that of the population II objects, i.e. spherical distribution with a concentration in the galactic nucleus. The variables with the longest periods are distributed more nearly like the population I objects, i.e. in a flat disk around the galactic plane.

The motions of the long-period variables confirm this representation. Using MERRILL'S [*31*] numerous observations of radial velocities combined with proper motions and other useful data, R. E. WILSON and P. W. MERRILL [*32*] studied the space motions of about 170 long-period variables of spectral classes *Me* and *Se*. They found a correlation of velocity with period which is shown in Table 6.

For the periods shorter than 150 days the scatter in velocity is very large, several stars of this group have velocities of about 300 km/sec. According to the velocities shown in Table 6, the variables with short periods belong to the spherical system, the bulk of the variables belongs to an intermediate system.

[1] B. V. KUKARKIN: Astronom. Journ., USSR. **31**, 489 (1954).

It is to be noted that the mean distances from the galactic plane, computed from the velocities, are systematically larger, according to OORT, than the values given above, for the long-period variables as well as for the RR Lyrae stars.

Table 6. *Correlation of space velocity with period.*

Period days	Number	Mean period days	Mean velocity km/sec
<150	9	132	105
<200	23	159	125
150–199	14	173	139
200–299	51	254	93
300–399	56	344	56
>400	16	440	40

Tabelle 7.

Period	M_{vis}
150	−2.2
175	−2.7
200	−2.2
250	−1.4
300	−0.7
350	−0.2
400	+0.3
450	+0.6

In their paper, WILSON and MERRILL described also the correlations of spectral type and absolute magnitude with period. With increasing period, the mean spectral type advances gradually from *M* 1 *e* to *M* 8 *e*, i.e. the variables become increasingly cooler. The correlation between the visual absolute magnitude and period is given in Table 7, which shows that the luminosity decreases with increasing period, that is just opposite to the behaviour of the Cepheids.

However, one must not forget that the visual brightness of the cool red stars does not give a good impression of the total radiated energy. When a bolometric correction is applied to the visual magnitude of Table 7, the resulting bolometric absolute magnitude is nearly independent of period and is about $M_{bol} = -6$. This means that, in terms of total radiated energy, the long-period variables must be placed among the most luminous stars.

The variables with carbon-spectra show more or less the same tendencies as those with *M*-type spectra. In passing from the *R*-type to the cooler *N*-type stars, the mean period increases whilst the luminosity and space-velocity decrease.

Recently CAMERON and NASSAU[1] made a spectral survey of the red variables in a belt 12° wide along the galactic equator. Fig. 26 shows how these variables are distributed over spectral types and periods.

It is seen that for the Mira-type variables the spectral type is statistically correlated with period and advances with increasing period.

The study of the distribution of the numbers of variables of various spectral types as a function of galactic longitude yielded an interesting result. Whereas the semi-regular and irregular variables were evenly distributed along the galactic equator, the Mira-type and *LP* variables showed a marked maximum near galactic longitude 40°. The most pronounced peak is shown by the variables with long periods and spectral classes *M* 7–*M* 10. At this galactic longitude we are looking in the direction of a local spiral arm of the galaxy and therefore CAMERON'S and NASSAU'S observations suggest that the Mira-type variables of late spectral types are located in the spiral arms and consequently belong to the flat system of our galaxy.

From the foregoing, it follows that statistically the long-period variables, like the Cepheids or the RR Lyrae stars, can be arranged in a sequence according to period, which appears to be also a sequence of spectral types, of luminosities and of other characteristics.

[1] D. CAMERON and J. J. NASSAU: Astrophys. Journ. **124**, 346 (1956).

However, an important difference is to be noted. The sequence of the classical Cepheids is formed entirely of stars with sharply defined Population I characteristics. The W Virginis stars are all pure Population II objects, as are also the sequences of RR Lyrae stars found in globular clusters.

The long-period variables on the other hand are primarily of a population type intermediate between I and II. The sequence of these stars according to period is also a transition from nearly pure population type II through the intermediary types to nearly Population I.

Only a few Mira-type variables have been found in globular clusters. The average period of these variables is 200 days. It is noteworthy that they occur preferably in clusters, such as 47 Tucanae, which contain no RR Lyrae stars.

In the Small Magellanic Cloud, SHAPLEY and MCKIBBEN NAIL[1] found eleven long period variables of which five could be assigned to the Mira-class. The periods of these variables are rather long, they vary between 532 and 741 days. They must be of a class different from the galactic long-period variables, because they are at least 3 magnitudes brighter than the galactic variables with corresponding periods.

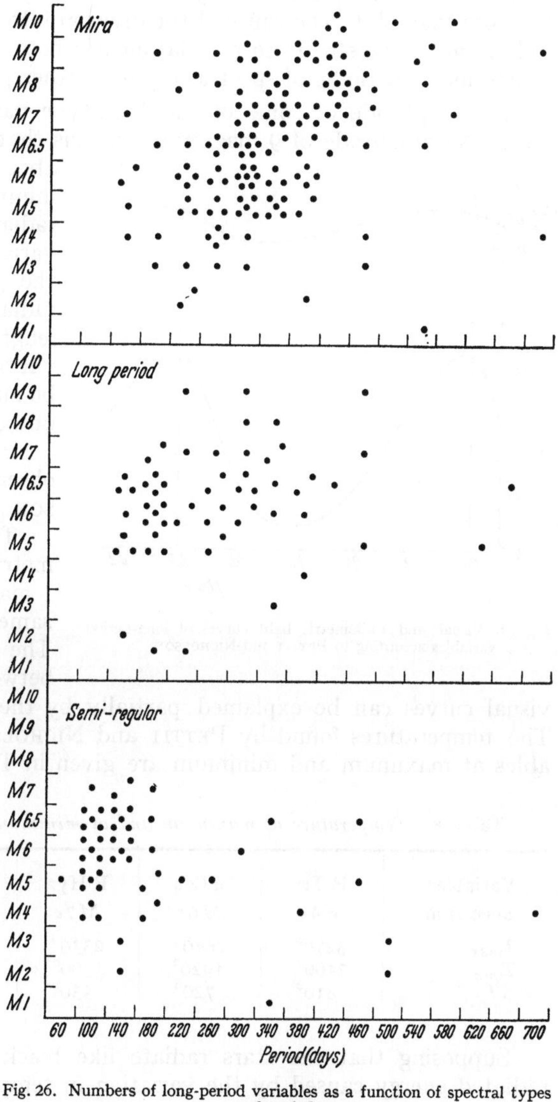

Fig. 26. Numbers of long-period variables as a function of spectral types and periods.

29. The form of the light curves and correlation with periods. The amplitudes of the light curves of the long-period variables vary within wide limits. For some variables as for example χ Cygni the visual range is 9 magnitudes, but any smaller amplitude may be found down to hardly observable variations in light.

The various amplitudes are not observed equally often. Variables with amplitudes of about 2.5 magnitudes are relatively rare, so that apparently the variables can be divided into two groups. The group with large amplitudes, or the Mira-

[1] H. SHAPLEY and V. MCKIBBEN NAIL: Proc. Nat. Acad. Sci. U.S.A. **37**, 138 (1951).

type variables, shows a maximum in number at the amplitude of 5 magnitudes. The number of variables with amplitudes smaller than 2.5 magnitudes increases strongly towards the smaller ranges. Probably a continuous transition towards the non-variable red supergiant stars takes place, as it was found by STEBBINS, HUFFER and WHITFORD[1] that 30 out of a sample of 32 red giants were variable by more than $0^m.1$. The cause of the division into groups according to the amplitude is not understood, no correlation has been found with the distribution as a function of periods, of spectral types or other properties of the stars.

The amplitude of some of the Mira-type variables may seem surprisingly large. An amplitude of 9 magnitudes means that the star at maximum light is about 4000 times brighter than at minimum. However, if the total radiated energy is considered the variation is much smaller. By far the greatest part of the energy is radiated in the infrared, at wavelengths of 1 or 2 microns. Observations in this spectral regions can be made by means of thermocouples or bolometers.

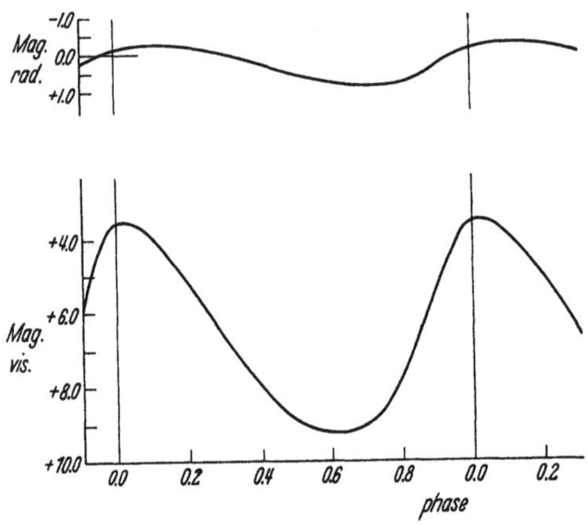

Fig. 27. Visual and radiometric light curves of long-period variables according to PETTIT and NICHOLSON.

Fig. 27 shows the results of such observations by PETTIT and NICHOLSON [33].

The amplitude of the radiometric light curve amounts to about 1 magnitude and is therefore of the same order as for the Cepheids. The large difference in amplitude between the radiometric and the visual curves can be explained partially by the laws of black-body radiation. The temperatures found by PETTIT and NICHOLSON for some long-period variables at maximum and minimum are given in Table 8.

Table 8. *Temperature at maximum and at minimum of some Mira-type variables.*

Variable	R Tri	o Cet	R Hya	R Aql	R Cnc	χ Cyg
Spectrum	*M4e*	*M6e*	*M7e*	*M7e*	*M7e*	
T_{max}	3270°	2640°	2330°	2320°	2320°	2240°
T_{min}	2460°	1920°	1900°	1920°	1700°	1640°
ΔT	810°	720°	430°	400°	620°	600°

Supposing that the stars radiate like black bodies the variation in total radiated energy caused by the variation in temperature can be computed with the STEFAN-BOLTZMANN'S law. For χ Cygni we find a variation of 1.35 magnitudes. The variations in the energy emitted by a black body at wavelength 0.6 μ, i.e. in the visual region, comes out at 4.3 magnitudes. Although the difference between these amplitudes is large, 2.75 magnitudes, it is decidedly smaller than the observed difference of 8 magnitudes and a mechanism has to be found in order to explain the discrepancy.

[1] J. STEBBINS, C. M. HUFFER and A. E. WHITFORD: Publ. Washburn Obs. **15**, Part. 3, 139 (1930).

The great sensitivity of the brightness of the red stars in the visual region to changes in temperature may explain perhaps the large number of variables found among the red stars.

The classification of the long-period variables according to the shape of the light curve is a difficult problem. The successive light curves are never exactly the same, not even for the most regular long-period variables. On the other hand the various types of mean light curves are not very different in shape. Consequently only a simple schematic system of classification is possible. If this system contains too many sub-classes the differences between the characteristic details of the standard light curves become smaller than the differences between successive cycles of one variable.

A good compromise is the scheme set up by LUDENDORFF [*4b*]. In this scheme three main classes α, β and γ are distinguished.

Type α: the rise of the light curve is steeper than the decline, the maximum is sharp and the minimum wide.

Type β: essentially symmetrical light curve.

Type γ: the light curve shows a hump or still-stand on the rising branch or has a double maximum.

For the cases where the light curve is known with somewhat greater precision the main classes are divided into respectively four, four and two sub-classes. The sub-classes are, like the main classes, arranged in order of decreasing asymmetry. For example, type α_1 is a curve with very steep rise and very flat and wide minimum resembling the light curve of RR Lyrae type *a*. The curves of types α_2, α_3 and α_4 in this order show an increasing symmetry.

The classification system developed by CAMPBELL is, except for the number of classes and the symbols used, essentially the same. CAMPBELL[1] first demonstrated that a statistical relationship exists between the shape of the light curve and the period.

Table 9 taken from LUDENDORFF, gives the numbers of variables of each class as a function of periods.

Table 9.

P days	α	β, $\alpha-\beta$	γ	pec	Σ
91−209	1	31	5	3	40
210−250	17	20	1	2	40
251−279	19	18	2	1	40
279−318	29	10	1	0	40
318−342	21	17	0	2	40
344−396	26	6	7	1	40
>398	30	2	8	0	40
All	143	104	24	9	280

The table shows that the relationship between the type of curve and the period has only a statistical significance. Towards the short periods the curves of type β or $\alpha-\beta$ are predominant while for the long periods, type α occurs more often. In other words the asymmetry of the light curves becomes gradually more pronounced as the period increases. Light curves with humps (type γ) are found either at long or at short periods.

The mean light curves of many long period variables can be found in CAMPBELL's book "Studies of Long Period Variable Stars" [*5*].

[1] L. CAMPBELL: Harvard. Repr. **1925**, No. 21.

30. Fluctuations and possible secular variations. The light variations of the long-period variables are subject to irregularities, both in the period and in the amplitude. The fluctuations in height of the maxima or of the minima take place in completely haphazard fashion. Possibly the only statistical rule is that the fluctuations in the height of the maxima are greater than those of the minima for variables with light curves presenting a sharp maximum as illustrated for V Delphini in Fig. 28a. In the case of the light curves with a sharp minimum the fluctuations of the minimum are stronger as shown in Fig. 28b for V Ophiuchi.

Some semi-regular variables show variations in amplitude which have a periodic character. According to PAYNE-GAPOSCHKIN[1] this periodic change takes

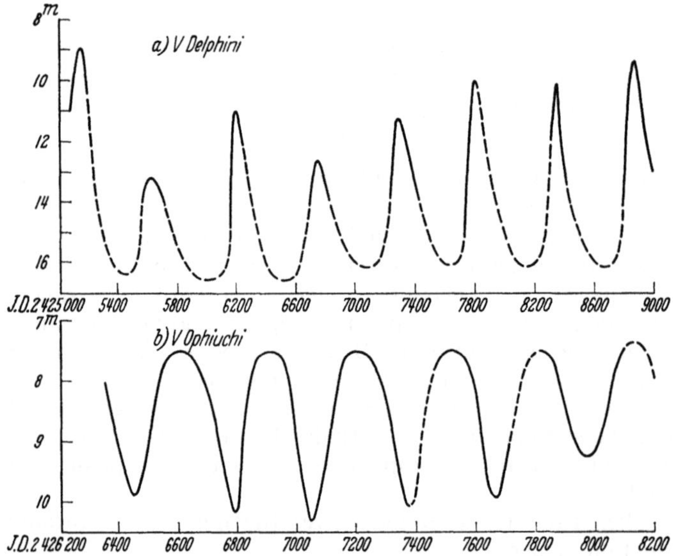

Fig. 28. Schematic light curve of V Delphini (a) and of V Ophiuchi (b) after CAMPBELL and JACCHIA [4e]. Full line: parts of the cycles actually observed.

place in about 9 cycles for the variables with spectral type M and in about 12 cycles for the N-type variables. FRITZOVÁ, PEKNY and SVESTKA[2] described periodic changes in amplitude accompanied by simultaneous changes in length of the period for two long-period variables (S Bootis and V Bootis).

These observations suggest that the long-period variables may show the phenomenon of double periodicity, found also for the Cepheids, RR Lyrae stars and the β Cephei stars. Unfortunately in the case of the long-period variables the effect is masked by the strong erratic fluctuations.

The variations in the length of the period have been investigated with great care by STERNE and CAMPBELL [34] for 377 well observed long-period variables. They found that most of the earlier reported "changes of period" were illusory and that, in nearly all cases, the variations in length of period are merely statistical fluctuations.

Real changes in periods were established with certainty only for five variables. Of these, two: R Hydrae and R Aquilae have diminishing periods and three: R Cancri, U Bootis and S Serpentis have sinusoidally changing periods.

[1] C. PAYNE-GAPOSCHKIN: Harvard Ann. **113**, 191 (1954).
[2] L. FRITZOVÁ, Z. PEKNY and Z. SVESTKA: Bull. Astr. Inst. Czechosl. **5**, 49 (1954).

31. Relations between light curve and radial velocity curve for absorption and emission lines. In contrast to the Cepheids which show a well defined variation in velocity, proportional to the variation in light, the long-period variables, notwithstanding their enormous variation in light exhibit only hardly measurable displacements of the spectral lines. In his classical study of *o* Ceti, Joy [*35*] gives a radial velocity curve derived from the absorption lines which has a total amplitude of 12 km/sec. This curve reproduced in the upper part of Fig. 29 is in phase with the light curve instead of being its mirror-image as in the case of the Cepheids. The radial velocities measured from the emission lines in the spectrum vary in a quite different way. When these lines appear first, some time after minimum light, they are displaced towards the violet with respect to the absorption lines by about 15 km/sec. During maximum light the lines shift rapidly towards the violet and shift backward again when the middle of the descending branch of the light curve has been reached. When the velocity of the emission lines has become nearly equal to that of the absorption lines the emission fades out, just before minimum light.

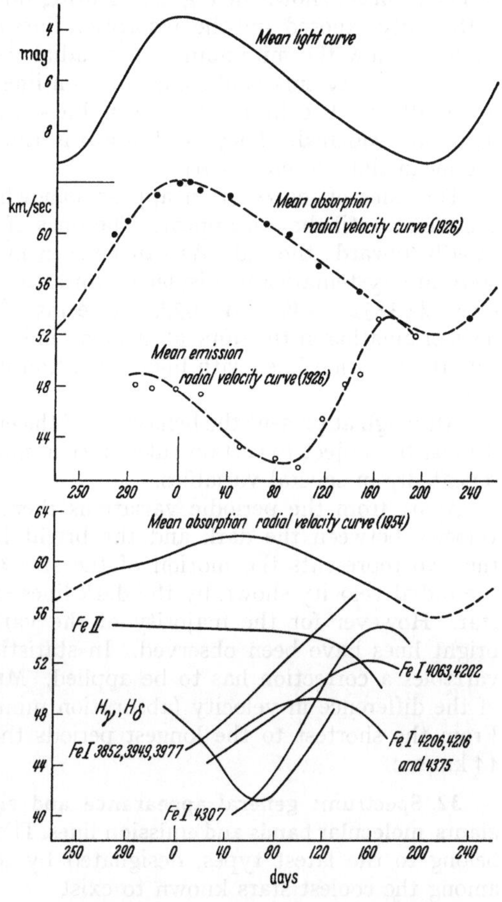

Fig. 29. Light curves and velocity curves of Mira Ceti. In the lower part, velocity curves of various emission lines are shown. The differences in behaviour escaped attention in the older observations shown in the upper part.

Merrill [*36*] in a series of investigations on Coudé-spectra of long-period variables, found that, if studied in detail, the behaviour is rather complicated. The radial velocity curve of the absorption lines sometimes, for example for R Hydrae and *o* Ceti, resembles the light curve, but in most cases the variation in velocity is very small and cannot be compared with the light curve.

The velocity curves of the emission lines are different for various elements, and actually they cannot be combined in a unique curve as in the upper part of Fig. 29. A bewildering variety of curves is shown by the lines of different elements, the neutral and the ionized atoms and even by lines of the same atom, but with different excitation potentials.

As an illustration some velocity curves for Mira Ceti are given in the lower part of Fig. 29. These curves are determined by Joy[1] from spectra obtained by P. W. Merrill, W. S. Adams and A. H. Joy during 16 years. During this

[1] A. H. Joy: Astrophys. Journ., Suppl. 1, 39 (1954).

interval which covers many cycles, the length of the period varied between 311 and 355 days, the maximum brightness varied between $2^{m}_{\cdot}5$ and $4^{m}_{\cdot}4$. The maximum radial velocity of the absorption lines varied correspondingly between 68.3 km/sec and +57.2 km/sec that is more than the total range of the mean velocity curve, shown in Fig. 29. During minimum light, the spectra cannot be sufficiently exposed and the absorption lines are not measurable. It is not known therefoıe how the minimum of the radial velocity varied.

The velocity curves of some emission lines shown in the lower part of Fig. 29 are baffling. The bright hydrogen lines, after their appearance at maximum light, are widened. They shift towards the violet and backward again and in the mean time become sharper.

The lines of ionized iron appear somewhat earlier and shift slowly towards the violet until they disappear. The lines of neutral iron appear later and move rapidly towards the red. As can be seen in Fig. 29, the FeI lines, $\lambda\lambda$ 4202 and 4063 are systematically displaced towards the violet with respect to the Fe I lines, $\lambda\lambda$ 3852, 3949 and 3977. It seems difficult to understand how spectral lines originating in the same atoms may indicate different velocities. As a general rule the emission lines with high excitation potential indicate higher radial velocities.

Although at present the behaviour of the emission lines seems entirely puzzling, it must be subject to certain rules since a more or less similar pattern is observed repeatedly in several variables.

Apart from the periodic variations there is a large systematic difference in velocity between the dark and the bright lines. The question arises which of the two represents the motion of the star as a body. According to MERRILL, the radial velocity shown by the dark lines should be identified with that of the star. However for the majority of the variables only radial velocities of the bright lines have been observed. In statistical studies of the motions of these variables a correction has to be applied. MERRILL[1] showed that the correction of the difference in velocity (absorption minus emission) depends on the period. From the shortest to the longest periods the correction increases from zero to 14 km/sec.

32. Spectrum: general appearance and similarities with non-variable supergiants, molecular bands and emission lines. The spectra of the long-period variables belong to the latest types, designated by *M*, *R*, *N* and *S*, which places them among the coolest stars known to exist.

As in the somewhat earlier types *G* and *K* the spectra of the latest types are filled with numerous metallic absorption lines, but they are characterized especially by absorption bands, produced by molecular compounds.

Depending on the types of molecules found in the spectra they can be divided into three groups. The *M*-type spectra are characterized by bands of TiO, the spectra of type *S* contain ZrO bands and the so-called carbon-spectra, or types *R* and *N*, show bands of CN and C_2.

With respect to the appearance of the metallic absorption lines and bands the spectra of the long-period variables are identical with the spectra of non-variable supergiants. The variable stars, however, often show a special feature, which is the presence of bright emission lines in the spectrum. Their spectral types are designated accordingly by the symbols *Me*, *Re*, *Ne* or *Se*. The presence of emission lines is so characteristic for the variable stars that it may be used as a criterium for detecting such stars.

[1] P. W. MERRILL: Astrophys. Journ. **58**, 215 (1923).

The brilliance of the emission lines is correlated with the amplitude of the variation in light. The Mira-type variables, with large amplitudes, without exception show strong emission lines in the spectrum during a large part of the cycle. The long-period variables with small amplitude show weaker lines and during a smaller part of the cycle. The spectra of the normal non-variable red giants may show weak emission only in the cores of strong absorption lines such as the *H* and *K* lines of ionized calcium.

It is obvious that the emission lines are formed under special circumstances brought forth by the cyclical variations which take place in the star and have no further bearing on the spectral type.

The classification of the spectra of the long-period variables and the non-variable red stars is based mainly on the intensities of the bands [*26*]. The *M*-type spectra are arranged in sub-classes *M* 0 to *M* 10 in a sequence along which the strength of the TiO bands increases.

Effective spectral classification of weak red stars is possible with low-dispersion objective prism spectra taken on infra-red sensitive plates[1].

In this method also the TiO bands are used. For the latest spectral types *M* 7 to *M* 10 the strength of some VO bands provides a sensitive test. It is interesting to note that according to NASSAU and CAMERON[2] nearly all the stars which have such *M* 7 to *M* 10 spectra are variables. Other compounds which have been identified in *M* type spectra are ScO, YO and AlO.

The *S*-type spectra are characterized by ZrO bands. The bands of TiO may be present, but they are much weaker than in the spectra of class *M*. Also bands of YO and LaO may occur. Relatively few stars belong to the class *S*, which was introduced in 1922 by MERRILL[3] to designate the spectra of π' Gruis and the long-period variables R Andromedae and R Cygni.

For a long time, differences between the individual spectra prevented a more detailed classification. Recently KEENAN[4] has developed a two-dimensional classification scheme for the *S*-type spectra based on the relative intensities of the TiO and ZrO bands in which the two variables are temperature and chemical composition.

The *R* and *N* type stars show several bands of carbon compounds, such as the Swan bands of C_2 and those of CN. Depending on the strength of the bands and the redness of the star, the spectral types were divided in sub-classes *R* 0 to *R* 9 continued by *N* 0 to *N* 7 which were considered as a sequence of decreasing temperature. This assumption proved to be not entirely correct. KEENAN and MORGAN[5] set up a new system of classification in sub-classes *C* 0 to *C* 9 based on atomic line ratios, which is more truly a temperature sequence. A comparison of the *R*-*N* types with the types in the new system showed that the *R* types form indeed a sequence of decreasing temperature. On the other hand the sequence of *N*-types is not related with temperature but more probably corresponds to a varying abundance of carbon.

The effect of chemical abundances on the latest spectral types is described by MERRILL[6] in a two-dimensional diagram. The position of the spectral type in the diagram depends on the two parameters which are the oxygen-carbon ratio and the Ti—Zr ratio. If carbon is more abundant than oxygen, all oxygen

[1] Cf. for instance, J. J. NASSAU and G. B. VAN ALBADA: Astrophys. Journ. **109**, 391 (1949).
[2] J. J. NASSAU and D. CAMERON: Astrophys. Journ. **120**, 468 (1954).
[3] P. W. MERRILL: Trans. Internat. Astronom. Union **1**, 98 (1922).
[4] P. C. KEENAN: Astrophys. Journ. **120**, 484 (1954).
[5] P. C. KEENAN and W.W. MORGAN: Astrophys. Journ. **94**, 501 (1941).
[6] P. W. MERRILL: Publ. Astronom Soc. Pacific **67**, 199 (1955).

is bound in CO molecules which produce no bands in the observable part of the spectrum. The excess of carbon forms molecules C_2 and CN which are observed in the *R-N* type spectra. In oxygen is more abundant an excess of oxygen is left over which produces oxides with various metals such as observed in the *M* and *S* type spectra.

If carbon and oxygen are equally abundant no excess is left of either element after the formation of the CO molecules. Perhaps this is the case in variables like R Canis Majoris or GP Orionis which show no marked bands.

In the second case, the Ti—Zr ratio determines whether we observe an *M*- or *S*-type spectrum. The hypothesis that the relative abundance of the elements determines the spectral type is supported by the fact that in the *S*-type spectra not only the ZrO bands are predominant but also the atomic lines of Zr and Zr^+ are stronger, relative to the Ti lines, than in the *M*-type spectra[1].

A similar strengthening is observed in the case of the bands and lines of Y and, as a general rule, the heavier elements are more abundant in the atmospheres of the *S*-type stars.

The numerous strong bands present in the late type spectra seriously impair the significance of the temperature determined from the colour of the stars. According to Merrill the entire spectrum of the *M*-type stars from green to red does not present any region completely free from bands. The remarkably deep red colour of the *N*-type stars is caused by strong absorption in the blue-violet part of the spectrum. Especially in the late type *N* spectra the ultraviolet is so weak that below wavelength λ 3800 it cannot be photographed. That this extreme weakness is not normal was first noticed by Shane[2] and confirmed by Shain and Struve[3].

The colour of the *N*-type stars corresponds to a temperature of only 1000° K but the atomic lines indicate the same temperature as for the *M*-type stars, which is about 2500° K.

The molecule responsible for the veiling of the violet part of the spectrum of the *N*-stars was long unidentified because it produces no sharp bandheads. Except for a few hazy bands near λ 4050, the so-called Swings' bands, no details can be seen in tbe continuous absorption. Swings, McKellar and Rao[4] suggested that the molecule C_3 might be responsible for it. Even larger conglomerates of carbon, or soot-like particles could play a role[5]. It was shown by McKellar and Richardson that the spectral intensity distribution of the cool *N*-stars resembles closely the spectrum of carbon vapour obtained by Phillips and Brewer with a King-furnace[6].

The effect of band-absorption or of the formation of clouds of solid particles increases strongly when the star becomes cooler, and it is possible that this mechanism will play an important rôle in the final explanation of the large amplitudes of light variations of the Mira-stars, especially that part which, as we have seen (cf. Sect. 29, discussion following Table 8), is not accounted for by black-body radiation.

---

[1] P. W. Merrill: Astrophys. Journ. **105**, 360 (1947).

[2] C. D. Shane: Lick Obs. Bull. **13**, 123 (1928).

[3] G. A. Shain and O. Struve: Astrophys. Journ. **106**, 86 (1947).

[4] P. Swings, A. McKellar and K. N. Rao: Monthly Notices Roy. Astronom. Soc. London **113**, 571 (1953).

[5] B. Rosen and P. Swings: Ann. d'Astrophys. **16**, 82 (1953). — P. Swings: Ann. d'Astrophys. **16**, 287 (1953).

[6] J. G. Phillips and L. Brewer: Les particules solides dans les astres. Coll. Liège, p. 341, 1954.

A specially interesting feature of the spectra of the long-period variables is the presence of the emission lines. Their appearance is somehow connected with the variability of the star and it is still a problem to explain how they can be generated in the atmospheres of cool stars. Most of our knowledge concerning the character and behaviour of the emission lines has been gathered by MERRILL, who published many detailed studies of spectra of long-period variables [*27*]. The lines are emitted by hydrogen and by metals, both neutral and ionized. They are observable during a part of the cycle. In general the hydrogen lines appear first while the brightness is increasing, they reach maximum brillance shortly after maximum light and fade out again before minimum light is reached. The metallic emission lines appear and also fade out somewhat later than the hydrogen lines, but the individual lines show appreciable differences in behaviour. Differences in the variation of the strength of the lines are also shown by individual variables and even by the same star in successive cycles.

At minimum light or just before minimum, sharp forbidden lines of ionized iron, Fe II, appear in emission in the spectra of all *Me*-variables investigated by MERRILL and JOY. These lines have not been noticed in the spectra of the *S*- or *C*-type variables. In one case (Mira Ceti), emission bands of AlO have been observed[1].

The intensities of the emission lines are entirely anomalous. Whereas normally, the hydrogen lines of the Balmer series, H_α, H_β, H_γ etc. form a sequence with regularly decreasing intensity, as is observed in the spectra of shell stars or of gaseous nebulae, the intensities in the spectra of the long-period variables show no relation at all with the series number. In the *Me*-type spectra H_α is relatively weak, H_β nearly invisible, H_δ very strong, H_ε is practically absent and so on. Similar vagaries are observed in the intensities of the metallic emission lines. The *Se*-type spectra show again quite different anomalies in the strength of the lines. The relative intensities of the lines change with phase. For example H_δ is considerably stronger than H_γ during the earlier stage of their appearance, but gradually the difference becomes smaller and finally, before they disappear, H_γ has become stronger than H_δ. Also in this respect the *Se*-type variables show a quite different behaviour. Many of these curious phenomena can be explained by the hypothesis that the emission lines are produced at a considerable depth in the atmosphere of the star. During the passage of the light through the overlying layers the lines are weakened in various degrees by absorption bands or lines. In this way several of the hydrogen lines are weakened by TiO bands. The extreme weakness of H_ε is caused by the strong absorption line of Ca II. The differences between the anomalies of the *Me*- and *Se*-stars can now be interpreted as due to the differences in the band absorptions.

The variations of the relative intensities of the emission lines with phase are connected with the variations in radial velocity of the layers emitting the lines relative to the absorbing layers.

Large dispersion spectrograms revealed that the hydrogen lines which are widened during the first stages of their appearance have a complex structure as they are split up into several components by sharp absorption lines. Due to changing displacements in wavelength and a gradual sharpening of the hydrogen lines the patterns change with phase. However, not all the observed pecularities can be explained in terms of overlying absorption. MERRILL pointed out long ago that the Fe I lines λ 4202 and λ 4308 are much too strong if compared to other lines of their multiplet, and in this case overlying absorption could not be

[1] A. H. JOY: Astrophys. Journn. **63**, 281 (1926).

responsible. The explanation was found by THACKERAY and MERRILL[1], who remarked that the two strong lines are the only ones of the multiplet $a^3F - z^3G^0$ which have the upper level $z^3G_4^0$ and that the energy necessary to bring the electrons in an iron atom from the ground level to this level $z^3G_4^0$ is just the energy of the strong line of MgII, λ 2795. This explanation rests on the same mechanism of fluorescence as that proposed by BOWEN for the mutilated multiplets of OIII in the spectra of gaseous nebulae. The hypothesis is supported by the fact that in the case of another multiplet of iron, $a^5F - z^3G^0$, again the only line present is the one originating from the upper level, which is excited by the MgII line. This MgII line, λ 2795 is not observable, but in the case of the line InI λ 4511, MERRILL was able to demonstrate the mechanism of fluorescence completely. This line forms a doublet with the line InI, λ 4101.8, i.e. they originate both from the same upper level 6^2S. In this case the exciting line is the strong emission line H_δ, in which the line InI, λ 4101.8 can be observed as an absorption. This absorption brings many electrons into the upper level from which they can emit the line InI, λ 4511. This line persists in the spectrum as long as H_δ does.

The emission lines make the spectra of the long-period variables most peculiar and interesting. Although some of the phenomena have been explained, the most fundamental problem, that of their origin is not yet solved. Their formation in cool extended atmospheres must be connected with strong disturbances of thermodynamical equilibrium, caused perhaps by shockwaves [*37*], perhaps by outbursts of far ultraviolet radiation from the hydrogen convection zone[2]. In addition, the spectra of the long-period variables offer many other interesting problems, such as the presence of forbidden lines near minimum light or the presence of several unidentified emission lines in the spectrum.

Also the absorption lines show several peculiar effects. For example the sodium *D* lines which are weak in the *M*-type spectra due to TiO absorption, are strong in the *S*-type stars but unusually intense in the *N*-type stars and show unexpectedly large variations in intensity with temperature.

One of the most spectacular peculiarities of the spectra of the long-period variables is the presence of technetium in the *S*-type spectra[3]. Several lines of this element were identified by MERRILL in the spectrum of *R* Andromedae and other *S*-type variables. The laboratory wavelengths were measured by MEGGERS and SCRIBNER at the U.S. Bureau of Standards from samples produced artificially by nuclear reactions. The element is unstable and its lifetime is less than half a million year. The presence of an unstable element in the atmosphere of the *S*-type stars means that either it is produced by processes involving very high energy, or that these stars were born not earlier than a few hundred thousand years ago in a process involving the formation of an appreciable quantity of technetium.

d) The RV Tauri stars and yellow semiregular variables.

33. The period interval between the Cepheids and the long-period and red semiregular variable stars is covered by a group of variables presenting a wide range of properties. In some respects, they may be considered as providing a more or less continuous transition between these classes. However they distinguish themselves from the Cepheids and to a smaller extent from the long-period variables by unstable light curves showing irregularities in shapes and periods

---

[1] A. D. THACKERAY and P. W. MERRILL: Publ. Astronom. Soc. Pacific **48**, 331 (1936).
[2] A. MCKELLAR and G. J. ODGERS: Publ. Astronom. Soc. Pacific **64**, 222 (1952).
[3] P. W. MERRILL: Trans. Internat. Astronom. Union **1952**.

so strong that they usually preclude the formation of a mean light curve. In some cases, the mean brightness is variable with a long period which may be 10 to 50 times the short period.

Confusing changes and irregularities in their light cycles render their identification by purely photometric criteria difficult so that, in recent years[1], the condition that their mean spectral types be *F*, *G* or *K* has usually been added. The number of known stars belonging definitely to the group is not large and significant lists may be found in the recent papers of L. ROSINO[2] and A. H. JOY[3] which have increased considerably our knowledge of the physical character of these stars.

Among them, some show a well marked tendency to alternation of deep and shallow minima with occasional interchange as illustrated on Fig. 30 for RV Tauri which gave its name to this group. In this case, the periods are usually defined (cf. for instance [*4b*] p. 174) as the time elapsing between two main minima and range from about 25 to 150 days. But the evidence of radial velocity curves when available suggests that the physically significant interval is the one between two successive minima whatever their depth, i.e. about half the above period.

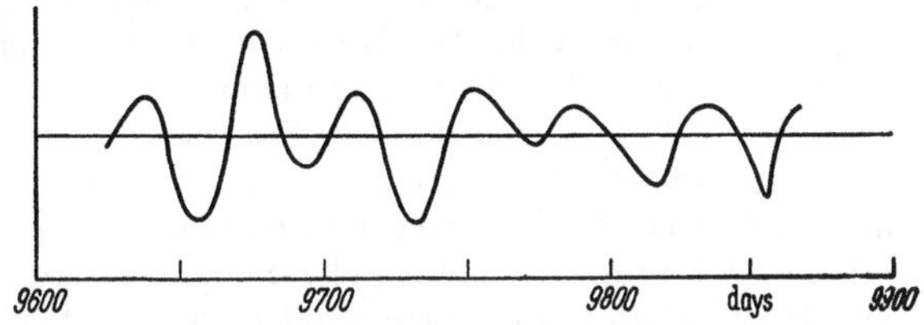

Fig. 30. Light curve of RV Tauri (after VAN DER BILT).

However not all variables falling in the gap between the Cepheids and the red semiregular variables can be included in the RV Tauri group although they are certainly related to it. In particular, the yellow semiregular variables sometimes show double maxima, sometimes, ill-defined alternation of deep and shallow minima or long intervals of erratic changes of light. Their periods defined as the mean interval between two successive maxima are definitely longer than the corresponding intervals in RV Tauri stars.

Altogether these stars exhibit a large range in luminosity probably spanning completely the gap in absolute magnitude between the long period Cepheids and the red variables. According to ROSINO, if the half-period is used for the RV Tauri stars, the whole group obeys a period-luminosity relation, with a slope opposite to that of the Cepheids. The absolute photographic magnitude decreases from a maximum $M_{pg} \approx -3$ for the RV Tauri stars with half-periods of the order of 35 to 40 days to a value of the order of $M_{pg} \approx -1$ for the yellow semiregular variables with periods of the order of 100 days. Furthermore, the brightest RV Tauri stars seem to fall on the extension of the period-luminosity relation for the W Virginis stars[4].

The spectra at maximum range from *F* 0 to late *G* and do not show any marked relation to period. Near minimum light, in many cases, the bands of TiO appear in spectra which otherwise would be classified as *G* or *K* and which again show no correlation with the period. If a classification on the basis of the TiO bands is adopted, then these minimal spectral types seem to obey a period-spectrum

---

[1] C. P. GAPOSCHKIN, V. K. BRENTON and S. GAPOSCHKIN: Harvard Ann. **113**, 1 (1942). — L. ROSINO: Astrophys. Journ. **113**, 60 (1951).

[2] L. ROSINO: Astrophys. Journ. **113**, 60 (1951).

[3] A. H. JOY: Astrophys. Journ. **115**, 25 (1952).

[4] Cf. however F. E. KAMENY: Astronom. J. **62**, 20 (1957), where appreciably higher absolute magnitudes are assigned to these stars.

relation nearly parallel to that valid for the maximal spectra of long-period variables[1]. Another fairly common characteristic is the appearance of bright-lines, notably of hydrogen, especially at times of increasing light and around secondary maxima if present. ROSINO finds that the maximum excitation and the earliest spectral types are reached halfway from deep minimum to maximum.

The radial velocities show variations associated with the light cycle with a mean range of about 35 km/sec, but here again irregularities are so large that mean velocity curves are impossible. The phase relationship between light and velocity variations resembles that of the Cepheids, maximum positive radial velocities occurring around light-minimum. In U Mon, R. F. SANFORD[2] discovered that the center of mass velocity (γ-velocity) varies with a long period of 2320 days which is also present in the light variations. According to TSESEVICH[3], the same phase relationship prevails for these slow velocity and light variations as for the more rapid ones. It would seem that this long period is also unstable.

As far as the γ-velocities are concerned there is a general tendency in the group toward high velocity. However JOY considers that they may be subdivided into two groups. The first group comprises those with γ-velocities smaller than 70 km/sec. These stars also have brighter absolute magnitude and greater maximum intensity of the *G* band of CH. Among the stars with γ-velocities higher than 70 km/sec, the occurrence of emission lines and TiO bands at deep minima is somewhat more frequent. These high-velocity stars belong definitely to the type II population and correspond closely with the semiregular variables of the globular clusters. In this respect, the situation is not quite so clear for the stars in the low-velocity group as they present a larger scatter with reference to the galactic plane than typical Population I objects and fail to show the effect of galactic rotation. Furthermore, a study by H. ABT[4] of high-dispersion spectrograms of three of these stars U Mon, R Sct and AC Her has revealed discontinuous velocity curves very similar to that of W Virginis, the prototype of Cepheids of type II. This together with the general shape of the light curve (narrow minima and flat maxima) supports the extrapolation by ROSINO of the period-luminosity relation of the W Virginis stars through the RV Tauri stars. It also suggests that JOY's subdivision on the basis of γ-velocities may not be very significant.

According to ABT, a weak set of lines displaced shortward with respect to the stronger lines appears just before light-maximum. These lines quickly strengthen, move longward and then fade during the next light-maximum (cf. Fig. 31 for U Mon). In this first announcement[4], ABT also mentions that, in two of these stars, there are times when lines from three distinct layers are visible at once. Displacements of the atmospheric layers computed from the radial velocity curves are very large and may reach values of the order of the stellar radius. The largest expansions are associated with the deepest light minima, a correlation first pointed out by SANFORD[5] and MCLAUGHLIN[6]. Moreover the detailed investigation of U Mon[7] shows that minimum light is a little earlier than maximum positive radial velocity and indeed follows rather closely ($\approx$ 10 days) maximum expansion.

[1] Cf. C. P. GAPOSCHKIN's article in "Astrophysics", A Topical Symposium 1951, Fig. 12.20 p. 523.

[2] R. F. SANFORD: Astrophys. Journ. **77**, 120 (1933).

[3] V. P. TSESEVICH: Astronom. Circ., Akad. Nauk **1952**, No. 131; cf. also V. P. TSESEVICH: Peremennye Zveody Bull. **8**, 121 (1951), and I. JDANOVA and V. P. TSESEVICH: Astronom. Circ. **1953**, No. 135, and [*38*], p. 116.

[4] H. ABT: Astronom. J. **58**, 210 (1953).

[5] R. F. SANFORD: Astrophys. Journ. **73**, 364 (1931).

[6] D. B. MCLAUGHLIN: Astrophys. Journ. **94**, 94 (1941).

[7] H. A. ABT: Astrophys. Journ. **122**, 72 (1955).

Data on colours are still meagre but they suggest that the colour variations are small and show little correlation with light; the star is often as blue during a light-minimum as during a maximum or even bluer. According to KAMENY[1] the RV Tauri stars are all too blue for their spectral classes while the yellow semiregular variables are all too red.

ABT's spectrophotometric analysis of U Mon has also brought to light many interesting results. A very striking one consists in large apparent changes in the abundances of Fe I and Fe II while the degree of ionization is practically constant. This is attributed by ABT to variations of the continuous opacity within the regions of line formation, a hypothesis which also brings a fair agreement between the effective temperatures derived from the spectral types and ionization temperatures.

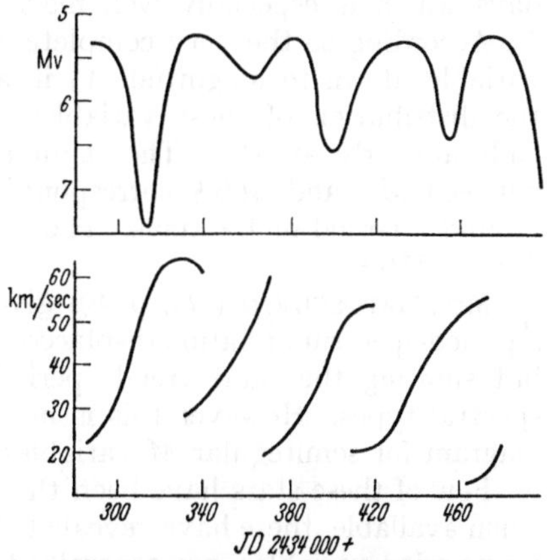

Fig. 31. Top: light curve of U Monocerotis. Bottom: radial velocity curves (after H. A. ABT).

The Doppler width of the lines belonging to a given set corresponds to velocities increasing from about 3 km/sec soon after the appearance of this set at the time of a light-maximum, to 7 km/sec at the time of its fading out at the following one. These velocities are too large to be due to thermal motions and must be attributed either to turbulence or to a dispersion in the pulsation velocity.

The effective gravity varies by an enormous factor (10^4) taking a value close to GM/R^2 at maximum contraction and decreasing to a very small value at maximum expansion.

Further progress in the knowledge of these stars will depend mainly on individual detailed studies. Apart from improved discussions of the problems already tackled by ABT, such studies may also provide new information on different effects such as those mentioned by JOY: veiling and absorption effects of titanium and carbon bands contributing to the dimming of the light at certain phases; flare effects contributing to the increase in light and accompanied by hydrogen emission and a marked diminution in the visibility of the absorption spectrum at times, in many of these stars.

e) The red semiregular and irregular variables.

34. As already mentioned in discussing the long-period variables, the majority of *M* stars are appreciably variable[2] and this is probably true also of other red stars especially those of advanced spectrum and high luminosity. Here, the separation between different types of variability is particularly ill-defined. The main distinctive characteristic of the red semiregular and irregular variables is their small amplitudes. The extreme range rarely exceeds two magnitudes and the mean range is appreciably smaller[3].

The same is true of the spectral type variations which, in general, do not exceed 1 or 2 spectral subdivisions[4].

[1] See footnote 4, p. 415.

[2] J. STEBBINS and C. M. HUFFER: Publ. Washburn Obs. **15**, 19 (1930). — J. J. NASSAU and V. M. BLANCO: Astrophys. Journ. **120**, 468 (1954).

[3] Cf. for instance, C. PAYNE-GAPOSCHKIN: Harvard Ann. **113**, No. 4 (1954).

[4] A. H. JOY: Astrophys. Journ. **96**, 344 (1942).

In general, the stars in the first group present large irregularities in the period although it comprises stars like W Cygni which is as regular as most long-period variables. The irregular type shows no periodicity altogether so that for these stars one of the most useful parameters in the discussion of the properties of variable stars and their classification, the period, is lacking.

Most of these stars belong to the spectral classes, *M*, *S*, *R* and *N* (rarely *K*) and their properties are often discussed in each spectral class separately.

Mrs. GAPOSCHKIN'S discussion[1] of about 500 red semiregular variables brighter photographically than the tenth magnitude has shown that the frequency distribution as a function of period of both *M* and *N* stars presents two maxima, one around 100 days which is most prominent for *M* stars and the other around 350 days which is especially well marked for *N* stars.

According to the very complete study by CAMERON and NASSAU[2] of the red variables down to magnitude 15 in a 12° belt centered on the galactic equator, the distribution of these variables covers all *M* subclasses through *M* 6.5 but ends abruptly at *M* 7. The semiregular *M* variables show a strong grouping between *M* 5 and *M* 6.5 corresponding to the previously mentioned maximum around a period of 100 days. The irregular variables are concentrated at classes *M* 6 to *M* 6.5.

Mrs. GAPOSCHKIN ([*1*], p. 49) gives for both *M* and *N* semiregular variables a period-spectrum relation displaced with respect to that for long-period variables but showing the same trend: period increasing in going from earlier to later spectral types. However this is not very apparent on CAMERON and NASSAU'S diagram for semiregular *M* variables.

Few of these stars have been the object of detailed spectroscopic studies but, when available, these have revealed the presence of bright lines which, as in the long-period variables, are generally the more intense, the greater the amplitude.

Little precise information on the luminosity of these stars is available but they seem to spread over a considerable range. According to Mrs. GAPOSCHKIN [*4h*], they can be separated into two groups, those of absolute visual magnitude about -1 (low-luminosity group) and those brighter than -3 (high-luminosity group) corresponding respectively to the maximum in period frequency around 100 days and 350 days. Thus the period increases with the luminosity when going from one group to the other although there is no apparent correlation between period and luminosity within each group.

As far as population types are concerned, definite conclusions have not been reached as yet but the general consensus of opinion (cf. [*1*], p. 48 and [*2b*], Sect. 11, p. 49) is that the *M* variables of the high-luminosity group which are also fairly abundant in the large Magellanic Cloud are probably typical Population I objects while the low-luminosity group is more heterogeneous containing a few high-velocity members which might be similar to the irregular variables reported in ω Centauri and Messier 22.

The semiregular and irregular carbon stars (*N* and *R* variables) show a disk distribution and no concentration towards the galactic center which make them likely members of Population I.

As far as physical properties susceptible to shed light on the variations are concerned, very little is known. The radial velocities of the semiregular *M* and *N* stars vary inappreciably during the cycle so that we do not know if and what kind of surface displacements are associated with the variations. In this respect, it is interesting to recall that in the case of μ Cephei, an irregular variable,

[1] See footnote 3, p. 417.

[2] D. CAMERON and J. J. NASSAU: Astrophys. Journ. **124**, 346 (1956).

J. ASHBROOK, R. L. DUNCOMBE and A. J. VAN WOERKOM[1] came to the conclusion that the light variation may be interpreted as arising from temporary, random surface disturbances on the star. As they mention, continuous photoelectric light curves might allow recognition of individual disturbances and help to unravel the physical mechanisms involved.

f) The explosive variable stars.

35. Some general comments. One usually groups under this title all variables which experience considerable and abrupt changes in luminosity. The most typical of these stars present strong similarities with novae and good reasons can be advocated to discuss them together[2]. In particular the nova-like variables such as Z Andromedae, AX Persei, CI Cygni, AG Pegasi which combine nova-like spectra with giant *M* character are very similar to the recurrent novae such as T Coronae Borealis or RS Ophiuchi. Outside of outbursts, they show essentially giant *M* spectra and could be included in the group of the red semiregular variables of small range. As in the case of the recurrent novae, the problem of their composite spectra often called "symbiotic" after MERRILL[3] has not yet been solved definitely.

Another interesting object of this type is the long-period variable R Aquarii. It presents, at long intervals, outbursts characteristic of a slow nova which practically suppress the normal variation.

In the same way, the U Geminorum and Z Camelopardalis stars which are intrinsically less luminous are sometimes interpreted as representing novae of small amplitudes and short cycles. Here, it is the only group which we shall discuss in some details.

36. The U Geminorum stars. These stars are characterized by large and very rapid increases in brightness occurring very suddenly and followed by a generally slower decline. The maxima last only a short time compared to the relatively long intervals of quietness separating them. These intervals vary irregularly but nevertheless significant mean values can be defined, the deviations obeying the normal law of errors[4]. These mean intervals fall in the range of, say 20 to 200 days.

Two of the best known members of this class are U Geminorum itself and SS Cygni. According to Mrs. MAYALL[5], the mean interval in the case of U Geminorum in the course of the last hundred years, is 101.8 days while its values in the first and second half of this period are 96.5 and 107.6 days respectively. This may reveal a real tendency to a slight increase. The extreme values of the individual intervals are 57 and 201 days. The mean amplitude is of the order of 5.5 magnitudes, the star reaching a magnitude of 8.8 at maximum. There is a correlation between the length of individual intervals and the intensity of the associated maxima: the smaller the interval, the smaller the range. Furthermore, the maxima are alternatively wide (10 to 20 days) and narrow (3 to 8 days) with temporary breaks in this alternation.

[1] J. ASHBROOK, R. L. DUNCOMBE and A. J. VAN WOERKOM: Astrophys. Journ. **59**, 12 (1954).

[2] Cf. [*4g*], part D which contains a wealth of information on individual stars; also [*4b*] and [*4e*]; for a discussion of the transition between these different groups, cf. especially H. SCHNELLER: Astronom. Nachr. **276**, 7 (1948). For the U Geminorum stars, cf. the very complete monograph by A. BRUN and M. PETIT: Bull. Assoc. Franc. Obs. Et. Var. **12** (1952).

[3] P.W. MERRILL: Astrophys. Journ. **111**, 484 (1950); cf. L. H. ALLER: Publ. Dom. Astrophys. Obs. **9**, 321 (1953).

[4] T. E. STERNE and L. CAMPBELL: Harvard Ann. **90**, 189 (1934).

[5] M. W. MAYALL: J. Roy. Astronom. Soc. Canada **51**, 165 (1957).

SS Cygni is probably the star whose light variations have been followed most continuously and L. CAMPBELL[1] has published a general discussion of all the light observations of this star since its discovery in 1896 up to 1933. The mean interval and the mean range in magnitude are smaller than for U Geminorum, the mean maximum and minimum brightness being 8.4 and 11.8 magnitude. SS Cygni differs further from U Geminorum by a greater irregularity in the form of the light curve, a portion of which is represented in Fig. 32. For instance although in the majority of cases, the rise to maximum is rapid (≈ 1 day), there are exceptions where the maximum is reached through a more regular and slower rise. This tendency to irregular spells reaching sometime franctic activity increases through the class towards shorter mean intervals.

The statistical study by T. E. STERNE and L. CAMPBELL[2] of all the available observations of SS Cygni has revealed many interesting correlations between different properties and confirmed namely the association between long inter-

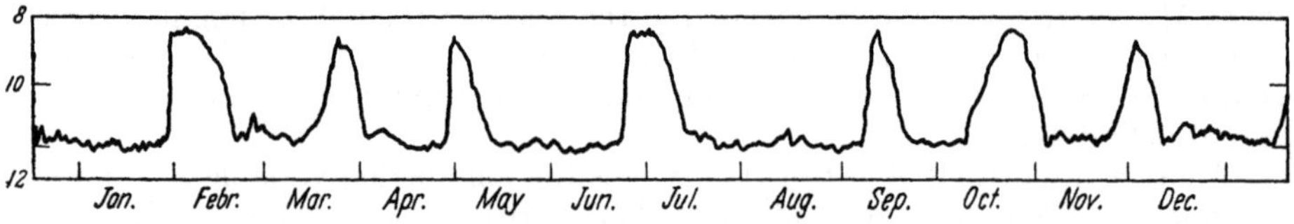

Fig. 32. Light curve of SS Cygni for the year 1922 (after L. CAMPBELL).

vals and important maxima. In this respect, one should note that the same type of relation exists also between the mean interval and the mean light range of different U Geminorum variables, the amplitude increasing with the period. This correlation has been emphasized by KUKARKIN and PARENAGO[3] who have proposed the following relation between the mean amplitude $\bar{A}$ expressed in magnitudes, and the length $\bar{P}$ of the cycle in days,

$$\bar{A} = 0.63 + 1.667 \operatorname{Log} \bar{P} \tag{36.1}$$

which according to the authors is also valid for the recurrent novae. KUKARKIN and PARENAGO insist also on the analogy between the form of the maxima in these stars and in novae. However according to KOPYLOV ([*38*], p. 71) the relation satisfied by the recurrent novae is quite distinct from (36.1), the amplitude increasing more rapidly with the period.

The question of the absolute magnitude of these stars is still controversial. In the case of SS Cygni, KUKARKIN and PARENAGO derived from the proper motion a parallax of 0".038 corresponding to an absolute magnitude at minimum of the order of $+10$. The mean trigonometric parallax was measured by VAN MAANEN and STRAND giving the discordant results: $-0".012$ and $+0".032$. STRAND's and KUKARKIN and PARENAGO's parallax would put SS Cygni among the white dwarfs. But this seems incompatible with its spectrum and recently HINDERER[4] and JOY[5] have proposed a parallax of 0".004. Furthermore, an analysis of the radial velocities at minimum derived on the one hand from the weak absorption lines and on the other hand from the wide emission lines led JOY to the conclusion that SS Cygni is a binary with components of type *dG*5 and *sdBe* of nearly

[1] L. CAMPBELL: Harvard Ann. **90**, 93 (1934).

[2] Cf. Ref. 4, p. 419 and also Harvard Bull. **1934**, No. 897, 9.

[3] B. V. KUKARKIN and P. P. PARENAGO: Verein Freunde d. Phys. u. Astron. in Gorki (Nishni-Novgorod), Veränderl. Sterne, Forsch.- u. Inform.-Bull., N.N.V.S. **4**, 249 (1934); cf. also L. ROSINO: Bologna Publ. **5**, Nr. 3 (1946), where a steeper slope is adopted for the relation (36.1).

[4] F. HINDERER: Astronom. Nachr. **277**, 193 (1949).

[5] A. H. JOY: Astrophys. Journn. **124**, 321 (1956).

equal masses ($M \leqq 0.4 M_\odot$) revolving around each other in 0.276 day on a nearly circular orbit of radius $\leqq 1.16 \cdot 10^6$ km. In that case, the last parallax brings the *G*-type star practically on the main sequence with an absolute magnitude of + 5.5, while the *B* component is a small subdwarf of about the same absolute magnitude surrounded by a gaseous envelope with large systematic and internal motions accounting for the wide (20 Å) emission lines of H (strong) and HeI, HeII and CaII (fainter) present at minimum. It is interesting to note that this places it somewhere in the region of the Hertzsprung-Russell diagram characteristic of the pre-novae which makes it likely that the hot component is really the seat of the instability leading to the outbursts.

A photoelectric study by G. GRANT[1] failed to show any eclipsing effects, a result which permitted to fix the upper limits to the masses and the dimensions of the system referred to above. However it revealed short-period fluctuations, sometime of a flare-like character during the minimum, a fact which strengthens the analogy with AE Aquarii in which LENOUVEL[2] discovered strong sharp flares lasting only a few minutes but sufficiently intense to double the total light of the star.

AE Aquarii was long considered as a long-period variable, but ZINNER[3] found that it presents occasional outbursts of several magnitudes and suggested that it belonged to the U Geminorum type, although not quite typical of the class. This was confirmed later by JOY[4] who showed that its spectrum resembled that of SS Cygni at minimum. Since then JOY[5] has established the binary nature of AE Aquarii by an analysis similar to the one recalled above for SS Cygni. The cool component shows a *dK*0 spectrum[6] and its absolute magnitude is +6. The small hot companion is probably ot type *B*0 and is somewhat fainter.

All the U Geminorum stars show spectra with the same characteristics. They are generally blueish-white in colour. At minimum, the prominent features of the spectrum are wide emission lines which become less and less distinguishable as the intensity of the continuous spectrum increases until finally at maximum, these lines appear in absorption. This suggests the same physical nature for all these objects and makes it probable that, as AE Aquarii and SS Cygni, they may all show duplicity when studied carefully. The older spectroscopic observations[7] should be rediscussed and extended with this in mind not only at minimum but also on the rising branch and at maximum. The same applies to colour[8] and spectrophotometric discussions such as HINDERER'S. In particular, informations on radial motions during the outbursts, if at all possible, would be very valuable.

37. The Z Camelopardalis stars. These stars have often been classified together with the U Geminorum stars and they are certainly related to them. Their ranges

[1] G. GRANT: Astrophys. Journ. **122**, 568 (1955).
[2] F. LENOUVEL: C. R. Acad. Sci., Paris **235**, 1282 (1952).
[3] E. ZINNER: Astronom. Nachr. **265**, 345 (1938).
[4] A. H. JOY: Publ. Astronom. Soc. Pacific **55**, 283 (1943).
[5] A.H. JOY: Astrophys. Journ. **120**. 377 (1954).
[6] Cf. also A. CRAWFORD and R. P. KRAFT: Astrophys. Journ. **123**, 44 (1956), who have discussed a possible interpretation of AE Aquarii in terms of material ejected from the *K* star through the inner Lagrangian point of the system some of which is collected by the hot companion. They classify the cool star, on the Morgan-Johnson system, as of type *K* 5 IV–V.
[7] C.T. ELVEY and H.W. BABCOCK: Publ. Amer. Astr. Soc. **10**, 51 (1940). — Astrophys. Journ. **97**, 412 (1943).
[8] Cf. M.-C. ZUCKERMANN: C.R. Acad. Sci., Paris **243**, 567 (1956), where colours computed for a system comprising a cool component constant in brightness and a hot component varying in luminosity due to variations of the radius are compared with observed colours.

of about two to three magnitudes and their cycles of some 10 to 20 days during their regular spells could be interpreted as those of U Geminorum stars of short "period". However, in general, they spend relatively much less time at minimum than the latter. Furthermore, as illustrated in Fig. 33, their light curves show much greater irregularities and a long period of regular variations may suddenly be succeeded by erratic fluctuations with a considerably reduced range of variation. In particular, from time to time, the descending branch of a maximum comes suddenly to a halt at about one third of the way from maximum to minimum, the brightness remaining then more or less constant at this level for periods of a few weeks to many months.

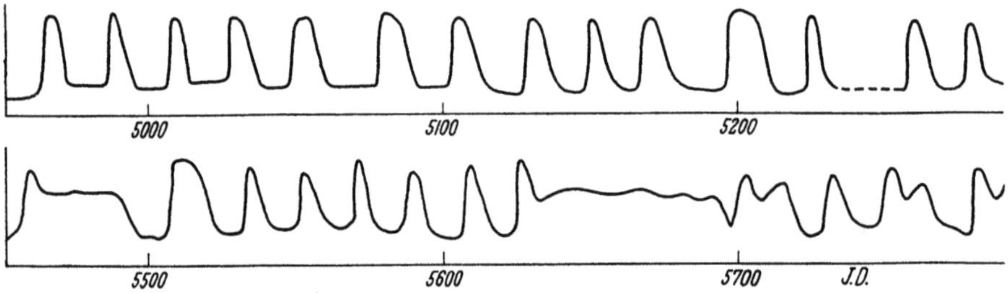

Fig. 33. Two portions of the light curve of Z Camelopardalis (after L. JACCHIA).

As far as their spectra are concerned, the observations of ELVEY and BABCOCK[1] have shown that, on the whole, they present very strong similarities with those of ordinary U Geminorum stars but detailed spectrophotometric discussions seem to be lacking as well as reliable informations on radial velocities.

g) The R Coronae Borealis stars.

38. As far as the form of their light curve is concerned the R Coronae Borealis stars show a behaviour which is just about the opposite of that of the U Geminorum stars. As shown on Fig. 34, they remain a very long time at "maximum"

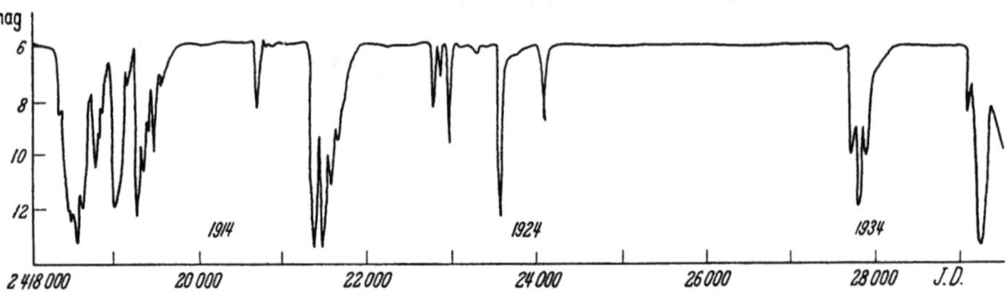

Fig. 34. Schematic light curve of R Coronae Borealis.

which is probably the normal state and then, without the least warning, they drop very rapidly to a deep minimum which corresponds to a decrease in brightness by 5 to 9 magnitudes.

In R Coronae Borealis itself, while it may remain at maximum for periods as long as ten years, the minimum may be reached in a period of the order of one month or two. The following recovery is usually slow and often interrupted by fluctuations and altogether, it may remain below its maximum brightness for periods of a few months to a few years.

[1] See footnote 7, p. 421.

Most of these stars are very steady in light at maximum. On the other hand, as shown by T.E. STERNE[1], the amplitudes of the light variations as well as the times of occurrence of the minima are distributed completely at random, following the laws of pure chance. S Apodis is somewhat of an exception to both rules, the luminosity at maximum fluctuating appreciably and the deep minima tending to recur at intervals around a thousand days. Another interesting object of this class, RY Sagittarii, shows a semi-regular variation of about half a magnitude with a mean period of 39 days in between its deep random minima. Not much is known about the absolute magnitude of these stars but all indications point to giant or even supergiant characteristics.

Detailed spectroscopic data and analysis are available for the type star[2]. At maximum, the spectrum resembles that of a supergiant of spectral class *F* 5 to *F* 8 although significant differences are present: weakness of the hydrogen absorption lines, presence of some bands of C_2 and high excitation C I lines. On the descent to the 1948 to 1949 minimum, HERBIG reports that changes in the spectrum did not occur until the star had fallen to magnitude 10.0. Emission cores appeared then in the *H* and *K* lines and went on increasing as the variable continued to fade while new emissions developed due mainly to NaI, ScII, TiII, SrII and OII. It is the only known case where the presence of CN in emission has ever been noted. The relative strength of these lines suggests a low temperature excitation. A very remarkable feature is the absence of hydrogen emission and the weakness of FeII.

The absorption lines of C_2 and C I did not show any appreciable changes during the decline but a curious "veiling" of the absorption spectrum was observed.

Changes in radial velocities, at least during the first 4.8 magnitudes of the descent, if present must have been very small. The emission lines were displaced shortward with respect to the absorption lines by a constant shift corresponding to 12 km/sec while, on previous occasions, JOY and HUMASON reported a 20 km/sec shift. HERBIG makes the interesting comment that perhaps the emission lines are there all the time but become only visible when the continuum fades sufficiently. What evidence is available for the rising branch suggests that the phenomena occur exactly in the reverse order, the emission features becoming unobservable around 4 magnitudes below maximum.

In his analysis, BERMAN came to the conclusion that carbon must be extremely abundant in the atmosphere of R Coronae Borealis (67% C, 27% H, ~6% light metal) and, as suggested by the relatively large number of *R* stars belonging to this class, it may well be that this is a fundamental characteristic of these variables. This, combined with a suggestion of LORETA[3], has been made the basis of a tentative interpretation of their variations by J. O'KEEFE[4] who showed that condensation of soot particles out of ejected carbon gas would occur at reasonable distances above the surface of R Coronae Borealis and in sufficient quantity to account fully for the observed decrease in brightness by 8 to 9 magnitudes, an extreme example of MERRILL'S veil theory for long-period variables. As pointed

---

[1] T. E. STERNE: Harvard Bull. **1935**, 896.

[2] (a) A. H. JOY and M. L. HUMASON: Publ. Astronom. Soc. Pacific **35**, 325 (1923). — (b) The spectrum at maximum has been discussed in great details by L. BERMAN: Astrophys. Journ. **81**, 369 (1935). — (c) The spectrum on the descending branch and at minimum has been the object of a careful investigation by G. H. HERBIG: Astrophys. Journ. **110**, 143 (1949).

[3] E. LORETA: Astronom. Nachr. **254**, 251 (1934).

[4] J. O'KEEFE: Astrophys. Journ. **90**, 294 (1939); cf. also an account of new computations by H. PILLANS based on more recent physico-chemical data in a paper by O. STRUVE: Sky and Telescope **12**, 261 (1953).

out by SWINGS[1], the recent identification of the absorption feature around λ 4050 in *N* stars as due to C_3 makes the presence of more complicated molecules and, under suitable conditions, of soot particles very likely. This certainly lends support to O'KEEFE's hypothesis which could perhaps now be refined and elaborated[2] on the basis of the new facts brought to light by HERBIG and in trying to follow quantitatively the development in time of the advocated processes: the dispersal of the clouds of particles as well as their formation.

h) RW Aurigae and T Tauri stars.

39. Some general comments. These denominations are fairly new and many of the stars assigned now to these classes can be found in older texts under other names, one of the most common being Nebular Variables or Orion Type Variables (cf. for instance [*4b*], [*4e*], and [*4g*]) emphasizing their frequent association with dark or bright nebulosities. But some of the members have also been classified with the R Coronae Borealis stars (cf. [*4b*]), others have been designated as RR Tauri stars and the general non-committal designation Main Sequence Variables has also been proposed.

The RW Aurigae type was introduced by P.P. PARENAGO[3] in 1932 mainly on the basis of photometric criteria. On the other hand, in JOY's definition[4] of the T Tauri variable stars, the spectral characteristics play a dominant rôle. It does not seem that general agreement either on terminology or on the correspondence of different classes or on their respective limits has been reached. Nor is it certain that all variables designated at one time or another by one of the various names recalled above will finally prove to be true members of the two classes considered here. As to these classes and their possible overlapping, we shall take a point of view similar to that adopted by P.N. KHOLOPOV at the Symposium on "Non-Stable Stars" held during the 1955 I.A.U. meeting in Dublin [*38*].

According to HOFFMEISTER's extended definition[5] of the RW Aurigae stars, these variables fall close to the main sequence and show, at times, rapid (0.5 to 1 magnitude per day or sometimes per hour) non-periodic fluctuations in brightness with ranges of 1 to 4 magnitudes which may be separated by short or long intervals of practically constant light or superposed on slower variations. While PARENAGO's original class was restricted to *G* type stars, this new definition permits to include in the class, stars of all spectral types from *B* to *M* and thus, the necessity for a special group, the Orion type variables, disappears.

On the other hand, according to JOY, the T Tauri variables are characterized by (a) irregular light variations with an amplitude of about 3 magnitudes, (b) a (*F*5 to *G*5 type spectrum with H_α and Ca II emissions, (c) low luminosity and d) connexion with a dark or bright nebula.

The emission spectrum mentioned under (b) and which had escaped the attention before JOY's investigation turned out to be a very important property and, since then, many stars picked out for this characteristic[6] were later found

[1] P. SWINGS: Ann. d'Astrophys. **16**, 287 (1953).

[2] Cf. HERBIG's discussion at the end of the paper referred to above and also the papers by O. STRUVE, in: Les particules solides dans les astres, Liège, 6th Symposium, Mém. Soc. Roy. Sci. Liège **15**, 193 (1955), and J. O'KEEFE: Mém. Soc. Roy. Sci Liège **15**, 223 (1955).

[3] P. P. PARENAGO: Variable Stars **4**, 222 (1933).

[4] A. H. JOY: Astrophys. Journ. **102**, 168 (1945).

[5] C. HOFFMEISTER: Astronom. Nachr. **278**, 24 (1949).

[6] A. H. JOY: Astrophys. Journ. **110**, 424 (1949). — P. N. KHOLOPOV: Variable Stars **8**, 83 (1951). — O. STRUVE and M. RUDKJOBING: Astrophys. Journ. **109**, 92 (1949). — G. HARO and A. MORENO: Bol. Obs. Tonantzintla y Tacubaya **1953**, No. 7, 11. — G. H. HERBIG: Astrophys. Journ. **111**, 15 (1950). — G. HARO: Astrophys. Journ. **117**, 73 (1953); etc.

to be variable with light fluctuations characteristic of the RW Aurigae type. Adopting the typical T Tauri type emission spectra as criterion, HERBIG extended the class to later spectral types down to *M* inclusively.

However this raises the question: all the stars classified as of RW Aurigae type according to the photometric criteria, do they present the spectral features characteristic of the T Tauri type and vice versa?

Exceptions are easily found. For instance R Monocerotis which has a typical T Tauri spectrum has never shown any rapid light variations and is not referred by HOFFMEISTER to the RW Aurigae type. T Tauri itself whose light curve

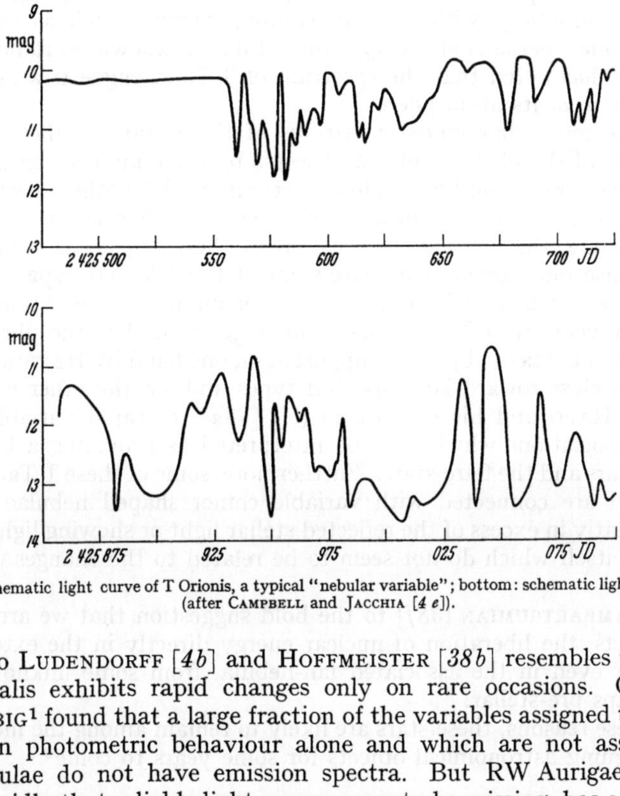

Fig. 35. Top: Schematic light curve of T Orionis, a typical "nebular variable"; bottom: schematic light curve of RR Tauri (after CAMPBELL and JACCHIA [4 e]).

according to LUDENDORFF [4b] and HOFFMEISTER [38b] resembles that of R Coronae Borealis exhibits rapid changes only on rare occasions. On the other hand, HERBIG[1] found that a large fraction of the variables assigned the RW Aurigae type on photometric behaviour alone and which are not associated with diffuse nebulae do not have emission spectra. But RW Aurigae itself which varies so rapidly that reliable light curves seem to be missing, has a typical emission spectrum but shows no appreciable relationship with nebular matter. Thus there is no doubt that the RW Aurigae class is rather heterogeneous and may, as suggested by HERBIG, have little physical meaning.

However, according to KHOLOPOV[2], the majority of RW Aurigae variables belong to *T*-associations and this is favourable to a unique cause for their variations connected some way or other with the fact that they must be fairly young stars. He even suggests that the presence or the absence of the emission may simply be an indication of an earlier or later stage of evolution of the same objects. The hypothesis of the relative youth of these stars is supported by the existence of the "Herbig-Haro objects" consisting in curious associations of a very small emission nebula with a stellar or semi-stellar nucleus presenting apart from the

[1] G. H. HERBIG: Trans. Internat. Astronom. Union **8**, 805 (1954).

[2] P. N. KHOLOPOV: Trans. of the Fourth Conf. on Problems of Cosmogony, Moscow, 1955.

nebular spectrum, a second spectrum, reminiscent of that of some T Tauri stars. The discovery by HERBIG [*38c*], that two objects which seem to be stellar have appeared in one of the brightest Herbig-Haro objects between January 1947 and December 1954, suggests that they are connected with the birth of new stars which may develop into new T Tauri stars.

In any case, the old explanation of their light variations in terms of obscuration suggested by their frequent association with nebulae, could not very well be upheld now that better information is available on the interstellar medium. Moreover, the spectroscopic investigations have revealed that the light variations are accompanied by spectral changes which point towards internal causes although complicated interactions with the surrounding medium such as suggested by GREENSTEIN[1] may perhaps also play a rôle. Little is known on radial velocities, but SANFORD[2] has noted that the spectrum of T Tauri suggests a possible ejection of matter from its atmosphere.

One of the most characteristic features of these stars is the enhancement at some phases of the blue end of their spectra by some kind of continuous emission, veiling the absorption lines. This is accompanied by other spectral changes (colour, emission lines) which can be interpreted as of thermal origin. According to AMBARTSUMIAN the continuous emission in these objects is very similar to the very intense one observed in flare-stars of the UV Ceti type, except that, while in this case it lasts only a few seconds or minutes, it may remain present for months or years in T Tauri stars. The suggestion that the phenomenon is the same in both types of objects is supported on one hand by HERBIG'S extension of the T Tauri class toward later spectral types and, on the other hand, by the discovery by HARO and his co-workers (cf. [*38d*]) of rapid variables (*dK*6 to *dM*6) in *T*-associations which may be interpreted as providing a link between the T Tauri stars and the flare-stars. Furthermore, some of these T Tauri variables like DD Tauri are connected with variable comet shaped nebulae reaching a brightness greatly in excess of the reflected stellar light or showing light variations as in T Tauri itself which do not seem to be related to the changes occurring in the star.

This led AMBARTSUMIAN [*38f*] to the bold suggestion that we are witnessing in these objects, the liberation of nuclear energy directly in the external layers of the star or even in the associated fan-nebula, from some unknown state of matter, perhaps pre-stellar.

For all these reasons, these stars are likely to remain among the most interesting and intriguing astronomical objects for some years to come[3].

i) The spectrum and magnetic variables.

40. There is a large group of stars in which the most obvious variation is that of their spectra. Outstanding among these is a group of high luminosity objects encountered most frequently in classes *B*0 to *B*3 and extending but with a much reduced frequency to classes *A*3 to *A*4. These objects have been catalogued by MERRILL and BURWELL[4] who distinguish the *Be* stars proper (with one or two bright components superposed centrally on a broad absorption line) and the

---

[1] J. L. GREENSTEIN: Publ. Astronom. Soc. Pacific **62**, 156 (1950).

[2] R. F. SANFORD: Publ. Astronom. Soc. Pacific **59**, 134 (1947).

[3] For further interesting observational data and new suggestions concerning the origin of the ultraviolet continuum cf. K. HUNGER and G. E. KRON: Publ. Astronom. Soc. Pacific **61**, 347 (1957) and K. H. BÖHM: Z. Astrophys. **43**, 245 (1957).

[4] P. W. MERRILL and C. G. BURWELL: Catalogue and Bibliophaphy of Stars of Classes *B* and *A* whose Spectra have bright Hydrogen lines. Mt. WILSON Contr. **1933**, No. 471.

P Cygni stars (with a bright component to the red of an absorption line). The bright lines show a variation of the relative intensity of their red and violet components as well as of their total intensity sometimes exhibiting a rough periodicity of a few years.

Changes in radial velocities may also occur but it is often difficult to disentangle this effect from the changes in the structure and in the distribution of intensity in the lines.

In a number of cases, the light is also variable usually with a very small amplitude and no doubt the number of these cases may increase as more powerful methods to detect light variations are applied.

The generally accepted inference is that, in most of these cases, the variations observed are the result of changes limited to the very external layers of these stars or in semi-detached shells. As a consequence, the interpretations proposed appeal usually to rotational instability of the external layers with formation of rings[1] or to the expansion of envelopes[2] repelled by radiation pressure. But as far as the origin of the observed phenomena is concerned, this class may be very heterogeneous as its characteristics are also encountered in eclipsing variables such as β Lyrae, in post-novae such as η Carinae and P Cygni or in irregular variables with slow and large or rapid and small fluctuations such as AG Carinae or XX Ophiuchi. Moreover, the gregarious tendency of the *Be* stars and their marked concentration in the galactic plane suggest the possibility of association with diffuse nebulosities and their interactions cannot be excluded as a cause of some of the spectral features observed.

But all this takes us rather far from the typical intrinsic variables which are the main object of this article. However a special group, the spectrum variables of type *A*, deserves a special mention. The stars included in this class are *A* stars with rather regular variations of fairly short period ($\frac{1}{2}$ to 20 days) in the intensity or appearance of some of their absorption lines but which do not show the *c*-characteristic[3]. DEUTSCH estimates that they lie about one magnitude above the main sequence. Recently, they have become the object of enhanced interest due to the fact that an appreciable fraction of the magnetic variables discovered by H.W. BABCOCK[4] belong to this class.

As a special article in this volume by A. J. DEUTSCH is devoted to the magnetic stars there is no need for us to linger on the subject here. Let us recall only that large magnetic fields have been recorded in stars covering a wide range of spectral types (*Be p* to *M* 2 *ep*). The largest periodic variations of the magnetic field have been found in conspicuous spectrum variables of type *A* and especially in α^2 Canum Venaticorum and HD 125248 where the magnetic field is reversed periodically with periods of 5.46939 and 9.298 days and amplitudes of over 4000 and 6000 gauss respectively. In those cases, while many lines show little variation, there are others which can be divided into two groups represented respectively by the lines of Eu II and by those of Cr I and Cr II which vary widely and in opposite phase.

The radial velocity is also variable with the same period as the magnetic field, but, in the case of the spectrum variables, different elements often show different

[1] O. STRUVE: Astrophys. Journ. **73**, 94 (1931). — O. STRUVE and P. SWINGS: Astrophys. Journ. **75**, 161 (1932).

[2] B. P. GERASIMOVIC: Monthly Notices Roy. Astronom. Soc. London **94**, 737 (1934).

[3] Cf. A. J. DEUTSCH: Astrophys. Journ. **105**, 283 (1947); Publ. Astronom. Soc. Pacific **68**, 92 (1956); for a detailed analysis of the spectrum variations cf. G. R. and E. M. BURBIDGE: Astrophys. Journ., Suppl. **1**, 431 (1955).

[4] For a general survey of this field cf. H. W. BABCOCK and T. G. COWLING: Monthly Notices Roy. Astronom. Soc. London **113**, 366 (1953).

velocity curves[1]. On the other hand, in the case of the peculiar A-type star HD 153882 [2] which is not a spectrum variable but has a magnetic field varying periodically between $+4500$ gauss and -4000 gauss in about 6 days, the radial velocities of different elements show practically no systematic differences with respect to a mean velocity curve with a total range of about 5 km/sec.

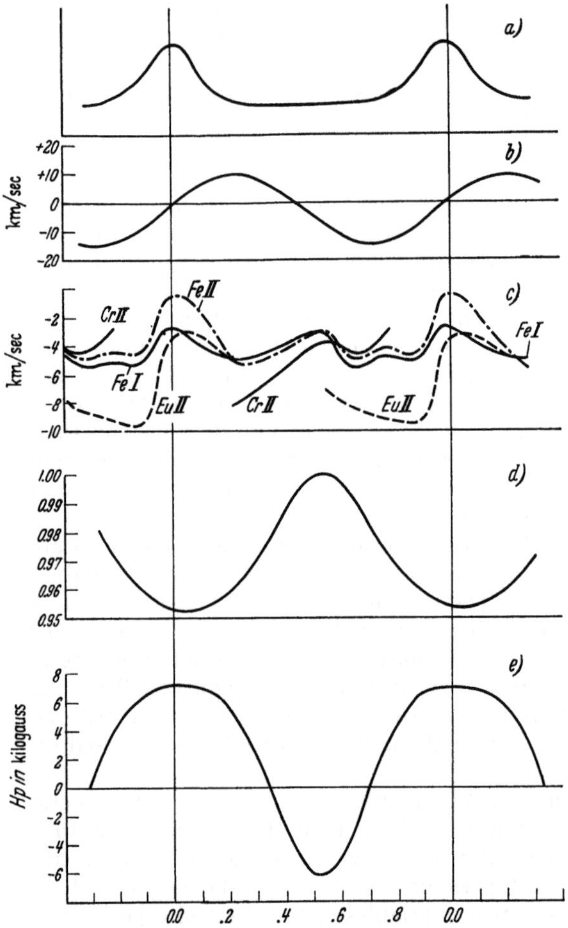

Fig. 36a–e. Variations in the magnetic variable HD 125248. (a) strength of Eu II lines; (b) "mean" radial velocity (Stibbs); (c) radial velocity of individual elements (Babcock); (d) light intensity; (e) polar magnetic field intensity (Babcock).

Deutsch[3] has shown that the line widths of relatively non-variable lines in many of these stars are nearly inversely proportional to the periods which suggests strongly that the observed period is essentially one of rotation. This favours an interpretation in terms of a magnetic field inclined with respect to the axis of rotation (oblique rotator) different types of elements accumulating in regions of opposite polarities. This type of theory would also provide a simple explanation of the "cross-over effect" discovered by Babcock which seems to indicate that the reversal of polarity propagates over the surface rather than occurring instantaneously at all points.

In some cases, photo-electric measurements have also revealed light variations[4] with amplitudes of a few hundredths of a magnitude. As an illustration, Fig. 36 summarizes the variations observed in the case of HD 125248. Curve b corresponds to an analysis by Stibbs of radial velocities measured in 1931 by W. W. Morgan and in the light of the more recent observations by Babcock (diagram c) it is probably not very significant.

Abundance determinations are difficult and very different results especially for the rare-earth elements may be obtained at different phases. According to

---

[1] O. Struve and P. Swings: Astrophys. Journ. **98**, 361 (1943). — H. W. Babcock and S. Burd: Astrophys. Journ. **116**, 8 (1952).

[2] G. Gjellestad and H. Babcock: Astrophys. Journ. **117**, 12 (1953).

[3] A. J. Deutsch: Trans. Internat. Astronom. Union **8**, 801 (1954).

[4] Cf. D. W. N. Stibbs: Monthly Notices Roy. Astronom Soc. London **110**, 395 (1950). — S. S. Provin: Astrophys. Journ. **117**, 21 (1953); **118**, 281, 489 (1953). — G. R. Miczaika: Trans. Internat. Astronom. Union **8**, 805 (1954).

the BURBIDGES[1] who have derived average abundances over the cycle, the rare-earth elements and other heavy elements are abnormally abundant. FOWLER and the BURBIDGES[2] have suggested that nuclear synthesis due to α-particle or neutron capture could explain the enrichment in heavy elements. But this requires accelerating nuclei to very high energies (1 to 100 Mev) and thus, strong magnetic fields, variable in a reference frame moving with the local gas velocity. This condition would not be realized in a rigid oblique rotator and would require a certain amount of true intrinsic variability.

As reported in the introduction, H. W. BABCOCK[3] has recently found appreciable variable magnetic fields in classical Cepheids and RR Lyrae stars. The correlation between these variations and the light and radial velocity cycle is not yet known but it will certainly raise interesting problems.

j) Stars with extremely rapid light variations.

41. We have already encountered incidentally some examples of these extremely rapid light variations in U Geminorum stars and T Tauri stars where they appear as a secondary phenomenon and also in discussing the possible connection between the T Tauri stars and the "flash" stars discovered by HARO. These last objects, dwarfs of late spectral type (dK), seem to be related[4] to the "flare-stars" recently discovered in the vicinity of the Sun among the dMe stars of the lowest known luminosity. On the other hand, M. F. WALKER[5] has discovered extremely rapid variations in many old novae and in a number of related stars. Despite some common characteristics these last variations distinguish themselves from the former by their more continuous character and also by the type of objects in which they occur.

α) The UV Ceti variables. This new class of variable stars was officially recognized by the I.A.U. in 1950 and was called after the most typical of the flare stars, UV Ceti, discovered by W. J. LUYTEN in 1948. Previously, the variability of only two other members of the class had been noted by A. VAN MAANEN. But after LUYTEN'S discovery a systematic search was started at different places and, in the following years, their number increased fairly rapidly so that about a dozen stars are now acknowledged true members of the class[6].

These stars are characterized by extremely sharp outbursts of radiation, the maximum being reached in a few minutes or seconds, the decline in brightness being somewhat slower although usually the normal brightness is resumed after some minutes. These flares occur in a haphazard way and, in the same star, their amplitudes may vary considerably. For instance in UV Ceti, while the most frequent outbursts are of the order of 1 to 2 magnitudes, two flares have been observed in September 1952 at a week interval, corresponding respectively to a brightening of 3.4 and ≈6 magnitudes in 15 and 20 seconds. In the first case, the decline lasted only 8 minutes but, in the second, it took nearly two hours.

[1] E. M. BURBIDGE and G. R. BURBIDGE: Astrophys. Journ. **122**, 396 (1955).

[2] W. A. FOWLER, G. R. BURBIDGE and E. M. BURBIDGE: Astrophys. Journ. Suppl. **2**, 167 (1955).

[3] H. W. BABCOCK: Publ. Astronom. Soc. Pacific **68**, 70 (1956) and communication at the Stockholm Symposium on "Electromagnetic Phenomena in cosmical Physics", August 1956.

[4] G. HARO and E. CHAVIRA: Bol. Obs. Tonantzintla y Tacubaya **1955**, No. 12; also Ref. [*38d*].

[5] M. F. WALKER: Publ. Astronom. Soc. Pacific **66**, 71 (1954); also Ref. [*38*], p. 146.

[6] Cf. the lists given by M. PETIT: Ciel et Terre **70**, 401 (1954), and P. E. ROQUES: Publ. Astronom. Soc. Pacific **67**, 34 (1955).

Although there is certainly no proper period there might be a significant mean interval between outbursts of the order of 1.5 day for UV Ceti, the larger part of it by far being spent at normal minimum brightness. In between flares in UV Ceti, V. OSKANJAN has noted secondary irregular but continuous variations with a total amplitude smaller than 0.8 magnitude and a mean interval of about 30 minutes.

All the known flare stars are of spectral type *dMe* and of extremely low luminosity ($M_v \approx 11$ to 16) so that the large increase in magnitude associated with the flares corresponds actually to an excess radiation which remains small compared to the luminosity of somewhat intrinsically brighter stars. This makes it difficult to decide whether the weak intrinsic luminosity of these objects is an active factor in the production of flares or if it is simply a condition for their visibility.

On the contrary, the presence of relatively strong emission lines of H and Ca II during the quiescent periods seems to be physically connected with the occurrence of flares.

Only a few spectra have been obtained at maximum and they show [*38e*] a strong reinforcement of the continuum in the ultraviolet, a considerable intensification of the hydrogen lines and, in the case of the flare of UV Ceti observed spectroscopically by JOY and HUMASON, even bright lines of He I and a faint line of He II. In the last case, it was estimated that the corresponding temperature is of the order of 10000° K or higher.

The rapidity of the whole phenomenon is usually taken to imply that only a small fraction of the star surface is affected, a point which strengthens the similarity with solar flares. However, since an increase in temperature from 3000° to 10000° multiplies the flux in the visual region by a factor of the order of 500, an ordinary flare ($\Delta M_v \approx 1.5$) corresponds already to an outburst affecting about 1% of the stellar surface; this might be appreciably larger in very strong flares. It is not obvious that such a thermal perturbation could die down in a few minutes. In this respect, it is worth keeping in mind AMBARTSUMIAN'S point of view [*38f*] recalled in Sect. 39. In any case, it seems pretty certain, that this energy must be liberated right in the atmosphere as the perturbation of the temperature gradient necessary to bring it from greater depths could hardly have such a short relaxation time.

Let us add that many of these flare stars are components of binaries, duplicity being possibly a favourable factor. In particular flaring has also been observed at maximum phase[1] in U Pegasi, an eclipsing variable of very short period and this circumstance has led HURUHATA to suggest that the responsible phenomena might be confined to the space between the components and associated with the tidal bulges.

If HARO'S "flash-stars" may be included in this group, there would seem to be a relation between the quiescent spectrum and the total duration of the outburst (cf. p. 429, Ref. 4).

β) Other cases of very rapid light-variations. The discovery[2] of very rapid and random light-variations reaching up to 0.44 magnitude and having cycles of 1 to 30 min in MacRae +43°1, a star with a spectrum resembling that of an old nova, led WALKER to search for similar changes in a series of objects related to novae ([*38*], p. 146). Of 13 old novae investigated, all showed some sign of variability of this type, T Coronae Borealis being the most active. In all cases,

[1] M. HURUHATA: Publ. Astronom. Soc. Pacific **64**, 200 (1952).
[2] M. F. WALKER: Publ. Astronom. Soc. Pacific **66**, 71 (1954).

the range of the variation increases with the frequency. For instance in MacRae $+43^\circ 1$, the amplitudes in the ultraviolet, the blue and the yellow are proportional respectively to 1, 0.8 and 0.7.

Are these variations related to the flares in the cool *dMe* stars? Because of the higher superficial temperatures of the stars considered here, one must expect that a larger fraction of the surface must be involved. For instance, WALKER adopting for MacRae $+43^\circ 1$ a temperature of 13000° K, finds that even if the material is heated up to 100000° K, the dimensions of the heated spot should be of the order of one tenth of the radius of the star. Similar variations were found in 3 recurrent and 2 possible novae while they seem to be absent or very weak in nova-like variables and in the white dwarf 40 Eri B. The four U Geminorum stars studied showed also the same type of extremely rapid fluctuations as well as the blue companions of *o* Ceti and R Aquarii.

It is difficult to be sure that all these variations although of the same type are physically related. But it is interesting to note that, as with flare stars, a large number of these objects are known to be binaries.

In particular, it was in the course of this investigation that WALKER discovered that Nova (DQ) Herculis 1934 is an eclipsing binary having the shortest known period, $4^h\,39^m$. This star kept another surprise in store since in addition to the eclipse and the erratic variability similar to that found in other old novae, its light curve revealed periodic variations between phases $0.110\,P$ and $0.325\,P$ in the eclipse cycle. These last fluctuations have a period of 1.18 minute and a range in the ultraviolet of 0.07 magnitude. This is the only known case where rapid variations are properly periodic and since they occur at special phases of the eclipse cycle, it is likely that they find their origin in the binary nature of the system.

C. Theory.

42. General remarks. Of all the attempts to explain intrinsic stellar variations, it is the hypothesis of stellar oscillations and more particularly of radial pulsations which has met with the greatest amount of approval among astronomers and it is the only one which has given rise to extensive theoretical developments. There exist, at least for radial pulsations, systematic accounts [*39*] of the successive stages reached by the theory and the latest one by S. ROSSELAND [*39d*] provides a fairly complete and detailed picture which, in many respects, is still up to date.

Of all the direct arguments in favour of the pulsation theory, the one which contributed most to its early success is the proportionality between the theoretical period of pulsation P and the inverse square-root of the mean density $\bar{\varrho}$ of the star

$$P \propto \bar{\varrho}^{-\frac{1}{2}} \tag{42.1}$$

a relation which is confirmed by the observations.

It should be noted however that, according to the principle of similitude[1], this relation holds for any type of periodic motion under the gravitational attraction of the star, provided that the significant length in the problem be of the order of the star radius. It remains true, in particular, for any non-radial oscillations and also for the revolution of a companion close to the surface of the star. It is only through the value of the *constant of proportionality* which, of course, can only be fixed by a complete theory in each particular case, that one may expect to distinguish between these different types of motion.

[1] Lord. RAYLEIGH: Nature, Lond. **95**, 66 (1915).

We may recall here a few results for orbital motion which will be useful later for purpose of comparison. According to DARWIN (cf. [14], Eqs. 201.4 and 360.1), the angular velocity of rotation of an incompressible mass at the moment when instability by fission[1] just sets in, is given by

$$\omega^2 = 0.071\, 4\pi G \bar{\varrho}.$$

But in this case, the period of the light variation is only half the period of rotation.

$$P = 1.29 \times 10^4 \bar{\varrho}^{-\frac{1}{2}} \sec \quad \text{or} \quad P \sqrt{\frac{\bar{\varrho}}{\bar{\varrho}_\odot}} = 0.125 \text{ day}. \tag{42.2}$$

While more generally, according to POINCARÉ, some form of rotational instability will set in if $\omega^2 > 2\pi G \bar{\varrho}$, i.e. if

$$P \sqrt{\frac{\bar{\varrho}}{\bar{\varrho}_\odot}} < 0.09 \text{ day}. \tag{42.3}$$

For two equal spherical masses rotating in contact, one has

$$\omega^2 = (0.0833)\, 4\pi G \bar{\varrho} \tag{42.4}$$

which corresponds to a period of light variation

$$P = 1.19 \times 10^4 \bar{\varrho}^{-\frac{1}{2}} \sec \quad \text{or} \quad P \sqrt{\frac{\bar{\varrho}}{\bar{\varrho}_\odot}} = 0.116 \text{ day}. \tag{42.5}$$

If the radius of the companion is small compared to that of the primary

$$\omega^2 = (0.3333)\, 4\pi G \bar{\varrho}. \tag{42.6}$$

In this case, due to the lack of symmetry, the period of light variation is equal to the period of revolution which is still given by formula (42.5).

I. General equations.

43. Remarks on notations. When discussing general deformations of the whole mass of a star, the most convenient language is that of hydrodynamics.

Two types of representation are then available, one using Eulerian and the other Lagrangian coordinates [40]. In both cases, the best suited system of coordinates will be dictated by the general symmetry of the problem. This will often lead to adopt curvilinear coordinates and then, the general tensorial notation [41] is of interest in some of the more complex aspects of the problem.

In general, we shall designate Cartesian coordinates in an inertial frame of reference by (x) and although the distinction between covariant and contravariant components has no meaning here, we shall keep to it to make the analogy with the case of general coordinates q^i clearer.

In the Lagrangian representation, the coordinates (x^1, x^2, x^3) are the components of the radius vector $\boldsymbol{r}$ bound to a given particle and are functions of the time t and three parameters (a^1, a^2, a^3) which serve to identify the particle considered. The a^i are often taken as the initial values of the coordinates $(a^i = x_0^i)$. Denoting differentiation with respect to t by a dot, the equation of motion

[1] For the exact type of instability concerned cf. R. A. LYTTLETON: The stability of rotating liquid Masses, especially Chap. X. Cambridge: Cambridge University Press 1953.

keeps the usual form

$$\gamma = \dot{\boldsymbol{v}} = \ddot{\boldsymbol{r}} = \boldsymbol{f} \tag{43.1}$$

if $\boldsymbol{f}$ is the total force acting per unit mass.

LAGRANGE'S equations allow one to apply many of the resources of classical mechanics to fluid dynamics[1]. Moreover, in this notation, it is easy to deal with the case where some parameters (for instance the molecular weight) vary from particle to particle. On the other hand, the explicit expressions of spatial differential operators in the independent variables a^i become, in general, very complicated. However, if the problem involves only *one degree of freedom* as for instance in the case of radial pulsations all these difficulties disappear and then, the use of Lagrangian coordinates is often advantageous.

In Eulerian notations, the x^i are purely geometrical coordinates which label the space in which the motion of the fluid, defined by a velocity field $\boldsymbol{v}(x^i, t)$, takes place. The time derivatives of Eulerian coordinates have no meaning and in the definitions

$$\dot{x}^i = v^i(x^i, t) \quad \text{or} \quad \dot{\boldsymbol{r}} = \boldsymbol{v}(\boldsymbol{r}, t) \tag{43.2}$$

the x^i or $\boldsymbol{r}$ must be treated as Lagrangian coordinates.

Eqs. (43.2) establish the connection between the two representations. If the Eulerian solution $\boldsymbol{v}(x^i, t)$ is known, the integration of (43.2), gives the Lagrangian solution

$$x^i = x^i(x_0^1, x_0^2, x_0^3, t) \quad \text{or} \quad x^i = x^i(a^1, a^2, a^3, t). \tag{43.3}$$

Remembering again that the x^i or $\boldsymbol{r}$ in (43.2) are functions of t, one finds

$$\left.\begin{aligned} \gamma^i = \ddot{x}^i = \dot{v}^i = \frac{\partial v^i}{\partial t} + \sum_k \frac{\partial v^i}{\partial x^k} v^k = \frac{\partial v^i}{\partial t} + \boldsymbol{v}\cdot\operatorname{grad} v^i \\ \text{or} \qquad \gamma = \ddot{\boldsymbol{r}} = \frac{\partial \boldsymbol{v}}{\partial t} + (\boldsymbol{v}\cdot\operatorname{grad})\,\boldsymbol{v}. \end{aligned}\right\} \tag{43.4}$$

The same type of relation exists between the local and total (or following the motion) time derivatives of any function F or $\boldsymbol{F}$ attached to the particles of the medium

$$\dot{F} = \frac{\partial F}{\partial t} + \boldsymbol{v}\cdot\operatorname{grad} F = \frac{\partial F}{\partial t} + \sum_k v^k \frac{\partial F}{\partial x^k}, \tag{43.5}$$

$$\dot{\boldsymbol{F}} = \frac{\partial \boldsymbol{F}}{\partial t} + (\boldsymbol{v}\cdot\operatorname{grad})\,\boldsymbol{F} \quad \text{or} \quad \dot{F}^i = \frac{\partial F^i}{\partial t} + \sum_k v^k \frac{\partial F^i}{\partial x^k}. \tag{43.6}$$

The Lagrangian solutions (43.3) can be interpreted as defining a continuous transformation in the course of time from the variables x_0^i (or a^i) to the x^i, and the Jacobian of the transformation

$$X = \frac{\partial(x^1, x^2, x^3)}{\partial(a^1, a^2, a^3)} \tag{43.7}$$

recurs often in the Lagrangian representation. The useful identity

$$\dot{X} = X \sum_k \frac{\partial v^k}{\partial x^k} = X \operatorname{div} \boldsymbol{v} \tag{43.8}$$

is easily verified.

[1] Cf. H. GOLDSTEIN: Classical Mechanics, Chap. XI. Cambridge, Mass.: Addison-Wesley Publ. Co. 1953.

In the case of generalized coordinates q^i introduced by

$$x^i = x^i(q^1, q^2, q^3, t) \tag{43.9}$$

the following generalizations hold

$$\dot{F} = \frac{\partial F}{\partial t} + \sum_k \frac{\partial F}{\partial q^k} \dot{q}^k \tag{43.5'}$$

and

$$\dot{Q} = Q \sum_k \frac{\partial \dot{q}^k}{\partial q^k} \tag{43.8'}$$

if

$$Q = \frac{\partial(q^1, q^2, q^3)}{\partial(a^1, a^2, a^3)}. \tag{43.7'}$$

a) Equation of continuity. (Conservation of mass.)

44. LAGRANGE's form. In going from an initial state (t_0, x_0^i) to a final state (t, x^i), the elementary volume occupied by a given amount of matter $(a_0^i \leqq a^i \leqq a_0^i + d a^i)$ changes in the ratio

$$\frac{X}{X_0} = \frac{\varrho_0}{\varrho} \tag{44.1}$$

if ϱ_0 and ϱ denote the initial and final densities (specific mass). If $a^i \equiv x_0^i$, this reduces to

$$\varrho X = \varrho_0. \tag{44.1'}$$

In generalized coordinates q^i, if we represent by J the Jacobian $\partial(x^i)/\partial(q^i)$, (44.1) becomes

$$\varrho J Q = \varrho_0 J_0 Q_0 \tag{44.2}$$

which for $a^i = q_0^i$ reduces to

$$\varrho J Q = \varrho_0 J_0. \tag{44.2'}$$

45. EULER's form. To obtain the Eulerian form, we take the time derivative of (44.1) using (43.8) and get

$$\dot{\varrho} + \varrho \sum_k \frac{\partial v^k}{\partial x^k} = 0 \quad \text{or} \quad \dot{\varrho} + \varrho \operatorname{div} \boldsymbol{v} = 0 \tag{45.1}$$

which according to (43.5) can also be written

$$\frac{\partial \varrho}{\partial t} + \sum_k \frac{\partial(\varrho v^k)}{\partial x^k} = 0 \quad \text{or} \quad \frac{\partial \varrho}{\partial t} + \operatorname{div}(\varrho \boldsymbol{v}) = 0. \tag{45.2}$$

The same procedure applied to (44.2) with the help of (43.8') gives

$$\frac{\dot{\varrho}}{\varrho} + \frac{\dot{J}}{J} + \sum_k \frac{\partial \dot{q}^k}{\partial q^k} = 0. \tag{45.3}$$

By (43.5'), Eq. (45.3) can also be written as

$$\frac{\partial \varrho}{\partial t} + \frac{\varrho}{J} \frac{\partial J}{\partial t} + \sum_k \frac{\partial J}{\partial q^k} \dot{q}^k + \sum_k \frac{\partial(\varrho \dot{q}^k)}{\partial q^k} = 0. \tag{45.4}$$

But in the generalized coordinates, the elementary geometrical distance ds is defined by

$$ds^2 = \sum_{i,k} g_{ik} dq^i dq^k \quad \text{with} \quad g_{ik} = \sum_j \frac{\partial x^j}{\partial q^i} \cdot \frac{\partial x^j}{\partial q^k} \tag{45.5}$$

and one verifies immediately that

$$g = |g_{ik}| = J^2. \tag{45.6}$$

On the other hand, if we denote the CHRISTOFFEL's symbols of the second kind by

$$\Gamma^j_{ik} = \sum_h g^{jh} \frac{1}{2}\left(\frac{\partial g_{ih}}{\partial q^k} + \frac{\partial g_{hk}}{\partial q^i} - \frac{\partial g_{ki}}{\partial q^h}\right) \tag{45.7}$$

we have

$$\sum_k \Gamma^k_{kh} = \frac{1}{\sqrt{g}} \frac{\partial \sqrt{g}}{\partial q^h} = \frac{1}{J} \frac{\partial J}{\partial q^h} \tag{45.8}$$

according to (45.6). Introducing (45.8) in (45.4), one gets

$$\frac{\partial \varrho}{\partial t} + \frac{\varrho}{J}\frac{\partial J}{\partial t} + \sum_k \left[\sum_h \Gamma^k_{kh}\, \varrho\, v^h + \frac{\partial(\varrho\, v^k)}{\partial q^k}\right] = 0$$

in denoting by v^i the generalized velocities $\dot{q}^i$.

Taking into account the definition of the covariant derivative Δ_k with respect to q^k

$$\Delta_k(\varrho\, v^i) = \frac{\partial(\varrho\, v^i)}{\partial q^k} + \sum_h \Gamma^i_{kh}\, \varrho\, v^h, \tag{45.9}$$

we can also write

$$\frac{\partial \varrho}{\partial t} + \frac{\varrho}{J}\frac{\partial J}{\partial t} + \sum_k \Delta_k(\varrho\, v^k) = 0 \tag{45.10}$$

or

$$\frac{\partial \varrho}{\partial t} + \frac{\varrho}{J}\frac{\partial J}{\partial t} + \operatorname{div}(\varrho\, \boldsymbol{v}) = 0. \tag{45.11}$$

If J does not depend explicitly of t and this may occur, even if the transformation (43.9) does depend of t, Eqs. (45.10) or (45.11) reduce to

$$\frac{\partial \varrho}{\partial t} + \sum_k \Delta_k(\varrho\, v^k) = 0 \quad \text{or} \quad \frac{\partial \varrho}{\partial t} + \operatorname{div}(\varrho\, \boldsymbol{v}) = 0\,. \tag{45.12}$$

b) Equation of motion. (Conservation of momentum.)

46. The stresses. To obtain the equation of motion, we have only to express Eq. (43.1) explicitly in the appropriate coordinates. A fluid is characterized by the existence of typical internal forces: the stresses or tensions which can be described by means of nine quantities P_G^{ik} which are the components of a symmetric tensor. Furthermore, $\sum P_G^{ii}$ is independent of the direction of the axes and we may write

$$\sum_i (P^i_i)_G = -3 p_G$$

where p_G denotes the mean pressure.

We shall admit that the gaseous pressure keeps always the form

$$p_G = \frac{R\varrho T}{\bar{\mu}} = \frac{1}{3} n\, \bar{\mu}\, m_H \overline{C_T^2} = n\, k T \tag{46.1}$$

where R is the gas constant, T the absolute temperature, $\bar{\mu}$ the mean molecular weight with respect to the mass m_H of hydrogen, $\overline{C_T^2}$, a mean square velocity of thermal motion, k Boltzmann constant and n is the number density of particles:

$$\mathsf{R} = \frac{\mathsf{k}}{m_H} \quad \text{and} \quad n = \frac{\varrho}{\bar{\mu}\, m_H}. \tag{46.2}$$

If as usual, it is assumed that the stresses are related linearly to the strains, one has

$$P_G^{ik} = P_G^{ki} = -p_G g^{ik} + \mathfrak{P}_G^{ik} \tag{46.3}$$

where, in Cartesian coordinates, $g^{ik} = \delta_{ik}$ (KRONECKER's symbol) and the $\mathfrak{P}_G^{ik}$ are defined by

$$\mathfrak{P}_G^{ik} = -\frac{2}{3}\eta_G g^{ik} \sum_j \frac{\partial v^j}{\partial x^j} + \eta_G \sum_l \left(g^{il}\frac{\partial v^k}{\partial x^l} + g^{lk}\frac{\partial v^i}{\partial x^l}\right) \tag{46.4}$$

or

$$\left.\begin{aligned} \mathfrak{P}_G^{ii} &= -\frac{2}{3}\eta_G \operatorname{div} \boldsymbol{v} + 2\eta_G \frac{\partial v^i}{\partial x^i}, \\ \mathfrak{P}_G^{ik} &= \eta_G\left(\frac{\partial v^k}{\partial x^i} + \frac{\partial v^i}{\partial x^k}\right), \qquad i \neq k \end{aligned}\right\} \tag{46.5}$$

which are valid even in the case of a variable coefficient of dynamical viscosity η_G defined by

$$\eta_G \approx \tfrac{1}{3}\varrho\, \bar{C}_T \bar{l} \tag{46.6}$$

where $\bar{l}$ is a mean free path.

The k component of the total stress acting on any internal volume limited by a surface S is

$$\iint\limits_S T^k dS = \iint\limits_S \sum_i P_G^{ik} dS_i = \iiint\limits_V \sum \frac{\partial P_G^{ik}}{\partial x^i} dV \tag{46.7}$$

by OSTROGRADSKI's theorem. Thus, these stresses correspond to a force per unit volume

$$f^k = \sum_i \frac{\partial P_G^{ik}}{\partial x^i}.$$

The k component of Eq. (43.1) can then be written, taking (46.3) into account,

$$\varrho \ddot{x}^k = \varrho F^k - \sum_i g^{ik}\frac{\partial p_G}{\partial x^i} + \sum_i \frac{\partial \mathfrak{P}_G^{ik}}{\partial x^i} \tag{46.8}$$

which becomes vectorially

$$\varrho \ddot{\boldsymbol{r}} = \varrho \boldsymbol{F} - \operatorname{grad} p_G + \operatorname{div} \mathfrak{P}_G \tag{46.9}$$

where $\boldsymbol{F}$ denotes all other forces, external or internal.

Introducing the values (46.4) of the $\mathfrak{P}_G^{ik}$ in (46.8), we obtain

$$\left.\begin{aligned} \varrho \ddot{x}^k = {} & \varrho F^k - \sum_i g^{ik}\frac{\partial p_G}{\partial x^i} - \frac{2}{3}\eta_G \sum_{i,j} g^{ik}\frac{\partial}{\partial x^i}\left(\frac{\partial v^j}{\partial x^j}\right) + \eta_G \sum_{i,l}\left(g^{il}\frac{\partial^2 v^k}{\partial x^l \partial x^i} + g^{lk}\frac{\partial^2 v^i}{\partial x^l \partial x^i}\right) - \\ & -\frac{2}{3}\sum_{i,j} g^{ik}\frac{\partial \eta_G}{\partial x^i}\left(\frac{\partial v^j}{\partial x^j}\right) + \sum_{i,l}\frac{\partial \eta_G}{\partial x^i}\left(g^{il}\frac{\partial v^k}{\partial x^l} + g^{lk}\frac{\partial v^i}{\partial x^l}\right). \end{aligned}\right\} \tag{46.10}$$

Taking into account explicitly the Cartesian character of the x^i, this reduces to

$$\left.\begin{aligned} \varrho \ddot{x}^k = {} & \varrho F_k - \frac{\partial p_G}{\partial x_k} + \frac{1}{3}\eta_G \frac{\partial}{\partial x_k}\left(\sum_j \frac{\partial v_j}{\partial x_j}\right) + \eta_G \sum_i \frac{\partial^2 v_k}{\partial x_i^2} - \\ & -\frac{2}{3}\frac{\partial \eta_G}{\partial x_k}\sum_j \frac{\partial v_j}{\partial x_j} + \sum_i \frac{\partial \eta_G}{\partial x_i}\left(\frac{\partial v_k}{\partial x_i} + \frac{\partial v_i}{\partial x_k}\right) \end{aligned}\right\} \tag{46.11}$$

where the position of the indices has no special meaning any more. If η_G is constant, it can further be simplified and written in the usual vectorial form

$$\varrho\,\ddot{\boldsymbol{r}} = \varrho\,\boldsymbol{F} - \operatorname{grad} p_G + \tfrac{1}{3}\eta_G \operatorname{grad}\operatorname{div}\boldsymbol{v} + \eta_G\,\nabla^2\boldsymbol{v}. \tag{46.12}$$

47. LAGRANGE's form. To obtain the Lagrangian equation, we must first replace everywhere $\boldsymbol{v}$ by $\dot{\boldsymbol{r}}$ or v^i by $\dot{x}^i$ and then introduce the identifying parameters a^i as independent variables. In the general case, as already stated, this leads to very complicated expressions (cf. for instance Ref. [*40b*], Vol. I, p. 121). As a simple illustration, let us consider the case where the viscosity is negligible. Then Eq. (46.11), becomes

$$\ddot{x}_k = F_k - \frac{1}{\varrho}\frac{\partial p_G}{\partial x_k}. \tag{47.1}$$

Multiplying by $\dfrac{\partial x_k}{\partial a_j}$ and summing over k gives

$$\sum_k \ddot{x}_k \frac{\partial x_k}{\partial a_j} = \sum_k F_k \frac{\partial x_k}{\partial a_j} - \frac{1}{\varrho}\frac{\partial p_G}{\partial a_j} \tag{47.2}$$

where the two terms on the right represent respectively the generalized component of the force and the gradient of pressure with respect to the a_j, these parameters now playing the rôle of independent variables.

The introduction of the generalized coordinates defined by (43.9) can proceed in exactly the same way in multiplying (47.1) by $(\partial x^k/\partial q^j)$ and summing over k. Transforming the left-hand member as in classical mechanics and introducing the kinetic energy per unit mass

$$K = \tfrac{1}{2}\sum_i \dot{x}_i^2 \tag{47.3}$$

one finds

$$\frac{d}{dt}\left(\frac{\partial K}{\partial \dot{q}^j}\right) - \frac{\partial K}{\partial q^j} = Q_j - \frac{1}{\varrho}\frac{\partial p_G}{\partial q^j}, \qquad j = 1, 2, 3 \tag{47.4}$$

where the generalized covariant component of the force in q^j is given by

$$Q_j = \sum_{i=1}^{3} F_i \frac{\partial x_i}{\partial q^j} \tag{47.5}$$

and K has to be expressed explicitly in terms of the $(q^i, \dot{q}^i, t)$:

$$\left.\begin{aligned} K &= \frac{1}{2}\sum_{i=1}^{3}\left(\frac{\partial x_i}{\partial t}\right)^2 + \sum_{i,j=1}^{3}\left(\frac{\partial x_i}{\partial t}\cdot\frac{\partial x_i}{\partial q^j}\right)\dot{q}^j + \frac{1}{2}\sum_{j,k=1}^{3}\sum_{i=1}^{3}\left(\frac{\partial x_i}{\partial q^j}\cdot\frac{\partial x_i}{\partial q^k}\right)\dot{q}^j\dot{q}^k \\ &= a_0 + \sum_{j=1}^{3} a_j\,\dot{q}^j + \sum_{j,k}\frac{1}{2}\,g_{ik}\,\dot{q}^j\,\dot{q}^k \end{aligned}\right\} \tag{47.6}$$

with the definition (45.5) of the g_{ik}.

Since the general transformation (43.9) involves the time, Eq. (47.4) is valid for *relative coordinates* as well as for inertial systems. Of course, we still have to introduce the independent variables a^i in (47.4). As previously, we multiply by $(\partial q^j/\partial a^i)$ and sum over j, obtaining

$$\sum_j \frac{d}{dt}\left(\frac{\partial K}{\partial \dot{q}^j}\right)\frac{\partial q^j}{\partial a^i} - \sum_j \frac{\partial K}{\partial q^j}\frac{\partial q^j}{\partial a^i} = \sum_j Q_j \frac{\partial q^j}{\partial a^i} - \frac{1}{\varrho}\sum_j \frac{\partial p_G}{\partial q^j}\frac{\partial q^j}{\partial a^i}. \tag{47.7}$$

In some cases (cf. Ref. [2], p. 123), it is possible to press the analogy with classical mechanics further and to construct a Lagrangian density

$$\mathcal{L}\left(q^i, \dot{q}^i, \frac{\partial q^i}{\partial a^j}, t\right) = \varrho_0[K - V] \tag{47.8}$$

where V is a generalized potential for the forces acting on unit mass and ϱ_0 is the density at the time $t=0$ when $a^i = x_0^i$, such that the total Lagrangian

$$L = \iiint \mathcal{L}\, d a^1\, d a^2\, d a^3 \tag{47.9}$$

obeys HAMILTON's principle

$$\delta \int L\, dt = \delta \int_{t_1}^{t_2} dt \iiint \mathcal{L}\, d a^1\, d a^2\, d a^3 = 0. \tag{47.10}$$

In that case, EULER's equations expressing the condition (47.10) become

$$\frac{d}{dt}\left(\frac{\partial \mathcal{L}}{\partial \dot{q}^j}\right) + \sum_{i=1}^{3} \frac{d}{d a^i}\left[\frac{\partial \mathcal{L}}{\partial\left(\frac{\partial q^j}{\partial a^i}\right)}\right] - \frac{\partial \mathcal{L}}{\partial q^j} = 0, \quad j = 1, 2, 3 \tag{47.11}$$

since the a^i are also independent variables. Of course (47.11) should reduce to the equation of motion (47.7).

48. EULER's form. In an inertial system of Cartesian coordinates, the Eulerian form of (46.10), (46.11) or (46.12) are obtained simply by replacing $\ddot{x}_k$ or $\ddot{\boldsymbol{r}}$ in the first member of these equations by their expressions (43.4). For instance, Eq. (46.12) becomes

$$\varrho \frac{d\boldsymbol{v}}{dt} = \varrho\left[\frac{\partial \boldsymbol{v}}{\partial t} + (\boldsymbol{v}\,\mathrm{grad})\,\boldsymbol{v}\right] = \varrho \boldsymbol{F} - \mathrm{grad}\, p_G + \frac{1}{3}\eta_G\, \mathrm{grad\, div}\, \boldsymbol{v} + \eta_G \nabla^2 \boldsymbol{v}. \tag{48.1}$$

In the case of an inertial system of arbitrary curvilinear coordinates q^i, the general form (46.9) or (46.10) remains valid provided that we interpret the v^i as $\dot{q}^i$, replace everywhere x^i by q^i, $\partial/\partial x^i$ by Δ_i with respect to q^i as defined by (45.9) and introduce for the accelerations $\ddot{x}^k$, their covariant definitions

$$\gamma^k = \Delta_t v^k = \frac{d v^k}{dt} + \sum_{i,h} \Gamma_{ih}^k v^i v^h. \tag{48.2}$$

Expanding dv^k/dt by formula (47.5) and grouping the terms, one can also write

$$\gamma^k = \frac{\partial v^k}{\partial t} + \sum_i v^i\left(\frac{\partial v^k}{\partial q^i} + \sum_h \Gamma_{ih}^k v^h\right) = \frac{\partial v^k}{\partial t} + \sum_i v^i \Delta_i v^k. \tag{48.3}$$

In this way, the general Eqs. (46.9) and (46.10) become

$$\varrho\left(\frac{\partial v^k}{\partial t} + \sum_i v^i \Delta_i v^k\right) = \varrho F^k - \sum_i g^{ik} \Delta_i p_G + \sum_i \Delta_i \mathfrak{P}_G^{ik} \tag{48.4}$$

and

$$\left.\begin{aligned} \varrho\left(\frac{\partial v^k}{\partial t} + \sum_i v^i \Delta_i v^k\right) &= \varrho F^k - \sum_i g^{ik} \Delta_i p_G - \frac{2}{3}\eta_G \sum_{i,j} g^{ik} \Delta_i(\Delta_j v^j) + \\ &+ \eta_G \sum_{i,l}\left[g^{il} \Delta_i(\Delta_l v^k) + g^{lk} \Delta_i(\Delta_l v^i)\right] - \frac{2}{3}\sum_{i,j} g^{ik} \Delta_i \eta_G \Delta_j v^j + \\ &+ \sum_{i,l} \Delta_i \eta_G\left(g^{il} \Delta_l v^k + g^{lk} \Delta_l v^i\right). \end{aligned}\right\} \tag{48.5}$$

In all this, it is useful to keep in mind the relations between covariant and contravariant components

$$g^{ik} = \frac{G_{ik}}{g}, \quad v_i = \sum_k g_{ik} v^k, \quad \mathfrak{P}^i_k = \sum_h g_{kh} \mathfrak{P}^{ih}, \quad \mathfrak{P}_{ik} = \sum_h g_{ih} \mathfrak{P}^h_k \tag{48.6}$$

where G_{ik} is the minor of g_{ik} in $g = |g_{ik}|$.

Very often the coordinates used are orthogonal

$$g_{ik} = g^{ik} = 0 \quad \text{if} \quad i \neq k, \quad g^{ii} = \frac{1}{g^{ii}}. \tag{48.7}$$

In that case, it is useful at some stage to introduce the natural components

$$\left.\begin{aligned} \hat{v}_i &= \sqrt{v^i v_i} = v^i \sqrt{g_{ii}} = v_i \sqrt{g^{ii}}, \\ \hat{T}_{ik} &= T_{ik} \sqrt{g^{ii} g^{kk}} = T^{ik} \sqrt{g_{ii} g_{kk}} = T^i_k \sqrt{g_{ii} g^{kk}}. \end{aligned}\right\} \tag{48.8}$$

In the case of relative Cartesian coordinates (O', x'_i) whose motion with respect to an inertial system is defined by a translation $\boldsymbol{v}_{O'}$ of the origin and an instantaneous rotation $\boldsymbol{\omega}$ at the origin, one can express the absolute velocity $\boldsymbol{v}$ and acceleration $d\boldsymbol{v}/dt$ for instance in (48.1) by the usual formulae

$$\left.\begin{aligned} \boldsymbol{v} &= \boldsymbol{v}' + \boldsymbol{v}_{O'} + (\boldsymbol{\omega} \times \boldsymbol{r}'), \\ \gamma_a &= \frac{d\boldsymbol{v}}{dt} = \gamma_r + \gamma_e + \gamma_c \\ &= \frac{d\boldsymbol{v}'}{dt} + \frac{d\boldsymbol{v}_{O'}}{dt} + \boldsymbol{\omega} \times (\boldsymbol{\omega} \times \boldsymbol{r}') + \frac{d\boldsymbol{\omega}}{dt} \times \boldsymbol{r}' + 2(\boldsymbol{\omega} \times \boldsymbol{r}') \end{aligned}\right\} \tag{48.9}$$

where $d\boldsymbol{v}'/dt$ can again be written $[\partial \boldsymbol{v}'/\partial t + (\boldsymbol{v}' \operatorname{grad}_{x'}) \boldsymbol{v}']$. In the case of most systems of generalized coordinates commonly used, the corresponding substitutions for the components v^i and dv^i/dt are not too complicated and can also be used in the general Eqs. (48.4) and (48.5).

However for a general transformation involving the time t, such as (43.9), the most direct way is to proceed as we did to obtain (47.4). This equation can then be put into the proper Eulerian form by replacing $\dot{q}^i$ by v^i in K which becomes

$$K = K(q^i, v^i, t).$$

The total time derivatives of $\partial K/\partial v^i$ must then be carried out in treating the q^i as function of t, so that in Eulerian notations (47.4) becomes

$$\frac{\partial}{\partial t}\left(\frac{\partial K}{\partial v^k}\right) + \sum_i v^i \frac{\partial}{\partial q^i}\left(\frac{\partial K}{\partial v^k}\right) - \frac{\partial K}{\partial q^k} = Q_k - \frac{1}{\varrho}\frac{\partial p}{\partial q^k}, \quad k = 1, 2, 3 \tag{48.10}$$

where the q^i can now be considered as purely geometrical variables.

Even if the transformation (43.9) includes t, very often the coefficients a_0, a_j and g_{jk} in (47.6) are independent of t and in that case, the left-hand member of (48.10) reduces again to the covariant component (48.3) plus terms which give automatically the complementary accelerations γ_e and γ_c. In most cases, the last method using the expression of K in terms of the relative coordinates is more convenient than the one based on the substitutions (48.9).

49. Special form of the Eulerian equation. Eq. (48.4) can be written according to (45.12) as

$$\frac{\partial(\varrho v^k)}{\partial t} + \sum_i \Delta_i(\varrho v^i v^k - P^{ik}_G) = \varrho F^k \tag{49.1}$$

which expresses explicitly the conservation of momentum.

α) *Turbulent stresses.* Eq. (49.1) is especially appropriate when the fluid is in turbulent motion, a case which must occur very frequently in cosmical masses due to their large dimensions. In fact, according to Eqs. (48.1) and (48.5), the constraint of molecular viscosity on the general motion will decrease as the dimensions increase. Finally, superposed on the mean motion, there will appear all kinds of irregularities, covering a whole range of sizes and speeds (spectrum of turbulence). In other words the Reynolds number

$$R = \frac{l \varrho v}{\eta} \tag{49.2}$$

will become larger and larger and overshoot the critical value R_C past which laminar motion is impossible. However in the case of irregularities on a small enough scale, molecular viscosity will again play a rôle or even control the motion tending to make it laminar in small enough regions of a size given in order of magnitude by

$$l_m \approx \frac{R_C \eta}{\varrho \bar{v}} \tag{49.3}$$

where $\bar{v}$ is an appropriate mean velocity. This implies a cut-off in the spectrum on the side of large wave-numbers (small size) as suggested by HEISENBERG and verified in a special case by CHAMBERLAIN and ROBERTS[1]. Dissipation of the energy of mean motion occurs in two stages, one in which the kinetic energy is transferred to smaller and smaller eddies until finally inside the smallest ones it is transformed into heat.

This complex situation has been the subject of two main approaches. One initiated by Taylor and von Karman relies essentially on correlations between values of the physical variables at two different points or two different times or both [*42*]. Unfortunately until now, this group of theories remain ill-adapted to the problems which we shall encounter here.

In the other attempt due to TAYLOR, PRANDTL and SCHMIDT, the turbulent elements (eddies) are supposed to detach themselves from a given layer with the average properties of that layer and move a certain distance (mixing length) keeping their individuality before being reabsorbed in another layer with different average properties [*43*]. The theory is then developed by analogy with the kinetic theory of gases assuming conservation of momentum (PRANDTL-SCHMIDT) or vorticity (TAYLOR), the average mixing length $\bar{l}$ playing the rôle of the mean free-path. No doubt this type of theory could also be improved in introducing the notion of a spectrum of turbulence perhaps in a more detailed form than usual.

In both types of theories, the fluid is nearly always treated as incompressible, an approximation which is reasonable in many physical applications where the turbulent velocities remain small compared to the local velocity of sound but which must fail utterly in the external layers of many stars. However we shall not worry for the time being over the resulting complications and, although we shall treat the fluid as compressible, we shall suppose that it is possible to define significant mean values of the physical variables $\bar{\varrho}$, $\bar{p}$, $\bar{T}$ etc. at each point. In those conditions, if we designate the turbulent velocities by capital letters,

$$v^i = \bar{v}^i + V^i \tag{49.4}$$

the mean velocity $\bar{\boldsymbol{v}}$ is defined by

$$\bar{\varrho}\, \bar{v}^i = \overline{\varrho\, v^i} \tag{49.5}$$

[1] J. W. CHAMBERLAIN and P. H. ROBERTS: Phys. Rev. **99**, 1674 (1955).

so that

$$\overline{\varrho V^i} = 0, \quad \overline{V^i} \neq 0. \tag{49.6}$$

This insures automatically that on the average there is no transfer of mass due to turbulence and the continuity equations (45.11) and (45.12) remain valid for the mean field. For instance (45.12) becomes

$$\frac{\partial \bar{\varrho}}{\partial t} + \operatorname{div}(\bar{\varrho}\,\bar{\boldsymbol{v}}) = 0. \tag{49.7}$$

On the other hand, if we introduce (49.4) into (49.1) and average, taking into account (49.5) and (49.6), we are left with

$$\frac{\partial}{\partial t}(\bar{\varrho}\,\bar{v}^k) + \sum_i \Delta_i(\bar{\varrho}\,\bar{v}^i\,\bar{v}^k + \overline{\varrho V^i V^k} - \overline{P_G^{ik}}) = \bar{\varrho}\,F^k \tag{49.8}$$

which is identical to (49.1) with all quantities replaced by their mean values, except for the appearance of supplementary stresses $\overline{\varrho V^i V^k}$ known as the turbulent Reynolds stresses.

Let us note that according to what was said earlier, if turbulence is present, the effect of molecular viscosity on the *mean motion* will be negligible so that the average viscous stresses will reduce to [cf. Eq. (46.4) and (46.5)]

$$\overline{P_G^{ik}} = -\bar{p}_G g^{ik} + \overline{\mathfrak{P}_G^{ik}}(\boldsymbol{V}) \tag{49.9}$$

with

$$\overline{\mathfrak{P}_G^{ik}} = -\tfrac{2}{3}\eta_G\, g^{ik} \sum_j \Delta_j \overline{V^j} + \eta_G \sum_l (g^{il}\Delta_l \overline{V^k} + g^{lk}\Delta_l \overline{V^i}) \tag{49.10}$$

where only the average turbulent velocities appear and which will be very small in general since η_G and $\overline{V^j}$ are small.

The difficult problem now is to evaluate explicitly the turbulent stresses which, in general, do not cancel out since the exact equation (49.1) introduces a relation between the V^i and the space derivatives of the v^i. Here, we shall adopt Boussinesq's suggestion that the turbulent stresses should be of the same form as the viscous stresses and write

$$P_t^{ik} = -\overline{\varrho V^i V^k} = -p_t g^{ik} + \mathfrak{P}_t^{ik} \tag{49.11}$$

with

$$\mathfrak{P}_t^{ik} = -\tfrac{2}{3}\eta_t g^{ik} \sum_j \Delta_j \overline{v^j} + \eta_t \sum_l (g^{il}\Delta_l \overline{v^k} + g^{kl}\Delta_l \overline{v^i}) \tag{49.12}$$

and

$$p_t = \overline{\varrho V^i V_i} = \tfrac{1}{3}\overline{\varrho\,|\boldsymbol{V}|^2} = \tfrac{1}{3}\varrho C^2 \tag{49.13}$$

which is analogous to (46.1). In simple cases, it is easy to show that the coefficient of turbulent viscosity η_t must be of the form

$$\eta_t \approx \tfrac{1}{3}\bar{\varrho}\,\bar{l}\,C \tag{49.14}$$

where C is a mean absolute velocity of turbulence of the same order as the root-mean square velocity.

Actually, the justification of these formulae presents serious difficulties which have been discussed by Wasiutynski (cf. [*43a*]). In the isotropic case, his formula (2.19) supports Boussinesq's hypothesis but the explicit calculation of the stresses in spherical coordinates carried out in Sect. 3.1 of his book leads to expressions of the $\mathfrak{P}_t^{ik}$ in which the first terms on the right of (49.12) have disappeared. This would imply that the sum of the natural values of the diagonal components would not reduce to three times the isotropic pressure as defined by (49.13). On the other hand, anisotropic turbulence might be fairly frequent in stars but a general treatment of such cases is lacking. Here, we shall keep to the above values of the stresses and the equation of motion (49.8) can then be written

$$\frac{\partial}{\partial t}(\bar{\varrho}\bar{v}^k) + \sum_i \Delta_i(\bar{\varrho}\bar{v}^i\bar{v}^k) = \bar{\varrho}F^k - \sum_i g^{ik}\Delta_i(\bar{p}_G + p_t) + \sum_i \Delta_i(\mathfrak{P}_G^{ik} + \mathfrak{P}_t^{ik}), \tag{49.15}$$

β) Radiative stresses. The detailed interaction between a radiation field and matter in relative motion has been the subject of many investigations[1], but the most complete discussion is due to L. H. THOMAS [*44*] who treated the problem from the relativistic point of view and cleared away the misconceptions which had arisen in some of the previous work. He gave general expressions including all terms in v/c where v is a material velocity but carried the actual calculations only to the first order terms. Here, we shall limit ourselves to these terms from the start.

If we designate by I_ν, the intensity of radiation of frequency ν at a fixed point $\boldsymbol{r}$ in a Cartesian system of coordinates (x_1, x_2, x_3) at time t, in a direction $\boldsymbol{l}(l_1, l_2, l_3)$, the number of quanta between ν and $\nu + d\nu$ flowing through the area $d\boldsymbol{S}$ in a direction within the solid angle $d\omega$ about $\boldsymbol{l}$ is

$$\frac{I_\nu(\boldsymbol{r}, \boldsymbol{l}, \nu, t)}{h\nu}(\boldsymbol{l}\cdot d\boldsymbol{S})\, dt\, d\omega\, d\nu. \tag{49.16}$$

Let us denote by a prime, the corresponding quantities measured in a system (x_i') having the velocity $\boldsymbol{v}$ with respect to the (x_i). Since the numbers of quanta counted in the two systems must be equal, one finds

$$I'_{\nu'} = I_\nu\left(1 - 3\frac{\boldsymbol{v}\cdot\boldsymbol{l}}{c}\right) \tag{49.17}$$

if one takes into account the relations

$$d\omega = d\omega'\left(1 - 2\frac{\boldsymbol{v}\cdot\boldsymbol{l}'}{c}\right), \quad d\nu = d\nu'\left(1 + \frac{\boldsymbol{v}\cdot\boldsymbol{l}'}{c}\right) \tag{49.18}$$

and the transformation-formula for direction-cosines of a light-track

$$\boldsymbol{l} = \boldsymbol{l}' + \frac{\boldsymbol{v}}{c} - \frac{(\boldsymbol{l}'\cdot\boldsymbol{v})}{c}\boldsymbol{l}'. \tag{49.19}$$

In any system, I_ν satisfies the same transfer equation

$$\frac{1}{c}\frac{\partial I_\nu}{\partial t} + (\boldsymbol{l}\cdot\nabla)\, I_\nu = -\varrho k_\nu I_\nu + \varrho j_\nu \tag{49.20}$$

where k_ν and j_ν are the coefficients of absorption and emission per unit-mass. This equation is established in *following the radiation* in any given system. If we take into account KIRCHHOFF's law

$$j_\nu/k_\nu = B_\nu \tag{49.21}$$

where B_ν is the black-body intensity, separate the spontaneous and induced emission in j_ν and define an effective coefficient of absorption by

$$\varkappa_\nu = k_\nu\left(1 - e^{-\frac{h\nu}{kT}}\right), \tag{49.22}$$

[1] J. JEANS: Monthly Notices Roy. Astronom Soc. London **85**, 917 (1925); **86**, 328, 444 (1926). — Cf. also S. ROSSELAND: Astrophys. Journ. **63**, 342 (1926). — H. VOGT: Astronom. Nachr. **232**, 1 (1928); E. A. MILNE: Quart. J. Math. **1**, 1 (1930). — Monthly Notices Roy. Astronom. Soc. London **89**, 518 (1929).

Eq. (49.20) becomes

$$\frac{1}{c}\frac{\partial I_\nu}{\partial t} + (\boldsymbol{l}\cdot\nabla)\, I_\nu = -\,\varrho\,\varkappa_\nu (I_\nu - B_\nu)\,. \tag{49.23}$$

As the emission and absorption are directly associated with the properties of matter, it is convenient to express the corresponding coefficients in terms of their proper values $\varrho_0\, k_{\nu_0}$ and $\varrho_0\, j_{\nu_0}$ as measured in a system $(x_{i,0})$ moving with the matter at the point considered, say with a velocity $\boldsymbol{v}$ with respect to the (x^i). In general, in this section and the following, the index zero will keep the same meaning. From the invariance of (49.20) or (49.23) and using the previous transformation formula with the prime replaced by an index zero, one finds

$$\varrho\, k_\nu = \varrho_0\, k_{\nu_0}\left(1 - \frac{\boldsymbol{v}\cdot\boldsymbol{l}_0}{c}\right), \quad \varrho\,\varkappa_\nu = \varrho_0\,\varkappa_{\nu_0}\left(1 - \frac{\boldsymbol{v}\cdot\boldsymbol{l}_0}{c}\right), \quad \varrho\, j_\nu = \varrho_0\, j_{\nu_0}\left(1 + 2\,\frac{\boldsymbol{v}\cdot\boldsymbol{l}_0}{c}\right) \tag{49.24}$$

and Eq. (49.23) becomes

$$\frac{1}{c}\frac{\partial I_\nu}{\partial t} + (\boldsymbol{l}\cdot\nabla)\, I_\nu = -\,\varrho_0\,\varkappa_{\nu_0}\left(1 - \frac{\boldsymbol{v}\cdot\boldsymbol{l}_0}{c}\right)(I_\nu - B_\nu)\,. \tag{49.25}$$

If we multiply (49.20) by $l_i\, d\omega$ and integrate over all directions, we find that the components of the flux-vector $\boldsymbol{F}_\nu$

$$F_{\nu,i} = \int I_\nu\, l_i\, d\omega \tag{49.26}$$

satisfy the equation

$$\frac{1}{c^2}\frac{\partial F_{\nu,i}}{\partial t} + \frac{1}{c}\int l_i \sum_j l_j \frac{\partial I_\nu}{\partial x_j}\, d\omega = \frac{1}{c}\int (\varrho\, j_\nu - \varrho\, k_\nu I_\nu)\, l_i\, d\omega\,. \tag{49.27}$$

The right-hand member represents the exchange of momentum along x_i between matter and radiation and thus either side of this equation expresses the i component of the resulting force. The second term in the first member can be written

$$\sum_j \frac{\partial}{\partial x_j}\left(\frac{1}{c}\int l_i\, l_j\, I_\nu\, d\omega\right)$$

and corresponds to the i component of the divergence of the radiation stress-tensor defined by

$$(P_{ij})_\nu = -\frac{1}{c}\int l_i\, l_j\, I_\nu\, d\omega\,. \tag{49.28}$$

If we integrate (49.26) and (49.28) with respect to ν, we obtain the more important quantities

$$(F_i)_R = \iint I_\nu\, l_i\, d\omega\, d\nu \quad \text{and} \quad (P_{ij})_R = -\frac{1}{c}\iint l_i\, l_j\, I_\nu\, d\omega\, d\nu \tag{49.29}$$

related by an equation similar to (49.27) which can be written vectorially

$$\frac{1}{c^2}\frac{\partial \boldsymbol{F}_R}{\partial t} - \operatorname{div}\mathfrak{P}_R = \frac{1}{c}\iint (\varrho\, j_\nu - \varrho\, k_\nu I_\nu)\,\boldsymbol{l}\, d\omega\, d\nu\,. \tag{49.30}$$

From the definition (49.29) for $\boldsymbol{F}_R$ and the relations (49.18) and (49.19), one verifies the transformation formula

$$\boldsymbol{F}_R = \boldsymbol{F}'_R + \boldsymbol{v}\, U'_R - \boldsymbol{v}\cdot\mathfrak{P}'_R \tag{49.31}$$

where we have used the general definition of the density of radiative energy

$$U_R = \frac{1}{c}\iint I_\nu\, d\nu\, d\omega\,. \tag{49.32}$$

To find explicitly U_R and $(P_{ij})_R$, we can solve (49.25) for I_ν by successive approximations, the second of which is sufficient here, giving

$$I_\nu = B_{\nu,0}\left(1+3\frac{\boldsymbol{v}\cdot\boldsymbol{l}_0}{c}\right) - \frac{1}{\varrho_0\varkappa_{\nu 0}}\left(1+\frac{\boldsymbol{v}\cdot\boldsymbol{l}_0}{c}\right)\left(\frac{1}{c}\frac{\partial}{\partial t}+\boldsymbol{l}\cdot\nabla\right)\left[\left(1+3\frac{\boldsymbol{v}\cdot\boldsymbol{l}_0}{c}\right)B_{\nu,0}\right] \tag{49.33}$$

taking into account the transformation of B_ν to $B_{\nu 0}$ according to formula (49.17) where the primes can be changed into zero indices. If we introduce (49.33) in (49.32) transform, to ν_0, ω_0 and $\boldsymbol{l}_0$ (in the operator) and take into account the relations

$$\frac{3}{4\pi}\int l_i l_j\, d\omega = \delta_{ij} \quad \text{and} \quad B_{\nu 0} \propto \nu_0^3 f\left(\frac{\nu_0}{T}\right)$$

we find

$$U_R = U_{R0} - \frac{1}{\varkappa_0\varrho_0}\frac{\partial U_{R0}}{\partial T}\left(\frac{1}{c}\frac{\partial T}{\partial t}+\frac{5}{3}\frac{\boldsymbol{v}}{c}\cdot\operatorname{grad} T+\frac{T}{3c}\operatorname{div}\boldsymbol{v}\right) \tag{49.34}$$

where $\varkappa_0$ is the usual ROSSELAND's mean absorption coefficient defined by

$$\frac{1}{\varkappa_0}\int\frac{\partial B_{\nu 0}}{\partial T}\,d\nu_0 = \int\frac{1}{\varkappa_{\nu 0}}\frac{\partial B_{\nu 0}}{\partial T}\,d\nu_0. \tag{49.35}$$

The density of black-body radiation U_{R0} is given by

$$U_{R0} = \frac{1}{c}\int B_{\nu 0}\,d\nu_0\,d\omega_0 = \frac{4\pi B_0}{c}, \qquad B_0 = \int B_{\nu 0}\,d\nu_0 \tag{49.36}$$

and is known to be equal to

$$U_{R0} = a\,T^4 \tag{49.37}$$

where a is the Stefan-Boltzmann constant. With this relation, one verifies easily that the second term in the right-hand member of (49.34) is very small compared to the first for all velocities and rate of variation in time that can be expected. We can proceed in the same manner to the evaluation of the $(F_i)_R$ starting from formula (49.29) and we obtain

$$\boldsymbol{F}_R = -\frac{c}{\varkappa_0\varrho_0}\operatorname{grad}\frac{U_{R0}}{3} + U_{R0}\,\boldsymbol{v} + \frac{U_{R0}}{3}\,\boldsymbol{v}. \tag{49.38}$$

Again, the radiation stresses can be computed from the second formula (49.29) and their general contravariant components can be written

$$\left.\begin{aligned}(P^{ij})_R = &-\frac{U_{R0}}{3}g^{ij} + \frac{1}{3c\varkappa_0\varrho_0}\Big(g^{ij}\dot{U}_{R0} + v^i\sum_h g^{jh}\Delta_h U_{R0} + v^j\sum_h g^{ih}\Delta_h U_{R0}\Big) + \\ &+\frac{4U_{R0}}{15c\varkappa_0\varrho_0}\Big(g^{ij}\sum_h\Delta_h v^h + \sum_h g^{ih}\Delta_h v^j + \sum_h g^{jh}\Delta_h v^i\Big).\end{aligned}\right\} \tag{49.39}$$

The transformation formula (49.31) in which we introduce this tensor and neglect all terms of the order of $(v/c)^2$ or smaller as compared to $(\boldsymbol{F}/c)$ becomes identical to (49.38) provided that

$$\boldsymbol{F}_{R0} = -\frac{c}{\varkappa_0\varrho_0}\operatorname{grad}\frac{U_{R0}}{3}. \tag{49.40}$$

In the case of static equilibrium the $(P^{ij})_R$ reduce to

$$P_R^{ii} = -p_{R0} = -\frac{U_{R0}}{3} = -\frac{a\,T^4}{3}, \qquad P^{ij} = 0\ (i \neq j). \tag{49.41}$$

The second group of terms in (49.39) represents stresses which subsist even in the absence of velocity-gradients and which can be written, taking (49.40) into account

$$\frac{1}{3c\varkappa_0\varrho_0}\left(g^{ij}\dot{U}_{R0}\right) - \frac{1}{c^2}\left(v^i F_{R0}^j + v^j F_{R0}^i\right) \tag{49.42}$$

the last part of which has been discussed recently by KIPPENHAHN[1]. The last group of terms in (49.39) corresponds to the radiative viscosity proper. In fact, defining an isotropic pressure p_R as the average of the diagonal terms

$$-p_R = \tfrac{1}{3}\sum (P^i_i)_R$$

expanding the time derivative and taking (49.34) and (49.37) into account, we get

$$p_R = \frac{U_{R0}}{3} - \frac{1}{3c\varkappa_0\varrho_0}\left(\frac{\partial U_{R0}}{\partial t} + \frac{5}{3}\boldsymbol{v}\cdot\operatorname{grad} U_{R0}\right) - \frac{4}{9c\varkappa_0\varrho_0}\operatorname{div}\boldsymbol{v}. \tag{49.43}$$

With this definition, the tensor (49.39) can be written in the form

$$P^{ij}_R = -p_R g^{ij} + \mathfrak{P}^{ij}_R, \tag{49.44}$$

with

$$\left.\begin{aligned}\mathfrak{P}^{ij}_R = \frac{1}{3c\varkappa_0\varrho_0}&\left(-\frac{2}{3}g^{ij}\sum_h v^h\Delta_h U_{R0} + v^i\sum_h g^{ih}\Delta_h U_{R0} + v^j\sum_h g^{ih}\Delta_h U_{R0}\right)\\ &+\frac{4U_{R0}}{15c\varkappa_0\varrho_0}\left(-\frac{2}{3}g^{ij}\sum_h\Delta_h v^h + \sum_h g^{ih}\Delta_h v^j + \sum_h g^{jh}\Delta_h v^i\right).\end{aligned}\right\} \tag{49.45}$$

This brings out the similarity with the ordinary viscous stresses provided the coefficient of radiative viscosity be defined by

$$\eta_R = \frac{4U_{R0}}{15c\varkappa_0\varrho_0}. \tag{49.46}$$

This is again of the form (46.6) or (49.14) with a density and a mean free path for the radiation of the order

$$\varrho_R = \frac{U_{R0}}{c^2},\quad \bar{l}_R = \frac{1}{\varkappa_0\varrho_0}. \tag{49.47}$$

In general, the radiative stresses except for p_R are negligible or at most of the same order as the molecular viscous stresses except perhaps in the very external layers.

c) Conservation of energy.

50. Conservation of mechanical energy. From the preceding discussion, p_G and $\mathfrak{P}^{ik}_G$ in Eq. (48.4) must be replaced in general by

$$p = p_G + p_R,\quad \mathfrak{P}^{ik} = \mathfrak{P}^{ik}_G + \mathfrak{P}^{ik}_R. \tag{50.1}$$

After this substitution, multiplying (48.4) by v_k and summing over k, we obtain

$$\frac{\partial}{\partial t}\left(\frac{1}{2}v^2\right) + \sum_i v^i\Delta_i\left(\frac{1}{2}v^2\right) = \sum_k v_k F^k - \frac{1}{\varrho}\sum_i v^i\Delta_i p + \frac{1}{\varrho}\sum_l v^l\sum_i\Delta_i\mathfrak{P}^i_l \tag{50.2}$$

or

$$\varrho\frac{d}{dt}\left(\frac{1}{2}v^2\right) = \varrho\,\boldsymbol{v}\cdot\boldsymbol{F} - \boldsymbol{v}\cdot\operatorname{grad}p + \boldsymbol{v}\cdot\operatorname{div}\mathfrak{P}(\boldsymbol{v}) \tag{50.3}$$

which express the conservation of mechanical energy per unit mass. To obtain it by unit volume, one can transform (50.2) by means of (45.12) or start directly from (49.1). One finds

$$\frac{\partial}{\partial t}\left(\frac{1}{2}\varrho v^2\right) + \operatorname{div}\left(\frac{1}{2}\varrho v^2\boldsymbol{v}\right) = \varrho\boldsymbol{F}\cdot\boldsymbol{v} - \boldsymbol{v}\cdot\operatorname{grad}p + \boldsymbol{v}\cdot\operatorname{div}\mathfrak{P}(\boldsymbol{v}). \tag{50.4}$$

In presence of turbulence, we can proceed as COWLING [45], introduce (49.4) in (50.4) and take mean values, remembering (49.6) and neglecting the effect of

[1] R. KIPPENHAHN: Z. Astrophys. 35, 165 (1954).

molecular and radiative viscosity on the mean motion as well as the first part of the $\mathfrak{P}_R^{ij}$ as defined by (49.45). If we use the notation

$$\left.\begin{aligned}\sum_{i,k} V_k \varDelta_i \{-\tfrac{2}{3}(\eta_G+\eta_R)\, g^{ik} \sum_j \varDelta_j V^j + (\eta_G+\eta_R)\times \\ \times \sum_l (g^{il}\varDelta_l V^k + g^{lk}\varDelta_l V^i)\} = \sum_{i,k} V_k \varDelta_i \mathfrak{P}^{ik}(\boldsymbol{V})\end{aligned}\right\} \tag{50.5}$$

this leads to

$$\left.\begin{aligned}\frac{\partial}{\partial t}\left(\frac{1}{2}\bar{\varrho}\,\bar{v}^2 + \frac{1}{2}\overline{\varrho V^2}\right) + \sum_i \varDelta_i\left(\frac{1}{2}\bar{\varrho}\,\bar{v}^2\bar{v}^i + \frac{1}{2}\overline{\varrho V^2}\,\bar{v}^i\right) \\ = \bar{\varrho}\sum_k F^k \bar{v}_k - \bar{v}^i\varDelta_i p - \overline{V^i\varDelta_i p} - \sum_i \varDelta_i\left(\frac{1}{2}\overline{\varrho V^2 V^i}\right) + \\ + \sum_{i,k}\varDelta_i(P_t^{ik}\bar{v}_k) + \sum_{i,k}\overline{V_k\varDelta_i\mathfrak{P}^{ik}(\boldsymbol{V})}\end{aligned}\right\} \tag{50.6}$$

where we have also neglected terms in $\overline{\boldsymbol{V}}$ in the effects of the stresses as well as purely viscous stresses. On the other hand, multiplying (49.15) by $\bar{v}_k$ and summing over k, we find, for the mean motion

$$\left.\begin{aligned}\bar{\varrho}\frac{d}{dt}\left(\frac{1}{2}\bar{v}^2\right) = \frac{\partial}{\partial t}\left(\frac{1}{2}\bar{\varrho}\,\bar{v}^2\right) + \sum_i \varDelta_i\left(\frac{\bar{\varrho}\,\bar{v}^2}{2}\bar{v}^i\right) \\ = \bar{\varrho}\sum_k F^k v_k - \sum_i \bar{v}^i\varDelta_i(p+p_t) + \sum_{i,k}\bar{v}_k\varDelta_i\mathfrak{P}_t^{ik}\end{aligned}\right\} \tag{50.7}$$

where, contrarily to COWLING, we do not neglect the turbulent viscosity. Subtracting (50.7) from (50.6) and using (49.14) we obtain

$$\left.\begin{aligned}\bar{\varrho}\frac{d}{dt}\left(\frac{1}{2}\frac{\overline{\varrho V^2}}{\bar{\varrho}}\right) = \frac{\partial}{\partial t}\left(\frac{1}{2}\overline{\varrho V^2}\right) + \sum_i \varDelta_i\left(\frac{1}{2}\overline{\varrho V^2}\,\bar{v}^i\right) \\ = -\sum_i \overline{V^i\varDelta_i p} - \sum_i \varDelta_i(\tfrac{1}{2}\overline{\varrho V^2 V^i}) + \sum_{i,k} P_t^{ik}\varDelta_i\bar{v}_k + \\ + \sum_{i,k}\overline{V_k\varDelta_i\mathfrak{P}^{ik}(\boldsymbol{V})}\end{aligned}\right\} \tag{50.8}$$

which expresses the conservation of kinetic energy of turbulence. The last term in (50.6) or (50.8) may be written

$$\overline{\sum_{ik} V_k\varDelta_i\mathfrak{P}^{ik}(\boldsymbol{V})} = \overline{\sum_{i,k}\varDelta_i[V_k\mathfrak{P}^{ik}(\boldsymbol{V})]} - \overline{\sum_{i,k}\mathfrak{P}^{ik}(\boldsymbol{V})\varDelta_i V_k} \tag{50.9}$$

and the last term in (50.9) which, in Cartesian coordinates is

$$\left.\begin{aligned}\bar{\varrho}\,\varepsilon_2 = \sum_{i,k}\overline{\mathfrak{P}^{ik}\varDelta_i V_k} \\ = \frac{2}{3}(\bar{\eta}_G+\bar{\eta}_R)\left[\overline{\left(\frac{\partial V_1}{\partial x_1}-\frac{\partial V_2}{\partial x_2}\right)^2} + \overline{\left(\frac{\partial V_1}{\partial x_1}-\frac{\partial V_3}{\partial x_3}\right)^2} + \overline{\left(\frac{\partial V_2}{\partial x_2}-\frac{\partial V_3}{\partial x_3}\right)^2}\right] + \\ + (\bar{\eta}_G+\bar{\eta}_R)\left[\overline{\left(\frac{\partial V_3}{\partial x_2}+\frac{\partial V_2}{\partial x_3}\right)^2} + \overline{\left(\frac{\partial V_1}{\partial x_3}+\frac{\partial V_3}{\partial x_1}\right)^2} + \overline{\left(\frac{\partial V_2}{\partial x_1}+\frac{\partial V_1}{\partial x_2}\right)^2}\right]\end{aligned}\right\} \tag{50.10}$$

corresponds to a pure dissipation of kinetic energy into heat at a rate ε_2 per unit mass. We can also introduce the rate ε_4 of transformation of kinetic energy of mean motion into turbulent kinetic energy, by

$$\bar{\varrho}\,\varepsilon_4 = \sum_{i,k}\mathfrak{P}_t^{ik}(\boldsymbol{v})\,\varDelta_i\bar{v}_k \tag{50.11}$$

and the convection vector of turbulent energy, by

$$F_2^i = \overline{\tfrac{1}{2}\varrho V^2 V^i}\,. \tag{50.12}$$

With these notations, Eq. (50.8) can be written

$$\left.\begin{aligned}\bar\varrho\frac{d}{dt}\left(\frac{1}{2}\frac{\overline{\varrho V^2}}{\bar\varrho}\right)+p_t\operatorname{div}\boldsymbol{v}=\bar\varrho\,\varepsilon_4-\bar\varrho\,\varepsilon_2-\operatorname{div}\boldsymbol{F}_2-\\ -\overline{\boldsymbol{V}\cdot\operatorname{grad}p}+\operatorname{div}\left[\overline{\boldsymbol{V}\,\mathfrak{P}(\boldsymbol{V})}\right]\end{aligned}\right\} \tag{50.13}$$

where the last term will be very small in general and the preceding term represents the work of the pressure gradient or essentially the transformation of gravitational potential energy into energy of turbulence.

51. Conservation of the total energy. Let us denote by U_0 the total thermal energy per unit mass

$$U_0 = U_{G0} + \frac{U_{R0}}{\varrho} \tag{51.1}$$

where U_G represents the energy associated with the particles (thermal kinetic energy plus any ionization or excitation energy) and, by E_S the subatomic energy associated with the proper mass of the particles. The variation following the motion of the total energy of a mass limited by a surface S bound to the particles is due to the work of the external forces $\boldsymbol{F}$, the radiative flux $\boldsymbol{F}_{R0}$ and the conductive flux $\boldsymbol{F}_{C0}$ with respect to the material and the work of the tensions P^{ik} [cf. (46.7)] on the surface S. This is expressed by the equation

$$\left.\begin{aligned}\iiint\frac{d}{dt}\left(U_0+E_S+\frac{1}{2}v^2\right)dm=\iiint_M\boldsymbol{F}\cdot\boldsymbol{v}\,dm-\iint_S(\boldsymbol{F}_{R0}+\boldsymbol{F}_{C0})\,d\boldsymbol{S}+\\ +\iint_S\sum_{i,k}v_k\,P^{ik}\,dS_i\end{aligned}\right\} \tag{51.2}$$

or, reducing it to unit volume and separating the pressure in the stress-tensor,

$$\left.\begin{aligned}\varrho\frac{d}{dt}\left(U_0+E_S+\frac{v^2}{2}\right)=\varrho\,\boldsymbol{F}\cdot\boldsymbol{v}-\operatorname{div}(\boldsymbol{F}_{R0}+\boldsymbol{F}_{C0})-\\ -\operatorname{div}(p\,\boldsymbol{v})+\operatorname{div}\left[\boldsymbol{v}\,\mathfrak{P}(\boldsymbol{v})\right].\end{aligned}\right\} \tag{51.3}$$

Let us note that in some cases, such as the emission of neutrinos, one might have to introduce explicitly a flux of subatomic energy. However up to now, the rôle of such processes in astrophysics remains negligible and the energy liberated by nuclear reactions can be considered as transformed instantaneously into thermal energy. We shall also neglect everywhere the corresponding loss of mass.

52. Conservation of thermal energy. If we substract (50.3) from (51.3), we obtain, taking (45.12) into account,

$$\left.\begin{aligned}\varrho\frac{dU_0}{dt}+p\operatorname{div}\boldsymbol{v}&=\frac{\partial(\varrho U_0)}{\partial t}+\operatorname{div}(\varrho U_0\,\boldsymbol{v})+p\operatorname{div}\boldsymbol{v}\\ &=\varrho\,\varepsilon_1-\operatorname{div}(\boldsymbol{F}_{R0}+\boldsymbol{F}_{C0})+\sum_{i,k}\mathfrak{P}^{ik}(\boldsymbol{v})\Delta_i v_k\end{aligned}\right\} \tag{52.1}$$

where we have introduced the rate of liberation of nuclear energy

$$\varepsilon_1 = -\frac{dE_s}{dt}\,. \tag{52.2}$$

This equation is nothing more than the explicit expression of the first principle of thermodynamics

$$dQ = dU - \frac{p}{\varrho^2}\,d\varrho \tag{52.3}$$

where the external heat dQ given to the system comprises here the generation of energy, the flux of energy in the system and the heat liberated by friction. In the presence of turbulence, we can introduce again (49.4) into (52.1) and take the mean values obtaining, to the same order of approximation as (50.6),

$$\left.\begin{aligned}&\frac{\partial(\overline{\varrho U_0})}{\partial t}+\operatorname{div}(\overline{\varrho U_0}\,\overline{\boldsymbol{v}})+\overline{p}\operatorname{div}\overline{\boldsymbol{v}}\\&\qquad=\overline{\varrho\,\varepsilon_1}-\operatorname{div}(\overline{\boldsymbol{F}}_{R0}+\overline{\boldsymbol{F}}_{C0}+\overline{\varrho U_0\boldsymbol{V}})-\overline{p\operatorname{div}\boldsymbol{V}}+\sum_{i,k}\overline{\mathfrak{P}^{ik}(\boldsymbol{V})\Delta_i V_k}\,.\end{aligned}\right\}\tag{52.4}$$

If we introduce the notation

$$\boldsymbol{F}_t=\overline{\varrho U_0\boldsymbol{V}}\tag{52.5}$$

and, for the rate of transformation of thermal energy into kinetic energy of turbulence,

$$\overline{\varrho}\,\varepsilon_3=\overline{p\operatorname{div}\boldsymbol{V}}\tag{52.6}$$

Eq. (52.4), taking (49.7) and (50.10) into account may be written

$$\overline{\varrho}\,\frac{d}{dt}\left(\frac{\overline{\varrho U_0}}{\overline{\varrho}}\right)+\overline{p}\operatorname{div}\overline{\boldsymbol{v}}=\overline{\varrho\,\varepsilon_1}+\overline{\varrho}\,\varepsilon_2-\overline{\varrho}\,\varepsilon_3-\operatorname{div}(\overline{\boldsymbol{F}}_{R0}+\overline{\boldsymbol{F}}_{C0}+\boldsymbol{F}_t)\,.\tag{52.7}$$

Remarks on notation. From now on, we shall, in general, omit the subscript zero as well as the bar above the symbols relating to the mean motion in the case of turbulence.

53. Adiabatic or isentropic transformations. Let us consider first the case where turbulence is absent. The transformation will be said to be adiabatic or, more correctly, isentropic if the second member of (52.1) cancels out. In that case, if radiation and ionization energy are negligible, using (45.1), (46.2) and

$$dU_G=C_v\,dT,\quad \bar{\mu}\,C_p=\bar{\mu}\,C_v+R\tag{53.1}$$

where C_v and C_p are the specific heats at constant volume and pressure per unit mass, Eq. (52.1) becomes

$$\frac{1}{p_G}\frac{dp_G}{dt}-\frac{\gamma}{\varrho}\frac{d\varrho}{dt}=0\tag{53.2}$$

from which it follows that

$$\frac{1}{T}\frac{dT}{dt}=(\gamma-1)\frac{1}{\varrho}\frac{d\varrho}{dt}=\frac{\gamma-1}{\gamma}\frac{1}{p_G}\frac{dp_G}{dt}\,.\tag{53.3}$$

In the same way, if U_R and p_R have to be taken into account, starting from the definitions (49.41) and (50.1) and the relation

$$dU=C_v\,dT+d\left(\frac{a\,T^4}{\varrho}\right)\tag{53.4}$$

Eq. (52.1) gives

$$\frac{1}{p}\frac{dp}{dt}=\Gamma_1\frac{1}{\varrho}\frac{d\varrho}{dt}\tag{53.5}$$

where

$$\Gamma_1=\beta+\frac{(4-3\beta)^2(\gamma-1)}{\beta+12(\gamma-1)(1-\beta)}\quad\text{and}\quad\beta=\frac{p_G}{p}\,.\tag{53.6}$$

But this time, the same Γ_1 cannot be used to relate the adiabatic variations of pressure and temperature and, following CHANDRASEKHAR [*46*], we shall write

$$\frac{1}{T}\frac{dT}{dt}=(\Gamma_3-1)\frac{1}{\varrho}\frac{d\varrho}{dt}=\frac{\Gamma_2-1}{\Gamma_2}\frac{1}{p}\frac{dp}{dt}\quad\text{and}\quad\frac{1}{\beta}\frac{d\beta}{dt}=(\Gamma_3-\Gamma_1)\frac{1}{\varrho}\frac{d\varrho}{dt}\tag{53.7}$$

with

$$\Gamma_3 = 1 + \frac{(4-3\beta)(\gamma-1)}{\beta + 12(\gamma-1)(1-\beta)} = 1 + \frac{\Gamma_1 - \beta}{4-3\beta} \quad \text{and} \quad \Gamma_2 = \frac{\Gamma_1}{1+(\Gamma_1-\Gamma_3)} \tag{53.8}$$

Furthermore, if an abundant element is changing its ionization appreciably in the range of temperature and density considered, we must include the ionization energy I in U. Apart from the effects of ionization, stellar material can be treated as a monatomic gas ($\gamma = \frac{5}{3}$) and we can write

$$\left.\begin{aligned} U &= \frac{3}{2}(N+M)\,kT + \frac{a\,T^4}{\varrho} + I = \frac{3p}{2\varrho} + \frac{a\,T^4}{2\varrho} + I, \\ p &= (N+M)\,k\,\varrho\,T + \frac{a\,T^4}{3} \end{aligned}\right\} \tag{53.9}$$

where N and M denote respectively the number of free electrons and ions per unit mass. This problem has been discussed in the more general case by FOWLER and GUGGENHEIM[1]. If only one kind of element of atomic number Z is present, practically the total number M_Z of atoms Z will be distributed between two consecutive states of ionization, say ions with $(Z-r)$ and $(Z-r-1)$ electrons such that the difference of energy between them χ_Z^r be of the order of χ defined by

$$\frac{\chi}{kT} = \ln \frac{2(2\pi\, m_e\, kT)^{\frac{3}{2}}}{h^3\,\varrho\, N}. \tag{53.10}$$

The corresponding numbers of ions M_Z^r and M_Z^{r+1} are related by SAHA's equation

$$\frac{M_Z^{r+1}}{M_Z^r} = \frac{2(2\pi\, m_e\, k\,T)^{\frac{3}{2}}}{h^3\,\varrho\,N} \cdot \frac{u_Z^{r+1}}{u_Z^r} \cdot \mathrm{e}^{-\frac{\chi_Z^r}{kT}} \tag{53.11}$$

with

$$M_Z^r + M_Z^{r+1} \approx M_Z\,. \tag{53.12}$$

If the modification considered displaces the equilibrium of ionization between ions $(Z-r)$ and $(Z-r-1)$ but does not introduce new states of ionization, the variation of the energy of ionization reduces to

$$d I_Z = -\chi_Z^r\, d M_Z^r = \chi_Z^r\, dN\,. \tag{53.13}$$

In that case, Eq. (52.1) with U defined by (53.9) leads to

$$\frac{1}{p}\frac{dp}{dt} = \Gamma_1 \frac{1}{\varrho}\frac{d\varrho}{dt} \tag{53.14}$$

with

$$\Gamma_1 = \frac{\{16 - 12\beta - \frac{3}{2}\beta^2 + \beta\,[4 - \frac{3}{2}\beta + \beta\,\chi_Z^r/kT]\,A_p\}\,(1+A_T)}{12 - \frac{21}{2}\beta + \beta\,(\frac{3}{2} + \chi_Z^r/kT)\,A_\varrho}, \tag{53.15}$$

where

$$A_p = \frac{N}{N+M_Z}\left(\frac{\partial \ln N}{\partial \ln T}\right)_p, \quad A_T = \frac{N}{N+M_Z}\left(\frac{\partial \ln N}{\partial \ln p}\right)_T, \quad A_\varrho = \frac{N}{N+M_Z}\left(\frac{\partial \ln N}{\partial \ln T}\right)_\varrho. \tag{53.16}$$

Similarly, one can write

$$\frac{1}{T}\frac{dT}{dt} = (\Gamma_3 - 1)\frac{1}{\varrho}\frac{d\varrho}{dt} = \frac{\Gamma_2 - 1}{\Gamma_2}\frac{1}{p}\frac{dp}{dt} \tag{53.17}$$

with

$$\Gamma_3 - 1 = \frac{4 - 3\beta + \beta\,A_\varrho}{12 - \frac{21}{2}\beta + \beta\,(\frac{3}{2} + \chi_Z^r/kT)\,A_\varrho} = \frac{\Gamma_1 - \beta\,(1+A_T)}{(1+A_T)\,(4 - 3\beta + \beta\,A_p)}. \tag{53.18}$$

[1] R. H. FOWLER and E. A. GUGGENHEIM: Monthly Notices Roy. Astronom. Soc. London **85**, 939, 961 (1925).

In the case of a mixture, a reasonable approximation can be obtained following BIERMANN[1] by noting that all the χ_Z^r of interest will be of the order of χ defined by (53.10). In that case, Eq. (53.13) and the definitions of Γ_1, Γ_2 and Γ_3 remain valid provided χ_Z^r be replaced everywhere by χ and M_Z by $M=\sum M_Z$. With the large abundances of hydrogen and helium adopted at present, these formulae are of interest only in the external layers since for T greater than some 50000° K the stellar material will approximate very closely to a monatomic gas. Furthermore since the ionization potentials of H and He are well separated, one can admit as a first approximation that in any physical conditions, only one element will be in a critical stage of ionization. In that case, the quantities A_p, A_T and A_ϱ with M substituted for M_Z can be evaluated easily. Let us denote by X_Z the relative abundance (by numbers) of element Z and by X_i that of the element $Z=i$ in a critical stage of ionization and write $x_Z^r = M_Z^r/M$. We then have

$$N = M \sum_Z X_Z (r + x_Z^{r+1}) = M x, \tag{53.19}$$

$$p_G = M(1+x)\, k\, \varrho\, T = \frac{R\, \varrho\, T}{\bar{\mu}}, \quad p = p_G/\beta = p_R/(1-\beta), \quad \bar{\mu} = \frac{\sum X_Z (m_Z/m_H)}{1+x}, \tag{53.20}$$

$$dx = X_i\, d(x_i^{r+1}), \qquad d\bar{\mu}/\bar{\mu} = -\, dx/(1+x), \tag{53.21}$$

$$dI = -\chi_i^r\, dM_i^r = \chi_i^r\, dN, \tag{53.22}$$

$$\frac{x_i^{r+1}}{1-x_i^{r+1}} \cdot \frac{x}{1+x}\, p_G = \frac{2(2\pi\, m_e\, k\vartheta)^{\frac{3}{2}}\, k}{h^3}\, \frac{u_i^{r+1}}{u_i^r}\, T^{\frac{5}{2}}\, e^{-\frac{\chi_i^r}{kT}}. \tag{53.23}$$

If we differentiate (53.23) logarithmically, we obtain

$$\frac{X_i}{(1+x)\, B}\, dx_i^{r+1} = \left(\frac{5}{2} + \frac{\chi_i^r}{kT}\right) \frac{dT}{T} - \frac{dp_G}{p_G} \tag{53.24}$$

where

$$B = \frac{X_i\, x\, x_i^{r+1} (1 - x_i^{r+1})}{x(1+x) + X_i\, x_i^{r+1} (1 - x_i^{r+1})}. \tag{53.25}$$

If using the relations (53.20) we explicitly express dp_G/p_G in (53.24) successively in terms of dT/T and $d\varrho/\varrho$ or dT/T and dp/p, we obtain

$$A_\Gamma = -\frac{B}{1+B}, \quad A_p = B\left[\frac{5}{2} + \frac{\chi_i^{r+1}}{kT} + \frac{4(1-\beta)}{\beta}\right], \quad A_\varrho = \frac{B}{1+B}\left(\frac{3}{2} + \frac{\chi_i^{r+1}}{kT}\right), \tag{53.26}$$

and our results become essentially equivalent to those of A.B. UNDERHILL[2]. A more refined calculation is however possible following a procedure introduced by A. UNSÖLD [47] and generalized by A.B. UNDERHILL[2]. This method consists in establishing numerically a nomogram of the entropy S in the plane of two of the variables ($\ln p$, $\ln \varrho$, $\ln T$) and then computing numerically along the lines of constant S, the desired derivatives $(d \ln p/d \ln \varrho) = \Gamma_1$ etc.

In presence of turbulence, one should start from Eq. (52.7) instead of (52.1) but the extra-terms in the right-hand member will partly cancel out and are very small anyway so that the previous values of Γ_1, Γ_2 and Γ_3 will remain valid.

In the same way, the right-hand member of Eq. (50.13) can be considered as small and we can write

$$\bar{\varrho}\, \frac{d}{dt}\left(\frac{1}{2}\, \frac{\overline{\varrho V^2}}{\bar{\varrho}}\right) + p_t \operatorname{div} \bar{\boldsymbol{v}} = 0. \tag{53.27}$$

[1] L. BIERMANN: Astronom. Nachr. **259**, 221 (1936).

[2] A. B. UNDERHILL: Monthly Notices Roy. Astronom. Soc. London **109**, 562 (1949). — Ann. d'Astrophys. **12**, 243 (1949).

If the turbulence is isotropic, using (49.13) and (49.7), this reduces to

$$\frac{1}{p_t}\frac{dp_t}{dt} = \frac{5}{3}\frac{1}{\bar{\varrho}}\frac{d\bar{\varrho}}{dt}. \tag{53.28}$$

In that case, the turbulent pressure behaves during an "adiabatic" change as the thermal pressure of a monatomic gas. If the turbulence is not isotropic but takes place for instance only along one direction ($V_y = V_z = 0$) and the mean motion is also limited to that direction ($\bar{v}_y = \bar{v}_z = 0$), one easily verifies, starting from (50.8) instead of (50.13) and writing $p_t = \overline{\varrho V_x^2}$, that

$$\frac{1}{p_t}\frac{dp_t}{dt} = \frac{3}{\bar{\varrho}}\frac{d\bar{\varrho}}{dt}. \tag{53.29}$$

But if the mean motion is similar in the three directions (symmetrical contraction or expansion), the adiabatic exponent reduces again to $\frac{5}{3}$. In the case of a spherical distribution of matter, if the turbulence has a preponderant radial component ($V_r \gg V_\vartheta$ or V_φ) and if the mean motion is radial, one gets writing $p_t = \overline{\varrho V_r^2}$

$$\frac{1}{p_t}\frac{dp_t}{dt} = \frac{3}{\bar{\varrho}}\frac{d\bar{\varrho}}{dt} + \frac{4}{r}\bar{v}_r \tag{53.30}$$

with

$$\frac{1}{\bar{\varrho}}\frac{d\bar{\varrho}}{dt} = -r\frac{\partial}{\partial r}\left(\frac{\bar{v}_r}{r}\right) - 3\frac{\bar{v}_r}{r}. \tag{53.31}$$

But if the mean motion is homologous, the adiabatic exponent again reduces to $\frac{5}{3}$. Only if a large fraction of the change in density is due to the variations of $(\bar{v}_r/r)$ would (53.30) deviate appreciably from (53.28). Furthermore, as BATCHELOR[1] has pointed out, the adiabatic exponents as defined here will also tend to be different depending on whether the time taken by the general distortion is long or short as compared to the relaxation time of turbulence.

Remark. It is necessary to distinguish carefully between *piezotropic* relations of the types (53.2) valid for a given element of mass during its isentropic transformations and the *barotropic* relations of the same form which are used to characterize a geometrical distribution of matter such as the polytropes.

54. The conditions at a surface of discontinuity. For the time being, we shall limit our discussion to the case of *contact surfaces* [*48*] through which no mass is flowing contrarily to what happens at a shock front. Across a contact surface, the density or the temperature may experience a discontinuous jump but the pressure and the normal velocities must be continuous. Let

$$f(x_1 x_2 x_3, t) = 0 \tag{54.1}$$

represent such a surface and q_1 and q_2 denote the values of any physical variable on both sides of the surface.

The kinematic conditions can be written

$$\frac{\partial f}{\partial t} + \dot{\boldsymbol{r}}_1 \cdot \operatorname{grad} f = 0, \qquad \frac{\partial f}{\partial t} + \dot{\boldsymbol{r}}_2 \cdot \operatorname{grad} f = 0, \qquad (\dot{\boldsymbol{r}}_1 - \dot{\boldsymbol{r}}_2)\cdot \operatorname{grad} f = 0 \tag{54.2}$$

or in Eulerian notation

$$\frac{\partial f}{\partial t} + \boldsymbol{v}_1 \cdot \operatorname{grad} f = 0, \qquad \frac{\partial f}{\partial t} + \boldsymbol{v}_2 \cdot \operatorname{grad} f = 0 \quad \text{or} \quad (\boldsymbol{v}_1 - \boldsymbol{v}_2)\cdot \operatorname{grad} f = 0. \tag{54.3}$$

[1] G. K. BATCHELOR: Vistas in Astronomy, Vol. 1, p. 290. London: Pergamon Press 1955.

On an external boundary surface, only one of the conditions (54.3) subsists

$$\frac{\partial f}{\partial t} + \boldsymbol{v} \cdot \operatorname{grad} f = 0. \tag{54.4}$$

In Lagrangian coordinates (cf. [40b], § 18), if one takes for the a_i the initial values of the x_i, (54.1) and (54.2) become

$$f(a_1 a_2 a_3, t) = 0 \tag{54.5}$$

and

$$\left.\begin{aligned} \dot{f}(a_1 a_2 a_3, t) &\equiv \frac{\partial f}{\partial t} = 0, \\ \sum_i [(\dot{x}_i)_1 - (\dot{x}_i)_2] \frac{\partial(x_{i-1}, f, x_{i+1})}{\partial(a_1, a_2, a_3)} \Big/ \frac{\partial(x_1, x_2, x_3)}{\partial(a_1, a_2, a_3)} &= 0. \end{aligned}\right\} \tag{54.6}$$

At a boundary, moving with the imposed velocity $\boldsymbol{u}$, one has

$$\dot{f}(a_1, a_2, a_3, t) = 0, \qquad \sum_i (\dot{x}_i - u_i) \frac{\partial(x_{i-1}, f, x_{i+1})}{\partial(a_1, a_2, a_3)} \Big/ \frac{\partial(x_1, x_2, x_3)}{\partial(a_1, a_2, a_3)} = 0. \tag{54.7}$$

The dynamical condition at the interface, is simply

$$p_1(x_1, x_2, x_3, t) - p_2(x_1, x_2, x_3, t) = 0 \tag{54.8}$$

if (x_1, x_2, x_3) satisfy Eq. (54.1). In fact (54.8) can be considered as defining the interface just as well as (54.1) and one can replace f by $p_1 - p_2$ in the kinematic conditions (54.2) which then read

$$\left.\begin{aligned} \frac{\partial p_1}{\partial t} - \frac{\partial p_2}{\partial t} + \dot{\boldsymbol{r}}_1 (\operatorname{grad} p_1 - \operatorname{grad} p_2) &= 0, \\ \frac{\partial p_1}{\partial t} - \frac{\partial p_2}{\partial t} + \dot{\boldsymbol{r}}_2 (\operatorname{grad} p_1 - \operatorname{grad} p_2) &= 0. \end{aligned}\right\} \tag{54.9}$$

To obtain the Eulerian form of these mixed conditions we must write $\dot{\boldsymbol{r}} = \boldsymbol{v}$ and interpret the x_i as geometrical coordinates. In particular, a free boundary ($p_2 \equiv 0$) is now defined by

$$p(x_1, x_2, x_3, t) = 0 \tag{54.10}$$

and the mixed conditions become

$$\frac{\partial p}{\partial t} + \boldsymbol{v} \cdot \operatorname{grad} p = 0. \tag{54.11}$$

These conditions can also be translated into Lagrangian coordinates (cf. [40b], Sect. 31) and, in particular, on a free boundary, one gets

$$\left.\begin{aligned} & p(a_1 a_2 a_3, t) = 0, \quad \frac{\partial p}{\partial t}(a_1, a_2, a_3, t) = 0 \\ \text{or} \qquad & p(a_1 a_2 a_3) = 0 \end{aligned}\right\} \tag{54.12}$$

which means that, following the particles forming the free boundary, the pressure always remains zero.

II. Linearized equations.

55. General method of linearization. The general equations established in the previous sections form a system of non-linear partial differential equations which can be solved only in very special cases. However if the dependent variables ψ_i deviate little from the values $\psi_{i,0}$ characterizing a state of equilibrium or a given motion, the general equations where one neglects all powers of $(\psi_i - \psi_{i,0})$ higher than the first and takes into account the equations satisfied by the $\psi_{i,0}$ become linear in the $(\psi_i - \psi_{i,0})$ and their solutions may provide useful approximations. Here, we shall always denote a perturbation of $\boldsymbol{r}$ by $\delta \boldsymbol{r}$ but for all dependent variables ψ we shall distinguish between the local or Eulerian perturbations ψ' and the Lagrangian perturbations $\delta\psi$ taken in following the motion. To obtain the linearized form of any of the general equations, one has simply to take the first variation of that equation with respect to the dependent variables. In doing this, one must take care that, in Lagrangian coordinates, the x_i must be treated as dependent variables. In Eulerian coordinates, $\boldsymbol{v}$ must be treated as a dependent variable but, once all d/dt have been explicited according to (43.5) or (43.6), the x^i are independent variables and all spatial differential operators are invariant. This facilitates greatly the process so that, in general, we shall use the Eulerian equations.

Chap. VII of Ref. [*40b*] contains a detailed discussion of the procedure and also the explicit linearized equations both in Lagrangian and Eulerian notations for different systems of coordinates. It should be noted however that the gravitational potential is always treated there as unaffected by the perturbation of the mass distribution, a reasonable approximation in an atmosphere for instance but, unsatisfactory in many of the cases of interest here.

56. The linearized equations. The equation of continuity (45.12) becomes

$$\left.\begin{aligned} &\frac{\partial \varrho'}{dt} + \sum_k \Delta_k (\varrho' v^k + \varrho v'^k) = 0 \\ \text{or}\quad &\frac{\partial \varrho'}{\partial t} + \operatorname{div}(\varrho' \boldsymbol{v} + \varrho \boldsymbol{v}') = 0. \end{aligned}\right\} \tag{56.1}$$

In the same way, the equation of motion (48.4) generalized to take the effects of radiation into account [cf. Eq. (50.1)] leads to

$$\left.\begin{aligned} \frac{\partial v'^k}{\partial t} + \sum_i (v'^i \Delta_i v^k + v^i \Delta_i v'^k) = F'^k + \frac{\varrho'}{\varrho^2} \sum_i g^{ik} \Delta_i p - \frac{1}{\varrho} \sum_i g^{ik} \Delta_i p' - \\ - \frac{\varrho'}{\varrho^2} \sum_i \Delta_i \mathfrak{P}^{ik}(\boldsymbol{v}) + \frac{1}{\varrho} \sum_i \Delta_i \mathfrak{P}'^{ik}(\boldsymbol{v}). \end{aligned}\right\} \tag{56.2}$$

In presence of turbulence, we have only to add the turbulent stress-tensor and, if we now neglect the effects of the molecular and radiative viscosity on the mean motion, we have

$$\left.\begin{aligned} \frac{\partial v'^k}{\partial t} + \sum_i (v'^i \Delta_i v^k + v^i \Delta_i v'^k) = F'^k + \frac{\varrho'}{\varrho^2} \sum_i g^{ik} \Delta_i (p + p_t) - \frac{1}{\varrho} \sum_i g^{ik} \Delta_i (p' + p_t') - \\ - \frac{\varrho'}{\varrho^2} \sum_i \Delta_i [\mathfrak{P}^{ik}(\boldsymbol{V}) + \mathfrak{P}_t^{ik}(\boldsymbol{v})] + \frac{1}{\varrho} \sum_i \Delta_i [\mathfrak{P}'^{ik}(\boldsymbol{V}) + \mathfrak{P}_t'^{ik}(\boldsymbol{v})]. \end{aligned}\right\} \tag{56.3}$$

In relative coordinates, the expression of the absolute acceleration for the perturbed motion can be obtained in taking the variation either of (48.9) or of the left-hand side of (48.10) written down explicitly in terms of the v^i.

Eq. (52.7) expressing the conservation of thermal energy can be written, using the definitions of Sect. 53,

$$\left.\begin{aligned} &\frac{\partial p}{\partial t} + \boldsymbol{v} \cdot \operatorname{grad} p - \frac{\Gamma_1 p}{\varrho} \left(\frac{\partial \varrho}{\partial t} + \boldsymbol{v} \cdot \operatorname{grad} \varrho \right) \\ &\qquad = (\Gamma_3 - 1)\, \varrho \left[\varepsilon_1 + \varepsilon_2 - \varepsilon_3 - \frac{1}{\varrho} \operatorname{div} (\boldsymbol{F}_R + \boldsymbol{F}_C + \boldsymbol{F}_t) \right] \end{aligned}\right\} \tag{56.4}$$

where if no turbulence is present the terms in $\boldsymbol{F}_t$ and ε_3 disappear and the terms in $\varrho\,\varepsilon_2$ may be interpreted as corresponding to the dissipation of energy $\sum_{i,k} \mathfrak{P}^{ik}(\boldsymbol{v}) \Delta_i v_k$

[cf. Eq. (52.1)] due to the effects of gaseous and radiative viscosity. Its first variation gives

$$\left.\begin{aligned}&\frac{\partial p'}{\partial t}+\boldsymbol{v}'\cdot\operatorname{grad}p+\boldsymbol{v}\cdot\operatorname{grad}p'-\left(\frac{\Gamma_1' p}{\varrho}+\frac{\Gamma_1 p'}{\varrho}-\frac{\Gamma_1 p}{\varrho^2}\varrho'\right)\left(\frac{\partial\varrho}{\partial t}+\boldsymbol{v}\cdot\operatorname{grad}\varrho\right)-\\&\qquad-\frac{\Gamma_1 p}{\varrho}\left(\frac{\partial\varrho'}{\partial t}+\boldsymbol{v}'\cdot\operatorname{grad}\varrho+\boldsymbol{v}\cdot\operatorname{grad}\varrho'\right)\\&\quad=\Gamma_3'\varrho\left[\varepsilon_1+\varepsilon_2-\varepsilon_3-\frac{1}{\varrho}\operatorname{div}(\boldsymbol{F}_R+\boldsymbol{F}_C+\boldsymbol{F}_t)\right]+\\&\qquad+(\Gamma_3-1)\left[\varrho\,\varepsilon_1+\varrho\,\varepsilon_2-\varrho\,\varepsilon_3-\operatorname{div}(\boldsymbol{F}_R+\boldsymbol{F}_C+\boldsymbol{F}_t)\right]'.\end{aligned}\right\}\qquad(56.5)$$

According to our discussion of Eq. (53.27), Eq. (50.13) can in many cases be written

$$\left.\begin{aligned}&\frac{\partial p_t}{\partial t}+\boldsymbol{v}\cdot\operatorname{grad}p_t-\frac{5}{3}\frac{p_t}{\varrho}\left(\frac{\partial\varrho}{\partial t}+\boldsymbol{v}\cdot\operatorname{grad}\varrho\right)\\&\quad=\frac{2}{3}\varrho\left(\varepsilon_4-\varepsilon_2-\frac{1}{\varrho}\operatorname{div}\boldsymbol{F}_2-\frac{1}{\varrho}\overline{\boldsymbol{V}\cdot\operatorname{grad}p}\right)\end{aligned}\right\}\qquad(56.6)$$

or in linearized form

$$\left.\begin{aligned}&\frac{\partial p_t'}{\partial t}+\boldsymbol{v}'\cdot\operatorname{grad}p_t+\boldsymbol{v}\cdot\operatorname{grad}p_t'-\frac{5}{3}\left(\frac{p_t'}{\varrho}-\frac{p_t}{\varrho^2}\varrho'\right)\left(\frac{\partial\varrho}{\partial t}+\boldsymbol{v}\cdot\operatorname{grad}\varrho\right)-\\&\qquad-\frac{5}{3}\frac{p_t}{\varrho}\left(\frac{\partial\varrho'}{\partial t}+\boldsymbol{v}'\cdot\operatorname{grad}\varrho+\boldsymbol{v}\cdot\operatorname{grad}\varrho'\right)\\&\quad=\frac{2}{3}\left(\varrho\,\varepsilon_4-\varrho\,\varepsilon_2-\operatorname{div}\boldsymbol{F}_2-\overline{\boldsymbol{V}\cdot\operatorname{grad}p}\right)'.\end{aligned}\right\}\qquad(56.7)$$

If the original state is stationary, all partial time derivatives of the non perturbed variables disappear. Furthermore, if the perturbation is applied to an equilibrium state, characterized by

$$\boldsymbol{v}=0,\qquad(56.8)$$

$$\varrho F^k=\sum_i g^{ik}\Delta_i(p+p_t),\qquad(56.9)$$

$$\varrho\,\varepsilon_1+\varrho\,\varepsilon_2-\varrho\,\varepsilon_3=\operatorname{div}(\boldsymbol{F}_R+\boldsymbol{F}_C+\boldsymbol{F}_t),\qquad(56.10)$$

$$\varrho\,\varepsilon_2+\operatorname{div}\boldsymbol{F}_2+\overline{\boldsymbol{V}\cdot\operatorname{grad}p}=0,\qquad(56.11)$$

the linearized equations take the simple form

$$\frac{\partial\varrho'}{\partial t}+\operatorname{div}(\varrho\,\boldsymbol{v}')=0,\qquad(56.12)$$

$$\frac{\partial v'^k}{\partial t}=F'^k+\frac{\varrho'}{\varrho^2}\sum_i g^{ik}\Delta_i(p)-\frac{1}{\varrho}\sum_i g^{ik}\Delta_i p'+\frac{1}{\varrho}\sum_i\Delta_i\mathfrak{P}^{ik}(\boldsymbol{v}'),\qquad(56.13)$$

or in the presence of turbulence

$$\left.\begin{aligned}&\frac{\partial v'^k}{\partial t}=F'^k+\frac{\varrho'}{\varrho^2}\sum_i g^{ik}\Delta_i(p+p_t)-\frac{1}{\varrho}\sum_i g^{ik}\Delta_i(p'+p_t')-\frac{\varrho'}{\varrho^2}\sum_i\Delta_i\mathfrak{P}^{ik}(\boldsymbol{V})+\\&\qquad+\frac{1}{\varrho}\sum_i\Delta_i[\mathfrak{P}^{ik}(\boldsymbol{V}')+\mathfrak{P}_t^{ik}(\boldsymbol{v}')],\end{aligned}\right\}\qquad(56.14)$$

$$\left.\begin{aligned}&\frac{\partial p'}{\partial t}+\boldsymbol{v}'\cdot\operatorname{grad}p-\frac{\Gamma_1 p}{\varrho}\left(\frac{\partial\varrho'}{\partial t}+\boldsymbol{v}'\cdot\operatorname{grad}\varrho\right)\\&\quad=(\Gamma_3-1)\,\varrho\left\{\varepsilon_1+\varepsilon_2-\varepsilon_3-\frac{1}{\varrho}\operatorname{div}(\boldsymbol{F}_R+\boldsymbol{F}_C+\boldsymbol{F}_t)\right\}'\end{aligned}\right\}\qquad(56.15)$$

and

$$\left.\begin{aligned}&\frac{\partial p_t'}{\partial t}+\boldsymbol{v}'\cdot\operatorname{grad}p_t-\frac{5}{3}\frac{p_t}{\varrho}\left(\frac{\partial\varrho'}{\partial t}+\boldsymbol{v}'\cdot\operatorname{grad}\varrho\right)\\&\quad=\frac{2}{3}\varrho\left[\varepsilon_4-\varepsilon_2-\frac{1}{\varrho}\operatorname{div}\boldsymbol{F}_2-\frac{1}{\varrho}\overline{\boldsymbol{V}\cdot\operatorname{grad}p}\right]'.\end{aligned}\right\}\qquad(56.16)$$

Here, we always have

$$\delta f = f' + \delta \boldsymbol{r} \cdot \operatorname{grad} f \tag{56.17}$$

and we can interchange at will f' and δf for any quantity f such that at equilibrium, $f=0$ or $\operatorname{grad} f=0$. This is the case for the second members of (56.15) and (56.16) [cf. Eqs. (56.10) and (56.11)] which take on their simplest form in Lagrangian notation

$$\frac{d\,\delta p}{dt} - \frac{\Gamma_1 p}{\varrho}\frac{d\,\delta\varrho}{dt} = (\Gamma_3 - 1)\,\varrho\,\delta\left\{\varepsilon_1 + \varepsilon_2 - \varepsilon_3 - \frac{1}{\varrho}\operatorname{div}(\boldsymbol{F}_R + \boldsymbol{F}_C + \boldsymbol{F}_t)\right\} \tag{56.18}$$

and

$$\frac{d\,\delta p_t}{dt} - \frac{5}{3}\frac{p_t}{\varrho}\frac{d\,\delta\varrho}{dt} = \frac{2}{3}\varrho\,\delta\left[\varepsilon_4 - \varepsilon_2 - \frac{1}{\varrho}\operatorname{div}\boldsymbol{F}_2 - \frac{1}{\varrho}\overline{\boldsymbol{V}\cdot\operatorname{grad} p}\right] \tag{56.19}$$

with

$$\frac{1}{\varrho}\frac{d\,\delta\varrho}{dt} = -\operatorname{div}\boldsymbol{v}' \tag{56.20}$$

according to (45.1).

The surface of discontinuity defined by (54.1) becomes

$$f(x_1, x_2, x_3, t) + f'(x_1, x_2, x_3, t) = 0$$

and the kinematic conditions (54.3) can be written

$$(\boldsymbol{v}_1 - \boldsymbol{v}_2)\operatorname{grad} f' + (\boldsymbol{v}_1' - \boldsymbol{v}_2')\cdot\operatorname{grad} f = 0 \tag{56.21}$$

while at an external boundary, we have

$$\frac{\partial f'}{\partial t} + \boldsymbol{v}\cdot\operatorname{grad} f' + \boldsymbol{v}'\cdot\operatorname{grad} f = 0. \tag{56.22}$$

The dynamical conditions (54.8) gives

$$p_1' - p_2' = 0, \tag{56.23}$$

and the mixed condition at a free boundary (54.11) can be written

$$\frac{\partial p'}{\partial t} + \boldsymbol{v}\cdot\operatorname{grad} p' + \boldsymbol{v}'\cdot\operatorname{grad} p = 0. \tag{56.24}$$

These equations combined with the appropriate equation of state, a definition of the external forces and, in some cases, auxiliary boundary conditions permit to treat the general linear oscillations of any compressible fluid.

III. Radial oscillations of a gaseous sphere under its own gravitation.

57. General equations. In this case, the external forces are defined by

$$\boldsymbol{F} = -\operatorname{grad}\Phi \tag{57.1}$$

where the gravitational potential Φ satisfies POISSON's equation

$$\nabla^2\Phi = 4\pi G\varrho.$$

The equilibrium conditions are determined by [cf. Eqs. (56.9) to (56.11)]

$$-\frac{\partial\Phi}{\partial r} = \frac{1}{\varrho}\frac{\partial(p+p_t)}{\partial r} = -\frac{G\,m(r)}{r^2}, \tag{57.2}$$

$$\varepsilon_1 + \varepsilon_2 - \varepsilon_3 = \frac{1}{\varrho r^2}\frac{\partial}{\partial r}\left[r^2(F_R + F_t)\right], \tag{57.3}$$

$$\varrho\,\varepsilon_2 + \frac{1}{r^2}\frac{\partial}{\partial r}(r^2 F_R) + \overline{V_r\frac{\partial p}{\partial r}} = 0, \tag{57.4}$$

where $m(r)$ is the mass of the sphere of radius r. The conduction flux has been neglected since it is very small in a non-degenerate gaseous configuration. Eq. (49.40) gives

$$F_R = -\frac{4acT^3}{3\varkappa\varrho}\frac{dT}{dr}. \tag{57.5}$$

On the other hand, using the definition (52.5) of F_t, a slight generalization of the procedure followed in Ref. [45a] and [45b] gives

$$F_t = \overline{\varrho U V_r} = -\varrho l C C_v T\left\{1 + \frac{2}{3}\frac{A_p^2}{B} - \frac{2}{3}\frac{A_p}{\beta} + \frac{2(1-\beta)}{\beta}\left[5 + \frac{4(1-\beta)}{\beta}\right]\right\} D \tag{57.6}$$

with the same notation as in Eqs. (49.14) and (53.26). Here C_v is the specific heat of a monatomic gas

$$C_v = \tfrac{3}{2}(N+M)\,k = \tfrac{3}{2}M(1+x)\,k \tag{57.7}$$

and D represents the excess of the actual logarithmic temperature gradient over its adiabatic value

$$D = \frac{1}{T}\frac{dT}{dr} - \frac{\Gamma_2 - 1}{\Gamma_2}\frac{1}{p}\frac{dp}{dr}. \tag{57.8}$$

In the same way, one obtains

$$\overline{V}_r = -lC\left(\frac{4-3\beta}{\beta} + A_p\right)D. \tag{57.9}$$

The generalized specific heats C_v' and C_p' take the values

$$C_v' = C_v\left[1 + \frac{2}{3}A_\varrho^2\frac{B+1}{B} + \frac{8(1-\beta)}{\beta}\right], \tag{57.10}$$

$$C_p' = C_v\left\{\frac{5}{3} + \frac{2}{3}\frac{A_p^2}{B} + \frac{8(1-\beta)}{\beta}\left[\frac{5}{3} + \frac{4(1-\beta)}{\beta}\right]\right\}, \tag{57.11}$$

and we notice that F_t given by (57.6) cannot be expressed simply in terms of C_v' or C_p'. However, if ε_2 is eliminated of (57.3) and (57.4) and $\varrho\varepsilon_3$ is replaced by its value (52.6), one obtains

$$\varrho\varepsilon_1 - \frac{1}{r^2}\frac{\partial}{\partial r}(r^2 F_2) - \frac{1}{r^2}\frac{\partial}{\partial r}\left[r^2(F_t + p\overline{V}_r)\right] - \frac{1}{r^2}\frac{\partial}{\partial r}(r^2 F_R) = 0$$

and if we denote the quantity $F_t + p\overline{V}_r$ by F_t^*, one can show that

$$F_t^* = F_t + p\overline{V}_r = -\varrho l C C_p' T D. \tag{57.12}$$

With this notation, Eq. (57.3) becomes

$$\varrho\varepsilon_1 - \frac{1}{r^2}\frac{\partial}{\partial r}\left[r^2(F_t^* + F_2 + F_R)\right] = 0 \tag{57.13}$$

which is the more usual form.

If we separate the time, writing

$$\delta r = \delta r\, e^{i\sigma t},\quad v' = i\sigma\,\delta r\, e^{i\sigma t},\quad p' = p'(r)\,e^{i\sigma t},\quad \varrho' = \varrho'(r)\,e^{i\sigma t},\quad \Phi' = \Phi'(r)\,e^{i\sigma t} \tag{57.14}$$

where, for simplicity, we have kept the same symbols for the amplitudes, the linearized Eqs. (56.12) to (56.14) become

$$\varrho' + \frac{1}{r^2}\frac{\partial}{\partial r}(r^2\varrho\,\delta r) = 0, \tag{57.15}$$

$$-\sigma^2\delta r = -\frac{\partial\Phi'}{\partial r} + \frac{\varrho'}{\varrho^2}\frac{\partial p}{\partial r} - \frac{1}{\varrho}\frac{\partial p'}{\partial r} + \frac{1}{\varrho}\sum_i \Delta_i\,\mathfrak{P}^{ir}(i\sigma\,\delta r) \tag{57.16}$$

with

$$\left.\begin{aligned}\sum_i \varDelta_i \mathfrak{P}^{ir}(i\sigma\,\delta r) &= i\sigma\Big\{\frac{4}{3}\frac{\partial\eta}{\partial r}r\frac{\partial}{\partial r}\Big(\frac{\delta r}{r}\Big)+\frac{4}{3}\eta\Big[r\frac{\partial^2}{\partial r^2}\Big(\frac{\delta r}{r}\Big)+4\frac{\partial}{\partial r}\Big(\frac{\delta r}{r}\Big)\Big]\Big\}\\ &= i\sigma\frac{4}{3}\frac{1}{r^3}\frac{d}{dr}\Big[r^4\eta\frac{d\xi}{dr}\Big]=i\sigma\psi(\eta,\xi)\end{aligned}\right\}\qquad(57.17)$$

where $\eta=\eta_R+\eta_G$ and the relative amplitude $\delta r/r$ is denoted by ξ. If turbulence is present, neglecting small terms in $\mathfrak{P}^{ir}(\overline{V})$ and $\mathfrak{P}^{ir}(\overline{V'})$ we have

$$-\sigma^2\,\delta r = -\frac{\partial\Phi'}{\partial r}+\frac{\varrho'}{\varrho^2}\frac{\partial(p+p_t)}{\partial r}-\frac{1}{\varrho}\frac{\partial(p'+p_t')}{\partial r}+\frac{1}{\varrho}\sum_i \varDelta_i\mathfrak{P}_t^{ir}(\boldsymbol{v}'),\qquad(57.18)$$

where $\sum_i \varDelta_i\mathfrak{P}_t^{ir}(\boldsymbol{v}')$ has the same form as (57.17) with η_t replacing η and will be denoted by $i\sigma\psi(\eta_t,\xi)$. The Eqs. (56.15) and (56.16) may be written

$$\left.\begin{aligned}&p'+\delta r\frac{\partial p}{\partial r}-\frac{\Gamma_1 p}{\varrho}\Big(\varrho'+\delta r\frac{\partial\varrho}{\partial r}\Big)\\ &\quad=\frac{(\Gamma_3-1)}{i\sigma}\varrho\,\delta\Big\{\varepsilon_1+\varepsilon_2-\varepsilon_3-\frac{1}{\varrho r^2}\frac{\partial}{\partial r}\big[r^2(F_R+F_t)\big]\Big\},\end{aligned}\right\}\qquad(57.19)$$

$$\left.\begin{aligned}&p_t'+\delta r\frac{\partial p_t}{\partial r}-\frac{5}{3}\frac{p_t}{\varrho}\Big(\varrho'+\delta r\frac{\partial\varrho}{\partial r}\Big)\\ &\quad=\frac{1}{i\sigma}\frac{2}{3}\varrho\,\delta\Big[\varepsilon_4-\varepsilon_2-\frac{1}{\varrho r^2}\frac{\partial}{\partial r}(r^2F_2)-\frac{1}{\varrho}\overline{V_r\frac{\partial p}{\partial r}}\Big].\end{aligned}\right\}\qquad(57.20)$$

On the other hand, POISSON's equation gives

$$\nabla^2\Phi'=4\pi G\varrho'\qquad(57.21)$$

or, using (57.15)

$$\frac{1}{r^2}\frac{\partial}{\partial r}\Big(r^2\frac{\partial\Phi'}{\partial r}\Big)=-\frac{1}{r^2}\frac{\partial}{\partial r}(4\pi Gr^2\varrho\,\delta r)$$

and thus

$$\frac{\partial\Phi'(r)}{\partial r}=-4\pi G\varrho\,\delta r.\qquad(57.22)$$

The elimination of p', p_t', ϱ', Φ' leads to

$$\left.\begin{aligned}&\frac{d}{dr}\Big[\Big(\Gamma_1 p+\frac{5}{3}p_t\Big)\frac{1}{r^2}\frac{\partial}{\partial r}(r^2\,\delta r)\Big]+\varrho\,\delta r\Big[\sigma^2+\frac{4Gm(r)}{r^3}\Big]\\ &\quad=\frac{\sigma}{i}\psi(\eta_t)+\frac{1}{i\sigma}\frac{d}{dr}\Big[(\Gamma_3-1)\varrho\,\delta\{\ldots\}+\frac{2}{3}\varrho\,\delta[\ldots]\Big],\end{aligned}\right\}\qquad(57.23)$$

with $\{\ldots\}$ and $[\ldots]$ meaning the brackets of Eqs. (57.19) and (57.20). In absence of turbulence, we may set $p_t=0$, $\varepsilon_4=0$, $\varepsilon_3=0$, $F_t'=0$, replace $\psi(\eta_t)$ by $\psi(\eta_G+\eta_R)$ and interpret ε_2 as due to the ordinary viscosity.

One may also proceed with these eliminations without previously separating the time, but this raises the order of the resulting equation which can be written

$$\left.\begin{aligned}\frac{\partial^3\delta r}{\partial t^3}&=\frac{1}{\varrho}\frac{\partial}{\partial r}\Big[\Big(\Gamma_1 p+\frac{5}{3}p_t\Big)\frac{1}{r^2}\frac{\partial}{\partial r}\Big(r^2\frac{\partial\,\delta r}{\partial t}\Big)\Big]-\frac{4}{\varrho r}\frac{\partial(p+p_t)}{\partial r}\cdot\frac{\partial\,\delta r}{\partial t}+\\ &+\frac{1}{\varrho}\frac{\partial}{\partial t}\sum\varDelta_i\mathfrak{P}_t^{ir}\Big(\frac{\partial\,\delta r}{\partial t}\Big)-\frac{1}{\varrho}\frac{\partial}{\partial r}\Big\{(\Gamma_3-1)(\varrho\{\ldots\})'+\frac{2}{3}(\varrho[\ldots])'\Big\}\end{aligned}\right\}\qquad(57.24)$$

where v' has been replaced by $\partial(\delta r)/\partial t$ which is correct to the first order of approximation. If we substitute now $\delta r=\delta r\,e^{i\sigma t}$, we recover Eq. (57.23) only after division by a factor $i\sigma$ eliminating a solution corresponding to a slow aperiodic motion of expansion or contraction due to a lasting lack of balance between energy generation and dissipation[1].

With the relative amplitude ξ, Eq. (57.23) may be written

$$\left.\begin{aligned}\frac{d}{dr}\left[r^4\left(\Gamma_1 p+\frac{5}{3}p_t\right)\frac{d\xi}{dr}\right]+\xi\left\{\sigma^2\varrho r^4+r^3\frac{d}{dr}\left[(3\Gamma_1-4)p+p_t\right]\right\}\\ =\frac{r^3\sigma}{i}\psi(\eta_t,\xi)+\frac{r^3}{i\sigma}\frac{\partial}{\partial r}\left[(\Gamma_3-1)\varrho\delta\{\ldots\}+\frac{2}{3}\varrho\delta[\ldots]\right].\end{aligned}\right\}\qquad(57.25)$$

If the spherical symmetry is to be conserved, the possible discontinuities occur on spheres

$$f(r,t)+f'(r,t)=0 \qquad (57.26)$$

where we have

$$\left.\begin{aligned}v_1'-v_2'=0 \quad\text{or}\quad \delta r_1=\delta r_2\\ p_1'-p_2'=0.\end{aligned}\right\}\qquad(57.27)$$

and

At the centre of the sphere, we must have

$$v'=0 \quad\text{or}\quad \delta r=0 \qquad (57.28)$$

and on its external surface [cf. Eq. (56.24)]

$$p'+\delta r\frac{\partial p}{\partial r}=0. \qquad (57.29)$$

Using the adiabatic part of (56.18) and (56.20) which here becomes

$$\frac{\delta\varrho}{\varrho}=-\frac{1}{r^2}\frac{d}{dr}(r^2\delta r)=-\left(3\xi+r\frac{d\xi}{dr}\right), \qquad (57.30)$$

Eq. (57.29) can also be written

$$\delta p=-\Gamma_1 p\left(3\xi+r\frac{d\xi}{dr}\right)=0. \qquad (57.31)$$

58. Adiabatic radial oscillations. Some mathematical aspects. Here the quantities in brackets, $\{\ldots\}$ and $[\ldots]$, cancel out and if we admit that friction of any type is also negligible, Eq. (57.25) becomes

$$\frac{d}{dr}\left(r^4\Gamma_1 p\frac{d\xi}{dr}\right)+\xi\left\{\sigma^2\varrho r^4+r^3\frac{d}{dr}\left[(3\Gamma_1-4)p\right]\right\}=0. \qquad (58.1)$$

We have also neglected the turbulent pressure since, except perhaps in the very external layers, it will always be very small compared to the gaseous pressure $(\overline{V^2}\ll\overline{C_T^2})$. Except for terms in the derivatives of Γ_1, this equation is identical to the equation established by EDDINGTON in his first papers [*13*] on the subject. It must be solved subject to the boundary conditions

$$\left.\begin{aligned}\delta r=r\xi=0 \quad\text{at}\quad r=0,\\ \delta p=-\Gamma_1 p\left(3\xi+r\frac{d\xi}{dr}\right)=0 \quad\text{at}\quad r=R.\end{aligned}\right\}\qquad(58.2)$$

As far as the interpretation of variable stars is concerned the only elements which this equation can provide, are the periods

$$P_j=\frac{2\pi}{\sigma_j} \qquad (58.3)$$

---

[1] Cf. J. JEANS [*14*], § 108 and S. ROSSELAND [*39d*], §§ 5.1 and 5.12 to 5.14.

of the different modes of oscillations. In this respect, the problem has been studied extensively and worked out numerically for many models.

From a mathematical point of view, Eq. (58.1) which can be written

$$\frac{d}{dr}\left[\alpha(r)\frac{d\xi}{dr}\right] - \beta(r)\,\xi + \sigma^2\mu(r)\,\xi = -L\xi + \sigma^2\mu(r)\,\xi = 0 \tag{58.4}$$

with

$$\left.\begin{aligned} L &= -\frac{d}{dr}\left[\alpha(r)\frac{d}{dr}\ldots\right] + \beta(r)\ldots, & \alpha(r) &= r^4\Gamma_1 p,\\ \beta(r) &= -r^3\frac{d}{dr}\left[(3\Gamma_1-4)\,p\right], & \mu(r) &= \varrho\, r^4 \end{aligned}\right\} \tag{58.5}$$

and the boundary conditions (58.2) form a typical self-adjoint, linear, homogeneous eigenvalue problem of the second order. Such problems have given rise to an extensive literature [*49*]. A simple starting point is provided by the comparison with a standard Sturm-Liouville problem in which the continuous functions $\alpha(r)$ and $\mu(r)$ are subject to the further conditions

$$\alpha(r) > 0, \quad \mu(r) > 0 \quad \text{in} \quad 0 \leqq r \leqq R, \tag{58.6}$$

and the boundary conditions are of the general form

$$h_0\,\xi(0) = \xi'(0) \quad \text{and} \quad -h_1\,\xi(R) = \xi'(R). \tag{58.7}$$

In that case, one knows that the problem admits a solution (eigenfunction) only for certain values (eigenvalues) of the parameter σ^2. The infinite discrete set of eigenvalues ordered by increasing values will be represented by

$$\sigma_0^2, \sigma_1^2, \sigma_2^2, \ldots, \sigma_i^2, \ldots. \tag{58.8}$$

To these correspond the complete set of eigenfunctions

$$\xi_0, \xi_1, \xi_2, \ldots, \xi_i, \ldots \tag{58.9}$$

which are orthogonal with respect to $\mu(r)$

$$\int_0^R \mu(r)\,\xi_i\,\xi_k\,dr = \delta_{ik}. \tag{58.10}$$

The eigenfunction ξ_0 corresponding to the smallest eigenvalue σ_0 has no node in $0 \leqq r \leqq R$ and will be called here the fundamental mode. The i-th mode ξ_i has i modes in $0 \leqq r \leqq R$ and the zeros of one eigenfunction divide the zeros of the preceding one. One shows also that all σ_i^2 are positive provided that

$$\beta(r) \geqq 0, \quad h_0 \geqq 0, \quad h_1 \geqq 0. \tag{58.11}$$

In our case however, $\alpha(r)$ and $\mu(r)$ tend to zero at both ends of the interval $0 \leqq r \leqq R$ and to study these singularities let us rewrite Eq. (58.1) as

$$\left.\begin{aligned} &\frac{d^2\xi}{dr^2} + \frac{d\xi}{dr}\left(\frac{4}{r} + \frac{1}{\Gamma_1}\frac{d\Gamma_1}{dr} - \frac{\varrho}{p}\frac{G\,m(r)}{r^2}\right) + \\ &\quad + \xi\left[\frac{\sigma^2\varrho}{\Gamma_1 p} + \frac{3}{r\Gamma_1}\frac{d\Gamma_1}{dr} - \frac{(3-4/\Gamma_1)}{r\,p}\frac{G\,m(r)}{r^2}\,\varrho\right] = 0. \end{aligned}\right\} \tag{58.12}$$

Physically Γ_1 and $d\Gamma_1/dr$ must remain finite everywhere although the last quantity might become very large in a region where an abundant element undergoes ionization. The point $r=0$ is a first-order pole of the coefficients of ξ and $d\xi/dr$. Close to the surface ϱ/p tends also towards infinity, and very often simply as T^{-1} or $z^{-1} = \left(\frac{r}{R} - 1\right)^{-1}$, in which case the point $r = R$ is also a first-order pole of the coefficients of ξ and $d\xi/dr$. Let us denote by β_{-1}, the value in $r=0$ (or $r=R$) of the coefficient of r^{-1} (or z^{-1}) in the series expansion of the coefficient of $d\xi/dr$ in the vicinity of $r=0$ (or $r=R$). In this case, both singularities are regular[1] and independent solutions of the form

$$\left.\begin{aligned} \xi_1 &= r^{\vartheta_1} \sum_{i=0} a_{1i} r^i \quad \text{or} \quad \xi_1 = z^{\vartheta_1} \sum_{i=0} b_{1i} z^i, \\ \xi_2 &= r^{\vartheta_2} \sum_{i=0} a_{2i} r^i \quad \text{or} \quad \xi_2 = z^{\vartheta_2} \sum_{i=0} b_{2i} z^i \end{aligned}\right\} \tag{58.13}$$

exist if the roots ϑ_1 and $\vartheta_2 (\vartheta_1 > \vartheta_2)$ of the indicial equation

$$\vartheta(\vartheta-1) + \beta_{-1}\vartheta = 0 \tag{58.14}$$

do not differ by an integer. If $\vartheta_1 - \vartheta_2 = q$ is zero or an integer, the two solutions (58.13) cease to be independent and ξ_2 must be replaced by

$$\xi_2 = \xi_1(r) \int \frac{dr\, e^{-\int \left(\frac{4}{r} + \frac{1}{\Gamma_1}\frac{d\Gamma_1}{dr} - \frac{\varrho}{p}\frac{Gm}{r^2}\right) dr}}{\xi_1^2(r)} = r^{\vartheta_1} \sum_{i=0} a_{1i} r^i \int \frac{dr\, e^{-\int \left(\frac{\beta_{-1}}{r} + \sum\limits_{i=0} \beta_i r^i\right) dr}}{r^{2\vartheta_1} \left(\sum\limits_{i=0} a_{1i} r^i\right)^2},$$

or

$$\xi_2 = z^{\vartheta_1} \sum b_{1i} z^i \int \frac{e^{-\int \left[\frac{\beta_{-1}}{z} + \sum\limits_{i=0} \beta_i z^i\right] dz}}{z^{2\vartheta_1} \left[\sum b_{1i} z^i\right]^2}.$$

Performing the integral in the exponential, one gets

$$\xi_2 = r^{\vartheta_1} \sum_{i=0} a_{1i} r^i \int \frac{S(r)}{r^{q+1}} dr \quad \text{or} \quad z^{\vartheta_1} \sum_{i=0} b_{1i} z^i \int \frac{S(z)}{z^{q+1}} dz$$

noting that $2\vartheta_1 = (q+1-\beta_{-1})$ and

$$S = \frac{e^{-\int \sum\limits_{i=0} \beta_i r^i dr}}{(\sum a_{1i} r^i)^2} \quad \text{or} \quad \frac{e^{-\int \sum \beta_i z^i dz}}{(\sum b_{1i} z^i)^2}$$

is regular in $0 \leq r \leq R$. Expressing S as a series expansion around $r=0$ or $r=R$

$$S = \sum_{i=0} A_i r^i \quad \text{or} \quad \sum_{i=0} A_i z^i$$

one finally gets

$$\xi_2 = r^{\vartheta_1 - q} \sum_{i=0} B_j r^i + \xi_1 A_q \ln r \quad \text{or} \quad z^{\vartheta_1 - q} \sum_{i=0} C_i z^i + \xi_1 A_q \ln z, \tag{58.15}$$

the logarithmic singularity disappearing only if A_q is zero, a case which never happens if $q=0$.

The general solution around $r=0$ or $r=R$ will be represented as a linear combination of the two independent solutions

$$\xi = A\xi_1 + B\xi_2. \tag{58.16}$$

More general situations could arise, but in all cases of physical interest the singularities in $r=0$ and $r=R$ remain regular [at most first or second order poles of the coefficients of $d\xi/dr\,(\beta^{-1}/r)$ and $\xi\,(\gamma_{-2}/r^2)$ respectively] and the construction of the independent solutions ξ_1 and ξ_2 proceeds exactly as before except that (58.14) may contain an independent term γ_{-2}. To satisfy the boundary

[1] Cf. for instance E. GOURSAT: Cours d'Analyse, Chap. XX, § III, p. 444. Paris: Gauthier-Villars 1905.

conditions (58.2), ξ as given by (58.16) must be regular and generally ξ_1 only is to be kept ($B=0$).

Note that in $r=R$, this implies

$$\left(\frac{d\xi}{dr}\right)_R = \frac{\xi_R}{R\,\Gamma_1}\left[\omega^2 - (3\Gamma_1 - 4)\right] \tag{58.17}$$

which is the degenerate form of (58.12), with

$$\omega^2 = \frac{3\sigma^2}{4\pi G\bar{\varrho}}\,. \tag{58.18}$$

In general, if the extremities of the interval of integration are singular points or if the interval becomes infinite, the spectrum is no longer purely discrete but may become continuous or comprise a continuous part. However if, as it is the case here, the singularities are regular and the boundary conditions are automatically satisfied by the regular part of the solution, the spectrum in general remains purely discrete[1] and the eigenvalues and eigenfunctions have the same properties as in the case of the Sturm-Liouville problem.

Some of those properties can easily be verified here and in such a way as to introduce a new point of view which has proved very fruitful in all those problems. First, two distinct solutions ξ_i and $\xi_k\,(i\neq k)$ are orthogonal with respect to $\mu(r)$ since one verifies easily that

$$\int_0^R \left[\xi_i(L\,\xi_k - \sigma_k^2\mu\,\xi_k) - \xi_k(L\,\xi_i - \sigma_i^2\mu\,\xi_i)\right] dr = (\sigma_i^2 - \sigma_k^2)\int_0^R \mu\,\xi_i\,\xi_k\,dr = 0 \tag{58.19}$$

if one takes the value of $\alpha(r)$ and the boundary conditions into account. In the same way, integrating $\xi_i(L\xi_i - \sigma_i^2\mu\xi_i)$ from 0 to R, any eigenvalue must be given by

$$\sigma_i^2 = \frac{\int_0^R \xi_i\,L\,\xi_i\,dr}{\int_0^R \mu(r)\,\xi_i^2\,dr}\,. \tag{58.20}$$

To prove the existence of these eigenvalues, let us suppose first that the problem is definite positive, i.e.

$$\int_0^R u\,L\,u\,dr > 0 \tag{58.21}$$

for any function u continuous and with piecewise continuous first and second derivatives in $0\leqq r\leqq R$ and satisfying the boundary conditions (58.2). Expliciting the operator L, condition (58.21) becomes, after an integration by parts and taking the boundary conditions into account

$$\int_0^R u\,L\,u\,dr = \int_0^R \left[\alpha(r)\left(\frac{du}{dr}\right)^2 + \beta(r)\,u^2\right] dr \geqq 0\,. \tag{58.22}$$

Since $\alpha(r) = r^4\,\Gamma_1 p \geqq 0$ everywhere, this condition will certainly be realized if

$$\beta(r) = -\,r^3\frac{d}{dr}\left[(3\Gamma_1 - 4)\,p\right] \geqq 0 \quad \text{in} \quad 0\leqq r\leqq R. \tag{58.23}$$

If Γ_1 is constant, this reduces to the well-known condition

$$\Gamma_1 > \tfrac{4}{3} \quad \text{in} \quad 0\leqq r\leqq R. \tag{58.24}$$

---

[1] Cf. for instance E. C. Kemble: The Fundamental Principles of Quantum Mechanics, § 23d and 23e. New-York: Mac Graw-Hill 1937.

If the pressure of turbulence is appreciable [compare the first member of (57.25) with (58.1)] the corresponding condition (58.23) could still be satisfied for values of Γ_1 a little smaller than $\frac{4}{3}$.

It is then possible to prove that, if for any function u of the class considered we define a number M_1 by

$$M_1 = \frac{\int_0^R u\,L\,u\,dr}{\int_0^R \mu\,u^2\,dr},$$

there is at least one eigenvalue comprised between 0 and M_1.

From this, it follows immediately that the lowest eigenvalue σ_0^2 is equal to the minimum value of M_1 when we vary the function u or

$$\sigma_0^2 = \min_{u(r)} \frac{\int_0^R u\,L\,u\,dr}{\int_0^R \mu\,u^2\,dr} \tag{58.25}$$

and, according to (58.20), ξ_0 is identical to the function $u = u_m$ for which the minimum is reached.

This can be generalized to the higher modes taking into account the orthogonality conditions which they satisfy and one gets

$$\sigma_i^2 = \min_{u(r)} \frac{\int_0^R u\,L\,u\,dr}{\int_0^R \mu\,u^2\,dr} \tag{58.26}$$

with the auxiliary conditions

$$\int_0^R \mu(r)\,u\,\xi_k\,dr = 0, \qquad k = 0, 1, 2, \ldots, i-1. \tag{58.27}$$

We shall admit here, that the set of eigenfunctions ξ_i is complete i.e. any arbitrary square integral function $f(r)$ can be expanded as

$$f(r) = \sum_{i=0}^{\infty} a_i\,\xi_i.$$

If the condition (58.21) is no longer verified but $\int_0^R u L u\,dr$ remains bounded negatively, the previous expressions for σ_i^2 remain valid and negative values of σ_i^2 may occur but only in finite number.

The expressions (58.25) and (58.26) for the eigenvalues can also be derived from the variational interpretation of the problem which is also known as RAYLEIGH's principle [*49e*]. Let us consider the integral

$$D(u) = \int_0^R \left[\alpha(r)\left(\frac{du}{dr}\right)^2 + \beta(r)\,u^2\right] dr. \tag{58.28}$$

To eliminate the degeneracy arising from the homogeneous character of the problem, we can impose a normalization condition

$$\int_0^R \mu(r)\,u^2\,dr = 1. \tag{58.29}$$

The condition

$$\delta \int_0^R \left[\alpha(r)\left(\frac{du}{dr}\right)^2 + \beta(r)\, u^2 - \lambda\, \mu(r)\, u^2\right] dr = \delta \int_0^R F(u, u', r)\, dr = 0 \tag{58.30}$$

where λ is an arbitrary parameter expresses the condition that $D(u)$ be a minimum when (58.29) is satisfied. This condition is equivalent to the well-known EULER's equation

$$\left[\frac{\partial F}{\partial u} - \frac{d}{dr}\left(\frac{\partial F}{\partial u'}\right)\right]_{u_m} = -\left\{\frac{d}{dr}\left[\alpha(r)\,\frac{du_m}{dr}\right] - \beta(r)\, u_m + \lambda\, \mu(r)\, u_m\right\} = 0 \tag{58.31}$$

with the boundary conditions

$$\left(\frac{\partial F}{\partial u'}\right)_0 = [\alpha(r)\, u']_0 = [\Gamma_1 p\, r^4 u_m']_0 = 0 \quad \text{and} \quad [\Gamma_1 p\, r^4 u_m']_R = 0 \tag{58.32}$$

which are identical to (58.4) with $\lambda = \sigma^2$ and (58.2). Thus the function u_m which makes $D(u)$ a minimum, is an eigenfunction of the problem. The corresponding value of λ is given by the minimum value $D(u_m)$ of $D(u)$ since, integrating (58.28) by parts and taking into account the boundary conditions and the fact that u_m satisfies (58.4), we find

$$D(u_m) = \sigma^2 \int_0^R \mu(r)\, u_m^2\, dr.$$

Thus according as to whether the u are normalized or not, we have

$$\left.\begin{aligned} \sigma^2 &= \min_{u(r)} D(u) = \min_{u(r)} \int_0^R \left[\alpha(r)\left(\frac{du}{dr}\right)^2 + \beta(r)\, u^2\right] dr \\ \text{or} \quad & \\ \sigma^2 &= \min_{u(r)} \frac{\int_0^R \left[\alpha(r)\left(\frac{du}{dr}\right)^2 + \beta(r)\, u^2\right] dr}{\int_0^R \mu(r)\, u^2\, dr} \end{aligned}\right\} \tag{58.33}$$

which is equivalent to (58.25) and thus defines the lowest eigenvalue σ_0 while $u_m = \xi_0$. In the same way, one can also recover (58.26) and (58.27).

This point of view can be generalized (cf. [49a], Chap. VI, § 1.4) so that the preliminary knowledge of the $(i-1)$ first eigenfunctions [conditions (58.27)] is no longer necessary for the determination of σ_i. If $(u_0, u_1, \ldots, u_{i-1})$ are any $(i-1)$ functions of the class u, one can show that

$$\sigma_i^2 = \operatorname*{Max}_{(u_0,\, u_1 \ldots u_{i-1})} \ \min_{u_i} \frac{\int_0^R \left[\alpha(r)\left(\frac{du_i}{dr}\right)^2 + \beta(r)\, u_i^2\right] dr}{\int_0^R \mu(r)\, u_i^2\, dr} \tag{58.34}$$

with the auxiliary conditions

$$\int_0^R \mu(r)\, u_i\, u_k\, dr = 0, \quad k = 0, 1, 2, \ldots \ (i-1) \tag{58.35}$$

This last formulation is very important because it permits to establish qualitative relations between the eigenvalues, the interval of integration, the boundary conditions and the coefficients of the differential equations (cf. [49a], Chap. VI, § 2, theorems 1 to 7). In particular, this provides a confirmation of the physically intuitive result that any increase in the external constraints leads to an increase of the eigenvalues. Again, it can be shown that if $\mu(r)$ in (58.4) is increased (or decreased) everywhere in $0 \leq r \leq R$, all eigenvalues decrease (or increase); on the contrary, if $\alpha(r)$ [or $\beta(r)$] is altered in the same sense everywhere, all eigenvalues vary in that sense too. This last theorem provides a convenient means to compare the eigenvalues of two problems differing only by the values of the coefficients.

59. Adiabatic oscillations. General physical principles. α) *Hamilton's principle.* Here we must use the Lagrangian notation which however is quite simple since we have only one spatial coordinate and since the mass $m(r)$ and its increment $dm(r)$ are invariant following the motion. During the oscillation, a unit mass is submitted to the following forces:

1. The variation of the gravitational attraction

$$\delta\left(-\frac{G\,m(r)}{r^2}\right) = \frac{2G\,m(r)}{r^3}\,\zeta \tag{59.1}$$

in denoting δr by ζ.

2. The variation of the pressure gradient

$$\delta\left(-\frac{1}{\varrho}\,\frac{dp}{dr}\right) = \delta\left(-4\pi r^2\,\frac{dp}{dm}\right) = \frac{2G\,m(r)}{r^3}\,\zeta + \frac{1}{\varrho}\,\frac{d}{dr}\left[\frac{\Gamma_1 p}{r^2}\,\frac{\partial}{\partial r}\,(r^2\zeta)\right] \tag{59.2}$$

in reverting to the spatial coordinate r which can be treated in the varied expressions as a Lagrangian coordinate. The first term in (59.2) can be combined with (59.1) to give

$$4\,\frac{G\,m(r)}{r^3}\,\zeta = \frac{\partial}{\partial\zeta}\left[2\,\frac{G\,m(r)}{r^3}\,\zeta^2\right] \tag{59.3}$$

which is associated with the change in the gravitational potential energy. The second term in (59.2) corresponds to the variation of the internal potential energy of the gas, the significant part of which is here

$$\frac{1}{\varrho}\,\frac{\Gamma_1 p}{2}\left[\frac{1}{r^2}\,\frac{\partial}{\partial r}\,(r^2\zeta)\right]^2. \tag{59.4}$$

Accordingly, the acting forces per unit volume can be considered as deriving from a potential density

$$\varrho V = -\frac{2G\,m(r)}{r^3}\,\varrho\zeta^2 + \frac{\Gamma_1 p}{2}\left[\frac{1}{r^2}\,\frac{\partial(r^2\zeta)}{\partial r}\right]^2,$$

and the Lagrangian density defined by (47.8) becomes

$$\mathscr{L} = \frac{1}{2}\,\varrho\dot{\zeta}^2 + \frac{2G\,m(r)}{r^3}\,\varrho\zeta^2 - \frac{\Gamma_1 p}{2}\,\frac{1}{r^4}\left[\frac{\partial(r^2\zeta)}{\partial r}\right]^2. \tag{59.5}$$

We then have

$$L = 4\pi\int_0^R\left\{\frac{1}{2}\,\varrho\dot{\zeta}^2 r^2 + 2\,\frac{G\,m(r)}{r^3}\,\varrho\zeta^2 r^2 - \frac{\Gamma_1 p}{2}\,\frac{1}{r^2}\left[\frac{\partial(r^2\zeta)}{\partial r}\right]^2\right\}dr. \tag{59.6}$$

Introducing $q = \zeta r^2$, the expression of HAMILTON's principle is

$$\delta\int_{t_1}^{t_2} L\,dt = 4\pi\,\delta\int_{t_1}^{t_2} dt\int_0^R\left[\frac{1}{2}\,\frac{\varrho}{r^2}\,\dot{q}^2 + \frac{2G\,m(r)}{r^3}\,\varrho\,\frac{q^2}{r^2} - \frac{\Gamma_1 p}{2}\,\frac{1}{r^2}\left(\frac{\partial q}{\partial r}\right)^2\right]dr = 0 \tag{59.7}$$

and EULER's equation (47.11) reduces to

$$\frac{\varrho}{r^2}\,\ddot{q} - \frac{4G\,m(r)}{r^3}\,\varrho\,\frac{q}{r^2} - \frac{d}{dr}\left[\frac{\Gamma_1 p}{r^2}\,\frac{dq}{dr}\right] = \varrho\ddot{\zeta} - \frac{4G\,m(r)}{r^3}\,\varrho\zeta - \frac{d}{dr}\left[\frac{\Gamma_1 p}{r^4}\,\frac{d(r^2\zeta)}{dr}\right] = 0$$

which is equivalent to the equation of motion (57.23) where one neglects all non-adiabatic terms and p_t. Writing as before $\zeta \propto e^{i\sigma t}$, taking care in evaluating squares that we have really to deal with the real part of ζ and introducing the variable $\xi = \delta r/r$, we obtain

$$\left.\begin{aligned} J &= \int_0^{2\pi/\sigma} L\,dt = \frac{4\pi^2}{\sigma} \times \\ &\times \left\{\sigma^2 \int_0^R \varrho r^4 \xi^2\,dr - \int_0^R \Gamma_1 p r^4 \left(\frac{d\xi}{dr}\right)^2 dr + \int_0^R r^3 \xi^2 \frac{d}{dr}\left[(3\Gamma_1 - 4)\,p\right] dr\right\}. \end{aligned}\right\} \tag{59.8}$$

One will readily verify that EULER's equation expressing the condition

$$\delta J = 0 \tag{59.9}$$

is identical to (58.1) or (58.4). Thus any function minimizing J will also satisfy (58.20) and in that case (59.8) yields for the fundamental mode,

$$\sigma_0^2 = \min \frac{\int_0^R \Gamma_1 p r^4 \left(\frac{d\xi}{dr}\right)^2 dr - \int_0^R r^3 \xi^2 \frac{d}{dr}\left[(3\Gamma_1 - 4)\,p\right] dr}{\int_0^R \varrho r^4 \xi^2\,dr} \tag{59.10}$$

which is simply the explicit expression of (58.33).

This of course provides the real physical basis of the variational interpretation.

β) The virial theorem. The use of the virial theorem to discuss the oscillations of a star has been introduced by P. LEDOUX [*50*] and has been extended since then in different directions as we shall see later.

Let us consider a cloud of particles interacting by collisions and in motion under their mutual gravitational attraction. In that case, the following Lagrangian identity is satisfied [*51*]

$$\frac{1}{2}\frac{d^2 I}{dt^2} = 2K^* + W \tag{59.11}$$

where I is the moment of inertia with respect to the origin, K^* the total kinetic energy of the particles, and W the gravitational potential energy. For a spherical mass composed of monatomic particles one has, taking the center as origin,

$$\left.\begin{aligned} I &= \int_0^M r^2\,dm(r) \\ K^* &= \int_0^M \frac{1}{2}\dot{r}^2\,dm + \int_0^R n\left(\frac{3}{2}kT\right) 4\pi r^2\,dr = \int_0^M \frac{1}{2}\dot{r}^2\,dm + \int_0^M \frac{3}{2}\frac{p_G}{\varrho}\,dm. \end{aligned}\right\} \tag{59.12}$$

If the pressure of radiation is appreciable p_G must be replaced by $p = p_G + p_R$ so that

$$K^* = \int_0^M \frac{1}{2}\dot{r}^2\,dm + \int_0^M \frac{3}{2}\frac{p}{\varrho}\,dm = K + \frac{3}{2}(\Gamma_1 - 1)\,U \tag{59.13}$$

where K is the kinetic energy of mass motion and U, the total thermal energy. Of course, the pressure of turbulence, if important, should also be included. Finally,

$$W = -\int_0^M \frac{G\,m(r)\,dm(r)}{r}. \tag{59.14}$$

For radial oscillations such as those considered before the first order Lagrangian variations of I and W are

$$\delta I = \int_0^M 2\xi\, dI \quad \text{and} \quad \delta W = -\int_0^M \xi\, dW. \tag{59.15}$$

To the same order of approximation, K is negligible and after some integrating by parts and using the boundary conditions, one finds

$$2\delta K^* = 3\int_0^M \left(\frac{\delta p}{p} - \frac{p}{\varrho}\frac{\delta\varrho}{\varrho}\right) dm = 3\int_0^M (\Gamma_1 - 1)\,\xi\, dW + 3\int_0^M \xi\frac{p}{\varrho}\, r\frac{d\Gamma_1}{dr}\, dm. \tag{59.16}$$

The substitution of (59.15) and (59.16) in the Lagrangian variation of (59.11), leads to

$$\sigma^2 = -\frac{\int_0^M (3\Gamma_1 - 4)\,\xi\, dW + 3\int_0^M \frac{p}{\varrho}\,\xi\, r\frac{d\Gamma_1}{dr}\, dm}{\int_0^M \xi\, dI} \tag{59.17}$$

which can also be written

$$\sigma^2 = -\frac{\int_0^R 4\pi r^3 \xi \frac{d}{dr}[(3\Gamma_1 - 4)\, p]\, dr}{\int_0^R \xi\, dI} \tag{59.18}$$

This expression could also be derived from (58.1) integrated from 0 to R, however the virial theorem makes the physical meaning of the result clearer and facilitates the extension to more complicated cases. In particular, Eq. (59.11) or its first variation may be considered as a kind of mean equation of motion for the whole star.

60. Some general properties and approximate solutions. We have already mentioned [cf. condition (58.24)] that if Γ_1 is constant and greater than $\frac{4}{3}$ all the frequencies σ will be real and will correspond to true oscillations. As shown by (59.10), this result subsists if $\Gamma_1 \geqq \frac{4}{3}$ varies, with $d\Gamma_1/dr$ everywhere negative. If $d\Gamma_1/dr$ is positive or changes sign in $0 \leqq r \leqq R$, the discussion is more difficult although as long as $d[(3\Gamma_1 - 4)\, p]/dr$ keeps the same sign as dp/dr, all eigenvalues σ^2 remain positive.

On the other hand, with the present abundances of H and He, Γ_1 remains fairly close to $\frac{5}{3}$ and a detailed discussion by P. LEDOUX[1] has failed to reveal any cases of physical significance where σ^2 could become negative and give rise to dynamical instability. Thus in the following, σ may be considered as real and we shall first discuss the main factors which affect its value.

For any homologous transformation corresponding to a general modification of the radius in the ratio α ($r \propto \alpha$, $p \propto \alpha^{-4}$, $\varrho \propto \alpha^{-3}$, $T \propto \alpha^{-1}$), the second member of (59.10) or (59.18) varies as α^{-3} or

$$\sigma^2 \propto \bar{\varrho}. \tag{60.1}$$

Let us now consider especially the fundamental mode which, in any case, is likely to be the most important one for the theory of variable stars. If we set $\xi = \text{const.}$ in (59.10) and note that

$$3\int_0^R p\, d\mathcal{V} = -W \tag{60.2}$$

[1] P. LEDOUX: Astrophys. Journ. **104**, 333 (1946) and [*50b*], Chap. IV.

we obtain

$$\sigma_0^2 \leq -\frac{(3\bar{\Gamma}_1 - 4)\,W}{I}, \tag{60.3}$$

a result derived by P. LEDOUX and C.L. PEKERIS [52]. Formula (60.3) reveals the analogy with the oscillations of simple mechanical systems such for instance as the pendulum of length l where $\sigma^2 = g/l = -W/I$ with $W = -mgl$ and $I = ml^2$. The extra-factor $(3\bar{\Gamma}_1 - 4)$ in (60.3) results from the combination of the pressure forces with gravity.

Using (60.2) and the definition of the velocity of sound c, Eq. (60.3) can also be written

$$\sigma_0^2 \leq 3\overline{\left(\frac{3\Gamma_1 - 4}{\Gamma_1}\right)} \frac{\int\limits_0^M \frac{\Gamma_1 p}{\varrho}\,dm}{\int\limits_0^M r^2\,dm} = 3\bar{\alpha}\frac{\bar{c}^2}{\bar{R}^2} \tag{60.4}$$

where $\bar{\alpha}$, $\bar{c}$ and $\bar{R}$ are appropriate mean values. The period given by

$$P_0 \geq \frac{2\pi}{\sqrt{3\bar{\alpha}}}\frac{\bar{R}}{\bar{c}} \tag{60.5}$$

is thus seen to be proportional to the time required for a sound wave to travel through the star, a result which would be expected if the restoring forces depended only on the compression of the material. The combined effects of the extra-forces of gravity and pressure appear again in the factor $\bar{\alpha}$.

On the other hand, if we use the definitions

$$q = \frac{m(r)}{M} \quad \text{and} \quad x = \frac{r}{R} \tag{60.6}$$

Eq. (60.3) becomes

$$\sigma_0^2 \leq (3\bar{\Gamma}_1 - 4)\frac{4\pi G\bar{\varrho}}{3}\frac{\int\limits_0^1 \frac{q\,dq}{x}}{\int\limits_0^1 x^2\,dq} \tag{60.7}$$

which shows that the frequency depends also on the relative distribution of mass in the model. In this respect, the ratio $\varrho_c/\bar{\varrho}$ is not always sufficient to characterize this distribution. But if we say that model 2 is more centrally condensed than model 1 if, for any value $x_2 = x_1$, we have $q_2 > q_1$, then (60.7) shows that, in general $\sigma_2 > \sigma_1$. In this sense, the fundamental frequency increases with the central condensation of the model.

If Γ_1 is constant, (59.10) can be written

$$\sigma_0^2 = \min\left\{\frac{\int\limits_0^R \Gamma_1 p r^4 \left(\frac{d\xi_0}{dr}\right)^2 dr}{\int\limits_0^R \varrho r^4 \xi_0^2\,dr} + (3\Gamma_1 - 4)\frac{4\pi G\bar{\varrho}(R)}{3}\frac{\int\limits_0^R \xi_0^2 \frac{\bar{\varrho}(r)}{\bar{\varrho}(R)}\varrho r^4\,dr}{\int\limits_0^R \varrho r^4 \xi_0^2\,dr}\right\} \tag{60.8}$$

In the case of the homogeneous model, $\varrho = \bar{\varrho}(r) = \bar{\varrho}(R)$ and we are left with

$$\sigma_0^2 = \min\left\{\frac{\int\limits_0^R \Gamma_1 p r^4 \left(\frac{d\xi_0}{dr}\right)^2 dr}{\int\limits_0^R \varrho r^4 \xi_0^2\,dr} + (3\Gamma_1 - 4)\frac{4\pi G\bar{\varrho}(R)}{3}\right\} \tag{60.9}$$

which shows that the minimum is realized for a constant ξ_0 corresponding to a frequency

$$\sigma_0^2 = (3\Gamma_1 - 4)\frac{4\pi G\varrho}{3}, \tag{60.10}$$

a well known result since the investigation of RITTER [10].

If the star is not homogeneous but $\bar{\varrho}(r)$ is monotonically decreasing so that $\bar{\varrho}(r) \geqq \bar{\varrho}(R)$, the minimum of (60.8) will always be larger than the minimum of the same expression with $\bar{\varrho}(r)/\bar{\varrho}(R) = 1$ and

$$\sigma_0^2 \geqq \min\left\{\frac{\int\limits_0^R \Gamma_1 p r^4 \left(\frac{d\xi}{dr}\right)^2 dr}{\int\limits_0^R \varrho r^4 \xi^2 dr} + (3\Gamma_1 - 4)\frac{4\pi G\bar{\varrho}(R)}{3}\right\} \geqq \tag{60.11}$$
$$\geqq (3\Gamma_1 - 4)\frac{4\pi G\bar{\varrho}(R)}{3}.$$

Thus the fundamental frequency of any physical model of mean density $\bar{\varrho}(R)$ has a lower bound equal to the frequency of the homogeneous model with $\varrho = \bar{\varrho}(R)$ a result recently pointed out by R. SIMON[1]. Eqs. (60.3) and (60.11) can be combined to give

$$(3\Gamma_1 - 4)\frac{4\pi G\bar{\varrho}(R)}{3} \leqq \sigma_0^2 \leqq -\frac{(3\Gamma_1 - 4)W}{I}. \tag{60.12}$$

Attempts to improve these limits have proved unsuccessful up to now. As we shall see later, for a given model, higher numerical approximation can be built up using RITZ's method. Also, if an approximating function has been constructed for the first mode, TEMPLE's procedure[2] can be applied to sharpen numerically the lower bound of σ_0.

As LEDOUX and PEKERIS have shown [52], if $\varrho_c/\bar{\varrho}$ is not too large, formula (60.3) with the sign of equality provides a reasonable approximation for σ_0. However if the central condensation becomes very large, the increase of ξ with r gives so much weight, in formula (59.10), to the external half of the star, that σ_0^2 becomes very sensitive to the type of density distribution there, a point first brought forcibly to the fore in a paper by I. EPSTEIN [53]. This can also be seen from a discussion of the coefficients of Eq. (58.1) according to the remarks at the end of Sect. 58. Taking $m(r)$ as independent variable and using the definitions (60.6), Eq. (58.1) may be written

$$\frac{d}{dq}\left[\Gamma_1 \varrho x^6 \left(\int\limits_q^1 \frac{q\,dq}{x^4}\right)\frac{d\xi}{dq}\right] + \xi\left[\frac{\sigma^2 x^2}{4\pi} - \frac{G\bar{\varrho}(R)}{3}\frac{q}{x}(3\Gamma_1 - 4)\right] = 0$$

or

$$\frac{d}{dq}\left[\alpha(q)\frac{d\xi}{dq}\right] - \beta(q)\,\xi + \sigma^2\mu(q)\,\xi = 0. \tag{60.13}$$

If $\bar{\varrho}(R)$ is kept constant while the central condensation is increased, $\mu(q)$ decreases and $\beta(q)$ increases, both effects contributing to the increase of all σ^2. On the other hand, if in $\alpha(q)$ the integral increases, the factor x^6 rapidly decreases while ϱ increases in the central part and decreases in the external layers. As long as ξ does not vary too rapidly, the effects of $\mu(q)$ and $\beta(q)$ predominate and the σ^2 increase with the central condensation. But if the latter goes on

[1] R. SIMON: Bull. Acad. Roy. Sci. Belg., Ser. V **42**, 950 (1956).

[2] Cf. Ref. [49e] and E. LYTKENS: Astrophys. Journ. **119**, 413 (1954) for an application to an astrophysical problem.

increasing, $d\xi/dq$ becomes larger and larger as can be seen from (60.13) integrated once, and the influence of $\alpha(q)$ becomes important. In general, it will counteract the effects of $\mu(q)$ and $\beta(q)$ and limit the increase of σ^2. In extreme cases, it might even reverse the trend of σ^2.

Attempts have been made to derive approximate formulae for σ_0^2 based on the expected dependence of ξ on the central condensation. In particular, assuming $\xi \propto \overline{\varrho(r)}^{-1}$, one finds[1], starting from (59.18), a simple formula

$$\sigma_0^2 = (3\Gamma_1 - 4)\,\frac{4\pi G\,\overline{\varrho}(R)}{3}\,\frac{\int_0^R \frac{\varrho}{\overline{\varrho}(R)}\, r^4\, dr}{\int_0^R \frac{\varrho}{\overline{\varrho}(r)}\, r^4\, dr} \tag{60.14}$$

which gives a fairly good approximation for a very large range of models.

61. The Rayleigh-Ritz method of successive approximations[2]. This method consists in expanding ξ into a series of linearly independent functions $\varphi_i(r)$

$$\xi = \sum_{i=0}^{n} a_i\,\varphi_i(r). \tag{61.1}$$

Introducing this into (59.8), the conditions that J be a minimum become

$$-\frac{\partial J}{\partial a_j} = \sum_{i=0}^{n} a_i\left[A_{ij} - \sigma^2 B_{ij}\right] = 0, \qquad j = 0, 1, \ldots, n \tag{61.2}$$

with

$$\left.\begin{aligned} A_{ij} &= \int_0^R \left\{\Gamma_1 p\, r^4 \varphi_i' \varphi_j' - r^3 \frac{d}{dr}\left[(3\Gamma_1 - 4)\, p\right] \varphi_i \varphi_j\right\} dr, \\ B_{ij} &= \int_0^R \varrho\, r^4\, \varphi_i \varphi_j\, dr. \end{aligned}\right\} \tag{61.3}$$

The system (61.2) admits solutions only if

$$|A_{ij} - \sigma^2 B_{ij}| = 0. \tag{61.4}$$

If the φ_i are ortho-normal with respect to ϱr^4

$$\int \varphi_i \varphi_j\, \varrho\, r^4\, dr = \delta_{ij} \tag{61.5}$$

Eq. (61.4) reduces to

$$|A'_{ij} - \sigma^2 \delta_{ij}| = 0. \tag{61.6}$$

The matrix (A'_{ij}) is always symmetrical [this corresponds to the self-adjoint character of the differential operator in (58.1)] and, its terms being real, the solutions σ^2 are all real. It is easy to show (cf. [*49c*], §§ 15.4 to 15.8) that those solutions ordered by increasing values $(\sigma_0^2)_n, (\sigma_1^2)_n \ldots (\sigma_n^2)_n$ define upper bounds for

---

[1] P. Ledoux, R. Simon and J. Bierlaire: Ann. d'Astrophys. **18**, 65 (1955).

[2] S. A. Zhevakin has introduced the interesting concept of "discrete model" in which the mass of the star is concentrated in discrete layers connected by forces replacing gravitation and the pressure gradient. For linear adiabatic oscillations, this is essentially equivalent to the Rayleigh-Ritz method. But Zhevakin's method is physically simple to extend to more general cases [cf. Z. A. Zhevakin: Dokl. Akad. Nauk USSR. **62**, 191 (1948)].

A direct application of this idea to the determination of the eigenvalues has been worked out by Ch. Whitney and P. Ledoux: Bull. Acad. Roy. Sci. Belg., Ser. V **43**, 622 (1957).

the $(n+1)$ first eigenvalues of our problem:

$$\sigma_i^2 \leq (\sigma_i^2)_n. \tag{61.7}$$

The same properties subsist for the solutions of (61.4). One can then solve (61.2) for the a_i and thus obtain corresponding approximations for the eigenfunctions.

Table 10. *Comparison between the exact and approached values of ω^2 for the standard model.*

$\alpha=3-4/\Gamma_1$	(ω_0^2)exact	$(\omega_0^2)_3$	(ω_1^2)exact	$(\omega_1^2)_3$
0,2	2.516	2.508	8.188	8.637
0.4	4.692	4.653	9.933	10.97
0.6	9.261	9.133	17.00	19.30

This method was applied by P. LEDOUX and C.L. PEKERIS [*52*] to the standard model with $\varphi_i = r^{2i}$. The third approximation $(n=2)$ already provides an excellent value for the fundamental frequency and a very reasonable one for the first mode as shown in Table 10 for ω^2 defined by (58.18). The exact values are taken from a paper by M. SCHWARZSCHILD [*54*]. The run of the amplitudes in the case of the fundamental mode are also compared in Fig. 37. Of course, the rate of convergence of the approximate solutions decreases when the central condensation increases as illustrated by Table 11[1]. D. LUCAS[2] has used the method with orthogonal functions. This simplifies the characteristic equation, but the building up of the orthogonal functions with respect to ϱr^4 is quite cumbersome. These two investigations have shown that the present day accuracy of stellar models limits the precision attainable. For instance, in most cases considered, the fourth approximation and, in some cases the third, correspond to the useful limit of the method as far as the fundamental mode is concerned.

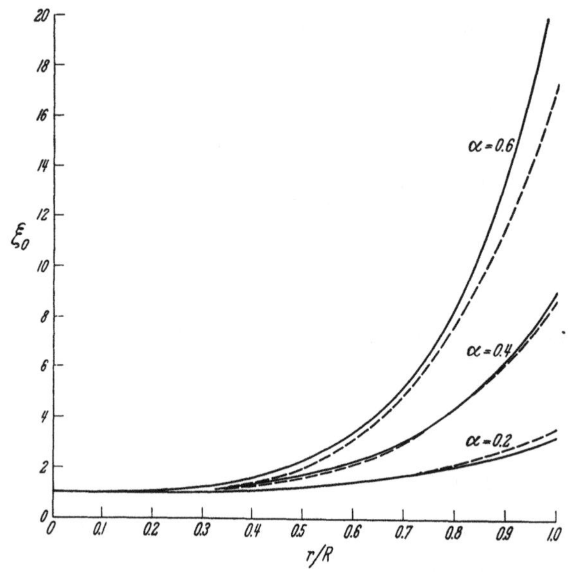

Fig. 37. Comparison of the exact (full line) and approached (dashed line) solutions ξ_0 for the standard model.

Table 11. *Successive approximations in the case of* EPSTEIN'S *model 4.*

Modes	First approx.	Second approx.	Third approx.	Fourth approx.	Exact
Fundamental .	429.4	24.50	15.09	14.14	14.08
First		783.90	50.76	27.99	24.065

[1] H. ELSEN and P. LEDOUX: Bull. Soc. Roy. Sci. Liège **24**, 239 (1955).
[2] D. LUCAS: Bull. Soc. Roy. Sci. Liège **25**, 585 (1956).

62. Analytical or numerical solutions. There are a few models for which Eq. (58.1) or (58.12) with the boundary conditions (58.2) admits analytical solutions in the form of polynomials. They correspond to the case where a solution ξ of the form (58.13): $\xi_1 = (r/R)^{\theta} \sum a_i (r/R)^i$ introduced into Eq. (58.12) leads to a two terms recurrence formula giving $a_{n+k}/a_n = f(\sigma^2, n)$, where k is a small constant integer (1 or 2 or 3 ...) depending on the powers which really subsist in the general expansion. If $a_{n+k}/a_n \to 1$ as $n \to \infty$, one has to limit the series to insure its convergence in $r = R$ by setting $a_{n+k} = 0$ or $f(\sigma^2, n) = 0$. This last equation defines σ_n^2 and the corresponding polynomial of degree n is the n-th eigenfunction. A review of such models and their properties has been given by S. ROSSELAND[1a]. More recently, Z. KOPAL[2] has generalized the discussion to all models for which the solutions of Eq. (58.12) can be expressed by means of the hypergeometric series. Apart from the homogeneous model and the different types of ROCHE's models already discussed by STERNE[1b] and PRASAD[1c] this investigation has revealed the existence of models admitting a continuous spectrum of frequencies. However, it is unlikely that the corresponding distribution of density could have any physical meaning.

In most cases of physical interest however, analytical solutions are not available and one must resort to numerical integration. Due to the singularities of (58.12) in $r = 0$ and $r = R$, the solution around these points has to be expressed in series expansions of the form (58.16) where the boundary conditions permit to set $B = 0$. As the computation of the series coefficients is laborious, only a small number of these are evaluated and the series expansion are used only up to a short distance from $r = 0$ and $r = R$. From there on, the solutions are continued inwards and outwards by numerical integration for a given value of σ^2, say $(\sigma^2)_1$. At the point where they meet, $\frac{1}{\xi}\frac{d\xi}{dr}$ should be continuous. If this condition is not satisfied, the process has to be renewed for another value of σ^2, $(\sigma^2)_2$ etc. until the continuity is satisfied or at least until two values of σ^2, $(\sigma^2)_i$ and $(\sigma^2)_{i+1}$ are obtained which differ by less than the desired precision on σ^2 and such that the continuity condition is violated in opposite directions for each of them. Let us note that usually the coefficients of the series expansion near $r = 0$ depend very little on σ^2 so that one does not have to recompute initial values for the outward integrations for each trial value of σ^2. This is not the case for the series expansion around $r = R$.

From a physical point of view, Eq. (58.12) ceases to be valid in the very outer layers of the star ($T \lesssim 50000°$ K) due to important deviations from isentropy in this region. However, as we shall see later, these changes have very little effect on the periods themselves.

In the static model studied, discontinuities in ϱ or in Γ_1 might occur at some point $r = r^*$. In the first case, the conditions (57.27) again imply the continuity of $\frac{1}{\xi}\frac{d\xi}{dr}$ at that point. In the second case, if we denote by indices i and e, the values on the internal and external side of $r = r^*$, the dynamical condition may

[1] a) S. ROSSELAND, cf. Ref. [*39d*], Chap. III; note that there is a misprint in Eq. (3.25) where the 2 should be replaced by a 3 and that, in (3.22), the admissible values of m should start with $m = 0$. — Cf. also b) T. E. STERNE: Monthly Notices Roy. Astronom. Soc. London **97**, 582 (1937). — c) C. PRASAD: Monthly Notices Roy. Astronom. Soc. London **108**, 414 (1948); **109**, 103 (1949). — d) H.K. SEN: Proc. Nat. Acad. India **13**, 44 (1943); cf. criticisms by T. G. COWLING: Math. Rev. **8**, 60 (1947), and a new discussion by Z. KOPAL: Astrophys. Journ. **111**, 395 (1950), which however does not appear to clear up completely the question.

[2] Z. KOPAL: Proc. Nat. Acad. Sci. U.S.A. **34**, 377 (1948).

be written

$$\left(\frac{d\xi}{dr}\right)_e^* = \frac{3\xi^*}{r^*}\left(\frac{\Gamma_{1i}}{\Gamma_{1e}} - 1\right) + \frac{\Gamma_{1i}}{\Gamma_{1e}}\left(\frac{d\xi}{dr}\right)_i^* \tag{62.1}$$

if the kinematical condition $\xi_i^* = \xi_e^*$ is already satisfied[1]. This implies a discontinuity of the first derivative at any discontinuity of Γ_1.

Radial oscillations have been studied numerically in the case of many models[2]. As an example, Fig. 38 reproduces the run of the amplitudes for the first four modes of the standard model computed by M. SCHWARZSCHILD [*54*] for $\Gamma_1 = \frac{5}{3}$. The most significant numerical integration of Eq. (58.12) in the last few years is due to EPSTEIN [*53*] in the case of a few models with high central condensations ($\varrho_c/\bar{\varrho} \approx 10^6$). These integrations, as already reported, showed that σ^2 increases more slowly with the central condensation than suggested by the first approximation (60.3) or (60.7) taken with the sign of equality. They led EPSTEIN to the suggestion that a lowering of the effective polytropic index n_e, and thus of the density gradient, in the external half of the star could even decrease σ^2 below its value for less centrally condensed models. This conclusion was checked by LEDOUX[3] and by LEDOUX, SIMON and BIERLAIRE[4] for two models with high central condensation but extensive homogeneous ($n_e = 0$) or convective ($n_e = 1.5$) external zones and it is verified again in the case of P. DUMÉZIL-CURIEN's model.

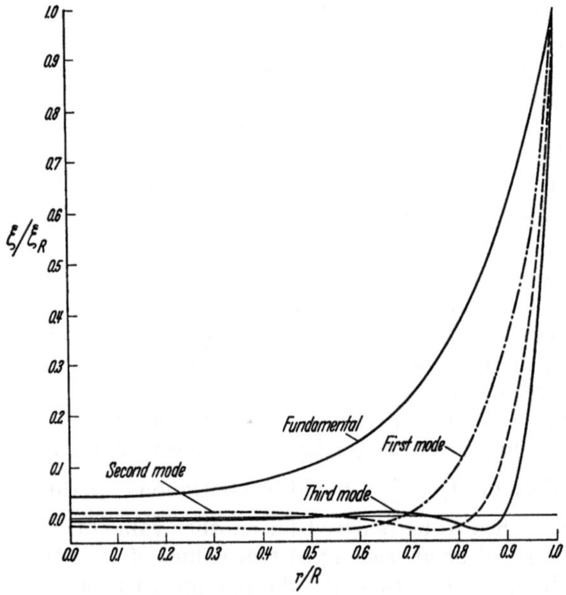

Fig. 38. Amplitudes of the first four modes of radial pulsation of the standard model after M. SCHWARZSCHILD.

Properties of some models are summarized in Table 12 while Fig. 39 gives q as a function of x for these models. One sees clearly here that $\varrho_c/\bar{\varrho}$ is not sufficient to characterize the effects of the density distribution on the periods. On the other hand, the slope of $q(x)$ is significant especially if, going from the less centrally to the more centrally condensed models, one shifts the emphasis from more internal to more external points.

All values of the periods P are given here for $\Gamma_1 = \frac{5}{3}$. According to (53.6), the pressure of radiation could lower Γ_1 somewhat, say to Γ_1' and this would lead to somewhat larger periods P' which for the fundamental mode are given in a first approximation by

$$P_0' \approx P_0 \sqrt{\frac{3\Gamma_1 - 4}{3\Gamma_1' - 4}}. \tag{62.2}$$

---

[1] Cf. Ref. 1, p. 466.

[2] Cf. a) S. ROSSELAND [*39d*], Chap. IV; b) G. KELLER: Ph. D. Thesis, Columbia University 1948; c) I. EPSTEIN: Astrophys. Journ. **112**, 6 (1950).

[3] P. LEDOUX: Bull. Soc. Roy. Sci. Liège **21**, 408 (1952).

[4] P. LEDOUX, R. SIMON and J. BIERLAIRE: Ann. d'Astrophys. **18**, 65 (1955).

Table 12. *Properties of the first two modes of radial oscillations for different stellar models with $\Gamma_1 = \frac{5}{3}$.*

Model	$\frac{\varrho_c}{\bar{\varrho}}$	ω_0^2	ω_1^2	$P_0\sqrt{\frac{\bar{\varrho}}{\bar{\varrho}_\odot}} = Q_0$	$\frac{P_1}{P_0}$	$(\xi_R/\xi_c)_0$	$(\xi_R/\xi_c)_1$	Node of ξ_1	Reference
1. Homogeneous	1	1	12.666	0.1158	0.28	1	−0.4	$x = 0.845$ $q = 0.603$	T. E. STERNE: Monthly Notices Roy. Astronom. Soc. London **97**, 582 (1937).
2. PRASAD (1) Point-mass+ homogeneous env.	$\to\infty$	1.442	15.658	0.0964	0.303	$+\infty$[2]	$-\infty$[2]	$x = 0.865$	C. PRASAD: Monthly Notices Roy. Astronom. Soc. London **108**, 414 (1948).
3. PRASAD (2) $\varrho = \varrho_c(1 - x^2)$	2.5	1.657	12.299	0.0900	0.365	1.1[1]	1.28[1]	$x = 0.77$[1] $q = 0.732$[1]	C. PRASAD: Monthly Notices Roy. Astronom. Soc. London **109**, 103 (1949).
4. STERNE (2) $\varrho = \bar{\varrho}/3x^2$	$\to\infty$	2.173	12.853	0.0786	0.411	$+\infty$[2]	$-\infty$[2]	$x = 0.820$ $q = 0.820$	T. E. STERNE: Monthly Notices Roy. Astronom. Soc. London **97**, 582 (1937).
5. Convective Model (Polytrope $n = \frac{3}{2}$)	6.00	2.712	12.516	0.0703	0.465	1.42	−3.39	$x = 0.72$ $q = 0.83$	D. LUCAS: Bull. Soc. Roy. Sci. Liège **25**, 585 (1956).
6. Modèle de ROCHE	$\to\infty$	3.237	17.71	0.0644	0.427	$+\infty$[2]	$-\infty$[2]	$x = 0.87$ $q = 1$	T. E. STERNE: Monthly Notices Roy. Astronom. Soc. London **97**, 582 (1937).
7. EPSTEIN's model 4 with external convection zone	$1.2 \cdot 10^6$	4.30	14.63	0.0558	0.543	7890	−14620	$x = 0.76$	P. LEDOUX et al.: Ann. d'Astrophys. **18**, 65, 232 (1955).
8. DUMÉZIL-CURIEN's model	7730	$\simeq 5.03$	$\simeq 14.04$	0.052	0.598			$x = 0.75$ $q = 0.964$	H. ELSEN and P. LEDOUX: Bull. Soc. Roy. Sci. Liège **24**, 239 (1955).
		4.70	14.96						E. SCHATZMAN: Ann. d'Astrophys. **19**, 51 (1955).
9. Standard model Polytrope = 3	54.2	9.261	17.00	0.0383	0.738	22.41	59.12	$x = 0.67$ $q = 0.993$	M. SCHWARZSCHILD: Astrophys. Journ. **94**, 245 (1941).
10. KELLER's model (R.M.)	79	13.125	20.739	0.0320	0.795	154.5	243.7	$x = 0.703$	G. KELLER: Dissertation, Delaware, Ohio, 1948.
11. EPSTEIN's model 7	$2.4 \cdot 10^7$	13.64	22.435	0.0314	0.780	$6.431 \cdot 10^5$	$9.51 \cdot 10^5$	$x = 0.74$ $q = 0.9990$	I. EPSTEIN: Astrophys. Journ. **112**, 6 (1950).
12. EPSTEIN's model 4	$1.9 \cdot 10^6$	14.082	24.065	0.0309	0.765	$6.124 \cdot 10^5$	$1.05 \cdot 10^6$	$x = 0.76$ $q = 0.9999$	I. EPSTEIN: Astrophys. Journ. **112**, 6 (1950).
13. KELLER's model (0)	230	15.72	25.20	0.0292	0.79	275.5	400.2	$x = 0.749$	G. KELLER: Dissertation, Delaware, Ohio, 1948.

[1] Approached values obtained from the series developments given by PRASAD for $\Gamma_1 = 1.538$ ($\alpha = 0.4$).
[2] These infinite ratios are due to the density becoming infinite at the centre so that δr tends to zero more rapidly than r and $\xi_c = (\delta r/r)_c$ tends also to zero.

Eq. (59.10) shows that the effects of a variation of $(3\Gamma_1 - 4)$ is the smaller the higher the mode. For instance P_1/P_0 decreases with Γ_1 as illustrated in Fig. 40 for the standard model.

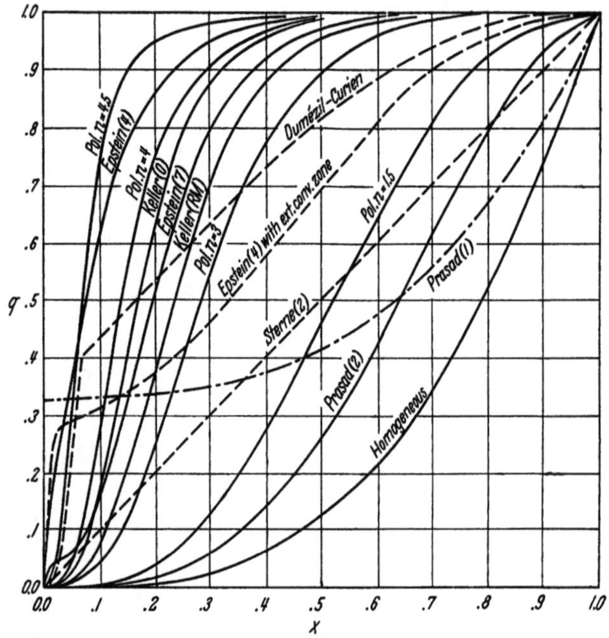

Fig. 39. Distribution of the relative mass $m(r)/M$ as a function of the relative radius r/R in various models.

For a constant Γ_1, factors that increase ω_0^2 also increase the ratio P_1/P_0 as illustrated in Fig. 41.

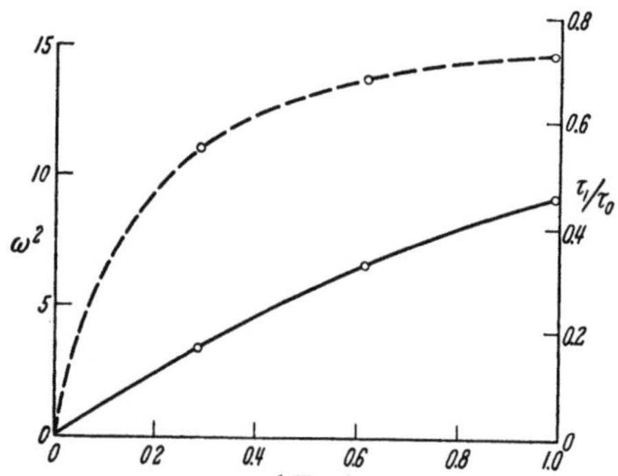

Fig. 40. Variation with $(3\Gamma_1 - 4)$ of ω_0^2 (full line) and P_1/P_0 (dashed line) for the standard model.

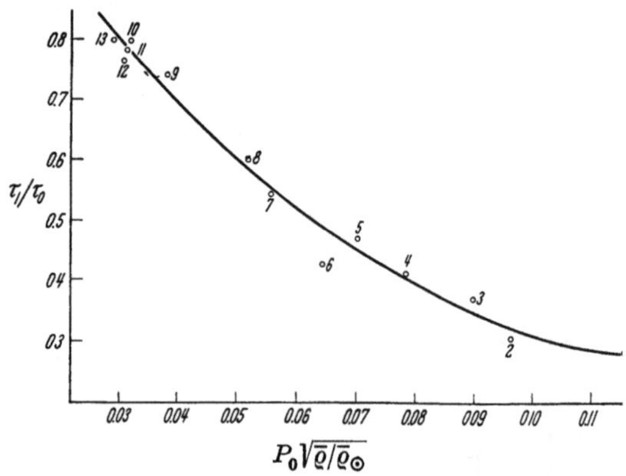

Fig. 41. Variation of the ratio P_1/P_0 as a function of $P_0\sqrt{\bar{\varrho}/\bar{\varrho}_\odot}$ (days). The numbers refer to the models in Table 12.

The quantity

$$P\sqrt{\frac{\bar{\varrho}}{\bar{\varrho}_\odot}} = Q \tag{62.3}$$

given in days is convenient for the discussion of observations.

63. Non-adiabatic radial oscillations and vibrational stability. Up to now, we have neglected the right-hand member of the complete Eq. (57.25) which however is not rigorously zero. From a physical point of view its effects may be discussed either by analogy with the simplest oscillatory systems obeying an equation of the form

$$\frac{d^2\xi}{dt^2} + n^2\xi = F \tag{63.1}$$

or with the help of the principle of conservation of energy.

α) Local discussion. In the case of (63.1) if F is in phase with ξ, the frequency is changed with respect to its value when $F=0$. If F is in phase with $v=r_0\, d\xi/dt$ the amplitude becomes a function of time. It will decrease (vibrational stability) or increase (instability) if F is in phase with $-v$ or $+v$ respectively. If we only consider standing oscillations ($\xi(r)$: real), Eq. (57.25) at a given point is essentially equivalent to (63.1) and can be written

$$\mu(r)\frac{d^2\xi}{dt^2} + L\xi = +i\sigma r^3\psi + \frac{i}{\sigma} r^3 \frac{d}{dr}\left\{(\Gamma_3-1)\,\varrho\,\delta\{\ldots\} + \frac{2}{3}\varrho\,\delta[\ldots]\right\}. \tag{63.2}$$

with the same notation as in (58.4). Since the coefficients of $i\sigma$ or i/σ are in phase with ξ, F will always be in phase with $\pm v = \pm i\sigma\xi$. To illustrate the situation let us consider simple cases.

If the terms in σ^{-1} are negligible on the right-hand side of Eq. (63.2), the only term left $i\sigma r^3\psi$ is in phase with $\pm v$ depending on whether ψ as defined by (57.17) is positive or negative. Thus, although the overall effect of the viscous dissipation is a positive damping, the amplitude tends to increase in regions where ψ is positive and decrease in regions where ψ is negative, momentum being transferred from one to the other.

On the other hand, if only the terms in i/σ are kept and if turbulence is neglected, the second member of (63.2) can be written

$$\frac{i r^3}{\sigma}\frac{d}{dr}\left\{(\Gamma_3-1)\left[\varrho\,\delta\varepsilon_1 - \frac{1}{r^2}\frac{d}{dr}\delta(r^2F_R)\right]\right\} = \frac{i r^3}{\sigma}\frac{d}{dr}\left\{(\Gamma_3-1)\,\varrho\left[\delta\varepsilon_1 - \frac{d\,\delta L(r)}{dm}\right]\right\}. \tag{63.3}$$

For the usual thermonuclear reactions, one has

$$\varepsilon_1 = \varepsilon_0\,\varrho\,T^\nu \tag{63.4}$$

and, neglecting any possible phase delays due to abundances variations,

$$\frac{\delta\varepsilon_1}{\varepsilon_1} = \frac{\delta\varrho}{\varrho} + \nu\frac{\delta T}{T} = \frac{\delta\varrho}{\varrho}\left[1+\nu(\Gamma_3-1)\right]. \tag{63.5}$$

The first part of (63.3) becomes, using (57.30),

$$-\frac{i r^3}{\sigma}(\Gamma_3-1)\left[1+\nu(\Gamma_3-1)\right]\frac{d}{dr}\left[\varrho\,\varepsilon_1\left(3\xi + r\frac{d\xi}{dr}\right)\right]. \tag{63.6}$$

Thus if $\varrho\,\varepsilon_1\left(3\xi + r\frac{d\xi}{dr}\right)$ decreases everywhere, this term is in phase with $+v$ and the generation of energy reinforces the oscillations. Developing the derivative in (63.6), one finds that this is the case in all regions where $d(\log\xi)/dr$ is positive provided

$$\nu > 1 \tag{63.7}$$

which is a very mild requirement.

According to (57.5), if

$$\varkappa = \varkappa_0\,\varrho^m\,T^{-n} \tag{63.8}$$

one has

$$\frac{\delta L_R}{L_R} = 4\frac{\delta r}{r} + (4+n)\frac{\delta T}{T} - m\frac{\delta\varrho}{\varrho} + \frac{\frac{d}{dm}\left(\frac{\delta T}{T}\right)}{\frac{1}{T}\frac{dT}{dm}}, \tag{63.9}$$

but the effects of the corresponding term in (63.3) are more difficult to estimate especially in the external layers where m and n as well as Γ_3 may vary widely. If one treats m, n and Γ_3 as constant all through the star, one finds in a very rough approximation, that if $d(\log\xi)/dr$ is positive, this term will be in phase with $-v$ provided that

$$n > \frac{3m+4}{3(\Gamma_3-1)} - 3. \tag{63.10}$$

In most cases of physical interest, this condition is well satisfied and the radiative conductivity, barring exceptional circumstances in the external layers, contributes to the vibrational stability of the star.

If the oscillation had a progressive character, the phase relationship between v and the different terms in (63.3) would become more complicated because of the phase shift introduced by the space derivatives.

Eq. (63.2) shows also that the relative importance of the viscous terms as compared to the radiative terms increases with the order of the mode approximately as σ^2.

β) Overall dissipation. To get a better idea of the overall effect of the non-adiabatic terms, let us rewrite Eq. (50.7) in the form

$$\left.\begin{aligned}\frac{d}{dt}\left(\frac{1}{2}v'^2\right) = &-\sum_i \frac{\partial\Phi'}{\partial x^i}v'^i - \sum_i \frac{1}{\varrho}\Delta_i[v'^i(p'+p_t')] + \frac{p'+p_t'}{\varrho}\sum_i \Delta_i v'^i + \\ &+\sum_i \frac{v'^i\varrho'}{\varrho^2}\Delta_i(p+p_t) + \frac{1}{\varrho}\sum_{i,m}\Delta_i[v'^m(\mathfrak{P}^i_m)_t] - \frac{1}{\varrho}\sum_{i,m}(\mathfrak{P}^i_m)_t\Delta_i v'^m\end{aligned}\right\} \tag{63.11}$$

limiting ourselves to terms of the second order. If we sum over the whole mass noting that all stresses vanish on a free surface and then integrate with respect to t over a period $P=2\pi/\sigma$, we obtain

$$\left.\begin{aligned}\left[\int_0^M \frac{1}{2}v'^2\,dm\right]_0^P = &-\int_0^P dt\int_0^M \sum_i \frac{\partial\Phi'}{\partial x^i}v'^i\,dm - \int_0^P dt\int_0^{\mathscr{V}} \frac{p'+p_t'}{\varrho}\left(\frac{\partial\varrho'}{\partial t} + \boldsymbol{v}'\cdot\operatorname{grad}\varrho\right)d\mathscr{V} + \\ &+\int_0^P dt\int_0^{\mathscr{V}} \frac{\varrho'}{\varrho}\sum_i v'^i\Delta_i(p+p_t)\,d\mathscr{V} - \int_0^P dt\int_0^{\mathscr{V}} \sum_{i,m}(\mathfrak{P}^i_m)_t\Delta_i(v'^m)\,d\mathscr{V}\end{aligned}\right\} \tag{63.}$$

where p' and p_t' are to be replaced by their values (57.19) and (57.20). As a first approximation, we may then introduce in the second member of (63.12), the corresponding adiabatic solution: δr, ϱ', $\Phi' \propto \pm\cos\sigma t$ and $v' = -\sigma\delta r\sin\sigma t$. Denoting the variation of the kinetic energy in the course of a period by $(\Delta K)_P$, one gets, after performing the time integrations,

$$\begin{aligned}(\Delta K)_P = &\frac{\pi}{\sigma}\int_0^M \frac{\delta T}{T}\,\delta\left\{\varepsilon_1+\varepsilon_2-\varepsilon_3-\frac{1}{\varrho}\operatorname{div}(\boldsymbol{F}_R+\boldsymbol{F}_c+\boldsymbol{F}_t)\right\}_a dm + \\ &+\frac{\pi}{\sigma}\int_0^M \frac{2}{3}\frac{\delta\varrho}{\varrho}\,\delta\left[\varepsilon_4-\varepsilon_2-\frac{1}{\varrho}\operatorname{div}\boldsymbol{F}_2-\frac{\boldsymbol{V}}{\varrho}\operatorname{grad}p\right]_a dm - \sigma\pi\int_0^{\mathscr{V}}\left[\sum_{i,m}\mathfrak{P}^i_m(\delta\boldsymbol{r})\,\Delta_i(\delta r^m)\right]_a d\mathscr{V}\end{aligned}$$

where the index a recalls that the second member must be evaluated for the corresponding adiabatic solution. On the other hand, if we denote by

$$K_M = \frac{\sigma^2}{2} \int_0^M (\delta r)_a^2 \, dm$$

the maximum kinetic energy of the system, we may introduce a damping coefficient σ'' defined by

$$\sigma'' = -\frac{1}{2} \frac{\sigma (\Delta K)_P}{2\pi K_M} \tag{63.13}$$

such that,

$$\delta r \propto \cos \sigma t \, \mathrm{e}^{-\sigma'' t} . \tag{63.14}$$

More explicitly, one has

$$\sigma'' = \sigma''_\eta + \sigma''_E = \frac{1}{2} \frac{\int_0^{\mathscr{V}} (\sum \mathfrak{P}^i_m (\delta \boldsymbol{r}) \, \Delta_i \, \delta r^m)_a \, d\mathscr{V}}{\int_0^M (\delta r)_a^2 \, dm} - \frac{1}{2\sigma^2} \frac{\int_0^M \left(\frac{\delta T}{T} \delta\{\cdots\} + \frac{2}{3} \frac{\delta \varrho}{\varrho} \delta[\cdots] \right)_a dm}{\int_0^M (\delta r^2)_a \, dm} \tag{63.15}$$

which is also known as the coefficient of vibrational stability (Eddington overstability).

This is physically equivalent to the evaluation of the work of all stresses during one cycle, a method favoured by EDDINGTON [*13*].

This discussion may be extended to cover the case of more general non-adiabatic solutions. For instance, if we consider purely radial oscillations with no viscous forces, the Lagrangian equation of motion (47.2) takes, in this case, the simple form

$$\frac{dv}{dt} \frac{d}{dr} (r + \delta r) = -\frac{d(\Phi + \delta \Phi)}{dr} - \frac{1}{\varrho + \delta \varrho} \cdot \frac{d(p + \delta p + p_t + \delta p_t)}{dr} .$$

If we neglect second order terms and use (57.2) and (44.2′) noting that

$$\left. \begin{aligned} \frac{d \, \delta \Phi}{dr} &= \delta \left(\frac{d\Phi}{dr} \right) - \frac{1}{\varrho} \frac{dp}{dr} \cdot \frac{d(\delta r)}{dr} \\ \text{and} \qquad \delta \left(\frac{d\Phi}{dr} \right) &= \delta \left(\frac{G \, m(r)}{r^2} \right) = -\frac{2G \, m(r)}{r^3} \delta r \end{aligned} \right\} \tag{63.16}$$

the equation of motion becomes

$$\frac{dv}{dt} = \frac{2G \, m(r)}{r^3} \delta r - 2 \frac{\delta r}{r} \frac{1}{\varrho} \frac{d(p + p_t)}{dr} - \frac{1}{\varrho} \frac{d(\delta p + \delta p_t)}{dr} \tag{63.17}$$

where r may be treated as an identifying Lagrangian parameter identical to its value at equilibrium.

Multiplying by $v \, dm$ and integrating over the whole mass gives

$$\left. \begin{aligned} \frac{d}{dt} \int_0^M \frac{1}{2} v^2 \, dm = \frac{d}{dt} \int_0^M \frac{G \, m(r)}{r^3} \delta r^2 \, dm - \frac{d}{dt} \int_0^M \frac{1}{\varrho} \frac{d(p + p_t)}{dr} \frac{\delta r^2}{r} \, dm - \\ - \int_0^{\mathscr{V}} \frac{d(\delta p + \delta p_t)}{dr} v \, 4\pi r^2 \, dr . \end{aligned} \right\} \tag{63.18}$$

Integrating the last term by parts and noting that $\delta p = \delta p_t = 0$ in $r = R$ and $v = 0$ in $r = 0$, we obtain, using (56.18) and (56.19),

$$
\begin{aligned}
-\int_0^V \frac{d(\delta p + \delta p_t)}{dr} v\, 4\pi r^2\, dr = &-\frac{d}{dt}\int_0^M \frac{\delta p + \delta p_t}{\varrho}\,\frac{\delta\varrho}{\varrho}\, dm + \\
&+\int_0^M \frac{\delta\varrho}{\varrho^2}\left[\frac{\Gamma_1 p}{\varrho}\,\frac{d\,\delta\varrho}{dt} + (\Gamma_3 - 1)\,\varrho\,\delta\{\ldots\}\right] dm + \\
&+\int_0^M \frac{\delta\varrho}{\varrho^2}\left\{\frac{5}{3}\,\frac{p_t}{\varrho}\,\frac{d\,\delta\varrho}{dt} + \frac{2}{3}\,\varrho\,\delta[\ldots]\right\} dm \\
= -\frac{d}{dt}\int_0^M \frac{\delta p + \delta p_t}{\varrho}\,\frac{\delta\varrho}{\varrho}\, dm &+ \frac{d}{dt}\int_0^M \frac{1}{2}\left(\frac{\Gamma_1 p}{\varrho} + \frac{5}{3}\,\frac{p_t}{\varrho}\right)\left(\frac{\delta\varrho}{\varrho}\right)^2 dm + \\
&+\int_0^M \left\{(\Gamma_3 - 1)\,\frac{\delta\varrho}{\varrho}\,\delta\{\ldots\} + \frac{2}{3}\,\frac{\delta\varrho}{\varrho}\,\delta[\ldots]\right\} dm.
\end{aligned}
$$

Reintroducing this value into (63.18) and integrating over a period we are led to

$$
\left.
\begin{aligned}
\left[\int_0^M \frac{1}{2} v^2\, dm\right]_0^P = &\left[\int_0^M \frac{G\,m(r)}{r^3}\,\delta r^2\, dm\right]_0^P - \left[\int_0^M \frac{1}{\varrho}\,\frac{d(p + p_t)}{dr}\,\frac{\delta r^2}{r}\, dm\right]_0^P - \\
&-\left[\int_0^M \frac{\delta p + \delta p_t}{\varrho}\,\frac{\delta\varrho}{\varrho}\, dm\right]_0^P + \left[\int_0^M \frac{1}{2}\left(\frac{\Gamma_1 p}{\varrho} + \frac{5}{3}\,\frac{p_t}{\varrho}\right)\left(\frac{\delta\varrho}{\varrho}\right)^2 dm\right]_0^P + \\
&+\int_0^P dt \int_0^M \left\{\frac{\delta T}{T}\,\delta\{\ldots\} + \frac{2}{3}\,\frac{\delta\varrho}{\varrho}\,\delta[\ldots]\right\} dm.
\end{aligned}
\right\} \qquad (63.19)
$$

If the motion is strictly periodic, the last integral on the right-hand side of (63.19) must vanish. On the other hand, if we know a first approximate solution which is periodic, introducing it in the right-hand member of (63.19), we get for the dissipation of kinetic energy

$$
(\Delta K)_P = \int_0^P dt \int_0^M \left(\frac{\delta T}{T}\,\delta\{\ldots\} + \frac{2}{3}\,\frac{\delta\varrho}{\varrho}\,\delta[\ldots]\right)_P dm \qquad (63.20)
$$

where the index P recalls that the integrand must be evaluated for a periodic solution. This form has been used by EDDINGTON to discuss the dissipation in the case of a non-adiabatic solution [*55*].

64. Perturbation theory. When the coefficients of Eq. (58.4) and (or) those of the boundary conditions (58.2) are subjected to slight modifications, how are the eigenvalues and eigenfunctions affected? This question has been the subject of extensive mathematical discussions [*49*] and the most relevant results will be summarized here as they seem susceptible of wider applications to our problem than have hitherto been attempted.

Let us first consider the case of a perturbation affecting *only* the coefficients of the differential equation. Our system (58.2) plus (58.4) can be written

$$
(L - \lambda M)\,\xi = 0; \quad U_1(\xi) = 0 \ \text{ in } \ r = 0, \quad U_2(\xi) = 0 \ \text{ in } \ r = R \qquad (64.1)
$$

where $\lambda=\sigma^2$ and $M=\mu(r)$. More generally, M could be a differential operator of order smaller than L, the ξ_i being then orthogonal with respect to M. When perturbed, the system (64.1) becomes

$$(L'-\lambda' M')\,\xi'=(L-\lambda M)\,\xi'+\Delta L\,\xi'=0;\qquad U_1(\xi')=0,\qquad U_2(\xi')=0 \tag{64.2}$$

where ΔL is a small bounded operator with bound ε. If we denote by (λ_k,ξ_k) one of the eigensolutions of (64.1), we may surmise that there is a solution of (64.2) close to it (λ_k',ξ_k') which varies continuously with ε and which can be represented for ε sufficiently small by the convergent series

$$\left.\begin{aligned}\lambda_k'&=\lambda_k+\sum_{i=1}^{\infty}\varepsilon^i\mu_i,\\ \xi_k'&=\xi_k+\sum_{i=1}^{\infty}\varepsilon^i u_i,\end{aligned}\right\} \tag{64.3}$$

where the μ_i are unknown constants and the u_i unknown functions satisfying the boundary conditions. Introducing (64.3) into (64.2) one obtains by identification a series of equations which can be solved successively for (μ_1,u_1), (μ_2,u_2) etc.

Very often, only the first power terms in ε are needed. One gets then

$$\lambda_k'-\lambda_k=\varepsilon\mu_1=\frac{\int_0^R\xi_k\,\Delta L\,\xi_k\,dr}{\int_0^R\xi_k\,M\,\xi_k\,dr} \tag{64.4}$$

and then εu_1 is the solution of

$$(L-\lambda_k M)\,\varepsilon u_1=\varepsilon\mu_1 M\,\xi_k-\Delta L\,\xi_k \tag{64.5}$$

where $\varepsilon\mu_1$ has its value (64.4). Using for u_1 a series expansion in terms of the complete set ξ_j, one finds

$$\xi_k'-\xi_k=\varepsilon u_1=\sum_{\substack{j=0\\ j\neq k}}^{\infty}\frac{\int_0^R\xi_j\,\Delta L\,\xi_k\,dr}{(\lambda_k-\lambda_j)\int_0^R\xi_j\,M\,\xi_j\,dr}\,\xi_j+C\,\xi_k \tag{64.6}$$

where the constant C may be put equal to zero.

Although up to now, only the first order terms have been considered in practical applications, in the future, it might be necessary to take into account the second order correction to λ_k which can be written

$$(\lambda_k'-\lambda_k-\varepsilon\mu_1)=\varepsilon^2\mu_2=-\frac{\int_0^R\{\xi_k\Delta L(\varepsilon u_1)-\varepsilon\mu_1\xi_k[M(\varepsilon u_1)+(M'-M)\,\xi_k]\}\,dr}{\int_0^R\xi_k\,M\,\xi_k\,dr}. \tag{64.7}$$

Its evaluation requires of course the computation of u_1 by (64.6).

Following S. ROSSELAND [56], an immediate application of (64.4) and (64.6) can be made to Eq. (57.25), treating the left-hand member as a small perturbing quantity

$$\Delta L\,\xi_k=\frac{r^3\sigma_k}{i}\,\psi(\eta,\xi_k)+\frac{r^3}{i\sigma_k}\,\frac{d}{dr}\Big\{(\Gamma_3-1)\,\varrho\,\delta\{\ldots\}+\frac{2}{3}\,\varrho\,\delta\,[\ldots]\Big\}_k.$$

We must introduce this in (64.4) where we can write $\lambda_k'=(\sigma_k+i\sigma_k'')^2$, $\lambda_k=\sigma_k^2$ and use for ψ its expression (57.17). We then integrate by parts noting that, according to (57.19) and (57.20), we have

$$\varrho\,\delta\{\ldots\}=0 \quad\text{and}\quad \varrho\,\delta[\ldots]=0 \quad\text{in}\quad r=R \tag{64.8}$$

since the total pressure must vanish on a free surface. In this way, we obtain

$$\left.\begin{aligned}\sigma_k'' &= (\sigma_\eta'')_k+(\sigma_E'')_k\\ &= \frac{2}{3}\,\frac{\int\limits_0^M \frac{\eta r^2}{\varrho}\left(\frac{d\xi_k}{dr}\right)^2 dm}{\int\limits_0^M r^2\xi_k^2\,dm}-\frac{1}{2\sigma_k^2}\,\frac{\int\limits_0^M\left(\frac{\delta T}{T}\,\delta\{\cdots\}+\frac{2}{3}\,\frac{\delta\varrho}{S}\,\delta[\cdots]\right)_k dm}{\int\limits_0^M r^2\xi_k^2\,dm}\end{aligned}\right\}\tag{64.9}$$

which is the appropriate form of (63.15) for the k-mode of adiabatic pulsation. According to the above value of $\Delta L\xi_k$, the coefficient of ξ_j in (64.6) can be written $C'_{jk}=iC_{jk}$ where C_{jk} is real and given by

$$\left.\begin{aligned}C_{jk}(\sigma_k^2-\sigma_j^2)\int\limits_0^M \xi_j^2 r^2\,dm &= \sigma_k\frac{4}{3}\int\limits_0^M\frac{\eta}{\varrho}\,\frac{d\xi_k}{dr}\cdot\frac{d\xi_j}{dr}\,r^2\,dm-\frac{1}{\sigma_k}\int\limits_0^M\left(\frac{\delta\varrho}{\varrho}\right)_j\times\\ &\quad\times\left((\Gamma_3-1)\,\delta\{\ldots\}+\frac{2}{3}\,\delta[\ldots]\right)_k dm.\end{aligned}\right\}\tag{64.10}$$

Reintroducing the time factor, the real part of the corrected solution can be written

$$\left.\begin{aligned}\xi_k' &= e^{-\sigma_k'' t}\left(\xi_k\cos\sigma_k t-\sum_{j\neq k}^{\infty}C_{jk}\xi_j\sin\sigma_k t\right),\\ &= e^{-\sigma_k'' t}\,\xi_k\sqrt{1+\tan^2\vartheta_k}\cos\left[\sigma_k t+\vartheta_k(r)\right]\end{aligned}\right\}\tag{64.11}$$

where the variable phase ϑ_k is defined by

$$\tan\vartheta_k=\frac{1}{\xi_k}\sum_{j\neq k}C_{jk}\,\xi_j. \tag{64.12}$$

Moreover, as ROSSELAND already pointed out, the method is not limited to the particular type of perturbation considered above in details. The effects of any small changes in the coefficients of (58.1) corresponding to a modification of the physical conditions in a part of the star could be estimated in the same way. The method could even be extended to the comparison of neighbouring models since very often it provides a good first approximation even if the operators L, M and L', M' differ appreciably.

The method could also be used to discuss the effects of small external forces (gravitational field of a companion) or extra-factors such as a rotation or a magnetic field provided the rotational or magnetic energy remains small compared to the gravitational energy. This is particularly interesting in the case of non-radial oscillations where the separation of the variables, say (r,ϑ,φ), leads to a differential operator in r of the form (64.1) (at least in a first approximation) with multiple eigenvalues each of which, say λ_k^l, is associated with l independent solutions $f(r,\vartheta,\varphi,m_i)$, $i=1\ldots l$, where m_i is a parameter. If the rotation or the magnetic field perturbs Eq. (64.1) into (64.2), ΔL will, in general, depend on m_i and the corrections to the degenerate λ_k^l given by (64.4) will be different so that the perturbation will lift (or decrease) the degeneracy of the eigenvalues.

Finally, the method can be extended to cover cases where the boundary conditions are also perturbed (cf. especially [49b], Sects. 4.4 and 15.2 to 15.4) either through a change of the coefficients in $U_1=0$, $U_2=0$ or through a modification of the limits of integration. If there is more than one space variable, the last type of perturbation may correspond to a modification of the form of the boundaries such as occurs, for instance, in a rotating star when terms in ω^2 are taken into account.

65. Effects on vibrational stability of the generation and transfer of energy in the main body of the star. We shall first consider the fundamental mode in which the amplitudes of all the variations δr, $\delta\varrho$, δT etc. keep the same sign all through the star. In the first investigations of the problem by JEANS and EDDINGTON (cf. for instance [*14*] and [*39a*], p. 196) the relative amplitude $\xi=\delta r/r$ was assumed constant. But as we shall see, this implies a gross overestimate of the effects of energy generation and a gross underestimate of the damping due to radiative conductivity.

Even after he had derived the correct criterion, ROSSELAND [*56a*] made the same approximation in evaluating it numerically and, of course, got similar results. Although in a later paper ([*56b*], p. 20) he pointed out the stabilizing influence of an increase of ξ with r, he failed to appreciate the generality and the strength of this effect.

COWLING [*45a*] was the first to fully realize its importance and to work out all its implications in a correct quantitative discussion of vibrational stability. He also took convection into account and derived a general expression of σ_E'' equivalent to (64.9) in the case where p_R is negligible with respect to p_G. For the quantitative discussion, it is advantageous to eliminate ε_3 using its definition (52.6) which can be written here

$$\varepsilon_3=\frac{d}{dm}(4\pi r^2\overline{V}_r p)-\frac{\overline{V}_r}{\varrho}\frac{dp}{dr}. \tag{65.1}$$

After some regrouping of the terms, the expression of σ_E'' becomes

$$\sigma_E''\int_0^M \xi_k^2 r^2\,dm=-\frac{1}{2\sigma_k^2}\left\{\begin{aligned}&\int_0^M\left(\frac{\delta\varrho}{\varrho}\right)_k(\Gamma_3-1)\left\{\delta\varepsilon_1-\frac{d}{dm}\left[\delta 4\pi r^2(F_R+F_t^*)\right]\right\}_k dm+\\&+\int_0^M\left(\frac{\delta\varrho}{\varrho}\right)_k\left(\Gamma_3-\frac{5}{3}\right)\left[\delta\varepsilon_2+\delta\left(\frac{\overline{V}_r}{\varrho}\frac{dp}{dr}\right)\right]_k dm-\\&-\int_0^M\frac{2}{3}\left(\frac{\delta\varrho}{\varrho}\right)_k\left[\frac{d}{dm}(\delta 4\pi r^2 F_2)\right]_k dm\end{aligned}\right\} \tag{65.2}$$

where F_t^* has its value (57.12). If convection is limited to the small central core where the generation of energy takes place, one verifies that, grouping all the terms due to convection, its contribution to σ_E'' can be written

$$(\sigma_E'')_C\int_0^M \xi_k^2 r^2\,dm=-\frac{1}{2\sigma_k^2}\left\{\begin{aligned}&\int_0^{m_C}\left(\frac{\delta\varrho}{\varrho}\right)_k\left(\Gamma_3-\frac{5}{3}\right)\left[\delta\varepsilon_2+\delta\left(\frac{\overline{V}_r}{\varrho}\frac{dp}{dr}\right)\right]_k dm+\\&+\int_0^{m_C}\delta\left[4\pi r^2 F_t^*\frac{d}{dm}\left[(\Gamma_3-1)\left(\frac{\delta\varrho}{\varrho}\right)_k\right]\right]dm+\\&+\int_0^{m_C}\delta(4\pi r^2 F_2)_k\frac{d}{dm}\left(\frac{2}{3}\frac{\delta\varrho}{\varrho}\right)_k dm\end{aligned}\right\} \tag{65.3}$$

after some integrations by parts and noting that $\overline{V}_r$, F_2 and F_t^* are zero in $m=0$ and $m=m_C$. In the core, ξ and $\delta\varrho/\varrho$ may be treated as constant at least for the first few modes and if $p_R \ll p_G$, $\Gamma_3 \approx \frac{5}{3}$ so that $(\sigma_E'')_C \approx 0$. Thus, in this case, convection does not affect appreciably the coefficient of vibrational stability which reduces to

$$\sigma_E'' = \sigma_\varepsilon'' + \sigma_R'' = -\frac{1}{2\sigma_k^2} \frac{\int\limits_0^M \left(\frac{\delta T}{T}\right)_k (\delta\varepsilon)_k \, dm}{\int\limits_0^M r^2 \xi_k^2 \, dm} + \frac{1}{2\sigma_k^2} \frac{\int\limits_0^M \left(\frac{d\,\delta L}{dm}\right)_k \left(\frac{\delta T}{T}\right)_k dm}{\int\limits_0^M r^2 \xi_k^2 \, dm} \tag{65.4}$$

where $L(r) = 4\pi r^2 F_R$.

Since in the central region where ε is appreciable, ξ does not vary much, σ_ε'' can be written, by aid of (60.3) and (63.5),

$$\sigma_\varepsilon'' \approx -\frac{9(\Gamma_3 - 1)\, L}{2(3\Gamma_1 - 4)(-W)} \frac{\xi_C^2}{\overline{\xi^2}} [1 + \nu(\Gamma_3 - 1)] \tag{65.5}$$

where $\overline{\xi^2}$ is a mean value with respect to $r^2\,dm$. Eq. (65.5) implies that the energy sources ε account for the whole radiation of the star. In the case of a special nuclear reaction generating only a fraction α of the total luminosity L, (65.5) should be multiplied by α.

As to σ_R'', its evaluation in general has to be made numerically. However, if we admit, as is usually the case, that, in (63.9), the dominating terms are those in $\delta T/T$ and $\delta\varrho/\varrho$, we obtain in a first approximation

$$\sigma_R'' \approx \frac{9(\Gamma_3 - 1)^2 \left(4 + n - \frac{m}{\Gamma_3 - 1}\right) L}{4(3\Gamma_1 - 4)(-W)\,\overline{\xi^2}} \left[\xi + \frac{r}{3}\frac{d\xi}{dr}\right]_R^2$$

or using (58.17)

$$\sigma_R'' \approx \frac{9(\Gamma_3 - 1)^2 \left(4 + n - \frac{m}{\Gamma_3 - 1}\right) L}{4(3\Gamma_1 - 4)(-W)} \cdot \frac{\xi_R^2}{\overline{\xi^2}} \cdot \left[1 + \frac{1}{3\Gamma_1}(\omega^2 - 3\Gamma_1 + 4)\right]^2. \tag{65.6}$$

Grouping (65.5) and (65.6), we find

$$\left.\begin{aligned} \sigma_E'' = -\frac{3}{2}\,\frac{3(\Gamma_3-1)}{3\Gamma_1-4}\left(\frac{L}{-W}\right) \times \\ \times\left\{\frac{\xi_C^2}{\overline{\xi^2}}[1+\nu(\Gamma_3-1)] - \frac{\xi_R^2}{\overline{\xi^2}}\,\frac{\Gamma_3-1}{2}\left(4+n-\frac{m}{\Gamma_3-1}\right)\left[1+\frac{1}{3\Gamma_1}(\omega^2-3\Gamma_1+4)\right]^2\right\} \end{aligned}\right\} \tag{65.7}$$

which shows that the time τ'' needed to modify the amplitude by a factor e is of the order of the gravitational time-scale (cf. [*46*], p. 454)

$$\tau_G = \frac{(3\Gamma_1 - 4)(-W)}{3(\Gamma_3 - 1)\, L} \tag{65.8}$$

multiplied by a factor depending on energy sources and energy transfer.

Assuming that this last factor remains constant and since $L \propto M^4$, $-W \propto M^2/R$, we obtain

$$\tau'' \propto \frac{1}{M^2 R} \quad \text{or} \quad \propto \frac{1}{L^{\frac{1}{2}} R}. \tag{65.9}$$

This shows that any instability will be the more effective the greater the radius and the luminosity, a conclusion already reached by ROSSELAND ([*56a*], p. 36) although on less general grounds. This result may have some real interest since practically all stars in the upper right-hand corner of the Hertzsprung-Russell diagram do present some kind of variability.

If ξ is constant which implies $\omega^2 = (3\Gamma_1 - 4)$, the critical value of ν beyond which instability sets in ($\sigma'' \leqq 0$) is given by

$$\nu_C = \frac{n+4}{2} - \left(1 + \frac{m}{2}\right) \Big/ (\Gamma_3 - 1)$$

which, in usual conditions ($n = 3.5$, $m = 1$, $\Gamma_1 = \Gamma_3 = \frac{5}{3}$), leads to the very mild instability condition $\nu > 1.5$. But we can now see why this result which was even used at some time as an argument against thermonuclear reactions is so misleading. For instance, in the standard model, SCHWARZSCHILD's solution for $\Gamma_1 = \frac{5}{3}$ gives $\xi_R/\xi_C \approx 20$, $\omega^2 = 9.26$ and using (65.7), we find

$$\nu_C \approx 4000 \tag{65.10}$$

which is correct in order of magnitude as shown by comparison with the results of the detailed discussions by COWLING [*45a*] and later by LEDOUX [*57*]. This shows that even a moderate degree of central condensation ($\varrho_C/\bar{\varrho} \approx 30$ to 80) is sufficient to insure, by a very large margin, the vibrational stability of stars with ordinary masses ($p_R \ll p_G$) since ν is at most of the order of 20 in normal conditions.

However, convection extending from the center to a large fraction of the radius, would decrease ν_C very much mainly because it lowers the central condensation of the model considerably ($\varrho_C/\bar{\varrho} \approx 6$) and reduces ξ_R/ξ_C to a value of the order of 1.4. In this case, assuming that close enough to the surface, the equilibrium becomes radiative again, formulae (65.7) gives $\nu_C \approx 9$ which is exactly the value found by COWLING[1].

The discussion was extended by LEDOUX [*57*] to stars with large masses. In that case, as M increases, p_R/p_G increases too and Γ_1 and Γ_3 decrease. This favours the instability mainly because a low value of Γ_1 lowers ξ_R/ξ_C. Thus since ν_C decreases as M increases, conversely, for a given model and a given ν, there is a critical mass M_C beyond which instability sets in. For instance, for the standard model and KRAMERS' law of opacity ($n = 3.5$, $m = 1$) if the generation of energy is due to the carbon-cycle ($\nu \approx 18$), M_C is of the order of $100\,(\bar{\mu})^{-2}\,M_\odot$.

Other factors susceptible to increase the vibrational instability were discussed by LEDOUX [*57*]. One possibility, would be to move the generation of energy away from the center into regions where the relative amplitude ξ is a larger fraction of the surface amplitude ξ_R. However for this to be effective, one has to push the shell containing the sources of energy much further out than allowed by the maximum extension of the isothermal core[2].

Since then, following a suggestion by ÖPIK and SALPETER[3], it has been recognized that, in the course of the contraction of a sufficiently massive star following the exhaustion of hydrogen in its core, central temperatures of the order of 2×10^8 °K or higher might be reached. At that stage, various processes involving the fusion of α-particles occur[4] capable of generating the energy radiated by the star during

[1] T. G. COWLING: Monthly Notices Roy. Astronom. Soc. London **98**, 528 (1938).

[2] Cf. [*57*] and S. CHANDRASEKHAR and M. SCHÖNBERG: Astrophys. Journ. **96**, 161 (1942); also J. P. COX: Astrophys. Journ. **122**, 286 (1955).

[3] E. J. ÖPIK: Proc. Roy. Irish Acad. A **54**, 49 (1951). — E. E. SALPETER: Nature, Lond. **169**, 304 (1952). — Astrophys. Journ. **115**, 326 (1952).

[4] Cf. for instance: a) J. GREENSTEIN: Les Processus Nucléaires dans les Astres, p. 307, 5th Symposium, Liège, 1953; b) E. M. BURBIDGE, G. R. BURBIDGE, W. A. FOWLER and F. HOYLE: Synthesis of the Elements in stars, Kellogg Radiation Lab. and Mount Wilson and Palomar Observatories, 1957; c) A. G. W. CAMERON: Publ. Astronom. Soc. Pacific **69**, 201 (1957).

a short fraction of its life. However, ε is still of the form (63.4) with a value of ν of the same order as for the carbon-cycle so that this particular stage of stellar evolution does not introduce new possibilities of vibrational instability.

Likely modifications of the law of opacity in the region of the star where the adiabatic approximation is valid do not lead either to appreciable changes in the coefficient of stability although one may note that, if electron scattering was the main source of opacity in a large external region, the chances for vibrational instability would be increased appreciably.

For higher modes (cf. columns 8 and 9, Table 12) the ratio of the square of the amplitudes in the external and central regions is even greater and since the stabilizing factors arise mainly in the external layers, the star will in general be even more stable towards these modes.

The main significance of these results is that ordinary stars are very stable. But we are mainly interested here in variable stars where, on the contrary, a rather strong vibrational instability must have been present to permit incipient oscillations to develop and reach the observed amplitudes in a time short compared to their life-time or to maintain these oscillations during a long period. Thus these variables must possess some rather exceptional internal properties. As the members of the most important classes among them are giants or supergiants, one might think that the requirements of the static models capable of accounting for their giant characteristics might lead the way to some of these properties. Although there is as yet no completely adequate model for these giants, they must certainly have a very high central condensation to account for their energy production by ordinary thermonuclear reactions[1]. But from the preceding discussion this is exactly the opposite of what is needed to bring about vibrational instability. Recently, COX[2] has computed σ_E'' for such a giant star model and found it extremely stable, as can be checked rapidly from (65.7). The generation of energy (σ_ε'') causes an extremely slow increase of the amplitude amounting to a factor e in about 2×10^9 years while the damping time due to radiative conductivity (σ_R'') is only of the order of 10 days.

On the other hand, he estimates that the adiabatic solution used to evaluate σ_E'' is certainly valid up to $r/R\approx 0.85$ and the value of σ_R'' for this part of the star alone corresponds to a damping time of the order of 160 days. To compensate this, there should be a rather strong source of vibrational instability in the very external layers where the adiabatic solution breaks down.

However, before tackling the problem of the external layers, some more remarks on the generation of energy might be made. First, it should be noted that a given nuclear reaction will have an appreciable effect on the vibrational stability only if, in the static model studied, it generates a sizable fraction of the energy radiated [cf. remark following Eq. (65.5)]. For instance, recently CAMERON[3] has pointed out that, at high temperatures (10^8 °K), apart from the fusion of α-particles into C^{12}, neutrons might become available for the building up of heavier elements.

[1] A. REIZ: Vgl. danske Vidensk. Selsk., mat. -fys. Medd. **25**, 1 (1948); Ark. Astron. **1**, 7 (1949). — LI HEN and M. SCHWARZSCHILD: Monthly Notices Roy. Astronom. Soc. London **109**, 631 (1949). — C. M. BONDI and H. BONDI: Monthly Notices Roy. Astronom. Soc. London **109**, 62 (1949); **110**, 287 (1950); **111**, 397 (1951). — J. G. GARDINER: Monthly Notices Roy. Astronom. Soc. London **111**, 102 (1951). — J. B. OKE and M. SCHWARZSCHILD: Astrophys. Journ. **116**, 317 (1952).

[2] J. P. COX: The Pulsational Stability of Models for Red Giant Stars. Astrophys. Journ. **122**, 286 (1955). — Cf. also I. N. RABINOWITZ [Astrophys. Journ. **126**, 386 (1957)] where it is shown that, even if the envelope is wholly convective, the usual thermonuclear reactions are hopelessly inadequate to maintain the pulsation.

[3] A. G. W. CAMERON: Astrophys. Journ. **121**, 144 (1955).

But all these subsidiary reactions liberate only a very small fraction of the total energy radiated and play a negligible rôle as far as vibrational stability is concerned. The same is true of all kinds of synchrotron-betatron effects in variable magnetic fields such as have been recently discussed[1] to explain the abnormally high abundances of some of the heavier elements in some stellar atmospheres.

If fission could become important in a star[2], it might provide a source of instability which, in some cases, might take the form of a vibrational instability.

The burning up of the light elements (D, Li^6, Li^7, Be^9, B^{10}, B^{11}) at low temperatures[3] in the early stages of stellar evolution provides another case which has never been studied. Although here the sensitivity of the reactions to the temperature would be of the same type as for the carbon cycle, the appropriate stellar models might be sufficiently different from current models to deserve a special discussion of their vibrational stability. But whatever interest these cases might present for the evolution of stars in general, it is unlikely that they could have much bearing on the problem of regular stellar variability since the time spent in the corresponding stages of energy generation would be short.

66. Phase delay in energy generation. In the preceding discussion, we have always assumed that the equilibrium expression (63.4) of ε remains valid during the pulsation. But, as Eddington pointed out, if the abundances of the different nuclei entering the reaction do not remain in phase with the pulsation, resulting delays in the generation of energy may affect considerably the stability of the star[4].

In a chain of nuclear reactions comprising n combinations and m desintegrations involving $\mathfrak{N}$ different types of nuclei, the rate of energy generation per unit mass can be written

$$\varepsilon = \frac{1}{\overline{m}}\left(\sum_{i=1}^{n} K^i_{l\to k}\, x_l\, x_k\, N\, \varepsilon_i + \sum_{j=1}^{m} \lambda^j\, x_j\, \varepsilon_j\right) \tag{66.1}$$

where $\overline{m}$ is the average mass of a particle and $N = \varrho/\overline{m}$ is the total number of particles per unit volume; x_l and x_k are the abundances by number of particles of kinds l and k involved in the i reaction transforming l into k and $K^i_{l\to k}\, x_l\, x_k\, N^2$ is the number of these reactions per unit volume and unit time. The K^i's depend only on the temperature and, during a pulsation, they will vary as

$$K^i = K^i_0\left(1 + \nu_i \frac{\delta T}{T}\, e^{i\sigma t}\right) \tag{66.2}$$

where ν_i is the temperature exponent for the i reaction. The desintegration probabilities λ^j are constants.

As the changes in $\overline{m}$ are negligible, the equation of continuity (45.2) can be written

$$\frac{\partial N}{\partial t} + \operatorname{div}(N\boldsymbol{v}) = 0 \tag{66.3}$$

where

$$N = N_0\left(1 + \frac{\delta\varrho}{\varrho}\, e^{i\sigma t}\right). \tag{66.4}$$

[1] Cf. Ref. 4, p. 483.

[2] D. Stanley-Jones: Nature, Lond. **162**, 627 (1948).

[3] E. E. Salpeter: Les processus nucléaires dans les Astres, 5th Symposium Liège, 1953.

[4] For general references on this subject and a detailed discussion cf. Ref. [*39d*], §§ 5.3 to 5.5.

We must add the equations governing the kinetics of the reactions which after simplification by (66.3) and assuming the material well mixed can be written

$$\left.\begin{aligned}\frac{\partial x_k}{\partial t} = \sum_i N K^i_{l\to k} x_l x_k - \sum_j N K^j_{k\to m} x_k x_m + \sum_p \lambda^p_{p\to k} x_p - \sum_q \lambda^k_{k\to q} x_k ,\\ k = 1, 2 \ldots \mathfrak{N}\end{aligned}\right\} \qquad (66.5)$$

where the sums must be extended to all reactions or desintegrations involving the element k.

Introducing (66.2) and (66.4) in (66.5), the resulting set of equations can be solved for the abundances x_l which will be of the form

$$x_l = x_{l,0}\left(1 + \alpha_l \frac{\delta\varrho}{\varrho} e^{i\sigma t} + \beta_e \frac{\delta T}{T} e^{i\sigma t}\right) \qquad (66.6)$$

where the α's and β's may be complex. Using these in (66.1) and grouping all terms proportional to δT and $\delta\varrho$, we obtain

$$\varepsilon = \varepsilon_0\left[1 + \left(\alpha \frac{\delta\varrho}{\varrho} + \beta \frac{\delta T}{T}\right) e^{i\sigma t}\right]$$

where α is a linear combination of the α_l and ν_i and β, a linear combination of the β_l.

The coefficient of stability σ''_E should really be computed with

$$\delta\varepsilon = \varepsilon_0\left(\alpha \frac{\delta\varrho}{\varrho} + \beta \frac{\delta T}{T}\right). \qquad (66.7)$$

However, only the real parts of $\delta\varepsilon$ will affect the stability while the imaginary part will correspond to a negligible modification of the frequency σ. Thus as far as the stability is concerned defining

$$\mu_{\text{eff}} = \operatorname{Re}\alpha \quad \text{and} \quad \nu_{\text{eff}} = \operatorname{Re}\beta \qquad (66.8)$$

we may write

$$\delta\varepsilon = \varepsilon_0\left(\mu_{\text{eff}} \frac{\delta\varrho}{\varrho} + \nu_{\text{eff}} \frac{\delta T}{T}\right) \qquad (66.9)$$

which should replace (63.5).

The whole procedure might look forbidding but, in practice, many obvious simplifications arise in any actual computation[1]. For instance, steps in the chain which evolve negligible amounts of energy may be neglected in (66.1). Also the abundances of the elements with mean-lifes long as compared to the period of the star will exhibit large phase-lags of the order of $\pm\pi/2$. In that case, the real parts of α_l and β_l will be small and, for these elements, one may write $x_l \approx x_{l,0}$ in (66.1). In such cases, the contribution to μ_{eff} and ν_{eff} comes essentially from (66.2) and (66.4). This happens for instance for all collision processes in the carbon cycle and since, in this case, the two desintegrations liberate very little energy, one may immediatly foresee that μ_{eff} and ν_{eff} will have values close to 1 and 18 as in the static case.

On the other hand, if the mean-life of an element is short compared to the period, the phase-lag of its abundance is small and the corresponding α_l and β_l will have an appreciable real part which will contribute to μ_{eff} and ν_{eff}. For instance, in the proton-proton reaction, the mean-life of D^2 is very short, much

[1] For examples of applications cf. a) P. LEDOUX and E. SAUVENIER-GOFFIN: Astrophys. Journ. **111**, 611 (1950); b) E. SCHATZMAN: Ann. d'Astrophys. **14**, 305 (1951) and "Processus nucléaires dans les Astres", 5th Symposium, Liège, p. 163, 1953; c) D. A. FRANK-KAMENETSKY: Dokl. Akad. Nauk USSR. **77**, 385, 817 (1951); d) J. P. COX: Astrophys. Journ. **122**, 286 (1955).

shorter in general than any pulsation period. Its abundance will vary in phase with $\delta\varrho$ and δT and ν_{eff} will be greater than ν, its maximum value being about 2ν.

However, except in very special circumstances[1], the substitution of (63.5) by (66.9) does not affect the order of magnitude of the results and for ordinary stars our previous conclusions remain valid.

67. Large departure from isentropy and the non-adiabatic region. In the external layers, the oscillation may depart drastically from adiabatic conditions and the second member of (57.25) may become as large or larger than the first one. More directly, we may compare the adiabatic and non-adiabatic parts of δT. Eq. (56.18) connecting δp and $\delta\varrho$ can be transformed in a relation between δT and $\delta\varrho$ using (53.24) and

$$\frac{dp}{p} = (4-3\beta)\frac{dT}{T} + \beta\frac{d\varrho}{\varrho} + \beta X_i \frac{dx_i}{1+x}.$$

One finds, since ε_1 is regligible here,

$$\frac{1}{T}\frac{d\,\delta T}{dt} = (\Gamma_3 - 1)\frac{1}{\varrho}\frac{d\,\delta\varrho}{dt} + \frac{(\Gamma_3-1)}{[4-3\beta+\beta A_\varrho](B+1)}\frac{\varrho}{p}\,\delta\left\{\varepsilon_2 - \varepsilon_3 - \frac{1}{\varrho}\operatorname{div}(\boldsymbol{F}_R + \boldsymbol{F}_t)\right\}. \tag{67.1}$$

We may fix the upper boundary of the adiabatic region by the condition that the second term (non-adiabatic) on the right of (67.1) evaluated with the help of the adiabatic solution be of the same order as the first one (adiabatic). Let us suppose also that p_R is negligible ($\beta \to 1$) and that the interface between the two regions occurs under the ionization zone of hydrogen (A_ϱ and $B \to 0$). Then, if we admit with COWLING [*45a*] that

$$\varepsilon_2 \approx \frac{A\,C^2}{l/C} = \frac{A\,C^3}{l}, \qquad \frac{\delta l}{l} \approx \frac{\delta r}{r} = \xi \tag{67.2}$$

and use (53.28) where p_t is replaced by its value (49.13), we find

$$\frac{\delta C}{C} = -\frac{1}{3}\left(3\xi + r\frac{d\xi}{dr}\right) \quad \text{and} \quad \delta\varepsilon_2 = -\varepsilon_2\left(4\xi + r\frac{d\xi}{dr}\right). \tag{67.3}$$

According to the definition (52.6) of ε_3, its variation depends on $\delta\overline{V}_r$ which according to (57.9) is given by

$$\frac{\delta\overline{V}_r}{\overline{V}_r} = \frac{\delta l}{l} + \frac{\delta C}{C} + \frac{\delta\left(\frac{4-3\beta}{\beta} + A_p\right)}{\left(\frac{4-3\beta}{\beta} + A_p\right)} + \frac{\delta D}{D}. \tag{67.4}$$

Noting that D can be written

$$D = 4\pi\varrho r^2\left(\frac{1}{T}\frac{dT}{dm} - \frac{\Gamma_2 - 1}{\Gamma_2}\frac{1}{p}\frac{dp}{dm}\right) = 4\pi\varrho r^2 D_m \tag{67.5}$$

it follows that

$$\frac{\delta D}{D} = \frac{\delta\varrho}{\varrho} + 2\frac{\delta r}{r} + \frac{1}{D_m}\left[\frac{\delta p}{p}\frac{d}{dm}\left(\frac{\Gamma_2-1}{\Gamma_2}\right) - \frac{1}{p}\frac{dp}{dm}\,\delta\left(\frac{\Gamma_2-1}{\Gamma_2}\right)\right]. \tag{67.6}$$

But outside the ionization zone of hydrogen $(\Gamma_2 - 1)/\Gamma_2$ changes very little either in the star or in the course of a pulsation and we may neglect the last term in (67.6) as well as the third term in (67.4). In that case, writing $\Gamma_1 = \Gamma_2 = \Gamma_3 = \gamma$, it follows that

$$\frac{\delta\overline{V}_r}{\overline{V}_r} = -\xi - \frac{4}{3}r\frac{d\xi}{dr} \tag{67.7}$$

---

[1] E. SCHATZMAN: Ann. d'Astrophys. **14**, 305 (1951).

and

$$\left.\begin{aligned}\frac{\delta\varepsilon_3}{\varepsilon_3} &= -\frac{d\log r}{d\log(r^2\overline{V}_r)}\left\{\frac{d\xi}{dr}\left(r+\frac{4}{3}\frac{r^2}{p}\frac{dp}{dr}\right)+\frac{4}{3}r^2\xi\left[\frac{\sigma^2\varrho}{\gamma p}+\left(3-\frac{4}{\gamma}\right)\frac{1}{rp}\frac{dp}{dr}\right]\right\}-\\ &\quad -\left(\gamma+\frac{4}{3}\right)r\frac{d\xi}{dr}-(3\gamma+1)\,\xi\end{aligned}\right\}\tag{67.8}$$

if $d^2\xi/dr^2$ is eliminated by (58.12).

To compute the contribution due to the fluxes, let us note that in the region considered here, the energy is still transported wholly either by radiation or by convection and that the total flux $L=4\pi r^2 F_R$ or $4\pi r^2 F_t$ is a constant. Then proceeding with the same general assumptions as before, one finds

$$\left.\begin{aligned}&\delta\left(\frac{1}{\varrho}\operatorname{div}\boldsymbol{F}_t\right)=\frac{d}{dm}\left[\delta(4\pi r^2 F_t)\right]\\ &=\frac{4\pi r^2 F_t}{4\pi\varrho r^2}\left\{\frac{d\xi}{dr}\left[s+\left(\frac{4}{3}+\gamma\right)\frac{r}{p}\frac{dp}{dr}\right]+\xi\left(\frac{4}{3}+\gamma\right)\left[\frac{\sigma^2\varrho r}{\gamma p}+\left(3-\frac{4}{\gamma}\right)\frac{1}{p}\frac{dp}{dr}\right]\right\}.\end{aligned}\right\}\tag{67.9}$$

Eq. (63.9) together with

$$\frac{d\,\delta p}{dr}=\left(\sigma^2 r\varrho-4\frac{dp}{dr}\right)\xi\tag{67.10}$$

which is the reduced form of (63.17), leads to

$$\left.\begin{aligned}&\frac{\delta L}{L}=\frac{\delta(4\pi r^2 F_R)}{4\pi r^2 F_R}\\ &=r^2\frac{d\log r}{d\log T}\left\{\frac{d^2\xi}{dr^2}+\frac{d\xi}{dr}\left[\frac{4}{r}+\frac{1}{p}\frac{dp}{dr}\left(\gamma-\left[(4+n)(\gamma-1)-m\right]\frac{d\log T}{d\log p}\right)\right]+\right.\\ &\left.\quad+\xi\left[\frac{\sigma^2\varrho}{p}+(3\gamma-4)\frac{1}{rp}\frac{dp}{dr}+(4-3\left[(4+n)(\gamma-1)-m\right])\frac{1}{rT}\frac{dT}{dr}\right]\right\}.\end{aligned}\right\}\tag{67.11}$$

With this expression, treating $4\pi r^2 F_R$ as a constant and eliminating $d^2\xi/dr^2$ as before by (58.12), we obtain

$$\left.\begin{aligned}\varrho\,\delta\left(\frac{1}{\varrho}\operatorname{div}\boldsymbol{F}_R\right)&=F_R\frac{d\xi}{dr}\left\{\begin{aligned}&\frac{\sigma^2\varrho r^2}{p}\left(\frac{\gamma-1}{\gamma}\right)\frac{d\log r}{d\log T}+4\left[1-\frac{\gamma-1}{\gamma}(n_e+1)\right]-\\ &-\frac{r}{p}\frac{dp}{dr}\left[(\gamma-1)(n_e+1)-(4+n)(\gamma-1)+m\right]\end{aligned}\right\}+\\ &\quad+F_R\xi\left\{\begin{aligned}&\frac{\sigma^2\varrho r}{\gamma p}(\gamma-1)\frac{d\log r}{d\log T}+\sigma^2 r\left(\frac{\gamma-1}{\gamma}\right)\frac{d}{dr}\left(\frac{\overline{\mu}}{\mathrm{R}}\frac{dr}{dT}\right)-\\ &-\left(\frac{\sigma^2\varrho r}{\gamma p}+\frac{3\gamma-4}{\gamma}\frac{1}{p}\frac{dp}{dr}\right)\left[(\gamma-1)(n_e+1)-(4+n)(\gamma-1)+m\right]\end{aligned}\right\}\end{aligned}\right\}\tag{67.12}$$

where we have treated the effective polytropic index n_e defined by

$$\frac{d\log p}{d\log T}=n_e+1=\frac{n+m+4}{m+1}\tag{67.13}$$

as a constant.

Expression (67.12) as well as some others which we shall derive from it should not in fact be used too close to the surface. Indeed the second member of (67.12) tends to a finite value in $r=R$ while according to (64.8) it should be zero there. This is due to the fact that the adiabatic solution used for ξ corresponds to the vanishing in $r=R$ of the *adiabatic* perturbation of the pressure only while (64.8) implies the vanishing of the total perturbation of pressure which is in fact the correct boundary condition.

If we separate the adiabatic and non-adiabatic parts in (67.1), we can write

$$\left(\frac{\delta T}{T}\right)_a = -(\gamma-1)\left(3\xi + r\frac{d\xi}{dr}\right) \tag{67.14}$$

and

$$\left(\frac{\delta T}{T}\right)_n = -\frac{i}{\sigma}(\gamma-1)\frac{\varrho}{p}\left[\delta\varepsilon_2 - \delta\varepsilon_3 - \frac{d\,\delta(4\pi r^2 F_R)}{dm} - \frac{d\,\delta(4\pi r^2 F_t)}{dm}\right] \tag{67.15}$$

which expresses simply the relative elevation of temperature due to the accumulation of heat per unit mass.

In the case of radiative equilibrium, according to (67.13) and the hydrostatic equation, we have

$$T = \frac{G M \bar{\mu}}{(n_e+1)\,\mathrm{R}}\left(\frac{1}{r} - \frac{1}{R}\right). \tag{67.16}$$

In that case, the term in F_R only is important in (67.15) and it can be evaluated with the help of (67.12). Using (67.16) and (58.17) we have

$$\left(\frac{\delta T}{T}\right)_n = -\frac{i}{\sigma}(\gamma-1)\,F_S\frac{1}{p}\left\{\left(\frac{3\xi}{R} + \frac{d\xi}{dr}\right)\frac{\gamma-1}{\gamma}\omega^2(n_e+1) + 4\frac{d\xi}{dr}\left[\frac{\gamma-1}{\gamma}(n_e+1) - 1\right]\right\} \tag{67.17}$$

where F_S is the surface flux. Thus $(\delta T/T)_n$ increases indefinitely as p tends towards zero.

The comparison of (67.14) and (67.17) for EPSTEIN's model 4 [*53*] with $\gamma = \frac{5}{3}$, $n_e = 4.25$, $F_S = 8\times 10^{11}$ ergs cm^{-2}, $R = 2.78\times 10^{12}$ cm, $\omega^2 \approx 14$, $\sigma = 2.4\times 10^{-5}$ sec^{-1} shows that they are of the same order around $x = r/R = 0.85$ where T is still very high of the order of 10^5 °K; $(\delta T/T)_n$ is already of the order 0.1 $(\delta T/T)_a$ when $T \approx 2\times 10^5$ °K. For the model of LI HEN and SCHWARZSCHILD also studied by EPSTEIN (model 15: $R = 3\times 10^{12}$ cm, $F_S = 4.4\times 10^{10}$ erg cm^{-2}, $\omega^2 \approx 14$), we find that $(\delta T)_n \approx (\delta T)_a$ at $x = 0.9$, $T = 4.5\times 10^4$ °K.

In the case of adiabatic equilibrium, we may proceed in the same way evaluating (67.3), (67.8) and (67.9). It is likely that (67.3) and (67.8) will not affect the order of magnitude of the result, and the term in F_t after simplification by (58.17) gives

$$\left(\frac{\delta T}{T}\right)_n = \frac{i}{\sigma}(\gamma-1)\,F_S\frac{1}{p}\,5\,\frac{d\xi}{dr}. \tag{67.18}$$

For the model of P. DUMÉZIL-CURIEN with $\gamma = \frac{5}{3}$, $F_S = 4\times 10^{10}$ erg cm^{-2}, $R = 5\times 10^{12}$ cm, $\sigma = 5.5\times 10^{-6}$ sec^{-1}, $(\delta T/T)_a$ and $(\delta T/T)_n$ become of the same order for a very small pressure p of the order of 7×10^3 dynes/cm^2. Even if one would admit that (67.3) and (67.8) could contribute terms of the same order as (67.18), the critical pressure would be at most of the order of 2×10^4 and this occurs in the ionization zone of hydrogen where our approximations break down. One may conclude that here the adiabatic approximation remains valid practically up to the base of this zone.

In the atmosphere proper, $(\delta T/T)_n$ becomes the dominating term and the adiabatic solution can no longer be used to evaluate the bracket in (67.15). Neglecting $(\delta T/T)_a$, we have according to (63.9)

$$\frac{d\,\delta T}{dt} = -\frac{1}{C_v\varrho}\frac{1}{r^2}\frac{d}{dr}\left\{r^2 F_R\left[4\frac{\delta r}{r} + (4+n)\frac{\delta T}{T} - m\frac{\delta\varrho}{\varrho}\right] - \frac{4acT^3}{3\varkappa\varrho}r^2\frac{d\,\delta T}{dr}\right\}, \tag{67.19}$$

If the last term on the right dominates, this equation reduces to the ordinary heat-conduction equation with a coefficient of conductivity

$$k = \frac{4acT^3}{3\varkappa\varrho}. \tag{67.20}$$

Neglecting the curvature and the variations of k and denoting by h the depth in the atmosphere, (67.19) becomes

$$\frac{d\,\delta T}{dt} = \frac{k}{C_v \varrho} \frac{d^2 \delta T}{d h^2} = m \frac{d^2 \delta T}{d h^2}. \tag{67.21}$$

If at a given depth h_0, $\delta T = \delta T_0 \, e^{i\sigma t}$, this fluctuation propagates towards the surface ($h < h_0$) according to

$$\delta T = \delta T_0 \, e^{-\sqrt{\frac{\sigma}{2m}}(h_0 - h)} \, e^{i\sigma\left(t - \frac{h_0 - h}{c'}\right)} \tag{67.22}$$

with a local velocity

$$c' = \sqrt{2 m \sigma} \tag{67.23}$$

describing in a period $P = 2\pi/\sigma$, a distance Δh

$$\Delta h = 2\pi \left(\frac{8 a c T^3}{3 \varkappa \varrho^2 C_v \sigma}\right)^{\frac{1}{2}}, \tag{67.24}$$

its amplitude decaying by a factor $e^{-2\pi}$.

At the point $r/R = 0.9$ in model 15 considered previously ($h = 3 \times 10^{11}$ cm, $T = 5 \times 10^4$ °K, $\varrho = 5 \times 10^{-9}$ g, $\varkappa \approx 1$ cm² g⁻¹, $\sigma = 2 \times 10^{-5}$ sec⁻¹) we find

$$c' = 2 \times 10^7 \text{ cm sec}^{-1}, \qquad \Delta h = 6 \times 10^{12} \text{ cm}$$

so that the surface would be reached in a time $\Delta t \approx P/20$. In these circumstances, the oscillation of the layers external to $r = 0.9R$ is certainly not adiabatic.

If we consider a point in the atmosphere of η Aquilae as derived by M. and B. SCHWARZSCHILD and ADAMS[1] with average characteristics, $T = 5000$ °K, $\varrho = 3 \times 10^{-9}$ g, $\varkappa \approx 2 \times 10^{-2}$ cm² g⁻¹, we still obtain for c' a value of the order of 10^7 cm/sec and the surface will still be reached in a time of the order of one hour.

Note that in model 15, at $r/R = 0.5$, $c' = 10^4$ cm/sec, $\Delta h = 3 \times 10^9$ cm $\approx 10^{-3}$ R and at the centre, $c' = 10^2$ cm/sec, $\Delta h = 3 \times 10^7$ cm $= 10^{-6}$ R so that energy released in the central regions and transmitted by radiation takes a much longer time to reach the surface than if transmitted in the form of a pressure wave[2].

A complementary aspect of the problem can be characterized by a relaxation time τ_r. Suppose that at a given instant $t = 0$,

$$\delta T_0 = A_0 \, e^{-\left(\frac{h - h_0}{\lambda_0}\right)^2}. \tag{67.25}$$

Eq. (67.21) admits a solution of the same form

$$\delta T = \frac{A_0 \lambda_0}{(\lambda_0^2 + 4 m t)} \, e^{-\frac{(h - h_0)^2}{(\lambda_0^2 + 4 m t)^2}}. \tag{67.26}$$

Thus if τ_r is the time taken to reduce the amplitude to one tenth of its initial value, we have

$$\tau_r = \frac{\lambda_0^2 \, 10^2}{4 m}. \tag{67.27}$$

Applying (67.27) at the point $r/R = 0.9$ in model 15, with $\lambda_0 = 10^{10}$ cm, we get

$$\tau_r = 2.5 \times 10^3 \text{ sec} \approx P/100 \tag{67.28}$$

and one can verify that it is of the same order in the atmosphere of η Aquilae.

---

[1] M. and B. SCHWARZSCHILD and W. S. ADAMS: Astrophys. Journ. **108**, 207 (1948).

[2] Cf. for instance, E. SCHATZMAN: Bull. Acad. Roy. Belg., Cl. Sci., Sér. V **34**, 828 (1948).

One should not forget however that these estimations rest on a very rough approximation. In particular, it is not at all sure that the last term in (67.19) is really prevalent. In many circumstances, the effect of the term in $(4+n)\,\delta T/T$ may be more important and Eq. (67.19) which then becomes

$$\frac{d\,\delta T}{dt} = -\frac{F_R(4+n)}{C_v\,\varrho\,T}\,\frac{d\,\delta T}{dr} \tag{67.29}$$

admits solutions of the form

$$\delta T = f(r \pm c''\,t) \tag{67.30}$$

with

$$c'' = \frac{(n+4)\,F_R}{C_v\,\varrho\,T}\,. \tag{67.31}$$

But numerical evaluation of c'' at $r/R=0.9$ in model 15 and in the atmosphere of η Aquilae gives respectively

$$c'' = 10^7\ \mathrm{cm\,sec^{-1}} \quad \text{and} \quad c'' = 10^8\ \mathrm{cm\,sec^{-1}}$$

which are of the same order as c'. However, in this case, the amplitude does not decay during the propagation.

A more refined treatment by C. WHITNEY[1] taking into account the equations of radiative transfer has given relaxation times in supergiant atmospheres of the same order as (67.28). As he points out, in RR Lyrae stars however, τ_r may become closer to the actual period. Finally, one may want to extend the discussion to include the adiabatic part $(\delta T)_a$ in (67.1) which means adding an energy source in Eq. (67.21) or (67.29) corresponding to the work of the pressure in the course of the oscillation. WHITNEY has also discussed this problem in his thesis and concluded that this work is too small to affect appreciably the flux which may be considered as imposed on the atmosphere by the changes occurring in the interior. This implies that the atmosphere does not depart appreciably at any phase from radiative equilibrium conditions. The same conclusion was reached by P. LEDOUX and J. GRANDJEAN[2] in testing static models used to represent the atmosphere of η Aquilae at different phases. In fact, Eq. (52.7) or (67.1) can be written, neglecting the curvature,

$$\left.\begin{aligned} &\int\limits_{M-\Delta M}^{M} C_v\,T\left(\frac{1}{T}\frac{dT}{dt} - \frac{\gamma-1}{\gamma}\frac{d\varrho}{dt}\right)dm \\ &\qquad = -\left[F_M - F_{M-\Delta M}\right] = -\left[\delta F_M - (\delta F)_{M-\Delta M}\right] \end{aligned}\right\} \tag{67.32}$$

where ΔM is the mass per cm² above say the ionization zone of hydrogen and is at most a few hundred grams. Each term in the first member is only of the order of 10^8 erg cm^{-2} sec^{-1} while F_M and δF_M are of the order of 10^{10}. Inversely, the conversion of a very small fraction of the flux into thermal energy can raise the temperature of the atmosphere considerably and for periods large as compared to τ_r.

The problem has also been discussed by GUREVITCH and LEBEDINSKY in two papers[3] in which some of the aspects are treated in great details. Their results concerning the relaxation time or the velocity c' (or c'') agree with our previous discussion, although they consider that the heat capacity of these layers is due mainly to the radiation [(cf. their second paper, discussion of formula (56)], an unlikely situation resulting from the adopted value for $\varkappa(\approx 20)$ which seems much too large.

[1] C. WHITNEY: Stellar Pulsation. Thesis, Harvard University 1955.
[2] P. LEDOUX and J. GRANDJEAN: Bull. Acad. Roy. Belg., Cl. Sci., Sér. V **41**, 1010 (1955).
[3] L. E. GUREVITCH and A. I. LEBEDINSKY: Astrophys. Journ. USSR. **26**, 97, 138 (1949).

On the other hand, they consider that the heat liberated by the work of the pressure in a layer of thickness $\Delta h/10$ is communicated practically immediately to the surface. Numerically their value of Δh is smaller than suggested by our evaluation of (67.25) because their $\varkappa$ and ϱ are larger. However, it is still sufficiently large, according to the authors, for the compression work in the whole layer to account for the variations of the flux which will then be in phase with $d\,\delta\varrho/dt$ and consequently with $\pm v$ according as whether the amplitude of v increases or decreases with depth.

They consider that these two cases can explain the phase relationship between δL and v observed respectively for the Cepheids and the Mira variables. It is difficult however to accept these conclusions. First, the region of thickness Δh comprises the ionization zone of hydrogen which will modify completely the results. Secondly, the velocity distribution derived especially for Cepheids can not be reconciled with the pulsation of the interior.

68. Behaviour and influence of the atmospheric layers. Let us first consider the atmospheric layers, say above the ionization zone of hydrogen, ignoring for the time being other factors such as non-linearity which complicate the problem further.

α) Atmospheres with an external boundary where $p=0$. Neglecting viscosity and turbulence as well as ε_1, Eq. (57.25) can be written

$$\left.\begin{aligned} c^2\frac{d^2\xi}{dr^2}+\frac{d\xi}{dr}\left[\frac{4c^2}{r}+\frac{1}{\varrho}\frac{d(\Gamma_1 p)}{dr}\right]+\xi\left\{\sigma^2+\frac{1}{rp}\frac{d}{dr}[(3\Gamma_1-4)p]\right\}\\ =\frac{i}{\sigma r\varrho}\frac{d}{dr}\left[\frac{(\Gamma_3-1)}{4\pi r^2}\frac{d\,\delta L_R}{dr}\right]\end{aligned}\right\}\tag{68.1}$$

where $c=\sqrt{\Gamma_1 p/\varrho}$ is the Laplacian velocity of sound.

The presence of the imaginary unit in the second member means that in general ξ must be complex or, in other words, that the phase of the pulsation will vary with r in these layers [cf. also Eq. (64.11)]. But on the other hand, when the density becomes very small, we must have approximatively

$$\frac{d}{dr}\left[(\Gamma_3-1)\frac{1}{r^2}\frac{d\,\delta L_R}{dr}\right]\approx 0\tag{68.2}$$

and close enough to the surface, we know from (64.8) that the first derivative of δL_R itself must tend toward zero. In those conditions, one may admit that the atmosphere is characterized by

$$\delta L_R=C=\text{const}\tag{68.3}$$

Physically, this agrees with our discussion in Sect. 67.

In that case, the "atmosphere" will, at least in the linear approximation considered here, pulsate with a uniform phase albeit a different one from the interior since a change of phase must necessarily occur where the two members of (68.1) are of the same order of magnitude.

Using (67.11), Eq. (68.3) becomes

$$\left.\begin{aligned}&\frac{d^2\xi}{dr^2}+\frac{d\xi}{dr}\left\{\frac{4}{r}+\frac{1}{p}\frac{dp}{dr}\left(\gamma-[(4+n)(\gamma-1)-m]\frac{d\log T}{d\log p}\right)\right\}+\\ &\quad+\xi\left\{\frac{\sigma^2\varrho}{p}+\frac{1}{rp}\frac{dp}{dr}\left[(3\gamma-4)+(4-3[(4+n)(\gamma-1)-m])\frac{d\log T}{d\log p}\right]\right\}\\ &=\frac{\delta L_R}{L_R}\frac{1}{T}\frac{dT}{dr}\frac{1}{r}.\end{aligned}\right\}\tag{68.4}$$

The complete equation corresponds to a forced oscillation of the atmosphere driven by the flux. The free oscillations are solutions of the homogeneous equation which is very similar to the ordinary adiabatic equation.

If the opacity is constant ($n=0$, $m=0$), it reduces to

$$c_N^2 \frac{d^2\xi}{dr^2} + \frac{d\xi}{dr}\left[\frac{4c_N^2}{r} - g\right] + \xi\sigma^2 = 0 \tag{68.5}$$

where g is the acceleration of gravity and c_N is now the Newtonian velocity of sound.

In the general case however, the coefficient of ξ does not reduce simply to the frequency. For instance if $m\approx 1$, $n\approx 3.5$, $\gamma=\frac{5}{3}$ we have approximately

$$c_N^2 \frac{d^2\xi}{dr^2} + \frac{d\xi}{dr}\left[\frac{4c_N^2}{r} - \frac{2}{3}g\right] + \xi\left[\sigma^2 + \frac{G\,m(r)}{r^3}\right] = 0. \tag{68.6}$$

If we neglect the curvature, Eqs. (68.5) and (68.6) become

$$c_N^2 \frac{d^2\zeta}{dh^2} + g\frac{d\zeta}{dh} + \sigma^2\zeta = 0, \tag{68.7}$$

$$c_N^2 \frac{d^2\zeta}{dh^2} + \frac{2}{3}g\frac{d\zeta}{dh} + \zeta\left[\sigma^2 + \frac{4\pi G\bar{\varrho}}{3}\right] = 0, \tag{68.8}$$

where ζ is now the displacement δr and h, the depth in the atmosphere.

In the same conditions, the adiabatic equation [left-hand member of (68.1)] gives

$$c^2 \frac{d^2\zeta}{dh^2} + \gamma g\frac{d\zeta}{dh} + \zeta\left[\sigma^2 - (3\gamma-4)\frac{4\pi G\bar{\varrho}}{3}\right] = 0. \tag{68.9}$$

If

$$\sigma^2 \gg \frac{4\pi G\bar{\varrho}}{3}. \tag{68.10}$$

both Eqs. (68.8) and (68.9) again reduce to the form (68.7). In particular, (68.9) becomes

$$c^2 \frac{d^2\zeta}{dh^2} + \gamma g\frac{d\zeta}{dh} + \sigma^2\zeta = 0. \tag{68.11}$$

Equations of this type have been the object of elaborate treatment in applications to the terrestrial atmosphere[1].

In the simplest case of an homogeneous compressible atmosphere in which $p=g\varrho h$ and $c^2=\gamma g h$, Eq. (68.11) may be written

$$\frac{d^2\zeta}{dy^2} + \frac{d\zeta}{dy}\frac{1}{y} + \zeta = 0$$

with

$$y = 2\sigma\sqrt{\frac{h}{\gamma g}}.$$

If the top of the atmosphere ($h=0$) is free ($\delta p=0$), the solution is

$$\zeta = A\,J_0(y)$$

[1] H. LAMB: On Atmospheric Oscillations. Proc. Roy. Soc. Lond. **1911**. Cf. also [*39d*], § 2.5, [*56a*], Sect. 7. One of the most systematic discussions of atmospheric oscillations is given in Chap. VIII of [*40b*]. In this book as in many publications directed towards meteorological applications [cf. also H. SOLBERG: Astrophys. Norv. **2**, No. 2 (1936)] the discussion is not limited as it is here, to oscillations of infinite horizontal wavelength.

where J_0 is a Bessel function of zero order. If we impose another condition at the bottom h_0 of the atmosphere such as

$$\zeta(h_0) = J_0(y_0) = 0, \tag{68.12}$$

there is an infinite number of modes with eigenvalues corresponding to the zero's of J_0. For instance, for the fundamental mode

$$(\sigma_0)_{\text{at}} \approx 1.2 \sqrt{\frac{\gamma g}{h_0}}. \tag{68.13}$$

In ordinary stars, $(\sigma_0)_{\text{at}}$ is much larger than the fundamental frequency σ_0 of the whole star but in supergiant variables, g, or at least its effective value g_e, may be very small and their atmospheres may have very considerable extent[1]. For instance, in the case of η Aquilae ($R = 4.5 \times 10^{12}$ cm) if we take $h_0 = 2 \times 10^{11}$ cm and $g \approx 10$ cm sec^{-2}, (68.13) gives a period $(P_0)_{\text{at}} \approx 7$ days which is very close to the fundamental period P_0 of the star.

Of course if $(\sigma_0)_{\text{at}}$ becomes of the same order as σ_0, the condition (68.10) can no longer be verified although for the most probable models (cf. values of ω_0^2 in Table 12), $4\pi G\bar{\varrho}/3$ remains small ($\approx \frac{1}{10}$) as compared to σ_0^2. But, if desirable, the general Eqs. (68.8) or (68.9) can be used without difficulty since they are again of type (68.11) with parameters

$$\sigma'^2 = \sigma^2 + \frac{4\pi G\bar{\varrho}}{3} \quad \text{or} \quad \sigma'^2 = \sigma^2 - (3\gamma - 4)\frac{4\pi G\bar{\varrho}}{3}, \tag{68.14}$$

Eq. (68.8) favouring somewhat longer periods.

For an atmosphere where $dT/dh = D$, (67.13) reduces to $(n_e + 1) = g\bar{\mu}/RD$ and Eq. (68.11) becomes

$$\frac{d^2\zeta}{dy^2} + \frac{1 + 2n_e}{y}\frac{d\zeta}{dy} + \zeta = 0 \tag{68.15}$$

with

$$y = 2\sigma\sqrt{\bar{\mu}\, h/\gamma R D}.$$

It has the solution

$$\zeta = A\, y^{-n_e} J_{n_e}(y).$$

For all positive values of n_e, the zero's of J_{n_e} are larger than those of J_0 and this could reduce $(P_0)_{\text{at}}$ appreciably.

Nevertheless, one should not neglect the possibility of resonance between the atmosphere and the interior. For instance, if we restore the time factor in (68.4) for the simple case ($m = 0$, $n = 0$), we find with the same assumptions as before

$$\frac{\partial}{\partial h}\left(p\frac{d\zeta}{dh}\right) - \varrho\frac{\partial^2\zeta}{\partial t^2} = \frac{(\delta L_R)_0}{L_R}\,\frac{R\varrho}{\bar{\mu}}\,\frac{dT}{dh}\,e^{i\Sigma_0 t} \tag{68.16}$$

where Σ_0 is the frequency of the adiabatic interior where δL_R is very generally in phase with $-\delta r$.

If we denote by ζ_i and σ_i, the solutions of the corresponding homogeneous equation

$$\frac{d}{dh}\left(p\frac{d\zeta}{dh}\right) + \sigma^2\varrho\zeta = 0, \tag{68.17}$$

[1] Cf. for instance A. PANNEKOEK: Bull. astronom. Inst. Netherl. **8**, 175 (1937); A. PANNEKOEK and F. B. VAN ALBADA: Publ. Astr. Inst. Univ. Amsterd. **1946**, No. 6, Part 2; D. KOELBLOED: Publ. Astr. Inst. Univ. Amsterd. **1953**, No. 10.

one verifies easily, following the procedure of Sect. 58, or using the properties of the Bessel functions, that the ζ_i are orthogonal

$$\int_0^{h_0} \varrho \zeta_i \zeta_k \, dh = 0 \quad \text{if} \quad i \neq k. \tag{68.18}$$

If we introduce in (68.16), the general solution

$$\zeta = \sum_i a_i \zeta_i \, e^{i \Sigma_0 t} \tag{68.19}$$

we obtain

$$\sum_i a_i (\Sigma_0^2 - \sigma_i^2) \varrho \zeta_i = \frac{(\delta L_R)_0}{L_R} \cdot \frac{g}{(n_e + 1)} \varrho = C \varrho$$

where dT/dh has been replaced by its value $g\bar{\mu}/R(n_e+1)$. Multiplying by $\zeta_k \, dh$ and integrating from 0 to h_0, we finally find

$$a_k = \frac{1}{(\Sigma_0^2 - \sigma_k^2)} \frac{\int_0^{h_0} C \varrho \zeta_k \, dh}{\int_0^{h_0} \varrho \zeta_k^2 \, dh} \tag{68.20}$$

so that in (68.19), the term corresponding to an atmospheric mode with a frequency σ_k close to Σ_0 would be strongly enhanced and would be in phase with $\pm \delta L_R$.

However, if dissipative forces, such for instance as turbulent viscosity, were at work in the atmosphere, the situation might be very different. In that case, to write Eq. (68.2) explicitly, we should use

$$\frac{\partial (\delta p)}{\partial r} = -\left(4 \frac{dp}{dr} \xi + r \varrho \frac{d^2 \xi}{dt^2}\right) + \frac{4}{3 r^3} \frac{\partial}{\partial r} \left[\eta r^4 \frac{\partial}{\partial r} \left(\frac{\partial \xi}{\partial t}\right)\right]$$

instead of (67.10) and, with the same hypothesis as before, Eq. (68.16) is replaced by

$$\frac{\partial}{\partial h} \left(p \frac{\partial \zeta}{\partial h}\right) - \varrho \frac{\partial^2 \zeta}{\partial t^2} + \frac{4}{3} \frac{\partial}{\partial h} \left[\eta \frac{\partial}{\partial h} \left(\frac{\partial \zeta}{\partial t}\right)\right] = C \varrho \, e^{i \Sigma_0 t}. \tag{68.21}$$

Introducing again the general solution (68.19), we obtain now

$$a_k (\Sigma_0^2 - \sigma_k^2) \int_0^{h} \varrho \zeta_k^2 \, dh - \frac{4 i \Sigma_0}{3} \sum_{j=1} a_j \int_0^{h_0} \eta \left(\frac{\partial \zeta_j}{\partial h} \cdot \frac{\partial \zeta_k}{\partial h}\right) dh = \int_0^{h_0} C \varrho \zeta_k \, dh \tag{68.22}$$

which shows that in general the a_k will be complex and the displacement (68.19) will show a more or less important phase-shift with respect to δL_R. In particular, if there is again a fairly close resonance for one of the modes ($\Sigma_0 \approx \sigma_k$), the corresponding term in (68.19) will become the most important and will essentially exhibit a phase-shift of one quarter period with respect to δL_R. This is particularly apparent in the formal case where η is taken proportional to p: $\eta = A p$. In that case, integrating by parts the term in η in (68.22) and using (68.17) to eliminate the second order derivative $d^2 \zeta_j / dh^2$, one finds, due to the orthogonality conditions, that

$$\sum_{j=1} a_j \int_0^{h_0} \eta \left(\frac{\partial \zeta_j}{\partial h} \cdot \frac{\partial \zeta_k}{\partial h}\right) dh = A \sigma_k^2 \int_0^{h_0} \varrho \zeta_k^2 \, dh.$$

Then

$$a_k = \frac{\int_0^{h_0} C\,\varrho\,\zeta_k\,dh}{\{(\Sigma_0^2 - \sigma_k^2) - \frac{4}{3} i\,\Sigma_0\,A\,\sigma_k^2\} \int_0^{h_0} \varrho\,\zeta_k^2\,dh} \tag{68.23}$$

and at resonance, the term in ζ_k will be large and its phase 90° late with respect to that of δL_R, a situation rather similar to the actual one in Cepheid variables.

It is difficult to judge how far non-linearity and other factors would affect these results, but they suggest that it would be worthwhile to discuss this question further.

FRANK-KAMENETSKY[1] has attempted to treat the problem more generally without using (68.2). In that case, extra-terms must be added to the explicit expression (67.11) of δL_R due to the fact that $\delta T/T$ in (63.9) must be replaced by its general expression (67.1) where the second term on the right can no longer be put equal to zero. Introducing the entropy as a variable, he obtains a system of the fourth order which he discusses briefly. Some of his inferences do not seem warranted and certainly some of his solutions with amplitudes increasing inwards cannot, as suggested elsewhere (cf. p. 486, footnote 1 c), have a general significance for the vibrational stability of the star. More recently, it seems that the author[2] has come back to a point of view closer to the one adopted here.

β) Unbounded atmosphere. Up to now, we have assumed that the star has a definite boundary characterized by $p=0$ and it is the corresponding boundary condition $\delta p=0$ which insures perfect reflection and leads to the establishment of standing oscillations. However the actual boundary of a star is ill defined as its atmosphere blends more or less continuously with the interstellar material. However, the drop in density between the reversing layers which are still mainly controlled by the gravitational attraction of the star and the interstellar medium is very large on any theory and this allows us to treat the star as a pulsating unit. In particular the mass in these external layers is extremely small and according to formula (64.4) or (59.10), the effect on the period will be negligible. As to the other consequences, it is very difficult to discuss them properly without, at least, a reasonable static or stationary model for the outermost layers of the star including the chromosphere and corona.

The case of a plane infinite isothermal atmosphere with a constant g in which $p=p_0\,e^{-x/H}$, $\varrho=\varrho_0\,e^{-x/H}$, H being the scale height: $H=RT/\bar{\mu}g$, is often used to illustrate the problem[3]. The adiabatic equation is then

$$\frac{d^2\zeta}{dx^2} - \frac{\gamma g}{c^2}\frac{d\zeta}{dx} + \zeta\frac{\sigma^2}{c^2} = 0$$

where c^2 is now a constant. It has solutions of the form

$$\zeta = A\,e^{ikx}$$

where

$$k = -\frac{i\gamma g}{2c^2} \pm \sqrt{\frac{\sigma^2}{c^2} - \frac{\gamma^2 g^2}{4c^4}} = -\frac{i}{2H} \pm \sqrt{\frac{\sigma^2}{\gamma g H} - \left(\frac{1}{2H}\right)^2}\,. \tag{68.24}$$

Two cases have to be distinguished

1. $\sigma > \frac{\gamma g}{2c}$, k: real, $\zeta = e^{\frac{x}{2H}}\left[A\,e^{i(\sigma t + kx)} + B\,e^{i(\sigma t - kx)}\right]$,

2. $\sigma < \frac{\gamma g}{2c}$, $k = i\,k'$: imaginary, $\zeta = e^{\frac{x}{2H}}\left[A\,e^{-k'x} + B\,e^{k'x}\right]$.

[1] D. A. FRANK-KAMENETSKY: Dokl. Akad. Nauk USSR. **80**, 185 (1951).
[2] D. A. FRANK-KAMENETSKY: Dokl. Akad. Nauk USSR. **99**, 41 (1954).
[3] Cf. Ref. [*39d*], § 6.4.

The corresponding standing waves are respectively of the form

$$1. \qquad \zeta = e^{\frac{x}{2H}} (A \cos k\, x + B \sin k\, x) \cos \sigma t,$$

$$2. \qquad \zeta = e^{\frac{x}{2H}} (A\, e^{-k' x} + B\, e^{k' x}) \cos \sigma t. \tag{68.25}$$

If the total energy of wave motion

$$E = \frac{\sigma^2}{2} \int_0^\infty \zeta^2 \varrho\, dx$$

is to remain finite, only standing waves of type (68.25) with $B=0$ are allowed and thus occur when

$$\sigma < \frac{\gamma g}{2c}. \tag{68.26}$$

In the other cases, only outgoing progressive waves of the type

$$\zeta = e^{\frac{x}{2H}} A \cos \sigma \left(t - \frac{k}{\sigma} x\right)$$

are possible.

In term of the wavelength $\lambda = 2\pi c/\sigma$, condition (68.26) becomes

$$\lambda > 4\pi H \tag{68.27}$$

which fits the general result that considerable reflection will occur only if conditions change markedly on a distance of the order of λ.

If we introduce numerical values in (68.26) taking $\gamma = \frac{5}{3}$, we find for the critical period P_c

$$P_c \approx \frac{\sqrt{T}}{g} \text{ days} \tag{68.28}$$

where g is in cm sec^{-2}.

If we again refer to η Aquilae ($R = 4.5 \times 10^{12}$ cm, $M = 1.35 \times 10^{34}$ g, $g = GM/R^2 = 43$ cm sec^{-2}) and if we take $T \approx T_e \approx 6000°$ we get

$$P_c \approx 2 \text{ days} \tag{68.29}$$

which is appreciably smaller than the observed period P_0 of about 7 days. However, if we take for g the same effective value $g_e \approx 10$ cm sec^{-2} [cf. discussion of Eq. [(68.13)], P_c becomes of the same order as P_0. If, in addition, one would consider that T in (68.28) should rather be of the order of the temperature in the high chromosphere or the corona, let us say 6×10^4 to 6×10^5 °K, P_c would become much larger

$$P_c \approx 20 \text{ to } 70 \text{ days}. \tag{68.30}$$

In this case, the fundamental oscillation ($P_0 \approx 7$ days) could definitely take a progressive character in these layers. In his thesis ,WHITNEY[1] has shown that non-adiabatic terms do not modify appreciably this result.

On the other hand, SCHATZMAN in a recent paper[2] has discussed in some details the case of an isothermal layer of thickness h and temperature T resting on a solid surface and surmounted by an infinite layer at temperature T'. In agreement with the previous results, he finds that if $T' \ll T$ standing oscillations are always established while if $T' \gg T$ only progressive waves are possible. In

[1] CH. WHITNEY: Stellar Pulsation. Harvard University 1955.

[2] E. SCHATZMAN: Ann. d'Astrophys. 19, 45 (1956); cf. also [29b].

this last case, if the origin of x is in the plane separating the two layers, the general solution in the lower layer is of the form

$$\zeta = e^{-\sigma'' t} e^{\frac{x}{2H}} [e^{b x}(A_1 + i A_2) e^{i(\sigma t - a x)} + e^{-b x}(C_1 + i C_2) e^{i(\sigma t + a x)}] \quad (68.31)$$

where a and b are the real and imaginary parts of the radical in (68.24). Considering the case $g = 10$ cm sec^{-2}, $T = 6000$ °K, $T' = 6 \times 10^5$ °K, $h \approx H = 5 \times 10^{10}$ cm, SCHATZMAN's formulae give

$$\left.\begin{aligned} a h &\approx 2 \text{ radians} \approx 116°, & b h &= 0.077, \\ \sigma &= 4 \times 10^{-5} \text{ sec}^{-1}, & \sigma'' &= 1.3 \times 10^{-6} \text{ sec}^{-1}, \end{aligned}\right\} \quad (68.32)$$

the last value corresponding to a large dissipation with a damping-time $\tau'' \approx 5 P$.

One might expect that, in such a case, the wave in the lower layer should show a rather strong progressive character. However, applying the boundary condition

$$(A_1 + i A_2) e^{-bh + ah} + (C_1 + i C_2) e^{bh - ah} = 0$$

in $x = -h$, one finds, if A_2 is set equal to zero,

$$C_1 = -A_1 e^{-2bh} \cos 2ah, \qquad C_2 = -A_1 e^{-2bh} \sin 2ah$$

and (68.31) becomes

$$\zeta = e^{-\sigma'' t} e^{\frac{x}{2H}} A_1 [e^{bx} \cos(\sigma t - a x) - e^{-b(2h + x)} \cos(\sigma t + a x + 2ah)]. \quad (68.33)$$

Using the numerical value $e^{-2bh} = 0.861$ and treating $|b x|$ as a small quantity, (68.33) can also be written

$$\zeta = e^{-\sigma'' t} e^{\frac{x}{2H}} A_1 \frac{1,861 + 0,139 b x}{\sin \varphi(x)} \sin[a(x + h)] \cos[\sigma t + a h - \varphi(x)] \quad (68.34)$$

where $\varphi(x)$ is defined by

$$\tan \varphi(x) = \tan[a(x + h)] \frac{1,861 + 0,139 b x}{0,139 + 1,861 b x}. \quad (68.35)$$

This is a progressive wave, but its phase $\varphi(x)$ changes very slowly. For instance in $x = 0$, $x = -h/2$, $\varphi(x)$ takes the following values

$$\varphi(0) = 87° 55, \quad \varphi\left(-\frac{h}{2}\right) = 88° 42', \quad (68.36)$$

which show that the wave departs very little from a pure oscillation.

The evaluation of the damping can be generalized easily. The energy $(\Delta K)_P$ lost per period is equal to the energy spread over a wavelength in the external layer. In the case which led to the larger one of the critical values (68.30), the outgoing wave is

$$\zeta = A e^{\frac{x}{2H}} e^{2\pi i \left(\frac{t}{P_0} - \frac{x}{\lambda}\right)} \quad (68.37)$$

with

$$\left.\begin{aligned} H &= 5 \times 10^{12} \text{ cm}, \quad P_0 \approx 7 \text{ days}, \\ \lambda &= 2\pi \left[\frac{\sigma_0^2}{\gamma g H} - \left(\frac{1}{2H}\right)^2\right]^{-\frac{1}{2}} \approx 6 \times 10^{12} \text{ cm}. \end{aligned}\right\} \quad (68.38)$$

Then

$$\begin{aligned} (\Delta K)_P &\approx -\frac{4\pi R^2}{2} \int_0^\lambda \varrho \dot{\zeta}^2 dx = -\frac{4\pi R^2 \varrho_0}{2} (\sigma_0 A)^2 \int_0^\lambda \sin^2 2\pi \left(\frac{t}{P_0} - \frac{x}{\lambda}\right) dx \\ &= -\pi R^2 \varrho_0 (\sigma_0 A)^2 \lambda \end{aligned}$$

where ϱ_0, the density at the base of the isothermal layer, must be very small. Assuming continuity of δr through the surface, formula (63.13) gives

$$\sigma'' = \frac{1}{\tau''} = \frac{\sigma_0}{8\pi} \frac{\varrho_0 \lambda}{R\int\limits_0^1 \varrho\,\xi^2 x^4\,dx} = \frac{1}{P_0} \frac{\varrho_0 \lambda}{4R\int\limits_0^1 \varrho\,\xi^2 x^4\,dx} \tag{68.39}$$

where the interior solution ξ is normalized to unity at the surface. If one uses EPSTEIN'S solution for model 4 [*53*] to evaluate the denominator of (68.39), the damping-time τ'' for η Aquilae is

$$\tau'' = 10^5\,P_0 \tag{68.40}$$

if the density ϱ_0 at the base of the high temperature region is taken equal to 10^{-14} g cm^{-3}. This shows that here too the dissipation could be important. However, the resulting progressive character will probably be slight as in the preceding example. This can also be inferred from (64.12) since the integrals in the C_{jk} [Eq. (64.10)] are in general much smaller than the integrals in the damping coefficient σ_k'' [Eq. (64.9)].

69. Influence of the non-adiabatic region below the atmosphere. We have seen in Sect. 67 that the adiabatic approximation remains valid up to a level where T reaches values from 10^5 to 10^4 °K according to the model and in Sect. 65 we concluded that, in the region below this level, the condition of vibrational stability was well satisfied. On the other hand, the discussion at the end of Sect. 67 and in Sect. 68 made it very plausible that, in the atmospheric layers, Eq. (68.2), is verified in which case the transfer of energy in these layers should not affect the vibrational stability. If dissipation occurs there, it can only be due to some form of viscosity or to progressive waves carrying the energy away in some kind of corona where it is dissipated into heat. This will correspond to a positive damping reinforcing the vibrational stability. Thus in the present state of our knowledge of nuclear reactions, the source of vibrational instability must be looked for in the intermediate layer between the adiabatic interior and the atmosphere proper.

With a small or moderate abundance of H or He, the heat capacity of this layer is very small and, accordingly, early investigations[1] led to the conclusion that its effects on vibrational instability or the character of the wave would be negligible. However, after the large predominance of H was recognized, EDDINGTON pointed out [*55*] that the ionization of H which occurs exactly in that region, might increase its heat capacity sufficiently to undo the stabilizing effect of radiative conductivity in the rest of the star, heat being accumulated in the form of ionization energy at compression and liberated by recombination during the expansion. The efficacity of the process depends on the total heat capacity of the region and hence on the total mass M_c of hydrogen atoms in a critical stage of ionization. If radiative equilibrium prevails in the external layers, M_c is much too small. To avoid this difficulty, EDDINGTON in his first paper advocated strong deviations from ionization equilibrium in the convective currents but, soon after, this was shown to be impossible by M. SCHWARZSCHILD[2]. In his second paper, EDDINGTON advanced a series of arguments to bolster up the convective transfer of energy and lower the radiative gradient, but it is

[1] J. J. REESINCK: Monthly Notices Roy. Astronom. Soc. London **87**, 414 (1927). — A. S. EDDINGTON: Monthly Notices Roy. Astronom. Soc. London **87**, 539 (1927). — M. SCHWARZSCHILD: Z. Astrophys. **11**, 152 (1935).

[2] M. SCHWARZSCHILD: Monthly Notices Roy. Astronom. Soc. London **102**, 152 (1942).

only recently that the first giant star model with an extensive external convection zone was actually constructed by P. DUMÉZIL-CURIEN. Since the vibrational stability of this model has been discussed by SCHATZMAN[1], we shall use it also to illustrate the situation. SCHATZMAN has neglected all effects of convection and turbulence except the convective flux F_t^* [cf. Eq. (57.12)]. Probably this does not affect his results too much since Γ_3 is still close to $\frac{5}{3}$ in a very large fraction of the mass while C and $\overline{V}_r$ are small compared to $\overline{C}_T$ so that ε_2, F_2 and p_t and their variations are generally small compared to the thermal energy and pressure. The most serious effect, as we shall see later, may arise from the neglect of the turbulent viscosity in the equations of motion. In those conditions, σ_E'' as given by (65.2) becomes

$$\sigma_E'' = -\frac{1}{2\sigma^2} \frac{\int\limits_0^M \frac{\delta T}{T}\left[\delta\varepsilon_1 - \frac{d(\delta L_R + \delta L_t^*)}{dm}\right] dm}{\int\limits_0^M \xi^2 r^2\, dm}. \tag{69.1}$$

In evaluating (69.1), SCHATZMAN has extended the integral right up into the very external layers where according to his discussion, $\varkappa$ increases strongly with T, $-n$ [cf. Eq. (63.8)] reaching values as large as 10 while m remains positive. This violates strongly the local condition (63.10) and contributes to the instability. Actually, the accumulation of energy in these layers at contraction is so large (cf. columns 7 and 8 of Table 13) that it renders σ_E'' negative as illustrated on Fig. 42. For the fundamental mode, extending the integral up to $\log p = 3.5$, SCHATZMAN finds

$$\sigma_E'' = -7.33 \times 10^{-8}\ \mathrm{sec}^{-1}, \tag{69.2}$$

and the amplitude would increase by a factor e every 13 periods since, for this model $P_0 = 1.14 \times 10^6$ sec.

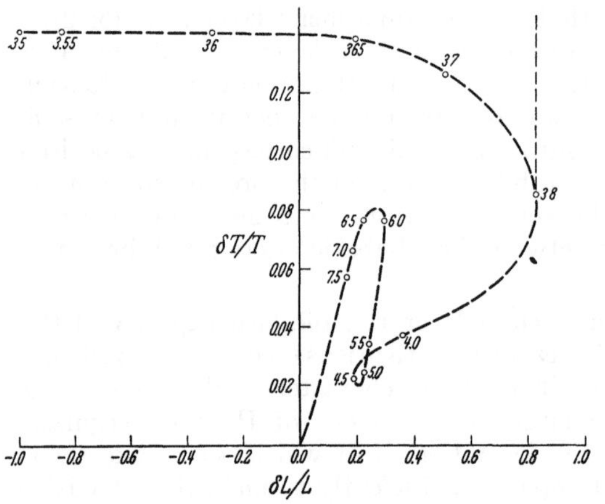

Fig. 42. Diagram $(\delta T/T, \delta L/L)$ in which areas are proportional to the integral in (69.1). Each point is labelled with the corresponding value of $\log p$.

However, the application of (69.1) in these external layers is very delicate as it appears already from some of SCHATZMAN'S results. For instance, a very formal evaluation of the effects of the atmosphere reduces already σ_E'' to -1.82×10^{-8}. From a physical point of view, we know (cf. Sect. 67) that, for this model, the adiabatic approximation remains valid very high, probably up to the base of the ionization zone. However, in this zone, it deteriorates rapidly as shown by column 9 of Table 13 where $(\delta T)_n$ has been computed by $(\delta T)_n = -\Delta(\delta L)/C_v' \Delta m$ with C_v' as given by (57.10).

High enough in the atmosphere, δL must remain constant due to the small heat capacity of the most external layers. To gain some idea of the level at which the variation of δL becomes negligible, we have computed the variation of the ionization energy between equilibrium condition and maximum compres-

[1] E. SCHATZMAN: Ann. d'Astrophys. **19**, 51 (1956).

sion, once for SCHATZMAN's adiabatic solution and once assuming that $(\delta T/T)$ is a constant equal to its normalized surface value $(\delta T/T)_R = 0.14$. In the first case, the extra-ionization energy is obtained from the compression work while, in the second, it is furnished in part by non-adiabatic processes at the expense of the flux.

The modification in the degree of ionization x_H has been computed from (53.24) putting $x_H \approx x$. Combining the result with (53.22), one gets for the change in ionization energy per unit mass

$$\delta I = \chi_H M \frac{x(1-x^2)}{x(1-x)+2}\left[\left(\frac{3}{2}+\frac{\chi_H}{kT}\right)\frac{\delta T}{T} - \frac{\delta \varrho}{\varrho}\right]. \tag{69.3}$$

The ionization energy accumulated in the layer above the level p, $m(r)$ is then

$$\delta \int_{m(r)}^{M} I\, dm \approx 4\pi R^2 \int_0^p \delta I \frac{dp}{g}. \tag{69.4}$$

This quantity is tabulated in columns 10 and 11 of Table 13 for the two cases mentioned above and the corresponding variations of x_H are illustrated in Fig. 43 together with some other typical quantities.

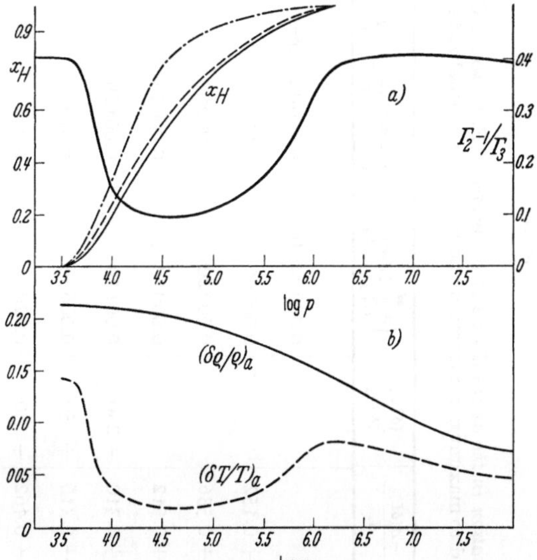

Fig. 43a and b. (a) Variation of x_H with depth at equilibrium (full line), at maximum adiabatic compression (dashed line) and at maximum compression assuming $\delta T/T$ constant and equal to $(\delta T/T)_R$ (interrupted line). (b) Adiabatic variation of the density and the temperature.

With the same normalization, the variation of the total energy radiated during the corresponding quarter period is approximately

$$\frac{LP}{2\pi} = 2 \times 10^{42}\,\text{ergs}. \tag{69.5}$$

As long as this quantity is large compared to the ionization energies in columns 10 and 11, δL cannot vary appreciably with depth and the corresponding part of the integral in σ_E'' should vanish. This comparison suggests that, for an appreciable range in the possible variation of $\delta T/T$, δL may be treated as a constant approximately from the level $\log_{10} p = 3.8$ upwards. But in that case, referring to Fig. 42, σ_E'' becomes positive of the order of $+1.4 \times 10^{-8}\,\text{sec}^{-1}$ and the amplitude would now *decrease* by a factor e every 60 periods. Even if one would admit as an extreme case that, at all levels, half of the neutral hydrogen could get ionized at compression, the base of the external zone with constant δL would be lifted only to $\log_{10} p \cong 3.7$ and this would just about bring the star into a state of indifferent equilibrium ($\sigma_E'' = 0$).

To arrive at a definite answer, it is necessary to study the non-adiabatic terms in the layer between the approximate limits

$$\left.\begin{aligned} &\log_{10} p_e \approx 3.8, \quad T_e \approx 8800\ ^\circ\text{K} \\ \text{and} \quad &\log_{10} p_i \approx 6.0, \quad T_i \approx 23000\ ^\circ\text{K}. \end{aligned}\right\} \tag{69.6}$$

Table 13. *Characteristics of the pulsation in the external part of the model of* P. DUMÉZIL-CURIEN.

Note: $L = L_R + L_t$; the amplitudes as given here correspond to maximum contraction and have been normalized in such a way that $\delta L/L = 1$ at the surface.

$\log_{10} p$	r/R	p	T	$(\delta T/T)_a$	x_H	$\left(\frac{\delta L}{L}\right)_a$	$\frac{-\Delta(\delta L)_a}{L}$	$\frac{(\delta T)_n}{(\delta T)_a}$	$\left(\delta \int_{m(r)}^{M} I\,dm\right)_a$	$\left(\delta \int_{m(r)}^{M} I\,dm\right)_{\frac{\delta T}{T} = \left(\frac{\delta T}{T}\right)_S}$	L_t/L	c	c'	c''
3.470	1	$2.95 \cdot 10^3$	5360	0.142	0.0003						0.000			
									$0.00002 \cdot 10^{42}$	$0.00002 \cdot 10^{42}$				
3.500		3.16	5530	0.141	0.0004	−1.0					0.017	$9 \cdot 10^5$	$2 \cdot 10^6$	10^8
							+0.152		0.00006	0.00006				
3.550		3.55	5940	0.141	0.001	−0.848					0.183			
							+0.536		0.00020	0.00020				
3.600	0.999	3.98	6590	0.141	0.0036	−0.312					0.576	$9 \cdot 10^5$	$6.4 \cdot 10^5$	$9 \cdot 10^7$
							+0.512		0.00070	0.00073				
3.650	0.999	4.47	7310	0.138	0.0123	+0.200					0.872			
							+0.310	+250	0.00179	0.00189				
3.700	0.998	5.01	7910	0.126	0.0288	+0.510					0.955			
							+0.316	+50	0.00686	0.010				
3.800	0.997	6.31	8770	0.085	0.0776	+0.826					0.987	10^6	$5 \cdot 10^4$	$6 \cdot 10^6$
							+0.462	−30	0.020	0.064				
4.000	0.995	$1.00 \cdot 10^4$	9930	0.037	0.200	+0.364					0.995			
							+0.170	−2	0.041	0.313				
4.500	0.988	3.16	12020	0.022	0.528	+0.194					1			
							+0.029	+0.14	0.159	2.278				
5.000	0.977	$1.00 \cdot 10^5$	14060	0.024	0.691	+0.223					1	$3 \cdot 10^6$	$2.6 \cdot 10^2$	$2.5 \cdot 10^4$
							+0.016	+0.016	0.466	4.200				
5.500	0.964	3.16	16900	0.034	0.888	+0.239					1			
							+0.061	+0.02	1.139	10.000				
6.000	0.945	$1.00 \cdot 10^6$	22910	0.076	0.977	+0.300					1			
							−0.070	−0.01	2.091	9 000				
6.500	0.919	3.16	27990	0.076	0.990	+0.230					1			
							−0.046	−0.002						
7.000	0.881	$1.00 \cdot 10^7$	55980	0.066	1.000	+0.184								
							−0,022							
7.500	0.826	3.16	88 500	0.057	1.000	+0.162								

70. EDDINGTON's treatment of the intermediate layers. The immediate effect of the non-adiabatic terms is a non-adiabatic modification $(\delta T)_n$ of the temperature. The resulting dynamical effects will be important unless c' or c'' as defined by (67.23) and (67.31) are large compared to the velocity of sound c. As shown in Table 13, c' and c'' are rather small in the ionization zone due to its high heat capacity and opacity and the exact evaluation of the dynamical effects there is rather difficult. In the following ξ, δp, $\delta\varrho$ and δT will designate total variations comprising an adiabatic part ξ_a, δp_a, etc. and a non-adiabatic part ξ_n, δp_n, etc. (Eq. 67.10), which is valid in general, can be written here with $m(r) = M$, $r \approx R$

$$\frac{d\,\delta p}{dp} = -(\omega^2 + 4)\,\xi \tag{70.1}$$

and gives

$$\delta p_n = (\delta p_a - \delta p_{a,i})\,\frac{\bar{\xi}_n}{\bar{\xi}_a} \tag{70.2}$$

where $\bar{\xi}$ denotes a mean value with respect to the pressure taken from p_i to p. If $\bar{\xi}_n/\bar{\xi}_a$ remains small, δp_n will always be small compared to the adiabatic increment of δp in the region.

On the contrary, the equation of continuity

$$\frac{\delta\varrho}{\varrho} = -\left(3\xi + r\frac{d\xi}{dr}\right) \approx -\left(3\xi + R\frac{\Delta\xi}{\Delta r}\right) \tag{70.3}$$

shows that in this region a small extra-change in ξ may give a large variation in $\delta\varrho/\varrho$. Let us suppose, for instance, that the effect of the non-adiabatic terms is to give to ξ a progressive character

$$\xi = \xi_0(r)\cos[\sigma t - \vartheta(r)], \quad \vartheta(r) = 0 \text{ in } r^* \text{ and } \vartheta(r) = \frac{\pi}{n} \text{ in } r = R$$

with n large. From Eq. (70.3), it is easy to verify that, if the variations of the amplitude ξ_0 are neglected,

$$\left(\frac{\delta\varrho}{\varrho}\right)_R = -3\xi_0\sqrt{1 + \frac{1}{9}\left(\frac{\pi}{n}\,\frac{R}{R - r^*}\right)^2}\,\cos(\sigma t - \varphi)$$

with

$$\tan\varphi = \frac{\dfrac{\pi}{n} + \dfrac{1}{3}\,\dfrac{\pi}{n}\,\dfrac{R}{R - r^*}}{1 - \dfrac{1}{3}\left(\dfrac{\pi}{n}\right)^2 \dfrac{R}{R - r^*}}.$$

If $n = 30$ and $R - r^* = R/30$, a phase-shift of 6° in ξ implies a phase-shift of 50° in $(\delta\varrho/\varrho)_R$ and an increase in amplitude by a factor 1.4. Thus a very small readjustment ξ_n and the resulting $(\delta\varrho/\varrho)_n$ will be able to compensate a fairly large $(\delta T/T)_n$ with very little change in $\delta p/p$. As a first approximation, EDDINGTON writes

$$\xi_n \approx 0 \qquad \left(\frac{\delta p}{p}\right)_n \approx 0. \tag{70.4}$$

In that case, (56.18) and (67.1) give

$$\frac{d}{dt}\left(\frac{\delta\varrho}{\varrho}\right)_n = \frac{\Gamma_3 - 1}{\Gamma_1}\,\frac{\varrho}{p}\,\frac{d\,\delta L}{dm} \tag{70.5}$$

and

$$\frac{d}{dt}\left(\frac{\delta T}{T}\right)_n = -\frac{\Gamma_3 - 1}{\Gamma_1}\,\frac{\varrho}{p}\left[\frac{\Gamma_1}{(1 + A_\varrho)(B + 1)} - (\Gamma_3 - 1)\right]\frac{d\,\delta L}{dm}. \tag{70.6}$$

From (63.13) and (63.20), neglecting small terms, the general coefficient of vibrational stability can be written

$$\left.\begin{aligned}\sigma'' = \sigma_a'' + \sigma_n'' = -\frac{1}{2\sigma^2}\,\frac{\int\limits_0^{M_i}\left(\frac{\delta T}{T}\right)_a\left(\delta\varepsilon - \frac{d\,\delta L}{dm}\right)_a dm}{\int\limits_0^M \xi^2 r^2\,dm} + \\ + \frac{1}{2\pi\sigma}\,\frac{\int\limits_{M_i}^{M_e} dm \int\limits_0^P \left[\left(\frac{\delta T}{T}\right)_a + \left(\frac{\delta T}{T}\right)_n\right]\frac{d\,\delta L}{dm}\,dt}{\int\limits_0^M \xi^2 r^2\,dm}\end{aligned}\right\} \tag{70.7}$$

where M_i and M_e refer to the limits defined by (69.6). Taking (70.6) into account, the time integral involving $(\delta T/T)_n$ cancels out and the last term reduces to

$$\sigma_n'' = \frac{1}{2\pi\sigma}\,\frac{\int\limits_{M_i}^{M_e} dm \int\limits_0^P \left(\frac{\delta T}{T}\right)_a \frac{d\,\delta L}{dm}\,dt}{\int\limits_0^M \xi^2 r^2\,dm}\,. \tag{70.8}$$

If the non-adiabatic phenomena in these layers introduce a progressive phase-shift $\varphi(m)$ of δL with respect to $(\delta T)_a$, the time integral in (70.8) becomes

$$\int\limits_0^P \left(\frac{\delta T}{T}\right)_a \cos\sigma t\,\frac{d}{dm}\{\delta L\cos[\sigma t + \varphi(m)]\}\,dt = \frac{\pi}{\sigma}\left(\frac{\delta T}{T}\right)_a \frac{d\,[\delta L\cos\varphi(m)]}{dm}\,. \tag{70.9}$$

Reintroducing this into (70.8) and integrating by parts, we obtain

$$\left.\begin{aligned}\sigma_n'' \int\limits_0^M \xi^2 r^2\,dm = \frac{1}{2\sigma^2}\Bigg\{\left[\left(\frac{\delta T}{T}\right)_a \delta L\right]_{M_e}\cos\varphi(M_e) - \left[\left(\frac{\delta T}{T}\right)_a (\delta L)_a\right]_{M_i} - \\ - \int\limits_{M_i}^{M_e} \delta L\cos\varphi(m)\,\frac{d}{dm}\left(\frac{\delta T}{T}\right)_a dm\Bigg\}\end{aligned}\right\} \tag{70.10}$$

with $\varphi(M_i) = 0$. The second term will cancel with a similar term in σ_a'' and we have

$$\left.\begin{aligned}\left(2\sigma^2\int\limits_0^M \xi^2 r^2\,dm\right)\sigma'' = -\int\limits_0^{M_i}\left(\frac{\delta T}{T}\right)_a \delta\varepsilon_a\,dm - \int\limits_0^{M_i}(\delta L)_a \frac{d}{dm}\left(\frac{\delta T}{T}\right)_a dm + \\ + \left[\left(\frac{\delta T}{T}\right)_a \delta L\right]_{M_e}\cos\varphi(M_e) - \int\limits_{M_i}^{M_e}\delta L\cos\varphi(m)\,\frac{d}{dm}\left(\frac{\delta T}{T}\right)_a dm.\end{aligned}\right\} \tag{70.11}$$

Since very generally $(\delta L)_a$ and the space derivative of $(\delta T/T)_a$ have the same sign in the adiabatic region (cf. for instance Table 13 columns 5 and 7), the first two terms will contribute to the instability. If as suggested by EDDINGTON, $\varphi(M_e)$ is close to $\pi/2$, the third term will cancel out. As to the fourth term, since $\cos\varphi$ tends to 0 when M tends to M_e, the main contribution comes from the region around the lower limit M_i. There the space derivative of $(\delta T/T)_a$ is negative (cf. Fig. 43b) while δL which does not yet deviate very much from (δL) is still positive. This term will thus contribute to the stability and EDDING-

TON, who did not want to rely on negative dissipation to excite the oscillation, thought that it might just about compensate the second term. In fact, this is practically realized in P. DUMÉZIL-CURIEN's model, if one assumes that

$$\varphi(m) = \frac{\pi}{2}\left(\frac{m - M_i}{M_e - M_i}\right). \tag{70.12}$$

In that case, the nuclear reactions (first term) deep down into the star should be responsible for the incipient instability. But we have seen in Sect. 65 [cf. Eq. (65.5) and COX's results] that, in all reasonable giant star models, this is very inefficient as the time needed to build up the amplitude is much too large. Of course, the instability increases if the phase-shift $\varphi(M_e)$ becomes larger than $\pi/2$ and it reaches a maximum for $\varphi(M_e) = \pi$. In that case, the last integral in (70.11) becomes positive and contributes about half as much to the instability as the third term $-(\delta T/T)_a \delta L$. A rough estimate based on (70.12) where $\pi/2$ is replaced by π gives for σ'' a value of about one third that found by SCHATZMAN [cf. (69.2)] and this would still correspond to a strong instability. Indeed a phase-shift of 100° would already lead to an appreciable instability.

It seems that EDDINGTON was fascinated by the observed lag of about a quarter period between luminosity and displacement. But it is not perhaps absolutely essential to recover this phase-lag just at the top of the ionization zone since factors such as those discussed in Sect. 68α, could still modify the phase relationship and bring it to the observed value, in consequence of some small extra-dissipation in the outermost layers.

A more serious question is whether the alleged phase-shift can really take place. The detailed argument of EDDINGTON was based on radiative transfer assuming that δL_R reduces to the term in δT. In that case, as we have seen in Sect. 67 [discussion of Eq. (67.29)], the main effect tends to be a change in phase of δT_n and of δL. But in the model considered here, the transfer of energy in the whole intermediate layer is still mainly by convection (cf. Table 13, column 12) and the necessary revision is rather tricky.

Column 9 of Table 13 shows that the tendency for the intermediate layers to remain cooler than their surroundings at compression is still reinforced in part of the layer by the first non-adiabatic approximation. This tendency could not persist indefinitely in successive approximations and it might lead to a complete breakdown of the normal modes of transfer and result in an inflow of energy in that region. If it is large enough to raise δT appreciably, especially towards the bottom of the region, enough energy could be accumulated there in form of ionization energy [cf. column 11 of Table 13 and Eq. (69.5)] to cause a large phase-shift in δL. The necessary analysis to settle this point definitely will certainly be very delicate. On should not forget either, that the model used here as an example is not the only possible one, in particular, as regards the external layers. In this respect, one may remark that, while the predominance of convective transport increases the extent and the heat capacity of the ionization zone, on the other hand a larger rôle of radiative conduction would favour the accumulation of energy in that layer since the low adiabatic temperature variation would permit a large increase of the opacity, even if n remains positive. In all this, any improvement in the treatment of the convective transport especially in the meaning and the evaluation of the mean free path would be very welcome.

In a series of papers, ZHEVAKIN[1] has extended the discussion to the case of helium ionization. He believes that the second ionization can play an important

[1] C. A. ZHEVAKIN: Astrophys. Journ. USSR. **30**, 161 (1953); **31**, 385 (1954); **32**, 124 (1955) in which references to earlier papers are given.

rôle in the problem. These papers contain many interesting remarks, but some of the criticisms of the rôle of the hydrogen ionization zone, as well as some of the positive arguments for the importance of the helium ionization, suffer from the very rough static model on which they are based. One of the advantages of the helium ionization zone is its occurrence at greater depths ($35000^\circ < T < 55000^\circ$). This increases its total heat-capacity which may remain adequate even in the case of radiative equilibrium. To some extent, the treatment is also simpler since the adiabatic approximation remains relatively better at least in the inner part of the zone. In the last paper referred to above, ZHEVAKIN comes to the conclusion that, to be effective in maintaining the oscillations, the helium ionization zone should really be in radiative equilibrium which is in agreement with his estimation of the relative importance of convective and radiative transfer in the static model used for this zone. Apart from the uncertainties in all evaluations of the convective flux F_t and the influence of the static model adopted, ZHEVAKIN'S discussion is also influenced to some extent by his desire to recover the observed phase-shifts between δr and δL for different types of variable stars, at the top of the ionization zone. As we have already remarked, this is perhaps not essential.

Negative dissipation in the external layers sets another type of problem because it will, in general, be at least as efficient for higher modes up to an appreciable order than for the fundamental one. For instance, SCHATZMAN finds that the first mode is about 10 times more unstable than the fundamental one. Of course, positive dissipation due to viscous forces [cf. (64.9)] will also increase with the order of the mode. Furthermore, the higher the mode, the easier it is for the wave to take a progressive character in an extended atmosphere [cf. condition (68.28)] and dissipate its energy according to (68.39) which however is not sensitive to the frequency. Nevertheless, to excite just one mode, as is required in many cases, would call for a very delicate balance.

The main conclusion is that with the emphasis being shifted from a search for a positive source of increasing amplitudes in the deep interior to that for a negative dissipation in the very external layers, a refined discussion of the static equilibrium of these layers as well as of their dynamics is urgently needed.

In particular, since the deviations from isentropy can be very large there, it will probably be necessary to use higher approximations in the computation of σ'' such for instance as provided by (64.7) which will require a fairly precise evaluation of the first order correction εu_1 to the displacement.

An essentially equivalent procedure would be to generalize to a physical model, the method of successive approximations introduced by WOLTJER[1]. From a physical point of view this method might present some advantage as it provides, at each step, the interesting quantity $\left(\frac{\delta T}{T}\right)_n$.

71. Effects of friction. Up to now, we have neglected the dissipation due to viscous forces. This is given here by σ''_η as defined in (64.9)

$$\sigma''_\eta = \frac{2}{3}\,\frac{\int\limits_0^M \eta\,\frac{r^2}{\varrho}\left(\frac{d\xi}{dr}\right)^2 dm}{\int\limits_0^M r^2\,\xi^2\,dm}\,. \tag{71.1}$$

[1] J. WOLTJER jr.: Bull. astronom. Inst. Netherl. **8**, 17 (1936); the corresponding first order approximation which is equivalent to the evaluation of σ'' in our notation has been applied by Mrs. PELS-KLUYVER in Bull. astronom. Inst. Netherl. **8**, 293 (1936). Cf. also a recent attempt by P. LAL and P. L. BHATNAGAR: Z. Astrophys. **41**, 21 (1956) which however avoids again the really significant part of the problem in the external layers.

Although it has been found recently[1] that the preponderance of hydrogen increases the coefficient of molecular viscosity η by a factor which may be as large as 200 with respect to the value used by Persico[2] the corresponding damping times $\tau''=1/\sigma''$ given in Table 14 (cf. footnote 1, b) are still so large that they present no real interest for the first 2 or 3 modes. However in models with high central condensation, the fourth and higher modes could be damped out completely by this cause alone in periods short as compared to the life-time of the star in his variable stage.

The effects of radiative viscosity will be at most of the same order but turbulent viscosity will be much more efficient as can be seen from a comparison of (46.6) and (49.14) since the mean free path is always much larger for the turbulent elements than for the molecules. For instance, in the convective zone of Dumézil-Curien's model taking on the average $l\approx 10^{11}$ cm $\approx R/50$, $C\approx 1$ km/sec, $\varrho=10^{-5}$ g, η_t comes out at least 10^{11} times greater than η. Applying (71.1) to this model with values of l and η_t computed according to the definition used by Dumézil-Curien, Longe[3] finds

$$\sigma''_{\eta_t}=1.2\times 10^{-8}\ \text{sec}^{-1}$$

which means that the amplitude is reduced by a factor e in about 3 years or 70 periods. This is so short a time that even a large reduction in l and C would still leave a very strong damping.

Table 14. *Damping times in years due to molecular viscosity.*

Models	Modes	
	Fundamental	First
Standard-model . . .	2.5×10^{13}	2×10^{12}
Model No. 4, Epstein .	10^{11}	10^{10}

But apart from the global dissipation of energy, momentum is continuously transferred from the exterior towards the interior and the relative amplitude ξ tends to become a constant, a stage in which friction would cease to act as a dissipative agency as shown by formula (71.1). The complete discussion of this aspect would require the solution of (57.25) keeping the first term on the right-hand side, but some ideas of the effect can be gained in introducing $\sigma=\sigma'+i\sigma''(r)$ in (57.25) and solving separately the real and imaginary parts of the equation. Neglecting terms of the order of $(\sigma''/\sigma)^2$, Longe finds that

$$\sigma''(r)=-\frac{2}{3}\frac{1}{\varrho\xi}\left[\frac{d}{dr}\left(\eta_t\frac{d\xi}{dr}\right)+\frac{4}{r}\eta_t\frac{d\xi}{dr}\right] \tag{71.2}$$

where ξ may be replaced by its adiabatic value. Note that the average of (71.2) with respect to $\xi^2 r^2\,dm$ gives back (71.1). The numerical evaluation of (71.2) for the model of Mme Dumézil-Curien shows that $\sigma''(r)$ changes sign around $r/R=0.66$, its mean values on the external and internal sides being of the order respectively of $+10^{-7}$ and -10^{-7} sec^{-1}. The ratio ξ_e/ξ_i would then be reduced by a factor e in about 4 periods. Since, in this model, the maximum ratio ξ_e/ξ_i is of the order of 5, it would be reduced practically to a constant in the convective region very rapidly. Of course, there will be dynamical interactions between the convection zone and the deep interior, and since the kinetic energy in the whole core is at most $^1/_{30}$ of the total kinetic energy, the amplitude in the core itself would be strongly affected. If the ratio ξ_R/ξ_C could be reduced sufficiently, perhaps to a value of the order of 2 or 3, the nuclear sources of energy close to

[1] a) S. Chapman: Astrophys. Journ. **120**, 151 (1954). — b) J. Counson, P. Ledoux and R. Simon: Bull. Soc. Roy. Sci. Liège **25**, 144 (1956).

[2] E. Persico: Monthly Notices Roy. Astronom. Soc. London **86**, 93, 98 (1926).

[3] P. Longe: Bull. Soc. Roy. Sci. Liège **26**, 541 (1956); also P. Ledoux in [*38*], p. 186.

the center could again play an important rôle in the problem and perhaps compensate the total dissipation by friction and by radiative conduction in the interior, both of which would be reduced very much in any case.

However, the simple picture of turbulent convection used here encounters serious difficulties in some parts of the star. For instance there are regions where l/C becomes of the same order as the period and the exchange of momentum cannot really be represented adequately by a coefficient of viscosity. On the other hand, the exact solution of (57.25) would exhibit phase-shifts which might be important.

72. Effects of a progressive character of the wave. We have already seen in Sect. 68β, Eq. (68.40), that if the wave takes a progressive character in the external layers, the resulting dissipation, even in the linear approximation, can be quite considerable. There is however another aspect which deserves some comments. To explain the phase shift between δL and δr, SCHWARZSCHILD [*58*] assumed that the solution of the adiabatic Eq. (58.1) could become progressive in the external layers of the star. This would violate the boundary conditions (58.2), but our purpose here is not to discuss the meaning of this type of solution but to estimate its effects on vibrational stability. If we write

$$\xi = \xi_0(r)\cos[\sigma t - \varphi(r)] \tag{72.1}$$

we have

$$\frac{\delta\varrho}{\varrho} = -\,\xi_0(r)\left\{\left(3+\frac{r}{\xi_0}\frac{d\xi_0}{dr}\right)\cos[\sigma t-\varphi(r)] + r\varphi'(r)\sin[\sigma t-\varphi(r)]\right\}, \tag{72.2}$$

δT and δp being in phase with $\delta\varrho$.

If the progressive character manifests itself only in the external layers, it will affect only the term in δL. Using (63.20), we can write, inverting the order of integration,

$$(\Delta K)_P = -\int_0^R dr_0 \int_0^P \frac{\delta T}{T}\,\frac{d\,\delta L}{dr_0}\,dt.$$

From the expression (63.9) of δL, its phase $\Theta(r_0)$ will not be exactly equal to the phase $\vartheta(r_0)$ of δT because of the terms in δr and $d(\delta T)/dr_0$. We have explicitly,

$$(\Delta K)_P = -\int_0^R dr_0 \int_0^P \left(\frac{\delta T}{T}\right)_0 \cos[\sigma t - \vartheta(r_0)]\,\frac{d}{dr_0}\{\delta L_0\cos[\sigma t - \Theta(r_0)]\}\,dt.$$

Developing and integrating, we find

$$(\Delta K)_P = -\frac{\pi}{\sigma}\int_0^R \left(\frac{\delta T}{T}\right)_0 \left[\frac{d\,\delta L_0}{dr_0}\cos(\Theta-\vartheta) + \delta L_0\sin(\Theta-\vartheta)\,\frac{d\Theta}{dr_0}\right] dr_0. \tag{72.3}$$

If $(\Theta-\vartheta)$ is exactly zero, the dissipation reduces to its usual expression in terms of the amplitudes. In general, $(\Theta-\vartheta)$ will be small since the dominating terms in δL are proportional to δT and $\delta\varrho$. In particular, this is true of the numerical solutions used by SCHWARZSCHILD. In that case, the progressive character of the adiabatic wave as such will have very little influence on the stability.

In this respect, ROSSELAND (cf. [*56a*], § 3) has discussed some interesting examples of purely progressive waves, such as the increase of the amplitude of an acoustic wave propagating in a radioactive gas. But the discussion has not been extended to the general case despite its obvious interest.

IV. Non-radial oscillations of a gaseous sphere under its own gravitation.

73. General introductory remarks. The general discussion of the non-radial oscillations of a spherical compressible mass presents a rather complex mathematical problem even if we ignore rotation or other perturbing factors such as magnetic fields or gravitational interaction with a companion.

As far as the most typical regular variable stars are concerned, the high symmetry of their observed properties (cf. Sects. 11 and 17) favours certainly the hypothesis of purely radial pulsations.

However *dynamical instability* leading to "explosions" might be easier to reach for some modes of non-radial oscillations than for purely radial pulsation and be of importance for the interpretation of some of the more violent intrinsic variable stars. On the other hand, a strong rotation or magnetic field could hinder considerably any spherically symmetrical pulsation. Interactions between the components of a binary will directly introduce a non-radial perturbation which, in the case of resonance, will lead to non-radial oscillations of one of the components. For all these reasons, it is likely that in the future the study of non-radial oscillations will gather interest. But up to the present, relatively little work has been devoted to it and few applications to stellar variability have been attempted.

In the case of an incompressible sphere, only non-radial oscillations are possible. If the mass is further homogeneous, the problem is considerably simplified as the perturbation Φ' of the gravitational potential reduces to the potential of a surface distribution of mass with density $\varrho(\delta r)_R$. If such a configuration is submitted to a perturbation represented by a series of spherical harmonics having for general term

$$\left.\begin{aligned} f'(r,\vartheta,\varphi) &= f'(r)\,P_l^m(\cos\vartheta)\,\mathrm{e}^{im\varphi} = f'(r)\,Y_l^m(\vartheta,\varphi)\,,\\ m &= -l\,,\ \ldots,\ -1\,,\ 0\,,\ 1\,,\ \ldots l \end{aligned}\right\} \tag{73.1}$$

where the P_l^m are the associated Legendre polynomials, KELVIN [*11*] found frequencies independent of m, given by

$$\sigma_l^2 = \frac{4\pi G\varrho}{3}\,l\,\frac{2l-2}{2l+1}\,. \tag{73.2}$$

We shall always consider $l \geq 2$ since for $l=0$, we fall back on radial pulsation and $l=1$ corresponds to a shifting of the centre of gravity.

ROSSELAND [*56b*] extended the discussion to the case of an incompressible heterogeneous sphere in which the density is a function of r. This case already reveals the main feature of the general problem, namely that the exact elimination of Φ' leads to a differential equation of the fourth order, the coefficients of which depend in a complicated way of the frequency σ. Following EMDEN[1], ROSSELAND remarks that the error introduced by the neglect of Φ' is not very large especially for higher spherical harmonics and that it is maximum in the homogeneous model where it corresponds to dropping $(2l-2)/(2l+1)$ in (73.2), a factor which tends rapidly to unity as l increases. This approximation leads to a second order differential equation which is more tractable.

For compressible masses, the problem becomes even more complicated except in a few special cases and the approximation $\Phi' \approx 0$ introduced by EMDEN has nearly always been used. EMDEN's own investigation in the particular case of a polytropic gas was vitiated by the use of the continuity equation in

[1] R. EMDEN: Gaskugeln, p. 448. Leipzig: Teubner 1907.

a form appropriate for a homogeneous incompressible fluid. Later work will be reviewed in the course of the following sections.

74. General equations. If the perturbation is applied to an equilibrium state, the discussion must rest on the basic Eqs. (56.12) to (56.16). The time may be separated again as in (57.14) and, neglecting all effects of turbulence, except perhaps the turbulent viscosity, we have in terms of the amplitudes:

a) the continuity Eq. (56.12):

$$\varrho' + \operatorname{div}(\varrho\,\delta\boldsymbol{r}) = 0, \tag{74.1}$$

or in spherical polar coordinates

$$\left.\begin{aligned} &\frac{\varrho'}{\varrho} + \frac{\delta r}{\varrho}\frac{\partial\varrho}{\partial r} + \frac{1}{r^2}\frac{\partial}{\partial r}(r^2\,\delta r) + \frac{1}{r\sin\vartheta}\frac{\partial}{\partial\vartheta}(\sin\vartheta\, r\,\delta\vartheta) + \\ &\qquad + \frac{1}{r\sin\vartheta}\frac{\partial}{\partial\varphi}(r\sin\vartheta\,\delta\varphi) = 0. \end{aligned}\right\} \tag{74.2}$$

b) the equation of motion (56.13):

$$\sigma^2\,\delta\boldsymbol{r} = \operatorname{grad}\Phi' - \frac{\varrho'}{\varrho^2}\operatorname{grad}p + \frac{1}{\varrho}\operatorname{grad}p' - \frac{i\sigma}{\varrho}\operatorname{div}\mathfrak{P}(\delta\boldsymbol{r}), \tag{74.3}$$

or in components

$$\sigma^2\,\delta r = \frac{\partial\Phi'}{\partial r} - \frac{\varrho'}{\varrho^2}\frac{\partial p}{\partial r} + \frac{1}{\varrho}\frac{\partial p'}{\partial r} - \frac{i\sigma}{\varrho}\sum_i \varDelta_i\,\mathfrak{P}^{ir}(\delta\boldsymbol{r}), \tag{74.4}$$

$$\sigma^2 r\,\delta\vartheta = \frac{\partial}{r\,\partial\vartheta}\left(\Phi' + \frac{p'}{\varrho}\right) - \frac{i\sigma r}{\varrho}\sum_i \varDelta_i\,\mathfrak{P}^{i\vartheta}(\delta\boldsymbol{r}), \tag{74.5}$$

$$\sigma^2 r\sin\vartheta\,\delta\varphi = \frac{1}{r\sin\vartheta}\frac{\partial}{\partial\varphi}\left(\Phi' + \frac{p'}{\varrho}\right) - \frac{i\sigma r\sin\vartheta}{\varrho}\sum_i \varDelta_i\,\mathfrak{P}^{i\varphi}(\delta\boldsymbol{r}), \tag{74.6}$$

using (57.1) and remembering that, at equilibrium, all variables depend on r only;

c) the equation of conservation of energy (56.15)

$$p' + \delta r\frac{\partial p}{\partial r} - \frac{\Gamma_1 p}{\varrho}\left(\varrho' + \delta r\frac{\partial\varrho}{\partial r}\right) = \frac{\Gamma_3 - 1}{i\sigma}\,\varrho\,\{\ldots\}' \tag{74.7}$$

d) POISSON's equation

$$\nabla^2\Phi' = 4\pi G\varrho'. \tag{74.8}$$

With the help of (74.7), Eq. (74.4) can be written

$$\left.\begin{aligned} \sigma^2\,\delta r &= \frac{\partial}{\partial r}\left(\Phi' + \frac{p'}{\varrho}\right) + A\left(\frac{p'}{\varrho} + \frac{\delta r}{\varrho}\frac{\partial p}{\partial r}\right) + \\ &\quad + \frac{\Gamma_3 - 1}{i\sigma}\frac{1}{\Gamma_1 p}\frac{\partial p}{\partial r}\{\ldots\}' - \frac{i\sigma}{\varrho}\sum_i \varDelta_i\,\mathfrak{P}^{ir}(\delta\boldsymbol{r}) \end{aligned}\right\} \tag{74.9}$$

or using (74.1)

$$\left.\begin{aligned} \sigma^2\,\delta r &= \frac{\partial}{\partial r}\left(\Phi' + \frac{p'}{\varrho}\right) - A\frac{\Gamma_1 p}{\varrho}\operatorname{div}\delta\boldsymbol{r} + \\ &\quad + \frac{\Gamma_3 - 1}{i\sigma}\frac{1}{\varrho}\frac{\partial\varrho}{\partial r}\{\ldots\}' - \frac{i\sigma}{\varrho}\sum_i \varDelta_i\,\mathfrak{P}^{ir}(\delta\boldsymbol{r}) \end{aligned}\right\} \tag{74.10}$$

where

$$A = \frac{1}{\varrho}\frac{\partial\varrho}{\partial r} - \frac{1}{\Gamma_1 p}\frac{\partial p}{\partial r}. \tag{74.11}$$

The corresponding vectorial equations can also be written down immediately

$$\left.\begin{aligned}\sigma^2\,\delta\boldsymbol{r} &= \operatorname{grad}\left(\Phi' + \frac{p'}{\varrho}\right) + \boldsymbol{A}\left(\frac{p'}{\varrho} + \frac{\delta\boldsymbol{r}}{\varrho}\operatorname{grad}p\right) + \\ &\quad + \frac{\Gamma_3-1}{i\sigma}\,\frac{1}{\Gamma_1 p}\operatorname{grad}p\{\ldots\}' - \frac{i\sigma}{\varrho}\operatorname{div}\mathfrak{P}(\delta\boldsymbol{r})\end{aligned}\right\}\tag{74.12}$$

or

$$\left.\begin{aligned}\sigma^2\,\delta\boldsymbol{r} &= \operatorname{grad}\left(\Phi' + \frac{p'}{\varrho}\right) - \boldsymbol{A}\,\frac{\Gamma_1 p}{\varrho}\operatorname{div}\delta\boldsymbol{r} + \\ &\quad + \frac{\Gamma_3-1}{i\sigma}\,\frac{1}{\varrho}\operatorname{grad}\varrho\{\cdots\}' - \frac{i\sigma}{\varrho}\operatorname{div}\mathfrak{P}(\delta\boldsymbol{r}),\end{aligned}\right\}\tag{74.13}$$

where $\boldsymbol{A}$ is a vector with a radial component only given by (74.11).

75. Non-radial adiabatic oscillations. In this approximation, we shall neglect the dissipation due to viscosity, energy sources and energy transfer. Let us assume further that $\delta\boldsymbol{r}$, ϱ', p' and Φ' are developed in series of spherical surface harmonics, satisfying the well-known equation

$$\frac{1}{\sin^2\vartheta}\,\frac{\partial^2 Y_l^m}{\partial\varphi^2} + \frac{1}{\sin\vartheta}\,\frac{\partial}{\partial\vartheta}\left(\sin\vartheta\,\frac{\partial Y_l^m}{\partial\vartheta}\right) + l(l+1)\,Y_l^m = 0.$$

In that case, the equation of motion

$$\sigma^2\,\delta\boldsymbol{r} = \operatorname{grad}\Phi' - \frac{\varrho'}{\varrho^2}\operatorname{grad}p + \frac{1}{\varrho}\operatorname{grad}p' \tag{75.1}$$

may be written

$$\sigma^2\,\delta\boldsymbol{r} = \operatorname{grad}\chi + \boldsymbol{A}\,(\chi - \Phi' - \delta r\, g) \tag{75.2}$$

or

$$\sigma^2\,\delta\boldsymbol{r} = \operatorname{grad}\chi - \boldsymbol{A}\,\frac{\Gamma_1 p}{\varrho}\,\alpha \tag{75.3}$$

where

$$\chi = \Phi' + \frac{p'}{\varrho};\qquad \alpha = \operatorname{div}\delta\boldsymbol{r}\quad\text{and}\quad g = -\frac{1}{\varrho}\,\frac{\partial p}{\partial r} = \frac{G\,m(r)}{r^2}. \tag{75.4}$$

They have the scalar components

$$\sigma^2\,\delta r = \frac{\partial\chi}{\partial r} + A(\chi - \Phi' - \delta r\, g), \tag{75.5}$$

or

$$\sigma^2\,\delta r = \frac{\partial\chi}{\partial r} - A\,\frac{\Gamma_1 p}{\varrho}\,\alpha, \tag{75.6}$$

$$\sigma^2\, r\,\delta\vartheta = \frac{1}{r}\,\frac{\partial\chi}{\partial\vartheta} = \frac{\chi(r)}{r}\,\frac{\partial Y_l^m}{\partial\vartheta}, \tag{75.7}$$

$$\sigma^2\, r\sin\vartheta\,\delta\varphi = \frac{1}{r\sin\vartheta}\,\frac{\partial\chi}{\partial\varphi} = \frac{\chi(r)}{r\sin\vartheta}\,\frac{\partial Y_l^m}{\partial\varphi}. \tag{75.8}$$

Using the definition of the Y_l^m and Eqs. (75.7) and (75.8), Eq. (74.2) gives

$$\alpha = \operatorname{div}\delta\boldsymbol{r} = -\left(\frac{\varrho'}{\varrho} + \frac{\delta r}{\varrho}\,\frac{\partial\varrho}{\partial r}\right) = \frac{1}{r^2}\,\frac{\partial}{\partial r}(r^2\,\delta r) - \frac{l(l+1)}{\sigma^2 r^2}\,\chi. \tag{75.9}$$

Eliminating δr by (75.6) or, more directly, taking the divergence of (75.3), one obtains also

$$\left.\begin{aligned}\alpha &= \frac{1}{\sigma^2}\left[\nabla^2\chi - \frac{1}{r}\,\frac{\partial}{\partial r}\left(r^2 A\,\frac{\Gamma_1 p}{\varrho}\,\alpha\right)\right] \\ &= \frac{1}{\sigma^2}\left[\frac{1}{r^2}\,\frac{\partial}{\partial r}\left(r^2\,\frac{\partial\chi}{\partial r}\right) - \frac{l(l+1)}{r^2}\,\chi - \frac{1}{r^2}\,\frac{\partial}{\partial r}\left(r^2 A\,\frac{\Gamma_1 p}{\varrho}\,\alpha\right)\right].\end{aligned}\right\}\tag{75.10}$$

Finally, the energy equation and POISSON's equation can be written

$$p' + \delta r \frac{\partial p}{\partial r} = -\frac{\Gamma_1 p}{\varrho} \alpha \tag{75.11}$$

or

$$\frac{p'}{\varrho} = \frac{\Gamma_1 p}{\varrho}\left(\frac{\varrho'}{\varrho} + \delta r\, A\right) \tag{75.12}$$

and

$$\left.\begin{aligned} \nabla^2 \Phi' &= \frac{1}{r}\frac{\partial}{\partial r}\left(r^2 \frac{\partial \Phi'}{\partial r}\right) - \frac{l(l+1)}{r^2}\Phi' \\ &= -4\pi G\left(\varrho\,\alpha + \delta r \frac{\partial \varrho}{\partial r}\right) = 4\pi G\,\varrho\left(\frac{p'}{\Gamma_1 p} - \delta r\, A\right). \end{aligned}\right\} \tag{75.13}$$

The form of these equations suggests that α or χ are likely to be the most convenient dependent variables to use in the general differential equation.

PEKERIS [59] who was the first to carry out completely the elimination, derived a fourth order differential equation in α which is however too complicated to be reproduced here.

One might as well proceed with the variable χ. Solving (75.5) with respect to δr gives

$$\delta r = \frac{\dfrac{\partial \chi}{\partial r} + A(\chi - \Phi')}{\sigma^2 + A\,g}. \tag{75.14}$$

On the other hand, solving (75.6) for α and eliminating δr by (75.14) leads to

$$\alpha = \frac{\varrho}{\Gamma_1 p}\,\frac{g\dfrac{\partial \chi}{\partial r} - \sigma^2 \chi + \sigma^2 \Phi'}{\sigma^2 + A\,g}. \tag{75.15}$$

Equating this to (75.9) where δr is again expressed by (75.14), we obtain

$$\left.\begin{aligned} &\frac{1}{r^2}\frac{\partial}{\partial r}\left[\frac{r^2}{\sigma^2 + A\,g}\left(\frac{\partial \chi}{\partial r} + A\,\chi\right)\right] - \frac{1}{r^2}\frac{\partial}{\partial r}\left(\frac{r^2 A\,\Phi'}{\sigma^2 + A\,g}\right) - \frac{l(l+1)}{\sigma^2 r^2}\chi \\ &\quad = \frac{\varrho}{\Gamma_1 p}\,\frac{\left(g\dfrac{\partial \chi}{\partial r} - \sigma^2 \chi + \sigma^2 \Phi'\right)}{\sigma^2 + A\,g} \end{aligned}\right\} \tag{75.16}$$

which may be solved for $\partial \Phi'/\partial r$ giving

$$\left.\begin{aligned} \frac{\partial \Phi'}{\partial r} &= \frac{1}{A}\nabla^2 \chi + \frac{\partial \chi}{\partial r}\,\frac{1}{A}\,\frac{\partial}{\partial r}\ln\left(\frac{\varrho}{\sigma^2 + A\,g}\right) + \\ &\quad + \chi\left[\frac{\partial}{\partial r}\ln\left(\frac{r^2 A}{\sigma^2 + A\,g}\right) - \frac{l(l+1)\,g}{r^2 \sigma^2} + \frac{\sigma^2 \varrho}{A\,\Gamma_1 p}\right] - \\ &\quad - \Phi'\left[\frac{\sigma^2 \varrho}{\Gamma_1 p\, A} + \frac{\partial}{\partial r}\ln\frac{r^2 A}{\sigma^2 + A\,g}\right]. \end{aligned}\right\} \tag{75.17}$$

From this, we may compute $\dfrac{\partial}{\partial r}\left(r^2 \dfrac{\partial \Phi'}{\partial r}\right)$ and introduce it in (75.13) where, in the second member δr and α are replaced by their values (75.14) and (75.15). If we write

$$m = \ln\left(\frac{\varrho}{\sigma^2 + A\,g}\right), \quad n = \ln\left(\frac{A\,r^2}{\sigma^2 + A\,g}\right)$$

and denote their derivatives by n', m' etc., this procedure leads to

$$\left.\begin{aligned}
&\Phi'\left[\frac{l(l+1)}{r^2}+\frac{2}{r}\left(\frac{\sigma^2\varrho}{\Gamma_1 p A}+n'\right)+\frac{\partial}{\partial r}\left(\frac{\sigma^2\varrho}{\Gamma_1 p A}\right)+n''-\left(\frac{\sigma^2\varrho}{\Gamma_1 p A}+n'\right)^2-\right.\\
&\left.\quad-\frac{4\pi G\varrho}{\sigma^2+Ag}\left(\frac{\sigma^2\varrho}{\Gamma_1 p}-\frac{A}{\varrho}\frac{\partial\varrho}{\partial r}\right)\right]\\
&=\frac{1}{r^2 A}\frac{\partial}{\partial r}(r^2 \Delta\chi)-\frac{\nabla^2\chi}{A}\left(\frac{1}{A}\frac{\partial A}{\partial r}+n'+\frac{\sigma^2\varrho}{\Gamma_1 p A}\right)+\\
&\quad+\frac{1}{r^2}\frac{\partial}{\partial r}\left(r^2\frac{\partial\chi}{\partial r}\right)\frac{m'}{A}+\frac{1}{A}\frac{\partial\chi}{\partial r}\left[\frac{4\pi G\varrho A^2}{\sigma^2+Ag}+m''-\frac{m'}{A}\frac{\partial A}{\partial r}+\right.\\
&\quad\left.+n' A-\frac{l(l+1)gA}{\sigma^2 r^2}+\frac{\sigma^2\varrho}{\Gamma_1 p}-m'\left(\frac{\sigma^2\varrho}{\Gamma_1 p A}+n'\right)\right]+\\
&\quad+\chi\left[\frac{2}{r}n'+n''-\frac{l(l+1)}{\sigma^2 r^2}\frac{\partial g}{\partial r}+\frac{\sigma^2}{r^2}\frac{\partial}{\partial r}\left(\frac{\varrho r^2}{\Gamma_1 p A}\right)-\right.\\
&\quad\left.-\left(\frac{\sigma^2\varrho}{\Gamma_1 p A}+n'\right)\left(n'+\frac{\sigma^2\varrho}{\Gamma_1 p A}-\frac{l(l+1)g}{\sigma^2 r^2}\right)-\frac{4\pi G\varrho}{\sigma^2+Ag}\left(\frac{\sigma^2\varrho}{\Gamma_1 p}-\frac{A}{\varrho}\frac{\partial\varrho}{\partial r}\right)\right].
\end{aligned}\right\}\qquad(75.18)$$

This may now be reintroduced into (75.17) giving finally a fourth order equation in χ. Once it is solved, Φ' can be derived from (75.18), p' from (75.4) and the displacements from the equations of motion (75.5) to (75.8).

These solutions must be determined, subject to the boundary condition

$$\delta r=0 \quad \text{in} \quad r=0 \qquad (75.19)$$

which according to (75.5) implies

$$\frac{\partial\chi}{\partial r}=0 \quad \text{in} \quad r=0 \qquad (75.20)$$

since $A\to 0$ as $r\to 0$. Furthermore, the Eqs. (75.7) and (75.8) show that we must also have

$$\chi=0 \quad \text{in} \quad r=0. \qquad (75.21)$$

But since Φ' on its own tends always towards zero at least as r^2, condition (75.21) is equivalent to

$$\Phi'=0, \quad \frac{p'}{\varrho}=0 \quad \text{in} \quad r=0. \qquad (75.22)$$

This is one of the important characteristics of non-radial oscillations which always have a loop as well as a node at the origin.

At the surface, we must have

$$\delta p=p'+\delta r\frac{\partial p}{\partial r}=0 \quad \text{in} \quad r=R \qquad (75.23)$$

but according to (75.11), (75.15) this is equivalent to

$$-\varrho\frac{\left[g\frac{\partial\chi}{\partial r}-\sigma^2\chi+\sigma^2\Phi'\right]}{(\sigma^2+Ag)}=0 \quad \text{in} \quad r=R. \qquad (75.24)$$

In physical models $\varrho\to 0$ and $A\to\infty$ in $r=R$ and since Φ' is given by (75.18), the condition (75.24) will certainly be satisfied, if χ and its first three derivatives remain finite.

On the other hand, if we denote by indices i and e the values of the potential on the internal and external sides of the surface, the continuity of Φ across $r=R$ gives

$$\Phi'_{i,R}=\Phi'_{e,R}=\frac{C}{R^{l+1}} \qquad (75.25)$$

since, corresponding to a surface harmonic Y_l^m, Φ'_e will be of the form C/r^{l+1}. Furthermore, we must also have

$$\left(\frac{\partial \Phi'_i}{\partial r} + \delta r \frac{\partial^2 \Phi'_i}{\partial r^2}\right)_R = \left(\frac{\partial \Phi'_e}{\partial r} + \delta r \frac{\partial^2 \Phi'_e}{\partial r^2}\right)_R$$

or, eliminating the second derivatives by the appropriate POISSON's equations and remembering the form of Φ'_e,

$$\left(\frac{\partial \Phi'_i}{\partial r}\right)_R + \frac{l+1}{R} \Phi'_{i,R} = -(4\pi G \varrho\, \delta r)_R. \tag{75.26}$$

Usually since $\varrho_R = 0$, the second member of (75.26) will vanish. Condition (75.26) may be transformed into a condition on χ and its derivatives by (75.18) and (75.17).

76. Special cases where the differential equation is of the second order only. *α) The homogeneous compressible model.* Apart from the incompressible sphere, the best known case is that of the homogeneous compressible model discussed by PEKERIS [*59*]. Here we have

$$\left.\begin{aligned} g &= -\frac{1}{\varrho}\frac{\partial p}{\partial r} = \frac{4\pi G \varrho r}{3}, \quad p = \frac{4\pi G \varrho^2}{6}(R^2 - r^2), \quad A = -\frac{1}{\Gamma_1 p}\frac{\partial p}{\partial r}, \\ \frac{\varrho'}{\varrho} &= -\alpha, \quad \frac{p'}{\varrho} = -\frac{\Gamma_1 p}{\varrho}\alpha - \frac{\delta r}{\varrho}\frac{\partial p}{\partial r} = \frac{4\pi G \varrho}{3} r\, \delta r - \frac{\Gamma_1 p}{\varrho}\alpha \end{aligned}\right\} \tag{76.1}$$

and

$$\nabla^2 \Phi' = -4\pi G \varrho \alpha. \tag{76.2}$$

Using these relations, Eq. (75.10) becomes

$$\omega^2 \alpha = \left\{\nabla^2 (r\, \delta r) - 3\alpha - \frac{1}{r^2}\frac{\partial}{\partial r}(\alpha r^3) - \nabla^2\left[\frac{\Gamma_1}{2}(R^2 - r^2)\alpha\right]\right\} \tag{76.3}$$

if, as before, we define

$$\omega^2 = \frac{3\sigma^2}{4\pi G \varrho}. \tag{76.4}$$

Solving (75.9) for χ and introducing it into (75.6) we get

$$\nabla^2 (r\, \delta r) = \frac{1}{r}\frac{\partial}{\partial r}(r^2 \alpha) + \frac{l(l+1)}{\omega^2}\alpha \tag{76.5}$$

if we note that

$$\nabla^2 (r\, \delta r) = \frac{1}{r}\frac{\partial^2 (r^2\, \delta r)}{\partial r^2} - \frac{l(l+1)}{r}\delta r.$$

With (76.5), Eq. (76.3), after some rearrangements, may be written

$$\left.\begin{aligned} &(1 - x^2)\frac{\partial^2 \alpha}{\partial x^2} + \frac{\partial \alpha}{\partial x}\frac{2 - 6x^2}{x} + \\ &\quad + \alpha\left[\frac{2\omega^2}{\Gamma_1} + \frac{8}{\Gamma_1} - 6 - \frac{2l(l+1)}{\Gamma_1 \omega^2} - l(l+1)\frac{1 - x^2}{x^2}\right] = 0 \end{aligned}\right\} \tag{76.6}$$

with $x = r/R$.

PEKERIS has shown that the eigenfunctions are polynomials in even powers of x [cf. condition (75.20)]

$$\alpha_{l,k} = x^l \sum_{j=0}^{k} C_{2j} x^{2j}, \quad k = 0, 1, 2, \ldots \tag{76.7}$$

with

$$C_{2j} + 2 = C_{2j}\left[\frac{2j(2j + 5 + 2l) - B}{2(j+1)(2j + 3 + 2l)}\right] \tag{76.8}$$

where

$$B = -6 - 4l + \frac{8}{\Gamma_1} + \frac{2\omega^2}{\Gamma_1} - \frac{2l(l+1)}{\Gamma_1 \omega^2}.$$

The eigenvalues $\omega_{l,k}^2$ are given by the condition

$$C_{2k+2} = 2k(2k+5+2l) - B = 0 \tag{76.9}$$

or explicitly

$$\omega^2 - \frac{l(l+1)}{\omega^2} = -4 + \Gamma_1[k(2k+5+2l) + 3 + 2l] \equiv 2D_k.$$

The solutions are

$$\omega_{l,k}^2 = D_k \pm [D_k^2 + l(l+1)]^{\frac{1}{2}}, \tag{76.10}$$

one of which is always negative and corresponds to a dynamically unstable mode. If k becomes large, the asymptotic values of the roots (76.10) are

$$\left.\begin{aligned} \omega_{l,k}^2 &= 2D_k + \frac{1}{2}\frac{l(l+1)}{D_k^2}, \\ \text{and} \qquad \omega_{l,k}^2 &= -\frac{1}{2}\frac{l(l+1)}{D_k^2}. \end{aligned}\right\} \tag{76.11}$$

Thus for any given l, we have two spectra, one of positive eigenvalues tending towards infinity as k increases and one of negative eigenvalues tending towards zero as k increases. We note that to one pair of associated positive and negative eigenvalues, there corresponds only one eigenfunction $\alpha_{l,k}$. The same is true of the $\Phi'_{l,k}$ but the displacements $\delta r_{l,k}$ are different as shown by their general expressions[1]:

$$\Phi'_{l,k} = \frac{4\pi G \varrho R^2}{B + 4l + 6}\alpha_{l,k}(x^2 - 1), \tag{76.12}$$

$$\delta r_{l,k} = \frac{1}{B + 4l + 6}\left\{\frac{d}{dx}[\alpha_{l,k}(x^2-1)] + \frac{l(l+1)}{\omega_{l,k}^2 x}\alpha_{l,k}(x^2-1)\right\}. \tag{76.13}$$

These solutions have been represented graphically by E. SAUVENIER-GOFFIN[1] for $k = 0, 1, 2$ and $l = 2, 3, 4$. The most significant characteristic is that for the stable modes ($\omega^2 > 0$), δr has one more node than α, while for unstable modes α and δr have the same number of nodes. To appreciate the errors introduced by the neglect of Φ', E. SAUVENIER-GOFFIN has also solved the problem in that case and Table 15 presents a comparison of the exact and approached eigenvalues.

Table 15. *Exact and approximate eigenvalues* $(\omega_l^2)_k$.

l	$k=0$			$k=1$			$k=2$		
	Exact	Approx.	$\Delta\omega^2/\omega^2$	Exact	Approx.	$\Delta\omega^2/\omega^2$	Exact	Approx.	$\Delta\omega^2/\omega^2$
2	8.39	11.20	0.335	26.23	29.205	0.113	51.12	54.111	0.058
	−0.73	−0.534	−0.268	−0.225	−0.205	−0.089	−0.117	−0.111	−0.051
3	12.00	14.810	0.234	33.03	36.0	0.090	61.20	64.187	0.049
	−1.00	−0.810	−0.190	−0.363	−0.333	−0.082	−0.196	−0.187	−0.046
4	15.61	18.419	0.179	39.84	42.801	0.074	71.28	74.269	0.042
	−1.281	−1.086	−0.152	−0.502	−0.467	−0.069	−0.281	−0.269	−0.041

The relative error is of the same order for stable and unstable modes and it decreases with l and k increasing. For the lowest mode ($k=0$) it implies an error

[1] E. SAUVENIER-GOFFIN: Bull. Soc. Roy. Sci. Liège **20**, 20 (1951).

of some 15% on the frequencies and this is already appreciably smaller than in the case of the incompressible sphere [cf. Eq. (73.2) and the discussion following]. The smallness of the error might seem somewhat startling at first sight since the term $\nabla^2\Phi'$ which is proportional to ϱ' is not strictly negligible. But simultaneously with $\nabla^2\Phi'$ other terms such as $\nabla^2\varrho'$ appear which are proportional to $l(l+1)\,\varrho'$ and apparently, even for $l=2$, the factor $l(l+1)$ is large enough to reduce $\nabla^2\Phi'$ to a subsidiary rôle.

If one adds to the compressibility an appreciable central condensation of the model one may expect that these errors will decrease further. As far as δr is concerned, the approximation $\Phi'=0$ is excellent in the cases $\omega^2>0$, but much poorer when $\omega^2<0$. The possibility of dynamical instability corresponds here to the fact that for a compressible fluid, an homogeneous configuration is not a state of minimum potential energy. It can also be interpreted from the point of view of SCHWARZSCHILD's criterion of convective stability

$$A<0. \tag{76.14}$$

Here A which reduces to $-(1/\Gamma_1 p)(dp/\partial r)$ is positive and whenever (76.14) is violated in the whole star, we shall see that unstable modes exist.

β) Roche's model. The differential equation is also of the second order in the case of ROCHE's model and can be written directly with $\xi=\delta r/r$ and $x=r/R$ as

$$\left.\begin{aligned} &\frac{d^2\xi}{dx^2}(1-x^3)+\frac{d\xi}{dx}\left\{\frac{1-4x^3}{x}-\frac{3\Gamma_1}{x}\,\frac{l(l+1)(1-x^3)}{[3\omega^2x^3-l(l+1)\Gamma_1(1-x^3)]}\right\}+\\ &\quad+\xi\left\{\left[3\omega^2x^3-l(l+1)(1-x^3)\Gamma_1\right]\frac{1}{\Gamma_1x^2}-\frac{3}{x^2}\left(3-\frac{4}{\Gamma_1}\right)-\right.\\ &\quad-\frac{3l(l+1)}{\omega^2\Gamma_1x^5}-\frac{3}{x^2}\,\frac{l(l+1)[3\Gamma_1(1-x^3)-5]}{[3\omega^2x^3-l(l+1)(1-x^3)\Gamma_1]}-\\ &\quad\left.-\frac{6l(l+1)}{3\omega^2x^5}\cdot\frac{3\omega^2x^6+l(l+1)(1-x^3)^2\Gamma_1}{3\omega^2x^3-l(l+1)(1-x^3)\Gamma_1}\right\}=0.\end{aligned}\right\} \tag{76.15}$$

Unfortunately, this equation has non-regular singularities in $x=0$, except if $\Gamma_1=\frac{3}{2}$. But, even in that case, it does not seem to admit simple solutions. It is rather a pity as this equation presents many characteristic features of the more general problem.

A numerical discussion of this equation is also lacking.

77. The heterogeneous incompressible model. Like the models treated in Sect. 76, the heterogeneous incompressible model has no direct application to stars, but like the others it is useful to illustrate some properties of the problem. We have recalled that ROSSELAND [*56b*] established the correct fourth order equation in this case. Neglecting Φ', he derived an approximate second order equation for p'. In this case, however, the equation in δr is easier to discuss, at least for the spectrum converging towards zero.

Since here $\alpha=\operatorname{div}\boldsymbol{\delta r}=0$, Eqs. (74.1) and (75.9) where Φ' is neglected $(\chi\to p'/\varrho)$ can be written

$$\varrho'=-\delta r\,\frac{\partial\varrho}{\partial r}, \tag{77.1}$$

$$\frac{1}{r^2}\,\frac{\partial}{\partial r}(r^2\,\delta r)-\frac{l(l+1)}{\sigma^2r^2}\,\frac{p'}{\varrho}=0. \tag{77.2}$$

With ϱ' and p' given by (77.1) and (77.2), the first equation of motion (74.4) when neglecting Φ' and friction becomes

$$\frac{d}{dr}\left(\varrho\,\frac{du}{dr}\right)-u\left[\frac{l(l+1)\,g}{\sigma^2r^2}\,\frac{d\varrho}{dr}+\frac{l(l+1)}{r^2}\,\varrho\right]=0 \tag{77.3}$$

with $u = r^2\,\delta r$ vanishing at the origin and remaining finite everywhere else. Multiplying (77.3) by u and integrating from 0 to R gives

$$\sigma^2 = -\frac{l(l+1)\int\limits_0^R g\,\frac{d\varrho}{dr}\,\frac{u^2}{r^2}\,dr}{\int\limits_0^R \varrho\left(\frac{du}{dr}\right)^2 dr + \int\limits_0^R \varrho\,\frac{u^2}{r^2}\,dr - \left(u\,\varrho\,\frac{du}{dr}\right)_R} \tag{77.4}$$

which is sufficient to discuss the sign of σ^2. In particular, if the density vanishes at the surface, we see that, as expected, the configuration will be stable ($\sigma^2 > 0$) or unstable ($\sigma^2 < 0$) according as whether an appropriate average of $d\varrho/dr$ is negative or positive.[1]

78. The convective model. α) *Perfect adiabatic equilibrium.* In that case, $A = 0$ everywhere and Eqs. (74.1), (75.2) and (75.12) reduce to

$$\varrho' = -\operatorname{div}(\varrho\,\delta\boldsymbol{r}), \tag{78.1}$$

$$\sigma^2\,\delta\boldsymbol{r} = \operatorname{grad}\chi, \tag{78.2}$$

$$\frac{p'}{\varrho} = \chi - \Phi' = \frac{\Gamma_1 p}{\varrho^2}\,\varrho'. \tag{78.3}$$

Multiplying (78.2) by ϱ and taking its divergence give

$$-\sigma^2\varrho' = \varrho\,\nabla^2\chi + \frac{\partial\varrho}{\partial r}\,\frac{\partial\chi}{\partial r} \tag{78.4}$$

or, using (78.3),

$$\sigma^2\Phi' = \frac{\Gamma_1 p}{\varrho}\left(\nabla^2\chi + \frac{1}{\varrho}\,\frac{\partial\varrho}{\partial r}\,\frac{\partial\chi}{\partial r}\right) + \sigma^2\chi. \tag{78.5}$$

If we take the Laplacian of both members, use Poisson's equation and eliminate again ϱ' with the help of (78.4), we finally obtain

$$\nabla^2\left(\frac{\Gamma_1 p}{\varrho}\,\nabla^2\chi\right) + \nabla^2\left(\frac{\Gamma_1 p}{\varrho}\,\frac{\partial\varrho}{\partial r}\,\frac{\partial\chi}{\partial r}\right) + \nabla^2\chi\,(\sigma^2 + 4\pi G\varrho) + 4\pi G\,\frac{\partial\varrho}{\partial r}\cdot\frac{\partial\chi}{\partial r} = 0 \tag{78.6}$$

which can also be written

$$\left.\begin{aligned} &\nabla^2\left(\frac{\Gamma_1 p}{\varrho}\,\nabla^2\chi + \frac{\Gamma_1 p}{\varrho^2}\,\frac{\partial\varrho}{\partial r}\,\frac{\partial\chi}{\partial r} + \sigma^2\chi\right) + \\ &\quad + \frac{4\pi G\varrho^2}{\Gamma_1 p}\left[\frac{\Gamma_1 p}{\varrho}\,\nabla^2\chi + \frac{\Gamma_1 p}{\varrho^2}\,\frac{\partial\varrho}{\partial r}\,\frac{\partial\chi}{\partial r} + \sigma^2\chi\right] - \frac{4\pi G\varrho^2}{\Gamma_1 p}\,\sigma^2\chi = 0. \end{aligned}\right\} \tag{78.7}$$

If one neglects Φ' completely and writes

$$\chi = \frac{p'}{\varrho} = y \tag{78.8}$$

Eq. (78.5) gives immediately

$$\frac{d^2y}{dr^2} + \frac{dy}{dr}\left(\frac{2}{r} + \frac{1}{\varrho}\,\frac{\partial\varrho}{\partial r}\right) + y\left[\frac{\sigma^2\varrho}{\Gamma_1 p} - \frac{l(l+1)}{r^2}\right] = 0. \tag{78.9}$$

If Φ' and $\partial\Phi'/\partial r$ are neglected, but not $\nabla^2\Phi'$, Eq. (78.5) leads to

$$\frac{d^2y}{dr^2} + \frac{dy}{dr}\left(\frac{2}{r} + \frac{1}{\varrho}\,\frac{\partial\varrho}{\partial r}\right) + y\left(\frac{\sigma^2\varrho}{\Gamma_1 p} - \frac{l(l+1)}{r^2} - \frac{4\pi G\varrho^2}{\Gamma_1 p}\right) = 0. \tag{78.10}$$

[1] Cf. also S. Chandrasokhar: Phil. Mag., Ser. VII, **43**, 1317 (1952); **44**, 233 (1953). — Quart. J. Mech. Appl. Math. **8**, 1 (1955).

The boundary condition (75.22) becomes

$$y = 0 \quad \text{in} \quad r = 0 \tag{78.11}$$

while Eq. (75.23) can be written

$$y = \frac{p'}{\varrho} = g\,\delta r \quad \text{in} \quad r = R \tag{78.12}$$

and the radial component of (78.2) shows that y and $\partial y/\partial r$ remain finite at the surface.

Eq. (78.9) or (78.10) are again of the type (58.4) and one may verify readily, taking the boundary conditions into account, that the eigenfunctions are orthogonal. Condition (58.21) applied to (78.9) shows that all eigenvalues σ^2 are positive.

In the case of (78.10), the same condition becomes

$$\int_0^R r^2 \left(\frac{dy}{dr}\right)^2 \varrho\, dr + \int_0^R \left[l(l+1) - \frac{4\pi G \varrho^2 r^2}{\Gamma_1 p}\right] y^2 \varrho\, dr > 0 \tag{78.13}$$

and it is certainly satisfied for high modes and large enough l. However, for small l and low modes, it is difficult to prove that the left-hand member remains always positive although it is certainly negatively bounded (cf. [*50b*], Chap. III, § 6).

β) Approach to adiabatic equilibrium. All the σ^2 found above belong to the family of eigenvalues tending towards infinity. But we may expect in general [cf. the discussion following Eqs. (76.11) and Sect. 77] the existence of another spectrum tending towards zero. In fact, it has been eliminated by setting A equal to zero.

Let us now suppose that A is not rigorously zero but can be represented by

$$A = \varepsilon\, a(r), \tag{78.14}$$

where ε is an arbitrarily small positive constant, and that eigenvalues exist that are also proportional to ε

$$\sigma^2 = \varepsilon \beta^2. \tag{78.15}$$

In that case, Eq. (75.5) where we neglect Φ' shows that we must have

$$y = \varepsilon z \tag{78.16}$$

and, neglecting second order terms in ε, it becomes

$$\beta^2 \delta r = \frac{\partial z}{\partial r} - a(r)\, \delta r\, g. \tag{78.17}$$

On the other hand, the piezotropy relation (75.12) implies

$$\frac{\varrho'}{\varrho} = \varepsilon\, q(r) \tag{78.18}$$

with

$$q(r) = \frac{\varrho}{\Gamma_1 p} z - a(r)\, \delta r \tag{78.19}$$

and (75.9) gives

$$-\frac{1}{r^2}\frac{\partial}{\partial r}(r^2 \delta r) + \frac{l(l+1)}{\beta^2 r^2} z - \frac{\delta r}{\varrho}\frac{\partial \varrho}{\partial r} = \varepsilon\, q(r) \to 0 \tag{78.20}$$

where the second member can be neglected. Solving for z and introducing in (78.17) leads to

$$\frac{d}{dr}\left(\varrho\frac{du}{dr}\right)+u\left[\varrho\frac{\partial}{\partial r}\left(\frac{1}{\varrho}\frac{\partial\varrho}{\partial r}\right)-\frac{a(r)\,g\,\varrho}{r^2}\,\frac{l(l+1)}{\beta^2}-\frac{l(l+1)\,\varrho}{r^2}\right]=0 \qquad (78.21)$$

with $u=r^2\,\delta r$ being zero or finite everywhere.

Once more the equation is of the type (58.4), with a parameter $\lambda=1/\beta^2$ which admits an infinite set of discrete values increasing indefinitely, i.e. β^2 decreases and tends towards zero.

It is easy to verify that the u_i are orthogonal and the sign of β^2 is most easily discussed from its expression

$$\beta^2=\frac{l(l+1)\int\limits_0^M a(r)\,g\,(\delta r)^2\,dm}{\int\limits_0^M r^2\frac{\partial}{\partial r}\left(\frac{1}{\varrho}\frac{\partial\varrho}{\partial r}\right)(\delta r)^2\,dm-\int\limits_0^M\left(\frac{1}{r}\frac{du}{dr}\right)^2 dm-l(l+1)\int\limits_0^M(\delta r)^2\,dm}\,. \qquad (78.22)$$

Let us first consider the case where the perturbation has small dimensions

$$l\gg 2\,,\qquad \frac{1}{r}\frac{du}{dr}\approx\frac{2r}{R}\,s\,\delta r\,,\qquad s\gg 1 \qquad (78.23)$$

where s is the number of times that δr changes sign along the radius $[\lambda\approx 2(R/s)]$. Then (78.22) reduces to

$$\beta^2=-\frac{l(l+1)\int\limits_0^M a(r)\,g\,(\delta r)^2\,dm}{s^2\int\limits_0^M\left(\frac{r}{R}\right)^2(\delta r)^2\,dm+l(l+1)\int\limits_0^M(\delta r)^2\,dm} \qquad (78.24)$$

which shows that dynamical stability or instability will prevail according as to whether the regions convectively stable $[a(r)<0$, cf. (76.14)$]$ or unstable $[a(r)>0]$ predominate in the star. If l is appreciably larger than s, (78.24) becomes

$$\beta^2=-\frac{\int\limits_0^M a(r)\,g\,(\delta r)^2\,dm}{\int\limits_0^M(\delta r)^2\,dm}\,, \qquad (78.25)$$

the numerator measuring simply the total gravitational energy transformable in kinetic energy due to convective instability.

On the other hand, as s increases with respect to l, β^2 decreases in absolute value so that the most favourable perturbation for dynamical instability is one of small horizontal extent and with a fairly large vertical wavelength, a result which has been emphasized by P. LEDOUX[1].

If neither l nor s is large, the discussion is more difficult because of the first term in the denominator of (78.22), but it would be interesting to follow it numerically. It would not be very difficult in that case to include terms of the order of ε^2 as well in the equations. As far as the possibilities of dynamical instability revealed by this analysis are concerned, the main question is whether,

[1] Cf. [*50b*], Chap. III, §§ 5 and 6; also, for the case of an atmosphere, A. SKUMANICH: Astrophys. Journ. **121**, 408 (1955).

as suggested at one time by BIERMANN[1], appreciable superdiabatic gradients can arise and persist for some time in fairly large regions inside a star. Of course, even where convective equilibrium is established, the actual gradient always remains somewhat superadiabatic [cf. Eq. (57.6)] but generally by an amount which is very small. Nevertheless, this question deserves to be explored thoroughly as it presents one of the rare instances of a dynamical instability with a physical meaning.

79. General models. A few years after PEKERIS' discussion, another important paper was published by COWLING [60] on the non-radial oscillations of polytropes. In this investigation, Φ' was again neglected in a first approximation, but its effects on the periods were later computed by the method of perturbation. The comparison between the exact and approached values for the homogeneous compressible model in Sect. 76 supports this procedure.

Here, we shall start directly with the general case[2] and particularize later to the polytropes. If in (75.9), we explicit χ and replace α by its value in terms of p' as given by (75.11), we obtain

$$\frac{du}{dr} - \frac{\varrho}{\Gamma_1 p} g u = \left[\frac{l(l+1)}{\sigma^2} - \frac{\varrho r^2}{\Gamma_1 p}\right] y + \frac{l(l+1)}{\sigma^2}\Phi' \tag{79.1}$$

with $u = r^2\,\delta r$ and $y = p'/\varrho$.

With the same notations, Eq. (75.5) can be written directly as

$$\frac{dy}{dr} + yA = \frac{1}{r^2}(\sigma^2 + A g)\,u - \frac{d\Phi'}{dr}. \tag{79.2}$$

If we add POISSON's equation (75.13) rewritten as

$$\frac{1}{r^2}\frac{d}{dr}\left(r^2\frac{d\Phi'}{dr}\right) - \frac{l(l+1)}{r^2}\Phi' = 4\pi G\varrho\left(\frac{\varrho}{\Gamma_1 p}y - \frac{uA}{r^2}\right) \tag{79.3}$$

we obtain what is probably the most practical fourth order system for a numerical integration of the complete problem.

If we neglect Φ', Eqs. (79.1) and (79.2) reduce to

$$\frac{du}{dr} - \frac{\varrho}{\Gamma_1 p} g u = p^{-1/\Gamma_1}\frac{d}{dr}\left(u\,p^{1/\Gamma_1}\right) = \left[\frac{l(l+1)}{\sigma^2} - \frac{\varrho r^2}{\Gamma_1 p}\right] y \tag{79.4}$$

and

$$\frac{dy}{dr} + yA = \left(p^{1/\Gamma_1}\varrho^{-1}\right)\frac{d}{dr}\left(y\,\varrho\,p^{-1/\Gamma_1}\right) = \frac{1}{r^2}(\sigma^2 + A g)\,u \tag{79.5}$$

with the boundary conditions

$$\left.\begin{array}{lll} \text{in } r=0: & u=0, & y=0, \\ \text{in } r=R: & u_R = R^2(\delta r)_R: \text{ finite}, & y_R = g(\delta r)_R. \end{array}\right\} \tag{79.6}$$

We may introduce the new variables

$$v = u\,p^{1/\Gamma_1} = r^2\,\delta r\,p^{1/\Gamma_1}, \qquad w = y\,\varrho\,p^{-1/\Gamma_1} = \frac{p'}{p^{1/\Gamma_1}} \tag{79.7}$$

which, in physical models $\left(\Gamma_1 > \frac{n_e+1}{n_e}\right.$ or $A<0$ close enough to the surface: radiative equilibrium$\Big)$ satisfy the boundary conditions

$$\left.\begin{array}{lll} \text{in } r=0: & v=0, & w=0, \\ \text{in } r=R: & v_R=0, & w_R=0. \end{array}\right\} \tag{79.8}$$

[1] L. BIERMANN: Z. Astrophys. **18**, 344 (1939).

[2] P. LEDOUX: III. Congrès National des Sciences, Bruxelles, Vol. II, Sect. 3, p. 133, 1950.

Note that in a model where $\Gamma_1 \geq \frac{n_e+1}{n_e}$ or $A \geq 0$ right up to the surface, w_R would either remain finite or become infinite.

With these notations, Eqs. (79.4) and (79.5) become

$$\frac{dv}{dr} = \left[\frac{l(l+1)}{\sigma^2} - \frac{\varrho r^2}{\Gamma_1 p}\right] \frac{p^{2/\Gamma_1}}{\varrho} w, \tag{79.9}$$

$$\frac{dw}{dr} = \frac{1}{r^2}(\sigma^2 + A g) \frac{\varrho}{p^{2/\Gamma_1}} v. \tag{79.10}$$

If we eliminate w or v, we obtain the equivalent second order equations

$$\frac{d}{dr}\left\{\frac{\varrho}{p^{2/\Gamma_1}\left[\frac{l(l+1)}{\sigma^2} - \frac{\varrho r^2}{\Gamma_1 p}\right]} \frac{dv}{dr}\right\} = \frac{1}{r^2}(\sigma^2 + A g) \frac{\varrho}{p^{2/\Gamma_1}} v, \tag{79.11}$$

$$\frac{d}{dr}\left[\frac{p^{2/\Gamma_1} r^2}{\varrho(\sigma^2 + A g)} \frac{dw}{dr}\right] = \left[\frac{l(l+1)}{\sigma^2} - \frac{\varrho r^2}{\Gamma_1 p}\right] \frac{p^{2/\Gamma_1}}{\varrho} w. \tag{79.12}$$

The singularities depending on σ^2 in

$$\left[\frac{l(l+1)}{\sigma^2} - \frac{\varrho r^2}{\Gamma_1 p}\right]_{r_1} = 0 \quad \text{for } v \tag{79.13}$$

and

$$(\sigma^2 + A g)_{r_2} = 0 \quad \text{for } w \tag{79.14}$$

are regular and, according (to 79.9) and (79.10), they correspond respectively to points where dv/dr or dw/dr vanish. In the same way, the nodes of v (or w) correspond to extrema of w (or v).

A rigorous discussion of these equations is lacking, but one may reproduce COWLING's reasoning for the polytropes. For a given l, if σ^2 is large, Eq. (79.11) tends towards the Sturm-Liouville type with a parameter $\lambda \equiv \sigma^2$. This suggests the existence of a spectrum of indefinitely increasing eigenvalues. The corresponding eigensolutions were denoted p-modes by Cowling because the motion which is chiefly radial is largely originated by the pressure variations. Conversely for σ^2 very small, Eq. (79.12) tends to define a Sturm-Liouville problem with $\lambda \equiv 1/\sigma^2$. This corresponds to a spectrum of eigenvalues decreasing towards zero as the order of the modes increases. The associated eigensolutions, called the g-modes by COWLING, correspond to a motion chiefly horizontal with small pressure variations.

COWLING also distinguishes a fundamental mode (f-mode) in which δr and ϱ' keep the same sign along any radius.

However, because of the singularities (79.13) and (79.14), this reasoning fails close enough to the surface or the centre. If we integrate (79.11) multiplied by v from 0 to R and take Eq. (79.8) into account, we obtain

$$\sigma^2 \int_0^R \frac{\varrho v^2}{r^2 p^{2/\Gamma_1}} dr + \int_0^R \frac{A g \varrho}{r^2 p^{2/\Gamma_1}} v^2 dr + \int_0^R \frac{\varrho\left(\frac{dv}{dr}\right)^2}{\left(\frac{l(l+1)}{\sigma^2} - \frac{\varrho r^2}{\Gamma_1 p}\right) p^{2/\Gamma_1}} dr = 0. \tag{79.15}$$

If for a given l, σ^2 is large, we may certainly neglect $l(l+1)/\sigma^2$ in the last term and we get for the p-modes

$$\sigma_p^2 = -\frac{\int\limits_0^R \frac{A\,g\,\varrho}{r^2\,p^{2/\Gamma_1}}\,v^2\,dr - \int\limits_0^R \frac{\Gamma_1 p^{1-2/\Gamma_1}}{r^2}\left(\frac{dv}{dr}\right)^2 dr}{\int\limits_0^R \frac{\varrho\,v^2}{r^2\,p^{2/\Gamma_1}}\,dr}. \tag{79.16}$$

For high enough modes, the second term in the numerator is the dominating one and σ_p^2 is certainly positive and increases indefinitely with the order of the mode. In the same way, if σ^2 is small, Eq. (79.12) gives

$$\frac{l(l+1)}{\sigma^2}\int\limits_0^R \frac{w^2}{\varrho}\,p^{2/\Gamma_1}\,dr - \int\limits_0^R \frac{r^2\,w^2}{\Gamma_1}\,p^{2/\Gamma_1-1}\,dr + \int\limits_0^R \frac{r^2\,p^{2/\Gamma_1}}{(\sigma^2+A\,g)\,\varrho}\left(\frac{dw}{dr}\right)^2 dr = 0 \tag{79.17}$$

where we may certainly neglect σ^2 in the denominator of the last term, obtaining

$$\sigma_g^2 = -\frac{l(l+1)\int\limits_0^R \frac{w^2}{\varrho}\,p^{2/\Gamma_1}\,dr}{\int\limits_0^R \frac{r^2\,p^{2/\Gamma_1}}{A\,g\,\varrho}\left(\frac{dw}{dr}\right)^2 dr - \int\limits_0^R \frac{r^2}{\Gamma_1}\,p^{2/\Gamma_1-1}\,w^2\,dr}. \tag{79.18}$$

As the order of the mode increases, the first term in the denominator predominates and

$$\sigma_g^2 = -\frac{l(l+1)\int_0^R w^2\,\varrho^{-1}\,p^{2/\Gamma_1}\,dr}{\int\limits_0^R r^2\,p^{2/\Gamma_1}\,(A\,g\,\varrho)^{-1}\left(\frac{dw}{dr}\right)^2 dr} \tag{79.19}$$

which shows that σ^2 decreases and tends towards zero with the order of the mode and that its sign is determined by the value of A. In particular, if A tends towards zero everywhere, (79.18) shows that all σ_g also tend towards zero. Again dynamical instability occurs if A is positive (convective instability) in a large enough region of the star and is stronger for modes of lower order. One should however be careful here when treating the relations obtained by integration of the differential equations as algebraic equations for the σ^2. For instance, multiplying (79.9) by w and (79.10) by v, adding and integrating from $r=0$ to $r=R$ gives

$$\sigma^4\int\limits_0^M \left(\frac{\delta r}{r}\right)^2 dI - \sigma^2\left[\int\limits_0^M A\,r\left(\frac{\delta r}{r}\right)^2 dW + \int\limits_0^{\mathcal{V}} \frac{1}{\Gamma_1}\left(\frac{p'}{p}\right)^2 p\,d\mathcal{V}\right] + $$
$$+ 4\pi\,l(l+1)\int\limits_0^R \left(\frac{p'}{\varrho}\right)^2 \varrho\,dr = 0$$

with the same notations as in (59.17). In fact for $l=0$, this equation reduces to the approximate form of (59.17) obtained in neglecting Φ'. But the two solutions for σ^2 do not correspond to the p and g modes. In fact, for each mode (one eigensolution) we seem to get two values of σ^2, one of which is spurious.

The discussion of (79.9) and (79.10) taking the singularities (79.13) and (79.14) and the boundary conditions (79.8) into account may help to draw qualitative informations on the two types of solutions. In particular, it shows that for the p-oscillations, the node of least radius occurs closer to the centre than the loop of least radius and vice-versa for the g-oscillations.

In this respect, one should note that if A is negative σ_p^2 and σ_g^2 are positive and both singularities (79.13) and (79.14) exist for both types of modes. In fact (79.14) may be satisfied in more than one point. However, if A is positive in a large enough region to render σ_g^2 negative, the singularity (79.13) disappears for the g-modes. All the previous equations can be adapted to a polytrope of index n immediately in setting

$$\left.\begin{aligned} &r = r_0\xi \quad \text{with} \quad r_0^2 = \frac{(n+1)\,p_c}{4\pi G\varrho_c^2}, \\ &\varrho = \varrho_c\vartheta^n, \quad p = p_c\vartheta^{n+1}, \quad A = \frac{\vartheta'}{\vartheta}\left(n - \frac{n+1}{\gamma}\right), \\ &v = \xi^2\frac{\delta r}{r_0}\vartheta^{(n+1)/\gamma}, \quad w = \frac{p'}{p_c}\vartheta^{-(n+1)/\gamma}, \end{aligned}\right\} \tag{79.20}$$

where ξ and ϑ are the usual EMDEN's variables and the subscript c denotes central values. In particular, if we further define

$$\left.\begin{aligned} &\alpha = \frac{\sigma^2(n+1)}{4\pi G\varrho_c} = \frac{n+1}{3}\frac{\bar\varrho}{\varrho_c}\omega^2, \quad Q = \frac{2(n+1)}{\gamma} - n \\ &f = \vartheta^{-Q}\left[\alpha + \frac{1}{2}(Q-n)(n+1)\frac{\vartheta'^2}{\vartheta}\right], \quad q = \vartheta^Q\left\{\frac{l(l+1)}{\alpha} - \frac{\xi^2}{\gamma\vartheta}\right\}, \end{aligned}\right\} \tag{79.21}$$

Eqs. (79.11) and (79.12) become

$$\frac{d}{d\xi}\left(\frac{\xi^2}{f}\cdot\frac{dw}{d\xi}\right) = q\,w, \tag{79.22}$$

$$\frac{d}{d\xi}\left(\frac{1}{q}\cdot\frac{dv}{d\xi}\right) = \frac{f\,w}{\xi^2}. \tag{79.23}$$

COWLING's numerical integrations[1] in the case of the standard model ($n=3$) for the f- and first g-modes of the second harmonic ($l=2$) were extended by

Table 16. *Characteristics values of ω^2 for non-radial oscillations of the polytrope $n=3$, $\gamma=\frac{5}{3}$.*

Modes	$l=2$		$l=3$		$l=4$	
	Approx.	Corr.	Approx.	Corr.	Approx.	Corr.
p_1	15.395		18.431		19.89	
f	9.416	8.67	9.833	9.15	10.234	9.60
g_1	5.297	4.74	7.027	6.48	8.205	7.69
g_2	2.928	2.51	4.337	3.93	5.455	5.06

KOPAL[2] and the results concerning the frequencies are summarized in Table 16 as well as some results for p-modes kindly communicated before publications by OWEN.

Apart from the possible dynamical instability arising when the σ_g^2 become negative, the g-spectrum is also very interesting when the σ_g^2 remain positive

[1] There appears to be a numerical error in the column giving the subsidiary quantity $\varphi \propto \varrho'$ in COWLING's table ($n=3$, $N=\frac{3}{2}$) which has been reproduced in KOPAL's Fig. 4a, curve α_2. The quantity φ should have a node in $\xi \approx 2.25$.

[2] Z. KOPAL: Astrophys. Journ. **109**, 509 (1949).

because it introduces the possibility of indefinitely long periods, a situation which should favour resonance in close binary stars. However, in the case of the standard model, comparison of Table 16 with formula (42.2) to (42.6) shows that, even in the most favourable case (42.6), resonance could only occur with a fairly high g-mode ($\approx g_4$) while for two equal masses in contact [cf. (42.4)] it would not occur before g_8 to g_{10}. A low value of A would lower the order of the interesting g-mode but in all cases, as COWLING [60] has shown, the excitation of the corresponding mode will be very small as it contributes relatively little to the equilibrium distortion of the star. Thus a very close resonance is necessary to give an appreciable effect and this resonance is likely to deteriorate as the amplitude increases.

On the other hand, the frequencies of the f- and p-modes are fairly close to those of the modes of radial pulsation and one may wonder if despite the difference of symmetry, they may not excite one another especially in presence of rotation or a magnetic field.

80. Corrections for the perturbation of the gravitational potential. COWLING [60] has estimated the effects of Φ' by the methods of perturbation theory. Let us designate by $\boldsymbol{h}_i$ one of the solutions $\delta \boldsymbol{r}_i$ of the previous problem so that, according to (74.1), (75.1) and (74.7), we have

$$\varrho \sigma_i^2 \boldsymbol{h}_i = \frac{1}{\varrho} \operatorname{grad} p \operatorname{div} (\varrho \boldsymbol{h}_i) - \operatorname{grad} (\Gamma_1 p \operatorname{div} \boldsymbol{h}_i + \boldsymbol{h}_i \cdot \operatorname{grad} p). \tag{80.1}$$

If we subtract this equation multiplied scalarly by $\boldsymbol{h}_k$ from the same equation written for $\boldsymbol{h}_k$ and multiplied by $\boldsymbol{h}_i$ and integrate the result over the volume $\mathscr{V}$ of the star, $d\mathscr{V} = r^2 \sin\vartheta \, d\vartheta \, d\varphi \, dr$, we get

$$\left.\begin{aligned} (\sigma_k^2 - \sigma_i^2) \int_0^{\mathscr{V}} \varrho \boldsymbol{h}_i \cdot \boldsymbol{h}_k \, d\mathscr{V} &= \int_0^{\mathscr{V}} \operatorname{div} [\boldsymbol{h}_k (\Gamma_1 p \operatorname{div} \boldsymbol{h}_i + \boldsymbol{h}_i \cdot \operatorname{grad} p)] \, d\mathscr{V} - \\ &\quad - \int_0^{\mathscr{V}} \operatorname{div} [\boldsymbol{h}_i (\Gamma_1 p \operatorname{div} \boldsymbol{h}_k + \boldsymbol{h}_k \cdot \operatorname{grad} p)] \, d\mathscr{V}. \end{aligned}\right\} \tag{80.2}$$

The two integrals in the right-hand member cancel since p and $\operatorname{grad} p$ vanish on the surface and we are left with

$$\int_0^{\mathscr{V}} \varrho \boldsymbol{h}_i \cdot \boldsymbol{h}_k \, d\mathscr{V} = 0 \qquad i \neq k. \tag{80.3}$$

If, due to Φ' the corrected solutions become $(\sigma_n)_c$, $(\boldsymbol{h}_n)_c$ which differ from σ_n, $\boldsymbol{h}_n$ by small quantities, we may write

$$(\boldsymbol{h}_n)_c = \boldsymbol{h}_n + \sum_i a_i \boldsymbol{h}_i \tag{80.4}$$

where the a_i are small of the first order. From (74.1) and (74.7), we have also

$$(\varrho_n')_c = \varrho_n' + \sum a_i \varrho_i', \qquad (p_n')_c = p_n' + \sum_i a_i p_i'. \tag{80.5}$$

If we introduce these expressions into (74.3) where Φ' is kept, we obtain, simplifying by (80.1) and neglecting second order terms,

$$[(\sigma_n)_c^2 - \sigma_n^2] \, \varrho \boldsymbol{h}_n = \varrho \sum_i a_i (\sigma_i^2 - \sigma_n^2) \, \boldsymbol{h}_i + \varrho \operatorname{grad} \Phi'. \tag{80.6}$$

If we take the scalar product of this into $\boldsymbol{h}_n$ and integrate throughout the volume $\mathcal{V}$, using (80.3), we obtain

$$[(\sigma_n^2)_c - \sigma_n^2] \int^{\mathcal{V}} \varrho \boldsymbol{h}_n \cdot \boldsymbol{h}_n \, d\mathcal{V} = \int^{\mathcal{V}} (\varrho \boldsymbol{h}_n) \operatorname{grad} \Phi' \, d\mathcal{V}$$
$$= \int^{\mathcal{V}} \operatorname{div}(\varrho \boldsymbol{h}_n \Phi') \, d\mathcal{V} - \int^{\mathcal{V}} \Phi' \operatorname{div}(\varrho \boldsymbol{h}_n) \, d\mathcal{V}$$

or, since ϱ vanishes on the surface and $\varrho_n' = -\operatorname{div}(\varrho \boldsymbol{h}_n)$,

$$(\sigma_n^2)_c - \sigma_n^2 = \frac{\int^{\mathcal{V}} \varrho_n' \Phi' \, d\mathcal{V}}{\int^{\mathcal{V}} \varrho \boldsymbol{h}_n \cdot \boldsymbol{h}_n \, d\mathcal{V}}. \tag{80.7}$$

To the same order of approximation, Φ' may be computed in terms of ϱ_n' by means of its expression as a solution of (74.8):

$$\Phi_n' = -\frac{4\pi G}{2l+1}\left[r^l \int_r^{\infty} \varrho_n' r^{1-l} \, dr + \frac{1}{r^{l+1}} \int_0^r \varrho_n' r^{l+2} \, dr\right]. \tag{80.8}$$

Cowling and Kopal have carried out the process for their solutions and the corrected values ω_c^2 are listed in Table 16.

81. Viscous damping and vibrational stability. The same method can be used to evaluate the effects of viscosity if, instead of Φ', we keep the term in div $\mathfrak{P}$ in (74.3) which becomes

$$\varrho \sigma^2 \delta \boldsymbol{r} = -\frac{\varrho'}{\varrho} \operatorname{grad} p + \operatorname{grad} p' - i\sigma \operatorname{div} \mathfrak{P}(\delta \boldsymbol{r}). \tag{81.1}$$

As before, introducing the solutions (80.4) and (80.5) into (81.1) and subsequently multiplying by $\boldsymbol{h}_n$ and integrating over the volume, we obtain

$$[(\sigma_n^2)_c - \sigma_n^2] \int^{\mathcal{V}} \varrho \boldsymbol{h}_n \cdot \boldsymbol{h}_n \, d\mathcal{V} = -i\sigma \int^{\mathcal{V}} \operatorname{div} \mathfrak{P}(\boldsymbol{h}_n) \cdot \boldsymbol{h}_n \, d\mathcal{V}, \tag{81.2}$$

taking (80.3) into account.

As in Sect. 64, we write

$$(\sigma_n)_c = \sigma_n + i\sigma_n'' \tag{81.3}$$

and transform the last term of (81.2) as follows:

$$\int^{\mathcal{V}} (\operatorname{div} \mathfrak{P}) \cdot \boldsymbol{h}_n \, d\mathcal{V} = -\int^{\mathcal{V}} \sum_{j,k} \mathfrak{P}_j^k \Delta_k (h_n)^j \, d\mathcal{V},$$

since, on the surface, all tensions vanish. Then (81.2) becomes

$$\sigma_n'' = \frac{1}{2} \frac{\int^{\mathcal{V}} \sum_{j,k} \mathfrak{P}_j^k \Delta_k (h_n)^j \, d\mathcal{V}}{\int^{\mathcal{V}} \varrho \boldsymbol{h}_n \cdot \boldsymbol{h}_n \, d\mathcal{V}} = \frac{1}{2} \frac{\int^{\mathcal{V}} \varepsilon_f \, d\mathcal{V}}{\int^{\mathcal{V}} \varrho \boldsymbol{h}_n \cdot \boldsymbol{h}_n \, d\mathcal{V}} \tag{81.4}$$

where

$$\mathfrak{P}_j^k(\boldsymbol{h}_n) = \sum_i g_{ij} \mathfrak{P}^{ik}(\boldsymbol{h}_n)$$

and $\mathfrak{P}^{ik}(\boldsymbol{h}_n)$ is given by an expression of the form (49.12) with $v^l \to (h_n)^l$.

Formula (81.4) is equivalent to that obtained by Counson, Ledoux and Simon[1] from the principle of conservation of energy (cf. Sect. 63). These authors

[1] J. Counson, P. Ledoux and R. Simon: Bull. Soc. Roy. Sci. Liège 25, 144 (1956).

have given the explicit expression of ε_f in spherical coordinates and have applied (81.4) in the case of molecular viscosity, to the modes g_1 with $l=2$, $m=2$ and g_2 with $l=4$, $m=4$ of the standard model. The damping times $\tau''=(\sigma'')^{-1}$ found are respectively $(\tau'')_{g_1}=8\cdot 10^{12}$ years and $(\tau'')_{g_2}=2\cdot 10^{12}$ years which are not much smaller than the damping times for purely radial oscillations (cf. Table 14, Sect. 71). This is due to the small velocity gradients in the g-oscillations. It is likely that the p-oscillations would be damped more rapidly.

Of course, if convection would prevail in a large fraction of the star, turbulent viscosity could decrease these damping times enormously just as in the case of radial pulsation (cf. Sect. 71).

If in (74.3) we neglect Φ' and $\mathfrak{P}$ but eliminate p' by means of the complete relation (74.7), which becomes here

$$p' = p'_n + \sum a_i p'_i + \frac{\Gamma_3 - 1}{i\sigma}\,\varrho\,\{\ldots\}',$$

and then proceed to the same multiplication and integration as before we obtain

$$\sigma''_n = -\frac{1}{2\sigma_n^2}\,\frac{\int\limits^{\mathscr{V}}\left(\frac{\delta T}{T}\right)_n\left\{\varepsilon - \frac{1}{\varrho}\operatorname{div}\boldsymbol{F}_R\right\}'_n\varrho\,d\mathscr{V}}{\int\limits^{\mathscr{V}}\boldsymbol{h}_n\cdot\boldsymbol{h}_n\,\varrho\,d\mathscr{V}}. \tag{81.5}$$

It should be noted that here the Eulerian perturbation of div $\boldsymbol{F}_R$ has to be computed from the general vectorial expression (49.40) since the perturbed flux will have non-vanishing components F'_ϑ and F'_φ as well as F'_r due to the variations of ϱ' and T' along a level surface.

Another difference with respect to radial pulsation is that, as long as generation of energy is limited to a small central region, its influence on the vibrational stability will be much smaller since here p'/p, ϱ'/ϱ, T'/T or $\delta p/p$, $\delta\varrho/\varrho$, $\delta T/T$ always tend towards zero with r [cf. Eq. (75.22)] so that the mean value of ε' (or $\delta\varepsilon$) will be very small.

It is unlikely that Eq. (81.5) could lead to conditions for vibrational stability very different from those encountered for radial pulsations but a detailed discussion would require the numerical evaluation of (81.5) which has not yet been attempted[1].

82. Influence of rotation. *α) General equations.* The effects of a uniform rotation $\boldsymbol{\Omega}$ on the small oscillations of an incompressible homogeneous globe have been discussed very completely by BRYAN [*61*]. Apart from the simplifications due to incompressibility, the great advantage in this case is that equilibrium configurations in uniform rotation do exist and that their level surfaces admit of analytical representations in finite terms of well-known functions.

In going over to physical compressible configurations, both of these features are lost. VON ZEIPEL's theorem[2] indicates the incompatibility of uniform rotation and generation of energy by nuclear reactions. However, the resulting meridional currents are very slow[3] so that, perhaps, in this respect uniform rotation remains an adequate working hypothesis. But, even in that case, infinite series expansions have to be used to represent the level surfaces[4] and the distribution

[1] R. SIMON: Bull. Acad. Roy. Belg., cl. Sci., Ser. V **43**, 610 (1957).

[2] Cf. for instance [*39a*], p. 282. — See also M. WRUBEL, p. 39, this volume.

[3] P. A. SWEET: Monthly Notices Roy. Astronom. Soc. London **110**, 548 (1950). — E. J. ÖPIK: Monthly Notices Roy. Astronom. Soc. London **111**, 94 (1951).

[4] Cf. for instance, S. CHANDRASEKHAR: Monthly Notices Roy. Astronom. Soc. London **93**, 390 (1933).

of ϱ, T and p:

$$\varrho = \varrho_0 + \sum_{i=1}^{\infty} \varrho_i(r)\, P_i(\cos\vartheta), \qquad \text{etc.} \tag{82.1}$$

where the P_i represent Legendre polynomials and the ϱ_i are proportional to the square and higher powers of Ω.

Up to now, these difficulties have confined the discussion mainly to the first order effects in Ω in which case, we have not to take into account the deviations from spherical symmetry.

If we refer the perturbed motion to axes rotating with the angular velocity Ω of the star at equilibrium, the kinetic energy K in spherical coordinates is

$$K = \tfrac{1}{2}\left[\dot r^2 + r^2 \dot\vartheta^2 + r^2 \sin^2\vartheta\,(\dot\varphi + \Omega)^2\right]. \tag{82.2}$$

Expliciting (48.10) and introducing the natural components of velocity $v_r = \dot r$, $v_\vartheta = r\dot\vartheta$, $v_\varphi = r\sin\vartheta\dot\varphi$, we obtain EULER's equations

$$\left.\begin{aligned}
&\frac{dv_r}{dt} - \frac{v_\vartheta^2 + v_\varphi^2}{r} - 2\Omega\sin\vartheta\, v_\varphi = -\frac{\partial\Phi}{\partial r} - \frac{1}{\varrho}\frac{\partial p}{\partial r} + \Omega^2 r\sin^2\vartheta,\\
&\frac{dv_\vartheta}{dt} + \frac{v_\vartheta v_r}{r} - \frac{v_\varphi^2}{r}\cot\vartheta - 2\Omega\cos\vartheta\, v_\varphi = -\frac{\partial\Phi}{r\,\partial\vartheta} - \frac{1}{\varrho r}\frac{\partial p}{\partial\vartheta} + \Omega^2 r\sin\vartheta\cos\vartheta,\\
&\frac{dv_\varphi}{dt} + \frac{v_\vartheta v_\varphi}{r}\cot\vartheta + \frac{v_r v_\varphi}{r} + 2\Omega\sin\vartheta\, v_r + 2\Omega\, v_\vartheta\cos\vartheta\\
&\qquad\qquad = -\frac{\partial\Phi}{r\sin\vartheta\,\partial\varphi} - \frac{1}{\varrho r\sin\vartheta}\frac{\partial p}{\partial\varphi}
\end{aligned}\right\} \tag{82.3}$$

where we have neglected all viscous forces. At equilibrium with respect to the rotating axes, we have

$$\operatorname{grad}\left(\Phi - \frac{1}{2}\Omega^2 r^2\sin^2\vartheta\right) + \frac{1}{\varrho}\operatorname{grad} p = 0. \tag{82.4}$$

For a perturbation of this equilibrium state proportional to $e^{i\sigma t}$ the velocities $v_r = i\sigma\,\delta r\, e^{i\sigma t}$ etc. may be treated as small quantities of the first order in (82.3) which become

$$\left.\begin{aligned}
-\sigma^2\delta r - 2i\sigma\Omega r\sin^2\vartheta\,\delta\varphi &= -\frac{\partial\Phi'}{\partial r} + \frac{\varrho'}{\varrho^2}\frac{\partial p}{\partial r} - \frac{1}{\varrho}\frac{\partial p'}{\partial r},\\
-\sigma^2 r\,\delta\vartheta - 2i\sigma\Omega r\sin\vartheta\cos\vartheta\,\delta\varphi &= -\frac{\partial\Phi'}{r\,\partial\vartheta} + \frac{\varrho'}{\varrho^2}\frac{\partial p}{r\,\partial\vartheta} - \frac{1}{\varrho}\frac{\partial p'}{r\,\partial\vartheta},\\
-\sigma^2 r\sin\vartheta\,\delta\varphi + 2i\sigma\Omega(\sin\vartheta\,\delta r + r\cos\vartheta\,\delta\vartheta) &= -\frac{\partial\Phi'}{r\sin\vartheta\,\partial\varphi} +\\
&\quad + \frac{\varrho'}{\varrho^2}\frac{\partial p}{r\sin\vartheta\,\partial\varphi} - \frac{1}{\varrho}\frac{\partial p'}{r\sin\vartheta\,\partial\varphi},
\end{aligned}\right\} \tag{82.5}$$

taking Eq. (82.4) into account.

β) The homogeneous compressible model. As before (cf. Sect. 76α), one may expect that this model will present the simplest conditions. The problem was tackled by SEVERNY[1] but his results were vitiated by a confusion between the coefficients of piezotropy and barotropy. Later, LEDOUX ([*50b*], Chap. V) reconsidered the problem using as SEVERNY a method derived from the theory of tides (cf. [*40a*], §§ 313 and 316). Solving Eqs. (82.5) where ϱ is treated as

[1] A. B. SEVERNY: Dokl. Akad. Nauk. USSR **46**, 53 (1945).

constant, with respect to δr, $r\,\delta\vartheta$ and $r\sin\vartheta\,\delta\varphi$ we obtain three scalar equations equivalent to

$$\left.\begin{aligned}\delta\boldsymbol{r}\,(\sigma^2-4\Omega^2) &= \operatorname{grad}\chi+\frac{\alpha}{\varrho}\operatorname{grad}p+\frac{4\boldsymbol{\Omega}}{\sigma^2}\left[\boldsymbol{\Omega}\cdot\left(\operatorname{grad}\chi+\frac{\alpha}{\varrho}\operatorname{grad}p\right)\right]-\\ &\quad-\frac{2}{i\sigma}\left[\left(\operatorname{grad}\chi+\frac{\alpha}{\varrho}\operatorname{grad}p\right)\times\boldsymbol{\Omega}\right]\end{aligned}\right\}\tag{82.6}$$

where

$$\chi=\Phi'+\frac{p'}{\varrho},\qquad \alpha=\operatorname{div}\delta\boldsymbol{r}=-\frac{\varrho'}{\varrho}\tag{82.7}$$

as in (75.4) and (76.1). Taking the divergence of (82.6) and using (82.7) leads to

$$\left.\begin{aligned}\alpha(\sigma^2-4\Omega^2) &= \nabla^2\chi+\frac{\alpha}{\varrho}\nabla^2 p+\frac{1}{\varrho}\operatorname{grad}\alpha\cdot\operatorname{grad}p+\frac{2\boldsymbol{\Omega}}{i\sigma\varrho}\cdot(\operatorname{grad}\alpha\times\operatorname{grad}p)-\\ &\quad-\frac{4\Omega^2}{\sigma^2}\left[L(\chi)+\frac{\alpha}{\varrho}L(p)\right]-\frac{4}{\sigma^2\varrho}(\boldsymbol{\Omega}\cdot\operatorname{grad}p)\,(\boldsymbol{\Omega}\cdot\operatorname{grad}\alpha)\end{aligned}\right\}\tag{82}$$

where L represents the operator

$$\left.\begin{aligned}L &= \frac{\boldsymbol{\Omega}}{\Omega^2}\cdot\operatorname{grad}\left(\Omega\cos\vartheta\frac{\partial}{\partial r}-\Omega\sin\vartheta\frac{1}{r}\frac{\partial}{\partial\vartheta}\right)\\ &=\cos^2\vartheta\frac{\partial^2}{\partial r^2}+\frac{\sin^2\vartheta}{r}\frac{\partial}{\partial r}+\frac{\sin 2\vartheta}{r^2}\frac{\partial}{\partial\vartheta}-\frac{\sin 2\vartheta}{r}\frac{\partial^2}{\partial r\,\partial\vartheta}+\frac{\sin^2\vartheta}{r^2}\frac{\partial^2}{\partial\vartheta^2}.\end{aligned}\right\}\tag{82.9}$$

To obtain a differential equation in α, we still have to eliminate χ i.e. Φ' and p' and this is not too easy in the general case. Furthermore, the form of L shows that, in general, the separation in spherical coordinates will no longer be possible.

However, Eq. (82.8) leads to the distinction of two main cases. If the configuration keeps its axial symmetry, i.e.

$$\frac{\partial f'}{\partial\varphi}=0,\qquad f'\text{: any perturbation,}\tag{82.10}$$

grad α is in the meridian plane and the term proportional to Ω drops out so that, in this case, the effects of rotation are always proportional to Ω^2.

If the perturbation is not axially symmetric, terms in Ω subsist and, if the rotation is sufficiently small, terms in Ω^2 may be neglected and the equation becomes again separable in spherical coordinates.

Among all the axially symmetric perturbations let us consider the particularly simple one corresponding to a mainly radial displacement (pseudo-radial oscillations). In that case, in addition to (82.10), we have

$$\frac{\partial f'}{\partial\vartheta}\propto\Omega^2\tag{82.11}$$

just as for the equilibrium variable f. Of course, the inertial forces will always induce some motion along ϑ and φ but, from the last two equations (82.5) and (82.10) and (82.11), the leading terms in $r\,\delta\vartheta$ and $r\sin\vartheta\,\delta\varphi$ are

$$r\,\delta\vartheta\propto\frac{\Omega^2\,\delta r}{4\,\Omega^2\cos^2\vartheta-\sigma^2},\qquad r\sin\vartheta\,\delta\varphi=\frac{2i\Omega}{\sigma}\,\delta r\sin\vartheta.\tag{82.12}$$

Thus, provided σ does not become of the order of Ω, our assumption of pseudo-radial motion will be verified.

If we further assume that $\delta r/r$ is constant as for the fundamental mode of radial pulsation of this model, α is also a constant and Eq. (82.8) keeping terms in Ω^2 only, can be reduced (cf. [50b], Chap. V, § 3) to

$$\sigma^2 = \frac{4\pi G \varrho}{3}(3\gamma - 4) + \frac{2\Omega^2}{3}(7 - 3\gamma) - \frac{4\Omega^2}{\sigma^2}\left[\frac{4\pi G\varrho}{9}(3\gamma - 4)\right] \tag{82.13}$$

which has the solution

$$\sigma^2 = \frac{4\pi G \varrho}{3}(3\gamma - 4) + \frac{2\Omega^2}{3}(5 - 3\gamma). \tag{82.14}$$

The other solution proportional to Ω is irrelevant here since it does not satisfy our hypothesis.

For perturbations destroying the axial symmetry keeping only terms in Ω (all equilibrium variables can be given their values in the sphere without rotation) Eq. (82.8) can be written

$$\left.\begin{aligned}\sigma^2\alpha = &-4\pi G\varrho\alpha - \frac{1}{\varrho}\nabla^2(\Gamma_1 p\alpha) - \frac{1}{\varrho}\nabla^2\left(\delta r\frac{\partial p}{\partial r}\right) + \frac{\alpha}{\varrho}\nabla^2 p + \frac{1}{\varrho}\frac{\partial\alpha}{\partial r}\frac{\partial p}{\partial r} - \\ &- \frac{2\Omega}{i\sigma\varrho}\sin\vartheta\frac{\partial\alpha}{r\sin\vartheta\,\partial\varphi}\frac{\partial p}{\partial r}\end{aligned}\right\} \tag{82.15}$$

if we explicit χ and use the relations (76.1) and (76.2). Considering again solutions of the form (73.1), the continuity equation then becomes

$$\alpha = \frac{1}{r^2}\frac{\partial}{\partial r}(r^2\delta r) - \frac{l(l+1)\chi}{\sigma^2 r^2} - \frac{2\Omega m}{\sigma r}\left(\delta r + \frac{\chi}{\sigma^2 r}\right) \tag{82.16}$$

instead of (75.9) and the radial component of (82.6) gives to the same order of approximation

$$\sigma^2\delta r = \frac{\partial\chi}{\partial r} + \frac{\alpha}{\varrho}\frac{\partial p}{\partial r} + \frac{2\Omega m\chi}{r\sigma}. \tag{82.17}$$

If one solves (82.16) for χ and introduces it into (82.17), the terms in δr, as in Sect. 76α, reduce to $\nabla^2(r\,\delta r)$ which is proportional to $\nabla^2\left(\delta r\frac{\partial p}{\partial r}\right)$ so that this last term can be eliminated from (82.15) giving

$$\left.\begin{aligned}(1-x^2)\frac{d^2\alpha}{dx^2} + \frac{d\alpha}{dx}\left(\frac{2-6x^2}{x}\right) + \alpha\bigg[\frac{6\sigma^2}{4\pi G\varrho\Gamma_1} + \frac{8}{\Gamma_1} - 6 - \\ - \frac{2l(l+1)}{3\sigma^2\Gamma_1}4\pi G\varrho - l(l+1)\frac{1-x^2}{x^2} - \frac{4\Omega m}{\Gamma_1\sigma}\left(2 + \frac{4\pi G\varrho}{3\sigma^2}\right)\bigg] = 0\end{aligned}\right\} \tag{82.18}$$

which reduces to (76.6) if $\Omega = 0$.

The eigensolutions are again polynomials of the form (76.7), the coefficients obeying the same recurrence formula (76.8) where the value of B however is different and equal to

$$B = -6 - 4l + \frac{8}{\Gamma_1} + \frac{6\sigma^2}{\Gamma_1 4\pi G\varrho} - \frac{2l(l+1)}{3\sigma^2\Gamma_1}4\pi G\varrho - \frac{4\Omega m}{\Gamma_1\sigma}\left(2 + \frac{4\pi G\varrho}{3\sigma^2}\right). \tag{82.19}$$

The eigenvalue equation corresponding to (76.9) is a little more complicated, but it can be shown ([50b], Chap. V, § 4) that, apart from a spurious solution proportional to Ω which here is contrary to the hypothesis, the eigenvalues can be written

$$\sigma_m = \sigma + \frac{m\Omega(2\omega^2+1)}{\omega^4 + l(l+1)} \qquad -l \leq m \leq l \tag{82.20}$$

where σ ($\omega^2 = 3\sigma^2/4\pi G\varrho$) is the corresponding eigenvalue of the non-rotating sphere. Thus in presence of rotation, the $(2m+1)$-fold degeneracy disappears completely. This result is similar to that obtained by BRYAN for an incompressible globe where the relation (82.20) is particularly simple:

$$\sigma_m = \sigma + \frac{m}{l}\Omega, \qquad -l \leq m \leq l. \tag{82.21}$$

BRYAN found also frequencies directly proportional to Ω and it is likely that such frequencies exist also for the compressible configurations corresponding essentially to motions of inertia.

γ) General models. The two types of oscillations considered in Sect. 82β have also been studied in the case of more general models. The problem of the pseudo-radial oscillations of a rotating star was approached by LEDOUX [*50a*] from the point of view of the virial theorem. With respect to the treatment in Sect. 59β, the only generalization concerns the kinetic energy of mass motion K which is now given by (82.2) so that the expression (59.13) of the total kinetic energy becomes

$$K^* = K + \frac{1}{2}\int_0^M \Omega^2 r^2 \sin^2\vartheta\, dm + \frac{1}{2}\int_0^M 2\Omega r^2 \sin^2\vartheta\, \dot{\varphi}\, dm + \int_0^M \frac{3}{2}\frac{p}{\varrho}\, dm \tag{82.22}$$

if we denote by K the relative kinetic energy of mass motion. The virial theorem (59.11) becomes explicitly

$$\frac{1}{2}\frac{d^2 I}{dt^2} = 2K + \int_0^M 2(\boldsymbol{\Omega}\times\boldsymbol{r})\cdot\dot{\boldsymbol{r}}\, dm + \int_0^M \Omega^2 r^2 \sin^2\vartheta\, dm + \int_0^M 3\frac{p}{\varrho}\, dm + W. \tag{82.23}$$

If we consider a pseudo-radial displacement satisfying the relations (82.12) and introduce it into the first variation of (82.22) neglecting all powers of Ω greater than the second, we obtain

$$\delta K^* = -\frac{1}{2}\int_0^M 2\Omega^2 r \sin^2\vartheta\, \delta r\, dm + \frac{3}{2}(\gamma - 1)\int_0^M \frac{\delta\varrho}{\varrho}\cdot\frac{p}{\varrho}\, dm \tag{82.24}$$

if γ is treated as constant.

One can transform the last integral in (82.24), writing

$$\int_0^M \frac{\delta\varrho}{\varrho}\frac{p}{\varrho}\, dm = -\int_0^{\mathscr{V}} p \operatorname{div}\delta\boldsymbol{r}\, d\mathscr{V} = +\int_0^{\mathscr{V}} \delta\boldsymbol{r}\cdot\operatorname{grad} p\, d\mathscr{V}$$

since p vanishes on a free surface.

Using (82.4) to eliminate grad p, leads to

$$\left.\begin{aligned} \int_0^M \frac{\delta\varrho}{\varrho}\frac{p}{\varrho}\, dm &= -\int_0^M \frac{\delta r}{r}(\boldsymbol{r}\cdot\operatorname{grad}\Phi)\, dm + \int_0^M \frac{\delta r}{r}\Omega^2 r^2 \sin^2\vartheta\, dm \\ &= +\int_0^M \frac{\delta r}{r}\, dW + \int_0^M 2\frac{\delta r}{r}\, dK_R \end{aligned}\right\} \tag{82.25}$$

where K_R represents the kinetic energy of rotation.

To the same order of approximation, one may use, together with (82.24) and (82.25), the expressions (59.15) of δI and δW in the first variation of (59.11) which gives

$$\sigma^2 = -\frac{(3\gamma - 4)\int_0^M \xi\, dW}{\int_0^M \xi\, dI} + (5 - 3\gamma)\frac{\int_0^M 2\xi\, dK_R}{\int_0^M \xi\, dI}. \tag{82.26}$$

This shows that in particular the value $\gamma = \frac{4}{3}$ is no longer critical. If $\xi = \delta r/r$ is assumed constant, (82.26) becomes

$$\sigma^2 = -(3\gamma - 4)\frac{W}{I} + (5 - 3\gamma)\frac{2K_R}{I} \tag{82.27}$$

which reduces to (82.14) if ϱ is constant.

As COWLING and NEWING [62] have shown, RAYLEIGH'S principle which normally yields better approximations than the previous method can also be extended to this problem. In the case of pseudo-radial motion, it leads to a result similar to (82.26). As these authors remark, however, (82.26) does not completely describe the dependence of σ^2 on Ω^2 since I and W depart from their values in the non-rotating case by terms proportional to Ω^2. Taking this into account they have made some numerical applications to polytropes using for ξ its value for the non-rotating polytrope. Their results show that for a star of given mass, although the dynamical effect of the rotation [second term in (82.26)] tends to increase σ^2, this is usually more than compensated by the decrease of the mean density due to the general expansion under the centrifugal force. For common stellar velocities of rotation, the effect on the frequencies remains small and does not exceed a few percent.

RAYLEIGH'S principle enabled COWLING and NEWING to discuss more general axial-symmetric oscillations, too. Solving the third Eq. (82.5) for $\delta\varphi$ noting that its right-hand member is zero in this case, and introducing the value of $\delta\varphi$ into the first two equations multiplied respectively by δr and $r\,\delta\vartheta$ one gets after addition and integration over the whole volume

$$\left.\begin{aligned}
&\sigma^2 \int^{\mathcal{V}} \varrho\,(\delta r^2 + r^2\,\delta\vartheta^2)\,d\mathcal{V} \\
&= \int^{\mathcal{V}} \varrho\,\delta r\left[\frac{\partial \Phi'}{\partial r} - \frac{\varrho'}{\varrho^2}\frac{\partial p}{\partial r} + \frac{1}{\varrho}\frac{\partial p'}{\partial r} - 4\Omega^2(\delta r\sin\vartheta + r\,\delta\vartheta\sin\vartheta\cos\vartheta)\right]d\mathcal{V} + \\
&+ \int^{\mathcal{V}} \varrho\, r\,\delta\vartheta\left[\frac{\partial \Phi'}{r\,\partial\vartheta} - \frac{\varrho'}{\varrho^2}\frac{\partial p}{r\,\partial\vartheta} + \frac{1}{\varrho}\frac{\partial p'}{r\,\partial\vartheta} - 4\Omega^2(\delta r\sin\vartheta\cos\vartheta + r\,\delta\vartheta\cos^2\vartheta)\right]d\mathcal{V}
\end{aligned}\right\} \tag{82.28}$$

which can be shown to be equivalent to COWLING and NEWING'S expression given in terms of Lagrangian variations. This formula permits, when using the known axial-symmetric oscillations [$m=0$ in (73.1)] of a non-rotating star, to evaluate the effects of rotation up to terms in Ω^2.

Finally, they extended RAYLEIGH'S principle to completely general oscillations and, in the particular case where only terms in Ω are retained, they obtained a convenient generalization of formula (82.20). This result can also be readily obtained by the method of perturbation[1]. In this case, since only the terms in Ω are kept, p, ϱ, Φ may be given their values in a non-rotating star, and the equation

$$\sigma^2\,\delta \boldsymbol{r} - 2i\sigma\,(\boldsymbol{\Omega}\times\delta\boldsymbol{r}) = \operatorname{grad}\Phi' + \frac{1}{\varrho}\operatorname{grad}p' - \frac{\varrho'}{\varrho^2}\operatorname{grad}p \tag{82.29}$$

is identical to (75.1) except for the perturbing term in Ω. From our previous discussion (Sect. 80) we could neglect Φ', but this is not essential since the $(\delta\boldsymbol{r})_i \equiv \boldsymbol{h}_i$, solutions of the complete equation (75.1) are also orthogonal. The proof runs exactly as in Sect. 80 except that the second member of (80.2) contains the extra-term

$$\int_0^{\mathcal{V}} \varrho\,(\operatorname{grad}\Phi_k'\cdot\boldsymbol{h}_i - \operatorname{grad}\Phi_i'\cdot\boldsymbol{h}_k)\,d\mathcal{V} = \int_0^{\mathcal{V}} (\varrho_i'\,\Phi_k' - \Phi_i'\,\varrho_k')\,d\mathcal{V}$$

[1] P. LEDOUX: Astrophys. Journ. **114**, 373 (1951).

if ϱ vanishes at the surface. Using POISSON's equation (74.8) and GREEN's theorem this becomes

$$\frac{1}{4\pi G}\iint\left(\Phi_k'\frac{\partial\Phi_i'}{\partial n}-\Phi_i'\frac{\partial\Phi_k'}{\partial n}\right)dS$$

which cancels according to the boundary conditions (75.26) so that the orthogonality conditions (80.3) subsist. Introducing in (82.29), the expressions (80.4) and (80.5) to which we may add according to POISSON's equation

$$(\Phi_n')_C=\Phi_n'+\Sigma a_i\Phi_i'$$

we obtain by the usual procedure

$$\sigma_n''=\frac{i\int\limits_0^M(\boldsymbol{\Omega}\times\boldsymbol{h}_n)\cdot\boldsymbol{h}_n\,dm}{\int\limits_0^M(\boldsymbol{h}_n\cdot\boldsymbol{h}_n)\,dm}\tag{82.30}$$

which is COWLING and NEWING's expression for the perturbation of the frequency of the n-th mode due to the work of the Coriolis force.

Note that the scalar product must be expended in terms of conjugate components if the complex notation is used.

For instance if the problem has been solved for the non-rotating star, neglecting Φ' and using the notation of Sect. 79, we have according to (75.14), (75.7) and (75.8)

$$\left.\begin{aligned}\delta r_n&=\frac{\frac{\partial y(r)}{\partial r}+y(r)\,A}{\sigma_n^2+A\,g}P_l^m\,e^{im\varphi}=a_n(r)\,P_l^m\,e^{im\varphi},\\ r\,\delta\vartheta_n&=\frac{y(r)}{\sigma_n^2\,r}\frac{\partial P_l^m}{\partial\vartheta}e^{im\varphi}=b_n(r)\frac{\partial P_l^m}{\partial\vartheta}e^{im\varphi},\\ r\sin\vartheta\,\delta\varphi_n&=\frac{i\,m\,y(r)}{\sigma_n^2\,r}\frac{P_l^m}{\sin\vartheta}e^{im\varphi}=i\,m\,b_n(r)\frac{P_l^m}{\sin\vartheta}e^{im\varphi}.\end{aligned}\right\}\tag{82.31}$$

Introducing these expressions into Eq. (82.30) and making use of well-known properties of the spherical harmonics, we obtain

$$\sigma_{n,m}''=\frac{m\,\Omega\int\limits_0^R\varrho\,(2ab+b^2)_n\,r^2\,dr}{\int\limits_0^R\varrho\,[a^2+l(l+1)\,b^2]_n\,r^2\,dr}=m\,\Omega\,C_n\tag{82.32}$$

where C_n is a constant depending on the mode considered. In the particular case of the homogeneous model, (82.32) gives back the correcting term in (82.20).

This method could be extended to cover the effects of the terms in Ω^2 but then the equilibrium variables p, ϱ and Φ should be expanded up to terms in Ω^2. The boundary conditions would also be perturbed (cf. the end of Sect. 64) but it is unlikely that this last effect could affect the frequencies very much.

83. Influence of a magnetic field. α) *General equations.* The discovery by H.W. BABCOCK [*63*] of stars with strong variable magnetic field led to different attempts to interpret their properties in terms of the oscillations of a star in hydrostatic equilibrium under the action of gravity and a general magnetic field. In this case, to the complexity of the hydrodynamical equations must be added that of MAXWELL's equations. However previous investigations relating mainly to the general field of the Sun or the interpretation of sunspots, by COWLING (origin and decay time of magnetic fields), FERRARO (interaction with rotation). ALFVÉN and WALÉN (magneto-hydrodynamic waves) had already prepared the ground in establishing the main simplifications admissible in stellar interiors. Generally, the conductivity can be treated as infinite and the displacement currents and the rate of change of charge density can be neglected so that in electromagnetic units, MAXWELL's equations can be simplified according to the following scheme [64]:

1. $$\operatorname{curl} \boldsymbol{E} = -\frac{\partial \boldsymbol{B}}{\partial t} \quad \text{gives} \quad \operatorname{curl} \boldsymbol{E} = -\frac{\partial \boldsymbol{H}}{\partial t} \tag{83.1}$$

where $\boldsymbol{E}$ and $\boldsymbol{H}$ denote the electric and magnetic field intensities, the permeability μ being taken equal to unity so that $\boldsymbol{B} = \boldsymbol{H}$;

2. $$\operatorname{curl} \boldsymbol{B} - \frac{1}{c^2}\frac{\partial \boldsymbol{E}}{\partial t} = 4\pi \boldsymbol{j} \quad \text{gives} \quad \operatorname{curl} \boldsymbol{H} = 4\pi \boldsymbol{j} \tag{83.2}$$

where $\boldsymbol{j}$ is the current density;

3. $$\operatorname{div} \boldsymbol{B} = 0 \quad \text{gives} \quad \operatorname{div} \boldsymbol{H} = 0; \tag{83.3}$$

4. $$\operatorname{div} \boldsymbol{E} = 4\pi \varrho_e \quad \text{gives} \quad \operatorname{div} \boldsymbol{E} = 0 \tag{83.4}$$

where ϱ_e is the space charge density.

In the same way, the generalized OHM's law

$$\boldsymbol{j} = \sigma(\boldsymbol{E} + \boldsymbol{v} \times \boldsymbol{B}) \quad \text{gives} \quad \boldsymbol{E} = -\boldsymbol{v} \times \boldsymbol{H} \tag{83.5}$$

if the conductivity σ becomes infinitely large, and the equation of charge conservation

$$\frac{\partial \varrho_e}{\partial t} + \operatorname{div} \boldsymbol{j} = 0 \quad \text{gives} \quad \operatorname{div} \boldsymbol{j} = 0. \tag{83.6}$$

The elimination of $\boldsymbol{E}$ between (83.1) and (83.5) yields the useful relation

$$\operatorname{curl}(\boldsymbol{v} \times \boldsymbol{H}) = \frac{\partial \boldsymbol{H}}{\partial t}. \tag{83.7}$$

Apart from the term in $\boldsymbol{v}$ in (83.5), the interaction between the motion and the magnetic field results in an extra body force $\boldsymbol{j} \times \boldsymbol{H}$ per unit volume. If we assume that the only other acting force is gravity and neglect viscosity, the general equation of motion (48.1) becomes

$$\varrho\left[\frac{\partial \boldsymbol{v}}{\partial t} + (\boldsymbol{v} \operatorname{grad}) \boldsymbol{v}\right] = -\varrho \operatorname{grad} \Phi - \operatorname{grad} p + (\boldsymbol{j} \times \boldsymbol{H}) \tag{83.8}$$

or

$$\varrho \frac{d\boldsymbol{v}}{dt} = -\varrho \operatorname{grad} \Phi - \operatorname{grad} p - \frac{1}{4\pi} \boldsymbol{H} \times \operatorname{curl} \boldsymbol{H} \tag{83.9}$$

if $\boldsymbol{j}$ is eliminated by (83.2). In hydrostatic equilibrium, (83.9) reduces to

$$\operatorname{grad} p = -\varrho \operatorname{grad} \Phi - \frac{1}{4\pi} \boldsymbol{H} \times \operatorname{curl} \boldsymbol{H}. \tag{83.10}$$

We can further transform the last term of (83.9) by the well-known identity

$$\boldsymbol{H} \times \operatorname{curl} \boldsymbol{H} = \operatorname{grad} \tfrac{1}{2} H^2 - (\boldsymbol{H} \cdot \operatorname{grad}) \boldsymbol{H}$$

which, taking (83.3) into account, can also be written

$$\boldsymbol{H} \times \operatorname{curl} \boldsymbol{H} = \operatorname{grad} \tfrac{1}{2} H^2 - \operatorname{div} \mathcal{H}$$

where $\mathcal{H}$ is the dyadic $(\boldsymbol{H}\boldsymbol{H})$ with components $\mathcal{H}_{ij} = H_i H_j$. With this notation (83.9) bccomes

$$\varrho \frac{d\boldsymbol{v}}{dt} = -\varrho \operatorname{grad} \Phi - \operatorname{grad}\left(p + \frac{H^2}{8\pi}\right) + \operatorname{div}\left(\frac{\mathcal{H}}{4\pi}\right) \tag{83.11}$$

where the usual decomposition of the magnetic stresses into a hydrostatic pressure $H^2/8\pi$ and a tension $H^2/4\pi$ along the lines of force is now apparent.

If we neglect all dissipation effects (conduction, viscosity and Joule heat) the equation of conservation of thermal energy leads to the same adiabatic relation (53.5) as before.

Let us now consider a finite mass having a free surface on which all stresses vanish. In that case, CHANDRASEKHAR and FERMI [*65*], taking the scalar product of (83.11) into $\boldsymbol{r}$ and integrating over the whole volume $\mathcal{V}$ of the configuration, have shown that the appropriate generalization of the virial theorem (59.11) reads here[1]

$$\frac{1}{2} \frac{d^2 I}{dt^2} = 2K + 3(\gamma - 1)\, U + W + \mathfrak{M} \tag{83.12}$$

where U and $\mathfrak{M}$ are respectively the internal heat energy and the magnetic energy

$$U = \frac{1}{\gamma - 1} \int^{M} \frac{p}{\varrho}\, dm, \qquad \mathfrak{M} = \int^{\mathcal{V}} \frac{H^2}{8\pi}\, d\mathcal{V}. \tag{83.13}$$

In hydrostatic equilibrium, the total energy $E = U + W + \mathfrak{M}$ must be negative and eliminating U by means of the corresponding form of (83.12), this condition can be written

$$\overline{H^2} R^3 < 6\,|W| \tag{83.14}$$

which, according to CHANDRASEKHAR and FERMI, means that the maximum possible magnetic field in a star varies from 10^8 gauss to 10^4 gauss in going from ordinary dwarfs to supergiants.

β) Oscillations around a state of equilibrium. The difficulties that beset the problem, apart from algebraic complexity, are of the same type as those encountered in the oscillations of rotating stars. First of all the hydrostatic model whose oscillations we are supposed to study is ill-defined[2]. Secondly, there are numerous boundary conditions which are not always simple to apply and their physical meaning is sometime elusive.

As before, the first variation of (83.8) [or (83.9) or (83.11)] taking (83.10) into account will give the linearized equation of motion which may be written, assuming that all perturbed quantities are proportional to $e^{i\sigma t}$,

$$\sigma^2 \varrho\, \delta\boldsymbol{r} = \varrho \operatorname{grad} \Phi' + \varrho' \operatorname{grad} \Phi + \operatorname{grad} p' - (\boldsymbol{j} \times \boldsymbol{H}') - (\boldsymbol{j}' \times \boldsymbol{H}). \tag{83.15}$$

Apart from the continuity Eq. (74.1) and the adiabatic relation (75.11), Eq. (83.2) gives

$$\operatorname{curl} \boldsymbol{H}' = 4\pi \boldsymbol{j}' \tag{83.16}$$

[1] This implies that the magnetic stresses vanish on the stellar surface. For a discussion of the limitations involved by this hypothesis and of the effects of an external magnetic field cf. J. H. PIDDINGTON: Austral. J. Phys. **10**, 530 (1957).

[2] Despite the numerous papers which have appeared during recent years, the question is still far from being completely elucidated: E. FERMI and S. CHANDRASEKHAR: Astrophys. Journ. **118**, 116 (1953). — V. C. A. FERRARO: Astrophys. Journ. **119**, 407 (1954). — G. GJELLESTAD: Astrophys. Journ. **119**, 14 (1954). — P. H. ROBERTS: Astrophys. Journ. **121**, 508 (1955). — S. CHANDRASEKHAR and K. H. PRENDERGAST: Proc. Roy. Soc. Lond. **42**, 1, (1956). — S. CHANDRASEKHAR: Astrophys. Journ. **124**, 232 (1956). — K. H. PRENDERGAST: Astrophys. Journ. **123**, 496 (1956). — L. MESTEL: Monthly Notices Roy. Astronom. Soc. London **116**, 324 (1956). — In this last paper, the author has adopted a more physical approach taking a possible rotation and energy generation into account. The effects of this last factor remind one of its rôle in v. ZEIPEL's theorem in the case of rotation.

and (83.7) leads to

$$\operatorname{curl}\left(\frac{\partial \delta \boldsymbol{r}}{\partial t} \times \boldsymbol{H}\right) = \frac{\partial \boldsymbol{H}'}{\partial t} \quad \text{or} \quad \operatorname{curl}(\delta \boldsymbol{r} \times \boldsymbol{H}) = \boldsymbol{H}'. \tag{83.17}$$

This last equation expresses that, in a medium of infinite conductivity, the change in the magnetic field is due to the pushing aside or the crowding together of the magnetic lines of force which are "frozen" in the material.

The complete equations are rather complicated and, in the first investigations, only very simplified models were considered.

M. SCHWARZSCHILD[1] who was the first to tackle the problem treated a homogeneous incompressible sphere pervaded by a uniform magnetic field ($\boldsymbol{j}=0$). Furthermore, he neglected the perturbation of the gravitational potential and considered only meridian motion. With these simplifications, the necessary eliminations can be carried out without undue complication.

Limiting the series representing the solutions to the first few terms, SCHWARZSCHILD was able to satisfy the continuity of the electric and magnetic fields across the external surface of the sphere $r=R$ at all points but the pressure boundary condition could only be imposed at a few discrete points. Nevertheless this procedure should have yielded periods of the correct order of magnitude. But instead of the complete condition (75.23), only the local pressure variation p' was set equal to zero. This explains why he obtained frequencies σ depending on H in the form

$$\sigma^2 = \frac{s^2 H^2}{4\pi \varrho R^2} \tag{83.18}$$

where s^2 is a numerical factor of the order of 1 to 5 in the cases treated by SCHWARZSCHILD. These frequencies correspond to periods of the order of the time taken by a magnetohydrodynamic wave moving with Alfvén velocity $a = H\sqrt{4\pi\varrho}$ to cross the star's radius. In order that these periods be equal to a few days as observed by BABCOCK, H should be of the order of 10^6 gauss.

Actually, in the case of an incompressible sphere without magnetic field, even if we neglect Φ', the frequencies (73.2) do not vanish but reduce to

$$\sigma_l^2 = \frac{4\pi G \varrho}{3} l. \tag{83.19}$$

Thus one might expect that the correct boundary condition would have given frequencies of the form

$$\sigma_l^2 = \alpha \frac{4\pi G \varrho}{3} \pm \frac{\beta H^2}{4\pi \varrho R^2} \tag{83.20}$$

where α and β for simple modes, are close to unity. One verifies immediately that for a star of the type of Sirius A with $\varrho \approx 1$, $R = 10^{11}$ cm, even if one takes $H = 10^6$ gauss, the second term in (83.20) is at most of the order of 10^{-4} times the first one. As suggested by (83.14), it is only for fields of the order of 10^8 gauss that the gravitational and magnetic frequencies of a main sequence star become comparable. This means that, in general, it is the pressure and gravitational field which will control the periods and, for main sequence stars, they will be much shorter than the observed periods. The only case, where a formula of type (83.18) may keep a meaning is the one of purely torsional oscillations with displacements parallel to the level surfaces. FERRARO and MEMORY[2] treated the

[1] M. SCHWARZSCHILD: Ann. d'Astrophys. **12**, 148 (1949).

[2] V. C. A. FERRARO and D. J. MEMORY: Monthly Notices Roy. Astronom. Soc. London **112**, 361 (1952).

case of an incompressible sphere with a non-uniform radial magnetic field (magnetic pole at the centre). Using spherical harmonic analysis and applying the same boundary condition as SCHWARZSCHILD, they found frequencies given again by a formula of type (83.18) in which, however, H represents the value H_s of the field in the external layers. In that case, for observed fields ($H_s \approx 10^3$ to 10^4 gauss) the periods are of the order of a few years rather than a few days. But as before, the correct boundary condition would lead to a formula of type (83.20) and again the gravitational forces would control the periods.

In a first publication[1], Miss GJELLESTAD tackled the same case as SCHWARZSCHILD trying to keep the treatment as general as possible but did not carry the analysis through completely. Later[2], she turned to a special case, where $\boldsymbol{j}$ always remains zero in the interior of the star although the magnetic field acquires a variable component, the motion and the frequencies being the same as when the magnetic field is absent. This implies surface currents parallel to circles of latitude and according to COWLING [66] such currents flowing across H in the tenuous external layers of actual stars would imply impossibly large dynamical effects.

As in the case of rotation, it is possible to generalize the method of Sect. 59β to this case. To this end, CHANDRASEKHAR and LIMBER[3] take the first variation of (83.12) and after some transformations obtain the following integral formula for σ^2:

$$\left.\begin{aligned}\sigma^2 \int_0^M (\boldsymbol{r}\cdot\delta\boldsymbol{r})\,dm = (3\gamma-4)\Big\{\int_0^M \delta\boldsymbol{r}\cdot\operatorname{grad}\Phi\,dm - \frac{1}{8\pi}\int^{\mathscr{V}} H^2 \operatorname{div}\delta\boldsymbol{r}\,d\mathscr{V} + \\ + \frac{1}{4\pi}\int^{\mathscr{V}} \boldsymbol{H}\cdot[(\boldsymbol{H}\cdot\operatorname{grad})\,\delta\boldsymbol{r}]\,d\mathscr{V}\Big\}\end{aligned}\right\} \tag{83.21}$$

which for a pseudo-radial oscillation $\delta\boldsymbol{r} = C\boldsymbol{r}$ reduces to

$$\sigma^2 = -(3\gamma-4)\,\frac{W+\mathfrak{M}}{I}. \tag{83.22}$$

One should note that in the second member the volume on which the integrals are extended is such that on its surface the total stresses (dynamic and magnetic) vanish. In general, such a surface will not coincide with the photosphere proper but, according to CHANDRASEKHAR, in most astrophysical contexts, $\boldsymbol{H}$ will become negligible before violating the assumption of infinite conductivity, on which the whole treatment rests. Formula (83.21) again suggests that, as in (83.20), the correct expression for σ^2 contains a gravitational part corrected by terms corresponding to the work against the magnetic pressure and the magnetic tensions. In an incompressible fluid, the first correction vanishes while the second would cancel if $\delta\boldsymbol{r}$ was constant along a line of force.

If $\mathfrak{M} \ll W$, the method of perturbation can also be applied. Let us suppose that the changes in the hydrostatic values of the variables due to the presence of the magnetic field are small and such that their squares or products are negligible as well as their products with the variations in the course of the oscillation. In that case, the equation of motion (83.9) becomes

$$\left.\begin{aligned}(\sigma_n^2)_c\,\varrho\,\boldsymbol{h}\,n = \varrho_0 \operatorname{grad}\Phi' + \varrho' \operatorname{grad}\Phi_0 + \operatorname{grad}p' + \\ + \frac{1}{4\pi}[\boldsymbol{H}\times\operatorname{curl}\boldsymbol{H}' + \boldsymbol{H}'\times\operatorname{curl}\boldsymbol{H}],\end{aligned}\right\} \tag{83.23}$$

[1] G. GJELLESTAD: Report No. 1, Inst. for Theor. Astrophys., Oslo, 1950.
[2] G. GJELLESTAD: Ann. d'Astrophys. 15, 276 (1952).
[3] S. CHANDRASEKHAR and D. N. LIMBER: Astrophys. Journ. 119, 10 (1954).

with the same notation as in Sect. 80, ϱ_0 and Φ_0 being the hydrostatic values in the absence of $\boldsymbol{H}$. Proceeding as before, one obtains

$$(\sigma_n^2)_c - \sigma_n^2 = 2\sigma_n \sigma_n'' = \frac{\frac{1}{4\pi}\int^{\mathcal{V}} (\boldsymbol{H}\times \operatorname{curl}\boldsymbol{H}' + \boldsymbol{H}'\times \operatorname{curl}\boldsymbol{H})\cdot \boldsymbol{h}_n\, d\mathcal{V}}{\int^{\mathcal{V}} \varrho_0 (\boldsymbol{h}_n\cdot\boldsymbol{h}_n)\, d\mathcal{V}}. \tag{83.24}$$

Since $\boldsymbol{H}'$ as given by (83.17) is proportional to $\boldsymbol{H}$, the difference $(\sigma_n^2)_c - \sigma_n^2$ will be proportional to H^2/ϱ as in (83.20).

With the same boundary conditions as Chandrasekhar and Limber, (83.24) can also be written

$$\left.\begin{aligned} 2\sigma_n\sigma_n''\int^{\mathcal{V}} \varrho_0(\boldsymbol{h}_n\cdot\boldsymbol{h}_n)\,d\mathcal{V} &= \frac{1}{4\pi}\int^{\mathcal{V}} H^2(\operatorname{div}\boldsymbol{h}_n)^2\,d\mathcal{V} - \\ &\quad - \frac{1}{4\pi}\int^{\mathcal{V}} \boldsymbol{H}\cdot[(\boldsymbol{H}\cdot\operatorname{grad})\boldsymbol{h}_n - (\boldsymbol{h}_n\cdot\operatorname{grad})\boldsymbol{H}]\operatorname{div}\boldsymbol{h}_n\,d\mathcal{V} - \\ &\quad - \frac{1}{4\pi}\int^{\mathcal{V}} [(\boldsymbol{H}'\cdot\operatorname{grad})\boldsymbol{H} + (\boldsymbol{H}\cdot\operatorname{grad})\boldsymbol{H}']\cdot\boldsymbol{h}_n\,d\mathcal{V}. \end{aligned}\right\} \tag{83.25}$$

Unfortunately these expressions are cumbersome to handle. If the components (82.31) of $\boldsymbol{h}_n$ are introduced in (83.25), the first term according to (75.9) will not depend explicitly on the rank m of the spherical harmonic. However, if $\boldsymbol{h}_n$ has a non-vanishing φ-component (non-axialsymmetric oscillations), some of the other terms will exhibit such a dependence. It is difficult to judge whether, in the general case, this will lift completely the degeneracy of the eigenvalues as the rotation does. Simple cases suggest that only the absolute values of m play a rôle so that the degenerate frequency $(\sigma_l)_n$ of the n-th mode is replaced by only $(l+1)$ frequencies $(\sigma_l^{|m|})_n$, $|m| = 0, 1, \ldots, l$. Generally, these frequencies differing by multiples of H^2/ϱ would be extremely close to each other.

In all cases, excepting torsional oscillations (very long periods) and the case $\mathfrak{M} \approx W$, the conclusion remains that the frequencies should be of the form

$$\sigma^2 = \sigma_G^2 + C\,\frac{H^2}{\varrho}, \tag{83.26}$$

where σ_G is the corresponding gravitational frequency which fixes the order of magnitude of σ^2. In that case, the only oscillations capable to give periods of a few days as observed are the g-oscillations discussed in Sect. 79. And even then, it would have to be a mode of very high order, i.e. one with many nodes along r, except if the average gradient of temperature in the star should happen to be very close to its adiabatic value [$A \approx 0$, cf. Eq. (74.11)]. As remarked by Cowling [66], this condition might seem to be exactly what is required to explain the large observed variations of the magnetic field. For instance, let us assume that the star has a dipole field and that an observer is looking at the pole while the star is undergoing an oscillation consisting of a predominantly horizontal motion in the meridian, reversing its sense at a small depth under the surface. If the lines of force can be considered as frozen in the material their orientations will be affected very strongly by this type of motion giving the observer the impression of a large increase or decrease in the magnetic field as they are tilted towards or away from him. However after a careful discussion, Cowling [66] came to the conclusion that the condition of continuity of the magnetic field at the surface would limit strongly this type of motion and that the resulting variation of the magnetic field would be at most some 13% of the undisturbed field which falls very short of the observed amplitudes, let alone the cases where the field changes sign.

V. Non-linear radial oscillations.

84. General equations. The general equations (45.12), (48.4) with, if necessary, the radiative and turbulent stresses added and (52.7) valid for finite oscillations form a very complex problem. The GEORGE DARWIN lecture delivered in 1943 by S. ROSSELAND[1] aroused a new interest in the subject but nevertheless up to now, research in this field has been limited to the case of purely radial pulsations. Even so, the energy Eq. (52.1) has been used mainly in its adiabatic form (53.5). This is rather unfortunate as, in physical models, the largest relative amplitudes $\delta r/r$ occur in the external layers where, as we have seen in Sects. 67, 69 and 70, the adiabatic approximation breaks down.

Actually $\xi = \delta r/r$, at least in classical Cepheids, remains small reaching a maximum value of the order of 5% at the surface. However, the observed material velocities in the external layers are often an appreciable multiple of the local sound velocity and in these circumstances the term in $(\boldsymbol{v}\cdot \operatorname{grad})\boldsymbol{v}$ ceases generally to be negligible with respect to $\partial \boldsymbol{v}/\partial t$.

Furthermore, $\delta p/p$, $\delta \varrho/\varrho$ and $\delta L/L$ become large in the external layers. For instance, even in the linear approximation, the limit of (70.1) as p tends towards zero, gives

$$\left(\frac{\delta p}{p}\right)_R = \lim \left(\frac{\delta p}{p}\right)_{p\to 0} = -(\omega^2+4)\,\xi \tag{84.1}$$

and since ω^2 is of the order of 5 to 10 for reasonable models (cf. Table 12) $(\delta p/p)_R \approx 0.5$ to 1 if $\xi_R = 0.05$. Under the same conditions $(\delta\varrho/\varrho)_R$ is also of the order of 0.5 for the linear adiabatic standing wave and, as we have seen in discussing (70.3), any small change in the phase of ξ increases this value considerably.

Also the observed shape of the velocity curves which, in many cases, departs strongly from being sinusoïdal, suggests that non-linear terms play an important rôle in the external layers. Furthermore, the amplitudes of linear oscillations either decrease or increase depending on the sign of σ'' (cf. Sect. 63). In the last case which must be the relevant one here, finite oscillations with a constant amplitude can only result from the fact that, as the amplitude increases, non-linear factors appear which for a given value of the amplitude balance the initial vibrational instability. Thus a complete explanation of the finite oscillations observed cannot avoid the study of the non-linear terms.

For radial oscillations, if v denotes the radial velocity, the three fundamental Eulerian equations can be written:

1. *Conservation of mass:*

$$\frac{d\varrho}{dt} = \frac{\partial\varrho}{\partial t} + v\,\frac{\partial\varrho}{\partial r} = -\frac{\varrho}{r^2}\,\frac{\partial}{\partial r}(r^2 v) \quad \text{or} \quad \frac{\partial m}{\partial t} + v\,\frac{\partial m}{\partial r} = 0. \tag{84.2}$$

2. *Conservation of momentum:*

$$\frac{dv}{dt} = \frac{\partial v}{\partial t} + v\,\frac{\partial v}{\partial r} = -\frac{\partial\Phi}{\partial r} - \frac{1}{\varrho}\,\frac{\partial p}{\partial r} + \frac{\psi(\eta, v)}{\varrho} \tag{84.3}$$

[1] S. ROSSELAND: Monthly Notices Roy. Astronom. Soc. London **103**, 239 (1943).

where $\psi(\eta, v)$ represents the friction force per unit volume which according to (57.17) is of the form

$$\psi(\eta, v) = \frac{4}{3r^3}\frac{\partial}{\partial r}\left[\eta r^4 \frac{\partial}{\partial r}\left(\frac{v}{r}\right)\right]$$

also,

$$\frac{\partial \Phi}{\partial r} = \frac{G\,m(r)}{r^2}, \qquad \frac{\partial^2 \Phi}{\partial r^2} = 4\pi G \varrho - 2\frac{G\,m(r)}{r^3}.$$

3. *Conservation of energy*:

$$\frac{dp}{dt} = \frac{\partial p}{\partial t} + v\frac{\partial p}{\partial r} = \frac{\Gamma_1 p}{\varrho}\frac{d\varrho}{dt} + (\Gamma_3 - 1)\,\varrho\left\{\varepsilon - \frac{1}{\varrho r^2}\frac{\partial}{\partial r}(r^2 F)\right\} \tag{84.4}$$

where $\varepsilon = \varepsilon_1 + \varepsilon_2 - \varepsilon_3$ represents the total source of heat energy and F, the total flux.

If we take the total time derivative of (84.3) applying the rule (43.5) and eliminate Φ, $\frac{d\varrho}{dt}$ and $\frac{dp}{dt}$, we obtain

$$\left.\begin{aligned}\frac{d^2 v}{dt^2} &= \frac{1}{\varrho}\frac{\partial}{\partial r}\left[\frac{\Gamma_1 p}{r^2}\frac{\partial}{\partial r}(r^2 v)\right] + \frac{1}{r}\frac{dv^2}{dt} + 4\frac{G\,m(r)}{r^3} v - \\ &\quad - \frac{1}{\varrho}\frac{\partial}{\partial r}\left[(\Gamma_3 - 1)\,\varrho\left\{\varepsilon - \frac{1}{\varrho r^2}\frac{\partial}{\partial r}(r^2 F)\right\}\right] - \frac{2v}{r}\frac{\psi(\eta, v)}{\varrho} + \frac{d}{dt}\left[\frac{\psi(\eta, v)}{\varrho}\right]\end{aligned}\right\} \tag{84.5}$$

which in the isentropic case, reduces to

$$\frac{d^2 v}{dt^2} = \frac{1}{\varrho}\frac{\partial}{\partial r}\left[\frac{\Gamma_1 p}{r^2}\frac{\partial}{\partial r}(r^2 v)\right] + \frac{1}{r}\frac{dv^2}{dt} + 4\frac{G\,m(r)}{r^3} v \tag{84.6}$$

where $m(r)$ as well as p and ϱ are still functions of t.

In some respects, the Lagrangian equations are simpler here because the continuity equation and, in some cases, the piezotropy equation can be written in integrated form. We have

1. *Conservation of mass* [cf. (44.2′)]:

$$\varrho r^2 \frac{\partial r}{\partial r_0} = \varrho_0 r_0^2 \quad \text{or} \quad m(r) = m(r_0) \tag{84.7}$$

where the index zero denotes values in some initial state.

2. *Conservation of momentum*:

If the viscosity is negligible, one has

$$\ddot{r} = -\frac{1}{\varrho(\partial r/\partial r_0)}\frac{\partial p}{\partial r_0} - \frac{G\,m(r_0)}{r^2} = -\frac{r^2}{\varrho_0 r_0^2}\frac{\partial p}{\partial r_0} - \frac{G\,m(r_0)}{r^2} \tag{84.8}$$

taking (84.7) into account.

3. *Conservation of energy*:

$$\dot{p} = \frac{\Gamma_1 p}{\varrho}\dot{\varrho} + (\Gamma_3 - 1)\,\varrho\left\{\varepsilon - \frac{\partial L(r)}{\partial m}\right\}. \tag{84.9}$$

If the motion is isentropic and Γ_1 constant in time, this last equation also can be written

$$\frac{p}{p_0} = \left(\frac{\varrho}{\varrho_0}\right)^{\Gamma_1}. \tag{84.10}$$

In this case, the elimination of p from (84.8) gives

$$\ddot{r} = -\frac{r^2}{\varrho_0 r_0^2}\frac{\partial}{\partial r_0}\left\{p_0\left[\frac{3r_0^2}{(\partial r^3/\partial r_0)}\right]^{\Gamma_1}\right\} - \frac{G\,m(r_0)}{r^2} \tag{84.11}$$

which takes even a simpler aspect if m is used as the independent variable

$$\ddot{r} = -r^2 \frac{\partial}{\partial m}\left[\frac{p_0}{\varrho_0^{\Gamma_1}}\left(\frac{3}{\partial r^3/\partial m}\right)^{\Gamma_1}\right] - \frac{G m}{r^2}. \tag{84.12}$$

In the case of barotropy, the factor $p_0/\varrho_0^{\Gamma_1}$ reduces to a constant K and we are left with

$$\ddot{r} = -K r^2 \frac{\partial}{\partial m}\left(\frac{3}{\partial r^3/\partial m}\right)^{\Gamma_1} - \frac{G m}{r^2}. \tag{84.13}$$

If Γ_1 varies in the course of the oscillation, the simple relation (84.10) is no longer valid. Of course, formally one could write

$$\frac{p}{p_0} = \exp\left[-\int_0^t \dot{\Gamma}_1 \log\frac{\varrho}{\varrho_0}\, dt\right]\left(\frac{\varrho}{\varrho_0}\right)^{\Gamma_1} = K(r_0, t)\left(\frac{\varrho}{\varrho_0}\right)^{\Gamma_1} \tag{84.14}$$

but, in general, the explicit expression of $K(r_0, t)$ will be very complicated. If Γ_1 is constant, but the non-adiabatic part of (84.9) is taken into account, one could also resort to an expression of the form (84.14) with

$$K(r_0, t) = \exp\left[\int_0^t (\Gamma_3 - 1)\frac{\varrho}{p}\left(\varepsilon - \frac{dL}{dm}\right) dt\right], \tag{84.15}$$

but again this is not very practical.

85. Steady state oscillations. If we consider isentropic oscillations with Γ_1 constant equal to γ, the simplest case occurs when the variables in (84.11) are separable so that

$$r = \varphi(r_0)\, w(t). \tag{85.1}$$

If $\gamma = \frac{4}{3}$, this is always possible and (84.11) can be written

$$\ddot{w}\, w^2 = -\frac{\varphi(r_0)}{\varrho_0 r_0^2}\frac{\partial}{\partial r_0}\left\{p_0\left[\frac{r_0^2}{\varphi^2(\partial\varphi/\partial r_0)}\right]^{\frac{4}{3}}\right\} - \frac{G\, m(r)}{\varphi^3} \tag{85.2}$$

and admits the homology transformation $r = r_0 t$, $\varrho = \varrho_0 t^{-3}$, $p = p_0 t^{-4}$ as a solution. Excluding this case, the last term in (84.12) implies that $\varphi(r_0)$ be of the form

$$\varphi(r_0) = A\, m^{\frac{1}{3}} \tag{85.3}$$

where A is a constant and (84.12) becomes

$$\ddot{w}(t)\, w^{3\gamma-2} + A^{1-3\gamma} 3^\gamma m^{\frac{4}{3}} \frac{\partial}{\partial m}\left(\frac{p_0}{\varrho_0^\gamma}\right) + G A^{-3} w^{3\gamma-4} = 0. \tag{85.4}$$

The second term must also reduce to a constant, say B, giving

$$\frac{p_0}{\varrho_0^\gamma} = B\, m^{\frac{2}{3}} + \frac{p_C}{\varrho_C^\gamma}. \tag{85.5}$$

This is a rather stringent condition for the hydrostatic model. For instance, in the case of barotropy, (85.5) implies $B = 0$, but then (85.4) reduces to

$$\ddot{w}(t) = -\frac{G}{A^3}\frac{1}{w^2}$$

which does not correspond to a periodic motion. In the case of polytropes, (85.5) expressed in terms of EMDEN'S variables ϑ and ξ defined by (79.20), yields

$$\vartheta^{n(1-\gamma)+1} = \frac{B(n+1)\,\varrho_C^{\gamma-\frac{4}{3}}}{(4\pi)^{\frac{1}{3}} G}\left(-\xi^2\frac{d\vartheta}{d\xi}\right)^{\frac{2}{3}} + 1 \tag{85.6}$$

which, except for the homogeneous model ($n=0$), is not compatible with the hydrostatic EMDEN'S equation. It is also very unlikely that any general physical model could satisfy (85.5).

In the case of the homogeneous compressible model, BATHNAGAR and KOTHARI[1] have shown that the radius varies according to

$$r = r_0 w(t) \tag{85.7}$$

fhe star going through its equilibrium state for $w=1$. Using (85.7) and (76.1) tor p_0, Eq. (84.11) becomes

$$\left.\begin{aligned} \ddot{w} &= \frac{4\pi G \varrho_0}{3} \frac{1}{w^2}\left(\frac{1}{w^{3\gamma-4}} - 1\right) \\ &= f(w,\gamma). \end{aligned}\right\} \tag{85.8}$$

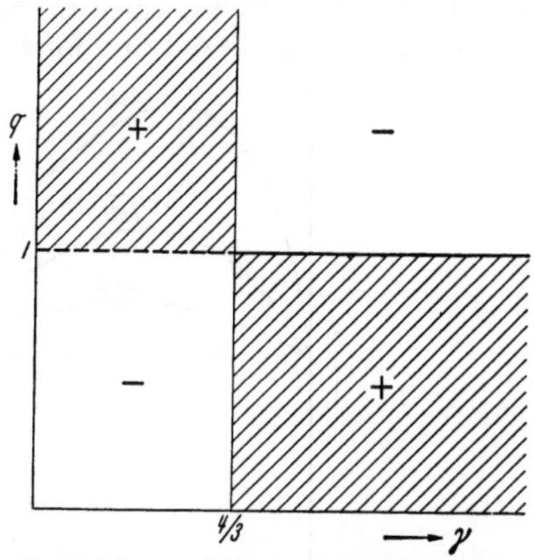

Fig. 44. Diagram of $f(q, \gamma)$. Among the equilibrium states ($q=1$), only those lying above a positive region (heavy line) are stable.

This shows that the dynamical instability encountered for infinitesimal oscillations when $\gamma < \frac{4}{3}$ subsists in this case for finite displacements since any motion goes on accelerating indefinitely. This can also be illustrated in the (w, γ) plane where the curve $f(w, \gamma) = 0$ degenerates into two straight lines (cf. Fig. 44) separating positive and negative regions. Of all the possible equilibrium states $w=1$, only those lying above positive regions are stable. The usefulness of this method due to POINCARÉ appears better when $f(w,\gamma)$ is more complicated.

Eq. (85.8) multiplied by $\dot{w}$ and integrated yields for the variation of the kinetic energy

$$\left.\begin{aligned} \frac{1}{2}\dot{w}^2 &= \frac{4\pi G \varrho_0}{3}\left[\frac{1}{w} - \frac{1}{3(\gamma-1)\, w^{3(\gamma-1)}} + C_1\right] \\ &= \frac{\sigma_0^2}{3\gamma-4}\left[\frac{1}{w} - \frac{1}{3(\gamma-1)\, w^{3(\gamma-1)}} + C_1\right] \end{aligned}\right\} \tag{85.9}$$

where σ_0 is the frequency of the fundamental mode of linear oscillations [cf. (60.10)], and the constant C_1 may be expressed in terms of the extreme values w_1 and w_2 ($w_2 < w_1$) taken by w when $\dot{w}=0$. This is particularly simple when $\gamma = \frac{5}{3}$ and gives

$$\frac{1}{C_1} = -(w_1 + w_2) = -2 w_1 w_2, \quad w_2 = \frac{w_1}{2w_1 - 1} \tag{85.10}$$

so that

$$\frac{1}{2}\dot{w}^2 = \sigma_0^2\left[\frac{1}{w} - \frac{1}{2w^2} - \frac{1}{2w_1 w_2}\right]. \tag{85.11}$$

Solving (85.11) for dt and integrating leads to

$$\sigma_0(t - t_0) = -(w_1 w_2)^{\frac{1}{2}}\left[(w_1 - w)(w - w_2)\right]^{\frac{1}{2}} + (w_1 w_2)^{\frac{3}{2}} \arcsin\frac{2w - (w_1 + w_2)}{w_1 - w_2} \tag{85.12}$$

so that the period P is given by

$$\frac{P}{P_0} = (w_1 w_2)^{\frac{3}{2}} = \left(\frac{w_1^2}{2w_1 - 1}\right)^{\frac{3}{2}}, \quad P_0 = \frac{2\pi}{\sigma_0}. \tag{85.13}$$

Thus P is somewhat lengthened with respect to P_0 but the effect is very small for usual amplitudes. For instance for $w_1 = 1.2$, $P = 1.043\, P_0$ and for w_1 of the order of 1.05 as observed, $P = 1.003\, P_0$.

[1] P. L. BATHNAGAR and D. S. KOTHARI: Monthly Notices Roy. Astronom. Soc. London **104**, 292 (1944).

The second effect of the non-linearity is to introduce some asymmetry into the variations. This can be illustrated in terms of the linear theory. In the course of a finite oscillation, the instantaneous free period being proportional to $\bar{\varrho}^{-\frac{1}{2}}$, all motions will tend to be slower during the expansion and quicker during the contraction so that the radius will tend to remain longer at values above the average than at values below it. For instance Fig. 45 represents the velocity curve $\dot{w}$ for the values 0.256, 0.133, 0.0477 of the average amplitude $\frac{1}{2}(w_1 - w_2)$. This asymmetry, however, remains relatively small. If t_1 and t_2 respectively denote the times during which the radius is greater or smaller than its equilibrium value, the ratio t_1/t_2 is only 1.2 for $\frac{1}{2}(w_1 - w_2) = 0.05$ and would reach the observed

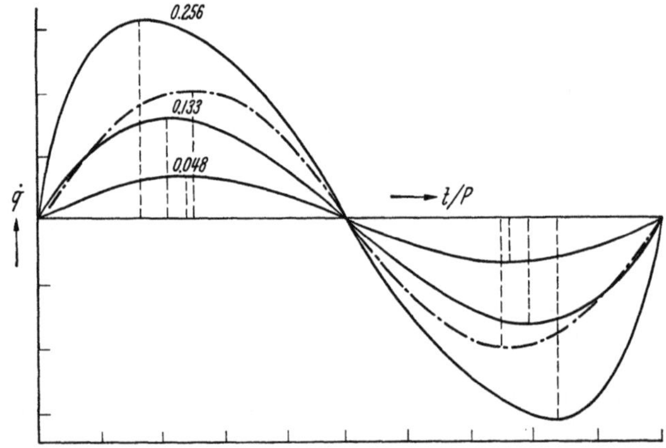

Fig. 45. Velocity curves $\dot{q}$ for the homogeneous model for different values of the semi-amplitude compared to a sinusoid.

values (2 to 4) only for amplitudes of the order of 0.3 to 0.4 which are much larger than the observed ones.

Of course the homogeneous model gives a very poor representation of an actual star and one might hope that more centrally condensed models would give a better agreement with the observations. Unfortunately, as we have just seen, a rigorous separation of the variables is unlikely in that case.

86. Successive approximations. Since it is impossible to find a general solution of (84.11), the natural procedure is to try a series development for the displacement writing for instance

$$r = r_0(1 + \xi), \tag{86.1}$$

$$\xi = \sum_{i=0} f_i(r_0)\, q_i(t), \tag{86.2}$$

where it is advantageous to choose either the $f_i(r_0)$ or the $q_i(t)$ as belonging to a known set of orthogonal functions. Both types of developments have been considered.

It has been the general practice, too, when (86.1) is introduced in (84.11), to limit the developed expression to the second order terms in ξ, obtaining

$$\left.\begin{aligned} &\frac{\varrho_0}{\gamma p_0}\ddot{\xi} - \xi'' - \left(\frac{4}{r_0} + \frac{p_0'}{p_0}\right)\xi' - \left(3 - \frac{4}{\gamma}\right)\frac{p_0'}{r_0 p_0}\xi + \\ &\quad + \frac{1}{2}\xi^2 \frac{p_0'}{r_0 p_0}\left[\left(3 - \frac{4}{\gamma}\right)(3\gamma + 1)\right] + \xi\xi'(3\gamma - 1)\left(\frac{4}{r_0} + \frac{p_0'}{p_0}\right) + \\ &\quad + \frac{1}{2}\xi'^2\left[(\gamma + 1)\frac{p_0'}{p_0} r_0 + 4(2\gamma + 1)\right] + \xi''[(\gamma+1) r_0 \xi' + (3\gamma - 1)\xi] = 0 \end{aligned}\right\} \tag{86.3}$$

where primes represent derivatives with respect to r_0.

In the case of the standard model, EDDINGTON[1] and Mrs. PELS-KLUYVER[2] have discussed solutions of (86.3) of the form

$$\xi = f_0(r_0) + f_1(r_0)\cos\sigma_0 t + f_2(r_0)\cos 2\sigma_0 t \qquad (86.4)$$

where $f_1(r_0) = \xi_0(r_0)$ and σ_0 correspond to the fundamental solution of the linear Eq. (58.12), f_0 and f_2 being found from (86.3). However, at present, the value $\gamma = 1.43$, adopted in this work, seems much too small and, since it brings near resonance between the fundamental and the first modes, the results obtained (large values of f_2, fairly large asymmetry) are not characteristic of the general case.

In more recent work[3], one has generally adopted for the f_i, the eigenfunctions of the linear problem, limiting (86.2) to the first two terms:

$$\xi = \xi_0(r_0)\,q_0(t) + \xi_1(r_0)\,q_1(t). \qquad (86.5)$$

This can be visualized as only a first step in a general method of integration of the complete problem initiated by WOLTJER [*67*] in an attempt to make available the resources of analytical mechanics. This method has been expounded at length and further developed by S. ROSSELAND ([*39d*], Chap. IV and VII).

Substituting (86.5) in (86.3) multiplied by $\gamma p_0 r_0^4$ and remembering that ξ_0 and ξ_1 are solutions of the linear equation, we obtain

$$\left.\begin{aligned}\ddot{q}_0\varrho_0\xi_0 r_0^4 + \ddot{q}_1\varrho_0\xi_1 r_0^4 + q_0\sigma_0^2\varrho_0\xi_0 r_0^4 + q_1\sigma_1^2\varrho_0\xi_1 r_0^4 +\\ + A_{00}q_0^2 + A_{11}q_1^2 + (A_{01}+A_{10})\,q_0 q_1 = 0\end{aligned}\right\} \qquad (86.6)$$

where

$$A_{ij} = \xi_i\xi_j m_j - \xi_i'\xi_j' l - \xi_i'\xi_j n_j$$

with

$$m_j = -\varrho_0\sigma_j^2(3\gamma-1)\,r_0^4 + \frac{3(1-\gamma)}{2}(3\gamma-4)\,p_0' r_0^3,$$

$$n_j = [\varrho_0\sigma_j^2 r_0^5 + (3\gamma-4)\,p_0' r_0^4]\,(\gamma+1),$$

$$l = \frac{\gamma+1}{2}\gamma p_0' r_0^5 + 2\gamma r_0^4 p_0.$$

If (86.6) is multiplied successively by $\xi_0\,dr_0$ and $\xi_1\,dr_0$ and integrated over the whole volume taking the orthogonality conditions (58.19) into account, we find the following two equations for q_0 and q_1

$$\left.\begin{aligned}\ddot{q}_0 N_0 + \sigma_0^2 q_0 N_0 + q_0^2\int_0^R A_{00}\xi_0\,dr_0 + q_0 q_1\int_0^{R_0}(A_{01}+A_{10})\,\xi_0\,dr_0 + q_1^2\int_0^R A_{11}\xi_0\,dr_0 = 0,\\ \ddot{q}_1 N_1 + \sigma_1^2 q_1 N_1 + q_0^2\int_0^{R_0} A_{00}\xi_1\,dr_0 + q_0 q_1\int_0^{R_0}(A_{01}+A_{10})\,\xi_1\,dr_0 + q_1^2\int_0^{R_0} A_{11}\xi_1\,dr_0 = 0\end{aligned}\right\} \qquad (86.7)$$

where

$$N_i = \int_0^R \varrho_0 r_0^4 \xi_i^2\,dr_0. \qquad (86.8)$$

[1] A. S. EDDINGTON: Monthly Notices Roy. Astronom. Soc. London **79**, 177 (1919).

[2] H. A. KLUYVER: Bull. astronom. Inst. Netherl. **7**, 265 (1935); cf. also [*39d*], § 7.2, where this work is summarized.

[3] a) H. K. SEN: Astrophys. Journ. **107**, 404 (1948). — b) C. PRASAD: Monthly Notices Roy. Astronom. Soc. London **109**, 529 (1949). — c) M. SCHWARZSCHILD and M. P. SAVEDOFF: Astrophys. Journ. **109**, 298 (1949).

SCHWARZSCHILD and SAVEDOFF solved equations equivalent to (86.7) numerically for initial conditions $q_1(0) = 0.06$, $\dot{q}_1(0) = 0$, $\dot{q}_2(0) = 0$, the other constant $q_2(0)$ being chosen so that $q_2(t)$ be periodic with the same period P as $q_1(t)$. They found that P is of the order of 1.002 P_0 which is very similar to the effect in the case of the homogeneous model [cf. Eq. (85.13)].

As far as the asymmetry is concerned t_1/t_2 is of the order of 1.5 which, although larger than in the homogeneous model, is still much smaller than the observed values. This suggests that the effect increases with the central condensation but the relation between the two is difficult to work out explicitly. If one considers only the term in q_0 in (86.5), the first Eq. (86.7) reduces to

$$\ddot{q}_0 + \sigma_0^2 q_0 = -\frac{\int_0^{R_0} A_{00}\,\xi_0\,dr_0}{\int_0^{R_0} \varrho_0 r_0^4 \xi_0^2\,dr_0}\, q_0^2 = D\,q_0^2$$

which has been discussed quite generally by S. ROSSELAND ([*39d*], Sect. 7.6). We may also solve it by successive approximations. As a first step, if we substitute the solution of the linear equation

$$q_{0,1} = A\cos\sigma_0 t + B\sin\sigma_0 t \tag{86.9}$$

and apply the method of variation of constants of Lagrange, we find

$$q_{0,2} = A\cos\sigma_0 t + B\sin\sigma_0 t + \int_0^t \frac{D}{\sigma_0}\, q_{0,1}^2(\tau)\sin\sigma_0(t-\tau)\,d\tau \tag{86.10}$$

or, if we express $q_{01}(\tau)$ explicitly,

$$\left.\begin{aligned} q_{0,2} = {} & \left(A - \frac{DA^2}{3\sigma_0^2} - \frac{2DB^2}{3\sigma_0^2}\right)\cos\sigma_0 t + \left(B + \frac{2DAB}{3\sigma_0^2}\right)\sin\sigma_0 t + \\ & + \frac{DA^2}{3\sigma_0^2}(1+\sin^2\sigma_0 t) + \frac{DB^2}{3\sigma_0^2}(1+\cos^2\sigma_0 t) - \frac{DAB}{3\sigma_0^2}\sin 2\sigma_0 t. \end{aligned}\right\} \tag{86.11}$$

In particular, if we suppose that B vanishes, the radial velocity is given by

$$\dot{q}_{0,2} = -A\left[\left(1 - \frac{DA}{3\sigma_0^2}\right)\sin\sigma_0 t - \frac{DA}{3\sigma_0^2}\sin 2\sigma_0 t\right]. \tag{86.12}$$

In this approximation, the period is not modified, but we see that the main factors affecting the asymmetry are of the order of $DA/3\sigma_0^2$. Of course, D depends on the model but in a complex fashion which would be most readily discussed numerically.

However Schwarzschild and Savedoff note that the term in q_1 which we have neglected here contributes appreciably to the asymmetry. In this respect, one should particularly keep in mind WOLTJER's suggestion [*67*] that a near-commensurability between the fundamental mode and some higher mode could affect considerably the shape of the radius variations. For instance, assuming that the mode (q_0, ξ_0) contributes predominantly to the oscillation, the second equation (86.7) can be approximated by

$$\ddot{q}_1 + \sigma_1^2 q_1 = -q_0^2\,\frac{\int_0^{R_0} A_{00}\,\xi_1\,dr_0}{\int_0^{R_0} \varrho_0 r_0^4 \xi_1^2\,dr_0} = d\,q_0^2. \tag{86.13}$$

If again we introduce the first approximation (86.9) into the second member, the solution is

$$\left.\begin{aligned} q_{1,2} = {} & \frac{(A^2+B^2)\,d}{2\sigma_1^2} + \cos\sigma_1 t\left(A_1 + \frac{A^2-B^2}{2}\,\frac{d}{4\sigma_0^2-\sigma_1^2} - \frac{A^2+B^2}{2}\,\frac{d}{\sigma_1^2}\right) + \\ & + \sin\sigma_1 t\left(B_1 + \frac{2\sigma_0 A B d}{\sigma_1[4\sigma_0^2-\sigma_1^2]}\right) - \\ & - \frac{(A^2-B^2)\,d}{2(4\sigma_0^2-\sigma_1^2)}\cos 2\sigma_0 t - \frac{A B d}{(4\sigma_0^2-\sigma_1^2)}\sin 2\sigma_0 t \end{aligned}\right\} \tag{86.14}$$

where we can always choose A_1 and B_1 to eliminate the periodicity in σ_1. We see that if σ_1 is close to $2\sigma_0$ the amplitude of some of the terms in q_1 could become large even if d, which again depends on the model, is small. This reasoning can be generalized to the case where the complete solution q_0 is expressed as a Fourier series of argument $(n\sigma_0' t)$ where σ_0' may be slightly different from σ_0 (cf. [*39d*], Sect. 7.6). In that case, resonance with any multiple n of σ_0' is possible (subharmonic resonance) and this would be especially important if in (86.5), higher modes than (ξ_1, q_1) would be taken into consideration.

Another important suggestion found in Woltjer's work [*67*] concerns the effects of the coupling between different modes. This can also be illustrated by means of Eqs. (86.7) where the coupling terms in $q_0 q_1$ are kept and which can be written

$$\left.\begin{aligned} \ddot{q}_0 + \sigma_0^2 q_0 &= D q_0^2 + E q_0 q_1, \\ \ddot{q}_1 + \sigma_1^2 q_1 &= d q_0^2 + e q_0 q_1. \end{aligned}\right\} \tag{86.15}$$

If, like Rosseland ([*39d*], Sect. 7.7), we introduce into the second members the first approximation

$$\left.\begin{aligned} q_{0,1} &= \cos\sigma_0 t, \\ q_{1,1} &= \cos(\sigma_1 t - \Phi), \end{aligned}\right\} \tag{86.16}$$

using the same method as before, we get a second approximation

$$\left.\begin{aligned} q_{0,2} = {} & \frac{D}{2\sigma_0^2} + A_0\cos(\sigma_0 t + \psi_0) - \boxed{\frac{E\cos(\sigma_0 t - \Phi)}{4\sigma_0(2\sigma_0-\sigma_1)}} - \frac{D}{6\sigma_0^2}\cos 2\sigma_0 t + \\ & + \boxed{\frac{E}{2}\,\frac{\cos[(\sigma_1-\sigma_0)t-\Phi]}{\sigma_1(2\sigma_0-\sigma_1)}} - \frac{E}{2}\,\frac{\cos[(\sigma_0+\sigma_1)t-\Phi]}{\sigma_1(2\sigma_0+\sigma_1)}, \end{aligned}\right\} \tag{86.17}$$

$$\left.\begin{aligned} q_{1,2} = {} & \frac{d}{2\sigma_1^2} + A_1\cos(\sigma_1 t + \psi_1) + \\ & + \boxed{\frac{d}{2}\,\frac{\cos\sigma_1 t}{(2\sigma_0-\sigma_1)(2\sigma_0+\sigma_1)}} - \boxed{\frac{d}{2}\,\frac{\cos 2\sigma_0 t}{(2\sigma_0-\sigma_1)(2\sigma_0+\sigma_1)}} + \\ & + \frac{e}{2}\,\frac{\cos[(\sigma_1-\sigma_0)t-\Phi]}{\sigma_0(2\sigma_1-\sigma_0)} - \frac{e}{2}\,\frac{\cos[(\sigma_0+\sigma_1)t-\Phi]}{\sigma_0(2\sigma_1+\sigma_0)}, \end{aligned}\right\} \tag{86.18}$$

where the constant terms shift slightly the position around which the oscillation takes place and where A_0, A_1, ψ_0, ψ_1 are new constants whose exact values have no importance in this context. If σ_1 is close to $2\sigma_0$, $\sigma_1 = 2\sigma_0 + \varepsilon$, some terms will be reinforced and may become important even if E and d are relatively small. Those in $q_{1,2}$ are the same as in (86.14) ($A=1$, $B=0$) and are due to resonance. They have frequencies $\sigma_1 = 2\sigma_0 + \varepsilon$ and $2\sigma_0$ and will modify the shape of the variation in the course of the fundamental period but, by interference between themselves and with the second term, they could also produce a beat of frequency ε corresponding to a very long period. The reinforced terms in $q_{0,2}$ are due to coupling and have frequencies σ_0 and $\sigma_1-\sigma_0=\sigma_0+\varepsilon$ and can give rise also to a beat of frequency ε. Of course, as in (86.14), one could choose Φ so that the second and third terms of (86.18) cancel out, but the beat phenomenon would subsist due to the interfering terms in $q_{0,2}$.

Of course the linear solution $q_{1,1}$ considered does not necessarily correspond to the first linear mode but may represent any mode which happens to have a frequency close to twice the fundamental frequency. However for higher modes the corresponding values of d and e would be appreciably smaller so that $2\sigma_0$ and σ_1 would have to be correspondingly closer. Also if higher harmonics had been included in (86.16), new terms reinforced by other commensurabilities between σ_0 and σ_1 would appear but their coefficients again would be much smaller.

One might also attempt to extend this discussion to the study of the simultaneous interactions between more than two modes (cf. [67], especially the fourth paper) but despite the simplifications introduced by WOLTJER's formalism and the use of angle-variables, the problem rapidly becomes very complex.

87. The non-adiabatic equation. If we want to take the deviations from isentropy into consideration the elimination of p from (84.8) should be carried out by means of a formula of type (84.14) with $K(r_0, t)$ given by (84.15). In all generality, this would lead to very complicated expressions, but admitting that in equilibrium $\varepsilon_0=(\partial L/\partial m)_0$, we may treat the integral in $K(r_0, t)$ as a very small quantity and write

$$K(r_0, t) = 1 + \int_0^t (\Gamma_3 - 1)\frac{\varrho}{p}\,\delta\left(\varepsilon - \frac{\partial L}{\partial m}\right)dt. \tag{87.1}$$

In that case, the elimination of p leads to

$$\left.\begin{aligned}\ddot{r} = &-\frac{r^2}{\varrho_0 r_0}\frac{\partial}{\partial r_0}\left\{p_0\left(\frac{3r_0^2}{\partial r^3/\partial r_0}\right)^{\Gamma_1}\right\} - \\ &-\frac{r^2}{\varrho_0 r_0}\frac{\partial}{\partial r_0}\left\{p_0\left(\frac{\varrho}{\varrho_0}\right)^{\Gamma_1}\int_0^t(\Gamma_3-1)\frac{\varrho}{p}\,\delta\left(\varepsilon-\frac{\partial L}{\partial m}\right)dt\right\} - \frac{G m}{r^2}.\end{aligned}\right\} \tag{87.2}$$

If we introduce, as before, $r=r_0(1+\xi)$ and expand up to the second order terms, we get back Eq. (86.3) with the additional terms

$$\left.\begin{aligned}\frac{\varrho_0}{\gamma p_0 r_0}\Bigg\{&\frac{1}{\varrho_0}\frac{\partial}{\partial r_0}\left[\int_0^t(\gamma-1)\,\varrho_0\,\delta\left(\varepsilon-\frac{\partial L}{\partial m}\right)dt\right] + \\ &+\frac{2\xi}{\varrho_0}\frac{\partial}{\partial r_0}\left[\int_0^t(\gamma-1)\,\varrho_0\,\delta\left(\varepsilon-\frac{\partial L}{\partial m}\right)dt\right] + \\ &+\frac{1}{\varrho_0}\frac{\partial}{\partial r_0}\left[\gamma\frac{\delta\varrho}{\varrho_0}\int_0^t(\gamma-1)\,\varrho_0\,\delta\left(\varepsilon-\frac{\partial L}{\partial m}\right)dt\right] - \\ &-\frac{1}{\varrho_0}\frac{\partial}{\partial r_0}\left[\int_0^t(\gamma-1)^2\,\delta\varrho\,\delta\left(\varepsilon-\frac{\partial L}{\partial m}\right)dt\right]\Bigg\}\end{aligned}\right\} \tag{87.3}$$

assuming $\Gamma_1=\Gamma_3=\gamma$.

But $\delta\left(\varepsilon-\frac{\partial L}{\partial m}\right)$ is very small in a large fraction of the star and as a first approximation we shall only keep the first term. If, as before, we substitute in Eq. (86.3) thus modified the expression (86.5), we obtain Eq. (86.6) with the additional term

$$r_0^3 \frac{\partial}{\partial r_0}\left\{(\gamma-1)\varrho_0\left(\delta\varepsilon-\frac{d\,\delta L}{dm}\right)_{\xi_0}\int_0^t q_0(t)\,dt + (\gamma-1)\varrho_0\left(\delta\varepsilon-\frac{d\,\delta L}{dm}\right)_{\xi_1}\int_0^t q_1(t)\,dt\right\}. \tag{87.4}$$

If we multiply by $\xi_0\,dr_0$ and integrate over the volume, we recover the first Eq. (86.7) plus the following terms

$$\int_0^R\left(\frac{\delta T}{T}\right)_{\xi_0}\left(\delta\varepsilon-\frac{d\,\delta L}{dm}\right)_{\xi_0}\varrho_0 r_0^2\,dr_0\int_0^t q_0(t)\,dt+\int_0^R\left(\frac{\delta T}{T}\right)_{\xi_0}\left(\delta\varepsilon-\frac{d\,\delta L}{dm}\right)_{\xi_1}\int_0^t q_1(t)\,dt. \tag{87.5}$$

Although the last term in (87.5) is not rigorously zero it is likely to be appreciably smaller than the first which is just equal to

$$-2\sigma_0^2\sigma_0''\int_0^R\varrho_0 r_0^4\xi_0^2\,dr_0\int_0^t q_0(t)\,dt$$

where σ_0'' is the damping constant of the fundamental mode [cf. (64.9)]. The first Eq. (86.7) divided by N_0 becomes then

$$\ddot q_0+\sigma_0^2 q_0-2\sigma_0^2\sigma_0''\int q_0\,dt=D\,q_0^2+E\,q_0 q_1 \tag{87.6}$$

if as in (86.15), we neglect the term in q_1^2.

Proceeding in the same manner, the second Eq. (86.7) becomes

$$\ddot q_1+\sigma_1^2 q_1-2\sigma_1^2\sigma_1''\int q_1\,dt=d\,q_0^2+e\,q_0 q_1. \tag{87.7}$$

88. Limitation of the amplitude by coupling. As we have already remarked, strictly periodic motion is impossible in a linear non-conservative system, the amplitude always decreasing to zero or increasing without limit. In the last case, as the amplitude increases, second order terms are becoming more and more important and may limit the amplitude to a finite value. Such a case may occur through the coupling of a mode with negative damping to a mode with positive damping. Introducing into the second member of (87.6) the simple solutions

$$q_{0,1}=\cos\sigma_0 t\,\mathrm{e}^{-\sigma_0'' t},\qquad q_{1,1}=\cos(\sigma_1 t-\Phi)\,\mathrm{e}^{-\sigma_1'' t}$$

of the corresponding linear equations, the second approximation is

$$q_{0,2}=\mathrm{e}^{-\sigma_0'' t}\cos\sigma_0 t+\mathrm{e}^{-\sigma_0'' t}\times\left\{\int_0^t\frac{1}{\sigma_0}\left[D\cos^2\sigma_0\tau\,\mathrm{e}^{-\sigma_0''\tau}+E\cos\sigma_0\tau\cos(\sigma_1\tau-\Phi)\,\mathrm{e}^{-\sigma_1''\tau}\right]\sin\sigma_0(t-\tau)\,d\tau\right\}.$$

If we develop the integral, neglecting terms proportional to σ_0'' or σ_1'' except if divided by $(2\sigma_0-\sigma_1)$ and also $\sigma_0''^2$ or $\sigma_1''^2$ as compared with σ_0^2 or σ_1^2 and their combinations except the particular one $(2\sigma_0-\sigma_1)$, we find

$$\left.\begin{aligned}q_{0,2}=&\frac{D}{2\sigma_0^2}\mathrm{e}^{-2\sigma_0'' t}+A_0\mathrm{e}^{-\sigma_0'' t}\cos(\sigma_0 t+\psi_0)-\frac{E\,\mathrm{e}^{-\sigma_0'' t}}{4\sigma_0}\frac{(2\sigma_0-\sigma_1)\cos(\sigma_0 t-\Phi)}{[(2\sigma_0-\sigma_1)^2+\sigma_1''^2]}-\\&-\frac{D}{6\sigma_0^2}\mathrm{e}^{-2\sigma_0'' t}\cos 2\sigma_0 t-\frac{E}{2}\mathrm{e}^{-(\sigma_0''+\sigma_1'')t}\frac{\cos[(\sigma_0+\sigma_1)t-\Phi]}{\sigma_1(2\sigma_0+\sigma_1)}+\\&+\frac{E}{2}\mathrm{e}^{-(\sigma_0''+\sigma_1'')t}\times\\&\times\left\{\frac{\sigma_1(2\sigma_0-\sigma_1)+\sigma_1''^2}{\sigma_1^2[(2\sigma_0-\sigma_1)^2+\sigma_1''^2]}\cos[(\sigma_0-\sigma_1)t+\Phi]-\frac{2(\sigma_0-\sigma_1)\sigma_1''\sin[(\sigma_0-\sigma_1)t+\Phi]}{\sigma_1^2[(2\sigma_0-\sigma_1)^2+\sigma_1''^2]}\right\}.\end{aligned}\right\}\tag{88.1}$$

In the same way, we obtain

$$\left.\begin{aligned}
q_{1,2} &= \frac{d}{2\sigma_1^2}\,e^{-2\sigma_0'' t} + A_1\cos(\sigma_1 t + \psi_1)\,e^{-\sigma_1'' t} + \frac{d}{4\sigma_1}\times\\
&\times e^{-\sigma_1'' t}\cos\sigma_1 t\left[\frac{2\sigma_0-\sigma_1}{(2\sigma_0-\sigma_1)^2+(2\sigma_0''-\sigma_1'')^2} - \frac{1}{2\sigma_0+\sigma_1}\right] - \frac{d}{4\sigma_1}\times\\
&\times e^{-2\sigma_0'' t}\frac{(2\sigma_0''-\sigma_1'')\sin 2\sigma_0 t}{(2\sigma_0-\sigma_1)^2+(2\sigma_0''-\sigma_1'')^2} + \frac{d}{4\sigma_1}\times\\
&\times e^{-2\sigma_0'' t}\cos 2\sigma_0 t\left[\frac{1}{2\sigma_0+\sigma_1} - \frac{2\sigma_0-\sigma_1}{(2\sigma_0-\sigma_1)^2+(2\sigma_0''-\sigma_1'')^2}\right] - \frac{e}{2}\times\\
&\times e^{-(\sigma_0''+\sigma_1'')t}\left\{\frac{\cos[(\sigma_0+\sigma_1)t-\Phi]}{\sigma_0(2\sigma_1+\sigma_0)} - \frac{\cos[(\sigma_1-\sigma_0)t-\Phi]}{\sigma_0(2\sigma_1-\sigma_0)}\right\}.
\end{aligned}\right\} \tag{88.2}$$

As expected, the dissipation excludes the tendency to infinite amplitudes which is otherwise present in case of perfect resonance ($\sigma_1 = 2\sigma_0$). Here the largest terms are proportional to $1/\sigma_1''$ and $1/(2\sigma_0''-\sigma_1'')$ and they are shifted a quarter period in phase with respect to the linear solution. On the other hand, the terms due to coupling have a damping coefficient equal to $(\sigma_0''+\sigma_1'')$ so that if one mode is unstable, say $\sigma_0''<0$, while the other is damped, say $\sigma_1''>0$, these two opposite influences may balance each other at least as far as these terms are concerned. In fact, the use of angle-variables has permitted to isolate a particular case ($\sigma_1 \approx 2\sigma_0$) admitting a finite solution q_0, q_1 (cf. [*39d*], Sect. 7.12). All the thermal energy fed into one of the modes is then used up completely to maintain the other mode through the coupling terms. It is difficult however to be sure that this process first pointed out by WOLTJER has a sufficiently general validity to explain the limitation of the amplitudes in actual variable stars. For instance, the above approximation will lead to finite solutions only if, apart from $\sigma_0'' \approx -\sigma_1'' < 0$ and $\sigma_1 \approx 2\sigma_0$, D and d in the selfcoupling terms are negligible and Φ is chosen so that A_0 be equal to zero.

Coupling with internal subharmonic resonance is not the only way to achieve finite amplitudes. Indeed, since the advent of electronics, the number of known systems capable of finite oscillations with amplitudes practically independent of the initial conditions has increased rapidly and their discussion and interpretation following the pioneering work of VAN DER POL[1], has made considerable progress thanks especially to the methods introduced earlier by POINCARÉ and LIAPOUNOFF [*68*].

89. Discussion of conservative systems in the phase plane. Among these methods, the use of the phase plane is particularly instructive at least as far as the qualitative discussion of the problems is concerned. To illustrate it, let us consider again the conservative case. Instead of using the processus of Sect. 86, we may as well derive an approximate mean equation from the virial theorem[2]. If we introduce in (59.11)

$$r = r_0\,\mu(r_0)\,w(t) \tag{89.1}$$

we readily obtain

$$\ddot{w}(t)\int_0^M \mu^2\,dI_0 = \frac{3}{w^{3\gamma-2}}\int_0^{\mathcal{V}} \frac{p_0\,d\mathcal{V}_0}{\mu^{2\gamma-2}[\mu + r_0(\partial\mu/\partial r_0)]^{\gamma-1}} + \frac{1}{w^2}\int_0^M \frac{dW_0}{\mu}. \tag{89.2}$$

This assumes separation of the space and time variables and, as we have seen, this is true only in very special cases but it is probably a reasonable

[1] B. VAN DER POL: Phil. Mag. **2** (1926).
[2] P. LEDOUX: Bull. Acad. Roy. Belg., Cl. Sci., Sér. V **38**, 352 (1952).

approximation when deriving a mean equation of motion. If we further assume that $\mu = 1$, Eq. (89.2) reduces to

$$\ddot{w} = -\frac{W_0}{I_0}\frac{1}{w^2}\left(\frac{1}{w^{3\gamma-4}} - 1\right) \tag{89.3}$$

which is similar to (85.8) for the homogeneous model. Introducing a new time unit such that $t = \tau(-W_0/I_0)^{-\frac{1}{2}}$, we have

$$\frac{d^2 w}{d\tau^2} = \frac{1}{w^2}\left(\frac{1}{w^{3\gamma-4}} - 1\right) \tag{89.4}$$

from which we derive as before

$$\left.\begin{aligned} &\frac{1}{2}\left(\frac{dw}{d\tau}\right)^2 \\ &= \left[\frac{1}{w} - \frac{1}{3(\gamma-1)\, w^{3(\gamma-1)}}\right] + C_1 \\ &= -V, \end{aligned}\right\} \tag{89.5}$$

where V can be considered as a generalized potential characterizing the forces acting on the system: Eq. (89.5) expresses simply the conservation of energy. Eq. (89.5) allows one to draw the phase picture of the pulsation in the phase plane $(w, dw/d\tau)$. This has been done in the lower part of Fig. 46 for $\gamma = \frac{5}{3}$ and for a given amplitude (full curve). In the upper part of the figure, the full line represents the corresponding potential V. The form of the V-curve shows immediately that appreciable deviations from an ellipse (linear oscillation) occur only for very large amplitudes. SCHWARZSCHILD and SAVEDOFF'S solution has also been included but does not show an appreciably larger asymmetry.

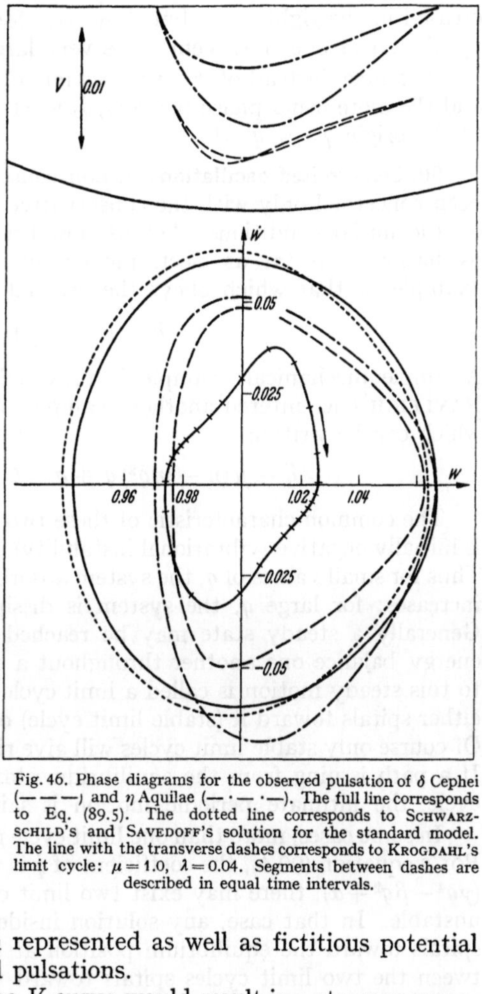

Fig. 46. Phase diagrams for the observed pulsation of δ Cephei (— — —) and η Aquilae (— · — · —). The full line corresponds to Eq. (89.5). The dotted line corresponds to SCHWARZSCHILD'S and SAVEDOFF'S solution for the standard model. The line with the transverse dashes corresponds to KROGDAHL'S limit cycle: $\mu = 1.0$, $\lambda = 0.04$. Segments between dashes are described in equal time intervals.

To illustrate the difference with the observations, the cycles of δ Cephei and η Aquilae have also been represented as well as fictitious potential curves capable of giving the observed pulsations.

Any increase in the skewness of the V-curve would result in a stronger asymmetry of the oscillation. But in the case of adiabatic oscillations, the only parameter which we have at our disposal is γ. Indeed, the observed asymmetry could be recovered with the observed amplitudes for a γ of the order of 10 as BHATNAGAR and KOTHARI have shown. However such values are excluded physically.

Another possibility is to treat γ as varying in the course of the oscillation. Such a case has been discussed by ZHEVAKIN[1] assuming that, in a small interval (w', w'') around $w = 1$, γ is smaller than $\frac{4}{3}$ while outside this interval $\gamma = \frac{5}{3}$. Of course, in this case, the ordinary equilibrium state $w = 1$, $\dot{w} = 0$, is unstable and

[1] Z. A. ZHEVAKIN: Dokl. Akad. Nauk. USSR. 62, 191 (1948).

for a positive value of C_1 in (89.5), the representative point in the phase plane describes a pear-shaped path around the points w' and w'' having horizontal tangents at the three points $w=1$, $w=w'$, $w=w''$. However this appeals to a very unlikely dynamical instability of the star and furthermore it does not increase the asymmetry very much by itself. If γ would take different values, say $\gamma_1<\frac{4}{3}$ for $w>1$ and $\gamma_2>\frac{4}{3}$ for $w<1$, the asymmetry would be increased in the right sense but, however, for physically possible values of γ_1 and γ_2, the effect does not seem to be very large either (cf. Ref. 2, p. 548).

Of course, instead of $w=r/r_0$, we might as well use the displacement $q=\delta r/r_0$ and the more usual phase plane $(q, \dot{q})$ where the equilibrium position corresponds to the origin $q=0$, $\dot{q}=0$.

90. Self-excited oscillations in non-conservative systems. Up to now we have been concerned only with the conservative case in which the amplitude depends on the initial conditions. Let us now briefly consider the case of self-excited oscillations. As far as electronic circuits are concerned, one of the simplest examples is that which obeys the original VAN DER POL's equation

$$\ddot{q}-\mu(1-q^2)\dot{q}+q=0. \tag{90.1}$$

A simple mechanical example is provided by FROUDE's pendulum and Lord RAYLEIGH encountered another example obeying an equation of the same type which can be written

$$\ddot{q}-\mu(n-m\dot{q}^2)\dot{q}+q=0, \quad m>0, \quad n>0. \tag{90.2}$$

The common characteristic of these two equations is that the term in $\dot{q}$ which is initially negative (vibrational instability) may change its sign as q or $\dot{q}$ increases. Thus for small values of q, the system absorbs energy and the amplitude gradually increases; for large q, the system is dissipative and the amplitude decreases. Generally a steady state may be reached when absorption and dissipation of energy balance one another throughout a cycle. The phase path corresponding to this steady motion is called a limit cycle since, in general, any nearby solution either spirals toward it (stable limit cycle) or away from it (unstable limit cycle). Of course only stable limit cycles will give rise to observable periodic phenomena. If a path issuing from the equilibrium state (origin) can reach the stable limit cycle, the ultimate periodic motion is said to be a *soft self-excited oscillation*.

In some cases, more than one limit cycle may occur. For instance if in VAN DER POL's equation (90.1), the coefficient of $\dot{q}$ is replaced by a more general expression $(\gamma q^4-\beta q^2+\alpha)$, there may exist two limit cycles the one nearer the origin being unstable. In that case, any solution inside the first limit cycle is damped and spirals toward the equilibrium position at the origin. Any solution initially between the two limit cycles spirals toward the stable external limit cycle. Thus to start the periodic motion, the initial amplitude has to exceed a minimum critical value which places the initial representative point outside the first limit cycle. Such cases are known as *hard self-excited oscillations*.

The discussion has been extended to more general equations of the form[1]

$$\ddot{q}+f(q)\dot{q}+g(q)=0 \quad \text{or} \quad \ddot{q}+f(q,\dot{q})\dot{q}+g(q)=0. \tag{90.3}$$

If we write

$$F(q)=\int_0^q f(q)\,dq \quad \text{and} \quad p=\dot{q}+F(q) \tag{90.4}$$

---

[1] A. LIÉNARD: Rev. gén. Electr. **23** (1928). — N. LEVINSON and O. K. SMITH: Duke Math J. **9** (1942).

the first Eq. (90.3) can be replaced by the equivalent system

$$\dot{q} = p - F(q), \qquad \dot{p} = -g(q). \tag{90.5}$$

This is of the general form

$$\dot{q} = Q(q, p), \qquad \dot{p} = P(q, p) \tag{90.6}$$

discussed by POINCARÉ and LIAPOUNOFF in terms of the critical points defined as the intersections (q_c, p_c) of the curves $P(q, p) = 0$ and $Q(q, p) = 0$ ($\dot{q} = 0, \dot{p} = 0$: stationary states). The nature of the critical point (q_c, p_c) can be ascertained by studying the Eqs. (90.6) in a small range around (q_c, p_c). In that range, it is generally sufficient to discuss the linear parts of (90.6), let us say

$$\dot{q} = a(q - q_c) + b(p - p_c), \quad \dot{p} = c(q - q_c) + d(p - p_c)$$

or the equivalent second order linear equation

$$\ddot{q} - (a + d)\dot{q} + q(ad - bc) = (ad - bc)\,q_c. \tag{90.7}$$

The roots S_1 and S_2 of its characteristic equation

$$S^2 - (a + d)\,S + (ad - bc) = 0 \tag{90.8}$$

provided their real parts do not vanish (LIAPOUNOFF), determine the way in which the critical point is approached by a general solution and its stability.

Furthermore, considering the vector field (Q, P), one may define the index of a closed curve in the phase plane (q, p) as the integral multiple m of 2π by which the vector (Q, P) turns in a chosen positive direction around this curve.

The index of a point (q, p) surrounded by this curve is the limiting value of m as the curve shrinks to the point (q, p). The index of an ordinary point is zero. One distinguishes three main types of critical points. A *node* corresponds to real roots S_1 and S_2 of (90.8). If S_1 and S_2 are negative (positive) it is stable (unstable) and all phase paths approach (leave) it along a common tangent and correspond to aperiodic damped motions. The index of a node is $+1$. At a *focus*, S_1 and S_2 are conjugate complex. If $\operatorname{Re} S_1$ and $\operatorname{Re} S_2$ are negative (one or both of them positive) it is stable (unstable) and all phase paths approach (leave) it along spirals corresponding to oscillatory damped motions. The index of a focus is $+1$. Finally, if S_1 and S_2 are real but of opposite sign, one has a *saddle-point* at the intersection of two phase paths. The index of a saddle-point is -1.

The index of a closed curve is the algebraic sum of those of the critical points inside it; as the index of a limit cycle is unity it must surround at least one critical point which cannot be a saddle-point. But, for instance, it can surround two foci and one saddle-point.

Two theorems of Bendixson give some information on the existence or otherwise of limit cycles. In particular, if the expression $\partial Q/\partial q + \partial P/\partial p$ does not change its sign in a domain D of the phase plane, no periodic motion can exist in that domain. But even if one has established the existence of a limit cycle $[q_1(t), p_1(t)]$, the stability of the motion along it has still to be investigated. In usual circumstances, this problem can be approached by the discussion of the linearized equations for small deviations from the periodic motion $[q_1(t), p_1(t)]$ [cf. Sects. 55 and 56]. This leads again to equations of the type (90.7) but where the coefficients a, b, c and d are known functions of the time. The solution will exhibit an exponential factor of the form e^{ht} where the characteristic exponent h is the average of

$$a(t) + d(t) = \frac{\partial Q(q_1, p_1)}{\partial q} + \frac{\partial P(q_1, p_1)}{\partial p}$$

over a cycle. Provided h does not vanish, the motion will be stable (unstable) if h is negative (positive).

However, more complicated cases may arise, for instance when the differential equations (90.3) depend on a parameter, the critical point changing from a stable one to an unstable one for some critical value of this parameter. Or, new unstable critical points may appear around which new limit cycles become possible. This is really what occurs around the points (w', $\gamma=\frac{4}{3}$; w'', $\gamma=\frac{4}{3}$) in the conservative stellar oscillations discussed by ZHEVAKIN (cf. Sect. 89) in the case of a variable parameter γ.

Returning to the first Eq. (90.3), one can show that if the following conditions are fulfilled

1. $f(q)$ is even thus $F(q)$ defined by (90.4) is odd,
2. $g(q)$ is odd,
3. $F(q)$ possesses a single positive zero: q_0, $F(q)<0$ if $q<q_0$, $F(q)>0$ and increases monotically with q, if $q>q_0$,

the trajectory is symmetrical with respect to the origin and there is one and only one limit cycle. Its position in the phase plane is determined by the condition

$$\int_{p_A}^{p_B} F(q)\,dq=0, \tag{90.9}$$

the curvilinear integral being taken along the half-cycle situated on one side of the p-axis. Physically this condition means that, for the periodic trajectory, the kinetic energy is equal at two consecutive passages through $q=0$. In other words, the total dissipation along such a half-cycle is exactly zero.

In more general cases, the determination of the limit cycles may be facilitated by the following considerations. If, in the phase plane, critical points whose sum of indices is $+1$ are surrounded by two closed curves along which the velocity vector (Q, P) of the representative point is nowhere directed outside the ring-like region D comprised between the two curves, there is at least one stable limit cycle within the region. If (Q, P) is always directed outside of D, there is at least one unstable limit cycle in D.

The order of magnitude of $f(q)$ is also important. For instance, in the simple case (90.1), if μ is small, the limit cycle is very close to a circle with radius $R=2$ and the motion deviates little from a harmonic oscillation. On the contrary, if μ is large, the limit cycle tends to take a rectangular shape. The phase path can be divided into four parts, two of which (opposite) are described very rapidly while the two others are described slowly so that as μ increases the oscillation departs more and more from the sinusoidal shape.

As μ becomes very large the oscillation tends finally to a square wave. This type of self-excited oscillation is known as a *relaxation oscillation*. It occurs essentially when, during one part of the cycle, the inertial force becomes negligible (very small accelerations) with respect to the restoring force ("spring" force) and the friction force which determine the motion while, in the other part of the cycle, the spring force becomes negligible with respect to the inertial force (large acceleration) and the friction force, the velocity experiencing practically an abrupt jump (sawtooth wave). The cycle may also comprise two portions of low acceleration and two portions corresponding to jump conditions (square wave). It also happens that the phenomenon is not strictly periodic, the interval between the phases of very fast accelerations becoming irregular as minor accidental variations in the system may retard or advance appreciably the instants at which the jumps occur.

Some very striking similarities between these relaxation oscillations and some types of stellar variability were pointed out by WESSELINK[1] as early as 1939. On the other hand, the general idea that the observed finite amplitudes must be due to the compensation of a vibrational instability (negative damping) for initially small amplitudes by a positive damping appearing when the amplitude increases has been in the air practically from the very beginning of the pulsation theory. Despite this, stellar self-excited oscillations have been the subject of very little work.

Recently KROGDAHL[2] has brought attention to some interesting points and has illustrated the problem by studying an equation of type (86.7) expanded up

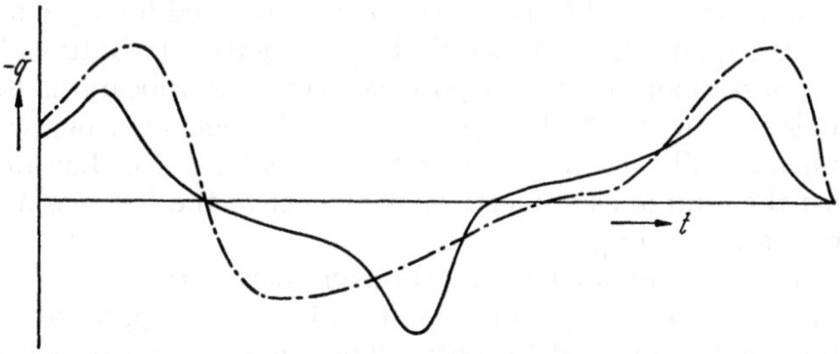

Fig. 47. Velocity curve of η Aquilae (dashed line) compared to the velocity curve (full line) corresponding to KROGDAHL's limit cycle represented on Fig. 46.

to the third order terms in q and with a term of type

$$\mu\left(1-\frac{q^2}{\lambda^2}\right)\dot{q} \tag{90.10}$$

added to the second member which makes it very similar to VAN DER POL's equation (90.1). If μ is positive and <2, the origin ($q=0$, $\dot{q}=0$) is an unstable focus and, an unstable node if $\mu>2$. KROGDAHL has discussed a few limit cycles corresponding to different combinations of values of μ between 0.01 and 1.0 and of λ between 0.04 and 0.15. As in the previous discussion, when μ increases, the asymmetry increases rapidly. For the purpose of illustration, the limit cycle corresponding to the largest value of μ considered by KROGDAHL ($\mu=1.0$, $\lambda=0.04$) is represented on Fig. 46. On the other hand, as λ decreases, the source of positive damping increases and the amplitude of the limit cycle decreases.

Because of the cubic terms, the symmetry with respect to the origin is lost in this case. As KROGDAHL points out, the non-conservative terms, when large enough, can increase the asymmetry considerably. However a comparison with the observed curves, at least in the case of the Cepheids, suggests that this type of asymmetry is not a dominating characteristic of stellar oscillations. This can be seen also on Fig. 47 where the observed velocity curve of η Aquilae is compared to the velocity curve corresponding to KROGDAHL's limit cycle.

To take a more physical point of view, let us go back to the complete Eq. (87.2). In our previous discussion (Sects. 87 and 88), we only kept the first of the non-adiabatic terms in (87.3). However, in some parts of the star at least, $\delta(\varepsilon-\partial L/\partial m)$ may become of the same order as $(\delta T/T)_a$ (cf. Sect. 69) and if we want an equation correct up to the second order terms we should add to (86.3) the whole expression (87.3). If instead of (86.5), we substitute

$$\xi=\xi_0(r_0)\,q(t) \tag{90.11}$$

[1] A. J. WESSELINK: Astrophys. Journ. **89**, 659 (1939).
[2] W. S. KROGDAHL: Astrophys. Journ. **122**, 43 (1955).

in (86.3) thus modified, multiply by $\xi_0\, dr_0$ and integrate over the star volume, we get instead of (87.6)

$$\ddot{q} + \sigma_0^2 q - 2\sigma_0^2 \sigma_0'' \int q\, dt + G q \int q\, dt + H \int q^2\, dt - D q^2 = 0 \qquad (90.12)$$

where G and H are constants depending on ξ_0 and on the stellar model. Changing from t to $\tau = \sigma_0 t$, (90.12) may be written

$$\frac{d^2 q}{d\tau^2} + q - \frac{2\sigma_0''}{\sigma_0}[1 + G' q] \int q\, d\tau + H' \int q^2\, d\tau - D' q^2 = 0 \qquad (90.13)$$

which unfortunately is more difficult to discuss than the previous equations[1]. But, physically the ordinary damping term $(2\sigma_0''/\sigma_0) \int q\, d\tau$ is now multiplied by $(1 + G' q)$ which means that if G' is positive it is reinforced for $q > 0$ and weakened for $q < 0$. Let us suppose for instance that σ_0'' is negative (vibrational instability). Then, in the phase plane $(q, \dot{q})$, no positive damping appears in the half-plane $q > 0$ while it is present in the half-plane $q < 0$. It seems reasonable to conclude that the amplitude will tend to reach larger values for $q > 0$ than for $q < 0$. This introduces just the kind of asymmetry with respect to the $\dot{q}$-axis which is revealed by the observations (cf. Fig. 46).

However, one should remember that the asymmetry arising in this way depends on the absolute value of σ_0''/σ_0 and that it will become appreciable only if this ratio is not too small compared to unity. Thus, while any case where the linear theory reveals a negative damping however small could lead to a self-excited oscillation of limited amplitude, only those cases where the negative damping is large could lead to large asymmetry. Furthermore if the effect is to be in the right direction, G' should be positive. But this is rather likely on general physical grounds. As we have seen in Sect. 69, at present, the best hope for a strong vibrational instability is that the opacity in the external layers increases at contraction and decreases at expansion with a corresponding accumulation or liberation of ionization energy. But both these factors have a more limited range at contraction than at expansion so that the reinforcement of vibrational instability at expansion and its weakening at contraction associated with a positive G' seems natural.

One last remark should be made. In all this, all the equations which we have used are essentially *mean* equations over the whole star. On the other hand, as far as we can see at present, all interesting effects arise only in the non-adiabatic layers under the atmosphere. It is likely that averaging minimizes these effects very much and it is urgent to try to discuss these external layers separately. A point of view somewhat similar to that adopted in Sect. 68 might be fruitful. The effect of the adiabatic interior could be reduced to that of a piston of variable temperature imposing a practically sinusoidal motion and a sinusoidal flux at the bottom of the interesting layer which should then be treated as a non-linear, non-conservative system undergoing forced oscillations. It would probably be possible to replace again the general partial differential equation by the differential equations with respect to time of a few coupled oscillators representing the average spatial properties of the layer. This approach might also teach us something concerning the phase relationship between the external layers and the interior.

VI. Progressive waves and shock waves.

91. Introductory remarks. Some observations especially in Cepheids of type II are most easily interpreted in terms of a wave front going through the atmospheric

[1] Cf. for instance the paper by L. L. RAUCH, in: Contributions to the Theory of Non-Linear oscillations. Annals of Math. Studies, No. 20. Princeton 1950.

layers. Earlier, SCHWARZSCHILD *[58]* had attempted to interpret the phase shift between light and velocity curves in Cepheids in terms of a progressive wave.

However from a theoretical point of view, very little work has been done which is of direct significance for ordinary variable stars. The semi-empirical attempts to interpret special features of their variations will be reviewed in the last part of this article.

The most direct origin to a progressive wave in a star is a sudden release of energy at some point inside it. It may be due to special nuclear reactions generating in a short time a quantity of energy at least of the same order as the internal energy of the region where they take place[1] or to a rapid transition between different types of equilibrium[2] or different states of matter[3]. Nearly all the possibilities suggested so far in this respect (accumulation of some special type of nuclei, building up of superadiabatic gradients through changes in opacity or mean molecular weight due to nuclear reactions, diffusion etc.) are associated with the evolution of the star and comprise phases of very slow and gradual changes which are supposed to lead to a state somewhat overcritical when the least perturbation will result in some violent turn over. The "explosion" might also be due to some powerful dynamical instability such as occurs for instance in HOYLE's theory[4] of the formation of heavy elements.

However the nature of all these processes is such that they are likely to occur only once or at least to repeat themselves only after long and somewhat irregular time intervals. It is also likely that their violence will create very strong shock waves which following ROSSELAND'S[5] suggestion have usually been advocated to explain novae and supernovae[6] whose study falls outside this article. In a mild form, they may have some significance for variables such as U Geminorum and Z Camelopardalis.

In a regular variable star, it is likely that the main phenomenon is of the nature of a standing wave on which a progressive wave may be superposed whose strength will depend on special circumstances. This progressive component may have its origin in a lack of perfect reflection at the surface of the star. Or it may be due to some phenomena qualitatively similar to those recalled above for novae, but in this case, the critical stage should be brought about periodically by the main oscillation instead of a one-way slow evolution. It is however difficult to advance any precise mechanism. Special nuclear reactions seem unlikely.

---

[1] Cf. for instance: a) L. E. GUREVIC and A. I. LEBEDINSKY: Zhourn. Theor. Exp. Phys. (Russian) **17**, 792 (1947); Dokl. Akad. Nauk USSR. **103**, 569 (1955). — b) P. DUMÉZIL-CURIEN and E. SCHATZMAN: Ann. d'Astrophys. **13**, 80 (1950). — c) E. SCHATZMAN: Ann. d'Astrophys. **14**, 294, 305 (1951). — d) Cf. also: Proceedings of the 4th Congress on Problems of Cosmogony. Acad. Sci., USSR., 1955, containing many papers on the subject some of which have been summarized in Astron. News Letter **1957**, No. 84.

[2] Cf. for instance: A. UNSÖLD: Z. Astrophys. **1**, 138 (1930); L. BIERMAN: Z. Astrophys. **18**, 344 (1939) (in this respect see also end of Sect. 78); E. J. ÖPIK: Armagh Obs. Contr. **1953**, No. 9; also Ref. 1d above, especially the papers by A. G. MASEVIC and by S. A. KAPLAN; J. WASIUTYNSKI: Astrophys. Norv. **4**, Chap. 3 (1946); P. LEDOUX: 5. Coll. Inst. d'Astrophys. Liège, p. 200, 1953.

[3] Cf. for instance: E. A. MILNE: Monthly Notices Roy. Astronom. Soc. London **91**, 4 (1930), and Observatory **54**, 140 (1931); S. CHANDRASEKHAR: Monthly Notices Roy. Astronom. Soc. London **95**, 258 (1935); E. SCHATZMAN: Ann. d'Astrophys. **9**, 199 (1946).

[4] F. HOYLE: Monthly Notices Roy. Astronom. Soc. London **106**, 343 (1946).

[5] S. ROSSELAND: Astrophys. Journ. **104**, 324 (1946).

[6] a) E. SCHATZMAN: Ann. d'Astrophys. **9**, 199 (1946); Bull. Acad. Roy. Belg., Cl. Sci., Sér. V **34**, 828 (1948); Ann. d'Astrophys. **14**, 294 (1951). — b) P. A. CARRUS, P. A. FOX, F. HAAS and Z. KOPAL: Astrophys. Journ. **113**, 193 (1951); **113**, 496 (1951); Z. KOPAL: Astrophys. Journ. **120**, 159 (1954). — c) Z. KOPAL and C. C. LIN: Proc. Nat. Acad. Sci. U.S.A. **37**, 495 (1951). — d) M. H. ROGERS: Astrophys. Journ. **125**, 478 (1957).

The general pulsation may alternatively increase or decrease the convective stability of the external layers and, apart from the first approximation effect on the vibrational stability (cf. Sect. 69), non-adiabatic terms which are difficult to take properly into account could perhaps give a sufficiently abrupt character to the phenomenon. Another possibility which has been tentatively discussed[1] is the building up of an excess pressure in the ionization zone of hydrogen.

However instead of treating the progressive component as a secondary phenomenon, a more extreme view might be taken. As expressed once by ROSSELAND in a lecture, one may consider the possibility of a pulse originating at the centre and propagating to the surface where it is partly reflected. The reflected wave, on reaching the centre, might reproduce the critical situation which caused the initial pulse, the phenomenon repeating itself with a period equal to the time taken by the wave to travel back and forth along the star radius which according to (60.5) is of the same order as the period of the fundamental mode of radial pulsation. Such a theory requires that the amplitude does not increase too rapidly towards the surface as otherwise violent shock conditions would arise which would be contrary to observation. On the other hand, the origin of the central pulse depends on the existence in that region of nuclear reactions critically sensitive to ϱ or T such as the resonant reactions once considered by GAMOW (T) or such as might be realized in case of fission (ϱ) in a critical pile.

A variant suggested by VAN HOOF[2] pushes the origin of the perturbation from the centre to a layer at some depth under the surface.

In any case, since the phenomena must not be too violent, their theoretical study might be approached on the basis of a linear theory supplemented by some indications on the behaviour of possible discontinuities.

92. Propagation of discontinuities. In the linear approximation, initial discontinuities in the dependent variables[3] propagate with the velocity of sound along the characteristics of the hyperbolic partial differential equation. But from a physical point of view, the linear equations really cease to be valid in the vicinity of such a discontinuity. For instance, assuming a solution

$$v = a(r)\, f[t - \vartheta(r)] \quad \text{with} \quad \vartheta(r) = \int \frac{dr}{c}, \tag{92.1}$$

we have

$$v\, \frac{\partial v}{\partial r} \Big/ \frac{\partial v}{\partial t} = \left(\frac{\partial a}{\partial r} f - \frac{a}{c} f'\right) \frac{f}{f'}$$

and this ratio becomes large there $\left(\frac{\partial v}{\partial r},\ \frac{\partial a}{\partial r} \to \infty\right)$. If the non-linear terms are taken into account no such sonic propagation of discontinuities of v (or ϱ) are deducible as limits of continuous solutions. But in this case, initial discontinuities may disappear in the course of time (rarefaction wave with characteristics spreading out as in the effects of an accelerated withdrawal of a piston) or new discontinuities may appear (compression wave with characteristics overtaking each other as in the accelerated forward motion of a piston).

This last type of discontinuity is known as a *shock front*. Material always flows through a shock front: it *enters* through the *front side* at supersonic speed with respect to the front and at lower pressure and density and *leaves* on the *back side* at subsonic speed but at higher pressure and density. One knows

[1] CH. WHITNEY: Astronom. J. **61**, 192 (1956).

[2] A. VAN HOOF: Trans. Internat. Astronom. Union **8**, 814 (1954).

[3] Cf. for instance [*49a*], Vol. II (1937), Chap. VI, § 2, p. 356; [*48*], § 51, p. 118; also [*68c*], the article by R. COURANT, pp. 92—109.

that across such a discontinuity the variations of the dependent variables are given by the Hugoniot-Rankine conditions ([*48*], §§ 54, 62) which express the conservation of mass, momentum and energy. In the case of spherical symmetry let us denote by $r^*(t)$ the position of the front which of course is not bound to any given particle. Consider a material shell bounded by the spheres $r_1(t)$ and $r_2(t)$ moving with the fluid particles and satisfying the inequalities

$$r_1 < r^* < r_2. \tag{92.2}$$

If $J(r, t)$ is a function discontinuous in $r = r^*$, one has

$$\left.\begin{aligned} \frac{d}{dt}\int_{r_1(t)}^{r_2(t)} J(r,t)\,dr &= \int_{r_1}^{r^*} \frac{dJ}{dt}\,dr + \int_{r^*}^{r_2} \frac{dJ}{dt}\,dr + \\ &+ J_1(r^*,t)\,\dot r^* - J(r_1,t)\,\dot r_1 - J_2(r^*,t)\,\dot r^* + J(r_2,t)\,\dot r_2 \end{aligned}\right\} \tag{92.3}$$

where the indices 1 and 2 to J denote values taken on the sides of the discontinuity facing respectively r_1 or r_2. If we let the limits r_1 and r_2 tend simultaneously toward r^* and write

$$\dot r_1 = v_1, \quad \dot r_2 = v_2 \text{ (material velocities)}; \quad \dot r^* = V \text{ (front velocity)} \tag{92.4}$$

we get

$$\lim_{r_1, r_2 \to r^*} \frac{d}{dt}\int_{r_1}^{r_2} J(r,t)\,dr = J_1(V - v_1) - J_2(V - v_2) \tag{92.5}$$

since the integrals on the right side of (92.3) vanish and $J(r_1, t) \to J_1(r^*, t)$.

α) Conservation of mass. The mass of the shell remaining constant, we have applying (92.5)

$$\lim \frac{d}{dt}\int_{r_1}^{r_2} \varrho\, r^2\,dr = \varrho_1(V - v_1) - \varrho_2(V - v_2) = 0. \tag{92.6}$$

Defining the material velocities with respect to the front

$$u_1 = v_1 - V, \quad u_2 = v_2 - V \tag{92.7}$$

the condition (92.6) becomes

$$\varrho_2 u_2 = \varrho_1 u_1 = \Delta m \tag{92.8}$$

which simply expresses that the mass flux Δm through the front is the same evaluated on one side or the other. Note that Δm, u_1 and u_2 have the same sign. According to our definitions, they are positive if the front is facing to the left and negative if the front is facing to the right.

β) Conservation of momentum. In the same way, we may multiply (84.3) by $\varrho r^2 dr$ and take the integral of both members from r_1 to r_2, integrating the term in $\partial p/\partial r$ by parts. If we then let r_1 and $r_2 \to r^*$, we find

$$\varrho_2 v_2 u_2 - \varrho_1 v_1 u_1 = p_1 - p_2 \tag{92.9}$$

if the viscosity is negligible ($\eta \to 0$). According to (92.7) and (92.8) this can be written

$$\Delta m (v_1 - v_2) = p_2 - p_1 \tag{92.10}$$

or

$$\varrho_1 u_1^2 + p_1 = \varrho_2 u_2^2 + p_2. \tag{92.11}$$

γ) Conservation of the total energy. In the same way, Eq. (51.3) yields

$$\varrho_2\left(U_2+\frac{v_2^2}{2}\right)u_2-\varrho_1\left(U_1+\frac{v_1^2}{2}\right)u_1=F_1-F_2+p_1v_1-p_2v_2 \tag{92.12}$$

where, as before, U and F respectively denote the internal energy and the flux.

δ) Variation of the entropy. Eq. (52.1) expressing the conservation of thermal energy may be written here

$$\frac{dS}{dt}=\frac{1}{T}\frac{dQ}{dt}=\frac{\varepsilon_1}{T}-\frac{1}{\varrho T}\frac{1}{r^2}\frac{\partial}{\partial r}(r^2F)+\frac{1}{\varrho T}\frac{4}{3}\eta\left(\frac{\partial v}{\partial r}-\frac{v}{r}\right)^2$$

or, after integration over the same shell as before,

$$\frac{d}{dt}\int_{r_1}^{r_2}S\varrho r^2dr=\int_{r_1}^{r_2}\frac{\varepsilon_1\varrho r^2dr}{T}-\left(\frac{Fr^2}{T}\right)_2+\left(\frac{Fr^2}{T}\right)_1-$$
$$-\int_{r_1}^{r_2}\frac{F}{T^2}\frac{\partial T}{\partial r}r^2dr+\int_{r_1}^{r_2}\frac{4}{3}\frac{\eta}{T}\left(\frac{\partial v}{\partial r}-\frac{v}{r}\right)^2r^2dr.$$

The total flux is

$$F=-(k_R+k_m)\frac{\partial T}{\partial r}=-k\frac{\partial T}{\partial r}$$

where k_R [cf. Eq. (67.20)] and k_m are the coefficients of radiative and molecular conductivity. If we substitute this value in the integral and perform the usual limiting process, we find

$$\left.\begin{aligned}S_2\varrho_2u_2-S_1\varrho_1u_1&=\left(\frac{F}{T}\right)_1-\left(\frac{F}{T}\right)_2+\lim_{r_1,r_2\to r^*}\left[\frac{1}{r_1^2}\int_{r_1}^{r_2}k\left(\frac{1}{T}\frac{\partial T}{\partial r}\right)^2r^2dr\right]+\\&+\lim_{r_1,r_2\to r^*}\left[\frac{1}{r_1^2}\int_{r_1}^{r_2}\frac{4}{3}\frac{\eta}{T}\left(\frac{\partial v}{\partial r}-\frac{v}{r}\right)^2\right]dr.\end{aligned}\right\} \tag{92.13}$$

If again we let η and k tend toward zero, the first two terms vanish, but this is not necessarily true for the two positive integrals in which $\partial T/\partial r$ and $\partial v/\partial r$ tend toward infinity. Thus, in the limit, we have for η and $k\to 0$

$$S_2\varrho_2u_2-S_1\varrho_1u_1\geqq 0 \quad\text{or}\quad \Delta m(S_2-S_1)\geqq 0 \tag{92.14}$$

so that, in this case, the entropy never decreases across a shock.

Actually, as shown by HUGONIOT, the equation of energy conservation implies a change in entropy across a shock. This can easily be checked for a perfect gas with constant $\bar{\mu}$ and C_v. In that case, the entropy variation assuming reversible processes is given by

$$S_1-S_2=C_v\ln\frac{p_1}{p_2}-C_p\ln\frac{\varrho_1}{\varrho_2}. \tag{92.15}$$

Introducing the strength of the shock

$$d=\frac{\varrho_1-\varrho_2}{\varrho_2}$$

and using the relation (92.31) which will be established later, (92.15) becomes

$$S_1-S_2=R\left\{\frac{1}{\gamma-1}\ln\left[\frac{2+d(\gamma+1)}{2-d(\gamma-1)}\right]-\frac{\gamma}{\gamma-1}\ln(1+d)\right\}. \tag{92.16}$$

It is easy to verify that, if the shock is weak ($d\to 0$), the increase in entropy across it is of the third order in d, a property which subsists in more general cases (cf. [48], § 65).

Physically η and k do not rigorously vanish and a more detailed discussion[1] will lead to replace the discontinuity by a zone of finite width where the dependent variables vary extremely rapidly. If only molecular conductivity k_m comes into play, the width of the transition zone is of the order of the mean free path l_m of the gas molecules and in a star l_m is always very small even in the atmosphere. However, in stellar conditions k_R is much more important than k_m and while the mean free path of a photon $l_R = 1/\varkappa\varrho$ [cf. Eq. (49.47)] is very small in the interior it becomes large in the external layers. For instance in the atmosphere of η Aquilae adopting the same conditions as in Sect. 67,

$$l_R \approx 2 \cdot 10^{10}\,\text{cm}, \tag{92.17}$$

i.e. of the order of the depth of the atmosphere itself. The same type of discussion as in Sect. 67 could also be applied and would lead to a similar result since in these layers a fluctuation in temperature, especially if sharp [cf. Eq. (67.23)], propagates much more rapidly than the shock. Thus one may expect that, the temperature discontinuity will flatten out very rapidly in that region while the discontinuity in pressure will remain sharp since it is still controlled by the molecular mean free path. The discontinuity in density will then increase to compensate smoothing out of the temperature, the conditions approaching those of an isothermal shock. However, this simple picture should be complemented by a detailed analysis of the transfer of energy from the kinetic to the internal degrees of freedom behind the front.

The jump conditions can also be expressed in Lagrangian coordinates. Let us denote by an index zero, values in an equilibrium state taken as reference, r_0 being used as the Lagrangian parameter labelling the particles. The position of the front is still given by $r^*(t)$ and its velocity by $V = \dot{r}^*(t)$. While it moves, it encounters particles which, in the equilibrium state, occupied positions $r_0^*(t)$ and $V_0 = \dot{r}_0^*(t)$ designates the fictitious velocity with which the "image" of the front moves across the equilibrium state. At the shock front, one has

$$r_1[r_0^*(t), t] = r_2[r_0^*(t), t] = r^*(t)$$

which, differentiated with respect to time, yields

$$v_1(r_0^*, t) + V_0\left(\frac{\partial r_1}{\partial r_0}\right)^* = v_2(r_0^*, t) + V_0\left(\frac{\partial r_2}{\partial r_0}\right)^* = V. \tag{92.18}$$

Furthermore, since the coordinates are bound to the particles, we have

$$\varrho\, r^2\, dr = \varrho_0\, r_0^2\, dr_0 \tag{92.19}$$

so that (92.18) becomes

$$V = v_1(r_0^*, t) + V_0\frac{\varrho_0(r_0^*)\, r_0^{*2}}{\varrho_1\, r^{*2}} = v_2(r_0^*, t) + V_0\frac{\varrho_0(r_0^*)\, r_0^{*2}}{\varrho_2\, r^{*2}} \tag{92.20}$$

or

$$u_1 = v_1 - V = -V_0\frac{\varrho_0\, r_0^{*2}}{\varrho_1\, r^{*2}}, \qquad u_2 = v_2 - V = -V_0\frac{\varrho_0\, r_0^{*2}}{\varrho_2\, r^{*2}}, \tag{92.21}$$

which satisfy (92.8) identically. If we represent by D the discontinuity across the front, (92.20) and 92.21) can also be written

$$D(v) = D(u) = -V_0\varrho_0\left(\frac{r_0^*}{r^*}\right)^2 D\left(\frac{1}{\varrho}\right). \tag{92.22}$$

With the help of (92.21), the momentum condition (92.9) becomes

$$D(p) = V_0\varrho_0\left(\frac{r_0^*}{r^*}\right)^2 D(v) \tag{92.23}$$

[1] Cf. for instance L. H. Thomas: J. Chem. Phys. **12**, 449 (1944).

and the energy condition (92.12) gives

$$D(p\,v) + D(F) = V_0\,\varrho_0 \left(\frac{r_0^*}{r^*}\right)^2 D\left(U + \frac{v^2}{2}\right). \tag{92.24}$$

All these relations hold also across a contact surface which however is distinguished from a shock by the fact that there is no flow of mass ($\Delta m = 0$) across it.

Multiplying (92.9) once by v_1 and once by v_2 and adding, we get

$$\Delta m\,(v_2^2 - v_1^2) = (p_1 - p_2)\,(v_2 + v_1). \tag{92.25}$$

If τ_1 and τ_2 are the specific volumes, Eq. (92.8) gives

$$u_1 = \Delta m\,\tau_1, \qquad u_2 = \Delta m\,\tau_2 \tag{92.26}$$

or

$$v_1 = \Delta m\,\tau_1 + V, \quad v_2 = \Delta m\,\tau_2 + V. \tag{92.27}$$

Substituting (92.25) and (92.27) into (92.12) we find

$$U_2 - U_1 = \frac{F_1 - F_2}{\Delta m} + \frac{p_1 + p_2}{2}\,(\tau_1 - \tau_2) \tag{92.28}$$

where the flux term is usually neglected. This equation is often referred to as the *Hugoniot relation*.

The first two conditions (92.8) and (92.11) do not depend on the thermodynamic nature of the fluid and any conclusion drawn from them alone is valid quite generally. If the fluid is barotropic [$p = f(\varrho)$], they determine the discontinuity completely. More generally, for a perfect gas, we have

$$p = \frac{R\,\varrho\,T}{\bar{\mu}}, \quad U = C_v\,T = \frac{p\,\tau}{\gamma - 1} \tag{92.29}$$

and if C_v (or γ) and $\bar{\mu}$ are constant, substituting (92.29) into (92.28) and neglecting F_1 and F_2, we find

$$\frac{\tau_2}{\tau_1} = \frac{\varrho_1}{\varrho_2} = \frac{(\gamma+1)\,p_1 + (\gamma-1)\,p_2}{(\gamma+1)\,p_2 + (\gamma-1)\,p_1} \quad \text{and} \quad \frac{T_1}{T_2} = \frac{(\gamma+1) + (\gamma-1)\,(p_1/p_2)}{(\gamma+1) + (\gamma-1)\,(p_2/p_1)} \tag{92.30}$$

which can also be inverted into

$$\frac{p_2}{p_1} = \frac{\varrho_2(\gamma+1) - \varrho_1(\gamma-1)}{\varrho_1(\gamma+1) - \varrho_2(\gamma-1)}. \tag{92.31}$$

The maximum jump in density ($p_1 \to \infty$) is

$$\frac{\varrho_1}{\varrho_2} = \frac{\gamma+1}{\gamma-1}$$

which is equal to 4 if $\gamma = \frac{5}{3}$. In presence of radiation, we have (cf. Sects. 8 and 12),

$$p = p_G + p_R = \frac{p_G}{\beta} = \frac{p_R}{1-\beta}, \quad U = \left[\frac{\beta}{\gamma - 1} + 3\,(1-\beta)\right] p\,\tau = \alpha\,p\,\tau \tag{92.32}$$

where β is related to p and τ by

$$\frac{1-\beta}{\beta^4} = \frac{a\,\bar{\mu}^4}{3\,R^4}\,p^3\,\tau^4 \tag{92.33}$$

so that (92.30) is replaced by

$$\frac{\tau_2}{\tau_1} = \frac{\varrho_1}{\varrho_2} = \frac{p_1\,(2\alpha_1 + 1) + p_2}{p_2\,(2\alpha_2 + 1) + p_1}. \tag{92.34}$$

If p_R becomes dominant, $\alpha \to 3$ and the maximum value of $\varrho_1/\varrho_2 \to 7$.

Neglecting again p_R, but keeping the terms in F_1 and F_2, we find

$$\frac{\tau_2}{\tau_1} = \frac{\varrho_1}{\varrho_2} = \frac{2(F_1 - F_2)(\gamma - 1) + [(\gamma - 1)\,p_2 + (\gamma + 1)\,p_1]\,u_1}{u_1[(\gamma - 1)\,p_1 + (\gamma + 1)\,p_2]} \tag{92.35}$$

and

$$\frac{T_1}{T_2} = \frac{2(F_2 - F_1)(\gamma - 1) + [(\gamma + 1)\,p_2 + (\gamma - 1)\,p_1]\,u_2}{u_2[(\gamma + 1)\,p_1 + (\gamma - 1)\,p_2]}\,\frac{p_1}{p_2}. \tag{92.36}$$

For a shock progressing in a medium at rest $u_2 = -V$ and u_1 are negative and for a given strength $(p_1 - p_2)/p_2$ of the shock, these formulae show that if $F_2 > F_1$ (shock radiating into the medium in front) the ratio ϱ_1/ϱ_2 is greater than when $F_2 - F_1 = 0$ and T_1/T_2 is smaller confirming our previous conclusions on the effects of radiation [cf. text accompanying Eq. (92.17)]. As $F_2 - F_1$ increases, the phenomenon approaches isothermal conditions which are reached when

$$\frac{F_2 - F_1}{\Delta m} = u_2\,\frac{p_2^2 - p_1^2}{2p_1} = \frac{p_1 + p_2}{2}\,(\tau_2 - \tau_1), \tag{92.37}$$

the whole energy liberated by the compression work on the mass crossing the front being radiated away.

If the ionization changes across the front, we must write according to (53.9)

$$U = \frac{3}{2}\,p\,\tau + I, \quad p = \frac{RT}{\bar{\mu}\,\tau}.$$

It is mainly in the external layers that the effect of I and the variations of $\bar{\mu}$ can be important so that we can use Eqs. (53.19) to (53.25). Let us assume further that hydrogen is so abundant that we can neglect all the other elements and write everywhere $X_H \approx 1$, $x_H \approx x$. Then

$$I_2 - I_1 = \frac{\chi_H}{m_H}\,(x_2 - x_1), \quad \bar{\mu} = \frac{1}{1 + x}.$$

Proceeding as before, we obtain

$$\frac{\tau_2}{\tau_1} = \frac{\varrho_1}{\varrho_2} = \frac{(2\chi_H/m_H\,\tau_1)(x_1 - x_2) + 4p_1 + p_2 + 2(F_1 - F_2)/\Delta m\,\tau_1}{4p_2 + p_1}, \tag{92.38}$$

$$\frac{T_1}{T_2} = \frac{p_1(1 + x_2)}{p_2(1 + x_1)}\,\frac{(2\chi_H/m_H\,\tau_2)(x_2 - x_1) + p_1 + 4p_2 + 2(F_2 - F_1)/\Delta m\,\tau_2}{4p_1 + p_2}. \tag{92.39}$$

For a shock progressing in the direction of increasing r, $x_1 > x_2$ and if p_1, p_2 and τ_2 are fixed and $(F_1 - F_2)$ is negligible, ϱ_1/ϱ_2 is greater and T_1/T_2 is smaller than in the case of constant ionization.

Reverting to the simple case (92.29) with C_v and $\bar{\mu}$ constant, the three jump conditions can also be written in term of the Mach number M characterizing the motion of the shock in the medium in front of it which, according to the above convention, is characterized by the index 2. With

$$M_2 = -\frac{u_2}{C_2} = -\frac{u_2\,\varrho_2}{\gamma\,p_2} \tag{92.40}$$

the three relations become

$$\frac{p_1}{p_2} = \frac{2\gamma}{\gamma + 1}\,M_2^2 - \frac{\gamma - 1}{\gamma + 1}, \tag{92.41}$$

$$\frac{\varrho_1}{\varrho_2} = \frac{\gamma + 1}{\gamma - 1 + 2M_2^{-2}}, \tag{92.42}$$

$$\frac{u_1 - u_2}{C_2} = \frac{v_1 - v_2}{C_2} = \frac{2}{\gamma + 1}\left(M_2 - \frac{1}{M_2}\right), \tag{92.43}$$

a form which is sometimes useful.

93. Solution of the wave equation. We may again start with the linear problem. If the conditions may be treated as adiabatic, the equation of propagation may be obtained from (84.6) neglecting second order terms

$$\frac{\partial^2 v}{\partial t^2} - 4\frac{G\,m(r)}{r_0^3}\,v - \frac{1}{\varrho_0}\frac{\partial}{\partial r_0}\left[\frac{\Gamma_1 p_0}{r_0^2}\frac{\partial}{\partial r_0}(r_0^2 v)\right] = 0. \tag{93.1}$$

Since here we may write $v = \partial\delta r/\partial t = \partial\zeta/\partial t$, Eq. (93.1) integrated once with respect to t, gives

$$\frac{\partial^2 \zeta}{\partial t^2} - 4\frac{G\,m(r)}{r^3}\,\zeta - \frac{1}{\varrho_0}\frac{\partial}{\partial r_0}\left[\frac{\Gamma_1 p_0}{r_0^2}\frac{\partial}{\partial r_0}(r_0^2 \zeta)\right] = 0 \tag{93.2}$$

or, introducing $\xi = \delta r/r_0$ and treating Γ_1 as a constant, equal to γ

$$\frac{\partial^2 \xi}{\partial t^2} + (3\gamma - 4)\frac{G\,m(r)}{r_0^3}\,\xi - \frac{\partial \xi}{\partial r_0}\left(\frac{\gamma p_0}{\varrho_0}\frac{4}{r_0} + \frac{\gamma}{\varrho_0}\frac{\partial p_0}{\partial r_0}\right) - \frac{\gamma p_0}{\varrho_0}\frac{\partial^2 \xi}{\partial r_0^2} = 0 \tag{93.3}$$

which can also be derived from (57.23) or (58.12) by substituting $\partial^2\delta r/\partial t^2$ for $\sigma^2\delta r$.

Here the boundary conditions to be satisfied by the solution at the wave front may be derived from the Hugoniot-Rankine conditions taking their limiting linear form. In particular, if the medium in front of the wave front $r^*(t)$ is at rest $[v_2(r_0, t) = 0,\ \varrho_2(r_0, t) = \varrho_2(r_0)$ for $r_0 \geq r_0^*]$, taking (92.19) into account, the conditions (92.22), (92.23) and (92.24) become

$$v_1 = V\frac{\delta\varrho_1}{\varrho_0}, \tag{93.4}$$

$$\delta p_1 = V\varrho_0 v_1 = V^2\,\delta\varrho_1, \tag{93.5}$$

$$p_0 v_1 = \varrho_0 V\,\delta U = \frac{V}{\gamma - 1}\left(\delta p_1 - p_0\frac{\delta\varrho_1}{\varrho_0}\right), \tag{93.6}$$

neglecting terms in δF.

But since in the linear approximation, the change in entropy across the front can be neglected [cf. (92.16)], the ordinary adiabatic relation

$$\frac{\delta p_1}{p_0} = \gamma\frac{\delta\varrho_1}{\varrho_0} \tag{93.7}$$

is still valid behind the front and the second member of (93.6) reduces to $p_0 V\,\delta\varrho_1$ so that this equation is equivalent to (93.4). Eqs. (93.5) and (93.7) show that the front advances with the sound velocity

$$V = \sqrt{\frac{\gamma p_0(r_0^*)}{\varrho_0(r_0^*)}} = C_0. \tag{93.8}$$

The problem now consists in finding solutions satisfying initial conditions, say at $t = 0$,

$$r(r_0, 0) = r_0 + \zeta(r_0),$$

$$v(r_0, 0) = \eta(r_0),$$

where $\zeta(r_0)$ and $\eta(r_0)$ are arbitrary functions and boundary conditions appropriate to the particular problem considered. For instance, if the wave is due to the motion of a spherical piston performing oscillations around an equilibrium position $r_0 = a$, according to a law $\delta a = Q(t)$, we must have

$$\delta r = \zeta(a, t) = Q(t). \tag{93.9}$$

In particular, if the piston starts oscillating with a finite velocity in a medium at rest, the wave will possess a definite front with discontinuities satisfying

(93.4) and (93.5). If on the other hand $Q(t)$ and $\dot{Q}(t)$ vanish simultaneously, the conditions will remain continuous everywhere.

Other boundary conditions may be imposed, such as reflection at the surface or at the centre.

In Eq. (93.3) the coefficient $(3\gamma-4)Gm(r)/r_0^3=(\gamma-\frac{4}{3})4\pi G\bar{\varrho}(r_0)$ is smallest in the external layers where it becomes of the order of σ_0^2 [cf. Eq. (60.12)]. The terms in $\partial\xi/\partial r_0$ are due to the curvature and the non-uniform distribution of pressure. For periodic solutions with $\sigma^2 \gg \sigma_0^2 (P \ll P_0)$ and wavelengths $\lambda \ll R$, the second order derivatives will be the dominating terms in (93.2) or (93.3) and, at least for the propagation over distances small as compared to the radius R, one may expect to find solutions of the form (92.1).

Apart from an early attempt by M. SCHWARZSCHILD [*58b*], it is only recently that the generalization of this solution for the complete Eq. (93.2) has been the object of systematic investigations. G. B. WHITHAM [*69*] and R. SIMON[1] have discussed the problem by means of the operational calculus, using a Laplace transform to eliminate the time. In special cases, this leads to explicit solutions. For instance, in the generalized Roche model with an envelope where $\varrho_0 \propto r_0^{-\alpha}$, the central homogeneous core acting as a piston, SIMON has found that ξ decreases or increases if $\alpha < 3.5$ or $\alpha > 3.5$ respectively. In his thesis he also discussed the reflection of those waves at a surface where $\delta P=0$ and verified that, in the case of perfect reflection, one may recover standing waves in the envelope provided the frequency of the piston be equal to one of the eigenvalues of the envelope[2]. In such cases, it would now be fairly simple to discuss the effects of imperfect reflection near the surface and make some progress in one of the problems mentioned in Sect. 91. It would also be easy in these cases to study the effects of the superposition of a small wave originating in a narrow layer on the standing oscillation. However, one disturbing feature of these models is that strictly, they extend to infinity and this may well affect some of the important physical characteristics of the solutions.

Unfortunately, it seems likely that all cases admitting of a fairly simple analytical treatment will be rather artificial and although their solutions may serve as useful models, methods directly applicable to physical models should be developed. In this respect, WHITHAM [*69*] has suggested an interesting approach to the problem of a wave propagating from $r_0=a$, in a medium at rest. Denoting by $\nu(r_0, t)$ the argument of the solution (92.1): $\nu=t-\Theta(r_0)$, we may, following a variant discussed by SIMON in his thesis, try to satisfy (93.2) by a solution of the form

$$\zeta=\sum_{n=0}^{\infty}\varphi_n(r_0)\,\psi_n(\nu) \tag{93.10}$$

valid behind the wave front whose position $r_0^*(t)$ is defined by

$$\nu(r_0^*, t)=0 \quad \text{or} \quad t=\int\limits_a^{r_0^*}\frac{dr_0}{C(r_0)},$$

assuming as WHITHAM the recurrence relations

$$\psi_n(\nu)=\int\limits_0^{\nu}\psi_{n-1}(\nu')\,d\nu' \tag{93.11}$$

with $\psi_0(0)=0$ since the displacement must vanish at the wave front. Substituting (93.10) into (93.2), taking (93.11) into account and equating the coefficients of $\psi_0', \psi_0, \psi_1 \ldots$ to zero, one obtains a series of differential equations

[1] R. SIMON: Ann. d'Astrophys. **18**, 92 (1955); **19**, 115 (1956).
[2] R. SIMON: Ann. d'Astrophys. **20**, 113 (1957).

admitting solutions

$$\left.\begin{aligned} \varphi_0(r_0) &= \frac{1}{r_0\sqrt{\varrho_0 C_0}}, \\ \varphi_{n+1}(r_0) &= \frac{1}{2r_0\sqrt{\varrho_0 C_0}} \int\limits_a^{r_0} \frac{f_n(r_0)}{r_0\sqrt{\varrho_0 C_0}}\, d r_0 \\ \text{with}\qquad f_n(r_0) &= r_0^2 \frac{d}{d r_0}\left\{\frac{\Gamma_1 p_0}{r_0^2}\frac{d}{d r_0}(r_0^2\varphi_n)\right\} + \frac{4G\,m(r)}{r_0}\varphi_n \end{aligned}\right\} \qquad (93.12)$$

and where the constants have been chosen so that $\varphi_1, \varphi_2 \ldots$ vanish in $r_0 = a$.

If we impose the motion in $r_0 = a$, $\delta a = Q(t)$, $\psi_0(\nu)$ is determined by

$$\varphi_0(a)\,\psi_0(\nu) = Q(\nu)$$

and, at the wave front, we have

$$\left.\begin{aligned} &v \propto \frac{1}{r_0^*}(\Gamma_1 p_0^* \varrho_0^*)^{-\frac{1}{4}}\dot{Q}(0), \qquad \delta\varrho \propto \frac{1}{r^*}(\Gamma_1 p_0^*)^{-\frac{3}{4}}\varrho_0^{*\frac{5}{4}}\dot{Q}(0), \\ &\delta p \propto \frac{1}{r_0^*}(\Gamma_1 p_0^* \varrho_0^*)^{\frac{1}{4}}\dot{Q}(0), \qquad \delta T \propto \frac{(\Gamma_1 - 1)\,T_0^*}{r_0^*}\varrho_0^{*\frac{1}{4}}(\Gamma_1 p_0^*)^{-\frac{3}{4}}\dot{Q}(0). \end{aligned}\right\} \qquad (93.13)$$

The complete solution behind the front could be worked out from (93.12), but formulae (93.13) already reveal some interesting characteristics of the wave. In particular, we see that δp always tends towards zero at the surface ($p_0, \varrho_0, T_0 \to 0$), but the strength of the discontinuity $\delta p/p_0$ tends to increase indefinitely as well as the material velocity v. Of course the linear approximation will cease to be valid well before that. According to WHITHAM [69], one may approximate the non-linear solution by the expansion

$$r(r_0, t) - r_0 = \sum_{n=0}^{\infty} \varphi_n(r)\,\psi_n(\eta) \qquad (93.14)$$

where $\eta(r, t)$ is a sufficiently good approximation to the exact characteristic variable which replaces the $\nu(_0, t)$ of the linear case. This modification corresponds to the fact that, on the exact theory, the velocity of propagation of wavelets along a characteristic η = constant, is equal to the sum of the local sound speed C and the local particle velocity v

$$\left(\frac{dr}{dt}\right)_\eta = C + v. \qquad (93.15)$$

As a result, the front of the compressed region steepens and if according to the linear theory v increases towards the surface, shock conditions are soon reached. As long as v remains smaller than C (weak shock), Eq. (93.15) where v is replaced by its linear approximation with η substituted for ν may be solved for dt and integrated giving a relation between t, r and η. However, close enough to the surface of a finite configuration where v becomes large a special investigation is necessary.

Even in simple cases like the polytropes treated by WHITHAM assuming barotropy, the discussion may become rather complicated but it points to conclusions similar to those already suggested by (93.13): the pressure discontinuity $(p_1 - p_2)$ tends to zero at the surface, but the strength of the shock $(p_1 - p_2)/p_0$ tends to increase indefinitely as well as the particle velocity behind the shock. As pointed out by WHITHAM, the first conclusion must be appraised with care. For instance, a shock which becomes very strong only when the mass ahead is negligible may not have any remarkable results. However the second conclusion concerning the velocity, if it remains valid in more general cases, suggests that even a moderate disturbance inside the star may lead to the ejection of particles from the stellar surface.

As mentioned before, the case where the initial perturbation is strong enough so that the wave takes the shock character from the very beginning seems of less direct interest here. Nevertheless the homology solutions (progressing waves) discussed[1] in that case may also prove to be useful at some stage in the present problem.

Of course one can always resort to straight numerical integration of Eqs. (84.2) to (84.4) or (84.7) to (84.10). However, as they contain a large number of dependent variables the step-by-step integration may be rather complicated. As KOPAL and LIN (Ref. 6c on p. 555) have pointed out it is possible to form a pair of equations where only two dependent variables appear in differential form and one may then expect to be able to associate with them two characteristic directions along which relations between the total differentials of these two dependent variables will hold. Introducing normalized variables

$$\left.\begin{aligned} &r = R\,x, \quad \varrho = \varrho_c\,\alpha, \quad p = (2\pi G\,\varrho_c^2 R^2)\,\pi, \quad m = (4\pi\,\varrho_c R^3)\,\mu, \\ &t = (2\pi G\,\varrho_c)^{-\frac{1}{2}}\,\tau, \qquad v = (2\pi G\,\varrho_c R^2)^{\frac{1}{2}}\,\nu, \end{aligned}\right\} \tag{93.16}$$

Eq. (84.2) becomes

$$\frac{\partial\alpha}{\partial\tau} + \nu\frac{\partial\alpha}{\partial x} + \alpha\left(\frac{\partial\nu}{\partial x} + \frac{2\nu}{x}\right) = 0 \tag{93.17}$$

which can also be written

$$\frac{\partial\mu}{\partial\tau} + \nu\frac{\partial\mu}{\partial x} = 0,$$

and Eqs. (84.3) and (84.4), where the non-adiabatic terms are neglected, yield

$$\frac{\partial\nu}{\partial\tau} + \nu\frac{\partial\nu}{\partial x} + \frac{1}{\alpha}\frac{\partial\pi}{\partial x} + 2\frac{\mu}{x^2} = 0, \tag{93.18}$$

$$\frac{\partial\pi}{\partial\tau} + \nu\frac{\partial\pi}{\partial x} - \frac{\gamma\pi}{\alpha}\left(\frac{\partial\alpha}{\partial\tau} + \nu\frac{\partial\alpha}{\partial x}\right) = 0. \tag{93.19}$$

If we eliminate α between (93.19) and (93.17), we obtain

$$\frac{\partial\pi}{\partial\tau} + \nu\frac{\partial\pi}{\partial x} + \gamma\pi\left(\frac{\partial\nu}{\partial x} + \frac{2\nu}{x}\right) = 0 \tag{93.20}$$

which as (93.18) contains only π and ν in differential form.

As an illustration, let us determine the characteristic directions of (93.18) and (93.20) by the general method ([48], § 22). Multiplying (93.20) by λ_1 and (93.18) by λ_2 and adding them, we obtain

$$\lambda_1\frac{\partial\pi}{\partial\tau} + \left(\lambda_1\nu + \frac{\lambda_2}{\alpha}\right)\frac{\partial\pi}{\partial x} + \lambda_2\frac{\partial\nu}{\partial\tau} + (\lambda_1\gamma\pi + \lambda_2\nu)\frac{\partial\nu}{\partial x} + \lambda_1\frac{2\gamma\pi}{x}\nu + \lambda_2\frac{2\mu}{x^2} = 0. \tag{93.21}$$

The combinations of π_τ, π_x and ν_τ, ν_x appearing in (93.21) will reduce to the total derivatives of π and ν along the same direction tangent to the curve $\tau(s)$, $x(s)$ if

$$\lambda_1 : \left(\lambda_1\nu + \frac{\lambda_2}{\alpha}\right) = \lambda_2 : (\lambda_1\gamma\pi + \lambda_2\nu) = \tau_s : x_s,$$

[1] Cf. Ref. 6b and d on p. 555; also [48], §§ 158 to 166; see also C. F. v. WEIZSÄCKER: Z. Naturforsch 9a, 269 (1954), and papers by collaborators in following issues.

that is, if λ_1 and λ_2 are solutions of

$$\lambda_1 (x_s - \nu \tau_s) - \frac{\lambda_2}{\alpha} \tau_s = 0,$$

$$\lambda_1 \gamma \pi \tau_s + \lambda_2 (\nu \tau_s - x_s) = 0.$$

The compatibility condition

$$x_s^2 - 2\nu \tau_s x_s + \tau_s^2 \left(\nu^2 - \frac{\gamma \pi}{\alpha}\right) = 0$$

defines two characteristic directions

$$x_s = (\nu \pm w)\, \tau_s \tag{93.22}$$

along which

$$\lambda_2 = \lambda_1 \alpha \frac{x_s - \nu \tau_s}{\tau_s} = \pm \lambda_1 \alpha w \tag{93.23}$$

with

$$w^2 = \frac{\gamma \pi}{\alpha} = \frac{C^2}{2\pi G \varrho_c R^2}. \tag{93.24}$$

Substituting this value of λ_2 in (93.21), we find

$$\frac{\partial \pi}{\partial \tau} + (\nu \pm w) \frac{\partial \pi}{\partial x} \pm \alpha w \left[\frac{\partial \nu}{\partial \tau} + (\nu \pm w) \frac{\partial \nu}{\partial x}\right] + \alpha w \left[\frac{2\nu}{x} w \pm \frac{2\mu}{x^2}\right] = 0. \tag{93.25}$$

If we eliminate π in terms of w and the entropy S which for a perfect gas, can be written according to (92.15)

$$dS = 2 C_p\, d \ln w - R\, d \ln \pi \tag{93.26}$$

we find

$$\left.\begin{aligned} &\frac{\partial}{\partial \tau}\left(\frac{2w}{\gamma - 1} \pm \nu\right) + (\nu \pm w) \frac{\partial}{\partial x}\left(\frac{2w}{\gamma - 1} \pm \nu\right) \mp \frac{w^2}{\gamma R} \frac{\partial S}{\partial x} - \\ &\qquad - \frac{w}{\gamma R}\left(\frac{\partial S}{\partial \tau} + \nu \frac{\partial S}{\partial x}\right) + \left(\frac{2\nu}{x} w \pm \frac{2\mu}{x^2}\right) = 0. \end{aligned}\right\} \tag{93.27}$$

But following the motion, Eq. (93.19) implies

$$\frac{\partial S}{\partial \tau} + \nu \frac{\partial S}{\partial x} = 0 \tag{93.28}$$

so that Eq. (93.27) becomes

$$\frac{\partial}{\partial \tau}\left(\frac{2w}{\gamma - 1} \pm \nu\right) + (\nu \pm w) \frac{\partial}{\partial x}\left(\frac{2w}{\gamma - 1} \pm \nu\right) \mp \frac{w^2}{\gamma R} \frac{\partial S}{\partial x} + \left(\frac{2\nu}{x} w \pm \frac{2\mu}{x^2}\right) = 0. \tag{93.29}$$

If the medium is isentropic ($\partial S/\partial x = 0$) and if the curvature and the gravity are neglected, this means that a given value of the function $[2w/(\gamma - 1) + \nu]$ is propagated with the speed $(w + \nu)$ while a given value of the function $[2w/(\gamma - 1) - \nu]$ is propagated in the opposite direction with the speed $(w - \nu)$. This justifies the inference leading to (93.15).

Eq. (93.25), where $(\nu \pm w)$ is replaced by x_s/τ_s and α is eliminated by (93.24), can also be written after multiplication by τ_s

$$\frac{d\pi}{\gamma \pi} \pm \frac{d\nu}{w} + \left(\frac{2\nu}{x} \pm \frac{2\mu}{w x^2}\right) d\tau = 0, \tag{93.30}$$

along the characteristics

$$dx = (\nu \pm w)\, d\tau. \tag{93.31}$$

On the other hand, (93.19) and (93.17) become

$$d\pi - w^2\, d\alpha = 0, \quad d\mu = 0 \tag{93.32}$$

along the path

$$dx = \nu\, d\tau \tag{93.33}$$

which plays the rôle of a double characteristic with an angular coefficient equal to the average of the angular coefficients of C^+ and C^- defined by (93.31).

For the purpose of numerical integration, the differentials in the above equations are replaced by finite differences. If, in the plane of the independent variables (x, τ), the values of the dependent variables (π, ν, α, μ) are given on a curve C distinct from C^+ or C^-, the characteristic directions through two points P_1 and P_2 of C may be determined from (93.31) as well as their intersection $P_3(x_3, \tau_3)$. The values of π and ν in P_3 are calculated from (93.30). The path line $P_0 P_3$ may then be determined by (93.33) and the values of α and μ, by (93.32). In practice, one may have to iterate, since the slope of the characteristics adopted in P_1 or P_2 should really be an average between the slopes in P_1 and P_3 or P_2 and P_3.

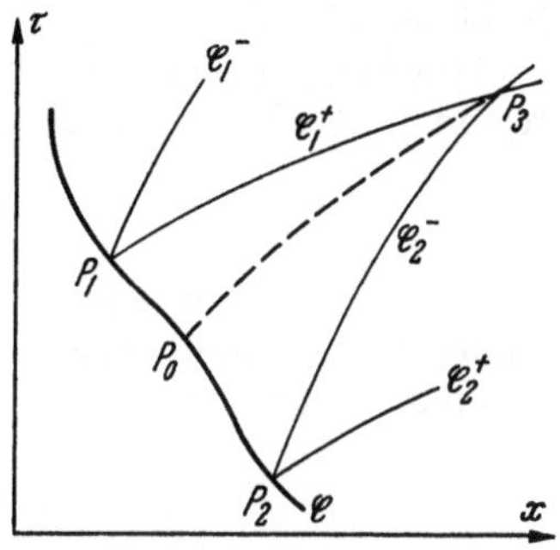

Fig. 48. Scheme for the numerical integration along the characteristics and the path, starting from boundary values given on C.

If two characteristics C_1^+ and C_2^+ (or C_1^- and C_2^-) intersect, the dependent variables become multivalued and a shock discontinuity should be introduced at that point, obeying the Hugoniot-Rankine conditions discussed in the previous section.

The discussion in Lagrangian coordinates, μ and τ being taken as independent variables (cf. Ref. 6c, p. 555), leads exactly to the same equations with a slight change in interpretation. One advantage, here, is that the path lines are directly integrable and are defined by $\mu =$ constant.

94. Some particular points concerning the propagation in the external layers. In this region, we may neglect the curvature and treat the acceleration of gravity g as a constant. If, furthermore, we keep the adiabatic approximation and suppose that the medium is isentropic, Eq. (93.29) can be written

$$\frac{\partial}{\partial\tau}\left(\frac{2w}{\gamma-1} \pm \nu \pm q\,\tau\right) + (\nu \pm w)\,\frac{\partial}{\partial x}\left(\frac{2w}{\gamma-1} \pm \nu \pm g\,\tau\right) = 0 \tag{94.1}$$

along

$$dx = (\nu \pm w)\, d\tau \tag{94.2}$$

where

$$q = \frac{2\mu}{x^2} \propto g = \text{constant}.$$

Considering an initial disturbance at $\tau = 0$, it gives rise to two waves, the function $[2w/(\gamma-1) + \nu + q\tau]$ propagating locally with the speed $(\nu + w)$ toward increasing x while the function $[2w/(\gamma-1) - \nu - q\tau]$ propagates in the opposite direction with the speed $(w - \nu)$. Applying the same reasoning as WALRAVEN[1], if the wavelength is not too large, the influence of the incoming wave will, after some time, be negligible in the region reached by the outgoing wave, or in other words the inward propagating function will reduce there to its equilibrium value, thus

$$\frac{2w}{\gamma-1} - \nu - q\,\tau \approx \frac{2w_0}{\gamma-1} \tag{94.3}$$

[1] TH. WALRAVEN: Bull. astronom. Inst. Netherl. **9**, 421 (1952).

which provides a relation between the material velocity ν and the pressure π or the density α since, with our assumptions, we have

$$\frac{\pi}{\pi_0} = \left(\frac{\alpha}{\alpha_0}\right)^{\gamma}, \quad w = \left(\frac{\gamma\,\pi_0}{\alpha}\right)^{\frac{1}{2}} \left(\frac{\alpha}{\alpha_0}\right)^{\frac{\gamma-1}{2}} = w_0 \left(\frac{\alpha}{\alpha_0}\right)^{\frac{\gamma-1}{2}}. \tag{94.4}$$

Taking this into account, Eq. (94.3) becomes

$$\frac{\varrho}{\varrho_0} \approx \left[1 + \frac{\gamma-1}{2}\,\frac{\nu + q\,\tau}{w_0}\right]^{\frac{2}{\gamma-1}} \tag{94.5}$$

which is very similar to SCHWARZSCHILD's solution [*58b*] except that the effect of gravity is apparent here[1]. This shows that, provided ν is large with respect to $q\tau$ (or $v > gt$) and the wavelength is small compared to R, ϱ will be practically in phase with v.

However, in that case, the time t_c needed to render the front of the compressed region practically vertical is of the order of

$$t_c = \frac{\lambda}{2}\,\frac{1}{2v} < \frac{P\,\overline{C}}{4\,\overline{g}\,t} \approx \frac{\overline{C}}{2\,\overline{g}} \tag{94.6}$$

where λ is the wavelength of the perturbation, P its period, $\overline{C}$ and $\overline{g}$ are mean values of the sound speed and gravity in the region considered and t has been taken of the order of $P/2$. Since the scale height H of the region is of the order of

$$H = \frac{\overline{RT}}{\overline{g}} = \frac{\overline{C^2}}{\overline{g}} = \frac{\overline{w^2}}{\overline{q}}\,R \tag{94.7}$$

we see that t_c will be smaller than the time t^* needed for the front to go through the region, say $t^* \approx H/\overline{C} \approx \overline{C}/\overline{g}$. One may thus expect that shock discontinuities will develop during the motion.

However, this approximation provides only some qualitative indications on the behaviour of the wave, since the adiabatic hypothesis breaks down in the very external layers. This is particularly troublesome, if we want to compare the results of the theory with the observations which refer to optically thin layers.

The isothermal case (T constant in space and time) provides another extreme approximation which according to our discussions following Eqs. (92.17) and (92.37) might even be better in some respects. In that case, if we introduce the new variables

$$\left.\begin{aligned} h &= \frac{x R}{H} = \frac{x q}{w^2}, \\ s &= \frac{\tau q}{w}, \quad n = \frac{\nu}{w} \end{aligned}\right\} \tag{94.8}$$

where $w = \sqrt{RT}$ is now a constant, Eq. (93.25), where we neglect again the curvature and use $\pi = \alpha w^2$, becomes

$$\frac{\partial(\ln\alpha \pm n \pm s)}{\partial s} + (n \pm 1)\,\frac{\partial(\ln\alpha \pm n \pm s)}{\partial h} = 0 \tag{94.9}$$

along

$$d h = (n \pm 1)\, d s. \tag{94.10}$$

These equations together with the jump conditions (92.41) to (92.43) have been integrated numerically by WHITNEY[2] using the method outlined at the end of

[1] For a detailed discussion of the effects of gravity in a somewhat similar case, cf. J. M. BURGERS: Proc. Kon. Ned. Akad. v. Wetensch. **51**, 145, 525 (1948).

[2] CH. WHITNEY: Thesis, Harvard 1955, and Ann. d'Astrophys. **19**, 34 (1956).

Sect. 93 and assuming that the bottom of the atmosphere has a sinusoïdal motion of frequency ω around its equilibrium position h_0, say,

$$h - h_0 = \frac{a}{\omega} \sin \omega t,$$

imposed by the stellar interior acting as a piston.

The computation was carried out for two frequencies but, in each case, it was assumed that the compression wave was headed from the start by a shock. Whitney found that the relaxation up to a few scale heights in the atmosphere was quite rapid so that after the second stroke of the piston the conditions repeated themselves nearly periodically. We reproduce in Fig. 49 the particle paths for one of the cases considered since it illustrates a state of motion which has been advocated on different occasions to explain the observations especially in Cepheids of type II[1]. The particle paths are labelled by the position of the corresponding particles at equilibrium. As pointed out by WHITNEY, the mean height of the particle is appreciably increased due to the upward flux of momentum associated with an outgoing compression wave of finite amplitude. He estimates that, in the cases treated, this may correspond to a reduction of the effective gravity by a factor of the order of 2 to 3, and an extrapolation based on the particle velocities led for W Virginis to a factor as large as 6. Although the influence of the frequency seems to be small a more detailed investigation of this question would be welcome[2].

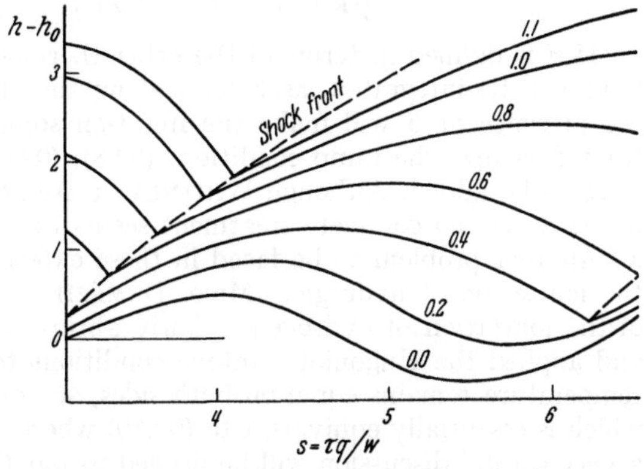

Fig. 49. Particle paths in the external layers after CH. WHITNEY.

But perhaps the most important points would be firstly to discuss numerically the establishment of the shock conditions in function of frequency or wavelength starting with a wave of small amplitude at some depth in the star and secondly to take into account the effects of radiative dissipation in the wave and at the shock front if it develops.

In the last case, Eq. (93.19) should be replaced by the complete Eq. (84.4) or, with the notation (93.16),

$$\frac{\partial \pi}{\partial \tau} + \nu \frac{\partial \pi}{\partial x} = \frac{\Gamma_1 \pi}{\alpha} \left(\frac{\partial \alpha}{\partial \tau} + \nu \frac{\partial \alpha}{\partial x} \right) + \frac{(\Gamma_3 - 1)\,\alpha}{(2\pi G \varrho_c)^{\frac{3}{2}} R^2} \left\{ \varepsilon - \frac{1}{R \varrho_c \alpha x^2} \frac{\partial (x^2 F)}{\partial x} \right\} \tag{94.11}$$

where ε and the curvature could generally be neglected in the external layers giving

$$\frac{\partial \pi}{\partial \tau} + \nu \frac{\partial \pi}{\partial x} = \frac{\gamma \pi}{\alpha} \left(\frac{\partial \alpha}{\partial \tau} + \nu \frac{\partial \alpha}{\partial x} \right) - \frac{(\gamma - 1)}{A} \frac{\partial F}{\partial x}, \tag{94.12}$$

if we neglect p_R before p_G and aT^4 before $C_v T$ and denote $(2\pi G \varrho_c)^{\frac{3}{2}} \varrho_c R^3$ by A. In the same conditions we also have along the path

$$\frac{\partial S}{\partial \tau} + \nu \frac{\partial S}{\partial x} = 2 C_p\, d \ln w - R\, d \ln \pi = - \frac{R}{\pi} \frac{1}{A} \frac{\partial F}{\partial x}. \tag{94.13}$$

[1] Cf. M. SCHWARZSCHILD: Trans. Internat. Astronom. Union **8**, 811 (1952).

[2] Cf. D. A. LAUTMAN: Astrophys. Journ. **126**, 537 (1957) where he concludes that for waves with frequencies of the order of that of the observed variations of the Cepheids, the effect is much smaller.

Replacing (93.19) by (94.12) and proceeding as before we find, instead of (93.25)

$$\frac{\partial \pi}{\partial \tau} + (\nu \pm w)\frac{\partial \pi}{\partial x} \pm \alpha w\left[\frac{\partial \nu}{\partial \tau} + (\nu \pm w)\frac{\partial \nu}{\partial x}\right] \pm \alpha w q + \frac{\gamma - 1}{A}\frac{\partial F}{\partial x} = 0 \qquad (94.14)$$

along the characteristics $dx = (\nu \pm w)\, d\tau$.

If we want to use w instead of π as dependent variable, an extra-term $[(\gamma-1)/A\alpha w](\partial F/\partial x)$ is added to the left-hand member of (93.27). Using (94.13), this last equation thus modified can be reduced to

$$\left.\begin{aligned} &\frac{\partial}{\partial \tau}\left(\frac{2w}{\gamma-1} \pm \nu \pm q\tau\right) + (\nu \pm w)\frac{\partial}{\partial x}\left(\frac{2w}{\gamma-1} \pm \nu \pm q\tau\right) - \\ &\quad - \frac{w}{\gamma R}\left[\frac{\partial S}{\partial \tau} + (\nu \pm w)\frac{\partial S}{\partial x}\right] - \frac{w}{C_p}\left(\frac{\partial S}{\partial \tau} + \nu\frac{\partial S}{\partial x}\right) = 0. \end{aligned}\right\} \qquad (94.15)$$

If F is defined in terms of the other thermodynamical variables, these equations can be integrated, as before, along the characteristics and the path, but the presence of S will make the iteration somewhat more delicate. In case a front develops, the jump conditions (92.8), (92.9), (92.12) or (92.35) and (92.16) have to be added. Although WHITNEY started discussing a problem of this type in his thesis, no complete treatment seems to exist up to now.

Another problem to be faced in these external layers concerns the effects of the ionization of hydrogen. Mme. DUMÉZIL-CURIEN[1] noting that the variation of the ionization of hydrogen is fairly abrupt, assimilated it to a discontinuity and applied the Hugoniot-Rankine conditions to it. Assuming further that the temperature remains equal on both sides, she derived an expression for $(F_2 - F_1)$ which is essentially equivalent to (92.39) where T_1/T_2 is set equal to 1. However, a very careful discussion will be needed to put this problem on a sound physical basis. Altogether, the propagation of finite waves in the external layers seems to constitute a rather complex question.

D. Interpretation and applications of the theory.

95. Fundamental data on intrinsic variable stars. In any theory attributing the observed variations to modifications in the interior of the stars, the fundamental stellar parameters are the mass M, the luminosity L, and the radius R. A general view of the distribution of these parameters among the variable stars is provided by their location in the Hertzsprung-Russell diagram. Fig. 50 reproduces such a diagram using as coordinates the absolute visual magnitude M_v and the spectral type. On this diagram a rough distinction between Populations I and II is made by horizontal and vertical hatchings. In Fig. 51, the coordinates have been changed to luminosity (absolute bolometric magnitude M_{bol}) and effective temperature T_e[2]. Loci of equal radius are represented by thin full lines.

Using the mass-luminosity relation [*70*]

$$\log_{10}\frac{L}{L_\odot} = 3.816\log_{10}\frac{M}{M_\odot} - 0.244 \qquad (95.1)$$

or

$$L \propto M^{3.82}, \qquad (95.2)$$

lines of equal mean density $\bar{\varrho}$ have also been drawn (dashed lines). Of course, the use of these mean densities is limited by the validity of the mass-luminosity

[1] Mme P. DUMÉZIL-CURIEN: C. R. Acad. Sci. Paris **233**, 1575 (1951).

[2] These diagrams have been constructed using mainly the data and the diagrams given in [*1*] and in STRUVE's paper in [*38*].

relation itself which rests entirely on the discussion of binary systems belonging to Population I. But even in this type of population there are, apart from an appreciable scatter, notable exceptions [70b]. In his discussion of the system BM Cas consisting of an $A\,5$ supergiant and a classical Cepheid of period $P=27$ days, THIESSEN[1] derives for this last component $M=14.3\,M_\odot$, $M_{\rm bol}=-6$. According to the author, the mass-luminosity relation would give a mass about

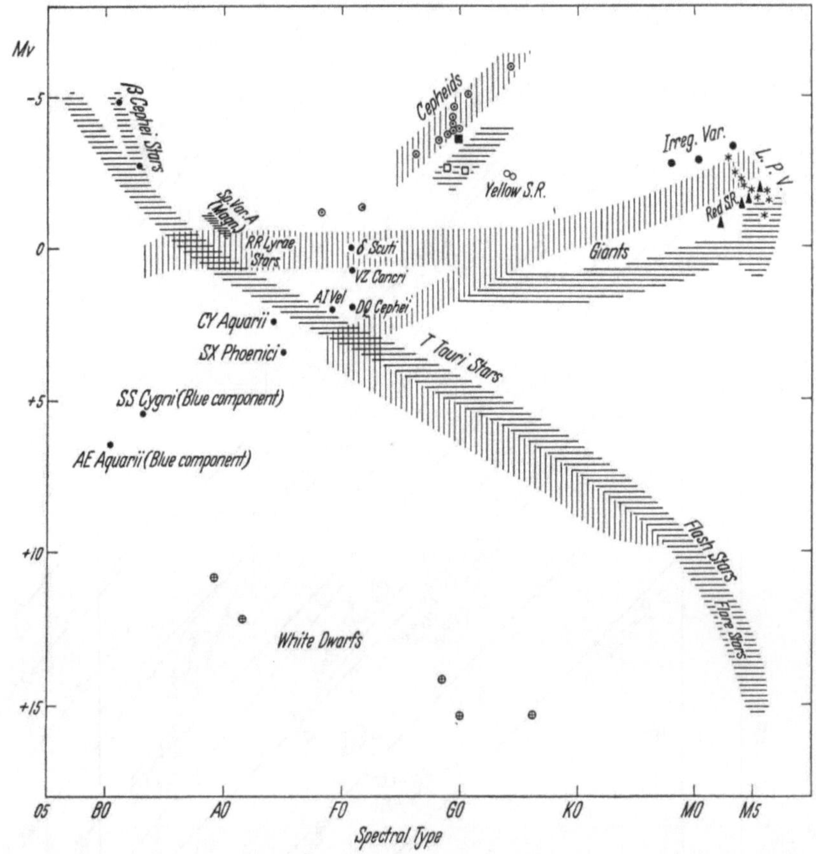

Fig. 50. Distribution of the intrinsic variable stars in the Hertzsprung-Russell diagram. Population I: horizontal hatchings; population II: vertical hatchings. A few individual stars or "mean" stars have been plotted: classical Cepheids: ⊙; W Virginis stars: □; RV Tauri: ■; yellow semiregular variables: ○; irregular variables: ●; long-period variables: ✱; red semiregular variables: ▲.

1.7 times larger. However this implies a low exponent of the order of 3 in (95.2) while the original relation (95.1) would give back practically the dynamical mass found by THIESSEN.

As far as Population II is concerned, we have no direct evidence but, according to the evolutionary considerations developed by SANDAGE, HOYLE and SCHWARZSCHILD and others[2], the masses of type II giants might be much smaller.

[1] G. THIESSEN: Z. Astrophys. **39**, 65 (1956).

[2] a) A. R. SANDAGE: 5th Symposium, Liège. Mém. Soc. Roy. Sci., Liège **14**, 254 (1953). — b) F. HOYLE and M. SCHWARZSCHILD: Astrophys. Journ., Suppl. **2**, 1 (1955); cf. also V. C. REDDISH: Monthly Notices Roy. Astronom. Soc. London **115**, 32, 480 (1955) as well as Monthly Notices Roy. Astronom. Soc. London **116**, 533, where the effects of small changes in chemical composition on the period-luminosity relation are considered.

The evolutionary track which they have estimated and which agrees fairly well with the colour-luminosity array of globular clusters corresponds to a mass of the order of $1.2\,M_{\odot}$.

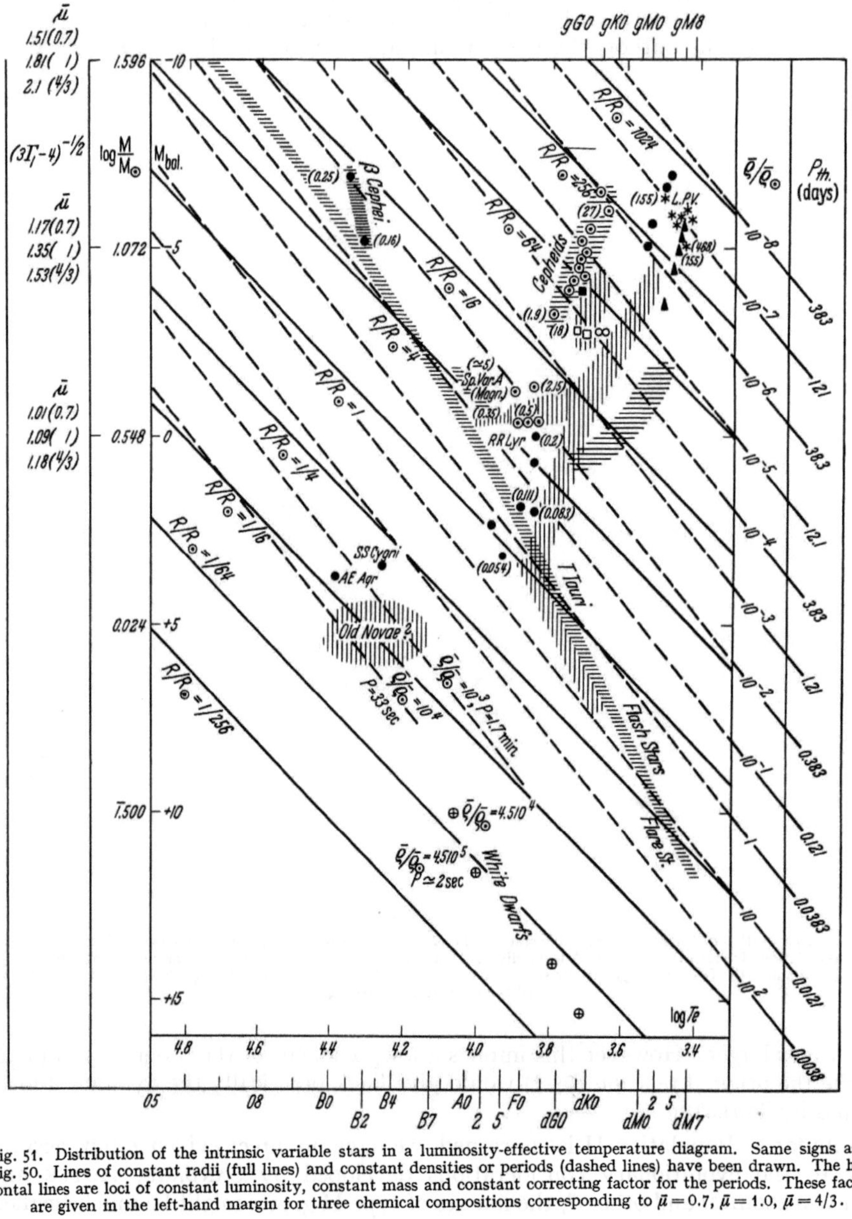

Fig. 51. Distribution of the intrinsic variable stars in a luminosity-effective temperature diagram. Same signs as in Fig. 50. Lines of constant radii (full lines) and constant densities or periods (dashed lines) have been drawn. The horizontal lines are loci of constant luminosity, constant mass and constant correcting factor for the periods. These factors are given in the left-hand margin for three chemical compositions corresponding to $\bar{\mu}=0.7$, $\bar{\mu}=1.0$, $\bar{\mu}=4/3$.

The same type of restrictions applies to the values of the periods indicated on the loci of equal mean density. Furthermore, these values have been computed with the relation

$$P_0 = 0.0383 \sqrt{\frac{\varrho_{\odot}}{\bar{\varrho}}} \tag{95.3}$$

valid for the fundamental mode of radial oscillation of the standard model with $\Gamma_1 = \frac{5}{3}$ (cf. Table 12, Sect. 62). There does not seem to be any reason at present to take for γ a value appreciably smaller than $\frac{5}{3}$. But if the pressure of radiation p_R is not negligible Γ_1 is smaller than $\frac{5}{3}$ [cf. Eq. (53.6)] and according to (62.2) the theoretical periods given here must be multiplied by the factor $(3\Gamma_1 - 4)^{-\frac{1}{2}}$ given in the left hand margin of Fig. 51 for three different chemical compositions corresponding to $\bar{\mu} = 0.7$, $\bar{\mu} = 1$, $\bar{\mu} = 1.33$.

Although, as shown in Table 12, the periods also depend on the adopted model, the values plotted on Fig. 51 provide the correct order of magnitude and are useful in a first survey of the problem. In particular, the comparison between the observed periods in brackets and the theoretical values reveals a fair amount of agreement for the more regular variables of Population I except for the spectrum variables of type *A* and the magnetic variables.

Apart from any other arguments, this is certainly in favour of pulsations rather than any form of revolution of contact binaries[1] [cf. Eq. (42.5)] or rotation of an asymmetric body [cf. Eqs. (42.2) and (42.3)].

On the other hand, as shown by comparing Table 12 for radial oscillations with Table 16, Sect. 79 or Table 19, Sect. 100 for non-radial oscillations, the differences in periods for some of the modes are not so large that non-radial oscillations could be excluded definitely on this ground alone. However, we are probably entitled to disregard them, at least in the more classical types of variables, on the basis of the high symmetry of the oscillations revealed by the observations and especially by the existence of a correlation between the period and the shape of the light curves and the indication of a period-amplitude relation discussed in Sect. 17.

Some of the fundamental data are again presented in a somewhat different form in Table 17. For the classical Cepheids and, to a lesser extent, for the other regular variables of Population I, the mean values in days of $Q_{\text{obs}} = P_{\text{obs}} \sqrt{\bar{\varrho}/\bar{\varrho}_\odot}$ are fairly reliable. But in the other cases, they might be subject to appreciable revisions. However a comparison with the corresponding theoretical values in Table 12 or in Table 19 immediately shows that, as far as their main cycle is concerned, the explosive variables (U Geminorum and UV Ceti stars) stand apart from the rest due to their large values of Q_{obs} as well as their large amplitudes ΔM.

In the case of the magnetic variables too, the large value of Q_{obs} raises difficulties which we have already discussed at the end of Sect. 83. We have indicated there that only high modes of non-radial oscillations of the *g*-type could give sufficiently long periods. In that discussion based on the linear approximation, we have neglected the effects of the amplitude on the period but, in these stars, the displacements derived from the radial velocity curves (cf. Fig. 36) are very large, of the order of the total radius. If these displacements were radial, we could apply the results of Sect. 85 and, in particular, formula (85.13) would give for the actual periods P_a

$$P_a = 1.6\, P_0$$

where P_0 is the period in the linear approximation. The non-radial oscillations have never been studied in this respect and we do not know whether the effect would be of the same order. In any case, it is too small to bridge the gap between the observed and the theoretical periods but it might help a little.

[1] Cf. the statistical studies of E. A. KREIKEN: Z. Astrophys. **31**, 256 (1952); **32**, 125 (1953), where a relation of the form (95.3) is established between the period and the inverse root square of the mean density of the most tenuous component of spectroscopic binaries. The constant of proportionality is of the order of 0.5 thus, much larger than in (95.3).

Table 17. *Some fundamental characteristics of variable stars.*

Class	$P_{obs}\sqrt{\bar{\varrho}/\bar{\varrho}_{\odot}} = Q_{obs}$	ΔM_{bol}	Type of cycle
Classical Cepheids (I)	~0.0365	~1	Regular
Dwarf Cepheids (I)	~0.045	~1	Regular, but multiple periods
Red semiregular variables (I + II ?)	~0.086	~1	Some irregularity
Long-period variables (I + II ?)	~0.096	~1	Some irregularity
RR Lyrae (II)	~0.075	~1	Regular, but multiple periods
Short period Cepheids in clusters (II)	~0.145	~1	Regular
W Virginis stars (II)	~0.160	~1	Regular with slow changes in periods
RV Tauri stars (prob. II)	~0.25	~1	Some irregularity
Yellow semiregular variables (prob. II)	~0.70	~1	Some irregularity
β Cephei stars (I)	~0.027	small	Regular, but multiple periods and perhaps secular changes in period
Magnetic variables (I)	1 to 2	small	Regular
Old novae, MacRae 43°.1 (II)	0.15—1.0	0.1 to 0.5	Erratic
Secondary periodic oscillations in Nova DQ Herculis	~0.07	~0.05	Regular
UV Ceti stars (I)			
main cycle (outbursts)	5—10 ?	1 to 3	Erratic
duration of outburst	~0.03		
secondary oscillations	~0.1	~0.5	Erratic
U Geminorum stars (I or II ?)			
main cycle (outbursts)	500—1000	~4	Erratic
duration of outburst	50—100		
secondary oscillations	0.05—0.5	0.1 to 1	Erratic

However, we have seen in Sect. 83 that, for these stars, another type of difficulty arises in the interpretation of the large variations of the magnetic field and it may be that, as suggested by DEUTSCH [cf. Sect. 40], the observed period is essentially one of rotation. Since such fundamental doubts subsist on the nature of the variations in the magnetic variables, we shall exclude them from the more detailed discussions in the following sections.

a) The periods.

96. The UV Ceti stars. In this case nothing is known on the mechanical effects associated with the light variations. If AMBARTSUMIAN's ideas recalled in Sect. 39 are right then the physical processes at work, those responsible for carrying up prestellar material in the external layers as well as those that accompany its modifications and the liberation of energy, escape for the time being any physical analysis.

As we have seen in Sect. 41, a more conservative point of view leads to the conclusion that only a small fraction of the star surface is affected in any one flare and, in this case, the theoretical periods of general oscillations of the star are not relevant for the phenomenon.

In this respect, one should note that although we are struck by the rapidity of the outburst, in fact, its duration is only of the order of the period of the fundamental mode of radial adiabatic oscillations and the time between outbursts is much longer than that period.

Generally speaking, these extremely short theoretical periods of adiabatic oscillations which, by the way, are not so much shorter than the observed erratic secondary cycle, raise a new problem because the non-adiabatic phenomena in the external layers might not be able to keep pace with these very rapid variations. For instance, the approximate formula (67.27) gives radiative relaxations times τ_R for the non-adiabatic temperature field in these dense stellar atmospheres of the same order as these periods. Thus the deviations from radiative equilibrium in the atmosphere may become very large and an appreciable part of the energy may be transferred through the atmosphere by the shock waves which would probably develop.

The conditions are even more extreme in the case of Nova DQ Herculis where τ_R might become appreciably larger than the period of oscillation. One may wonder whether, in such cases, the successive compressions could not accumulate heat at some depth in the atmosphere causing more violent breakdowns of the equilibrium at irregular intervals.

97. The U Geminorum stars. In this case, the main cycle and even the duration of the outburst are so long with respect to the theoretical periods of oscillation that, at first sight, they do not seem to be relevant for the main phenomena although they may be so for the secondary oscillations.

The main cycle has often been compared to that of the recurrent novae (cf. Sect. 35) and the same type of hypothesis briefly recalled at the beginning of Sect. 91 have often been advocated to explain both. The existence of a relation of the form (36.1) in both cases or the more or less equivalent fact noted by SCHATZMAN[1] that the quantity E/LP, where E is the total energy emitted during the outburst and L the luminosity at minimum, is of the order of unity for both types of stars supports this analogy. Such relations between E, L and P are qualitatively compatible with many of the suggested hypotheses where the energy accumulates in some potential form at a rate proportional to L. However in many cases, this rate is much too slow and could only explain explosions occurring once or at very long intervals. In other cases, it is difficult to specify the process determining the length of the cycle which, although variable, has a definite mean value in U Geminorum stars.

In the specific possibility discussed by SCHATZMAN[1], the accumulation of He^3 leads to vibrational instability and the main outburst should be announced by oscillations of increasing amplitudes which have never been observed. This could perhaps be explained if, as in the discussion at the end of the preceding section, the relaxation time τ_R of the atmosphere would still be much larger than the proper period of linear oscillation. But this is not very likely due to the much higher surface temperature of these stars.

Another possibility is provided at least for the U Geminorum stars by a revision of SCHATZMAN'S line of arguments. He assumes that at each explosion the He^3 content is reduced below the critical value necessary for vibrational instability. In that case, the length of the cycle is fixed by the time necessary to build up the concentration of He^3 to the required critical value. But let us imagine that, at some time, this critical value is slightly exceeded and that small oscillations of very short period are started. Due to the vibrational instability, the amplitude will grow and according to Eqs. (85.12) and (85.13), the period and the asymmetry will increase. Perhaps the oscillation will only get stabilized, through a balance of the effects of the non-linear terms in the external layers and the variations of the He^3 content in the interior, at such a very large amplitude that the length of the cycle is then mainly controlled by this amplitude and no longer by the period of the linear oscillation.

[1] E. SCHATZMAN: Ann. d'Astrophys. **14**, 905 (1951).

In that case, instead of a succession of outbursts, what we observe is really an extreme case of non-linear oscillations to which the computations of FRANK-KAMENETSKY[1] are applicable. He assumes that the amplitude of the displacement like that of the light curve is very large. Then, the actual period P_a is given in a first approximation [cf. Eqs. (89.3) and (85.8)] by (85.13) which can be written

$$P_a = P_0 \left(\frac{1}{1-A^2}\right)^{\frac{3}{2}} \tag{97.1}$$

if P_0 denotes the corresponding period in the linear approximation and A is related to the extreme values w_1 and w_2 of $w = r/r_0$ by

$$w_1 = \frac{1}{1-A}, \qquad w_2 = \frac{1}{1+A}. \tag{97.2}$$

With the same notations, formula (85.12) gives for the inverse ratio of the period P_a to the time t_1 during which the radius is greater than its equilibrium value,

$$\frac{t_1}{P_a} = \left(\frac{1}{2} + \frac{\operatorname{arc\,sin} A}{\pi}\right). \tag{97.3}$$

Frank-Kamenetsky assumes

$$L \propto (w)^{-\mu} \tag{97.4}$$

so that

$$\frac{L_M}{L_m} = \left(\frac{w_2}{w_1}\right)^{-\mu}. \tag{97.5}$$

If A is very close to 1 $(1 + A \approx 2)$, the elimination of $(1 - A)$ between (97.5) and (97.1) yields for the total range in magnitude

$$\Delta M_{\text{bol}} = 2.5 \log_{10} \frac{L_M}{L_m} = 5\mu \log_{10} 2 - \frac{5}{3}\mu \log_{10} P_0 + \frac{5}{3}\mu \log_{10} P_a \tag{97.6}$$

which is similar to the relation (36.1) of KUKARKIN and PARENAGO. If $\mu = 1$, the coefficients of $\log_{10} P_a$ in both relations become identical. In that case, to recover also the value of the constant in (36.1), FRANK-KAMENETSKY takes $P_0 \approx 3$ days. Unfortunately, this is about a thousand times larger than the theoretical period of linear pulsation of the blue component of SS Cygni which we believe is the seat of these large variations.

Does this mean that FRANK-KAMENETSKY's idea must be rejected completely? The weakest link in the whole reasoning is (97.4) and it would be interesting to discuss the effects of a better physical relation between L and w.

However as this is not too easy, one might, for the time being, abandon the attempt to recover the quantitative relation of KUKARKIN and PARENAGO and discuss some other consequences of FRANK-KAMENETSKY's hypothesis. For instance accepting for P_0 a value of the order of 0.002 days, one sees from (97.1) that A must be of the order of 0.9994 to give $P_a \approx 50$ days. This corresponds to an enormous variation of the radius from $R_2 \approx R_0/2$ to $R_1 \approx 1.7 \times 10^3 R_0$ and one may wonder whether such extreme conditions can ever be realized. Another aspect of the problem concerns the asymmetry associated with these enormous amplitudes. According to (97.3) if $A = 0.9994$,

$$\frac{t_1}{P_a} = 0.988. \tag{97.7}$$

This would correspond to an outburst lasting about one to two hundreds of a cycle. This is smaller than the observed length of the outburst by a factor of the order of 10 but it compares rather favourably with the time of rise to maximum

[1] D. A. FRANK-KAMENETSKY: Dokl. Akad. Nauk USSR 77, 385 (1951).

which, from the theoretical point of view adopted here, may be more significant, the time of decline being determined rather by the relaxation time of the external layers.

Moreover when, as in this case, P_a is mainly determined by an "amplitude" A which gets close to unity small irregularities in A will imply large erratic variations in the length of the cycle and in the asymmetry which recall the observed characteristics of these stars. Finally, this would provide, at least qualitatively, an immediate interpretation of the fact recalled in Sect. 36 that longer cycles are associated with larger amplitudes. Thus, despite the extreme character of this hypothesis, it may be worthwhile to discuss it further.

98. The classical Cepheids. For quite a long time, the theory yielded consistently periods about a factor 2 too small. The revision of the zero point of the period-luminosity relation implying an increase of the radii by about a factor 2 while the masses change very little provided just about the needed correction to the mean density to bring agreement between the theoretical and the observed periods. This point has been the object of numerous detailed discussions[1] in the case of Cepheids. Here we cannot go into such details but the mean value $Q_{\text{obs}} = 0.0365$ given in Table 17, shows a good agreement with the theoretical values in Table 12 for different models of fairly high central condensation.

It is perhaps premature to try to choose among these models on the basis of a strict agreement because the uncertainty on the masses and the radii is still considerable. However, if such an attempt is made, one should not forget the correcting factor [cf. Eq. (62.2)] due to radiation pressure. As indicated by the figures given in the left-hand margin of Fig. 51 this correction is not at all negligible especially for the brighter Cepheids even if a high hydrogen abundance is adopted. In fact, it should affect the slope of the period-luminosity relation. Here, since we consider only the average properties of the group, it may amount to a factor of the order of 1.15 by which the quantity Q_{obs} should be divided to be compared directly to the theoretical values in Table 12. This would reduce its mean value to about 0.0320 which would then indicate that the model should have a very high central condensation of the same order as the models discussed by EPSTEIN. Furthermore, the external layers should be essentially in radiative equilibrium since, as we have seen in Sect. 62, a large external zone in convective equilibrium increases the period appreciably.

The first conclusion above is very satisfactory since it agrees with the requirements of the energy generation in these stars.

We have neglected here the correction to the periods due to the finite amplitudes. In fact, these are so small that, as we have already seen in Sects. 85 and 86, the correction which is only of the order of $0.002 P_0$ is quite negligible.

On the whole, we may conclude that, for this class, the situation with regard to the periods is quite satisfactory. This means also that the period-luminosity relation is compatible with the mass-luminosity relation and the usual definition of the effective temperature in terms of the radius and the luminosity. However, it does not explain the *existence* of a period-luminosity relation. This last aspect is really connected with the origin of the pulsation and its maintenance.

99. The dwarf Cepheids. Here again, the values of Q_{obs} are certainly subject to revision[2], but the mean value 0.045 given in Table 17 probably denotes a real

[1] Cf. for instance M. SAVEDOFF: Bull. astronom. Inst. Netherl. **12**, 58 (1953); R. P. KRAFT: Publ. Astronom. Soc. Pacific **65**, 146 (1953); K. FERRARI D'OCCHIAPPO: Wien. mitt. **6**, No. 14 (1953).

[2] For a detailed discussion of some individual stars, cf. L. WOLTJER: Bull. astronom. Inst. Netherl. **13**, 62 (1957).

tendency to somewhat higher values than for the classical Cepheids. Here the correction due to p_R is negligible in all cases as well as the correction for the finite amplitude of the displacement[1] which is at most of the order of 0.1.

Our ignorance of the extent of the hydrogen convection zone in these stars makes the discussion somewhat difficult. However it is not likely to be very deep so that the higher value of Q_{obs} may be taken as indicating a lower central condensation than in the classical Cepheids. This is rather fortunate because, in these stars, such large values of $\varrho_c/\bar{\varrho}$ ($\approx 10^6$) would probably lead to a much too large generation of energy.

On the other hand, many of these stars exhibit multiple periods (cf. Sect. 22) and it is rather remarkable that in all known cases, the modulation of period P_M can be interpreted as resulting from the interference of two oscillations of periods P_0 and P_1, the ratio P_1/P_0 being nearly constant as shown in Table 18 taken from a recent paper by L. DETRE[2].

Table 18. *Values of the ratio P_1/P_0 for dwarf Cepheids.*

Stars	P_0 days	P_1/P_0
SX Phe	0.055	0.778
DQ Cep	0.0789	0.826
RV Ari	0.0931	0.773
AI Vel	0.1116	0.773
VZ Cnr	0.1784	0.801
δ Sct	0.1938	0.812

The simplest interpretation is that the first mode of radial pulsation is also excited in these stars. However although for the standard model, $(P_1/P_0)_{th}$ reaches already the value 0.738, Table 12 does not contain any model for which it reaches quite as high a value as 0.8. Moreover, according to Fig. 41, the highest values of $(P_1/P_0)_{th}$ are associated with values of Q_{th} which are definitely smaller than the observed value.

The numerical difference $(P_1/P_0)_{obs} - (P_1/P_0)_{th}$ is not very large even for the standard model but it might prove difficult to reduce it to zero because the effects of p_R (cf. Fig. 40) or of finite amplitudes (amplitude of mode P_0 greater than that of mode P_1) would only increase it. Perhaps a model with a not too high central condensation and a very steep density gradient in the external layers would offer the best chance for a solution of this difficulty. The same interpretation could also be applied to the two Cepheids at the head of Table 2, with a ratio P_1/P_0 a little smaller of the order of 0.71 to 0.73 which is easier to justify and corresponds according to Fig. 41 to a value of Q_{th} close to the observed value.

On the other hand, although the relative displacement $\delta R/R$ is small[3], on the theory of finite pulsation (cf. Sect. 86), one should expect a certain amount of coupling and the appearance of frequencies $2\sigma_0$, $\sigma_0+\sigma_1$ and $\sigma_1-\sigma_0$ even if there is no close resonance. In fact, in AI Velorum, WALRAVEN [*23*] has reported the presence of other frequencies σ_2 and σ_3 with $\sigma_3=\sigma_2+\sigma_0$ and $\sigma_0/\sigma_2=P_2/P_0=0.39795$ as well as another one $\sigma_4=\sigma_2-\sigma_1$. But these are not the coupling frequencies expected and seem rather difficult to interpret in terms of higher modes although, except for the standard model [*54*], we know little about these modes. Other schemes can be suggested (cf. for instance [*23*], p. 251) but they are all open to serious objections.

Despite the remarkable agreement between the main observed features of the variation and the sum of the two waves P_0 and P_1, one should not perhaps

[1] Cf. L. WOLTJER: Bull. astronom. Inst. Netherl. **13**, 58 (1957) where the known data on the velocity curves for these stars are summarized.

[2] L. DETRE: Mitt. Sternwarte Budapest **1956**, No. 40.

[3] For AI Velorum, cf. the radial velocity curves by L. GRATTON and C. J. LAVAGNINO: Z. Astrophys. 32, 69 (1953).

exclude completely the possibility that the modulation period may have a more physical meaning.

100. The β Cephei stars. According to Table 17, it is for these stars, that Q_{obs} reaches its smallest value, 0.027. Furthermore as shown by the correcting factors in the left-hand margin, the pressure of radiation plays a rather important rôle here. Even with a large abundance of hydrogen ($\bar{\mu}=0.7$), the average correcting factor is of the order of 1.22 so that the actual value of Q_{obs} to be compared to the corresponding theoretical quantity in Table 12 is of the order of 0.022. If we consider only the fundamental mode, this is appreciably smaller than any values in that table and, according to Fig. 39 and the accompanying discussion, this would point to a high value of the effective polytropic index (4 or greater) throughout the star. Physically this is probably not too easy to justify. If one is ready to admit that the first mode can be excited while the fundamental one remains absent, agreement between the observed and theoretical periods can be reached easily. On the other hand, because of subsisting uncertainties on M and R, too much weight should not be given to these conclusions.

Moreover, as recalled in Sects. 25 to 27, these stars exhibit many peculiar properties. First of all, many of them present a long period which suggests very strongly a beat phenomenon between two oscillations of very close periods, P_0 and P_0'. In this case, the fact that a variable broadening of the lines is associated with the second period P_0' strongly supports the physical reality of the two periods.

The type of coupling between two modes of radial pulsation of frequencies σ_0 and σ_1 with $\sigma_1 \approx 2\sigma_0$ proposed by WOLTJER (cf. Sect. 86) and often advocated to explain a somewhat similar beat phenomenon in some RR Lyrae stars does not seem to be very helpful here since it is difficult to understand how the resulting frequency $\sigma_0' = \sigma_1 - \sigma_0$ could be associated with a broadening of the lines[1]. Moreover, the observations do not suggest the same degree of symmetry as in the case of the Cepheids and RR Lyrae stars. Other systematic differences are summarized in Table 5 and the need for a distinct explanation has been felt very generally. This has led to different hypotheses which have been summarized by STRUVE[2].

The latest one, advanced by STRUVE [*30*] and ODGERS[3] has been devised especially to account for the recently observed splitting of the lines and the van Hoof effect (cf. Sect. 26). It makes use of radial pulsations and assumes that a thin layer of the atmosphere is ejected at regular intervals determined by the internal period P_0. This layer then falls back under the action of surface gravity. If the time of fall P_f is smaller than P_0 there is a standstill on the velocity curve as in BW Vulpeculae. If P_f is very slightly greater than P_0, the interfering motions of the two layers might give the appearance of a beat phenomenon.

It is difficult to see how the generally high symmetry of the light and velocity curves can be reconciled with this hypothesis which must imply very strong shocks in the external layers. Moreover, it is practically impossible that P_f could be fixed with the necessary precision to give rise to the regular beat phenomena observed, in some cases, over very long periods of time. The only possibility would be that P_f be strictly controlled as, for instance, in a proper oscillation of the atmosphere.

---

[1] A theory of a similar type has recently been proposed by A VAN HOOF: Publ. Astronom. Soc. Pacific **69**, 308 (1957). Unfortunately one of the modes P_0 or P_0' considered by VAN HOOF has no physical reality.

[2] Cf. O. STRUVE: Ann. d'Astrophys. **15**, 157 (1952) and [*30*].

[3] G. J. ODGERS: Publ. Dom. Astrophys. Obs. **10**, 215 (1956).

Although, in this form, the idea might be fruitful, the difficulty is to explain how the interior and the atmosphere can be made sufficiently independent so that their respective proper periods, say P_i and P_e, keep their significance. In Sect. 68, we have discussed the linear aspects of a somewhat similar problem where the atmosphere experiences a forced oscillation under the influence of the interior. If non-linear terms were taken into account, coupling between the internal and the atmospheric oscillations would be added and it may be that beat phenomena could result provided the P_i's and P_e's be in the proper ratios. However the whole idea would need a good deal of elaboration from the physical as well as from the mathematical point of view before any reliable conclusions could be reached.

Another objection has been raised by HUANG who remarked that the total equivalent width of the lines remains constant through the stages of broadening or doubling, suggesting that these phenomena cannot be due to superposed layers but rather to different parts of the star surface.

One should also attempt to disentangle properly the question of the broadening of the lines and that of their doubling. Much emphasis has been laid on the fact that the broadening is intimately associated with the oscillation of period P_0' and this should be true also of the doubling if, as often suggested, the broadening is only a non-resolved doubling. But STRUVE states that in 12 Lacertae, the double lines appear particularly clearly when the total amplitude K is large, that is, when K_0 and K_0' are in phase in the course of the beat period. This suggests that duplicity is really a global effect to which both oscillations contribute.

Finally, looking at Table 4, it is hard to believe that rotation does not play a rôle in the phenomenon. For instance, small rotational velocities are systematically associated with the absence of beat phenomena. This could be due either to the fact that this aspect depends directly on the rotation or that it does not show up when the line of sight makes a small angle with the axis of rotation or both.

A theory in which rotation enters as a fundamental factor has been proposed by LEDOUX[1] assuming non-radial oscillations. In that case, considering an oscillation corresponding to a spherical harmonic of the second degree ($l=2$) which is the most likely to be excited, the frequencies with respect to axes rotating with the angular velocity Ω are, according to (82.32)

$$\sigma_{2,0}, \quad \sigma_{2,0}+2C_2\Omega, \quad \sigma_{2,0}-2C_2\Omega, \quad C_2<1$$

where $\sigma_{2,0}$ is the frequency of the corresponding mode of non-radial oscillation of the non-rotating star.

The first one is the frequency of a stationary oscillation symmetrical with respect to the axis of rotation, while the other two correspond respectively to waves travelling in the opposite or the same direction as the rotation of the star.

For observers at rest, these frequencies or the corresponding periods become

$$\sigma_{2,0}, \quad \sigma_{2,0}-2\beta\Omega, \quad \sigma_{2,0}+2\beta\Omega \quad \text{with} \quad \beta=1-C_2$$

or

$$P_{2,0}, \quad P_{2,0}+2\beta'\Omega, \quad P_{2,0}-2\beta'\Omega \quad \text{with} \quad \beta'>0.$$

In Table 19 below, we have converted the values $\sigma_{2,0}$ of the standard model given in Table 16, Sect. 79 into the more practical quantities $P\sqrt{\bar{\varrho}/\bar{\varrho}_\odot}$.

[1] P. LEDOUX: Astrophys. Journ. **114**, 373 (1951).

Unfortunately, we have no information on more centrally condensed models[1], but it is very likely that here also agreement with the observed periods will require the excitation of at least a p_1-mode. If this is possible, the interference of the stationary oscillation ($P_0 = P_{2,0}$) and one of the travelling waves ($P_0' = P_{2,0} + 2\beta'\Omega$ or $P_0' = P_{2,0} - 2\beta'\Omega$) would give a beat frequency proportional to 2Ω so that the faster the star rotates, the shorter the beat period or the larger the difference between P_0 and P_0'. Excepting β CMa, this is in rather good agreement with the observations.

In this theory, the broadening is attributed to the differential velocities on the surface of the star resulting from the combination of rotation and the three components of the non-radial oscillations and is important mainly for the travelling wave whose amplitude is supposed to be of the same order or smaller than the linear rotational velocity.

Table 19. *Values of $P\sqrt{\bar{\varrho}/\bar{\varrho}_\odot}$ for different modes of non-radial pulsation ($l = 2$) of the standard model.*

Modes	$P\sqrt{\bar{\varrho}/\bar{\varrho}_\odot}$
p_1	0.0295
f	0.0393
g_1	0.0532
g_2	0.0731

The broadening is constant when viewed from the poles. It increases and becomes more and more variable as the line of sight approaches the equator. Its variations are shifted 90° in phase with respect to those of the radial velocity V_0' of the travelling waves. This phase shift corresponds to an advance (maximum broadening occurring when the radial velocity V_0' goes through zero from negative to positive values) for the wave travelling in the opposite direction to the rotation ($P_0' = P_{2,0} + 2\beta'\Omega$) and to a lag (maximum broadening occurring when V_0' goes through zero from positive to negative values) for the wave travelling in the same direction as the rotation ($P_0' = P_{2,0} - 2\beta'\Omega$).

In the case of β CMa to which the theory was originally applied, the broadening being associated with the longest period, P_0' was to be identified with the value $P_{2,0} + 2\beta'\Omega$ corresponding to a wave travelling in the opposite direction to the rotation and this implied the wrong sign for the phase-shift of the broadening. This difficulty subsists for β CMa but one should note that, in all the other cases (cf. Table 4), P_0' is smaller than P_0 so that P_0' should be associated with the wave travelling in the same direction as the rotation giving the right phase-shift. Again, the larger amplitudes K_0' in all these stars would facilitate the quantitative interpretation of the broadening as compared to the case of β CMa where K_0' is small.

In this simplified and linear approach, the greatest difficulty comes from the fact that we have to choose one of the travelling wave, the other remaining unexcited. This may be connected with the cause of the oscillation itself. The simplest one, in the case of non-radial oscillations, is the gravitational interaction of a companion. Unfortunately, since the period of revolution would probably have to be of the same order as P_0, this companion should be well inside the main star.

Other aspects such as the duplicity of the lines or the van Hoof effect do not receive direct explanation on the basis of this hypothesis. But, in any case, they are probably connected with non-linear aspects of the phenomenon which in any theory will be difficult to work out satisfactorily. Despite the above difficulties, LEDOUX's hypothesis has some definite advantages and it is probably

[1] A recent paper by J. W. OWEN: Monthly Notices Roy. Astronom. Soc. London **117**, 384 (1957) indicates that, as the central condensation increases, some of the lower modes vanish. Thus a polytrope $n = 3.25$ has no f-mode and the p_1-mode corresponds to a period $P\sqrt{\bar{\varrho}/\bar{\varrho}_\odot} \approx 0.026$. In the polytrope $n = 3.5$, the lowest p-mode is p_2 with a period $P\sqrt{\bar{\varrho}/\bar{\varrho}_\odot} \approx 0.019$.

worthwhile to keep it in mind in a more general and thorough approach of the problem.

101. The RR Lyrae stars. Let us consider first the class as a whole. The average value $Q_{\text{obs}} \approx 0.075$ given in Table 17 is larger than the theoretical value for any physically reasonable stellar model. Even, if we assume a liberal abundance of helium, p_R could at most reduce Q_{obs} to about 0.064 which is still large. If the mass given by the mass-luminosity relation ($\approx 3.5 M_{\odot}$) is too high, a reduction to the value $1.2 M_{\odot}$ suggested by evolutionary considerations (cf. Sect. 95) would lead to $Q_{\text{obs}} \approx 0.045$. Values comprised between 0.045 and 0.064 could be explained on the basis of a model with a fairly high central condensation provided an appreciable fraction of the external layers be in convective equilibrium. Although the relative amplitude of the displacement is fairly large in these stars ($\delta R/R \approx 0.1$) it is still too small, according to (97.1), to affect appreciably the period.

Actually, one should distinguish two groups of RR Lyrae stars, those of type *c* and those of type *a* and *b* whose respective properties have been discussed in Sects. 16 and 17. A study of the short-period variables in Messier 3 led SCHWARZSCHILD[1] to the conclusion that these variables fall along two distinct parallel lines in a colour-log *P* diagram, the *c*-type variables corresponding to periods systematically shorter by a factor 0.63.

SCHWARZSCHILD[1] suggested that these stars were oscillating in the first mode, but it is difficult to understand how this mode can be excited while no trace of the fundamental mode appears.

If this factor 0.63 is applied to the preceding average value $Q_{\text{obs}} \approx 0.075$, it is reduced to 0.047 and, in the most favourable case, it could be further decreased by the effect of p_R to about 0.04. Contrarily to the case of the other RR Lyrae variables, this value could then be interpreted as corresponding to the fundamental mode of pulsation of a reasonable model without any appreciable reduction in mass with respect to the value deduced from the ordinary mass-luminosity relation. But of course this implies some rather fundamental difference in structure and composition between two groups of stars occurring in the same cluster.

The existence in some members of the class of very long periods of modulation (cf. Table 2, Sect. 22) raises another problem. We have already discussed a somewhat similar question for the β Cephei stars and recalled two of the suggestions made to explain it.

In one, due to ODGERS, the effects of more or less independent "oscillations" of the interior and of the atmosphere with very close "periods" interfere in the external layers. The difficulties encountered by this hypothesis would still be larger here than in the case of the β Cephei stars.

In the other suggestion, non-radial oscillations are considered in the presence of rotation, the beat frequency being proportional to the angular velocity of rotation. According to formula (83.24) and the following discussion, the presence of a weak magnetic field would also lift, at least partially, the degeneracy of the frequencies of non-radial oscillations and provide a possibility of long beat periods. However, non-radial oscillations imply a general asymmetry of the phenomena which appear differently depending on the angle made by the line of sight with some privileged axis in the star. But, in the case of the RR Lyrae stars, the observations point out to a rather high symmetry of the oscillation (cf. especially Sect. 17).

[1] M. SCHWARZSCHILD: Harward Circ. **1940**, 437.

If non-radial oscillations are eliminated, the coupling theory of WOLTJER and Mrs. PELS-KLUYVER[1] offers another possibility for purely radial oscillations. We have studied some aspects of the possible coupling between different modes in Sect. 86. We have seen there in discussing a solution of the form (86.5) with q_0 and q_1 defined by (86.17) and (86.18) that, when the frequency σ_1 is close to $2\sigma_0$, say $\sigma_1 = 2\sigma_0 + \varepsilon$, coupling gives rise to a beat of frequency ε by interference between terms of frequencies σ_0 and $\sigma_0' = \sigma_1 - \sigma_0 = \sigma_0 + \varepsilon$ reinforced by the resonance $\sigma_1 \approx 2\sigma_0$.

The pulsations of frequencies σ_0 and σ_1 are not necessarily the fundamental and the first modes and Mrs. PELS-KLUYVER originally suggested a case where the commensurability 2 to 1 was realized between the fundamental and the second mode. However, since it is likely that the damping of the modes increases rather rapidly with the order (cf. Sect. 71), this theory would inspire more confidence if, actually, the commensurability 2 to 1 (i.e. $P_1/P_0 \approx \frac{1}{2}$) was found between the fundamental and the first modes. In the present state of our knowledge, Table 12 and Fig. 41 indicate that it is only in models with fairly extensive external convection zones that P_1/P_0 can approach the value $\frac{1}{2}$. As our previous discussion of the values of Q_{obs} led to a similar conclusion, it may still be possible, in the future, to find a model where both conditions be simultaneously satisfied.

Formula (86.17) and especially (86.18) suggest that a frequency $2\sigma_0$ should also appear in the observations. In this respect, it is rather satisfactory that, in the case of RR Lyrae, the very precise analysis by WALRAVEN [23] has indeed isolated a term with this frequency. Of course, it is always delicate to be sure that such harmonics have a true physical meaning.

Finally, in this theory, one should expect that the longer the beat period (the smaller $\sigma_1 - 2\sigma_0$), the larger the amplitudes of the terms excited by resonance in (86.17) and (86.18). However, according to DETRE[2], the observations indicate just the opposite a fact which seems incompatible with the coupling theory.

102. The short period Cepheids in clusters, the W Virginis stars and the RV Tauri stars. For this group, the values of Q_{obs} are definitely much larger than the corresponding theoretical values for any models in Table 12. If the usual mass-luminosity relation is valid, the correction due to p_R may amount to a factor 1.2 to 1.3 for the W Virginis and RV Tauri stars respectively reducing Q_{obs} to 0.133 and 0.192 which are still very large.

If as before, we adopt for the mass of these type II variables, the value 1.2 $M_\odot$, the reduced values of Q_{obs} become respectively 0.074, 0.067, 0.087 which certainly come already much closer to the possible theoretical values.

In the case of the last two classes, the relative displacement $\delta R/R$ is probably very large[3], the radius being practically doubled at maximum expansion in stars with the longest periods. By (97.2), this would correspond to $A \approx 0.5$ and according to (97.1), this would increase the theoretical periods by a factor 1.5. If this applies for instance to the RV Tauri stars, Q_{obs} should further be reduced by this factor, giving $Q_{\text{obs}} \approx 0.058$, before being compared to the theoretical values in Table 12. For the average W Virginis star considered here, A will be a little smaller, say ≈ 0.4 which gives a reduction factor of the order of 1.3 so that $Q_{\text{obs}} \approx 0.052$. These last two values are compatible with the theoretical

[1] H. A. KLUYVER: Bull. astronom. Inst. Netherl. **7**, 313 (1936).

[2] Cf. the article by L. DETRE, p. 1156 (Fig. 6) in: Vistas in Astronomy, Vol. 2 (Ed. by A. BEER). London: Pergamon Press 1956.

[3] Cf. for instance A. W. RODGERS: Monthly Notices Roy. Astronom. Soc. London **117**, 85 (1957); also Sect. 33.

periods of the fundamental mode of radial oscillations of models with fairly high central condensation provided the effective polytropic index be small (convection?) in a large external region.

However, in the case of the short period Cepheids in clusters, the reduction factor due to the finite amplitudes is probably much smaller and one does not see any obvious means of reducing appreciably the value 0.074 unless through a further revision of the fundamental data M and R, which of course cannot be excluded. This last possibility subsists also for the two other classes so that not too much weight should be given to the present discussion. Nevertheless, together with the evidence from the RR Lyrae stars, one may conclude that the observed periods in the type II variables definitely favour masses considerably smaller than those derived from the ordinary mass-luminosity relation and models with low values of the effective polytropic index in the external layers. Finally the correction for finite amplitudes is likely to be important for the variables of type II with fairly long periods.

103. The long-period variables and red semiregular variables. The values of Q_{obs} given in Table 17, are perhaps not very significant in this case as they vary considerably from one star to the other in both classes. This means that in a diagram $\log P$-$\log\sqrt{\bar{\varrho}}$ these stars fall along a line with a slope appreciably different from -1 (cf. for instance [*39d*], Fig. 7, p. 43). Is this to be taken as an indication that straight pulsation theory does not apply? Of course, large errors may still affect the bolometric magnitudes and the radii of these stars, perhaps in a systematic fashion which distorts the relation $\log P$-$\log\sqrt{\bar{\varrho}}$. Also, the ordinary mass-luminosity relation may be at fault especially since these classes may not be homogeneous as far as population types are concerned.

Keeping for the time being, to the mean values in Table 17, they should be corrected for the effect of p_R by a factor of the order of 1.2 to 1.6 depending on the abundance of hydrogen. For instance, if we take $\bar{\mu}=1$, this effect reduces the values of Q_{obs} respectively to 0.07 and 0.06 for the long-period variables and the red semiregular variables.

Because of the complicated pattern of the radial velocity curves (cf. Sect. 31), little can be said with certainty concerning $\delta R/R$. On the other hand, as important deviations from the law of black body radiation are probable (cf. Sect. 29), the interpretation of the large phase shifts between the light maxima at different wavelengths ([*37a*], pp. 47 to 50) is not straightforward. They suggest, however, that the radius variations are rather important and, according to formula (97.1), this would mean another reduction in the values of Q_{obs} before comparing them to the theoretical values. As they stand, these values are still a little high for any reasonable known model but nevertheless they suggest that, here too, the observed variations are connected with the fundamental mode of radial pulsation.

104. Conclusions. The discussion in the previous sections has shown that in many of the intrinsic variables considered, the length of the cycle is of the same order as the period of pulsation of the interior. Furthermore, in the most regular variables, this period is very stable suggesting that it is controlled by very steady forces applied to a system of large inertia which again suggests that the bulk of the star is concerned in these oscillations.

It is more difficult, on this basis alone, to distinguish between radial and non-radial pulsations since the periods are not so different. Up to now, our main argument in favour of purely radial pulsations in most of these variables is the impression derived from the observations and especially from the existing correlations between periods and amplitudes, that the phenomenon responsible for the variations must have a very high geometrical symmetry.

b) Origin and maintenance of finite oscillations.

105. Vibrational instability as a cause of pulsation. In addition to a reasonable approximation for the periods, one usually expects that the linear theory should give some idea on the possible instabilities. Since we are not interested here in dynamical instability, we may assume that the linear adiabatic approximation gives real values for the frequencies σ. Taking into account the non-adiabatic terms, we may then evaluate a damping coefficient σ'' so that the factor $e^{-\sigma'' t}$ will make the amplitude decrease or increase according as σ'' is positive or negative. In the second case, we say that the star is *vibrationally unstable* and we expect that the infinitely small original oscillation will grow indefinitely unless non-linear terms which become important after some time, are able to limit the amplitude to some finite value.

The evaluation of σ'' has been the object of a detailed study in Sects. 63 to 72. Unfortunately the conclusion remains mainly negative: up to now we have failed to discover the origin of the vibrational instability which is probably at the basis of the observed variations.

In the present state of our knowledge of thermonuclear reactions and of the sources of stellar opacity and for all physically acceptable models, the main bulk of the star, in which the adiabatic approximation is excellent, contributes a positive dissipation (cf. especially Sect. 65). Moreover, in the very external layers, the heat capacity is so small that $d\,\delta L/dr$ tends to vanish [cf. end of Sect. 67 and Eq. (68.3)] and their contribution to σ'' is negligible.

Thus the whole question depends on what happens in the difficult intermediate region where the ionization of hydrogen increases the heat capacity considerably. At a somewhat greater depth, the second ionization of helium could also play a rôle according to ZHEVAKIN. To bring about vibrational instability, the negative dissipation in this region should be larger than the net positive dissipation in the interior. Recent investigations have not yet led to definite conclusions (cf. Sects. 69 and 70). It seems possible to construct models in which the intermediate layers have a sufficient heat storage capacity due to ionization, but it is not certain that the corresponding changes in opacity will permit full use of this to introduce a sufficient phase shift in δL to render the star unstable. For instance, in SCHATZMAN'S paper, the instability is due only to changes in the opacity occurring above the ionization zone of hydrogen in a region where, according to our previous remark, the actual δL should more nearly be constant.

Nevertheless, at present our best hope to find a source of vibrational instability still lies in these intermediate layers especially for the highly condensed models which the requirements of energy generation and the interpretation of the periods (cf. Sect. 98) have led to adopt for the Cepheids.

In all this, the effects of turbulent viscosity have generally been neglected although some of the models considered have extensive convective zones. As shown in Sect. 71, this type of viscosity is much more effective than molecular or radiative viscosity and its influence is complex. In a first approximation, it gives a very strong damping as long as $\xi = \delta r/r$ varies appreciably in the convective region. But on the other hand, it may also lower very much the rate of increase of ξ through the whole star and thus reduce considerably ξ_R/ξ_C and the conductive damping. This aspect deserves further study especially in centrally condensed models with an isothermal core followed by an energy generating shell in radiative equilibrium surrounded by a large envelope in convective equilibrium. Strictly, all this is true only for the fundamental mode of radial oscillation. However, according to Sects. 71 and 81 modes of fairly high orders can be neglected as even ordinary molecular viscosity would have an appreciable damping effect on them. For the first few modes of radial pulsation, the situation is fairly similar to that for the fundamental mode except that the ratio ξ_R/ξ_C increases with the order of the mode. As a consequence, if the source of instability is fairly deep, the star will be appreciably more stable towards these higher modes. On the contrary, if the source of instability is pushed out into the intermediate layer under the atmosphere, some of the first higher modes may become more unstable than the fundamental one. This may prove to be a source of difficulties if this layer is the seat of vibrational instability since, in general, there is little evidence that the higher modes are excited.

For the first few modes of non-radial oscillations no detailed evaluation of σ'' exists (cf. Sect. 81) but since $\delta\varrho/\varrho$ and $\delta T/T$ tend towards zero at the center, the conventional energy sources will have very little influence[1]. As to the effects of the conductivity, one may expect that the total dissipation will be somewhat larger than for radial oscillations due to the extra components of the flux along the level surfaces during the oscillations. Thus, on the whole, the star should be more stable towards this type of perturbation. However, the situation in the intermediate non-adiabatic layer will be complex and only a detailed analysis could lead to reliable conclusions.

In this case, one may also think of external excitation but to be effective it should be accompanied by resonance and, as we have recalled at the end of Sect. 79, this is only possible for fairly high *g*-modes and even then, it is not likely to be very efficient. There is also the difficult case of a small dense companion moving well inside the extended atmosphere of the primary or that of two similar stars revolving around each other inside a common envelope as proposed by HOYLE and LYTTLETON[2] but they have received very little theoretical attention up to now.

Nearly all discussions of vibrational stability as a cause of pulsation have been directed towards the interpretation of the classical Cepheids. In that case, as often pointed out by EDDINGTON, the condition for vibrational instability should, at the same time, explain the existence of the period-luminosity relation or, in other words, the position of the Cepheids in the Hertzsprung-Russell diagram.

The variations of the RR Lyrae stars and of the W Virginis stars may perhaps be considered as the effect, in Population II, of the same cause that leads to the pulsation of the classical Cepheids. There are differences especially in the observed radial velocity curves (discontinuities and emission lines in type II variables) and in the form of the light curve of W Virginis as compared to classical Cepheids of the same period, but they may arise entirely in the external layers due, for instance, to larger relative amplitudes or, as suggested by WHITNEY[3], to a different ratio of the wavelength to the atmosphere scale height.

However the concentration of the cluster variables in the gap of the horizontal branch of the colour-luminosity array of these clusters and the absence

[1] Cf. however a recent investigation by R. SIMON: Bull. Acad. Roy. Belg., Cl. Sci., Sér. V **43**, 610 (1957).

[2] F. HOYLE and R. A. LYTTLETON: Monthly Notices Roy. Astronom. Soc. London **103**, 21 (1943); cf. also P. LEDOUX: Astrophys. Journ. **114**, 373 (1951).

[3] CH. WHITNEY: Ann. d'Astrophys. **18**, 375 (1955).

of any stable stars in this gap has certainly a very important meaning in this respect. If we want to keep up the analogy, what is the equivalent of this property in the younger Population I Cepheids? In this connection, what is the meaning of the Hertzsprung gap?

The long-period variables and the semiregular variables may perhaps be connected to the preceding classes as far as the origin of their variations is concerned. In these objects, the analysis in the external layers will probably be particularly difficult especially in those stars where molecules and perhaps solid particles can form at some phases in considerable numbers. In the latter case, it would be interesting to have a theoretical discussion of the effects of clouds of solid particles (veil theory) on the stability of the star.

In the case of the dwarf Cepheids, despite obvious similarities with the classical Cepheids, it is already more difficult to understand how the same factor of instability could be at work and this becomes definitely unlikely in the β Cephei stars. Groups such as T Tauri stars, the flare stars or, among the giants, the R Coronae Borealis stars or the S variables may each set its own distinct problem.

To a certain extent, the same may be said of the U Geminorum stars or of the old novae although if, as suggested by their location in the Hertzsprung-Russell diagram, they are stars on their way to the white dwarf stage[1], they may be the objects in which the condition of vibrational instability may be most likely to be realized. P. LEDOUX and E. SAUVENIER-GOFFIN[2] have shown that in white dwarfs, due mainly to the slow variations of ξ from centre to surface and to the pushing out of the energy generation zone, most nuclear reactions if present would lead to vibrational instability. But as pointed out by MESTEL[3] later, there really is a deeper incompatibility between the white dwarf stage and nuclear reactions since in the presence of the latter, the high degree of degeneracy of the electron gas prevents the reaching of any steady state configuration. But LEDOUX and SAUVENIER-GOFFIN'S conclusions remain applicable to partially degenerate configurations such as might be reached in the evolution towards the white dwarf stage. In some cases, the periodic accumulation of He^3 as proposed by SCHATZMAN (cf. Sect. 97) might be necessary to bring about the instability, or it might reinforce it. We would then be witnessing in these stars the last spasms before the dying out of the nuclear energy sources.

In any case, the variety of properties exhibited by variable stars as well as their differences in location in the Hertzsprung-Russell diagram and their belonging to different types of population suggest that, before we can explain all the cases of stellar variability, several distinct sources of vibrational instability will have to be discovered.

106. Other possible causes of pulsation. The conventional approach through the question of vibrational instability may not reveal all the possibilities of instability and of finite oscillations. For instance, the case of hard self-excited oscillations discussed in Sect. 90 would escape all enquiries along this line, since they occur only if the system is originally displaced from its equilibrium position by a finite amount.

This opens a new line of approach which up to now has received little attention. For instance, it may be that certain forms of secular instability could really bring the star to a state in which it could start finite oscillations. This possibility

[1] Cf. however the photometric study of the eclipsing system of Nova Herculis by M. F. WALKER: Publ. Astronom. Soc. Pacific **66**, 230 (1954); Astrophys. Journ. **123**, 68 (1956), which suggests such a very small value for its mass and its mean density that it should still be very far from the white dwarf stage. It becomes even questionable whether it could ever become degenerate.

[2] P. LEDOUX and E. SAUVENIER-GOFFIN: Astrophys. Journ. **111**, 611 (1950).

[3] L. MESTEL: Monthly Notices Roy. Astronom. Soc. London **112**, 583 (1952).

is especially interesting because it would connect the occurrence of pulsations rather directly with some phase in the evolution of the star. Recently, SU-SHU HUANG[1] has suggested a scheme in which the novae, the U Geminorum stars, the RR Lyrae variables and the Population II Cepheids would all be the result of the same type of instability developing in the course of evolution. It is suggested that the difference in behaviour can be accounted for by the fact (assumed to be a general property) that objects of the first two classes are components of binaries which have lost a large fraction of their mass through the inner Lagrangian point, a circumstance which has hastened considerably their evolution. The cause of the instability however is left very vague and is referred to only as an excess energy generated in the core. Without having in mind necessarily as ambitious a program as HUANG's, it may be worthwhile keeping in mind that, if such finite deviations from a true state of equilibrium can be brought about (and the loss of mass may be especially efficient in this respect), there may perhaps result finite oscillations of sufficient asymmetry to enable the star to get rid of the excess energy. In other words, apart from the purely static stellar models usually considered, a state of finite oscillations may provide, under special circumstances, another type of steady state for a star[2].

Another possibility is the existence, in the static configuration, of some mild form of dynamical instability. Such a case suggested by ZHEVAKIN has been referred to and summarized in Sects. 89 and 90.

107. Limitation of the amplitude to some finite value. Whatever the precise form of the origin of the pulsation, it is always some form of instability either for infinitely small oscillations around the equilibrium state or for a finite perturbation whose amplitude exceeds a certain critical value. In both cases, the amplitude will tend to increase. To limit it to a finite value, a new dissipative factor must come into play. This may also occur by coupling of the unstable mode with another one which is positively damped.

Both cases have been discussed in Sects. 88 and 90. Unfortunately, this discussion had to remain mainly formal because the significant factors in the case of stellar oscillations have not yet been isolated. It is a difficult but important problem which no doubt will receive a deserved attention in the future.

c) The correlation between the amplitudes and the phases of the velocity and light curves.

108. The amplitude correlation. The best known variables in that respect are the classical Cepheids, the RR Lyrae stars and the β Cephei stars. The corresponding empirical relations between the total range in photographic magnitude ΔM_{pg} and in velocity $2K = \Delta V_R$ measured in km/sec have been discussed in Sects. 18 and 25.

L. WOLTJER[3], using as coordinates ΔV_R and the range in bolometric magnitude ΔM_{bol} which is more significant, found for the RR Lyrae stars and the dwarf Cepheids

$$\Delta V_R \approx 65\, \Delta M_{bol}, \tag{108.1}$$

or

$$\Delta V_R \approx 160 \log_{10}\left(\frac{L_M}{L_m}\right). \tag{108.2}$$

---

[1] SU-SHU HUANG: Astronom. J. **61**, 49 (1956); cf. also O. STRUVE and SU-SHU HUANG: Occ. Notes of the Roy. Astronom. Soc. **3**, 161 (1957).

[2] Cf. also P. LEDOUX: Stellar Stability, Sect. 15, p. 657, this volume.

[3] L. WOLTJER: Bull. astronom. Inst. Netherl. **13**, 58 (1957).

The classical Cepheids satisfy a very similar relation with perhaps a slightly higher value of the constant of proportionality. W Virginis, the best known representative of the type II Cepheids falls also very closely on the same line.

From the theoretical point of view, it might even be better to eliminate ΔV_R in (108.1) in favour of $\Delta R/R$ where ΔR is the total range of the radius. In a first approximation, we have

$$\frac{\Delta R}{R} \approx \frac{\Delta V_R}{2\pi R} P \approx \Delta V_R \, 0.02 \, Q \sqrt{\frac{R/R_\odot}{M/M_\odot}} \tag{108.3}$$

where $Q = P\sqrt{\bar{\varrho}/\bar{\varrho}_\odot}$ which is measured in days is very nearly constant for all stars oscillating in the same mode while the last factor varies only slowly.

As long as $\Delta L = L_M - L_m$ is relatively small, (108.2) and (108.3) together imply a practically linear relation between $\Delta R/R$ and $\Delta L/L$ and, in those cases, the linear theory should already provide a reasonable approximation for the relation between ΔL and ΔR. However, when the range of variation reaches medium values ($\Delta L/L \approx 1$), according to (108.1), $\Delta L/L$ increases already more rapidly than $\Delta R/R$, and this tendency is still reinforced by the non-linearity of (108.1) itself for large amplitudes (cf. Sect. 18). This is also in qualitative agreement with the behaviour of the light and velocity curves in the course of the modulation period in the RR Lyrae and dwarf Cepheids as analysed by WALRAVEN (cf. Sect. 22) since the light curve shows a strong distortion in amplitude (vertical asymmetry) while the velocity curve shows only the phase distortion.

Although in the case of the Cepheids and the RR Lyrae stars, these relations provide an interesting basis for theoretical investigations, little progress in their interpretation has been done, mainly because of the complexity of ΔL which is, of course, a non-adiabatic effect and may be determined largely by the non-adiabatic external layers (cf. Sects. 65, 69 and 70).

In this respect also, the β Cephei stars (cf. Sect. 25) differ strongly from the previous classes, the coefficient of proportionality in (108.1) being nearly 10 times larger. This means that for a given range in velocity, the light-variations are very small. This may be taken as favourable to non-radial oscillations since, in that case, the alternation, on the star surface, of hotter and cooler patches will reduce very much the mean value of ΔL.

As far as the long-period variables are concerned, the complexity of the radial velocity curves (cf. Sect. 31) renders all comparison with the light curves practically impossible at the present time.

109. The phase correlation. Here again, the Cepheids and the RR Lyrae stars form a well defined group characterized by a 90° phase lag between the radius and the luminosity (cf. Sect. 18), maximum luminosity occurring approximately a quarter period after maximum compression. On the simple picture of a linear adiabatic standing oscillation assumed valid up to the surface, maximum temperature should occur at maximum compression and one would rather expect the star to appear brighter at that moment. In any case, on that picture the phases of mean radius should also coincide with those of mean luminosity and not, as observed, once with maximum and once with minimum luminosity.

This discrepancy between elementary theory and the observations has been the main argument of the opponents to the pulsation theory. Actually, there is little reason why this simple picture should be true. However, as shown early by EDDINGTON and others, towards the top of the adiabatic region where δL can still be computed by a formula of the type (63.9), it is indeed maximum at maximum compression. Now, EDDINGTON always considered that the phase of δr could not change appreciably in the external layers (cf. first paragraphs of Sect. 70) and, as long as the abundances of H and He were kept very low, the heat capacity of these layers was too small (cf. first paragraphs of Sect. 69) to affect the phase of δL. In those circumstances, δL and δr keep the same phase relationship up to the surface and the simple argument outlined above is justified.

Later, EDDINGTON realized that the ionization of hydrogen now considered as very abundant increases the heat capacity of the non-adiabatic layers very much, and he suggested that this could retard δL by a quarter period while δr would still keep the same phase up to the surface. His argument was very attractive because he intimated that, at the same time, this could bring about the needed vibrational instability. EDDINGTON'S theory with its possible extension to the helium ionization zone has been discussed in details especially in Sect. 70. Although some of its quantitative aspects remain very important, it is doubtful whether the more qualitative characteristics and especially the exact phase shift of a quarter period between $(\delta L)_R$ and $(\delta r)_R$ can be preserved.

In another attempt to explain this phase relationship, M. SCHWARZSCHILD [58] assumed that the oscillation takes the character of a progessive wave in the outer parts of the adiabatic region. In that case, Eqs. (72.1) and (72.2) show that, if the curvature is negligible and the amplitude $\xi_0(r)$ does not change too fast while the phase $\varphi(r)$ varies rapidly, $|\delta\varrho/\varrho|$ is in phase with $|\dot{\xi}|$ i.e. with the velocity $|v|$. Since the adiabatic relations (53.7) are supposed to hold, this is true also of $\delta T/T$ and $\delta p/p$ and, according to (63.9), $|\delta L|$ is also in phase with $|v|$. In fact, SCHWARZSCHILD uses a non-linear solution of the type (94.5) but the general hypotheses are the same as above. Finally, the relation between δL and v depends on five parameters, and, by adjusting them, SCHWARZSCHILD was able to reproduce with great accuracy the observed light curve from the observed velocity curve. However, all the adopted values of the parameters are not physically acceptable and in particular, one of them (K) implies that the radiative equilibrium in the region considered is violently unstable while δL was computed assuming radiative equilibrium.

However, this is not the most serious difficulty. As we have seen repeatedly, the ionization of hydrogen in the non-adiabatic layers through which SCHWARZSCHILD'S wave must travel, can affect the phase of δL considerably. Furthermore, the basic assumptions imply that the wavelength of the perturbation is small compared to R and thus the phase of v (or δr) may also change appreciably. Thus even if one admits the progressive character of the wave, the actual phase relationship between light and velocity in the atmosphere will not necessarily resemble that at the top of the adiabatic region.

Finally, it would remain to justify the progressive character of the wave. The most likely reason (cf. Sect. 91) is some dissipation in the external layers but the example in Sect. 68, β suggests that, in this case, the progressive character is rather weak and the phases of $|\delta\varrho|$ and $|v|$ may be quite different. On the other hand, if the progressive character is strong enough to give phase equality between $|\delta\varrho|$ and $|v|$, it is rather likely that shock conditions will rapidly develop and the propagation of such waves in the non-adiabatic layers raises difficult problems (cf. Sect. 94).

Any progress in the solution of this question would be welcome, even in idealized cases such as the isothermal shock discussed at the end of Sect. 94. For the present problem, it would be interesting to test whether the conclusion reached in Sect. 67, namely that the work of the pressure in the atmospheric layers is too small to affect the flux, remains valid. In the case of an isothermal shock, Eq. (92.37) gives as a rough approximation for the maximum extra flux due to the pressure work across the shock

$$F_1 - F_2 = \frac{p_2}{2}\left(\frac{\varrho_1}{\varrho_2} + 1\right)(v_1 - v_2). \tag{109.1}$$

If we apply this to the kinematic model of W Virginis worked out by CH. WHITNEY[1], with $v_1 - v_2 \approx 50$ km/sec, $\varrho_1/\varrho_2 \approx 50$ and $p_2 \approx 2 \cdot 10^2$ dynes cm^{-2}, we find $F_1 - F_2 \approx 2.5 \times 10^{10}$ ergs which is of the same order as the average flux in the atmosphere. Thus, in this case, this effect could play an important rôle in shaping the light curve. Indeed if it was dominant and the layers in front of the shock practically transparent, it would give the right phase relationship since $F_1 - F_2$ would be in phase with v_1.

However without a detailed discussion of the excitation conditions behind the front and of the radiative processes through the whole atmosphere, it is difficult to judge how close the actual situation may come to this extreme case. The observations give the impression that the excitation temperature changes appreciably and this could reduce very much the jump in density and in flux across the shock.

On the other hand, the discussion in Sect. 68 of the behaviour of the atmospheric layers submitted to a flux variable in time but constant in space has revealed that, contrarily to EDDINGTON'S expectation, the displacement may experience an important phase-shift with respect to its value in the deeper layers. Actually, in the case of near-resonance with some dissipation present, we found [cf. Eq. (68.23)] that the dominating terms in the displacement lags 90° with respect to the luminosity exactly as observed. But again the general possibility of resonance is far from being established and, even then, this result would need to be substantiated by a careful study of kinematical and dynamical factors which have been neglected here and of their continuous transition through the non-adiabatic layer.

Apart from the phase shift, there is generally a close correspondence between the light and velocity curves even in details, which renders rather attractive any hypothesis implying a direct relation between δL and v (or δr) as in some of the examples considered above.

Another line of approach which, at first sight, might appear promising is based on the application of the principle of conservation of energy to the whole star. As recalled in Sect. 11, Moulton already tried to derive the light-variation from it and more recently E. A. MILNE[2] and W. KÜHN[3] have taken up the attempt again in different forms. MILNE'S paper has been criticized by S. ROSSELAND[4] and P. LEDOUX[5] who have shown that due to the smallness of δL (or L) as compared to the total internal energy of the star, such attempts are doomed from the beginning.

It seems that, to many people, the case of the β Cephei stars where, in the hypothesis of radial pulsations, maximum and minimum luminosity are associated respectively with minimum and maximum radius appears simpler because it agrees with the elementary theory recalled at the beginning of this section. However, here also a precise study of the non-adiabatic layers is necessary before definite conclusions can be reached. It is possible that due to the much higher

---

[1] CH. WHITNEY: Ann. d'Astrophys. **19**, 142 (1956).
[2] E. A. MILNE: Monthly Notices Roy. Astronom. Soc. London **109**, 517 (1949).
[3] W. KÜHN: Astronom. Nachr. **280**, 177 (1952).
[4] S. ROSSELAND: Monthly Notices Roy. Astronom. Soc. London **110**, 440 (1950).
[5] P. LEDOUX: III. Congr. Nat. Sci., Bruxelles, Vol. II, p. 137, 1950.

temperatures and the greater atmospheric densities, this study will actually be simpler than in the Cepheids. On the other hand, if non-radial oscillations are the cause of the variations, new aspects of the problem will have to be taken into account.

As for the long-period variables, the complex character of the radial velocity curves renders any discussion difficult. Identifying the level of formation of the bright hydrogen lines with that of the photosphere, SCOTT[1] found a good agreement between the radius variations derived from the radial velocities of these lines and the radiometric measurements of PETTIT and NICHOLSON. He also applied SCHWARZSCHILD'S progressive wave theory to Mira Ceti and, starting from the hydrogen emission lines velocity curve, he was able to recover the observed luminosity curve by an appropriate choice of the different parameters. However, in view of the restrictions to the applicability of SCHWARZSCHILD'S theory, which should even be stronger here than in the case of the Cepheids, and of the new data on the radial velocities (cf. Sect. 31), it is probable that SCOTT'S results are not very significant.

d) The asymmetry.

110. If we exclude the β Cephei stars where the variations in light and velocity are both very symmetrical and the U Geminorum stars where the strong asymmetry in light has been discussed in Sect. 97, we are left with a well defined problem only in the case of the Cepheids and the RR Lyrae stars (cf. Sects. 16 and 17).

In the RR Lyrae stars of types *a* and *b* there is a fairly close correlation between the asymmetry and the amplitude. This is supported by the fact that in stars presenting the phenomenon of modulation (cf. Sect. 22), the asymmetry reaches its largest values at the same time as the total amplitude. In this case, the analysis by WALRAVEN, GRATTON and WOLTJER of the light and velocity curves has revealed the interesting fact, that while the asymmetry of the velocity curve can be interpreted mainly as a phase distortion, the asymmetry of the light curve requires in addition a distortion in magnitude.

In the case of the Cepheids, the relation between asymmetry and amplitude is not so apparent. Perhaps is it masked in part by the rather heterogeneous character of the observations. For instance, D. JEHOULET[2] using JOY'S velocity curves of northern Cepheids, found a very large scatter in the distribution of the asymmetry as a function of $\delta R/R$. This work has been repeated recently and extended to include the southern Cepheids studied by STIBBS. Again the scatter is very large for the class as a whole but, if the stars are subdivided into groups corresponding to the different sequences which according to CAMPBELL and JACCHIA present a continuous progression of the form of the light curve, the correlation in each group is fairly well marked. However the slope of the relation varies with the group, the asymmetry increasing most rapidly in the group of smallest periods.

While, qualitatively, all this confirms the theoretical expectation that the asymmetry is a non-linear effect increasing with amplitude, we have seen in Sects. 85 and 86 that, for the observed amplitudes, the ordinary non-linear theory treating the star as a conservative system gives an asymmetry much smaller than observed. In Sect. 90, we have suggested that non-conservative terms apart from providing the most natural way to limit the amplitude, can also increase the asymmetry appreciably.

---

[1] R. M. SCOTT: Astrophys. Journ. **95**, 58 (1942).

[2] D. JEHOULET: Bull. Soc. Roy. Sci. Liège **20**, 561 (1950).

Finally, as illustrated by the discussion of Eqs. (86.13) and (86.14) subharmonic resonance also favours large asymmetry.

If the wave takes a progressive character in the external layers, the well-known steepening of the front of a finite compression wave can also explain the observed asymmetry. Indeed this was the meaning given by WALRAVEN to the phase distortion discovered in AI Velorum, SX Phoenicis, RR Lyrae etc; the waves, assumed to be sinusoïdal at some depth, were supposed to get distorted while progressing through the external layers. But it should be realized that the progressive character is not essential since a standing oscillation can also be distorted in an analogous way by the non-linear terms.

Actually a difficulty of the progressive wave is that it tends to become rapidly a shock wave especially with the high material velocities observed (greater than the sound velocity) while, in many cases, the direct evidence for a discontinuity is lacking. It is true that in other cases, like RR Lyrae or W Virginis, a suggestive discontinuity does appear in the velocity.

As to the difference between the distortion of the light and velocity curves discussed by WALRAVEN, it is another proof of the importance of the non-linear terms in the external layers. It also suggests that the flux is more sensitive to them, perhaps through its non-adiabatic part, than the velocity. For instance, in the progressive wave or shock wave picture, WALRAVEN has suggested that the extra energy needed to explain the magnitude distortion may be provided by the pressure work across the front of the wave which, as shown by Eq. (109.1), may be considerable if the shock is strong enough.

In these last few sections, we had to mention side by side the standing and progressive wave hypothesis without being able to choose among them. Such a choice will perhaps become possible in the future thanks to better observations and important developments in the non-linear theory. But it may be also that future progress will reveal that the two aspects are really intimately mixed, the progressive wave character taking more and more importance in the outermost layers leading finally to the ejection of tenuous shells of matter unless, before that, the particles' mean free-path has become large enough to smooth out the conditions and dissipate the remaining energy.

E. Atmospheric phenomena.

111. Introduction. Many of the problems raised by the interpretation of the spectra of variable stars are common also to non-variable stars and we shall not discuss them here.

The problem can be roughly divided into two parts: the study of the continuous spectrum and that of the line spectrum. In the first case, a few years ago, comparatively little was known, which was directly related to the atmosphere pulsation outside the variation of the total flux. In the case of the line spectrum, the main significant data for the pulsation were the variable radial velocities deduced from the position of the lines. Some early comments on the form of the lines or the relative displacements of lines of different excitation (differential velocities in the atmosphere) had to be drastically revised as better and higher dispersion spectra became available.

In recent years, the amount of information in both domains has increased considerably although perhaps in a little haphazard fashion. It would seem however that, at least for some classes of variables, the time has come for a concentrated study of typical objects. But, for instance, good *simultaneous* photoelectric and spectroscopic observations which may turn out to be extremely significant from a physical point of view are still mainly lacking.

On the other hand, important as this direct information may be, we must remember in all interpretative attempts that it consists only of what filters through the atmospheric layers and that it can be appreciably modified by these layers themselves which represent only a very small fraction of the total mass of the star. In this respect, the most urgent task of the theorist is to devise reliable methods to extrapolate the observed conditions down to levels where the theory of internal structure can take over, and this means crossing the difficult non-adiabatic region under the atmosphere proper.

a) The continuous spectrum.

112. Phase shifts between the light curves in different colours. The six-colour photoelectric observations by STEBBINS and collaborators summarized in Sect. 23 have shown that, in δ Cephei, the interval between maximum and minimum is the same for all colours but that the corresponding light curves show a progressive phase shift, the maximum in the infrared occurring about 0.05 P later than in the ultraviolet. Similar results have been obtained for η Aquilae and RR Lyrae. An effect of the same type was already apparent in A. KOHLSCHÜTTER'S discussion (1909) of the visual and photographic light curves of η Aquilae[1] but not much attention was paid to it. In 1943, A. VAN HOOF[2] concluded from a theoretical discussion of the respective effects of the variations of the radius and of the temperature on the luminosity that, in Cepheids, the photographic maximum should always occur a little earlier than the visual maximum.

It soon became clear that the effect discovered by STEBBINS was due to the interaction of the two factors[3] discussed by VAN HOOF. Even on the simple picture of a star radiating as a black body.

$$L_\nu = 4\pi^2 R^2 \frac{2h\nu^3}{c^2} \frac{1}{e^{h\nu/kT}-1} \tag{112.1}$$

one may verify easily by differentiation of (112.1) that, due to the fact that, in a Cepheid, R increases while T goes through its maximum, the maximum of L_ν will occur later and later as ν decreases. In the case of δ Cephei, if the numerical values of dT/dt and dR/dt are taken for instance from the data gathered by SCHWARZSCHILD[4], one finds a phase shift of 0.02 to 0.03 between the ultraviolet and infrared maxima. A. F. WESSELINK[5] using BECKER'S empirical relation between colour index and radiation temperature (cf. Sect. 23) confirmed that the theoretical effect was in the same sense as the observed one. To go beyond this qualitative interpretation, the variations of the radius and of the "temperature" should be known with a very high accuracy. Formally, this can be provided by analytical representations of R and T as used by DAMBARA[6] but the test loses then some of its physical significance.

In the case of β Cephei, J. STEBBINS and G. E. KRON (cf. last paragraphs of Sect. 25) found a similar result and could also explain it by using the radius variations obtained by integration of the radial velocity curve, although here the phase relation between radius and temperature is quite different from that

[1] A. KOHLSCHÜTTER: Astronom. Nachr. **183**, 265 (1910).
[2] A. VAN HOOF: Kon. Vlaam. Acad. v. Wetensch. **5**, No. 12 (1943).
[3] J. STEBBINS: Publ. Astronom. Soc. Pacific **65**, 118 (1953).
[4] M. SCHWARZSCHILD: Harvard. Circ. **1938**, No. 431.
[5] A. F. WESSELINK: Bull. astronom. Inst. Netherl. **10**, 256 (1947).
[6] T. DAMBARA: Publ. Astronom. Soc. Japan **3**, 135 (1951).

in the Cepheids, and the changes in radius account for a much larger fraction of the total luminosity changes.

We have already recalled (cf. Sect. 103) that in the long-period variables, the phase shift between the visual maximum and the infrared maximum can also be interpreted in a similar way.

On the whole, this effect certainly supports the pulsation theory and the ordinary interpretation of the radial velocity curve as being practically identical to that of the photosphere.

113. BAADE's criterion and its modern developments. In 1926, BAADE[1] remarked that, if a Cepheid radiates like a black body, at all phases

$$L(t) = 4\pi R^2(t)\, T_e^4(t) \tag{113.1}$$

and if $T_e(t)$ is known throughout the cycle (from spectral type or colour index), the relative variation of the radius $R(t)/R_0$ may be computed from (113.1). If the pulsation theory is true, the curve $R(t)/R_0$ should agree in phase and amplitude with the displacement $(R-R_0)$ derived from the radial velocity curve for a value of R_0 compatible with the average absolute luminosity L_0 of the star. Of course, BAADE's criterion may be formulated in terms of monochromatic luminosities as well as in terms of the total luminosity.

BOTTLINGER[2] encountered difficulties in the application of this test but remarked that they could be due to deviations from the laws of black body radiation. BECKER[3] using his newly established relation between colour index and radiation temperature (cf. Sect. 23) found that BAADE's criterion was well verified on the average. This already suggests that the fundamental step in this problem is the adoption of a given relation between the colour-index CI $= m_{\lambda_2} - m_{\lambda_1}$ and the surface brightness m_{λ_1} in a small wavelength interval around λ_1, say

$$m_{\lambda_1}(T) = f(\text{CI}). \tag{113.2}$$

In BECKER's analysis, this relation rests on the assumption that in a Cepheid the radii at maximum and minimum luminosity are equal. As this is a consequence of the radial velocity curve, it implies already to a certain extent, that the pulsation theory is correct and that the variation of the photospheric radius R_p is parallel to that of the radius R_V characteristic of the level in the reversing layers where the absorption lines are formed. The ratios between Δ CI and $\Delta m_{\lambda_1}(T)$ thus obtained between these two phases in a series of Cepheids can then be used to build a relation of type (113.2) after the zero point of the relation has been fixed with the help of a known star.

In 1943, VAN HOOF (cf. Ref. 2, p. 594) extended BECKER's method to all pairs of phases in a given Cepheid which, according to the velocity curve, have the same radius. This is sufficient to establish a relation

$$\Delta m_{\lambda_1}(T) = a_{\lambda_1} \Delta\, \text{CI}. \tag{113.3}$$

If this relation can be extrapolated to all phases throughout the cycle and if the continuous variation of CI is known, one may then determine the variations of R_p since the total variation of magnitude $(m_{\lambda_1})_2 - (m_{\lambda_1})_1 = (\Delta m_{\lambda_1})_{1,2}$ between any two phases 1 and 2, is given by

$$(\Delta m_{\lambda_1})_{1,2} = -5.0\, (\Delta \log R_p)_{1,2} + (\Delta m_{\lambda_1}(T))_{1,2} \tag{113.4}$$

[1] W. BAADE: Astronom. Nachr. **228**, 359 (1926).
[2] K. F. BOTTLINGER: Astronom. Nachr. **232**, 3 (1928).
[3] W. BECKER: Z. Astrophys. **19**, 249 (1940).

or

$$(\varDelta m_{\lambda_1})_{1,2} = -5.0 \log \frac{R_{p,2}}{R_{p,1}} + a_{\lambda_1} (\varDelta\,\mathrm{CI})_{1,2}. \tag{113.5}$$

If the radius variations are small, this can also be written approximately as

$$(\varDelta m_{\lambda_1})_{1,2} = -2.151 \frac{(\varDelta R_p)_{1,2}}{(R_p)_0} + a_{\lambda_1} (\varDelta\,\mathrm{CI})_{1,2} \tag{113.6}$$

where $(R_p)_0$ is the radius at some phase, say the mean radius. Identifying the $\varDelta R_p$ given by (113.6) to the corresponding displacements $\varDelta R_V$ deduced by integration of the radial velocity curve corrected for limb darkening, one can determine the values of $(R_p)_0$ for each pair of phases. Since the method is rather sensitive to small errors in the empirically determined values of $(\varDelta m_{\lambda_1})_{1,2}$ and $(\varDelta\,\mathrm{CI})_{1,2}$ an agreement in order of magnitude between the different values of $(R_p)_0$ is generally considered satisfactory and the average value is adopted for the radius of the star.

WESSELINK[1] more or less inverted the method by considering pairs of phases having the same CI and liberating himself from the necessity to establish explicitly a relation of type (113.3), although, of course, he still postulates the existence of such a relation since he assumes that $\varDelta m_{\lambda_1}$ vanishes with $\varDelta$ CI. In that case, if now the indices 1 and 2 denote phases characterized by the same value of CI, Eq. (113.6) gives immediately

$$(\varDelta m_{\lambda_1})_{1,2} = -2.151 \frac{(\varDelta R_p)_{1,2}}{(R_p)_0}. \tag{113.7}$$

If again $(\varDelta R_p)_{1,2}$ is replaced by $(\varDelta R_V)_{1,2}$, the same considerations as before apply and an average value of the radius $(R_p)_0$ may be determined.

The six-colour photometry gave a new impetus to these methods[2]. In particular, the corresponding phases in WESSELINK's method could be determined by a much better match in colours. On the whole, in the case of the Cepheids, the values of the mean radius derived by this differential method show a reasonable agreement with other evidence concerning the radius.

WESSELINK's method has also been applied to some of the β Cephei stars [*30*]. It yields reasonable results in the case of β Cephei itself which has only one period and does not present any variable broadening of the lines. In other cases where these effects are present, it does not work at all or gives aberrant results which led WALKER to conclude that, if these results mean anything at all, they might be taken as strengthening the supposition that these stars undergo some sort of non-radial pulsation.

The next step is to come back to the original method and determine the absolute values of the radius as a function of phase. This can be done semi-empirically in combining VAN HOOF's and WESSELINK's methods as OPOLSKI and KRAVIECKA[3] have done for δ Cephei and η Aquilae with consistent results. But this implies again fixing empirically the constants a_{λ_1} in the different relations (113.3) corresponding to STEBBINS' six colour observations.

However, with the progress in the study of stellar atmospheres, it was natural that a more theoretical approach be attempted. At the end of Sect. 67, we reached the conclusion that deviations from ordinary radiative equilibrium remain very small in the atmosphere which essentially adapts itself at each

---

[1] A. F. WESSELINK: Bull. astronom. Inst. Netherl. **10**, 91 (1946).

[2] A. F. WESSELINK: Bull. astronom. Inst. Netherl. **10**, 256, 330 (1947). — J. STEBBINS, G. E. KRON and J. L. SMITH: Astrophys. Journ. **115**, 292 (1952).

[3] A. OPOLSKI and J. KRAVIECKA: Trav. Soc. Sci. Lett. Wroclaw, Ser. B **1956**, No. 81.

phase to the total flux of energy to be transported. On the other hand, the effective gravity g_e may vary appreciably due to the motion especially at phases between minimum and maximum light. It may also be affected at all phases by factors such as the finite transfer of momentum through the atmosphere (cf. Sect. 94) or turbulence and its variations[1]. But provided none of these factors imply a rapid spatial variation of g_e, it seems appropriate to represent the successive states of the atmosphere during a cycle by a series of quasi-static configurations which can be computed by the ordinary methods, in use in the theory of model atmospheres.

Having at one's disposal such a series of models for a range of values of T_e and g_e (or p_e) around likely values of these parameters, one can then pick out the models that give back the relative distribution of the fluxes observed by STEBBINS and at the same time the value of the Balmer discontinuity[2] which is rather sensitive to g_e. The comparison of the magnitudes at two phases, one of which is taken as reference, by means of the formula

$$(m_\lambda)_i - (m_\lambda)_0 = 2.5 \log\left(\frac{F_{\lambda,0}}{F_{\lambda,i}}\right) - 5 \log \frac{R_{p,i}}{R_{p,0}} \tag{113.8}$$

where the F_λ represent the theoretical values of the flux given by the respective models, allows one to determine the relative variation of the radius. This can be repeated for a series of colours, providing an internal test on the values of R/R_0. As a rule, the observations in the ultraviolet are not used because they are strongly affected by line absorption and the Balmer discontinuity. Moreover, the theoretical value of F_λ should really refer not to a given wavelength but to the total flux in the wavelength interval $\Delta\lambda$ covered by STEBBINS' filter.

This method has been applied by a number of authors[3] with slightly different assumptions and different ranges of T_e and g_e. In all cases, it led to a serious disagreement between the variations of R_p and R_V. It did seem that one was falling back on the early difficulties of BOTTLINGER. Indeed, CAVANAGGIA and PECKER showed that the method was essentially equivalent to a theoretical determination of the a_{λ_1} in (113.3) for the different wavelengths used and that the models, as the black body, gave too large values for these a_{λ_1}.

Ad hoc reductions of the a_λ can be found which are consistent all through the spectrum and bring back agreement between R_p and R_V. GRANDJEAN and LEDOUX suggested that the necessary reductions could be brought about by a change in chemical composition. But more detailed discussions did not support this point of view and furthermore they brought to light other factors which had been neglected[4]. Among these, the correction for line absorption was at least partially taken into account by WHITNEY[5] in a paper where the previous method was put on an absolute basis thanks to a calibration of the six-colour observations by reference to the solar continuum. In that case, the quantities relative to the phase chosen as reference in formula (113.8) can be replaced by the corresponding quantities for the sun. But this led again to a similar discrepancy between R_p and R_V. In another paper[6], WHITNEY took into consideration the

[1] Cf. for instance [*29a, b*]; also A. PANNEKOEK: Zeeman Congress, Physica, Haag **12**, 761 (1946), and D. H. MENZEL: Physica, Haag **12**, 768 (1946); P. LEDOUX and J. GRANDJEAN: Bull. Acad. Roy. Belg., Cl. Sci., Sér. V **41**, 1010 (1955).

[2] R. CANAVAGGIA: Ann. d'Astrophys. **12**, 21 (1949).

[3] Cf. Z. HITOTUYANAGI: Sci. Rep. Tôhoku Univ., Ser. I **36**, No. 4 (1952); R. CAVANAGGIA and J. C. PECKER: C. R. Acad. Sci. Paris **234**, 1739 (1952); Ann. d'Astrophys. **15**, 260 (1952); **16**, 47 (1953); P. LEDOUX and J. GRANDJEAN: Ann. d'Astrophys. **17**, 161 (1954).

[4] Cf. for instance, CH. WHITNEY: Thesis, Harvard 1955; Astrophys. Journ. **122**, 385 (1955).

[5] CH. WHITNEY: Astrophys. Journ. **121**, 682 (1955).

[6] CH. WHITNEY: Astrophys. Journ. **122**, 385 (1955).

variations of the weak lines' absorption with phase and, although the data at his disposal were not sufficient for a complete discussion, he concluded that this effect works in the proper direction and may completely remove the discrepancy.

In the case of η Aquilae, WHITNEY applied also the necessary correction for interstellar reddening which according to HARRIS[1] is particularly important for this star. This effect has also been discussed by Miss CAVANAGGIA and PECKER[2]. After reduction of δ Cephei and η Aquilae to the same degree of interstellar reddening, Miss CAVANAGGIA[3] has developed a comparative study of their radii variations which, according to the author, favours the radius variations derived from the models rather than those obtained from the velocity curve.

Z. HITOTUYANAGI and K. VJI-IYE[4] have tackled the problem again with model atmospheres whose characteristics (T_e, g_e) were chosen on the basis of spectro-photometric data which they found preferable for this purpose to those obtained by wide-band filter photometry. They found a good agreement in phase between R_p and R_V but a fairly large discrepancy in amplitude subsists.

With the new correction factors lately introduced into the problem and the somewhat discordant results mentioned, it is difficult to draw any definite conclusions.

Even, if the difficulties raised by the models persist after these corrections have been taken properly into account, it will be hazardous to interpret them as an indication that the photosphere has a motion appreciably different from that of the reversing layers since, as we have seen in Sect. 24, differential velocities, if present in the atmosphere, are very small.

In any case, new precise computations of model atmospheres with different chemical abundances and a careful discussion of all possible sources of opacity would be advisable. A detailed comparison of their properties with those observed in the case of non-variable supergiants may also provide an interesting check. Finally, the conversion of the measured displacements of the lines into true velocities of pulsation may be affected by factors which have been neglected up to now.

b) The line spectrum.

114. Line displacements and velocities. Pulsational line profile. Up to the present, the measured displacements have been converted directly into a mean radial velocity V_R over the visible stellar hemisphere. If this disk was uniformly illuminated, the true velocity v along the radius, assuming radial pulsation, would be simply equal to $^3/_2\, V_R$. But, as SHAPLEY and NICHOLSON[5] have shown, the limb darkening if proportional to $l-k+k\cos\vartheta$ modifies the geometric mean and one finds

$$v = \frac{6-2k}{4-k}\, V_R$$

or with the grey law of darkening ($k = {}^3/_5$)

$$v = \frac{24}{17}\, V_R \tag{114.1}$$

which is the relation generally used.

[1] D. L. HARRIS, III: Astrophys. Journ. **119**, 297 (1954).
[2] R. CAVANAGGIA and J. C. PECKER: Ann. d'Astrophys. **18**, 151 (1955).
[3] R. CAVANAGGIA: Ann. d'Astrophys. **18**, 431 (1955).
[4] Z. HITOTUYANAGI and K. VJI-IYE: Sci. Rep. Tôhoku Univ. **40**, 54 (1956).
[5] H. SHAPLEY and S. D. NICHOLSON: Proc. Nat. Acad. Sci. U.S.A. **5**, 417 (1919).

Recently, in part because of the difficulties recalled in the previous section, the problem has been discussed anew[1] taking into account the effects of the combination with the pulsation velocity v of the other possible velocity fields C due for instance to macro-turbulence or rotation. If actual measurements of line position refer to the deepest point in the line profile, the schematic investigations reported above suggest that the velocities computed from the measured displacements should be corrected by a factor varying rapidly with the ratio C/v. The practical consequences of this however remain to be worked out, and it is a little doubtful whether they could really affect BAADE's criterion in a decisive manner.

SHAPLEY and NICHOLSON already noted that the pulsation should also distort the line giving it a certain asymmetry. This effect has been studied by A. VAN HOOF and R. DEURINCK[2] in the spectrum of η Aquilae. They succeeded in reproducing the periodic asymmetries of weak Fe lines by considering a limb-darkened reversing layer assuming that the velocity remains constant with depth. How much would this result be affected by even small differential velocities in the atmosphere or by the presence of extra velocity fields such as discussed above?

As reported in Sect. 24, differential velocities in the main body of the atmospheres of classical Cepheids are very small. From a careful study of high dispersion spectra (2.9 Å/mm) of ξ Geminorum and η Aquilae, J. GRANDJEAN[3] concludes that there is a slight but definite variation of the velocity with depth in both stars, amounting to 3 to 4 km/sec between the lines forming at the lowest and highest levels. One should note however that the measures have been restricted to relatively sharp simple lines forming in the main body of the atmosphere. These differences could be interpreted as due to a small lag of the higher levels with respect to the lower ones. However these results relate to only two phases in both stars and slight variations of the amplitudes with height could also explain them. However, one should note that, in the case of η Aquilae, if the first interpretation is adopted, a straight extrapolation of the phase lag to very high levels is compatible with the shift in phase of the radial velocity curve derived by JACOBSEN from the cores of the H and K lines[4]. This may be an indication in favour of the running wave picture. In this respect, a repetition of GRANDJEAN's work would be very welcome.

The evidence concerning the occurrence of emission lines and double absorption lines and their relative displacements has been summarized in Sect. 24 for the Cepheids and RR Lyrae stars and in Sect. 26 for the β Cephei stars. The general inferences drawn from their behaviour have also been summarized in these sections and have been referred to again on different occasions in the course of this article. In cases, where these phenomena are well marked as in W Virginis, the most straightforward interpretation is in terms of a shock going through the atmosphere. But although the kinematic model of WHITNEY[5] has met with a certain amount of success, a great deal of theoretical work is still needed (cf. Sect. 94) to consolidate the basis of this interpretation. In other

[1] a) J. DUQUESNE and E. SCHATZMAN: Ann. d'Astrophys. **18**, 279 (1955). — b) G. THIESSEN: Z. Astrophys. **41**, 259 (1957).

[2] A. VAN HOOF and R. DEURINCK: Astrophys. Journ. **115**, 166 (1952). — M. P. SAVEDOFF: Astronom. J. **57**, 25 (1952), has studied the same problem in the case of RR Lyrae and has found agreement with the pulsation theory.

[3] J. GRANDJEAN: Mém. Acad. Roy. Sci. Belg. **29**, 3 (1956).

[4] Cf. for instance P. LEDOUX and J. GRANDJEAN: Bull. Acad. Roy. Sci. Belg., Sér. V **41**, 1025 (1955).

[5] CH. WHITNEY: Ann. d'Astrophys. **19**, 142 (1956).

cases, when these phenomena are much weaker as in the classical Cepheids, they still favour the idea that, at least in the very external layers, the wave takes a progressive character. In this respect, the long-period variables present probably the most complex problem and up to now, the attempts at interpretation are no more than suggestions which together with the most relevant facts have been summarized in Sects. 31 and 32.

115. Physical analysis of the structure and the chemical composition of the atmospheres of variable stars. Since the atmospheres of the most regular variables can be considered as in equilibrium at each phase (cf. Sect. 113), the theory of stellar atmospheres and of the formation of stellar absorption lines so successful for the stars with static atmospheres can also be applied to these variables. Up to now, only some Cepheids and RR Lyrae have been the object of detailed investigations along these lines. In these cases, many details in the appearance of the line spectrum have been interpreted satisfactorily (cf. Sect. 24).

This approach has the further advantage that, at the same time, it provides independent information on the conditions of density, temperature and pressure in the atmosphere and their variations in the course of the cycle providing a test for the nature of the oscillation in the external layers. As far as possible, these studies should be based on the techniques of model atmospheres which permit a much more detailed comparison with the observations and are essential for the interpretation of any level effects. In Sect. 113, we already encountered model atmospheres built on the basis of the distribution of the energy in the continuous spectrum. There are no reasons why the two procedures should not be used simultaneously increasing considerably the obtainable precision and the reliability of the results. In the case of η Aquilae for which both the line spectrum [*29b*] and the continuous spectrum have been used quite independently to determine the properties of the atmosphere, it is satisfactory to find that there is, at least, qualitative agreement for the variations of these properties with phase (cf. Ref. 4, p. 599).

One should note that, in this method, the atmospheric changes are essentially determined by the values of the effective gravity g_e and of the flux L to be transported through the atmosphere, L being taken as imposed by the stellar interior. From the point of view of the pulsation theory, the main task is the dynamical interpretation of the values of g_e which one is led to adopt. In principle, they should yield the most interesting and useful information on the wave but up to now the nature of different factors which affect not only the variations of g_e but also its mean value has not been completely elucidated (cf. Sect. 113).

In any case, one of the consequences of these g_e is that the density ϱ varies practically in phase with v in the atmosphere. Qualitatively, this suggests again a predominating progressive character for the wave. However, as shown in [*29b*] difficulties are encountered in the quantitative development of this interpretation. One may also check the general consistency of the changes in ϱ, T and p from the hydrodynamical and energetic points of view and derive the required differential velocity field [cf. Ref. 4, p. 599). In the case of η Aquilae, this field turned out to be compatible with the actual velocity observations which unfortunately were available only for two phases. On the other hand, the theoretically derived velocity field is not too precise and it does not permit to distinguish clearly between the progressive or standing wave hypothesis.

116. Conclusions. At many places in this article and especially at the end, in the short account of some of the aspects of the pulsation theory in the external layers of the best known variable stars, the reader must have gathered the impression that we are still very far from an adequate picture. We hope that this will be a strong incentive to workers on both the observational and theoretical sides and that, in the course of the next few years a coherent scheme will emerge from their concerted efforts. Let us hope that it will provide the necessary basis not only for the direct interpretation of the observational data but also for a sound extrapolation toward the interior which will permit to tackle with better chances of success, the fundamental problem of the origin and the maintenance of the pulsations.

We have mentioned incidentally the possible interactions of the theories of stellar variability and of stellar evolution[1] but we have not discussed them in any details because, at the present time, the information both observational and theoretical remains confined to limited domains and is still often indefinite. But it is likely that, in the future, the mutual impact of the two theories will become fundamental to the deep understanding of the messages which these enigmatic variable stars try to convey to us from the depths of space.

Bibliography.

[1] PAYNE-GAPOSCHKIN, C.: Variable Stars and Galactic Structure, Table 1.5, p. 11. London 1954.

[2] a) SHAPLEY, H.: Galaxies, Harvard Books on Astronomy. Philadelphia 1945.
b) KUKARKIN, B. W.: Erforschung der Struktur und Entwicklung der Sternsysteme auf der Grundlage des Studiums veränderlicher Sterne. Berlin: Akademie-Verlag 1954.
c) Also Ref. [1].

[3] EDDINGTON, A. S.: The Internal Constitution of the Stars, p. 397. Cambridge 1926.

[4] The readers interested in historical and technical details can be referred to the exhaustive work by:
a) HAGEN (S. J.), J. G., and J. STEIN (S. J.): Die veränderlichen Sterne, 2 Bde. Freiburg i. Brsg.: Herder & Co. 1921—1924. — The first volume (being also vol. 5 of the Pubblicazioni della Specola Vaticana) is especially devoted to the discoveries of variable stars and their observations while the second (being also vol. 6 of the same publications) contains a detailed account of the different theories proposed up to 1923.
The articles by:
b) LUDENDORFF, H.: Die veränderlichen Sterne, Handbuch der Astrophysik, Bd. VI, pp. 49—250, 1928, and Bd. VII, pp. 614—670, 1936, present a very complete survey of the literature up to 1934.
The review by:
c) TEN BRUGGENCATE, P.: Die veränderlichen Sterne. Ergebn. exakt. Naturw. **10**, 2—82 (1931), has a more physical orientation and contains many interesting comments on the major attempts at interpretation.
Written in a lighter vein, the following books provide in addition to an integrated view on the subject, many interesting historical details:
d) MERRILL, P. W.: The Nature of Variable Stars. New York: MacMillan Company 1938.
e) CAMPBELL, L., and L. JACCHIA: The Story of Variable Stars. Philadelphia and Toronto: Blackiston Co. 1946.
As far as the history of individual stars is concerned the most useful bibliography remains the
f) Geschichte und Literatur des Lichtwechsels der bis Ende 1915 als sicher veränderlich anerkannten Sterne nebst einem Katalog der Elemente ihres Lichtwechsels. Leipzig: Poeschel & Trepte, 3 volumes (1918, 1920, 1922) by G. MÜLLER and E. HARTWIG.
This monumental work has been continued for the period 1916—1933 by:
PRAGER, R.: Geschichte und Literatur des Lichtwechsels der veränderlichen Sterne. Bd. 1: Andromeda-Crux. Berlin u. Bonn: Ferd. Dümmler 1934; Bd. 2: Cygnus-Ophiuchus. Berlin u. Bonn: Ferd. Dümmler 1936.
and by:
SCHNELLER, H.: Bd. 3: Orion-Vulpecula. Berlin: Akademie-Verlag 1952.
The last volume covers the literature from 1916—1950.

[1] Cf. for instance O. STRUVE: Publ. Astronom. Soc. Pacific **67**, 29 (1955); H. SHAPLEY: Proc. Nat. Acad. Sci. U.S.A. **41**, 824 (1955).

A great deal of information together with an attempt at systematization is also contained in:
g) PAYNE-GAPOSCHKIN, C., and S. GAPOSCHKIN: Variable Stars. Harvard Obs. Monogr. **1938**, No. 5.
Cf. also the short but compact article by
h) PAYNE-GAPOSCHKIN, C. in: Astrophysics, a Topical Symposium, ed. by J. A. HYNEK. New York-Toronto-London: McGraw Hill 1951.
[5] CAMPBELL, L.: Harvard. Ann. **79**, Part 2 and Studies of Long Period Variable Stars. Cambridge (Mass.) 1955.
[6] KUKARKIN, B. V. and P. P. PARENAGO: The General Catalogue of Variable Stars. Moscow 1948. Seven supplements, the latest dating from 1955.
[7] BAILEY, S. I.: Astrophys. Journ. **10**, 255 (1899) and Harvard Ann. **38**, 132, 208 (1902).
[8] LEAVITT, H. S.: Harvard. Circ. **1912**, 173.
[9] BAADE, W.: Trans. Internat. Astronom. Union. **8**, 397 (1954); for a systematic account by the same author see:
a) Galaxies: Symposium on Astrophysics, Univ. Michigan 1953, and
b) Publ. Astronom. Soc. Pacific **68**, 1 (1956).
[10] RITTER, A.: Wiedemanns Ann. **5—20** (1878—1883); see especially **8**, 179 (1879) and **13**, 366 (1881).
[11] THOMSON, W.: Phil. Trans. Roy. Soc. Lond. **153**, 612 (1863).
[12] EGGEN, O. J.: Astrophys. Journ. **113**, 367 (1951).
[13] EDDINGTON, A. S.: Monthly Notices Roy. Astronom. Soc. London **79**, 2, 177 (1919).
[14] JEANS, J.: Astronomy and Cosmogony, p. 361. Cambridge 1928.
[15] a) JOY, A. H.: Astrophys. Journ. **110**, 105 (1949).
b) SAWYER, H. B.: David Dunlap Obs. Publ. **2**, No. 2 (1955).
[16] MARTIN, N. C.: Ann. Leiden Obs. **17**, Part 2 (1938).
[17] ARP, H. C.: Astronom. J. **60**, 1 (1955).
[18] JOY, A. H.: Astrophys. Journ. **86**, 363 (1937).
[19] STIBBS, D. W. N.: Monthly Notices Roy. Astronom. Soc. London **115**, 363 (1955).
[20] PARENAGO, P. P.: Variable Stars **10**, 193 (1955).
[21] SHAPLEY, H.: Star Clusters, p. 135. 1930.
[22] Papers of the 4th Conference on Cosmogonical Problems, p. 389, Moscou 1955.
[23] WALRAVEN, TH.:Bull astronom. Inst. Netherl. **12**, 223 (1955).
[24] a) STEBBINS, J.: Astrophys. Journ. **101**, 47 (1945).
b) STEBBINS, J., G. E. KRON and J. L. SMITH: Astrophys. Journ. **115**, 292 (1952).
c) STEBBINS, J.: Publ. Astronom. Soc. Pacific **65**, 118 (1953).
[25] STRUVE, O.: Observatory **65**, 257 (1944).
[26] MORGAN, W. W., P. C. KEENAN and E. KELLY: An Atlas of Stellar Spectra. Chicago, Ill.: Chicago University Press 1943.
[27] UNSÖLD, A.: Physik der Sternatmosphären. Berlin: Springer 1955.
[28] ALLER, L. H.: Astrophysics, I. New York: Ronald Press 1953.
[29] a) WALRAVEN, TH.: Amsterdam Publ. **1948**, No. 8.
b) SCHWARZSCHILD, M. and B., and W. S. ADAMS: Astrophys. Journ. **108**, 207 (1948).
c) LUST-KULKA, RH.: Z. Astrophys. **33**, 211 (1954).
d) PELS-KLUYVER, H. A.: Bull. astronom. Inst. Netherl. **12**, 151 (1954).
e) ABT, H. A.: Astrophys. Journ., Suppl. **1**, 63 (1954).
[30] STRUVE, O.: Publ. Astronom. Soc. Pacific **67**, 135 (1955).
[31] MERRILL, P. W.: Astrophys. Journ. **94**, 171 (1941).
[32] WILSON, R. E., and P. W. MERRILL: Astrophys. Journ. **95**, 248 (1942).
[33] PETTIT, E., and S. B. NICHOLSON: Astrophys. Journ. **78**, 320 (1933).
[34] CAMPBELL, L., and T. E. STERNE: Harvard Ann. **105**, 459 (1937).
[35] JOY, A. H.: Astrophys. Journ. **63**, 281 (1926).
[36] MERRILL, P.W.: Astrophys. Journ. **102**, 347 (1945) (V. Orionis); **103**, 6 (1946) (R Hydrae); **103**, 275 (1946) (R Serpentis); **105**, 360 (1947) (R Andromedae); **106**, 274 (1947) (X Cygni).
[37] MERRILL, P. W.: The Spectra of Long Period Variable Stars. Chicago, Ill.: Chicago University Press 1940. — Emission Lines in the Spectra of Long Period Variables. J. Roy. Astronom. Soc. Canada **46**, 181 (1952).
[38] Non-Stable Stars. Internat. Astronom. Union Symposium No. 3, Cambridge 1957. Apart from a) P. N. KHOLOPOV's article, p. 11, on RW Aurigae and T Tauri Stars, this volume contains also (b) a paper by C. HOFFMEISTER, p. 22, on the general problem of the classification of these stars as well as different papers on their physical properties and tentative interpretations: (c) G. H. HERBIG, p. 3; (d) G. HARO, p. 26; (e) A. H. JOY, p. 31, (f) V. A. AMBARTSUMIAN, p. 177.
[39] a) EDDINGTON, A. S.: The Internal Constitution of the Stars, Chap. 8. Cambridge 1926.
b) MILNE, E. A.: Theory of Pulsating Stars, Handbuch der Astrophysik, Vol. III/2,

p. 804. 1930. — Unfortunately, the expression of the general principle of conservation of energy for a mixture of matter and radiation used in this paper is not correct.
c) STRÖMGREN, B.: Thermodynamik der Sterne und Pulsationstheorie, Kap. 8, Handbuch der Astrophysik, Vol. VII, p. 192—202. 1936. — This paper presents a short but very clear summary of the state reached by the theory around 1935 but it is directed more towards the discussion of questions of stability than towards the detailed interpretation of variable stars.
d) ROSSELAND, S.: The Pulsation Theory of Variable Stars. Oxford 1949.
e) SEVERNY, A. B.: On the Stability and Oscillations of gaseous Spheres and Stars. Izvest. Crim. Astrophys. Obs. **1**, 3 (1948).
[40] a) LAMB, H.: Hydrodynamics, Chap. I, 6th ed. Cambridge 1932.
b) BJERKNES, V., J. BJERKNES, H. SOLBERG and T. BERGERON: Hydrodynamique Physique, Chap. II. Paris: Presses Universitaires de France 1934.
[41] Cf. for instance A. LICHNEROWICZ: Eléments de Calcul Tensoriel, especially Chap. VI, § III. Paris: Coll. Armand Colin 1955.
[42] For a short but excellent summary, cf. S. CHANDRASEKHAR: Turbulence—A Physical Theory of Astrophysical Interest. Astrophys. Journ. **110**, 329 (1949).
For a more elaborated discussion, cf.:
a) BATCHELOR, G. K.: The Theory of Homogeneous Turbulence. Cambridge 1953.
b) AGOSTINI, L., et J. BASS: Les Théories de la Turbulence. Service de Documentation et d'Information technique de l'Aéronautique, Paris 1950.
c) Also for an interesting new point of view:
CHANDRASEKHAR, S.: A Theory of Turbulence. Proc. Roy. Soc. Lond., Ser. A **229**, (1955).
[43] For a brief discussion of this type of theory cf.:
a) WASIUTYNSKI, J.: Studies in Hydrodynamics and Structure of Stars and Planets, Chap. I. Astrophys. Norv. **4** (1946).
b) BRUNT, D.: Physical and Dynamical Meteorology. Cambridge 1941.
[44] THOMAS, L. H.: The Radiation Field in a Fluid in Motion. Quart. J. Math. **1**, 239 (1930).
[45] a) COWLING, T. G.: Monthly Notices Roy. Astronom. Soc. London **96**, 42 (1936).
For a generalisation to the case where the effects of radiation must be taken into account, cf.:
b) LEDOUX, P.: Astrophys. Norv. **3**, 193 (1940).
[46] CHANDRASEKHAR, S.: An Introduction to the Study of Stellar Structure. Chicago, Ill.: Chicago University Press 1939, cf. Chap. II, § 12, p. 55.
[47] UNSÖLD, A.: Physik der Sternatmosphären. Berlin: Springer 1955.
[48] COURANT, R., and K. O. FRIEDRICHS: Supersonic Flow and Shock Waves, §§ 52, 56. New York: Interscience Publishers 1948.
[49] For general summaries cf. for instance:
a) COURANT, R., u. D. HILBERT: Methoden der Mathematischen Physik, Vol. I, Berlin: Springer 1931. Chap. V.
b) VOGEL, T.: Les Fonctions orthogonales dans les problèmes aux limites de la physique mathématique, especially Chap. I. Centre Nation. Rech. Sci. Paris **1953**.
c) COLLATZ, L.: Eigenwertaufgaben mit technischen Anwendungen, Chap. III, § 8 and Chap. V, § 15, 1949.
d) FRIEDMAN, B.: Principles and Techniques of Applied Mathematics, especially Chap. IV. New York: J. Wiley & Sons, Inc. 1956.
e) For a very interesting account of RAYLEIGH's principle which never loses sight of the physical meaning, cf. G. TEMPLE and W. G. BICKLEY: RAYLEIGH's Principle and its Applications to Engineering. Dover Publ. 1956.
[50] a) LEDOUX, P.: Astrophys. Journ. **102**, 143 (1945).
b) LEDOUX, P.: Contribution à l'étude de la structure interne des étoiles et de leur stabilité. Mém. Soc. Roy. Sci. Liège **9**, Chap. IV et V (1949).
[51] Cf. for instance H. POINCARÉ: Hypothèses cosmogoniques, p. 90—95. 1911.
[52] LEDOUX, P., and C. L. PEKERIS: Astrophys. Journ. **94**, 124 (1941). — For early applications of the variational method to this and similar problems cf. also A. B. SEVERNY: Dokl. Akad. Nauk USSR. **30**, 405 (1941). — Publ. Sternberg State Astronom. Inst. **13**, 95 (1940).
[53] EPSTEIN, I.: Astrophys. Journ. **112**, 6 (1950).
[54] SCHWARZSCHILD, M.: Astrophys. Journ. **94**, 245 (1941).
[55] EDDINGTON, A. S.: Monthly Notices Roy. Astronom. Soc. London **101**, 182 (1941); **102**, 154 (1942).
[56] ROSSELAND, S.: Univ. Obs. Oslo: a) Publ. **1931**, No. 1; b) **1932**, No. 2.
[57] LEDOUX, P.: Astrophys. Journ. **94**, 537 (1941).
[58] a) SCHWARZSCHILD, M.: Z. Astrophys. **15**, 14 (1938).
b) Harvard Circ. **1938**, 429.
c) Harvard Circ. **1938**, 431.

[59] PEKERIS, C. L.: Astrophys. Journ. **88** (1938).
[60] COWLING, T. G.: Monthly Notices Roy. Astronom. Soc. London **101**, 367 (1941).
[61] BRYAN, G. H.: Phil. Mag. (5) **27**, 254 (1889).
[62] COWLING, T. G., and R. A. NEWING: Astrophys. Journ. **109**, 149 (1949).
[63] BABCOCK, H. W.: Publ. Astronom. Soc. Pacific **59**, 260 (1947).
[64] Two recent books in the collection "Interscience Tracts on Physics and Astronomy" have brought considerable progress in the systematization of the field. The macroscopic equations are given in: T. G. COWLING: Magnetohydrodynamics, § 1.2. New York 1957. For a detailed and instructive discussion of the passage from the equations for ions and electrons to macroscopic equations cf. L. SPITZER jr.: Physics of Fully Ionized Gases, § 2.2 and Appendix. New York 1956.
[65] CHANDRASEKHAR, S., and E. FERMI: Astrophys. Journ. **118**, 116 (1953).
[66] COWLING, T. G.: Monthly Notices Roy. Astronom. Soc. London **112**, 527 (1952).
[67] WOLTJER, jr. J.: Monthly Notices Roy. Astronom. Soc. London **95**, 260 (1935). — Bull. astronom. Inst. Netherl. **8**, 193(1937); **9**, 435, 441 (1943); **10**, 125, 130, 135 (1946).
[68] For a general survey of these methods and their applications and new developments in this field cf. for instance:
a) MINORSKY, N.: Nonlinear mechanics. Ann Arbor: Edwards Bros 1947.
b) ANDRONOV, A., and C. CHAIKIN: Theory of Oscillations. Princeton: Univ. Press 1949.
c) For a short summary, cf. the article by S. LEFSCHETZ, pp. 7—30, in: Modern Mathematics for the Engineer, ed. by BECKENBACH. New York-Toronto-London: McGraw Hill 1956.
d) For an extension of the Ritz averaging method to non-linear problems especially interesting in the case of forced oscillations, cf. K. KLOTTER: Proceedings of the First National Congress of Applied Mechanics, p. 125, Amer. Soc. of Mech. Eng., New York; also: Symposium on Non-Linear Circuit Analysis, p. 234, Polytechnic Institute of Brooklyn, 1953.
[69] WHITHAM, G. B.: Comm. Pure a. Appl. Math. **6**, 397 (1953).
[70] a) RUSSELL, H. N., and C. E. MOORE: The Masses of the Stars, p. 112. Chicago, Ill.: Chicago University Press 1940.
b) Cf. also O. STRUVE: Stellar Evolution, p. 24. Princeton 1950.
c) For recent detailed discussions cf. R. M. PETRIE: Publ. Dom. Astr. Obs. **8**, 341 (1950). and Z. KOPAL and C. TREUENFELS: Harvard Circ. **1951**, No. 457.

Stellar Stability.

By

P. LEDOUX.

With 6 Figures.

1. Introduction. The most immediate purpose of the study of stellar stability is to discover the sources of the incipient instabilities which must be responsible for the observed variability of a great number of stars. In that respect, the observations already suggest that different types of instability must be at work.

In supernovae and at least in some of the novae, the phenomena are characteristic of a kind of explosion involving a part of the star which may vary from a large fraction of the total mass to a fairly small external fringe. This seems to be directly connected with the type of instability generally known as *dynamical instability* which arises when an infinitesimal perturbation starts increasing exponentially.

On the other hand, the more regular intrinsic variables such as the Cepheids seem to imply the existence of a tendency for the amplitude of any small linear oscillation around the state of equilibrium of these stars to increase until non-linear factors become important enough to limit the oscillation to some finite amplitude. In this case, the basic instability is most often referred to as *vibrational instability* although other names have been used such as overstability (EDDINGTON) and thermodynamical instability (JEANS).

Moreover, irregular fluctuations such as occur in the flare-stars may be due to limited *local instability* in the external layers of these stars, being in some way a magnified version of the instability which, in the Sun, gives rise to the protuberances and the flares.

But there are intermediary types of variable stars which have sometimes been interpreted as a link between the typical cases referred to above. For instance, the U Geminorum stars have been considered alternatively as very weak recurrent novae or as intrinsic variables of very large amplitude so that, in some cases, justified doubts may subsist as to the kind of instability responsible for the observed phenomena.

On the other hand, at least for perturbations which are not purely radial, there may be a continuous transition between some forms of dynamical and local instability as in the presence of a superadiabatic temperature gradient. In that case, provided the difference between the actual and the corresponding adiabatic gradient be larger in absolute value than a quantity which, in the interior of a star, is indeed very small, there are always many types of irregular perturbations of small wavelengths capable of initiating more or less irregular motions of limited amplitudes leading to the establishment of a new regime in which a continual mixing is maintained and which is usually known as convective equilibrium. This is the milder manifestation of local instability and it does not endanger the existence of the star.

If the superadiabatic layer is narrow, this is probably the only possible course of events, but if the layer covers a sufficiently large fraction of the star, there are special types of perturbations capable of leading to dynamical instability of the whole star. This provides an example in which the type of instability encountered depends on the form of the perturbation.

There is another type of relation between stability and the perturbation, the important factor this time being the size of the perturbation. Examples of static *metastable states* of equilibrium such as the rectangular box resting on one of its smallest faces have long been known. In that case, the system is stable toward any perturbation small enough but violently unstable for a perturbation of amplitude greater than a certain amount which will, however, lead to a new state of equilibrium corresponding to a lower potential energy.

No well-established cases of metastability have been discovered as yet in the study of stellar structure. If convection was somewhat harder to initiate, the sudden substitution of convective equilibrium to radiative equilibrium in a large region with an appreciable superadiabatic gradient, a possibility envisaged originally by UNSÖLD and BIERMANN to explain ordinary novae, would provide an important example of such instability. But it is difficult to understand why the starting of convection should be subject to any appreciable delays in time and space.

On the other hand, in the case of planets, an interesting possibility of this type has been discovered by RAMSEY and further discussed by RAMSEY, LIGHTHILL and COLLEAU[1]. They have shown that, if the equation of state is of the form $\varrho = \varrho(p)$ and is continuous on both sides of a critical pressure where a phase change causes a jump in density by a factor exceeding $\frac{3}{2}$, there will be three possible static configurations for values of the mass M comprised in a range $M_1 \leqq M \leqq M_0$, two with a central core in which the matter is in the second phase and one without a core.

The configuration with the smallest core is always unstable while, among the two others, one will be more stable than the other depending on the mass. In this case, a finite perturbation is needed to pass from the less stable to the more stable configuration. The critical masses M_0 and M_1 correspond respectively to the cases where the configuration with the smallest core degenerates either into the one without a core or into the one with the largest core so that only two equilibrium states are left, the degenerate one being always just unstable. An infinitely small perturbation of one of these last states is then sufficient to push the planet on an accelerating course leading in a finite time to the stable state of equilibrium.

As remarked by LIGHTHILL, the results of the necessary finite readjustments may depend strongly on the sense in which the transition takes place. If the planet goes from an unstable state without a core ($M = M_0$) to a stable one with a core, the difference of energy is released all through the mass and it is likely to lead to some kind of radial oscillations with very little, if any, release of matter from the planet's gravitational field. On the other hand, if the transition is from an unstable state with a core to a stable one without a core, the energy will be primarily released by the change of phase in the core and may be sufficiently concentrated locally to start blast waves which could lead to an appreciable ejection of matter from the planet's surface.

It is not impossible that such situations may arise also in composite stellar configurations and it may be worthwhile keeping this in mind in the study of evolutionary sequences of stellar models[2].

---

[1] W.H. RAMSEY: Monthly Notices Roy. Astronom. Soc. London **108**, 406 (1948); **110**, 325 (1950). — M.J. LIGHTHILL: Monthly Notices Roy. Astronom. Soc. London **110**, 338 (1950). — J. COLLEAU: Bull. Astron. **18**, 193 (1954).

[2] Cf. for instance G. GAMOW: Nature, Lond. **168**, 72 (1951).

In recent years, thanks to the study of non-linear systems especially in electronics, more complicated cases have become familiar in which a system displaced by a finite amount from an equilibrium state, stable toward sufficiently small perturbations, may evolve toward a stationary state of stable periodic motion known as a hard self-excited oscillation (cf. [*1*], Sect. 90). For instance, if our failure to discover a source of vibrational instability in Cepheids persists, we might have to investigate this possibility more carefully to explain their oscillations.

Apart from the interpretation of variable stars, the study of stability has a more general meaning for the problem of stellar structure since any stability criterion may be looked upon as a necessary condition to be satisfied by any hydrostatic model. For instance, any gaseous configuration in which the generalized ratio of specific heats becomes smaller than $\frac{4}{3}$ is dynamically unstable and, if for no other reasons, a cold gaseous mass composed of polyatomic molecules could hardly exist. In the same way, the condition of vibrational stability applied to massive stars has shown that, in presence of generation of energy by the carbon-cycle, stellar masses larger than about $100\,M_{\odot}$ are impossible. No doubts, many other applications remain to be discovered.

Up to now, we have neglected factors such as stellar rotation or the presence of a general magnetic field or of an appreciable external gravitational field as occurs in close double stars. These conditions may introduce new forms of instability such as the shedding of material through the equator or through the Lagrangian points or they may facilitate or hinder the occurrence of dynamical or local instabilities or introduce privileged classes of displacements.

One might also think of the interaction between a star and the interstellar medium which, in some cases, may give rise to unstable phenomena affecting at least the surface of the star. For instance, it is quite possible, as often suggested, that this can explain some of the features of the T Tauri stars.

All the previous types of instability may be characterized by the fact that, if present, they lead to motions which, although not necessarily fatal to the existence of the star itself, are always very fast on a cosmic scale.

Direct evidence in the case of the Earth and the solar system indicates that a star like the Sun can remain in practically the same state for a very long time of the order of 5×10^9 years despite the fact that the gravitational time-scale which is fixed by the time needed by the star to contract from a state of infinite dilution to the present state or to reduce its radius further by a factor 2 is only of the order of some 2×10^7 years. This can arise only through a proper balance of the energy generation and the opacity in the star and is known as *secular stability*.

Here we touch the very important and difficult problem of stellar evolution for, if one may admit that any hydrostatic model must instantaneously satisfy the condition of secular stability, it is obvious that it is the extremely slow departure from this condition caused by the progressive depletion of energy sources that permits the star to evolve. Furthermore, in the early stages of star formation and evolution when the stabilizing factor due to the generation of subatomic energy is lacking, the star changes more rapidly at a rate determined by the gravitational time-scale. Up to now, these early phases of a star life have hardly been studied from the point of view of the stability criteria.

We know also very little of the transition stages when the star switches from one type of nuclear reaction to another especially as far as the burning up, at fairly low temperatures, of the light elements D, Li, Be, B is concerned.

Moreover, the study of sequences of stellar models representing a star at different stages of evolution has already revealed hydrostatic incompatibilities such, for instance, as the existence of a maximum extension for the isothermal core which develops in the central regions of the star when the energy sources become exhausted there[1]. Another example is provided by the evolution of models with a convective core and a radiative envelope as the ratio $\bar{\mu}_c/\bar{\mu}_e$ of the mean molecular weight in the core and in the envelope increases due to the nuclear reactions in the former. At some phase, the mass of the radiative envelope which has kept its original chemical composition starts increasing again which is clearly impossible.

In some of these cases, the solution may be found in a revision of some of the hypothesis used in the hydrostatic model[2] but, in other cases, this type of hydrostatic difficulty may indicate an important change in the evolutionary trend. Although, sometimes, general principles and intuition may help to find the most likely direction in which the evolution will proceed past the critical point, it would be very interesting to base the discussion of the star's behaviour around this point on stability considerations.

If secular stability is so intimately connected with stellar evolution, one may wonder if the first question which it should answer is not: when can a very dilute and cold mass of gas or gas and other material start contracting under its own gravitation? This is also the fundamental question of the origin of the stars if one thinks that they are generally formed by condensation of interstellar matter. However, in this case, the mass considered is so far from hydrostatic equilibrium that the usual procedure adopted to discuss secular stability is of no avail and, furthermore, the influence of the surrounding medium cannot be discarded completely. As a result, a special name has been reserved for this type of questions which are referred to as problems of *gravitational stability*. Since they depend more on the properties of the medium in which stars may form than on the properties of the stars themselves, these problems do not really come under the heading of this article.

2. Available methods for the study of stability. First of all, it should be noted, as it may already have been inferred from the great variety of names (local, dynamical, vibrational, secular, gravitational stability, metastability, etc.) encountered in the introduction, that it is difficult to formulate an entirely general and consistent definition of stability. Even in the case of purely mechanical systems, carefully worded definitions often admit of counter-examples in which they lead to frustrating conclusions[3]. However the most serious of these difficulties arise really in discussing the stability of motions, and here we shall be mainly concerned with the stability of states of equilibrium which is a somewhat simpler problem. Indeed for a conservative mechanical system with a finite number of degrees of freedom, it is well known that a necessary and sufficient condition for an absolute equilibrium to be stable is that the potential energy be an absolute minimum (theorem of LEJEUNE-DIRICHLET). This means that provided the initial displacement of the system be sufficiently small, the distances of the different points of the system to their equilibrium positions will always remain small.

---

[1] M. SCHÖNBERG and S. CHANDRASEKHAR: Astrophys. Journ. **96**, 161 (1942).

[2] For such discussions of these difficulties cf. for instance P. LEDOUX: Mém. Soc. Roy. Sci. Liège **9**, Chap. II, § 10 (1949); A. R. SANDAGE and M. SCHWARZSCHILD: Astrophys. Journ. **116**, 463 (1952); F. HOYLE and M. SCHWARZSCHILD: Astrophys. Journ. Suppl. Ser. **2**, 1 (1955).

[3] Cf. for instance J. J. STOKER: On the Stability of Mechanical Systems. Comm. Pure Appl. Math. **8**, 133 (1955.)

The generalizations which we have to consider here are of different types. First of all, stars must be treated as systems with an infinite number of degrees of freedom. From a rigorous point of view, the passage from a discrete system to a continuous one may raise delicate questions [2], [3]. For instance, WAVRE (cf. [2], Sects. 94, 95) has proposed to generalize the definition of stability in that case, as follows. Let dm be a mass element of the system and E the volume of space swept by this element in the course of time. It is understood that a given region of space is counted as many times as the element covers it. The measure of the displacement of the system is then defined by the integral

$$\int E\, dm \tag{2.1}$$

extended over the whole system. The *distance* between two configurations S and S_0 which will be denoted by $|S-S_0|$ is then taken as the lower bound of the measures, as defined by (2.1), of all the virtual displacements leading from one configuration to the other. An equilibrium state S_0 will be stable if, ε_1 being an arbitrary small number, it is always possible to define another small number ε_2 such that any modification of the system, bringing it to a new state S at a distance from S_0 smaller than ε_2 and with a kinetic energy also smaller than ε_2, will only lead to motions such that all subsequent positions of S will be at distances from S_0 smaller than ε_1.

Of course, this definition is not the only possible one but it is fairly general and it has some definite advantages. For instance, it prevents automatically the inclusion among unstable states of those in which a small perturbation leads to a periodic motion for, although, in that case, the *displacement* (2.1) may increase continually (for instance when it is measured by some non-palpable coordinate such as an angle), the *distance* will reach a maximum during the first period which will never be exceeded in the subsequent motion.

With this definition, finite or even infinitely large displacements E of individual mass elements may occur provided that the corresponding set of elements be sufficiently small (of measure zero).

Note that, in this respect, the total kinetic energy plays a special rôle since, although it may always remain small, the distance travelled by the system may nevertheless increase continuously in the course of time.

Very generally, one may then define a functional $F(S)$ of the configuration S which will be continuous in S_0 provided

$$|F(S)-F(S_0)|<\varepsilon(r) \tag{2.2}$$

if

$$|S-S_0|\leqq r \tag{2.3}$$

$\varepsilon(r)$ tending towards zero with r.

In the same way, $F(S)$ may be said to be a minimum in a strong sense in $S=S_0$, if

$$F(S)-F(S_0)<\eta^2(r) \tag{2.4}$$

provided (2.3) be verified and $\eta^2(r)$ be a positive function decreasing to zero with r. Note that the restrictive condition (2.3) which expresses that the state S as a whole belongs to the set of states at distances from S_0 smaller than r is less strict than the corresponding condition for the individual elements of the system so that if F is a minimum in the sense implied by (2.4) and (2.3) then it is, a fortiori, a minimum when the condition (2.3) is replaced by the usual one that every elements of the system remain at distances from their equilibrium positions smaller than a given small quantity.

With the help of these definitions, taking for the functional F which must be a minimum in a strong sense, the potential energy V, the extension of the theorem of LEJEUNE-DIRICHLET to continuous systems is immediate.

The state of systems which are not purely mechanical may also depend on other variables than the position, but all the preceding definitions may be extended to them, the notion of "displacement" and "distance" as defined above then being applied to the variations of any of these variables, and, in some cases, appropriate generalizations of the theorem of LEJEUNE-DIRICHLET can also be formulated for these systems (cf. for instance [*3*], Chap. XVI, Sect. 6).

Although we do not intend to study the stability of motions, another necessary generalization concerns the case of ***relative equilibrium*** in which, at any time, the points of the system are at rest with respect to axes bound to the system as, for instance, in the case of a solid-body rotation. Obviously, in this case, the distances and kinetic energy will have to be measured with respect to the relative axes. Let us suppose that the relative kinetic energy K_R can be written

$$K_R = E - V_R \tag{2.5}$$

where E is a constant if there is no dissipation of energy in the system and can only decrease if such a dissipation is present and where V_R is a functional depending only on the configuration S of the system. Then a sufficient condition for stability of the state S_0 is that V_R be minimum in a strong sense in $S = S_0$. In presence of dissipation (i.e. E decreasing) it is also a necessary condition.

For simple systems the main features of which remain essentially mechanical, this generalized form of the principle of the minimum of the potential energy is fairly simple to apply. In particular, one knows [*4*] to what advantageous uses it has been put in the discussion of the stability of the equilibrium configurations of gravitating masses of incompressible fluid in rotation by POINCARÉ, LIAPOUNOFF, DARWIN, JEANS, SCHWARZSCHILD, etc.

In the case of gaseous configurations, the thermodynamical factors are often important and have to be taken into account in any generalization of this principle to actual stars. This point of view has found expression in a few papers either as a deliberate attempt [*5*] to put such a generalized principle at the basis of stability discussions in general or as a means to reach an answer to some particular question of stability [*6*].

This principle can also be expressed in the following way: the state of the system is stable (unstable), if the work of the external forces needed to displace the system from its equilibrium position or to maintain it in a state of harmonic oscillations is positive (negative). In this form, it was used for instance by EDDINGTON[1] to discuss the vibrational stability of stars.

However up to now, this method has met with only limited successes which, in the main [*5b*], consisted in recovering conditions of stability already well-known. Nevertheless, it possesses a degree of generality which may justify its further study.

The other method which has been much more extensively used in this context is the well-known method of perturbations in which it is assumed that, if the perturbation is sufficiently small, the solution of the *linearized* equations of motion will approximate closely enough to the actual solution of the complete non-linear problem to reveal the trend of the motion in a small region around the equilibrium position.

[1] A. S. EDDINGTON: Monthly Notices Roy. Astronom. Soc. London **101**, 182 (1941); cf. [*1*], Sect. 63, β.

LIAPOUNOFF [7] has shown this to be rigorously true in special cases where it can be expressed explicitly in the following form: if the real parts of the characteristic exponents[1] are different from zero, the equations of the linear approximation always give a correct answer to the question of stability. The equilibrium is stable if the real parts of all the characteristic exponents are negative but if at least one exponent has a positive real part, the equilibrium is unstable. However if some or all of the characteristic exponents have vanishing real parts, the question of stability cannot be decided by this linear approach alone.

It seems likely that, at least in the immediate future, the perturbation method will remain the most useful one for many of the problems mentioned in the introduction. To a large extent it reduces essentially to the study of the small oscillations around the state of equilibrium, a subject which has been discussed in great details in the theoretical part of the article on "Intrinsic Variable Stars" in this volume [1]. Thus, in many cases, we shall be content to summarize the main results, referring the reader to that article for detailed equations and discussions.

A. Incompressible masses.

3. Introduction and general principles. Although some of the authors who contributed most to the theory of liquid configurations in rotation were mainly concerned with the mathematics of the problem, most of the general interest in this question was aroused by the cosmogonical implications stressed especially by DARWIN and JEANS. However, as soon as the gaseous nature of the stars was recognized following EDDINGTON'S work, it should have been clear that, at the same time, the domain of application of this theory had shrinked enormously and, at best, was reduced essentially to the problem of the evolution of primitive planets once they had reached more or less uniform densities and, perhaps, to the problem of the subsequent formation of satellites (cf. [4a], Chap. X).

On that basis, one might consider that this question is irrelevant to the present article. However, it provides an extremely interesting illustration of general principles which are fundamental in all studies of stability. Furthermore, there are still lingering misconceptions as to the actual significance of the results and their possible applications.

The study of stability of liquid rotating masses is generally introduced by the consideration of the same problem for a mechanical system with a finite number of degrees of freedom. Indeed, if it has been possible to treat the stability of these configurations so exhaustively, it is mainly because an incompressible fluid has essentially purely mechanical properties. The second reason which facilitated the research in this case, is that the possible configurations of equilibrium are ellipsoids so that the theory of ellipsoidal harmonics introduced by LAMÉ provides an adequate mathematical tool. There remains of course the question of going from a system with a finite number of degrees of freedom to one with an infinite number but, although particular results may be affected by this, the methods remain essentially the same.

Let us consider first a *conservative* mechanical system possessing f degrees of freedom and obeying the Lagrangian equations

$$\frac{d}{dt}\left(\frac{\partial L}{\partial \dot{q}_i}\right) - \frac{\partial L}{\partial q_i} = 0, \quad i = 1, 2, \ldots, f \tag{3.1}$$

[1] Cf. [1], Sect. 90, where a brief mention of LIAPOUNOFF'S characteristic equation and exponents and further references will be found.

with

$$L = K(q_i, \dot{q}_i) - V(q_i), \tag{3.2}$$

where K is the kinetic energy which is a positive definite quadratic form in the $\dot{q}_i$, and V is the potential energy.

At equilibrium, Eqs. (3.1) reduce to

$$\frac{\partial V}{\partial q_i} = 0, \quad i = 1, 2, \ldots, f \tag{3.3}$$

which express the condition that V be stationary and their solutions $(q_{i,0})$ define the equilibrium configuration of the system.

Since, the total energy of the system is constant, we have also

$$K + V = E. \tag{3.4}$$

From this relation, we conclude immediately that, if V goes through an absolute minimum for $q_i = q_{i,0}$, the equilibrium state $q_{i,0}$ is stable in the sense that, following any small perturbation of the equilibrium state, V and consequently all the q_i's can only deviate from their equilibrium values by correspondingly small amounts since K is always positive. On the other hand, if there is at least one generalized coordinate q_i following which V is not a minimum, then, for an appropriate perturbation, the system can deviate more and more from the equilibrium state.

In the case of very small perturbations δq_i, these results are recovered easily from the equations of motion (3.1) which, in the vicinity of $q_{i,0}$ become, assuming a time-dependence of the form

$$\delta q_i = a_i \, e^{i\omega t},$$

$$\sum_j (V_{ij} - \omega^2 K_{ij}) \, a_j = 0, \quad i = 1, 2, \ldots, f \tag{3.5}$$

where the V_{ij} and K_{ij} represent respectively the values, in $q_i = q_{i,0}$, of $\partial^2 V/\partial q_i \, \partial q_j$ and those of the coefficients of the quadratic form K. The compatibility condition

$$|V_{ij} - \omega^2 K_{ij}| = 0 \tag{3.6}$$

determines f eigenvalues $\lambda^k = (\omega^k)^2$ which are always real since the real matrices $\boldsymbol{V}$ and $\boldsymbol{K}$ are symetric. Introducing these values in turn in (3.5), these equations can be solved for the corresponding f eigenvectors a^k which are orthogonal and can further be normalized since the system (3.5) leaves one of their components arbitrary. Then, in the general configuration space of the q_i's, we have

$$\sum_{i,j} K_{ij} \, a_i^k \, a_j^l = \delta_{k,l}. \tag{3.7}$$

If we multiply Eq. (3.5) written for the eigenvector a^k by a_i^k and sum over i and j, we find that λ^k can be expressed as

$$\lambda^k = (\omega^k)^2 = \sum_{i,j} V_{ij} \, a_i^k \, a_j^k. \tag{3.8}$$

Here the condition of stability is of course that all λ^k be positive but, according to (3.8), this is equivalent to the condition that $V(q_{i,0})$ be an absolute minimum since the second member represents simply the variation of the potential energy when the system is displaced from $(q_{i,0})$ to $(q_{i,0} + a_i^k)$. In that case, all possible motions of the system around its equilibrium position reduce to a linear combination of the different modes of oscillation

$$\delta q_i = \sum_k C_k \, a_i^k \, e^{-i\omega^k t}. \tag{3.9}$$

Some aspects of the problem are simplified by the simultaneous diagonalization of the matrices $\boldsymbol{V}$ and $\boldsymbol{K}$ or of the corresponding quadratic forms by means of a principal axis transformation of matrix $\boldsymbol{A}$. In that case, the relevant parts of the kinetic and potential energy can be written

$$K = \tfrac{1}{2}\sum_i k_i \dot{\eta}_i^2, \qquad V = \tfrac{1}{2}\sum v_i \eta_i^2 \tag{3.10}$$

where the η_i's are new generalized coordinates of the type of the δq_i's. If the system is stable, V is positive definite and all the v_i's which are proportional to the λ^i's must be positive. They are sometimes referred to as POINCARÉ's "coefficients of stability".

In particular, if the matrix of the transformation $\boldsymbol{A}$ is formed with the eigenvectors, all the k_i's reduce further to unity so that the new axes are Cartesian and

$$v_i = \lambda^i, \tag{3.11}$$

$$|v_i\,\delta_{ij}| = |\lambda^i\,\delta_{ij}| = |a_{ij}| \times |V_{ij}|. \tag{3.12}$$

Let us now suppose that the state of the system depends on a parameter μ. If the system is free or if the constraints are independent of μ, this dependence will appear only in the potential $V = V(q_i, \mu)$. For a given value of μ, all the preceding considerations remain valid, but the solutions of (3.3) defining the equilibrium state will depend on μ. In fact, if we differentiate (3.3) with respect to μ we obtain f linear equations

$$\sum_j \left(\frac{\partial^2 V}{\partial q_i\,\partial q_j}\right)_0 \frac{d q_j}{d\mu} + \left(\frac{\partial^2 V}{\partial q_i\,\partial \mu}\right)_0 = \sum_j V_{ij}\frac{\partial q_j}{\partial \mu} + V_{i,\mu} = 0 \tag{3.13}$$

which are simply the new conditions (3.3) that the modified potential

$$\left.\begin{aligned} V = V(q_{i,0}, \mu_0) &+ \frac{1}{2}\sum_{i,j}\left(\frac{\partial^2 V}{\partial q_i\,\partial q_j}\right)_0 d q_i\, d q_j + \left(\frac{\partial V}{\partial \mu}\right)_0 d\mu + \\ &+ \frac{1}{2}\left(\frac{\partial^2 V}{\partial \mu^2}\right)_0 (d\mu)^2 + \sum_i \left(\frac{\partial^2 V}{\partial \mu\,\partial q_i}\right)_0 d q_i\, d\mu \end{aligned}\right\} \tag{3.14}$$

be stationary at the new equilibrium position $(q_{i,0} + d q_i;\ \mu_0 + d\mu)$.

If $|V_{ij}|$ does not vanish, these Eqs. (3.13) yield well determined values for all the $d q_j/d\mu$ and the q_j are uniform functions of μ around $(q_{i,0}, \mu_0)$. Thus as long as $|V_{ij}| \neq 0$, if we let μ vary continuously, we shall obtain a continuous set of successive equilibrium states known as a *linear series*.

However when we reach a point C where

$$|V_{ij}|_C = 0 \tag{3.15}$$

Eqs. (3.13) no longer determine uniquely the continuation of the series past this point. To lift the indeterminacy we should have to go to a higher approximation in the development (3.14) of V. But then the corresponding Eqs. (3.13) would become non-linear and might have either no real solutions or yield more than one set of values for the $d q_j/d\mu$. Thus in a point such as C, the linear series may stop abruptly, turn around or branch off into different series. Following POINCARÉ, points of this type are known as *bifurcation points*.

Moreover, the condition (3.15) implies that a change in the stability of the system must take place at such a point since according to (3.12) or (3.6) at least one of the λ^k or $(\omega^k)^2$ must vanish when $|V_{ij}| = 0$.

4. Graphical discussions of stability. All these results are susceptible to be illustrated and, to some extent, completed by diagrammatic means. Adopting q_i and μ as rectangular coordinates, the equations

$$V(q_i, \mu) = V \tag{4.1}$$

represent, for different values of the constant V, non-intersecting hyper-surfaces in a $(f+1)$-dimensional space. Equilibrium states of the system for a given value of μ will correspond to points where these surfaces have tangent planes normal

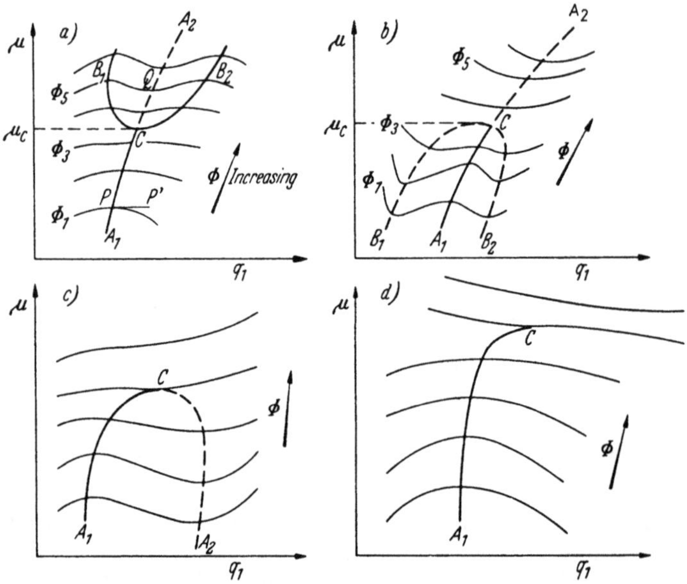

Fig. 1 a—d. Examples of linear series of equilibrium configurations and of exchange of stabilities. The thin lines represent the potential $\Phi(\mu, q_1)$; the heavy full lines represent stable linear series while the dashed lines correspond to unstable equilibrium configurations.

to the μ-axis. The case $f=1$, is illustrated on Fig. 1 where the heavy lines represent linear series of equilibrium configurations.

If at an equilibrium point P, the concavity of the surface (4.1) along all coordinates is turned toward decreasing values of V, this point corresponds to a stable configuration since any small displacement of the system at constant μ such as PP' will lead to an increase of the potential energy which will thus be an absolute minimum in P. On the other hand, if the concavity is turned toward increasing values of V as in Q, the corresponding equilibrium state will be unstable.

The two types of points considered on the linear series $A_1 A_2$ will be separated by a point C where the curvature goes through zero. This will be the point of bifurcation where the stability of the linear series $A_1 C$ vanishes. If the change in curvature of V at the point C is accompanied by the appearance of new concavities turned toward decreasing V as illustrated at the points B_1 and B_2 in Fig. 1, there will be a new linear series $B_1 C B_2$ crossing the primitive one in C and "inheriting" the stability of $A_1 C$. For $\mu < \mu_C$ there is one stable configuration while for $\mu > \mu_C$, there are two such configurations.

Other possible cases are also illustrated. In Fig. 1 b the stability of the linear series $A_1 C$ disappears completely in C, no stable configurations existing for $\mu > \mu_C$. In Figs. 1 c and 1 d no equilibrium configuration at all exists for $\mu > \mu_C$.

According to POINCARÉ ([4d], pp. 162—172), the situation can also be described as follows. Let us suppose that, around an equilibrium state (q_i, μ), all the first order minors of $|V_{ij}|$ do not vanish. In particular, let

$$|V_{r,s}| \neq 0 \text{ for } r \text{ and } s = 2, 3, \ldots, f. \tag{4.2}$$

Then Eqs. (3.3) for $i = 2, 3, \ldots, f$ can be solved yielding $q_2, q_3, \ldots, q_f$ as uniform functions of q_1 and μ since, in that case, the last $(f-1)$ Eqs. (3.13) give well determined values of $(dq_i/d\mu)$, $i = 2, 3, \ldots, f$. Introducing these values in $V(q_i, \mu)$, we get a function $\psi(q_1, \mu)$ with a derivative with respect to q_1 for a fixed value of μ, given by

$$\frac{\partial \psi}{\partial q_1} = \frac{\partial V}{\partial q_1} + \sum_{i=2}^{f} \frac{\partial V}{\partial q_i} \frac{dq_i}{dq_1} = \frac{\partial V}{\partial q_1}. \tag{4.3}$$

Since $(q_1, q_2, \ldots, q_f)$ is an equilibrium state, we must have

$$\frac{\partial \psi(q_1, \mu)}{\partial q_1} = 0 \tag{4.4}$$

which defines a relation between q_1 and μ. But q_1 will be a uniform function of μ only if

$$\frac{\partial^2 \psi}{\partial q_1^2} \neq 0 \tag{4.5}$$

and the condition (3.15) for a bifurcation point is equivalent here to

$$\frac{\partial^2 \psi}{\partial q_1^2} = 0. \tag{4.6}$$

When (4.5) is satisfied, the equilibrium will be stable if

$$\frac{\partial^2 \psi}{\partial q_1^2} > 0. \tag{4.7}$$

These relations reduce the general discussion to one in the (q_1, μ) plane or, for that matter, in any (q_k, μ) plane provided that the minor of $|V_{ij}|$ corresponding to the element V_{kk} does not vanish.

The condition of equilibrium (4.4) defines a curve in the (q_1, μ) plane which represents a linear series of equilibrium states and separates in this plane regions where $\partial\psi/\partial q_1$ is respectively positive or negative. Any point where (4.6) is verified but $\partial^2\psi/\partial q_1\, \partial\mu \neq 0$ has a horizontal tangent and the curve is all on one side of this tangent. On the other hand, wherever $\partial^2\psi/\partial q_1\, \partial\mu = 0$ at the same time as $\partial^2\psi/\partial q_1^2$, the curve has a double point with two tangents which may be distinct or superposed. One of these tangents may be horizontal or not so that the main cases of interest are as illustrated on Fig. 2 where the regions with $\partial\psi/\partial q_1 > 0$ are covered with hatching. Any portions of these linear series with a negative region on the left and a positive region on the right represent stable states of equilibrium since, in that case, the condition (4.7) is satisfied at any of their points.

The preceding discussion and the present representation are very appropriate for systems depending on a parameter μ. Physically, this parameter has to vary slowly so that the evolution of the system can be represented by a succession of equilibrium states. This is a situation which arises quite often in astronomy.

The method was developed originally for rotating liquid masses letting the angular momentum which, in this case, plays the rôle of the parameter μ increase continuously. But, in principle, it could be used as well in the case, already recalled in the introduction, of a planet in which a discontinuous phase change occurs at a critical pressure p_c, the mean temperature or the mean density playing the rôle of the parameter μ. Again, in the case of a star, it would be very interesting to be able to follow its evolution by these methods using for instance the amount of nuclear fuel as the slowly varying parameter.

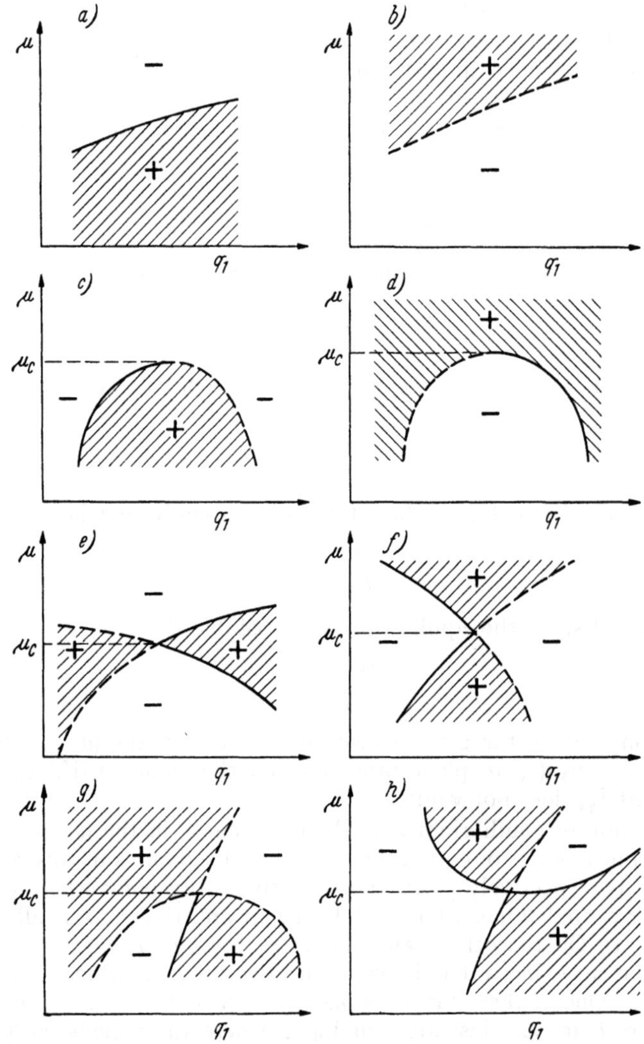

Fig. 2 a—h. Some examples of stable (full lines) and unstable (dashed lines) linear series represented by the curves $\partial \psi/\partial q_1 = 0$.

5. Effects of rotation. Dynamical and secular stability. The only type of instability which we have encountered in the previous discussion of a conservative system without rotation is the one which we have called dynamical instability and in which the coordinates increase exponentially. In the discussion of a rotating liquid mass this is more often referred to as ordinary instability. However for the sake of consistency with the rest of this article we shall keep the first name.

As far as rotating liquid masses are concerned, the case of interest is the one where the whole configuration rotates as a solid body with a constant angular velocity Ω around a fixed axis through the centre of mass. In that case, it is indicated to use uniformly rotating axes with respect to which the Lagrange equations (3.1) become

$$\frac{d}{dt}\left(\frac{\partial K_R}{\partial \dot{q}_i}\right) - \frac{\partial K_R}{\partial q_i} - \Omega \sum_j \beta_{ij} \dot{q}_j = -\frac{\partial}{\partial q_i}\left(V - \frac{1}{2}\Omega^2 I\right) \tag{5.1}$$

where K_R is the kinetic energy relative to the rotating frame, I is the moment of inertia round the axis of rotation, and the β_{ij} are geometrical quantities depending on the relations between the generalized coordinates and the ordinary Cartesian coordinates (x_k, y_k, z_k) with respect to the rotating axes,

$$\beta_{ij} = 2\sum_k m_k \left(\frac{\partial x_k}{\partial q_i}\frac{\partial y_k}{\partial q_j} - \frac{\partial y_k}{\partial q_i}\frac{\partial x_k}{\partial q_j}\right) = -\beta_{ji}, \quad \beta_{ii} = 0.$$

Eq. (5.1) shows that a constant rotation influences the motion in two ways, first, by the appearance of centrifugal forces proportional to Ω^2 and admitting a potential and secondly, by the appearance, in presence of relative motion, of Coriolis forces proportional to Ω.

The conditions of relative equilibrium ($K_R=0$, $\dot{q}_i=0$) which replace (3.3) are

$$\frac{\partial V_R}{\partial q_i} = 0, \quad i = 1, 2, \ldots, f, \tag{5.2}$$

denoting by V_R the total relative potential

$$V_R = V - \tfrac{1}{2}\Omega^2 I \tag{5.3}$$

while, after diagonalization of K_R and V_R as in (3.10), the equations of small motion around such a state of equilibrium become

$$k_{R,i}\ddot{\eta}_i + v_{R,i}\eta_i - \Omega \sum_j \beta'_{ij}\dot{\eta}_j = 0, \quad i = 1, 2, \ldots, f. \tag{5.4}$$

They admit solutions of the form

$$\eta_i = a_i \, e^{i\omega t}$$

provided that the compatibility condition

$$\begin{vmatrix} v_{R,1} - \omega^2 k_{R,1} & -i\Omega\omega\beta'_{12} & \cdots & -i\Omega\omega\beta'_{1f} \\ -i\Omega\omega\beta'_{21} & v_{R,2} - \omega^2 k_{R,2} & \cdots & -i\Omega\omega\beta'_{2f} \\ \cdots & \cdots & \cdots & \cdots \\ -i\Omega\omega\beta'_{f1} & -i\Omega\omega\beta'_{f2} & & v_{R,f} - \omega^2 k_{R,f} \end{vmatrix} = 0 \tag{5.5}$$

be satisfied. Since $\beta'_{ij} = -\beta'_{ji}$, this is really an equation in ω^2 for it is unchanged when ω is replaced by $-\omega$ but, contrarily to the case of (3.6), its eigenvalues ω^2 may not in general be all real but may contain pairs of complex conjugate roots giving frequencies

$$\omega^k = \pm(\alpha^k \pm i\beta^k)$$

which would always lead to an exponential increase of the corresponding perturbation.

In absence of rotation, all non-diagonal terms vanish, $v_{R,i} \to v_i$, $k_{R,i} \to k_i$ and, as we have seen previously, the condition of dynamical stability is that all the v_i be positive. Here, the condition for the system to be dynamically stable is still that all the roots ω^2 be real and positive, but this does not depend only on the sign of the $v_{R,i}$ but also on the values of the β'_{ij} or β_{ij} i.e. on the gyroscopic terms proportional to Ω.

On the other hand, if we multiply (5.1) by $\dot{q}_i$ and sum over i, we obtain remembering the antisymmetry of the β_{ij}

$$\frac{d}{dt}(K_R + V_R) = 0, \tag{5.6}$$

which yields the energy integral

$$K_R + V_R = E. \tag{5.7}$$

Let us now consider an equilibrium configuration $(q_{i,0})$ and let us suppose that, apart from being stationary [conditions (5.2)], $V_R = V - \frac{1}{2}\Omega^2 I$ is an absolute minimum for $q_i = q_{i,0}$ whose value can always be taken equal to zero. In that case, if a slight perturbation is applied to the equilibrium state, (5.7) will be valid with the second member equal to some small positive constant, say e. K_R will be maximum and equal to e when the motion goes through the equilibrium state and will vanish for some small displacement from equilibrium when $V_R = \sum_{i,j} (V_{i,j})_R\, \delta q_i\, \delta q_j = e$ so that the motion must consist of small oscillations around the state $(q_{i,0})$.

But if the total potential energy is an absolute minimum, its value out of equilibrium, $\sum_{i,j} (V_{i,j})_R\, \delta q_i\, \delta q_j$, must be positive definite. This means that, if it is diagonalized as above, all the coefficients $v_{R,i}$ must be positive. Thus, we may conclude that, if all the $v_{R,i}$ are positive, the system is certainly dynamically stable. But, as we have seen in discussing directly the frequencies of the possible oscillations, because of the presence of the Coriolis terms, the argument cannot be reversed to show that the system is necessarily dynamically unstable if V_R is not an absolute minimum i.e. if some of the $v_{R,i}$ are not positive.

In other words, the condition

$$v_{R,i} > 0, \quad i = 1, 2, \ldots, f \tag{5.8}$$

is always a sufficient condition for dynamical stability, but it is not a necessary condition. In particular, we deduce from (5.5) that

$$\prod_{k=1}^{f} (\omega^k)^2 = \frac{\prod_{i=1}^{f} v_{R,i}}{\prod_{i=1}^{f} k_{R,i}} \tag{5.9}$$

which shows that the $(\omega^k)^2$ can all remain positive if an even number of $v_{R,i}$ become negative. But if one or an odd number of $v_{R,i}$ become negative then at least one of the ω^2 becomes negative too and the system is dynamically unstable.

However, if in a rotating system the dynamical stability is not necessarily lost when V_R ceases to be an absolute minimum, could it not be that a milder form of instability appears in that case? To discuss this distinction introduced by Lord KELVIN, let us consider with him a non-conservative system submitted to friction forces which, for the small motions contemplated here, can always be taken as proportional to the velocities and as deriving from a dissipation function $\mathfrak{F}$ which is a positive definite quadratic form in the $\dot{q}_i$ and equal to half the absolute value of the rate of energy dissipation due to these forces. In that case, the second member of (5.1) will contain an extra-term $-\partial\mathfrak{F}/\partial\dot{q}_i$ and proceeding as before, we obtain instead of (5.6)

$$\frac{d}{dt}(K_R + V_R) = -2\,\mathfrak{F}(\dot{q}_i^2). \tag{5.10}$$

If V_R is still an absolute minimum which can be put equal to zero in the equilibrium state $(q_{i,0})$, the total energy (K_R+V_R) after a small displacement will still take a small positive value but, because of the negative character of the second member in (5.10), this positive value will decrease continuously until all $\dot{q}_i$ and $(q_i-q_{i,0})$ vanish simultaneously. Thus, in this case, any amount of friction however small will always bring the system back to equilibrium after a sufficiently long time. The system is said to be *secularly stable* as well as dynamically stable.

On the other hand, suppose that V_R is no longer an absolute minimum, i.e. some of the $v_{R,i}$ are negative. In that case, it is always possible to choose a displacement along the coordinates with respect to which V_R is a maximum such that (K_R+V_R) takes a small negative value, say $-e$. But, in that case, because of the negative character of the second member of (5.10), e will increase continuously, the system moving farther and farther away from the equilibrium position. The system is said to be *secularly unstable* but, as we have seen above, it may be dynamically stable or unstable.

Thus this distinction between dynamical and secular stability which has no meaning for a system at rest becomes important for a rotating system.

If we want to apply these considerations to a linear series of equilibrium states of a system with a variable moment of inertia which plays the rôle of the parameter μ of Sects. 3 and 4, our assumption of a constant angular velocity Ω requires the introduction of an external couple capable of modifying the azimuthal angle φ which fixes the position of our rotating system with respect to some given direction. But φ is a cyclic variable and the corresponding Lagrangian equation reduces to

$$\frac{dH}{dt}=G \tag{5.11}$$

where $H=\Omega I$ is the absolute angular momentum of the system, the other equations of motion being unaffected.

However, this is a rather artificial situation and, as SCHWARZSCHILD has remarked, for a free system $G=0$ and it will evolve with H constant. In that case, if we still refer the motion to axes rotating with a constant angular velocity Ω, the actual rotation of the system may lag or advance continuously with respect to these axes when its moment of inertia varies and this may imply that some of the q_i may increase indefinitely although the system is stable. To avoid this, let us suppose that, at any time we choose Ω in such a way that the relative angular momentum H_R is always zero. Then the general relations

$$K=K_R+\Omega H_R+\tfrac{1}{2}\Omega^2 I,$$
$$H=H_R+\Omega I,$$

which are always valid whether Ω be variable or constant in time, become

$$K=K_R+\tfrac{1}{2}\Omega^2 I, \qquad H=\Omega I,$$

and the elimination of Ω yields

$$K=K_R+\frac{H^2}{2I}. \tag{5.12}$$

The energy equations (5.7) and (5.10) now assume the form

$$K+V=K_R+V+\frac{H^2}{2I}=\text{const} \tag{5.13}$$

or

$$\frac{d}{dt}\left(K_R+V+\frac{H^2}{2I}\right)=-2\mathfrak{F}(\dot{q}_i^2) \tag{5.14}$$

and the same arguments as before will lead to the conclusion that systems with constant angular momentum will be secularly stable if $V_R' = V + H^2/2I$ is an absolute minimum, this being also a sufficient condition for the system to be at the same time, dynamically stable. If V_R' ceases to be a minimum with respect to some coordinates, then secular stability is lost, but the system may or may not be dynamically stable.

Summing up, the relation between secular stability and dynamical stability may be stated as follows:

(1) If an equilibrium configuration is secularly stable, it is also dynamically stable.

(2) If an equilibrium configuration is secularly unstable, it may or may not be dynamically unstable.

Thus although dynamical stability cannot vanish before secular stability and while it may subsist after secular stability is lost, it may also happen that both types of stability vanish *together*. The disregard of this last possibility, has led JEANS to some unreliable conclusions in his discussion of the cosmogonical implications of the theory.

6. Applications to rotating liquid masses. To a large extent, the preceding theory of exchange of stability at the bifurcation points of linear series has reduced the problem of determining a sequence of stable configurations to the much simpler problem of mapping out all configurations of equilibrium. One should note however that, in the case of a rotating system, this exchange of stability concerns only the secular stability since it is the one which is determined entirely by the behaviour of the total potential $(V + H^2/2I)$, dynamical stability being able to persist on the original series past such a point. However as we have already recalled it may also vanish simultaneously with secular stability. This is an important point to check because the behaviour of the system will be very different according as to whether secular instability only sets in or whether both types of instability set in simultaneously.

In the absence of rotation, the stable equilibrium state of a liquid mass is that one which makes the gravitational potential energy V an absolute minimum. It is easy to show if the fluid is homogeneous that, whatever its shape, its potential Φ at any point satisfies the condition

$$-\Phi < 2\pi G \varrho R^2 \tag{6.1}$$

if R is the radius of the sphere having the same total volume $\mathcal{V}$ as the mass considered. From this and the definition of the potential energy

$$V = \tfrac{1}{2} \int^{\mathcal{V}} \Phi \varrho \, d\mathcal{V} \tag{6.2}$$

one concludes that

$$V > -\tfrac{4}{3} \pi^2 G \varrho^2 R^5 . \tag{6.3}$$

Thus V has a minimum and LIAPOUNOFF (cf. [4d], p. 17) has shown that this minimum which is equal to

$$V_{\min} = -\tfrac{16}{15} \pi^2 G \varrho^2 R^5 \tag{6.4}$$

is actually reached only for the sphere.

The same result has also been proved and extended to heterogeneous configurations by LICHTENSTEIN (cf. [4a], Chap. III and [2], Sects. 23, 24) following a simpler line of arguments showing that, in absence of rotation, the stable equilibrium configuration must be symmetrical with respect to all planes going through its mass centre. This is a particular application of the general result that, in any rotating configuration (homogeneous or not) in relative equilibrium, there exists an equatorial plane of symmetry perpendicular to the axis of rotation.

If we superpose a solid-body rotation of the whole mass, intuition and common experience tell us that, at least for small angular velocities, the body will flatten along the axis of rotation keeping a symmetry of revolution. If we increase the angular velocity, the subsequent forms of equilibrium are less evident but we know that, at some time centrifugal force will overcome any cohesion force present and disrupt the mass.

In the case of gravitational forces deriving from a potential Φ, solution of the Poisson equation

$$\nabla^2 \Phi = 4\pi G \varrho, \tag{6.5}$$

it is fairly simple, following POINCARÉ, to translate this into an analytical condition. The total force with respect to Cartesian axes (x, y, z) rotating around z with the instantaneous angular velocity Ω of the mass admits a generalized potential

$$\Phi_R = \Phi - \tfrac{1}{2}\Omega^2 (x^2 + y^2). \tag{6.6}$$

For equilibrium, the corresponding force must always be directed towards the interior at each point on the surface so that

$$-\iint_S \operatorname{grad} \Phi_R \cdot d\boldsymbol{S} < 0. \tag{6.7}$$

If the surface integral is transformed into a volume integral and (6.5) is used, condition (6.7) becomes

$$\Omega^2 < 2\pi G \frac{M}{\mathscr{V}} = 2\pi G \bar{\varrho} \tag{6.8}$$

where M is the total mass and $\mathscr{V}$ the total volume. For a homogeneous configuration, this condition becomes

$$\Omega^2 < 2\pi G \varrho. \tag{6.9}$$

It fixes an upper limit to the speed of rotation beyond which, whatever the state of the fluid, no equilibrium is possible. If the external surface is everywhere convex, CRUDELI[1] was able to sharpen the inequality (6.8) into

$$\Omega^2 < \pi G \bar{\varrho}. \tag{6.10}$$

Another general property of rotating masses in equilibrium is that the smaller axis of the inertia ellipsoid always coincides with the axis of rotation.

The problem now is to find all the possible equilibrium configurations. First of all, when an homogeneous incompressible sphere is given a very small uniform rotation, it can be proved as suggested above that it assumes the shape of an ellipsoid of revolution and so it turns out that ellipsoids will play a major rôle in the evolution with increasing Ω. This is a rather fortunate circumstance as the gravitational potential Φ of a homogeneous ellipsoid whose external surface is represented by

$$\frac{x^2}{a^2} + \frac{y^2}{b^2} + \frac{z^2}{c^2} = 1 \tag{6.11}$$

is well known. At any internal point, it can be written

$$\left.\begin{aligned}\Phi &= \pi G \varrho\, a\, b\, c \int_0^\infty \left(\frac{x^2}{a^2+\lambda} + \frac{y^2}{b^2+\lambda} + \frac{z^2}{c^2+\lambda} - 1\right)\frac{d\lambda}{\Delta} \\ &= \pi G \varrho (\alpha x^2 + \beta y^2 + \gamma z^2 - \delta)\end{aligned}\right\} \tag{6.12}$$

[1] U. CRUDELI: Velocità angolare e schiacciamento delle figure di equilibro dei fluidi rotanti, Coop. Tipografica Azzoguidi, Bologna, 1930. — W. NIKLIBORC: Math. Z. **30** (1929).

where

$$\Delta^2 = (a^2 + \lambda)(b^2 + \lambda)(c^2 + \lambda)$$

and λ is the parameter defining the family of confocal ellipsoids

$$\frac{x^2}{a^2 + \lambda} + \frac{y^2}{b^2 + \lambda} + \frac{z^2}{c^2 + \lambda} = 1.$$

On the other hand, the pressure must vanish everywhere on a free surface so that the external surface is the isobaric surface $p = 0$. But according to the hydrostatic equation

$$\frac{1}{\varrho} \operatorname{grad} p = -\operatorname{grad} \Phi_R = -\operatorname{grad}\left[\Phi - \frac{1}{2}\Omega^2 (x^2 + y^2)\right] \tag{6.13}$$

it must also be an equipotential surface

$$\Phi_R = \pi G \varrho \left[\left(\alpha - \frac{\Omega^2}{2\pi G \varrho}\right) x^2 + \left(\beta - \frac{\Omega^2}{2\pi G \varrho}\right) y^2 + \gamma z^2\right] = \text{const.} \tag{6.14}$$

The identification of (6.14) with (6.11) provides two relations between the axes and Ω. One of them can be written

$$\frac{\Omega^2}{2\pi G \varrho} = \frac{a c}{b} (b^2 - c^2) \int_0^\infty \frac{\lambda \, d\lambda}{(b^2 + \lambda)(c^2 + \lambda) \Delta}$$

and, in particular, it shows that b is always larger than c or, in other words, since we can always take $a \geq b$, that the smallest axis always coincides with the axis of rotation. In the case $a = b$, this reduces to

$$\frac{\Omega^2}{2\pi G \varrho} = \frac{3 - 2e^2}{e^3} \sqrt{1 - e^2} \arcsin e - \frac{3}{e^2} (1 - e^2) \tag{6.15}$$

where e is the excentricity defined by

$$c^2 = a^2 (1 - e^2).$$

The other relation

$$(a^2 - b^2) \int_0^\infty \left[\frac{a^2 b^2}{(a^2 + \lambda)(b^2 + \lambda)} - \frac{c^2}{c^2 + \lambda}\right] \frac{d\lambda}{\Delta} = 0 \tag{6.16}$$

can be verified in two ways, either in choosing $a = b$ or in having the integral vanish. In the first case, the resulting configurations form a linear series of ellipsoids of revolution also known as the Maclaurin spheroids while, in the second case, we obtain the Jacobi ellipsoids with three unequal axes.

According to (6.2) and (6.12) the gravitational potential energy is given by

$$V = -\frac{8}{15} \pi^2 G \varrho^2 a^2 b^2 c^2 \int_0^\infty \frac{d\lambda}{\Delta} \tag{6.17}$$

where the last integral can be evaluated explicitly for a Maclaurin spheroid $(a = b)$ giving

$$V = -\frac{16}{15} \pi^2 G \varrho^2 (a b c)^{\frac{5}{3}} (1 - e^2)^{\frac{1}{6}} \frac{\arcsin e}{e}. \tag{6.18}$$

For the Jacobi ellipsoids, the integral in (6.17) must be computed numerically.

The angular momentum H is defined by

$$H = I\Omega = \tfrac{1}{5} M (a^2 + b^2) \Omega \tag{6.19}$$

where the total mass M is given by

$$M = \tfrac{4}{3}\pi\, a\, b\, c\, \varrho = \tfrac{4}{3}\pi R^3 \varrho \quad \text{if} \quad R = (a\, b\, c)^{\frac{1}{3}} \tag{6.20}$$

and where I is the moment of inertia with respect to the z-axis.

Table 1. *Maclaurin spheroids.*
The lines in heavy numbers correspond to the setting in of some kind of instability.

e	$\frac{a}{R}$	$\frac{c}{R}$	$\frac{\Omega^2}{2\pi G\varrho}$	$\frac{H}{(GM^3R)^{\frac{1}{2}}}$	$\frac{V}{GM^2R^{-1}}$	$\frac{H^2}{2I}\cdot\frac{R}{GM^2}$
0.0	1.0	1.0	0.0	0.0	− 0.6	0.0
0.2	1.0068	0.9865	0.0107	0.0514	− 0.6	0.0033
0.4	1.0295	0.9435	0.0436	0.1085	− 0.5997	0.0139
0.6	1.0772	0.8618	0.1007	0.1804	− 0.5974	0.0351
0.8127	**1.1973**	**0.6976**	**0.1868**	**0.3035**	**− 0.5850**	**0.0803**
0.85	1.236	0.655	0.2028	0.3382	− 0.5792	0.0935
0.90	1.3189	0.5749	0.2203	0.4000	− 0.5660	0.1150
0.91	1.3411	0.5560	0.2225	0.4156	− 0.5621	0.1200
0.9299	**1.3957**	**0.5134**	**0.2247**	**0.4524**	**− 0.5520**	**0.1313**
0.94	1.4311	0.4883	0.2239	0.4748	− 0.5453	0.1376
0.95	1.4740	0.4603	0.2213	0.5008	− 0.5370	0.1443
0.9529	**1.4884**	**0.4514**	**0.2201**	**0.5092**	**− 0.5342**	**0.1463**
0.96	1.5286	0.4280	0.2160	0.5319	− 0.5262	0.1514
0.97	1.6023	0.3895	0.2063	0.5692	− 0.5116	0.1577
0.98	1.7128	0.3409	0.1890	0.6249	− 0.4898	0.1664
0.99	1.9210	0.2710	0.1551	0.7121	− 0.4509	0.1772
1.0	∞	0.0	0.0	∞	0.0	0.0

Table 2. *Jacobi elllipsoids.*

$\frac{a}{R}$	$\frac{b}{R}$	$\frac{c}{R}$	$\frac{\Omega^2}{2\pi G\varrho}$	$\frac{H}{(GM^3R)^{\frac{1}{2}}}$	$\frac{V}{GM^2R^{-1}}$	$\frac{H^2}{2I}\,\frac{R}{GM^2}$
1.197	1.197	0.698	0.187	0.304	− 0.585	0.0806
1.279	1.123	0.696	0.186	0.306	− 0.584	0.0808
1.383	1.045	0.692	0.181	0.313	− 0.581	0.0815
1.59	0.93	0.679	0.167	0.338	− 0.563	0.0841
1.601	0.924	0.677	0.166	0.341	− 0.561	0.0851
1.8858	**0.8150**	**0.6507**	**0.1420**	**0.3896**	**− 0.552**	**0.0899**
1.899	0.811	0.649	0.141	0.392	− 0.551	0.0901
2.346	0.702	0.607	0.107	0.481	− 0.519	0.0965
3.129	0.588	0.543	0.066	0.639	− 0.467	0.1007
5.041	0.452	0.439	0.026	1.009	− 0.355	0.0994
∞	0.0	0.0	0.0	∞	0.0	0.0

The preceding formulae enable one to compute the characteristics of ellipsoids belonging to the two series and some examples are reproduced in the following tables where only non-dimensional ratios are given, an actual configuration being fixed by the values of R and M (or ϱ).

To obtain information on the stability of these configurations, we may use the procedure discussed in Sect. 4 and map the linear series contained in Tables 1 and 2 in an appropriate plane (μ, q_1). Physically, we want to know at which stages in the course of the contraction of a given mass of constant angular momentum H, instability sets in. Then the obvious parameter of the problem is the density ϱ. But one verifies easily that one may as well keep ϱ a constant and let H increase.

In presence of constraints allowing only configurations of revolution, the Maclaurin spheroids may be represented in the plane $(\mu = H, q_1 = e)$. In that case, the diagram fails to reveal any instability, the configuration evolving from a sphere for $H=0$ to an infinite flat disk for $H=\infty$. In this respect, let us note that if Ω had been chosen as parameter, ϱ remaining constant (PLATEAU'S experiences), then, as shown on Fig. 3, secular instability would set in at the point where Ω reaches its maximum ($\Omega^2/2\pi G\varrho=0.2247$). For higher values of $\Omega^2/2\pi G\varrho$, there are no equilibrium states at all. One notices that this maximum value is appreciably smaller than the upper limit given by (6.9) or even (6.10).

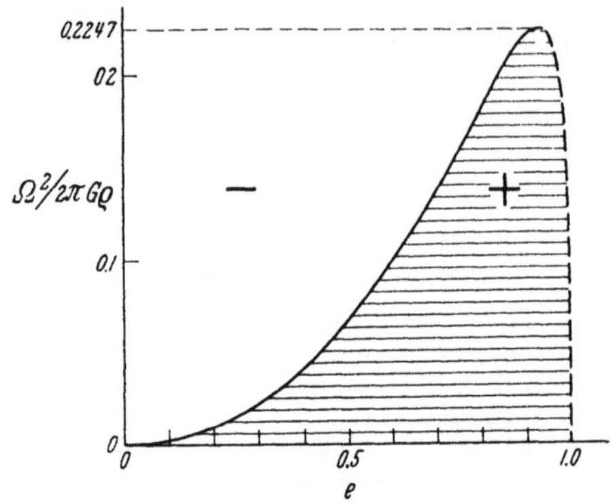

Fig. 3. Representation of the Maclaurin spheroids in a diagram ($\Omega^2/2\pi G\varrho$, e). The linear series starts in (0, 0) with the sphere which is stable. Thus the first part of the linear series is also stable and this fixes the sign of $\partial\psi_R/\partial e$ on both sides of the curve. Instability sets in after the maximum $\Omega^2/2\pi G\varrho=0.2247$.

If we lift some of the constraints so as to allow general ellipsoidal configurations with three unequal axes, the Jacobi ellipsoids come into competition with the Maclaurin spheroids and Fig. 4 shows that they form a separate branch which cuts the Maclaurin series in $e=0.8127$, the corresponding point $a/b=1$ on the Jacobi series being a critical point through which H goes through a minimum $H=0.304\,G^{\frac{1}{2}}M^{\frac{3}{2}}R^{\frac{1}{2}}$. The two symmetrical parts of the Jacobi series with respect to the line $a/b=1$ correspond simply to the same ellipsoid turned by $\pi/2$.

At this bifurcation point, the secular stability of the Maclaurin spheroids is transferred to the Jacobi ellipsoids which, in presence of the appropriate constraints, remain stable from the minimum value of H just given to $H=\infty$, the configuration evolving along both branches from a spheroid to an infinite cylinder.

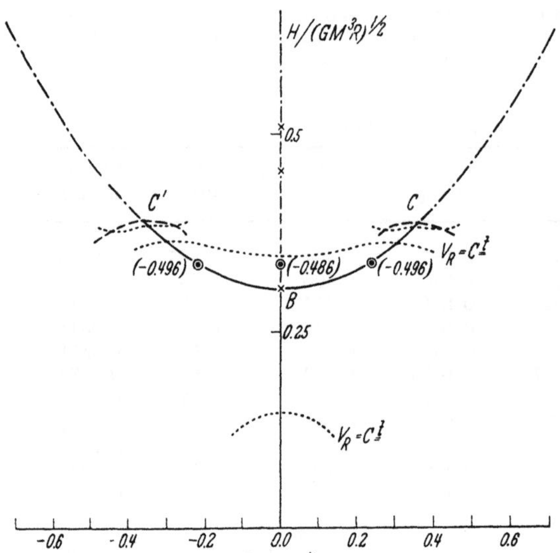

Fig. 4. Qualitative illustration of the linear series of equilibrium configurations in the plane (H, log a/b). The stable configurations are represented by full lines, the secularly unstable ones, by dashed lines and those both secularly and dynamically unstable by a mixed line (—•—). The numbers in brackets give the value of $V_R=\left(V+\frac{H^2}{2I}\right)$ at the corresponding points (cf. lines $H=0.338$ in Tables 1 and 2) so that the curves $V_R=c^t$ in this plane are as illustrated by the dotted lines, confirming the exchange of stability at the bifurcation point B where $H=0.304\cdot(GM^3R)^{\frac{1}{2}}$. New bifurcation points appear in C and C' and correspond to pear-shaped configuration series which however are secularly unstable from the start.

However, if the constraints are further relaxed so that other forms of equilibrium than ellipsoids are allowed, new bifurcation points appear in C and C' corresponding to the pear-shaped configurations. But according to LIAPOUNOFF and JEANS these series, as shown qualitatively on the figure, are secularly unstable from the beginning as they start deviating from C towards regions of lower potential energy.

This qualitative discussion is confirmed by the detailed investigation of POINCARÉ's stability coefficients $v_{R,i}$ defined in Sect. 5.

7. Detailed discussion of the stability of rotating liquid masses and fission-theory. This implies rather complicated algebra, but we shall be content to underline the argument which is most adequately developed in terms of ellipsoidal harmonics. These are special solutions of LAPLACE's equation separated in ellipsoidal coordinates (λ, μ, ν) whose values for a point (x, y, z) are the three real roots $\lambda = \vartheta_1$, $\mu = \vartheta_2$, $\nu = \vartheta_3$ of

$$\frac{x^2}{a^2+\vartheta} + \frac{y^2}{b^2+\vartheta} + \frac{z^2}{c^2+\vartheta} = 1. \tag{7.1}$$

One verifies easily that

$$\lambda > -c^2 > \mu > -b^2 > \nu > -a^2, \tag{7.2}$$

so that the λ-surfaces are ellipsoids, the μ-surfaces are hyperboloids of one sheet and the ν-surfaces are hyperboloids of two sheets.

If LAPLACE's equation is expressed in these coordinates and a solution of the form

$$\Phi = L(\lambda)\, M(\mu)\, N(\nu) \tag{7.3}$$

is sought for, one finds that all three functions must verify LAMÉ's differential equation

$$A\frac{d}{ds}\left(A\frac{dY}{ds}\right) - (B s + C)\, Y = 0 \tag{7.4}$$

where B and C are separation constants and

$$A(s) = [(a^2+s)(b^2+s)(c^2+s)]^{\frac{1}{2}}.$$

The variables Y and s may successively take the values L and λ, M and μ, N and ν. When B is of the form

$$B = \tfrac{1}{4} n(n+1) \qquad \text{with } n \text{ an integer}, \tag{7.5}$$

Eq. (7.4) admits as solutions, polynomials $f(s)$ of degrees related to n multiplied by a factor

$$(s+a)^{r_1}(s+b)^{r_2}(s+c)^{r_3} \tag{7.6}$$

where the exponents r_1, r_2 and r_3 are each independently equal to 0 or $\frac{1}{2}$ provided C takes a certain number of permissible values, this number depending also on n.

Altogether, counting the different solutions arising from the different combinations of r_1, r_2 and r_3 and the number of possible values of C, it is found that Eq. (7.4) where B is given by (7.5) admits $(2n+1)$ independent solutions of the form considered and known as LAMÉ's functions of order n. They could each be represented by Y_n^m where m is a formal superscript taking $(2n+1)$ values corresponding to the different combinations referred to above. The product (7.3) of the corresponding functions L, M, N is a polynomial of degree n in (x, y, z) known as the LAMÉ polynomial of order n or the solid ellipsoidal harmonic of degree n.

Among the ellipsoidal coordinates, $\lambda = \text{const}$ defines an ellipsoid going through the point (x, y, z), and its value increases with the distance r from this point to the origin tending to $\lambda \propto r^2$ as $\lambda \to \infty$. On the other hand, μ and ν are always bounded [cf. Eq. (7.2)] and for a given λ, they describe a point moving on the surface of the corresponding ellipsoid. Thus, there is an analogy with spherical coordinates, λ corresponding to r and μ and ν corresponding to ϑ and φ. In the same way, there is an analogy between the ellipsoidal harmonics and spherical harmonics, $L(\lambda)$ corresponding to r^n and the product MN corresponding to $P_n^m(\cos\vartheta)\, e^{\pm i m \varphi}$.

In fact, Eq. (7.4) being of the second order admits a second independent solution $S(s)$ which is related to $Y(s)$ by

$$S(s) = Y(s) \int\limits_s \frac{(2n+1)\,ds}{2Y^2(s)\sqrt{(a^2+s)(b^2+s)(c^2+s)}}$$

which, in the case $s=\lambda$, has the asymptotic form

$$S(\lambda) = r^{-(n+1)}, \qquad \lambda \to \infty.$$

Thus the general solution of LAPLACE's equation in ellipsoidal coordinates (λ, μ, ν) may be written

$$\sum_{n,m} \{\alpha_{1,n}^m L_n^m(\lambda) + \beta_{1,n}^m S_n^m(\lambda)\}\{\alpha_{2,n}^m M_n^m(\mu) + \beta_{2,n}^m S_n^m(\mu)\}\{\alpha_{3,n}^m N_n^m(\nu) + \beta_{3,n}^m S_n^m(\nu)\}$$

in complete analogy with the corresponding solution in spherical coordinates

$$\sum_{n,m} \{\alpha_{1,n} r^n + \beta_{1,n} r^{-(n+1)}\}\{\alpha_{2,n}^m P_n^m(\cos\vartheta) + \beta_{2,n}^m Q_n^m(\cos\vartheta)\}\{\alpha_3^m \cos m\varphi + \beta_3^m \sin m\varphi\}.$$

Just as surface spherical harmonics are orthogonal on the surface of the sphere, so are the ellipsoidal surface harmonics MN on the surface of an ellipsoid

$$\iint\limits_E D(MN)_n^m (MN)_{n_1}^{m_1}\, dS = 0 \qquad \text{for } n \neq n_1 \text{ or } m \neq m_1 \tag{7.7}$$

where

$$a\,b\,c\,D = \frac{a\,b\,c}{\sqrt{\mu\,\nu}}$$

is the distance from the origin to the tangent plane in (μ, ν).

Any continuous function F of (μ, ν) on the surface of an ellipsoid can be expressed as a linear combination of the surface harmonics MN

$$F = \sum_{n,m} A_n^m M_n^m N_n^m \tag{7.8}$$

where the A_n^m are defined by

$$A_n^m \iint\limits_E D(M_n^m N_n^m)^2\, dS = \iint\limits_E D F M_n^m N_n^m\, dS. \tag{7.9}$$

A function $F(\mu, \nu)$ being given on the ellipsoid $\lambda=0$, the harmonic function which reduces to F on this surface is given by

$$\Phi = \sum \alpha\, S(0)\, LMN \qquad \text{inside the ellipsoid}$$

and

$$\Phi = \sum \alpha\, L(0)\, SMN \qquad \text{outside}$$

where indices have been dropped to facilitate the notation, but it must be understood that the summation is extended to ellipsoidal harmonics of all orders and ranks. The constants α are related to the coefficients A of the expansion (7.8) of F by

$$\alpha = \frac{A}{L(0)\, S(0)}.$$

This enables one to write down the potential of a surface distribution of unit density and variable depth ξ given by

$$\xi = D \sum \beta MN, \tag{7.10}$$

as

$$\left.\begin{aligned} \Phi &= \sum \frac{4\pi\beta}{2n+1} L(0)\, SMN \quad \text{outside}, \\ \Phi &= \sum \frac{4\pi\beta}{2n+1} S(0)\, LMN \quad \text{inside}. \end{aligned}\right\} \tag{7.11}$$

As any infinitesimal perturbation of an ellipsoidal homogeneous configuration can be thought of as resulting from such a surface distribution of matter, it is now possible to evaluate the change in the total potential energy $V_R = V - \frac{1}{2}\Omega^2 I$ of the configuration due to such a perturbation. One finds, taking $abc = 1$

$$\left.\begin{aligned} \delta V_R &= \delta\left(V - \frac{\Omega^2 I}{2}\right) \\ &= 2\pi \sum_{n,m} \left(\frac{1}{3} L_1^3 S_1^3 - \frac{1}{2n+1} L_n^m S_n^m\right)_{\lambda=0} (\beta_n^m)^2 \iint_E D\,(M_n^m N_n^m)^2\, dS \end{aligned}\right\} \tag{7.12}$$

where the β_n^m are the coefficients in the expansion (7.10) of the perturbation and where L_1^3 and S_1^3 represent the Lamé functions of order 1 $[f(\lambda) = 1]$ associated with $r_1 = 0$, $r_2 = 0$, $r_3 = \frac{1}{2}$ in (7.6).

The β_n^m can be interpreted as the general coordinates q_i associated with the infinite number of degrees of freedom of our continuous system and we see that δV_R is already reduced to a sum of squares so that POINCARÉ's coefficients of stability, now in infinite number, can be written down immediately

$$(v_R)_n^m = \frac{1}{3} L_1^3 S_1^3 - \frac{1}{2n+1} L_n^m S_n^m. \tag{7.13}$$

It may seem somewhat surprising that the total energy V_R corresponding to axes rotating with constant angular velocity is still used here while we have insisted that physically the more significant quantity is $V_R' = \left(V + \frac{H^2}{2I}\right)$. However, it can be shown that, for all harmonics of order higher than two in the perturbation (7.10), the corresponding variation δI is of the second order of smallness and, in that case, one verifies immediately that $\delta\Omega$ is also of the second order and

$$\delta\left(V + \frac{H^2}{2I}\right)_H = \delta\left(V - \frac{\Omega^2 I}{2}\right)_\Omega.$$

Moreover, the particular second order harmonics of importance here either have the same property or correspond to a deformation which keeps to the configuration the form of an ellipsoid with the same principal axes as the original one and this case has been covered thoroughly by our previous graphical discussions.

Finally, the three harmonics of order 1 correspond to displacements of the whole mass along one of the axes and they have no meaning here and can be excluded from the expansion (7.10) of the perturbation.

Altogether, we see that, for all significant harmonics in (7.10), the stability discussion may rest on the study of V_R without risking to introduce any spurious instability corresponding to a first order change in Ω (if such changes were possible, their effects viewed from axes rotating with a constant velocity might, through the continuous increase of some non-palpable coordinate, give the impression of instability). Thus, true secular instability will occur each time that one of the $(v_R)_n^m$ given by (7.13) vanishes.

The study of this condition for the Maclaurin spheroids shows that the first stability coefficient to vanish corresponds to the second order ellipsoidal harmonics which in the case $a = b$, can be written

$$L_2^2 M_2^2 N_2^2 = (a^2 + \lambda)\,(a^2 + \mu)\, e^{\pm i m \varphi}.$$

This implies a deformation of the spheroid into an ellipsoid with unequal axes so that this instability corresponds to the bifurcation point occurring at $e = 0.8127$ where the stability is transferred to the Jacobi ellipsoids.

The same study can now be repeated for the Jacobi ellipsoids and it is found that, as the Jacobi series is described in the direction of increasing angular momentum, the coefficients of stability corresponding to harmonics of order 3, 4, 5, ... vanish successively and that, for a given order n, only one of the $(2n+1)$ coefficients of stability vanishes fixing completely the ratios $a:b:c$. As before we are only interested in the first one to vanish (i.e. one of the harmonics of order 3) and DARWIN has shown that this occurs for

$$a:b:c = 1.8858:0.8150:0.6507$$

which corresponds to the bifurcation point of the Jacobi series indicated in Table 2.

It can also be shown that the implicated third order harmonic is associated with the Lamé function

$$L = \sqrt{\lambda + a^2}\,\{\lambda + \tfrac{1}{5}(a^2 + 2b^2 + 2c^2) + \tfrac{1}{5}(a^4 + 4b^4 + 4c^4 - 7b^2c^2 - c^2a^2 - a^2b^2)^{\frac{1}{2}}\}$$

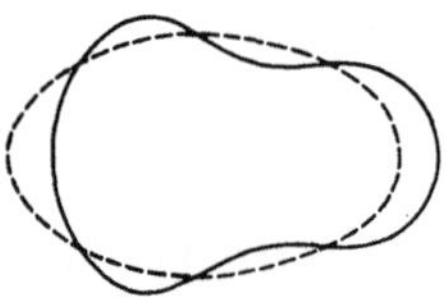

Fig. 5. The dashed line represents a section through the longest axis of the last stable Jacobi ellipsoid and the full line, its deformation by the third zonal ellipsoidal harmonic through which secular instability enters first.

and corresponds to a deformation of the Jacobi ellipsoid as indicated on Fig. 5. This is the famous pear-shaped configuration, on which the furrow was supposed to deepen continuously leading, in the end, to a new series of stable configurations in which the primitive mass is divided into two detached bodies.

However we have not mentioned up to now the question of dynamical stability. As we have seen, in the case of rotating bodies, this type of stability does not coincide with secular stability and its existence or otherwise, after secular stability is lost, can only be decided by the study of the frequencies of the possible infinitesimal oscillations of the system. In case of a system with a finite number of degrees of freedom this means determining the eigenvalues of a matrix as in (5.5). But for continuous systems (infinite number of degrees of freedom) such as those considered here, the eigenvalues are generally defined by linear differential equations whose solutions must satisfy given boundary conditions. Here however as shown by CARTAN[1], the use of ellipsoidal harmonics analysis can still reduce the problem to the resolution of an infinite number of determinants, one of order $(2n+1)$ for each component harmonic of order n.

In the case of the Maclaurin spheroids, the problem had already been solved by BRYAN[2] who showed that they remain dynamically stable until their excentricity reaches the value $e = 0.9529$ corresponding to the last line in heavy type in Table 1. But, as far as the Jacobi ellipsoids are concerned, the answer was only provided more recently by CARTAN's work referred to above in which it was established that dynamical instability sets in at the same time as secular instability for the same harmonic of the third order.

This implies exponentially increasing displacements of quite a different nature as those corresponding to secular instability. They will lead to a state of dynamical motions very difficult to follow in details, but its consequences will certainly be extremely different from those contemplated by DARWIN and JEANS. In particular, one may conclude that the mass will certainly not split itself into two bodies of the same order of magnitude gravitating around each other on stable orbits. Thus, even if the liquid state would bear some resemblance to the

[1] E. CARTAN: Bull. Sci. math. **46**, 332 (1922). — Proc. Int. Math. Congress at Toronto 1924, II, 9 (1928).

[2] G.H. BRYAN: Phil. Trans. roy. Soc. Lond., Ser. A **190**, 187 (1887).

actual state of stellar matter (which is out of the question) these purely dynamical results would be sufficient to discard definitely the "fission" theory as an explanation for the formation of double stars (cf. [4a] for further detailed comments).

8. Some effects of compressibility. α) *General considerations.* If the incompressible case has been given so much attention in the preceding sections it is because it affords a clear illustration of the methods involved in stability discussions. But, as we have already insisted, it is hopelessly inadequate to represent actual stars. Even if we neglect many of the physical aspects of stellar structure such as the generation of energy inside the star and its transfer through the star and all forms of viscosity (molecular, radiative, turbulent) the mere compressibility of the material is sufficient to render the problem very complex.

One may distinguish two aspects in the effects of compressibility, the heterogeneity which results from the increase of the density ϱ with the pressure p in going from the surface toward the centre of any equilibrium configuration and the existence of variations of density $\delta\varrho$ associated with those of pressure δp in the course of any small deformation. For a physical fluid and more particularly for a perfect gas, the resulting "geometrical" relation between ϱ and p characterizing the heterogeneity may be quite different from the physical relation between $\delta\varrho$ and δp in a modification of the fluid. The simplest case, when these two relations are identical, corresponds to a barotropic fluid in which

$$p = f(\varrho). \tag{8.1}$$

As far as rotation is concerned, one may still show[1] that, just as in the case of a homogeneous incompressible mass, a fluid obeying the relation (8.1) can reach a state of relative equilibrium only if it rotates around a fixed axis. This remains true even if it is submitted to the gravitational attraction of other bodies having fixed or variable positions.

In the case of a more general fluid, such as a gas in which the equation of state depends also on the temperature T

$$p = \frac{R}{\bar{\mu}}\,\varrho T \tag{8.2}$$

where $\bar{\mu}$ is the mean molecular weight, the situation is more complex. In a state of relative equilibrium, taking the curl of (6.13), we find that the surfaces of constant density must again coincide with the isobaric and isothermal surfaces if the chemical composition is constant.

However, as was shown by VON ZEIPEL [8], this is incompatible with any plausible generation of energy inside the star. Since then, many investigations[2] have been devoted to elucidating the possible states of stationary internal motions resulting from this incompatibility. The general conclusion seems to be that

---

[1] P. APPELL: Acta math. **47**, 15 (1926); cf. also [2], Sect. 96.

[2] A. S. EDDINGTON: Observatory **48**, 73 (1925); also The Internal Constitution of the Stars, p. 285. Cambridge: Cambridge University Press 1926. — S. ROSSELAND: Astrophys. Norv. **2**, 173, 249 (1937). — G. RANDERS: Astrophys. Norv. **3**, 98 (1939). — I. K. CSADA: Contr. from the Konkoly Obs. Budapest, No. 19, 1948 and No. 22, 1949. — P. A. SWEET: Monthly Notices Roy. Astronom. Soc. London **110**, 548 (1950). — E. J. ÖPIK: Monthly Notices Roy. Astronom. Soc. London **111**, 94 (1951). — L. MESTEL: Monthly Notices Roy. Astronom. Soc. London **113**, 716 (1953). — Astrophys. Journ. **126**, 550 (1957). — ST. TEMESVARY: Z. Naturforsch. **7a**, 105 (1952). Cf. also H. FABRE: Ann. Obs. Toulouse **17**, 91 (1945); **19**, 16 (1949). These papers contain an interesting discussion of the steady state of a rotating mass based on the principle of minimum dissipation.

the equipotential (and isobaric) surfaces will cease to coincide with the isothermal and isosteric surfaces and the resulting solenoidal field will maintain meridional circulations. These in turn will be deflected adding a component to the primitive zonal circulation, the process leading in the end to a definite stationary state if viscosity is taken into account. However, a complete theory is still lacking and the main reliable results concern only the case of very slow rotation and consist mainly in an evaluation of the speed of the meridional currents which are very slow. But this is far from what we would actually need to extend the previous discussion of stability in presence of fast rotation.

Instead of trying to solve the problem of the stationary state of an actual rotating star, one might tackle the problem more formally looking for instance for the possible states of stationary motion when only compressibility is taken into account. But, in that case, the problem remains very indefinite[1] and it is difficult to pick out among the possible formal solutions those that are physically significant.

This difficulty persists even if one limits oneself to barotropic fluids verifying (8.1) and to permanent rotations in which a given particle describes always the same circle around an axis of rotation common to all particles. Moreover such configurations may depend on more than one parameter since, alongside the value of the mean rotation, other parameters are needed to describe the variation of the angular velocity in the star. Finally, there are no axes with respect to which the configuration may be considered as in relative equilibrium so that the stability discussion will exhibit some of the typical features appearing in the case of essentially dynamical systems, making it more delicate.

Altogether, it would seem that the best approach would still be to assume relative equilibrium, imposing a uniform rotation to the whole mass and trying to clarify first the effects of the heterogeneity. But we encounter immediately a new difficulty concerning the representation of the equipotential surfaces which, as we have seen before, coincides with the isobaric and isosteric surfaces in the case of relative equilibrium.

To see this, let us transform the Laplacian of the generalized potential (6.6) according to

$$\nabla^2 \Phi_R = -\frac{dg}{dn} + c\,g \tag{8.3}$$

where g is the value of gravity (i.e. gravitational plus centrifugal forces), n is the normal to the level surface $\Phi_R = \text{const}$, and c represents twice the mean curvature of this surface. In this relation, all normals are counted positively in the direction of g. POISSON's equation (6.5) now becomes

$$\frac{dg}{dn} - c\,g = -4\pi G \varrho + 2\Omega^2 \tag{8.4}$$

which is BRUNS' relation.

On the other hand, if we consider two lines of forces denoted ϑ_1 and ϑ_2 orthogonal to a set of level surfaces $\Phi(\lambda)$ where λ is a parameter, we have on each surface

$$(g\,N)_{\vartheta_1} = (g\,N)_{\vartheta_2} \tag{8.5}$$

where

$$N(\lambda, \vartheta) = \frac{dn}{d\lambda}$$

with dn positive in the direction of λ increasing.

[1] Cf. for instance (a) Ref. [2]. — (b) P. DIVE: Rotations internes des astres fluides. Paris: Blanchard 1930. — (c) V. VOLTERRA: Rotation des corps dans lesquels existent des mouvements internes. Paris: Gauthier-Villars 1938.

Taking the derivative of (8.5) with respect to λ and eliminating dg/dn with the help of (8.4), WAVRE ([2], Sects. 31 and 34) finds the following condition for the stratification in the fluid

$$\frac{4\pi G\varrho - 2\Omega^2}{d\Phi(\lambda)/d\lambda} = \frac{[c\,N + d\ln N/d\lambda]_{\vartheta_2}^{\vartheta_1}}{[N^2]_{\vartheta_2}^{\vartheta_1}} \tag{8.6}$$

and he shows, using POINCARÉ's condition (6.8), that (8.6) is never verified for a stratification along homothetic surfaces whatever their form [except, in absence of rotation, the case of spheres for which the second member of (8.6) becomes indeterminate]. In the case of ellipsoids this had already been shown by VOLTERRA as early as 1903.

On the other hand, suppose that we write the equation of the surfaces as

$$F(x, y, z, \lambda) = 0 \tag{8.7}$$

yielding

$$\frac{\partial\lambda}{\partial x} = -\frac{\frac{\partial F}{\partial x}}{\frac{\partial F}{\partial\lambda}} \quad \text{and} \quad \frac{\partial^2\lambda}{\partial x^2} = -\frac{\frac{\partial^2 F}{\partial x^2} + 2\frac{\partial^2 F}{\partial x\,\partial\lambda}\frac{\partial\lambda}{\partial x} + \frac{\partial^2 F}{\partial\lambda^2}\left(\frac{\partial\lambda}{\partial x}\right)^2}{\frac{\partial F}{\partial\lambda}} \tag{8.8}$$

etc. Since the equipotential surface $\Phi(\lambda)$ must coincide with the level surfaces, the derivatives (8.8) keep the same values on both types of surfaces so that POISSON's equation may be transformed in the following condition (cf. [2], Sects. 35, 36)

$$\nabla^2 F + Q\frac{\partial F}{\partial\lambda} - P R - \frac{\partial R}{\partial\lambda} = 0 \tag{8.9}$$

where

$$P = \frac{d^2\Phi/d\lambda^2}{d\Phi/d\lambda}, \qquad Q = \frac{(-4\pi G\varrho + 2\Omega^2)}{d\Phi/d\lambda},$$

$$R = \left[\left(\frac{\partial F}{\partial x}\right)^2 + \left(\frac{\partial F}{\partial y}\right)^2 + \left(\frac{\partial F}{\partial z}\right)^2\right] \Big/ \frac{\partial F}{\partial\lambda}.$$

If it is assumed that F is the equation of an ellipsoid, it is easy to verify that (8.9) implies that the three axes be proportional to a common parameter which can be identified with λ. But this leads to homothetic surfaces which are excluded by the condition (8.6). Thus ellipsoidal level surfaces are impossible in a heterogeneous configuration in relative equilibrium. This result was first proved by HAMY in 1887 for the case of a mass composed of n ellipsoidal homogeneous shells of different densities and generalized later by VOLTERRA, VERONNET and DIVE.

Of course, if the rotation is very weak so that terms in Ω^2 can be considered as small quantities of the first order, all those of higher orders being neglected, it is well known that ellipsoidal configurations with a stratification along ellipsoidal level surfaces are possible (cf. for instance [*4d*], Chap. IV). The theory of these approximate equilibrium configurations has received considerable developments because of its interest for the Earth and it may also find applications in the discussion of the first order effects of rotation on stellar structure, but it is not of much help where the effects of rotation on the overall stability of the star are concerned.

β) Roche's model and the adiabatic model. There is however a very simple model, ROCHE's model, where all the mass is practically concentrated at the centre, which admits a complete analytical treatment and is generally considered as giving at least qualitative indications as to the behaviour of compressible masses. In a certain sense, this model represents the opposite extreme to the homogeneous model.

Indeed, if for the present we exclude rotation and external gravitational fields, the equilibrium configuration being then spherical, the mean density $\bar{\varrho}(r)$ in the sphere of radius r is always related to the density ϱ by

$$3\varrho(r) = 3\bar{\varrho}(r) + r\bar{\varrho}'(r), \tag{8.10}$$

$$0 \geq \frac{r\bar{\varrho}'(r)}{\bar{\varrho}(r)} \geq -3. \tag{8.11}$$

The upper limit in (8.11) corresponds to the homogeneous model, and the lower limit to a model in which all the mass is concentrated at the centre. One could of course approach the last model by setting

$$\frac{r\bar{\varrho}'(r)}{\bar{\varrho}(r)} = -3 + \varepsilon \tag{8.12}$$

where ε is a small positive quantity.

If ε is constant, one gets laws of density distribution leading to infinite density but no rest mass at the centre. But if ε is variable with r, a central rest mass M_c may occur. For instance when

$$\varepsilon = \frac{4\pi a r}{M_c + 4\pi a r}, \tag{8.13}$$

where a is a small number, (8.10) and (8.12) yield

$$\varrho(r) = \frac{a}{r^2} \tag{8.14}$$

and

$$\bar{\varrho}(r) = \frac{M_c + 4\pi a r}{(4\pi r^3/3)}, \tag{8.15}$$

giving for the whole star

$$\bar{\varrho}(R) = \frac{M_c + 4\pi a R}{(4\pi R^3/3)} = \frac{M}{\mathcal{V}} \tag{8.16}$$

where M is the total mass and $\mathcal{V}$ the total volume. This corresponds to a pressure p varying according to

$$p(r) = G M \left\{\frac{a}{3}\left(\frac{1}{r^3} - \frac{1}{R^3}\right) - \frac{3}{2}\,\frac{a^2}{\bar{\varrho}(R)}\left(\frac{1}{r} - \frac{1}{R}\right)^2\right\}. \tag{8.17}$$

If a tends toward zero, the mass of the envelope $M_e = 4\pi a R$ tends also toward zero, while M_c tends toward the total mass M. According to (8.16), one may however consider that the total volume remains constant. If a is sufficiently small, the contribution of the envelope to the gravitational potential at any point may be neglected. Physically, ROCHE's model may be interpreted as referring to a small and intensively dense nucleus surrounded by an atmosphere of negligible density.

If a solid-body rotation of angular velocity Ω is now applied to this model, the envelope will get distorted, but the gravitational potential will remain the same since we neglect the mass in the envelope and, with respect to rotating axes bound to the fluid, we always have

$$\Phi_R = -\frac{G M}{r} - \frac{1}{2}\Omega^2 (x^2 + y^2). \tag{8.18}$$

The equipotential and level surfaces which are symmetrical with respect to the rotation axis and the (x, y) plane, can be represented by

$$\frac{GM}{r} + \frac{\Omega^2}{2}(x^2+y^2) = \frac{GM}{R_P} \tag{8.19}$$

where R_P is the polar radius $(x=y=0)$. For R_P very small, this equation defines a small closed surface differing little from a sphere and one open surface at large distance from the axis of rotation as shown in Fig. 6. As R_P increases, the first surface flattens, its equatorial radius increasing, while the second surface intersects the (x, y) plane along a circle of decreasing radius. In the meridian plane

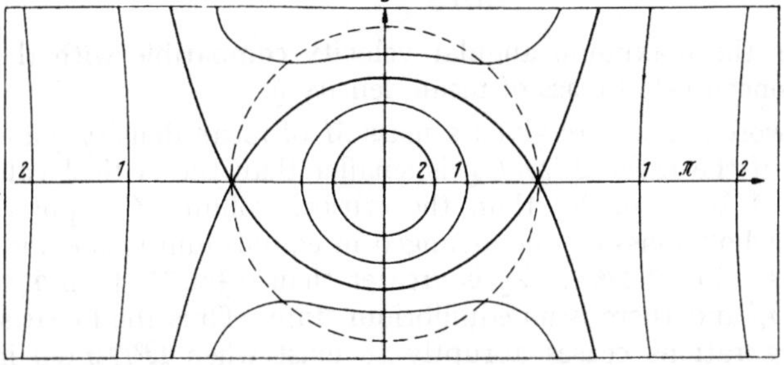

Fig. 6. Equipotentials for Roche's model. The dashed circle is the locus of the points where the centrifugal force balances exactly the horizontal component of the gravitational attraction. The heavy line represents the critical equipotential.

$(z, \tilde{\omega} = \sqrt{x^2+y^2})$, the equation of the section may be written

$$\frac{1}{(z^2+\tilde{\omega}^2)^{\frac{1}{2}}} + \frac{\Omega^2}{2GM}\tilde{\omega}^2 = \frac{1}{R_P} \tag{8.20}$$

or, if we take care to exclude the branch corresponding to a negative value of the radical in (8.20),

$$z^2 = \frac{1}{(\alpha - \beta\,\tilde{\omega}^2)^2} - \tilde{\omega}^2 \quad \text{with} \quad \alpha = \frac{1}{R_P} \quad \text{and} \quad \beta = \frac{\Omega^2}{2GM}. \tag{8.21}$$

We notice that the derivative

$$\frac{dz}{d\tilde{\omega}} = \frac{\tilde{\omega}}{z}\left[2\beta(z^2+\tilde{\omega}^2)^{\frac{3}{2}} - 1\right] \tag{8.22}$$

is always infinite in $z=0$, except if

$$2\beta\,\tilde{\omega}_E^3 = 1 \quad \text{or} \quad \Omega^2 R_E = GM/R_E^2 \tag{8.23}$$

where $R_E = \tilde{\omega}_E$ is the equatorial radius. This defines the critical surface separating the closed equipotentials from the open ones. Physically Eq. (8.23) expresses the equality in absolute value of the centrifugal force and the gravitational attraction along the equatorial edge where the critical equipotential intersects itself.

If we apply (8.20) at the equator and use (8.23), we find that on this critical surface

$$R_E = \tfrac{3}{2} R_P.$$

We notice also that $dz/d\tilde{\omega}$ is always zero on the z-axis and on the circle of radius $r = \sqrt{z^2+\tilde{\omega}^2}$ defined by

$$r_c^3 = \frac{GM}{\Omega^2} = \frac{1}{2\beta}$$

along which the centrifugal force balances exactly the horizontal component of the gravitational attraction.

If M is fixed, the larger Ω^2, the smaller the value of R_P and R_E corresponding to the critical surface and the smaller the volume $\mathcal{V}_C$ enclosed by this surface. The latter can be evaluated fairly easily (cf. [4b], Sect. 152) and one finds

$$\mathcal{V}_C = 4\pi R_E^3 \times 0.180373. \tag{8.24}$$

If the mean density $\bar{\varrho}$ is known, the corresponding critical mass $M_C = \bar{\varrho}\mathcal{V}_C$ introduced in (8.23) yields

$$\frac{\Omega^2}{2\pi G\bar{\varrho}} = 0.36075 \tag{8.25}$$

which defines the maximum angular velocity compatible with the equilibrium of a highly concentrated mass of mean density $\bar{\varrho}$.

In other words, if we are given a mass M of mean density $\bar{\varrho}$ and angular velocity Ω and if the ratio $\Omega^2/2\pi G\bar{\varrho}$ is smaller than the critical value (8.25), the total volume $M/\bar{\varrho}$ is smaller than the critical volume $\mathcal{V}_C$ appropriate to this case and the whole mass can be arranged in equilibrium inside one of the closed equipotentials. But if $\Omega^2/2\pi G\bar{\varrho}$ is greater than 0.36075, then $\mathcal{V} = M/\bar{\varrho}$ will be greater than $\mathcal{V}_C$ and there is no equilibrium state. Thus the linear series of equilibrium configurations ceases abruptly to exist when $\Omega^2/2\pi G\bar{\varrho}$ increases past the limit (8.25).

Although for such configurations, at least in the zero order approximation considered here, it is impossible to show rigorously that $\Omega^2/\bar{\varrho}$ increases when the mass contracts keeping its angular momentum, it is likely to do so for any reasonable physical model resembling the ideal ROCHE's model treated here. In that case, when, in the course of the contraction, $\Omega^2/2\pi G\bar{\varrho}$ reaches the value (8.25), molecules on the equatorial edge cease to belong really to the star and start revolving on circular orbits with an angular velocity equal to the critical value at which they were "liberated" from the star. As the latter shrinks further, $\Omega^2/2\pi G\bar{\varrho}$ for the mass controlled by gravitation becomes larger than its critical value (8.25) which means that the new critical surface does not enclose the whole mass. The excess mass drifts equatorward along equipotential surfaces and is shed through the unstable edge in the equatorial plane.

Whether this will occur continuously or by discrete steps leaving more or less distinct rings of material behind depends on the physical processes underlying the contraction[1].

These considerations can be extended to the case where the central mass instead of reducing to a point is a homogeneous incompressible nucleus (cf. [4b], Sect. 154) of density ϱ_N and volume $\mathcal{V}_N$. Since we neglect the gravitational attraction of the envelope, the equilibrium configuration of the nucleus will be either a Maclaurin spheroid or a Jacobi ellipsoid and the gravitational potential will take the corresponding value. Adding to the latter, the centrifugal potential, the form of the envelope can be discussed.

It is found that as long as the nucleus remains spheroidal there is again a critical surface with a sharp equatorial edge where centrifugal force balances gravity.

If the central nucleus becomes a Jacobi ellipsoid then the critical surface in the envelope takes the form of a pseudo-ellipsoid with pointed ends at the extremities of the long axis.

---

[1] Cf. for instance H. POINCARÉ: Hypothèses cosmogoniques, Chap. III. Paris 1911.

The evolution of a given mass M of mean density $\bar{\varrho}$ can be discussed as before. The instability in the equatorial plane will manifest itself by ejection all along the equator if the critical volume $\mathcal{V}_C$ becomes smaller than the actual volume $\mathcal{V}=M/\bar{\varrho}$ before the nucleus ceases to be spheroidal.

If the equality $\mathcal{V}=\mathcal{V}_C$ is not reached until the nucleus has become a Jacobi ellipsoid then the instability of the envelope will lead to the shedding of streams of material in the equatorial plane through the pointed ends of the critical surface.

One must note however that, if the volume of the envelope $(\mathcal{V}-\mathcal{V}_N)$ is larger than about $\mathcal{V}/3$, the instability always takes the first form.

It is not likely that these considerations will ever be directly applicable to actual stars, however, they suggest that a fairly thin compressible envelope is sufficient for rotational instability to appear first as a shedding of material in the equatorial plane. JEANS has also studied another somewhat more realistic model (cf. [4b], Sects. 166 to 183 and [4c], Sect. 235) often referred to as the adiabatic model because an equation of state of the form

$$p = K\,\varrho^{\gamma} - p_0 \tag{8.26}$$

is assumed. The introduction of a constant p_0 allows one to consider the case where the density does not vanish with the pressure at the external boundary of the configuration.

JEANS again imposed a solid-body rotation to the whole mass and, starting with a series expansion of the density from the centre outward, he was led to adopt for the distribution of density a law of the form

$$\varrho = \varrho_C - (\varrho_C - \varrho_S)\left(\frac{x^2}{a^2} + \frac{y^2}{b^2} + \frac{z^2}{c^2} + e P_0\right) \tag{8.27}$$

where ϱ_C is the central density and ϱ_S the density on the free surface defined by

$$\frac{x^2}{a^2} + \frac{y^2}{b^2} + \frac{z^2}{c^2} - 1 + e P_0 = 0. \tag{8.28}$$

In these equations, eP_0 will be small if the mass is only slightly compressible since in the limit of incompressibility (8.28) must reduce to an ellipsoid ($eP_0=0$) while if the matter is strongly compressible eP_0 may become of the same order as the other terms.

After writing an adequate expression for the potential at a point inside a configuration stratified according to (8.27), JEANS was able to set up a process of successive approximations which, in principle, would permit to take compressibility into account up to any order of accuracy desired. In practice, however, the second step is already very laborious.

Nevertheless, these first two approximations show that, compared to the incompressible case, compressibility increases the critical value of $\Omega^2/2\pi G\bar{\varrho}$ at which the pseudo-spheroids (symmetry of revolution) give place to pseudo-ellipsoids. This increase depends both on the quantity $(\varrho_C-\varrho_S)/\varrho_C$ and the adiabatic exponent γ.

At the same time, the equator is drawn out further from the axis of rotation and, in general, both factors tend to increase the relative importance of the centrifugal force with respect to the gravitational attraction. This suggests that, in some cases, the centrifugal force may balance gravitation, leading to the shedding of material in the equatorial plane, before the configuration has a chance to reach the critical shape through which the pseudo-ellipsoids enter. Supposing that this balance occurs exactly for the critical configuration, we get a condition in which Ω^2 may be given its critical value, so that we are left with a relation between $(\varrho_C - \varrho_S)/\varrho_C$ and γ.

If ϱ_S is supposed to vanish, which is the most significant case, we obtain an equation defining a critical value of γ, say γ_C, such that for $\gamma > \gamma_C$ the centrifugal force is still inferior to the gravitational attraction on the equator of the critical pseudo-spheroid and further evolution will introduce pseudo-ellipsoidal configurations which might finally lead to an instability of the type encountered in the pear-shaped incompressible configurations.

On the other hand, if $\gamma < \gamma_C$, loss of matter through the equator will occur before this critical pseudo-ellipsoid is reached and the further evolution will be of the type encountered in the discussion of ROCHE's model.

Thus it would seem that this value of γ_C estimated by JEANS to be

$$\gamma_C \approx 2.2 \tag{8.29}$$

separates the compressible configurations into two groups following two essentially different patterns of evolution modelled respectively on the behaviour of an incompressible mass $(\gamma > \gamma_C)$ or of ROCHE's model $(\gamma < \gamma_C)$.

However, at present, all the evidence indicates that in stellar conditions γ never exceeds $\frac{5}{3}$ and, qualitatively, we may safely conclude that instability due to rotation will manifest itself in stars by the shedding of matter in the equatorial plane. A quantitative analysis of this process however remains difficult and it may have to take into account the non-uniform rotation and the meridional ur rents of actual compressible stars.

B. Compressible masses.

9. Introduction and general principles. If the fluid is compressible, the thermodynamical aspects of the problem become important since, in the course of a small perturbation, part of the internal thermal energy can be made available in the form of work to increase or decrease the kinetic energy and thus modify the motion. In that case, the simple mechanical idea of minimum potential energy can no longer be used as the criterion for stability. In fact, the potential energy of a fluid takes an ambiguous meaning when the work accompanying changes in density may depend on the rate of the modification and on the heat flow during the process. What we need really if we want to formulate a principle of minimum is a potential characterizing the "available" energy which can be delivered in the form of work in the course of a fluctuation in the distribution of the fluid.

Of course, whether this is possible or not, we can always use the small perturbation method applied directly to the equations of motion. We have already commented on this method in Sect. 2 and recalled there some of its limitations: linearization of the equations, assumption of an exponential time dependence for the displacement $\delta r \propto e^{\sigma^* t}$. However, it would seem that the criticisms of this method have sometimes been exaggerated. We may, with LIN (cf. [*9*], Introduction), say that if all the characteristic values of σ^* have negative real parts, the non-perturbed state is certainly stable with respect to sufficiently small disturbances. If some of the characteristic values σ^* have positive real parts, the initial state is certainly unstable since, if a perturbation is to die out, it must first become so small that its further behaviour would be determined by the linearized equations.

In any case, if the small perturbation method has its drawbacks, it is doubtful whether, in the general conditions contemplated here, the other methods, based on energy considerations, present any real advantages. They consist essentially in trying to estimate whether the total kinetic energy of the perturbing motion

$$K = \iiint \tfrac{1}{2} v^2 \varrho \, d\mathcal{V} \tag{9.1}$$

increases or decreases. If the system is stable in the sense of the small perturbation method, the displacement $|\delta\boldsymbol{r}|$ and the absolute velocity $|\boldsymbol{v}|$ of any part of the system can always be taken smaller than some arbitrary small positive number

$$|\delta\boldsymbol{r}| < \varepsilon, \quad |\boldsymbol{v}| < |\sigma^*| \, \varepsilon, \quad t > t_0 \tag{9.2}$$

so that

$$K < \eta, \quad t > t_0 \tag{9.3}$$

where η is some other small positive number. However the reverse is not necessarily true, since $|\boldsymbol{v}|$ could very well become infinite while K keeps a meaning or even continues to satisfy (9.3). Moreover, if in applying conditions of the type (9.3), one does not want to make any hypothesis on the form of the time dependence of $\delta\boldsymbol{r}$ or $\boldsymbol{v}$ then, even if $|\boldsymbol{v}|$ remains bounded, the same will not necessarily be true of $|\delta\boldsymbol{r}|$.

Of course if K increases beyond any limit or even if it admits a lower positive bound, it is clear that (9.2) can no longer be satisfied everywhere in the system so that (9.3) may be taken as a necessary but not sufficient condition for stability. However it is not a very convenient condition to use, and very often it is replaced by

$$\frac{dK}{dt} \leqq 0, \quad t > t_0 \tag{9.4}$$

which may be considered as a sufficient condition for the necessary condition (9.3) to be verified. But it is not itself a necessary condition since stability could very well be insured without (9.4) being verified for all values $t > t_0$.

Moreover, (9.4) cannot rigorously be considered as a sufficient condition of stability in the sense given to negative values of σ^* in the small perturbation method.

Before condition (9.4) can be used, it is necessary to express the change in kinetic energy in terms of the generalized forces acting on the system. If they admit a generalized potential whose variations are equal to those of $-K$, a given state will be stable, if the generalized potential is an absolute minimum for this state. When this is a necessary as well as a sufficient condition, it is sufficient, to prove that the system is unstable, to show that there exists at least one type of perturbation for which the associated variation of the generalized potential is negative.

It is sometimes easier to establish the existence of such a perturbation than to solve the equations of small motion to determine the actual values of the frequencies. Of course, the latter method yields more information on the details of the motion and on its time-scale but, if one is interested only in knowing, whether the system is stable or not, the first method may be simpler. However, one should note that many of its advantages may also be recovered in the variational interpretation (cf. [*1*], end of Sect. 58) of the linear equation of small motion. Nevertheless, a greater freedom subsits in the energy method as far as the normalization of the perturbation is concerned. A judicious use of this freedom may lead to important analytical simplifications although, of course, the exact values of the frequencies are then lost.

Finally, actual applications generally require some a-priori specification of the form of the perturbation and this is not without danger in boundary value problems. In such cases, there does not seem to be much reason to prefer the energy method to the small perturbation method. According to Villat (cf. [10], Chap. XIV, especially §§6, 7, 35 and 36), the latter may actually be the most reliable.

In the compressible case, only non-rotating masses will be considered so that the shape of the equilibrium configuration is always spherical (cf. Sect. 6). This compensates somewhat for the greater complexity introduced by the compressibility and, anyway as already pointed in Sect. 8, we know too little about the effects of a general rotation on the equilibrium state itself to improve on the qualitative conclusions reached there. The same may be said of the influence of a general magnetic field and, in both cases, we shall limit the discussion to a few remarks at the end.

10. The energy method. As the small perturbation method has been developed at great length and used extensively in [1], there is no need to go into it again here. Instead, we shall expound in some details the energy method which was only touched upon in [1] (cf. Sect. 63β).

The equation governing the variations of the mechanical energy per unit mass may be written (cf. [1] Sect. 50, Eq. (50.3)), if we note that the forces derive from a gravitational potential Φ,

$$\left.\begin{aligned}\frac{d}{dt}\left(\frac{1}{2}v^2\right) &= -\boldsymbol{v}\cdot\operatorname{grad}\Phi - \frac{1}{\varrho}\operatorname{div}(p\,\boldsymbol{v}) + \frac{1}{\varrho}p\operatorname{div}\boldsymbol{v} + \\ &\quad + \frac{1}{\varrho}\operatorname{div}(\boldsymbol{v}\cdot\mathfrak{P}) - \frac{1}{\varrho}\sum \mathfrak{P}^{ik}\Delta_i v_k\end{aligned}\right\} \tag{10.1}$$

where the last term, as in Eq. (50.10) of [1], is always negative and expresses the mechanical energy dissipated into heat by viscosity. In this equation, the pressure p and the tensions $\mathfrak{P}^{ik}$ represent the contributions of the gas and the radiation together. Turbulence, if present, could also be taken into account by the addition of a turbulent pressure p_t and Reynolds stresses $\mathfrak{P}_t^{ik}$ as in Eq. (50.7) in [1]. But in this case, a further equation (50.8) is required to determine the rate of change of the kinetic energy of turbulence and, for the sake of simplicity, we shall assume that turbulence is absent.

If (10.1) is multiplied by dm and integrated over the whole mass, we obtain, noting that $dm = \varrho\, d\mathcal{V}$ is invariant following the motion and that all stresses vanish on the free surface of the star

$$\frac{d}{dt}\int_M \frac{1}{2}v^2\,dm = -\int_M \boldsymbol{v}\cdot\operatorname{grad}\Phi\,dm + \int_M \frac{p}{\varrho}\operatorname{div}\boldsymbol{v}\,dm - \int_{\mathcal{V}} \sum \mathfrak{P}^{ik}\Delta_i v_k\,d\mathcal{V}. \tag{10.2}$$

If V denotes the total gravitational energy, the first integral on the right gives its variation with time dV/dt. The second may be transformed with the help of the equation of continuity (cf. [1], Eq. (45.1)) so that (10.2) becomes

$$\frac{dK}{dt} = -\frac{dV}{dt} - \int_M \frac{p}{\varrho^2}\frac{d\varrho}{dt}\,dm - \int_{\mathcal{V}} \sum \mathfrak{P}^{ik}\Delta_i v_k\,d\mathcal{V}. \tag{10.3}$$

When p is a known function of ϱ (barotropy), the second term on the right can further be reduced directly to the time derivative of

$$\mathcal{U}' = \int_M dm \int_{t_0}^{t} \frac{p}{\varrho^2}\frac{d\varrho}{dt}\,dt = -\int_M dm \int_{\varrho=0}^{\varrho} p\, d\left(\frac{1}{\varrho}\right) \tag{10.4}$$

if the zero of energy is taken to correspond to a state of infinite dilution ($\varrho \to 0$).

However, in general, p depends also on the temperature T and it is necessary to add to (10.3) another equation relating the variations of T to those of the other variables in the course of a real modification. This "supplementary relation" in the sense of P. DUHEM ([3], Vol. II, pp. 26, 80, 216) is provided by the equation of conservation of thermal energy ([1], Sect. 52) which may be written, per unit mass,

$$\left.\begin{aligned}\frac{dQ}{dt} = T\frac{dS}{dt} &= \varepsilon - \frac{1}{\varrho}\operatorname{div}\boldsymbol{F} + \frac{1}{\varrho}\sum_{i,k}\mathfrak{P}^{ik}\Delta_i v_k\\ &= \frac{dU}{dt} + p\frac{d\tau}{dt} = \frac{dU}{dt} - \frac{p}{\varrho^2}\frac{d\varrho}{dt}\end{aligned}\right\} \tag{10.5}$$

where the external heat furnished to the system per unit time comprises here the rate of liberation of subatomic energy ε, the total flux of energy $\boldsymbol{F}$ in the system and the heat set free by internal friction. It may of course be expressed in terms of the rate of change of the entropy S and it must be equal to the variation of the internal thermal energy U plus the work of the pressure p in the course of a modification $d\tau = -d\varrho/\varrho^2$ of the specific volume τ.

In particular, this shows that

$$-p\frac{d\tau}{dt} = \frac{dU}{dt} - \frac{dQ}{dt} = \frac{dU}{dt} - T\frac{dS}{dt} \tag{10.6}$$

is not, in general, an exact total differential since dQ/dt is not. This may however be realized in special cases such as an isothermal transformation when a thermodynamical potential (free energy) may be defined by

$$F = U - TS \tag{10.7}$$

so that the external work against the pressure may be written

$$p\frac{d\varrho}{\varrho^2} = dF. \tag{10.8}$$

In the case of a "normal" system (cf. [3], Vol. I, Chap. VI) each part of which may be characterized by a temperature T and a set of variables (e) defining, independently of T, the "mechanical" state of the system, any transformation may be replaced by an isothermal change of state $(e)_1 \to (e)_2$ followed by a transformation $T_1 \to T_2$ without any necessary change of state and which does not imply any external work. Here, the mechanical state of the unit mass may be defined simply by the density ϱ and U and F may be treated as functions of ϱ and T. In that case, a general transformation will correspond to an amount of work

$$-p\,d\tau = p\frac{d\varrho}{\varrho^2} = dF(\varrho, T) - \left(\frac{\partial F}{\partial T}\right)_\varrho dT \tag{10.9}$$

with

$$\left[\frac{\partial F(\varrho, T)}{\partial T}\right]_\varrho = -S, \quad \left[\frac{\partial F(\varrho, T)}{\partial \varrho}\right]_T = \frac{p}{\varrho^2}. \tag{10.10}$$

Eq. (10.3) may then be written

$$\frac{dK}{dt} = -\frac{d(V+\mathfrak{F})}{dt} + \int\limits_M \left(\frac{\partial F}{\partial T}\right)_\varrho \frac{dT}{dt}\,dm - \int\limits_{\mathscr{V}} \sum \mathfrak{P}^{ik}\Delta_i v_k\,d\mathscr{V} \tag{10.11}$$

where $\mathfrak{F}$ is the total thermodynamical potential defined by

$$\mathfrak{F} = \int\limits_M F\,dm. \tag{10.12}$$

If, taking the supplementary condition (10.5) into account, $(\partial F/\partial T)_\varrho$ becomes a function of T only i.e.

$$\left(\frac{\partial F}{\partial T}\right)_\varrho = - S(\varrho, T) = - \varphi(T), \tag{10.13}$$

we may introduce a new function $\Phi(T)$ defined by

$$\frac{d\Phi(T)}{dT} = \varphi(T)$$

so that

$$\int_M \left(\frac{\partial F}{\partial T}\right)_\varrho \frac{dT}{dt}\, dm = -\frac{d}{dt}\int_M \Phi(T)\, dm = -\frac{d\Psi}{dt}. \tag{10.14}$$

Introducing a function

$$\Lambda = \mathfrak{F} + \Psi \tag{10.15}$$

which may be called the available thermal energy, Eq. (10.11) becomes

$$\frac{d}{dt}(K + V + \Lambda) = -\int_{\mathcal{V}} \mathfrak{P}^{ik} \Delta_i v_k\, d\mathcal{V} < 0 \tag{10.16}$$

or, if viscosity is absent,

$$K + V + \Lambda = E. \tag{10.17}$$

When the equation of conservation of energy reduces to this form which is exactly similar to the purely mechanical energy integral [cf. for instance (3.4), (5.7), and (5.10)] except for the addition of a thermodynamical part Λ to the gravitational potential V, the extension of the methods developed for the mechanical problem is straightforward. In particular, an equilibrium state will certainly be stable towards a sufficiently small perturbation provided that $V + \Lambda$ goes through an absolute minimum for this state.

A simple and fairly common case verifying the condition (10.13) occurs when the entropy of each element of the system remains constant in the course of the motion so that $\varphi(T)$ reduces to a constant say φ which may depend of course on the element considered. In that case,

$$\Lambda = \mathfrak{F} + \int_M \varphi\, T\, dm = \int_M (F + \varphi\, T)\, dm.$$

Using (10.7) and remembering that φ is the value (constant in the course of time) taken by the entropy S in each mass element, this becomes

$$\Lambda = \int_M U\, dm = \mathfrak{U}, \tag{10.18}$$

the available energy being equal, in this case, to the total internal energy.

This suggests that it may be interesting to substitute S to T as an independent variable and write

$$\Lambda = \mathfrak{U}(\varrho, S) = \int_M U(\varrho, S)\, dm. \tag{10.19}$$

It is obvious that if S is constant as above, we have

$$\int_M \frac{p}{\varrho^2}\frac{d\varrho}{dt}\, dm = \frac{d\Lambda}{dt} = \int_M \left(\frac{\partial U}{\partial \varrho}\right)_S \frac{d\varrho}{dt}\, dm = \frac{d}{dt}\int_M (U\, dm)_S = \left(\frac{d\mathfrak{U}}{dt}\right)_S.$$

But taking the relations

$$T = \left(\frac{\partial U(\varrho, S)}{\partial S}\right)_\varrho, \qquad \frac{p}{\varrho^2} = \left(\frac{\partial U(\varrho, S)}{\partial \varrho}\right)_S \tag{10.20}$$

into account, it is easy to verify, either from (10.6) or (10.9), that this is always true. Thus we may always identify Λ with the total internal energy $\mathfrak{U}$ expressed as a function of ϱ and S provided the variation of U be taken for constant S.

Up to now, we have considered only real modifications of the system and we have evaluated the variation of the kinetic energy from the equations of motion considered as given a priori. But one may, in a sense, reverse the procedure and try to derive the equations of motion from some general principle which will, at the same time, embody or, at least, render obvious some of the previous conclusions.

One may, for instance, try to formulate a generalized form of d'ALEMBERT's principle which, by the introduction of the virtual work of the inertial forces per unit mass say

$$\delta\mathfrak{T}_i = -\boldsymbol{\gamma}\cdot\delta\boldsymbol{r} \tag{10.21}$$

where γ is the acceleration, reduces the dynamical problem to a static one. DUHEM [3] has shown in great details that, if $\delta\mathfrak{T}_a$ denotes the virtual work of the applied forces, $\delta\mathfrak{R}$ that of the friction, which is always negative, one has for any virtual modification of the system (i.e. a virtual displacement plus a virtual variation of temperature)

$$\int_M \delta\mathfrak{T}_i\,dm = \delta\mathfrak{F} - \int_M \left(\frac{\partial F}{\partial T}\right)_\varrho \delta T\,dm - \int_M \delta\mathfrak{T}_a\,dm - \int \delta\mathfrak{R}\,dm \tag{10.22}$$

where F and $\mathfrak{F}$ have the same meaning as in (10.11). If the internal friction is not taken into account, (10.22) becomes

$$\int_M \delta\mathfrak{T}_i\,dm \geq \delta\mathfrak{F} - \int_M \left(\frac{\partial F}{\partial T}\right)_\varrho \delta T\,dm - \int_M \delta\mathfrak{T}_a\,dm \tag{10.23}$$

since, in any case,

$$\delta\mathfrak{R} \leq 0.$$

If the applied forces admit a potential, (10.22) and (10.23) can be written

$$\int_M \delta\mathfrak{T}_i\,dm = \delta(\mathfrak{F}+V) - \int_M \left(\frac{\partial F}{\partial T}\right)_\varrho \delta T\,dm - \int_M \delta\mathfrak{R}\,dm \tag{10.24}$$

or

$$\int_M \delta\mathfrak{T}_i\,dm \geq \delta(\mathfrak{F}+V) - \int_M \left(\frac{\partial F}{\partial T}\right)_\varrho \delta T\,dm. \tag{10.25}$$

As in mechanics, the virtual modifications represented by the symbol δ are instantaneous and must be compatible with the definition of the system but, apart from this, they are submitted to no restriction whatsoever and some of them may very well violate the physical laws governing the system. However, the real change occurring in time dt will coincide with one of them and it is easy to verify that, in this particular case, (10.24) becomes equivalent to (10.11) since the work of the inertial forces [cf. Eq. (10.21)] is equal to the decrease of the kinetic energy, $\delta\mathfrak{T}_i = -\delta K$.

If a function Λ representing the available energy may be defined for the system [cf. Eqs. (10.13) and (10.14)], Eqs. (10.22) and (10.23) become

$$\int_M \delta\mathfrak{T}_i\,dm = \delta\Lambda - \int_M \delta\mathfrak{T}_a\,dm - \int_M \delta\mathfrak{R}\,dm \tag{10.26}$$

and

$$\int_M \delta\mathfrak{T}_i\,dm \geq \delta\Lambda - \int_M \delta\mathfrak{T}_a\,dm \tag{10.27}$$

or, if the system admits a mechanical potential energy, V,

$$\int_M \delta \mathfrak{T}_i \, dm = \delta(\Lambda + V) - \int_M \delta \mathfrak{R} \, dm \tag{10.28}$$

and

$$\int_M \delta \mathfrak{T}_i \, dm \geqq \delta(\Lambda + V). \tag{10.29}$$

Any one of these expressions may be used to formulate at least a sufficient condition of stability. For instance, an equilibrium state of a system admitting an available energy Λ is certainly stable provided that, for any small virtual modification of this state, the total work of the applied forces be smaller than the corresponding change in the available energy. Indeed, according to (10.26) or (10.27), this would imply a positive work of the inertial forces for all virtual displacements and thus, for that one which coincides with the real modification, a decrease of the kinetic energy which, being originally zero, should then become negative which is clearly impossible. This may also be interpreted in terms of the work of the fictitious extra-external forces which have to be brought in to realize these virtual displacements. If this work is always positive, the equilibrium state is certainly stable.

If a mechanical potential energy V may be defined as well, this result means according to (10.28) and (10.29), that the equilibrium state is certainly stable if $(\Lambda + V)$ is an absolute minimum for this state.

Since, in the course of the motion of the system from t_0 to t_1, Eq. (10.22) is valid at each instant t, we may also write

$$\int_{t_0}^{t_1} dt \left\{ \int_M \left[\delta \mathfrak{T}_a + \delta \mathfrak{T}_i + \delta \mathfrak{R} + \left(\frac{\partial F}{\partial T}\right)_\varrho \delta T \right] dm - \delta \mathfrak{F} \right\} = 0. \tag{10.30}$$

If, furthermore, we assume that the virtual modification vanishes identically in t_0 and t_1, the virtual work of the inertial forces may be transformed by a partial integration with respect to the time giving

$$\int_{t_0}^{t_1} dt \int_M \delta \mathfrak{T}_i \, dm = \int_{t_0}^{t_1} \delta K \, dt .$$

Let us suppose also that the applied forces admit a potential, then (10.30) becomes

$$\int_{t_0}^{t_1} \delta \left(K - V - \mathfrak{F} + \int_M \mathfrak{R} \, dm \right) dt + \int_{t_0}^{t_1} dt \int_M \left(\frac{\partial F}{\partial T}\right)_\varrho \delta T \, dm = 0 \tag{10.31}$$

or, denoting by δ_T a virtual variation in which the temperature does not change

$$\delta_T \int_{t_0}^{t_1} \left(K - V - \mathfrak{F} + \int_M \mathfrak{R} \, dm \right) dt = 0 . \tag{10.32}$$

If the available energy Λ exists, (10.31) may be written directly

$$\delta \int_{t_0}^{t_1} \left(K - V - \Lambda + \int_M \mathfrak{R} \, dm \right) dt = 0 \tag{10.33}$$

or in absence of viscous forces

$$\delta \int_{t_0}^{t_1} (K - V - \Lambda) \, dt = 0 . \tag{10.34}$$

Eqs. (10.31) to (10.34) represent the general forms of Hamilton's principle for this problem (cf. [*3*], Chap. XIII, Sect. 5 and also [*11*], Chap. IX).

A certain care is necessary in writing the explicit expression of the available energy Λ by means of the supplementary relation. While the general expression (10.15) of Λ with the definitions (10.12) and (10.14) is always allowed, all the forms in which it may be transformed with the help of the supplementary relation are not necessarily equivalent when a perfectly general virtual variation is considered. In fact, if the supplementary relation must certainly be verified in any real displacement it is not always possible to recognize, from the mere specification of the state to which the virtual modification is applied, whether the latter verifies the same supplementary relation or not. However if this is possible, then the virtual modifications may be limited to those obeying the supplementary condition and then all forms of Λ are equivalent.

As before, Eqs. (10.33) and (10.34) imply that, at equilibrium, the Lagrangian which reduces to $(V+\Lambda)$ must be an extremum. However it does not seem that, in this connection, HAMILTON's principle presents any real advantages on d'ALEMBERT's principle.

11. The energy method in the stellar problem. As already mentioned in Sect. 2, the only detailed stellar applications of the energy method are due to L. H. THOMAS and R. C. TOLMAN ([5], a and b) who discussed explicitly only the case of purely radial motions. As an illustration, we shall summarize here some of their results following more closely THOMAS' treatment which is slightly more general.

Let us consider a spherical star the state of which varies very slowly in time and let us take the entropy as the parameter whose variations are associated with these slow changes. At any time, the entropy S per unit mass is a definite function of the position inside the star of the element considered. If we have in mind purely radial motions, the mass inside the sphere of radius r is a convenient Lagrangian variable to characterize this position. Then at each instant, S is a definite function of m, but its rate of variation given by (10.5) depends on the state of the system at the time considered and perhaps also on the time explicitly. Thus m and t may be taken as independent variables and r and S as dependent variables while the other thermodynamical variable ϱ is related to m and r by

$$\varrho = \frac{1}{4\pi r^2}\frac{dm}{dr} = \frac{1}{4\pi r^2 r'} \tag{11.1}$$

which is the simple form taken by the equation of continuity in this case.

We have here

$$K = \int\limits_M \frac{1}{2}\dot{r}^2\,dm, \qquad \Lambda = \int\limits_M U(\varrho, S)\,dm, \qquad V = -\int \frac{G m\,dm}{r} \tag{11.2}$$

so that, neglecting viscosity, HAMILTON's principle (10.34) gives

$$\delta \int\limits_{t_0}^{t_1}\int\limits_0^M \left[\frac{1}{2}\dot{r}^2 - U(\varrho, S) + \frac{Gm}{r}\right] dm\,dt = 0 \tag{11.3}$$

it being understood that in taking the variation of U, S must be treated as a constant [cf. Eq. (10.19) and the following discussion] while ϱ must be considered as a function of r and m as defined by (11.1). In these conditions, one verifies that the Euler condition for an extremum, i.e.

$$\frac{d}{dt}\left(\frac{\partial L}{\partial \dot{r}}\right) + \frac{d}{dm}\left(\frac{\partial L}{\partial r'}\right) - \frac{\partial L}{\partial r} = 0$$

where L represents the integrand in (11.3) and $r' = dr/dm$, gives

$$\ddot{r} + 4\pi r^2 \frac{d}{dm}\left[\left(\frac{\partial U}{\partial \varrho}\right)_S \varrho^2\right] + \frac{Gm}{r^2} = 0 \tag{11.4}$$

which, as may be checked immediately taking the second relation (10.20) into account, is the ordinary Lagrangian equation for radial motions.

Moreover it is not a fixed end point problem since the variation of the dependent variable r does not necessarily vanish in $m = M$ and we must add to the Euler equation the end conditions[1]

$$\left[\frac{\partial L}{\partial r'}\right]_{m=0}^{m=M} = \left[-4\pi r^2 \varrho^2 \left(\frac{\partial U}{\partial \varrho}\right)_S\right]_{m=0}^{m=M} = 0 \tag{11.5}$$

which coincide with the physical conditions that r should always be zero in $m = 0$ and that p should vanish at the free surface $m = M$.

The supplementary condition (10.5) may be written, using the first relation (10.20)

$$\left(\frac{\partial U}{\partial S}\right)_\varrho \dot{S} = \varepsilon - \frac{d}{dm}\left[4\pi r^2 F(r)\right] \tag{11.6}$$

if, as above, we neglect the effects of viscosity. Of course, we must add the equation of state

$$p = p_G + p_R = \frac{\mathrm{R}\varrho T}{\bar{\mu}} + \frac{1}{3} a T^4 \tag{11.7}$$

and the definition of the internal energy and the entropy per unit mass for a mixture of gas and radiation

$$U = \frac{a T^4}{\varrho} + \int C_v \, dT, \tag{11.8}$$

$$S = \frac{4}{3} a T^4 + \int \frac{C_v}{T} dT - \frac{\mathrm{R}}{\bar{\mu}} \ln \varrho \tag{11.9}$$

where C_v is the specific heat of the gas at constant volume.

Since we want mainly to illustrate the method, we shall limit ourselves here to the case of radiative equilibrium when

$$F(r) = -\frac{4\pi r^2 c}{3\varkappa} \frac{d}{dm}(a T^4) = -\frac{4\pi r^2 c}{3\varkappa} \frac{d}{dm}\left[a\left(\frac{\partial U}{\partial S}\right)_\varrho^4\right] \tag{11.10}$$

where $\varkappa$ is the coefficient of opacity by unit mass and c the velocity of light.

If the motion starts from an equilibrium configuration ($\ddot{r}_0 = 0$), three types of *initial* motions satisfying the boundary conditions are possible:

1. A motion during which S remains constant (adiabatic perturbation) which may take the form of an oscillation (dynamical stability) or of a displacement increasing exponentially with the time (dynamical instability). As can be seen directly from (11.4), the square of the period or of the time needed for an increase of the displacement by a factor e is proportional to the ratio of the moment of inertia to the total potential energy or inversely proportional to the mean density.

2. If the variation of S due to the small motion is taken into account, this will modify the previous solution and, in the case of dynamical stability, the

[1] For these end conditions and other results of the calculus of variation, cf. for instance Forsythe: Calculus of Variations. Cambridge 1927 or G. A. BLISS: Lectures on the Calculus of Variations. Chicago, Ill.: Chicago University Press 1946.

effect may be either an exponential decrease (vibrational stability, positive damping) or an exponential increase (vibrational instability, negative damping) of the amplitude of the adiabatic oscillation. The absolute value of the damping constant will be proportional to $\dot{S}$, i.e. to the rate at which the internal energy is modified by the generation of energy and the flux. But it is well known that, in a normal star, this is fairly small compared to the total potential energy so that, in general, the damping time will be fairly long compared to the adiabatic period. In other words this second type of motion will always be slow compared to the first one.

3. Finally, if we suppose that the conditions of dynamical and vibrational stability are verified, i.e. all small velocities—large compared with those necessitated by the changes of S between successive equilibrium configurations—tend to die out, the system will pass slowly through a set of states departing from equilibrium only by a very small amount proportional to $\dot{S}$. In this case, it remains to be shown whether this slow motion will be stable or unstable or, in other words, whether any small perturbation of it will be damped or whether there are perturbations which might lead to an accelerated change in S and consequently in all the state variables. This is the problem of secular stability.

12. The conditions of dynamical stability towards a radial perturbation. According to our previous discussion, the condition required for a dynamically stable equilibrium state is that the total available energy

$$J = V + \Lambda = \int_0^M \left[U(\varrho, S) - \frac{Gm}{r}\right] dm = \int_0^M L(m, r, r')\, dm$$

be a minimum. In the integrand, S at each point, must be treated as a constant in the course of any perturbation of the integrand and ϱ must be considered as defined in terms of m, r and $r' = dr/dm$ by (11.1).

The calculus of variations yields different well-known criteria for the minimizing of the integral J. The conditions that the first variation δJ be equal to zero are provided by the Euler equation and, in the case of mobile end points, by appropriate end conditions. They become here

$$\frac{d}{dm}\left(\frac{\partial L}{\partial r'}\right) - \frac{\partial L}{\partial r} = -\left\{4\pi r^2 \frac{d}{dm}\left[\varrho^2\left(\frac{\partial U}{\partial \varrho}\right)_S\right] + \frac{Gm}{r^2}\right\} = 0 \tag{12.1}$$

which, according to (10.20), reduces simply to the ordinary hydrostatic equation and

$$\left[\frac{\partial L}{\partial r'}\right]_0^M = \left[-4\pi r^2 \varrho^2 \left(\frac{\partial U}{\partial \varrho}\right)_S\right]_0^M = 0. \tag{12.2}$$

However, this insures only the stationarity of J and, if J is to be a minimum, the second variation $\delta^2 J$ should verify the condition

$$\delta^2 J > 0. \tag{12.3}$$

This implies two types of tests, one which concerns the whole configuration and one which must be verified locally for each value of m.

α) Dynamical stability of the star as a whole. The evaluation of $\delta^2 J$, taking (12.2) into account yields

$$\delta^2 J = \int_0^M 2\Omega(\eta, \eta')\, dm \tag{12.4}$$

where

$$2\Omega(\eta, \eta') = L_{r'r'}\, \eta'^2 + 2L_{rr'}\, \eta\eta' + L_{rr}\, \eta^2 \tag{12.5}$$

and

$$\eta = \delta r. \tag{12.6}$$

The condition (12.3) will certainly be verified if the minimum value of $\delta^2 J$ is greater than zero. Thus it is normal to associate the discussion of (12.3) with the Euler condition

$$\frac{d}{dm}\left(\frac{d\Omega}{d\eta'}\right) - \frac{\partial \Omega}{\partial \eta} = 0 \tag{12.7}$$

for a minimum of (12.4). Eq. (12.6) is called the Jacobi equation. If η is a solution of (12.7) which vanishes at a point $m = m_1$, the next point $m = m_2$ where η vanishes again is termed conjugate to the point m_1 and the integral (12.4) extended from m_1 to m_2 vanishes. It can then be seen that the condition (12.3) will be satisfied if no conjugate point to $m = 0$ occur for $0 < m \leqq M$.

But this condition can be transformed in the following way[1]. One may associate with (12.7) the following linear homogeneous problem

$$\frac{d}{dm}\left(\frac{\partial \Omega}{\partial \eta'}\right) - \frac{\partial \Omega}{\partial \eta} + \lambda \eta = 0 \tag{12.8}$$

with boundary conditions which here reduce to

$$[\Omega_{\eta'}]_{m=0}^{m=M} = 0. \tag{12.9}$$

It can then be shown that a necessary and sufficient condition for (12.3) is that all the eigenvalues λ_i of this problem be positive.

To apply this to our problem, we note that

$$L_{r'r'} = 16\pi^2 r^4 \varrho^2 \left[\varrho^2 \left(\frac{\partial^2 U}{\partial \varrho^2}\right)_S + 2\varrho \left(\frac{\partial U}{\partial \varrho}\right)_S\right],$$

$$L_{rr'} = 8\pi r \varrho \left[\varrho^2 \left(\frac{\partial^2 U}{\partial \varrho^2}\right)_S + 2\varrho \left(\frac{\partial U}{\partial \varrho}\right)_S - \varrho \left(\frac{\partial U}{\partial \varrho}\right)_S\right],$$

$$L_{rr} = \frac{4}{r^2}\left[\varrho^2 \left(\frac{\partial^2 U}{\partial \varrho^2}\right)_S + 2\varrho \left(\frac{\partial U}{\partial \varrho}\right)_S - \frac{1}{2}\left(\frac{\partial U}{\partial \varrho}\right)_S \varrho\right] - 2\frac{Gm}{r^3}.$$

These expressions can be simplified using the definitions (11.7) to (11.9) which yield

$$\left(\frac{\partial U}{\partial \varrho}\right)_S = \frac{p}{\varrho^2}, \quad \varrho^2 \left(\frac{\partial^2 U}{\partial \varrho^2}\right)_S + 2\varrho \left(\frac{\partial U}{\partial \varrho}\right)_S = \left(\frac{\partial p}{\partial \varrho}\right)_S = \Gamma_1 \frac{p}{\varrho} \tag{12.10}$$

where Γ_1 is the usual generalized adiabatic exponent (cf. [*1*], Sect. 53) for a mixture of gas and radiation characterized by $p_G/p = \beta$, i.e.

$$\Gamma_1 = \beta + \frac{(4-3\beta)^2 (\gamma - 1)}{12(1-\beta)(\gamma-1) + \beta}. \tag{12.11}$$

It is then easy to verify that (12.8) may be written, setting $\delta r = \eta = r\xi$ and using (11.1),

$$\frac{d^2\xi}{dr^2} + \frac{d\xi}{dr}\left[\frac{4}{r} + \frac{1}{\Gamma_1 p}\frac{d(\Gamma_1 p)}{dr}\right] + \xi\left[\frac{\lambda\varrho}{\Gamma_1 p} + \frac{3}{r\Gamma_1 p}\frac{d(\Gamma_1 p)}{dr} - \frac{4}{r\Gamma_1 p}\frac{dp}{dr}\right] = 0. \tag{12.12}$$

But this is nothing else than the usual equation for radial adiabatic pulsations with the square of the frequency σ^2 set equal to λ [cf. [*1*], Eqs. (58.1) or (58.12)].

---

[1] Cf. M. MORSE: The Calculus of Variations in the Large, Chap. II. The Amer. Mat. Soc. Coll. Publ., Vol. XVIII, 1934.

The boundary condition (12.9) can also be transformed into

$$[r^2\,\delta p + 2r^2 p\,\xi]_{m=0}^{m=M} = 0, \qquad \delta p = \Gamma_1 p\,\frac{\delta\varrho}{\varrho} = -\,\Gamma_1 p\left(r\,\frac{d\xi}{dr} + 3\xi\right)$$

which will certainly be satisfied by all physical solutions for which the Lagrangian pressure perturbation δp must remain zero at the free surface $m = M$ at all times while ξ remains bounded everywhere.

The application of the Jacobi condition is then equivalent to the requirement that all frequencies σ of radial pulsation be real which is also the form given to the condition of dynamical stability when the method of small perturbations is used. Thus, since the hydrostatic conditions (12.1) and (12.2) are automatically satisfied by any hydrostatic stellar model, the studies of its dynamical stability by the energy method or by the small perturbation method are essentially equivalent. One then understands better the exact correspondence between the results reached by TOLMAN [5b] on the basis of (12.7) and the location of the conjugate points and those yielded by the small perturbation method.

β) Local dynamical stability. The Weierstrass-Legendre conditions may be considered as the form taken by (12.3) for a special class of perturbations $\eta = \delta r$ which vanish everywhere except in the immediate neighbourhood of a given value of m where their derivatives η' do not tend towards zero with η. If these derivatives are not bound to be very small, we obtain the Weierstrass condition

$$L[m, r, (r')^*] - L(m, r, r') - [(r')^* - r']\,\frac{\partial L(m, r, r')}{\partial r'} > 0 \tag{12.13}$$

which can also be written using ROLLE's theorem and noting that L depends on r' only through ϱ in $U(\varrho, S)$

$$\left(\frac{\partial U}{\partial\varrho}\right)_S + \frac{1}{2}\,\varrho^*\left(\frac{\partial^2 U\,[\varrho + \alpha(\varrho^* - \varrho)]}{\partial\varrho^2}\right)_S > 0 \tag{12.14}$$

if α is a number between 0 and 1.

If, on the contrary, $(r')^*$ or ϱ^* are as close to r' or ϱ as one wants, condition (12.14) becomes

$$\left(\frac{\partial U}{\partial\varrho}\right)_S + \frac{\varrho}{2}\left(\frac{\partial^2 U}{\partial\varrho^2}\right)_S > 0 \tag{12.15}$$

which is the Legendre condition. Taking (12.10) into account, these conditions reduce to

$$\left(\frac{\partial p}{\partial\varrho}\right)_S > 0 \tag{12.16}$$

both for a finite or an infinitesimal change in ϱ or r'. This condition that ϱ must change in the same sense as p in any adiabatic perturbation is certainly a necessary condition for local stability, but it will be satisfied for any fluid of physical interest and thus is not very significant. But it is the limitation to purely radial perturbations which prevents us from recovering more useful local conditions such as the Schwarzschild criterion of stability for radiative equilibrium.

13. The condition of vibrational stability towards radial oscillations. Let us suppose that the conditions of dynamical stability are verified so that the possible small *isentropic* motions reduce to linear harmonic oscillations around an instantaneous state of equilibrium. In the general case, when the entropy varies, the equilibrium state of the system itself changes and the subscript 1 will denote one of the equilibrium configurations through which it goes. Each one of these verifies Eq. (12.1) which defines the dependence of (r_1, ϱ_1) on S. In general, however, the rate of change of the entropy which, in this case, depends on the state of the system and is given by [cf. (10.5)]

$$\dot{S} = \frac{1}{T}\left[\varepsilon - \frac{d}{dm}\left(4\pi r^2 F(r)\right)\right] = \frac{1}{T}\left[\varepsilon - \frac{dL(r)}{dm}\right] \tag{13.1}$$

can be taken as small, compared to the rate of change of r or ϱ in an adiabatic oscillation: $\dot{S} = \alpha f(\varrho, S)$ where α is a small quantity.

The condition of vibrational stability is that the changes in S do not cause the dynamical energy $\mathscr{E}$ of any normal oscillation to increase. $\mathscr{E}$ may be defined as the difference between the total dynamical energy of the oscillating system at a given time and that of an equilibrium configuration having the same entropy. Its rate of change is given by

$$\frac{d\mathscr{E}}{dt} = \frac{d}{dt}\int_0^M \left(\frac{1}{2}\dot{r}^2 + U - \frac{Gm}{r} - U_1 + \frac{Gm}{r_1}\right) dm$$

where ϱ_1 and r_1 must be treated as functions of S as recalled above. After differentiation under the integral sign and taking (11.4) and (12.1) into account, this reduces to

$$\left.\begin{aligned} \frac{d\mathscr{E}}{dt} &= \int_0^M \left(\frac{\partial U}{\partial S} - \frac{\partial U_1}{\partial S}\right)\dot{S}\, dm \\ &\approx \int_0^M \left[(\varrho - \varrho_1)\left(\frac{\partial^2 U}{\partial S\, \partial \varrho}\right)_1 + \frac{1}{2}(\varrho - \varrho_1)^2\left(\frac{\partial^3 U}{\partial S\, \partial \varrho^2}\right)_1\right](\dot{S}_1 + \delta\dot{S})\, dm \end{aligned}\right\} \tag{13.2}$$

if only small oscillations are considered.

Neglecting all terms containing products or powers higher than the first of $\delta\dot{S}$, $\dot{S}_1$, δS, Eq. (13.2) may be written

$$\frac{d\mathscr{E}}{dt} = \int_0^M \left\{\frac{\partial^2 U}{\partial S\, \partial \varrho}\left[(\varrho - \varrho_1)\,\dot{S}_1 + \delta\varrho\,\delta\dot{S}\right] + \frac{1}{2}\,\frac{\partial^3 U}{\partial S\, \partial \varrho^2}\left(\delta\varrho^2\,\dot{S}_1 + \delta\varrho^2\,\delta\dot{S}\right)\right\} dm \tag{13.3}$$

where all finite coefficients except $\dot{S}_1$ in the first term are to be evaluated in the equilibrium state (not distinguishing between ϱ and ϱ_1 in them) and $\delta\dot{S}$ may be computed from (13.1) in terms of the adiabatic variations $\delta\varrho$, δT etc.

In the average of Eq. (13.3) over a period P of the harmonic oscillation, the second factor in the coefficient of $\partial^3 U/\partial S\, \partial\varrho^2$ may be safely neglected, but the term in $(\varrho - \varrho_1)\,\dot{S}_1$ is more troublesome. Although its first order contribution will vanish in the mean, there may subsist a second order contribution which may be comparable to that coming from the term in $\delta\varrho^2\,\dot{S}_1$. Since we may assume that all perturbations are proportional to $\sin\left(\frac{2\pi}{P}t\right)$, this average yields

$$\left.\begin{aligned} 2\,\overline{\frac{d\mathscr{E}}{dt}} &= \frac{2(\Delta\mathscr{E})_P}{P} = \int_0^M \frac{\partial^2 U}{\partial S\, \partial\varrho}\,\delta\varrho\,\delta\dot{S}\, dm + \int_0^M \frac{1}{2}\,\frac{\partial^3 U}{\partial S\, \partial\varrho^2}\,\dot{S}_1\,\delta\varrho^2\, dm + \\ &\quad + \int_0^M \overline{(\varrho - \varrho_1)\,\dot{S}_1\,\frac{\partial^2 U}{\partial S\, \partial\varrho}}\, dm \end{aligned}\right\} \tag{13.4}$$

where $\delta\varrho$, $\delta\dot{S}$ etc. now represent the amplitudes of the corresponding perturbations.

THOMAS ([*5a*], second paper, Sect. 5) remarked that Eq. (13.4) where he completely neglected the last term, can be interpreted as the second variation

$$\delta^2 J = \int_0^M \Omega(m, \eta, \eta', \eta'')\, dm \tag{13.5}$$

of an appropriate integral in which η denotes the variation δv of the volume $4\pi r^3/3$ and the primes represent again differentiation with respect to m. Applying the same procedure as in Sect. 12α, the discussion of Eq. (13.5) may be associated with the linear differential equation

$$\frac{d^2}{dm^2}\left(\frac{\partial\Omega}{\partial\eta''}\right) - \frac{d}{dm}\left(\frac{\partial\Omega}{\partial\eta'}\right) + \frac{\partial\Omega}{\partial\eta} = 0 \tag{13.6}$$

which, as Eq. (12.7), may be used to define conjugate points. A *sufficient* condition for vibrational stability, i.e. for $\delta^2 J$ or $\overline{(d\mathscr{E}/dt)}$ to be negative, is that no conjugate points to the origin exist in the interval $0 < m \leqq M$.

One might of course proceed one step further and discuss the eigenvalue problem [cf. Eqs. (12.8) and (12.9)] associated with Eq. (13.5), but the final expressions are rather cumbersome to handle and the direct discussion of Eq. (13.4) is really simpler. Of course, it requires the solution of the equation of small motion (12.12), but the discussion of the zeros of the solutions of Eq. (13.6), necessary to fix the position of the conjugate points to the origin, will involve just as much work and, in the general case, will also have to proceed by numerical integration.

If we use Eq. (13.1) and note that, from Eqs. (11.8) and (11.9),

$$\frac{\partial^2 U}{\partial S\,\partial\varrho} = (\Gamma_3 - 1)\frac{T}{\varrho}, \tag{13.7}$$

with the usual definition of Γ_3 for a mixture of gas and radiation

$$\Gamma_3 = 1 + \frac{(4-3\beta)(\gamma-1)}{\beta + 12(\gamma-1)(1-\beta)}, \tag{13.8}$$

Eq. (13.4) becomes

$$\begin{aligned}\frac{2(\Delta\mathscr{E})_P}{P} = &\int_0^M \frac{\delta T}{T}\left[\delta\varepsilon - \frac{d\,\delta L(r)}{dm}\right] dm + \int_0^M \left[\frac{\varrho^2}{2T}\,\frac{\partial^3 U}{\partial S\,\partial\varrho^2}\left(\frac{\delta\varrho}{\varrho}\right)^2 - \left(\frac{\delta T}{T}\right)^2\right] \times \\ &\times \left[\varepsilon - \frac{dL(r)}{dm}\right] dm + \int_0^M \frac{\partial^2 U}{\partial S\,\partial\varrho}(\varrho - \varrho_1)\left[\varepsilon - \frac{dL(r)}{dm}\right] dm.\end{aligned} \tag{13.9}$$

If we assume that, in the equilibrium state, $\dot{S}_1$ is totally negligible, as is certainly the case during long intervals in the course of stellar evolution when the nuclear reactions exactly balance the energy radiated, Eq. (13.9) reduces to

$$\frac{2(\Delta\mathscr{E})_P}{P} = \int_0^M \frac{\delta T}{T}\left[\delta\varepsilon - \frac{d\,\delta L(r)}{dm}\right] dm.$$

As in Sect. 63β of [*1*], we can define a damping coefficient by

$$\sigma'' = -\frac{1}{2}\sigma\,\frac{(\Delta\mathscr{E})_P}{2\pi K_M}$$

where $\sigma = 2\pi/P$ and K_M is the maximum kinetic energy. This gives

$$\sigma'' = -\frac{1}{2\sigma^2}\,\frac{\int_0^M \frac{\delta T}{T}\left[\delta\varepsilon - \frac{d\,\delta L}{dm}\right] dm}{\int_0^M \delta r^2\, dm}. \tag{13.10}$$

which is the usual expression of the coefficient of vibrational stability. Of course, it could be generalized, as in [1], Sect. 63 and 64, to take into account viscosity, convection and turbulence. The condition of vibrational stability is that σ'' be positive or $(\Delta\mathscr{E})_P$ negative.

The extra-terms in Eq. (13.9) may be of interest during phases of stellar evolution when the exact balance between ε and $dL(r)/dm$ is no longer realized and may perhaps serve to excite oscillations in such cases. Up to now, they have not been the object of any detailed discussion.

14. The condition of secular stability. If the star is dynamically and vibrationaly stable, it will tend to evolve slowly through a set of states departing from equilibrium only by very small amounts. The rate of evolution will depend on the balance between the generation of energy and the outflow of radiation both varying continually as a consequence of the changes in chemical composition due to the nuclear reactions. In most stars, these changes are very slow during the greater part of their lives in which we may distinguish different stages corresponding to different nuclear reactions. The time-scale characteristic of one of them, say stage i, is fixed essentially by the ratio

$$(\tau_N)_i = \frac{(E_N)_i}{\overline{L}_i} \tag{14.1}$$

where $\overline{L}_i$ is the average luminosity of the star in this stage and $(E_N)_i$ is the total energy liberable by the corresponding nuclear reaction i.e. approximately the mass excess Δm of the elements entering the reaction over the final products multiplied by a fraction of the star mass proportional to the initial abundance of the transformed elements.

These stages will be separated by phases of more rapid evolution during which the star readjusts its internal temperature to the level necessary to initiate the next nuclear reaction. In some cases, this will be realized by a general contraction of the star and the gravitational time-scale

$$\tau_G \approx -\frac{V}{2L} \tag{14.2}$$

provides an adequate order of magnitude for the time taken by this contraction.

In other cases, however, these transition stages may be more complex. For instance, after the formation of an isothermal core, the zone of energy generation shifts progressively towards the external layers where nuclear fuel is still available. If the resulting generation of energy is sufficient, this may appreciably lower the rate of evolution or it may even modify the latter considerably leading to an increase in the radius of the star while the core gets more and more centrally condensed. Here, the properties of the star and, in particular, its luminosity may vary considerably and fairly rapidly.

In the last case, it is difficult to characterize the time-scale of the corresponding evolutionary phases. In this respect, one should note that, even in the other cases, formulae such as (14.1) or (14.2) have a meaning only if the luminosity L does not vary too rapidly. This depends on the way in which the opacity, or the opacity and the energy generation react to the changes in stellar structure brought about by the evolution. The aim of the study of secular stability is to establish the conditions under which the resulting motion remains always very slow.

In this context, the restriction to purely radial motion is probably better justified than in the preceding sections since one may expect that, at least as long as the motion remains slow, each instantaneous configuration will make the gravitational potential a minimum and thus assumes the spherical shape (cf. Sect. 6).

Let us apply, to one of these instantaneous equilibrium states verifying Eqs. (12.1) and (12.2), a small radial perturbation δr, $\delta\varrho$, δS during which the changes in chemical composition are considered as negligible. These variations must satisfy the following linearized equations:

$$\left.\begin{aligned}\frac{2Gm}{r^3}\,\delta r - 8\pi r\,\delta r\,\frac{d}{dm}\left(\varrho^2\frac{\partial U}{\partial\varrho}\right) - 4\pi r^2\frac{d}{dm}\left[\left(2\varrho\frac{\partial U}{\partial\varrho} + \varrho^2\frac{\partial^2 U}{\partial\varrho^2}\right)\delta\varrho\right]\\ = 4\pi r^2\frac{d}{dm}\left(\varrho^2\frac{\partial^2 U}{\partial\varrho\,\partial S}\,\delta S\right)\end{aligned}\right\} \tag{14.3}$$

where we have neglected the inertial terms which may be expected to be very small, in any case, much smaller than the first order time-derivative in

$$\frac{d\,\delta S}{dt} = \delta\left\{\frac{1}{\left(\frac{\partial U}{\partial S}\right)}\left[\varepsilon - \frac{dL(r)}{dm}\right]\right\} \tag{14.4}$$

which is obtained by linearization of (13.1). The solutions must verify the boundary conditions

$$\left\{2\,\delta r\,\varrho^2\frac{\partial U}{\partial\varrho} + r\left(2\varrho\frac{\partial U}{\partial\varrho} + \varrho^2\frac{\partial^2 U}{\partial\varrho^2}\right)\delta\varrho + r\varrho^2\frac{\partial^2 U}{\partial\varrho\,\partial S}\,\delta S\right\}_0^M = 0.$$

All coefficients are to be evaluated in the equilibrium configuration considered. Taking Eqs. (12.10) and (13.7) into account and noting that p, ϱ and T are assumed to vanish at the surface, these conditions can be expressed explicitly as

$$\delta r = 0 \quad \text{in} \quad m = 0$$

and

$$\delta p = \frac{\Gamma_1 p}{\varrho}\,\delta\varrho + \varrho\,T(\Gamma_3 - 1)\,\delta S = 0 \quad \text{in} \quad m = M \tag{14.5}$$

Using the continuity equation (11.1), which yields here

$$\frac{\delta\varrho}{\varrho} = -4\pi\varrho\frac{d}{dm}(r^2\,\delta r) = -\left(2\frac{\delta r}{r} + \frac{d\,\delta r}{dr}\right), \tag{14.6}$$

one may associate with the preceding problem a simpler one in which the entropy S is treated as constant in time and in which δr satisfies the following equation

$$\left.\begin{aligned}\frac{d}{dm}\left[\left(2\varrho\frac{\partial U}{\partial\varrho} + \varrho^2\frac{\partial^2 U}{\partial\varrho^2}\right)4\pi\varrho^2\frac{d}{dm}(r^2\,\delta r)\right] -\\ -2\,\frac{\delta r}{r}\,\frac{d}{dm}\left(\varrho^2\frac{\partial U}{\partial\varrho}\right) + \frac{2Gm}{4\pi r^5}\,\delta r = 0\end{aligned}\right\} \tag{14.7}$$

with the boundary conditions

$$\delta r = 0 \quad \text{in} \quad r = 0 \quad \text{and} \quad \delta p = \Gamma_1 p\,\frac{\partial\varrho}{\varrho} = 0 \quad \text{in} \quad m = M. \tag{14.8}$$

If two independent solutions of this problem, say $\delta r = \eta$ and $\delta r = \zeta$ can be found, the general solution of Eq. (14.3) may be written

$$\left.\begin{aligned}\delta r = C_1\eta + C_2\zeta + \eta\int_m^M \frac{\varrho^2}{W_0}\,\frac{\partial^2 U}{\partial\varrho\,\partial S}\,\delta S\,\frac{d}{dm}(r^2\zeta)\,dm +\\ +\zeta\int_0^m \frac{\varrho^2}{W_0}\,\frac{\partial^2 U}{\partial\varrho\,\partial S}\,\delta S\,\frac{d}{dm}(r^2\eta)\,dm\end{aligned}\right\} \tag{14.9}$$

where W_0 is a constant. Note that if one solution, say η, of Eq. (14.7) is known, a second independent one, ζ, may be obtained by quadrature from the expression of the Wronskian

$$W = W_0 \left[\frac{\varrho}{r'} r^2 \left(2\varrho \frac{\partial U}{\partial \varrho} + \varrho^2 \frac{\partial^2 U}{\partial \varrho^2}\right)\right]^{-1} = \left(\zeta \frac{\partial \eta}{dm} - \eta \frac{\partial \zeta}{dm}\right). \tag{14.10}$$

The expression (14.9) of δr should now be introduced in the second member of Eq. (14.4) after the perturbations of ε and $L(r)$ have been expressed explicitly in terms of δr and its derivatives. This will lead to a complex integro-differential problem, the rate of change of δS, at any point, depending on its value at all other points in the star and not only on its values at neighbouring points.

In that respect, the following alternative approach may be of interest. First, note that, using Eqs. (12.10), Eq. (14.7) can be expressed explicitly as

$$\frac{d}{dr}\left(r^4 \Gamma_1 p \frac{d\xi}{dr}\right) + \xi \left\{r^3 \frac{d}{dr}\left[(3\Gamma_1 - 4)\, p\right]\right\} = 0$$

where $\xi = \delta r/r$. Except for the absence of the inertial term in λ (or σ^2), this is identical with the equation of radial adiabatic pulsation (12.12) which admits a complete set of orthogonal eigensolutions ξ_i (cf. [1], Sect. 58) which may be used to expand the general solution ξ of the present problem.

This solution must satisfy the differential equation (14.3) which, if treated as Eq. (14.7) above, becomes

$$\frac{d}{dr}\left(\Gamma_1 p r^4 \frac{d\xi}{dr}\right) + \xi \left\{r^3 \frac{d}{dr}\left[(3\Gamma_1 - 4)\, p\right]\right\} = r^3 \frac{d}{dr}\left[(\Gamma_3 - 1)\,\varrho\, T\, \delta S\right]. \tag{14.11}$$

Provided that none of the eigenvalues $\sigma_i^2 = \lambda_i$ of Eq. (12.12) vanish, we may write

$$\xi = \Sigma \alpha_i \xi_i \tag{14.12}$$

which, substituted in Eq. (14.11), yields

$$\sum_j \alpha_j \xi_j \sigma_j^2 \varrho r^4 = r^3 \frac{d}{dr}\left[(\Gamma_3 - 1)\,\varrho\, T\, \delta S\right].$$

Multiplying this equation by ξ_i and integrating from 0 to R, keeping in mind the orthogonality properties of the ξ_i [cf. [1], Eq. (58.19)] we obtain

$$\alpha_i = -\frac{\int_0^R r^3 \xi_i \frac{d}{dr}\left[(\Gamma_3 - 1)\,\varrho T\, \delta S\right] dr}{\sigma_i^2 \int_0^R \varrho r^4 \xi_i^2\, dr}.$$

If the numerator is integrated by parts, using the boundary condition (14.5) and the definition (14.6), α_i can finally be written

$$\alpha_i = -\frac{\int_0^M \left(\frac{\delta T}{T}\right)_i T\, \delta S\, dm}{\sigma_i^2 \int_0^M r^2 \xi_i^2\, dm}, \tag{14.13}$$

where

$$\left(\frac{\delta T}{T}\right)_i = (\Gamma_3 - 1)\left(\frac{\delta \varrho}{\varrho}\right)_i \tag{14.14}$$

is the relative adiabatic temperature deviation corresponding to the eigensolution ξ_i.

With this definition, ξ as given by Eq. (14.12) becomes an explicit function of δS and can be used to expand the second member of Eq. (14.4). In carrying this through, one will need expressions of $\delta\varrho/\varrho$ and $\delta T/T$. In the case of the former, Eq. (14.6) gives immediately

$$\frac{\delta\varrho}{\varrho} = \sum_i \alpha_i \left(\frac{\delta\varrho}{\varrho}\right)_i. \tag{14.15}$$

However, here, $\delta T/T$ cannot be replaced by its adiabatic approximation (14.14) but must be evaluated from Eq. (11.9) in terms of $\delta\varrho$ and δS

$$\frac{\delta T}{T} = (\Gamma_3 - 1)\frac{\delta\varrho}{\varrho} + \frac{\bar{\mu}}{\mathsf{R}}\,\frac{\beta(\gamma-1)\,\delta S}{[12(1-\beta)(\gamma-1)+\beta]} = (\Gamma_3 - 1)\frac{\delta\varrho}{\varrho} + \Theta\,\delta S. \tag{14.16}$$

As usual, let us assume laws of energy generation and opacity of the form

$$\varepsilon = \varepsilon_0\,\varrho^{\mu}\,T^{\nu}, \qquad \varkappa = \varkappa_0\,\varrho^{m}\,T^{-n} \tag{14.17}$$

where ε_0 und $\varkappa_0$ depend only on the chemical composition. If furthermore, the equilibrium is radiative everywhere in the star, i.e.

$$L(r) = 4\pi r^2 F(r) = -\frac{16\pi^2 r^4 c}{3\varkappa}\,\frac{d}{dm}\,(a\,T^4), \tag{14.18}$$

Eq. (14.4) may be written

$$\left.\begin{aligned} T\frac{d\,\delta S}{dt} &= \varepsilon\left(\mu\frac{\delta\varrho}{\varrho} + \nu\frac{\delta T}{T}\right) - \\ &\quad - \frac{d}{dm}\left\{L(r)\left[4\xi + (4+n)\frac{\delta T}{T} - m\frac{\delta\varrho}{\varrho} + \frac{\frac{d}{dm}\left(\frac{\delta T}{T}\right)}{\frac{1}{T}\frac{dT}{dm}}\right]\right\} \end{aligned}\right\} \tag{14.19}$$

if we suppose that $(\varepsilon - dL/dm)$ vanishes in the instantaneous equilibrium configuration considered. If in Eq. (14.19), we substitute for ξ, $\delta\varrho/\varrho$ and $\delta T/T$ their expressions (14.12), (14.15) and (14.16) with the definition (14.13) of the α_i's we obtain, as before, a linear integro-differential equation in δS

$$\left.\begin{aligned} T\frac{d\,\delta S}{dt} &= \varepsilon\left\{\nu\Theta\,\delta S - [\mu + \nu(\Gamma_3 - 1)]\sum_i \frac{\int_0^M \left(\frac{\delta T}{T}\right)_i T\,\delta S\,dm}{\sigma_i^2 \int_0^M r^2\,\xi_i^2\,dm}\left(\frac{\delta\varrho}{\varrho}\right)_i\right\} \\ &\quad - \frac{d}{dm}\left\{L(r)\,(4+n)\,\Theta\,\delta S - L(r)\sum_i \frac{\int_0^M \left(\frac{\delta T}{T}\right)_i T\,\delta S\,dm}{\sigma_i^2 \int_0^M r^2\,\xi_i^2\,dm}\times\right. \\ &\qquad \times\left[4\xi_i + [(4+n)(\Gamma_3 - 1) - m]\left(\frac{\delta\varrho}{\varrho}\right)_i\right] + \\ &\qquad \left. + \frac{L(r)\,T}{\frac{dT}{dm}}\,\frac{d}{dm}\left[\Theta\,\delta S - (\Gamma_3 - 1)\sum_i \frac{\int_0^M \left(\frac{\delta T}{T}\right)_i T\,\delta S\,dm}{\sigma_i^2 \int_0^M r^2\,\xi_i^2\,dm}\left(\frac{\delta\varrho}{\varrho}\right)_i\right]\right\} \end{aligned}\right\} \tag{14.20}$$

which looks forbidding.

It may however be possible to make some progress in its solution by assuming for δS a series expansion of the form

$$\delta S = \sum_k \left[\beta_k(t)\,\xi_k(r) + \gamma_k(t)\,\frac{d\xi_k}{dr}\right] \tag{14.21}$$

and identifying the coefficients of $\xi_i(r)$ and $d\xi_i/dr$ in Eq. (14.20) after all $d^2\xi_i/dr^2$ have been eliminated with the help of Eq. (12.12). In practice, it is also likely that after substituting Eq. (14.21) in the numerator of α_i, the term corresponding to $k=i$ in the expression of δS will be the dominating one, many or all of the others being negligible.

However, up to now, these general aspects of the problem have received very little attention[1] and the condition of secular stability has been worked out explicitly only in the simplest case when Eqs. (12.1) and (13.1) where S is put equal to zero, i.e.

$$\frac{d}{dm}\left[\varrho^2\,\frac{\partial U}{\partial \varrho}\right] + \frac{G m}{4\pi r^4} = 0, \tag{14.22}$$

$$\varepsilon - \frac{dL(r)}{dm} = 0, \tag{14.23}$$

admit homologous solutions satisfying the boundary conditions (12.2) together with the following conditions for the temperature

$$\frac{dT}{dr} = 0 \quad \text{in} \quad m = 0, \qquad T = 0 \quad \text{in} \quad m = M. \tag{14.24}$$

Denoting one of these solutions by a subscript zero, say $r_0(m)$, a family of homologous solutions will be given by

$$r(m) = \lambda\, r_0(m), \quad \varrho(m) = \lambda^{n_1}\varrho_0(m), \quad T(m) = \lambda^{n_2}\, T_0(m), \quad p(m) = \lambda^{n_3} p_0(m). \tag{14.25}$$

Using the definitions

$$\varrho^2\,\frac{\partial U}{\partial\varrho} = p, \qquad \frac{\partial U}{\partial S} = T, \qquad \varrho = \frac{1}{4\pi r^2 r'}, \qquad p = \frac{R\varrho T}{\bar\mu} + \frac{1}{3}\,a\,T^4 \tag{14.26}$$

together with Eqs. (14.17) and (14.18), one finds that Eqs. (14.22) and (14.23) are compatible with the transformation (14.25) where

$$n_1 = -3, \quad n_2 = -1, \quad n_3 = -4 \quad (\text{Lane's law}) \tag{14.27}$$

provided the condition

$$3\mu + \nu = n - 3m \tag{14.28}$$

be satisfied. Then, the transformation (14.25) associates to any equilibrium state, a continuous series of other configurations which has often been interpreted in the past as a sequence of evolution.

One may try to generalize the transformation (14.25) so that it will provide a solution of Eq. (14.22) and the general time-dependent form (13.1) of (14.23). To this end, we let λ be a function of the time t, say $\lambda(t)$, and proceeding as before we recover again, on one hand, the values (14.27) of the exponents while, on the other hand, the definition of the entropy (11.9) gives

$$\dot S = \frac{R}{\bar\mu}\,\frac{3\gamma-4}{\gamma-1}\,\frac{\dot\lambda}{\lambda} \tag{14.29}$$

---

[1] For another line of approach cf. [12], Sects. 5.12 to 5.14.

where γ is the ratio of the specific heats of the gas. If $\gamma > \frac{4}{3}$, we may have a general homologous expansion if $\dot{S}$ is positive, i.e. if more energy is generated than radiated, or a general contraction if $\dot{S}$ is negative. If $\gamma < \frac{4}{3}$, we know (cf. [*1*], Sects. 58 and 59) that the star is dynamically unstable and the question of secular stability and slow evolution looses all meaning.

With the help of (14.29), Eq. (13.1) becomes

$$\varepsilon - \frac{R}{\bar{\mu}} \frac{(3\gamma - 4)}{(\gamma - 1)} T \frac{\dot{\lambda}}{\lambda} - \frac{dL(r)}{dm} = 0, \tag{14.30}$$

the second term corresponding to the fraction of the gravitational energy radiated away in the course of the motion. However, the last relation imposes restrictions on the form of ε and $\varkappa$ which should be such that they get transformed according to

$$\frac{\varepsilon}{(\varepsilon)_0} = \frac{\dot{\lambda}}{\lambda^2 \dot{\lambda}_0} \quad \text{and} \quad \frac{\varkappa}{\varkappa_0} = \frac{\dot{\lambda}_0 \lambda^2}{\dot{\lambda}}.$$

If ε and $\varkappa$ are given by Eqs. (14.17), this implies again the equality (14.28) between their exponents. Thus, a star in which (14.28) is verified may evolve along the series of homologous configurations considered previously, but of course it is not the only possible path of evolution.

In the absence of nuclear reactions[1], Eq. (14.30) becomes, substituting for $R/\bar{\mu}$ its value $(\gamma - 1)\, C_v$,

$$(3\gamma - 4)\, C_v T \frac{\dot{\lambda}}{\lambda} = -\frac{dL(r)}{dm} = -\frac{d}{dm}\left[\lambda^{3m-n} L_0(r)\right]$$

or, if m and n are constant,

$$(3\gamma - 4)\, C_v T_0 \frac{\dot{\lambda}}{\lambda^2} = -\lambda^{3m-n} \frac{dL_0(r)}{dm}.$$

In any case, integrating over the whole star and denoting surface values by a subscript R, we obtain

$$\frac{\dot{\lambda}}{\lambda^{2+(3m-n)_R}} = -\frac{L_0(R)}{\int\limits_0^M (3\gamma - 4)\, C_v T_0\, dm} \tag{14.31}$$

so that, if we admit that $\gamma > \frac{4}{3}$, the rate of the contraction will depend on the ratio of the total luminosity to the internal thermal energy. Since the latter is of the same order as the gravitational potential energy, this confirms that an appropriate time-scale, in this case, is defined by Eq. (14.2). On the other hand, the type of motion is determined by the value of $n - 3m$: it is exponential, if $n - 3m = 1$, and given by a power-law in $(t - t_0)$, in the other cases.

If Eq. (14.28) is not verified, we may nevertheless gain some information on the stability of the possible slow motion by assuming that the small perturbation δr of the instantaneous equilibrium state corresponds to an infinitesimal Lane's transformation (14.25) with

$$\lambda(t) = 1 + \alpha(t), \quad \alpha(t) \text{ very small},$$

and substitute this particular solution for δr, $\delta\varrho$, δT in Eq. (14.4) instead of the general solutions (14.9) or (14.12). According to Eq. (14.23) and the hypothesis of radiative equilibrium (14.18) this yields

$$\frac{d\,\delta S}{dt} = -\frac{\alpha(t)}{T}\left\{\varepsilon(3\mu + \nu) - \frac{d}{dm}\left[L(r)(n - 3m)\right]\right\} \tag{14.32}$$

[1] For recent and, in some respects, more general discussions of the evolution in the early stages when nuclear sources of energy are absent cf. for instance R. D. Levee: Astrophys. Journ. **117**, 200 (1953). — G. C. McVittie: Astronom. J. **61**, 453 (1956).

or, if n and m are assumed constant,

$$\frac{d\,\delta S}{dt} = -\frac{\alpha(t)}{T}\,\varepsilon\,[3\mu + \nu - (n - 3m)]. \tag{14.33}$$

But the transformation considered must be compatible with the physical laws of the problem and, in particular, the temperature variation $\delta T = -\alpha T$ must be equal to that given by Eq. (14.16) so that

$$\alpha = \frac{\Theta}{3\Gamma_3 - 4}\,\delta S. \tag{14.34}$$

Using the definitions (14.16) of Θ and (13.8) of Γ_3, Eq. (14.33), where δS is eliminated with the help of Eq. (14.34), becomes

$$\frac{d\alpha}{dt} = -\frac{\varepsilon\,[3\mu + \nu - (n - 3m)]}{(3\gamma - 4)\,C_v\,T}.$$

This is a local equation which shows that the motion will tend to die out at each point provided that

$$3\mu + \nu > n - 3m, \tag{14.35}$$

an inequality which is usually referred to as the *condition of secular stability*. In that case, the "half-life" τ of any such perturbation,

$$\tau = \frac{(3\gamma - 4)\,C_v\,T}{\varepsilon}\,\frac{1}{3\mu + \nu - (n - 3m)}, \tag{14.36}$$

is proportional to the ratio of the internal energy to the rate of generation of energy at the point considered.

An average condition for the whole star may also be derived simply by multiplying Eq. (14.32) by dm and integrating over the whole mass. The result is

$$\frac{d\alpha}{dt} = -\frac{L(R)}{\int_0^M (3\gamma - 4)\,C_v\,T\,dm}\,\{\overline{(3\mu + \nu)} - (n - 3m)_R\}\,\alpha \tag{14.37}$$

where $\overline{(3\mu + \nu)}$ is an average with respect to ε and will differ very little from the value that $(3\mu + \nu)$ takes at the centre while $(n - 3m)_R$ is the value taken by the exponents in the opacity law at the surface. Since, close enough to the surface, there will always be a layer where the transfer of energy is mainly due to the radiation, Eq. (14.37) will always be valid independently of the type of the energy transfer inside the star[1].

If $\gamma > \frac{4}{3}$, the star as a whole will be secularly stable if

$$\overline{3\mu + \nu} > (n - 3m)_R. \tag{14.38}$$

In that case, one may expect the evolutionary motion accompanying the modifications of the chemical composition due to the nuclear reactions to remain very slow, any small homologous perturbation of this motion decaying with a mean half-life

$$\tau = \frac{\int_0^M (3\gamma - 4)\,C_v\,T\,dm}{L(R)\,[\overline{3\mu + \nu} - (n - m)_R]} \tag{14.39}$$

which is proportional to the ratio of the total internal energy to the luminosity $L(R)$ and inversely proportional to the difference $\overline{3\mu + \nu} - (n - m)_R$. When this

[1] This has not always been recognized in the past, but the problem was solved in a way essentially equivalent to the one above by T. G. COWLING: Monthly Notices Roy. Astronom. Soc. London **98**, 528 (1938).

difference vanishes, τ tends towards infinity and, except for the effects of the chemical changes, any homologous transformation would again become a possible path of evolution.

However, as m and n are respectively of the order of 1 and 3.5, the condition (14.38) will be amply satisfied in all stars deriving their energy from nuclear reactions, since, in that case, μ is at least equal to 1 and ν is often very large, and, in any case, is greater or equal to about 4. In the absence of nuclear reactions, the time-scale of the perturbation given by Eq. (14.39) where $\overline{3\mu+\nu}=0$ is of the same order as that of the slow gravitational contraction itself [cf. Eq. (14.31)] which, in this case, is the source of energy. Because of our homology assumptions, the perturbation is in this case essentially tangent to the slow motion itself, at any time.

Of course these conclusions rest on the consideration of a rather special type of perturbation but, unfortunately as we saw above, a more general treatment of Eq. (14.3) is still lacking. Our discussion has also been limited to the case of a mixture of a perfect gas and radiation and it needs revision before it could be applied, for instance, to white dwarfs. In their case, it is now generally admitted[1] that if the degenerate interior is devoid of any nuclear fuel, no instabilities could arise and, in particular, they can live for a very long time on a very slow stable gravitational contraction.

Less extreme deviations from a perfect gas may be introduced by the ionization of some abundant element and the present discussion could easily be extended to cover such cases. However as long as dynamical stability is preserved, it is unlikely that essentially new features should appear in the discussion of secular stability.

Finally, one may wonder whether it is always permissible, as we have done above, to neglect the chemical changes in the course of the perturbation of the slow motion. Of course the half-life of the perturbation defined by Eq. (14.39) is ordinarily very small on a time-scale defined by Eq. (14.1) which characterizes the rate of the chemical changes. However there may be special phases in stellar evolution where this is no longer true. Unfortunately it is likely that, in such circumstances, the chemical changes should be taken into account *explicitly* not only in the perturbation but in the slow motion as well and this may well lead to a very difficult problem.

15. Concluding remarks on the energy method and its extension to finite oscillation. In the preceding sections, the energy method has enabled us to distinguish the main aspects in the problem of stellar stability and to recover the corresponding criteria which, in the end, are essentially equivalent to those obtained directly by the small perturbation method.

Actually, this has been established only for purely radial perturbations but there are no reasons to doubt that this would also be generally true in the case of non-radial perturbations. Nevertheless, the energy method may provide additional interesting information on special questions. For instance, as far as local dynamical stability is concerned (cf. Sect. 12β) it may be worthwhile in the case of non-radial motions to develop explicitly the Weierstrass condition.

The energy method may also prove useful in tackling more general aspects of the problem such as appear when non-linear terms are taken into account. For instance, in Sect. 13, we have seen [cf. Eq. (13.9)] that, if a star reaches a state in which there is a lack of balance between the total energy generated, say E_0, and the energy radiated, say L_0, the difference between the two plays a rôle in the excitation of oscillations. As long as the amplitude of these oscillations

[1] L. Mestel: Monthly Notices Roy. Astronom. Soc. Lond. **112**, 583, 598 (1952).

is very small, the fluctuations of E and L around E_0 and L_0 may be treated as harmonic and, on the average, $\bar{L}=L_0, \bar{E}=E_0$ and the initial lack of equilibrium between L_0 and E_0 subsists feeding more and more energy into the oscillation. But as the amplitude increases, the non-linear terms will become more and more important and may affect differently E and L.

Since the energy generation takes place in the central region where the amplitudes are likely to remain very small, δE may well remain practically harmonic while δL, which is determined essentially by the behaviour of the external layers where the amplitude is largest, will be strongly anharmonic. Its average value $\bar{L}$ may be quite different from L_0 and together with whatever energy is dissipated in the course of the oscillation it may exactly equilibrate the total energy generated, $\bar{E} \approx E_0$. In the language of non-linear systems (cf. [1], Sects. 89, 90), the resulting finite oscillation would correspond to a stable limit-cycle.

Long ago, MILNE[1] discussed in great mathematical details a somewhat similar case where E was treated as a constant. Let us take for the origin of time, the instant when the lack of balance between E and L was first brought about and let us assume with MILNE that, at any subsequent time t, it is possible to associate a later time $t+T(t)$ such that, apart from a constant, all the energy generated from $t=0$ to t, i.e. Et, has been radiated away by the time $t+T$. Denoting by $f(t)$ the total amount of energy lost at the time t, we have

$$f(t+T) = \int_{t=0}^{t+T} L(t)\,dt = E(t+C) \tag{15.1}$$

while in a steady state, whatever the origin of time, one always has $f(t)=Et$.

At any instant t, the total internal energy $I(t)$ is given by

$$I(t) = Et - f(t) + I_0. \tag{15.2}$$

Using (15.1), Eq. (15.2) becomes

$$I(t) = Et - [E(t-T+C)] + I_0 = E(T-C) + I_0$$

which, applied at $t=0$, shows that

$$C = T(0) \equiv T_0.$$

On the other hand, for $t=0$, (15.1) yields

$$f(T_0) = EC = ET_0$$

which means that, on the average, from $t=0$ to $t=T_0$ the system behaves formally as if in a steady state. In other words, T_0 may be interpreted approximately as the time which elapses before a perturbation of energy production at the centre manifests itself in the surface emission and inversely.

If $T(t)$ may be treated as a one-valued function of $I(t)$ and $f(t)-Et$ is small compared to I_0, one may write

$$T = T_0 + \frac{k}{E}[f(t) - Et] = T_0 + \Phi(t) \tag{15.3}$$

where

$$k = -E\left(\frac{\partial T}{\partial I}\right)_0 \tag{15.4}$$

and

$$f(t) = E\left[t + \frac{1}{k}\Phi(t)\right]. \tag{15.5}$$

---

[1] E. A. MILNE: The Energetics of non-steady States, with Applications to Cepheid Variations. Quart. J. Math. 3, 258 (1933).

Expressing $f(t+T)$ in Eq. (15.1) by means of Eqs. (15.3) and (15.5), we obtain the following functional equation

$$k\,\Phi(t) + \Phi[t + T_0 + \Phi(t)] = 0. \tag{15.6}$$

One will notice that the establishment of (15.6) is somewhat hybrid in character since, to obtain (15.3), $\Phi(t)$ has been treated as a small quantity whose square and higher powers may be neglected while in Eq. (15.6) it is considered as finite.

If (15.6) is treated as (15.3), it becomes

$$k\,\Phi(t) = -\,\Phi(t + T_0) \tag{15.7}$$

and the whole problem is now linear. In particular, from Eqs. (15.1) and (15.5), we have

$$L(t) = f'(t) = E\left(1 + \frac{\Phi'(t)}{k}\right). \tag{15.8}$$

If $k=1$, Eq. (15.7) corresponds to a stable periodic phenomenon

$$\Phi(t) = \sin\left(\frac{2\pi}{2T_0}\,t\right)$$

with period $P=2T_0$ and constant amplitude. As far as the interpretation of P is concerned, T_0 is not necessarily the time that the energy would take to reach the surface if transmitted by radiation alone. This is much too long (cf. [1], Sect. 67) to have any significance for the mechanical oscillations of a star. But the release of extra-energy at the centre will create a pressure perturbation which will reach the surface in a time of the order of $R/\bar{c}$ where $\bar{c}$ is an appropriate mean value of the sound velocity in the star and, with this interpretation, $P=2T_0$ is of the right order of magnitude [cf. [1], Eq. (60.5)].

If $k \neq 1$, the solution of (15.7) is of the form

$$\Phi = k^{t/T_0} \sin\left(\frac{2\pi t}{2T_0}\right)$$

and the motion is damped if $k<1$ while, if $k>1$, the motion is unstable and the amplitude increases continuously.

In the non-linear case, the discussion of Eq. (15.6) is more difficult but MILNE's results show that, qualitatively, the situation remains similar except that the variations of Φ are anharmonic. In particular, in that case, Eq. (15.8) gives for $L(t)$, if $k=1$, an asymmetric variation with a pronounced peak and a flat minimum reminiscent of the observed light-curves of some Cepheids.

Of course, MILNE's investigation is much too formal for direct application to actual stars, but it provides a good example of the power of the energy method to analyse more complex situations including the effects of non-linear terms. Of course, other more detailed methods have been developed, at least in the case of fairly simple systems, to take the influence of such terms into account and discuss the special types of instability or of steady motions to which they may give rise. For further details we shall refer the reader to Ref. [1], Sects. 84 to 90.

16. Main results concerning the dynamical stability of stars. Since, in this case, the energy method leads to results essentially equivalent to those obtained by the small perturbation method which has been discussed in great details in Ref. [1], Sects. 57 to 62 and 73 to 80, we shall limit the discussion here to a short summary of the main results.

α) Radial perturbations. In the case of a purely radial perturbation, the problem reduces to the discussion of the eigenvalues of a linear equation of the Sturm-Liouville type where the square of the frequency σ^2 plays the rôle of the parameter [cf. Eq. (12.12)].

The eigenvalues σ_i^2 $(i=0, 1, 2, \ldots)$ ordered by increasing values correspond to eigenfunctions ξ_i defining the relative displacements $(\delta r/r)_i$ for successive modes with $0, 1, 2, \ldots, i$ nodes between the centre and the surface. Thus any possible instability ($\sigma^2<0$) will enter first through the fundamental mode (σ_0, ξ_0). But for this mode the frequency is given by the very simple expression [cf. [*1*], Eq. (59.18)]

$$\sigma_0^2 = -\frac{\int\limits_0^R 4\pi\xi_0 r^3 \frac{d}{dr}[(3\Gamma_1-4)\,p]\,dr}{\int\limits_0^R 4\pi\varrho r^4 \xi_0\,dr} = \frac{\int\limits_0^{\mathcal{V}} \left(3-\frac{4}{\Gamma_1}\right)(-\delta p)_0\,d\mathcal{V}}{\int\limits_0^{\mathcal{V}} \varrho r^2\xi_0\,d\mathcal{V}} \tag{16.1}$$

where ξ_0 and $-(\delta p)_0$ are everywhere positive and Γ_1 is a generalized ratio of specific heats for a mixture of radiation and ionized gas. If Γ_1 is constant, one recovers immediately the well-known condition of dynamical stability,

$$\Gamma_1 > \tfrac{4}{3}, \tag{16.2}$$

since dp/dr is always negative.

Strictly speaking Γ_1 is never a constant as the ionization of an electronic shell of an abundant element can lower its value appreciably and even render it smaller than $\frac{4}{3}$ in the region of the star where it takes place. The effects of the variations of Γ_1 have been discussed by TOLMAN (cf. [*5b*] second paper) and by LEDOUX (cf. [*13*], Chap. IV) with very similar conclusions. In general, the variations of Γ_1 occurring somewhere around the middle of the star radius have the largest effects. All conditions in this region being equal, the most favourable case for instability occurs when $(3\Gamma_1-4)$ has the deepest possible minimum both at the centre and a little below the surface. However, in actual stars with the accepted predominance of H and He, only the last minimum is likely to be present, Γ_1 being practically a constant close to $\frac{5}{3}$ in the rest of the star. In those conditions, the superficial layers of low $\Gamma_1<\frac{4}{3}$ due to the ionization of H and He are much too narrow and of too little weight in (16.1) (δp very small) to affect appreciably the dynamical stability of the star.

However, if we do not limit ourselves to actual stars but also want to consider the configurations of very large radii through which a star must presumably pass in the course of its condensation from interstellar material, the investigation of BIERMANN and COWLING [*6*] has shown that, for masses of the order of 1 to $10\,M_\odot$, there are phases corresponding to values of the radius between 30 to $100\,R_\odot$ when dynamical stability requires the presence of a fairly high hydrogen content.

In presence of a large amount of H, the same difficulty is encountered at an earlier stage ($R\approx 100$ to $1000\,R_\odot$) unless one assumes that the contracting hydrogen has been previously ionized by the ultraviolet radiation of pre-existing hot stars of early spectral types.

If the contracting mass was primitively endowed with a fairly large amount of kinetic energy, for instance in the form of turbulent velocities, one might also think of collisional ionization[1] transforming the turbulent energy into radiation and ionization energy. But this process will be efficient only if the contracting mass is much larger than the critical mass M_c below which no contraction is

[1] Cf. for instance F. HOYLE: Astrophys. Journ **118**, 513 (1953).

possible and this condition is not likely to be realized at the stage considered here. Note that in computing M_c either by means of JEANS' criterion or the condition that, in absolute value, the potential energy of a particle at the surface be greater than some mean value of its kinetic energy, the energy of turbulence should be included in the latter[1].

In their work, BIERMANN and COWLING have used essentially the energy method applied to neighbouring homologous configurations. This is equivalent to setting ξ equal to a constant in (16.1) and the result could be stated also in saying that, in the conditions recalled above, the mean value of Γ_1 with respect to $p\,d\mathscr{V}$ is smaller than $\frac{4}{3}$. As these conclusions may have an important bearing on cosmological problems, a more detailed investigation might be worthwhile.

Since Γ_1 is the important parameter in this problem, a little more attention may be devoted to it. By definition (cf. [*1*], Sect. 53) Γ_1 is the coefficient relating the Lagrangian variations of pressure δp and density $\delta\varrho$ in the course of an isentropic modification. This implies that the energy generation ε and the divergence of the total flux $\boldsymbol{F}$, if they do not cancel out exactly in the course of the motion, are nevertheless very small and contribute only a very small correction represented by the second member of the general relation between δp and $\delta\varrho$ which can be written [cf. [*1*], Eq. (56.18)]

$$\frac{d\,\delta p}{dt} - \frac{\Gamma_1 p}{\varrho}\,\frac{d\,\delta\varrho}{dt} = (\Gamma_3 - 1)\,\varrho\,\delta\left[\varepsilon - \frac{1}{\varrho}\operatorname{div}\boldsymbol{F}\right]. \tag{16.3}$$

If the time has been separated by the introduction of a factor $e^{i\sigma t}$ in all perturbations, Eq. (16.3) becomes

$$e^{i\sigma t}\left(\delta p_0 - \Gamma_1\frac{p}{\varrho}\,\delta\varrho_0\right) = \frac{(\Gamma_3 - 1)}{i\sigma}\,\varrho\left[\delta\varepsilon - \frac{d\,\delta L}{dm}\right] \tag{16.4}$$

where a subscript zero denotes the amplitudes.

In all cases, δL is expressible in terms of δT, $\delta\varrho$ etc. and their spatial derivatives, so that its contribution to (16.4) is always of the form

$$\frac{d\,\delta L}{dm} = e^{i\sigma t}\,\frac{d\,\delta L_0}{dm}. \tag{16.5}$$

If the energy generation is due to thermonuclear reactions [cf. Eq. (14.17)], $\delta\varepsilon$ also is a linear function of δT and $\delta\varrho$ and its contribution to (16.4) can be written

$$\delta\varepsilon = e^{i\sigma t}\,\delta\varepsilon_0 = e^{i\sigma t}\,\varepsilon\left(\mu\frac{\delta\varrho_0}{\varrho} + \nu\frac{\delta T_0}{T}\right). \tag{16.6}$$

Substituting (16.5) and (16.6) into (16.4) we have

$$\delta p_0 - \frac{\Gamma_1 p}{\varrho}\,\delta\varrho_0 = \frac{\Gamma_3 - 1}{i\sigma}\,\varrho\left[\delta\varepsilon_0 - \frac{d\,\delta L_0}{dm}\right], \tag{16.7}$$

and we see that the second member of (16.7) is imaginary. Expressing $\delta\varepsilon_0$ and δL_0 explicitly in terms of $\delta\varrho_0$ and δp_0, it is easy to define a generalized coefficient Γ_1' such that (16.7) could be written as

$$\delta p_0 = \Gamma_1'\,\frac{p}{\varrho}\,\delta\varrho_0 \tag{16.8}$$

[1] S. CHANDRASEKHAR: Proc. Roy. Soc. Lond., Ser. A **210**, 26 (1951).

and Eq. (12.12) would still be valid with Γ_1' substituted for Γ_1 as can also be seen from the general equation (57.25) in Ref. [1]. But, in this case, since $(\Gamma_1'-\Gamma_1)$ is imaginary the solution of the generalized equation (12.12) will be complex. The imaginary parts of ξ and σ^2, which in general are very small in absolute value, correspond respectively to a phase shift and to a damping, which are the special object of the study of vibrational stability, while the real part of the equation depending on Γ_1 is used to determine the dynamical stability of the star.

However subatomic energy may also be liberated or absorbed otherwise than by thermonuclear reactions. In particular in the course of its evolution, a star may, in certain circumstances, reach a state of very high ϱ and T in which a true nuclear equilibrium is established in its interior such, for instance, as that contemplated in HOYLE's theory[1] of the formation of heavy elements. In that case, ε depends on the *rate* of variation of the state variables, say ϱ and p, and instead of (16.6), we have

$$\left.\begin{aligned}\delta\varepsilon &= \varepsilon\left[f_1(\varrho,p)\frac{d\delta\varrho}{dt}+f_2(\varrho,p)\frac{d\delta p}{dt}\right]\\ &= i\sigma\, e^{i\sigma t}\left[f_1(\varrho,p)\,\delta\varrho_0+f_2(\varrho,p)\,\delta p_0\right]\end{aligned}\right\}\tag{16.9}$$

which substituted in (16.4) brings this time a real contribution to Γ_1. The exact evaluation of Γ_1 would have to take into account the accompanying changes in number density of particles or in mean molecular weight just as in the case of ionization (cf. [1], Sect. 53). However, considering the large energy made available in such displacements of a nuclear equilibrium, the energy term given by (16.9) s likely to be dominant.

If the equilibrium between nuclei and elementary particles is such that any increase in density leads on the whole to recombination with liberation of energy, Γ_1 may reach values appreciably higher than $\frac{5}{3}$ corresponding to an increased incompressibility of the stellar fluid and increased stability.

On the other hand, if any further increase in ϱ favours on the whole endothermic reactions with dissociation of complex nuclei, Γ_1 may be reduced to a value close to 1 in a large part of the star causing a violent instability. In HOYLE's theory, the collapse under gravity may be interpreted as the first phase of this instability. Due to the large kinetic energy acquired by the stellar material, the contraction will certainly overshoot the point where the centrifugal force in a large fraction of the star balances exactly gravity. After this, the centrifugal force increasing more and more will eventually stop the inward motion and transform it in a general expansion which, due to the same instability as before, will now accelerate very rapidly and may very well lead to the disruption of the star in some kind of supernova explosion.

Of course, it must be very difficult to follow this process in details and, when $d\varrho/dt$ becomes large, even the study of the stability of the motion would require the use of more complete perturbation equations [cf. [1], Eqs. (56.1) to (56.7)].

Up to now, we have treated stellar matter as a perfect gas but we know that, in white dwarfs, the hydrostatic support is essentially provided by the pressure of a degenerate electron gas obeying Fermi statistics. In that case, the equation of state reduces in a first approximation (complete degeneracy) to a relation between p and ϱ only, which can be expressed in a parametric form (cf. [14], Chap. X or [15], Chap. 4) by the following relations

$$p = A\,f(x), \qquad \varrho = n\bar{\mu}_e m_H = B\,x^3 \tag{16.10}$$

[1] F. HOYLE: Monthly Notices Roy. Astronom. Soc. London **106**, 343 (1946).

where the parameter x is related to the maximum value p_0 of the electron momentum in the conditions considered and where $\bar{\mu}_e$ is the mean molecular weight per electron

$$\bar{\mu}_e = \left(\sum_z \frac{X_z z}{A_z}\right)^{-1} \tag{16.11}$$

X_z being the abundance by mass of the atomic species of atomic number z and atomic weight A_z.

For small values of x ($p_0 \ll m_e c$: non-relativistic degeneracy) or large values of x ($p_0 \gg m_e c$: relativistic degeneracy) the elimination of x in (16.10) is simple and leads respectively to the following limiting equations of state:

$$p = K \varrho^{\frac{5}{3}} \quad (x \to 0) \quad \text{or} \quad p = K \varrho^{\frac{4}{3}} \quad (x \to \infty).$$

To the same order of approximation, the adiabatic exponent Γ_1' relating the variations of pressure and density, in the course of an isentropic modification[1] can be written

$$\Gamma_1' = \frac{8x^5}{3(x^2+1)^{\frac{1}{2}} f(x)} \tag{16.12}$$

and one verifies easily that

$$\lim_{x\to\infty} \Gamma_1' = \tfrac{4}{3}. \tag{16.13}$$

From the preceding discussion, one might think that this implies a tendency towards dynamical instability when the configuration tends towards a state of complete relativistic degeneracy. However, this is not true because the radius of the configuration tends towards zero simultaneously with $(3\Gamma_1' - 4)$ while the mass remains finite and tends towards the critical mass of CHANDRASEKHAR. As a consequence, the gravitational energy [numerator of (16.1)] increases indefinitely and the moment of inertia [denominator of (16.1)] decreases and tends towards zero. This factor more than compensates the decrease of $(3\Gamma_1' - 4)$ and, actually σ^2 tends to increase indefinitely.

The question has been discussed quantitatively by E. SAUVENIER-GOFFIN[2] on the basis of the explicit expression (16.1) of σ^2 which can be written here

$$\sigma^2 = \frac{A}{B} \frac{\int\limits_0^{x_c} \frac{8x^4 r^3}{(x^2+1)^{\frac{3}{2}}} \xi \, dx}{\int\limits_0^{x_c} \xi r^4 x^3 \, dx}. \tag{16.14}$$

As $(3\Gamma_1' - 4) \to 0$, ξ tends to become a constant and we have

$$\lim_{(3\Gamma_1'-4)\to 0} \sigma^2 = \lim_{x_c\to\infty} 4\pi G B (1+x_c^2)^{\frac{1}{2}} \frac{\int\limits_0^{\eta_R} \vartheta_3 \vartheta_3' \eta^3 \, d\eta}{\int\limits_0^{\eta_R} \vartheta_3^3 \eta^4 \, d\eta} \to \infty$$

where the integrals can be evaluated for the polytrope of index $n=3$ and remain finite. Thus on that approximation, dynamical instability is never reached.

However, SCHATZMAN ([*15*], Chap. 4, § 3 and Chap. 5, § 4) has recently remarked that, when the density becomes extremely large, the capture of electrons

---

[1] E. SAUVENIER-GOFFIN: Mém. Soc. Roy. Sci. Liège, Sér. IV **10**, 1 (1949). — Bull. Soc. Roy. Sci. Liège **19**, 47 (1950).

[2] E. SAUVENIER-GOFFIN: Mém. Soc. Roy. Sci. Liège, Sér. IV **10**, 1, Chap. 2, § 9 (1949). There is an error on p. 35 in the derivation of the general expression for σ^2 which in fact is always given by formula (29) with an extrafactor ξ in both integrals. This has been corrected by E. SCHATZMAN (cf. [*15*], Chap. 5, Eq. 115).

by nuclei can no longer be neglected. If one treats the effect as an equilibrium between capture and radioactivity there arises a correction to Γ_1' as defined by (16.12) which is essentially due here to the change in $\bar{\mu}_e$ or, in other words, to the change in the number of free electrons. As the latter decreases when p increases, Γ_1' is decreased below its value (16.12) and may reach values smaller than $\frac{4}{3}$ for sufficiently large concentration (cf. [*15*], Table 21). This in turn will affect (16.14) adding a negative part to the numerator and, according to SCHATZMAN, this will lead to dynamical instability for a critical value of the radius of the order of $2.7\times 10^{-3} R_\odot$.

Thus we may conclude that, while any agregate of matter of normal stellar dimensions is very likely to be dynamically stable, there are however possibilities of dynamical instability for very large or very small radii, which, in both cases, may have interesting cosmological consequences.

β) Non-radial perturbations and interactions between local and dynamical stability. In this case, if we expand the perturbation δr, ϱ', p' and Φ' in series of spherical harmonics

$$f(r)\, P_l^m(\cos\vartheta)\, e^{\pm i m \varphi}, \qquad m = 0, 1, \ldots l \tag{16.15}$$

the small perturbation method leads to a more complicated eigenvalue problem in which the exact differential equation is of the fourth order with some of the coefficients depending in a complicated manner on the parameter $\lambda = \sigma^2$ (cf. [*1*], Sect. 75).

Except special cases ([*1*], Sect. 76) where the order of the equation reduces to 2, the complete problem has never been treated. However, the hypothesis that the perturbation of the gravitational potential is negligible seems reasonable even for harmonics of small degree l, especially in stars with fairly high central condensation (cf. Tables 15 and 16 in [*1*] and Sect. 80) and it simplifies greatly the problem which becomes one of the second order in the coefficients of which the parameter appears as σ^2 and $1/\sigma^2$ [cf. [*1*], Eqs. (79.9) to (79.12)].

As pointed out by COWLING [*16*], this implies the existence of two spectra, one comprising discrete eigenvalues denoted by $(\sigma_{l,k}^2)_p$ which increase indefinitely with the degree l of the spherical harmonics and the order k of the mode (number of nodes from $r=0$ to $r=R$) and another one in which the discrete eigenvalues $(\sigma_{l,k}^2)_g$, taken in absolute value, decrease toward zero as k increases while they increase indefinitely with l. The modes of oscillations associated respectively with these two spectra are known as the p- and g-modes. The first ones are actuated mainly by pressure variations and possess important radial components, the second correspond to motions having comparatively larger components along the level surfaces and are mainly due to the action of gravity tending to smooth out any density irregularities on a level surface.

It is very likely that these general inferences remain true for the complete problem including the effects of the perturbations Φ' of the gravitational potential. Anyway, they are certainly confirmed by the exact analytical solutions which the complete equation admits in the case of the homogeneous compressible model (cf. [*1*], Sect. 76α). Nevertheless, a somewhat detailed mathematical analysis of the general case, even for the incomplete problem ($\Phi' = 0$), would be welcome.

In that respect, one of the difficulties arises from the mobile singularities in the second order equations which have been written down in Ref. [*1*], Eqs. (79.11) and (79.12), for the variables $v = r^2 p^{1/\Gamma_1}\,\delta r$ and $w = p'/p^{1/\Gamma_1}$. One may note however that these singularities are always regular. The general solution around such a singular point is always free of a logarithmic part and remains finite, its derivative tending towards zero.

This can also be verified directly on the equations expressed in terms of the variables $\zeta = \delta r$ and $y = p'/\varrho$ which are sometimes more directly useful:

$$\left.\begin{aligned} &\frac{d^2\zeta}{dr^2} + \frac{d\zeta}{dr}\left\{\frac{4}{r} + \frac{1}{\varrho}\frac{d\varrho}{dr} + \frac{1}{\left[\frac{l(l+1)}{\sigma^2} - \frac{\varrho r^2}{\Gamma_1 p}\right]}\frac{d}{dr}\left(\frac{\varrho r^2}{\Gamma_1 p}\right)\right\} + \\ &+\zeta\left\{\frac{\sigma^2\varrho}{\Gamma_1 p} - \frac{l(l+1)}{r^2}\frac{\sigma^2 + Ag}{\sigma^2} + \frac{2}{r^2} - \frac{1}{r^2}\frac{d}{dr}\left(\frac{g\varrho r^2}{\Gamma_1 p}\right) + \frac{2A}{r} + \right.\\ &\left. + \frac{\left(\frac{2}{r} - \frac{g\varrho}{\Gamma_1 p}\right)}{\left[\frac{l(l+1)}{\sigma^2} - \frac{\varrho r^2}{\Gamma_1 p}\right]}\frac{d}{dr}\left(\frac{\varrho r^2}{\Gamma_1 p}\right)\right\} = 0 \end{aligned}\right\} \tag{16.16}$$

and

$$\left.\begin{aligned} &\frac{d^2y}{dr^2} + \frac{dy}{dr}\left[\frac{2}{r} + \frac{1}{\varrho}\frac{d\varrho}{dr} - \frac{d(Ag)}{dr}\frac{1}{\sigma^2 + Ag}\right] + \\ + y&\left[\frac{\sigma^2\varrho}{\Gamma_1 p} - \frac{l(l+1)}{r^2}\frac{\sigma^2 + Ag}{\sigma^2} + \frac{1}{r^2}\frac{d}{dr}(r^2 A) - \frac{A}{\sigma^2 + Ag}\frac{d(Ag)}{dr}\right] = 0. \end{aligned}\right\} \tag{16.17}$$

In these equations $g = GM/r^2$ is the gravity, and

$$A = \frac{1}{\varrho}\frac{d\varrho}{dr} - \frac{1}{\Gamma_1 p}\frac{dp}{dr} \tag{16.18}$$

is the quantity which appears in SCHWARZSCHILD'S criterion of convective stability which may be written

$$A < 0. \tag{16.19}$$

In the first equation, the mobile singularity vanishes if σ^2 is negative i.e. for unstable modes. If σ^2 is positive there is always one singularity as $\varrho r^2/\Gamma_1 p$ varies from zero at the centre to infinity at the surface. In the second equation, if A is positive (local instability), there are no mobile singularities when σ^2 is positive (i.e. for stable modes) while, if A is negative (local stability), the singularity vanishes for σ^2 negative (i.e. for unstable modes, if such exist). The number of singularities of the second equation when present, depends critically on the behaviour of Ag.

Even in the case of polytropes, the variation of Ag is rather complicated: it vanishes at the centre and becomes infinite at the surface but, in between, it may present a maximum M and a minimum m whose intensities and positions depend on the model. Thus for values of σ^2 [or α: cf. [1], Eqs. (79.21) to (79.23)] comprised in the range $M - m$, there will be three such singularities considerably affecting the shape of the corresponding modes (y or w). For instance, in the standard model, this occurs for the f-mode (no node or loop in the star) and explains the existence of a minimum close to zero in w_f. According to OWEN[1], as the central condensation increases, the minimum of w_f deepens and finally becomes negative and the f-mode vanishes. As the concentration of the model increases further g and p-modes of higher and higher orders disappear in their turn.

In the same way, the difference in the behaviour of the mobile singularities for stable and unstable modes implies marked differences between them. In

[1] J. W. OWEN: Monthly Notices Roy. Astronom. Soc. London **117**, 384 (1957).

that respect, the p-modes behave very much like the radial oscillations discussed previously, i.e. instability can only enter through the mode of lowest degree l and lowest order k. In fact, it is doubtful whether any instability at all could manifest itself through these modes. For instance, in the case of the homogeneous compressible model, Eq. (76.10) in Ref. [1] taken with the positive sign before the radical shows that $(\sigma_{2,0}^2)_p$ is always positive even if $\Gamma_1 < \frac{4}{3}$, although this is a critical value for stability towards radial perturbations in any model.

As far as the g-modes are concerned, the situation is very different since the sign of σ_g^2 depends primarily on the sign of A as shown by the discussion following Eqs. (77.4), (78.22) and (79.18) in Ref. [1]. This means that dynamical instability towards non-radial perturbations is intimately related to the existence of superadiabatic gradients. In fact, the discussions recalled above suggest that no dynamical instability will arise unless the condition (16.19) is violated in an appreciable portion of the star.

This is understandable since SCHWARZSCHILD's criterion (16.19) expresses simply the condition that σ^2 be positive locally when the perturbation of pressure p' is neglected as well as Φ'. As was shown by LEDOUX ([13], Chap. III, §§ 5 and 6), this implies harmonics of very high degrees ($l \gg 1$), i.e. perturbations of very small horizontal wavelengths and such that the associated vertical displacement does not vary too rapidly with r (low order modes), i.e. fairly large vertical wavelengths. He also showed that if p' becomes appreciable, instability will still manifest itself most strongly in the same conditions (cf. [1], discussions below Eqs. (78.25) and (79.19)].

Of course, the main problem here is to distinguish between local and general dynamical instability. If a region violating (16.19) develops in a star, it probably starts as a very narrow layer and we generally assume that convective equilibrium rapidly replaces radiative equilibrium in it and that convection extends progressively as the unstable layer grows reducing A to a very small positive value. This is in reasonable agreement with the previous results since the most unstable perturbations are those of small horizontal extent which are also likely to be the most common and since, in any case, as long as that region is fairly narrow, it could not, whatever the initial value of A, endanger the dynamical stability of the star as a whole. In those conditions, one may wonder whether any true dynamical instability could ever arise from this cause.

There may, however, be a flaw in this argument because initial perturbation inside a star may not at all be as common as we have assumed and, furthermore, there are factors such as a rotation or a magnetic field (cf. the end of Sect. 17α and β) which tend to delay the establishement of a convective regime with closed circulations while they would not probably oppose a true dynamical instability leading to ejection of matter, for instance, along the lines of force of the magnetic field or parallel to the axis of rotation. There is no doubt that if A could become appreciable in a large external zone as was once proposed by BIERMANN[1], a violent instability could arise for a fairly low harmonic which would lead to the ejection of material in the form of a few separated jets recalling the conditions in some novae[2].

But even, if such extreme cases are never realized, the question deserves further study. First of all, our previous result, that instability increases as the

---

[1] L. BIERMANN: Z. Astrophys. **18**, 344 (1939).

[2] Cf. H. VAN DE HULST: Problems of Cosmical Aerodynamics. Central Air Documents Office III, p. 116, 1951.

horizontal size of the perturbation decreases, has been reached in neglecting all dissipation due to friction (ordinary or turbulent) and heat transfer (conduction and radiation). For very small sizes, the damping effect of those factors predominates so that actually there exists a finite optimum wavelength for the perturbation[1].

This manifests itself also in the more realistic condition of convective stability of the type of RAYLEIGH's criterion (cf. Sect. 17α) which, in a layer of given depth, imposes a minimum positive value of A under which convection stops. This is true as well of thermal turbulence if the conservation of kinetic energy of turbulence [cf. [*1*], Eq. (50.8)] is properly taken into account.

In a deep convective region this minimum positive value is very small but, in any case, if the star was wholly convective there would certainly be unstable g-oscillations (cf. [*1*], Sect. 78β) however small the value of A and it would be very interesting to compute the lowest degree l of the harmonic which can render the star dynamically unstable, let us say for the first few modes. Of course, the characteristics of these unstable perturbations may be so particular that the chances of them being realized may be small but, on the other hand, a star has plenty of time for experimenting.

Finally, there are other aspects which, although less fundamental in the sense that they do not endanger the stability of the star as a whole, may nevertheless be important for the interpretation of many minor and erratic changes occurring on the surface of a star like the Sun or, perhaps, even in the case of flare stars.

Let us assume, for instance, the existence of a superficial layer in convective equilibrium. In that case, because of the lower density, the minimum positive value of A may be appreciably larger than deeper down in the star. Actually, in some cases, convection may be so inefficient, that large superadiabatic gradients may subsist in these layers. In these circumstances, it may happen that, from time to time, conditions particularly favourable to a strong instability become realized locally due either to an increase in A or through the fortuitous building up of a suitable perturbation (upward displacement through the whole layer, inside a fairly narrow column). This could easily lead to the ejection above the stellar surface of narrow jets of material.

Since this last type of phenomena is of interest only close to the stellar surface adequate simplifications of the corresponding equations may facilitate its study appreciably. For instance, the curvature may be neglected and, if we consider again an adiabatic perturbation, we may in Eqs. (16.16) introduce, as independent variable, the depth $z=R-r$ and neglect all terms proportional to positive powers of $1/r$ and $1/R$ except $l(l+1)/r^2$ which tends toward $k^2=(\mu^2+\nu^2)$ with $\mu=2\pi/\lambda_x$ and $\nu=2\pi/\lambda_y$ where λ_x and λ_y are the horizontal wavelengths of the perturbation. This yields

$$\left.\begin{aligned}&\frac{d^2\zeta}{dz^2}+\frac{d\zeta}{dz}\left[\frac{1}{\varrho}\frac{d\varrho}{dz}+\frac{1}{\left(\frac{k^2}{\sigma^2}-\frac{\varrho}{\Gamma_1 p}\right)}\frac{d}{dz}\left(\frac{\varrho}{\Gamma_1 p}\right)\right]+\\&+\zeta\left[\frac{\sigma^2\varrho}{\Gamma_1 p}-\frac{k^2}{\sigma^2}(\sigma^2+Ag)+\frac{d}{dz}\left(\frac{\varrho g}{\Gamma_1 p}\right)+\frac{g\varrho/\Gamma_1 p}{\left(\frac{k^2}{\sigma^2}-\frac{\varrho}{\Gamma_1 p}\right)}\frac{d}{dz}\left(\frac{\varrho}{\Gamma_1 p}\right)\right]=0\end{aligned}\right\}\tag{16.20}$$

[1] Cf. for instance S. CHANDRASEKHAR: Proc. Cambridge Phil. Soc. **51**, 162 (1955) where a somewhat similar problem in the physically simple case of two superposed layers of uniform incompressible fluids of different densities is solved in details.

and

$$\left.\begin{aligned}&\frac{d^2y}{dz^2}+\frac{dy}{dz}\left[\frac{1}{\varrho}\frac{d\varrho}{dz}-\frac{1}{\sigma^2+Ag}\frac{d(Ag)}{dz}\right]+\\&+y\left[\frac{\sigma^2\varrho}{\Gamma_1 p}-\frac{k^2}{\sigma^2}(\sigma^2+Ag)-\frac{dA}{dz}+\frac{A}{\sigma^2+Ag}\frac{d(Ag)}{dz}\right]=0.\end{aligned}\right\}\quad(16.21)$$

If furthermore, g may be considered as constant and if the atmosphere is polytropic with $p \propto \varrho^{\Gamma}$, the equilibrium conditions are given by

$$\left.\begin{aligned}&p=\frac{\Gamma-1}{\Gamma}gKz^{\frac{\Gamma}{\Gamma-1}}, && \varrho=Kz^{\frac{1}{\Gamma-1}},\\&T=\frac{\Gamma-1}{\Gamma}\frac{\bar{\mu}}{\mathsf{R}}gz, && A=\frac{\Gamma-\Gamma_1}{\Gamma_1(\Gamma-1)}\frac{1}{z}=\frac{\varepsilon}{z},\end{aligned}\right\}\quad(16.22)$$

and Eqs. (16.20) and (16.21), where we write $\beta=\Gamma/\Gamma_1(\Gamma-1)$, reduce to

$$\frac{d^2\zeta}{dz^2}+\frac{d\zeta}{dz}\left(\frac{\beta\Gamma_1}{z}-\frac{1}{z-\frac{\beta\sigma^2}{gk^2}}\right)+\zeta\left[\frac{\sigma^2\beta}{gz}-k^2-\varepsilon\frac{k^2g}{\sigma^2z}-\frac{\beta}{z}\frac{1}{\left(z-\frac{\beta\sigma^2}{gk^2}\right)}\right]=0,\quad(16.23)$$

$$\frac{d^2y}{dz^2}+\frac{dy}{dz}\left[\frac{\beta\Gamma_1}{z}-\frac{1}{\left(z+\frac{\varepsilon g}{\sigma^2}\right)}\right]+y\left[\frac{\sigma^2\beta}{gz}-k^2-\varepsilon\frac{k^2g}{\sigma^2z}+\frac{\varepsilon}{z}\left(\frac{1}{z+\frac{\varepsilon g}{\sigma^2}}\right)\right]=0.\quad(16.24)$$

These equations or rather their equivalent form in Lagrangian coordinates have been discussed in some details mainly in view of meteorological applications[1]. It has been shown, in particular, that their general solutions may be expressed in terms of hypergeometric functions but few results are directly available for the application which we have in mind here. However recently, A. SKUMANICH[2] has tackled the problem numerically and has obtained results which confirm completely our conclusions concerning the most unstable type of perturbation.

To reach useful results one might have to give up some of the simplifying assumptions since the external layers of a star depart appreciably from a polytrope and, instead of considering g as a constant, one might have to treat it as inversely proportional to r^2:

$$g\approx g_0\left(1+2\frac{z}{R}\right).$$

More important perhaps, one should take into account the sources of dissipation and consider in that case, a few special forms of the perturbation such, for instance, as a narrow rising column surrounded by a large sinking region which may be the most efficient and which could be analysed in Fourier components in terms of the elementary solutions. No doubt, this would complicate the problem but it may remain tractable in these external layers and its solution would present a real interest.

17. Some remarks on the influence of extra-factors such as a rotation, a magnetic field or an external gravitational field. Up to now, we have always considered the star as submitted to its own gravitational attraction only. But in many cases, this field of forces is modified by centrifugal forces (rotating stars), Lorentz-forces (magnetic stars) or an external gravitational field (binaries).

[1] Cf. for instance, V. BJERKNES, J. BJERKNES, H. SOLBERG and T. BERGERON: Hydrodynamique physique, Vol. 3, Chap. VIII. Paris: Presses Universitaires de France, 1934. Also H. SOLBERG: Astrophys. Norv. 2, No. 2 (1936).

[2] A. SKUMANICH: Astrophys. Journ. 121, 408 (1955).

Often, these extra-forces are very small compared to the star's own gravitational attraction and they may, at most, slightly modify the frequencies of oscillations of the star. However, in each case, if the extra-force increases, there is a limit past which equilibrium becomes impossible and it is interesting to know the corresponding critical values. Furthermore, the presence of these extra-fields of forces may affect the local stability in some regions of the star. As far as the influence of rotation and magnetic field are concerned, a compact summary of many of the most important aspects has been given by S. CHANDRASEKHAR [17] in the George Darwin Lecture delivered in 1953.

α) *Effects of a rotation.* Let us start with a slowly rotating star of angular velocity Ω. As long as the terms in Ω^2 are negligible, a straightforward application of the method of perturbations (cf. [1], Sect. 82) yields a reasonable approximation for the frequencies. For a perturbation corresponding to a spherical harmonic of degree l and rank m [cf. Eq. (16.15)], the frequencies of the mode of order k are given by [cf. [1] Eqs. (82.30) and (82.32)]

$$\sigma_{k,m} = \sigma_k + \sigma''_{k,m} = \sigma_k \pm m\, C_k \Omega \tag{17.1}$$

where σ_k is the frequency of the non-rotating star. The plus and minus signs in (17.1) correspond respectively to a wave travelling in the opposite or the same direction as the rotation of the star.

If terms in Ω^2 have to be taken into account, the application of the perturbation method becomes more difficult, due mainly to the perturbation of the boundary conditions. The only case which has been considered in some detail (cf. [1], Sect. 82) corresponds to pseudo-radial oscillations for which a fairly simple expression of σ^2 can be given [cf. [1], Eq. (82.27)]:

$$\sigma^2 = -(3\gamma - 4)\frac{V}{I} + (5 - 3\gamma)\frac{2K_R}{I} \tag{17.2}$$

where V is the potential gravitational energy and I the moment of inertia about the centre.

But both formula (17.1) and (17.2) have been established specifically for the case where σ^2 is large compared to its perturbation due to the rotation. They are useful to estimate the effects of rotation on the period of oscillations of stable stars, but they cannot be directly applied to cases of marginal stability ($\sigma^2 \to 0$) which should be the object of a special discussion. Anyway, in such cases, the main source of instability is not the rotation which could only exercise a trigger action.

Only large values of Ω can on their own, endanger the stability of a star, but in that case (cf. Sect. 8), the lack of an adequate model for the steady state of the star must be added to the dynamical difficulties. Up to now, we have only general indications often drawn from highly idealized models, such for instance, as upper limits to the value of Ω [cf. Eqs. (6.9), (6.10) and (8.25)] past which secular instability sets in. As far as the form taken by this instability is concerned, the discussion in Sect. 8 has shown that, for a gaseous mass, the most likely course is a shedding of material in the equatorial plane. However, many important questions are still awaiting an answer. For instance, it would be very interesting to know in the case of a compressible mass, whether dynamical instability could also occur and whether it may set in at the same time as the secular instability, as in the case of the Jacobi ellipsoids, or whether it occurs only later as for the Maclaurin spheroids (cf. end of Sect. 7).

Finally, rotation may also affect local stability, modifying SCHWARZSCHILD'S criterion (16.19). In this respect, the simplest illustration of the influence of rotation is provided by a rotating incompressible fluid. If the law of rotation is such that ΩR^2 increases with the distance R from the axis, an element displaced outward arrives in its new surroundings with less than the local angular velocity, due to the conservation of angular momentum. Hence it is subject to less centrifugal force and tends to fall back. This leads to a very simple stability criterion due to Lord RAYLEIGH[1].

$$\frac{d(\Omega R^2)}{dR} > 0. \tag{17.3}$$

If the medium is compressible, the buoyancy of the displaced element must be composed with the defect of centrifugal force to obtain the net force acting on it and one may expect that the global condition will involve a combination of the terms on the left of (16.19) and (17.3). In particular, if the rotation satisfies a condition of type (17.3) it is likely that it could have an appreciable inhibiting effect on the outset of convection at least in certain directions[2].

For instance, in the case of uniform rotation, RANDERS[3] obtained the following stability criterion for radial displacements

$$g A < 4\Omega^2 \sin^2 \vartheta \tag{17.4}$$

where $g = -\frac{1}{\varrho}\frac{\partial p}{\partial r}$ and ϑ is the angle which $\boldsymbol{r}$ makes with $\boldsymbol{\Omega}$. It reduces to (16.19) if $\Omega = 0$ and to (17.3) if the fluid is incompressible. We notice that for displacements parallel to the axis, instability will always occur as soon as SCHWARZSCHILD'S criterion is violated. On the other hand, in a direction perpendicular to $\boldsymbol{\Omega}$, the absolute value of the actual temperature gradient has to overshoot its adiabatic value by a finite amount before instability sets in. Randers concluded that, in a rotating star, convection would be hindered in directions parallel to the equator creating an asymmetry in the energy transfer by convection with an excess energy flowing towards the poles. For displacements perpendicular to $\boldsymbol{\Omega}$, the criterion can be generalized fairly easily to the case of a general law of rotation[4].

The difference between motions parallel or perpendicular to the axis of rotation is a significant one which subsists generally, but it also suggests that the criterion may be fairly sensitive to the type of displacements considered. In particular, in the cases referred to above the criteria were established for displacements symmetric about the axis and conservation of angular momentum for each element. But as pointed out by COWLING[5], in more general types of displacements, azimuthal pressure gradients may develop, for instance between contiguous ascending and descending currents, which may be able to destroy locally the conservation of angular momentum.

The influence of these terms has been discussed by COWLING[6] using the small perturbation method and assuming the dimensions of the convected elements to be small as compared to the scale of the variations of pressure, density etc. in the equilibrium state. In that case and for uniform rotation (surfaces of equal pressure and equal density coinciding), he finds using cylindrical coordinates (R, z, φ) that if l_R, l_z, l_φ are wave numbers inversely proportional to the wavelengths of the perturbation along the coordinate axes, *instability* will occur, provided

$$\boldsymbol{g} \cdot \boldsymbol{A}[l_\varphi^2 + (l_R \cos\alpha - l_z \sin\alpha)^2] < -4\Omega^2 l_z^2 \tag{17.5}$$

---

[1] Lord RAYLEIGH: Proc. Roy. Soc. Lond. Ser. A **93**, 148 (1916).
[2] H. JEFFREYS: Proc. Roy. Soc. Lond., Ser. A **118**, 195 (1928).
[3] G. RANDERS: Astrophys. Journ. **95**, 454 (1942).
[4] Cf. C. WALÉN: Ark. Mat. Fys. **33**a, No. 18 (1946).
[5] T. G. COWLING: Monthly Notices Roy. Astronom. Soc. London **105**, 166 (1945).
[6] T. G. COWLING: Astrophys. Journ. **114**, 272 (1951).

where $\boldsymbol{g}$ and $\boldsymbol{A}$ are the obvious vectorial generalizations of our previous notation and α is the angle between z and $-\boldsymbol{g}$ which, in general, differs very little from ϑ.

For radial displacements ($l_\varphi = 0$, $l_R \sin\vartheta + l_z \cos\vartheta = 0$), this reduces essentially to RANDERS' criterion and again, motion parallel to the axis becomes unstable as soon as SCHWARZSCHILD's condition (16.19) is violated. But this time, (17.5) shows that, provided l_z be small compared to l_R and l_φ, there are also displacements with non-vanishing components along R and φ which lead to instability for arbitrary small positive values of A. The resulting motion however possesses considerable regularity and is organized in whirl tubes perpendicular to the equatorial plane, the plane of whirling being comprised within the acute angle between the tube axis and the direction of local gravity. However its efficacity, as far as transfer of energy normal to the axis of rotation is concerned, is difficult to estimate. Nevertheless, we must conclude that, as far as incipient instability is concerned, it always appears (for some types of displacements) as soon as the gradient becomes superadiabatic.

The case of a non-uniform rotation is more complex. If (17.3) is verified, it can stabilize some types of displacements but, for others, its influence is negligible or it may even exercise a destabilizing action because of the shear that it produces.

The same problem had already been discussed previously independently of its relation with stellar structure by H. SOLBERG in 1936 and later by E. HØILAND[1] who endeavoured to interpret the problem in terms of the circulation theorems of V. BJERKNES. The approach of these authors is based more or less explicitly on the energy method and on the consideration of the simultaneous displacements of elements in a closed stream tube which, however, is reduced in most cases to an elementary tube with two parallel branches (δR, δz) and ($-\delta R$, $-\delta z$) in the meridian plane joined by two others which are considered infinitesimal as compared to the first ones. This method again postulates axial symmetry in the displacements but, with this limitation, it leads to a very general criterion which can be reduced to the following necessary and sufficient conditions of stability

$$\frac{2\Omega}{R}\frac{\partial(\Omega R^2)}{\partial R} + \boldsymbol{g}\cdot\boldsymbol{A} > 0, \quad g_z A_z > 0, \tag{17.6}$$

$$g_z\left[\frac{\partial(\Omega R^2)}{\partial R} A_z - \frac{\partial(\Omega R^2)}{\partial z}\cdot A_R\right] > 0. \tag{17.7}$$

If rotation is absent, one recovers easily SCHWARZSCHILD's criterion while, if the fluid is in adiabatic equilibrium ($A_R = 0$, $A_z = 0$), (17.7) reduces to RAYLEIGH's condition. The criteria (17.6) and (17.7) have been critically discussed by WASIUTYNSKI[2] who pointed out that some of the possibilities of instability which they imply have probably no physical meaning and that they would disappear if the discussion was limited to displacements along those stream-tubes of steady motion which are actually possible in the circumstances considered. In that case, dissipation should be taken into account as well as appropriate boundary conditions, as was done for instance, by RAYLEIGH, JEFFREYS, LOW and others[3] in discussing cellular convection of the Bénard type in absence of rotation.

The simplest case, treated by RAYLEIGH, concerns a horizontal layer of fluid of thickness d in presence of gravity g and of a vertical gradient of temperature

[1] Cf. E. HØILAND: Arch. Mat. Naturvidensk., Oslo **42**, No. 5 (1939). — Avhandl. Norske Vidensk. Akad., Mat.-Naturv. Kl., Oslo, No. 11, 1941.

[2] J. WASIUTYNSKI: Astrophys. Norv. **4**, Sects. 6.2, 6.3 and 26.1 (1946).

[3] J. W. STRUTT (Lord RAYLEIGH): Phil. Mag. **32**, 529 (1919). — H. JEFFREYS: Proc. Cambridge Phil. Soc. **26**, 170 (1930). — A. R. LOW: Proc. Roy. Soc. Lond. Ser. A **125**, 180 (1929). — Proc. of the 3rd Intern. Congr. for Appl. Mechanics, Vol. 1, p. 109, Stockholm 1930.

(temperature T decreasing upward). If the fluid can be treated as incompressible except for the variation of the *weight* $g\varrho$ due to the varying temperature, i.e. $g\varrho = g\varrho_0(1 - \alpha\Delta T)$ where α denotes the coefficient of thermal expansion, RAYLEIGH found that convective instability occurs only if

$$\left|\frac{dT}{dh}\right| \geq \frac{27\pi^4 \varkappa \nu}{4g d^4 C_p \varrho \alpha} \tag{17.8}$$

where ν is the coefficient of kinematic viscosity, $\varkappa$ the effective conductivity ($\varkappa = k\varrho C_p$, if k is the thermometric conductivity) and C_p the specific heat at constant pressure.

If we denote by β the absolute value of dT/dh, this condition is often expressed in terms of the values taken by the Rayleigh number

$$\mathcal{R} = \frac{g\alpha\beta}{k\nu} d^4 \tag{17.9}$$

which must become larger than a definite critical value $\mathcal{R}_c$ for convection to be possible. $\mathcal{R}_c$ depends on the type of boundary conditions adopted and on the wave number a defined by

$$a^2 = 4\pi^2 d^2 \left(\frac{1}{\lambda_x^2} + \frac{1}{\lambda_y^2}\right)$$

where λ_x and λ_y are the wavelengths of the perturbation along two horizontal directions at right angle. For a given set of boundary conditions, the values of $\mathcal{R}$ allowing convection vary with a and admit, in general, a single minimum, say $\mathcal{R}_c$, for a certain value of a, say a_c. It is clear that as $\mathcal{R}$ increases, $\mathcal{R}_c$ corresponds to the earliest possibility for the setting in of convection and a_c fixes the most favourable horizontal dimensions of the convective cells, which, in this case, extend vertically through the whole layer. Note that this is in agreement with previous results that instability should be greatest for a perturbation with the least possible number of nodes along the vertical.

For instance, the value of $\mathcal{R}_c$ defined by (17.8), $\mathcal{R}_c = 657$, corresponds to the case where the fluid is bounded by two "free" surfaces (meaning free slip) on which, in any case, the vertical component of velocity and the perturbation of temperature vanish. The corresponding value of a_c is $a_c = 2.23$ or, for a square perturbation, $\lambda = 4d$. Of course, other modes of convection are possible corresponding to superposed layers of cells of smaller dimensions requiring larger values of $\mathcal{R}$ but, in moderately thick plane layers, convection will certainly establish itself through the first mode.

JEFFREYS showed that the criterion (17.8) holds good for compressible fluids as well, provided the density does not vary too much within the system and provided the temperature gradient is replaced by the excess of the temperature gradient over its adiabatic value

$$\left|\frac{dT}{dh}\right| - \left(\frac{\gamma - 1}{\gamma}\right)\frac{T}{p}\left|\frac{dp}{dh}\right| > \frac{27\pi^4 \varkappa \nu T}{4g d^4 C_p \varrho}$$

or

$$A > \frac{27\pi^4 \varkappa \nu}{4g d^4 C_p \varrho}. \tag{17.10}$$

In most cases of astrophysical interest, the second member is so small that (17.10) is not significantly different from SCHWARZSCHILD's criterion (16.19). One must keep in mind however that (17.8) or (17.10) are strictly valid for horizontal layers and a constant g only and, for some astrophysical and geophysical problems (convective stability close to the centre or in a thick layer), it would be interesting to discuss the effects of curvature and of the variation of g. Up to now, a few attempts have been made in that direction for simple fluids[1].

In recent years, CHANDRASEKHAR[2] has extended the discussion to the case of a uniformly rotating layer of angular velocity Ω using the same general assumptions as adopted by RAYLEIGH and recalled above. If the axis of rotation is vertical, the problem can be discussed very completely and CHANDRASEKHAR has shown that, in this case, the Coriolis force inhibits the onset of convection, $\mathcal{R}_c$ and a_c increasing by an amount which depends on the dimensionless parameter (Taylor number).

$$\mathcal{T} = \frac{4\, d^4 \Omega^2}{\nu^2}. \tag{17.11}$$

The increasing value of a_c means that the cells get more and more elongated along the vertical.

CHANDRASEKHAR also found that, in presence of sufficiently strong Coriolis forces ($\mathcal{T} > \mathcal{T}^*$) and if k/ν is greater than a certain critical value (1.478 in the case of two "free" surfaces), the instability manifests itself first as oscillations of increasing amplitudes (overstability or vibrational instability). This means that, for $k/\nu > 1.478$, $\mathcal{T} > \mathcal{T}^*$ the critical value for overstability $\mathcal{R}_0$ is always such that

$$\mathcal{R}_c < \mathcal{R}_0 < \mathcal{R}_c' \tag{17.12}$$

if $\mathcal{R}_c$ and $\mathcal{R}_c'$ denote respectively the critical values of the Rayleigh number when rotation is absent and when it is present.

This provides another interesting example where the introduction of dissipation induces a new type of instability. This can be understood qualitatively as follows: in the present case, if we neglect all dissipation, the restoring forces (buoyancy and defect of centrifugal force) acting on unit mass after a displacement (ζ_x, ζ_y, ζ_z) have components in the plane of the motion

$$g\, A\, \zeta_z \quad \text{and} \quad -4\,\Omega^2 (\zeta_x^2 + \zeta_y^2)^{\frac{1}{2}}$$

since $\boldsymbol{\Omega}$ and $-\boldsymbol{g}$ are along the z-axis. We say that instability occurs, if the distance from the displaced element from its place of origin tends to increase continually, i.e., if the projection of the forces on the displacement is positive or if

$$g\, A\, \zeta_z^2 - 4\,\Omega^2 (\zeta_x^2 + \zeta_y^2) > 0$$

which is essentially equivalent to COWLING's condition (17.5) with $\alpha = 0$.

Let us now compare the case of two stable layers without and with rotation, the respective conditions of stability being

$$A < 0 \quad \text{or} \quad g\, A\, \zeta_z^2 - 4\,\Omega^2 (\zeta_x^2 + \zeta_y^2) < 0. \tag{17.13}$$

In absence of dissipation, the displaced element will in both cases, oscillate indefinitely around its position of equilibrium.

If dissipation due to heat conduction and viscosity is present, we expect that, in the first case at least, the motion will be damped. Indeed, heat conduction reduces the difference between the temperature (or density) of the displaced

[1] J. WASIUTYNSKI: Astrophys. Norv. **4**, Sect. 22.2 (1946). — H. JEFFREYS and M. E. M. BLAND: Montly Notices Roy. Astronom. Soc. London, Geophys. Suppl. **6**, 148 (1951). — S. CHANDRASEKHAR: Phil. Mag., Ser. VII **43**, 1317 (1952); **44**, 233 (1953) also Quart. J. Mech. Appl. Math. **8**, 1 (1955).

[2] S. CHANDRASEKHAR: Proc. Roy. Soc. Lond., Ser. A **217**, 306, (1953); **231**, 198 (1955); also D. FULTZ and Y. NAKAGAWA: Proc. Roy. Soc. Lond., Ser. A **231**, 211 (1955).

element and that of the surrounding medium and the buoyancy is decreased by a fraction proportional to k, so that the restoring force becomes

$$g A \zeta_z (1 - \alpha k). \tag{17.14}$$

This will already yield a positive damping which viscosity will further reinforce.

When rotation is present, heat conduction will still have the same effect as before, but as far as viscosity is concerned, the situation is modified. Apart from a general dissipation of kinetic energy affecting about equally the different components of motion, viscosity will tend to equalize the velocity of rotation of the convected element and that of its surrounding, reducing the defect of centrifugal force by a fraction proportional to ν. Thus the force components now become

$$g A \zeta_z (1 - \alpha k) \quad \text{and} \quad - 4\Omega^2 (\zeta_x^2 + \zeta_y^2)^{\frac{1}{2}} (1 - \alpha \nu)$$

where we let the constant of proportionality α be equal. Evaluating the projection along the displacement, we obtain

$$|\zeta|^{-1} \{[g A \zeta_z^2 - 4\Omega^2 (\zeta_x^2 + \zeta_y^2)] - \alpha [g A \zeta_z^2 k - 4\Omega^2 (\zeta_x^2 + \zeta_y^2) \nu]\} \tag{17.15}$$

which must, of course, still be negative. If, in absolute value, (17.15) is larger than the left-hand member of the second condition (17.13), the restoring force is increased and vibrational instability (overstability) will result. But this requires

$$g A \zeta_z^2 k > 4\Omega^2 (\zeta_x^2 + \zeta_y^2) \nu$$

which is only possible if

$$0 < g A \cdot \zeta_z^2 < 4\Omega^2 (\zeta_x^2 + \zeta_y^2) \quad \text{and} \quad \frac{k}{\nu} > 1. \tag{17.16}$$

Thus while, in the purely thermal case, overstability is impossible[1] it may occur in presence of rotation if the conditions (17.16) are satisfied. Of course, the above treatment is incomplete especially as far as the effects of viscosity are concerned but, qualitatively, it agrees well with CHANDRASEKHAR'S conclusions and it clarifies the rôle of the different physical factors. As is apparent from the exact condition (17.12) and the accompanying discussion, the main effect of the viscosity terms neglected here, is to modify the limits in (17.16).

As far as astrophysical applications are concerned, it is probably safer, considering the different assumptions at the basis of the theory, to limit them to the external layers of stars. In that case, if the only form of viscosity to be taken into account is of molecular or radiative origin, k/ν becomes so large that, according to CHANDRASEKHAR, we should expect local instability due to superdiabatic gradients to manifest itself much more frequently as oscillations of increasing amplitudes than as cellular circulations. However, if turbulence is present, it may well reverse this conclusion.

When the axis of rotation makes an angle ϑ with the vertical, the problem remains unchanged for convective instability starting as elongated rolls parallel to the plane $(\boldsymbol{\Omega}, \boldsymbol{g})$ provided Ω is replaced by $\Omega \cos \vartheta$. In other cases, however, the problem becomes more complex and the solutions have not been worked out. The same is true of the effects of a non-uniform rotation. There is no doubt that this is a field where much difficult work remains to be done before completely general conclusions can be reached.

---

[1] A. PELLEW and R. V. SOUTHWELL: Proc. Roy. Soc. Lond., Ser. A **176**, 312 (1940).

β) Effects of a magnetic field. In many respects, the situation is similar to that encountered in the case of rotation and it has been summarized in [1], Sect. 83. As long as the total magnetic energy

$$\mathfrak{M} = \int^{\mathcal{V}} \frac{H^2}{8\pi} d\mathcal{V} \tag{17.17}$$

is small compared to the total gravitational energy V, the perturbation method yields a reasonable approximation for the correction σ_k'' to the frequency σ_k of any stable mode of oscillation. It is of the form (cf. [1], Sect. 83)

$$\sigma_k'' = C \frac{H^2}{\bar{\varrho}}. \tag{17.18}$$

However while, in the case of rotation, the method is straightforward if one limits oneself to the Coriolis forces neglecting terms in Ω^2 which affect the equilibrium distribution of density and pressure and the form of the boundary, in the case of a magnetic field, all the perturbing terms, in the equilibrium state as in the dynamical motion, are proportional to H^2 and all should be taken into account simultaneously[1]. This is particularly troublesome as far as the perturbation of the form of the boundary is concerned. Furthermore, in the case of a non-radial oscillation represented by (16.15), while the rotation lifts completely the degeneracy of the frequencies of the non-rotating star ($2l+1$ values of σ_k'', one for each value of $\pm m$, $0 \leqq m \leqq l$) the situation is more complex in presence of a magnetic field.

When the field strength increases, the situation becomes more and more difficult because of the lack of adequate static models. However, as in the case of rotation, one may resort to the general theorems which at least set an upper limit to the maximum possible field inside a star. Assuming that all stresses, magnetic as well as material vanish on the surface, S. CHANDRASEKHAR and E. FERMI[2] have derived such a limit from the virial theorem (cf. [1], Eq. (83.14)), which can be written in order of magnitude,

$$\overline{H^2} < 4 \cdot 10^{16} \frac{M^2}{R^4} \; (H \text{ in gauss}) \tag{17.19}$$

if M and R are expressed in units of the solar mass and the solar radius.

Recently, PIDDINGTON[3] has remarked that due to the special boundary conditions adopted by CHANDRASEKHAR and FERMI, some care must be exercised in the application of their results. As he points out, the necessary generalization is however fairly simple. Indeed, taking the scalar product of Eq. (83.11) in [1] with $\boldsymbol{r}$ and integrating over the volume, we have for the general expression of the virial theorem

$$\frac{1}{2}\frac{d^2 I}{dt^2} = 2K - \iiint_{\mathcal{V}} \boldsymbol{r}\cdot \operatorname{grad}\left(p + \frac{H^2}{8\pi}\right) d\mathcal{V} + \iiint_{\mathcal{V}} \boldsymbol{r}\cdot \operatorname{div}\left(\frac{\mathfrak{H}}{4\pi}\right) d\mathcal{V} + \iiint_{\mathcal{V}} \boldsymbol{r}\cdot \operatorname{grad}\Phi\, \varrho\, d\mathcal{V} \tag{17.20}$$

where $\mathfrak{H}$ is the dyadic ($\boldsymbol{H}\,\boldsymbol{H}$) and I and K represent respectively the moment of inertia and the kinetic energy of mass motion. In a steady state d^2I/dt^2 is zero but $2K$ does not necessarily vanish [cf. [1], Eq. (82.22)]. However, here we

[1] P. LEDOUX and R. SIMON: Ann. d'Astrophys. **20**, 185 (1957).
[2] S. CHANDRASEKHAR and E. FERMI: Astrophys. Journ. **118**, 116 (1953).
[3] J. H. PIDDINGTON: Austral. J. Phys. **10**, 530 (1957).

shall limit ourselves to the static case, when (17.20) becomes

$$\iiint_{\mathscr{V}} \boldsymbol{r}\cdot\operatorname{grad}\left(p+\frac{H^2}{8\pi}\right)d\mathscr{V}+\iiint_{\mathscr{V}} \boldsymbol{r}\cdot\operatorname{div}\left(\frac{\mathfrak{H}}{4\pi}\right)d\mathscr{V}+\iiint_{\mathscr{V}} \boldsymbol{r}\cdot\operatorname{grad}\Phi\,\varrho\,d\mathscr{V}. \tag{17.21}$$

If we are interested in the equilibrium of a given star which may be considered as surrounded by empty space, we may take for $\mathscr{V}$ the volume contained inside the stellar surface S on which ϱ and p vanish and which must be free of surface currents (continuity of the vertical and horizontal components of $\boldsymbol{H}$). Integrating (17.21) by parts as CHANDRASEKHAR and FERMI, we obtain

$$3(\gamma-1)\,U_i+\mathfrak{M}_i+V_i-\iint_S\left[\frac{H^2}{8\pi}\boldsymbol{r}-\left(\boldsymbol{r}\cdot\frac{\mathfrak{H}}{4\pi}\right)\right]d\boldsymbol{S}=0 \tag{17.22}$$

where U_i, $\mathfrak{M}_i$, V_i represent respectively the thermal energy, the magnetic energy and the gravitational potential energy of the star. The surface integral can be transformed so that (17.22) becomes

$$3(\gamma-1)\,U_i+\mathfrak{M}_i+V_i+\frac{1}{8\pi}\iint_S\left[(\boldsymbol{H}\cdot\boldsymbol{n})(\boldsymbol{H}\cdot\boldsymbol{r})-(\boldsymbol{n}\times\boldsymbol{H})\cdot(\boldsymbol{r}\times\boldsymbol{H})\right]dS=0. \tag{17.23}$$

Of course, we might just as well have extended the volume integrals in (17.21) to the whole of space ($S\to\infty$). Integrating by parts as before and, since in the conditions specified above H decreases as r^{-3} at sufficiently large distances, the surface integral vanishes and we are left with

$$3(\gamma-1)\,U_i+\mathfrak{M}_i+V_i+\mathfrak{M}_e=0 \tag{17.24}$$

with

$$\mathfrak{M}_e=\iiint_{\mathscr{V}_e}\frac{H^2}{8\pi}\,d\mathscr{V} \tag{17.25}$$

where the subscript e denotes values corresponding to the space outside the star. Eq. (17.24) is exactly equivalent to (17.23) since, in this case, $\mathfrak{M}_e$ can be evaluated indifferently by the surface integral in (17.23) or by (17.25).

For equilibrium to be possible, the total energy of the system must be negative, as otherwise there would be enough energy available to expand the star to infinity against the gravitational attraction of its parts. Combining this condition

$$U_i+\mathfrak{M}_i+\mathfrak{M}_e+V_i<0$$

with (17.24), we find

$$\frac{(3\gamma-4)}{3(\gamma-1)}\left(\mathfrak{M}_i+\mathfrak{M}_e+V_i\right)<0 \tag{17.26}$$

or, if $\gamma>\frac{4}{3}$,

$$\mathfrak{M}_i+\mathfrak{M}_e<-V_i.$$

If the field approximates to a dipole field at the surface of the star and in the empty space around it, this condition can be written numerically to the same order of approximation as (17.19)

$$\left(\overline{H^2}+\frac{H_P^2}{2}\right)<4\cdot10^{16}\,\frac{M^2}{R^4}. \tag{17.27}$$

Since H_P^2 is likely to be appreciably smaller than the average square value $\overline{H^2}$, the exact boundary conditions, in this case, do not affect appreciably the upper limit of CHANDRASEKHAR and FERMI.

For more general external fields, provided $H^2 r^3 \to 0$ as $r \to \infty$, the condition (17.27) will be replaced by

$$\overline{H^2}(1+\eta) < 4 \cdot 10^{16} \frac{M^2}{R^4} \tag{17.28}$$

where η is unlikely to become larger than about $\frac{1}{3}$ unless there is matter outside the star which may be the seat of electrical currents.

In that case, as pointed out by PIDDINGTON, although the virial theorem may still be used yielding, for the system as a whole, the generalized condition

$$\mathfrak{M}_i + \mathfrak{M}_e < -(V_i + V_e) \tag{17.29}$$

if $U_e \ll U_i$, some caution is necessary in its interpretation. On the one hand, in evaluating the destabilizing influence of a magnetic field on a given mass, only the magnetic energy corresponding to the lines of force actually crossing this mass must be taken into account. In that case, large external fields for which the star presents a very small cross section will have very little influence on its stability, so that the star itself may remain stable even if (17.29) is not satisfied.

On the other hand, the dilute gas cloud in which the external currents are flowing may very well get disrupted by the external field even if (17.29) is verified for the system as a whole, because the stabilizing influence of the star gravitational field on the cloud and the gravitational potential of the latter may be very small.

In many respects, the effects of a magnetic field on local stability is also similar to that of a rotation. As was suggested by BIERMANN[1], if convection results in motions characterized by closed stream lines, it must be hindered by the presence of a general magnetic field since a highly conducting fluid tends to stick to the magnetic lines of force (frozen in fields). In the same way, in a fluid of low viscosity, the vortex lines tend to be attached to the fluid and this explains the general similarity between the inhibiting influence of a rotation or of a magnetic field on the onset of convection.

As this type of effect is likely to play an important rôle in the explanation of the sunspots, it has been discussed by different authors[2] from a more or less qualitative point of view but the mathematical discussion has been developed in details by THOMPSON[3] and especially by CHANDRASEKHAR[4]. A very interesting account may also be found in COWLING's recent book [*18*].

The typical effect of a magnetic field, which here may be taken as uniform since the scale of the perturbation to be considered is relatively small, may be visualized following a simple argument due to WALÉN. Let us consider, for instance, a horizontal layer of gas presenting a vertical superadiabatic gradient ($A > 0$) and submitted to a horizontal magnetic field $\boldsymbol{H}$. In the absence of the latter, the layer would be convectively unstable according to Eq. (16.19) due to the volume force $g\varrho A\zeta$ acting on any element submitted to a small vertical displacement ζ. But in presence of the magnetic field, the same displacement assumed sinusoidal and of wavelength λ in the direction of $\boldsymbol{H}$ imparts to the lines

---

[1] L. BIERMANN: Vjschr. Astronom. Ges. **76**, 194 (1941).

[2] C. WALÉN: On the Vibratory Rotation of the Sun. H. LINDSTAHLS Bokhandel. Stockholm 1949. — T. G. COWLING: Monthly Notices Roy. Astronom. Soc. London **106**, 446 (1946). — F. HOYLE: Recent researches in Solar Physics. Cambridge-New York: Cambridge University Press 1949.

[3] W. B. THOMPSON: Phil. Mag. **42**, 1417 (1951).

[4] S. CHANDRASEKHAR: Phil. Mag. **43**, 501 (1952); **45**, 1117 (1954). — Proc. Roy. Soc. Lond., Ser. A **216**, 293 (1953).

of force a curvature $4\pi^2\zeta^2/\lambda^2$ which results in a volume force [cf. [1], Eq. (83.11)] $-\pi H^2\zeta/\lambda^2$. We have assumed as usual in these considerations, that the fluid pressure at equilibrium balances exactly the magnetic hydrostatic pressure $H^2/8\pi$. Thus the displaced element is submitted to a total force

$$g\varrho A\zeta - \frac{\pi H^2\zeta}{\lambda^2}$$

and convective instability will subsist only if the first term in absolute value is larger than the second, i.e. if

$$A > \frac{\pi H^2}{\lambda^2 g\varrho}. \tag{17.30}$$

This illustrates the stabilizing effect of the magnetic field and also the influence of the wavelength, the instability being the greater, for a given A, the larger the wavelength along the magnetic field[1].

It is interesting to compare further the two cases when local dynamical stability prevails, i.e.

$$A < 0 \quad \text{or} \quad A < \frac{\pi H^2}{\lambda^2 g\varrho}.$$

In those cases, the element will oscillate around its position of equilibrium with frequencies given in order of magnitudes respectively by

$$\sigma^2 = -gA \quad \text{or} \quad \sigma^2 = \frac{\pi H^2}{\lambda^2\varrho} - gA.$$

To investigate the behaviour of the amplitudes in the course of time, we have to take the dissipation into account. The effect of heat conduction has already been discussed in the case of rotation [cf. Eq. (17.14)] and, in the absence of a magnetic field, the motion is always damped.

If a magnetic field is present, we must also evaluate the effect of the finite electrical resistance which may be characterized by $\eta = (4\pi\sigma)^{-1}$. With this notation Eqs. (83.1), (83.2) and the first Eq. (83.5) in [1] yield in the present conditions

$$\frac{\partial \boldsymbol{H}'}{\partial t} = (\boldsymbol{H}\cdot\operatorname{grad})\,\boldsymbol{v} + \eta\,\nabla^2\boldsymbol{H}'$$

which shows that the extra-variation of the local perturbation $\boldsymbol{H}'$ due to the electrical resistance, corresponds to a kind of "diffusion" of $\boldsymbol{H}'$ in the surrounding medium. This reduces the magnetic restoring force by a fraction proportional to η so that the total force becomes

$$g\varrho A\zeta - \frac{\pi H^2}{\lambda^2}\zeta - \alpha\left[g\varrho A\zeta k - \pi\frac{H^2\zeta}{\lambda^2}\eta\right]$$

where α is the positive factor of proportionality which may be taken as equal for both types of forces. If

$$g\varrho A k > \frac{\pi H^2}{\lambda^2}\eta,$$

i.e. if

$$0 < A < \frac{\pi H^2}{\lambda^2 g\varrho}, \qquad k \gg \eta \tag{17.31}$$

the total restoring force increases leading to oscillations of increasing amplitudes. Thus in the presence of a magnetic field (as in the presence of rotation), vibrational

[1] In this respect cf. especially M. KRUSKAL and M. SCHWARZSCHILD: Proc. Roy. Soc. Lond. Ser. A 223, 348 (1954).

instability (or overstability) is possible, as was first shown quantitatively by THOMPSON and CHANDRASEKHAR.

Of course, viscosity should also be included in the dissipation. This was done by CHANDRASEKHAR in a very complete discussion of the case where $\boldsymbol{H}$ and $\boldsymbol{g}$ are parallel and for the "ideal" fluid commonly adopted in these discussions [cf. comments preceding Eq. (17.8)]. As suggested by the qualitative results above, he found that the presence of a magnetic field always inhibits the setting in of instability. In particular, the critical value $\mathcal{R}_c'$ of the Rayleigh number past which cellular convection sets in is always larger than its value $\mathcal{R}_c$ when the magnetic field is absent and $\mathcal{R}_c'$ increases with the non-dimensional number (Hartmann number)

$$\mathcal{M} = \frac{H^2 d^2}{4\pi \varrho \nu \eta}. \tag{17.32}$$

CHANDRASEKHAR'S discussion also confirmed the possibility of vibrational instability if the second condition (17.31) is satisfied. As, in the case of rotation, the effect of viscosity is to modify the limits in the first relation (17.31) which can then be written

$$\mathcal{R}_c < \mathcal{R}_0 < \mathcal{R}_c'$$

where $\mathcal{R}_0$ is the critical Rayleigh number at which overstability manifests itself provided $\mathcal{M}$ is greater than a certain value $\mathcal{M}^*$. In stellar atmospheres $k \gg \eta$ and, one may expect that instability due to superadiabatic gradients, often takes the form of overstability.

One should remark however that this does not necessarily imply catastrophic consequences for the layer in which it occurs. First of all, it is likely that second order terms can provide the necessary damping except perhaps for exceptional types of perturbations. Secondly, in a star, the layer considered has no fixed boundaries as in the theoretical case, so that, as remarked by COWLING, purely standing oscillations of wavelength comparable to the thickness of the unstable layer are unlikely and the effect of overstability, in this case, is more likely to be a limited increase of the amplitude of progressive waves travelling through the unstable region than the building up, at any fixed point, of a motion dangerous for the existence of the layer.

CHANDRASEKHAR has also shown that, as in the case of rotation, if $\boldsymbol{H}$ makes an angle ϑ with $-\boldsymbol{g}$, the previous results remain true for convection starting in elongated rolls parallel to the plane $(\boldsymbol{H}, \boldsymbol{g})$, provided, in the Hartmann number defined by (17.32), H be replaced by $H\cos\vartheta$. On the contrary, convection in transverse rolls is more difficult to excite ($\mathcal{R}_c$ much larger). Finally, he has also discussed the case of a layer submitted simultaneously to a rotation and a magnetic field[1], a case which is characterized by the fact that the curve $\mathcal{R}(a)$ representing the variation of the Rayleigh number at which convection sets in as the horizontal wave number a is varied, has now two minima in a certain range of values of the Taylor and Hartmann numbers $\mathcal{T}$ and $\mathcal{M}$. This implies some unexpected results as to the shape of the most favoured cells and its variations with $\mathcal{T}$ and $\mathcal{M}$.

Many other stability problems in magnetohydrodynamics have been formulated[2] (stability of flow, pinch-effect etc.) and with the present interest in the physics of plasma it is likely that important developments, some of which may have considerable astrophysical interest, will take place in this field in the next few years.

[1] S. CHANDRASEKHAR: Proc. Roy. Soc. Lond., Ser. A **225**, 173 (1954).

[2] Cf. for instance [*18*]; also Ref. 1, p. 678; D. H. MENZEL: Report of Conference on the Dynamics of Ionized Media, London 1951; J. W. DUNGEY: Monthly Notices Roy. Astronom. Soc. London **113**, 180 (1953).

In the investigations reported in this section, the small perturbation method has been used nearly exclusively although, CHANDRASEKHAR has made extensive use of the variational interpretation of the differential equations and we know that, from a physical point of view, this always corresponds to some aspect of the energy method. Furthermore, the detailed discussions of convective instability nearly always resort to the method of marginal stability which consists in determining the characteristics of the motion at vanishing stability and the condition for the setting in of instability [here $\mathcal{R}(a)$ and $\mathcal{R}_c$] by neglecting all acceleration terms in the equations of the problem. If as usual the dependance with respect to the time t is taken to be of the form $e^{\omega t}$, this means letting $\omega\to 0$ everywhere.

This procedure is only valid if it can be shown that either ω is always real in the neighbourhood of the critical state or, if complex, that $\operatorname{Im}\omega$ tends towards zero simultaneously with $\operatorname{Re}\omega$ when the latter tends towards zero by positive values. On the other hand, if in the same conditions, $\operatorname{Im}\omega$ tends towards a finite limit, the principle of marginal stability ceases to be valid and instability will develop in the form of oscillations of increasing amplitudes (vibrational in stability). The papers referred to in this section present many discussions of this distinction which is an important one to keep in mind in any stability problem.

Let us note also that the energy principle in presence of a magnetic field has recently been the object of a detailed analysis by BERNSTEIN[1] and others which may promote its use in the future.

γ) Effects of an external gravitational field. Again, in many respects, the problems arising in this case can be compared to those encountered in presence of a rotation. For instance, let us consider two masses M_1 and M_2, their centres being a distance d apart, and use spherical coordinates r, ϑ, φ taking the centre of gravity of M_1 as the origin of r, ϑ being measured from the z-axis oriented along the lines of centres from M_1 to M_2. If we suppose that the mass of M_2 can be considered as collected at its centre, the potential Φ_{21} of the instantaneous gravitational force exerted by M_2 on the point (r, ϑ, φ) of M_1 is

$$\left.\begin{aligned}\Phi_{21} &= -\frac{GM_2}{(d^2+r^2-2dr\cos\vartheta)^{\frac{1}{2}}}\\ &= -\frac{GM_2}{d}\left(1+\frac{r}{d}P_1(\cos\vartheta)+\frac{r^2}{d^2}P_2(\cos\vartheta)+\frac{r^3}{d^3}P_3(\cos\vartheta)+\cdots\right).\end{aligned}\right\}\tag{17.33}$$

The first term is a constant which can be neglected. The second term gives an acceleration GM_2/d^2 to the centre of M_1 and can be neutralized by supposing the axes to move with the same acceleration, so that the tide-generating potential reduces to

$$-\frac{GM_2}{d}\left(\frac{r^2}{d^2}P_2+\frac{r^3}{d^3}P_3+\cdots\right)\tag{17.34}$$

where, if d is sufficiently large, we may keep only the first term.

In that case, the hydrostatic equation of equilibrium for M_1 is

$$\frac{1}{\varrho_1}\operatorname{grad}p_1 = -\operatorname{grad}\left(\Phi_1-\frac{GM_2}{d^3}r^2+\frac{3}{2}\frac{GM_2}{d^3}r^2\sin^2\vartheta\right).\tag{17.35}$$

If we compare this equation with (6.13), we see that, apart from a uniform decrease of the density ϱ_1 by a quantity ϱ_2' equal to the mean density of M_2

[1] I. B. BERNSTEIN, E. A. FRIEMAN, M. D. KRUSKAL and R. M. KULSRUD: Proc. Roy. Soc. Lond., Ser. A 244, 17 (1958).

spread over a sphere of radius d, the effect of the external body is comparable to that of an "imaginary" rotation of angular velocity $\Omega = i\sqrt{3GM_2/d^3}$. Then if M_1 is composed of a homogeneous incompressible fluid, the stability problem can be treated along similar lines as in Sect. 6 (cf. [4], b §§ 46–49). However this time, no Jacobi-ellipsoidal configurations are possible and, in the plane $\left(\frac{3M_2}{2\pi\varrho_1 d^3}, e\right)$ where e is the excentricity, the linear series of possible configurations for M_1 is represented by a curve similar to that of Fig. 3, the maximum of $3M_2/(2\pi\varrho_1 d^3)$ equal to 0.18825 occurring for $e = 0.8826$. Past this maximum, the configurations become unstable and thus for a value of d smaller than d_c defined by

$$d_c = 2.1984\left(\frac{M_2}{M_1}\right)^{\frac{1}{3}} R_1 = 2.1984\left(\frac{\varrho_2}{\varrho_1}\right)^{\frac{1}{3}} R_2 \tag{17.36}$$

where R_i is the "mean" radius of M_i ($M_i = 4\pi\varrho_i R_i^3/3$), the tidal pull of M_2 on M_1 will disrupt the latter provided the cohesion forces in the latter are due only to gravity.

This discussion can easily be extended to the case where the two bodies are revolving on circular orbits about their common centre of gravity G with angular velocity

$$\Omega^2 = \frac{G(M_1 + M_2)}{d^3}. \tag{17.37}$$

If both masses rotate about themselves in the (z, x) plane with the same angular velocity the problem reduces to a statical one with respect to the same axes as before, provided one introduces the potential

$$\Omega^2 d_1 r\cos\vartheta - \tfrac{1}{2}\Omega^2 r^2 (1 - \sin^2\vartheta\sin^2\varphi) \tag{17.38}$$

of the centrifugal forces due respectively to the revolution of the origin and the rotation of the axes about themselves, d_1 being the distance M_1G given by

$$d_1 = \frac{M_2 d}{M_1 + M_2}, \tag{17.39}$$

and φ an azimuthal angle measured from the x-axis in the (x, y) plane.

If M_2 can again be assimilated to a mass-point and (17.38) is combined with (17.33) where the constant term is neglected, we find using (17.37) and (17.39)

$$\Phi'_{21} = -\frac{GM_2}{d}\left\{\frac{r^2}{d^2}\left[1 + \frac{1}{2}\left(\frac{M_1+M_2}{M_2}\right)\right] - \frac{3}{2}\frac{r^2}{d^2}\sin^2\vartheta\left[1 + \frac{1}{3}\frac{M_1+M_2}{M_2}\sin^2\varphi\right]\right\}. \tag{17.40}$$

This leads to the hydrostatic equation

$$\left.\begin{aligned} \frac{1}{\varrho_1}\operatorname{grad} p_1 = -\operatorname{grad}\Big\{\Phi_1 - \frac{GM_2}{d^3} r^2\left(1 + \frac{1}{2}\frac{M_1+M_2}{M_2}\right) + \\ + \frac{3}{2}\frac{GM_2}{d^3}\left[1 + \frac{1}{3}\frac{M_1+M_2}{M_2}\sin^2\varphi\right] r^2\sin^2\vartheta\Big\}. \end{aligned}\right\} \tag{17.41}$$

Comparing (17.41) to (17.35), we see that formally the rotation increases the second term in the total potential and decreases ϱ_1 by a further quantity equal to $\varrho'_2(M_1+M_2)/2M_2$ so that we may expect the instability to occur for a somewhat larger value of d_c. In a first approximation, we may substitute for $\sin^2\varphi$ its mean value $\frac{1}{2}$ and introduce a fictitious distance d' and mean density ϱ'_1

$$d' = d\left[1 + \frac{1}{6}\frac{M_1+M_2}{M_2}\right]^{-\frac{1}{3}}, \qquad \varrho'_1 = \varrho_1 - \varrho'_2\frac{M_1+M_2}{M_2}$$

to render Eq. (17.41) identical to (17.35). Then the critical value of d' is still given by (17.36) with $\varrho_1 = \varrho_1'$ and corresponds to a critical value of the actual distance d given by

$$d_c = 2.1984 \left(\frac{\varrho_2}{\varrho_1}\right)^{\frac{1}{3}} \left(1 + \frac{1}{5} \frac{M_1 + M_2}{M_2}\right)^{\frac{1}{3}} R_2 \tag{17.42}$$

provided M_1 remains smaller than M_2. If it is much smaller, (17.42) yields

$$d_c = 2.34 \left(\frac{\varrho_2}{\varrho_1}\right)^{\frac{1}{3}} R_2$$

instead of the exact value, known as *Roche's limit*

$$d_c = 2.4554 \left(\frac{\varrho_2}{\varrho_1}\right)^{\frac{1}{3}} R_2 .$$

Thus, for instance, a small satellite S in which the only force of cohesion is gravity will get disrupted if its distance to the centre of its planet P becomes smaller than $2.4554\,(\varrho_P/\varrho_S)^{\frac{1}{3}}$ times the planet radius.

G. DARWIN[1] has pursued the discussion for the general case of two incompressible fluid masses of the same density and both of finite volumes so that each is subject to distortion from the tidal forces generated by the other. The problem can be tackled by the methods of Part A, building first the equilibrium configurations and discussing then the stability of the linear series which they form. This however involves much numerical work and the main result may be summarized as follows:

for $\frac{M_2}{M_1} =$ 1 0.8 0.5 0.4 0.0

the critical distances corresponding to the limit of secular stability are

$d_c =$ 2.638 ? 2.574 2.59 ∞

which in general are very close to the distance d_a of closest approach (unstable configurations)

$d_a =$ 2.343 2.36 ? ? 2.457

where the unit of length is $[3\,(M_1 + M_2)/4\pi\varrho]^{\frac{1}{3}}$.

The fact that, if one of the masses becomes infinitesimal ($M_2/M_1 \to 0$) $d_c \to \infty$ corresponds really to orbital instability but not to instability of the masses themselves. If this effect of tidal friction on the orbits is excluded ("partial stability"), the critical distance d_c' becomes very insensitive to the ratio M_2/M_1 and remains always of the order of 2.5 times the radius of the combined mass assumed spherical.

As in Sect. 8α, an idea of the effects of compressibility on these results can be formed by replacing, in the preceding discussion, the incompressible masses by two configurations built according to ROCHE's model. This appreciably simplifies the algebra since the total potential may be written immediately in a closed form. Adopting as origin of the axes the centre of gravity G of the two masses M_1 and M_2, taking the x-axis along GM_2 and the z-axis along the vector $\boldsymbol{\Omega}$, we have

$$\Phi = -\frac{1}{2}\Omega^2(x^2 + y^2) - \frac{G M_1}{[(x - x_1)^2 + y^2 + z^2]^{\frac{1}{2}}} - \frac{G M_2}{[(x - x_2)^2 + y^2 + z^2]^{\frac{1}{2}}} .$$

If we choose as unit mass $(M_1 + M_2)$ and as unit length the distance d between

[1] G. DARWIN: Phil. Trans. Roy. Soc. Lond., Ser. A **206**, 161 (1906); cf. [4], b §§ 60 to 65.

the two masses, the potential becomes using (17.37)

$$\Phi = -\frac{G}{2}\left\{(x^2+y^2) + \frac{2(1-\mu)}{[(x-x_1)^2+y^2+z^2]^{\frac{1}{2}}} + \frac{2\mu}{[(x-x_2)^2+y^2+z^2]^{\frac{1}{2}}}\right\} \tag{17.43}$$

where

$$\mu = M_2/(M_1+M_2).$$

If a small body whose gravitational attraction may be neglected moves in this field with velocity $\boldsymbol{v}$, its motion which, as a special case of the three-body problem has been the subject of extensive studies in the past [*19*], admits the integral

$$v^2 + 2\Phi = -C. \tag{17.44}$$

This gives the velocity v at any permissible point, once C has been determined by the initial conditions. For a particular value of C, the motion will be limited to the inside of the equipotential surface, often known as the zero-velocity surface, defined by

$$-2\Phi = C. \tag{17.45}$$

These surfaces are symmetrical with respect to the (x, y) and the (x, z) planes and also if $\mu = \frac{1}{2}$ with respect to the (y, z) plane. They are often illustrated by their intersections in these different planes. Examples are given in [*19*], Chap. VIII which however correspond to rather a small value of μ for stellar applications. A more appropriate example is given by STRUVE and HUANG[1]. For large values of C, the equipotential surfaces degenerate into two closed surfaces around M_1 and M_2 which approach more and more the spherical shape as C increases or, due to the term in (x^2+y^2), in pseudo-cylinders at large distances from the masses. As C decreases, the first of these surfaces get elongated towards each other until they fuse together at a single point L_1 on the x-axis forming a sort of dumb-bell. At any such points as L_1 where a surface intersects itself, the acceleration vanishes and they are often known as *Lagrangian points*. As C is further decreased, two more Lagrangian points appear on the x-axis, the first one L_2 on the outside of the smaller mass and the second one L_3 on the outside of the larger mass. Two more Lagrangian points L_4 and L_5 occur at the summits of the equilateral triangles built on d as basis. All these points correspond to well known particular solutions of the three-body problem in which the particle, brought at any of these points with zero-velocity, remains indefinitely at rest in these positions with respect to the rotating axis. However the straight line solutions L_1, L_2, L_3 are generally unstable while the equilateral triangle solutions L_4, L_5 are stable or unstable according as μ is smaller or greater than 0.0385. This is so small that, if M_1 and M_2 are stars, stability at L_4 and L_5 must not be very frequent.

As far as the stability of the individual masses M_1 and M_2 is concerned, each of them will remain stable as long as its volume is smaller than that of the lobe of the first critical surface which contains it. If for some reasons, for instance because of an evolutionary change, the volume of M_1, say, increases, instability will arise as soon as it reaches the critical value and matter will be lost to the second star through the interior Lagrangian point L_1. Eventually, the system as a whole can loose mass through the external Lagrangian points.

Since actual stars have fairly high central condensations, this model is believed to be useful for the discussion of the behaviour of close binary stars whose observations often reveal, indeed, material streams circulating around and between

[1] O. STRUVE and SU SHU HUANG: Occasional Notes, Roy. Astronom. Soc. London 3, 161 (1957); also in Vol. L of this Encyclopedia, p. 243. Cf. also S. GAPOSCHKIN: Vol. L, p. 234.

the components or even drifting away from them. This has brought a new life to the subject as may be judged by the very interesting account published recently by STRUVE and HUANG (footnote 1, p. 683) which contains many references to recent work in this field by KUIPER, KOPAL and others.

Thus, as in the case of rotation, we have some informations, for extreme models, on the conditions in which instability occurs in a mass acted upon by a very large external gravitational field. However, the distinction between secular and dynamical instability is not always very clear and we know very little on the exact form taken by dynamical instability when it arises in this case. It may be that general principles such as those involved in the virial theorem or in POINCARÉ'S theorem on rotating mass (cf. Sect. 6) could also be useful here to extend the previous results to more physical models.

On the other hand, when the external field is very weak, one could study its effects on any physical stellar model by the perturbation method although very little work has been done on this subject. For instance, the square of the frequencies of stable non-radial oscillations of a mass subjected to the attraction of a second body will certainly be perturbed by quantities proportional to the gravitational potential energy of the first one in the field of the second and this combined with the tidal interaction of the two bodies may increase the chance of resonance between the oscillations and the orbital revolution. SCHATZMAN[1] has recently suggested that nova phenomena may be associated with such a resonance.

Unfortunately, the intermediate case which is probably fairly often realized especially in eclipsing binaries sets a very difficult problem, the dynamical aspects of which are still little understood (for a general survey, cf. [*20*]).

As far as the effects on local stability are concerned, we may distinguish two aspects. Following a recent remark by MESTEL[2], the deformation of the level surfaces due to the external field will, in case of purely static equilibrium, lead to a paradox similar to von ZEIPEL'S for a rotating star and will enforce meridional circulations. This can be seen directly on the hydrostatic equations (17.35) and (17.41) where the first spherically symmetrical perturbing term may be absorbed in Φ_1. The remaining perturbing term is then essentially equivalent, as already noted, to the effect of an imaginary rotation so that, formally, the two problems are similar and one may expect that, here also, the meridional currents will be very slow.

As to convection, it does not seem that an external field will have much local influence on it. Of course, the general departure of the level surfaces from the spherical shape implies a decrease in A and g along the polar axis directed towards the perturbing body and an increase in the equatorial plane of the prolate spheroid. In the case of an unstable region ($A>0$), this will intensify radial motions in the equatorial belt as compared to the polar axis, but this corresponds to a similar effect as that connected with von ZEIPEL'S paradox in a stable layer and will involve a compensation by superposed slow meridional circulations.

18. Main results concerning the vibrational stability of stars. This question has been discussed very fully in [*1*], Sects. 63 to 72, for radial pulsations and the little we know in the case of non-radial oscillation has been recalled in Sect. 81 of [*1*]. To sum up the results very briefly we may say that, with the presently accepted laws of energy generation ($\varepsilon=\varepsilon_0 \varrho T^\nu$) and reasonable stellar models, ordinary stars are thoroughly stable because of the large increase in the relative

---

[1] E. SCHATZMAN: Ann. d'Astrophys. (in press).

[2] L. MESTEL: Astrophys. Journ. **126**, 550 (1957).

amplitude $\delta r/r = \xi$ from the centre (destablizing factors) to the surface (stabilizing factors). Barring exceptional circumstances in the external layers, the only factor which can reduce the vibrational stability of gaseous stars mainly in radiative equilibrium, is the increase of the pressure of radiation which decreases the effective value of the ratio of specific heats Γ_1 which, in turn, reduces the variation of ξ. Thus vibrational stability decreases as the mass increases, but for stars built on the standard model and with the carbon cycle as source of energy, instability occurs only for masses larger than about 100 $M_\odot$ [1].

Another possible cause would be an extreme form of viscosity equalizing the amplitude ξ in a very large external fraction of the star but this is more speculative and a correct analysis of the physical factors involved is difficult.

In a wholly convective star (polytrope $n = \frac{3}{2}$), even if the pressure of radiation is negligible, the variation of ξ is already so small that vibrational instability arises as soon as ν becomes larger than about 9.

A star possessing a large internal region in which the electron gas is degenerated and in which the energy generation by nuclear reactions is going on in an external shell, would easily become vibrationally unstable [2] again because the variation of ξ through the star is very small in this case. This would probably remain true, in case of extensive partially degenerated cores. As pointed out by SCHATZMAN (cf. [*15*]) in the case of the proton-proton reaction, the effective exponents (cf. [*1*], Sect. 66) of ϱ and T in ε may be strongly affected by the abundance of He^3 present in the energy-generating region and this may lead to the possibility of temporary phases of vibrational instability during which the abundance of He^3 is reduced sufficiently to restore stability.

In the case of the oscillations observed in intrinsic variable stars, it remains uncertain, up to the present, whether, as has usually been assumed, they find their origin in vibrational instability. If they do, all the present evidence indicates that the source of instability must be looked for in the very external layers of the star where it may be connected with the ionization of hydrogen (EDDINGTON) or helium (ZHEVAKIN) or both. Some recent investigations such as that by SCHATZMAN, reported in detail in [*1*], Sect. 69, suggest that this is quite possible for Cepheids possessing a large external convection zone. But one should be aware of the dangers of extending the integral in the numerator of the criterion (13.10) of vibrational instability too close to the surface, using for δL the usual expression valid for the interior. It is certainly better to stop the integral at some level below the surface such that the heat capacity of the layers above is small compared to δL since, in such a region, the spatial variations of δL must be vanishingly small.

This is a very important problem but it should be kept in mind that if vibrational instability is a likely source for the observed oscillations, it is not absolutely the only possible one and, perhaps, the case of *hard self-excited oscillations* (Cf. [*1*], Sect. 90) deserves a little more attention than it has received up to now.

19. Explosions and shock-waves. There is another side to the problem of stability, namely the case where a certain amount of energy is suddenly released in a narrow region inside the star, which we have neglected so far, although some of the possibilities of local instability reviewed above could have similar consequences. Relatively few significant results have been obtained up to now in this field and some of them have been summarized and discussed in [*1*], Sects. 91

---

[1] P. LEDOUX: Astrophys. Journ. **94**, 537 (1941).

[2] P. LEDOUX and E. SAUVENIER-GOFFIN: Astrophys. Journ. **111**, 611 (1950); also [*15*], Chap. V, § 4 and Chap. VII, § 3.

to 94, where the general equations of the problem have been recalled and numerous references have been given.

If the energy release is small, its consequences can be discussed in terms of the linear progressive waves or the weak shocks to which it gives rise. In some respects, this raises problems somewhat similar to those encountered in the discussion of standing linear oscillations except for typical boundary conditions at the wave-front. A beginning in their study has been made by WHITHAM, WHITNEY, SIMON and others mainly for spherically symmetrical isentropic perturbations. In many cases, the non-perturbed models used are rather artificial and it would be worth while solving the problem numerically for some models with a more direct physical meaning[1]. It may also be interesting to extend the investigation to more general types of perturbation which do not preserve the spherical symmetry of the star especially when local instability prevails in some region. One should also consider the damping affecting the amplitude of the wave as it progresses. The relevant factors will be the same as those encountered in the discussion of vibrational stability. For non-radial waves, the possibility of overstability in a convectively unstable layer in presence of rotation or a magnetic field (cf. Sect. 17α and β) should also be taken into account.

When the energy released is large, of the order of the internal energy of the layer in which it occurs or larger, conditions will resemble more closely those of an explosion, a case which is usually considered significant for the interpretation of novae and supernovae. We have recalled in [*1*], Sect. 91, the main suggestions as to the origin of such finite perturbations. The problem in this case is certainly very difficult and, up to now, it has been approached only for models with simple laws of density distribution.

MCVITTIE and ROGERS[2], using the general solution of gas dynamics established by MCVITTIE[3] for radial motions, have studied an explosion in which the velocity is continuous everywhere (no shocks) and is, at each instant, proportional to the distance to the centre. Assuming an initial density distribution of the form

$$\varrho = \varrho_c \left\{1 - \left(\frac{r}{R}\right)^n\right\}$$

they find that, for a given value of n, i.e. a given degree of central condensation $\varrho_c/\bar{\varrho} = (n+3)/n$, there is emission of energy only if the central temperature T_c is greater than a critical value which decreases as the central condensation increases (or n decreases). For instance, with $T_c = 2 \cdot 10^7$ °K, $\varrho_c/\bar{\varrho} = 50$ and a surface ejection velocity of 1000 km/sec, the energy released is of the order of 10^{50} erg. Although the exact physical meaning of this "explosion" is difficult to ascertain, it is satisfactory that all these figures are compatible with average stellar conditions and supernova explosions.

But if the motion is due to the sudden release of energy in a limited region of the star, there must be, at least in the initial phases of the phenomenon, an external unperturbed region separated from the central core by a strong shock front. The study of the propagation of such fronts through the star and of its immediate consequences has been tackled by SCHATZMAN, KOPAL and ROGERS[4].

---

[1] Cf. for instance D. A. LAUTMAN: Astrophys. Journ. **126**, 537 (1957).

[2] G. C. MCVITTIE and M. H. ROGERS: Astronom. J. **60**, 374 (1955).

[3] G. C. MCVITTIE: Astronom. J. **59**, 173 (1954).

[4] E. SCHATZMAN: Ann. d'Astrophys. **13**, 384 (1950); **14**, 294 (1951). — P. A. CARRUS, P. A. FOX, F. HAAS and Z. KOPAL: Astrophys. Journ. **113**, 193, 496 (1951). — Z. KOPAL: Astrophys. Journ. **120**, 159 (1954). — M. H. ROGERS: Proc. Roy. Soc. Lond., Ser. A **235**, 120 (1956). — Astrophys. Journ. **125**, 478 (1957).

Most of the solutions discussed so far are particular cases of the so-called "progressing wave"[1] corresponding to homology solutions obtained by separation of the variables in the general problem, a type of solutions introduced by TAYLOR[2] and GUDERLEY[3] and which have been further elaborated by different authors[4]. They have also been extensively studied in relation with another astrophysical problem, i.e. that of the propagation of shock waves in the interstellar medium, by von WEIZSÄCKER and his collaborators[5] who, in particular, have discussed the stability of these homology solutions and found that as the time increases, non-homology solutions converge towards the homology type, at least, for plane waves and, in the case of spherical waves, in a sufficiently narrow region about the centre. Although this work is concerned with conditions somewhat simpler than those encountered in a star (no gravitational field, uniform density), some of the results may be useful in particular cases of the stellar problem.

At present, the connection between the theoretical results of those investigations directly aimed at the stellar problem (cf. footnote 4, p. 686) and actual phenomena such as nova or supernova explosions remains rather uncertain. Nevertheless, they have brought to the attention different points such as the possible appearance of cavitation at some distance behind the front when the density decreases rapidly, or relations between the Mach number of the shock, the energy release, the thickness of the blast wave and the possible ejection of surface material, which will help in orienting further researches in this field.

Bibliography.

[1] The discussion of stellar stability, just as the interpretation of intrinsic stellar variability, rests for a good part on the theory of small oscillations so that, we shall often have the occasion to refer to: LEDOUX, P., and TH. WALRAVEN: Variable Stars. In this volume.

[2] WAVRE, R.: Figures Planétaires et Géodésie. Paris: Gauthier-Villars 1932.

[3] DUHEM, P.: Traité d'énergétique, 2 Vol. Paris: Gauthier-Villars 1911. (The transition from a discrete to a continuous system is discussed in Vol. II, Chap. XVI, § 6, p. 304).

[4] For a recent and generally excellent summary of the most significant results including those of E. CARTAN on the dynamical stability of the Jacobi series, cf.:
a) LYTTLETON, R. A.: The Stability of Rotating Liquid Masses. Cambridge 1953.
b) JEANS, J.: Problems of Cosmosgony and Stellar Dynamics. Cambridge 1919.
c) JEANS, J.: Astronomy and Cosmogony. Cambridge 1929. The general accounts of the theory given in these books remain valuable even if, as pointed out by LYTTLETON, they suffer here and there from some confusion concerning the relation between secular and dynamical (or ordinary) stability. In particular, they contain an interesting attempt to take into account at least partially, the effects of compressibility.
d) POINCARÉ, H.: Figures d'équilibre d'une masse fluide. Paris: Gauthier-Villars 1902. This book still provides an interesting introduction to the fundamentals of the problem.
e) LICHTENSTEIN, L.: Gleichgewichtsfiguren rotierender Flüssigkeiten. Berlin: Springer 1933.
f) APPELL, P.: Mécanique rationnelle, Vol. IV. Paris: Gauthier-Villars 1921.

[5] a) THOMAS, L. H.: Monthly Notices Roy. Astronom. Soc. London **91**, 122, 619 (1930).
b) TOLMAN, R. C.: Astrophys. Journ. **90**, 541, 568 (1939).

---

[1] Cf. R. COURANT and K. O. FRIEDRICHS: Supersonic Flow and Shock Waves, Chap. VI New York: Interscience Publishers Inc. 1948.

[2] G. I. TAYLOR: Proc. Roy. Soc. Lond. Ser. A **201**, 159 (1950).

[3] G. GUDERLEY: Luftf.-Forschg. **19**, 302 (1942).

[4] Cf. G. J. KYNCH: Modern Development in Fluid Dynamics: High Speed Flow, ed. by L. HOWARTH: Oxford: Clarendon Press 1953. — L. I. SEDOV: Similarity and Dimensional Methods in Mechanics. Moscow: State Publishing House 1954.

[5] C. F. v. WEIZSÄCKER: Z. Naturforschg. **9**a, 269 (1954). — K. HAIN and S. v. HOERNER: Z. Naturforschg. **9**a, 993 (1954). — W. HÄFELE: Z. Naturforschg. **10**a, 1006, 1017 (1955). **11**a, 183 (1956).

[6] BIERMANN, L., and T. G. COWLING: Z. Astrophys. **19**, 1 (1939).
[7] LIAPOUNOFF, M. A.: Problème général de la stabilité du mouvement. Ann. Toulouse, Sér. II **9** (1907).
[8] VON ZEIPEL, H.: Monthly Notices Roy. Astronom. Soc. London **84**, 665 (1924).
[9] LIN, C. C.: The Theory of hydrodynamic Stability. Cambridge: Cambridge University Press 1955.
[10] VILLAT, H.: Leçons sur les fluides visqueux. Paris: Gauthier-Villars 1943.
[11] LICHTENSTEIN, L.: Grundlagen der Hydromechanik. Berlin: Springer 1929.
[12] ROSSELAND, S.: The Pulsation Theory of Variable Stars. Oxford: Clarendon Press 1949.
[13] LEDOUX, P.: Contribution à l'étude de la structure interne des étoiles et de leur stabilité. Mém. Soc. Roy. Sci. Liège **9** (1949).
[14] CHANDRASEKHAR, S.: An Introduction to the Study of stellar Structure. Chicago, Ill.: Chicago University Press 1939.
[15] SCHATZMAN, E.: White Dwarfs. Amsterdam: North-Holland Publishing Co. 1958.
[16] COWLING, T. G.: Monthly Notices Roy. Astronom. Soc. London **101**, 367 (1941).
[17] CHANDRASEKHAR, S.: Monthly Notices Roy. Astronom. Soc. London **113**, 667 (1953).
[18] COWLING, T. G.: Magnetohydrodynamics, Chap. IV. Interscience Tracts on Physics and Astronomy No. 4. New York and London 1957.
[19] MOULTON, F. R.: An Introduction to Celestial Mechanics, 2nd ed., Chap. 8. New York: MacMillan 1914.
[20] KOPAL, Z.: An Introduction to the Study of Eclipsing Variables. Cambridge U.S.A.: Harvard Univers. Press 1946.

Magnetic Fields of Stars.

By

ARMIN J. DEUTSCH.

With 15 Figures.

I. Introduction.

1. General magnetic fields of the order of one gauss are known to exist at the surfaces of the Earth and the Sun, and general fields of the order of a few thousand gauss have been observed in the reversing layers of a certain small proportion of stars. In sunspots, fields of order two thousand gauss are commonly observed. In addition, a magnetic field of order 10^{-4} gauss has been inferred to exist in the Crab nebula, and it is thought that comparable fields occur in many or all of the non-thermal radio sources of radio astronomy. The properties of cosmic rays, and the observed polarization in the light of distant stars, both seem to require the existence of a field of order 10^{-6} gauss throughout large parts of the galactic plane.

Our understanding of such cosmical magnetic fields is still in its infancy. In particular, we cannot say how the known astronomical fields are genetically related to each other, if at all. Wherever fields are known to occur, it is clear that the associated magnetic forces may play a major role in the dynamics of the system. This has led to the conjecture that such forces are essentially involved in many astronomical phenomena that have not otherwise been adequately explained: stellar pulsation, mass loss, mass accretion, the nova phenomenon, planetary nebulae, spiral structure of galaxies, etc., etc. While sometimes fruitful, such conjectures are necessarily speculative in the present state of knowledge.

In this article, we shall be principally concerned with the stars that are known to have large general magnetic fields in their reversing layers. This criterion excludes the Sun, about which we have considerable information relating to magnetic phenomena. There are theoretical reasons for believing that many, or even all, stars may have strong fields somewhere in their atmospheres or interiors; but these reasons are not compelling, and for want of information we shall not discuss these possibilities here.

As compared with magnetic phenomena in the laboratory, those we observe in the stars are usually qualitatively distinctive. To understand this distinction, we write the Maxwell equations for a slowly changing field in an isotropic conducting fluid, with internal motions that are small compared with the velocity of light. In e.m.u., the equations are, in a familiar notation,

$$\operatorname{curl} \boldsymbol{H} = 4\pi \boldsymbol{j}, \tag{1.1}$$

$$\operatorname{curl} \boldsymbol{E} = -\mu \frac{\partial \boldsymbol{H}}{\partial t}, \tag{1.2}$$

$$\operatorname{div} \boldsymbol{H} = 0, \tag{1.3}$$

$$\boldsymbol{j} = \sigma(\boldsymbol{E} + \mu \boldsymbol{V} \times \boldsymbol{H}). \tag{1.4}$$

From these, it can be readily shown that

$$\frac{\partial \boldsymbol{H}}{\partial t} = \operatorname{curl}(\boldsymbol{V} \times \boldsymbol{H}) + \eta \nabla^2 \boldsymbol{H}, \tag{1.5}$$

where

$$\frac{1}{\eta} = 4\pi \mu \sigma. \tag{1.6}$$

In order of magnitude, the relative sizes of the two terms in the right member of Eq. (1.5) depend on the ratio

$$R_M = \frac{L V}{\eta}, \tag{1.7}$$

where L and V are, respectively, a length and a velocity characteristic of the system. In most laboratory situations, $R_M \ll 1$, and the velocity term is negligible in Eq. (1.5). But in most stellar fields, $R_M \gg 1$, and the velocity term is dominant.

The consequences of this distinction have been extensively developed by COWLING in his monograph [4]. The conditions in which Eqs. (1.1) to (1.4) are valid are carefully prescribed by ELSASSER[1], and by SPITZER[2]. For the application of these equations to problems of stellar magnetism, it is necessary to have a microscopic theory for the calculation of the conductivity σ of highly ionized gases. SPITZER'S monograph reviews this subject in detail, and it elaborates the theory of the other transport coefficients which may be needed in discussions of ionized gases. In sufficiently strong magnetic fields, these transport coefficients will become anisotropic; however, in the fields thought to exist in stellar reversing layers and interiors, σ remains isotropic.

II. Observations of magnetic stars.

a) Zeeman effect in stellar spectra.

2. Instrumentation. Virtually all existing observations of the Zeeman effect in stellar spectra have been made photographically by H. W. BABCOCK, at the coudé spectrographs of the 100-inch and 200-inch telescopes [1]. BABCOCK utilizes a double analyzer for circularly polarized light, which produces two analyzed spectra side by side on the photographic plate (see Fig. 1). Into one of these spectra goes all light elliptically polarized in one sense, into the other goes all light elliptically polarized in the opposite sense. Any light which is unpolarized or plane polarized is equally divided between the two spectra.

BABCOCK'S analyzer (Fig. 2) is mounted directly in front of the slit of the spectrograph. It consists of a mica quarter-wave plate followed by a plane-parallel crystal of calcite. The calcite produces two analyzed images side by side on the slit. For compensation of the small phase shift introduced by the oblique reflection at the flat coudé mirrors, it is sometimes necessary to insert a pair of crossed retardation plates with adjustable azimuths.

For a given spectrum line, an effective magnetic field H_e may be defined in the formula [1]

$$H_e = 52.7 \left(\frac{4500}{\lambda}\right)^2 F \left(\frac{\Delta s}{z}\right) \text{gauss}, \tag{2.1}$$

where F is the dispersion in angstroms per millimeter, z is the weighted mean displacement of the σ-components of the Zeeman pattern of the line as compared

[1] W. M. ELSASSER: Rev. Mod. Phys. **28**, 135 (1956).

[2] L. SPITZER jr.: Physics of Fully Ionized Gases. New York: Interscience Publ. 1956.

to the normal triplet, and Δs is the measured displacement in microns between the line images in the dextrogyrate and levogyrate spectra. The effective field is then just that uniform field which would produce the observed Δs in the pure

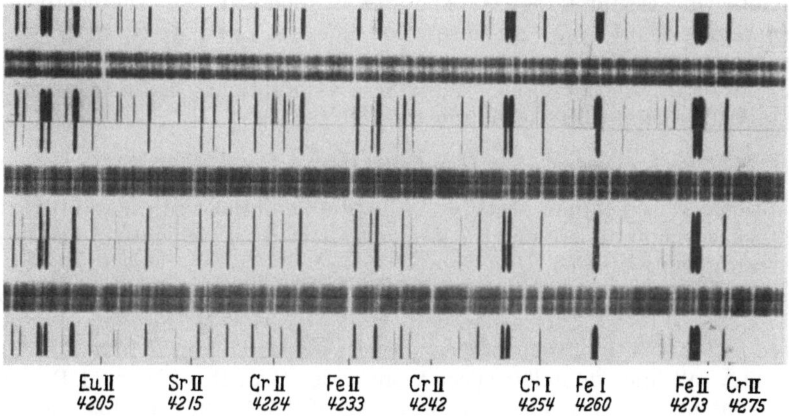

Fig. 1. [Reproduced from H. W. BABCOCK: Astrophys. Journ. **124**, 489 (1956).] The spectrum of HD 125248, $\lambda\lambda$ 4200 to 4280, taken with the double polarizing analyzer at phases 0.11 (top), 0.77 (middle), and 0.55 (bottom).

longitudinal Zeeman effect. The sign of H_e is taken as positive when the magnetic vector is directed outwards from the surface of the star, or towards the observer. On any given spectrogram of a magnetic star, measures of a number of lines usually yield a correlation between $(4500/\lambda)^2 \Delta s$ and z (Fig. 3), such that all

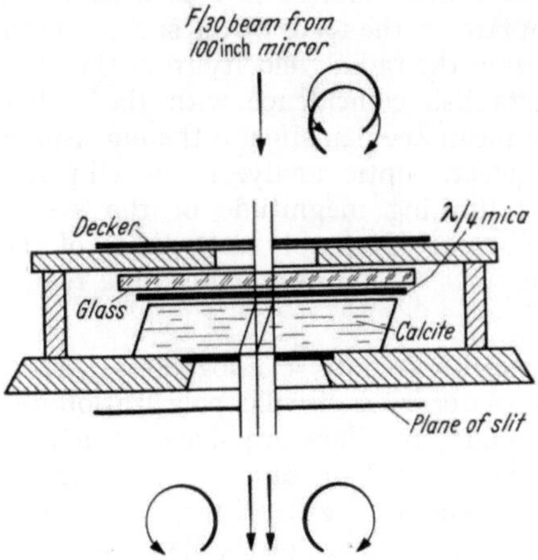

Fig. 2.

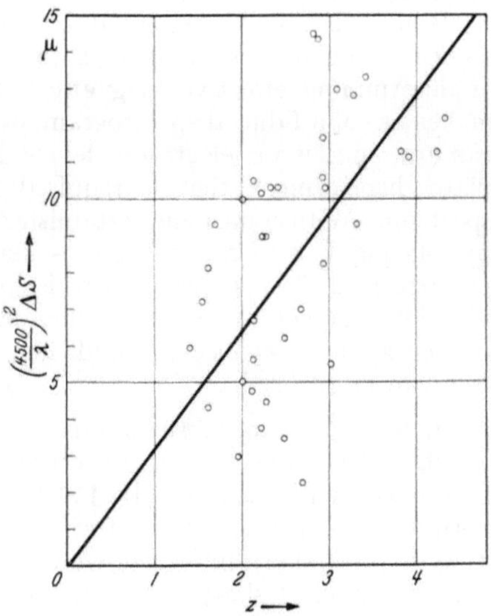

Fig. 3.

Fig. 2. (Adapted from Ref. [1].) BABCOCK's differential analyzer for longitudinal Zeeman effect.

Fig. 3. Correlation of theoretical Zeeman splitting z with measured displacements between analyzed spectra of the magnetic star HD 111133. The field obtained from these measures, of plate Pb 3146 (4.5 Å/mm), is -1.538 ± 0.064 kgauss. The line width is $w' = 0.14$ Å.

lines in the group yield comparable values of H_e. The published values of H_e are usually obtained from the measurements by taking the ratio of mean values, $\langle \Delta s \rangle / \langle z \rangle$, instead of the mean ratio $\langle (\Delta s / \Delta z) \rangle$; this is equivalent to weighting more heavily the lines of larger Δs and z.

In most of Babcock's early work on stellar magnetism, the results of the measurements of Zeeman effect were published in the form of "polar fields", H_p, instead of effective fields as defined above. H_p was obtained from H_e on the basis of the following assumptions: (1) The star is spherical. (2) The coefficient of limb-darkening is 0.45. (3) An absorption line has constant strength over the visible hemisphere. (4) The surface field is that of a central dipole with zero inclination to the line of sight. With these assumptions; one can integrate H_e over the visible hemisphere to obtain the result[1]

$$H_e = 0.303\, H_p. \tag{2.2}$$

Assumptions (3) and (4) are probably unjustifiable in most cases, and it is therefore customary now to present the Zeeman effect measurements in the form of the derived effective fields, H_e.

When $H_e = 1$ kilogauss (kgs), Eq. (2.2) indicates that the Zeeman displacement $\Delta\lambda$ will be 0.017 Å for a line with $z = 1$ at λ 4500 Å. For comparison, the width to half-intensity of the Doppler core of a weak iron line at 10000° is 0.043 Å at λ 4500 Å; all lines in stellar spectra are wider than this. Most of Babcock's observations have been made at a dispersion of 4.5 Å/mm. At this scale, a $\Delta\lambda$ of 0.017 Å corresponds to a Δs of 3.8 μ. The probable error of measurement of Δs is only 1 or 2 microns on the best plates of very sharp-lined stars; from measures of about 60 lines, Babcock can then obtain H_e with a probable error of about 0.15 kgs. Where the lines are intrinsically wide, or the star faint enough to require a lower dispersion of 10 Å/mm, the probable error of H_e may increase to 1.0 kgs or more.

Babcock has experimented at the 200-inch telescope with a photoelectric device which may be able to supplement the ordinary photographic methods of determining effective magnetic field and radial velocity[2]. This instrument makes use of a fiducial spectrogram, or template, in the focus of the spectrograph camera, and an a.c. electronic device, to obtain the radial velocity from the template displacement that is required to establish coincidence with the stellar spectrum. With coincidence established, the mean Zeeman effect is then measured by placing before the slit an oscillating electro-optic analyzer for elliptical polarization. The expected precision and limiting magnitude of the stellar template magnetometer are estimated to compare favorably with those of the photographic procedures; in addition, an important gain in speed of data reduction can be realized.

3. Survey of the observations. Babcock has recently catalogued 338 stars which he has observed with his differential analyzer for circular polarization due to the Zeeman effect [*1*]. In 150 of these stars, the lines are sharp enough to permit the ready detection of effective fields larger than, say, 0.5 kgs. Eighty-nine of these sharp-line stars show definite evidence of general magnetic fields; the remaining 61 sharp-line stars give little or no evidence of the Zeeman effect. Among the stars with broader lines, 66 are listed as probably showing some evidence of the Zeeman effect, and 122 are listed as having "lines too broad to permit measurements".

The 338 stars in Babcock's catalogue obviously do not constitute a random sample, but rather a highly selected one. A wide variety of stars have been sampled for Zeeman effect, and this has been most commonly found in the

[1] M. Schwarzschild: Astrophys. Journ. **112**, 222 (1950).

[2] H. W. Babcock: Ann. Report of the Director 1954-55, Mount Wilson and Palomar Observatories.

peculiar $A(Ap)$ stars. These stars have therefore been observed the most intensively. No attempt can be made to discuss the statistics of BABCOCK's magnetic stars without taking account of the strong selection factors of his catalogue.

The 61 sharp-line stars that show no evidence of the Zeeman effect include a small number of normal A stars, but no Ap stars. Examples of apparently non-magnetic sharp-line stars are o Peg (A1 V), α CMi (F5 IV—V), ε Peg (K2 Ib), and α Equ (F8 III + A3).

The 84 sharp-line stars that definitely show the Zeeman effect include 64 peculiar A stars, seven metallic-line stars, three abnormal M giants, two S stars, one subdwarf, and one cluster-type variable. It appears that all these stars have general magnetic fields that vary with time, and with amplitudes of the order of a few kilogauss. Some of the details of these magnetic stars will be discussed below. BABCOCK summarizes the well-observed Ap stars in his catalogue as comprising five regular magnetic variables of large amplitude and reversing polarity; 22 irregular variables with reversing polarity; and 15 irregular variables with non-reversing polarity. He also notes that "the irregular variations prove that large-scale intrinsic hydromagnetic fluctuations occur at the surfaces of these stars" [1].

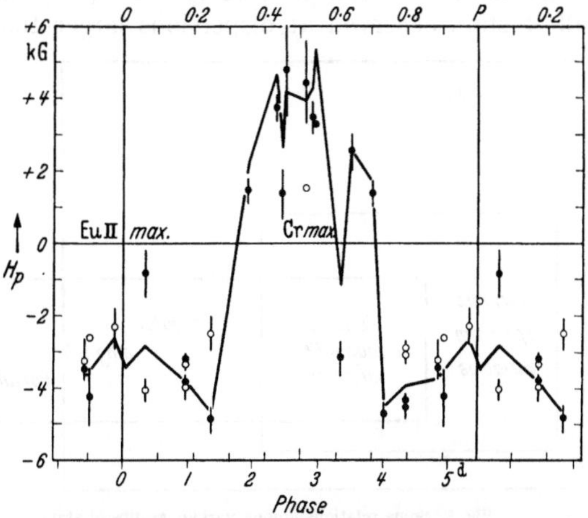

Fig. 4. Polar fields H_p in α^2 CVn. The lines give the curve from the mean of all absorption lines. The lines of Eu II give the open circles; of Cr II, the full circles. [Adapted from H. W. BABCOCK and S. BURD: Astrophys. Journ. 116, 8 (1952).]

It is a characteristic feature of many magnetic Ap stars that on a given plate lines of different elements may indicate systematically different values for the effective field H_e. The magnetic observations of the periodic star α^2 CVn, for example, show this effect (Fig. 4). This can only mean that lines of different elements are sometimes produced preferentially in different volumes of gas. Moreover, the observed rates of change of H_e, as indicated by lines of any one element, can easily be shown to imply the occurrence of large differential velocities in the stellar atmosphere. If we measure the electric field $\boldsymbol{E}$ and the magnetic field $\boldsymbol{H}$ relative to the inertial frame that is fixed at the center of the star, then the law of induction requires that, at a representative point on the observed hemisphere,

$$\operatorname{curl} \boldsymbol{E} = -\frac{\partial \boldsymbol{H}}{\partial t},$$

in electromagnetic units. If R is the radius of the star, this implies that, in order of magnitude,

$$\frac{|\boldsymbol{E}|}{R} \approx \frac{|\boldsymbol{H}|}{P},$$

where P is the time for $\boldsymbol{H}$ to vary through its full range. But since the conductivity is very high, the electric field must vanish when it is measured in a reference frame that moves with the gas. If the material velocity across the lines of force is $V_\perp$, therefore,

$$|\boldsymbol{E}| \approx V_\perp |\boldsymbol{H}|.$$

Eliminating $|\boldsymbol{E}|$ and $|\boldsymbol{H}|$, we find that $V_\perp \approx R/P$; i.e., in the time required for $\boldsymbol{H}$ to vary through its full range, the gas must move perpendicular to $\boldsymbol{H}$ a distance of the order of the stellar radius. In moving, of course, the gas will convect the magnetic field with it. The computed value of $V_\perp$ is of the same order as the velocities observed in many magnetic variables. For the periodic variables, it will be of the same order as the equatorial speed of rotation, if the observed period is that of rotation.

b) The peculiar *A* stars.

4. Introduction: Spectroscopic "patches". The characteristic spectroscopic abnormalities of the peculiar *A* stars (*Ap*) can be recognized in about 12% of all *A* stars observed at dispersions larger than, say, 125 Å/mm. Since general magnetic-fields have been most commonly observed among the *Ap* stars, and since none of these objects is known to be without a large field, it is appropriate to consider this group in some detail [2]. The hottest *Ap* stars show enhancement of the Mn II lines. As the excitation diminishes, the *Ap* stars tend to show, in turn, abnormally strong lines of Si II; Eu II and certain other rare earths; Cr II; and Sr II. The lines of Ca II and O I are usually too weak for the Draper type; the lines of H, Fe I, Fe II, Ti II, and Mg II are most nearly normal and form the basis for the Draper classification. A conventional interpretation of the anomalous line strengths leads to abundance anomalies ranging up to factors of the order of 10^3.

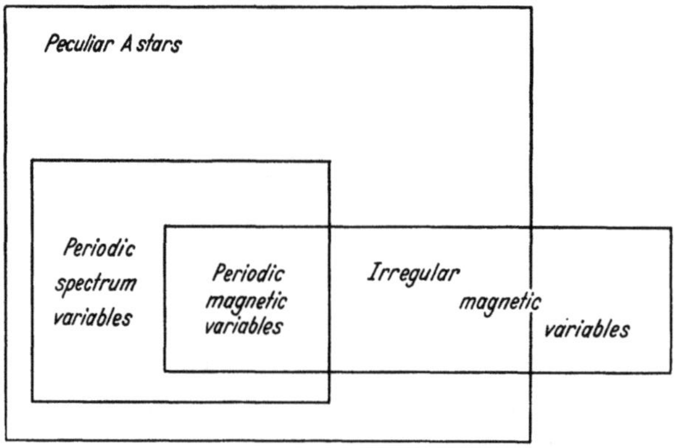

Fig. 5. Some relations among various groups of stars.

In most of the *Ap* stars, the spectroscopic anomalies are not observed to change appreciably with time. In the ones that are spectrum variables, however, the line-strengths show a regular variation (Fig. 1). The mean period of the known spectrum variables is about four days, which is in the range of rotational periods of *A* stars, as these have been inferred from line-width studies. It is possible, then, that in the spectrum variables we have simply a group of spectroscopically "patchy" *A* stars that are viewed at high inclination to their axes of rotation. This is the so-called "oblique rotator" model for a spectrum variable. It is still controversial, for reasons which will appear below. In the present discussion, however, we shall tentatively adopt it as a convenient working hypothesis.

The spectrum variables among the sharp-line *Ap* stars constitute the group of periodic magnetic variables, the magnetic cycles being synchronous with the line-intensity variations. This fact suggests that the surface magnetic field rotates rigidly with the star, the spectroscopic patches being in some way related to the field. Some of the periodic spectrum-magnetic variables seem to exhibit certain irregular changes, superimposed on the aspect changes due to rotation. These irregularities, together with the more pronounced irregular variations of most magnetic stars, must be attributed to an intrinsic variation of the magnetic field.

It is important to note that in most *Ap* stars, including many spectrum variables, the lines are too wide to permit the ready detection of the Zeeman

Table 1. *Periodic magnetic variables and/or spectrum variables.*

No.	Star	HD	Period (days)	Line-width (Å)	Principal variable lines	Magnetic field H_e (kilogauss)	Remarks
1	HD 124224	*A* 0*p*	0.52	4	Si II, He I, λ 4201	Unknown	a
2	56 Ari	*A* 0*p*	0.73	3	Si II, He I, λ 4201	Unknown	a
3	21 Com	*A* 3*p*	1.03:	1:	Ca II, Sr II	Unknown	
4	χ Ser	*A* 0*p*	1.60	1.2:	Ca II, Sr II	Unknown	
5	ι Cas	*A* 5*p*	1.74	1.3:	Ca II, Sr II	Unknown	
6	Hd 34452	*A* 0*p*	2.47	1	He I, λ 4201	Probably present	
7	HD 224801	*A* 0*p*	3.74:	0.8:	Eu II	< 0 to + 2.3	
8	ADS 16252B	(*A* 2*p*)	3.77	Unknown	Ca II, Eu II	Unknown	
9	ε UMa	*A* 0*p*	5.09	0.6	Ca II, Cr II	Probably present	
10	α^2 CVn	*A* 0*p*	5.47	0.36(v)	Cr II, Eu II	− 1.4 to + 1.6	a, b, c
11	HD 98088	*F* 0	5.90	0.38:	Ti II, Sr II, Ba II	− 1.2 to + 1.2	a, b, d
12	HD 153882	*A* 0*p*	6.00	0.40:	Sr II	− 1.2 to + 1.4	c
13	HD 71866	*A* 0*p*	6.80	0.26	Eu II, Gd II	− 2 to + 2	b, c
14	53 Cam	*A* 2*p*	7.80	0.15 to 1	Ti II, Mg II	− 4 to + 4	c
15	HD 125248	*A* 0*p*	9.30	0.18	Cr II, Eu II	− 1.9 to + 2.1	a, b, c, e
16	73 Dra	*A* 2*p*	20.27:	0.13:	Ca II, Eu II	− 0.6 to > 0	
17	HD 188041,2	*F* 0*p*, *A*	226	0.11	Eu II, Sr II	− 0.2 to + 1.5	

Remarks.

(a) Lines of different elements yield different radial velocities.
(b) Lines of different elements yield different effective fields.
(c) Crossover effect.
(d) Spectroscopic binary with period of 5ᵈ90.
(e) Spectroscopic binary with period of 4ʸ40.

effect. Since the spectroscopic peculiarities of these relatively wide-line *Ap* stars are essentially the same as those found in their sharp-line counterparts, which always do show the Zeeman effect, it is reasonable to assume that all the *Ap* stars have large general magnetic fields. The wide-line *Ap* stars may then be regarded as magnetic stars that are in rapid rotation and that are viewed at high inclinations. Fig. 5 subsumes the observed relationships among these partially overlapping groups of stars. Table 1 summarizes some of the known data pertaining to the periodic spectrum variables.

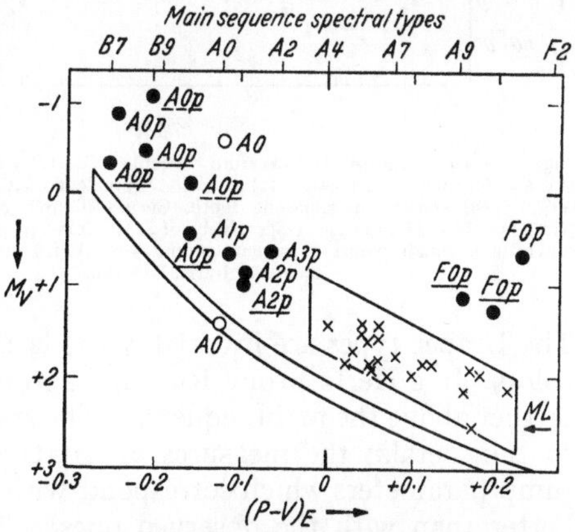

Fig. 6. The color-absolute-magnitude-spectral type diagram in the region of the peculiar *A* stars (*Ap*) and metallic-line stars (*ML*). [Adapted from O. EGGEN: Astronom. J. **62**, 45 (1957).] See text.

5. Photometric characteristics. Except for their marked spectroscopic abnormalities, the *Ap* stars are not distinguishable from the normal *A* stars of BAADE's Population I. They frequently occur in open clusters, and in binary systems with normal or metallic-line stars. Their photometric characteristics appear to be quite ordinary[1]. They lie slightly above the standard main sequence in the color-magnitude diagram, in a region which is also populated by stars with

[1] S. S. PROVIN: Astrophys. Journ. **118**, 489 (1953).

normal spectra (Fig. 6). Since the *Ap* stars are often so closely associated with normal stars of very similar radii and luminosities, there is no reason to expect that they are essentially different in structure, origin, or composition.

The color-magnitude diagram of Fig. 6 is adapted from the work of EGGEN on stars with well-determined parallaxes[1]. The band represents the main sequence observed in the Pleiades, which is very nearly the same as the standard main sequence in this region of the color-magnitude diagram. The spectral types are from the Draper catalogue. Underlining indicates that the star shows the Zeeman effect, or probably shows it. The *Ap* stars without underlined types have not yet been tested for the effect, or they have lines too wide to show it. The open circles represent α CMa and (at higher M_v) γ Gem; these are non-peculiar *A* stars with relatively sharp lines which show no Zeeman effect. The crosses represent metallic-line stars; these will be discussed in a later section.

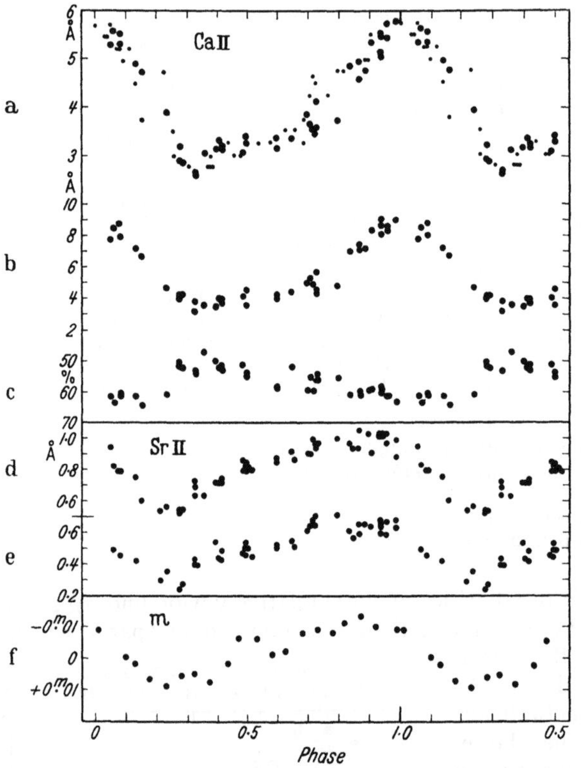

Fig. 7. Observations of the spectrum variable ι Cas ($P = 1^d.74$). Curves (a), (b), and (c) refer to Ca II 3934: (a) gives the equivalent width; (b) the width at 30% depth; and (c) the apparent depth. Curves (d) and (e) give equivalent widths of λλ 4078 and 4215, respectively, of Sr II. Curve (f) gives the apparent magnitude. Small points represent estimates by A. J. DEUTSCH; the others, measures by K. BAHNER.

The scattering of the stars above the main sequence of Fig. 6 is probably to be understood in terms of ordinary evolutionary effects. The only real anomaly shown by Fig. 6 is the relation between spectral type and color for the *Ap* stars. The Draper types are invariably too late for main-sequence stars of the observed colors. In a Hertzsprung-Russell diagram, the *Ap* stars would obviously lie much farther above the main sequence. The spectroscopic peculiarity is mainly confined to the metals; the measures of CHALONGE and his collaborators give Balmer-jump parameters which correspond with the indicated main-sequence types much better than with the observed ones[2]. The *A 0p* stars, in particular, appear to have all the gross characteristics of main-sequence *B* stars, although their spectra show most of the metals too strong for such stars, and He too weak. The most notable anomaly of this kind occurs in HD 34452. This star, which is too distant for inclusion in Fig. 6, has the Balmer-jump appropriate to a *B*3 V star; its Draper type, however, is *A* 0*p*. The anomaly is almost as great in α And (*B*5 V, *A*0*p*).

[1] O. J. EGGEN: Astron. J. **62**, 45 (1957).
[2] D. CHALONGE and L. DIVAN: Ann. d'Astrophys. **15**, 201 (1952).

Small variations in brightness seem to occur in all the spectrum variables [2]. The photometric changes are synchronous with the spectrum and/or magnetic variation. The amplitude is always less than 0.2 magnitude, and is usually least at longer wavelengths. The light-curves are generally anharmonic, and in some cases they resemble the curves for line-intensity variation. Fig. 7 shows these phenomena in ι Cas ($P=1^{\mathrm{d}}.74$); the figure was kindly communicated by the author, K. BAHNER, in advance of publication. On the oblique-rotator model, the light variations are to be understood in terms of a variation of brightness over the photosphere of the star.

6. Line profiles: rotational effects. While some *Ap* stars exhibit wide lines, SLETTEBAK[1] has shown that the average line-width is distinctly less than in non-peculiar stars in the same part of the color-magnitude diagram. On the usua assumption that line-widening in most *A* stars is attributable to rotation, SLETTEBAK has derived the mean line-of-sight component of equatorial rotational velocity V_e in several groups of stars. Table 2 gives some of SLETTEBAK'S results, from measurements of profiles of Mg II 4481. The mean velocities $\langle V_e\rangle$ have been derived from the measured quantity, $V_e \sin i$, on the assumption of a random distribution of the inclination i.

Table 2.

Type	Number of stars	$\langle V_e \sin i\rangle$ (km/sec)	$\langle V_e\rangle$ (km/sec)
B 8 to *A* 2 V	87	139	177
B 8 to *A* 2 III–IV	24	73	93
Ap	16	41	52
ML	6	40	51

Peculiarities will be discovered somewhat more easily in sharp-line spectra than in others, but this selection effect is not strong enough to produce the marked differences of Table 2. A real effect is certainly present. It cannot be explained by supposing that all *A* stars have the *Ap* characteristic if they rotate more slowly than a certain limit. The widest-line *Ap* stars would then require so high a limit for V_e that a much larger proportion of *A* stars would fall below the limit than is actually observed.

A possible alternative explanation of SLETTEBAK'S result is that the *Ap* stars represent the selection of rotating *A* stars that are viewed at low inclination. This hypothesis implies the occurrence of zonal stratification in the reversing layers of rotating *A* stars; they must present normal (but wide-line) *A*-type spectra at high inclinations, and *Ap*-type spectra at low inclinations. The few known very sharp-line stars that are "non-peculiar" and non-magnetic would, on this view, be the stars with very low V_e. The statistics of this situation have not yet been satisfactorily worked out; however, a rough argument that tends to support this interpretation may be given as follows. Let us assume that $\langle V_e\rangle$ is really about the same for the *Ap* stars as for the *B* 8 to *A* 2 V group. If we define a mean inclination $\langle i\rangle$ for the group in the relation

$$\sin\langle i\rangle = \langle V_e \sin i\rangle / \langle V_e\rangle ,$$

we then find $\langle i\rangle = 13^\circ$ for the *Ap* stars. The proportion of all *A* stars with $\langle i\rangle \leq 13^\circ$ would then be 11%, which is very nearly the observed frequency of *Ap* stars relative to normal *A* stars[2].

These arguments imply that a large proportion of all *A* stars develop zonally stratified atmospheres, and that the zones produce pronounced aspect effects in *A*-type spectra. If this model is correct, it requires that the *Ap* stars represent only the most conspicuous of these aspect effects, and that all lesser degrees of

[1] A. SLETTEBAK: Astrophys. Journ. **119**, 146 (1954).
[2] M. E. WALTHER: Astrophys. Journ. **110**, 68 (1949).

the aspect peculiarities must occur among stars considered "normal". A few stars are known to be "border-line" cases; they have been classified among the *Ap*'s by some investigators, and among the *ML* or normal stars by others. MORGAN's estimates of line-strengths in a group of 125 *A* stars, which include 45 *Ap* stars, suggest that the characteristic peculiarities do, in fact, occur in all degrees, and that the Draper *Ap* stars do not really represent a discrete group[1]. The color-magnitude diagram testifies to the same conclusion.

BABCOCK's observations at dispersions higher then 10 Å/mm have provided an important addition to our knowledge of line-widths in *A* stars. In particular, his catalogue of magnetic stars gives an index of line-width for most of these objects [*1*]. He defines his index *w* as the extreme width in angstroms of typical unblended lines of moderate intensity, in the region near λ 4200, as measured directly on the plate with a comparator, minus the projected slit-width (usually 0.07 Å). The sharpest lines known in stellar spectra occur among these *A* stars; in several, $w < 0.10$. In some of these "ultra-sharp-line" stars, however, the field is small (ι CrB, $H_e \approx 0$ to -0.30) or undetectable (o Peg).

BABCOCK finds that *w* is often variable in the magnetic stars, even among those that show no appreciable changes in line-intensity. In addition, the line-widths sometimes vary from element to element, or in some more complicated way. Apart from the line-width itself, one commonly observes that at high dispersion the line profiles are diverse in an *Ap* spectrum, and often variable with time. Simple rotational broadening is obviously incapable of accounting for all the observed effects. These include relatively "square" profiles for some lines, weak emission wings, extended but shallow absorption wings, asymmetries, etc. In addition, the profiles sometimes differ systematically between the two spectra in canonical states of polarization. Since this phenomenon is most often observed near the phases where H_e changes sign, it is called the "crossover effect". A full discussion of the crossover effect will be deferred to another section.

DEUTSCH has noted that among the spectrum variables with known periods, the widths of relatively non-variable lines are strongly correlated with the observed periods [*2*]. Over the range of periods from $0^d.52$ to $9^d.30$, the widths of such lines are approximately in inverse ratio to the periods. The period-line-width relation for spectrum variables is clearly consistent with the oblique rotator model and, indeed, constitutes one of its chief supports. DEUTSCH assumed that the spectrum variables have radii normal for an *A* 0 dwarf; that the observed periods of spectrum variation are the periods of rotation; and that each star is viewed in the equatorial plane. For each of the seven spectrum variables known in 1951 to have periods less than six days, he then computed the rotational disturbance, and applied it to the observed profile of Mg II 4481 in the *F* 0*p* star γ Equ. The result, in each case, was a profile in reasonably close agreement with the observations[2]. For the four spectrum variables known to have periods longer than five days, the same procedure was carried out for the weaker and intrinsically sharper line Fe II 4576, with similar results.

Subsequent work has added a number of spectrum variables of known period, as shown in Table 1. The profiles of relatively non-variable lines in these stars indicate that there is, indeed, some scatter about the mean relation between period and line-width. This would be expected, of course, if only as the result of inclinations that deviate from 90°. BABCOCK notes that in HD 153882, the measured line widths yield an approximate upper limit of 10 km/sec for $V_e \sin i$; this is to be compared with a predicted V_e of 17 km/sec on the oblique rotator

[1] W. W. MORGAN: Publ. Yerkes Obs. **7**, Part 3 (1935).

[2] A. J. DEUTSCH: Trans. Internat. Astronom. Union **8**, 801 (1952).

hypothesis. For HD 125248, he finds $V_e \sin i \leqq 4$ km/sec; the predicted V_e is 11 km/sec. For HD 71866, he finds $V_e \sin i \lesssim 9$ km/sec; the prediction is 15 km/sec. Limb-darkening and irregular abundances on the stellar surfaces may partly account for these differences. In any case, the period-line-width relation remains a strong piece of evidence in support of the oblique-rotator hypothesis for the periodic spectrum variables. It is noteworthy that the mean period of the stars in Table 1 is about four times greater than the mean rotation period for *B* 8 to *A* 2 V stars. The oblique-rotator hypothesis therefore implies that slowly-rotating *A* stars preponderate among the spectrum variables.

7. Line profiles: magnetic effects. In some of the sharp-line *Ap* stars, where the Zeeman effect can be detected, the Zeeman pattern may become so wide that it determines the width of the observed line profile. Thus, for Eu II 4205 in a field of 3 kgs, the extreme σ-components are separated by 0.14 Å; for Fe I 4210, by 0.15 Å. At $T = 10000°$, the Doppler widths of these lines are, respectively, 0.014 and 0.025 Å. These lines, and others with comparably wide Zeeman patterns, have frequently been noted in *Ap* spectra as uncommonly wide for their strengths.

Magnetic intensification of lines is a related phenomenon. A simple manifestation would be the following. Consider an absorption line originating from an atomic state whose population is N_0 cm^{-2} over the stellar surface. Suppose that its equivalent width is W_0 in the absence of a magnetic field, and that it lies well out on the flat part of the curve of growth. In the presence of a sufficiently strong magnetic field, the Zeeman components of this line will be separated by distances large compared with their Doppler width. If the magnetic field is everywhere purely transverse, each of the Zeeman components will then produce an absorption line in one of the two canonical states of polarization. The equivalent width of each such line will be W_0 when measured with respect to the corresponding polarized continuum, and $W_0/2$ with respect to the whole unpolarized continuum. If the total number of Zeeman components is n, the equivalent width of the whole Zeeman pattern will then be $W = (n/2)\, W_0$.

The magnetic intensification W/W_0 will be less than $n/2$ whenever the Doppler width exceeds the Zeeman splitting; when N_0 is small enough so that the less intense Zeeman components slide off the flat part of the curve of growth; or when the field $\boldsymbol{H}$ is not purely transverse. Babcock has taken into account the second and third of these effects in calculations for fully resolved triplet patterns observed in the flux from a stellar atmosphere in a dipole field[1]. For an Fe line at $T = 8000°$, with Doppler width 0.22 Å, he finds the magnetic intensification W/W_0 to be 1.30 at an inclination $i = 0°$ relative to the dipole axis, and 1.32 at $i = 90°$; the field at the pole is taken as $H_p = 8$ kgs. The intensification will more closely approach the theoretical limit of 1.5 as N_0 increases, or as the Doppler width decreases.

For most lines with anomalous Zeeman patterns, the fields in stellar atmospheres are never strong enough to prevent overlapping of the absorption coefficients of the various components. The complications due to blending effects of this kind will become particularly disagreeable in the oblique Zeeman effect, when π- and σ-components overlap and compete with each other in absorption from non-canonical states of polarization. Detailed calculations have not been made in the context of stellar spectroscopy, but Babcock has attempted to estimate the magnetic intensification factor for a variety of Zeeman patterns in a stellar dipole field with $H_p = 8$ kgs. His estimate is "based on the number of components

[1] H. W. Babcock: Astrophys. Journ. **110**, 126 (1949).

in the pattern and on their spacing in the π and σ groups in relation to the half-width of the Doppler profile". He finds that these estimated magnetic intensification factors correlate with the intensity variations estimated by STRUVE and SWINGS for a number of lines in the spectrum-magnetic variable α^2 CVn. In particular, the line Eu II 4205 has the largest intensity change and the largest magnetic intensification factor (7); the number of Zeeman components for this line is 21.

The effects of magnetic intensification must indeed be present in the spectra of all magnetic stars. However, they are obviously incapable of explaining spectrum variation of large amplitude. Magnetic intensification can clearly not depend on the sign of the field, and by itself, cannot produce the observed phase differences between the intensity curves for different elements. In some cases the observed intensity variation exceeds the theoretical limiting value, $n/2$, of the magnetic intensification factor. In their spectrophotometric study of α^2 CVn, the BURBIDGES found no correlation between the observed intensity variation of lines and their Zeeman splitting[1].

Probably the complicated effects of magnetic intensification are still further complicated in an *Ap* star by abundance irregularities over the observed surface, and by gradients with depth. HUBENET has worked out the effects of a vertical gradient of $\boldsymbol{H}$ upon the contour of a Zeeman triplet observed in the direction of $\pm\boldsymbol{H}$, as in a sunspot at the center of the solar disk[2]. HUBENET's results yield both an intensification of the line and an asymmetry, although a uniform longitudinal field obviously produces no effect at all.

8. Line-profiles: crossover effect. In BABCOCK's analyzed spectra of some magnetic stars, at the epochs when $H_e \approx 0$, some lines show profiles systematically wider in one state of polarization than in the conjugate state. This "crossover effect" appears most conspicuously in the spectrum of HD 71866, which is a magnetic variable with $P = 6.80$ days and a spectrum variable of moderate amplitude[3]. Fig. 8 illustrates the crossover effect in this star at two phases. The figure shows the effect to differ strongly among different lines; BABCOCK notes that the effect may be specific with respect to element, state of ionization, or even multiplet, and that it also depends on the Zeeman pattern of the line. The sharper lines may occur either in the levogyrate or dextrogyrate spectrum; in the latter case, the sign of the effect is taken as positive.

BABCOCK's explanation of the crossover effect supposes that the affected lines arise predominantly in two areas A and B on the disk which have different magnetic fields and different radial velocities. Fig. 9 illustrates BABCOCK's hypothesis. If we neglect the intrinsic (Doppler) width of the line, for example, in Fe I 4233.6 the differential velocity ΔV between A and B will produce two lines separated by the corresponding Doppler shift $\Delta\lambda$. If $\boldsymbol{H}$ in A is equal and opposite to $\boldsymbol{H}$ in B, then the Zeeman patterns of the two Doppler-shifted lines will be reversed, as shown. In a purely longitudinal $\boldsymbol{H}$, only the σ-components will appear. The dextrogyrate spectrum will then show a narrow pattern in the illustrated case, or, in the actual case, a narrow line. The levogyrate spectrum, on the other hand, will show a wide, double pattern, or, in the actual case, a widened line. If the field is oblique instead of purely longitudinal, the π-components will be shared equally between the two analyzed spectra.

[1] G. R. BURBIDGE and E. M. BURBIDGE: Astrophys. Journ. Suppl. **1**, 431 (1955).
[2] H. HUBENET: Z. Astrophys. **34**, 110 (1954).
[3] H. W. BABCOCK: Astrophys. Journ. **124**, 489 (1956).

BABCOCK's measures of λ 4233.6 indicate a maximum H_e of 4.4 kgs. If, as in Fig. 9, the velocity ΔV is adjusted to obtain coincidence of the dextrogyrate σ-components in a purely longitudinal $\boldsymbol{H}$ of 4.4 kgs, then BABCOCK finds $\Delta V=$ 10.4 km/sec. If the intensity-centroids of these components are matched, ΔV

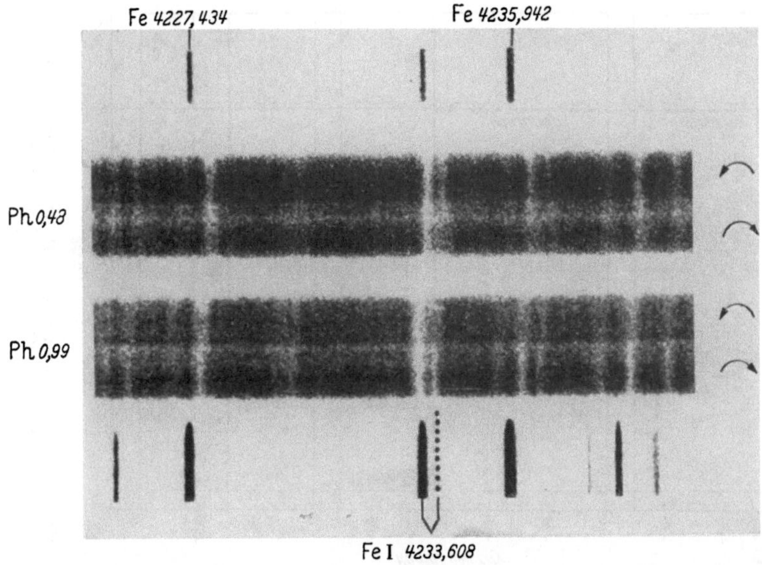

Fig. 8. Crossover effect in the spectrum of HD 71866 ($P=6^{\mathrm{d}}.80$). The line profiles differ conspicuously between the two states of polarization in Fe I 4233.61, but they are nearly the same in Fe II 4233.12. The effective field H_e was nearly zero, and increasing, at the phase 0.99 of the lower pair of analyzed spectra; in the upper pair, it was nearly zero and decreasing. [Reproduced from H. W. BABCOCK: Astrophys. Journ. **124**, 489 (1956).]

must be 7.9 km/sec. Similar calculations for other lines sensitive to the crossover effect yield a mean "matching velocity" ΔV of 5 km/sec. Presumably any other velocity ΔV between A and B will yield a less conspicuous crossover effect than is observed.

It is possible that this differential velocity between two parts of the visible hemisphere arises simply as the result of the stellar rotation. The oblique rotator can clearly produce a crossover effect, but only with positive sign when H_e is increasing to more positive values. This condition is satisfied in the five regularly reversing magnetic variables that show the crossover effect; in HD 98088, the effect is not seen. If ΔV is simply the result of rotation, BABCOCK's mean value of 5 km/sec implies an equatorial speed of rotation of $V_e \approx 3.5$ km/sec. However, the period of an A0 star, with $R=2R_{\odot}$, and $V_e=3.5$ km/sec, is 29 days. The discrepancy between this and the observed magnetic period of 6.8 days militates against this explanation. On the other hand, it is possible that in computing the matching velocities with the maximum effective field for each line, BABCOCK has systematically underestimated the Zeeman splitting that actually occurs in

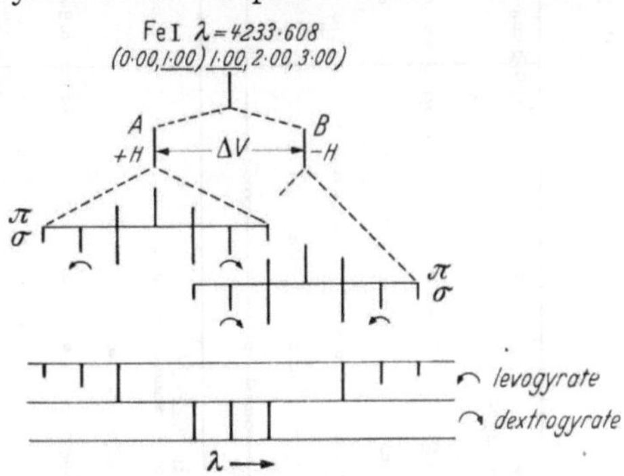

Fig. 9. Schematic representation of BABCOCK's hypothesis for the crossover effect. See text. [Adapted from H. W. BABCOCK: Astrophys. Journ. **124**, 489 (1956).]

the two areas that dominate the line profile at the crossover phases. It will always be true, of course, that in some parts of the observed hemisphere the local field must be numerically greater than the effective field, which is always an average over the disk.

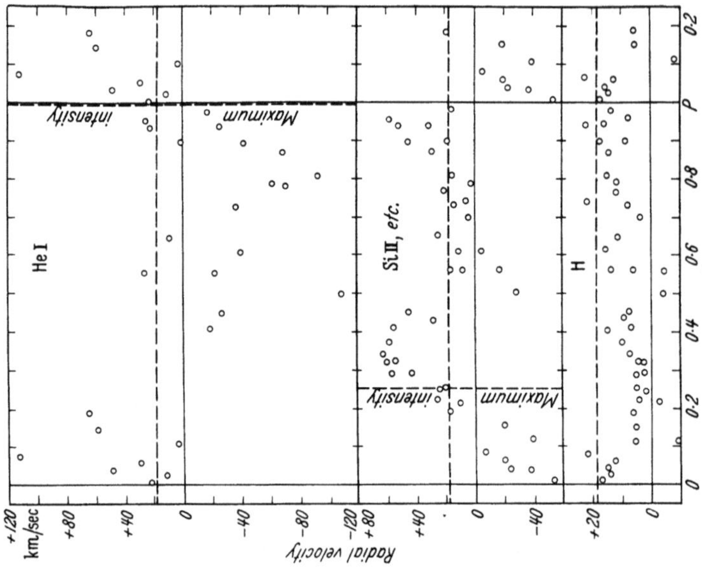

Fig. 11. Velocity variation in 56 Ari ($P = 0^{\mathrm{d}}.73$). [Adapted from W. BONSACK: Publ. Astronom. Soc. Pacific **70**, 90 (1958)].

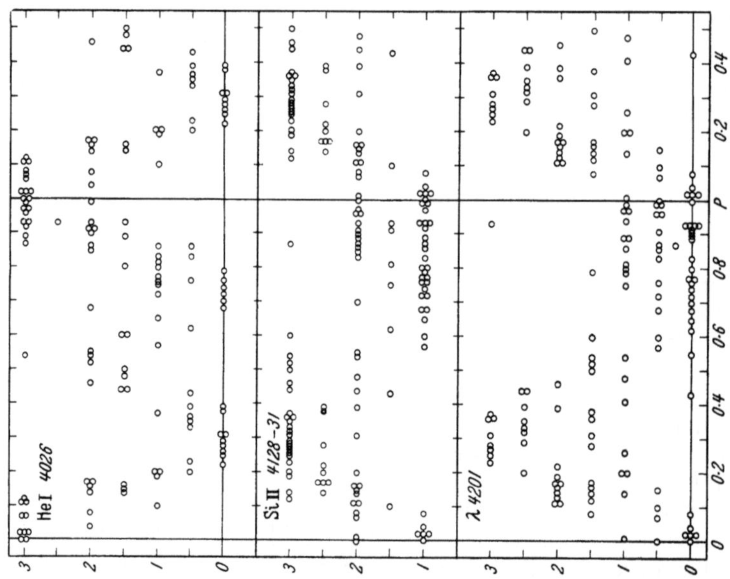

Fig. 10. Spectrum variation in 56 Ari ($P = 0^{\mathrm{d}}.73$). (Adapted from Ref. [2].) The intensities are in arbitrary units.

9. Radial velocities. Of the *70 A* stars in which BABCOCK has observed the Zeeman effect, ten are known to be spectroscopic binaries. One of these, HD 98088, shows magnetic and spectrum variation with the same period ($5^{\mathrm{d}}.9$) as the orbital motion. This star is undoubtedly an oblique rotator. However, "it is the only spectroscopic binary known to show magnetic variations that are related to the orbital motion" [1]. BABCOCK also notes that the stars in this system are separated by only about ten stellar radii, and that tidal interactions are probably important.

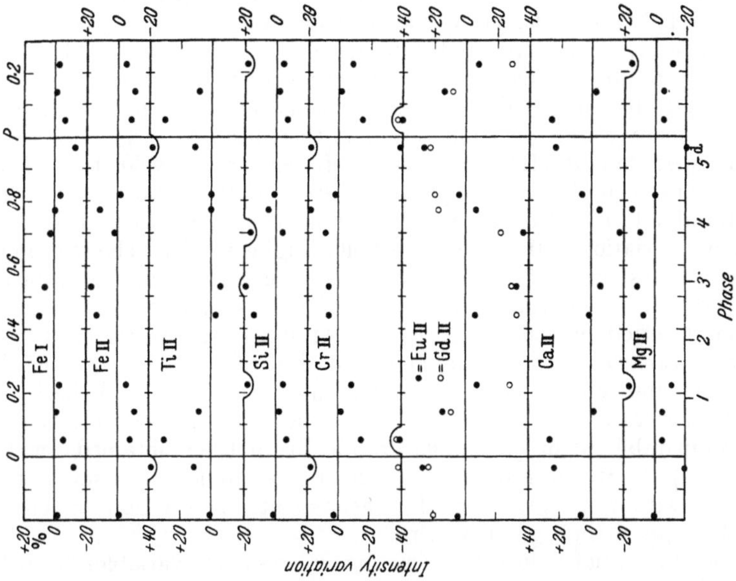

Fig. 13. Intensity variations in α^2CVn ($P = 5^{d}.47$). (Adapted from Ref. [2].)

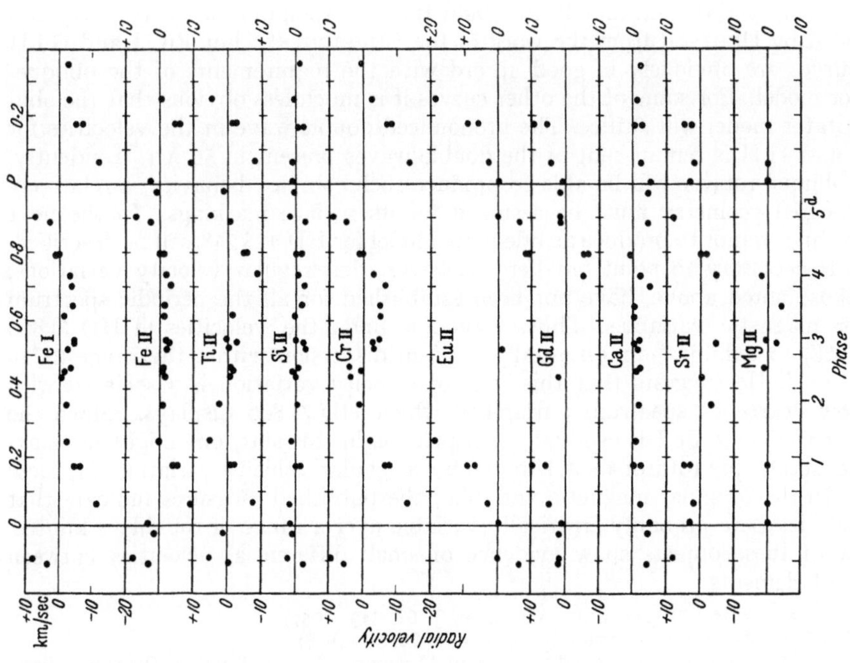

Fig. 12. Velocity variation in α^2CVn ($P = 5^{d}.47$). [Adapted from H. W. BABCOCK and S. BURD: Astrophys. Journ. 116, 8 (1952).]

HD 98088 has been classified as *gF0p*, but its spectrum is very closely similar to that of a number of *Ap* stars, including, for example, the spectrum-magnetic variable 53 Cam ($P = 8^{d}.0$).

The statistics of binary incidence among *Ap* stars, as compared with normal *A* stars, has been studied recently by C. and M. JASCHEK[1]. They find a normal incidence of visual binaries among the *Ap*'s, viz., 25%; but a significantly low incidence (~15%) of spectroscopic binaries as compared with normal *A* stars, of which 43% are indicated as spectroscopic binaries. The JASCHEKS note that this deficiency of spectroscopic binaries among *Ap* stars would be expected if the *Ap* stars are selected for low inclination of their rotational axes, for these are likely to be correlated with the orbital axes in short-period systems.

As in many of the spectrum variables, there are times when different elements in the spectrum of HD 98088 yield different radial velocities. The oblique-rotator model obviously can produce effects of this kind. On this model, the equatorial velocity of rotation in km/sec is $V_e = 100/P$, where P is the period of spectrum variation in days, and the stellar radius is $2R_\odot$. The motion of spectroscopic patches across the visible hemisphere, as a result of rigid rotation, should then produce differential velocities, and these should be especially notable in the short-period objects. Figs. 10 and 11 illustrate the relations between the intensity-variations observed by DEUTSCH [2] in 56 Ari ($P = 0\overset{d}{.}73$) and the velocity variations observed by BONSACK[2]. The intensity-curves lead the corresponding velocity-curves in phase by about 0.25 cycles, as expected, and the amplitudes are comparable with V_e. Similar results have been obtained for HD 124224 ($P = 0\overset{d}{.}52$)[3,4]. In both stars, the Balmer lines indicate constant strength and velocity, within the precision of measurement. This is low, of course, for all lines because of their great width.

The differential velocities occurring in spectrum variables have been most exhaustively studied in α^2 CVn ($P = 5\overset{d}{.}47$). Figs. 12 and 13 give the velocities determined by BABCOCK and BURD[5], and the corresponding intensity variations compiled by DEUTSCH from the work of the BURBIDGES[6]. For Eu II and Gd II the curves are obviously in good accord with the requirements of the oblique-rotator model. For some of the other curves it is much less obvious that the oblique rotator model will suffice. The pronounced double-wave in the velocities for Cr II and Ti II is reminiscent of the double-waves present in 56 Ari. Evidently, if an oblique rotator will be able to produce such complex behavior, a relatively complicated geometry must be assumed for its surface markings. In the next section an attempt to produce an adequate model for HD 125248 will be described.

It is necessary to point out here, however, that regular velocity variations, like those noted above, have not been established for all the periodic spectrum and/or magnetic variables. Thus, BABCOCK finds the velocities in HD 71866 ($P = 6\overset{d}{.}8$) to exhibit "an essentially random dispersion within the range 25 to 33 km/sec". He suggests that this type of velocity variation is associated with the low degree of spectrum variability which HD 71866 displays. Since the crossover effect in Fe I is especially conspicuous in this star, one might have expected an oblique-rotator model to exhibit a regular velocity-variation in these lines. In the irregular magnetic variables, the published measures indicate that the velocities are probably erratically variable over a range of roughly 3 km/sec. A few of these objects show evidence of small differential velocities between different elements.

---

1 C. JASCHEK and M. JASCHEK: Astronom. J. **62**, 343 (1957).

2 W. BONSACK: Publ. Astronom. Soc. Pacific **70**, 90 (1958).

3 A. J. DEUTSCH, in A. BEER (ed.): Vistas in Astronomy, Vol. 2. London: Pergamon Press 1956.

4 A. J. DEUTSCH: Astrophys. Journ. **116**, 536 (1952).

5 H. W. BABCOCK and S. BURD: Astrophys. Journ. **116**, 8 (1952).

6 G. R. BURBIDGE and E. M. BURBIDGE: Astrophys. Journ. Suppl. **1**, 431 (1955).

10. Harmonic analysis of spectrum variables. If one specifies the distribution of line strength w over the surface of an oblique rotator, and the inclination i of the rotator's axis, he can then integrate over the visible hemisphere to obtain the line strength W that will be observed at the Earth. The result of this integration will give W as a function of phase Φ. If one also knows the equatorial velocity of rotation V_e, and the stellar velocity V_0, a similar integration gives the radial velocity $V(\Phi)$ that will be indicated by the line. Again, if one also knows $\boldsymbol{H}$ as a function of position over the stellar surface, one can by integration obtain the effective field H_e that will be indicated by the line.

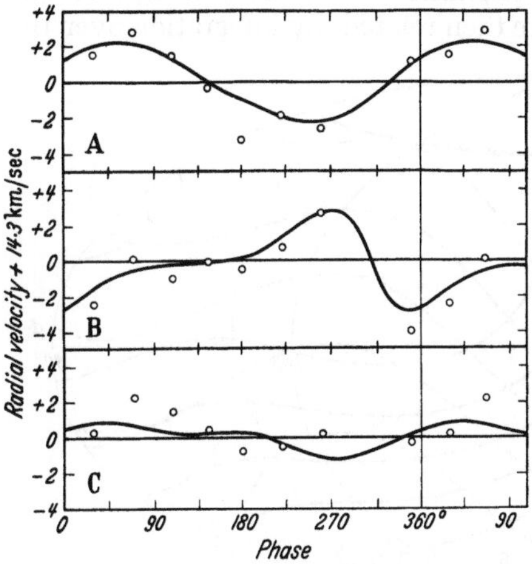

a

DEUTSCH has attempted to invert this argument and derive the distribution function w, the field $\boldsymbol{H}$, and the parameters i, V_e, and V_0 from the obser-

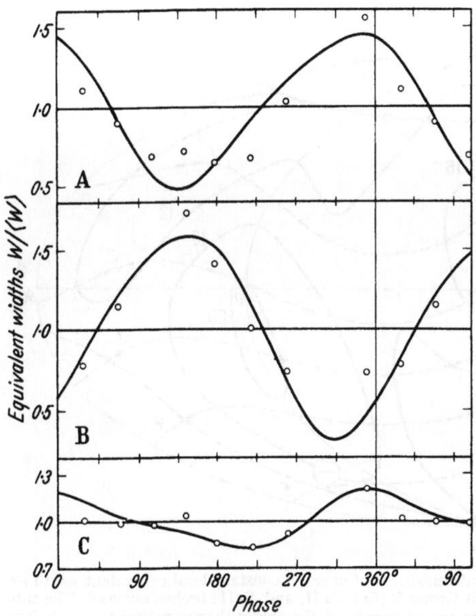

b

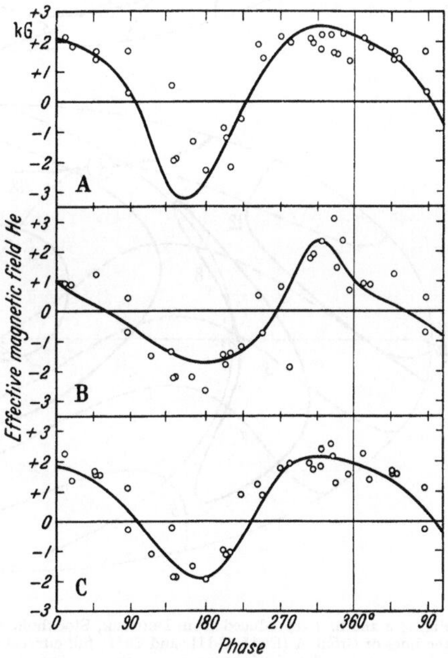

c

Fig. 14 a—c. (Adapted from A. J. DEUTSCH: Proceedings of the Stockholm (1956) Conference on Electromagnetic Phenomena in Cosmical Physics [in press].) Observations and theoretical representations of HD 125248 ($P = 9\overset{d}{.}30$). See text. (a) Velocity variations. (b) Intensity variations. (c) Effective field variations.

vations[1]. These show that, to a sufficient approximation, the ratio $W/\langle W\rangle$ is the same function of Φ for all lines of a given element, where $\langle W\rangle$ is the average equivalent width observed over a whole cycle. In the magnetic-spectrum variable HD 125 248 ($P = 9\overset{d}{.}30$), DEUTSCH finds that most lines can be assigned to one of

[1] A. J. DEUTSCH: Proceedings of the Stockholm Conference (1956) on "Electromagnetic Phenomena in Cosmical Physics" (in press).

three groups, in each of which all the lines indicate the same variation with Φ of $W/\langle W\rangle$, V, and H_e. Group A comprises Gd II, Ce II, and Eu II; Group B, Cr I, Cr II, and Sr II; and Group C, Fe I, Fe II, and Ti II. For the distribution of Group A lines, $(w/\langle W\rangle)_A$, Deutsch uses a Laplace expansion of the second degree. The Laplace coefficients L_n^m of this expansion are then related, by integration over the

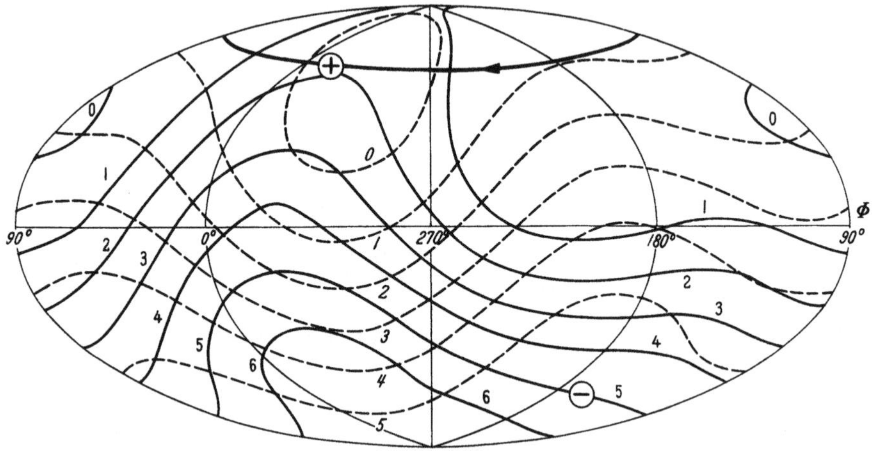

a

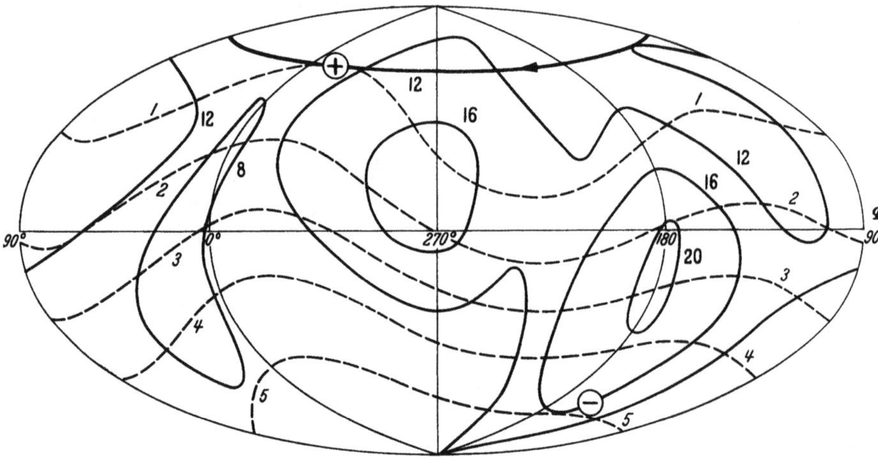

b

Fig. 15 a and b. (Reproduced from Deutsch, Stockholm Conference). (a) Curves of constant local equivalent width for the lines of Group A (Eu II, Gd II, and Ce II; full curves) and Group B (Cr I, Cr II, and Sr II; broken curves). The subsolar point describes the heavy curve as shown. The plus and minus signs mark the axis of the magnetic dipole. (b) The dashed curves are curves of constant local equivalent width for the lines of Group C (Fe I, Fe II, Ti II). The full curves are the contours of $|\boldsymbol{H}|$. The heavy curve and the plus and minus signs have the same significance as in Fig. 15(a).

visible hemisphere, to the Fourier coefficients F_m of the expansion representing the observed equivalent widths $(W/\langle W\rangle)_A$. Another integral relates the L_n^m to the F_m of the Fourier expansion representing the observed radial velocities V_A.

To derive an expression for $(H_e)_A$, Deutsch assumes that $\boldsymbol{H}$ can be derived from a scalar potential S, and this function S is also represented by a second-degree Laplace expansion, with coefficients M_n^m. An integral then relates the M_n^m to the F_m of the Fourier expansion representing the observed effective field

$(H_e)_A$. The distribution functions for Groups B and C yield a similar set of relations giving the Fourier coefficients of the observed curves in terms of the Laplace coefficients of the functions $(w/\langle W\rangle)_B$, $(w/\langle W\rangle)_C$, S, and the parameters i, V_e, and V_0.

The nine observed curves yield 45 Fourier coefficients, and these are now expressed in terms of 35 Laplace coefficients and the three additional parameters i, V_e, and V_0. It should therefore be possible to proceed from the Fourier coefficients back to the Laplace coefficients. In practice it appears that this solution can be made in a way that yields results which are significant, if rough. DEUTSCH's preliminary solution for HD 125248 is illustrated in Figs. 14 and 15. The smooth curves of Fig. 14 are the Fourier representations of the indicated observations. Fig. 15 gives the corresponding contours of the distribution functions and of $|\boldsymbol{H}|$. Additional integrations over the derived distribution functions have also been made to predict the variation of line-widths due to rotation. These predictions are in reasonable accord with the line-width variations actually observed.

Recent observations have established the period of the Kepler motion in HD 125248 as 4.40 years, and have made it possible to combine the velocity observations over many years to strengthen the velocity-curves of Fig. 15. A new solution is now under way, based on these improved velocity curves. Probably it will change the derived maps somewhat, but the principal large-scale features of the maps are not likely to be affected. Attempts are also being made to map other suitable periodic magnetic variables, including α^2 CVn.

An interesting feature of the solution for HD 125248 is the situation of all three abundance maxima in the unobserved zone of the star. A calculation has shown that if this configuration were observed under an inclination of 150° instead of 30°, all three groups of lines would be about four times stronger when averaged over the cycle. The amplitudes of the curves giving H_e and V as a function of phase would be comparable with the present values. But the amplitudes of the curves for $W/\langle W\rangle$ would be less than one third as great, because of the increase in $\langle W\rangle$. It is possible that aspect effects of this kind could account for some of the magnetic variable stars that do not show conspicuous spectrum variation.

11. Abundance determinations. Abundance determinations in peculiar A stars have been attempted by several investigators, but usually from spectrograms of relatively low dispersion. The earlier work has now been largely supplanted by the recent studies of the BURBIDGES[1]. They have discussed 4.8 Å/mm spectrograms of α^2 CVn, and 18 Å/mm spectrograms of HD 133029 and HD 151199. In all cases, abundances relative to those of a standard star have been obtained by a curve-of-growth analysis. The abundances are summarized in Table 3. Table 4 gives the adopted values of certain other important parameters.

Such abundance determinations always suffer from certain difficulties. Photometric errors may be important, and uncertainties due to blending are unavoidable in spectra as rich as these. Such effects become particularly invidious when, as is often the case, the only accessible lines of an interesting element are very weak. Ions dominate the spectra of these stars, and for these lines neither laboratory nor theoretical line-strengths are available. Recourse has been had to some solar line-strengths, and to some stellar line-strengths from BUSCOMBE's

[1] G. R. BURBIDGE and E. M. BURBIDGE: Astrophys. Journ. Suppl. **1**, 431 (1955). — Astrophys. Journ. **122**, 396 (1955); **124**, 130, 655 (1956).

Table 3. *Atomic abundances (by number) in three peculiar A stars.*

Element	Abundance ratio			Element	Abundance ratio		
	α^2 CVn: γ Gem	HD 133029: α^2 CVn	HD 151199: 95 Leo		α^2 CVn: γ Gem	HD 133029: α^2 CVn	HD 151199: 95 Leo
Mg	0.4	3.5	1.2	Y	20:	—	—
Al	1.1	2.0	—	Zr	30	1.3	—
Si	10	2.5	1.3	Ba	0.9:	—	0.6
Ca	0.02	2.6	2.6	La	1020	0.2:	—
Sc	0.7	—	—	Ce	400	0.6	—
Ti	2.6	1.0	—	Pr	1070	0.6	—
V	1.3	2.3	—	Nd	250	0.6	—
Cr	5.2	1.8	1.8	Sm	410	0.6	—
Mn	16	1.0	9	Eu	1910	0.5	130
Fe	2.9	1.5	1.1	Gd	810	0.4	—
Ni	3.0	0.8	—	Dy	760	0.6	—
Sr	14	1.1	65				

Table 4. *Atmospheric parameters for three peculiar A stars.*

Star	α^2 CVn	HD 133029	HD 151199
Observed range of H_p (kilogauss)	$+5$ to -4	$+10$ to $+4$	—
Excitation temperature T_e	8000°	8000°	7400°
Ionization temperature T_i	8400°	8000°	7800°
Electron pressure P_e (dynes/cm²)	100	100	130
Velocity parameter (Thermal + turbulent + magnetic; km/sec)	2 to 4	3 to 6	4

work[1] on the *A* 0 IV star γ Gem and from GREENSTEIN's[2] on the *ML* star τ UMa. The derived abundances are therefore subject to errors of the curves of growth adopted for these standard stars, as well as to errors of the curves adopted for the peculiar stars. For the latter especially, it is dangerous to assume that all lines of all elements will fall on a single curve of growth that is computed on an approximate theory for a normal and homogeneous stellar atmosphere. Effects of magnetic intensification may cause other uncertainties.

The BURBIDGES have also had to contend with two additional difficulties. For the elements Ce, Pr, Nd, Gd, and Dy, the second ionization potentials are still unknown. The authors have assumed a value of 11.5 ev, which is the average for the other rare earth elements. Since these do not differ much among themselves, it is unlikely that the assumed values are greatly in error. On the other hand, the ionization is such that about 80% of the rare earth elements will be in the second state of ionization, which, in general, is not observed. A small error in the observed abundance of Gd II, for example, can therefore be magnified five times or more in the total abundance for Gd. Then, in α^2 CVn, the lines of most elements change with phase, as we have seen. The results of Table 3 are derived from mean line strengths, and therefore represent an average abundance over the cycle. However, in Ti, the abundance at $\Phi = 0.0\,P$ is 2.6 times the abundance at $\Phi = 0.05\,P$; for the rare earth elements, this factor may lie anywhere in the range 4.5 to 30!

With these difficulties in mind, the BURBIDGES have summarized their own abundance determinations for α^2 CVn and HD 133029 in the following words [3]:

1. O is probably underabundant, although the factor could not be determined.
2. Ca is underabundant by a factor of about 30.

[1] W. BUSCOMBE: Astrophys. Journ. **114**, 73 (1951).

[2] J. L. GREENSTEIN: Astrophys. Journ. **107**, 151 (1947).

3. Si is overabundant by a factor of about 15.

4. The metallic group is overabundant by a factor of about 6; the effect is largest for Mn and Cr; the factor for Fe, which is the element contributing the major part of the metal peak, is only about 2, and this value is still subject to observational uncertainties.

5. Sr, Y, and Zr are overabundant by a factor of about 25.

6. La, Ce, Pr, Nd, Sm, Eu, Gd, and Dy have a mean excessive abundance ratio of the order of 600, while Ba is the only member of this group to have an apparently normal (i.e. solar) abundance.

7. Pb is probably overabundant by a factor of about 1500.

8. The presence of U cannot be ruled out. In addition, there are several unidentified lines in the spectra of these stars which may be attributable to other heavy elements.

In addition to these conclusions, the BURBIDGES believe that at least some of the apparent abundance differences between α^2 CVn and HD 133029 are probably real. In particular the rare earth elements are only about half as abundant in the latter star as in the former. HD 151199 shows no significant anomalies in Ca and Si; in this star, the only notable effects are the overabundances in Mn, Sr, and Eu.

Some of the abundance anomalies observed in the peculiar *A* stars are reminiscent of those in the *S* stars; in these, the rare earth elements and other heavy elements appear to be abnormally abundant. The *S*-type stars are red giants, and they may well contain hydrogen-exhausted central cores in which helium-burning takes place. CAMERON has indicated that at the temperature ($T \approx 10^{8°}$) and density ($\varrho \approx 10^3$ gm/cm^3) of such a core, nuclear reactions with neutrons will synthesize heavy nuclei from the relatively abundant metal nuclei of intermediate weight[1]. In collaboration with FOWLER and HOYLE, the BURBIDGES have recently elaborated upon the arguments of CAMERON and have shown that most or all of the heavy nuclei now present in stars may actually have been synthesized in past generations of red giants[2].

However, for the reasons that have been outlined in Sect. 5, it seems probable that the peculiar *A* stars have originated relatively recently from matter with the normal cosmic abundances, and that the physical conditions in their interiors are not sufficiently extreme for the type of element synthesis that occurs in the cores of red giants. Accordingly, FOWLER and the BURBIDGES have recently examined the possibility of explaining the anomalous abundances of a peculiar *A* star in terms of nuclear synthesis right in the atmosphere itself [*3*]. This is an attractive kind of synthesis for an *A* star, because the atmosphere is probably mixed only very slowly into the interior. In an *A* star, the radiative gradient becomes unstable against convection at optical depth $\tau \approx 0.25$. However, the convective energy transport is likely to remain small, particularly under the influence of a strong magnetic field. In this circumstance, the temperature gradient will closely approximate the radiative gradient, and the whole thickness of the hydrogen convective zone will be only of the order of 1000 km. FOWLER and the BURBIDGES conservatively estimate the thickness of the mixing zone as 10^4 km; with a mean density of 10^{-4} gm/cm^3, it comprises 2×10^{28} gm, or 3×10^{-6} of the total mass of the star.

12. Nuclear reactions in stellar atmospheres. Since it seems necessary to explain the observed abundance anomalies in terms of nuclear reactions occurring

[1] A. G. W. CAMERON: Astrophys. Journ. **121**, 144 (1955).

[2] E. M. BURBIDGE, G. R. BURBIDGE, W. A. FOWLER, and F. HOYLE: Rev. Mod. Phys. **29**, 547 (1957).

within the "mixing layer" near the photosphere, FOWLER and the BURBIDGES have proposed a model in which the following sequence of events occurs[1].

(a) In volumes measuring $\sim 10^9$ cm in diameter, the protons are accelerated into a Maxwell-Boltzmann distribution with energies $kT \sim 6$ Mev. These energetic proton fluxes presumably result from unspecified disturbances of the magnetic field. At the known densities and field strengths, the acceleration processes must occur in times of the order of one second or less; otherwise, radiation from electrons in bremsstrahlung and/or the synchrotron process will prevent the protons from achieving energies >1 Mev. The optical radiation from these "hot spots" will correspond to energies $kT_a \sim 1$ kev; this radiation is responsible for the color-spectral-type anomaly of the *Ap* stars.

(b) Neutrons are liberated within the hot spots by (p, n) and (a, n) reactions on the light nuclei, and especially by $N^{14}(p, n) O^{14}$. Depending on the time available, the neutron-proton ratio will build up to $\sim 10^{-1}$ in the hot spots. The diffusion length for neutrons being of the order of the spot radius, or larger, they will diffuse out of the hot spot, be thermalized to $kT \sim 2$ ev, and be captured by protons.

(c) The deuterium so formed will be largely decomposed by the $H(d, \gamma) He^3$ reaction in "cooler" hot spots, where the proton flux has energies distributed in a flat power law, $N(E) \propto E^{-1}$, with a sharp cutoff near 6 Mev. D will also be partly decomposed by the $H(d, n)\, 2\, H$ reaction in these areas. When the latter reaction has provided about 20 neutrons per Fe nucleus in the mixing zone, the observed overabundances of heavy elements will have been (roughly) reproduced. Nuclei which have magic-number isotopes, and are therefore particularly stable against neutron capture, should be enriched 10 or 100 times more than other nuclei of comparable Z. The observed abundances of Sr, Y, and Zr, which possess magic-number nuclei, seem to show this effect. However, Eu, which has the highest overabundance, has no magic-number isotope; and Ba, which has a nearly normal abundance, does have such an isotope.

(d) The $H(d, \gamma) He^3$ reaction will produce a He^3 concentration approximating that of He^4. The expected equilibrium values will be $He^3/H \sim D/H \sim 10^{-1}$. The BURBIDGES have given some evidence for an isotope shift in the HeI lines of the magnetic star 21 Aql.[2]. Their observations are consistent with the predicted enrichment of He^3, but they may be vitiated by intermolecular Stark effect.

(e) Silicon may be enriched by the reaction $S^{33}(n, \alpha) Si^{30}$, and oxygen depleted by $O^{16}(p, \gamma) F^{17}(\beta^+) O^{17}(p, \alpha) N^{14}$. The anomalies in Ca, Cr, and Mn are still unexplained.

If we assume that $\sim 10^9$ years are available for producing the observed enrichment of heavy elements in the mixing zone, this proposal requires the development of a "hot" hot spot every 11 hours, on the average. In these spots, the total number of neutrons produced by (p, n) reactions on light nuclei is about 5×10^{51}. The "cool" hot spots responsible for heavy-element synthesis would have to develop every 10^3 seconds, on the average.

This theory bears some resemblance to the proposals of PARKER[3], who has shown that in very large solar flares roughly 1% of the energy dissipated will appear as protons with energies averaging ~ 1 Bev. PARKER's theory requires that hydromagnetic shock waves raise to thermal energies of the order of a few kev a small fraction—say 10^{-7}—of the protons in a collapsing magnetic field.

[1] E. M. BURBIDGE, *et al*: Proceedings of the Stockholm Conference (1956) on "Electromagnetic Phenomena in Cosmical Physics" (in press).

[2] E. M. BURBIDGE and G. R. BURBIDGE: Astrophys. Journ. **124**, 655 (1956).

[3] E. N. PARKER: Phys. Rev. **107**, 830 (1957).

Before escaping, these protons are further accelerated in a Fermi mechanism within the collapsing field, the processes of energy loss being negligible at this injection energy and at particle densities of the order of 10^{11} cm^{-3}, as in the chromosphere. In smaller flares, where the Fermi process operates for shorter intervals, the escaping protons might be expected to have energies of the order of only a few Mev, and to produce a local enrichment of D by interaction with the thermal protons around the flare. Actually, SEVERNY[1] has found spectroscopic evidence for a considerable enrichment of D near the short-lived bright grains first studied by ELLERMAN[2] and called by him "hydrogen bombs".

PARKER notes that the energy generated per cm^3 in a large flare is of the same order as the energy density in a magnetic field of order 300 gauss. The self-induction of the required field is so large that the decay time is of order 10^6 years. But the flare dissipates its energy in some 15 minutes! This indicates the occurrence of dissipative processes many orders of magnitude more efficient than simple Ohmic currents. The macroscopic theory of a disturbed magnetic field leads us to expect fluid velocities of the order of the Alfvén velocity. Where this is highly supersonic, we must expect strong local heating which is essentially compressive, as in an ordinary shock wave, and not Ohmic. The most energetic ions may then be free to move through the fluid along the lines of force, and be subjected to further accelerations by Fermi reflections. While the details of such processes remain to be worked out, that the processes must occur is clearly indicated by the evidence from solar flares.

c) Other magnetic stars.

13. The metallic-line stars. These objects lie above the main sequence in the color-magnitude diagram at lower absolute magnitudes than the *Ap* stars (see Fig. 6, p. 695). The Balmer lines indicate spectral types in the range *A* 5 to *F* 5; the Ca II *K* line always yields an earlier type, the other metallic lines usually a later type. High-dispersion studies of three *ML* stars have shown that their atmospheres simulate those of supergiants, with high turbulence, low electron pressure, and low opacity. In addition, there is a tendency for elements to be deficient when they have second ionization potentials near the Lyman limit[3].

The *ML* stars have nearly the same average rotational velocity as the *Ap* stars (Table 2, p. 697); relative to main-sequence stars of the same color, which have types in the range *A* 5 to *F* 0, they have significantly low rotational velocities. The *ML* stars may well form an extension of the *Ap* group to fainter absolute magnitudes. The incidence of spectroscopic binaries among *ML* stars seems to be higher than among normal stars, however. The *ML* stars are therefore not likely to be a selection of rapidly-rotating stars at low inclinations.

BABCOCK lists seven *ML* stars that definitely show the Zeeman effect. He also lists a larger number of sharp-line *ML* stars in which the Zeeman effect is not seen. Where a field is observed, it appears to be irregularly variable, and in some cases it reverses sign.

14. RR Lyrae. This is the prototype of a populous group of intrinsic variable stars characteristic of BAADE's Population II. The period is 0.567 day; long-period terms of smaller amplitude also occur in both the light and velocity curves. The spectral type varies between *A* 2 and *F* 0, as judged from the metallic lines,

[1] A. B. SEVERNY: Astronom. J. USSR. **34**, 684 (1957).
[2] F. ELLERMAN: Astrophys. Journ. **46**, 298 (1917).
[3] G. R. MICZAIKA, F. A. FRANKLIN, A. J. DEUTSCH and J. L. GREENSTEIN: Astrophys. Journ. **124**, 134 (1956).

but the spectrum is always peculiar in that the Balmer lines are too weak. The star apparently executes radial pulsations. At some phases absorption lines are formed in two separate layers with different velocities[1].

BABCOCK finds a large reversing field; his catalogue gives -1.6 kgs $\leqq H_e \leqq +1.2$ kgs. The field does not seem to change smoothly with phase in the light-cycle. Significant changes in H_e occur within a few hours; long-period changes in H_e also seem to be present.

This star is a highly strategic object for the study of hydromagnetic oscillations. G. R. BURBIDGE has shown that, depending on the central condensation of the field, it may be expected to change the period by a few percent. WOLTJER has inferred that an RR Lyrae star, with a field comparable to that observed in the prototype, was the ancestor of the Crab nebula (see Sect. 18, below).

15. The subdwarf HD 19445. This star is a subdwarf with high space velocity. Like RR Lyrae, it is a typical member of BAADE's Population II. The spectral type is ambiguous, but in the neighborhood of $F0$. CHAMBERLAIN and ALLER have found the Fe abundance per gram of matter to be only $\frac{1}{6}$ that in the Sun, and the Ca abundance only $\frac{1}{25}$. Different elements indicate significantly different values for H_e.

16. Late-type giants. Three M-type giants are known to show the Zeeman effect. All are abnormal in showing forbidden emission lines, and one (HD 4174) shows nebular lines of [Ne III] and [O III]. The most luminous of the group is the $M2$ Iab supergiant VV Cep. It has a peculiar B-type companion which suffers a chromospheric eclipse, and then a total eclipse, every 20 years. From detailed observations of the Zeeman effect during the 20-day chromospheric eclipse, it should be possible to obtain unique information on the chromospheric field and its gradients. Another supergiant (α Ori, $M2$ Ib) shows no Zeeman effect. A field is probably present, however, in the long-period variable star R Leo ($gM\,8e$, $P=313^d$). The S-type long-period variable R Gem ($P=370^d$) also shows a general field, with significant differences in H_e between elements. A nonvariable S star (HR 1105) also shows a field.

Depending on the relative central concentrations of gas density and magnetic energy in a late-type giant star, its total magnetic energy may be comparable with the magnitude of its gravitational potential energy. This can lead to dynamical instability, as will be shown in a later section. Another point of particular interest in these stars is the unexpectedly fast changes sometimes observed in H_e. In VV Cep for example, H_e has been observed to vary through about 1 kgs in about 10^2 days. But since the radius of this star is 10^3 solar radii, or greater, we find by the argument of Sect. 3 that differential gas motions must occur with velocities of order 10^2 km/sec. The line profiles, however, cannot be reconciled with velocities much larger than 10 km/sec. To remove this discrepancy, we require large changes in surface brightness or line strength over the disk.

17. *Be* and *Ae* stars. BABCOCK finds that among the *Be*, *Ae*, and shell stars with sharp lines, some show the Zeeman effect and some do not. Among the objects with definitely established fields is HD 190073. Weak absorption lines of the metals are superposed on weak, wide emission lines. The absorption lines indicate different H_e's for different atoms and ions; e.g., on one plate, 19 lines of Fe I give $H_e=+0.27$ kgs, and 7 lines of Fe II give -0.03. In AG Peg ($Bep+M$), the metallic emission lines are often sharp enough to show the Zeeman effect, and indicate a large reversing field. Observations in the visual spectrum have

[1] R. F. SANFORD: Astrophys. Journ. **109**, 210 (1949).

shown that the absorption lines of the *M* star indicate a field of the same order as in the *Be* star. The BURBIDGES concluded from a spectrophotometric study of the emission lines that the emission lines arise in an envelope of radius about 400 stellar radii, or 2000 solar radii. They point out that $\boldsymbol{H}$ is likely to be several orders of magnitude greater at the photosphere than in the shell; the measured fields "have their origin in large spot regions on the stellar surface".

The star HD 45677 (*B2e*) is perhaps the most surprising object in this group, for in its spectrum the nebular lines [SII] 4069 and 4076 indicate a field of order -1.6 kgs. Other emission lines indicate somewhat different H_e's.

18. The Crab nebula. This unique object is the expanding remnant of the supernova of 1054. It radiates principally in the continuous spectrum. Recent work has shown that the optical continuum is strongly plane-polarized, and that the non-thermal radio continuum is very intense. Following a suggestion by SHKLOVSKY, OORT and his collaborators have attempted to explain this continuum as synchrotron radiation from relativistic electrons. WOLTJER[1] has proposed a model in which a force-free magnetic field fills the spheroidal volume of the nebula, with a maximum strength of order 10^{-4} gauss near the center of the nebula. The bounding shell of the nebula contains the order of 10^{-1} to 1 solar mass. The filaments that are observed to emit a fluorescence (line) spectrum are low-temperature condensations in this shell, analogous to the solar prominences in the corona; the filamentary structure is attributed to the pinch effect in the surface currents.

The adopted distance of the Crab nebula is 1030 parsec, but there is some evidence that its actual distance is twice as large. At the adopted distance, the average radius is 0.8 parsec, and the total magnetic energy $\mathcal{M}$ of order 10^{46} erg. The gravitational potential energy $|\Omega|$ is several orders of magnitude smaller than this, and the system must be unstable. If the magnetic field is the result of diluting an initial, stellar, field over the volume of the nebula, then $\mathcal{M}$ has been inversely proportional to the nebular radius R throughout the expansion. But $|\Omega|$ also is proportional to R^{-1}, and at all stages of the expansion $|\Omega| \ll \mathcal{M}$. The nebula then would always have been unstable, and could never have originated. The alternative is to acknowledge that the present field is not a diluted stellar field, but rather one which was generated after the beginning of the expansion.

The present kinetic energy of the shell is 10^{49} erg. WOLTJER associates this with the magnetic energy $\mathcal{M}_0$ at a time when R was 10^{-3} its present value. He argues that this magnetic field could have been created by turbulent amplification of the stellar field in and behind the expanding shell. The pre-supernova is identified with a condensed stage of Population II stars, intermediate between the RR Lyrae and white dwarf stages. The radius of the pre-supernova is of order 10^{10} cm, the surface field of order 10^5 kgs, and the magnetic energy $\mathcal{M}$ of order 10^{47} erg. The magnetic energy is within the range of what BABCOCK'S observed field of RR Lyrae itself might yield as the result of evolutionary contraction, and is still much smaller than the limit set by the virial theorem.

WOLTJER also discusses the processes by which the relativistic electrons are replenished in the nebula, to compensate for their rapid degradation by synchrotron radiation. He invokes a Fermi-type mechanism in which electrons collide with hydromagnetic waves within the nebula. The waves are identified with the light-ripples, which BAADE has observed to move through the nebula at speeds comparable with the velocity of light. The waves result from hydromagnetic disturbances at or near the stellar remnant. The electrons may be

[1] L. WOLTJER: Bull. astronom. Inst. Netherl. **14**, 40 (1958).

injected into the acceleration mechanism by strong radioactivity in this core, which has been enriched in radioactive nuclei by the implosive process that set off the supernova explosion, according to the theory of BURBIDGE *et al.*[1].

III. Theory of magnetic stars.

a) The generalized dynamo problem.

19. It is well-known that the gravest mode of rigid decay for a general magnetic field in the Sun has a half-life of order 10^{10} years [*4*]. In a recent discussion, WRUBEL[2] has taken account of the variation of conductivity σ with depth, and has found that the longest half-life is $\tau = 4 \times 10^9$ years; for an $A\,0$ star, it will be an order of magnitude larger. Stellar fields may therefore possibly be relics from the time when the stars were formed out of interstellar matter.

On the other hand, the longest time of rigid decay for the Earth's magnetic field is only about 1.5×10^4 years, and there is evidently no possibility that this field can be regarded as a relic in rigid decay. Evidently fluid motions within the Earth have enormously retarded the time of decay, and there is even the possibility that they have reversed the decay process and have amplified the field initially present. But if fluid motions can retard the decay of a field, it would seem that they also could accelerate it. SWEET[3] has concluded that turbulent motions, in particular, will increase the Ohmic dissipation in an axisymmetric field and considerably hasten the field decay. Since turbulent motions probably occur within the Sun and the A stars, it is possible that the rigid-decay time is really of little relevance.

If fluid motions within a conducting sphere can either retard or accelerate the decay time of a magnetic field, it is probable that they can also exactly maintain it. The classical dynamo problem addresses itself to this problem: Can there exist within a bounded fluid of uniform conductivity σ a solenoidal flow $\varrho \boldsymbol{V}$ and a stationary magnetic field $\boldsymbol{H}$, such that $\boldsymbol{V}$ and $\boldsymbol{H}$ are regular and satisfy appropriate boundary conditions at the origin, at the boundary, and at infinity? The equations to be satisfied are, in e.m.u.,

$$\operatorname{curl} \boldsymbol{E} = 0, \tag{19.1}$$

$$\operatorname{curl} \boldsymbol{H} = 4\pi\sigma(\boldsymbol{E} + \mu \boldsymbol{V} \times \boldsymbol{H}), \tag{19.2}$$

$$\operatorname{div} \boldsymbol{H} = 0, \tag{19.3}$$

$$\operatorname{div} (\varrho \boldsymbol{V}) = 0. \tag{19.4}$$

The boundary conditions require that ϱ, $\boldsymbol{V}$, and $\boldsymbol{H}$ be finite at the origin; that $\boldsymbol{V} \cdot \boldsymbol{N}$ vanish over the boundary Σ, where $\boldsymbol{N}$ is the normal to Σ; and that $\boldsymbol{H}$ be continuous over Σ and vanish at infinity.

The author knows of no rigorous solution of this problem. Solutions have been found that are not regular, e.g., by DAVIS[4]. If regular solutions exist at all, they are probably geometrically complex. COWLING has proved, in a celebrated theorem, that no solution exists when $\boldsymbol{H}$ and $\boldsymbol{V}$ are axisymmetric and have no toroidal components [*4*]; the theorem has recently been extended to the axi-

[1] E. M. BURBIDGE, G. R. BURBIDGE, W. A. FOWLER and F. HOYLE: Rev. Mod. Phys. **29**, 547 (1957).

[2] M. H. WRUBEL: Astrophys. Journ. **116**, 291 (1952).

[3] P. A. SWEET: Monthly Notices Roy. Astronom. Soc. London **110**, 69 (1950).

[4] L. DAVIS jr.: Phys. Rev. **102**, 939 (1956).

symmetric case where toroidal components are present[1]. Fields of still more general kinds have also been shown to be incapable of dynamo maintenance. By a lengthy numerical argument, however, BULLARD appears to have found a solution; his method is briefly summarized by COWLING [4]. BULLARD'S solution, which was obtained in the context of the problem of the Earth's magnetic field, relies upon a combination of differential rotation with convective circulation in a dipole field. PARKER has described a similar set of motions which could be responsible for the Sun's general field, and for the solar cycle.

In the formulation we have given of the classical dynamo problem, no account has been taken of the requirements that energy and momentum be conserved. The energy dissipated as Ohmic heat must evidently be transferred back to the magnetic field from the kinetic energy of the fluid. This can be accomplished only by the action of forces that will drive the fluid across the lines of force at the required rate. In BULLARD'S model, the ultimate energy source is radioactive heating; the forces are those which drive the convection, plus the Coriolis force. In the idealized dynamo problem, however, we may suppose that some unspecified process will provide whatever body forces may be required by the equation of motion.

Since the magnetic fields we observe in nature are not truly stationary, but at most stationary in the mean, an obviously relevant modification of the dynamo problem is to seek solutions which have this property. Several authors have advanced arguments for believing that such solutions exist when the motions are turbulent and the field tangled. The conjecture is that a stationary state will exist, in the mean, when equipartition of energy is established between the magnetic field and the velocity field. Qualitatively, the proposal is that non-reentrant fluid motions will stretch the lines of force until the tension in them is great enough to resist further stretching. COWLING summarizes these theories by remarking that "nearly every argument for or against equipartition between magnetic and turbulent energy can be strongly criticized". He adds that "At present most workers believe in equipartition in some sense" [4].

In an obvious generalization of the classical dynamo problem, we may replace Eq. (19.1) by

$$\operatorname{curl} \boldsymbol{E} = -\mu \frac{\partial \boldsymbol{H}}{\partial t}, \tag{19.5}$$

and inquire to what degree a suitable velocity field $\boldsymbol{V}$ can change the decay time for $\boldsymbol{H}$ when $\boldsymbol{V}$ vanishes. In the static case, we obtain the well-known diffusion equation

$$\nabla^2 \boldsymbol{H} = 4\pi \mu \sigma \frac{\partial \boldsymbol{H}}{\partial t}. \tag{19.6}$$

The solutions of Eq. (19.6) are known to compose a complete set provided that they include the solutions that are singular at the origin[2]. The latter solutions would be admissable, for example, in the discussion of the field within a rigid conducting shell and its surroundings, the outer surface of the shell being spherical and its inner surface excluding the origin. It follows that the most general field at the surface of a fluid sphere, and outside it, is a superposition of the rigid decay

[1] G. BACKUS and S. CHANDRASEKHAR: Proc. Nat. Acad. Sci. U.S.A. **42**, 105 (1956).

[2] Strictly speaking, a complete set is constituted by the solutions of the vector wave equation,

$$\nabla^2 \boldsymbol{H} - 4\pi \mu \sigma \frac{\partial \boldsymbol{H}}{\partial t} - \frac{\mu}{c^2} \frac{\partial^2 \boldsymbol{H}}{\partial t^2} = 0.$$

See, e.g., J. A. STRATTON: Electromagnetic Theory, Chap. VII. New York 1941.

fields of a sphere and a spherical shell. But it is readily shown that the latter solutions are not limited to a set of discrete eigenfrequencies; in them, any decay time is possible. For example, a small bar magnet inside a spherical conducting shell obviously yields a stationary dipole field on the surface of the shell and outside it.

In view of these arguments, it is somewhat surprising that axisymmetric motions in a fluid sphere are not merely incapable of supporting a classical dynamo, but that they are also incapable of appreciably changing the decay of the external field from its value for a static sphere. BACKUS[1] has shown in particular, that such motions "cannot lengthen the decay time of the dipole moment by a factor of more than about 4". SPITZER[2] has obtained essentially the same result, and has shown in addition that strong convective motions transfer energy from an initial large-scale field to a localized small-scale field of greater intensity.

The present position would seem to be that in astronomical bodies where internal motions occur, the time of rigid decay is probably not a parameter of great significance. Depending on the character of the motions that occur, the energy initially present in the magnetic field may be either amplified or diminished, and the external moments of the current distribution may either increase or decrease. However, complicated motions are certainly essential to obtain deviations from rigid decay conditions. The greatest deviation to be expected from these conditions will consist of an equipartition, in the mean, between magnetic energy and kinetic energy. But the possibility of such an equilibrium has not been actually demonstrated, and the time required to achieve it is unknown. The possibility cannot be ruled out that still larger deviations from rigid decay may occur, including configurations in which the magnetic energy increases far beyond the kinetic energy.

Upon elimination of $\boldsymbol{E}$ between Eqs. (19.2) and (19.5), we obtain the general result

$$\frac{\partial \boldsymbol{H}}{\partial t} = \operatorname{curl}\,(\boldsymbol{V}\times\boldsymbol{H}) + \frac{1}{4\pi\mu\sigma}\,\nabla^2\boldsymbol{H}, \qquad (19.7)$$

of which Eq. (19.6) is the special case when $\boldsymbol{V}\equiv 0$. In most astronomical problems we are interested in describing a field over a time interval Δt short compared to the decay time τ of the gravest mode of rigid decay. In these problems, we may always neglect the last term of Eq. (19.7), which is equivalent to the assumption that σ is infinite. We shall limit our further discussion of stellar fields to this case. In the first instance, we shall also assume that $\boldsymbol{V}\equiv 0$. Then Eq. (19.7) indicates that $\boldsymbol{H}$ does not change with time.

b) Magnetohydrostatic equilibrium of stars (σ infinite).

20. The equilibrium configuration. In the magnetohydrostatic equilibrium of an infinitely conducting fluid, the pressure gradients and the gravitational potential $\mathcal{V}$ must balance the magnetic forces; therefore

$$\operatorname{grad} p - \varrho \operatorname{grad} \mathcal{V} = \frac{\mu}{4\pi}\,(\operatorname{curl}\boldsymbol{H})\times\boldsymbol{H}. \qquad (20.1)$$

Solutions of this equation for a fluid sphere are known only for certain axisymmetric fields; not every magnetic field can be accommodated in a static fluid

[1] G. BACKUS: Astrophys. Journ. **125**, 500 (1957).
[2] L. SPITZER jr.: Astrophys. Journ. **125**, 525 (1957).

sphere. For example, if ϱ is constant, then Eq. (20.1) clearly requires that $\boldsymbol{H}$ be subject to the equation

$$\operatorname{curl}[(\operatorname{curl}\boldsymbol{H})\times\boldsymbol{H}]=0. \tag{20.2}$$

The consequences of this relation have been studied by CHANDRASEKHAR under conditions of axisymmetry[1]. An important special case arises when the field is force-free; then

$$(\operatorname{curl}\boldsymbol{H})\times\boldsymbol{H}=0. \tag{20.3}$$

WOLTJER has shown that at the bounding surface Σ of a region containing a force-free field, $H_r\equiv 0$. Furthermore, if no currents flow outside Σ, then $\boldsymbol{H}$ vanishes outside Σ, and a surface current flows in Σ. For the case where

$$\operatorname{curl}\boldsymbol{H}=\alpha\boldsymbol{H}, \tag{20.4}$$

with α a constant, the general solution for $\boldsymbol{H}$ can be obtained in spherical coordinates from the complete set of solutions of the vector wave equation. The axisymmetric case has been studied by LÜST and SCHLÜTER[2], who have shown how to meet the boundary conditions when currents do flow outside Σ. The same authors have also given a useful decomposition of any axisymmetric $\boldsymbol{H}$ into a poloidal field depending on a single scalar function P, and a toroidal field depending on another scalar function T. CHANDRASEKHAR has derived the conditions on T and P for the magnetohydrostatic equilibrium of an incompressible fluid.

A number of investigators have attempted to impose fields of arbitrary geometry upon incompressible static spheres. In general, magnetohydrostatic equilibrium is not possible in these cases; the equilibrium configurations, if they exist at all, are aspheric. Thus, it has been shown that a uniform magnetic field tends to produce an oblate spheroid by contracting the sphere along the direction of the field[3], and that another simple type of field tends to produce a prolate spheroid by expanding the sphere along the direction of the field[4]. In none of these investigations is the assumed magnetic field found to be compatible with a spherical boundary.

PRENDERGAST[5] has succeeded in finding an axisymmetric solution of Eq. (20.2) which is not force-free, and which is strictly compatible with a spherical boundary. He characterizes the geometry of this field by noting that "a typical line of force resembles a helix wrapped on the surface of a torus" around the symmetry axis. The field vanishes on the surface of the sphere.

21. Stability considerations. Configurations of magnetohydrostatic equilibrium will generally be of physical interest only when they are stable against small perturbations. LUNDQUIST[6] has given a stability criterion for an incompressible fluid in the absence of a gravitational field. For the axisymmetric case of a field with no toroidal component, BERNSTEIN *et al.*[7] have generalized LUNDQUIST's results to adiabatically compressible fluids in a gravitational field. BERNSTEIN's stability criterion reduces to the consideration of the eigenvalues of an ordinary Sturm-Liouville differential equation of the second order. One general conclusion

[1] S. CHANDRASEKHAR: Proc. Nat. Acad. Sci. U.S.A. **42**, 5 (1956).
[2] R. LÜST and A. SCHLÜTER: Z. Astrophys. **34**, 263 (1954).
[3] S. CHANDRASEKHAR and E. FERMI: Astrophys. Journ. **118**, 116 (1953).
[4] F. C. AULUCK and D. S. KOTHARI: Z. Astrophys. **38**, 237 (1956).
[5] K. H. PRENDERGAST: Astrophys. Journ. **123**, 498 (1956).
[6] S. LUNDQUIST: Phys. Rev. **83**, 307 (1951). — Ark. Mat. Astronom. Fys. **5**, 297 (1952).
[7] I. B. BERNSTEIN, E. A. FRIEMAN, M. D. KRUSKAL and R. M. KULSRUD: Proc. Roy. Soc. London, Ser. A **244**, 17 (1958).

is that instability is possible "only if somewhere a line of force is concave toward the side of the larger gas pressure". These stability criteria appear not to have been applied to the few true equilibrium configurations that have been obtained for a sphere.

In an unpublished communication, PRENDERGAST reports that he has established the stability of his incompressible model with magnetic forces. For axisymmetric perturbations of the Lüst and Schlüter force-free field in a uniform sphere, TREHAN[1] has given the criterion for stability against adiabatic perturbations. TREHAN notes, e.g., that for such a sphere of interstellar gas at $T = 10000°$ and number density of one hydrogen atom per cm^3, his criterion indicates stability or instability in a uniform ambient field of 10^{-5} gauss depending on whether the cloud radius is less than or greater than 1.7 kpc.

To remain in a steady state, a static system must satisfy the condition of thermal equilibrium as well as the condition of magnetohydrostatic equilibrium. For a prescribed rate of energy transport, this may not be possible; convection may be required to accomodate the energy transport. A number of authors have investigated the transition to convective instability in the presence of a magnetic field; COWLING's monograph gives a survey of this work. The conclusion is that an incompressible fluid in a magnetic field may support without convection a much steeper temperature gradient than it could support without a field. However, for σ sufficiently large, the field may equally well diminish the largest stable temperature gradient. When this occurs, it comes about through "overstability": a perturbation is accompanied by an excessive restoring force, which produces an oscillation of increasing amplitude. The effects of overstability are thought to occur in the photosphere of the Sun and stars. Nevertheless, several authors have proposed theories to account for the darkness of sunspots in terms of a magnetic suppression of convection, the energy flow being effectively dammed up below the spot and escaping around the edges of the strong field in the spot. Evidently the effect of a strong field is not always to produce a dark spot; the observations of the *Ap* stars will not permit such a model.

A necessary condition for stability of any magnetohydrostatic equilibrium has been obtained from a generalization of the virial theorem[2]. The condition is

$$(3\gamma - 4)\,(|\Omega| - \mathcal{M}) > 0, \tag{21.1}$$

where γ is the ratio of specific heats, Ω the gravitational potential energy, and $\mathcal{M}$ the magnetic energy. For the lowest modes of free decay in a Schwarzschild model for solar-type stars, BURBIDGE[3] has found the limiting surface fields given by this equation. The gravest mode becomes unstable in the Sun when $H_p = 8.5 \times 10^4$ kgs. In an incompressible sphere, Eq. (21.1) yields for the root-mean-square field

$$\langle H^2 \rangle^{\frac{1}{2}} < 2.0 \times 10^5 \frac{M}{R^2} \text{ kgs},$$

where M and R are mass and radius in solar units. For the *Ap* stars, $M/R^2 \approx \frac{3}{4}$, and the limiting value of $\langle H^2 \rangle^{\frac{1}{2}}$ is much less than BABCOCK's observed fields in these objects. For the M-type supergiant VV Cep, with R at least 1000 and $M \approx 50$, $\langle H^2 \rangle^{\frac{1}{2}} < 10$ kgs. This is greater than BABCOCK's observed fields, which are in the range $-0.36 \text{ kgs} \leqq H_e \leqq +0.85$ kgs. However, it is likely that in the above calculation R has been underestimated and M overestimated, and that

[1] S. K. TREHAN: Astrophys. Journ. **126**, 429 (1957).
[2] S. CHANDRASEKHAR and E. FERMI: Astrophys. Journ. **118**, 116 (1953).
[3] G. R. BURBIDGE: Astrophys. Journ. **120**, 589 (1954).

the star would indeed be perilously close to dynamical instability if not strongly condensed toward the center.

22. Oscillations. If a fluid in a stable state of magnetohydrostatic equilibrium is perturbed slightly, it will execute small oscillations about the equilibrium state. CHANDRASEKHAR and his collaborators have given an argument that sometimes enables an estimate to be made of the largest possible oscillation period[1]. An elaboration of the virial theorem gives for the total energy ε of a stable, static, spherical star

$$\varepsilon = -\frac{3\gamma-4}{3(\gamma-1)}\left(|\Omega| - \mathcal{M}\right), \tag{22.1}$$

where Ω is the gravitational potential energy, $\mathcal{M}$ the magnetic energy, and γ the ratio of specific heats. Now, a stable star of any Ω may have an arbitrarily large $\mathcal{M}<|\Omega|$; for LÜST and SCHLÜTER'S force-free fields, at least, will leave $|\Omega|$ unaffected. Therefore the total energy may be made arbitrarily near zero. On general grounds, then, the largest oscillation period may be made arbitrarily long. An approximate formula for the period of the slowest radial pulsation is

$$P = -\frac{2\pi}{(3\gamma-4)}\,\frac{I}{(\Omega+\mathcal{M})}, \tag{22.2}$$

where I is the moment of inertia.

If $\mathcal{M}\ll|\Omega|$, the fundamental period P_0 of a star is known to be of the same order as the time for a sound wave to traverse the diameter; for an $A\,0$ star, $P_0\approx 3$ hours. Since the observed periods of the Ap stars average some 4 days, these may be the fundamental periods of radial pulsation only if $\mathcal{M}\approx|\Omega|$, which is equivalent to $\langle H^2\rangle^{\frac{1}{2}}\approx 300$ kgs.

Probably one can rule out the possibility that internal fields exist of the order required by the observed periods in Eq. (22.2). If a periodic magnetic variable has the same density law $\varrho(r)$ as a nonmagnetic star of the same mass, then it follows from the extended virial theorem that

$$U_1 = \frac{P_0}{P_1}\,U_0$$

where P_0 and U_0 are, respectively, the longest period and the total internal (heat) energy of the nonmagnetic star; and P_1, U_1 are the corresponding quantities for the magnetic star. Since we observe P_1 to be of the order $30\,P_0$, this implies much lower temperatures throughout the magnetic star. But the magnetic A stars are known to lie very near the main sequence in the color-magnitude diagram. One can doubt that $\varrho(r)$ could adjust itself to compensate so nicely for the large temperature changes between magnetic and nonmagnetic stars. In other words, it remains to be shown that stars, in which the magnetic pressure is comparable with the gas pressure, will populate the main sequence in the color-magnitude diagram.

It now remains to describe briefly the oscillations that can occur about a stable state of magnetostatic equilibrium. The simplest such oscillations were found by ALFVÉN for the case of a uniform field in an infinitely conducting inviscid liquid. In an Alfvén wave, the fluid oscillates with a velocity $v_y(x,t)$, say, perpendicular to the equilibrium field H_x, and produces a field perturbation $h_y(x,t)$ which is in phase with v_y. This oscillation propagates in the directions

[1] S. CHANDRASEKHAR and E. FERMI: Astrophys. Journ. **118**, 116 (1952). — S. CHANDRASEKHAR and D. N. LIMBER: Astrophys. Journ. **119**, 10 (1954).

$\pm H_x$, then, with the phase velocity

$$V_A = \frac{\mu H_x}{\sqrt{4\pi\varrho_0}}\,. \tag{22.3}$$

The oscillation may be regarded as a vibration of the lines of force, with the fluid along each line of force moving like the particles in a vibrating string. With appropriate boundary conditions in an incompressible fluid, Alfvén waves can occur with finite amplitudes.

VAN DE HULST[1] has shown the existence of three propagation modes for a plane wave arbitrarily inclined to a uniform field $\boldsymbol{H}_0$ in a compressible fluid. One is essentially an Alfvén wave; the perturbations are transverse to $\boldsymbol{H}_0$ and are propagated with the speed V_A. Another mode is essentially a retarded sound wave, with perturbations parallel to $\boldsymbol{H}_0$. The third mode is a modified Alfvén wave, with perturbations transverse to $\boldsymbol{H}_0$. Additional modes also appear in an ionized gas when the frequency is comparable with or greater than the cyclotron frequency of the positive ions. These modes, which ordinarily produce no observable fields or motions on the macroscopic scale, require a detailed microscopic theory for their description[2]. The effect of small gradients in ϱ_0 and $\boldsymbol{H}_0$ have been discussed by WALÉN, who has shown that the fluid velocity $v \propto \varrho^{-\frac{1}{4}}$ and that $h \propto \varrho^{\frac{1}{4}}$.

Several authors have attempted to discuss the standing hydromagnetic waves that could be set up in a fluid sphere. For incompressible poloidal oscillations, SCHWARZSCHILD formulated and solved the eigenvalue problem. The longest period is essentially the transit time of an Alfvén wave across the stellar diameter. When account was taken of the probable increase of $\boldsymbol{H}_0$ with depth, this period was found by FERRARO and MEMORY to be of the order of a year for an A star. But COWLING has shown that mechanical restoring forces will, in fact, preponderate in virtually all oscillations, and will lead to short periods, of the order of the transit-time of a sound wave. Only when the motions are nearly horizontal will the period be governed by the magnetic field. FERRARO and PLUMPTON have described torsional oscillations in an axisymmetric field; but in this case the eigenvalues degenerate into a continuous set, with successive shells of the star in uncoupled motion.

The chief obstacle to the application of these oscillation theories to the periodic magnetic variables is that the theories seem quite incapable of yielding the large changes in H_e that are observed. Since the theories necessarily start from linearized equations, they can adequately discuss only small perturbations from equilibrium. Even if one abandons the dynamical analysis, and makes an *ad hoc* specification of a flow pattern calculated to produce large changes in H_e, the results indicate that, for a large class of field geometries, field reversal would be much less common than it is observed to be; and that the non-reversing fields should show much larger effective fields, relative to the reversers, than observed[3].

c) Magnetohydrodynamical steady states (σ infinite).

23. Laminar flow. There is no *a priori* reason why we must suppose a magnetic star to be in or near a state of magnetohydrostatic equilibrium. These states are limited, as we have already seen, to configurations in which the magnetic

[1] H. VAN DE HULST: In "Problems of Cosmical Aerodynamics". Central Air Documents office 1951.

[2] L. SPITZER jr.: Physics of Fully Ionized Gases, Chap. 4. New York 1956.

[3] P. A. SWEET: Monthly Notices Roy. Astronom. Soc. London **114**, 549 (1954).

field is of a special kind, viz.,

$$\operatorname{curl}\left[(\operatorname{curl}\boldsymbol{H})\times\boldsymbol{H}\right]=0. \tag{23.1}$$

In virtually all known cases, the observations admit the possibility of a stationary but non-static configuration. If we neglect viscosity, the equation of motion is

$$\varrho\frac{\partial\boldsymbol{V}}{\partial t}+\boldsymbol{V}\cdot\nabla\boldsymbol{V}=-\nabla p-\varrho\nabla\mathscr{V}+\frac{\mu}{4\pi}(\operatorname{curl}\boldsymbol{H})\times\boldsymbol{H}. \tag{23.2}$$

In addition, Eq. (19.1) and (19.2) require that

$$\operatorname{curl}(\boldsymbol{V}\times\boldsymbol{H})=0, \tag{23.3}$$

which is an expression of the conservation of flux through any closed curve that moves with the fluid.

Stationary solutions of these equations exist for fields of arbitrary geometry. For the incompressible case, CHANDRASEKHAR has found the solution

$$\boldsymbol{V}=\pm\frac{\mu}{\sqrt{4\pi\varrho}}\boldsymbol{H},\qquad p+\varrho\mathscr{V}+\frac{1}{2}\varrho\boldsymbol{V}^2=\text{const}, \tag{23.4}$$

and he has demonstrated the stability of this solution. MENZEL has shown that Eqs. (23.4) give a solution in compressible flow as well, provided that ϱ is constant along any line of force.

If a system is initially in a stable magnetohydrodynamical steady state and is then perturbed slightly, it will execute small oscillations about this state. Eqs. (23.4) show that an Alfvén wave moves with the fluid; in the case of toroidal motion in a sphere, the period of the matter at any (r,ϑ) is the same as the period of an Alfvén wave at that point. CHANDRASEKHAR has pointed out that this kind of argument removes the sharp distinction previously maintained between rotational motions and periods, on the one hand, and hydromagnetic motions and periods, on the other[1].

The interaction between rotational and elastic motions has been discussed by several authors. If the stationary velocity field is one of pure rotation, then Eq. (23.3) can be shown to imply that $\boldsymbol{H}$ is axisymmetric, and that the angular velocity ω is constant along any line of force. This is FERRARO'S law of isorotation. LAYZER, KROOK, and MENZEL[2] have discussed the torsional oscillations of an idealized star that rotates according to FERRARO'S law, but with gravitational energy large compared with its rotational and magnetic energy. They have ignored the poloidal components of $\boldsymbol{V}$. By taking these into account, COWLING[3] has shown that the resulting coupling of adjacent magnetic shells selects a discrete set of eigenfrequencies from the continuous set found by FERRARO in his study of the torsional oscillations of a static star. The motions, however, must have an appreciable vertical component, and the periods are therefore still determined, in a real star, by the gravitational forces rather than by electromagnetic or Coriolis forces.

SPITZER[4] has noted the probability that hydromagnetic oscillations will necessarily occur in a rotating star where the magnetic field is not symmetric about the rotation axis. His argument is that the magnetic forces will distort the

[1] S. CHANDRASEKHAR: Astrophys. Journ. **124**, 571 (1956).
[2] D. LAYZER, M. KROOK and D. H. MENZEL: Proc. Roy. Soc. Lond., Ser. A **233**, 302 (1955).
[3] T. G. COWLING: Proc. Roy. Soc. Lond., Ser. A **233**, 319 (1955).
[4] L. SPITZER jr.: Proc. Stockholm Conference 1956 on "Electromagnetic Phenomena in Cosmical Physics" (in press).

surfaces of constant density so that the rotation axis no longer coincides with a principal axis of inertia. A rigid star, in these circumstances, would wobble, and stresses would be set up in its interior. In a fluid sphere, these stresses will produce motions that perturb the original stationary state. Eventually, the magnetic field should be changed by dissipative processes into one that is symmetric around the rotation axis. The argument is equivalent to the assertion that no stationary solution of Eq. (23.2) exists in which $\boldsymbol{V}$ and $\boldsymbol{H}$ have different symmetry axes. Possibly the appreciable inclination of the Earth's magnetic axis is an indication that this conjecture is incorrect.

24. Turbulent flow. If a system is initially in an unstable magnetohydrodynamical steady state and is then perturbed slightly, the original disturbance will be amplified and will not remain synchronous with the perturbation. In ordinary hydrodynamics, the criterion for the occurrence of this kind of instability is that the REYNOLDS' number $\mathfrak{R}$ should be larger than, say, 5000. If L and V are numbers of the order of the characteristic length and velocity of the steady state, and ν is the viscosity of the fluid, then

$$\mathfrak{R} = \frac{L V}{\nu}.$$

Experimental and theoretical studies on incompressible fluids in simple geometries have shown that a sufficiently strong magnetic field can stabilize laminar flow to much larger values of $\mathfrak{R}$ than would be possible without a field. The field produces a kind of magnetic viscosity which may considerably exceed ν. This effect is additional to the one which causes the transport coefficient ν, itself, to change in an ionized gas under a magnetic field.

The phenomenon of magnetic stabilization of laminar flow may be of importance in some astronomical processes; e.g., gas motions in nebulae and in the interstellar medium. In stars, however, the requirements of heat transport normally determine the stability of flow patterns. When there is a transition from a static state or from a stationary flow, the new state is normally one of convective equilibrium (Sect. 21). Considerable intrinsic interest attaches, however, to the condition of stationary, isotropic, hydromagnetic turbulence. The problem has been discussed by CHANDRASEKHAR and others.

References.

[1] BABCOCK, H. W.: A Catalogue of Magnetic Stars. Astrophys. Journ. Suppl. **3**, 141 (1958).
[2] DEUTSCH, A. J.: The Spectrum Variables of Type A. Publ. Astronom. Soc. Pacific **68**, 92 (1956).
[3] FOWLER, W. A., G. R. BURBIDGE and E. M. BURBIDGE: Nuclear Reactions and Element Synthesis in the Surfaces of Stars. Astrophys. Journ. Suppl. **2**, 167 (1955).
[4] COWLING, T. G.: Magnetohydrodynamics. New York and London: Interscience Publ. 1957.

Théorie des naines blanches.

Par

E. SCHATZMAN.

Avec 4 Figures.

Introduction.

La théorie des naines blanches fait appel aux propriétés physiques de la matière dense: propriétés quantiques des gaz dégénérés d'électrons, propriétés analogues à celles des métaux, propriétés nucléaires. Des développements théoriques importants ne s'appliquent en définitive qu'à un très petit nombre de naines blanches. Bien que le nombre de naines blanches détectées à l'heure actuelle soit de l'ordre de deux cents, trois seulement, composantes d'étoiles doubles Sirius B, 40 Eridani B, Procyon B ont à la fois leur masse, leur rayon et leur luminosité connus. On peut espérer vérifier quantitativement la théorie des naines blanches en en comparant les résultats à l'observation de ces trois étoiles. L'ensemble des autres naines blanches devant, par ailleurs, servir à cette vérification au moyen particulièrement de leur position dans le diagramme de HERTZSPRUNG-RUSSELL. Le diagramme tracé par LUYTEN[1] résume de façon cohérente, bien que dans un système d'indices de couleur difficilement relié au système international, notre connaissance des naines blanches en 1952 (Fig. 1).

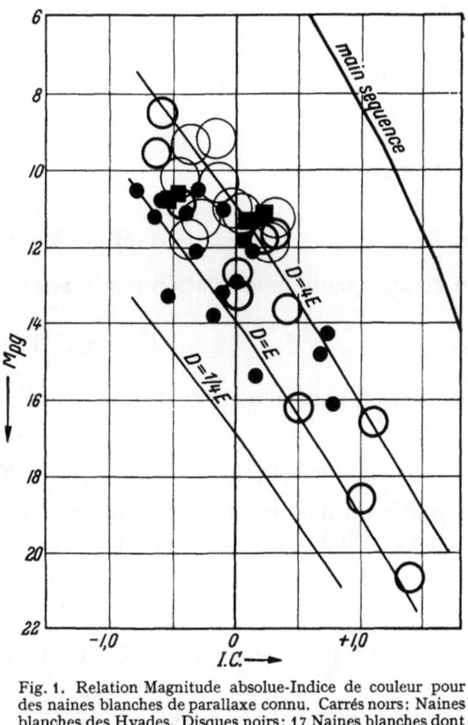

Fig. 1. Relation Magnitude absolue-Indice de couleur pour des naines blanches de parallaxe connu. Carrés noirs: Naines blanches des Hyades. Disques noirs: 17 Naines blanches dont la parallaxe est bien connue. Cercles épais: Naines blanches de parallaxe incertaine. Cercles fins: Naines blanches de parallaxe très incertaine.

Dans l'ensemble, celles-ci sont des étoiles de masse comparable à celle du Soleil, de rayon comparable à celui de la Terre, et dont la luminosité est cent à dix mille fois plus faible que celle du Soleil. Ce sont ces données physiques de base qu'il convient d'expliquer tant en ce qui concerne l'état actuel des naines blanches que le chemin évolutif suivi pour y parvenir.

[1] W. J. LUYTEN: Astrophys. Journ. **116**, 283 (1952).

A. Physique de la matière dense.

I. Equation d'état.

1. Gaz dégénérés d'électrons. Les formules reliant la pression à la densité ont été rassemblées par S. CHANDRASEKHAR [1]. En raison de leur grande importance, nous les rassemblerons ici, nous réservant de discuter Sect. 3 de leur légitimité.

A l'intérieur d'une boîte rectangulaire de côtés l_i un électron peut avoir les quantités de mouvement p_i

$$p_i = \frac{h k_i}{l_i} \tag{1.1}$$

où k_i ($i = 1, 2, 3$) sont des nombres entiers positifs ou négatifs. L'énergie d'un électron est donnée par

$$\left(\frac{E}{c}\right)^2 = m^2 c^2 + h^2 \sum_i \left(\frac{k_i}{l_i}\right)^2. \tag{1.2}$$

Deux états quantiques correspondent à chaque point de coordonnées k_i. Nous calculons le nombre d'états quantiques d'énergie inférieure à E. Il est égal à 2 fois le nombre de points représentatifs à l'intérieur de l'ellipsoïde d'axes

$$a_i = \frac{l_i}{h}\left[\left(\frac{E}{c}\right)^2 - m_e^2 c^2\right]^{\frac{1}{2}}, \tag{1.3}$$

et par conséquent, le nombre d'états quantiques $J(E)$ d'énergie inférieure à E est

$$J(E) = \frac{8\pi}{3} \frac{l_1 l_2 l_3}{h^3}\left[\left(\frac{E}{c}\right)^2 - m_e^2 c^2\right]^{\frac{3}{2}}. \tag{1.4}$$

Si l'on pose

$$\left.\begin{aligned} l_1 l_2 l_3 &= V, \\ p_1^2 + p_2^2 + p_3^2 &= p^2 \end{aligned}\right\} \tag{1.5}$$

et compte tenu de la relation relativiste

$$E = m_e c^2 \left[1 + \left(\frac{p}{m_e c^2}\right)^2\right]^{\frac{1}{2}}, \tag{1.6}$$

on obtient, pour le nombre d'états quantiques $Z(E)\, dE$ dans l'intervalle d'énergie E, $E + dE$

$$Z(E)\, dE = \frac{8\pi V}{h^3} p^2 \cdot dp. \tag{1.7}$$

A partir du nombre d'états quantiques on peut aisément calculer la répartition la plus probable des électrons entre ces différents états. Si l'on pose

$$\frac{1}{kT} = \vartheta \tag{1.8}$$

on obtient pour le nombre d'électrons d'énergie comprise entre E et $E + dE$:

$$dN = \frac{8\pi p^2\, dp}{h^3} V \frac{1}{1 + \frac{1}{A} e^{\vartheta E}} \tag{1.9}$$

et l'on peut aisément calculer la pression, l'énergie interne et le nombre d'électrons dans le volume V en fonction du paramètre A. Si l'on pose

$$\frac{p}{mc} = \operatorname{Sin} \xi, \tag{1.10}$$

$$E = m_e c^2 \operatorname{Cos} \xi \tag{1.11}$$

on trouve

$$N = \frac{8\pi V m_e^3 c^3}{h^3} \int_0^\infty \frac{\operatorname{Sin}^2 \xi \operatorname{Cos} \xi \, d\xi}{\frac{1}{A} e^{\vartheta m_e c^2 \operatorname{Cos} \xi} + 1}, \tag{1.12}$$

$$U = \frac{8\pi V m_e^4 c^5}{h^3} \int_0^\infty \frac{\operatorname{Sin}^2 \xi \operatorname{Cos} \xi (\operatorname{Cos} \xi - 1) \, d\xi}{\frac{1}{A} e^{\vartheta m_e c^2 \operatorname{Cos} \xi} + 1}, \tag{1.13}$$

$$P V = \frac{8\pi V m_e^4 c^5}{3 h^3} \int_0^\infty \frac{\operatorname{Sin}^4 \xi \, d\xi}{\frac{1}{A} e^{\vartheta m_e c^2 \operatorname{Cos} \xi} + 1}. \tag{1.14}$$

De ces formules très générales différentes approximations peuvent être tirées suivant les cas considérés. Lorsque la dégénérescence est très forte, on peut obtenir un développement asymptotique des formules (1.12) à (1.14) au moyen de la relation due à SOMMERFELD:

$$\int_0^\infty \frac{du}{\frac{1}{\lambda} e^u + 1} \frac{d\varphi(u)}{du} = \varphi(u_0) + 2\left[c_2 \varphi''(u_0) + c_4 \varphi^{\mathrm{IV}}(u_0) + \cdots\right] \tag{1.15}$$

où l'on a:

$$u_0 = \log \lambda,$$

$$c_\nu = \sum_n (-1)^{n+1} n^{-\nu},$$

$$c_2 = \frac{\pi^2}{12}, \quad c_4 = \frac{7\pi^4}{720}, \quad c_6 = \frac{31\pi^6}{30240}.$$

On peut alors introduire un argument ξ_0

$$\vartheta \, m_e \, c^2 \operatorname{Cos} \xi_0 = \log A \tag{1.16}$$

tel que

$$\frac{p_0}{mc} = \operatorname{Sin} \xi_0 = x \tag{1.17}$$

définisse le maximum p_0 de la quantité de mouvement des électrons lorsque le gaz d'électrons est complètement dégénéré.

On introduit alors les fonctions

$$f(x) = x(2x^2 - 3)(x^2 + 1)^{\frac{1}{2}} + 3 \operatorname{Ar Sin} x, \tag{1.18}$$

$$g(x) = 8x^3\left[(x^2 + 1)^{\frac{1}{2}} - 1\right] - f(x) \tag{1.19}$$

qui permettent de donner les premiers termes du développement asymptotique de N, U, P

$$N = \frac{8\pi V m_e^3 c^3}{3h^3} x^3 \left[1 + \frac{\pi^2}{(\vartheta m_e c^2)^2} \frac{2x^2 + 1}{2x^4} + \frac{7\pi^4}{40(\vartheta m_e c^2)^4} \frac{1}{x^8} + \cdots\right], \tag{1.20}$$

$$P = \frac{\pi m_e^4 c^5}{3h^3} f(x) \left[1 + \frac{4\pi^2}{(\vartheta m_e c^2)^2} \frac{x(x^2 + 1)^{\frac{1}{2}}}{f(x)} + \cdots\right], \tag{1.21}$$

$$U = \frac{\pi V m_e^4 c^5}{3h^3} g(x) \left[1 + \frac{4\pi^2}{(\vartheta m_e c^2)^2} \frac{(3x^2 + 1)(x^2 + 1)^{\frac{1}{2}} - (2x^2 + 1)}{x \, g(x)} + \cdots\right]. \tag{1.22}$$

Ces formules sont à la base du calcul de la structure interne des naines blanches. Dans le cas des dégénérescences complètes, on a

$$\varrho = B\,\mu_e\,x^3 \quad \text{avec} \quad B = 10^{5,993}. \tag{1.23}$$

Pour x petit (dégénérescences non relativistes), on peut écrire

$$P = K_1 \left(\frac{\varrho}{\mu_e}\right)^{\frac{5}{3}} \tag{1.24}$$

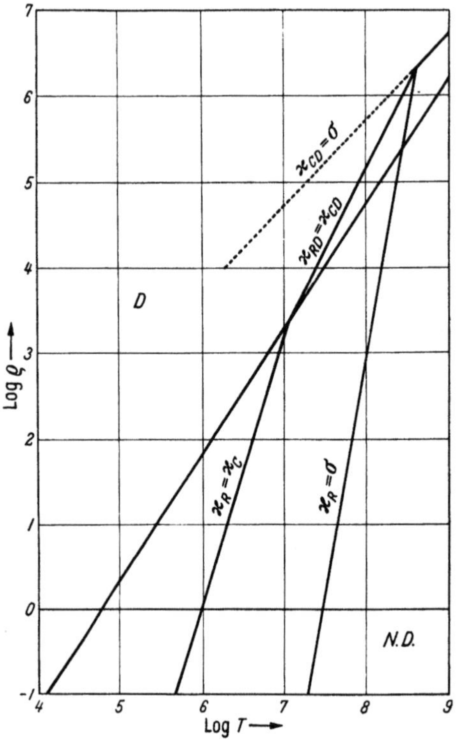

Fig. 2.

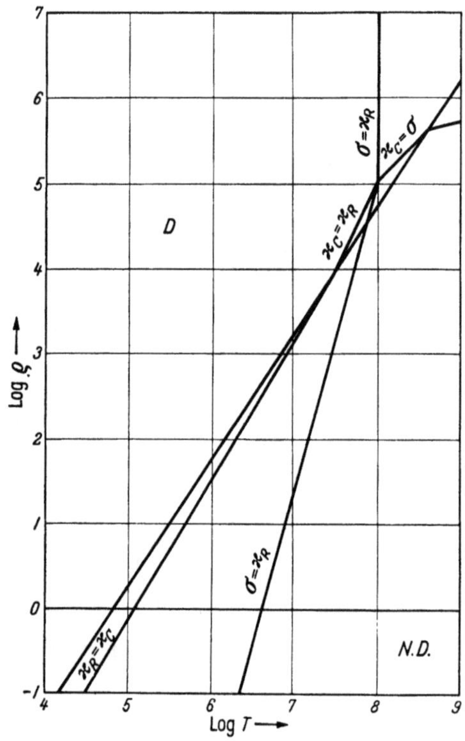

Fig. 3.

Fig. 2. Propriétés physiques de la matière dense pour le mélange de RUSSELL. Régions dégénérée et non dégénérée; régions où prédominent l'absorption photoélectrique, la diffusion et la conductibilité thermique.

Fig. 3. Propriétés physiques de la matière dense pour l'hydrogène pur.

où μ_e est le poids atomique moyen par électron libre et

$$\operatorname{Log} K_1 = 12{,}996.$$

Pour x grand (dégénérescences relativistes), on peut écrire

$$P = K_2 \left(\frac{\varrho}{\mu_e}\right)^{\frac{4}{3}} \tag{1.25}$$

avec

$$\operatorname{Log} K_2 = 15.090.$$

La relation (1.25) est connue sous le nom de relation de STONER-ANDERSON.

On peut délimiter, dans le plan $\operatorname{Log} T$, $\operatorname{Log} \varrho$ les différentes régions où s'appliquent les différentes approximations physiques. On trouvera (Fig. 2 et 3) la courbe séparant la région des gaz dégénérés des gaz non dégénérés d'électrons respectivement pour $\mu_e = 1$ (hydrogène) et $\mu_e = 2$ (éléments moyens et lourds)

2. Effet des interactions noyaux-électrons. Les réactions du type

$$X_Z^A + e_{-1}^0 \rightleftharpoons X_{Z-1}^A + \nu \tag{2.1}$$

déplacent le poids atomique moyen par électron libre μ_e vers des valeurs supérieures à 2. Diverses conséquences importantes en résultent pour les propriétés thermodynamiques de la matière dense (Sect. 5), pour la structure des naines blanches (Sect. 13) et pour leur stabilité (Sect. 17).

Ces réactions ne commencent à jouer un rôle important qu'à des densités élevées, mais qui néanmoins se rencontrent à l'intérieur des naines blanches. Deux méthodes s'offrent à nous pour étudier ces réactions. L'une consisterait à les étudier individuellement. L'autre consiste à calculer l'énergie totale électrons plus nucléons et à admettre que statistiquement, la charge Z des noyaux de masse atomique A correspond à l'état d'énergie minimum par nucléon. Nous suivrons ici cette méthode, étudiée par VAN ALBADA[1].

L'énergie moyenne par nucléon comprend:

(1) l'énergie au repos par nucléon. Exprimée en millièmes d'unité de masse (mMU) elle vaut

$$\mathcal{E}_m = 1000 + f \tag{2.2}$$

où f, le défaut de masse par nucléon, est

$$f = 83\left(\frac{A-2Z}{2A}\right)^2 + 0{,}8\,\frac{A-2Z}{2A} - 6{,}65 + 0{,}63\,Z^2 A^{-\frac{4}{3}} + 15\,A^{-\frac{1}{3}}. \tag{2.3}$$

(2) l'énergie par nucléon du gaz dégénéré d'électrons, donnée dans les mêmes unités par

$$\mathcal{E}_d = \frac{10^3 m_0}{\varrho} \cdot \frac{8\pi m_e^3 c^3}{3h^3}\left[x^3 + \frac{1}{8}\, g(x)\right] \tag{2.4}$$

où m_e est la masse de l'électron, m_0 l'unité de masse atomique, x et $g(x)$ la variable et la fonction définies par les équations (1.17) et (1.19).

(3) l'énergie électrostatique par nucléon, donnée dans les mêmes unités par

$$\mathcal{E}_e = -\frac{9}{10}\,\frac{10^3}{c^2}\, e^2 \left(\frac{4\pi}{3}\right)^{\frac{1}{3}} m_0^{-\frac{4}{3}}\, \varrho^{\frac{1}{3}}\, Z^2 A^{-\frac{4}{3}} \tag{2.5}$$

où l'on a négligé la variation du défaut de masse et où l'on a supposé la charge d'espace constante. On a supposé que chaque atome de charge Z était en interaction avec Z électrons seulement. La condition d'énergie minimum $\partial\mathcal{E}/\partial Z = 0$ entraîne

$$\frac{(-83{,}8\,A + 166\,Z + 1{,}26\,Z A^{\frac{2}{3}})}{A^2} = -\frac{10^3 m_e}{m_0 A}\,(1 + x^2)^{\frac{1}{2}}. \tag{2.6}$$

La pression est donnée par le premier terme du développement (1.21) et la densité par

$$\varrho = \frac{8\pi m_e^3 c^3}{3h^3} \cdot \frac{A m_0}{Z}\, x^3. \tag{2.7}$$

A titre d'exemple, nous donnons ici la relation entre x et Z pour un élément moyen $A = 16$, $Z = 8$ (Tableau 1).

[1] G. B. VAN ALBADA: Bull. astronom. Inst. Netherl. **10**, 161 (1946). — Astrophys. Journ. **105**, 393 (1947).

Tableau 1. *Equilibre de capture des électrons pour un élément moyen $A = 16$; $Z = 8$.*

x	Z	Log ϱ	x	Z	Log ϱ	x	Z	Log ϱ
1	7,63	6,314	2,0	7,600	7,218	5,0	7,450	8,421
1,2	7,625	6,351	2,2	7,585	7,343	7,0	7,350	8,865
1,4	7,620	6,753	2,6	7,555	7,563	10,0	7,19	9,268
1,6	7,615	6,927	3,0	7,550	7,750	20,0	6,59	10,281
1,8	7,610	7,080	4,0	7,500	8,127			

3. Comportement collectif ou individuel des électrons? Dans une série d'articles, EDDINGTON[1] a contesté la validité de la formule de STONER et ANDERSON (1.25). Selon EDDINGTON, il n'est pas légitime de dénombrer les états quantiques que peuvent occuper les électrons sans tenir compte des interactions nucléons-électrons; il n'est pas non plus légitime de dénombrer ces états sans tenir compte de l'ensemble des électrons présents.

La question essentielle est en réalité de savoir si dans le comportement des électrons, le comportement collectif ou le comportement individuel domine. La théorie du plasma [*2*] permet de répondre à cette question. Si la longueur de DEBYE est plus grande que le rayon de la sphère moyenne d'action d'un électron, le comportement individuel des électrons domine; si de plus, le libre parcours moyen est plus grand que la longueur de DEBYE, les électrons peuvent être considérés comme libres et le dénombrement classique des états quantiques des électrons est valable.

En adaptant la procédure de DEBYE-HÜCKEL[2], on peut estimer la longueur de DEBYE devant être associée aux électrons dégénérés[3]:

$$\lambda_D = \left(\frac{m_e c^2}{12\pi e^2}\right)^{\frac{1}{2}} (1 + x^2)^{-\frac{1}{4}}\, x\, N_e^{-\frac{1}{2}} \tag{3.1}$$

où N_e est le nombre moyen d'électrons par centimètre cube. Si l'on appelle d le rayon de la sphère contenant 1 électron, on a, numériquement

$$\frac{\lambda_D}{d} = 3{,}4\, x^{\frac{1}{2}} (1 + x^2)^{-\frac{1}{4}}. \tag{3.2}$$

De même, on peut estimer le libre parcours moyen en se servant des formules relatives aux collisions de 2 électrons de grande énergie[4]

$$l_{el} = \frac{1}{4\pi}\left[1 + \frac{1}{4}\,x\right] N_e^{-1} \left(\frac{e^2}{m_e c^2}\right)^{-2} \tag{3.3}$$

d'où l'on tire le rapport

$$\frac{l_{el}}{\lambda_D} = 10^{3{,}93}\left(1 + \frac{1}{4}\,x\right)(1 + x^2)^{\frac{1}{4}}\, x^{-\frac{5}{2}}. \tag{3.4}$$

On voit donc qu'aux fortes dégénérescences, l'électron peut être considéré comme *libre*, ce qui légitime le dénombrement d'états qui conduit à la formule de STONER et ANDERSON.

[1] EDDINGTON: Monthly Notices Roy. Astronom. Soc. London **95**, 194 (1935); **100**, 582 (1940). — Novae and white dwarfs. Actualités scientifiques et industrielles. Paris: Hermann 1941. White dwarfs, p. 249.

[2] Voir, par exemple, tome XX de cette Encyclopédie, article de M. DARMOIS, Sect. 17.

[3] Voir également R. P. Singh, Astrophys. Journ. **126**, 213 (1957).

[4] CH. MÖLLER: Ann. d. Phys. **14**, 568 (1932).

II. Propriétés thermodynamiques de la matière dense.

4. Gaz dégénérés d'électrons. Il est aisé de calculer à partir des formules (1.20), à (1.22) les coefficients[1]

$$\Gamma_P = (d \log P / d \log \varrho)_{ad}, \tag{4.1}$$

$$\Gamma_T - 1 = (d \log T / d \log \varrho)_{ad}, \tag{4.2}$$

$$\Gamma' / (\Gamma' - 1) = (d \log P / d \log T)_{ad}. \tag{4.3}$$

Le calcul, simple mais un peu long, conduit aux expressions suivantes:

$$\Gamma_T - 1 = \frac{x^2 + 2}{3(x^2+1)}, \tag{4.4}$$

$$\Gamma_P = \frac{8x^5}{3(1+x^2)^{\frac{1}{2}} f(x)}, \tag{4.5}$$

$$\frac{\Gamma' - 1}{\Gamma'} = \frac{(x^2+2) f(x)}{8x^5 (x^2+1)^{\frac{1}{2}}}. \tag{4.6}$$

On trouvera dans le Tableau 2 quelques valeurs de ces trois coefficients, en fonction de x.

Tableau 2. *Coefficients adiabatiques pour le gaz dégénéré.*

x	Γ_P	Γ_T	Γ'	Γ'_P
0	$\frac{5}{3}$	$\frac{5}{3}$	$\frac{5}{3}$	$\frac{5}{3}$
1	1,5333	1,5000	1,4841	1,530
1,2	1,5048	1,4700	1,4541	1,502
1,4	1,4810	1,4464	1,4311	1,477
1,6	1,4608	1,4269	1,4130	1,456
1,8	1,4440	1,4119	1,3993	1,439
2,0	1,4299	1,4000	1,3885	1,424
2,2	1,4179	1,3904	1,3798	1,412
2,6	1,3997	1,3763	1,3675	1,392
3,0	1,3869	1,3667	1,3595	1,380
4,0	1,3669	1,3529	1,3481	1,355
5,0	1,3560	1,3461	1,3428	1,340
7,0	1,3456	1,3400	1,3381	1,326
10,0	1,3397	1,3367	1,3357	1,308
20.0	—	—	—	1,267
∞	$\frac{4}{3}$	$\frac{4}{3}$	$\frac{4}{3}$	—

5. Rôle de l'interaction nucléons-électrons. Si l'on tient compte, dans l'expression de l'énergie interne, de l'énergie au repos des noyaux, et de l'énergie électrostatique, on obtient en tenant compte des expressions (2.3) à (2.5), pour la compressibilité adiabatique l'expression

$$\Gamma'_P = \frac{8x^5}{3(x^2+1)^{\frac{1}{2}} f(x)} \cdot \frac{1}{1 + \frac{0{,}1822\, x^2 (1+x^2)^{-\frac{1}{2}}}{83{,}8 - 0{,}546 (1+x^2)^{\frac{1}{2}}}}. \tag{5.1}$$

L'effet de l'interaction nucléons-électrons, analogue à celui de l'interaction ions-électrons et de l'équilibre de dissociation qui en résulte, est de faire tomber, aux grandes densités, la compressibilité adiabatique au-dessous de $\frac{4}{3}$. Le Tableau 2 contient également les valeurs de Γ'_P.

III. Conductibilité thermique et opacité.

6. Conductibilité thermique. Lorsque la densité s'élève, la conductibilité thermique s'élève et le transfert par conduction finit par l'emporter sur le transfert par rayonnement. Depuis les calculs déjà anciens de KOTHARI[2], des perfectionnements notables ont été apportés à la théorie de la conductibilité thermique

[1] Le calcul des trois coefficients a été fait par Mme SAUVENIER GOFFIN, Mémoires in-8° de la Société Royale des Sciences de Liège, Quatrième Série, Tome X, Fascicule I; Bulletin de la Société Royale des Sciences de Liège, No. 1, Janvier 1950. Le calcul du troisième a été également fait par E. SCHATZMAN: Ann. d'Astrophysique **9**, 199 (1946).

[2] D. S. KOTHARI: Phil. Mag. **13**, 361 (1932). — Monthly Notices Roy. Astronom. Soc. London **93**, 61 (1932).

de la matière dense, perfectionnements inspires de la théorie des métaux[1]. Il nous suffira ici de donner les résultats.

Si l'on appelle G_3, G_4, G_5, K, L les fonctions de Λ suivantes

$$G_3 = \int_0^\infty \frac{1}{1+\frac{1}{\Lambda}\,e^{\vartheta m_e c^2 \operatorname{Cos}\xi}}\, d\left(\frac{\operatorname{Sin}^6 \xi}{\operatorname{Cos}^2 \xi}\right), \tag{6.1}$$

$$G_4 = \int_0^\infty \frac{1}{1+\frac{1}{\Lambda}\,e^{\vartheta m_e c^2 \operatorname{Cos}\xi}}\, d\left(\frac{\operatorname{Sin}^6 \xi\,(\operatorname{Cos}\xi - 1)}{\operatorname{Cos}^2 \xi}\right), \tag{6.2}$$

$$G_5 = \int_0^\infty \frac{1}{1+\frac{1}{\Lambda}\,e^{\vartheta m_e c^2 \operatorname{Cos}\xi}}\, d\left(\frac{\operatorname{Sin}^6 \xi\,(\operatorname{Cos}\xi - 1)^2}{\operatorname{Cos}^2 \xi}\right), \tag{6.3}$$

$$K = \int_0^\infty \frac{1}{1+\frac{1}{\Lambda}\,e^{\vartheta m_e c^2 \operatorname{Cos}\xi}}\, \frac{1}{5}\, d\,(\operatorname{Sin}^5 \xi)\,, \tag{6.4}$$

$$L = \int_0^\infty \frac{1}{1+\frac{1}{\Lambda}\,e^{\vartheta m_e c^2 \operatorname{Cos}\xi}}\, \frac{1}{3}\, d\,(\operatorname{Sin}^3 \xi) \tag{6.5}$$

on obtient pour la conductibilité thermique d'un mélange d'atomes (A_i, Z_i) dans la proportion en poids X_i l'expression

$$\lambda = \frac{2}{3}\,\frac{m_e^5 c^8}{h^3 e^4 I}\,\frac{m_e c^2}{T}\,\frac{G_3 G_5 - G_4^2}{G_3} \tag{6.6}$$

avec

$$I = \frac{\varrho}{m_0} \sum_i \frac{X_i Z_i^2}{A_i} \left\{\frac{1}{3}\log 6\pi + \frac{1}{3}\log Z_i + \frac{1}{2}\log K - \frac{5}{6}\log L\right\}. \tag{6.7}$$

Dans le cas non relativiste, en introduisant les fonctions

$$F_n = \int_0^\infty \frac{u^n\,du}{1+\frac{1}{\Lambda}\,e^u} = \Gamma(n+1)\,U_n \tag{6.8}$$

on obtient

$$\lambda = \frac{64}{3}\,\frac{m_e k^5 T^4}{h^3 e^4 I}\,\frac{\left(\frac{15}{4}\right) F_2 F_4 - 4F_3^2}{3F_2} \tag{6.9}$$

avec

$$I = \frac{\varrho}{m_0} \sum_i \frac{X_i Z_i^2}{A_i} \left\{\frac{1}{3}\log 12\pi + \frac{1}{3}\log Z_i + \frac{1}{2}\log F_{\frac{3}{2}} - \frac{5}{6}\log F_{\frac{1}{2}}\right\}. \tag{6.10}$$

Dans le cas non dégénéré, (6.9) se simplifie en

$$\lambda = 128\,\frac{m_e k^5 T^4}{h^3 e^4 I}\,\Lambda\,. \tag{6.11}$$

[1] R. E. MARSHAK: Ann. N.Y. Acad. Sci. **41**, Art. 1, 49 (1940). — T. D. LEE: Astrophys. Journ. **111**, 625 (1950). —L. MESTEL: Proc. Cambridge Phil. Soc. **46**, 331 (1950).

Dans le cas des dégénérescences fortes:

$$\lambda = \frac{2\pi^2}{9} \frac{m_e^4 c^6 k^2 T}{h^3 e^4 I} \frac{x^6}{x^2+1}. \tag{6.12}$$

La comparaison ultérieure (Sect. 7) de la conductibilité thermique et de la conductibilité radiative permet de délimiter dans le plan $\operatorname{Log} T$, $\operatorname{Log} \varrho$ (Fig. 2 et 3), les régions où dominent soit le transfert radiatif soit la conductibilité. En réalité, on introduit le concept d'opacité conductive

$$\varkappa_c = \frac{4acT^3}{3\varrho\lambda} \tag{6.13}$$

si bien que l'opacité $\varkappa$ résultant de l'opacité radiative $\varkappa_R$ et de l'opacité conductive $\varkappa_c$ est

$$\frac{1}{\varkappa} = \frac{1}{\varkappa_R} + \frac{1}{\varkappa_c}. \tag{6.14}$$

7. Opacité. L'opacité radiative est donnée par la moyenne de ROSSELAND:

$$\frac{1}{\varkappa_R} = \frac{\int\limits_0^\infty \frac{1}{\varkappa_\nu} \frac{1}{1-\exp[-(h\nu/kT)]} \frac{\partial B_\nu}{\partial T} d\nu}{\int\limits_0^\infty \frac{\partial B_\nu}{\partial T} d\nu} \tag{7.1}$$

où B_ν est la fonction de PLANCK à la température T. Le facteur $(1 - e^{-h\nu/kT})$ tient compte de l'émission induite.

L'absorption du rayonnement est due à la photo-ionisation des atomes fortement ionisés (transitions libre-libre et lié-libre).

En théorie classique, si tous les états d'énergie dans l'intervalle dE sont occupés, la section de choc totale d'absorption (libre-libre) est

$$\frac{32\pi^2 e^6 Z^2}{3\sqrt{3}\, h^4 c \nu^3} dE. \tag{7.2}$$

Pour obtenir le coefficient d'absorption par gramme, il faut multiplier (7.2): (1) par le nombre d'atomes par gramme, $1/(A m_0)$, (2) par la probabilité de trouver un état d'énergie E occupé, c'est-à-dire par

$$f(E) = \left(1 + \frac{1}{\Lambda} e^{E/kT}\right)^{-1}, \tag{7.3}$$

(3) par la probabilité de trouver un état d'énergie $(h\nu + E)$ libre, c'est-à-dire par

$$1 - f(E + h\nu) = (1 + \Lambda\, e^{-(E+h\nu)/kT})^{-1}, \tag{7.4}$$

et intégrer le produit sur E. On obtient alors le coefficient d'absorption par gramme, dû aux transitions libre-libre

$$a_\nu = \frac{32\pi^2 e^6 Z^2 k T}{3\sqrt{3}\, c h^4 A m_0} \frac{1}{\nu^3} \frac{e^{h\nu/kT}}{e^{h\nu/kT}-1} \log\left(\frac{1+\Lambda}{1+\Lambda\, e^{-(h\nu)/(kT)}}\right). \tag{7.5}$$

Dans certaines conditions, principalement dans le cas des fortes dégénérescences, la majeure contribution à l'absorption est due aux transitions «libre-libre». Dans ce cas, il suffit d'insérer l'expression (7.5) dans la moyenne de ROSSELAND. L'intégration est faite de façon approchée et l'on peut montrer que le coefficient d'absorption est, dans le cas de la matière dégénérée:

$$\varkappa_{RD} = \frac{4\pi^4}{15\cdot 7!\cdot\zeta(7)}\,\frac{32\pi^2 Z^2 e^6 kT}{3\sqrt{3}\,c\,h^4 A\,m_0}\left(\frac{h}{kT}\right)^3 \log\left(\frac{A+1}{1+A\,e^{-7}}\right) \tag{7.6}$$

où $\zeta(7) = \sum_1^\infty q^{-7} = 1{,}0083$. A la limite, $A \to 0$, l'équation (7.6) tend vers la loi habituelle de KRAMERS; à la limite $A \to \infty$ elle tend vers l'expression usuelle pour le cas complètement dégénéré. Ce dernier cas offre d'ailleurs peu d'importance pratique car aux densités élevées, l'opacité conductive l'emporte largement sur l'opacité radiative.

Pour tenir compte des transitions d'état lié à un état libre, il faut introduire un facteur de guillotine:

$$\varkappa_R = \frac{4\pi^4}{15\cdot 7!\,\zeta(7)}\,\frac{32\pi^2 Z^2 e^6 kT}{3\sqrt{3}\,c\,h^4 A\,m_0}\left(\frac{h}{kT}\right)^3 \frac{g\cdot 196{,}5}{\tau} \tag{7.7}$$

où g est le facteur de GAUNT. Le facteur de guillotine, τ, augmente avec la population des états liés, ce qui compense l'augmentation de l'opacité avec ϱ. MARSHAK[1] a calculé le facteur de guillotine suivant la méthode indiquée par B. STRÖMGREN pour le mélange suivant:

O (8/16), Na, Mg (4/16),

Si (1/16), K, Ca (1/16),

Fe (2/16)

dit mélange de RUSSELL. Lorsque la densité est faible, $\tau = 1$, et $196{,}5 \times g = 172$. La table de MARSHAK est reproduite ci-contre.

Tableau 3. *Facteur de Guillotine.*

T	$\log A$					
	−2	−1	0	1	2	3
$2\cdot 10^6$	3,8					
4	6,0	11	27	45		
6	6,5	14	29	59	100	
8	8,6	15	35	68	108	
10	10	17	37	71	113	
12	11	18	38	72	116	150
14	10	18	39	74	118	153
16	11	18	40	75	119	155
18	11	18	40	75	119	156

Pour un mélange contenant une proportion X_R de mélange de RUSSELL le coefficient d'absorption est

$$\varkappa_R = 6{,}00\cdot 10^{17}\, X_R\, \tau^{-1}\, T^{-2} \log\frac{1+A}{1+A\,e^{-7}} \tag{7.8}$$

et pour l'hydrogène pur

$$\varkappa_R = 6{,}02\cdot 10^{14}\, T^{-2} \log\frac{1+A}{1+A\,e^{-7}}\,. \tag{7.9}$$

Il est intéressant de comparer l'importance des différents processus contribuant à l'opacité: photoionisation, diffusion, opacité conductive. Pour le mélange de RUSSELL et pour l'hydrogène, on trouvera sur les Fig. 2 et 3 la délimitation des différentes régions où dominent ces différents processus.

IV. Production d'énergie.

8. Barrière de potentiel. Les réactions thermonucléaires sont susceptibles d'être à l'origine du débit d'énergie des naines blanches. La discussion de cette possibilité sera faite à la Sect. 20. Pour le moment, nous nous contentons d'examiner les réactions thermonucléaires dans les conditions spéciales des naines blanches.

[1] R. E. MARSHAK: Ann. N.Y. Acad. Sci. **41**, Art. 1, 49 (1940).

Le nombre total de réactions par centimètre cube et par seconde est donné par l'intégrale suivante:

$$p = N_1 N_2 \int_0^\infty [E^{\frac{1}{2}} e^{-(E/kT)}] \, T(E) \, \sigma_{\text{nucl}}(E) \, dE \tag{8.1}$$

où N_1 et N_2 sont les nombres de noyaux d'espèces (1) et (2) par centimètre cube.

Le premier terme sous le signe somme est le facteur de MAXWELL-BOLTZMANN, donnant la probabilité pour que l'énergie des deux noyaux, dans le système du centre de gravité, soit comprise entre E et $E + dE$. Le second facteur T est le facteur de pénétration de la barrière de potentiel. Le dernier facteur $\sigma_{\text{nucl}}(E)$ est un facteur purement nucléaire qui dépend de l'interaction après pénétration et habituellement varie assez lentement avec E [*3*].

Le facteur de pénétration de la barrière de potentiel commande largement la vitesse des réactions thermonucléaires entre particules chargées. Exprimé dans le système du centre de gravité le facteur de pénétration est

$$T = \exp\left\{-2 \frac{(2\,\overline{m})^{\frac{1}{2}}}{\hbar} \int_{r^*}^{r_E} (U - E)^{\frac{1}{2}} \, dr\right\} \tag{8.2}$$

où $\overline{m}$ est la masse réduite

$$\overline{m} = \frac{m_1 m_2}{m_1 + m_2}, \tag{8.3}$$

r_E la distance classique d'approche, r^* le rayon du noyau composé.

Aux grandes densités, il est impossible de supposer le potential $U(E)$ coulombien. L'effet d'écran des électrons libres modifie la barrière de potentiel, ce qui augmente le facteur de pénétration. Le nombre de réactions thermonucléaires par centimètre cube et par seconde se trouve accru par un facteur A que nous appellerons facteur d'accélération des réactions thermonucléaires.

Etant donné les hautes densités régnant dans les naines blanches, il suffit de remplacer les électrons libres par un fluide d'électricité négative de densité de charge uniforme. Une approximation suffisante est alors la suivante[1]:

$$T = \exp\left\{-2 \frac{(2\,\overline{m})^{\frac{1}{2}}}{\hbar} \frac{\pi}{2} (Z_1 Z_2 e^2 r_E)^{\frac{1}{2}}\right\}. \tag{8.4}$$

On peut alors aisément calculer r_E et, par la méthode du col qui sert habituellement, obtenir le taux des réactions thermonucléaires.

Pour deux particules de charge Z_1 et $Z_2 (Z_1 > Z_2)$, si l'on pose

$$\beta = \left(\frac{8 h^2 e^2 Z_2}{3 \pi^3 \bar{A} \, k^2 m_0^2}\right)^{\frac{1}{3}} \left(\frac{\varrho}{\mu_e T^2}\right)^{\frac{1}{3}} \tag{8.5}$$

on obtient pour le facteur d'accélération A

$$A = f_2(\beta) \exp[\tau f_1(\beta)] \tag{8.6}$$

où τ est classiquement

$$\tau = 3 \left[\frac{\pi^2 e^4 m_0 Z_1^2 Z_2^2 A_1 A_2}{2 \hbar^2 k T (A_1 + A_2)}\right]^{\frac{1}{3}}. \tag{8.7}$$

Les fonctions $f_1(\beta)$ et $f_2(\beta)$ sont assez compliquées. Pour β petit, on peut écrire

$$\left.\begin{aligned} f_1(\beta) &= 2^{-\frac{2}{3}} \beta, \\ f_2(\beta) &= 1 - \tfrac{1}{6} \beta^3. \end{aligned}\right\} \tag{8.8}$$

[1] E. SCHATZMAN: J. Phys. Radium **9**, 46 (1948).

Pour β grand, on peut écrire

$$\left.\begin{aligned} f_1(\beta) &= 1 - \frac{2^{\frac{1}{3}}}{3}\beta^{-\frac{1}{2}}, \\ f_2(\beta) &= 2^{\frac{1}{3}}\beta^{-1}. \end{aligned}\right\} \tag{8.9}$$

Les quantités f_1 et f_2 sont tabulées ci-après.

9. Revue des réactions thermonucléaires [4]. Le cycle proton-proton

$$\mathrm{I}\left\{\begin{array}{ll} H^1 + H^1 \rightarrow H^2 + \beta^+ + \nu, & (1) \\ H^2 + H^1 \rightarrow He^3 + \gamma, & (2) \\ He^3 + He^3 \rightarrow He^4 + H^1 + H^1 & (3) \end{array}\right.$$

se produit tant que la densité n'est pas trop élevée. La réaction la plus lente est la réaction proton-proton. Si l'on pose

$$\tau = 3380\, T^{-\frac{1}{3}} \tag{9.1}$$

le débit d'énergie par gramme est

$$\varepsilon_I = 2{,}18 \cdot 10^3\, \varrho\, x_H^2\, \tau^2\, e^{-\tau} A. \tag{9.2}$$

Lorsque le nombre d'électrons libres par centimètre cube est supérieur à $1{,}7 \cdot 10^{28}$, le cycle I devient

$$\mathrm{II}\left\{\begin{array}{ll} H^1 + H^1 \rightarrow H^2 + \beta^+ + \nu, & \\ H^2 + H^1 \rightarrow He^3 + \gamma, & \\ He^3 + e^- \rightarrow H^3 + \nu, & (4) \\ H^3 + H^1 \rightarrow He^4 + \gamma. & (5) \end{array}\right.$$

Tableau 4. *Fonctions pour le facteur d'accélération.*

$\log \beta^3$	$\log f_1$	$\log f_2$
−5	−2,1684	0
−4	−1,8351	0
−3	−1,5017	0
−2	−1,1684	0
−1	−0,8485	−0,012
0	−0,5551	−0,089
1	−0,3404	−0,316
2	−0,0942	−0,616

Chaque réaction proton-proton conduit à la fonction d'un atome d'hélium (contre $\frac{1}{2}$ pour le cycle I). Le débit d'énergie par gramme est alors

$$\varepsilon_{II} = 4{,}36 \cdot 10^3\, \varrho\, x_H^2\, \tau^2\, e^{-\tau} A. \tag{9.3}$$

La variable β définie par l'équation (8.2) est

$$\beta = 3{,}2 \cdot 10^2 \left(\frac{\varrho}{\mu_e T^2}\right)^{\frac{1}{3}}, \tag{9.4}$$

et le facteur d'accélération A est

$$A = \exp\left(\frac{3{,}38 \cdot 10^5\, \varrho^{\frac{1}{3}}}{T \mu_e^{\frac{1}{3}}}\right). \tag{9.5}$$

Pour des conditions typiques dans les naines blanches, $\varrho = 10^6$, $T = 10^7$, $\mu_e = 2$, on a $A = 14{,}6$. La réaction proton-proton peut se trouver modifiée par la capture d'électrons libres[1]:

$$H^1 + H^1 + e^- \rightarrow D^2 + \nu. \tag{6}$$

Le nombre de réactions N_6 par gramme et par seconde peut s'écrire

$$N_6 = N_1 \cdot 2{,}5\, x^3 \left[2 + x_1^2 + \frac{3x^2}{5} + \frac{3}{4}(1 + x^2)^{\frac{1}{2}} \frac{x(1+x^2)^{\frac{1}{2}}(1+2x^2) - \operatorname{Ar\,Sin} x}{x^3}\right] \tag{9.6}$$

où x est lié à la densité par la relation (1.20) et x_1 est défini par l'énergie du neutrino émis:

$$\sqrt{1 + x_1^2}\, m_e c^2 = 1{,}81\, m_e c^2;$$

[1] C. L. CRITCHFIELD: Astrophys. Journ. **96**, 1 (1942).

on a donc $x_1 = 1{,}508$. Le nombre de réactions N_6 devient supérieur au nombre de réactions N_1 lorsque $x > 0{,}335$, ou encore lorsque le nombre d'électrons libres par centimètre cube est supérieur à $2{,}2 \cdot 10^{28}$. La densité correspondante, pour de l'hydrogène pur, est $3{,}7 \cdot 10^4$ grammes par centimètre cube. Pour des densités plus élevées, le nombre de réactions N_6, au facteur A près, est proportionnel à la puissance $\frac{8}{3}$ de la densité.

Aux densités très élevées, notamment celles qui règnent dans les régions centrales des naines blanches, les réactions de l'hélium, favorisées par un facteur d'accélération énorme, deviennent importantes:

$$\text{III}\begin{cases} He^4 + He^4 \rightleftharpoons Be^8\,, & (7)\\ Be^8 + He^4 \rightarrow C^{12}\,, & (8)\\ C^{12} + He^4 \rightarrow O^{16}\,. & (9)\end{cases}$$

En raison de l'importance du facteur d'accélération, la résonance qui favorise le cycle III joue un rôle négligeable aux densités élevées.

Le nombre des réactions de formation du carbone C^{12} sera donc, par centimètre cube et par seconde[1]

$$N_8 = (3\pi)^{\frac{1}{2}} N_\alpha^3 h^6 (2\pi m_{He} kT)^{-3}\, 6{,}5 \cdot 10^{-9} \frac{kT}{D} \frac{kT}{h} \tau^{\frac{1}{2}} e^{-\tau} A \tag{9.7}$$

où N_α est le nombre de particules α par centimètre cube, D l'espacement des niveaux du noyau composé, τ la quantité

$$\tau = 23\,640\, T^{-\frac{1}{3}} \tag{9.8}$$

et A le facteur d'accélération. Numériquement, on a

$$p = 1{,}41 \cdot 10^{-50} N_\alpha^3\, T^{-1} \tau^{\frac{1}{2}} e^{-\tau} A \tag{9.9}$$

et le débit d'énergie par gramme est

$$\varepsilon_{III} = 5{,}58 \cdot 10^{14} X_\alpha^3 \varrho^2\, T^{-1} \tau^{\frac{1}{2}} e^{-\tau} A \tag{9.10}$$

où X_α est la concentration en poids en hélium.

La variable β définie par l'équation (8.5) vaut

$$\beta = 2{,}31 \cdot 10^2 \left(\frac{\varrho}{\mu_e T^2}\right)^{\frac{1}{3}}. \tag{9.11}$$

Pour $T = 10^7$, $\varrho = 2 \cdot 10^9$ on a $\beta = 4{,}98$, $\tau = 109{,}7$. L'expression du facteur d'accélération A est donnée par les équations approchées (8.9):

$$A = 2^{\frac{1}{3}} \beta^{-1} \exp(0{,}812\,\tau) = 1{,}13 \cdot 10^{38},$$

et le débit d'énergie par gramme serait alors

$$\varepsilon = 5{,}2 \cdot 10^{17} X_\alpha^3. \tag{9.12}$$

Un tel débit d'énergie est incompatible avec l'existence des naines blanches. L'hélium ne peut exister dans les régions centrales des naines blanches denses qu'à de très faibles concentrations.

Nous complétons ces renseignements par la donnée de la durée de vie des éléments réagissant dans les différentes réactions.

---

[1] E. E. Salpeter: Symposium on Astrophysics. Ann. Arbor. Michigan, 1953.

La réaction proton-proton est très lente. Le nombre de réactions par proton et par seconde est

$$p_1 = 1{,}73 \cdot 10^{-16}\, \varrho\, x_H\, \tau_1^2\, e^{-\tau_1} A_1 \tag{9.13}$$

avec

$$\tau_1 = 3380\, T^{-\frac{1}{3}}. \tag{9.14}$$

La réaction proton-deutérium est extrêmement rapide. Le nombre de réactions par atome de deutérium et par seconde est

$$p_2 = 39\, \varrho\, x_H\, \tau_2^2\, e^{-\tau_2} A_2 \tag{9.15}$$

avec

$$\tau_2 = 3710\, T^{-\frac{1}{3}}. \tag{9.16}$$

La réaction de formation de l'hélium est plus lente:

$$p_3 = 6 \cdot 10^7\, \varrho\, X_{He^3}\, \tau_3^2\, e^{-\tau_3} A_3 \tag{9.17}$$

avec

$$\tau_3 = 12270\, T^{-\frac{1}{3}}. \tag{9.18}$$

Dans le cas du cycle I, l'abondance de l'hélium est donnée par

$$\frac{X_{He^3}}{X_H} = 2\left(\frac{2{,}4 \cdot 10^8}{1{,}73 \cdot 10^{-16}}\, \frac{\tau_3^2}{\tau_1^2}\, e^{-\tau_3+\tau_1}\, \frac{A_3}{A_1}\right)^{-\frac{1}{2}} \tag{9.19}$$

et la durée de vie P_{He^3} est donnée par

$$P_{He^3}^{-1} = \varrho\, X_H \left[2 \cdot 1{,}73 \cdot 10^{-16} \cdot 2{,}4 \cdot 10^8\, \tau_3^2\, \tau_1^2\, e^{-\tau_3-\tau_1} A_3 A_1\right]^{\frac{1}{2}}. \tag{9.20}$$

Pour $T = 10^7$, $\varrho = 2 \cdot 10^4$, on obtient $A_1 = 1{,}97$; $A_3 = 18{,}4$; et $P_{He^3} = 2{,}05 \cdot 10^{11}$ sec.

Le nombre de réactions de capture des électrons par atome He^3 est donné par

$$P_4 = K_4.$$

Lorsque le gaz n'est pas dégénéré, on obtient[1]

$$K_4 = \frac{\pi^2 g^2 m_e^2 c}{h^4} \left(\frac{8}{\pi}\right)^{\frac{1}{2}} \operatorname{Sin} 2\xi_0\, z^{-\frac{3}{2}} N_e \exp\left[-(\operatorname{Cos}\xi_0 - 1)\, z\right] \tag{9.21}$$

où g est la constante d'interaction des réactions bêta. $\operatorname{Cos}\xi_0$ est lié à l'énergie de réaction $\varepsilon_0 = 0{,}0185$ Mev par

$$\varepsilon_0 / m_e c^2 = \operatorname{Cos}\xi_0 - 1,$$

$$\operatorname{Cos}\xi_0 = 1{,}0362$$

et z est lié à la température par

$$z = m_e c^2 / kT = 5{,}9 \cdot 10^9\, T^{-1}.$$

Numériquement:

$$K_4 = 1{,}036 \cdot 10^{-37} \times 10^{-0.028/T_8} \times T_8^{\frac{3}{2}} N_e \tag{9.22}$$

avec $T_8 = 10^{-8}\, T$.

Pour le gaz dégénéré d'électrons, la constante de la réaction K_4 est

$$K_4 = 0{,}00562\, H_1 \tag{9.23}$$

avec

$$H_1 = \int_{\operatorname{Cos}\xi_0}^{\infty} \frac{e^{-z(\operatorname{Cos}\xi - \operatorname{Cos}\xi_1)}}{1 + e^{-z(\operatorname{Cos}\xi - \operatorname{Cos}\xi_1)}} (\operatorname{Cos}\xi - \operatorname{Cos}\xi_0)^2 \operatorname{Sin}^2\xi \operatorname{Cos}\xi\, d\xi \tag{9.24}$$

---

[1] E. SCHATZMAN: Ann. d'Astrophysique **16**, 162 (1953).

où $z \operatorname{Cos} \xi_1 = \log \Lambda$. Le développement (1.15) permet d'obtenir une expression approchée de H_1:

$$H_1 = H_{10} + \frac{\pi^2}{6} H_{10}'' + \frac{7\pi^4}{360} H_{10}^{\mathrm{IV}}. \tag{9.25}$$

H_{10} est

$$H_{10} = \int\limits_{\operatorname{Cos}\xi_0}^{\operatorname{Cos}\xi_1} (\operatorname{Cos}\xi - \operatorname{Cos}\xi_0)^2 \operatorname{Sin}^2\xi \operatorname{Cos}\xi \, d\xi. \tag{9.26}$$

La différentiation est faite par rapport à $z \operatorname{Cos}\xi_1$. On trouvera cidessous des valeurs des H_{10} en fonction de ξ et de H_1 en fonction de T lorsque $\xi_1 = \xi_0$

Tableau 5.

ξ	H_{10}	ξ	H_{10}
0,268	0	0,328	$6,62 \cdot 10^{-6}$
0,278	$2,05 \cdot 10^{-9}$	0,338	$1,15 \cdot 10^{-6}$
0,288	$1,79 \cdot 10^{-8}$	0,348	$1,85 \cdot 10^{-6}$
0,298	$6,56 \cdot 10^{-8}$	0,358	$2,83 \cdot 10^{-6}$
0,308	$1,59 \cdot 10^{-7}$	0,368	$4,17 \cdot 10^{-6}$
0,318	$3,59 \cdot 10^{-7}$		

Tableau 6.

T	H_1
$4 \cdot 10^6$	$1,63 \cdot 10^{-10}$
$6 \cdot 10^6$	$5,59 \cdot 10^{-10}$
$8 \cdot 10^6$	$1,34 \cdot 10^{-9}$
$10 \cdot 10^6$	$2,66 \cdot 10^{-9}$

On en déduit la durée de vie de He^3 pour la capture des électrons. Dans l'hydrogène pur, pour $T = 10^7$, $\varrho/\mu_e = 2 \cdot 10^4$, on a

$$P = 6,7 \cdot 10^{10} \text{ sec},$$

plus court donc que la durée de vie de He^3 pour la formation de l'hélium He^4.

10. Influence des variations de température et de densité. La stabilité vibrationnelle des étoiles fait intervenir la sensibilité des réactions thermonucléaires aux variations de température et de densité.

Le débit d'énergie par gramme et par seconde pour la réaction d'un cycle peut être écrite sous la forme

$$\varepsilon_i = \varepsilon_{0i}\, \varrho^{\mu_i} T^{\nu_i}. \tag{10.1}$$

Les exposants effectifs μ_{eff} et ν_{eff} pour un cycle sont une combinaison des μ_i et ν_i faisant intervenir les débits d'énergie ε_i et la période de pulsation [*5*].

Lorsque la durée de vie d'un élément est longue comparée à la période de pulsation, le nombre par seconde des réactions subies par cet élément varie en phase avec la pulsation. Au contraire, lorsque la durée de vie est très courte, le nombre de réactions varie en quadrature avec la pulsation et la réaction ne contribue pas à la stabilité de la pulsation.

Nous donnons ici les exposants μ_i, ν_i des différentes réactions et les exposants μ_{eff}, ν_{eff} des différents cycles[1].

Cycle I. Réaction (1): pour les faibles densités

$$\left.\begin{aligned} \mu_1 &= 1 + \tfrac{1}{3}\log_e \Lambda_1, \\ \nu_1 &= -\tfrac{2}{3} + \tfrac{1}{3}\tau_1 - \log_e \Lambda_1. \end{aligned}\right\} \tag{10.2}$$

Pour les grandes densités ($\varrho/\mu_e T^2 > 3,2 \cdot 10^{-8}$)

$$\mu_1 = \tfrac{2}{3}; \quad \nu_1 = 0. \tag{10.3}$$

[1] E. Schatzman: Ann. d'Astrophysique **16**, 162 (1953).

Les exposants de la réaction (2) n'interviennent pas dans les exposants μ_{eff}, ν_{eff} on raison de la très courte durée de vie du deutérium. Les exposants de la réaction (3) sont donnés par des formules semblables aux formules (10.2) et (10.3) où l'indice 1 est remplacé par l'indice 3.

Pour la réaction (4), les choses sont plus compliquées car elles impliquent une étude des formules (9.24) et (9.25). La densité $\varrho_0 = B\,\text{Sin}^3\xi_0$ est critique. On distingue trois cas:

gaz non dégénéré:

$$\left.\begin{aligned} \mu_4 &= 1\,, \\ \nu_4 &= \tfrac{3}{2} + 0{,}0362\,z \end{aligned}\right\} \qquad (10.4)$$

où z a le même sens que dans les formules (9.23) et suivantes;

gaz dégénéré: $\varrho < \varrho_0$

$$\left.\begin{aligned} \mu_4 &= z\,\text{Sin}^2\xi/\text{Cos}\,\xi\,, \\ \nu_4 &= 3 + z\,(\text{Cos}\,\xi_0 - \text{Cos}\,\xi)\,; \end{aligned}\right\} \qquad (10.5)$$

gaz dégénéré $\varrho > \varrho_0$

$$\left.\begin{aligned} \mu_4 &= \frac{\text{Sin}\,\xi}{3\,\text{Cos}\,\xi}\;\frac{\dfrac{dH_{10}}{d\xi} + \dfrac{\pi^2}{6}\,\dfrac{dH_{10}^{\text{II}}}{d\xi}}{H_1}\,, \\ \nu_4 &= \frac{\dfrac{\pi^2}{3}\,H_{10}^{\text{II}} + \dfrac{7\pi^4}{90}\,H_{10}^{\text{IV}}}{H_1}\,. \end{aligned}\right\} \qquad (10.6)$$

Pour $\varrho = \varrho_0$, un calcul spécial de μ_4 est nécessaire. Aux densités élevées, μ_4 tend vers $\frac{5}{3}$ et ν_4 vers zéro. Le maximum de μ_4 est atteint pour $\varrho = \varrho_0\,(\xi = \xi_0)$. Le tableau donne quelques valeurs de μ_4 pour $\varrho = \varrho_0$ (Tableau 7).

Tableau 7. *Exposant de la densité pour la réaction de capture bêta de l'hélium.*

T	μ_4	T	μ_4
$4\cdot 10^6$	34,9	$8\cdot 10^6$	17,4
$6\cdot 10^6$	23,3	10^7	$13{,}9_5$

Appelons q_i le nombre de réactions par gramme et par seconde, pour la $i^{\text{ème}}$ réaction, dont le bilan d'énergie est χ_i. L'exposant effectif ν pour le cycle I est

$$\nu_{\text{eff I}} = \frac{\nu_1(\chi_1 + \chi_2) + \nu_3\chi_3}{\chi_1 + \chi_2 + \chi_3}\,.$$

Pour le cycle II, l'exposant effectif $\mu_{\text{eff II}}$ est:

$$\mu_{\text{eff II}} = \frac{\mu_1(\chi_1 + \chi_2) + \mu_4(q_4/q_1)(\chi_4 + \chi_5)}{\chi_1 + (q_2/q_1)\,\chi_2 + (q_5/q_1)\,\chi_5 + (q_4/q_1)\,\chi_4}$$

où q_i est le nombre de réactions par centimètre cube et par seconde.

Dans le régime stationaire

$$\mu_{\text{eff II}} = \frac{\mu_1(\chi_1 + \chi_2) + \mu_4(\chi_4 + \chi_5)}{\chi_1 + \chi_2 + \chi_4 + \chi_5}\,.$$

Au voisinage de $\varrho = \varrho_0$ l'exposant effectif peut être appréciablement plus grand que 1, en raison de la grande valeur de μ_4 et malgré le fait que $\chi_4 < 0$.

Pour le cycle III, l'exposant ν_{eff} est élevé, sauf dans les régions centrales où, en raison de l'énorme valeur du facteur d'accélération, ν_{eff} se trouve ramené au voisinage de zéro et μ_{eff} se trouve prendre des valeurs assez faibles. On a en effet,

$$\mu_i = \frac{2}{3} + \frac{2^{\frac{1}{3}}}{18}\,\tau\,\beta^{-\frac{1}{3}}.$$

Pour la réaction de formation de C^{12}, pour $T = 10^7$, $\varrho = 2\cdot 10^9$, on trouve $\mu_4 = 4{,}11$.

B. Constitution interne des naines blanches.

I. Configurations complètement dégénérées.

11. Théorie de CHANDRASEKHAR. La théorie de CHANRASEKHAR des configurations complètement dégénérées reste la base de la théorie de la structure interne des naines blanches [*1*].

On part des équations (1.21) et (1.23) écrites sous la forme

$$P = A\,f(x); \qquad \varrho = B\,\mu_e\,x^3 \tag{11.1}$$

avec

$$A = 6{,}01 \cdot 10^{22}; \quad B = 9{,}82 \cdot 10^{5}. \tag{11.2}$$

L'équation d'équilibre hydrostatique est

$$\frac{A}{B}\,\frac{1}{r^2}\,\frac{d}{dr}\left(\frac{r^2}{x^3}\,\frac{df}{dr}\right) = -\,4\pi\,G\,B\,\mu_e\,x^3. \tag{11.3}$$

On pose

$$y = x^2 + 1. \tag{11.4}$$

Si y_0 est la valeur de y au centre de l'étoile, on pose

$$y = y_0\,\Phi. \tag{11.5}$$

Φ prend la valeur 1, au centre et la valeur $1/y_0$ à la surface. Après quelques réductions, l'équation d'équilibre hydrostatique s'écrit:

$$\frac{1}{\eta^2}\,\frac{d}{d\eta}\,\eta^2\,\frac{d\Phi}{d\eta} = -\left(\Phi^2 - \frac{1}{y_0^2}\right)^{\frac{3}{2}} \tag{11.6}$$

avec

$$r = \alpha\,\eta \tag{11.7}$$

et

$$\alpha = (2A/\pi G)^{\frac{1}{2}}\,(B\,\mu_e\,y_0)^{-1} = 7{,}71 \cdot 10^{8}\,(\mu_e\,y_0)^{-1}. \tag{11.8}$$

L'équation (11.6) a été intégrée par CHANDRASEKHAR pour 10 valeurs de $1/y_0^2$ définissant dix valeurs différentes de la densité centrale. Dans le cas d'une discontinuité du poids atomique par électron libre μ_e à travers une certaine surface de rayon r, on doit satisfaire aux conditions de continuité de la pression et de la masse. SCHATZMAN[1] a intégré l'équation (11.6) pour dix neuf modèles différents combinant différentes valeurs de y_0 au centre de l'étoile et du rayon de la surface de discontinuité; le poids atomique par électron libre prenant les valeurs 2 et 1 de part et d'autre de la surface de discontinuité.

La masse de la configuration est

$$\mathcal{M} = -\,4\pi\left(\frac{2A}{\pi G}\right)^{\frac{3}{2}}\frac{1}{(B\,\mu_e)^2}\left(\eta^2\,\frac{d\Phi}{d\eta}\right)_s. \tag{11.9}$$

Quand $1/y_0^2$ décroît, la configuration tend vers un modèle polytropique d'indice 3, le rayon décroît comme $1/y_0$ et la masse atteint une limite

$$M_3 = -\,4\pi\left(\frac{2A}{\pi G}\right)^{\frac{3}{2}}\frac{1}{(B\,\mu_e)^2}\left(\eta^2\,\frac{d\Theta_3}{d\eta}\right)_s \tag{11.10}$$

où Θ_3 est la fonction polytropique d'indice 3.

Entre $1/y_0^2 = 0{,}01$ et $1/y_0^2 = 0$, SCHATZMAN a calculé une solution approchée et A. REIZ, par une méthode de perturbation, a donné la solution exacte pour

[1] E. SCHATZMAN: Ann d'Astrophysique **10**, 19 (1947). — Contrib. Inst. d'Astrophysique, Sér. A **1951**, Nr. 61.

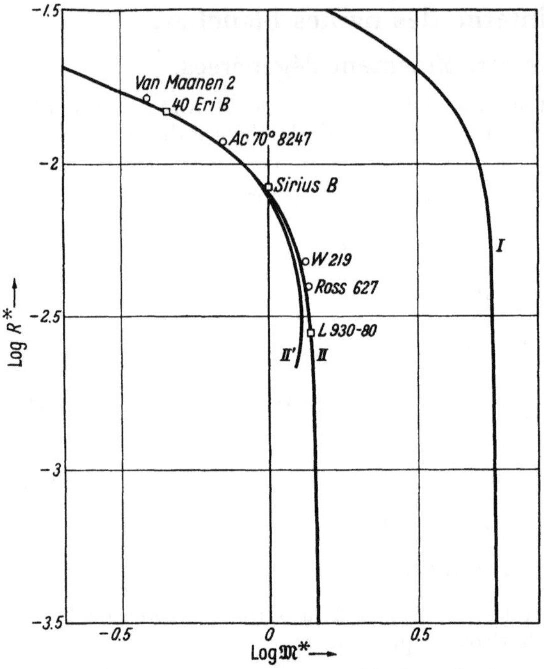

Fig. 4. Relation masse rayon pour les naines blanches: *I* Courbe pour $\mu_e = 1$ (hydrogène pur); *II* Courbe pour $\mu_e = 2$ (mélange de RUSSELL); *II'* Courbe pour $\mu_e = 2$, compte tenu de l'effet des captures beta. L'emplacement approximatif de 7 naines blanches est donné. $R^* = R/\odot$, $\mathfrak{M}^* = \mathcal{M}/\odot$.

Tableau 8. *Les caractéristiques physiques des configurations complètement dégénérées.*

$1/y_0^2$	$\mathcal{M}/\odot$	ϱ_0 g cm^{-3}	$\bar{\varrho}$ g cm^{-3}	R cm
0,0001	5,753	$9{,}82 \cdot 10^{11}$	$2{,}00 \cdot 10^{10}$	$5{,}14 \cdot 10^{7}$
0,00025	5,748	$2{,}48 \cdot 10^{11}$	$5{,}33 \cdot 10^{9}$	$7{,}99 \cdot 10^{7}$
0,0005	5,740	$8{,}78 \cdot 10^{10}$	$2{,}00 \cdot 10^{9}$	$1{,}11 \cdot 10^{8}$
0,001	5,725	$3{,}10 \cdot 10^{10}$	$7{,}62 \cdot 10^{8}$	$1{,}53 \cdot 10^{8}$
0,0025	5,682	$7{,}83 \cdot 10^{9}$	$2{,}21 \cdot 10^{8}$	$2{,}30 \cdot 10^{8}$
0,005	5,615	$2{,}76 \cdot 10^{9}$	$8{,}92 \cdot 10^{7}$	$3{,}10 \cdot 10^{8}$
0,01	5,51	$9{,}85 \cdot 10^{8}$	$3{,}70 \cdot 10^{7}$	$4{,}13 \cdot 10^{8}$
0,02	5,32	$3{,}37 \cdot 10^{8}$	$1{,}57 \cdot 10^{7}$	$5{,}44 \cdot 10^{8}$
0,05	4,87	$8{,}13 \cdot 10^{7}$	$5{,}08 \cdot 10^{6}$	$7{,}69 \cdot 10^{8}$
0,1	4,33	$2{,}65 \cdot 10^{7}$	$2{,}10 \cdot 10^{6}$	$9{,}92 \cdot 10^{8}$
0,2	3,54	$7{,}85 \cdot 10^{6}$	$7{,}9 \cdot 10^{5}$	$1{,}29 \cdot 10^{9}$
0,3	2,95	$3{,}50 \cdot 10^{6}$	$4{,}04 \cdot 10^{5}$	$1{,}51 \cdot 10^{9}$
0,4	2,45	$1{,}80 \cdot 10^{6}$	$2{,}29 \cdot 10^{5}$	$1{,}72 \cdot 10^{9}$
0,5	2,02	$9{,}82 \cdot 10^{5}$	$1{,}34 \cdot 10^{5}$	$1{,}93 \cdot 10^{9}$
0,6	1,62	$5{,}34 \cdot 10^{5}$	$7{,}7 \cdot 10^{4}$	$2{,}15 \cdot 10^{9}$
0,8	0,88	$1{,}23 \cdot 10^{5}$	$1{,}92 \cdot 10^{4}$	$2{,}79 \cdot 10^{9}$
1,0	0	0	0	

$\mathcal{M}$ et R. On trouvera dans le tableau les caractéristiques physiques des configurations complètement dégénérées. Le point représentatif d'une naine blanche dans le plan Log $\mathcal{M}$, Log R est situé entre deux courbes, la courbe supérieure correspondant à l'hydrogène pur ($\mu_e = 1$), la courbe inférieure aux éléments moyens et lourds ($\mu_e = 2$). On trouvera Fig. 4 ces deux courbes ainsi que les points représentatifs des naines blanches dont la masse et le rayon sont connus. En fait, des conditions physiques très strictes viennent limiter dans ce diagramme le domaine d'existence possible des naines blanches. C'est particulièrement à propos de Sirius B que se pose un difficile problème d'interprétation. D'après les résultats d'observation, Sirius B serait entièrement dégénérée [conditions (15.5)] mais son rayon est incompatible avec les structures théoriques. De nouvelles mesures photométriques et spectroscopiques de Sirius B sont nécessaires pour trancher la question.

12. Modèle de RUDKJÖBING[1]. Sous l'effet de la gravité, une séparation partielle des charges a lieu dans les naines blanches. Il en résulte un champ électrique radial. Ce champ électrique radial produit un effet relativiste analogue à l'interaction spin-orbite. La solution du problème peut être obtenue aisément, car il est possible de considérer des domaines à la fois petits devant les dimensions de l'étoile et grands devant la longueur d'onde.

On trouve alors que le nombre d'électrons par centimètre cube est

$$N_e = \frac{8\pi m_e^3 c^3}{3h^3}\, x^3 \left[1 - \frac{1}{x^2+1}\left(r\,\frac{dx}{dr}\right)^2\right]^{\frac{3}{2}} \tag{12.1}$$

[1] M. RUDKJÖBING: Publ. Obs. Copenhague **1952**, Nr. 160.

et l'équation (11.6) des structures complètement dégénérées se trouve modifiée en

$$\frac{1}{\eta^2}\frac{d}{d\eta}\left(\eta^2\frac{d\varphi}{d\eta}\right) = -\left(\varphi^2 - \eta\frac{d\varphi}{d\eta} - \frac{1}{y_0^2}\right)^{\frac{3}{2}}. \tag{12.2}$$

Les solutions de (12.2) donnent une nouvelle relation masse-rayon pour les naines blanches. Si $1/y_0^2$ tend vers l'unité, la solution tend vers le polytrope d'indice $\frac{3}{2}$ comme dans la solution de CHANDRASEKHAR. Si $1/y_0^2$ tend vers zéro, la solution tend vers le polytrope d'indice 3, le rayon tend vers zéro et la masse tend vers une masse limite de l'ordre de 84% de la masse limite de CHANDRASEKHAR.

On trouvera dans le Tableau 9 les caractéristiques physiques du modèle de RUDKJÖBING.

Tableau 9.

$1/y_0^2$	R cm	$\mathcal{M}/\odot$
0,5	$9{,}93\cdot 10^8$	0,426
0,2	7,58	0,661
0,1	5,97	0,779
0	0	1,205

13. Influence des modifications nucléaires. Ainsi que nous l'avons vu à la Sect. 2, la modification du rapport (Z/A) des noyaux aux grandes densités entraîne une altération de la relation pression-densité qui peut, en réalité, être traitée comme une altération du poids atomique par électron libre. L'équation de la structure interne des naines blanches est alors modifiée en

$$\frac{1}{\eta^2}\frac{d}{d\eta}\eta^2\frac{d\Phi}{d\eta} - \frac{1}{\mu_e}\frac{d\mu_e}{dy}y_0\left(\frac{d\Phi}{d\eta}\right)^2 = -\left(\Phi^2 - \frac{1}{y_0^2}\right)^{\frac{3}{2}} + \left(1 - \frac{\mu_e^2}{(\mu_e)_0^2}\right)\left(\Phi^2 - \frac{1}{y_0^2}\right)^{\frac{3}{2}} \tag{13.1}$$

où $(\mu_e)_0$ est le poids atomique par électron libre à pression nulle. Les modifications nucléaires ne se produisant qu'aux densités très élevées, on les traite comme une perturbation. La fonction Φ étant voisine du polytrope d'ordre 3, on pose

$$\Phi = \Theta_3 + \frac{1}{y_0^2}\Psi + \lambda_1 u \tag{13.2}$$

avec

$$\lambda_1 = \frac{2\cdot 10^3 (m_e/m_0)\, y_0}{83{,}8} = 0{,}01298\, y_0 \tag{13.3}$$

et l'on suppose à la fois $(1/y_0^2)$ et λ_1 petits.

Tableau 10. *Caractéristiques physiques des naines blanches (compte tenu des modifications nucléaires).*

y_0^{-2}	$\mathcal{M}/\odot$	$R/R_\odot$
0,01562	1,270	$3{,}49\cdot 10^{-3}$
0,01	1,285	$2{,}92\cdot 10^{-3}$
0,00695	1,267	$2{,}52\cdot 10^{-3}$
0,00500	1,245	$2{,}21\cdot 10^{-3}$

On calcule alors la fonction u, ce qui permet d'obtenir la masse et le rayon des configurations. Le rayon est donné par la condition

$$\Theta_3 + \frac{1}{y_0^2}\Psi + \lambda_1 u = \frac{1}{y_0}. \tag{13.4}$$

Les résultats sont rassemblés dans le Tableau 10, qui donne les caractéristiques physiques des naines blanches, compte tenu des modifications nucleaires.

On notera que la masse passe par un maximum pour une valeur finie du rayon. La branche inférieure de la courbe $R(\mathcal{M})$ correspond manifestement à des configurations instables, car une contraction s'accompagne d'une diminution du nombre d'électrons libres par gramme de matière donc à une diminution de pression.

14. Modifications relativistes[1]. Lorsque le rayon des naines blanches est petit, il devient nécessaire de modifier l'équation de POISSON pour tenir compte des

[1] S. A. KAPLAN: Notes Savantes Université Lwow **15**, 110 (1949).

corrections de relativité générale. Il convient alors de remplacer l'équation d'équilibre hydrostatique par

$$\frac{dP}{dr} = -\frac{GM}{r^2}\varrho\,\frac{\left(1+\frac{P}{\varrho c^2}\right)\left(1+\frac{4\pi P r^3}{GM}\right)}{1-\frac{2GM}{c^2 r}}. \tag{14.1}$$

Le second membre peut être réduit au premier terme de son développement. On peut poser

$$\frac{p}{\varrho c^2} + \frac{4\pi P r^3}{GM} + \frac{2GM}{c^2 r} = \frac{P_c}{\varrho_c c^2}\,\varphi\left(\frac{M}{\mathcal{M}}\right) \tag{14.2}$$

où $\mathcal{M}$ est la masse totale de la configuration et M la masse de la configuration de rayon $r \cdot \varphi(M)$ est sensiblement constant et voisin de 5. La masse de la configuration est alors

$$\mathcal{M} = M_3\left[1 - 7{,}5\cdot\frac{K_2\varrho_c^{\frac{1}{3}}}{c^2} - 0{,}61\left(\frac{m_e}{\mu_e m}\right)^2\left(\frac{c^2}{K_2\varrho_c^{\frac{1}{3}}}\right)^2 + \cdots\right]. \tag{14.3}$$

Elle présente un maximum pour une densité centrale $\varrho_c = 2{,}5\cdot 10^{10}$ g cm^{-3} ce qui correspond à une masse $\mathcal{M} = 0{,}96 M_3$. Le rayon correspondant est $R = 1{,}1\cdot 10^8$ cm.

II. Structure des couches superficielles.

15. Equilibre radiatif et convectif. Selon toute vraisemblance, les régions extérieures des naines blanches sont constituées d'hydrogène pur. L'atmosphère est en équilibre radiatif et dans l'atmosphère commence une zone convective dont l'extension est plus ou moins grande suivant les conditions à la surface. Si la zone convective superficielle est étendue, elle peut limiter les effets du triage des éléments (Sect. 16) et permettre aux éléments lourds d'apparaître dans le spectre.

La question n'est que partiellement débrouillée. Dans le plan Log T_{eff}, Log g, la région des zones convectives étendues se trouve à gauche d'une ligne Log $g = 29{,}1$, Log $T_{\text{eff}} - 111{,}5$[1]. Dans le cas des naines blanches, cela correspond à des températures superficielles assez basses (disons: inférieures à 10000°).

Dans un cas particulier, le calcul a été fait avec plus de détail. On choisit Log $g = 10$ et pour trois valeurs de la température superficielle définie par $\vartheta = 5040/T$, on calcule la température à la base de la zone convective. La base de la zone convective est définie par la condition

$$\frac{3}{8}\,\frac{\bar{\chi} P}{g}\left(\frac{\vartheta}{\vartheta_0}\right)^4 = 0{,}4 \tag{15.1}$$

avec

$$\bar{\chi} = [a X + b(1-X)](1+X)\,\varrho\, T^{-3{,}5}, \tag{15.2}$$

$$a = 3{,}7\cdot 10^{22}, \tag{15.3}$$

$$b = 1000\,a. \tag{15.4}$$

La zone convective est supposé en équilibre adiabatique. Les résultats sont rassemblés dans le Tableau 11.

La température à la base de la zone convective n'est pas encore la température centrale. La température continue à croître dans les couches extérieures.

[1] CH. PECKER: Ann. d'Astrophysique **16**, 321 (1953).

Elle devient pratiquement constante lorsque la dégénérescence est atteinte, en raison de la haute conductibilité thermique de la matière dense.

Tableau 11. *Zone convective dans une atmosphère de gravité* $g = 10^{10}$.

ϑ_0	profondeur optique à la limite supérieure	ϑ	Log P	Log P_e	Log T à la base de la zone convective	
					$1-x=0$	$1-x=0{,}01$
0,45	0,40	0,404	6,83	5,53	zone convective mince	
0,50	0,64	0,436	7,13	5,45	5,13	5,42
0,55	0,63	0,466	7,38	5,34	5,23	5,53

Le cas le plus simple est celui où la région extérieure, en équilibre radiatif sauf dans une mince zone superficielle, est faite d'hydrogène pur[1]. Dans ce cas, le coefficient d'absorption étant donné par

$$\varkappa = \varkappa_0 \varrho \, T^{-3,5} \tag{15.5}$$

l'on a

$$T = \frac{4}{17} \frac{\bar{\mu}}{R} \frac{GM}{R} \left(\frac{R}{r} - 1\right) = 5{,}44 \cdot 10^6 \bar{\mu} \frac{\mathcal{M}^*}{R^*} \left(\frac{R}{r} - 1\right), \tag{15.6}$$

$$\varrho = \left(\frac{3\varkappa_0 L}{16\pi a c R}\right)^{-\frac{1}{2}} \left(\frac{4}{17} \frac{\bar{\mu}}{R} \frac{GM}{R}\right)^{3,75} \left(\frac{R}{r} - 1\right)^{3,25}, \tag{15.7}$$

$$= 4{,}76 \cdot 10^{12} \varkappa_0^{-\frac{1}{2}} \bar{\mu}^{3,75} \mathcal{M}^{*\,3,75} R^{*\,-3,25} L^{*\,-0,5} \left(\frac{R}{r} - 1\right)^{3,25}. \tag{15.8}$$

L'examen de la condition de dégénérescence confirme, comme il est bien connu, que la dégénérescence est atteinte à faible profondeur dans les naines blanches, lorsque

$$3{,}97 \cdot 10^{13} \varkappa_0^{-\frac{2}{3}} \mu_e^{-\frac{4}{3}} \left(\frac{\mu \mathcal{M}^*}{R^*}\right)^3 \left(\frac{R^*}{L^*}\right)^{\frac{2}{3}} \left(\frac{R}{r} - 1\right)^{\frac{7}{2}} \gg 1. \tag{15.9}$$

D'autre part, compte tenu des valeurs exactes du coefficient d'absorption, il est possible de calculer la température interne des naines blanches à enveloppe d'hydrogène pur. On trouve alors que la relation entre le température interne, la luminosité et la masse peut être représentée par la formule d'interpolation

$$T = 10^{7,567} \left(\frac{L^*}{M^*}\right)^{0,349}. \tag{15.10}$$

Lorsque l'atmosphère n'est pas constituée d'hydrogène pur, d'autres processus (triage des éléments) viennent accroître considérablement la complexité des couches extérieures des naines blanches.

16. Triage des éléments. En l'absence de courants de convection, une séparation des éléments s'effectue dans les naines blanches, les éléments lourds tombent dans le champ de gravitation, l'hydrogène flottant à la surface.

Le problème général du triage dans un champ de gravitation se complique du fait que le nombre de particules libres par unité de volume varie, non seulement en raison de la présence d'un champ de gravitation et d'un champ électrique, mais aussi en raison de la variation de degré d'ionisation. Considérons un mélange d'hydrogène et d'éléments 1, 2, ... *n*. L'indice H se rapporte à l'hydrogène, l'indice *i* aux autres éléments. On peut montrer que le nombre $p_i^{(n)}$ d'atomes

[1] E. Schatzman: Publ. Obs. Copenhague **1950**, Nr. 149.

d'espèce i, ionisés n fois, le nombre r d'électrons sont donnés par les expressions[1]

$$\left.\begin{aligned} p_i^{(n)} &= g_i^n \frac{(2\pi m_i^{(n)} kT)^{\frac{3}{2}}}{h^3} A\, B^{Z_i-n}\, U^{Z_i-n+1} \times \\ &\times \exp\left[-\vartheta \sum \chi_i^n - \int \frac{1}{kT} d\,(m_i^{(n)} \varphi - n e \psi)\right], \end{aligned}\right\} \tag{16.1}$$

$$r = 2 \cdot \frac{(2\pi m_e kT)^{\frac{3}{2}}}{h^3} U_{\frac{1}{2}}, \tag{16.2}$$

$$U_{\frac{1}{2}} = \frac{1}{\Gamma(\frac{3}{2})} \int\limits_0^\infty \frac{u^{\frac{1}{2}}\, du}{\frac{1}{\Lambda} e^u + 1}, \tag{16.3}$$

$$\log \Lambda = \log B + \log U - \int \vartheta\, d\,(m_e \varphi + e\psi). \tag{16.4}$$

A et B sont des constantes. χ_i^n est l'énergie nécessaire pour ioniser le $n^{\text{ième}}$ électron. φ est le potentiel gravifique et ψ le potentiel électrostatique. La fonction U et la relation entre les potentiels φ et ψ sont déterminées par l'équation d'équilibre hydrostatique et la condition de neutralité électrique:

$$\frac{d}{dr} \sum (p_i^{(n)} + r)\, kT = -g \sum (m_i^{(n)} p_i^{(n)} + 2 m_e), \tag{16.5}$$

$$\sum n\, p_i^{(n)} = r. \tag{16.6}$$

Différentes méthodes de représentation paramétrique peuvent être utilisées pour l'étude des principaux cas intéressants. Nous noterons les cas suivants:

α) Zone de mélange dans un gaz non dégénéré. On introduit la variable

$$U = \exp\left(-\int \frac{1}{kT} m_0\, d\varphi\right). \tag{16.7}$$

On prend comme niveau de référence le point $U = 1$ où la concentration en poids de l'hydrogène égale la concentration de tous les autres éléments. On introduit le paramètre t lié à la fonction U par la relation:

$$U^3 = \frac{t^{2/Z_1}}{\frac{3}{2} - \frac{1}{2} \sum c_i t^{Z_i/Z_1}} \tag{16.8}$$

où c_i est la concentration de l'élément i au sein du mélange d'éléments lourds au niveau de réference. Les concentrations x_{H} et x_i sont données par:

$$x_{\mathrm{H}} = \frac{U^{-1} t^{1/Z_1}}{U^{-1} t^{1/Z_1} + \sum c_i t^{Z_i/Z_1}}, \tag{16.9}$$

$$x_i = \frac{c_i t^{Z_i/Z_1}}{U^{-1} t^{1/Z_1} + \sum c_i t^{Z_i/Z_1}}. \tag{16.10}$$

On met ainsi en évidence l'existence d'une région où la composition chimique varie très rapidement avec l'altitude, que l'on appelle zone de mélange.

β) Zone de mélange dans la zone dégénérée. Le problème est considérablement simplifié si l'on suppose le degré de dégénérescence constant dans la zone de mélange. Si l'on réduit le mélange à 1 élément et l'hydrogène, on peut introduire la *hauteur de mélange*, épaisseur de la région où à lieu la transition de l'hydrogène

[1] Voir notes 1, p. 743.

au mélange d'éléments lourds. Si l'on pose [U_n défini par (6.8)],

$$\zeta = \frac{Z+\xi}{1+\xi}, \tag{16.11}$$

$$\xi = \frac{U_{-\frac{1}{2}}(A)}{U_{\frac{1}{2}}(A)}, \tag{16.12}$$

on obtient pour la hauteur de mélange en milieu dégénéré:

$$h = \frac{kT}{m_0 g} \frac{9}{4} \frac{\zeta+1}{A-\zeta}, \tag{16.13}$$

où A et Z sont le poids et le numéro atomique de l'élément lourd supposé mélangé à l'hydrogène. Pour les dégénérescences complètes,

$$\left.\begin{aligned} &\xi = 0, \quad \zeta = Z, \quad \text{et } A/Z \approx A/(Z+1) \approx \tfrac{1}{2}, \quad \text{on a} \\ &h = 6{,}95 \cdot 10^7 \frac{T}{g}. \end{aligned}\right\} \tag{16.14}$$

En équilibre, la zone de mélange a une épaisseur très petite comparée au rayon de l'étoile.

γ) Zone extérieure non dégénérée en équilibre radiatif. Il faut alors intégrer à la fois l'équation de transfert et les équations d'équilibre statistique.

Une complication supplémentaire provient de ce qu'en raison du changement de composition chimique, la condition d'équilibre radiatif peut n'être plus satisfaite. L'expression (15.2) du coefficient d'absorption montre qu'au voisinage de $1-x=0{,}001$, une très faible augmentation de l'abondance des éléments lourds suffit pour que la condition d'équilibre radiatif:

$$\frac{d\log T}{d\log P} \leqq 0{,}4 + \frac{d\log\varrho}{d\log P} \tag{16.15}$$

ne soit pas vérifiée.

Le calcul a été fait par SCHATZMAN[1] en assimilant l'élément lourd moyen à l'oxygène. On trouve alors qu'un grand nombre de zones, alternativement en équilibre radiatif et convectif, peuvent exister dans les régions extérieures des naines blanches.

Négligeant l'apparition de l'instabilité, LEE a étudié l'équilibre radiatif des zones extérieures de naines blanches en supposant présent seulement un mélange hydrogène-hélium.

δ) Zone superficielle. Dans une région en équilibre radiatif où la concentration en métaux est faible, on peut admettre que la température varie avec la distance à la surface suivant la loi

$$T = \frac{2}{17} \frac{m_{\mathrm{H}} g z}{k}. \tag{16.16}$$

Le nombre, par centimètre cube, des atomes d'espèce i et de charge Z varie avec la distance à la surface suivant la loi

$$p_i^Z = p_{i0}^Z T^\alpha. \tag{16.17}$$

L'exposant α pour un gaz complètement ionisé est donné exactement par

$$\alpha = -1 - \tfrac{17}{4} Z + \tfrac{17}{2} A_i. \tag{16.18}$$

[1] E. SCHATZMAN: Astrophys. Journ. **110**, 261 (1949).

Pour un gaz partiellement ionisé, l'exposant est approximativement

$$\alpha = -1 - \tfrac{7}{4} n_i - \tfrac{5}{2} Z_i + \tfrac{17}{2} A_i . \tag{16.19}$$

Si à la base de l'enveloppe d'hydrogène des naines blanches l'abondance des métaux est encore notable, elle est évannescente dans l'atmosphère, sauf dans deux cas: (1) si une structure complexe de zones convectives existe dans les couches extérieures, (2) si une zone convective étendue a son sommet dans l'atmosphère. (Application possible notamment à van Maanen 2, naine blanche présentant dans son spectre des raies des métaux.)

III. Stabilité.

17. Stabilité dynamique. Une configuration stellaire est dynamiquement stable si la période de pulsation σ est réelle. On trouve par la méthode du viriel

$$\sigma^2 = \frac{A}{B} \frac{\int\limits_0^x r^3 \xi \left(\frac{8 x^4}{(x^2+1)^{\frac{3}{2}}} - \frac{d}{d x} \frac{\mu'}{3\mu} \frac{8 x^6}{(x^2+1)^{\frac{1}{2}}} \right) d x}{\int\limits_0^R \xi r^4 x^3 \, d x} \tag{17.1}$$

où ξ est l'amplitude de la pulsation, et A et B sauf donnés par les Eqs. (1.21) et (1.23) ou (11.1) et (11.2).

En première approximation, en supposant ξ constant et y_0 grand,

$$\sigma^2 = \frac{8A}{B} \frac{1}{\alpha^2 y_0} \frac{\int\limits_0^{\vartheta_c} \eta^3 \vartheta_3 \, d\vartheta_3 - \left(\frac{1}{\mu} \frac{d\mu}{d y}\right)_0 \int\limits_0^{\eta_s} y_0^{\frac{3}{2}} \vartheta_3^{\frac{5}{2}} \eta^2 \, d\eta}{\int\limits_0^{\eta_s} \vartheta_3^{\frac{3}{2}} \eta^4 \, d\eta} \tag{17.2}$$

σ^2 change de signe pour

$$\frac{1}{y_0^2} = 0{,}00833 . \tag{17.3}$$

Les naines blanches deviennent dynamiquement instables au voisinage du maximum de masse. Il ne peut donc exister aucune naine blanche de rayon inférieur à $2{,}7 \cdot 10^{-3} R_\odot$.

18. Stabilité séculaire. Tout changement de température interne sera stable si le changement correspondant dans le débit d'énergie est inférieur au changement dans l'énergie rayonnée. Dans le cas d'une couche épaisse d'hydrogène,

$$L \propto T^{2.87} . \tag{18.1}$$

Par conséquent, les naines blanches seront séculairement stables si l'exposant de la température dans la loi de débit d'énergie est inférieur à 2,87.

Cette conclusion est en opposition avec le résultat de MESTEL[1] qui semble ne pas avoir tenu des variations de l'énergie rayonnée. En effet, lorsque la température diminue, le débit d'énergie diminue également. Il suffit que le débit d'énergie diminue moins vite que l'énergie rayonnée pour que l'étoile puisse ultérieurement se réchauffer et par conséquent soit séculairement stable.

Les conséquences d'une structure plus complexe sur le débit d'énergie n'ont pas été étudiées.

---

[1] L. MESTEL: Monthly Notices Roy. Astronom. Soc. London **112**, 583 (1952).

19. Stabilité vibrationnelle. Nous suivons ici la solution donnée par LEDOUX et SAUVENIER GOFFIN[1]. Le problème consiste à chercher l'amplitude ξ de la pulsation et la période. La solution pour le mode fondamental est obtenue en minimisant l'expression

$$\sigma^2 = \min \frac{\int\limits_0^R P\,\Gamma_P r^4 \xi'^2 - r^3 \xi^2 d\,[(3\Gamma_P - 4)\,P]}{\int\limits_0^R \xi^2 \varrho\, r^4\, dr}. \tag{19.1}$$

On remplace ξ par un développement suivant les puissances paires du rayon dont on détermine les coefficients au moyen de la condition d'extremum.

Pour $1/y_0^2 = 0{,}2$, $R = 1{,}29 \cdot 10^9$ cm, Mme SAUVENIER GOFFIN[2] trouve les résultats suivants:

Première approximation	Mode fondamental	$\mu_e \sigma_0^2 = 0{,}5732$
Deuxième approximation	Mode fondamental	0,5724
	Second mode	1,5785
Troistéme approximation	Mode fondamental	0,5720
	Second mode	1,5379
	Troisième mode	2,0098

Quand on tient compte de l'écart à l'adiabaticité, un terme imaginaire s'ajoute à la pulsation σ: $\sigma + i\sigma'$. La pulsation σ' est donnée par

$$\sigma' = \frac{1}{2i\sigma^2} \int\limits_0^R \frac{d}{dr} \big\{(\Gamma_T - 1)\big(\delta(\varrho\,\varepsilon) - \delta \operatorname{div} F\big)\big\} r^3 \xi\, dr. \tag{19.2}$$

La condition de stabilité vibrationnelle est $i\sigma' < 0$, ou encore, après intégration par parties

$$\int\limits_0^{\mathfrak{M}} (\Gamma_T - 1) \frac{\delta\varrho}{\varrho}\, \delta \left[\varepsilon - \frac{d(4\pi r^2 F)}{dm}\right] dm < 0. \tag{19.3}$$

Si le débit d'énergie a lieu dans le noyau dégénéré, la condition de stabilité s'écrit

$$\left.\begin{aligned} &\mu\,\overline{(\Gamma_T - 1)} + \nu\,\overline{(\Gamma_T - 1)^2} + \frac{2}{9}[4 + 3n - 2(4 + s)] + \\ &+ \frac{4}{3}\int\limits_0^{\mathfrak{M}} \frac{L_r}{L}\, \frac{x\,dx}{3(x^2+1)^2} \left[\frac{T x}{(x^2+1)^2}\, \frac{dx}{dT} - 2 - \frac{1}{x^2+1} + \frac{1}{\log Q}\right] \leq 0 \end{aligned}\right\} \tag{19.4}$$

où μ et ν sont les exposants de ϱ et T dans la loi de débit d'énergie, n et s les exposants de ϱ et T dans l'expression du coefficient d'absorption, et $Q = 1 + \vartheta \varrho^{\frac{2}{3}}$, ϑ étant une constante. Les moyennes $\overline{(\Gamma_T - 1)}$ et $\overline{(\Gamma_T - 1)^2}$ sont calculées en pondérant à l'aide du débit d'énergie par gramme.

Si le débit d'énergie a lieu dans l'enveloppe, la condition de stabilité s'écrit

$$\xi^2 L \left\{6n - 4s + 6\mu + 4\nu - 8 - \int\limits_0^{\mathfrak{M}_a} \frac{48\pi^2 c\, a}{L}\, \frac{T^4 r^2}{\varkappa\varrho} \left(\frac{d\Gamma_T}{dr}\right)^2 dr\right\} \leq 0. \tag{19.5}$$

[1] P. J. LEDOUX et E. SAUVENIER GOFFIN: Astrophys. Journ. **111**, 611 (1950).
[2] E. SAUVENIER GOFFIN: Mém. Soc. Roy. Sci., Liège **10**, Fasc. 1.

LEDOUX et SAUVENIER GOFFIN ont calculé ces intégrales pour un modèle stellaire correspondant approximativement à Sirius B. Dans l'équation (19.4), l'intégrale prend la valeur $-0{,}17$, et dans l'équation (19.5) elle prend la valeur $-0{,}03$. Ces intégrales jouent donc un rôle secondaire par rapport aux autres termes. Si les sources d'énergie sont régulièrement réparties, on trouve la condition.

$$0{,}42\,\mu + 0{,}162\,\nu \leqq 2{,}01\,. \tag{19.6}$$

Si, au contraire, elles sont concentrées dans une zone voisine de la surface, on trouve la condition

$$\tfrac{2}{3}\,\mu + \tfrac{4}{9}\,\nu \leqq 1{,}78\,. \tag{19.7}$$

La condition d'instabilité est intermédiaire entre ces deux conditions si les sources d'énergie sont un peu plus profondes. $\overline{(\Gamma_T - 1)}$ et $\overline{(\Gamma_T - 1)^2}$ au lieu de prendre les valeurs 2/3 et 4/9 comme dans les régions superficielles, prennent alors des valeurs plus petites, rendant ainsi la condition de stabilité moins stricte.

IV. Origine du débit d'énergie des naines blanches.

20. Différentes hypothèses peuvent être faites sur l'origine du débit d'énergie des naines blanches. Une fois ces hypothèses faites, il reste encore à examiner dans quelle mesure elles sont compatibles avec les propriétés générales des naines blanches.

On peut supposer la production d'énergie avoir lieu dans l'enveloppe d'hydrogène[1].

Si ΔR est la distance comptée à partir de la surface, la température et la densité dans l'enveloppe d'hydrogène sont données par

$$T = 10^{6{,}434}\,\frac{\mathscr{M}^*}{R^*}\,\frac{\Delta R}{R}\,, \tag{20.1}$$

$$\varrho = 10^{0{,}116}\,R^{*-\frac{13}{4}}\,L^{*-\frac{1}{2}}\,\mathscr{M}^{*\frac{15}{4}}\left(\frac{\Delta R}{R}\right)^{\frac{13}{4}}. \tag{20.2}$$

Tableau 12.

$1/y_0^2$	Log λ	μ
0		
0,01	6,82	1,56
0,02	6,60	1,73
0,05	6,24	2,00
0,1	5,92	2,27
0,2	5,46	2,63
0,3	5,09	2,96
0,4	4,73	3,29
0,5	4,37	3,65
0,6	3,97	4,57
0,8	2,93	5,44

Le débit d'énergie est alors donné par

$$\left.\begin{aligned} L^* = 10^{2{,}76}\,R^{*-\frac{17}{12}}\,\mathscr{M}^{*\frac{41}{12}}\left(\frac{\Delta R}{R}\right)^{\frac{41}{12}} \times \\ \times \exp\left[-12{,}1\left(\frac{R^*}{\mathscr{M}^*}\right)^{\frac{1}{3}}\left(\frac{\Delta R}{R}\right)^{-\frac{1}{3}}\right], \end{aligned}\right\} \tag{20.3}$$

ou encore par

$$L^* = \lambda\left(\frac{\Delta R}{R}\right)^{\frac{41}{12}} \mathrm{e}^{-\mu\left(\frac{\Delta R}{R}\right)^{-\frac{1}{3}}}\,. \tag{20.4}$$

λ et μ sont donnés dans le Tableau 12.

Cette théorie s'applique si la dégénérescence n'est pas atteinte dans la couche d'hydrogène. Dans le cas de 40 Eridani B,

$$\operatorname{Log} L^* = -2{,}27, \quad \operatorname{Log} R^* = -1{,}85, \quad \operatorname{Log} \mathscr{M}^* = -0{,}35.$$

On en tire l'épaisseur de la couche d'hydrogène:

$$\frac{\Delta R}{R} = 0{,}122\,.$$

On en déduit la température et la densité à la base de la couche d'hydrogène:

$$T = 10{,}5 \cdot 10^6 \text{ degrés}$$

[1] E. SCHATZMAN: Publ. Obs. Copenhague **1950**, Nr. 149.

la formule (15.6) aurait donné $8 \cdot 10^6$ degrés

$$\operatorname{Log} \varrho = 2{,}993.$$

Or, des formules (10,5) on tire $\mu_4 = 5{,}62$, $\nu_4 = 20{,}5$. Si l'on calcule les exposants $\nu_1\,\nu_3$ et le nombre de réactions par seconde, on obtient les valeurs des exposants effectifs:

$$\mu_{\text{eff}} = 2{,}26, \quad \nu_{\text{eff}} = 10{,}4$$

qui ne satisfont évidemment pas à la condition de stabilité vibrationnelle (19.7). Cette conclusion peut être généralisée aux autres naines blanches. Deux possibilités se présentent alors: (1) les naines blanches ne tirent pas leur énergie de la réaction proton-proton ou celle-ci ne se produit pas dans les régions extérieures; (2) elles sont vibrationnellement instables.

Examinons tout d'abord cette deuxième conclusion. On peut imaginer que les naines blanches sont effectivement des étoiles pulsantes. On peut imaginer également que, de façon récurrente, ces étoiles subissent une explosion de faible amplitude, qui supprime une partie de l'hélium He^3 dans la réaction $He^3 + He^3 \rightarrow He^4 + H^1 + H^1$. Par suite de la diminution de l'abondance de He^3 résultant de l'explosion, les exposants μ_{eff} et ν_{eff} peuvent se trouver ramenés au-dessous des valeurs critiques pour lesquelles l'étoile devient vibrationnellement instable ou explosive. Ces explosions récurrentes pourraient être de nature à expliquer la présence d'enveloppes gazeuses en expansion observées par GREENSTEIN autour de certaines naines blanches.

On peut supposer que le débit d'énergie est dû aux réactions proton-proton se produisant dans des couches plus profondes de l'étoile. Toutefois, l'hydrogène ne saurait subsister à des densités électroniques supérieures à $6{,}9 \cdot 10^{30}$ électrons par centimètre cube (ou, pour le mélange de RUSSELL, à des densités supérieures à $2{,}28 \cdot 10^7$) car, à ces densités, la réaction $H^1 + e^- \rightarrow n + \nu$ se déroule avec une rapidité considérable.

A l'aide des relations (9.3), (9.5), (9.6), on peut calculer le débit d'énergie par gramme. Pour $\varrho = 2 \cdot 10^6$, $\mu_e = 2$, on trouve

$$\varepsilon_{\text{II}} = 10^{8{,}353}\, X_{\text{H}}^2.$$

Si, par exemple, la zone productrice d'énergie occupe 1% de la masse de l'étoile une concentration d'hydrogène $X_{\text{H}} = 10^{-4{,}18}$ suffit à produire le débit d'énergie observé dans 40 Eridani B. L'abondance de l'hydrogène dans l'étoile serait alors de l'ordre de $X = 10^{-6}$. La durée de vie de l'étoile serait alors de l'ordre de 10^7 ans seulement. En fait, il est possible de placer dans l'étoile la zone productrice d'énergie de telle façon que les conditions relatives au débit d'énergie, à la stabilité et à la durée de vie soient simultanément satisfaites.

Il resterait alors à expliquer comment, au cours de l'évolution des étoiles vers l'état de naine blanche, une quantité d'hydrogène, même si minime, a pu subsister dans les régions internes.

On peut alors abandonner l'idée que les réactions proton-proton sont à l'origine du débit d'énergie des naines blanches. Deux éventualités sont à considérer: (1) d'autres réactions thermonucléaires sont à l'origine du débit d'énergie des naines blanches, (2) les naines blanches ne tirent pas leur énergie des réactions thermonucléaires.

Ainsi que nous l'avons vu, la réaction de formation du carbone C^{12} est suffisamment rapide aux très hautes densités pour pouvoir être la source d'énergie des naines blanches très denses; mais cette éventualité n'a de chances d'être réalisée que dans un très petit intervalle de masse, voisin de la masse limite.

Les réactions du cycle du carbone, se produisant grâce à la présence de traces d'hydrogène, pourraient être à l'origine du débit d'énergie des naines blanches. Il paraît possible, en raison du rôle important de l'effet d'écran aux grandes densités, que les conditions de stabilité soient satisfaites par l'exposant de la température et de la densité dans la loi de débit d'énergie.

Il reste enfin possible que l'énergie rayonnée par les naines blanches soit due à leur refroidissement. L'énergie thermique disponible est de l'ordre de

$$W = \mathcal{M} R T \mu_A^{-1} \tag{20.5}$$

où μ_A est le poids atomique moyen des noyaux. Compte tenu de la relation (15.6), on obtient:

$$10^{6,33} \frac{dT}{dt} = -10^{-26,4}\, T^{2,87} \mathcal{M}^* \tag{20.6}$$

d'où l'on tire l'échelle de temps du refroidissement

$$t \approx 10^{11} \quad \text{années}$$

pour une naine blanche typique. Le refroidissement peut donc être la source de l'énergie rayonnée par les naines blanches pendant une très longue durée.

C. Conclusion.

21. L'évolution des étoiles et les naines blanches. Il a été démontré à plusieurs reprises[1] que la température centrale d'une étoile en contraction s'élève considérablement, disons aux environs de deux cent millions de degrés. Cette élévation de température se produit une fois achevée la combustion de l'hydrogène des régions centrales. Elle s'accompagne des réactions de formation du carbone, de l'oxygène et d'autres éléments moyens. Il paraît donc raisonable de supposer que les naines blanches résultent de la contraction d'étoiles normales, au terme de leur évolution.

Cependant, la masse maximum des naines blanches, compte tenu des différents effets (Sects. 11 à 14) est de l'ordre de la masse solaire. Une étoile de masse supérieures à celle du soleil ne peut devenir naine blanche qu'en se débarrassant de l'excès de masse. Les conditions dans lesquelles cette masse est rejetée ne nous sont pas connues avec certitude. Il paraît alors raisonnable de supposer que le noyau des étoiles WOLF-RAYET et le noyau des nébuleuses planétaires sont des étoiles en évolution vers l'état de naine blanche.

Cependant, le nombre d'étoiles déclinant ainsi vers l'état de naine blanche est grandement insuffisant pour expliquer l'abondance spatiale actuelle des naines blanches. Supposons alors que toutes les étoiles brillantes déclinent vers l'état de naine blanche, après passage par d'autres formes (telles que celle de supergéante rouge). Le nombre d'étoiles brillantes par unité d'intervalle de masse peut être écrit

$$\frac{dn}{d\mathcal{M}^*} = \frac{N}{\mathcal{M}^*} 10^{-kM} \times 2{,}5\, \alpha \operatorname{Log}_{10} e$$

où M est la magnitude bolométrique absolue. Avec

$$L = l\, \mathcal{M}^{*\alpha}$$

[1] Sir A. EDDINGTON: Novae et Naines Blanches. Paris: Hermann 1941. — E. J. ÖPIK: Proc. Roy. Ir. Acad. A **54**, Nr. 4 49 (1951). — M. SCHWARZSCHILD *et al.*: Premier article: Astrophys. Journ. **116**, 463 (1952).

on a pour la durée de vie des étoiles

$$T = K \mathcal{M}^{*1-\alpha}.$$

D'après la discussion de SALPETER[1]

$$N \approx 10^{-4}\, 10^{-0,42\, M_{\mathrm{Bol}\odot}}$$

pour $M_{\mathrm{Bol}} < 0$. D'autre part, K est de l'ordre de 10^{10} ans. Le nombre d'étoiles par unité d'intervalle de masse, déclinant en naine blanche par unité d'intervalle de temps est alors

$$\frac{dn/d\mathcal{M}}{T} = 2{,}5\,\alpha \cdot \mathrm{Log}_{10}\, e\, \frac{N}{K}\, 10^{-kM_{\mathrm{Bol}\odot}}\, \mathcal{M}^{*(2,5k-1)\alpha}.$$

Le nombre total d'étoiles déclinant en naine blanche par unité d'intervalle de temps, $d\mathcal{N}/dt$, s'obtient alors en intégrant sur la masse (par exemple dans l'intervalle allant de $1\odot$ à $100\odot$). On trouve alors

$$\frac{d\mathcal{N}}{dt} = 2{,}5\,\alpha\, \mathrm{Log}_{10}\, e\, \frac{N}{K}\, \frac{1}{(2{,}6k-1)\,\alpha+1} \left[\mathcal{M}^{*(2,5k-1)\alpha+1}\right]_{\mathcal{M}_1^*}^{\mathcal{M}_2^*}$$

soit

$$\frac{d\mathcal{N}}{dt} \approx 8 \cdot 10^{-13} \text{ étoiles psc}^{-3}\text{ an}^{-1}.$$

Avec une échelle de temps galactique de 10^{10} ans, on trouve $8 \cdot 10^{-3}$ naines blanches par parsec cube à l'époque actuelle, alors que l'observation indique une densité comprise entre $3 \cdot 10^{-3}$ et 10^{-2} naines blanches par parsec cube[2]. L'évolution des étoiles de la séquence principale paraît donc capable d'expliquer le nombre actuel de naines blanches observées.

Addenda au paragraphe 9. Il conviendrait également de tenir compte de la réaction

$$He^3 + He^4 \rightarrow Be^7 + \gamma$$

dont l'importance a été récemment reconnue par W. A. FOWLER, Astrophys. Journ. **127**, 550, (1958).

Bibliographie.

[*1*] CHANDRASEKHAR, S.: An Introduction to the study of stellar Structure. Chicago 1939.
[*2*] BOHM, D., and D. PINES: Phys. Rev. **85**, 338 (1952).
[*3*] BLATT, J. M., et V. F. WEISSKOPF: Theoretical nuclear physics. New York 1952.
[*4*] SALPETER, E. E.: Phys. Rev. **88**, 547 (1952); **97**, 1237 (1955).
[*5*] ROSSELAND, S., et G. RANDERS: Astrophys. Novegica **3**, 71 (1938).
[*6*] SCHATZMANN, E.: Withe Dwarfs, Amsterdam: North-Holland Publishing Company 1958.

[1] E. E. SALPETER: Astrophys. Journ. **121**, 161 (1955).
[2] W. J. LUYTEN: Publ. Astr. Obs. Univ. Minnesota **2**, Nı. 11 (1939).

The Novae.

By

Cecilia Payne-Gaposchkin.

With 5 Figures.

A nova, or new star, is the result of a stellar explosion. The nova phenomenon can be regarded as including the supernovae (where the star appears to undergo a radical change, see following article by F. Zwicky), the ordinary novae (where the changes are superficial), the *U* Geminorum stars and the shell stars. The ordinary novae are discussed in the present article.

I. Statistical information.

1. Occurrence of novae. About 150 novae have been recorded in the galactic system, but for many of them the records are too fragmentary to convey physical information. A list of those best observed is given in Table 1.

Successive columns contain:

Name assigned to the star.

Designation, which describes the position approximately by six digits, of which the first four give the Right Ascension in hours and minutes, the last two, the Declination in degrees. Negative Declinations are denoted by italics.

Galactic longitude, l, and latitude, b.

Apparent photographic magnitude at maximum, as observed (Obs.) and as extrapolated (Extr.) for incompletely observed maxima.

Apparent photographic magnitude at minimum of the pre-nova (Before) and of the post-nova (After). Entries placed between the two columns (usually upper limits) are based on all minimal observations.

Type: VF, F, S, VS denote very fast, fast, slow, and very slow decline after maximum: more than 0.20, 0.19 to 0.08, 0.06 to 0.01, less than 0.009 magnitudes per day respectively.

Absolute photographic magnitude (M) at maximum, deduced with an adopted modulus of $24^{\mathrm{m}}.2$ from the relationship established by Arp [*1*] for the relationship between maximal magnitude and rate of decline of the novae in Messier 31.

Distance (corrected for interstellar absorption): $r \sin b$ perpendicular to the galactic plane; $r \cos b$, projected along the plane.

Table 1. *Selected galatic novae.*

Star	Des.	l	b	Maximum Obs.	Maximum Extr.	Minimum Before	Minimum After	Type	Rate of Decline	M	$r \sin b$	$r \cos b$
				m	m	m	m		m/day		kpc	kpc
Do Aql	192*606*	0	−13	8.6	8.6	16.5		S	0.007	−5.9	−0.70	3.0
EL Aql	185*003*	358	−4	6.4	6.4	...	19.0	F	0.105	−7.35	−0.16	2.3
V356 Aql	191201	5	−6	7.0	7.0	16.0	...	S	0.013	−6.0	−0.19	1.9
V368 Aql	192107	11	−6	6.6	6.6	[15.0		F	0.13	−7.8	−0.38	2.7
V500 Aql	194708	15	−11	6.5	6.5	[17		F	0.13	−7.8	−0.49	2.5
V528 Aql	191400	4	−7	7.4	7.4	[18		F	0.12	−7.7	−0.38	3.1
V603 Aql	184300	1	−1	−1.1	−1.1	10.5*v*	10.5*v*	VF	0.57	−8.35	−0.01	0.2
V604 Aql	185*604*	358	−6	8.2	8.2	[16.8		F	0.25	−8.25	−0.41	4.0
V606 Aql	191*500*	4	−8	6.7	4.4	[15.4		S	...	−7.05*	−0.17	1.2
OY Ara	163*352*	302	−5	6.2	5.1	17.5	...	S	0.045	−6.5	−0.11	1.3
T Aur	052530	145	0	4.2	4.2	...	14.8*v*	S	0.025	(−5.3)	0.00	0.8
RS Car	110*361*	259	−1	7.2	5.0	[15.8		VF	...	−8.3*	−0.04	2.1
N Car	104*159*	255	−1	−0.8	−0.8		(7.9)	VF	0.001	(−11:)	0.00	0.2

Table 1. (Continued.)

Star	Des.	*l*	*b*	Maximum Obs.	Maximum Extr.	Minimum Before	Minimum After	Type	Rate of decline	M	*r* sin *b*	*r* cos *b*
				m	m	m	m		m/day		kpc	kpc
MT Cen	113960	262	+ 1	8.5	8.2	[14.0		VF	0.33	−8.3	+0.07	4.1
X Cir	143464	282	− 5	6.5	6.5	16.7		S	0.040	−6.45	−0.16	1.9
AR Cir	144059	285	− 2	10.6	10.6	(14.9)	(14.8)	S	0.010	−5.95	−0.14	4.1
T Cir B	155526	9	+47	2.0	2.0	10.6*v*		VF	0.52	−8.4	+0.56	0.6
Q Cyg	213742	58	− 8	3.0	3.0	...	14.8*v*	VF	0.33	−8.3	−0.17	1.2
V450 Cyg	205435	47	− 7	7.8	7.8	...	16.3	S	0.030	(−5.5)	−0.30	2.5
V476 Cyg	195553	55	+12	2.0	2.0	...	16.15*c*	VF	0.29	−8.3	+0.17	0.8
DM Gem	063730	153	+13	5.0	5.0	...	16.5*c*	VF	0.33	−8.3	+0.47	2.0
DN Gem	064832	152	+16	3.5	3.5	15:	14.8*v*	F	0.12	−7.7	+0.31	1.0
DQ Her	180445	40	+26	1.4	1.4	14−15*v*	13.8*v*	S	0.030	(−5.5)	+0.13	0.3
CP Lac	221255	70	− 1	2.1	2.1	15.3	(14.85)	VF	0.40	−8.3	−0.02	0.8
DI Lac	223152	71	− 5	4.6	4.6	14.0	14.4*v*	F	0.100	−7.3	−0.12	1.4
DK Lac	224552	72	− 6	5.4	5.4	[13.4		F	0.14	−7.95	−0.21	2.1
HR Lyr	184929	27	+11	6.5	6.5:	16.0	15.3	S	0.064	−6.75	+0.40	2.0
GI Mon	072106	191	+ 6	5.6	5.6	15.1*c*	...	F	0.15	−8.1	+0.24	2.3
KT Mon	061905	173	− 2	10.2	10.2	[14.0		F	0.100	−7.3	−0.17	4.7
RS Oph	174406	348	+ 9	4.3	4.3	11.7		VF	0.30	−8.3	+0.26	1.7
V840 Oph	164829	321	+ 7	6.5	6.5	[17.0		F	0.100	−7.3	+0.30	2.4
V841 Oph	165312	336	+16	5.0	2:	...	12.6*v*	F	...	−7.8*	+0.28	1.0
V849 Oph	180911	7	+12	7.4	7.2	[15.0		S	0.017	−6.05	+0.42	1.9
FU Ori	053909	165	+ 9	9.7	9.7	16:		S	0.001	−5.6:	+0.52	3.3
GK Per	032443	119	− 9	0.2	0.12	13.5*v*	13.0*v*	VF	0.33	−8.3	−0.06	0.4
RR Pic	063462	239	−25	1.2	1.2	12.7*c*	...	S	0.025	−6.2	−0.08	0.2
CP Pup	080735	221	0	0.2	0.2	17.0	...	VF	0.40	(−11)	0.00	0.4
DY Pup	080926	214	+ 5	7.0	7.0	[16.0		S	0.017	−6.05	+0.17	2.0
T Pyx	090031	225	+10	6.6	6.6	13.6		S	0.032	−6.4	+0.32	1.9
WZ Sge	200317	25	− 9	7.0	7.0	16.1		F	0.100	−7.3	−0.42	2.7
AT Sgr	175726	332	− 4	11.0	8.7:	[16.5		VF	...	−8.35*	−0.41	5.8
BS Sgr	182027	332	− 8	9.2	9.2	[16.0		S	0.006	−5.9	−0.36	3.1
FL Sgr	175334	324	− 7	8.3		[15.0		F	0.17	−8.25	−0.50	4.1
FM Sgr	181123	336	− 5	8.6	8.6	16.5		F	0.091	−7.15	−0.31	3.6
KY Sgr	175526	332	− 3	10.6	7.2	[16.5		VF	...	−8.35*	−0.18	3.4
V363 Sgr	190530	335	−18	8.8	7.9	[16.0		F	...	−7.3*	−1.93	5.9
V522 Sgr	184125	336	−12	12.9	12.9	[15.3		VF	...	−8.3*	−8.30	39.0
V630 Sgr	180234	325	− 8	4.5	4.5	15:	...	VF	0.50	−8.35	−0.25	1.8
V726 Sgr	181326	333	− 7	10.8	10.8	[16.5		F	0.14	−7.95	−0.71	5.8
V732 Sgr	174937	330	− 3	6.5	6.5	12.7		S	0.027	−6.2	−0.09	1.8
V737 Sgr	180028	330	− 5	10.3	10.0	[12.5		S	...	−6.2*	−0.33	3.8
V787 Sgr	175330	328	− 4	9.4	9.0	[16.5		F	0.12	−7.7	−0.30	4.2
V909 Sgr	181935	326	−12	6.8	6.8	[16.0		VF	0.50	−8.35	−0.67	3.1
V927 Sgr	180133	326	− 7	8.0	7.3	[16.5		VF	...	−8.3*	−0.42	3.4
V999 Sgr	175327	327	− 3	8.0	8.0	...	16.5*v*	VS	0.009	−5.95	−0.13	2.4
V1012 Sgr	175931	327	− 7	8.0	8.0	[17.0		VF	...	−8.3*	−0.53	4.4
V1014 Sgr	180027	331	− 5	10.3	10.3	[16.1		S	...	(−5.5*)	−0.35	4.1
V1015 Sgr	180232	327	− 8	7.1	6.5	[12.0		F	0.12	−7.7	−0.36	2.6
V1016 Sgr	181325	335	− 8	8.5	6.9	14.9	14.9*c*	F	0.17	−8.25	−0.34	3.2
V1017 Sgr	182529	332	−11	7.2	7.2	14.3		S	0.034	−6.4	−0.12	2.3
V1059 Sgr	185613	350	−10	4.9	2.0:	...	16.5*c*	VF	0.20	−8.25	−0.14	0.8
T Sco	161122	321	+18	6.7	6.7	[11.0		VF	...	−8.3*	+2.16	6.6
U Sco	161617	326	+20	8.8	8.8	[17.6		VF	0.67	−8.4	+4.65	12.8
KP Sco	173735	322	− 5	9.4	9.4	[16.5		F	0.091	−7.15	−0.42	4.8
V382 Sco	174535	323	− 6	9.4	9.4	[16.5		VF	...	−8.3*	−0.51	4.9
V384 Sco	175435	324	− 8	12.3	9.2	[16.5		S	...	−6.6*	−0.50	3.5
V697 Sco	174437	321	− 7	10.2	10.0	[16.5		S	0.024	−6.2	−0.46	3.8
V707 Sco	174136	321	− 6	9.9	9.6	[15.0		MF	0.050	−6.6	−0.40	3.8
V711 Sco	174734	324	− 6	9.8	9.8	[15.5		S	0.017	−6.05	−0.44	4.2
V719 Sco	173933	323	− 4	9.8	9.8	[13.5		F	0.13	−7.8	−0.34	4.8
V720 Sco	174535	322	− 5	7.8	7.8	[16		F	0.118	−7.55	−0.29	3.3
V723 Sco	174335	322	− 5	9.8	9.0	[15		VF	0.29	−8.3	−0.53	4.6
EU Sct	185004	357	− 4	8.4	8.4	17		F	0.14	−7.95	−0.28	4.0
FS Sct	185205	357	− 6	10.1	10.1	[15		F	0.14	−7.95	−0.53	5.1
X Ser	161402	339	+30	8.9	8.9	14.5*v*	15.4	VS	0.005	−5.9	+3.40	5.9
RT Ser	173411	322	+29	9.0	9.0	[16.0		VS	0.002	−5.9	+3.44	6.2
XX Tau	051316	155	+10	6.0	6.0	[15.0		F	0.083	−7.05	−0.35	2.0
RR Tel	195656	309	−33	6.8	6.8	13.5*v*		VS	0.003	−5.85	−0.90	1.4
CN Vel	105853	255	+ 5	10.2	10.2	[16.5		S	0.013	−6.0	+0.33	3.8
CQ Vel	085552	239	+ 4	9.2	8.9	[15		MF	0.054	−6.65	+0.23	3.4

Data for these and the less well-observed galactic novae are given in the author's monograph [*8*].

Novae in nearby galaxies are summarized in Table 2.

Table 2. *Novae in nearby galaxies.*

System	Type	Novae observed	Novae per year (estimated)	Reference
Galaxy	*Sb*	150	50	a, b, c
Messier 31	*Sb*	30	26 ± 4	d
Messier 33	*Sc*	4	1 ?	c
Large Magellanic Cloud.	*I*–*SB*	6	2 ?	e
Small Magellanic Cloud.	*I*	4	1 ?	e

(a) C. PAYNE-GAPOSCHKIN: The Galactic Novae. Amsterdam: North Holland Publishing Company 1957.
(b) S. I. BAILEY: Publ. Astronom. Soc. Pacific **24**, 554 (1921).
(c) C. PAYNE-GAPOSCHKIN: Variable Stars and Galactic Structure. London: Athlone Press 1954. p. 61
(d) H. C. ARP: Astrophys. Journ. **61**, 15 (1956).
(e) W. BUSCOMBE, S. C. B. GASCOIGNE and G. DE VAUCOULEURS: Austral. J. Sci. **17**, No. 3 1955.

Limiting seasonal frequencies ($\frac{1}{2}$ in every case) are given by ARP [*1*] for the elliptical companions of Messier 31 (Messier 32, NGC 205, NGC 147 and NGC 185). An approximate estimate for the three spirals (see PAYNE-GAPOSCHKIN [*8*]) suggests that roughly one Population II star in 10^9 becomes a nova every year. Possibly the fraction is even smaller in the elliptical systems.

Table 3. *Novae with expanding nebular shells.*

Star	Absolute maximal magnitude		References
	Arp relation	Nebular	
V603 Aql	−8.35	−8.5	a
V476 Cyg	−8.3	−8.9	b
CP Lac	−8.3	−8.6	b, c
GK Per	−8.3	−8.4	d
RR Pic	−6.2	−7.3	e
DQ Her	−6.25	−5.5	f
T Aur	−6.15	−5.3	g

(a) H. F. WEAVER: Private communication 1955.
(b) W. BAADE: Publ. Astronom. Soc. Pacific **56**, 218 (1944).
(c) D. B. MCLAUGHLIN: Publ. Astronom. Soc. Pacific **57**, 69 (1945).
(d) W. BAADE and M. L. HUMASON: Publ. Astronom. Soc. Pacific **55**, 260 (1943).
(e) D. B. MCLAUGHLIN: Astrophys. Journ. **45**, 149 (1936).
(f) W. BAADE: Publ. Astronom. Soc. Pacific **52**, 386 (1940).
(g) W. BAADE: Publ. Astronom. Soc. Pacific **55**, 261 (1943).

2. Luminosities of novae. The survey of novae in Messier 31 by ARP [*1*] has established a relationship between apparent maximal luminosity and rate of decline. With an adopted modulus of 24.2 for the system, the observed absolute magnitudes range from −8.5 for the brightest to −6.1 for the faintest nova observed. The average VF, F, S novae have absolute magnitudes −8.3, −7.3 and −6.1 respectively.

The luminosities thus derived are in accord with those derived from the observed rate of tangential expansion of ejected nebular "shells" and associated radial velocities (Table 3).

The agreement is less good for the three last entries. The nebula for RR Pic is not strictly an expanding disc; DQ Her and T Aur have light curves unlike anything observed by ARP, and his relationship may not apply to them.

Many novae are too faint at minimum to be observed. Those with reliable ranges lead to the minimal absolute magnitudes of Table 4.

Table 4. *Minimal absolute magnitudes.*

Speed class	Mean range m	Mean absolute minimal magnitude m
VF	12.54	+4.14
F	12.25	+3.90
S (except DQ Her, T Aur)	10.25	+3.70
DQ Her, T Aur[1]	11.85	+6.45

Table 5. *Dimensions at maximum and minimum.*

Speed class	Log radius at maximum (solar units)	Log radius at minimum (solar units)
VF	2.01	−0.64
F	1.97	−0.59
S	1.89	−0.55
DQ Her, T Aur	1.75	−1.10

At maximum, most novae have spectra and colors that correspond to temperatures between 7000 and 15000°. The spectra at minimum have been shown, principally by HUMASON [*5*], to be those of very blue stars, whose temperatures may be between 20000 and 40000°. These data permit a rough estimate of the dimensions of the stellar photosphere. The star has about 100 times the solar radius at maximum, about $\frac{1}{3}$ solar radius at minimum.

3. Distribution of novae. The deduced absolute magnitudes, when combined with the observed apparent magnitudes, show that the galactic novae are concentrated in the direction of the galactic center, and that they populate a flattened spheroidal system. The distribution has been discussed by McLAUGHLIN [*7*], by KOPYLOV [*6*], and by PAYNE-GAPOSCHKIN [*8*]. It identifies the novae with an "intermediate" polupation, between the extremes of "Populations I and II". Two novae have been observed in typical Population II systems, the globular clusters Messier 80 and NGC 6553.

II. Physical behavior.

The physical changes that are observed at a nova outburst are exceedingly complicated. Detailed summaries of the information for individual stars have been collected by the writer [*8*]. For the purposes of a short survey it is necessary to treat the processes schematically.

4. Changes of brightness. The typical nova brightens suddenly, probably within a few hours, by between 12 and 13 magnitudes (Table 4). Light curves of a very fast, a fast and a slow nova are compared in Fig. 1. Fig. 2 shows a schematic light curve that represents the course of most novae, adapted from McLAUGHLIN ([*7*], 1936). The curve passes from the pre-nova stage through the rapid initial rise, the pre-maximum halt, the slower final rise, the early decline, a "transition" stage marked by one or more fluctuations of brightness, and a smooth final decline, to the post-nova stage which is similar to the pre-nova stage. The whole process occupies a few years for a fast nova, and may last a century or more for a very slow one.

5. Changes of spectrum. The spectrum undergoes remarkable changes during the outburst. Only one pre-outburst spectrum has been recorded for a normal nova (V 603 Aquilae). However, as the post-nova brightness reverts, on the average, to the pre-nova value, we can assume that the very blue star that is typical of post-novae is characteristic before the outburst. Pre-maximum spectrum has rarely been recorded more than two magnitudes below the peak of

[1] "Nebular" parallax adopted.

brightness. When observed, they are predominantly *absorption* spectra, often with bright lines, and give evidence in their excitation and ionization of fairly high temperatures, from 10000 to 20000° K. In general, the temperatures are highest for the fastest novae.

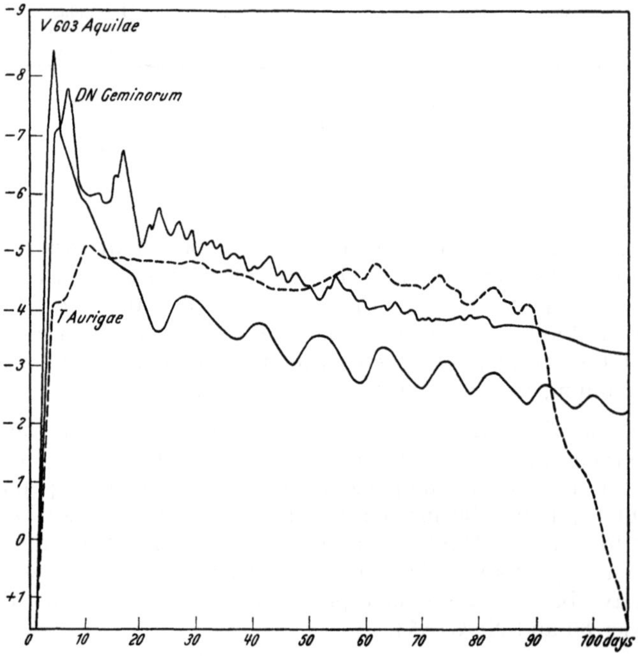

Fig. 1. Light curves of a very fast, a fast, and a slow nova. Ordinates are absolute visual magnitudes; abscissae are marked at intervals of ten days.

At maximum light the spectrum changes abruptly. The absorption lines, previously single, become double or complex, and the bright lines, which have tended to lose intensity at maximum light, become more conspicous. The ab-

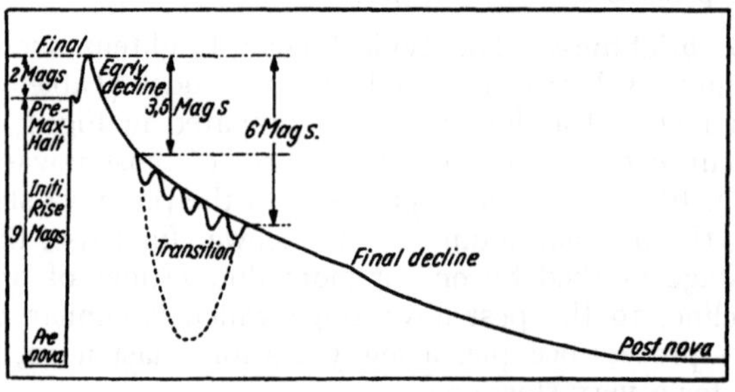

Fig. 2. Schematic light curve of a nova, showing typical stages. The time scale is not uniform throughout; it has been magnified in the early stages. After MCLAUGHLIN.

sorption spectrum is characteristic of a lower temperature than before maximum. The complexity of the absorption lines increases during the decline, and the bright lines increase in intensity relative to the continuum, though not usually in absolute value. When there are fluctuations of brightness, the temperature, as indicated by the excitation and ionization of absorption and emission lines, is highest at the minima of brightness, lowest at the maxima.

The bright lines first seen are permitted radiations of atoms already present in the absorption spectra. As the decline progresses, forbidden lines appear (roughly in order of increasing ionization). The spectrum gives evidence of rising ionization during the decline and transition. By the time the final decline begins, the absorption spectra have disappeared, the permitted bright lines weaken progressively and finally the forbidden bright lines of the atoms of highest ionization are the only observable features. When the star has declined to the post-nova stage even the forbidden lines are gone, and the spectrum may be featureless or may show very weak bright or dark lines of hydrogen and helium.

6. Radial velocities. The spectral changes are clarified by a study of the radial velocities of absorption lines and of the edges of the bright lines.

The pre-maximum absorption spectrum is displaced to the violet (greatest displacements being associated with greatest speeds of development). The bright lines present at this stage lie at the redward edges of the corresponding absorption lines, and have widths that indicate similar velocities.

At maximum light the second system of absorption lines is always further to the violet than the pre-maximum system, and the associated emissions (at their redward edges) are proportionately broader. The increasing complexity of the absorption spectra is a consequence of the appearance of yet more absorption systems, almost always displaced further to the violet than the systems that appeared previously, and usually accompanied by the corresponding bright lines.

The pre-maximum spectrum lasts only a day or two after maximum. The second, or *principal*, spectrum, which appears at maximum, usually survives almost as long as absorptions are visible, though its intensity decreases steadily. The so-called *diffuse enhanced* spectrum, which appears during the decline, also lasts for some time. The violetward shifts, and therefore the outward velocities, associated with these three spectral systems tend to be constant, or, if variable to increase uniformly.

The absorption spectrum of highest excitation, known as the *Orion* spectrum, may appear soon after, or together with, the diffuse enhanced spectrum. It tends to show large changes of radial velocity, which are associated with the fluctuations of the light curve and the changes of temperature indicated by the spectrum. Some absorption lines of the Orion spectrum, notably those of N III and N V, tend to persist after all other absorptions have disappeared. When they finally fade away, the spectrum enters the bright-line *nebular* stage, during which the permitted bright lines weaken progressively in comparison to the forbidden lines.

7. Geometrical interpretation. The light received from the nova clearly consists of a number of superimposed spectra. During the initial rise we observe an envelope which may be in spherically symmetrical expansion with constant velocity, but is more probably spheroidal. The level of the photosphere will be defined by the velocity and the local density, as suggested by GROTRIAN [3]. The observed radial velocity is not, of course, the velocity of expansion of the effective photosphere, which must be smaller; atoms are continually passing outward through the photosphere.

The principal spectrum represents a sudden increase of the velocity with which atoms are passing through the photosphere. The records are not complete enough to show how abrupt the change of velocity is. The short-lived pre-maximum absorption consists of atoms that outran the photosphere during the rise, and is quickly overtaken in its turn by the principal absorption of higher velocity. The effective photosphere at maximum is formed in an envelope expanding with the velocity of the principal spectrum. We do not know whether this envelope is spherically symmetrical; perhaps it is so for some novae and not for others. Neither do we know what causes the initial ejection and determines its velocity, or what causes the velocity to increase and to lead to the formation of the principal absorption.

If GROTRIAN'S idea is correct, an increase of the velocity of ejection will lead to a decrease in the dimensions of the effective photosphere, and will thus be responsible for the maximum of the light curve. If the velocity of ejection is not spherically symmetrical, neither will the resultant photosphere be so. Many novae in the later stages show definite *axial* symmetry; if the velocity distribution has axial symmetry, the maximal photosphere may, for example, be spheroidal. It is tempting to think that some of the differences between novae may stem from presentation effects. A slow nova at maximum might, for instance, have a flattened spheroidal photosphere viewed edgewise, and thus appear fainter than a fast nova, which presented the larger endwise area to observation.

The quality of the lines of the principal spectrum suggests "turbulence", and a "turbulent velocity" comparable to that observed for yellow supergiants like γ Cygni has been deduced by the writer for the principal spectrum of DQ Herculis.

The principal spectrum may stem from a single expanding shell, but the source of the later complex spectrum is clearly multiple. The lines of the diffuse enhanced and Orion spectra have considerable equivalent widths—greater than those of lines in the principal spectrum. Even if (as is very likely) these spectra are more affected by "turbulence" than earlier spectra, they must involve a comparable number of atoms. If the sources of the diffuse enhanced and Orion spectra are spherically (or spheroidally) symmetrical, each should produce its own continuum, and the simultaneous appearance of several absorption systems is difficult to visualize, especially as in some novae the lines of many systems are themselves multiple.

These facts, and the rate of fading of the principal spectrum, suggest that the later spectra, associated with higher velocities, represent *localized directional ejections* of material from below, which penetrate the photosphere formed by the principal spectrum and shred it gradually to pieces. The spectrum of the expanding nebula about V 603 Aquilae has been analyzed by Weaver into a series of axially symmetrical, conical emitting volumes, with their apices directed toward the nova. The ejections would in this case have been directed radially outward from a number of axially symmetrical rings on the star's surface. There is evidence also for two axially directed bursts with higher velocities, nearly in the line of sight. Other novae, notably CP Puppis, clearly show a similar structure, and perhaps DQ Herculis represents a comparable formation viewed from a direction differing by 90°.

The observed spectrum in decline and transition can thus be analyzed into several components. The principal absorption is associated with the main part of the continuum, and is seen as both absorption and emission spectra. The diffuse enhanced and Orion spectra make small (but increasing) contributions to the continuum as they gradually supplant the "principal" continuum, often in axially symmetrical zones. Their absorption lines are seen against the resulting composite continuum. Their bright lines overlie this composite continuum, and furnish a pseudo-continuum against which the absorption of less-displaced systems is observed.

The spectrophotometric problems presented by such a complex of lines and continua are wellnigh insuperable. A beginning may be made by determining the total magnitude of the continuum (freed from the effect of bright lines). Before, at, and directly after maximum, the course of the continuum can be traced with some confidence, and significant equivalent widths can be measured. During the decline and transition the complex continuum becomes increasingly difficult to trace, and equivalent widths of absorption lines grow indeterminate. The bright lines stem additively from all the velocity systems, and their redward halves (undisturbed by the absorptions) can be analyzed in favorable cases.

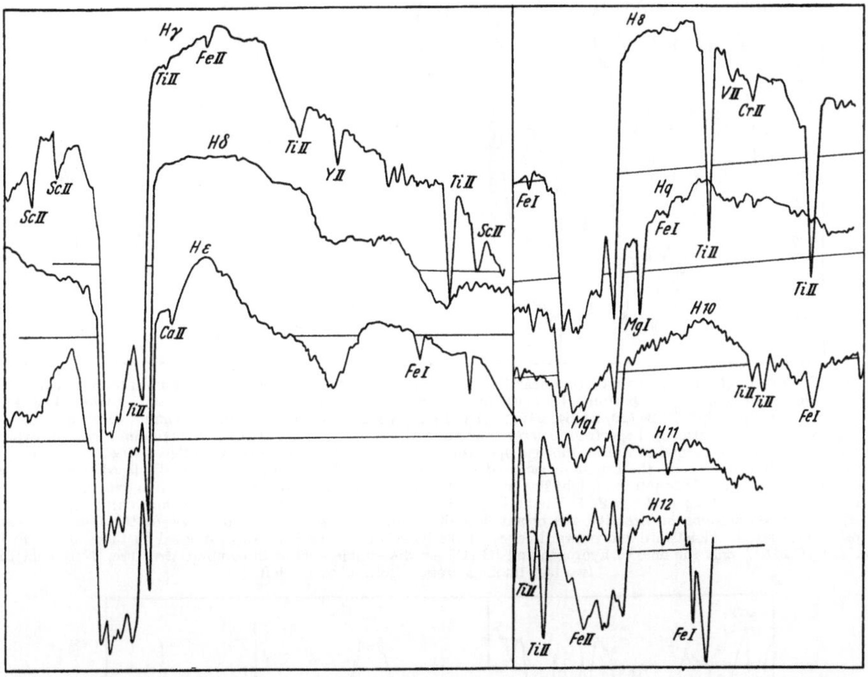

Fig. 3. DQ Herculis, profiles of hydrogen lines from a Mount Wilson Coudé spectrum of JD 27873, 79 days after maximum, not reduced to intensities. The left section covers about 90 Å, the right section, about 45 Å; the dispersion is prismatic. Each Balmer line shows a strong narrow absorption (the principal spectrum), complex strong absorptions to the violet (mainly Orion spectrum), and broad emission to the red. The spectrum is cut up by the strong sharp metallic lines of the principal spectrum, many of which are marked. Note the strength of the lines of Ti II and the common structure of the Balmer lines. Violet is to the left. The approximate course of the continuum is marked by light lines.

The spectral complexities are illustrated in Figs. 3, 4 and 5, which show profiles of hydrogen lines of DQ Herculis and RR Pictoris.

The last absorptions to disappear are those of N III and N V. This is largely a spectroscopic accident; the first are seen against the pseudocontinuum provided by the broad emission lines at H δ, the second, against the similar emission of He II at 4686.

A striking feature of the decline and transition, particularly in novae that show fluctuations in this interval, is so-called "nitrogen flaring". The lines of N III near λ 4640 grow intermittently broad and intense; for DN Geminorum they attained an overall width of 100 Å, far greater than twice the displacement of any observed absorption. The "Bowen mechanism", which gives a good account of anomalous intensities in spectra of planetary nebulae, has been invoked to interpret nitrogen flaring. In DQ Herculis, at least, the mechanism is not a major effect, for the O III multiplets, which are mutilated in planetary nebulae, were shown by Wyse [*11*] to be completely represented. It seems likely that

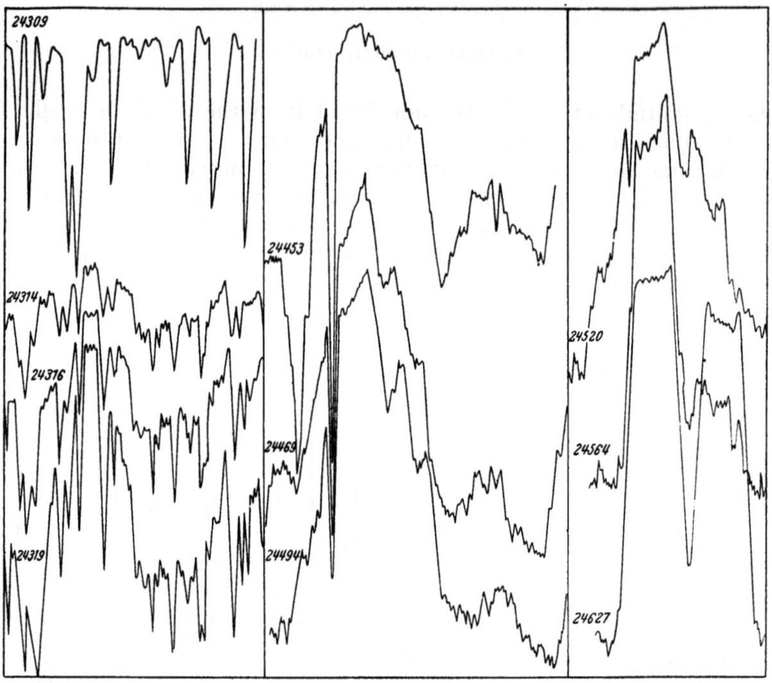

Fig. 4. RR Pictoris, changes in the spectrum near Hγ 4340 over about 320 days; tracings from Lick spectra, not reduced to intensities. Left strip, top to bottom: JD 24309 (date of maximum), 24314, 24316, 24319; middlestrip: JD 24453, 24469, 24474; right strip: JD 24520, 24564, 24627. The first tracing shows only the pre-maximum spectrum. In the subsequent tracings, the principal spectrum emerges and strengthens to the violet of the pre-maximum spectrum, which gradually fades. In the second strip the principal spectrum, and the intense, more highly-displaced spectra of hydrogen dominate the absorptions, and the bright redward edge, associated principally with the highly-displaced spectrum, becomes conspicuous. By JD 24469 the bright line has developed a distinctive structure, with a strong redward edge; the violetward and redward edges of the Fe II line at λ 4351 have also become prominent. In the third strip, the hydrogen absorptions are diminishing in intensity, the bright lines displaying more structure. On JD 24564 the absorptions are almost gone, and the violetward and redward edges of the [O III] line λ 4363 are superimposed on those of the Fe II line. On JD 24627, only the lines of hydrogen and [O III] are discernible, each with complex structure. Note that the two last tracings cross. Violet is to the left.

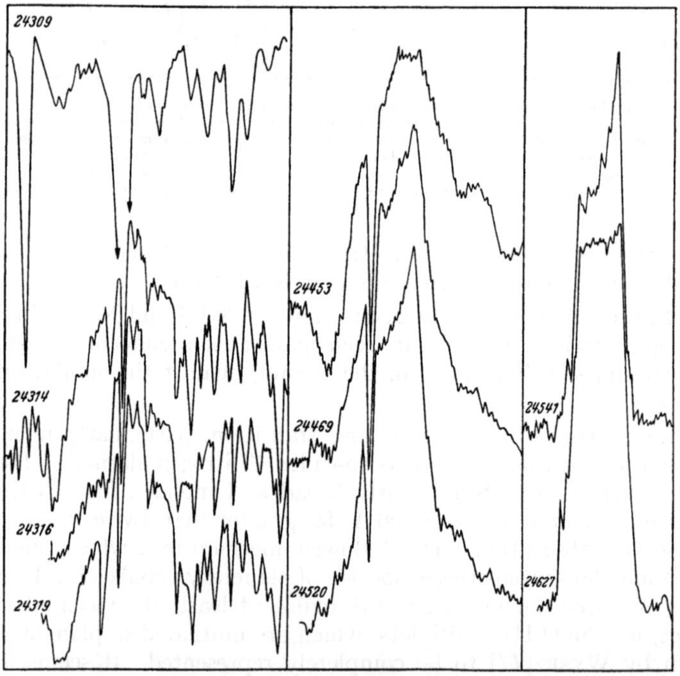

Fig. 5. RR Pictoris, changes in the spectrum near H δ 4101 over about 320 days; tracings from Lick spectra, not reduced to intensities. Left strip, top to bottom: JD 24309 (date of maximum; for clarity the deep center of the hydrogen line is omitted), 24314, 24316, 24319; center strip: JD 24453, 24469, 24520; right strip: JD 24541, 24627. The development is similar to that shown in Fig. 4, but the later spectra are less complicated by emissions other than that of hydrogen. Note that all the dates on the two figures are not identical. Violet is to the left.

the directional ejections that were suggested previously to account for the spectral peculiarities may also stimulate "nitrogen flaring".

As the spectrum enters the "nebular stage" and the absorptions die away, the bright lines can be more readily analyzed, though the location of the continuum is still a problem. The forbidden lines are the best subjects for quantitative study. Centrally-depressed, "saddle-shaped" profiles (often multiple) become conspicuous at this stage in the spectra of most novae. They seem to be best interpreted in terms of a ring or rings of emitting material. Such profiles often have their inception at earlier stages. Examples are shown in Figs. 4 and 5.

The interpretation sketched above is primarily *geometrical*. The *physical* problems presented by the nova spectrum are far from solution. This is partly a result of lack of quantitative data, such as spectrophotometry of the continuum, measures of equivalent widths of absorption lines, profiles of bright lines, Balmer decrements, Balmer discontinuities, etc. Also it results from our ignorance as to the excitation processes. Most treatments of the subject have assumed that the excitation is radiative. It now seems probable that in later stages the effects of collisions may preponderate, not only for excitation and ionization, but also by producing the accelerations of which the highly-displaced components of the spectrum give evidence. If this is so, localized zones on the central star must be the source of high-energy particles; we know nothing about the cause and circumstances of their ejection except that it must be a sequel to the original outburst.

Numerical data, averages from all available observations, are summarized in Table 6. Spectrum, radial velocity, interval dt from maximum light in days, and drop dm from maximum light in magnitudes, are given separately for the various absorption systems, and for novae of three speed classes. Although the novae develop at very different rates, the magnitude drops at which the features appear and disappear are rather similar for all. In terms of absolute magnitudes, the diffuse enhanced spectrum disappears at about −4.0, the Orion spectrum

Table 6. *Summary of velocity systems.*

Spectrum	Very fast novae				Fast novae				Slow novae			
	Spectrum	V km/sec	dt days	dm mag.	Spectrum	V km/sec	dt days	dm mag.	Spectrum	V km/sec	dt days	dm mag.
Pre-maximum												
First appearance	B 9.3	− 600			A 2.5	− 425			F2	− 72		
Maximum light	A 2.3	− 880	0	0.0	A 6	− 440	0	0.0	F4	− 120	0	0.0
Post-maximum	A 4	− 1020			F 0	− 460			F 6.5	− 230		
Principal												
First appearance	A 4.2	− 1420	0	0.0	A 7.5	− 640	0	0.0	A 8.5	− 570	0	0.0
Last appearance		− 1870	32	4.7		− 1120	40	4.6		− 660	184	3.1
Diffuse enhanced												
First appearance		− 2230	1	0.4		− 1620	4	0.7		− 710	16	2.3
Last appearance			16	3.8			21	3.0			89	2.6
Orion												
First appearance		− 2700	3	1.3		− 1900	14	2.6		− 1090	70	1.9
Last appearance		− 3280	27	4.2		− 2140	41	3.8		− 1270	132	2.4
N III												
First appearance		− 3100	4	1.7		− 1840	22	2.5		− 1450		3.4
Last appearance		− 3660	38	4.5		− 2150	69	3.3		...		...
N V												
First appearance		− 4010	21	4.2		− 2510	30	3.4		...		...
			63	4.8			42	4.3		...		...

at -3.8, and the principal spectrum at -3.1. The N III lines disappear at approximately the same time as the Orion spectrum (of which they are actually part), and the N V lines about at the same time as the principal spectrum.

III. Physical parameters.

8. Maximal and minimal luminosities of novae were summarized in Part I, and Table 5 shows their probable dimensions. The prenovae lie several magnitudes below the Main Sequence, and must be classed as Intermediates rather than as White Dwarfs.

There has been a tendency to assume that a typical nova has about the same mass as the Sun, in conformity with the mass-luminosity relation. The one direct determination is that made by WALKER [*10*] for the eclipsing star DQ Herculis. If it is legitimate to treat the light curve of this star by conventional methods, the deduced mass of the blue component is 0.006 suns, and its density 6.6 times the solar density—large for an Intermediate. Possibly the results are affected by the nebulous envelope, but the mass seems to be much lower than those of novae have been thought to be. We cannot at present make confident assumptions about the internal constitution of such a star.

The composition of novae is an important and difficult problem. The absorption spectra of many, though not all, show anomalies when compared with those of other stars of similar dimensions, such as the yellow supergiants. Absorption lines of the atoms of carbon, nitrogen and oxygen are sometimes abnormally prominent. The bright-line spectra show particularly striking differences in the intensities of the forbidden lines. The spectra of GK Persei and a few other novae displayed extremely strong [Ne III] and [Ne V]; in others, [O I] and [O III] are more prominent. The slow novae tend to display a series of iron spectra, [Fe II], [Fe III], [Fe IV]? [Fe V], [Fe VI], [Fe VII] occurring in succession as the brightness declines. The recurrent novae (see below) are unique in developing strong coronal lines of [Fe X] and [Fe XIII], but some fast novae show these lines weakly.

It would be premature to deduce differences of composition from the intensities of these forbidden lines, when the details of the excitation process are not fully understood. The absorption spectra are more promising, though here again we cannot rule out peculiar excitation mechanisms. The author has made a quantitative study of DQ Herculis, and reaches the conclusion that (if conventional curve-of-growth procedures are valid) the lighter elements are present in high abundance. Relative to the solar abundances, the numbers of atoms that contribute to the absorption spectra of DQ Herculis appear to fall off rather steadily with increasing atomic number. This result is no more than suggestive until analyses of other novae have been made. The spectrum of RR Pictoris does not show such high relative abundances of carbon, nitrogen and oxygen as that of DQ Herculis, but the latter star does indeed appear to be somewhat metal-poor.

IV. Relation of novae to other stars.

9. The explosive stars range from the supernovae to the U Geminorum stars. Table 7 contains a comparative summary of some of the physical parameters.

References to the sources of these data will be found in the author's monograph [*8*]. The suggestion that ordinary novae may be recurrent at very long intervals must be amplified by discussion of the recurrent novae.

Six recurrent novae are known. The average cycle is about 35 years, the average range about 8 magnitudes-sensibly less than for the novae not known to be recurrent.

Table 7. *Properties of explosive stars.*

Property	Supernovae Type I	Supernovae Type II	Novae	Recurrent novae	U Geminorum stars
Absolute photographic magnitude .	-16.1	-13.9	-7.8	(-7.8)	$+5.5$
Energy emitted in outburst (ergs)	10^{49}	10^{47}	6×10^{44}	10^{44}	6×10^{38}
Ejected mass (suns)	0.1	?	10^{-3}	5×10^{-6}	10^{-9}
Number per year in our galaxy . . .	0.005	0.025 ?	50	—	—
Interval of recurrence (years) . .	—	—	10^6 ?	20	0.2
Pre-nova: Mass (suns)	20 ?	—	0.01 ?	?	?
Absolute photographic magnitude .	?	?	$+3.5$	0	$+9.5$
Radius	?	?	0.2	6	0.03
Spectrum	?	?	*O* to *B*	*A* to *F*	*B* to *A*

The red giant spectrum observed for T Coronae Borealis at minimum has a regularly variable radial velocity (period 230.5 days) and the system is almost certainly a binary. The minimal spectra of RS Ophuichi and V 1017 Sagittarii suggest that these also are members of binary systems. If the maximal absolute magnitudes of recurrent novae can legitimately be deduced from the Arp relation-ship (an unverified assumption), their mean observed minimal absolute magnitude is $+0.54$, much brighter than for ordinary novae. It is likely that this is the absolute magnitude of a giant or subgiant companion, and that the nova itself is much fainter between outbursts. We note that even DQ Herculis, a nova not known to be recurrent, is an eclipsing binary.

Table 8. *Recurrent novae.*

Star	Range m	Cycle years	Minimal spectrum
T Coronae Borealis	10:	80	*gM*
RS Ophiuchi . .	7.5	35	*G* ?
T Pyxidis . . .	7.2	18	?
WZ Sagittae . .	9.1	33	*Be* ? very blue
V 1017 Sagittarii	7.1	33	*G* 5 ?
U Scorpii	over 8.5	35	not observed

The U Geminorum stars, sometimes known as dwarf novae or "novulae", are superficially similar to the recurrent novae. Their luminosities are much lower, their ranges smaller, and their cycles shorter (see Table 7). All of the eleven observed minimal spectra of U Geminorum stars show the characteristics of *G* dwarfs or subdwarfs. Two of these stars (SS Cygni and AE Aquarii) are spectroscopic binaries of short period, and RU Pegasi, which has variable radial velocity, is probably another. It seems very likely that all U Geminorum stars are close binaries, and their behavior may be plausibly attributed to their duplicity.

The spectroscopic behavior of the U Geminorum stars is only remotely like that of the novae. Greatly broadened, undisplaced, symmetrical absorption lines are observed at maximum, and are gradually replaced by undisplaced central emission lines as the brightness declines. Forbidden lines have never been observed in the spectra of U Geminorum stars. These features are incompletely understood; they may result from rapid rotation and involve violent mass motions.

The periods of the U Geminorum stars known to be binary are less than a day, and the irregular outburst cycles of about a month are therefore not directly related to them, but the tendency to outbursts may well be associated with the duplicity of these stars. The ratio between their binary periods and their cycles is the same order as for T Coronae Borealis, between 100 and 200.

The cycles and ranges for U Geminorum stars were shown by Kukarkin and Parenago[1] to be correlated. They deduced the relationship:

$$A = 0.63 + 1.667 \operatorname{Log} P$$

where A is the mean amplitude, P the mean cycle. They extended the relation to include the recurrent novae.

If the U Geminorum stars and recurrent novae are binaries, the true ranges of the nova-like components are larger than the observed ranges. After correcting for this effect, the writer has obtained the revised relationship:

$$A = 2.00 + 1.78 \operatorname{Log} P,$$

which is found to represent the recurrent novae as well as the U Geminorum stars.

This empirical information does not necessarily imply that the processes are identical; indeed, the spectroscopic differences suggest that they are not. The whole system must differ in scale; the companions of the U Geminorum stars seem to be subdwarfs, those of the recurrent novae, giants or subgiants. However, it seems probable that the presence of a companion stimulates the outbursts; the symbiotic variable stars such as Z Andromedae show analogous tendencies.

A very speculative extrapolation of the range-cycle relationship to the novae would predict outbursts in cycles of the order of a million years. If recurrent outbursts are stimulated by duplicity, such an extrapolation would imply that all novae are double stars. While not impossible, this has not been demonstrated; DQ Herculis is certainly binary, and a test of the binary character of other novae by spectroscopic and photometric means might lead to important results.

A comparison of the pre-novae with the general pattern of nonvariable stars places them definitely between the Main Sequence and the white dwarfs. They can thus be regarded as stars of rather advanced development, but current theoretical studies of stellar evolution have not been extended in detail to this stage of the stellar lifetime. Galactic distribution and the occurrence of two novae in globular clusters confirm the impression that is conveyed by the position of the pre-novae in the Hertzsprung-Russell array.

V. Theories of the nova outburst.

10. The constitution of the pre-nova must form a starting-point for theories of the cause of the outburst. As we have seen, they are blue Intermediates probably of low mass (incompatible with the mass-luminosity relation) and moderately low density. If their composition differs from the average, it is if anything, relatively richer in light elements.

The first modern theory of the nova process, developed by Biermann 1939 [*2*], required pre-novae to be hydrogen-poor subdwarfs of about solar mass, with contraction their main source of energy, and concluded that 10% of the star's mass would be ejected. The stellar mass, the ejected mass, and the composition are at variance with the facts as we see them today.

---

[1] Kukarkin and Parenago. Verein. Freunde d. Phys. u. Astron. in Gorki (Nishni-Novgorod), Veränd. Sterne, Forsch.- u. Inform.-Bull., N. N. V. S. (1934).

The theory developed by HOYLE [4] relied on the collapse of a hydrogen-exhausted star as a result of rotational instability. While it may be applicable to some supernovae, this theory is not compatible with the observed return of the nova to the pre-outburst brightness, and its probable reversion to the pre-outburst condition.

The long series of papers by SCHATZMAN (1946 to 1954) [9] gives what seems to be the best account, up to the present, of the nova process. He examines the structure, composition and energy sources of stars in the Intermediate area, and discusses the effects of a shock wave propagated within the star. The process of explosion is essentially that of a hydrogen bomb with He^3 acting as detonator.

Schatzman links the novae to the periodic variables by the idea that stars of both types oscillate because they are vibrationally unstable; conditions limit the amplitude of the pulsations in the latter case, but not in the former. We note that the light of DQ Herculis has been found by WALKER [10] to display a regular oscillation with a period of 1.180 minutes, presumably the free period of the post-nova.

SCHATZMAN's theory leads to an evaluation of the relationship between range and cycle for recurrent outbursts. The range-cycle relationship was shown by ZUCKERMANN [12] to be compatible with this theory. The revision of the range-cycle relationship, described above, alters the numerical results, but leaves the conclusions unaltered.

A plausible interpretation of the nova outburst seems, therefore, to have been reached. However, many problems are still unsolved. The propagation of shock waves in the interior depends on the still unknown constitution of the pre-nova. We have no clear idea of what determines the velocities of ejection from the star, the distribution of ejection over its surface, or the long-continued activity, often intermittent, at intervals that are great in comparison to the star's free period.

Bibliography.

[1] ARP, H. C.: Astrophys. Journ. **61**, 15 (1956).
[2] BIERMANN, L.: Z. Astrophys. **18**, 344 (1939).
[3] GROTRIAN, W.: Z. Astrophys. **13**, 238 (1937).
[4] HOYLE, F.: Monthly Notices, Roy. Astronom. Soc. London **106**, 343 (1946); **107**, 231 (1947).
[5] HUMASON, M. L.: Astrophys. Journ. **88**, 238 (1938).
[6] KOPYLOV, I. M.: Comparison of the Large Scale Structure of our Galaxy with Those of Other Galaxies. Int. Astr. Union, Dublin **41** (1955).
[7] MCLAUGHLIN, D. B.: Astrophys. Journ. **45**, 149 (1936); **51**, 136 (1945).
[8] PAYNE-GAPOSCHKIN, C. H.: The Galactic Novae. Amsterdam: North Holland Publishing Company 1957.
[9] SCHATZMAN, E.: Ann. d'Astrophys. **9**, 199 (1946); **10**, 14, 101 (1947); **12**, 161, 281 (1949); **13**, 384 (1950); **14**, 278, 294, 305 (1951); **17**, 152, 377, 398 (1954). — J. de Phys. **9**, No. 2 (1948). — Observatory **68**, 66 (1948). — Bull. Acad. Roy. Belg., Cl. Sci. **34**, 828 (1948). — C. R. Acad. Sci., Paris **232**, 1740 (1951). — Problems of Cosmical Aerodynamics, p. 96. Paris 1951. — Mém. Soc. Roy. Sci. Liège **14**, 168, 235 (1953); **15**, 163 (1954). — Contr. Inst. d'Astrophys. **1953**, No. 181.
[10] WALKER, M. F.: Astrophys. Journ. **123**, 68 (1956).
[11] WYSE, A. B.: Publ. Astronom. Soc. Pacific **47**, 204 (1936).
[12] ZUCKERMANN, M. C.: Ann. d'Astrophys. **17**, 243 (1954).

Supernovae.

By

F. Zwicky.

With 9 Figures.

I. The history of supernovae.

1. Introduction. For convenience of this review the history of supernovae may be divided into the following periods:

(i) Early history until 1885. This latter date marks a significant turning point since in this year the first supernova was discovered in an extragalactic nebula (Sect. 2).

(ii) From 1885 until about 1920 several supernovae were found in extragalactic systems. Their absolute brightness as well as other characteristics were not known to the astronomers of that period since the nature and the distances of the extragalactic nebulae remained as yet to be determined. Significant progress had, however, already been made toward the solution of these problems, mainly because of the work of K. Lundmark and H. D. Curtis (Sect. 3).

(iii) In the period from 1920 until 1933 the extragalactic nebulae were clearly identified as stellar systems and distances were derived for them which are correct in order of magnitude. From this knowledge it also followed that supernovae are on the average thousands of times brighter than common novae. An independent investigation of Tycho's nova of 1572 showed that this star must have been a supernova in our own galaxy rather than a common nova (Sect. 4α).

(iv) The systematic search for supernovae during the period from 1933 to 1942 was decided upon as a consequence of the knowledge gained in the previous period. This search netted about twenty supernovae and made possible the first detailed analysis of their physical characteristics (Sect. 4β).

(v) From 1942 until 1950 a few supernovae were casually discovered, but no systematic investigations were conducted (Sect. 4γ).

(vi) In the years since 1950 not only a new search was initiated which we hope to get into full swing by 1957, but several novel types of investigations were started which particularly concern themselves with the properties of the remnants of supernovae (Sect. 4δ).

For purposes of identification the following differences between common novae and supernovae may be noted. While the most luminous ordinary novae at maximum become perhaps one million times as bright as the Sun, the visual luminosity $L_{\max}$ of supernovae lies in the range $10^7 L_\odot$ to $10^{10} L_\odot$. The range of absolute magnitudes $\Delta M = M_{\min} - M_{\max}$ from maximum brightness to the final stage is smaller than 17.0 for common novae and probably larger for supernovae. This criterion can, however, be used only in the case of relatively nearby galactic objects. The spectra of some supernovae (type I) are clearly distinguishable from those of common novae. On the other hand, supernovae of the type II seem to have spectra similar to those of ordinary novae, except that the velocities of expansion of their gaseous shells are greater and lie in the range from about 5000 to 10000 km/sec. There are also certain features in the light curves of supernovae which may make them distinguishable from common novae. Whether or not there occur exploding stars with characteristics intermediate between those of common novae and of supernovae remains as yet to be determined.

2. Supernovae before 1885. The names of several historians and astronomers are prominently related to the investigation of the historical records on the appearance of new objects in the sky. Among these we mention BIOT[1] who translated many of the ancient Chinese astronomical books. More recently DUYVENDAK[2] occupied himself with the same task. Outstanding work of interpretation was done in particular by VON HUMBOLDT[3], ZINNER[4] and LUNDMARK[5]. The success of these men will no doubt stimulate the search for more historical records and the correlation of data on past supernovae with their still observable remnants. At the present time we have definite knowledge only on three supernovae in our own galaxy which we discuss in the following.

α) The Supernova of 1054 A. D. in Taurus. This star which appeared in the constellation of Taurus on July 4, 1054 A.D. was observed by both Chinese[2] and Japanese astronomers. At maximum it was visible in daytime and was probably somewhat brighter than Venus. At night it could be followed until April 17, 1056.

LUNDMARK[5] suggested in 1921 that the position of the Nova as given by the Chinese is near the present day Crab Nebula (Messier 1 or NGC 1952). The spectrum of the Crab Nebula had already previously been photographed by SLIPHER[6] who originally thought that the multiplicity in the emission lines was due to some sort of Stark effect. Shortly afterward LAMPLAND[7] observed changes in the appearance of the nebula and DUNCAN[8] showed that it is expanding at such a rate that, according to HUBBLE[9] "it must have required 900 years to reach its present dimensions". This circumstance greatly strengthened LUNDMARK's suggestion that the Crab Nebula is somehow related to the Nova of 1054. In 1937 MAYALL[10] carefully analyzed the spectrum of the nebula and determined its distance as 1250 parsecs by relating its radial expansion of 1300 km/sec, as measured by the Doppler effect of the emission lines, with its angular expansion of 0''.21 per year. He showed that the true apparent magnitude of the star of 1054, as corrected for galactic absorption was probably $m_{\max} = -6$. This with a distance modulus $m - M = +10.5$ gives an absolute magnitude $M_{\max} = -16.5$, corresponding to a supernova. For a most stimulating account of all of these developments the article by LUNDMARK[11] entitled "Was the Crab Nebula formed by a Supernova in 1054 A.D.?" should be consulted. In the present review we shall return to the discussion of some of the properties of this star and of its present day remnant later on in this article.

---

1 E. C. BIOT: Connaissance des Temps 1846, Additions, p. 67; and C. R. Acad. Sci., Paris **1843**.
2 J. J. L. DUYVENDAK: Publ. Astronom. Soc. Pacific **54**, 91–94 (1942).
3 A. v. HUMBOLDT: Kosmos, Part III. Stuttgart und Tübingen 1845–1858.
4 E. ZINNER: Geschichte und Literatur der veränderlichen Sterne, Part 2. 1920.
5 K. LUNDMARK: Publ. Astronom. Soc. Pacific **33**, 225–238 (1921).
6 V. M. SLIPHER: Publ. Astronom. Soc. Pacific **28**, 191 (1916).
7 C. O. LAMPLAND: Publ. Astronom. Soc. Pacific **33**, 79 (1921).
8 J. C. DUNCAN: Proc. Nat. Acad. Sci. U.S.A. **7**, 179 (1921).
9 E. P. HUBBLE: Astronom. Soc. Pacific Leaflets **1**, 58 (1934).
10 N. U. MAYALL: Publ. Astronom. Soc. Pacific **49**, 101–105 (1937).
11 K. LUNDMARK: Festskrift tillagnad Osten Bergstrand, p. 89. 1938.

β) Tycho's Star of 1572 (B Cassiopeiae). This new star, which was discovered in November 1572 by Tycho Brahe and other observers, represents a landmark in the history of astronomy inasmuch as it made Tycho and his successors realize that the stars are not eternal and unchangeable. His own words[1] in his book "On a New Star" are:

"I was moved by so great astonishment at this sight, that I was not ashamed to doubt the evidence which had been furnished by my own eyes. When, in truth, I noticed that the star could be seen by others in the place designated to them, and that the star really appeared there, no further doubts disturbed me. A miracle surely, either the greatest of all which have occurred in the course of nature, since the beginning of the world or one, certainly, which must be likened to those to which the Sacred Oracles testify in recording the holding back of the sun in its course, in answer to the progress of Joshua, and in the darkening of the sun which occurred at the time of the Crucifixion of Christ. Indeed, all philosophers agree, and the fact itself proves clearly, that in the ethereal region of the celestial world no change of generation or of corruption occurs; but that the heavens, and those celestial bodies which are contained in the heavens, do not increase or decrease; nor do they vary either in number or in size or in light or in any other way whatsoever; but remain always the same, like unto themselves in every way, no length of time wasting them away." TYCHO concludes that "It remains, therefore, for us to decide that this is a wonderful portent, contrary to all order of nature, created by God the Maker of the whole world machine and placed by Him in the beginning, but not until now to have appeared to the evening world. Furthermore, the divine power acts very freely and is not bound by any chains of nature, but, when He wishes, He causes the water in flowing streams to stand still, and turns back the stars in their courses."

While modern astronomers are convinced that supernovae are quite bound within the chains of nature they are nevertheless duly impressed by these stupendous cosmic explosions and much of the mystery remains. In the case of TYCHO's nova no remnant has as yet been found and our conviction that we here deal with a supernova is based on some rather indirect reasoning.

γ) Kepler's nova of 1604 in Ophiuchus. K. LUNDMARK[2] identified this star as a supernova. BAADE[3] who has investigated possible remnants of KEPLER's nova writes "The nova appeared at the beginning of October 1604 far down in the southwestern sky, close to Jupiter and Mars. Notwithstanding its unfavorable position, it was discovered promptly, since numerous observers were watching the approaching conjunction of the two planets which took place on October 9. The nova reached its maximum brightness—somewhat brighter than Jupiter—near the middle of October and was still about as bright as Jupiter when it disappeared in the rays of the sun in November. Reappearing in the eastern sky at the beginning of January, 1605, it has become much fainter, equalling Antares in brightness. The decrease in brightness continued throughout the summer, and, when last seen by KEPLER in October, 1605, the nova had reached the fifth magnitude. By the spring of the following year it had become invisible to the naked eye.". Estimates of the brightness of the nova relative to the brightness of some of the surrounding stars were given by both European and Chinese observers, in excellent agreement with one another.

BAADE states that no stellar remnant of KEPLER's nova brighter than the 18th magnitude has been found. This would make $\Delta M > 20.0$, which, as we have stated, seems characteristic for supernovae. The second reason why the star of 1604 is not considered as a common nova is that its light curve very closely coincides with those of the supernovae of the type I. Nevertheless both of these

[1] See excerpts from Tycho's writings in ALBERT EINSTEIN, Builders of the Universe. The U.S. Library Association, Inc. At Westwood Village, Los Angeles, California, 1932. Original Latin text in the introductory chapter of Tycho Brahe: De Nova Stella, Hafniae 1573, reprinted in Tychonis Brahe Opera omnia, Tomus I, Hauniae 1913, De Nova Stella, p. 16—19.

[2] K. LUNDMARK: Lund Medd. Ser. 2, No. 74 (1935).

[3] W. BAADE: Astrophys. Journ. **97**, 119—127 (1943).

arguments may be in error and a final decision is being awaited from the study of the spectrum of the nebulosity which BAADE has discovered near the indicated position of KEPLER'S star. A preliminary analysis of the spectrum of the said nebulosity by MINKOWSKI[1] seems to support the conclusion that the star was a supernova, but more work will be necessary to come to a final decision.

3. The period from 1885 until 1920. The real character of supernovae revealed itself first from the study of exploding stars in extragalactic nebulae rather than from the observations of new stars in our own Milky Way. S Andromedae of 1885 is the first nova ever to be discovered in an extragalactic stellar system. AGNES M. CLERKE writes in her book[2] that "This time the great nebula in the girdle of Andromeda was the scene of the outbreak. The unlooked for addition to it of a star-like nucleus was announced by Dr. HARTWIG at Dorpat, August 31, 1885; but it turned out that the change had already been perceived by Mr. ISAAC W. WARD of Belfast, August 19, and two nights earlier at Rouen by M. LUDOVIC GULLY, who set it down as an effect of bad definition. Concordant observations by TEMPEL at Florence, MAX WOLF at Heidelberg, and ENGELMANN at Leipzig showed decisively that the strange object made no show down to 10 P.M. on August 16; and a photograph taken by the late Dr. COMMON in August 1884 gave positive assurance that a year earlier its place had no stellar occupant as bright as the fifteenth magnitude. What were virtually the first rays reached the earth August 17, 1885."

"Between that date and August 31 it mounted from the ninth to the seventh magnitude; then without delay entered upon nearly as swift a downward course, checked by only one decided pause. Even the largest telescopes failed to keep it in view after March 1886."

"The spectrum of Nova Andromedae was of a dubious character. It bore witness to a completely different order of incandescence from that of the 'blaze stars' in the Northern Crown and the Swan. The bright rays which it perhaps included were inconspicuous. None were definitely determined, though the presence of several in the green and yellow was strongly suggested to Sir WILLIAM HIGGINS on September 9; and Dr. COPELAND succeeded, on September 30, in getting rough measures of three vaguely discernible accessions of brightness. The light, however, was mainly continuous."

In view of our present knowledge of the spectra of supernovae these observations are highly significant and strengthen the conclusion that S Andromedae was a supernova within the great galaxy Messier 31 and not a foreground star.

The second new star to be discovered in a nebula was Z Centauri, found by Mrs. FLEMING in NGC 5253 in July 1895. Agnes M. CLERKE writes in her book that "the spectrum of the star, fortunately recorded by a casual exposure at the Harvard Observatory (with an objective prism) in July 1895, showed the same irregularly continuous character with that of Nova Andromedae". This first spectrum ever to be photographed of a supernova was originally classified as an *R*-spectrum. This incorrect identification was maintained until 1936 and caused much confusion.

In the following years, from 1895 until 1921 about one dozen novae were discovered associated with nebulae in high galactic latitudes (see Table 1), but no clear realization was achieved as to the character of these new stars. It is of great interest, however, to quote a few of the opinions expressed by various prominent astronomers. For instance, CURTIS, in a discussion in 1917 of novae

[1] R. MINKOWSKI: Astrophys. Journ. **97**, 128–129 (1943).
[2] A. M. CLERKE: The Systems of Stars. London: Adam & Black 1905.

discovered in spirals, says[1] "The occurrence of these new stars in spirals must be regarded as having a definite bearing on the 'island universe' theory of the constitution of spiral nebulae".

LUNDMARK'S[2] basic opinions were equally sound. He had in fact determined in 1919 the distance of the Andromeda nebula as being of the order of 650000 light years. From this it followed naturally that the absolute magnitude at maximum of the nova of 1885 is of the order of $M = -15$. SHAPLEY on the other hand inclined toward the opinion that both the nebulae and the novae appearing in them are parts of our own galaxy. He writes[3]: "Moreover, if in real dimensions spiral nebulae were analogous to our galactic system, the absolute magnitude of the novae in spirals would far transcend any luminosity with which we are acquainted, and would be at direct variance with present results on intrinsic stellar brightness. For at the distance computed above the absolute magnitude of a nova of the sixteenth apparent magnitude would be -16, nearly two hundred thousand times as bright as the novae of the galactic system for which VAN MAANEN has determined the absolute luminosities. An upper limit to the intrinsic brightness attainable by stars is suggested by recent observational and theoretical work and this limit is much fainter than -16. The study of globular clusters, for example, has yielded sufficient knowledge of the luminosity of more than a million stars to show that not one is within ten magnitudes of this enormous brightness ... Hence stellar luminosities of this order [-16] seem out of the question, and accordingly the close comparability of spirals containing such novae to our galaxy appears inadmissible."

In the paper mentioned, SHAPLEY summarizes his conclusions thus: "Observation and discussion of the radial velocities, internal motions, and distribution of spiral nebulae, of the real and apparent brightness of novae, of the maximum luminosity of galactic and cluster stars, and finally the dimensions of our own galactic system, all seem definitely to oppose the 'island universe' hypothesis of the spiral nebulae."

In his detailed discussion of "the scale of the Universe" with H. D. CURTIS[4], SHAPLEY still held the opinion expressed in the preceding quotation from his article of 1919. H. D. CURTIS, on the other hand, arrived at a conclusion which in its essentials coincides with the view now taken with regard to the nature of spiral nebulae, although he considerably underestimated the dimensions of our own galaxy. His final statement reads "I hold, therefore, to the belief that the galaxy is probably not more than 30000 light years in diameter, that the spirals are not intragalactic objects but island universes, like our own galaxy, and that the spirals, as external galaxies indicate to us a greater universe into which we may penetrate to distances of ten million to a hundred million light years". There is also a first inkling of the truth about the possible existence of very bright novae in CURTIS' statement, "It seems certain, for instance, that the dispersion of the novae in the spirals, and probably also in our galaxy may reach at least ten absolute magnitudes, as is evidenced by a comparison of S Andromedae with the faint novae found recently in this spiral. A division into two magnitude classes is not impossible".

4. Since 1920. *α) The period from 1920 until 1933.* The analysis of the plates obtained with the 100-inch telescope on Mount Wilson once and for all showed that spiral nebulae as well as many of the irregular nebulae like NGC 6822 are extra-

[1] H. D. CURTIS: Publ. Astronom. Soc. Pacific **29**, 180 (1917); **29**, 206 (1917).
[2] K. LUNDMARK: Kgl. Svensk. Vetensk. Handl. **60**, No. 8 (1920).
[3] SHAPLEY: Publ. Astronom. Soc. Pacific **31**, 266 (1919).
[4] SHAPLEY and H. D. CURTIS: Bull. Nat. Res. Counc. **2**, Part 3, 171 (1921)

galactic stellar systems[1]. From this result it followed that the brightest novae appearing in these nebulae are not of the common type, but are supernovae. The discovery of about four of these (listed in Table 1) in the years from 1920 to 1933 did not lead to any systematic knowledge on supernovae, no spectra were taken and no determined search program was initiated. This rather surprising lack of interest may perhaps be ascribed to a failure to realize the importance of supernovae, to occupation with other important investigations, to the uncertainty about the frequency of occurrence of these giant explosions and to the unavailibility of suitable fast wide angle telescopes. The present author, however, felt that these difficulties did not constitute a sufficient deterrent and he therefore organized a systematic search program.

β) The extended search program from 1933 to 1942. The search was started with the 10 inch refractor on Mount Wilson and with a 3-inch objective mounted

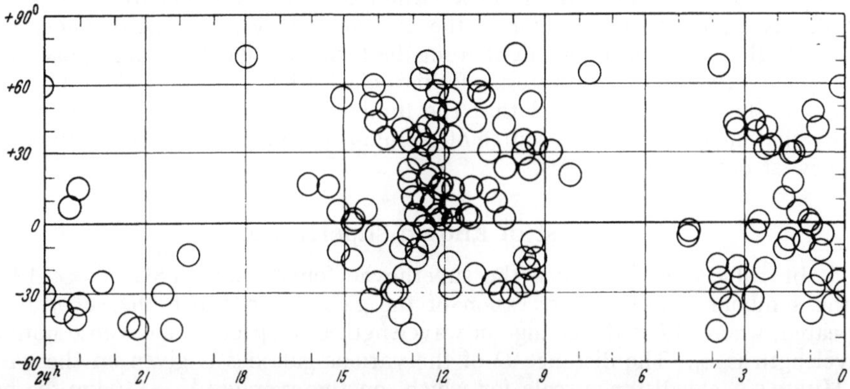

Fig. 1. Distribution of the fields controlled in the first Palomar supernova search 1936 to 1942. Abscissa = right ascension; ordinate = declination.

on the roof of the California Institute of Technology. During the preliminary attempts it was decided to build an 18-inch Schmidt telescope which was put in operation on September 5, 1936 and with which the 175 fields shown in Fig. 1 were searched for supernovae as often as possible.

The fields shown were chosen with a view to covering most of the neighboring galaxies such as Messier 31, 32, 33, 51, 81, 82, 101, IC 1613 etc. as well as many of the nearest clusters of galaxies including those in Virgo, Ursa Major, Coma, Leo, Cancer, Perseus, Hydra (I), Fornax and Centaurus. About 3000 galaxies brighter than apparent photographic magnitude $m_p = 15.0$ and perhaps 700 brighter than $m_p = 13.0$ were kept under surveillance[2]. While the present author occupied himself with the search program, some of the Mount Wilson astronomers had agreed to follow the light curves and the spectra of the supernovae found with the 18-inch Schmidt on Palomar Mountain. The whole program was highly successful and contributed by far the major part of our present knowledge on the light curves, the spectra and the frequency of occurrence of supernovae.

γ) The period from 1942 until 1950. World War II almost completely interrupted all investigations, and only a few supernovae were found, without any gain in basic knowledge.

---

[1] See for instance E. P. Hubble: The Realm of the Nebulae. New Haven, Conn.: Yale Univ. Press 1936.

[2] F. Zwicky: Astrophys. Journ. **88**, 529—541 (1938).

δ) The years since 1950 and the future. During the past few years a program was worked out for a survey of nearby galaxies at the observatories on Mount Hamilton, Tucson, Flagstaff, Palomar and Berne in Switzerland. A determined effort is to be made during a five year period to discover a few very bright supernovae and to construct their light curves in several colour ranges, to analyse their spectra, check on possible polarisation, observe the stars photoelectrically for flare characteristics and also watch for any possible emission of cosmic rays and of radio noise. During the preparations for this program three interesting supernovae were discovered by P. WILD. As a parallel effort it is intended to improve on the statistics of supernovae through an attempt to accumulate data on a few dozen supernovae which we may expect to spot in distant clusters of galaxies.

Important additions to our knowledge of the remnants of supernovae have recently been made through the observation of radio emission from the Crab Nebula. A most significant result is the discovery by Russian astronomers and by OORT[1] that the continuous light from the Crab Nebula is strongly polarised, and by ZWICKY[2] that the pattern of this polarisation is a most puzzling one. Finally it should be mentioned that the fundamental predictions made by the theory that supernovae are a source of cosmic rays[3,4] have now been verified by the newest data on these rays.

II. List of known supernovae.

5. In Table 1 we introduce the supernovae found since 1885. $V_s = c\,\Delta\lambda/\lambda_0$ is the symbolic velocity of recession of the galaxies in which the supernovae appeared, where $\Delta\lambda$ is the change in wavelength of a spectral line whose normal wavelength is λ_0. The distance D of the various galaxies is given on the basis of HUBBLE's old distance scale for which, on the average, $V_s = 550$ km/sec per million parsecs[5]. The distance modulus $m - M$ and the corresponding absolute photographic magnitudes M_{max} of the supernovae at maximum brightness are calculated on the basis of this old distance scale. Much work will have to be done to correct this old scale in order to derive an absolute luminosity function for supernovae.

The list of supernovae presented here must in no way be considered as complete. For the discussion of possible old supernovae in our own galaxy the work by K. LUNDMARK in particular should be consulted. LUNDMARK also discusses a number of possible supernovae which have appeared in extragalactic nebulae during the past century and which we have not included in our list[6].

III. The properties of supernovae.

6. The frequency of supernovae. The following questions may be asked.

α) What is the absolute frequency of supernovae either when referred to a given unit of cosmic space or in a given number of stellar systems?

β) How many types of supernovae exist and what are their relative frequencies of occurrence?

---

[1] J. H. OORT and TH. WALRAVEN: Bull. Astr. Inst. Netherl. **12**, No. 462 (1956).
[2] F. ZWICKY: Publ. Astronom. Soc. Pacific **68**, 121—124 (1956).
[3] W. BAADE and F. ZWICKY: Proc. Nat. Acad. Sci. **20**, 254—259, 259—263 (1934).
[4] F. ZWICKY: Proc. Nat. Acad. Sci. U.S.A. **25**, 604—609 (1939).
[5] See footnote 1, p. 771.
[6] K. LUNDMARK: Medd. Lund. Astr. Obs. Ser. I **1939**, No. 155.

γ) What is the luminosity function of one specific type of supernovae?

δ) Is there any correlation between the frequency of occurrence of various types of supernovae and the type of the stellar system?

ε) What is the probability of occurrence of supernovae within different parts of a stellar system?

ζ) What is the frequency of supernovae in "intergalactic space"?

α) The absolute frequency of supernovae. The data available are entirely insufficient to estimate accurately the number of supernovae per year and per unit volume of space (per cubic megaparsec for instance). Also, since it was found[1] that the number of stellar systems increases monotonely with decreasing brightness, the question of the frequency of supernovae per stellar system has obviously no meaning. We may, however, ask what is the frequency f of supernovae in galaxies which are brighter than those of a given absolute magnitude M or which are members of any collection of individually specified galaxies. Fair answers have been obtained to these questions as a result of the Palomar Search Program described in Sect. 4β.

Between September 6, 1936 and January 31, 1938 about 3000 galaxies brighter than the apparent photographic magnitude $m_p = 15$ were searched as often as possible with the 18-inch Schmidt telescope on Palomar Mountain and four supernovae were found. From these findings a frequency f_1,

$$f_1 = \text{one supernova per average galaxy per 457 years} \tag{6.1}$$

results[2], where an "average" galaxy refers to the 3000 galaxies surveyed. Among the 3000 galaxies about 840 are listed in the Shapley-Ames catalogue, and within this partial collection one supernova appeared (in NGC 4157) in the period mentioned. The calculation of the frequency of supernovae based on the search of these selected galaxies gives

$$f_2 = \text{one supernova per Shapley-Ames Galaxy per 460 years.} \tag{6.2}$$

Because of the small numbers involved the agreement between (6.1) and (6.2) is of course entirely fortuitous.

During the period from September 6, 1936 to December 31, 1939 about 3000 galaxies brighter than the apparent magnitude $m_p = 15$ were searched and 1625 photographs of 30 min exposure time were taken. It was found that the control value of the average photograph taken with the 18-inch Schmidt corresponds to about one month. The control value of the whole search mentioned therefore is equivalent to the search for supernovae in an average galaxy of our collection during a period of 61804 months or 5150 years. Since twelve supernovae were discovered we arrive at the result that, per average of our 3000 galaxies, one supernova makes its appearance in average intervals of

$$\tau = 430 \text{ years} \tag{6.3}$$

in satisfactory agreement with (6.1) and (6.2). Considering again only the galaxies listed in the Shapley-Ames catalogue, all of which are brighter than $m_p = 13.0$, we note that in the period mentioned five supernovae were discovered in 840 galaxies. The statistical evaluation of these findings leads to the result that in any average individual Shapley-Ames galaxy supernovae make their appearance at average intervals of

$$\tau_s = 359 \text{ years.} \tag{6.4}$$

τ_s must be considered as the best value so far available.

[1] F. Zwicky: Phys. Rev. **61**, 489–503 (1942). — Helv. phys. Acta **26**, 241–254 (1953).

[2] F. Zwicky: Astrophys. Journ. **88**, 529–541 (1938); **96**, 28–36 (1942).

Table 1. *List of*

No.	Galaxy	Type	(1950) R.A.	(1950) Decl.	m_{neb}	V_s km/sec	D in 10^6 L.Y.	Date of maximum brightness	m_{max}	M_{max}
1	NGC 224	*Sb*	$0^h40^m.0$	+41°0′	4.5	−266	0.8	August 1885	7.2	−15.0
2	NGC 4424	*SBb* pec	$12^h24^m.6$	+ 9°42′	12.5	—	7.0	March 1895	11.1	−15.6
3	NGC 5253	Irr	$13^h37^m.1$	−31°23.′2	11.0	+432	—	July 1895	8.0	—
4	NGC 2535	*SBc*	$8^h\ 8^m.2$	+25°21.′6	12.7	+4243	—	Jan. 1901	14.7	—
5	NGC 4321	*Sc*	$12^h20^m.4$	+16°6′	10.5	+1617	7.0	March 1901	11.9	−14.8
6	NGC 4674	*S*	$12^h30^m.8$	+20°12.′7	—	—	—	May 1907	13.5	—
7	NGC 5457	*Sc*	$14^h\ 1^m.4$	+54°35′	8.9	+247	—	Febr. 1909	(12.1)	—
8	NGC 2841	*Sb*	$9^h18^m.6$	+51°12′	10.6	+584	—	Febr. 1912	—	—
9	NGC 4321	*Sc*	$10^h20^m.4$	+16°6′	10.5	+1617	7.0	March 1914	< 14.0	< −12.7
10	NGC 4527	*Sc*	$12^h31^m.6$	+ 2°56′	11.3	+1727	7.0	March 1915	14.0	−12.7
11	NGC 6946	*Sc*	$20^h33^m.9$	+59°58′	11.1	+38	3.6	July 1917	12.9	−12.4
12	NGC 4486	*Eo*	$12^h28^m.3$	+12°40′	10.1	+1290	7.0	Febr. 1919	11.5	−15.2
13	NGC 2608	*SBc*	$8^h32^m.2$	+28°38′	13.6	+2119	—	— 1920	11.0	—
14	NGC 4038/39	*S* pec	$11^h59^m.3$	−18°35′	11.0	+1673	—	March 1921	—	—
15	NGC 3184	*Sc*	$10^h15^m.2$	+41°40′	11.8	+443	3.9	April 1921	13.9	−11.5
16	NGC 3184	*Sc*	$10^h15^m.2$	+41°40′	11.8	+443	3.9	Dec. 1921	11.0	−14.4
17	NGC 5236	*Sc*	$13^h34^m.3$	−29°37′	8.0	+491	3.2	May 1923	14.0	−10.8
18	NGC 4303	*SBc*	$12^h19^m.4$	+ 4°45′	10.4	+1671	7.0	May 1926	12.8	−13.9
19	NGC 6181	*Sc*	$16^h30^m.1$	+19°56′	12.6	+2158	—	June 1926	(14.8)	—
20	IC 4719	*S*	$18^h29^m.0$	−56°46′	13.9	—	—	Oct. 1934	(13.6)	—
21	NGC 4273	*Sc*	$12^h17^m.4$	+ 5°37′	12.4	+2302	7.0	Jan. 1936	14.2	−12.5
22	Anon (1)	*SBc*	$1^h18^m26^s$	+15°25.′9	—	—	—	Sept. 1936	< 15.1	—
23	NGC 4157	*Sc*	$12^h\ 8^m36^s$	+50°46′	12.0	—	—	Febr. 1937	(14.4)	—
24	Anon (2)	*SBc*	$22^h\ 7^m52^s$	−22°56.′3	—	—	—	August 1937	(15.3)	—
25	IC 4182	*Sc*	$13^h\ 3^m28^s$	+37°52.′1	13.5	—	2.9	August 1937	8.2	−16.6
26	NGC 1003	*Sc*	$2^h36^m\ 6^s$	+40°39′	13.1	+585	7.4	Sept. 1937	12.8	−14.0
27	NGC 3184	*Sc*	$10^h15^m.2$	+41°40′	11.8	+443	3.9	Dec. 1937	13.5	−11.9
28	NGC 1482	*Sa* pec	$3^h52^m25^s$	−20°38.′9	14.4	—	—	Nov. 1937	(14.0)	—
29	Anon (3)	*SBc*	$2^h34^m48^s$	+34°10′	15.0	+4800	29.8	Oct. 1938	14.5	−15.3
30	NGC 4636	E_1	$12^h40^m.3$	+ 2°57′	10.8	+973	7.0	Jan. 1939	12.5	−14.2
31	NGC 4621	E_5	$12^h39^m.5$	+11°55′	11.4	+414	7.0	May 1939	11.8	−14.9
32	NGC 6946	*Sc*	$20^h33^m.9$	+59°58′	11.1	+38	3.6	June 1939	13.2	−12.1
33	Anon (4)	*Sb*	$0^h54^m36^s$	− 4°16′	—	—	—	Nov. 1939	16.0	—
34	NGC 5907	*Sb*	$15^h14^m.6$	+56°31′	11.8	+553	4.2	Febr. 1940	< 13.0	—
35	NGC 4725	*SBb*	$12^h48^m.1$	+25°46′	10.8	+1114	6.1	May 1940	12.6	−13.5
36	IC 1099		$15^h\ 5^m.8$	+56°40.′9	15.0	—	—	May 1940	< 16.3	—
37	NGC 4545		$12^h32^m.4$	+63°47.′6	—	—	—	July 1940	< 15.0	—
38	NGC 253	*Sc*	$0^h45^m.1$	−25°34′	7.0	−81	3.2	Nov. 1940	< 14.0	< −11.0
39	NGC 4559	*Sc*	$12^h33^m.5$	+28°14′	10.7	+856	—	Febr. 1941	13.5	—
40	NGC 3254	*Sc*	$10^h26^m.5$	+29°45′	12.6	+1228	—	March 1941	15.0	—
41	NGC 4136	*Sc*	$12^h\ 6^m.7$	+30°12′	12.1	+445	—	April 1941	16.8	—
42	NGC 5195	Irr	$13^h27^m.9$	+47°31′	11.1	+542	—	April 1945	—	−12.8
43	NGC 3977	*Sb*	$11^h53^m.6$	+55°39.′6	—	—	—	May 1946	< 18.0	—
44	NGC 4632	*Sc*	$12^n40^m.0$	+ 0°11′	12.1	—	7.0	May 1946	< 15.7	< −11.0
45	NGC 3177	*Sb*	$10^h13^m.9$	+21°22′	12.8	+1220	7.0	March 1947	< 16.5	< −10.3
46	NGC 4699	*Sb*	$12^h46^m.5$	− 8°24′	10.5	+1511	—	March 1948	< 17.0	—
47	NGC 6946	*Sc*	$20^h33^m.9$	+59°58′	11.1	+38	3.6	July 1948	< 14.9	—
48	IC 4051	E_1	$12^h58^m.5$	+28°16.′6	—	+4932	45	March 1950	(14.5)	(−16.0)
49	NGC 5236	*Sc*	$13^h34^m.3$	−29°37′	8.0	+491	3.2	March 1950	(14.5)	—
50	NGC 4214	Irr	$12^h13^m\ 6^s$	+36°36′	10.7	+295	3.5	April 1954	9.0	−16.0
51	NGC 5668	*S*	$14^h30^m54^s$	+ 4°40′	12.4	+1770	10.5	May 1954	12.0	−15.6
52	NGC 5879	*Sb*	$15^h\ 8^m24^s$	+57°12′	12.1	+876	5.5	Sept. 1954	14.9	−11.2
53	Anon (5)	*SBa* pec	$1^h\ 5^m\ 0^s$	−13°29′30″	—	—	100	Sept. 1955	15.8	(−16.6)
54	NGC 3992	*SBc*	11^h55^m	+53°39′	11.2	+1059	—	March 1956	(11.0)	(−15.1)

Assuming that every cubic megaparsec of cosmic space contains about twenty galaxies of the same character as those listed in the Shapley-Ames catalogue we should expect on the average, the appearance of one supernova per cubic *known supernovae.*

No.	Type	Discoverer	Year of discovery	Cluster member	$m-M$	Literature and remarks
1	I	Hartwig and others	1885	Local Group	22.2	Lundmark: Kgl. Svensk. Vetens. Handl. **60**, 55 (1920). Light curve.
2	—	M. Wolf	1925	Virgo Cluster	26.7	Astronom. Nachr. **226**, 76 (1925).
3	—	Mrs. Fleming	1895	—	—	Pickering: H.C. 4 (1897). Z. Centauri.
4	—	K. Reinmuth	1923	—	—	Astronom. Nachr. **221**, 47 (1924).
5	—	H. D. Curtis	1917	Virgo Cluster	26.7	Lick Obs. Bull. 300 (1917).
6	—	W. L. Luyten	1936	—	—	H.A.C. 399 (1936).
7	—	M. Wolf	1909	—	23.8	SS Urs. Maj. in Messier 101. Astronom. Nachr. **180**, 375 (1909).
8	—	Pease, Curtis	1917	—	—	Pease: Publ. Astronom. Soc. Pacific **29**, 213 (1917). Only one observation
9	—	H. D. Curtis	1917	Virgo Cluster (M 100)	26.7	Lick Obs. Bull. 300 (1917). Only one observation.
10	—	H. D. Curtis	1917	Virgo Cluster	26.7	Lick Obs. Bull. 300 (1917).
11	—	G. Ritchey	1917	—	25.3	Publ. Astronom. Soc. Pacific **29**, 211 (1917).
12	—	I. Balanowski	1922	Virgo Cluster	26.7	Astronom. Nachr. **215**, 215 (1922). M 87.
13	—	M. Wolf	1920	—	—	K. Lundmark: Publ. Astronom. Soc. Pacific **35**, 116 (1923).
14	—	Hubble, Duncan	1939	—	—	Medd. Lunds Astr. Obs. Ser. 1, No. 155 (1939).
15	—	F. Zwicky	1939	—	25.4	H. Shapley: Proc. Nat. Acad. Sci. U.S.A. **25**, 569 (1939).
16	—	Miss R. Jones	1939	—	25.4	H. Shapley: Proc. Nat. Acad. Sci. U.S.A. **25**, 569 (1939).
17	—	C. Lampland	1923	—	24.8	M 83 MWC 671 H.B. 786 and 787 (1923). (Similar to Nova Puppis?)
18	—	Wolf, Reinmuth	1926	Virgo Cluster	26.7	M 61 H.B. 836 (1926).
19	—	A. van Maanen	1941	—	—	Publ. Astronom. Soc. Pacific **53**, 125 (1941).
20	—	Miss C. D. Boyd	1938	—	—	H.B. 907 (1938).
21	II	E. P. Hubble, G. Moore	1936	Virgo Cluster	26.7	Baade: Publ. Astronom. Soc. Pacific **48**, 226 (1936).
22	—	F. Zwicky	1938	—	—	Publ. Astronom. Soc. Pacific **51**, 36 (1939).
23	II	F. Zwicky	1937	Urs. Maj. Cloud	—	Publ. Astronom. Soc. Pacific **49**, 204 (1937).
24	—	F. Zwicky	1938	—	—	Publ. Astronom. Soc. Pacific **51**, 36 (1939).
25	I	F. Zwicky	1937	—	24.8	Astrophys. Journ. **88**, 411 (1938); **89**, 156 (1939).
26	I	F. Zwicky	1937	—	26.8	Astrophys. Journ. **88**, 411 (1938); **89**, 156 (1939).
27	—	F. Zwicky	1938	—	25.4	Astrophys. Journ. **96**, 28 (1942).
28	—	F. Zwicky	1937	—	—	Rev. Mod. Phys. **12**, 66 (1940).
29	—	F. Zwicky	1938	—	29.8	Publ. Astronom. Soc. Pacific **51**, 36 (1939).
30	—	F. Zwicky	1939	Virgo Cluster	26.7	Astrophys. Journ. **96**, 28 (1942).
31	—	F. Zwicky	1939	Virgo Cluster	26.7	Astrophys. Journ. **96**, 28 (1942).
32	—	F. Zwicky	1939	—	25.3	Astrophys. Journ. **96**, 28 (1942).
33	—	F. Zwicky	1939	—	—	Astrophys. Journ. **96**, 28 (1942).
34	II	J. J. Johnson	1940	—	—	H. A. C. 519.
35	II	J. J. Johnson	1940	Urs. Maj. Cloud	26.1	H. A. C. 522.
36	—	F. Zwicky	1940	—	—	H. A. C. 524.
37	—	J. J. Johnson	1940	—	—	H. A. C. 530.
38	I	F. Zwicky	1940	—	—	H. A. C. 552.
39	II	Miss R. Jones	1941	—	—	H. A. C. 576.
40	—	J. J. Johnson	1941	—	—	H. A. C. 578.
41	II	Miss R. Jones	1911	—	—	H. A. C. 581.
42	—	M. L. Humason	1945	—	—	Companion of M 51. Publ. Astronom. Soc. Pacific **57**, 174 (1945).
43	—	E. P. Hubble	1946	Urs. Maj. Cloud	—	Ann. Rep. Mt. Wilson Obs. **45**, 19 (1945 to 1946).
44	II?	E. P. Hubble	1946	Virgo Cluster	26.7	Ann. Rep. Mt. Wilson Obs. **45**, 19 (1945 to 1946).
45	II	E. P. Hubble	1947	—	26.7	Ann. Rep. Mt. Wilson Obs. **46**, 20 (1946 to 1947).
46	—	E. P. Hubble	1948	Virgo Cluster	26.7	Ann. Rep. Mt. Wilson Obs. **47**, 20 (1947 to 1948).
47	II?	N. U. Mayall	1948	—	25.3	Publ. Astronom. Soc. Pacific **60**, 266 (1948); **61**, 97 (1949).
48	I	M. L. Humason	1950	Coma Cluster	30.5	Publ. Astronom. Soc. Pacific **62**, 117 (1950).
49	—	G. Haro	1950	—	—	H. A. C. 1074 (1950).
50	(I)	P. Wild	1954	—	25.0	H. A. C. 1250 (1954).
51	I	P. Wild	1954	—	27.6	H. A. C. 1245 (1954).
52	II	P. Wild	1954	Urs. Maj. Cloud	26.1	H. A. C. 1275 (1954).
53	—	F. Zwicky	1955	A-cluster?	(32.4)	Publ. Astronom. Soc. Pacific **68**, 271 (1956).
54	—	H. S. Gates	1956	Urs. Maj. Cloud	26.1	—

megaparsec per 18 years. In a sphere of 10^9 parsecs radius which is about the distance reached by the 200-inch telescope there exist therefore at any given time about 200 million supernovae at stages near maximum light.

β) Types of supernovae. Detailed studies of supernovae by R. MINKOWSKI[1] led to the recognition of two groups (I and II) which differ radically in their spectra. ZWICKY found in particular that the continuum in the near ultraviolet is weak for supernovae of the type I and strong for type II. This serves as a means to distinguish between different types of supernovae when distant objects are in question and no spectra are available. Type I is intrinsically brighter than type II and seems to favor elliptical galaxies and the central parts of spirals while type II is found mostly in the spiral arms. The light curves of the two classes also seem to be different. Preliminary data indicate that type II is more frequent than type I. It must be emphasized, however, that far more data are needed to properly classify supernovae. In the author's opinion it is certain that many more types than just two will be found eventually, a prediction which is suggested by many anomalies observed in the past supernovae.

γ) Luminosity function. From the data available it appears as if the luminosity function of supernovae has a maximum which, for type I, corresponds to an absolute photographic magnitude $M_q \approx -14.0$ (on the old distance scale). This conclusion, however, will most likely be found to be erroneous for the following reasons. The probability of discovery of supernovae rapidly decreases with decreasing intrinsic brightness. Also, the distances for the nearer galaxies are so uncertain that it is quite impossible to derive the absolute luminosity function of the known supernovae. Finally, the newest findings by the author on Nova Puppis of 1942 indicate that within our own galaxy there may have been far more novae in the range $-10 > M_p > -13$ than in the range $M_p < -13$.

δ) Correlation with type of stellar system. The following fragmentary data are available so far. Between September 1936 and December 1941, 18 supernovae were discovered by ZWICKY and JOHNSON as a result of their systematic search with the 18-inch Schmidt on Palomar Mountain. Among these were 8 supernovae which appeared in the Shapley-Ames Galaxies NGC 4157, 4636, 3184, 4621, 6946, 5907, 4725 and 253 of which two are ellipticals and six are spirals. The ratio of elliptical galaxies to spirals among 721 Shapley-Ames galaxies, according to a private communication by Dr. HUBBLE is 139 to 416; that is exactly 1 to 3. The frequency of supernovae does not seem to depend markedly on the type of the galaxy involved. On the other hand there seem to be certain spirals which are exceptional, NGC 3184 and 6469 having had three supernovae between 1900 and 1950 while two appeared in NGC 4321 and 5236 respectively.

ε) The relative frequency of various types of supernovae and their appearance in various types of galaxies is as yet uncertain. Most of them, however, seem to favor the outskirts of stellar systems; the frequency does not seem to depend very markedly on the number of stars per unit volume in various localities within a galaxy.

ζ) Although supernovae were discovered within the tenuous outskirts of galaxies, no strictly intergalactic supernova has as yet been found. The search for objects of this type is of course most difficult.

7. The light curves of supernovae. Only fragmentary data are available on extragalactic supernovae which were found prior to the extended search with the 18-inch Schmidt telescope on Palomar Mountain. These data were reviewed in a paper by BAADE[2].

[1] R. MINKOWSKI: Publ. Astronom. Soc. Pacific **52**, 206–207 (1940); **53**, 224–225 (1941).
[2] W. BAADE: Astrophys. Journ. **88**, 285–304 (1938).

Of the eighteen supernovae discovered by ZWICKY and JOHNSON during their Palomar search from 1936 until 1941 several hundred films were turned over to Dr. BAADE of the Mount Wilson Observatory with the understanding that he would use these films, in conjunction with plates available to him from the Mount Wilson 60″ and 100″ reflectors, to determine the detailed light curves of the mentioned supernovae. Although it seems that Dr. BAADE has carried out the necessary reductions, only a few of his results have been published so far.

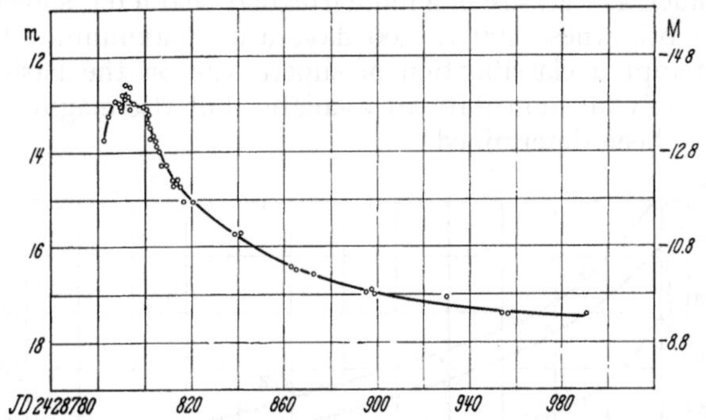

Fig. 2. Photographic light curve of the supernova in NGC 1003. Abscissa = Julian days; ordinate m = apparent photographic magnitude; M = absolute photographic magnitude (HUBBLE's old distance scale).

Some of the published results are as follows. The light curves for the first 200 days of the supernovae in NGC 1003 and IC 4182 were given in a paper by BAADE and ZWICKY[1]. We reproduce in Fig. 2 the partial light curve for the supernova in NGC 1003, which was probably of the type I.

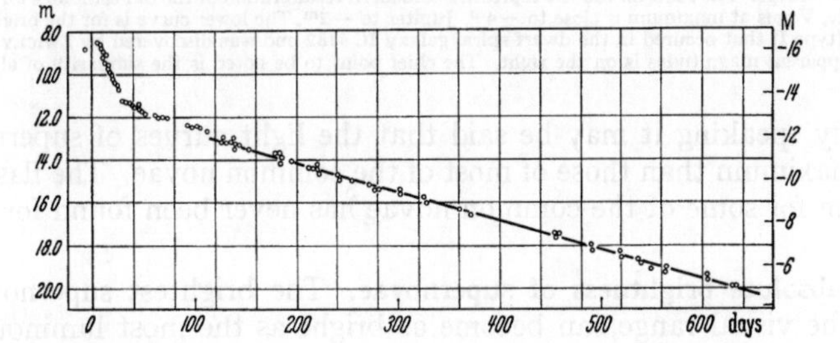

Fig. 3. Photographic light curve of the supernova in IC 4182. Abscissa = days from maximum brightness. Ordinates = apparent and absolute photographic magnitudes (on HUBBLE's old distance scale).

Fig. 3 shows a revised light curve by BAADE[2] for the supernova in IC 4182, extending over about 600 days.

BAADE[2] states that "about 100 days after maximum the decrease in brightness (meaning rather more correctly the increase in photographic magnitude) becomes linear with a gradient of 0.0137 magnitudes per day".

Of the several additional light curves of supernovae of the type I a fragmentary curve for the supernova in NGC 4636 has been published[3]. This object

[1] W. BAADE and ZWICKY: Astrophys. Journ. **88**, 411—421 (1938).

[2] See N. U. MAYALL: Sci. Monthly **66**, 17—24 (1948). — W. BAADE et al.: Publ. Astronom. Soc. Pacific **68**, 296—300 (1956).

[3] W. BAADE: Les Supernovae, article in: Théories des Novae, Supernovae. Paris: Hermann & Cie. 1941.

is of great interest because it was discovered by Zwicky several days before maximum brightness and spectra were obtained by Minkowski at this early stage of the development (data as yet unpublished).

Baade has also compared[1] the visual light curves of Tycho's nova of 1572 and of Kepler's nova of 1604 with the light curves of the supernovae of the type I and has found that they are very similar, as Fig. 4 shows.

The light curves of supernovae of the type II seem to differ from those of the type I inasmuch as they are of a more irregular character, showing a relatively steep decline in brightness 100 to 200 days after maximum. It is dangerous, however, to attempt a classification of supernovae on the basis of their light curves, until not only far more data are available, but also magnitudes in different color ranges have been determined.

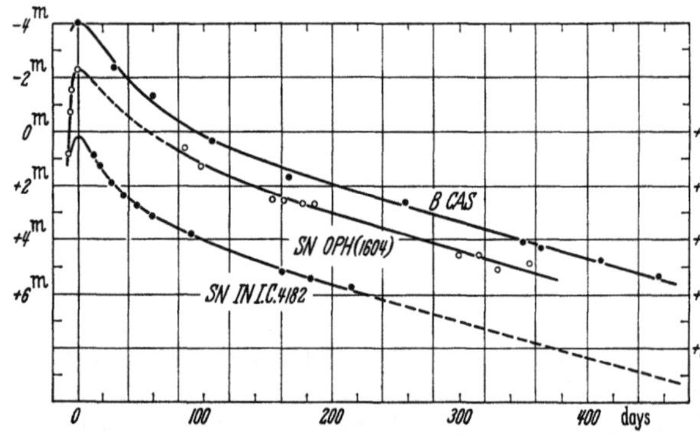

Fig. 4. Light curves of three supernovae. The upper two curves are for Tycho's star of 1572 (B Cassiopeiae) and for Kepler's star of 1604 (the supernova in Ophiuchus). The observations in both cases (dots and circles) were made before the invention of the telescope. The scale on the left represents a modern recalibration of the old estimates for the magnitudes. On the same scale, Venus at maximum is close to -4^m, Jupiter to -2^m. The lower curve is for the brightest known modern supernova (type I) that occured in the dwarf spiral galaxy IC 4182 and was discovered by Zwicky in August 1937. Its scale of apparent magnitudes is on the right. The chief point to be noted is the similarity of all three curves.

Generally speaking it may be said that the light curves of supernovae have a broader maximum than those of most of the common novae. The flash observed at maximum for some of the common novae has never been found for any supernova.

8. The absolute brightness of supernovae. The brightest supernovae of the type I, in the visual range can become as bright as the most luminous galaxies, that is, their absolute photographic magnitude can be as much as $M_p = -16$ to -17, on the old distance scale, or $M_p = -18$ to -19 on the basis of the more recent estimates of the distances of extragalactic nebulae. Supernovae of the type II and of other possible types do not seem to become quite as bright as supernovae of the type I. The values of M_p at the maximum brightness for different supernovae cover at least a range of five to six magnitudes. It is, however, not known at the present time, whether or not there are continuous transitions between the supernovae of various types and of various luminosities as well as continuous transitions between supernovae and common novae. Nova Puppis of 1942, for instance, appears to have had some of the characteristics of supernovae, such as the broad and rounded out maximum of the light curve, while its absolute magnitude[2] at maximum was about $M_{vis} \sim -10.0$.

[1] See footnote 2, p. 777.

[2] The present author has succeeded in the fall of 1955 in photographing the expanding gas shells from Nova Puppis, from whose diameter and spectral characteristics the quoted value of M_{vis} was determined.

A renewed systematic search for supernovae is badly needed in order to obtain more knowledge on the statistical distribution of supernovae in dependence on type and absolute magnitude. Also, the distances of all of the resolved galaxies in which supernovae have appeared in the past must be redetermined on the basis of new and more reliable distance criteria. In the present state of turmoil concerning the distances of individual nearby galaxies and clusters of galaxies it would be entirely premature to make any definite statement about the distribution of supernovae in dependence on absolute magnitude.

As far as the total energy liberated is concerned, it is important to notice that a bright supernova, during the first few hundred days, radiates a total of as much as 10^{49} to 10^{50} ergs in the visual range[1]. This estimate is made on the assumption that the old distance scale originally established by HUBBLE must be expanded by a factor 2 to 4. This means that if a mass of the size of the Sun, that is 2×10^{33} grams is involved, containing about 10^{57} protons, the energy liberated per proton in form of visual radiation alone is of the order of 10^{-7} ergs, or 10^{16} ergs/gram = 240000 kilocalories/gram. Since this energy is about 5000 times greater than that liberated in the most powerful chemical reactions, such as in $H + H = H_2$, the present author concluded[1,2] that in supernovae nuclear chain reactions were involved, for instance a rapid collapse into a neutron star. The observations of supernovae thus furnished the first proof, long before the processes of nuclear fission were discovered, of the *occurrence of nuclear chain reactions.*

9. The spectra of supernovae. Prior to the systematic search of 1936 to 1941 only a very few photographs of spectra of supernovae were available in addition to a number of visual observations. With the discovery in 1937 of the two bright (type I) supernovae in IC 4182 and NGC 1003 HUMASON and MINKOWSKI obtained a long series of beautiful spectra, tracings of three of which we reproduce in Figs. 5 to 7.

The remarkable aspect of these spectra is that for various supernovae of the type I, even if their absolute luminosity is quite different, they show on an absolute time scale the same sequence of changes. The various features in these spectra and their respective changes are so characteristic that Dr. MINKOWSKI, from one spectrum could always tell, how many days after maximum a supernova had been discovered, and his deduction could often be checked through the subsequent reconstruction of the light curve with the aid of prediscovery plates which were available at other observatories.

Unfortunately, in spite of much work during the 19 years which have elapsed since the discovery of the very bright supernova in IC 4182 we do not yet have the slightest inkling as to the interpretation of the spectra of supernovae of the type I. On the other hand, the supernovae of the type II seem to have spectra analogous to those of ordinary novae with velocities of expansion of the order of 5000 km/sec to 10000 km/sec. In Fig. 8 we reproduce three typical spectra of a common nova and of two supernovae of types I and II, for which I am indebted to Dr. R. MINKOWSKI.

The general status of our knowledge of the spectra of supernovae is still adequately described by a statement made by MINKOWSKI[3] as far back as 1941 which reads as follows:

---

[1] See footnote 1, p. 777.

[2] F. ZWICKY: Astrophys. Journ. **88**, 522–525 (1938). — Phys. Rev. **55**, 726–743 (1939). — Rev. Mod. Phys. **12**, 66–85 (1940). — Observatory **68**, 121–143 (1948).

[3] R. MINKOWSKI: Publ. Astronom. Soc. Pacific **53**, 224–225 (1941).

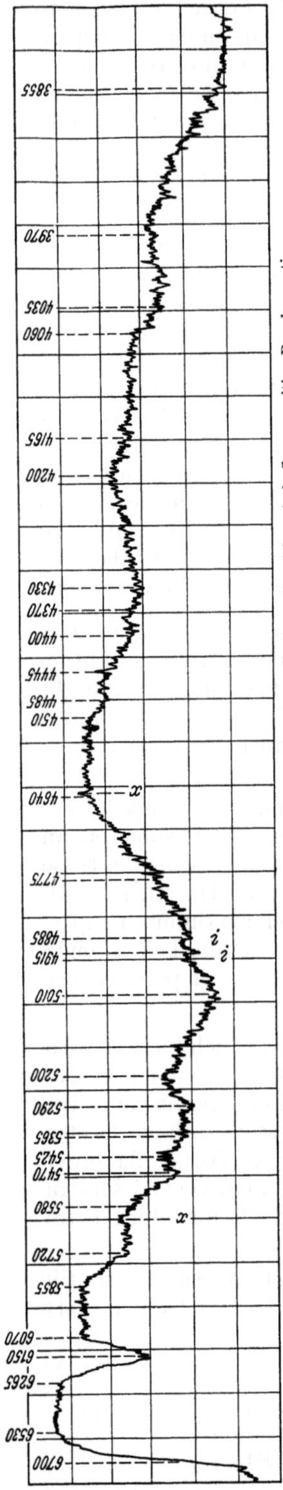

Fig. 5. Tracing of the spectrum of the supernova IC 4182, 9 days after maximum. Emulsion, Agfa Supersensitive Panchromatic.

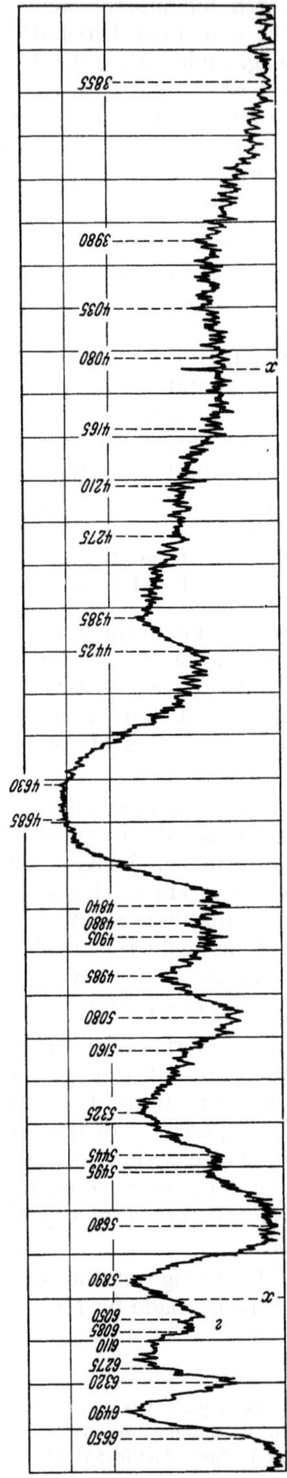

Fig. 6. Supernova in IC 4182, 136 day afters maximum. Emulsion, Agfa Supersensitive Panchromatic.

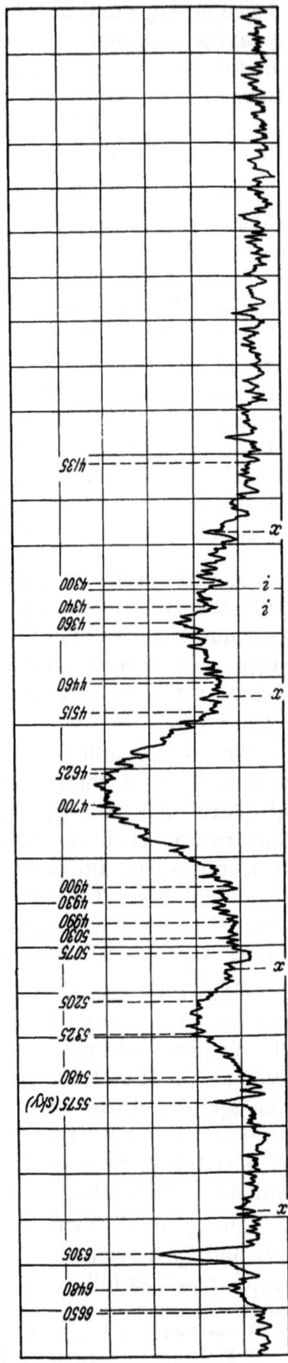

Fig. 7. Supernova in IC 4182, 312 days after maximum. Emulsion, Agfa Superpan Press.

"Spectroscopic observations indicate at least two types of supernovae. Nine objects (represented by the supernovae in IC 4182 and in NGC 4636) form an extremely homogeneous group provisionally called "type I". The remaining five objects (represented by the supernova in NGC 4725) are distinctly different; they are provisionally designated as "type II". The individual differences in this group are large; at least one object, the supernova in NGC 4559, may represent a third type or, possibly, an unusually bright ordinary nova.

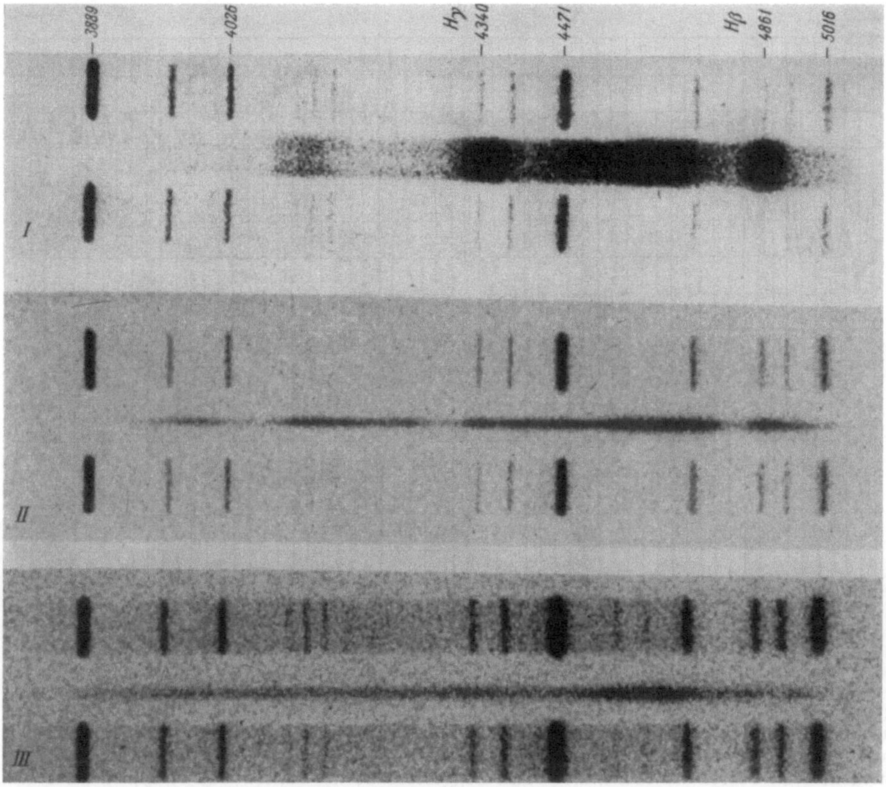

Fig. 8. Spectra of: I. Common nova Sagittarii, October 13, 1936. II. Type II supernova in NGC 5907, April 6, 1940. III. Type I supernova in NGC 1003, November 3, 1937. The wavelengths of some of the lines of the comparison spectrum (He + H) are indicated.

Spectra of supernovae of type I have been observed from 7 days before maximum until 339 days after. Except for minor differences, the spectrograms of all objects of type I are closely comparable at corresponding times after maxima. Even at the earliest premaximum stage hitherto observed, the spectrum consists of very wide emission bands. No significant transformation of the spectrum occurs near maximum. Spectra of type II have been observed from maximum until 115 days after. Up to about a week after maximum, the spectrum is continuous and extends far into the ultraviolet, indicating a very high color temperature. Faint emission is suspected near Hα. Thereafter, the continuous spectrum fades and becomes redder. Simultaneously, absorptions and broad emission bands are developed. The spectrum as a whole resembles that of normal novae in the transition stage, although the hydrogen bands are relatively faint and forbidden lines are either extremely faint or missing. The supernova in NGC 4559, while generally similar to the other objects in this group, shows multiple absorptions of H and Ca II; the emission bands are fainter than in the other objects.

No satisfactory explanation for the spectra of type I has been proposed. Two O I bands of moderate width in the later spectra of the supernova in IC 4182 are the only features satisfactorily identified in any spectrum of type I. They are, at the same time, the only indication of the development of a nebular spectrum for any supernova. The synthetic spectra by Gaposchkin and Whipple disagree in many details with the observed spectra of type I. However, these synthetic spectra agree better with spectra of type II and provide a very satisfactory confirmation of the identifications which, in this case, are already suggested by the pronounced similarity to the spectra of ordinary novae. As compared with normal novae, supernovae of type II show a considerably earlier type of spectrum at maximum hence a higher surface temperature (order of 40000°), and the later spectrum indicates

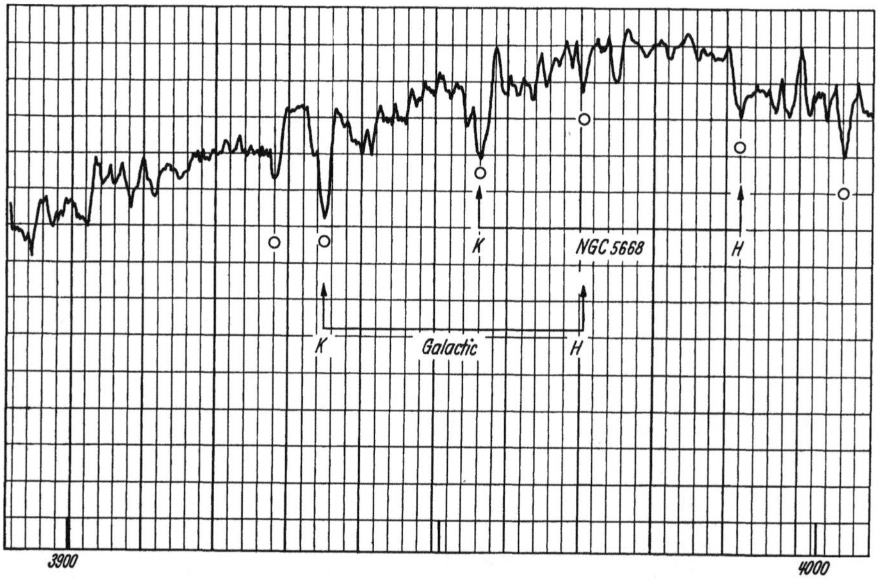

Fig. 9. Tracing of the spectrum of the supernova in NGC 5668 as obtained by A. Deutsch on May 8/9, 1954 with the Coudé spectrograph of the 200-inch telescope, dispersion 4 Å/mm. Wavelengths 3900 to 4000 Å.

greater velocities of expansion (5000 km/sec or more) and higher levels of excitation. Supernovae of type II differ from those of type I in the presence of a continuous spectrum at maximum and in the subsequent transformation to an emission spectrum whose main constituents can be readily identified. This suggests that the supernovae of type I have still higher surface temperature and higher level of excitation than either ordinary novae or supernovae of type II."

No significant discoveries about the spectra of supernovae[1] have been made during the past 15 years except perhaps the observation by A. Deutsch that in the spectrum of Wild's supernova in NGC 5668 both the H and K lines due to interstellar calcium within our own galaxy and within the spiral NGC 5668 are discernible, as shown in the tracing reproduced in Fig. 9.

10. The initial and the final stages of supernovae. Of the initial bodies in which supernovae occur we have no observational knowledge whatsoever. We do not even know if these bodies are stars. Theoretically it seems for instance possible that a gas and dust cloud becomes unstable, collapses and as a consequence results in the formation of a star with simultaneous ejection of much radiation and the re-expulsion of parts of the collapsing gases.

---

[1] C. P. and S. Gaposchkin: Variable Stars, p. 274—276, Harvard Obs. Monogr. No. 5, 1938.

As to the final stages of a supernova there is only one remnant about which we are quite certain, namely, the Crab Nebula (Messier 1), which is made up of the gases expelled by the supernova of the year 1054 A.D. These gases still expand with a mean radial rate of $V_r = 1116$ km/sec or a mean angular rate of 0".201/year, corresponding to a distance modulus $m - M = 10.35$. Much of the information available on the Crab Nebula can be found in two papers by BAADE[1] and MINKOWSKI[2], to which we herewith refer the readers. About the possible remnant star of the supernova of 1054 A.D. there has been some speculation that it is identical with one of the components of the faint double star ($m_p = 15.94$ and 15.96) at the center of the Crab nebula. BAADE[1], however, writes "The star that excites the nebula cannot be conclusively identified at present. The north following component of the central double star is definitely ruled out. Proof that the south preceding component is the exciting star can be expected only after the proper motion of the nebula has been better determined."—It should be added that a much more thorough investigation of the spectrum and of the polarisation of the star in question is most desirable in order to clear up some of the important problems related to the Crab Nebula.

Since those earlier papers there have taken place two most significant developments. It was found in the first place that the Crab Nebula is a fairly strong *cosmic radio source*[3]. In the second place VASHAKIDZE and DOMBROVSKY[4] discovered that the continuous radiation of the nebula is due to electrons of very high energy (2×10^{11} ev) moving in a magnetic field of about $H = 10^{-3}$ gauss (synchrotron mechanism). It is significant to note that on this theory the total integrated energy content of the magnetic field ($H^2/8\pi$ per cm^3) over the continuous mass of the Crab Nebula would be of the order of 10^{49} ergs, or comparable to the total visual energy emitted by a supernova.

Most recently ZWICKY has succeeded[5] with the aid of his newly developed *composite photography* to demonstrate that the directions of polarisation within the diffuse mass of the Crab Nebula are arranged in a most amazing array, resembling a rectangular basket weave. No rational explanation has as yet been given for this discovery.

In the case of KEPLER's nova of 1604 no remnant star has been found, and, whether or not the gas clouds photographed by BAADE[6] have any relation to the nova remains to be checked. For TYCHO's nova of 1572 neither any stellar remnant nor any of the expelled gases have ever been identified. It is therefore doubtful if the supernovae of 1054, 1572, and 1604, whose remnants so radically differ from one another, can be all of the same type I as Baade claims.

The fact that the Crab Nebula is a radio source has led to the suggestion that other radio sources might be remnants of old supernovae. It was thus found[7] that the strong radio source in Cassiopeia is associated with diffuse condensations showing radial velocities from -4000 to $+4900$ km/sec. This might indicate

---

[1] W. BAADE: Astrophys. Journ. **96**, 188—198 (1942).

[2] R. MINKOWSKI: Astrophys. Journ. **96**, 199—213 (1942).

[3] J. G. BOLTON and G. J. STANLEY: Austral. J. Sci. Res., Ser. A, **2**, 139—148 (1949). — Cf. also Sect. 17 of the contribution of R. H. BROWN in vol. LIII of this Encyclopedia.

[4] Dokl. Akad. Nauk. USSR. **91**, 475 (1953). For details see J. H. OORT and TH. WALRAVEN: Bull. Astr. Inst. Netherl. **12**, 285—308 (1956).

[5] F. ZWICKY: Publ. Astronom. Soc. Pacific **68**, 121—124 (1956).

[6] See footnote 3, p. 768.

[7] Year Book No. 54, (1954—1955) of the Carnegie Institution of Washington, p. 24—25.

that we here deal with the remnants of a supernova of type II. The Cassiopeia source, however, presents many mysteries, the elucidation of which must be awaited before we can draw any final conclusion as to its real character.

11. Theory of supernovae. In the course of the last twenty years many theories of supernovae have been proposed, most of which, however, are of little interest to-day. Clearly, no theory has as yet been advanced which can explain any of the features of the spectra of the supernovae of the type I. Neither is there any explanation for the astounding fact that the light curves and the spectra of the supernovae of the type I are so remarkably similar although the absolute magnitudes may considerably vary.

As to the problem of the suddenness and the magnitude of the energy generation in supernovae, the theory of ZWICKY[1,2] of the partial gravitational collapse of a star or of a large gas cloud into a neutron core, with simultaneous expulsion of the remaining peripheral more or less normal gases, still bids fair to provide the explanation for many if not for all supernovae. One of the reasons for a collapse of this type could be the fact that the ordinary exothermic reaction Neutron = Proton + Positive Electron becomes endothermic under high pressure or very low gravitational potential. In reversing itself under these conditions the said reaction results in the mass production of neutrons and in the rapid gravitational collapse of a star or a gas cloud, with tremendous liberation of energy.

ZWICKY[3] also showed that, by two processes which take place during a supernova outburst, electric potential differences of the order of 10^{11} and 10^{19} volts are built up respectively. The collapse of these electric fields leads to the ejection of cosmic ray particles, in particular of all of the heavy particles now found in the primary cosmic rays. The theory that the origin of the cosmic rays would be found in the supernovae led in the period from 1932 to 1939 to a number of predictions concerning the total energy of the cosmic rays, the maximum energy of individual corpuscles (10^{19} ev) and the material composition of the rays (nuclei of all masses present) which during the subsequent 15 years have found their observational confirmation.

In conclusion it should perhaps be mentioned that the almost exact exponential decline of the photographic brightness of supernovae of the type I has prompted some investigators[4] to propose that this decline corresponds to the radioactive decay (half lives equal or near 55 days) of some unstable nuclei such as Be^7, Sr^{89} or Californium 254, whose energy of disintegration is supposed to be transformed into light. The difficulty with any such theory is that the blue light from a supernova can only be a small part of the total energy liberated and that the intensity of the other components of this total energy are certainly not declining in the same exponential way as the blue light. But even if the "light curves" in all other colours (as well as in particle emission, creation of electric and magnetic field energies and heat content of the remnant masses) were similar to the photographic light curve, it would be difficult to prove why all of these energies, which are made up of low energy quanta, should under conditions of constantly varying density show the same decay constant as the

[1] See footnote 2, p. 779.

[2] See footnote 3, p. 772.

[3] F. ZWICKY: Phys. Rev. **55**, 986 (1939). — Proc. Nat. Acad. Sci. U.S.A. **25**, 338–344 (1939).

[4] W. BAADE and G. R. BURBIDGE et al.: Publ. Astronom. Soc. Pacific **68**, 296–300 (1956).

radioactive disintegration of the elements like Californium 254 which liberate quanta of enormously higher energy.

It appears therefore that for the time being the original theory of the author remains the only one which is to be taken seriously, namely, that most if not all supernovae are caused by the collapse of parts of stars or of other large aggregations of matter into neutron cores whose end states approach nuclear density[1,2].

Note added in proof. Since the present manuscript was submitted in September 1956, the following noteworthy progress has been made.

1. The combined search for supernovae which is being carried out by the observatories mentioned in Sect. 4δ of Chap. I has resulted in the discovery of four additional supernovae in NGC 2841 by M. SCHÜRER in Berne, in NGC 4374 (Messier 84) by G. ROMANO in Treviso, Italy and by H. S. GATES at the Palomar Observatory and in NGC 1365 and NGC 5236 (Messier 83), both by H. S. GATES. If the star in NGC 5236 is not just a very bright common nova, the galaxy NGC 5236 will be the third one which, in a period of less than fifty years had three supernovae.

2. R. MINKOWSKI in the Year Book 1955/56 of the Carnegie Institution of Washington, p. 51, announced that he has discovered a nebulosity near the position of TYCHO's nova of 1572 which he considers undoubtedly to be a remnant of the nova. The nebulosity which shows motion is under more detailed investigation.

[1] See footnote 2, p. 779.
[2] See footnote 3, p. 772.

Sachverzeichnis.

(Deutsch-Englisch.)

Bei gleicher Schreibweise in beiden Sprachen sind die Stichwörter nur einmal aufgeführt.

Subject Index.

(English-German.)

Where English and German spelling of a word is identical the German version is omitted.

Table des matières

pour la contribution écrite en français:

E. Schatzman: Théorie des naines blanches.

MIX
Papier aus verantwortungsvollen Quellen
Paper from responsible sources
FSC® C105338

If you have any concerns about our products,
you can contact us on
ProductSafety@springernature.com

In case Publisher is established outside the EU,
the EU authorized representative is:
Springer Nature Customer Service Center GmbH
Europaplatz 3, 69115 Heidelberg, Germany

Printed by Libri Plureos GmbH
in Hamburg, Germany